5th International Conference on Mathematical Modeling in Physical Sciences (IC-MSquare 2016)

AA006730

Journal of Physics: Conference Series Volume 738

Athens, Greece
23 – 26 May 2016

Editors:

Elias C. Vagenas
Dimitrios S. Vlachos

ISBN: 978-1-5108-2994-7
ISSN: 1742-6588

Printed from e-media with permission by:

Curran Associates, Inc.
57 Morehouse Lane
Red Hook, NY 12571

CURRAN ASSOCIATES INC. proceedings .com

Some format issues inherent in the e-media version may also appear in this print version.

Copyright© (2016) by the Institute of Physics
All rights reserved. The material featured in this book is subject to
IOP copyright protection, unless otherwise indicated.

Printed by Curran Associates, Inc. (2016)

For permission requests, please contact the Institute of Physics
at the address below.

Institute of Physics
Dirac House, Temple Back
Bristol BS1 6BE UK

Phone: 44 1 17 929 7481
Fax: 44 1 17 920 0979

techtracking@iop.org

Additional copies of this publication are available from:

Curran Associates, Inc.
57 Morehouse Lane
Red Hook, NY 12571 USA
Phone: 845-758-0400
Fax: 845-758-2633
Email: curran@proceedings.com
Web: www.proceedings.com

Table of contents

Volume 738

5th International Conference on Mathematical Modeling in Physical Sciences (IC-MSquare 2016)

23–26 May 2016, Athens, Greece

Accepted papers received: 27 July 2016
Published online: 5 September 2016

Preface

Papers

Editors:

Elias C. Vagenas, Kuwait University, Kuwait
Dimitrios S. Vlachos, University of Peloponnese, Greece

Content from this work may be used under the terms of the Creative Commons Attribution 3.0 licence. Any further distribution of this work must maintain attribution to the author(s) and the title of the work, journal citation and DOI.
Published under licence by IOP Publishing Ltd

Journal of Physics: Conference Series **738** (2016) 011001 doi:10.1088/1742-6596/738/1/011001

Preface

The 5th International Conference on Mathematical Modeling in Physical Sciences (IC-MSQUARE) took place at Athens, Greece, from Monday, 23th of May, to Thursday, 26th of May 2016.

The Conference was attended by more than 130 participants and hosted about 170 oral, poster, and virtual presentations while counted more than 500 pre-registered authors. The 5th IC-MSQUARE consisted of different and diverging workshops and thus covered various research fields where Mathematical Modeling is used, such as Theoretical/Mathematical Physics, Neutrino Physics, Non-Integrable Systems, Dynamical Systems, Computational Nanoscience, Biological Physics, Computational Biomechanics, Complex Networks, Stochastic Modeling, Fractional Statistics, DNA Dynamics, Macroeconomics etc.

The scientific program was rather heavy since after the Keynote and Invited Talks in the morning, three parallel oral and one poster session were running every day. However, according to all attendees, the program was excellent with high level talks and the scientific environment was fruitful, thus all attendees had a creative time.

We would like to thank the Keynote Speaker and the Invited Speakers for their significant contribution to IC-MSQUARE. We also would like to thank the Members of the International Advisory and Scientific Committees as well as the Members of the Organizing Committee.

The Conference Chairmen

Elias Vagenas
Kuwait University

Dimitrios Vlachos
University of Peloponnese

International Advisory Committee

- **Professor Constantino Tsallis,** Centro Brasileiro de Perquisas Fisicas, Brazil
- **Professor Athanasios S. Fokas,** University of Cambridge, UK
- **Professor Henrik J. Jensen** , Imperial College London, UK
- **Professor George E. Karniadakis,** Brown University, USA
- **Professor Nikolaos E. Mavromatos**, King's College London, UK
- **Professor John E. Straub** , Boston University, USA
- **Professor Jorge Zanelli** , Centro de Estudios Científicos (CECs), Chile

International Scientific Committee

- **Bari Ferenc,** University of Szeged, Hungary
- **Mark Cross,** Swansea University, UK
- **Morten Hjorth-Jensen,** University of Oslo, Norway
- **Nathan A Baker,** Pacific Northwest National Laboratory, Richland, WA, USA
- **Nikolas Markatos,** National Technical University of Athens, Greece
- **Shiyi Chen,** Peking University, People's Republic of China
- **Kostas Kokkotas,** Eberhard Karls University of Tübingen, Germany
- **Sergey Nazarenko,** University of Warwick, UK
- **Rho Shin Myong,** Gyeongsang National University, Republic of Korea
- **Michel A. Van Hove,** City University of Hong Kong, Hong Kong
- **Jacob Barhen,** Oak Ridge Natl. Lab, Oak Ridge, USA
- **Javier Hernández-Andrés,** University of Granada, Spain
- **Maia Martcheva,** University of Florida, USA
- **Emilio Elizalde,** CSIC, Barcelona, Spain
- **Angelo Tartaglia,** Politecnico di Torino, Italy
- **Yoshiharu Yamamoto,** The University of Tokyo, Japan
- **Amaury Mouchet,** Université de Tours, France
- **Mairi Sakellariadou,** King's College of London, UK
- **Kostas Skenderis,** University of Amsterdam, The Netherlands
- **Orfeu Bertolami,** University of Porto, Portugal
- **Saurya Das,** University of Lethbridge, Canada
- **Theocharis Kosmas,** University of Ioannina
- **Georgios Loukes-Gerakopoulos,** TPI, FSU Jena, Germany

Keynote Speaker

Constantino Tsallis, Centro Brasileiro de Perquisas Fisicas, Brazil

Invited Speakers

- **Professor Iannis Kominis,** University of Crete, Greece
- **Professor Yannis Kominis,** National Technical University of Athens, Greece
- **Professor Nicolas-Chris Markatos,** National Technical University of Athens, Greece
- **Professor Nikolaos E. Mavromatos,** King's College London, UK
- **Professor Fabrizio Pinto,** Jazan University, Saudi Arabia
- **Dr George Spyrou,** Biomedical Research Foundation, Academy of Athens, Greece
- **Professor Kalman Varga,** Vanderbilt University, USA

Peer review statement

All papers published in this volume of *Journal of Physics: Conference Series* have been peer reviewed through processes administered by the proceedings Editors. Reviews were conducted by expert referees to the professional and scientific standards expected of a proceedings journal published by IOP Publishing.

Content from this work may be used under the terms of the Creative Commons Attribution 3.0 licence. Any further distribution of this work must maintain attribution to the author(s) and the title of the work, journal citation and DOI.
Published under licence by IOP Publishing Ltd

The stochastic modeling of the short-time variations of the galactic cosmic rays

A. Wawrzynczak[1], **R. Modzelewska**[2]
[1]Institute of Computer Sciences, Siedlce University, Poland,
[2]Institute of Mathematics and Physics, Siedlce University, Poland.
E-mail: awawrzynczak@uph.edu.pl, renatam@uph.edu.pl

Abstract. We present the stochastic model of the galactic cosmic ray (GCR) particles transport in the heliosphere. The model is created based on the numerical solution of the Parker transport equation (PTE) describing the non-stationary transport of charged particles in the turbulent medium. We present the numerical schemes for the strong order integration of the set of the stochastic differential equations (SDEs) corresponding to the non-stationary PTE. Among the employed methods are the strong order Euler-Maruyama, Milstein and stochastic Runge-Kutta methods. We perform the selection of the method resulting in the highest agreement of the model of the 27-day variation of the GCR intensity with the experimental observations.

1. Introduction

The stochastic differential equations (SDEs), in conjunction with various Monte Carlo technics, are broadly used in many fields. Nowadays, in the era of high-performance computing, efficient parallel algorithms to realize large-scale simulations are crucial in solving realistic problems inspired by physical sciences and engineering. One of these problems is the non-stationary stochastic motion of charged particles in three-dimensional space, described by diffusion equation of Fokker-Planck type. The main statistical characteristic is the representation of the solution of the Fokker-Planck equation (FPE) as a probability density function. Unfortunately, most of the problems existing in nature are unsolvable analytically, so numerical methods must be applied. The difficulty of the numerical solution of this type of equations increases with the problem dimension. The reason is the instability of the numerical schemes like finite-differences and finite-volume in the higher dimensions. To ensure the scheme stability and convergence the density of numerical grid must be improved, with the computational complexity enlargement, at the same time. To overcome this problem, the stochastic methods can be applied (e.g. [1], [2]). In this approach, the individual particle motion is described as a Markov stochastic process, and the system evolves probabilistically. Employment of probabilistic description with Monte Carlo simulations allow reducing the solution of the partial differential equation (PDE) describing the analyzed phenomena to the integration of SDEs.

The aim of this paper is threefold: (1) to compose a reliable mathematical model of the galactic cosmic ray (GCR) particle transport in the heliosphere using the SDEs; (2) to solve Parker transport equation (PTE) by applying to the SDEs the strong order integration Euler-Maruyama, Milstein and stochastic Runge-Kutta numerical methods and then, (3) to employ the developed methods to simulate the 27-day variation of the GCR intensity being in an agreement

Content from this work may be used under the terms of the Creative Commons Attribution 3.0 licence. Any further distribution of this work must maintain attribution to the author(s) and the title of the work, journal citation and DOI.
Published under licence by IOP Publishing Ltd

with the experimental data.

We employ the stochastic methodology to model the GCR transport in the heliosphere. Modulation of the GCR is a result of an action of four primary processes: convection by the solar wind, diffusion on irregularities of the heliospheric magnetic field (HMF), particles drifts in the non-uniform magnetic field and adiabatic cooling. Transport of the GCR particles in the heliosphere can be described by the Parker transport equation [3]:

$$\frac{\partial f}{\partial t} = \vec{\nabla} \cdot (K_{ij}^S \cdot \vec{\nabla} f) - (\vec{v}_d + \vec{U}) \cdot \vec{\nabla} f + \frac{R}{3}(\vec{\nabla} \cdot \vec{U})\frac{\partial f}{\partial R}, \tag{1}$$

where $f = f(\vec{r}, R, t)$ is an omnidirectional distribution function of three spatial coordinates $\vec{r} = (r, \theta, \varphi)$, particles rigidity R and time t; $\vec{U}$ is the solar wind velocity, $\vec{v}_d$ the drift velocity, and K_{ij}^S is the symmetric part of the diffusion tensor of the GCR particles.

This paper is an extension of our previous results presented in [4, 5]. We have presented [4] that the GCR transport can be effectively modeled based on the solution of the set of SDEs corresponding to the PTE (Eq. 1) by Euler-Maruyama method. In [5] we have increased the accuracy of the SDEs solution by applying the higher order methods i.e. Milstein and stochastic Runge-Kutta. In this paper, we employ the higher order methods i.e. Milstein and stochastic Runge-Kutta to simulate the short-time variations of the GCR intensity. We perform the selection of the numerical integration method resulting in the highest agreement of the short-time variations of the GCR intensity with the experimental observations.

2. Stochastic differential equation corresponding to the Parker transport equation

The PTE (Eq. 1) in the 3-D heliocentric spherical coordinate system (r, θ, φ) written as time-backward FPE diffusion equation has the form:

$$\frac{\partial f}{\partial t} = A_1\frac{\partial^2 f}{\partial r^2} + A_2\frac{\partial^2 f}{\partial \theta^2} + A_3\frac{\partial^2 f}{\partial \varphi^2} + A_4\frac{\partial^2 f}{\partial r \partial \theta} + A_5\frac{\partial^2 f}{\partial r \partial \varphi} + A_6\frac{\partial^2 f}{\partial \theta \partial \varphi} + A_7\frac{\partial f}{\partial r} + A_8\frac{\partial f}{\partial \theta} + A_9\frac{\partial f}{\partial \varphi} + A_{10}\frac{\partial f}{\partial R} \tag{2}$$

the coefficients A_1,...,A_{10} are presented in detail in [4]. Depending on the choice of the numerical approximation the corresponding to Eq. 2 set of SDEs has a form:

$$\begin{aligned}
dr &= \underbrace{\underbrace{\underbrace{A_7 \cdot dt + B_{11} \cdot dW_r}_{Euler-Maruyama} + \frac{1}{2}B_{11}\frac{\partial B_{11}}{\partial r}(dW_r^2 - dt)}_{Milstein} + \Phi_1}_{Stochastic\ Runge-Kutta} \\
d\theta &= \underbrace{\underbrace{\underbrace{A_8 \cdot dt + B_{21} \cdot dW_r + B_{22} \cdot dW_\theta}_{Euler-Maruyama} + \frac{1}{2}B_{21}\frac{\partial B_{21}}{\partial \theta}(dW_r^2 - dt) + \frac{1}{2}B_{22}\frac{\partial B_{22}}{\partial \theta}(dW_\theta^2 - dt)}_{Milstein} + \Phi_2}_{Stochastic\ Runge-Kutta} \\
d\varphi &= \underbrace{\underbrace{\underbrace{A_9 \cdot dt + B_{31} \cdot dW_r + B_{32} \cdot dW_\theta + B_{33} \cdot dW_\varphi}_{Euler-Maruyama} + \Phi_3}_{Milstein} + \Phi_4}_{Stochastic\ Runge-Kutta} \\
dR &= A_{10} \cdot dt.
\end{aligned} \tag{3}$$

The B_{ij}, $(i, j = r, \theta, \varphi)$, is the lower triangular matrix presented in [4], and the coefficients Φ_1, Φ_2, Φ_3 and Φ_4 are presented in detail in [5].

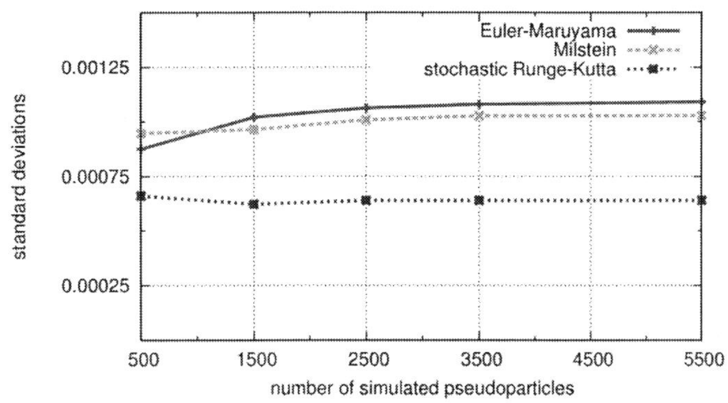

Figure 1. Standard deviations of the particle distribution function $f(\vec{r}, R)$ calculated for the Euler-Maruyama, Milstein and stochastic Runge-Kutta methods vs. a number of simulated pseudoparticles with the rigidity of 10 GV.

3. The simulations results and discussion

The code for the numerical solution of the set of Eqs. 3 is realized in MatLab environment. The code is easy to parallelize versus the number od simulated pseudoparticles using Matlab Parallel Toolbox. This approach fallouts from the assumption that any random process is independent of the other realization, accordingly each pseudoparticle's trajectory is independent on another. We performed the simulations applying all three numerical methods given by Eqs. 3. The trajectory of $n = 2000$ pseudoparticles was traced backward in time in the spherical heliocentric coordinate system. The pseudoparticles were initialized in the point representing the Earth's orbit (i.e. $r = 1AU$, $\theta = 90°$, $\varphi = 180°$) and traced backward in time until crossing the heliosphere boundary assumed at 100 AU.

The value of the particle distribution function $f(\vec{r}, R)$ for the starting point was found as an average of $f_{LIS}(R)$ value for pseudoparticles characteristics at the entry positions, $f(\vec{r}, R) = \frac{1}{N}\sum_{n=1}^{N} f_{LIS}(R)$, where $f_{LIS}(R)$ is the cosmic ray local interstellar spectrum (LIS) taken as in [6] for rigidity R of the nth particle at the entrance point.

To solve Eqs. 3 in spherical coordinates system we used the boundary conditions of form [7]: $\varphi_i < 0 \rightarrow \varphi_i = \varphi_i + 2\pi$; $\varphi_i > 2\pi \rightarrow \varphi_i = \varphi_i - 2\pi$; $\theta_i < 0 \rightarrow \theta_i = -\theta_i$ & $\varphi_i = \varphi_i \pm \pi$ and $\theta_i > \pi \rightarrow \theta_i = 2\pi - \theta_i$ & $\varphi_i = \varphi_i \pm \pi$. The reflecting boundary is considered as the inner radial boundary, $\frac{\partial f}{\partial r} = 0$ at $r = 0.001$ AU. An empty heliosphere constitutes an initial condition, $f_i(0.01AU < r < 100AU, \theta, \varphi, R, 0) = 0$, as was shown in [8].

In the case, when we know the analytical solution of the equation that we want to solve, it is easy to compare the efficiency of each applied numerical method. However, in the case of GCR particles transport in the heliosphere, we do not have such a possibility. In general, using the higher order numerical method should result in the increase of the accuracy of the solution. Particularly, the smaller number of simulated particles for higher order method should give better or comparable results to those provided by a weaker method employing more pseudoparticles. Thus, we have calculated the values of the particle distribution function, $f(\vec{r}, R)$ and its standard deviations for models employing all presented numerical methods (Eqs. 3) with number of simulated particles increasing from $n = 500$ to $n = 5500$. We have analyzed the changes of the standard deviations of the particle distribution function $f(\vec{r}, R)$ calculated for the Euler-Maruyama, Milstein and stochastic Runge-Kutta methods vs. the number of simulated pseudoparticles with the rigidity of 10 GV. Results of calculations are presented in Fig. 1. Fig. 1 shows that the standard deviation of the numerical solution of the Eqs. 3 is the largest for Euler-

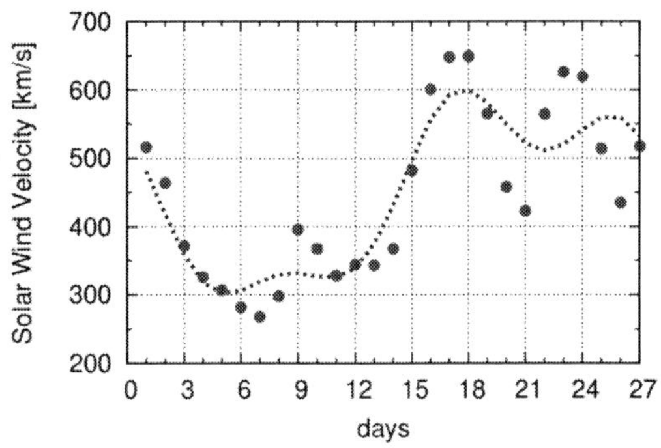

Figure 2. Temporal changes of the observed daily solar wind velocity (points) and dashed curve represent its approximation – sum of three harmonics (27, 14 and 9 days) wave during 7 September – 3 October 2007.

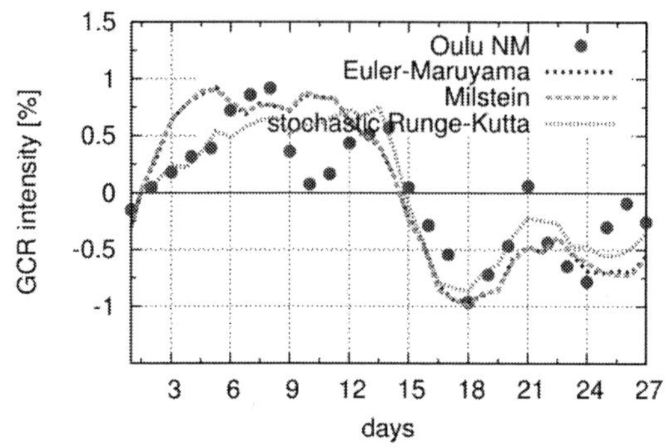

Figure 3. The expected 27-day variation of the GCR intensity at the Earth orbit for the particles with rigidity of 10 GV in comparison with the GCR intensity registered by Oulu neutron monitor during 7 September - 3 October 2007.

Maruyama method and the lowest for stochastic Runge-Kutta. So, by applying the higher order method (especially stochastic Runge-Kutta), the statistical accuracy of the numerical solution is significantly increased. The obtained results suggest that all applied numerical methods give reliable results. Performed tests proved that the Milstein and stochastic Runge-Kutta methods are more stable and return the same results of differential spectra when we decrease the number of simulated particles by a factor of three.

4. Model of the 27-day variation of the GCR intensity

The recurrence of the GCR intensity connected with the solar rotation period ($\sim$ 27 days) is commonly called the 27-day variation, although the duration can slightly differ. The 27-day variation of the GCR intensity is connected with the heliolongitudinal asymmetry of the electromagnetic conditions in the heliosphere. The recent minimum of solar activity between solar cycles No. 23 and 24 was quite exceptional. Recurrent variations connected with corotating structures ($\sim$ 27 days), at the end of 2007 and almost for the whole year 2008 were clearly established in all solar wind and interplanetary parameters. Accordingly, the 27-day variation of cosmic ray intensity was clearly visible in a variety of cosmic ray counts of neutron monitors (e.g., [9, 10]) and space probes (e.g., [11]).

We present the model of the 27-day variation of the GCR intensity considering a particular period of the solar rotation starting at 2007.09.07. As it was stated by [9, 12], the 27-day variation of the GCR intensity in the minimum epochs of solar activity is preferentially related to the heliolongitudinal asymmetry of the solar wind velocity. In the model, we apply approximation of in situ measurements of the solar wind speed as the source of the 27-day variation of the GCR intensity approximated by the formula: $U = U_0(1 - 0.31\sin(\varphi + 6.10) + 0.06\sin(2\varphi + 0.82) - 0.10\sin(3\varphi - 1.04))$, where $U_0 = 400km/s$. Fig. 2 presents temporal changes of the observed daily solar wind velocity (points) and dashed curve represents its aproximation being the sum of three harmonics (27, 14 and 9 days) wave during 7 September – 3 October 2007. The diffusion coefficient $K_{\|}$ of cosmic ray particles has the form: $K_{\|} = K_0 \cdot K(r) \cdot K(R)$, where $K_0 = 10^{22}cm^2/s$, $K(r) = 1 + 0.5 \cdot (r/1AU)$ and $K(R) = R^{0.5}$.

The expected changes of the 27-day variation of the GCR intensity for the rigidity of 10 GV in comparison with the profile of the daily GCR intensity recorded by Oulu neutron monitor during 7 September - 3 October 2007 presents Fig. 3. Fig. 3 compares the results of the solutions of the PTE with SDEs by the Euler-Maruyama, Milstein and stochastic Runge-Kutta methods. One can see that model utilizing all presented methods can reconstruct the 27-day wave of the GCR intensity at the Earth orbit (1 AU, $\theta = 90^0$). However, in some periods, one can see that the model results do not coincide one to one with experiment. The model indeed does not describe all heliolongitudinal variations in the GCR registration. The reason is the assumed approximation. We incorporated in the model of the 27-day variation only the approximation of the first three harmonics of the solar wind velocity (Fig. 2). Comparing the results of the SDEs solutions by Euler-Maruyama, Milstein and stochastic Runge-Kutta methods, one can see that the stochastic Runge-Kutta method reproduced the data with the highest statistical precision. The Euler-Maruyama and Milstein methods give almost the same results in the scope of the accuracy. However, short period fluctuations of the GCR intensity could not be completely reproduced by the proposed model.

Presented model results (Fig. 3) show an acceptable agreement between experimental data of Oulu neutron monitor and modeling results dealing mainly with the shape and average amplitude of the 27-day variation of the GCR intensity. It is hardly expected that modeling could completely explain any short period fluctuation of the GCR intensity. We demonstrated that proposed approach, with an implementation of the solar wind velocity in situ measurements from one hand, and application of the higher order numerical methods from the other, makes the proposed model more realistic and statistically significant. Moreover, the model of the 27-day variation obtained based on the solution of the SDEs allows to reflect the stochastic character of the GCR particles distribution in the heliosphere. The presented model visualizes the pseudoparticle trajectory throughout the 3D heliosphere, which is not possible based on the solution of the Parker transport equation by e.g. the finite difference method [10, 12].

5. Conclusions

(i) We presented the solution of the Parker transport equation using a numerical solution of the set of stochastic differential equations driven by a Wiener process with the strong order Euler-Maruyama, Milstein, and stochastic Runge-Kutta methods. We showed that application of the higher order methods (especially stochastic Runge-Kutta) significantly increased the statistical accuracy of the numerical solution.

(ii) The SDEs were integrated backward in time in the heliocentric spherical coordinates applying the full 3D anisotropic diffusion tensor. The models obtained based on the solution of the SDEs reflect the stochastic character of the pseudoparticles distribution in the heliosphere. Additionally, this technique allows observing the statistically possible changes of the pseudoparticles trajectories, which is not feasible based on the solution of the Parker transport equation by e.g. finite difference methods.

(iii) We presented the model of the 27-day variation of the GCR intensity obtained based on the stochastic approach to the solution of the Parker transport equation. The modeling results are in good agreement with the experimental data.

6. Acknowledgments

This work is supported by The Polish National Science Centre grant awarded by decision number DEC-2012/07/D/ST6/02488. We thank the principal investigators of Oulu neutron monitor and OMNIweb for the ability to use their data.

7. References

[1] Zhang M 1999 *Astrophys. J.* **513** 409-420
[2] Gervasi M, Rancoita P G, Usoskin I G and Kovaltsov G A 1999 *Nuclear Physics B (Proc. Suppl.)* **78** 26-31
[3] Parker E N 1965 *Planetary and Space Science* **13** 9-49
[4] Wawrzynczak A, Modzelewska R and Gil A 2015 *Journal of Physics: Conference Series* **574** 012078
[5] Wawrzynczak A, Modzelewska R and Kluczek M 2015 *Journal of Physics: Conference Series* **633** 012058

[6] Webber W R and Lockwood J A 2001 *Journal of Geophysical Research* **106** 29323-29332
[7] Kopp A, Busching I, Strauss R D and Potgieter M S 2012 *Computer Physics Communications* **183** 530-542
[8] Pei C, Bieber J W, Burger R A and Clem J 2010 *J. Geophys. Res.* **115** A12107
[9] Alania M V, Modzelewska R and Wawrzynczak A 2010 *Adv. Sp. Res.* **45** 421-428
[10] Modzelewska R and Alania M V 2013 *Solar Physics* **286** 593-607
[11] Leske, R.A., Cummings A.C.,Mewaldt R.A., Stone E.C. 2013 *Space Sci Rev.*, **176**, 1-4, 253-263
[12] Gil A and Alania M V 2013 *Solar Physics* **283** 565-578

Optimized design of thermo-mechanically loaded non-uniform bars by using a variational method

P Nayak, K N Saha[1]
Department of Mechanical Engineering, Jadavpur University, Kolkata-32, India

[1]E-mail: kashinathsaha@gmail.com

Abstract. The present paper evaluates the axial strain and stress of a thermo-mechanically loaded non-uniform bar by using a numerical method based on a variational principle. The solutions are obtained up to the elastic limit of the material based on the assumptions that material properties are independent of temperature variation and plane cross-sections remain plane maintaining axisymmetry. This approximation is carried out by Galerkin's principle, using a linear combination of sets of orthogonal co-ordinate functions which satisfy prescribed boundary conditions. The solution algorithm is implemented with the help of MATLAB® computational simulation software. Some numerical results of thermoelastic field are presented and discussed for different bar materials such as mild steel, copper, aluminium alloy 6061 (Al alloy 6061), aluminium alloy 7075 (Al alloy 7075) and diamond. The effect of geometry parameters like aspect ratio, slenderness ratio and the type of taperness is investigated and the relevant results are obtained in dimensional form. The term bar used in this paper is in generic sense and hence the formulation is applicable for all one dimensional elements, e.g., rods, pipes, truss members, etc.

1. Introduction

The post-elastic analyses of thermo-mechanically loaded non-uniform bars are important for designing mechanical, aerospace and civil structures. The solutions of thermal stress problem available in textbook are valid for uniform cross-section bar under uniform temperature field only. In general the solution fails because the assumption of equality of thermal and mechanical strains at every point within the domain is not true. However, an overall equilibrium of the bar system is achieved by balancing forces coming from the strain differentials, existing within the domain. Hence a detail analysis of the generalized system yields a different strain field, and subsequently a different stress field, which are often much more critical. The simple analytical method also fails when the induced thermal stress is in post-elastic region due to insufficiency in capturing the physics of material behaviour.

The solution of thermal stress problems for uniform bar subjected to a uniform temperature is found in the textbook of Timoshenko [1]. The prediction of the elasto-plastic behaviour of solid slender bars of various types of geometry as well as loading is an interesting area of work for the designers (Hill [2]). Niknam et al. [3] studied the non-linear bending of tapered functionally graded (FG) beam subjected to thermal and mechanical load with general boundary condition. They presented a closed form solution for the classical beam problem and employed Galerkin technique for the general

[1] Corresponding author

Content from this work may be used under the terms of the Creative Commons Attribution 3.0 licence. Any further distribution of this work must maintain attribution to the author(s) and the title of the work, journal citation and DOI.
Published under licence by IOP Publishing Ltd

case with axial force. Das et al. [4] presented a simulation study of the dynamic behaviour of non-uniform taper bars of circular and rectangular cross-section under body force loading due to gravity. The loading is controlled statically to take the bar to its post-elastic state so as to predict its dynamic behaviour in the presence of plastic deformation. Noda [5] presented an extensive review that covered a wide range of topics from thermo-elastic to thermo-inelastic problems. Grysa and Kozlowski [6] presented an approximate analytical and exact solution of the problems of heat flux and temperature determination on slabs. The interior thermal and mechanical responses are used in determining the unknown functions describing heat flux and temperature on the surface of the slab. Nayak and Saha [7] investigated the elastic limit speed of non-uniform rotating disks considering thermal effect on elasticity modulus.

The literature review in this area reveals that textbooks address analysis of thermal stress problem of uniform cross-section bar under uniform temperature field in elastic domain only and literatures explaining non-uniform bars under thermo-mechanical loading are rare. This paper attempts to address the thermo-mechanical behaviour of non-uniform bars upto the elastic limit of the material by using a numerical method based on variational principle. The effect of geometry parameters like aspect ratio, slenderness ratio and the type of taperness on the thermo-elastic performance of the bar is investigated and the relevant results are obtained in dimensional form.

2. Problem formulation

The present paper employs an energy approach to get the appropriate governing equations for the thermo-mechanically loaded non-uniform bars. The formulation is displacement based and the unknown displacement field is approximated by finite linear combination of admissible orthogonal functions. The mathematical model is framed on the assumption that the material of bar is homogeneous, isotropic and linear elastic. It is also assumed that the bar geometry is stub enough to exclude buckling failure from the scope of analysis. Further, the analysis is carried out based on the assumptions that material properties are independent of temperature variation and plane cross-sections remain plane maintaining axisymmetry. However, although the present analysis is carried out for an elastic material, extension of the analysis in the elasto-plastic region needs special mention.

The geometry taken for the present analysis is taper bar of solid circular cross-section with linear variation in diameter given by $d = d_0 - \xi(d_0 - d_1)$, where d_0 is the largest diameter and d_1 is the smallest diameter of the bar. The diameter of the bar (ϕd) at any axial location x is expressed in terms of the normalized axial coordinate $\xi(=x/L)$, where L is the total length of the bar. Furthermore, the geometry of the bar is defined by the slenderness ratio (ratio of length to radius of gyration corresponding to the minimum radius of the bar) and aspect ratio (ratio of difference in radii of two ends to length). The mathematical expression of slenderness ratio is given by $S_R = L/k(r_1)$, where $k(r_1)=\sqrt{I_1/A_1}$ is the radius of gyration of the bar, corresponding to the minimum cross-section of the bar. The expression for aspect ratio is $A_R = (r_0 - r_1)/L$.

Displacement will occur due to thermal expansion resulting in thermal load. The magnitude of this displacement field is also governed by the boundary conditions of the bar. The solution for the displacement field is obtained from the minimization of total potential energy principle $\delta(U+V)=0$ where, U is the strain energy stored in the bar and V is the potential energy developed by the external forces. When the bar material is subjected to a temperature field, it experiences strain due to an expansion proportional to the temperature rise ΔT, and the linear constitutive thermoelastic equations take the form $\varepsilon_x = \frac{\sigma_x}{E} + \alpha\Delta T$, where ε_x is the axial strain, σ_x is stress, E is the modulus of elasticity and α is the coefficient of linear thermal expansion. The total strain energy of the bar comes from the stress and strain field, and expressed as $\Pi = \frac{1}{2}\int_{Vol}(\sigma\varepsilon)dv = \frac{1}{2}\int_0^L(\sigma_x\varepsilon_x)A(x)dx$, where $A(x)$ is the area of cross-section at location x. Using the linear-strain displacement relation, $\varepsilon_x = du/dx$, where u is the axial displacement field, we get

$$\Pi = \frac{1}{2}\int_0^L \left\{E\left(u'\right)^2 - 2E\alpha\Delta T u'\right\}A(x)dx. \tag{1}$$

In equation (1), $(\ )'$ indicate first derivative with respect to the coordinate variable. Here Π is the total potential energy in which one part is similar to the strain energy U stored in the bar and the other part is work function due to temperature gradient loading. So, the expression for U is given by,

$$U = \frac{1}{2}\int_0^L \left\{E\left(u'\right)^2\right\}A(x)dx \tag{2}$$

and the expression for potential energy V, arising from the thermal loading is given by,

$$V = -\frac{1}{2}\int_0^L \left\{2E\alpha\Delta T u'\right\}A(x)dx. \tag{3}$$

Substituting equations (2) and (3) in the energy principle $\delta(U+V)=0$, the governing equilibrium equation becomes

$$\delta\left[\frac{1}{2}\int_0^L \left\{E\left(u'\right)^2 - 2E\alpha\Delta T u'\right\}A(x)dx\right] = 0. \tag{4}$$

Equation (4) is expressed in normalized co-ordinate ξ to facilitate the numerical computation work and the governing equation takes the form

$$\frac{E}{L}\int_0^1 \left\{u'\delta\left(u'\right)\right\}A(\xi)d\xi = E\alpha\Delta T\int_0^1 \delta\left(u'\right)A(\xi)d\xi. \tag{5}$$

The displacement function $u(\xi)$ in equation (5) is approximated by a linear combination of sets of orthogonal coordinate functions as $u(\xi)=\sum c_i\phi_i$, $i=1, 2,\ldots, n_f$, where ϕ_i is the set of orthogonal functions developed through Gram–Schmidt scheme, c_i is the set of unknown coefficients and n_f is number of functions. The necessary start function ϕ_0 is given by $\phi_0 = \xi(1-\xi)$ which satisfies the geometric boundary conditions of the bar which are, i.e. $u=0$ at $\xi=0$ and $u=0$ at $\xi=1$.

Now substituting the series approximation of $u(\xi)$ in Eq. (5) and replacing the operator 'δ' by $\partial/\partial c_j$, $j=1,2,\ldots,n$, according to Galerkin error minimization principle, we obtain the governing differential equation in matrix form.

$$\left(\frac{E}{L}\right)\left[\sum_{i=1}^{n}\sum_{j=1}^{n}\int_0^1\left(A(\xi)\phi_i'\phi_j'\right)d\xi\right]\{c_i\} = E\alpha\Delta T\int_0^1\left(A(\xi)\phi_j'\right)d\xi \tag{6}$$

Solution of equation (6) yields the solution vector $\{c_i\}$, obtained through a single step matrix inversion process. The axial displacement field (u) can be found out for any prescribed value of ΔT which in turn gives strain and stress fields. The numerical integration, differentiation and all other associated mathematical operations are carried out in the computational platform of MATLAB software.

3. Results and discussions

In this paper, a thermal stress analysis is carried out on a clamped taper bar for different materials such as mild steel, copper, Al alloy 6061, Al alloy 7075 and diamond. Mechanical properties of the bar materials such as elasticity modulus (E), Poisson's ratio (ν), coefficient of thermal expansion (α) and yield stress (σ_y) are listed in table 1.

The theoretical value of temperature rise at the point of yielding (ΔT_y) of a taper bar of 1.2 m long, having aspect ratio $A_R = 0.025$, and slenderness ratio $S_R = 20$ is given in table 2 for various bar materials. Table 2 also furnish the respective melting temperatures (T_m) of the materials.

Table 1. Material properties.

Bar material	E (GPa)	ν	α (10^{-6}/°C)	σ_y (MPa)
Mild steel	210	0.3	11.7	350
Copper	129	0.34	17	211
Al alloy 6061	69	0.33	23.6	276
Al alloy 7075	71	0.33	23.4	505
Diamond	1220	0.2	1.18	60000

Table 2. Temperature rise at yield and melting temperature.

Material	Mild steel	Copper	Al alloy 6061	Al alloy 7075	Diamond
ΔT_y (°C)	98.9	66.8	117.7	211.1	28936.9
T_m (°C)	1427	1085	582	477	-

The dimensionless parameter $T_R = \sigma_y / \alpha E \Delta T$, termed as thermal load resistance, is used to characterize the bar material upto the elastic limit. As the values of ΔT_y and T_m are different for different materials as shown in table 2, an operating temperature range of (20-120) °C has been selected. Such a selection is made because beyond this range, the plot of thermal load resistance cannot be significantly represented in a single plot for comparison. The plot of thermal load resistance $(T_R = \sigma_y / \alpha E \Delta T)$ with temperature rise (ΔT) is provided in figure 1(a) for mild steel, copper, Al alloy 6061 and Al alloy 7075 from (20-120) °C. Since the thermal load resistance of diamond is very much higher as compared to other bar materials, so a separate figure for diamond is provided (figure 1(b)).

The stress and strain fields induced in a clamped taper bar for five different materials under three different uniform thermal loads (ΔT=*30 °C, 50 °C* and *66.8 °C*) are shown in figure 2. The geometry of the bar is assumed to remain same as before ($A_R = 0.025$, and $S_R = 20$). Since the temperature rise at yield of copper is the lowest among all materials, so for comparison purpose three temperatures upto *66.8 °C* is considered for all other materials in the present paper.

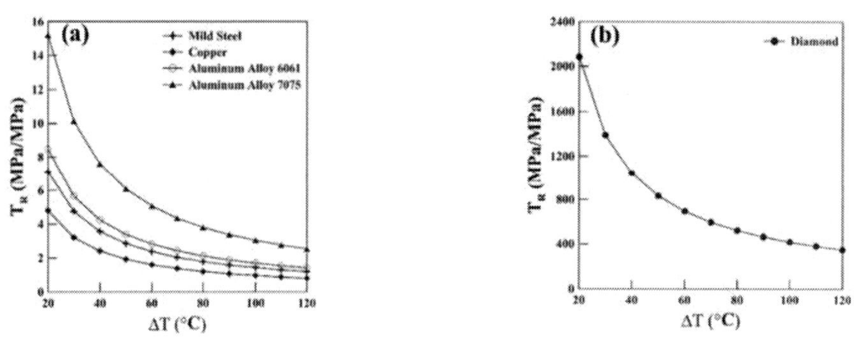

Figure 1(a) & (b). Variation of thermal load resistance (T_R) with rise in temperature (ΔT).

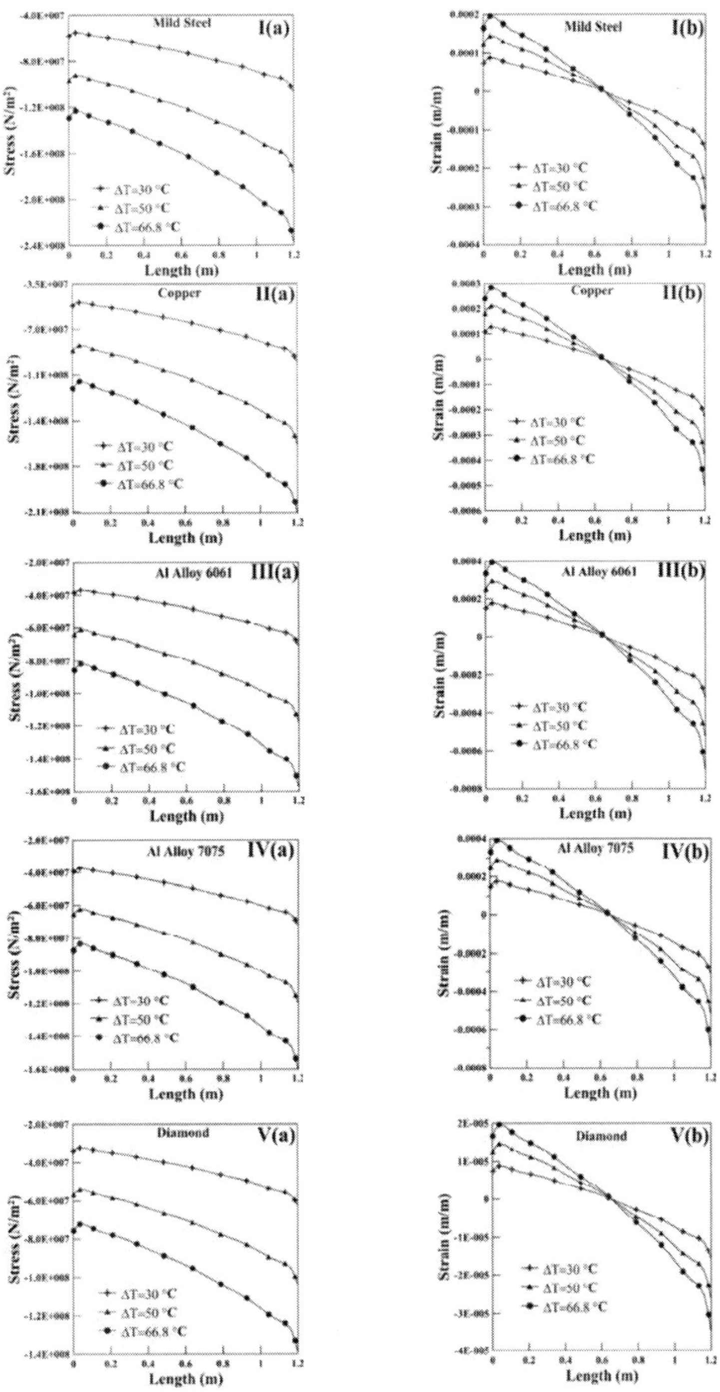

Figure 2. Stress and strain fields induced in different materials as indicated in plots I(a,b), II(a,b), III(a,b), IV(a,b) and V(a,b).

Although the mathematical formulation and the results of figure 2 refer to uniform temperature field, the present analysis can be carried out for various other temperature distributions also. In the present study, we consider linear and parabolic temperature distributions and the temperature boundary conditions at the largest and smallest ends of the bar are assumed as $T(0) = 20\,°C$ and $T(L) = 100\,°C$ respectively. The mathematical relations for linear and parabolic temperature distributions are taken as $T(\xi) = T(0) + (\{T(L) - T(0)\}\xi)$ and $T(\xi) = T(0) + (\{T(L) - T(0)\}\xi^2)$.

The stress and strain fields induced in a clamped mild steel taper bar for linear and parabolic temperature distributions are shown in figure 3 and the plots of elastic, thermal and total strain for uniform, linear and parabolic temperature distribution is shown in figure 4.

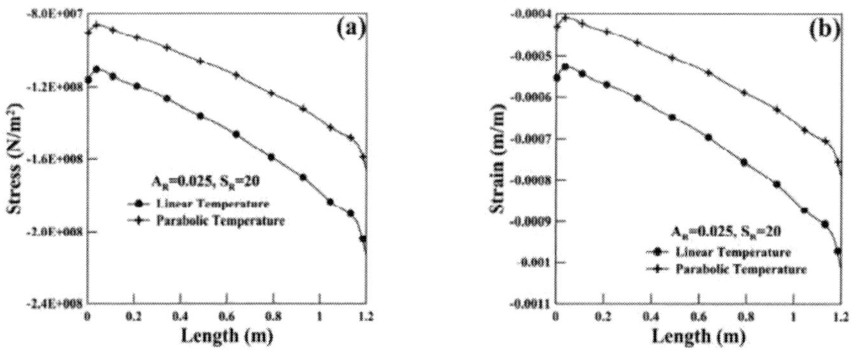

Figure 3. Stress and strain fields induced under linear and parabolic temperature distribution in a clamped mild steel taper bar.

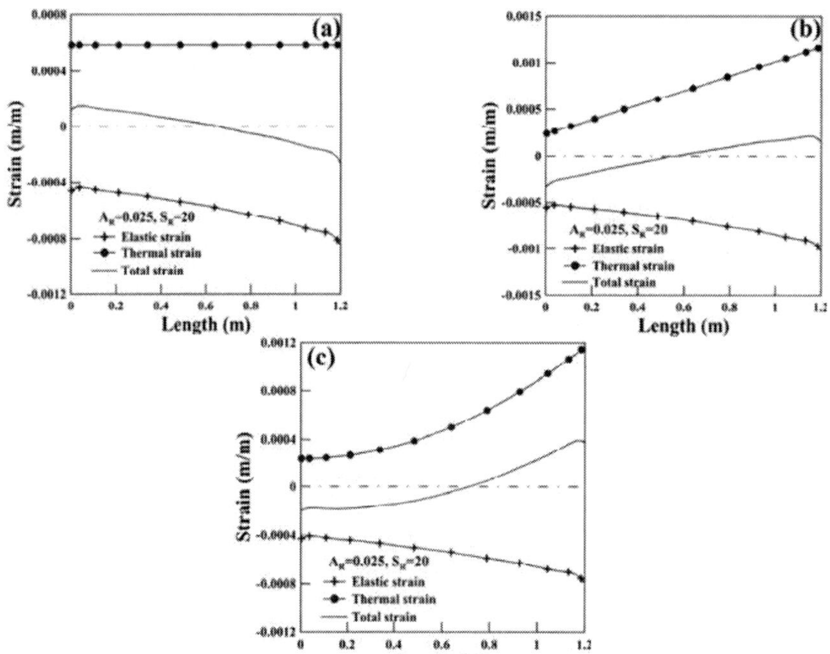

Figure 4. Elastic, thermal and total strain induced in a clamped mild steel taper bar for (a) uniform, (b) linear and (c) parabolic temperature distribution.

The stress and strain field as mentioned here is upto the elastic limit of the bar material. However, the present method has the potential of application in structural mechanics problems involving material non-linearity. A sample result obtained from the solution of such a problem is given in figure. 5 for the mild steel bar, assuming bilinear post-elastic material behaviour. It is observed that the stress and strain field remains similar but changes in their pattern becomes more prominent when the bar material enters in post-elastic domain with increase in thermal load.

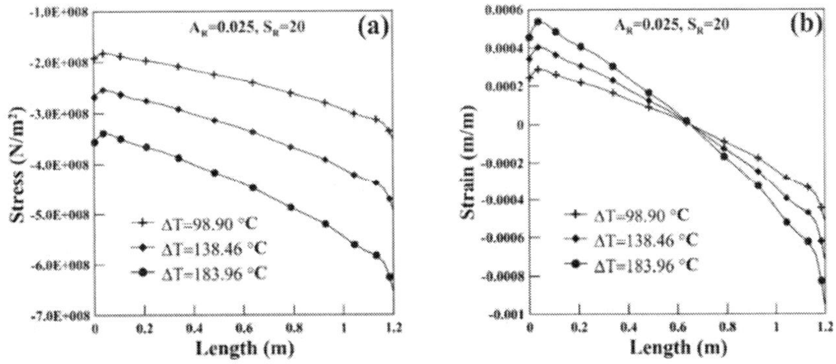

Figure 5. Stress and strain fields induced in a clamped mild steel taper bar at and above yield temperature.

4. Conclusion

The investigation of thermoelastic stress and strain fields of clamped taper bar is formulated through a variational method for different bar materials. The effect of uniform thermal load on different bar materials has been reported. Assuming a series approximation following Galerkin's principle, the solution of the governing partial differential equation is obtained. The solution algorithm is implemented with the help of MATLAB computational simulation software. The effect of non-uniform thermal load and material non-linearity in a clamped mild steel taper bar has also been reported. Finally, the paper paves the way towards design optimization, where a uniform stress field is sought in the multi-parameter domain of the problem.

References

[1] Timoshenko S 1930 Strength of Materials: Part I Elementary Theory and Problems (New York, D. Van Nostrand Company, Inc).

[2] Hill R 1950 The Mathematical Theory of Plasticity (Oxford University Press, Oxford).

[3] Niknam H, Fallah A and Aghdam M M 2014 Nonlinear bending of functionally graded tapered beams subjected to thermal and mechanical loading *Int. J. Nonlinear Mech.* **65** 141-147.

[4] Das D, Sahoo P and Saha K N 2009 Dynamic analysis of non-uniform taper bars in post-elastic regime under body force loading *Appl. Math. Model.* **33** 4163-4183.

[5] Noda N 1991 Thermal stress in materials with temperature-dependent properties *Appl. Mech. Rev.* **44** 383-397.

[6] Grysa K and Kozlowski Z 1982 One dimensional problems of temperature and heat flux determination at the surfaces of a thermoelastic slab: Part I The analytical solutions *Nucl. Engrg. Des.* **74** 1-14.

[7] Nayak P and Saha K N 2014 Analytical study on elastic limit speed of non-uniform rotating disks considering thermal effect on elasticity modulus *Appl. Mech. Mater.* 592-594 (DMMIR) 1001-1005.

On the pertinence to Physics of random walks induced by random dynamical systems: a survey

Dimitri Petritis

Institut de recherche mathématique, Université de Rennes 1, Campus de Beaulieu, 35042 Rennes Cedex, France

E-mail: `dimitri.petritis@univ-rennes1.fr`

Abstract.
Let $\mathbb{X}$ be an abstract space and $\mathbb{A}$ a denumerable (finite or infinite) alphabet. Suppose that $(p_a)_{a\in\mathbb{A}}$ is a family of functions $p_a : \mathbb{X} \to \mathbb{R}_+$ such that for all $x \in \mathbb{X}$ we have $\sum_{a\in\mathbb{A}} p_a(x) = 1$ and $(S_a)_{a\in\mathbb{A}}$ a family of transformations $S_a : \mathbb{X} \to \mathbb{X}$. The pair $((S_a)_a, (p_a)_a)$ is termed an *iterated function system with place dependent probabilities*. Such systems can be thought as generalisations of random dynamical systems. As a matter of fact, suppose we start from a given $x \in \mathbb{X}$; we pick then randomly, with probability $p_a(x)$, the transformation S_a and evolve to $S_a(x)$. We are interested in the behaviour of the system when the iteration continues indefinitely.

Random walks of the above type are omnipresent in both classical and quantum Physics. To give a small sample of occurrences we mention: random walks on the affine group, random walks on Penrose lattices, random walks on partially directed lattices, evolution of density matrices induced by repeated quantum measurements, quantum channels, quantum random walks, etc.

In this article, we review some basic properties of such systems and provide with a pathfinder in the extensive bibliography (both on mathematical and physical sides) where the main results have been originally published.

1. Introduction and motivation

The (discrete) time evolution of several phenomena in (classical and quantum) physics, biology, ecology, economics, etc. are modelled by *dynamical systems*. Mathematically, a dynamical system is a map $S : \mathbb{X} \to \mathbb{X}$ from some measurable space $(\mathbb{X}, \mathcal{X})$ into itself. The system starts at time $n = 0$ at some initial state $x_0 \in \mathbb{X}$; the next moment $n = 1$, the system gets at state $x_1 = S(x_0)$ and recursively for all (integer) instants $n \in \mathbb{N}$, we have $x_{n+1} = S(x_n)$. At this level, we don't specify further either the nature of the space $\mathbb{X}$ beyond its measurability properties (very often it is a topological or metric space) or the nature of the transformation S (very often it exhibits some non-linearity). Nevertheless, we mention that even very innocent looking examples on gentle spaces — like iteration of the logistic map $S(x) = 4x(1-x)$ on $\mathbb{X} = [0,1]$ — can exhibit a tremendously complicated behaviour (see [1] for instance, for a user-friendly introduction to the topic).

When dealing with realistic systems, it is natural to suppose that the evolution is influenced by external parameters, collectively denoted by some symbol $a \in \mathbb{A}$, i.e. instead of having a single evolution transformation S, we have a collection $(S_a)_{a\in\mathbb{A}}$, parametrised by $\mathbb{A}$. Now, on a realistic system the control of external parameters can never be complete; to model such imperfect knowledge, we introduce a randomness on the parameters a. Thus, the evolution

Content from this work may be used under the terms of the Creative Commons Attribution 3.0 licence. Any further distribution of this work must maintain attribution to the author(s) and the title of the work, journal citation and DOI.
Published under licence by IOP Publishing Ltd

is described by a random dynamical system, when the system at moment n is in state x, the applied transformation S_a is chosen among the possible ones with probability $p_a(x)$; the evolution is described by the Markovian kernel

$$P(x, B) := \mathbb{P}(X_{n+1} \in B | X_n = x) = \sum_{a \in \mathbb{A}} p_a(x) \delta_{S_a(x)}(B), \quad \text{for } B \in \mathcal{X}. \tag{1}$$

Such systems, in their simplest form with place independent probabilities (i.e. with p_a constants), have been introduced — and termed iterated function systems — in [18] where the conditions of the existence of an attracting set supporting their invariant measure have been given. The full-fledged place-dependent variant has been introduced in [2] when $\mathbb{X}$ is a metric space; the fractal dimension of the attractor has been obtained under the condition that the place-dependent probabilities are Lipschitz-continuous. Conditions of asymptotic stability of such systems are obtained in [19] (extending previous results in [24]).

Although these systems are very versatile, can model numerous phenomena, and exhibit very rich a structure in numerical simulations, rigorous mathematical results are lacking when the place-dependent probabilities are not continuous. (A very partial case is treated in [20]).

It is worth noting that the joint process $(A_n, X_n)_{n \in \mathbb{N}}$ taking values in $\mathbb{A} \times \mathbb{X}$ is also a Markov chain with transition probability kernel $Q((a, x), (b, B)) := \mathbb{P}((A_{n+1} = b, X_{n+1} \in B | A_n = a, X_n = x) = p_b(x) \delta_{S_b(x)}(B)$ for $a, b \in \mathbb{A}$ and $B \in \mathcal{X}$; the kernel P, introduced in equation (1), is obtained from the kernel Q by marginalisation when the $\mathbb{A}$-component is not observed. On the contrary, the marginalisation of Q where the $\mathbb{X}$-component is not observed does not — in general — give rise to a Markov chain but only to what is called a *hidden Markov chain*. A hidden Markov chain can become an ordinary Markov chain under some specific circumstances (for instance when the p_a's are constant). Let us mention finally that the restriction of $\mathbb{A}$ being denumerable can be removed is some cases.

In the sequel, we present some models fitting this formalism.

2. Random walks on lattices

2.1. Simple random walks on directed sublattices of $\mathbb{Z}^d$

Let $\mathbb{X} = \mathbb{Z}^d$, $\mathbb{A} = \{\pm e_1, \ldots, \pm e_d\} \subset \mathbb{X}$, and for any $a \in \mathbb{A}$ and any $x \in \mathbb{X}$, define $S_a(x) = x + a$. If the functions $p_a(x) = (2d)^{-1}$ for all a and x, the process (X_n) governed by (1) is equivalent to the simple symmetric random walk on $\mathbb{Z}^d$, introduced, almost a century ago, in the celebrated paper of Pólya [33]. A lot of properties of simple random walks can be subsumed under their recurrence/transience behaviour; already in [33] it has been established that the simple symmetric random walk on $\mathbb{Z}^d$ returns almost surely infinitely often (it is recurrent) to its starting point in $d \leq 2$ but almost surely only a finite number of times (it is transient) in $d \geq 3$. It is worth noting that interesting connections of random walks on $\mathbb{Z}^d$ with computational trajectories of finite automata or with properties of (Abelian) groups can be made. Let us mention also that simple random walks serve in modelling a plethora of physical phenomena: every time a physical phenomenon is described by a differential equation involving a Laplacian, one can coin a random walk representation of the phenomenon.

It is quite surprising that although random walks on $\mathbb{Z}^d$ have a centennial history, random walks on directed sublattices of $\mathbb{Z}^2$ had not been mathematically studied until 2003. These walks are more realistic if one wishes to model information flow on internet or vehicle traffic in urban networks for instance; they have been considered by hydrologists [28] and a related model has been heuristically studied by computer simulations in [34] but no mathematical result was known before [6, 7]. They are obtained with $\mathbb{A}$ and $\mathbb{X}$ as in the simple random walk on $\mathbb{Z}^d$ case but with different choices of place-dependent p_a's. Recall that $\mathbb{X} = \mathbb{Z}^2 = \mathbb{Z} \times \mathbb{Z}$, hence every $x \in \mathbb{Z}^2$ can be written as a pair $\mathbf{x} = (x_1, x_2) \in \mathbb{Z} \times \mathbb{Z}$. Three different models of partially directed lattice, depicted in the figure 1, have been studied in [6]. These directed lattices arise as sublattices of $\mathbb{Z}^2$ by rendering their horizontal lines unidirectional.

Alternate lattice	Half-plane one-way	Randomly horizontally directed
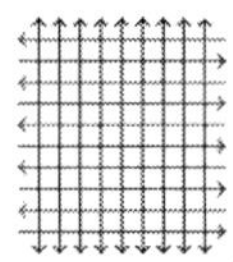	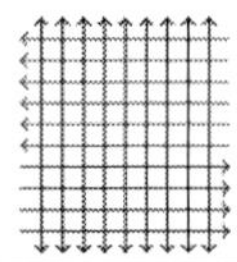	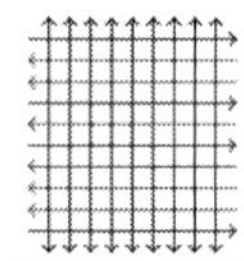
$p_a(\mathbf{x}) = \begin{cases} 1/3 & \text{if } a = \pm e_2 \\ 1/3 & \text{if } a = e_1 \text{ and } x_2 \in 2\mathbb{Z} \\ 1/3 & \text{if } a = -e_1 \text{ and } x_2 \in 2\mathbb{Z}+1 \\ 0 & \text{otherwise.} \end{cases}$	$p_a(\mathbf{x}) = \begin{cases} 1/3 & \text{if } a = \pm e_2 \\ 1/3 & \text{if } a = e_1 \text{ and } x_2 < 0 \\ 1/3 & \text{if } a = -e_1 \text{ and } x_2 \geq 0 \\ 0 & \text{otherwise.} \end{cases}$	$p_a(\mathbf{x}) = \begin{cases} 1/3 & \text{if } a = \pm e_2 \\ 1/3 & \text{if } a = \sigma_{x_2} e_1 \\ 0 & \text{otherwise.} \end{cases}$

Figure 1: Three different lattices having one-way horizontal lines and the corresponding probability functions p_a. (i) In the alternate lattice the horizontal lines are going eastward or westward according to the parity of the ordinate. (ii) In the half-plane lattice, all upper half-plane lines are westward and all lower half-plane eastward. (iii) In the random horizontal lattice a honest coin is flipped to decide the way each line is going and this choice is encoded into $\{-1, 1\}$-valued random variable σ_{x_2}.

We proved that the random walk on the alternate lattice is recurrent, on the half-plane one-way is transient, and on the randomly directed lattice almost surely transient. It is worth noting that all the above lattice have zero net drift; the dramatic change of behaviour in the recurrence/transience properties is due solely to the directedness of the lattice.

As expected, several new results have been triggered by this work (see [14, 15, 30, 32, 11, 13, 8] for instance). It is worth noting that from the above models only (i) can be obtained by the computational trajectories of a finite automaton; model (ii) requires a push-down automaton while model (iii) a genuine Turing machine with bi-infinite external tape. Additionally, the set of all possible trajectories looses the natural group structure to remain merely a semi-groupoid.

2.2. *Random walks on Penrose tilings*

Another non-trivial model that fits the aforementioned description is the random walk on the complexes obtained by the edges of Penrose tilings. Introduced first as a mathematical recreational problem, such tilings proved essential in the crystallographic description of some alloys (like Al-Mn, Ho-Mg-Zn) that exhibit 5-fold and icosahedral symmetries — forbidden by classical crystallography — in their X-ray diffraction patterns[1]. Recall that Penrose tiling of $\mathbb{R}^d$ (see figure 2) is an aperiodic covering of the space by finitely many types of prototiles. Although the initial construction of Penrose [31] for $d = 2$ was made by use of local matching rules, a cut-and-project method giving a systematic construction for every d has been introduced in [21, 29].

The idea of this cut-and-project method is to decompose $\mathbb{R}^N = E \oplus E'$ into a direct sum of mutually orthogonal Euclidean spaces E and E' with $N > d$, $\dim E = d$ and $\dim E' = N - d$. When the unit cube $C = \{y \in \mathbb{R}^N : y = \sum_{i=1}^{N} y_i e_i, y_i \in [0,1]\}$ is translated parallel to E it forms the "strip" $\mathcal{S} = C + E$. All points of the integer lattice $\mathbb{Z}^N$ within the strip $\mathcal{S}$ are then orthogonally projected on E. When the principal directions of E are incommensurate with $\mathbb{Z}^N$, the points in the strip are in bijection with their orthogonal projections on E or E'. When projected on E, they produce an aperiodic tiling of E by finitely many types of prototiles. To illustrate this method we give in the figure 2 below this construction for the aperiodic tiling

[1] This experimental observation [35] was essential for awarding the 2011 Nobel prize in Chemistry to Dan Shechtman.

of the line by short and long segements, obtained by projecting a strip in $\mathbb{Z}^2$ on a line with irrational slope and the tiling of the plane by projecting the strip in $\mathbb{Z}^5$ with respect to a given irrationally oriented plane.

Figure 2: *Left:* The aperiodic tiling of the one-dimensional space E by short and long segments. The shaded region within the two parallel oblique lines is the strip $\mathcal{S}$. The tiling is obtained by projecting on E the horizontal and vertical segments of the line zigzagging inside the strip. *Centre:* The aperiodic tiling of $\mathbb{R}^2$ by 2 different types of lozenges, each coming into 5 different orientations to give rise in 10 types of prototiles arranged in a pattern having locally a 5-fold symmetry (Public domain image, source: Wikipedia) *Right:* X-ray experimental diffraction pattern on a Zn-Mg-Ho quasicrystal showing a clear local 5-fold symmetry. (Image under Creative Commons BY-SA 3.0 licence, source: Wikipedia).

The projected figure constitutes the Penrose tiling of the plane. The corners and the edges of the lozenges define a lattice in $\mathbb{R}^2$. The simple random walk on a given vertex at time n, jumps with equal probability to a nearest-neighbour (in the graph distance defined by the edges) at moment $n+1$. It is worth noting that in the cut-and-project method, the parameter N can be chosen arbitrarily large (but finite): some vertices of the lattice can have thus arbitrarily large coordination number. In spite of that, in [37], results on the heat kernel of the simple random walk on a Penrose lattice have been obtained; in [10] the random walk is shown to be generically recurrent for $d \leq 2$ and generically transient for $d \geq 3$.

Remark also that the trajectories of the random walk can be shown to be in bijection with the computational trajectories of Turing machine; additionally, they have a natural groupoid structure but not that of a group. The random walk on the Penrose lattice is isomorphic to a random walk fitting the iterated function systems with place-dependent probabilities formalism by choosing $\mathbb{A} = \{\pm e_1, \ldots, \pm e_N\}, \mathbb{X} = \mathbb{Z}^N$ and

$$p_a(x) = \begin{cases} \frac{1}{d(x)} & \text{if } x + a \in \mathcal{S} \\ 0 & \text{otherwise,} \end{cases}$$

where $d(x) := \sharp\{b \in \mathbb{A} : x + b \in \mathcal{S}\}$, for all $x \in \mathcal{S}$. Since the integer points in the strip are in bijection with the corners of tiles in E, any trajectory of the random walk in the strip is bijectively projected to a trajectory on edges of the tiling.

3. Random walks on (semi-)groups

3.1. *The "$ax+b$" group*

In [36, 23] problems stemming from random walks in random envrionment have been studied. It appeared then that a random process $(X_n)_{n\in\mathbb{N}}$ on the set $\mathbb{X}$ of d-dimensional vectors occurred naturally. If $\mathbb{A}$ is a fixed set of $d \times d$ matrices and of d-dimensional vectors, the evolution of the process consists in choosing randomly a pair $a = (m, v) \in \mathbb{A}$ (i.e. a pair of a matrix m and a vector v and define $X_{n+1} = S_a(X_n)$ where $S_a = mx + v$. Obviously, this evolution fits the general formalism given in the introduction. In case the matrices m and vectors v have positive components, the asymptotic behaviour of X_n can be obtained in a more straightforward manner [22]. Connection of this process with a one-dimensional *disordered model of statistical mechanics* has been established in [12], while in [9], the asymptotic behaviour of the model has been shown be connected with the tail behaviour of the invariant measure of the process (X_n) by using Mellin transform. Same ideas have been developed in [16, 5], while in [17], a highly non trivial generalisation to the case of not necessarily positive matrices has been made.

3.2. Random dynamical systems on $\mathbb{R}_+$

In [3], we studied systems with $\mathbb{A} \simeq \mathbb{Z}$, $\mathbb{X} = \mathbb{R}_+$ and $S_a(x) = (a + x \pm x^\gamma)_+$ where γ is some fixed power in $(0, 1)$. When the random variable $A \in \mathbb{A}$ has moments, nothing very interesting occurs since the asymptotic behaviour of random dynamical system is governed by the systematic drift term x^γ. An interesting situation occurs when the random variable A is distributed according to a law having heavy tails, i.e. the distribution of A does not decay sufficiently fast for large values of $|A|$ for it to have expectation. More precisely, we suppose that $\mathbb{P}(|A| > y) \asymp \frac{C}{y^\alpha}$, with $\alpha \in (0, 1)$. A complete characterisation of the type diagram is obtained exhibiting a competition between the heavy-tail behaviour and the systematic drift, i.e. we obtain a non trivial type diagram in the (α, γ) plane.

4. Evolution of density matrices under repeated unsharp quantum measurements

The state of a quantum system is described by a *density operator*, i.e. a positive, trace-class and unit-trace operator x, acting on a separable Hilbert space $\mathbb{H}$; this set is denoted $\mathfrak{D}(\mathbb{H})$ and plays the *rle* of $\mathbb{X}$ in this quantum setting. When the system is isolated, a standard result in quantum mechanics states that the evolution is through a unitary operator U acting on x by U^*xU. When a sharp measurement of a quantity Q (represented by a self-adjoint operator acting on $\mathbb{H}$ and supposed — for simplicity — with discrete spectrum in this survey) is performed, the quantum mechanics postulates establish that the value a occurs with probability $\text{tr}(xE[a])$ where $E[a]$ denotes the spectral measure of Q at a; the family of $(E[a])_{a\in\mathbb{A}}$ is a family of orthogonal orthoprojections. After having obtained a value $a \in \mathbb{A}$, the state of the system evolves irreversibly to a new density operator $x' = S_a(x) = \frac{E[a]xE[a]}{\text{tr}(xE[a])}$. In this setting, the set $\mathbb{A}$ can be chosen to be the spectrum of Q and the probability of occurrence of every a is $p_a(x) = \text{tr}(xE[a]) = \text{tr}(E[a]xE[a]) = \text{tr}(E[a]^*xE[a])$.

Unsharp quantum measurements are families of positive operators $(E[a])_{a\in\mathbb{A}}$, verifying $\sum_{a\in\mathbb{A}} E[a] = I$ but the members $E[a]$ of the family are not orthogonal orthoprojections. Since they are positive however, there exists a family of operators $(Z[a])$ such that $E[a] = Z[a]^*Z[a]$ (they are called *Kraus operators* in the literature). The probability of getting the value a in this unsharp measurement is $p_a(x) = \text{tr}(xE[a]) = \text{tr}(xE[a]) = \text{tr}(Z[a]xZ[a]^*)$ while the state of the system evolves into the new state $x' = S_a(x) = \frac{Z[a]xZ[a]^*}{\text{tr}(Z[a]xZ[a]^*)}$. Therefore, this problem also fits the general formalism introduced above (see [4, 26] for instance) with $\mathbb{X} = \mathfrak{D}(\mathbb{H})$. It is remarkable that repeated quantum measurements induce a classical Markov chain on the space $\mathbb{X} = \mathfrak{D}(\mathbb{H})$.

In [27], the asymptotic behaviour of the thus induced Markov chain has been obtained for the case where $\mathbb{A}$ is finite while in [25] these results have been extended to the case of denumerably infinite $\mathbb{A}$.

5. Conclusion and open problems

We have demonstrated that there exists a unified formalism encompassing very diverse models that have been previously introduced in various areas of physics (statistical mechanics, disordered media, quantum mechanics, etc.). This formalism puts together mathematical methods developed somehow separately in the context of dynamical systems and/or Markov processes and as such has a very rich mathematical structure. In spite of the vast applicability of the formalism, very little is known about existence and uniqueness of the invariant measure for the random dynamical system described by the Markov kernel of equation (1) when the probabilities $(p_a)_{a \in \mathbb{A}}$ are genuinely place-dependent and this dependence is not continuous. This constitutes a challenge in the theoretical understanding of these systems. On the physical side, the existence of an underlying unified framework can be used to transfer methods developed ad-hoc in each area to the other. Within every model presented above, only some partial aspects are understood. We expect that the transfer of methods from one area to the other will help answering some of the remaining questions.

References

[1] Kathleen T. Alligood, Tim D. Sauer, and James A. Yorke. *Chaos*. Textbooks in Mathematical Sciences. Springer-Verlag, New York, 1997. An introduction to dynamical systems.

[2] M. F. Barnsley, S. G. Demko, J. H. Elton, and J. S. Geronimo. Invariant measures for Markov processes arising from iterated function systems with place-dependent probabilities. *Ann. Inst. H. Poincaré Probab. Statist.*, 24(3):367–394, 1988.

[3] Vladimir Belitsky, Mikhail Menshikov, Dimitri Petritis, and Marina Vachkovskaia. Random dynamical systems with systematic drift competing with heavy-tailed randomness. 2015. Accepted for publication 2016; ArXiv 1407.3475.

[4] Ph. Blanchard, A. Jadczyk, and R. Olkiewicz. Completely mixing quantum open systems and quantum fractals. *Phys. D*, 148(3-4):227–241, 2001.

[5] Dariusz Buraczewski, Ewa Damek, Yves Guivarc'h, Andrzej Hulanicki, and Roman Urban. Tail-homogeneity of stationary measures for some multidimensional stochastic recursions. *Probab. Theory Related Fields*, 145(3-4):385–420, 2009.

[6] Massimo Campanino and Dimitri Petritis. Random walks on randomly oriented lattices. *Markov Process. Related Fields*, 9(3):391–412, 2003.

[7] Massimo Campanino and Dimitri Petritis. On the physical relevance of random walks: an example of random walks on a randomly oriented lattice. In *Random walks and geometry*, pages 393–411. Walter de Gruyter, Berlin, 2004.

[8] Massimo Campanino and Dimitri Petritis. Type transition of simple random walks on randomly directed regular lattices. *J. Appl. Probab.*, 51(4):1065–1080, 2014.

[9] Claude de Calan, Jean-Marc Luck, Theo M. Nieuwenhuizen, and Dimitri Petritis. On the distribution of a random variable occurring in 1D disordered systems. *J. Phys. A*, 18(3):501–523, 1985.

[10] Basile de Loynes. *Graphes et marches aléatoires*. Thèse, Université Rennes 1, July 2012. tel-00726483.

[11] Basile de Loynes. Marche aléatoire sur un di-graphe et frontière de Martin. *C. R. Math. Acad. Sci. Paris*, 350(1-2):87–90, 2012.

[12] B. Derrida and H. J. Hilhorst. Singular behaviour of certain infinite products of random 2×2 matrices. *J. Phys. A*, 16(12):2641–2654, 1983.

[13] Alexis Devulder and Françoise Pène. Random walk in random environment in a two-dimensional stratified medium with orientations. *Electron. J. Probab.*, 18:no. 18, 23, 2013.

[14] N. Guillotin-Plantard and A. Le Ny. Transient random walks on 2D-oriented lattices. *Teor. Veroyatn. Primen.*, 52(4):815–826, 2007.

[15] Nadine Guillotin-Plantard and Arnaud Le Ny. A functional limit theorem for a 2D-random walk with dependent marginals. *Electron. Commun. Probab.*, 13:337–351, 2008.

[16] Yves Guivarc'h. Sur une extension de la notion de loi semi-stable. *Ann. Inst. H. Poincaré Probab. Statist.*, 26(2):261–285, 1990.

[17] Yves Guivarc'h and Émile Le Page. Spectral gap properties and Pareto's asymptotics for affine random walks. Acccepted for publication in Ann. Inst. Henri Poicar'e (2016), 2016.

[18] John E. Hutchinson. Fractals and self-similarity. *Indiana Univ. Math. J.*, 30(5):713–747, 1981.

[19] Joanna Jaroszewska. How to construct asymptotically stable iterated function systems. *Statist. Probab. Lett.*, 78(12):1570–1576, 2008.

[20] Joanna Jaroszewska. A note on iterated function systems with discontinuous probabilities. *Chaos Solitons Fractals*, 49:28–31, 2013.

[21] André Katz and Michel Duneau. Quasiperiodic patterns. *J. Physique*, 46(12, Suppl. Colloq. C8):C8–31–C8–40, 1985.

[22] H. Kesten, M. V. Kozlov, and F. Spitzer. A limit law for random walk in a random environment. *Compositio Math.*, 30:145–168, 1975.

[23] Harry Kesten. Random difference equations and renewal theory for products of random matrices. *Acta Math.*, 131:207–248, 1973.

[24] Andrzej Lasota and Michael C. Mackey. *Chaos, fractals, and noise*, volume 97 of *Applied Mathematical Sciences*. Springer-Verlag, New York, second edition, 1994. Stochastic aspects of dynamics.

[25] Bunrith Jacques Lim. *Poisson boundaries of quantum operations and quantum trajectories*. Thèse, Université Rennes 1, November 2010. tel-20111102.

[26] Artur Łoziński, Karol Życzkowski, and Wojciech Słomczyński. Quantum iterated function systems. *Phys. Rev. E (3)*, 68(4):046110, 9, 2003.

[27] Hans Maassen and Burkhard Kümmerer. Purification of quantum trajectories. In *Dynamics & stochastics*, volume 48 of *IMS Lecture Notes Monogr. Ser.*, pages 252–261. Inst. Math. Statist., Beachwood, OH, 2006.

[28] G. Matheron and G. de Marsily. Is transport in porous media always diffusive? a counter-example. *Water Resour. Res.*, 16:901–917, 1980.

[29] Christophe Oguey, Michel Duneau, and André Katz. A geometrical approach of quasiperiodic tilings. *Comm. Math. Phys.*, 118(1):99–118, 1988.

[30] Françoise Pène. Transient random walk in $\mathbb{Z}^2$ with stationary orientations. *ESAIM Probab. Stat.*, 13:417–436, 2009.

[31] Roger Penrose. Pentaplexity: a class of nonperiodic tilings of the plane. *Math. Intelligencer*, 2(1):32–37, 1979/80.

[32] Gábor Pete. Corner percolation on $\mathbb{Z}^2$ and the square root of 17. *Ann. Probab.*, 36(5):1711–1747, 2008.

[33] Georg Pólya. Über eine Aufgabe der Wahrscheinlichkeitsrechnung betreffend die Irrfahrt im Straßennetz. *Math. Ann.*, 84(1-2):149–160, 1921.

[34] Sidney Redner. Survival probability in a random velocity field. *Phys. Rev. E*, 56:4967–4972, 1997.

[35] D. Shechtman, I. Blech, D. Gratias, and J. W. Cahn. Metallic phase with long-range orientational order and no translational symmetry. *Phys. Rev. Lett.*, 53:1951–1953, Nov 1984.

[36] Fred Solomon. Random walks in a random environment. *Ann. Probability*, 3:1–31, 1975.

[37] András Telcs. Diffusive limits on the Penrose tiling. *J. Stat. Phys.*, 141(4):661–668, 2010.

Statistical Research for Probabilistic Model of Distortions of Remote Sensing

Iskakova Ayman
Mechanic and Mathematic Department, L.N. Gumilyov Eurasian Nathional University, Mirzoyan st, 2, Astana, Kazakhstan
E-mail: ayman.astana@gmail.com

Abstract. In this work the new multivariate discrete probability model of distribution of processes distortion of radiation from remote sensing data is proposed and studied. Research was performed on a full cycle adopted in mathematical statistics, namely, the model was constructed and investigated, various methods for estimating the parameters was proposed and test the hypothesis that the model adequacy observations, was considered.

1. Introduction

On the Earth's near-Earth orbit for the purpose of remote sensing of the surface today launched a variety of satellites (Terra, Aqua, Suomi NPP, etc.) with the sensors, which transmit to the monitoring station huge data flows (see [1], p. 28). Problems with their storage and their processing in real time makes very urgent to study the probability of information about distortions.

In general, the processing of data from the satellite may be viewed as a conveyor. This technology can be shown as partitioning of data by the time, by the sensor type and by range. It is known that incoming files are passing sequential processing of.

International Satellite System based on geostationary spacecraft is designed to meet the challenges of the global meteorological support and for consumers in the European , Asian and African regions. The management and maintenance work on the program is implemented by Eumetsat.

The remote sensing (RS) is an observation of the surface of the Earth aviation and space vehicles equipped with different types of imaging equipment. Wavelength range, which is receiving apparatus, is constitutes from a fraction of micrometer (visible optical radiation) to m (radio waves). The methods can be passive, that is to use a natural reflection or secondary heat radiation of objects on the Earth's surface due to solar activity, and active, these are using stimulated emission objects initiated by artificial light source directed action. The datas of RS, which is received from the spacecraft, is characterized by a large degree of depending on the transparency of the atmosphere. So on spacecraft equipment the multichannel passive and active types are used, which have recording of electromagnetic radiation in different ranges.

RS can be represented as a process by which information is gathered about an object, area or phenomenon without direct contact with him. Part of RS are entering in the digital form that allows you to use them for processing modern computer technology. Pictures on photo can be

Content from this work may be used under the terms of the Creative Commons Attribution 3.0 licence. Any further distribution of this work must maintain attribution to the author(s) and the title of the work, journal citation and DOI.
Published under licence by IOP Publishing Ltd

converted into a digital format for a bitmap using special scanning devices (scanners) . Digital image in the form of a raster is a matrix of numbers.

RS contain a series of random, systemic and systematic distortions due to the influence of the atmosphere, the Earth's curvature, motion shooting the vehicle with respect to its surface at the time of the shooting, the physical characteristics of sensors used and communication channels. To eliminate the mentioned quite numerous distortions, taking into account their specific uses several types of correction: radiation, radiometric, and geometric calibration.

Objectives of the study of: a) submission of a probabilistic model processes distortion RS; b) the construction of unbiased estimates of the probability distribution of the model under investigation. All major results are new and are as follows: a) proposed and studied by multivariate discrete probability model for the distribution of sums of unobservable independent identically distributed random matrices; b) it is shown that the best unbiased estimate of the probability of this distribution exist only if the uniqueness of the partition of the observed matrix of all possible components. c) if some element of the implementation of the sample has more than one partition on the possible components, then we construct a set of unbiased estimators studied for distribution; d) introduce the concept of the most appropriate estimate of the set of unbiased, which has good asymptotic properties.

The results and methods presented in the paper can be used in research specialists in Mathematical Statistics and in Space Monitoring .

Telemetry data transmitted from the Board of the satellite at a frequency of 1675.928 MHz. To send telemetry (see [2]) using directional antenna S-band high-gain, and at the stage of withdrawal satellite into orbit - directional antenna S-range frequencies. Command information is taken at a frequency of 2098.0 MHz using a directional antenna S-band high-gain.

On the energy characteristics of the radio transmission of data in HR with the IRS series of Meteosat satellites (see [3], p. 205) affect the EIIM, loss of distribution losses in the antenna pointing (± 1), gain receiving antenna with a mirror diameter of 4 m , losses in the feeder lines, return loss, received power $A+2+3+4+5+6$, modulation loss, the effective received signal $G+8$, the noise temperature of antenna elevation angle 30°, the noise temperature, system noise temperature $A0+11$, noise spectral density $198.6+A2$, data transmission speed 166 kbit/second, the total noise power of the $A3+A14$ and so on. In other words, the distortion of the influence factor of 23, ie, $d = 23$.

2. Multivariate discrete probability distribution processes remote sensing

Consider a probabilistic model of the processes of energy characteristics of radio satellite IRS Meteosat. In [4], a probabilistic-statistical model of weather prediction.

Assume that the true image can be represented as a matrix $\mathbf{l}_0 = \|l_{0_{i,j}}\|_{m\times q}$, which imposed distortion, consisting of four factors (matrices) of losses $\mathbf{u} = \|u_{i,j}\|_{m\times q}$, taking values from the set of $\mathbf{l}_1, \mathbf{l}_2, \ldots,, \mathbf{l}_d$.

Obviously, the factors (the matrix), the loss $\mathbf{l}_1, \mathbf{l}_2, \ldots,, \mathbf{l}_d$ are realizations of random matrices $\mathbf{L}_1, \mathbf{L}_2, \ldots,, \mathbf{L}_d$, which appear relevant probabilities $\mathbf{p} = (p_1, \ldots, p_d)$, where

$$\sum_{\alpha=1}^{d} p_\alpha = 1.$$

Assume that $V_\mathbf{u}$ is the number of possible combinations $r_{1_{v_\mathbf{u}}}\mathbf{L}_1, \ldots, r_{d_{v_\mathbf{u}}}\mathbf{L}_d$, which together form a matrix of $\mathbf{u}$, where $r_{1_{v_\mathbf{u}}}, \ldots, r_{d_{v_\mathbf{u}}}$ determine the possible number of balls taken out, which marked the relevant matrices $\mathbf{L}_1, \mathbf{L}_2, \ldots,, \mathbf{L}_d$. In other words, from [5] it follows that $V_\mathbf{u}$ is the number of partitions on the part of the matrix $\mathbf{u}$ on $\mathbf{L}_1, \mathbf{L}_2, \ldots,, \mathbf{L}_d$.

Theorem 1. *The probability that the distortion takes the value u is determined by the formula*

$$P(\mathbf{U}=\mathbf{u}) = \sum_{v_{\mathbf{u}}=1}^{V_{\mathbf{u}}} n! \prod_{\alpha=1}^{d} \frac{p_\alpha^{r_{\alpha v_{\mathbf{u}}}}}{r_{\alpha v_{\mathbf{u}}}!}. \tag{1}$$

3. Unbiased estimation of the probability distribution of the proposed model

In practice, as a rule, elements of the vector $\mathbf{p} = (p_1, \ldots, p_d)$ are not known. It is also not known matrix $\mathbf{L}_1$, $\mathbf{L}_2$, …,, $\mathbf{L}_d$. Consequently, formula (1) does not find the actual application.

Assume that there are photos in the number of k particular locality with the distortions $\mathbf{x} = \{\mathbf{x}_1, ..., \mathbf{x}_k\}$. In other words, a number of evidence-x can be interpreted as a realization of a sample of k, whose elements are subject to distribution (1). We denote $\mathbf{r}_{v_\beta}$ vector $(r_{1_{v_\beta}}, \ldots, r_{d_{v_\beta}})$, which defines v_β - th solution of equation

$$\begin{cases} \sum\limits_{\alpha=1}^{d} L_\alpha r_{\alpha_{v_\beta}} = \mathbf{u}, \\ \sum\limits_{\alpha=1}^{d} r_{\alpha_{v_\beta}} = n, \end{cases} \tag{2}$$

, where $v_\beta = 1, \ldots, V_\beta, V_\beta$ - the number of partitions of the matrix $\mathbf{x}_\beta$ on the matrices $\mathbf{L}_1$, $\mathbf{L}_2$, …,, $\mathbf{L}_d$. Using the system of equations (3), the matrices $\mathbf{L}_1$, $\mathbf{L}_2$, …,, $\mathbf{L}_d$, and the actual data $\mathbf{x}$, we define for each $\beta = 1, \ldots, k$ the number of partitions V_β matrix $\mathbf{x}_\beta$ at $\mathbf{L}_1$, $\mathbf{L}_2$, …,, $\mathbf{L}_d$, and vectors $\mathbf{r}_{1_\beta}$, …, $\mathbf{r}_{V_\beta}$.

Suppose that for each $j = 1, \ldots, \mu$, where

$$\mu = \prod_{\beta=1}^{k} V_\beta,$$

there is a vector $\mathbf{z}_j = (z_{1_j}, \ldots, z_{d_j})$, defined as

$$\mathbf{z}_j = \sum_{\beta=1}^{k} \mathbf{r}_{v_\beta}, \tag{3}$$

and the indices on the right and left side are linked one-to-one correspondence, which is not unique.

Thus, from the above lemma that if some element of the implementation of the sample $\mathbf{x} = (\mathbf{x}_1, \ldots, \mathbf{x}_k)$ of the distribution (1) has more than one partition on the submitted part, it is impossible, using the theorem Rao - Blackwell-Kolmogorov construct an unbiased estimate with minimum variance for the probability distribution (1).

Theorem 2. *The elements of* $W(\mathbf{u}, \mathbf{z}) = W(\mathbf{u}, \mathbf{z}_1), \ldots, W(\mathbf{u}, \mathbf{z}_\mu)$ *is an unbiased estimate for the probability* $P(\mathbf{U} = \mathbf{u})$ *distribution* (1) *that for* $j = 1, \ldots, \mu$ *is defined as*

$$W(\mathbf{u}, \mathbf{z}_j) = \frac{\sum\limits_{v_{\mathbf{u}}=1}^{V_{\mathbf{u}}} \prod\limits_{\alpha=1}^{d} \binom{z_{\alpha_j}}{r_{\alpha v_{\mathbf{u}}}}}{\binom{nk}{n}}, \tag{4}$$

where $V_{\mathbf{u}}$ *- number of partitions on the part of the matrix* $\mathbf{u}$ $\mathbf{L}_1$, $\mathbf{L}_2$, …,, $\mathbf{L}_d$; *for each partition* $r_{1_{v_{\mathbf{u}}}}, \ldots, r_{d_{v_{\mathbf{u}}}}$ *determine the possible number of matrices* $\mathbf{L}_1$, $\mathbf{L}_2$, …,, $\mathbf{L}_d$; $k \geq 1$ *and* $z_{\alpha_j} \geq r_{\alpha_{v_{\mathbf{u}}}}$, *when* $\alpha = 1, \ldots, d$, $v_{\mathbf{u}} = 1, \ldots, V_{\mathbf{u}}$.

4. The most suitable unbiased estimates for the probability distribution of the proposed model and their properties

Thus, we have a lot of unbiased estimates of the probability of distortion.

Definition 1. *Decision* $\mathbf{z}_g$*, based on observation, is the most appropriate set of* $\mathbf{z} = \{\mathbf{z}_1, \ldots, \mathbf{z}_m\}$*, if*

$$\prod_{\beta=1}^{k} W(\mathbf{x}_\beta, \mathbf{z}_g) = \max_{j=1,\ldots,\mu} \prod_{\beta=1}^{k} W(\mathbf{x}_\beta, \mathbf{z}_j), \tag{5}$$

where for $\beta = 1, \ldots, k$ elements of $W(\mathbf{x}_\beta, \mathbf{z}) = \{W(\mathbf{x}_\beta, \mathbf{z}_1), \ldots, W(\mathbf{x}_\beta, \mathbf{z}_\mu)\}$ *is an unbiased estimate for the probability* $P(\mathbf{U} = \mathbf{u})$ *distribution* (1) *defined in* (5).

Definition 2. *Unbiased estimate of* $W(\mathbf{x}_\beta, \mathbf{z}_g)$ *for the probability* $P(\mathbf{U} = \mathbf{u})$ *distribution* (1) *is the most suitable from the entire set of unbiased estimates of* $W(\mathbf{x}_\beta, \mathbf{z}) = \{W(\mathbf{x}_\beta, \mathbf{z}_1), \ldots, W(\mathbf{x}_\beta, \mathbf{z}_\mu)\}$ *defined in* (5)*, if* $\mathbf{z}_g$ *- the most appropriate solution, based on observation.*

Theorem 3. *The most suitable unbiased estimate of* $W(\mathbf{x}_\beta, \mathbf{z}_g)$ *for the probability* $P(\mathbf{U} = \mathbf{u})$ *model* (1) *is consistent, asymptotically normal and asymptotically efficient.*

5. Statistical hypothesis testing

Consider construction of the chi-square test of the hypothesis

$$H_0 = \{\mathbf{U} \sim P(\mathbf{U} = \mathbf{u})\}.$$

Using the simple's realization

$$\mathbf{x} = (\mathbf{x}_1, \ldots, \mathbf{x}_k)$$

of volume k for any

$$\mathbf{u} \in \Omega = \{\mathbf{u}\},$$

where Ω is the space of elementary events of model (1), define the followiing random variables for $k = 1, \ldots, k$

$$\delta_{\mathbf{u}}(\mathbf{x}_\beta) = \begin{cases} 1, & \text{if } \mathbf{u} = \mathbf{x}_\beta, \\ 0, & \text{otherwise.} \end{cases}$$

Then

$$\delta_{\mathbf{u}}(\mathbf{x}) = \sum_{\beta=1}^{k} \delta_{\mathbf{u}}(\mathbf{x}_\beta)$$

is the frequency of the matrix

$$\mathbf{u} \in \Omega = \{bfu\}$$

from simple's realization

$$\mathbf{x} = (\mathbf{x}_1, \ldots, \mathbf{x}_k).$$

Divide the set Ω on

$$\lambda \geq 2$$

sets by the following form

$$\Omega = \{\Omega_1, \ldots, \Omega_\lambda\},$$

where

$$\bigcap_{i=1}^{\lambda} \Omega_i = \Omega$$

Journal of Physics: Conference Series **738** (2016) 012004 doi:10.1088/1742-6596/738/1/012004

and for any $i_1 = 1, \ldots, \lambda$, $i_2 = 1, \ldots, \lambda$, if $i_1 \neq i_2$ then

$$\Omega_1 \Omega_2 = \oslash.$$

Define the vector

$$\zeta = (\zeta_1, \ldots, \zeta_\lambda)^T,$$

where for $i = 1, \ldots, \lambda$

$$\zeta_i = \sum_{\mathbf{u} \in \Omega_i} \delta_{\mathbf{u}}.$$

Introduce the vector

$$\mathbf{B} = (B_1, \ldots, B_\lambda)^T,$$

where for $i = 1, \ldots, \lambda$

$$B_i = \sum_{\mathbf{u} \in \Omega_i} P(\mathbf{U} = \mathbf{u})$$

and

$$P(\mathbf{U} = \mathbf{u})$$

is the probability of distributuion (1). Define the vector

$$\mathbf{Y} = \frac{1}{\sqrt{k}}(\zeta_1 - A_1, \ldots, \zeta_\lambda - A_\lambda),$$

where for $i = 1, \ldots . \lambda$

$$A_i = kB_i.$$

From the construction in accordance with the central limit Theorem Lindeberg - Levy (see, for example, [13])it follows that in the case of validity of the hypothesis

$$H_0 = \{\mathbf{u} \sim P(\mathbf{U} = \mathbf{u})\}$$

we have for $k \to \infty$ thr vector $\mathbf{\Upsilon}$ has the asymptotically normal distribution

$$N(\mathbf{0}_\lambda, \Sigma),$$

where $\mathbf{0}_\lambda$ is the zero vector of dimension λ and the matrix $\mathbf{\Sigma}$ is defined as

$$\mathbf{\Sigma} = \mathbf{D} - \mathbf{B}\mathbf{B}^T,$$

$$\mathbf{D} = \begin{vmatrix} B_1 & 0 & \ldots & 0 \\ 0 & B_2 & \ldots & 0 \\ \ldots & \ldots & \ldots & \ldots \\ 0 & 0 & \ldots & B_\lambda \end{vmatrix}_{\lambda \times \lambda}$$

and

$$\mathbf{B}\mathbf{B}^T = \begin{vmatrix} B_1B_1 & B_1B_2 & \ldots & B_1B_\lambda \\ B_2B_1 & B_2B_2 & \ldots & B_2B_\lambda \\ \ldots & \ldots & \ldots & \ldots \\ B_\lambda B_1 & B_\lambda B_2 & \ldots & B_\lambda B_\lambda \end{vmatrix}_{\lambda \times \lambda}.$$

Thus, we have following Theorem.

Theorem 4. *If* $k \to \infty$, *quadratic form*

$$\Phi_k^2 = \mathbf{\Upsilon}^T \mathbf{\Upsilon}$$

Journal of Physics: Conference Series **738** (2016) 012004 doi:10.1088/1742-6596/738/1/012004

is seek to chi-square distribution with $\lambda - d + 1$ degrees of freedom, ie

$$\lim_{k\to\infty} P\{\Phi_k^2 \geq x | H_0\} = P\{\chi^2_{\lambda-d+1} \geq x\}.$$

Let's

$$\Upsilon = \frac{1}{\sqrt{k}(\zeta_1 - A_1, \ldots, \zeta_{\lambda-1} - A_{\lambda-1})}$$

and the matrix Σ_0 is obtained if the matrix Σ strike last row and column. Based on the work of the [14] we have the following Theorem.

Theorem 5. *The quadratic form*

$$\mathbf{I} = \boldsymbol{\Upsilon}_0^T \Sigma_0^{-1} \boldsymbol{\Upsilon}_0$$

for $k \to \infty$ has limits distribution Chi-square with $\lambda - d + 1$ degrees of freedom.

Let's

$$\mathbf{N} = \mathbf{K}(\mathbf{K}^T\mathbf{K})^{-}\mathbf{K}^T,$$

where

$$\mathbf{K} = \|a_{ij}\|_{\lambda\times d}|_{\tilde{\mathbf{p}}}$$

or

$$\mathbf{K} = \left\| \frac{\partial B_i}{\sqrt{B_i}\partial p_j} \right\|_{\lambda\times d}\Bigg|_{\tilde{\mathbf{p}}},$$

$(\mathbf{K}^T\mathbf{K})^{-}$ is the generalized inverse of the matrix for martix $\mathbf{K}^T\mathbf{K}$, that is

$$\mathbf{K}^T\mathbf{K}(\mathbf{K}^T\mathbf{K})^{-}\mathbf{K}^T\mathbf{K} = \mathbf{K}^T\mathbf{K},$$

the vector

$$\tilde{\mathbf{p}} = \{\tilde{\mathbf{p}}_1, \tilde{\mathbf{p}}_2, \ldots, \tilde{\mathbf{p}}_\mu\}$$

is the vector of the unibiased estimations for parameters

$$\mathbf{p} = (p_1, p_2, \ldots, p_d)$$

defined in (3).

Theorem 6. *If the vector of paramrters*

$$\mathbf{p} = (p_1, p_2, \ldots, p_d)$$

isn't known and for $k \to \infty$, then quadratic form

$$\boldsymbol{\Theta}_k^2 = \tilde{\Upsilon}^T\tilde{\Upsilon} - \tilde{\Upsilon}^T N \tilde{\Upsilon}$$

has limits distribution Chi-square with $\lambda - R(K) + 1$ degrees of freedom where

$$\tilde{\Upsilon}$$

is estimatiom for Υ, which obtained using a vector of estimates for $\mathbf{p}$ and the most appropriate solutions $\mathbf{z}_g$, based on observation.

The validity of this Theorem follows from [8].

Let's for $\beta = 1, \ldots, \lambda$

$$S_\beta = \sum_{\mathbf{u}\in\Omega_\beta} W(\mathbf{u}, \mathbf{z}_g),$$

where $W(\mathbf{u}, \mathbf{z}_g)$ is most suitable unbiased estimation for distribution (1), and for $\alpha = 1, \ldots, d$

$$Q_{\alpha_\beta} = \sum_{\mathbf{u} \in \Omega_\beta} r_{\alpha_{v_{bfu}}} W(\mathbf{u}, \mathbf{z}_g),$$

where $r_{\alpha_{v_{bfu}}}$ is defined from z_α by using Lemma 1. Define

$$\mathbf{\Sigma}_1 = \mathbf{D}_1 - \mathbf{S}\mathbf{S}^T,$$

$$\mathbf{D}_1 = \begin{vmatrix} S_1 & 0 & \ldots & 0 \\ 0 & S_2 & \ldots & 0 \\ \ldots & \ldots & \ldots & \ldots \\ 0 & 0 & \ldots & S_{\lambda-1} \end{vmatrix}_{(\lambda-1)\times(\lambda-1)},$$

$$\mathbf{S}\mathbf{S}^T = \begin{vmatrix} S_1S_1 & S_1S_2 & \ldots & S_1S_{\lambda-1} \\ S_2S_1 & B_2S_2 & \ldots & S_2S_{\lambda-1} \\ \ldots & \ldots & \ldots & \ldots \\ S_\lambda S_1 & S_\lambda S_2 & \ldots & S_{\lambda-1}S_{\lambda-1} \end{vmatrix}_{(\lambda-1)\times(\lambda-1)},$$

$$\mathbf{\Sigma}_2 = \begin{vmatrix} np_1(1-p_1) & -np_1p_2 & \ldots & -np_1p_d \\ -np_1p_2 & np_2(1-p_1) & \ldots & -np_2p_d \\ \ldots & \ldots & \ldots & \ldots \\ -np_1p_d & -np_2p_d & \ldots & np_d(1-p_d) \end{vmatrix}_{d\times d}$$

and

$$\mathbf{\Sigma}_3 = \begin{vmatrix} Q_{11} - S_1np_1 & Q_{12} - S_2np_1 & \ldots & Q_{1\lambda-1} - S_{\lambda-1}np_1 \\ Q_{21} - S_1np_2 & Q_{22} - S_2np_2 & \ldots & Q_{2\lambda-1} - S_{\lambda-1}np_2 \\ \ldots & \ldots & \ldots & \ldots \\ Q_{d1} - S_1np_d & Q_{d2} - S_2np_d & \ldots & Q_{d\lambda-1} - S_{\lambda-1}np_d \end{vmatrix}_{d\times(\lambda-1)},$$

the elements of vector

$$\mathbf{p} = (p_1, p_2, \ldots, p_d)$$

are parameters of distribution (3). Define the matrix

$$\mathbf{C} = \Sigma_1 - \Sigma_3\Sigma_2^{-1}\Sigma_3$$

and vector

$$\mathbf{M} = \frac{1}{\sqrt{k}(\zeta_1 - kS_1, \ldots, \zeta_\lambda - S_{\lambda-d+1})}.$$

From results [15] - [16] we have following Theorem.
Theorem 7. *If $k \to \infty$ then the quadratic form*

$$\Phi_k^2 = M^T C^{-1} M$$

has limits distribution Chi-square with $\lambda - d + 1$ degrees of freedom.

Conclusion

Analysis conducted in this paper, allows us to formulate the following main results.

- Proposed and studied a new probability distribution of distortions of radiation processes from remote sensing data.
- Define the generating function for the distribution of the proposed model.
- The set of unbiased estimates for the probability distribution of the proposed model and the variance of these estimates.
- Introduced a new concept of the most appropriate evaluation of the set of unbiased estimates, with asymptotic properties.
- From the results of a method comparison of the estimates for the probability distribution of this model implies that the most suitable unbiased estimation is the best.
- Chi-square test for the adequacy of the model to experimental data in the case, where the distributions of the parameters of this model are known and unknown, is built.

References

[1] Poida A., Polykov A., Novikov A. Processing of Big Data in the clouds // Open systems , 10, 2013, P.28-31.
[2] Lisov I. Satellites. // Space News, 01, 1996.
[3] Garbuk S.V., Gershenzon VE Space remote sensing. Moscow, 1997, 297 p.
[4] Iskakova A.S. Identifying the most suitable unbiased estimate for a weather forecast. // Siberian Journal of Industrial Mathematics. 2002.Tom V, 1 (9). P.79-84.
[5] Andrews G.E. The theory of partitions, Encyclopedia of Mathematics and Its Applications (Rota, et.). G.-C. Addison - Wesley, Reading. 1976. Vol.2. 256p
[6] Cramer G., Methods of Mathematical Statistics. - M. 1975. - 648 p.
[7] Nikulin M.S., Smirnov T.I., Voinov V.G. Multivariate distributions induced by an urn scheme, linear diophantine equations, unbiased estimating end testing. // Rapports Internes de l'Unit Math. Applique'es Bordeaux. 1998. No. 98 023 .- 11 p.
[8] Mirvaliev M. Chi-Square Goodness-of-Fit Tests for a Family of Multidimensional Discrete Distributions //Theory Probab. Appl., 23(3.1), 461474.
[9] Riordan J. Introduction to Combinatorial Analysis. Courier Dover Publications, Dec 13, 2002 - Mathematics - 256 p.
[10] Smith K.T. Power series from a computational point of view. Springer - Verlag. 1987.- 132 p.
[11] Panaretos J., Xekalaki E. On generalized binomial and multinomial dis-tributions and their relation to generalized Poisson distributions. // Ann. Inst. Math. 1986.V.38.Part A. pp. 223 - 231.
[12] Haight F.A. The generalized Poisson distribution. Annals of the Institute of Statistical Mathematics 1959, Volume 11, Issue 1 Supplement, pp 101-105.
[13] Loeve M. Probability Theory. D. Van Nostrand Company INC, 710 p.
[14] Nikulin M.S. Chi-Square Test for Continuous Distributions with Shift and Scale Parameters. // Theory Probab. Appl., 18(3), 559568.
[15] Nikulin M.S., Smirnov T.I., Voinov V.G. Multivariate distributions induced by an urn scheme, linear diophantine equations, unbiased estimating end testing. // Rapports Internes de l'Unit Math. Appliqu'es Bordeaux. 1998. No. 98023.- 11 .
[16] Nikulin M.S., Smirnov T.I.,Voinov V.G. Multivariate discrete distributions induced by an urn scheme, linear diophantine equations, unbiased estimating, testing and applications.// J. of Stat. Plan. & Inf. 2002.Vol.101. Issue 1-2. 229-253pp.

Probing the hue of the stochastic magnetization dynamics

Stam Nicolis[1], Julien Tranchida[1,2] and Pascal Thibaudeau[2]

[1]CNRS-Laboratoire de Mathématiques et Physique Théorique (UMR 7350), Fédération de Recherche "Denis Poisson" (FR2964), Département de Physique, Université de Tours, Parc de Grandmont, F-37200, Tours, FRANCE

[2]CEA DAM/Le Ripault, BP 16, F-37260, Monts, FRANCE

E-mail: stam.nicolis@lmpt.univ-tours.fr; julien.tranchida@cea.fr; pascal.thibaudeau@cea.fr

Abstract. The Fokker–Planck equation describes the evolution of a probability distribution towards equilibrium–the flow parameter is the equilibration time. Assuming the distribution remains normalizable for all times, it is equivalent to an open hierarchy of equations for the moments. Ways of closing this hierarchy have been proposed; ways of explicitly solving the hierarchy equations have received much less attention. In this paper we show that much insight can be gained by mapping the Fokker–Planck equation to a Schrödinger equation, where Planck's constant is identified with the diffusion coefficient.

Release 1.5 on May 20, 2016

1. Introduction

Consider a stochastic process, described by a differential equation on a one-dimensional variable x, driven by an additive Gaussian noise $\eta(t)$ of intensity D,

$$\frac{dx}{dt} = -\frac{dV(x)}{dx} + \eta(t). \tag{1}$$

This induces a probability measure, $P(x,t)$, on the space of configurations,$\{x\}$, that satisfies a master equation:

$$\frac{\partial P(x,t)}{\partial t} = -\frac{\partial J(x,t)}{\partial x} \tag{2}$$

where $J(x,t)$ is the corresponding probability current. In this equation, both $P(x,t)$ and $J(x,t)$ depend on the potential $V(x)$. Under certain assumptions it's possible to show that the current, J, is a local function of the density P, in particular and it is given by the expression

$$J(x,t) = -D\frac{\partial P(x,t)}{\partial x} - \frac{dV(x)}{dx}P(x,t), \tag{3}$$

in which case the master equation is called the "Fokker–Planck equation"[1].

Content from this work may be used under the terms of the Creative Commons Attribution 3.0 licence. Any further distribution of this work must maintain attribution to the author(s) and the title of the work, journal citation and DOI.
Published under licence by IOP Publishing Ltd

This equation describes the approach to an equilibrium $P_{\rm eq}(x)$ with $t \geq 0$, provided $V(x)$ satisfies the condition

$$Z = \int_{-\infty}^{\infty} dx\, e^{-\frac{V(x)}{D}} < \infty \tag{4}$$

in which case the equilibrium distribution is, indeed, given by the expression

$$P_{\rm eq}(x) = \frac{1}{Z} e^{-\frac{V(x)}{D}} \tag{5}$$

Now, we shall work in units, where the coefficient D in front of $\partial^2 P/\partial x^2$ is set to 1.

The boundary conditions, for this to hold is that $\lim_{x\to\pm\infty} P(x,t) = 0$, i.e. that $V(x)$ confines "sufficiently strongly". For different boundary conditions the equilibrium distribution can, of course, be completely different, or, even, not exist at all.

In many practical situations, $P(x,t)$ and $P_{\rm eq}(x)$ can be characterized by their moments, $\langle x^m\rangle(t)$ and $\langle x^m\rangle_{\rm eq}$ respectively. Additionally, the moments are, typically, obtained, directly, by measurements, in either real or computer experiments and the probability densities are not known beforehand, but rather their properties are to be deduced from the properties of the moments. In such situations it is useful to write the evolution equations for the moments of $P(x,t)$ with the following recursion relation:

$$\begin{aligned} \frac{d}{dt}\left\langle x^m\right\rangle(t) \equiv \int dx\, x^m \frac{\partial P(x,t)}{\partial t} = -m\int dx\, x^{m-1}\left(\frac{\partial P(x,t)}{\partial x} + \frac{dV(x)}{dx}P(x,t)\right) = \\ m(m-1)\left\langle x^{m-2}\right\rangle - m\left\langle x^{m-1}\frac{dV(x)}{dx}\right\rangle \end{aligned} \tag{6}$$

for $m = 2, 3, \ldots$. For $m = 1$ one must be a bit more careful and go back to the definition of $\langle x\rangle(t)$ itself to obtain

$$\frac{d}{dt}\langle x\rangle = -\left\langle\frac{\partial V}{\partial x}\right\rangle. \tag{7}$$

The equations (6) and (7) provide an infinite set of ordinary differential equations for the moments $\langle x^m\rangle(t)$. In these equations, also, boundary terms have been assumed not to contribute and must be studied separately, once provided an infinite set of initial conditions $\langle x^m\rangle(0) \equiv u_m$.

It is possible to generalize this calculation to higher dimensions: Suffice to interpret the terms in the Fokker–Planck equation appropriately, at first, for flat manifolds, by replacing derivatives by gradients and products by scalar products:

$$\begin{aligned} \frac{\partial^2 P}{\partial x^2} &\to \Delta P(\boldsymbol{x},t) \\ \frac{\partial}{\partial x}\left(\frac{\partial V}{\partial x}P\right) &\to \nabla\cdot(P(\boldsymbol{x},t)\nabla V(\boldsymbol{x})) \end{aligned} \tag{8}$$

For more general manifolds one simply has to introduce the appropriate metric and all expressions can be written in covariant form. In this case the moments become tensors, e.g.

$$\left\langle x_{I_1}^{m_1} x_{I_2}^{m_2}\cdots x_{I_k}^{m_k}\right\rangle(t), \tag{9}$$

and are subject to coupled differential equations as well.

For $V(\boldsymbol{x})$ a polynomial, whatever the number of dimensions, equations (6) with proper initial conditions, are linear equations in the moments and describe their flow towards their equilibrium values. The existence of a normalizable equilibrium distribution implies that the equilibrium values should be independent of the initial conditions, provided the moments do exist[1].

[1] The normalizability of the density is a necessary, but not sufficient, condition for the existence of the moments.

For $V(x)$ not a polynomial, the equations are non–linear in the moments. However it's not this property that makes them hard to solve; the polynomial case is hard to solve, also.

The real problem for solving eqs. (6) and (7), resp. eq. (9), is that the equations are coupled and the matrix of the linear system is not of finite rank. Only two cases allow the hierarchy to be solved explicitly: the case $V(x) = 0$ and the case where $V(x) = kx^2/2$, with $k > 0$, because this second case can be reduced to the first one, by a change of variables. In fact these two cases illustrate the nature of the problem: each moment decays at a different time scale, set by its order. It is this property that provides the real obstruction to "closing the hierarchy". And the reason is the absence of "mixing", that depends on the interplay between diffusion and the interactions described by the potential. In the two cases at hand, this is due to the fact that the potential can, in fact, be eliminated, which eliminates mixing.

This was noted many years ago [2] and it was pointed out that the interplay between the mixing properties of the diffusion term of Fokker–Planck and the ergodic properties of the potential, can make certain moments, of different orders, tend to decay at the same rate. This is the theoretical foundation of the attempts for "closing" the hierarchy and thus reducing the system of moment equations to finite rank. In recent work [3] and work in progress, this approach was shown to be relevant, in practice, also, for closing the hierarchy of moments of the magnetization of a nanomagnet described by the Landau–Lifshitz–Gilbert equation. That this is possible at all in this system isn't a straightforward extension of the formalism.

The reason is that, for this system, noise does not appear additively, like in eq. (1), but multiplicatively; also that, instead of being white–with ultra–local two–point correlation function–it's colored, with a finite auto–correlation time. These properties mean that it is not true that the master equation for the probability density, $P(x,t)$, is of Fokker–Planck form–the expression for the current, $J(x,t)$ must be found and the finite autocorrelation time means that it won't be a local function of $P(x,t)$. Therefore the hierarchy for the moments is the only way of computing them and thus of reconstructing $P(x,t)$. Since the Fokker–Planck equation is a partial differential equation, it is possible to solve it directly–numerically if need be–to machine precision and compute the moments directly. If the master equation is not known explicitly at all, attempting to close the hierarchy seems to be a way to reliably compute the moments.

Another way, however, is possible: to try and parametrize density $P(x,t)$ and current $J(x,t)$ on the basis of the symmetries of the problem and deduce an equation that can be solved–if need be numerically. From its solution, the moments can then be directly obtained, thereby eliminating having to deal with the hierarchy altogether. As a bonus, such a scheme would provide a way of describing the approach to synchronization of the moments involved.

In this contribution we shall show that a practical way of computing the moments of the master equation is by associating it to a Schrödinger equation, where the role of Planck's constant is played by the diffusion coefficient.

2. Effective map

If the master equation is of Fokker–Planck form, and $P_{\mathrm{eq}}(x) \equiv \exp(-V(x))$ is normalizable, it is possible to perform a change of variables and rewrite the Fokker–Planck equation as a Schrödinger equation in imaginary time [4]

$$P(x,t) \equiv e^{-V(x)/2}\phi(x,t) \Rightarrow \frac{\partial\phi}{\partial t} = \frac{\partial^2\phi}{\partial x^2} + U(x)\phi \tag{10}$$

with boundary conditions $\lim_{|x|\to\infty} \phi(x,t) = 0$ and the potential $U(x)$ is given by the expression

$$U(x) = -\frac{1}{4}\left(\frac{\partial V}{\partial x}\right)^2 + \frac{1}{2}\frac{\partial^2 V}{\partial x^2} \tag{11}$$

We deduce that $\langle x^m\rangle(t)$ is given by the explicit expressions

$$\langle x^m\rangle(t) = \frac{\int_{-\infty}^{\infty} dx\, e^{-V(x)/2}\phi(x,t)x^m}{\int_{-\infty}^{\infty} dx\, e^{-V(x)/2}\phi(x,t)} \Rightarrow \langle x^m\rangle_{\text{eq}} = \frac{\int_{-\infty}^{\infty} dx\, e^{-V(x)/2}\phi_0(x)x^m}{\int_{-\infty}^{\infty} dx\, e^{-V(x)/2}\phi_0(x)} \tag{12}$$

where $\phi_0(x) = \exp(-V(x)/2)$ is the ground state wave-function. This is the solution to the problem, provided that $\phi_0(x)$, indeed exists, i.e. that the fluctuations do not affect the potential. Similarly, for higher dimensional target spaces, the equal time moments are given by the expressions

$$\langle x_{I_1}^{m_1}\cdots x_{I_k}^{m_k}\rangle(t) = \frac{\int d\boldsymbol{x}\, e^{-V(\boldsymbol{x})/2}\phi(\boldsymbol{x},t)x_{I_1}^{m_1}\cdots x_{I_k}^{m_k}}{\int d\boldsymbol{x}\, e^{-V(\boldsymbol{x})/2}\phi(\boldsymbol{x},t)}. \tag{13}$$

These expressions can then be readily evaluated, thereby obviating the technical need for closing the hierarchy. What the framework of closing the hierarchy provides is the motivation to understand which groups of moments synchronize towards equilibrium.

For the case of a nanomagnet in a magnetic field, however, what the appropriate expression for the master equation should be is not obvious at all, since the force term does not derive from a potential. In previous work on the subject we have shown [3, 5, 6] that it is possible to derive the hierarchy of equations for the moments directly from the Landau–Lifshitz–Gilbert equation and, using the techniques of ref. [2], to close the hierarchy and solve the equations numerically. Obtaining the master equation, however, is considerably harder, since it is not a local equation for the probability density, even for vanishing autocorrelation time. Therefore we need to understand what are the physical properties that such a master equation must have and attempt to parametrize the space of its solutions in another way, beyond Fokker–Planck.

For the case of a vector potential, as in the Landau–Lifshitz–Gilbert equation

$$\frac{d\boldsymbol{x}}{dt} = \boldsymbol{\omega}\times\boldsymbol{x} - \lambda\left(\boldsymbol{\omega}\times\boldsymbol{x}\right)\times\boldsymbol{x} \tag{14}$$

things become more complicated, because the RHS cannot be written as the gradient of a potential. Furthermore, the noise term doesn't appear explicitly. So this doesn't seem to be a stochastic equation at all. Stochastic effects could appear, were this system chaotic [7]; this isn't the case, however, since, for fixed precession frequency, the component of the magnetic moment parallel to the frequency vector is conserved and the transverse components define an integrable system.

If the precession frequency can fluctuate, however, then stochastic effects can appear and the master equation must be found by appealing to the symmetries of the problem. The key symmetry is gauge invariance.

The precession frequency $\boldsymbol{\omega}\equiv\boldsymbol{\omega}(t)$, can, in fact, be identified with the magnetic field, that can be expressed in terms of a vector potential $\boldsymbol{A}(t)$, such as $\boldsymbol{\omega}(t)\equiv\nabla\times\boldsymbol{A}(t)$. Gauge invariance is the symmetry $\delta_\Lambda\boldsymbol{A} = \nabla\Lambda$. Stochastic effects are described by assuming that $\boldsymbol{\omega}(t)\equiv\boldsymbol{\omega}_0+\boldsymbol{\Omega}(t)$, with $\boldsymbol{\Omega}(t)$ drawn from a Gaussian process with color defined by a finite auto-correlation time,

$$\langle\Omega_I(t)\Omega_J(t')\rangle = \frac{D}{\tau}\delta_{IJ}\exp\left(-\frac{|t-t'|}{\tau}\right) \tag{15}$$

This disorder induces a probability measure on the space of configurations for the magnetization, $P(\boldsymbol{x},t)$, that satisfies a master equation–that is not a Fokker–Planck equation for finite autocorrelation time. We would like to explore the possibility of deducing it as the continuity equation of an appropriate Schrödinger equation. This time there is not any scalar potential, however. But it is known how to write the Schrödinger equation for a particle in a magnetic field, in terms of its vector potential, including the anisotropy relevant for the Zeeman effect:

one introduces the covariant derivatives and the anisotropy involves the coupling to angular momentum. The mixing effects are captured, once more, by the diffusion coefficient, while the effects of the finite autocorrelation time are captured by averaging the moments over the distribution of the vector potential.

The advantage of this formulation is that it allows us to describe what happens, when the vector potential, $\boldsymbol{A}(\boldsymbol{x},t)$ is drawn from some distribution. This distribution must be invariant under gauge transformations, therefore, depend only on the precession field, derived from the vector potential. For a Gaussian distribution this means that it is, necessarily, of the form

$$\rho(\boldsymbol{A}) = \exp\left(-\frac{1}{2}\int dt\,dt'\,(\boldsymbol{\omega}[\boldsymbol{A}](t)-\boldsymbol{\omega}_0)\,G(t-t')\cdot(\boldsymbol{\omega}[\boldsymbol{A}](t')-\boldsymbol{\omega}_0)\right) \tag{16}$$

with $\omega_I[\boldsymbol{A}] = \varepsilon_{IJK}\nabla_J A_K$ and $G(t-t')$ the functional inverse of the 2–point function $C(t-t')$ of the precession field:

$$\langle(\omega_I(t)-\omega_{0I})\,(\omega_J(t')-\omega_{0J})\rangle = \delta_{IJ}C(t-t') \tag{17}$$

with the other correlation functions defined by Wick's theorem, if we assume that the precession field fluctuates about the value $\langle\boldsymbol{\omega}[\boldsymbol{A}](t)\rangle = \boldsymbol{\omega}_0$.

Since the probability density for the precession field is Gaussian, it is, in principle, possible to integrate out the vector potential and obtain an effective action for the magnetic moment. If the noise is "colored", then the effective action will, indeed, be non-local and the equation satisfied by the probability density, in the saddle point approximation will not be of Fokker–Planck type, for this reason.

In this way we obtain the effective action for the averaged wavefunction $\psi(\boldsymbol{x},t)$, from which we can compute the density, $P(\boldsymbol{x},t)$, whose equal time moments can then be computed. This is clearly a formulation consistent with all symmetries of the problem. While analytical expressions are difficult to obtain, numerical techniques, that are reliable to machine precision, are known.

The details of this approach will be presented in future work. It is possible to establish a benchmark for it, by studying the approach to equilibrium of the equal–time moments, for "small" , as well as "large" values of τ. The technical details can be found in ref. [5]; they are illustrated in fig. 1 where s_i stands for the i-th component of the average magnetization (i.e. the first moment).

3. Conclusions

We have described several, *a priori*, different ways for computing the equal time moments of the magnetization, that determine its probability distribution, when stochastic perturbations must be described with a finite auto–correlation time. On the one hand we have implemented a numerical method, based on an efficient integration of the Langevin equation and averaging over the realizations of the noise; on the other hand, we have implemented a method for testing a theoretical approach, based on chaotic dynamical systems and have found that the two approaches, while providing complementary information, are in agreement; and the framework put in place allows to improve the approximations in a systematic way. We have, also, set up a theoretical framework, based on a Schrödinger equation, that exploits the gauge invariance of the system and where the diffusion coefficient plays the role of Planck's constant. While, in quantum mechanics, this approach is, of course, well known, the fact that, from a mathematical viewpoint, it is of much more general validity, hasn't been, yet, appreciated.

The finite auto–correlation time of the noise implies, in general, that the probability current is a non–local functional of the density, which is natural in the Schrödinger approach. This leads to new numerical and conceptual tools for studying non–Markovian dynamics (cf. also [8]), beyond the vicinity of the white–noise limit [9] and can lead to new insights for further developing the functional integral approach [10]) and will be presented more fully in future work.

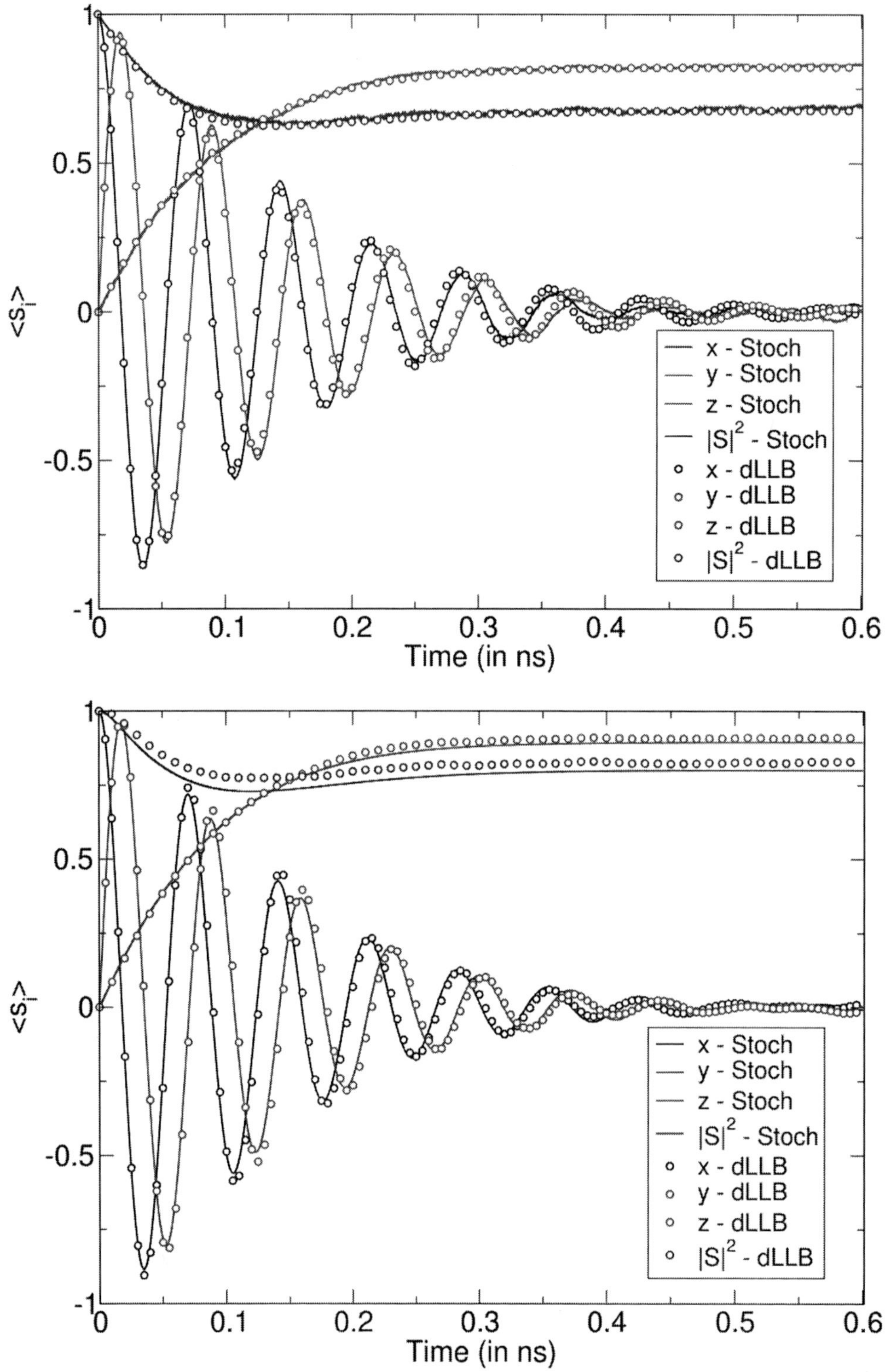

Figure 1. $\langle s\rangle(t)$ for weakly colored noise, $\tau = 10^{-7}$ ns (upper panel) vs. strongly colored noise, $\tau = 10^{-2}$ ns (lower panel). Taken from [5].

References

[1] Risken H 1989 *The Fokker-Planck Equation* (*Springer Series in Synergetics* vol 18) (Berlin, Heidelberg: Springer-Verlag) ISBN 978-3-540-61530-9 978-3-642-61544-3 URL `http://link.springer.com/10.1007/978-3-642-61544-3`

[2] Nicolis C and Nicolis G 1998 *Physical Review E* **58** 4391–4400 ISSN 1063-651X, 1095-3787 URL `http://link.aps.org/doi/10.1103/PhysRevE.58.4391`

[3] Tranchida J, Thibaudeau P and Nicolis S 2016 *Physica B: Condensed Matter* **486** 57–59 ISSN 0921-4526 URL `http://www.sciencedirect.com/science/article/pii/S0921452615302787`

[4] Nelson E 1966 *Phys. Rev.* **150**(4) 1079–1085 URL `http://link.aps.org/doi/10.1103/PhysRev.150.1079`

[5] Tranchida J, Thibaudeau P and Nicolis S 2016 *IEEE Transactions on Magnetics* 1–1 ISSN 0018-9464, 1941-0069 URL `http://ieeexplore.ieee.org/lpdocs/epic03/wrapper.htm?arnumber=7401053`

[6] Thibaudeau P, Tranchida J and Nicolis S 2016 *IEEE Transactions on Magnetics* 1–1 ISSN 0018-9464, 1941-0069 URL `http://ieeexplore.ieee.org/lpdocs/epic03/wrapper.htm?arnumber=7393845`

[7] Pérez L M, Bragard J, Mancini H, Gallas J A, Cabanas A M, Suarez O J and Laroze D 2015 *Networks and Heterogeneous Media* **10** 209–221

[8] Romá F, Cugliandolo L F and Lozano G S 2014 (*Preprint* 1403.4822) URL `http://arxiv.org/abs/1403.4822`

[9] Fox R F 1986 *Physical Review A* **33** 467–476 URL `http://link.aps.org/doi/10.1103/PhysRevA.33.467`

[10] Aron C, Barci D G, Cugliandolo L F, Arenas Z G and Lozano G S 2014 (*Preprint* 1402.1200) URL `http://arxiv.org/abs/1402.1200`

A drift-diffusion model based on the fractional exclusion statistics

G. A. Nemnes[1,2] and D. V. Anghel[1]

[1]Horia Hulubei National Institute for Physics and Nuclear Engineering, 077126, Magurele-Ilfov, Romania

[2]University of Bucharest, Faculty of Physics, MDEO Research Center, 077125 Magurele-Ilfov, Romania

E-mail: alexandru.nemnes@nipne.ro

Abstract. We propose a drift-diffusion model for systems which obey fractional exclusion statistics (FES), in a framework where the species include classical degrees of freedom such as positions. The transition rates are calculated and the relation between the step and acceptance probabilities on one hand and the diffusion and drift processes on the other hand are established. A Monte Carlo scheme is implemented on a prototypical double-junction system of particles with screened Coulomb interactions. In our approach the properties of interacting quantum gases are locally included using the FES methodology. The model is suitable to describe transient as well as stationary regimes.

1. Introduction

During the past decades the drift-diffusion models have been invaluable tools for describing the transport in semiconductor devices and the physics of cold plasmas and gas discharges. Efficient solvers based on Boltzmann transport equation implementing Monte Carlo schemes have been devised. However, in recent years, the quantum effects in modern semiconductor devices became increasingly important.

Solving microscopic quantum models–although they are most accurate–is becoming an increasingly difficult task as the size of the analyzed system is getting larger in the attempt to reach the experimentally attainable length scales. Therefore hybrid microscopic-macroscopic models are usually employed, which combine a microscopic description for short length scales with macroscopic fluid-type models, e.g. quantum drift diffusion model (QDD) [1], quantum corrected drift diffusion model (QCDD) [2], Schrödinger-Poisson-Drift-Diffusion model (SPDD) [3].

We propose here a method to locally incorporate in a transport model the behavior of interacting quantum gases, using fractional exclusion statistics (FES) [4–6]. The method was shown to be a tool for the study of the thermodynamic properties of interacting particle systems, regarded as ideal FES gases. The transition rates for homogeneous FES gases were established [7–9], which opened the possibility of analyzing the non-equilibrium dynamics of interacting bosonic and fermionic systems via Monte Carlo simulations [9]. The extension of this kinetic model to non-homogeneous FES systems with mutual exclusion parameters was realized in Refs. [10–12]. We make here a step further and introduce a transport model based on FES, using Monte Carlo simulations.

Content from this work may be used under the terms of the Creative Commons Attribution 3.0 licence. Any further distribution of this work must maintain attribution to the author(s) and the title of the work, journal citation and DOI.
Published under licence by IOP Publishing Ltd

2. Model and formalism

The system under investigation is a generic device made up by an active region and contacts. The transport properties are dominated by hopping between states localized at the sites $\mathbf{r}_i$. The contacts have the role of particle reservoirs which fix the chemical potentials according to the applied bias. The key elements of the active region, which determine properties of the system, are the local density of states $\sigma(\mathbf{r}, \epsilon)$ and the interacting potentials which, in general, may be position dependent. The Hamiltonian of the system is [11, 12]

$$E = \sum_I \epsilon_{\mathbf{r}_I} n_{\mathbf{r}_I} + \frac{1}{2} \sum_{I,J} V(|\mathbf{r}_I - \mathbf{r}_J|) n_{\mathbf{r}_I} n_{\mathbf{r}_J}. \tag{1}$$

where $\epsilon_{\mathbf{r}_I}$ is the energy and $n_{\mathbf{r}_I}$ is the occupation number of the site I; the total particle number is $N = \sum_I n_{\mathbf{r}_I}$. The quasiparticle energies are defined as $\tilde{\epsilon}_{\mathbf{r}_i} = \epsilon_{\mathbf{r}_i} + \sum_{\tilde{\epsilon}_{\mathbf{r}_j} < \tilde{\epsilon}_{\mathbf{r}_i}} V(|\mathbf{r}_i - \mathbf{r}_j|) n_{\mathbf{r}_j}$, such that $E = \sum_I \tilde{\epsilon}_{\mathbf{r}_I} n_{\mathbf{r}_I}$.

In order to apply the FES formalism we work in the quasicontinuous limit and we define the species as in Refs. [11, 12] by dividing the real space and the quasiparticle energy axis into elementary volumes and segments $\delta\mathbf{r}$ and $\delta\epsilon$. If the space is s-dimensional (sD), then the species will be the $(s+1)$D volumes $\delta\mathbf{r}_\xi \times \delta\epsilon_i$, where by ξ (Greek) and i (Latin) we index the volumes and energy intervals, respectively. Each species (ξ, i) contains $N_{\xi i}$ particles and $G_{\xi i}$ (for bosons) or $T_{\xi i}$ (for fermions) single-particle states. Then the number of micro-configurations corresponding to a distribution $\{G_{\xi i}, N_{\xi i}\}$ (bosons) or $\{T_{\xi i}, N_{\xi i}\}$ (fermions) of states and particles is $W^{B/F} = \prod_{(\xi,i)} W_{\xi i}^{B/F}$, where

$$W_{\xi i}^B = \frac{(G_{\xi i} + N_{\xi i} - 1)!}{N_{\xi i}!(G_{\xi i} - 1)!}, \quad \text{and} \quad W_{\xi i}^F = \frac{T_{\xi i}!}{N_{\xi i}!(G_{\xi i} - T_{\xi i})!}, \tag{2}$$

The FES parameters are defined by the change of the number of states in one species (ξi) due to the change of the number of particles in any (other or the same) species (ηj), which is $\delta G_{\xi i} = -\alpha_{\xi i,\eta j} \delta N_{\eta j}$ for bosons and $\delta T_{\xi i} = -\alpha_{\xi i,\eta j} \delta N_{\eta j}$ for fermions. For the Hamiltonian (1) and our choice of quasiparticle energies the FES parameters are $\alpha_{\xi i;\eta j} = [\delta_{ij}\sigma_\xi(\epsilon_i) + \theta(i - j)\delta\epsilon_i(d\sigma_\xi(\epsilon_i)/d\epsilon_i)]V(|\mathbf{r}_\xi - \mathbf{r}_\eta|)$, where $\sigma_\xi(\epsilon_i) \equiv \delta\mathbf{r}_\xi \sigma(\mathbf{r}_\xi, \epsilon_i)$ [11, 12]. Note that if the $\sigma_\xi(\epsilon_i)$ does not depend on the energy, the FES parameters simplify to $\alpha_{\xi i;\eta j} = \delta_{ij}\sigma_\xi(\epsilon_i)V(|\mathbf{r}_\xi - \mathbf{r}_\eta|)$.

In FES the description of bosons and fermions may be unified by the redefinition of the FES parameters and of the number of states obtaining what we may call *bosonic* or *fermionic descriptions*. In the bosonic description–the one that we shall use in the following–the number of states is $G_{\xi i}$ for bosons and $T_{\xi i} - N_{\xi i} + 1$ for fermions. Similarly, the FES parameters are $\alpha_{\xi i,\eta j}$ for bosons and $\alpha_{\xi i,\eta j} + \delta_{\xi\eta}\delta_{ij}$ for fermions. If we make these transformations then the number of microconfigurations is given by $W_{\xi i}^B$ for both, bosons and fermions.

If we assume that the system is in thermal equilibrium, then the probability to find the system in the state Υ, defined by the set $\{N_{\xi i}, G_{\xi i}\}_{\xi i}$, is

$$p_\Upsilon^{\rm eq} \equiv p^{\rm eq}(\{N_{\xi i}, G_{\xi i}\}_{\xi i}) = \frac{1}{\mathcal{Z}} \prod_{(\xi,i)} W_{\xi i} \exp\left(-\frac{E_\Upsilon}{k_B T}\right), \tag{3}$$

where $\mathcal{Z}$ is the partition function and $E_\Upsilon = \sum_{\xi,i} N_{\xi i}\epsilon_{\xi i}$ is the total energy–we assume that all the particles that belong to one species (ξi) have the same energy $\epsilon_{\xi i}$. Two states of the FES system, Υ and Φ, are considered neighbors if they differ by one particle jump from one species (ξ, i) to another (η, j),

$$\Upsilon = \{N_{\xi i}, G_{\xi i}\ ;\ N_{\eta j}, G_{\eta j}\ ;\ \{N_{\zeta k}, G_{\zeta k}\}\} \tag{4}$$

$$\Phi = \quad N_{\xi i} - 1, G'_{\xi i}\ ;\ N_{\eta j} + 1, G'_{\eta j}\ ;\ \{N_{\zeta k}, G'_{\zeta k}\}\Big\}, \tag{5}$$

(where $(\zeta,k) \neq (\xi,i),(\eta,j)$). Upon extracting one particle from species (ξ,i) and inserting it into species (η,j), the number of available states may change in any species (ζk) by $G'_{\zeta k} = G_{\zeta k} + \alpha_{\zeta k,\xi i} - \alpha_{\zeta k,\eta j}$. Introducing a discrete-time reversible Markov chain one can employ the detailed balance equation,

$$p_{\Upsilon}^{\rm eq}\Gamma_{\Phi\Upsilon} = p_{\Phi}^{\rm eq}\Gamma_{\Upsilon\Phi}, \tag{6}$$

where $\Gamma_{\Phi\Upsilon}$, $\Gamma_{\Upsilon\Phi}$ are the transition rates from Υ to Φ and vice-versa, respectively. From the detailed balance equation we obtain the ratio between the transition rates

$$\log\left(\frac{\Gamma_{\Upsilon\Phi}}{\Gamma_{\Phi\Upsilon}}\right) = \log(p_{\Upsilon}^{\rm eq}) - \log(p_{\Phi}^{\rm eq}) \simeq \frac{\partial(\log p_{\Upsilon}^{\rm eq})}{\partial N_{\xi i}} - \frac{\partial(\log p_{\Upsilon}^{\rm eq})}{\partial N_{\eta j}}, \tag{7}$$

where, in last step, we used a first order expansion for $\log(p_{\Phi}^{\rm eq})$ around Υ. In the thermodynamic limit, i.e. for large $N_{\xi i}$ and $G_{\xi i}$, we apply the Stirling formula to calculate $\log(p_{\Upsilon}^{\rm eq}) \simeq \sum_{(\xi,i)}[(G_{\xi i}+N_{\xi i})\log(G_{\xi i}+N_{\xi i}) - G_{\xi i}\log G_{\xi i} - N_{\xi i}\log N_{\xi i}]$ and by using the identity $\alpha_{\xi i,\eta j} = -\partial G_{\eta j}/\partial N_{\xi i}$ we find

$$\begin{aligned}\frac{\partial(\log p_{\Upsilon}^{\rm eq})}{\partial N_{\xi i}} &= (1-\alpha_{\xi i,\xi i})\log(G_{\xi i}+N_{\xi i}) - \log N_{\xi i} + \alpha_{\xi i,\xi i}\log G_{\xi i} \\ &- \alpha_{\xi i,\eta j}\log(G_{\xi i}+N_{\xi i}) + \alpha_{\xi i,\eta j}\log G_{\eta j} \\ &+ \sum_{(\zeta,k)}[-\alpha_{\xi i,\zeta k}\log(G_{\zeta k}+N_{\zeta k}) + \alpha_{\xi i,\zeta k}\log G_{\zeta k}],\end{aligned} \tag{8}$$

where $(\zeta,k) \neq (\xi,i),(\eta,j)$ and a similar expression for $\partial(\log p_{\Upsilon}^{\rm eq})/\partial N_{\eta j}$ by interchanging (ξ,i) and (η,j). Defining the number of available states in the absence of particles [5] by $G_{\xi i}^0 = G_{\xi i} + \sum_{\eta,j}\alpha_{\xi i,\eta j}(N_{\eta j} - \delta_{\xi i,\eta j})$ and the occupation numbers as $n_{\xi i} = N_{\xi i}/G_{\xi i}^0$, from Eq. (7) we find the transition rate for the $\Upsilon \to \Phi$ process

$$\Gamma_{\Phi\Upsilon} \sim n_{\xi i}(1+n_{\eta j}-A_{\eta j})\prod_{(\zeta,k)}[1+n_{\zeta k}-A_{\zeta k}]^{-\alpha_{\eta j,\zeta k}}\prod_{(\zeta,k)}[1-A_{\zeta k}]^{\alpha_{\eta j,\zeta k}}, \tag{9}$$

where $A_{\xi i} = \sum_{\eta,j}\alpha_{\xi i,\eta j}n_{\eta j}$. One should note that in the case of a homogeneous system with diagonal statistical parameters, $\alpha_{\xi i,\eta j} = \alpha\delta_{\xi\eta}\delta_{ij}$, one recovers the transition rates reported in Ref. [9]:

$$\Gamma_{\Phi\Upsilon}(\alpha) \sim n_{\xi i}(1-\alpha n_{\eta j})^{\alpha}[1+(1-\alpha)n_{\eta j}]^{1-\alpha}. \tag{10}$$

The transition rate in equation (9) represents the *step* probability for one particle in species (ξ,i) to select (η,j) as target species. Once the destination species is obtained, the move is *accepted* with a probability proportional with the Boltzmann factor, i.e. $\sim \min[1,\exp(-(E_\Phi - E_\Upsilon)/k_BT)]$. The step probability depends on the occupancies of both source and target species and can be associated with a diffusion process. The acceptance probability depends on the energy diference between the two species and, in the case of external fields, is influenced and may be associated with the drift process.

3. Application – Transport in a double junction system with screened Coulomb interactions

We apply the drift-diffusion model on a one-dimensional test-case system. We consider a two terminal device partitioned into three regions: the left contact, the active region and the right contact. The particles are interacting via a screened Coulomb interaction, which is parametrized by:

$$V(r,\lambda) = V_0\frac{1}{r/R_0}\exp\left(-\frac{r/R_0-1}{\lambda/R_0}\right), \tag{11}$$

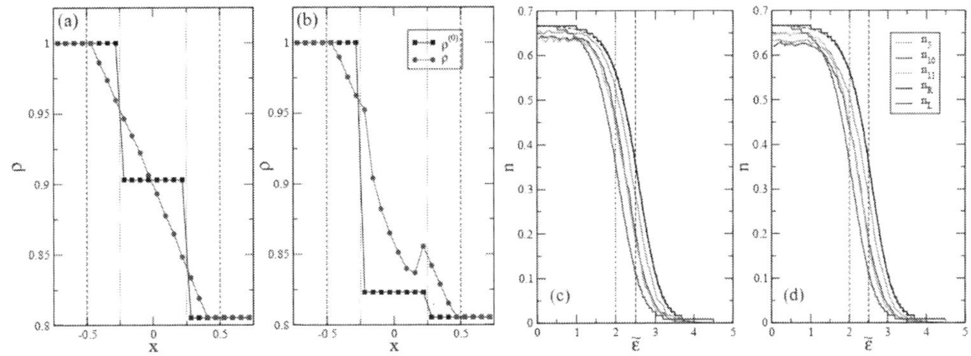

Figure 1. (Color online) (a) and (b) Particle densities across the two terminal device, for $\lambda_{AR} = \lambda_C$ and $\lambda_{AR} \to \infty$, respectively. The dashed lines mark the device system with N_x sites and the dotted lines show the interface between the contacts and the active region. Both initial (squares) and final (circles) particle densities, $\rho^{(0)}$ and ρ, are represented. (c) and (d) Populations vs. energy for the left (n_L) and the right (n_R) contact and three other sites, $\xi = 5, 10, 11$, for the two screening lengths considered. The vertical dashed lines mark the chemical potentials.

where r is the distance between the particles, R_0 is a cut-off radius which removes the singularity at the origin, $V_0 \equiv V(R_0, \lambda)$, and λ is the screening length. We assume the interaction inside the contacts is strongly screened, while in the active region we consider both limiting situations which correspond to short and large screening lengths.

Our prototypical system consists of $N_x \times N_\epsilon$ species, with $N_x = 16$ and $N_\epsilon = 50$, and constant density of states. The species are equally spaced in the interval $[-L_x/2, L_x/2]$, where $L_x = 1$. The length of each of the two contact regions is $L_x/4$ and it is assumed they have a short screening length $\lambda_C = L_x/N_x$, while for the active region, of length $L_x/2$, we compare two screening lengths, namely, $\lambda_{AR} = \lambda_C$ and $\lambda_{AR} \to \infty$. In the following, all energies are scaled with respect to E_F, the Fermi energy of the non-interacting system. The chemical potentials in the left and the right contacts are set to $\mu_L = 2.5$ and $\mu_R = 2$, respectively, their difference reflecting the applied bias. The cutoff radius is set to $R_0 = 0.05$ and $V_0 = 8$. The temperature of the system is $k_B T = 0.25$.

The species $\xi = 0$ and $\xi = N_x - 1$ have fixed populations, as imposed by the chemical potentials in the contacts. Whenever a particle enters / leaves these species, the particle is removed / added back, respectively.

We first performed a number of 10^{10} Monte Carlo steps in order to equilibrate the system, and then using an equal number of steps the thermal averages are performed. The particle densities, $\rho_\xi = \sum_i n_{\xi i}$, are presented in Fig. 1 (a) and (b) for the two cases – short and long screening lengths in the active region. For the homogeneous system, $\lambda_{AR} = \lambda_C$, one obtains a linear dependence of the particle density, as particles are moving from the left to the right contact. By contrast, for pure Coulombian interaction one observes a decrease in the particle density in the active region. The depletion region is skewed by the finite potential difference between the two contacts and a spike-like feature appears in the particle density in the vecinity of the right contact interface, in the species $\xi = 11$. However, the regions of the non-interacting contacts embedded in the active region still retain a linear dependence in the particle distribution. Correspondingly, Figs. 1 (c) and (d) show the populations in the two contacts, as well as for the species at three other sites. For $\xi = 10$ one obtains a minimum in the particle density for $\lambda_{AR} \to \infty$ and the populations of the quasiparticle energy levels are comparatively smaller. This charge depletion effect is due to the unscreened Coulombian repulsion and it may be found, e.g., in a n++ /

intrinsic / n++ semiconductor structure.

Within the proposed method one may extract the current of particles flowing between terminals, from the number of particles inserted or extracted from the boundary species, $\xi = 0$ and $N_x - 1$, respectively. Although in stationary regime the two currents are equal up to some finite system fluctuations, generally, in non-stationary regime, they may differ and one can describe in real time transient processes, e.g. charge accumulation or depletion.

4. Conclusions

We proposed a drift-diffusion model for systems which obey fractional exclusion statistics. We determined the transition rates, pointing out the relation between the step and acceptance probabilities and the diffusion and drift processes.

The model is employed on a prototypical double-junction system of particles with screened Coulomb interactions. These types of systems are relevant for device applications in semiconductor physics, e.g. field effect transistors. Depletion effects are analyzed with respect to the screening length in the device active region.

Our approach takes advantage of a local description of the system in terms of ideal FES gases. This enables us to handle larger systems, in the range where exact quantum mechanical calculations are difficult to be performed and yet incorporating locally the quantum behavior of the interacting particle systems by means of the FES methodology.

Acknowledgements

This work was supported by the National Authority for Scientific Research and Innovation (ANCSI) under grants PN-II-ID-PCE-2011-3-0960, PN 16420101 and PN 16420202.

References

[1] Degond P, Gallego S S and Mehats F 2007 *J. Comp. Phys.* **221** 226.
[2] de Falco C, Jerome J W and Sacco R 2009 *J. Comp. Phys.* **228** 1770.
[3] Lacaita A L, Pirovano A and Spinelli A S 2002 *IEEE Trans. Electron Devices* **1** 25.
[4] Haldane F D M 1991 *Phys. Rev. Lett.* **67** 937.
[5] Wu Y S 1994 *Phys. Rev. Lett.* **73** 922.
[6] Isakov S B 1994 *Phys. Rev. Lett.* **73**(16) 2150.
[7] Kaniadakis G, Lavagno A and Quarati P 1996 *Mod. Phys. Lett. B* **10** 497.
[8] Kaniadakis G, Lavagno A and Quarati P 1996 *Nucl. Phys. B* **466** 527.
[9] Nemnes G A and Anghel D V 2010 *J. Stat. Mech.* page P09011.
[10] Nemnes G A and Anghel D V 2013 *J. Phys.: Conf. Ser.* **410** 012120.
[11] Anghel D V, Nemnes G A and Gulminelli F 2013 *Phys. Rev. E* **88** 042150.
[12] Nemnes G A and Anghel D V 2014 *Rom. Rep. Phys.* **66** 336.

A Generalized Variational Principle for Gaussian Random Fields

Flora Koukiou

Laboratoire de physique théorique et modélisation, Université de Cergy-Pontoise, 2, av. A. Chauvin, 95302 Cergy-Pontoise, France

E-mail: flora.koukiou@u-cergy.fr

Abstract.

The freezing property for a family of models involving Gaussian Random Fields is studied in a unified manner and it is related to the behaviour of the entropy of the Gibbs measure in terms of the variational principle. The analysis includes the Gaussian Multiplicative Chaos, Random Energy Model, Spin Glass Mean-Field, and the Random Polymers models.

1. Introduction and Motivation

The freezing phenomenon, expected to occur in a wide class of log-correlated Gaussian fields and Branching Brownian Motion has been recently (see [?] for instance) the subject of many studies. Initially defined for free Gaussian fields, it was extensively studied by physicists in case of random polymers on trees. In physics literature freezing is related to the behaviour of the free energy at low temperatures. The freezing phenomenon can be characterized either as the fact that the free energy becomes linear for temperatures beyond some "freezing temperature", or as the concentration of the Gibbs measure on the extreme values of fields.

In the following, we study the freezing property in relation with the entropy; in particular, for a class of Gaussian random models we give a generalized variational principle allowing to show that the freezing property is equivalent to the vanishing of the entropy of the corresponding Gibbs measure. We expect that the same is valid for the log-correlated Gaussian fields and this will be given in a future publication.

2. Random Gaussian Models

In order to have an unified description of the models to be considered, we first introduce a generic Gaussian model (GGM) as follows. Consider the dyadic tree defined from the alphabet: $\mathbb{A} = \{a, b\}$. The words (or paths) of length n define the natural configuration space denoted by $\mathbb{A}^n$, e.g. $\alpha = abaa$ with $n = |\alpha| = 4$. To each configuration $\alpha \in \mathbb{A}^n$ of the GGM, we associate the random hamiltonian $H_n(\alpha, \omega)$ whose explicit form is depending on specific model.

For $\beta = \frac{1}{T} > 0$ we define the partition function

$$Z_n(\beta, \omega) = \sum_\alpha \frac{e^{-\beta H_n(\alpha,\omega)}}{e^{\frac{\beta n}{2}}} := \sum_\alpha W_n^\beta(\alpha, \omega).$$

Content from this work may be used under the terms of the Creative Commons Attribution 3.0 licence. Any further distribution of this work must maintain attribution to the author(s) and the title of the work, journal citation and DOI.
Published under licence by IOP Publishing Ltd

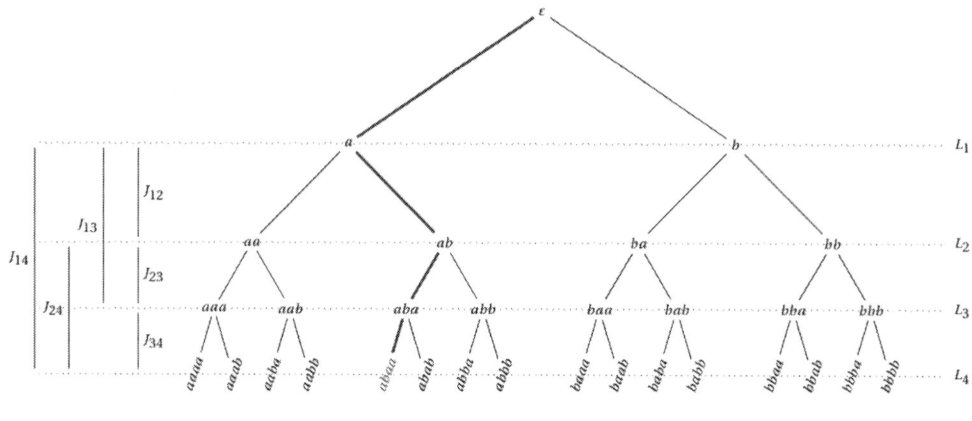

Figure 1: The tree corresponding to the Generalized Gaussian Model.

When the randomness J is fixed, the corresponding conditional Gibbs probability measure is denoted by

$$\mu_{n,\beta,\omega}(\alpha) = \frac{W_n^{\beta}(\alpha,\omega)}{Z_n(\beta,\omega)}.$$

Moreover, if E_J denotes the expectation with respect to the randomness J, the annealed free energy is given by

$$\overline{g}_n(\beta,\omega) = \frac{1}{n}\log E_J Z_n(\beta,\omega).$$

We recall moreover the definitions of the quenched free energy

$$g_n(\beta,\omega) = \frac{1}{n}\log Z_n(\beta,\omega),$$

and the entropy of $\mu_{n,\beta,\omega}(\alpha)$,

$$S(\mu_{n,\beta,\omega}) = -\sum_{\alpha} \mu_{n,\beta,\omega}(\alpha)\log \mu_{n,\beta,\omega}(\alpha).$$

It is well known that, $\forall \beta > 0$, the $n \to \infty$ (thermodynamic) limits exist *a.s.* and are non random

$$g_\infty(\beta) = \lim_{n\to\infty} g_n(\beta,\omega)$$

$$S(\mu_\beta) := \lim_{n\to\infty} \frac{1}{n} S(\mu_{n,\beta,\omega}).$$

From the variational principle one has also that

$$g_\infty(\beta) = S(\mu_\beta) + \lim_{n\to\infty} \frac{1}{n}\sum_{\alpha} \mu_{n,\beta,\omega}(\alpha)\log\frac{W_n^{\beta}(\alpha,\omega)}{\mu_{n,\beta,\omega}(\alpha)}.$$

The reader can now easily verify that the following models are included in the GGM:

2.1. The Random Energy Model

This model is indeed defined [?] by associating to each vertex of the nth level of the tree the i.i.d. centered Gaussian weights “energies” $(J_\alpha)_\alpha$ of variance $n/2$ indexed by the nodes. The hamiltonian is given by

$$H_n(\alpha,\omega) = \sum_{\alpha\in\mathbb{A}^n} J_\alpha(\omega),$$

and one can easily check that $EZ_n(\beta) = 2^n e^{n(\frac{\beta^2}{4}-\frac{\beta}{2})}$.

2.2. Random Polymers on Trees and Multiplicative Chaos
At each configuration $\alpha \in \mathbb{A}^n$, we associate [?] a family of i.i.d. Gaussian random "weights" $(J_{\text{parent, child}})$, indexed by the edges,

$$H_n(\alpha,\omega) = \sum_{\alpha\in\mathbb{A}^n}\sum_{i=0}^{n} J_{\alpha_1\alpha_{i+1}}(\omega).$$

2.3. Mean-Field Spin Glass Model
The Sherrington-Kirkpatrick (SK) mean-field model is defined in a complete graph [?, ?, ?, ?]. For this, we consider the family of i.i.d. centered Gaussian random interactions J_{ij} indexed by pairs of levels

$$H_n(\alpha,\omega) = \frac{1}{\sqrt{n-1}}\sum_{1\le i<j\le n} J_{ij}(\omega)\sigma(\alpha_i)\sigma(\alpha_j) \text{ (where } \sigma(a)=-1, \sigma(b)=1),$$

and it is very simple to show that $E_J Z_n(\beta, J) = 2^n e^{(\frac{\beta^2}{4}-\frac{\beta}{2})n}$.

These models are studied extensively and in different levels of rigor. There is a rich physics and, more recently, mathematical literature and studies of the phase transitions and properties of the Gibbs measures [?, ?, ?, ?, ?, ?]. Here we only focus on the relationship of the freezing property with the entropy of the Gibbs measureand show that vanishing of the entropy defines freezing as well.

All models covered by the GGM undergo a phase transition at a critical temperature β_c such that the high temperature region is characterized by

$$\beta_c = \sup\{\beta : g_\infty(\beta) = \overline{g}_\infty(\beta)\}.$$

Hence, at low temperature, from the concavity of logarithm, we always have

$$g_\infty(\beta) < \overline{g}(\beta), \text{ for } \beta > \beta_c.$$

We remark moreover that, for $\beta \le 1$, all previously defined models have similar thermodynamic behaviour and as we can see in the next section they have the same entropy $S(\mu_\beta)$.

3. Entropy versus Freezing

Returning to the GMM model, we define moreover the freezing temperature β_f by

$$\beta_f = \min\{\beta \ge \beta_c : S(\mu_\beta) = 0\}.$$

One can verify that $\beta_c = \beta_f$ for the REM, Random Polymers on Trees and Multiplicative Chaos models. In the mean-field spin glass (SK) model, the freezing temperature is bigger than the critical temperature. In particular, the SK model undergoes a phase transition at $\beta = 1$ but its entropy remains positive for small values of $\beta > 1$.

In the following we shall prove the

Theorem 3.1. *Almost surely, the freezing temperature for the Generic Gaussian Model occurs in the interval* $[\beta_c, \beta_*]$ *where* $\beta_* = 4\log 4 = 2.77258\cdots$. *Moreover, the specific entropy* $S(\mu_{\beta_*})$ *of the Gibbs measure vanishes.*

$$S(\mu_{\beta_*}) := \lim_{n\to\infty}\frac{1}{n}S(\mu_{n,\beta_*}) = -\lim_{n\to\infty}\frac{1}{n}\sum_{\sigma}\mu_{n,\beta_*}(\sigma)\log\mu_{n,\beta_*}(\sigma) = 0.$$

Journal of Physics: Conference Series **738** (2016) 012007 doi:10.1088/1742-6596/738/1/012007

Thus, for the Gaussian models we have considered, the value of β_* gives the maximum inverse temperature of positive entropy. As a consequence, the freezing phenomenon is common for all models covered by the GGM in the region $\beta \geq \beta_*$ and coincides with the fact that the entropy of the Gibbs measures vanishes. Since the GGM includes the SK model our result provides with the low temperature region where the freezing property is valid for the SK model as well.

For the proof of the theorem we first make the following remark concerning the inverse temperature β_*: the particular value of β_* is defined by the intersection of the graph of $\overline{g}_\infty(\beta)$ and the straight line $\frac{\beta}{\beta_1}\overline{g}_\infty(\beta_1) = \frac{\beta}{\beta_1}g_\infty(\beta_1)$. As one can see in figure 2, the annealed free energy $\overline{g}_\infty(\beta) = \log 2 + \frac{\beta^2}{4} - \frac{\beta}{2}$ is plotted as a function of β and the straight line is defined by $\frac{\beta}{\beta_1}g_\infty(\beta_1) \equiv \beta g_\infty(\beta_1)$. The two graphs intersect at $\beta_1 \equiv 1$ and $\beta_* \equiv 4\log 2 = 2{,}77258\ldots$ We can now readily check that the annealed limit, at $\beta = \beta_*$, is simply related to the limit at β_1 by

$$\overline{g}_\infty(\beta_*) = \frac{\beta_*}{\beta_1}g_\infty(\beta_1).$$

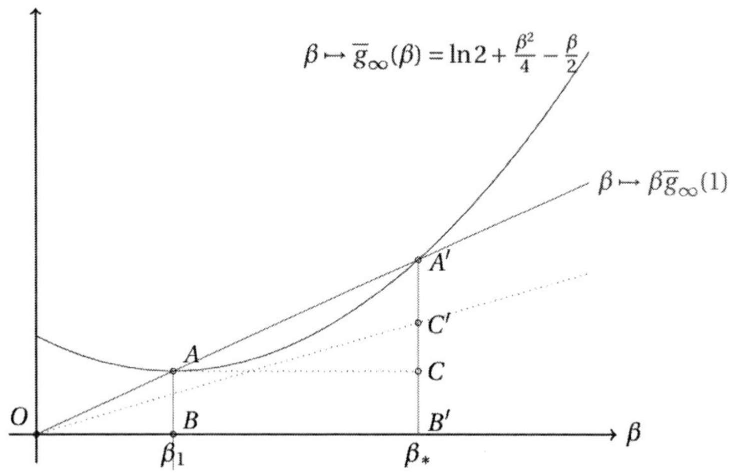

Figure 2: The value $\beta_* = 4\log 2$, is given by the intersection of the graph of the annealed free energy $\overline{g}_\infty(\beta)$ with the straight line $\beta\overline{g}_\infty(1)$.

In figure 2, the graph of the convex function $\overline{g}_\infty(\beta)$ is plotted and it is easy to check that the minimum is reached at $\beta = 1$. The same holds for the minimum of the limit $g_\infty(\beta)$ because $\overline{g}'_\infty(\beta_1) = \overline{g}'_\infty(\beta_1 - 0) = g'_\infty(\beta_1 - 0) = 0$ and the convexity of g_∞.

Introducing now the relative entropy entropy density $s(\lambda|\lambda')$ of the probability measure λ w.r.t. the probability measure λ', one can show easily that for all $\beta \geq \beta_1$, the limit $g_\infty(\beta)$ is related to the relative entropy by

$$g_\infty(\beta) = g_\infty(\beta_1) + s(\mu_{\beta_1}|\mu_\beta).$$

At the particular value $\beta = \beta_*$, one can show by a simple geometrical argument that

$$s(\mu_{\beta_1}|\mu_{\beta_*}) \geq (\beta_* - 2)g_\infty(\beta_1).$$

Using now the facts that at each point of the graph of a convex function there is a tangent line which never goes above the graph and the positivity of the entropy, we obtain, for $\beta_1 \leq \beta \leq \beta_*$, the convexity boundary of the limit $g_\infty(\beta)$ (represented by dashed lines AC and OC' in the figure).

In order to show that the entropy $S(\mu_{\beta_*})$ of the Gibbs measure vanishes at β_*, we assume that the relative entropy reaches its lower bound $s(\mu_{\beta_1}|\mu_{\beta_*})$. It follows, by the variational principle, that the entropy indeed vanishes. We conclude by remarking that, from convexity, if the relative entropy is bigger than its lower bound, the entropy of the Gibbs measure had already vanished at a value of β smaller than β_*.

4. Conclusion

In this note we showed the relationship between the entropy of Gibbs measure and the freezing property for a class of Gaussian Free Models. Another interesting question concerns the Hausdorff dimension of the support of the Gibbs measure. One can show using the theorem of this note that this dimension vanishes at β_*. We expect the same relationship for log-correlated Gaussian Fields and Brownian Branching Motion.

References

[1] M. Aizenman, J. L. Lebowitz, and D. Ruelle. Some rigorous results on the Sherrington-Kirkpatrick spin glass model. *Comm. Math. Phys.*, 112(1):3–20, 1987.

[2] P. Collet and F. Koukiou. Large deviations for multiplicative chaos. *Comm. Math. Phys.*, 147(2):329–342, 1992.

[3] Bernard Derrida. Random-energy model: an exactly solvable model of disordered systems. *Phys. Rev. B (3)*, 24(5):2613–2626, 1981.

[4] Francesco Guerra and Fabio Lucio Toninelli. The thermodynamic limit in mean field spin glass models. *Comm. Math. Phys.*, 230(1):71–79, 2002.

[5] F. Koukiou. The low-temperature free energy of the Sherrington-Kirkpatrick spin glass model. *Europhys. Lett.*, 33(2):95–98, 1996.

[6] F. Koukiou. Low temperature behaviour of the entropy for mean-field spin-glass models, 2016. Submitted for publication to *Reviews of Mathematical Physics.*

[7] G. Parisi. A sequence of approximated solutions to the S-K model for spin glasses. *J. Phys. A*, 13(3):L115–L121, 1980.

[8] L. A. Pastur and M. V. Shcherbina. Absence of self-averaging of the order parameter in the Sherrington-Kirkpatrick model. *J. Statist. Phys.*, 62(1-2):1–19, 1991.

[9] David Sherrington and Scott Kirkpatrick. Solvable model of a spin-glass. *Phys. Rev. Lett.*, 35:1792–1796, Dec 1975.

[10] Eliran Subag and Ofer Zeitouni. Freezing and decorated Poisson point processes. *Comm. Math. Phys.*, 337(1):55–92, 2015.

[11] Michel Talagrand. The Parisi formula. *Ann. of Math. (2)*, 163(1):221–263, 2006.

Risk management model in road transport systems

R L Sakhapov[1,2,3], R V Nikolaeva[1], M H Gatiyatullin[1] and M M Makhmutov[1]

[1]Kazan State University of Architecture and Engineering, 420043, Kazan, Zelenaya str., 1, Russia

[3]E-mail: rustem@sakhapov.ru

Abstract. The article presents the results of a study of road safety indicators that influence the development and operation of the transport system. Road safety is considered as a continuous process of risk management. Authors constructed a model that relates the social risks of a major road safety indicator - the level of motorization. The model gives a fairly accurate assessment of the level of social risk for any given level of motorization. Authors calculated the dependence of the level of socio-economic costs of accidents and injured people in them. The applicability of the concept of socio-economic damage is caused by the presence of a linear relationship between the natural and economic indicators damage from accidents. The optimization of social risk is reduced to finding the extremum of the objective function that characterizes the economic effect of the implementation of measures to improve safety. The calculations make it possible to maximize the net present value, depending on the costs of improving road safety, taking into account socio-economic damage caused by accidents. The proposed econometric models make it possible to quantify the efficiency of the transportation system, allow to simulate the change in road safety indicators.
Keywords: methods of mathematical modeling, motor roads, the risk assessment of the accident factors, the socio-economic damage, road safety performance management.

1. Introduction

The availability of reliable and efficient transport system is an important factor in the economic development of any country. Today transport is an important part of the global economy, as it is the material carrier between states. Area's well-developed transport system is one of the important factors to attract population and production and is an important advantage for the distribution of productive forces, which makes the integration effect.

The dominant type of transport in the world is an automobile that is growing rapidly every year. The importance of the road network is also reflected in the share of the total volume of passenger and cargo transportations by cars. Studies of foreign scientists have shown that road transport has one of the largest contributions to the budget revenues [1, 3]. For example, taxes and road user charges in the US totaled $78 billion in 1994 (6.2 percent of federal government revenue) and $33 billion in the UK in 1995-96 (of which only $10 billion was spent on the roads). The share of the transport system in the main production assets in Russia is 27%, and a significant share of transport services in GDP is 8%. Net financial flows from the road sector are generally positively correlated with economic development [8, 10].

---

[2] To whom any correspondence should be addressed.

Content from this work may be used under the terms of the Creative Commons Attribution 3.0 licence. Any further distribution of this work must maintain attribution to the author(s) and the title of the work, journal citation and DOI.
Published under licence by IOP Publishing Ltd

Efficient motor transportations are crucial to economic and social welfare of the country. However, in cities around the world, the existing transport network suffers from a policy of authorities responsible for traffic jams, unsafe conditions, high costs, pollution and poor service. As a result countries have huge economic losses, so the damage caused by congestion on the roads in European countries may be from 0.47% of GDP in Ireland to 2.95% in the Netherlands [1, 9]. In Russia, the loss caused by traffic congestion is 7-9% of GDP. A total annual loss from transport congestions in the US is estimated to be worth about $108 billion [4].

Therefore the question about the need to maintain and improve the existing road network and the construction of new infrastructure to increase the national wealth arises [11].

The economic losses of the country in which the third of major highways will work in overload mode, can reach tens of billions of rubles annually, and these costs will be included in the price of goods and services, which means the deterioration of living standards and reducing the competitiveness of the domestic economy, which in its turn, will reduce budget revenues [4, 6].

The researches of traffic flows that go through the Republic of Tatarstan have shown that they are growing faster than transport infrastructure develops. The need for roads is increasing annually, a large percentage of roads work in overload mode, and the entrances to major cities of the republic have become a permanent "bottleneck" in the movement of passengers and cargo. In this case, the increasing of the traffic, due to the steady growth of the number of cars, leads to the rapid destruction of road surfaces. Lag in the development of the roads limits population mobility, inhibits the development of the economy of the republic.

Inconsistency between the level of development of roads and the level of demand for road transport leads to unnecessarily huge budget expenditures. Thus, the annual damage from underdevelopment of the transport system of the Republic of Tatarstan is from 3.4 to 11.9 billion rubles, which corresponds to 1.2% and 4.1% of GRP. This compares with annual expenditure on transport that means the losses are equal or exceed the level of funding for the industry. From the point of view of the government (the ratio of "public investments" and return) the transport complex is very effective. Effectiveness of financing is 137%. This is one of the highest rates among the subjects of the Russian Federation [12].

The damage from undeveloped transport network occurs as a result of a combination of such factors as:

- Inefficient traffic management (social risk);
- Unrepaired road fund, failure of reserve maintenance periods of work on the roads (risk of severity of consequences);
- Existing traffic intensity exceeds the designed data of load on the road network of the Republic of Tatarstan, etc.

Setting of objectives for sustainable development of transport systems in terms of structural reforms of the economy should be carried out with the safety position. Achievement of road safety becomes the main organizational purpose of the functioning of the transport system and is considered as a continuous process of risk management.

In order to prevent road traffic accidents (RTA) it is necessary to evaluate the factors of accidents and manage road safety indicators. The possibilities of modeling help to understand better the mechanisms of accident, its development options and ways to prevent it. The present work is based on the main provisions of the theory of risk, analysis and assessment of the negative effects of an accident.

2. The impact of the transport complex on the economy

Due to its geographical position, a unique position for transport in the European part of Russia, the Republic of Tatarstan became one of the major economic regions of Russia. The Republic of Tatarstan is a competitive region, its competitive advantages is determined by its contribution to the Russian and global economy, comparative characteristics of socio-economic development, as well as the existing preconditions for sustainable growth of the Republican economic and social spheres [5].

One of the most important indicators of economic development of the Republic of Tatarstan is the gross regional product (GRP). GRP of the Republic of Tatarstan in 2014 was amounted to 1.631 trillion rubles; the growth rate was 1.3%. The growth of such indicator as the GRP and the development of the economy as a whole contribute to the fact that the republic steadily increases its macroeconomic indicators, which is important for investors.

The basic elements of the republic's economy are industry, construction, transport and communications, agriculture, wholesale and retail trade and other activities. The most important indicator of the importance of an industry to the economy and social sphere is the contribution of the sector to the formation of GRP.

Considering the basic spheres of the economy in the Republic of Tatarstan, we can say that transport is an essential part of the industrial and social infrastructure of the republic. The system of transport networks is a prerequisite for the region's territorial integrity and unity of its economic space. Transportation network connects the Republic of Tatarstan with other regions of Russia, with the international community, and appears as the base of the region's integration into the global economic system. Because of its geographical position, Tatarstan is one of the main transport arteries of the Russian Federation. In fact, the transport sector is the backbone sphere rather than simply serving infrastructure sector of the economy of the Republic of Tatarstan. The place and importance of transport is evidenced by its significant share in the basic production assets of the Republic of Tatarstan (13.2%) and significant share of transport services in the gross domestic product (6.2%), investments in the development of the sectors of the economy (10.1) and in the number of employed workers (6%) [12].

Among all types of transport the special role is played by automobile transport. Enterprises of the almost all industries give preference to it when shipping.

On average, the share of Tatarstan's trucking expenses is 6.02% of the cost of material production, and rail and inland waterway transport's shares are much less (0.63% and 0.35%, respectively). This clearly demonstrates the higher competitiveness of road transport. The percentage of cargo shipping provided by automobile transport in the Republic of Tatarstan is 60-70% of the total cargo traffic, and the percentage of passenger traffic is 50-60%.

Republic of Tatarstan is one of the leading regions in Russia in terms of the density of the road network - 406 kilometers per 1000 square kilometers (13th place of 83 subjects of the Russian Federation). For comparison, Moscow region has 695 kilometers per 1000 square kilometers (3rd place), Leningrad Oblast has 180 kilometers per 1000 square kilometers (42nd place) [12].

3. The theory of risks and assessment of socio-economic cost of road accidents

Indicators of safety in road transport should reflect the degree of safety of the transport system. They should take into account the impact of traffic safety on the timeliness and efficiency of road transport and allow to evaluate the effectiveness of road safety management.

For an integrated assessment of safety in road transport and the adoption of organizational and economic solutions for its improvement there can be used mathematical expectation of the socio-economic damage caused by the accident.

Applying the concept of socio-economic damage in this case is caused by the presence of a linear relationship between the natural (loss of life, loss of property) and economic (monetary) indicators of the damage caused by the accident.

Professor R. Smeed proposed a simple, but as shown by subsequent experience, a successful model that links social risks with the level of motorization. Based on the analysis of data on accidents of European twenty countries since 1938, he has developed a regression model (log-linear model) and set inversely proportional relationship between social risk (the number of deaths (injured) to the amount of the population) and the level of motorization (total fleet of vehicles to the amount of the population). This regression equation allows us to give a reasonably accurate assessment of the level of social risk for any given level of motorization [7].

In order to construct the model he used the heuristic considerations in the spirit of "the theory of dimension". In particular, he suggested that:

- the total number of road deaths (*D*) should be a power function from *P2* × *N*, i.e. *D~(P2* × *N) γ*;
- the specific number of people died in road accidents per unit of the car park (D / N) must be a power function of the level of motorization *(N / P)*, i.e. *D / N ~ (N / P) β*.

Calculations show that these assumptions imply that $2\gamma = -\beta$, $\gamma - 1 = \beta$ and, accordingly, $\gamma = 1/3$, $\beta = -2/3$.

From this we can get "Smeed formula" for transport risks:

$$\frac{D}{N} = \alpha \cdot \left(\frac{N}{P}\right)^{-\frac{2}{3}}, \tag{1}$$

Its obvious consequences are the formulas for the social risks and the total number of fatalities in road accidents:

$$\frac{D}{P} = \alpha \cdot \left(\frac{N}{P}\right)^{\frac{1}{3}}, \tag{2}$$

$$D = \alpha \cdot N^{\frac{1}{3}} \cdot P^{\frac{2}{3}}, \tag{3}$$

The coefficient α is the only adjustable parameter, which was used by the author of the model. The value of $\alpha = 30$ was calculated by twenty empirical points. Statistical tests showed a very high compliance of the model with the empirical data: the correlation coefficient $r = 0.87$ and $R2 > 0,7$.

To determine the damage caused by social risk in the Republic of Tatarstan there was built Smeed model in "motorization - social risk" coordinates, Figure 1.

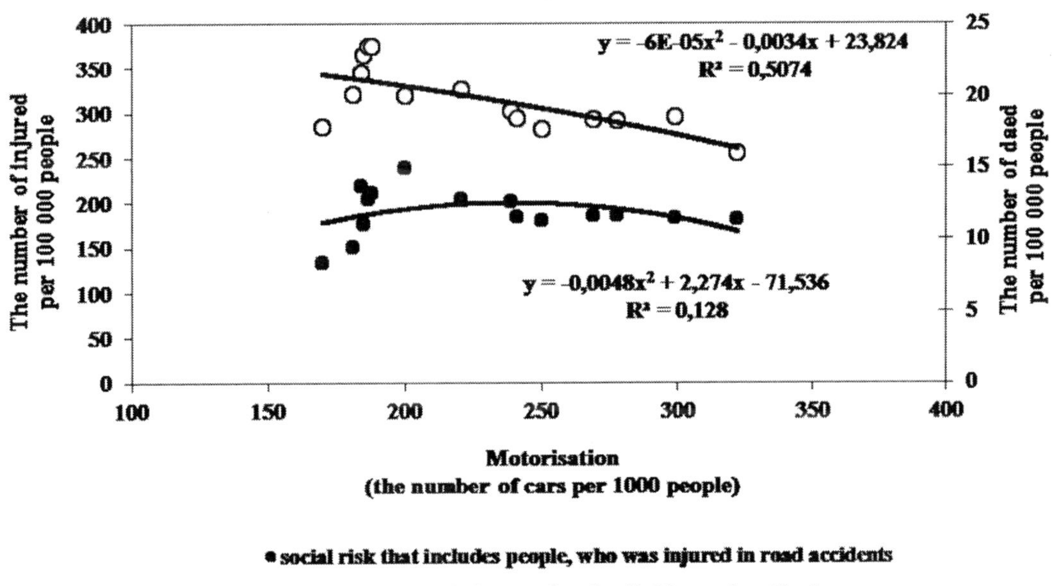

Figure 1. Smeed model in "motorization - social risk" coordinates of the Republic of Tatarstan in 2000-2014

The data presented in figure 1 shows that with the growth of car ownership in the country there is an increase in social risks, but after reaching a critical point (about 200 cars per 1,000 inhabitants) we can see a decrease of social risks as the further growth of motorization. This change in social risks can be attributed to the fact that the level of development of highways does not match the growth of motorization and the demand for road transport and it is resulted in a reduction in the average speed of the vehicle, appearing of traffic jams and overloads on the transport network and rising of accidents number.

Economic losses due to death and injuries in road accidents include losses due to the disposal of man from the sphere of production. The method of total revenues is used to estimate the losses due to the disposal of man from the sphere of material production. The basis of this method is expressed in monetary terms of the economic benefits that the society will receive due to the preventing of the death or injury of a person in an accident. In this approach human consumption is considered as an integral part of the state profits from the production and socio-economic activities of individuals.

The assessment of the socio-economic damage caused by the death and injuries in road accidents in the Republic of Tatarstan was held under the simplified method technique R-03112199-0502-00 "Methods of assessing and calculation of norms of social and economic damage caused by road traffic accidents (RTA)." This technique allows you to make an exact calculation of losses value from accidents for each year, because the assessment depends on the size of the GRP as the main indicator of the economic potential of the Republic of Tatarstan.

Dynamics of damage from road accident victims in the Republic of Tatarstan is shown in Figure 2.

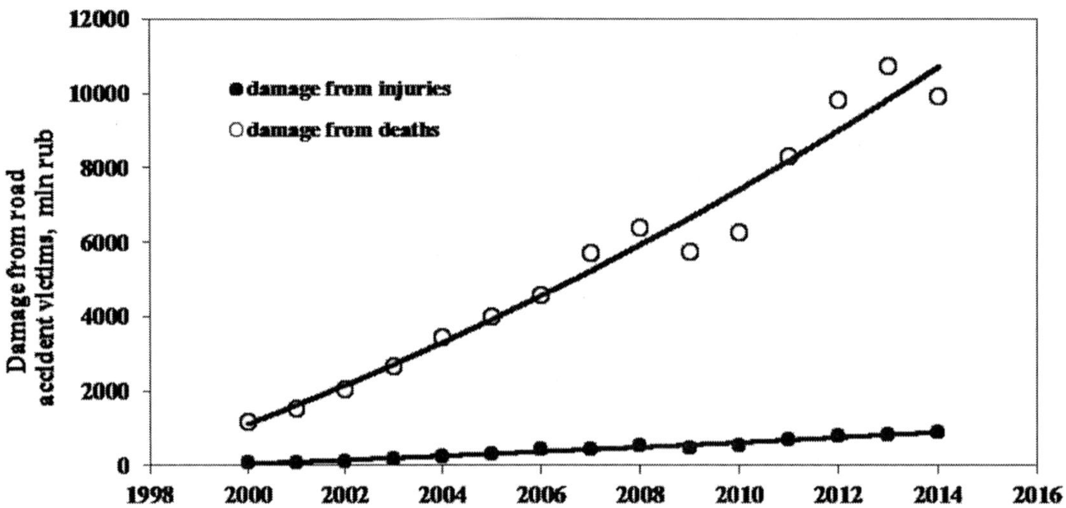

Figure 2. Dynamics of damage from road accident victims in the Republic of Tatarstan in 2000-2014

Calculations show that the damage from the people involved in road accidents in the Republic of Tatarstan increases every year, despite the fact of a decline in the accident rate in some years (the number of people affected in an accident). So, in 2014 the damage from the accident affected people in the Republic of Tatarstan was 0.65% of GRP or about 10.8 billion rubles, which could be used in the economy. If we consider that the social and economic damage is counted excluding damage from vehicles involved in an accident, without damage to the roads, this figure would be much higher.

Exploring this dependence we can assume that with a further increase of vehicles, number of people injured in road accidents will be reduced, but the damage will be bigger, while the growth of vehicles will increase other negative effects of growing motorization and the damage caused by the underdevelopment of the transport system will rise [10].

If we assume that the Z is a socio-economic damage (economic losses) from a traffic accident, P - probability of accident with victims (dead), then, according to the concept of $Z=\alpha \cdot P$, where the α is coefficient of proportionality, or the price of risk measured in units of currency per one additional death (cost of losses to the state in connection with the loss of health).

The other indicators required to assess the safety level of traffic in road transport can be:

- the probability of at least one traffic accident within a certain time;
- expectation of passengers delivery time delay and cargo safety violations;
- expectation of costs of preventing and reducing accidents consequences.

The investigation of these indicators makes it possible to quantify the effectiveness of road transport safety management.

Besides the frequency, it's necessary to take into account the extent of adverse events. The consequences of an adverse event can be expressed in many different ways. The most common way of expressing the consequences of the accident is the death of a person. On this basis, for the mathematical formalization, the risk R is nothing else but a function of two variables. These are the frequency of accidents (β) and the amount of damage (Z), i.e. $P = f(\beta, Z)$.

Individual risk is the probability of damage to health and (or) property as a result of accident for one individual. This indicator can be used to assess the safety of road transport in a particular region and the comparison of regions for this indicator with each other.

It should be noted that the particular person is primarily interested in the extent of its own security. The level of personal safety is an individual risk (p_i), or the probability of death of a person (harm to his health or property) as a result of an accident during a certain time period:

$$p_i = \sum_{j=1}^{m} \beta_j \mu_{ij} z_{ij} \ , \tag{4}$$

where β_j is the probability of occurrence of j-type; μ_{ij} is the probability that the i-th person is a party to the accident of j-type; z_{ij} is the probability that the i-th person in the accident of j-type die (get personal injury, property etc.).

Social risk determines the ratio between the probability of occurrence of damage and its size. This type of risk characterizes the severity of the consequences of the hazards associated with the occurrence of the accident.

Social or collective risk Po reflects the position of society and focused on the number of victims, which can be expected at this level of safety in road transport for a certain period of time. The value of social risk is determined by the statistics on accidents in road transport:

$$P_0 = \sum_{j=1}^{m} \beta_j G_j = \sum_{i=1}^{n} p_i \ , \tag{5}$$

where G_j - the number of died in the accident of j-type.

As can be seen from the above formula, the social risk is the expectation value of damage equal to the sum of the individual risks.

This indicator is important for assessing the impact of traffic safety in road transport on the economic security of the state as a whole.

In contrast with collective risk, social risk Pe additionally takes into account the magnitude of the risks γ:

$$P_e = \sum_{j=1}^{m} \beta_j G_j \gamma_j \ , \tag{6}$$

The inclusion of the scale factor γj in the calculations allows to assess accidents with a large number of victims.

For the public, as well as for the transport organization, collective risk is more significant. The probability of car accidents is small, so individual risks here are small and are characterized by a moderate value of the mathematical expectation of damage.

If the accident occurred and its consequences are huge, further work of transport organization may be associated with large losses due to the loss of competitive advantages. For the state the accident may also have large and long-term economic consequences, even an increase in social tension. In this regard, the government is interested in reducing the number of accidents.

4. Socio-economic benefits from improved road conditions in the republic

Economic aspects of road transport security management on the basis of the theory of risks require solutions of three groups of interrelated tasks:

1. The economic evaluation of risk indicators and the damage caused by the accident;
2. The development of economic mechanisms of road transport security control;
3. Optimization of an acceptable level of risk of an accident and the choice of the most effective measures to improve traffic safety in road transport.

Management of the risk of dangerous situations in road transport can be carried out at the particular accident source level, local, regional and federal levels. At the same time economic and mathematical formulations of risk optimization problems will be similar. There can be a probability of a modification of the objective functions at different levels of governance.

In every case, an important part of the control and risk optimization of traffic safety is the social and economic damage. This optimization procedure will be to search for the extremum of the objective function that characterizes the economic effect of the implementation of measures to improve safety. The meaning of this optimization is to maximize net present income CF, depending on the cost of increasing the security C, taking into account the socio-economic damage Z, caused by an accident. This is equal to minimization thc generalized reduced costs or expenses, including the cost of measures to improve road safety and the costs of compensation for damage (Figure 3).

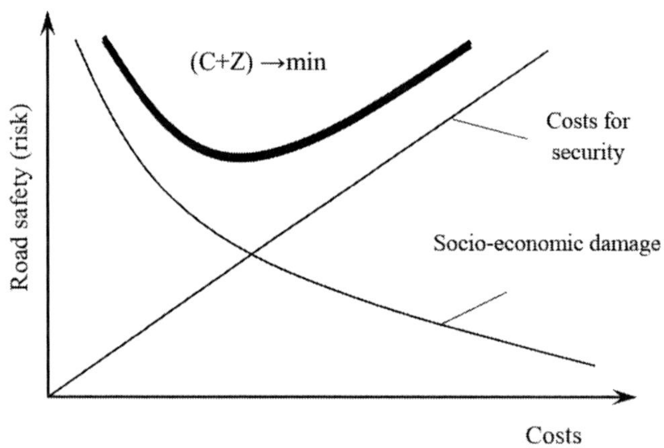

Figure 3. Dependence of the socio-economic costs of road accidents and the cost of measures to improve road safety

Optimality criterion of the cost of accident occurrence risk reduction with such an approach means that the limited economic costs of risk reduction must be equal to the unit cost of risk α:

$$MC = \frac{dC}{dP} = \alpha \ , \tag{7}$$

In this case, the level of economic costs must take into account which the maximum costs are allowed for the reduction of social risk, based on the death of one person. This is not the price of a human life, but rather the expenses which are society and the state ready to have to reduce the risk of fatalities on the roads.

In order to remedy this situation it is necessary to put Tatarstan roads in proper conditions. Currently part of the roads in the country is in a dangerous condition and requires a lot of money that would bring it into proper condition. The relevance of the question is caused by necessity of improving road safety, particularly by ensuring the efficiency of budget expenditures allocated for this purpose. The mechanism of formation of the positive effect that will be applied on the economy of the Republic of Tatarstan by the improving of road conditions is shown in Figure 4 [11].

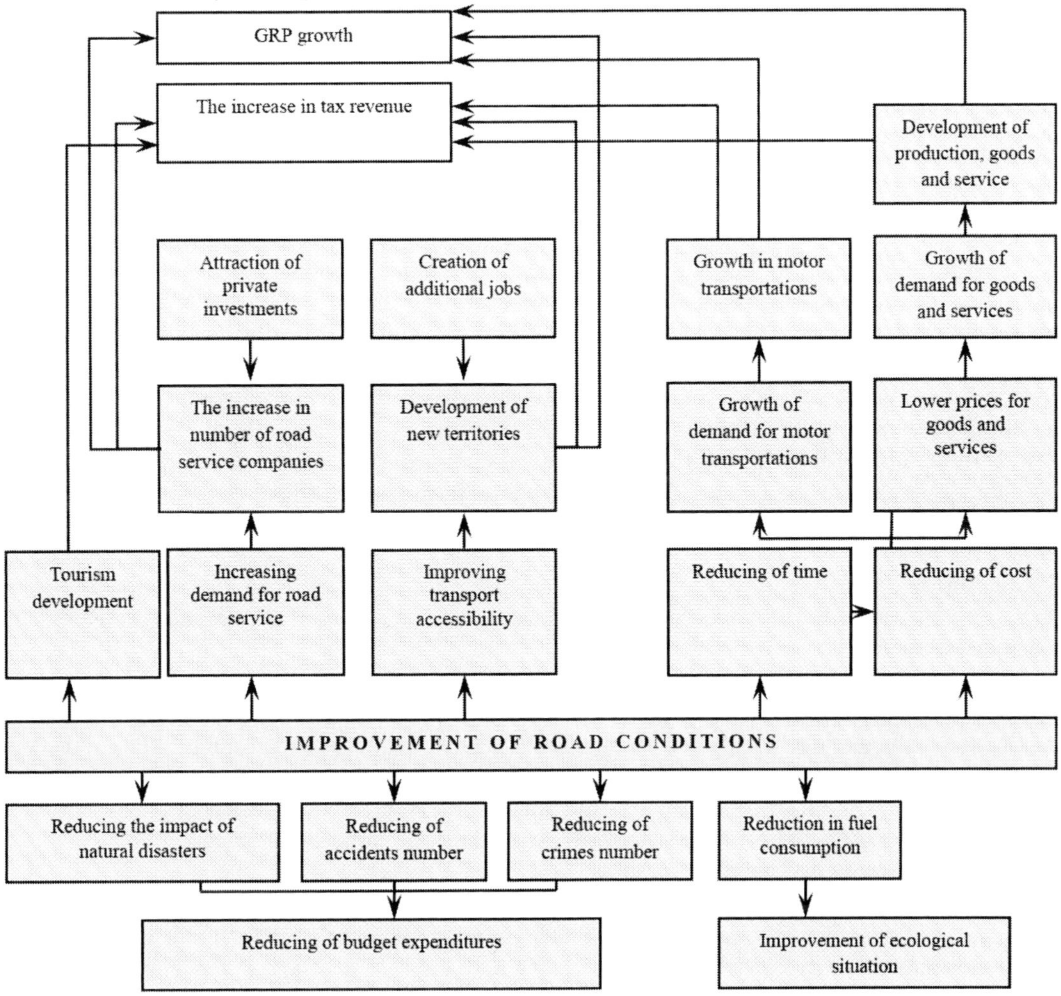

Figure 4. The impact of improved road conditions on the economy of the Republic of Tatarstan

Socio-economic benefits of improving the road conditions are determined by the influence of modernization and development of the road network on the costs of road transport and on the demand for road transport. Also there are the effects associated with increasing of vehicle speed and reliability of year-round connections across the network of roads [2].

Assessment of the socio-economic benefits from the implementation of measures for improvement the road conditions can be carried out in accordance with the Methodology for phasic improvement of transport and road performance, which is based on the principle of the calculation of the transport effects.

Transport effect is defined in terms of value on the basis of changes in the cost of passenger and cargo transport, the cost of road maintenance works, savings of costs of road traffic accidents and reduction of repair of vehicles:

$$E = C^{transport}_{pass+cargo} + C_{\substack{road\\ main-ce}} - C_{lands} + C_{\substack{repair\\ sav.}} + C_{accidents}, \quad (8)$$

where:

$C^{transport}_{pass+cargo}$ – savings from reducing the cost of passenger and cargo transport;

$C^{road}_{main-ce}$ – savings from reducing the cost of road maintenance works;

C_{lands} – the cost of lands for motorways;

$C_{\substack{repair\\sav.}}$ – savings from reduced repair of vehicles;

$C_{accidents}$ – savings from the reduction of road accidents.

Reducing the costs of road transport makes it possible to decrease the tariffs for transportation of goods and passengers, which affects the cost of goods, works and services in the intermediate and final consumption and leads to lower production costs or reduces producer prices, which leads to an increase of the demand for manufactured goods and services. These factors lead to an increase in GRP of Republic of Tatarstan. The reduction of transport margins on manufactured goods and services leads to additional (beyond reduction in producer prices) decrease in purchase prices of goods and services (customer prices) that even more increases the demand for them and value of GRP republic.

Studies have shown that 1 ruble invested in the construction of roads will give 3 rubles after 5 years and 1 ruble invested in the repair and reconstruction of roads - 6 rubles. The most effective investments are those in measures to ensure road safety: signposting gives 27 rubles on 1 ruble, lighting sections of roads gives 12 rubles, installation of barrier fencing – 11 rubles etc. [11].

5. Conclusion

A variety of methods of modeling and forecasting are being used to prevent dangerous traffic situations using.

Mathematical methods help to simulate a situation, assess it, to find ways to reduce the number of accidents.

Created mathematical models help to plan the measures to improve road safety. Marginal economic costs include the ratio of security and the cost of its maintenance; they cannot be defined solely on the basis of objective criteria. It shall enter into force some subjective assessments. The definition of marginal costs in order to prevent an accident is the subject of numerous studies.

They boil down to two main conclusions:

- procedures for the selection of measures of road safety improvement always contain assumptions about the availability of their funding. A large cost variation is possible, which indicates volatility of assessments;
- there is no universal marginal costs value for all occasions. Determining the limits of risk price, we should take into account the risk category.

The marginal cost of safety improvement can be determined on the basis of empirical research, the study of similar systems and processes, based on expert assessments and other ways.

Establishing the cost of risk unit price will allow for an economic optimization analysis of measures to improve safety in road transport.

The use of the methods of mathematical modeling is the possibility to assess factors of accidents and management of road safety.

References

[1] Aghadiayc E and Olamigoke' Adehayo A E, The Role of Road Transportation in Local Economic Development: *A Focus* 011 *Nigeria Transportation System*

[2] Gabriel R 2006 Street Smart: Competition, Entrepreneurship and the Future of Roads. *The Independent Institute*

[3] Heggie I G 1998 Commercial management and Financing of roads. *World Bank Technical Paper*. № 409

[4] Krepysheva N V 2008 Methodical approaches to the estimation of budget losses from traffic accidents. *Audit and financial analysis* №3

[5] Sakhapov R L and Absalyamova S G 2015 The use of telecommunication technologies in education network. *Proceedings of 2015 12th International Conference on Remote Engineering and Virtual Instrumentation, Bangkok, REV* pp 14-17

[6] Skvortsov O V 2012 Motorways as the country's economic development factor. *Science and Transportation Magazine. Transport building* №4 (4) pp 10-14

[7] Smeed R J 1949 Some statistical aspects of road safety research. (*Journal of the Royal Statistical Society. Series A (General),* Vol. 112: 1–34

[8] Zietlow G J and Bull A 2004 Reform of Financing and Management of Road Maintenance A New Generation of Road Funds in Latin America

[9] Long-term strategy of development of the transport complex of the Republic of Tatarstan (*from the standpoint of sustainable development*)

[10] R-03112199-0502-00 "Methods of assessing the calculation of norms of social and economic damage caused by road traffic accidents (RTA)"

[11] State program "Development of the transport system of the Republic of Tatarstan for 2014 - 2020", approved by the Cabinet of Ministers of the Republic of Tatarstan №1012 from 20.12.2013

[12] The official website of the Federal State Statistics Service, http://www.gks.ru/

The impact of corruption on the sustainable development of human capital

Svetlana Absalyamova[134], **Timur Absalyamov**[1], **Asiya Khusnullova**[2] **and Chulpan Mukhametgalieva**[1]

[1]Kazan Federal (Volga region) University, 18 Kremlyovskaya street, Kazan 420008 Russia
[2]Kazan Cooperation Institute (branch) of the Russian University of Cooperation, 58 Ershova street, Kazan 420081 Russia
[3]E-mail: s.absalyamova@yandex.ru

Abstract. The article explains the use of the human capital sustainable development index (HCSDI) to assess the quality of the reproduction of human capital. The paper provides the algorithm for calculating HCSDI and its components. Authors estimated cross-country differences of HCSDI and developed econometric model of the impact of corruption on HCSDI. The use of this model has allowed to reveal the mechanism and assess the impact of corruption on HCSDI and its components. The results of econometric analysis revealed a negative multiplier effect: an increase in the corruption of the socio-economic system of the state by 1% caused HCSDI reduce by more than 1%. The results and conclusions may be proxy-assessments of the socio-economic consequences of violations of the stability of reproduction of human capital in the conditions of the growth of corruption in the country.
Keywords: human capital, sustainable development, corruption, multiplier effect of corruption

1. Introduction

Nowadays the problem of increasing the efficiency of human capital in Russia is becoming particularly urgent. The possibility and pace of withdrawal from the crisis of the Russian economy depend on the correct chosen strategy of human capital sustainable development. In order to overcome the present crisis it is not enough to take measures to save the banking system, financial markets resuscitation and improve the investment climate. They should be complemented by programs of stimulation the development of country's human capital and improvement of the efficiency of its use.

Unfortunately, the high level of corruption in Russia, and especially in the social sphere, has a negative impact on the reproduction of the human capital. Corruption in the social sphere in Russia has reached a level at which the country has a serious loss in the economy, governance efficiency, quality of life, trust in public institutions. According to the degree of corruption, the lead areas are education and health care that directly impairs the quality of human capital.

Corruption, as a phenomenon, is widely studied by both foreign and Russian scientists and practitioners. The effectiveness of the proposed methods of combating corruption has been investigated by Elizabeth Harrrison [9], Mushtaq Khan [12], Graham Harrison (10), Dan Hough [11],

[4] To whom any correspondence should be addressed.

Content from this work may be used under the terms of the Creative Commons Attribution 3.0 licence. Any further distribution of this work must maintain attribution to the author(s) and the title of the work, journal citation and DOI.
Published under licence by IOP Publishing Ltd

and others. The last tries to track progress in the fight against corruption on specific country examples. The attempt to combine the description of corrupt practices in the broadest number of sectors of the economy has been made in the work under edition of Edgardo Campos and Pradhan Sanjay [5] under the auspices of the World Bank.
However, in our opinion, the impact of corruption on the stability of the development of human capital remains hardly explored. Insufficient degree of the problem development and the presence of a large range of controversial issues have defined the purpose and objectives of this study.

2. Materials and methods

The analysis of approaches to the evaluation of human capital allows us to choose a number of distinctive features in the selection of evaluation criteria. The classics of the human capital theory Gary Becker, L. Thurow, T. Shultz, M. Friedman [3,7,16,17] see the assessment of human capital in the calculation of the economic effects of its use and the revenue that will be generated in the future. M. Friedman considers human capital as a form of assets, alternative to money. A feature of this method is that it takes into account the total income of the individual property. Nevertheless, it does not reflect many of the indicators used for the analysis of human capital, a number of additional costs for human capital is not considered. G. Agiomirgianakis, D. Asteriou, V. Monastiriotis, E. Dinopoulos, P. Thomson [2,5] propose an approach to the evaluation of human capital, based on the accumulated knowledge during the personnel education, S. Cabrilo, X. Zhu, J..Huang develop their research [4,18].
The advantage of these approaches is the possibility of determining the income from the use of human capital, but it does not take into account the cost of its reproduction.
The other approach is the index method. The index method is used to bring the various elements of human capital to a single coordinate system.
The most famous index that is used in the practice of evaluation of human capital is the Human Development Index (HDI).
In our opinion, the drawback of HDI, as well as of other composite indices, is that the algorithm for their calculation does not include the degree of consistency with each other private indicator included in the calculation of the index. Meanwhile, if the values of the partial indicators of the index are relatively close to each other, then this index is more stable than in the case of private index figures are significantly different from each other.
To eliminate this drawback of HDI, we propose to use the human capital sustainable development index (HCSDI).

$$HCSDI = \frac{HDI}{\sigma} / \max\left\{\frac{HDI}{\sigma}\right\} \qquad (1),$$

where HCSDI – human capital sustainable development index;
HDI – human development index;
σ - standard deviation of private indicators included in the calculation of the HDI, from their average value;
max {HDI /σ} – the maximum value of the test indicator in the population.

$$\sigma = \left(\frac{\Sigma(X_i - HDI)^2}{n}\right)^{1/2} \qquad (2),$$

where X_i - the value of the i-th private indicator;
n - total number of individual parameters included in the composite index (in the case of HDI n = 3).
Division by max {HDI / σ} in proposed formula (2) is necessary in order to bring all values of HCSDI to comparable measurements scale - from 0 to 1 (by analogy with HDI scale).

The interest question is the mutual influence of the HDI and HCSDI on each other. In order to identify the nature of the influence we have constructed the regression model (Figure 1). The study went to the analysis of modern trends in the formation of human capital, carried out with using a statistical framework and methodological tools of UNDP [20], the Federal State Statistics Service of the Russian Federation and some selected surveys.

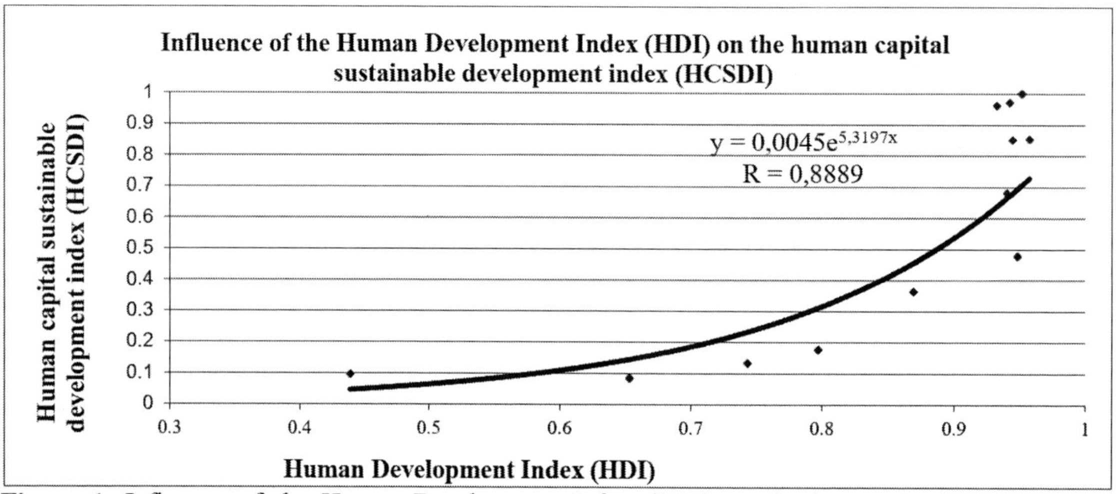

Figure 1. Influence of the Human Development Index (HDI) on the human capital sustainable development index (HCSDI)

As it shown in Figure 1, there is a stable relationship between the studied indexes, and it's evidenced by the rather high, close to 1, value of the correlation coefficient (R). So, there is a direct relationship of these factors. Therefore, ceteris paribus, the higher the value of HDI, the higher the value of HCSDI. The smaller the spread between the proportions of the individual elements of the reproduction of human capital (i.e. the higher the HCSDI), the higher the efficiency of the system as a whole (HDI). The conclusion is that the state, society and individual businesses should strive for sustainable, coordinated development of education, health and economic development in general. In this case the additional synergy effect will be provided for sustainable development of human capital.

3. Investigation of the influence of corruption on the sustainable development of human capital

The efficiency of the reproduction of the human capital is closely linked to an institutional trap of corruption. Views on corruption range from perceptions of corruption as a deviation from the law to a broader concept, including corrupt behavior that violates social and moral norms [8,13].

The most complicated forms of corruption acquired in countries with economies in transition.

Thus, according to The Corruption Perceptions Index Russia in 2015, Russia ranked 119 out of 167 countries [19].

Corruption in Russia is the result of the unformed institutional system. It gradually transformed from anomalies to the social norm, the most important regulator of social and economic relations. Corruption generates destructive economic, social and political effects such as reduce the rate of economic growth, reduce the public expenditure on education and health care, social stratification of the population, informal employment development [11], etc. All the listed effects of corruption negatively affect the development of human capital, but the most destructive factor is a reduction of costs on development of education and health care as on the basic spheres of formation and development of human capital. If in the size of bribes there are such leaders as lobbying legislative initiatives, corruption in the courts and state registration authorities, then in the intensity (frequency) of bribes in Russia leading spheres are education and health. [15]

Integrated model of the impact of corruption on the sustainable development of human capital can be represented as follows:

The impact of corruption on HCSDI = f(AQH, AQE, DI) (3),

where AQH -Availability and quality of health care;
AQE - Availability and quality of education;
DI - Differentiation of incomes of the population.

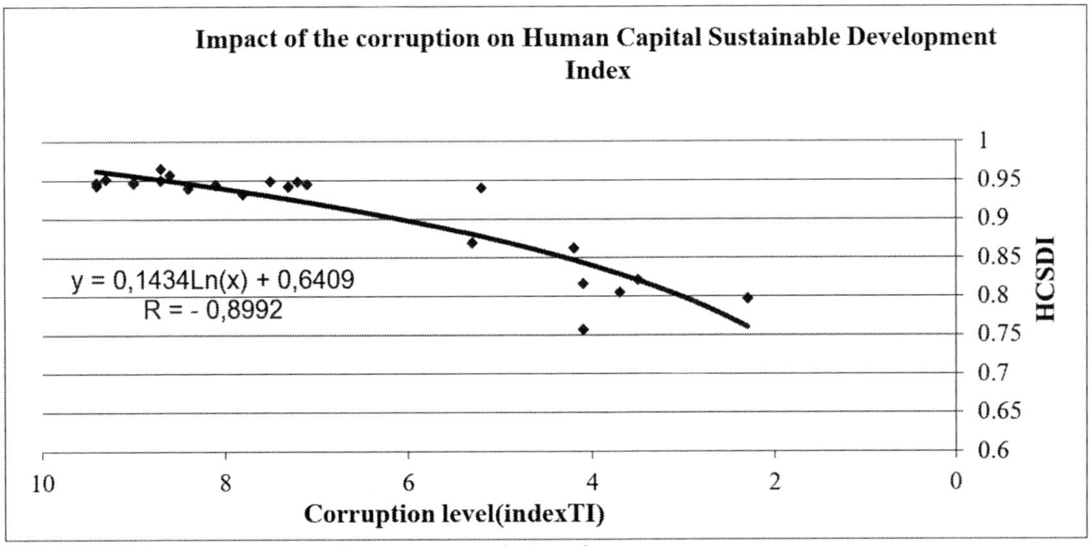

Figure 2. Impact of the corruption on HCSDI in Russia

The study has revealed the relationship between level of corruption in the state and the HCSDI (Figure 2). The resulting function is rapidly decreasing. This suggests that an increase in the level of corruption in the state by 1% leads to HCSDI reduced by more than 1%, i.e. there is a negative multiplier effect.

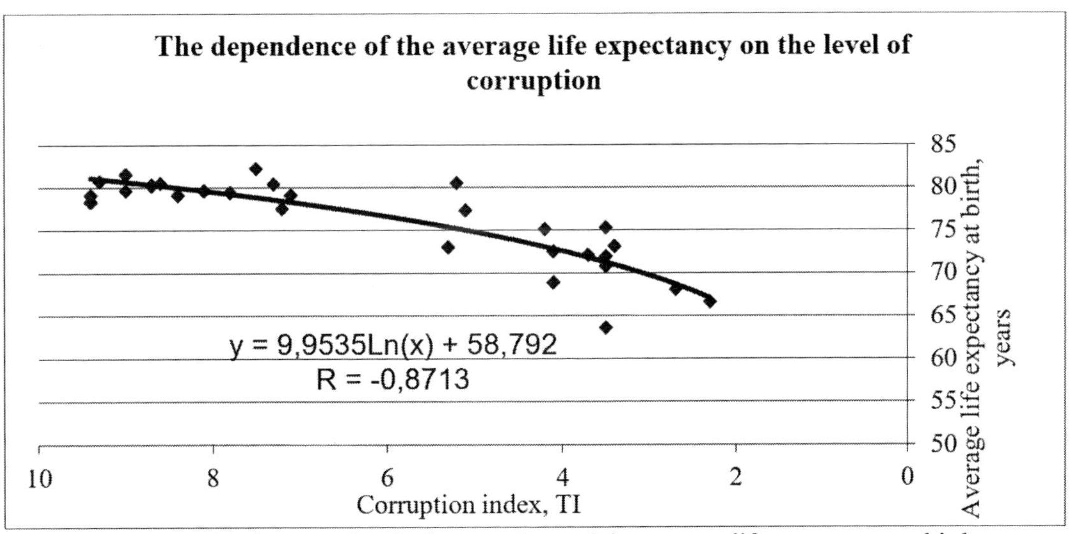

Figure 3. Relationship between level of corruption and the average life expectancy at birth

The negative impact of corruption on education and health is reflected in the following key trends:
- Corruption increases the costs of providing education [14] and health services, although formally the many types of services in a large part funded by the state;
- Corruption directly reduces the amount and quality of education and health services.

Effect of level of corruption in the socio-economic system of the state on the average life expectancy is shown in Figure 3.
The level of corruption has a negative effect on the differentiation of incomes of the population (Figure 4), which, in turn, undermines the stability of the reproduction of the human capital.

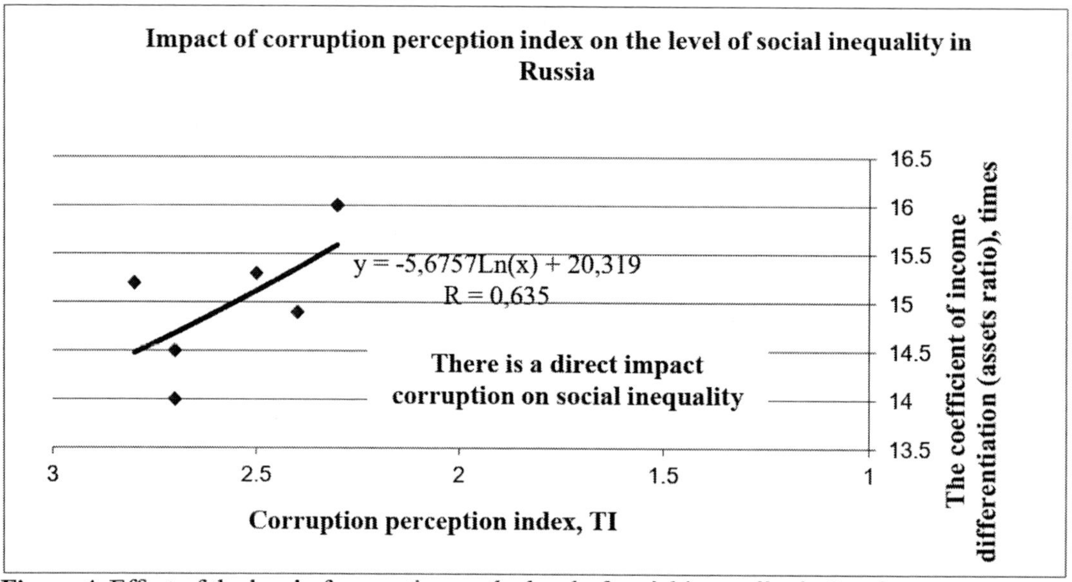

Figure 4. Effect of the level of corruption on the level of social inequality in Russia

Corruption has a greater influence on the income of the poorest segment of society, in other words it is reflected in the first place on the poor. The tax burden is being shifted on the poor, social programs funds aimed at reducing the income differentiation are being stolen.

4. Conclusions

To sum up, this study suggests the following conclusions:
1. Nowadays human capital has become, both in theory and in practice, an intense leading factor in the development, the foundation of the GDP growth, combined with innovations and new technologies.
2. The study proposes to calculate the human capital sustainable development index (HCSDI) for the assessment of the degree of consistency of particular indicators that are included in the human development index. HCSDI allows not only to assess the level of human capital of the State or the company, but also the degree of balance of its individual components. The reproduction of human capital is significantly affected by the cyclical economic development.
3. In Russia, the stability of human capital development is greatly influenced by the high level of corruption, which was transformed from an anomaly in a social institution. Reducing corruption level can significantly improve the quality of human capital and reduce income differentiation, and accelerate economic growth.
4. If the high corruption level in the system of socio-economic relations in Russia stays the same, then progressive reforms in the social sectors that have the greatest impact on the sustainable development of human capital will not be sufficiently effective.

References

[1] Absalyamova, S.G. , Absalyamov, T.B. Remote employment as a form of labor mobility of today's youth // Mediterranean Journal of Social Sciences Volume 6, Issue 1S3, 2015, pp. 227-231.
[2] Agiomirgianakis, G.,Asteriou,D.,Monastiriotis, V. (2002) 'Human capital and economic growth. Revisited: a dynamic panel data study'. International Advances in Economic Research. 3(8).pp.177-187.
[3] Becker, Gary, S. (2007) Health as Human Capital: Synthesis and Extensions. Oxford Economic Papers. 59(3). pp. 379-410.
[4] Cabrilo, S. , Nesic, L.G., Mitrovic, S. Study on human capital gaps for effective innovation strategies in the knowledge era // Journal of Intellectual Capital, Volume 15, Issue 3, 8 July 2014, pp. 411-429.
[5] Campos J.E., Pradhan S. The Many Faces of Corruption. Tracking Vulnerabilities at the Sector Level. World Bank. 2007.
[6] Dinopoulos, E.,Thompson, P. (1999) Reassessing the empirical validity of the humancapital augmented neoclassical growth model. Journal of Evolutionary Economics. 1(9). pp. 135-154.
[7] Friedman, M., Hayek,F. (2003) O svobode. Moscow: Socium.
[8] Haque M.E., Kneller R. Corruption Clubs: Endogenous Thresholds in Corruption and Developmen// Economics of Governance. 2009.
[9] Harrison E. Unpacking the Anti-corruption Agenda: Dilemm as for Anthropologists // Oxford Development Studies. 03/2006. (1360–0818). No34 (1). P. 15.
[10] Harrison G. The World Bank and Africa. The c onstruction of governance states. Routledge, 2004.
[11] Hough D. Corruption, Anti-Corruption and Governance. Palgrave Macmillan. April 2013.
[12] Khan M.H. Governance and Anti-Corruption. Reforms in Developing Countries: Policies, Evidence and Ways Forward. No. 42, November 2006. United Nations.
[13] Nielsen M., Haugaard J.Democracy, Corruption, and Human Development. Denmark, 2000.
[14] Sakhapov R., Absalyamova S. Integration of Universities and Business As the Condition of Formation of the Innovative Economy // 16th International Conference on Interactive Collaborative Learning & 42nd IGIP International Conference on Engineering Pedagogy, Kazan, Russia, 2013, pp. 189-191.
[15] Satarov G., Levin M. et al. Monitoring of Russian corruption // www.indem.ru
[16] Shultz, T.(1971) Investment in Human Capital. N.Y., London.pp. 26-28.
[17] Thurow, L. (1986) Zero sum solution: An economic and political agenda for the 80-s. N.Y.
[18] Zhu, X., Huang, J. Analysis of Chinese human capital investment strategy of scientific outlook on development // Engineering and Manufacturing Science - Proceedings of the 2014 International Conference on Future Information Engineering and Manufacturing Science, FIEMS 2014, 2015, pp. 319-322.
[19] The Corruption Perceptions Index//URL: http://gtmarket.ru/ratings/corruption-perceptions-index/info
[20] United Nations e-Government Survey 2012: From e-Government to Connected Governance //URL: hllp:// unpanI. un. org/intradoc/groups/public/documcnls/un/unpan 028607. pdf.

Cumulative sums for edge determination of a single object in PET and SPECT images

Nicholas E Protonotarios[1,2], George M Spyrou[3] and George A Kastis[1]

[1] Research Center of Mathematics, Academy of Athens, Athens 11527, Greece
[2] Department of Mathematics, National Technical University of Athens, Athens 15780, Greece
[3] Center of Systems Biology, Biomedical Research Foundation of the Academy of Athens, Athens 11527, Greece

E-mail: protost@hotmail.com

Abstract. The issue of edge determination of a single object in reconstructed nuclear medicine images has been examined thoroughly in the past, nevertheless most of the investigation has focused on the concepts of either numerical sinogram differentiation or segmentation. This work aims to develop an automated method for determining the contour of a single convex object in PET and SPECT reconstructed images, which can be used for computing body edges for attenuation correction, as well as for eliminating streak artifacts outside the specific object. This was accomplished by implementing a modified cumulative sums (CUSUM) scheme in the sinogram. Our method can automatically detect the object's boundary in the reconstructed image. This approach has been tested in simulated as well as real phantoms and it performed efficiently for all convex objects. We were able to detect the contour of a single object in the image space, which in turn enabled us to eliminate streak artifacts outside and thus to obtain body edges necessary for attenuation correction.

1. Introduction

Emission computed tomography (ECT) and its clinical applications can be employed to various medical fields such as cardiology, oncology and neurology. The two predominant nuclear imaging ECT modalities are positron emission tomography (PET) and single photon emission computed tomography (SPECT); they may be applied in either clinical or preclinical cases. PET is based on the detection of annihilation photons produced by decaying radiopharmaceuticals (such as fludeoxyglucose ^{18}F) injected into the patient and SPECT is based on the radiation of single photons detected by agent-labeled tracers (such as technetium ^{99}Tc).

Tomography is based on image reconstruction, specifically on the reconstruction from projection data such as sinograms. There exist two types of reconstruction algorithms, namely analytic and iterative. The main analytic reconstruction algorithm is *filtered back-projection* (FBP). FBP is fast and simple, however it usually generates "streak artifacts" in the reconstructions.

For the past three decades, edge determination through the sinogram of a single object for attenuation purposes in ECT has received considerable attention in the literature. Several methods have been suggested, [1–4]. These methods are applied to single objects being imaged and the first two of these methods emphasize on the skull.

Content from this work may be used under the terms of the Creative Commons Attribution 3.0 licence. Any further distribution of this work must maintain attribution to the author(s) and the title of the work, journal citation and DOI.
Published under licence by IOP Publishing Ltd

In the current work, we developed an edge-determination method for nuclear medicine reconstructed images. This method can be used in order to eliminate FBP streak artifacts outside the single object under investigation, as well as to obtain object edges for attenuation correction. Our approach applies a simplified cumulative sums (CUSUM) scheme in the sinogram, as originally proposed by E. S. Page [5].

2. Materials and methods

2.1. The Radon transform and its attenuated generalization

Let $g(x, y)$ denote the distribution of radioactivity. It is safe to assume that the function g is adequately smooth with compact support. If we introduce a set of so-called *local coordinates*, then every point (x_1, x_2) on a line can be expressed in terms of the line's (i) angle θ with the x_1-axis, (ii) distance from the origin ρ, as well as (iii) local arc length τ:

$$x_1 = \tau\cos\theta - \rho\sin\theta \quad \text{and} \quad x_2 = \tau\sin\theta + \rho\cos\theta. \tag{1}$$

We shall denote the Radon transform of g with $\widehat{g}$, defined as follows

$$\widehat{g}(\rho,\theta) = \int_{-\infty}^{\infty} g(\tau\cos\theta - \rho\sin\theta, \tau\sin\theta + \rho\cos\theta)\mathrm{d}\tau, \tag{2}$$

where

$$\rho = x_2\cos\theta - x_1\sin\theta \quad \text{and} \quad \tau = x_1\cos\theta + x_2\sin\theta. \tag{3}$$

The *sinogram* of $g(x_1, x_2)$ is none other than its Radon transform, $\widehat{g}(\rho,\theta)$, in matrix format and its elements $\widehat{g}_{ij}$ are given by

$$\widehat{g}_{ij} = \widehat{g}(\rho_i, \theta_j), \quad i = 1, \ldots, P, \quad j = 1, \ldots, \Theta. \tag{4}$$

Similarly, the attenuated Radon transform of g with linear attenuation coefficient $\mu(x_1, x_2)$, denoted by $\widehat{g}_\mu$, is defined as its weighted line integral, [6]:

$$\widehat{g}_\mu(\rho,\theta) = \int_{-\infty}^{\infty} e^{-\int_\tau^\infty \mu(s,\rho,\theta)\mathrm{d}s} g(\tau\cos\theta - \rho\sin\theta, \tau\sin\theta + \rho\cos\theta)\mathrm{d}\tau, \tag{5}$$

The *attenuated sinogram* of $g(x_1, x_2)$ with attenuation $\mu(x_1, x_2)$ is its attenuated Radon transform, $\widehat{g}_\mu(\rho,\theta)$, in matrix format and its elements $\{\widehat{g}_\mu\}_{ij}$ are given by

$$\{\widehat{g}_\mu\}_{ij} = \widehat{g}_\mu(\rho_i, \theta_j), \quad i = 1, \ldots, P, \quad j = 1, \ldots, \Theta. \tag{6}$$

Henceforth, we drop the subscript and we will denote either the sinogram or the attenuated sinogram simply with $\widehat{g}_{ij}$.

2.2. The CUSUM algorithm for edge determination

In medical applications such as PET and SPECT, both the sinogram and the attenuated sinogram have finite supports and hence we may assume that for every row of the sinogram there are two (left and right) "essentially zero-count" intervals. We also assume that these intervals' lengths are equal to L pixels and we treat L as a user-defined parameter. Due to the assumed left-right symmetry and without loss of generality, we will focus on finding the abrupt changes at the left side of the sinogram in a row-by-row basis. We introduce the CUSUM (cumulative sums) statistic $C_n^{(j)}$:

$$C_0^{(j)} = 0 \tag{7a}$$

$$C_n^{(j)} = \max\{0, C_{n-1}^{(j)} + g_{n,j} - (\mu_C^{(j)} + \lambda\sigma_C^{(j)})\}, \quad n = 1, \ldots P \tag{7b}$$

where $\mu_C^{(j)}$ and $\sigma_C^{(j)}$ are the mean and standard deviation of the "essentially zero-count" interval, i.e.:

$$\mu_C^{(j)} = \frac{1}{L}\sum_{i=1}^{L} \widehat{g}_{ij}, \qquad \sigma_C^{(j)} = \sqrt{\frac{1}{L}\sum_{i=1}^{L}\left(\widehat{g}_{ij} - \mu_C^{(j)}\right)^2}, \tag{8}$$

and λ is (a user-defined) positive, usually fixed at $\lambda = 3$, the value of which measures the size of the shift we aim to detect. It is obvious that, until it detects the abrupt change in question, the CUSUM statistic is non-decreasing, i.e. $C_k^{(j)} \geq C_l^{(j)}$ for all $k > l$.

In order to determine the left edge, we must find the index $i_L^{*(j)}$ of ρ such that the CUSUM statistic $C_i^{(j)}$ stops being zero for all $i \geq i_L^{*(j)}$, i.e.

$$i_L^{*(j)} = \max_{1<n<P} \ker C_n^{(j)}. \tag{9}$$

Similar considerations are valid for the right index $i_R^{*(j)}$, only this time we need to find the minimum of the kernel of the right CUSUM statistic $D_n^{(j)}$

$$i_R^{*(j)} = \min_{1<n<P} \ker D_n^{(j)}. \tag{10}$$

Given the above indices, we denote the non-zero interval of the variable ρ in row j with $A^{(j)}$:

$$A^{(j)} = \left[\rho_{i_L^*}^{(j)}, \rho_{i_R^*}^{(j)}\right]. \tag{11}$$

Given the information of $A^{(j)}$ for all rows, we are able to construct both the sinogram mask as well as the image mask.

2.3. Edge determination through sinogram and image masking

It is now evident that the sinogram mask $\mathbf{S}$ will be created as follows

$$\mathbf{S}(\rho_i, \theta_j) = \begin{cases} 1, & \text{if } \rho_i \in A^{(j)} \\ 0, & \text{otherwise} \end{cases}. \tag{12}$$

The next step is to calculate the image mask, $\mathbf{M}$. If we consider one of the results of Kastis et al. [7], namely that a pixel which is outside the boundary spanned by an object and hence has a value of 0, can be singled out from the sinogram by first identifying the detector locations $\rho_i = \rho_i(\theta_j)$ for all angles θ_j that receive contribution from this pixel. If this is the case, then for every (x_1, x_2) if there is even one θ_j such that $\widehat{g}(\rho_i, \theta_j) = 0$ it follows that $g(x_1, x_2)$ must be zero. This condition emerges from the nature of the sinogram, since it is a sum of positive terms (counts) and thus equals zero only when *all* terms are zero. In this manner, we are able to "reconstruct" the sinogram mask and calculate the resulting image mask,

$$\mathbf{M}(x_1^{(m)}, x_2^{(n)}) = \begin{cases} 0, & \text{if } \exists\ \theta^* \text{ such that } \mathbf{S}(\rho(\theta^*), \theta^*) = 0 \\ 1, & \text{otherwise} \end{cases}. \tag{13}$$

The resulting cleared image will then just be the element-by-element multiplication of the original image with the above image mask. This way the edges of the object are determined.

We implemented our algorithm in GNU Octave, which is a freely redistributable high-level interpreted language and software.

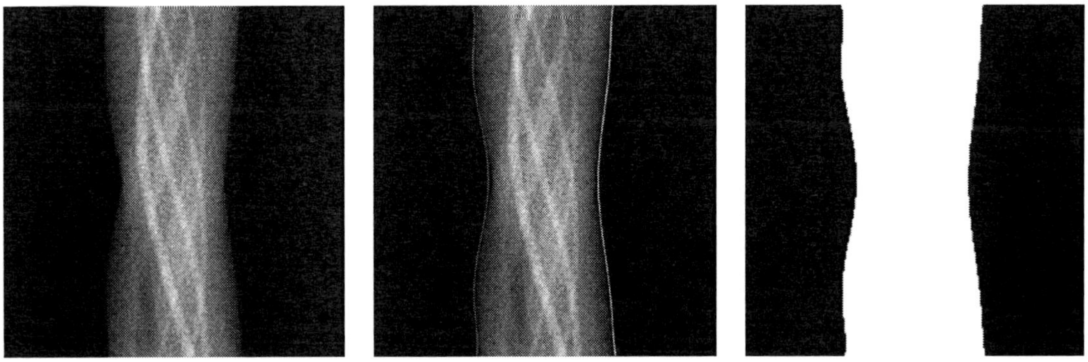

(a) (b) (c)

Figure 1: Sinogram masking for the simulated IQ phantom: (a) Initial sinogram, (b) Sinogram edge detection ($\lambda = 3$, $L = 60$) and (c) Sinogram mask

2.4. Simulation study

We performed a simulation study by employing the simulated sinogram of a simulated image-quality (IQ) phantom under Poisson noise of 10% of the total counts. Software for Tomographic Image Reconstruction (STIR) [8] was used to simulate the GE Discovery ST PET/computer tomography (CT) scanner. This specific scanner consists of 24 detector rings with diameter of 886.2 mm each. Every detector ring consists of 70 blocks, each of which consisting of an array of 6×6 crystals, i.e. a total of 420 crystals per ring. The sinograms' dimensions are $P = 221$ detectors and $\Theta = 210$ angles, corresponding to a detector bin size of 3.195 mm, see [9]. We used an 128×128 square image grid. We also set the parameter values $\lambda = 3$ and $L = 60$ pixels. The reconstruction was performed in STIR by employing 2D FBP with a ramp filter and a cutoff at Nyquist frequency.

3. Results

The algorithm's execution time was measured less than 5 seconds. The study was executed on a TOSHIBA laptop with Intel® Core™ i7-4710 HQ Processor, running on a 64-bit Windows® 10 environment.

The initial sinogram, the sinogram edge determination and the sinogram mask are shown in Fig. 1. The reconstructed image, image mask and masked image are shown in Fig. 2. This specific digital phantom simulates the human torso. It is clear from Fig. 2c, that the object is preserved intact and that FBP artifacts outside the object have been removed. Our algorithm has significantly eliminated the streak artifacts outside the object. The cleared image consists of 5471 nonzero pixels instead of the uncleared image, which consists of $128^2 = 16{,}384$ pixels, i.e. a percentage of less than 12% of the original number of nonzero pixels of the uncleared image.

4. Discussion and conclusions

Our main goal is to automatically determine the edges of single-object nuclear medicine images. In order to accomplish this goal, we implemented a CUSUM-based algorithm and tested it on an image quality simulated phantom. We performed sinogram edge determination and hence created a sinogram mask. Then we mapped the masked sinogram onto the image space and therefore we performed image masking and finally we determined the edges of the object under investigation.

The determination of the boundaries of a single object from its sinogram projections is essential for the determination of the attenuation correction map, if unavailable. Our method

(a) (b) (c)

Figure 2: Image masking for the simulated IQ phantom: (a) Image mask, (b) FBP (ramp filter) reconstructed image and (c) Masked image

is accurate and efficient and can be proven to be extremely useful, since it can be employed for other applications as well, such as computing body contours for attenuation correction in PET/SPECT systems that do not have a CT. Furthermore, in reconstruction techniques other than FBP (where the reconstruction is performed on a pixel-by-pixel basis), our method can substantially reduce reconstruction time, since it essentially cancels out the masked, zero-count pixels.

A concise method for the determination of the edges of single-object ECT images has been developed and tested. With our CUSUM-based method, we are able to systematically detect the boundaries of an object via its sinogram. Edge determination is essential for eliminating the streak artifacts outside the object under investigation, as well as for performing attenuation correction.

Acknowledgments
This work was partially supported by the research programme "Inverse Problems and Medical Imaging" (200/842) of the Research Committee of the Academy of Athens. The authors would like to thank Dr. Tsiamyrtzis of Athens University of Economics and Business (AUEB) for his brilliant suggestions.

References
[1] M. Bergström, J. Litton, L. Eriksson, C. Bohm, and G. Blomqvist. Determination of object contour from projections for attenuation correction in cranial positron emission tomography. *Journal of Computer Assisted Tomography*, 6(2):365–372, 1982.

[2] M. Hosoba, H. Wani, H. Toyama, H. Murata, and E. Tanaka. Automated body contour detection in SPECT: effects on quantitative studies. *Journal of Nuclear Medicine*, 27(7):1184–1191, July 1986.

[3] J. Case, M. King, D.-S. Luo, E. J. Soares, M. S. Z. Rabin, et al. Determination of concave body outlines from spect projection data. In *IEEE Nuclear Science Symposium and Medical Imaging Conference Record*, volume 2, pages 944–948. 1, 1995.

[4] L. R. Barnden, J. Dickson, and B. F. Hutton. Detection and validation of the body edge in low count emission tomography images. *Computer Methods and Programs in Biomedicine*, 84(23):153 – 161, 2006. Medical Image Segmentation Special Issue.

[5] E. S. Page. Continuous inspection schemes. *Biometrika*, 41(1/2):100–115, 1954.

[6] A. S. Fokas, A. Iserles, and V. Marinakis. Reconstruction algorithm for single photon emission computed tomography and its numerical implementation. *Journal of The Royal Society Interface*, 3(6):45–54, 2006.

[7] G. A. Kastis, D. Kyriakopoulou, A. Gaitanis, Y. Fernández, B. F. Hutton, and A. S. Fokas. Evaluation of the spline reconstruction technique for PET. *Medical physics*, 41(4):042501, 2014.

[8] K. Thielemans, C. Tsoumpas, S. Mustafovic, T. Beisel, P. Aguiar, N. Dikaios, and M. W. Jacobson. Stir: software for tomographic image reconstruction release 2. *Physics in Medicine and Biology*, 57(4):867, 2012.

[9] V. Bettinardi, M. Danna, A. Savi, M. Lecchi, I. Castiglioni, M. Gilardi, H. Bammer, G. Lucignani, and F. Fazio. Performance evaluation of the new whole-body PET/CT scanner: Discovery st. *European Journal of Nuclear Medicine and Molecular Imaging*, 31(6):867–881, 2004.

Network Analysis Using Spatio-Temporal Patterns

Gisele H B Miranda[1], **Jeaneth Machicao**[2] **and Odemir M Bruno**[2]

[1] Institute of Mathematics and Computer Science, University of São Paulo. Av. Trabalhador São Carlense, 400, São Carlos, SP - Brazil

[2] São Carlos Institute of Physics, University of São Paulo. Av. Trabalhador São Carlense, 400, São Carlos, SP - Brazil

E-mail: gimiranda@usp.br

Abstract. Different network models have been proposed along the last years inspired by real-world topologies. The characterization of these models implies the understanding of the underlying network phenomena, which accounts structural and dynamic properties. Several mathematical tools can be employed to characterize such properties as Cellular Automata (CA), which can be defined as dynamical systems of discrete nature composed by spatially distributed units governed by deterministic rules. In this paper, we proposed a method based on the modeling of one specific CA over distinct network topologies in order to perform the classification of the network model. The proposed methodology consists in the modeling of a binary totalistic CA over a network. The transition function that governs each CA cell is based on the density of living neighbors. Secondly, the distribution of the Shannon entropy is obtained from the evolved spatio-temporal pattern of the referred CA and used as a network descriptor. The experiments were performed using a dataset composed of four different types of networks: random, small-world, scale-free and geographical. We also used cross-validation for training purposes. We evaluated the accuracy of classification as a function of the initial number of living neighbors, and, also, as a function of a threshold parameter related to the density of living neighbors. The results show high accuracy values in distinguishing among the network models which demonstrates the feasibility of the proposed method.

1. Introduction

In the last decade, there has been a growing interest in Complex Networks, which is an area directly related to the study of complex systems [1, 2, 3]. The possibility of modeling different systems through the interactions between their constituent elements motivates the representation of discrete systems by means of a network structure. The focus of analysis becomes the connectivity among the elements of the system and not only the elements themselves. Many real world networks became available and significant studies have been conducted on the analysis of network structures with a large number of vertices. These networks have nontrivial topology and present communities and hierarchies [3, 4]. Consequently, statistical tools combined with graph theory have provided a more robust study of the topology of networks. The importance of network characterization is becoming more evident due to the vast amount of applications that can be handled by this approach. Different areas comprising biology, medicine, epidemic spreading, engineering, transportation, economy, scientific collaboration and social interactions are a few examples that can be benefited from network modeling. Many studies accounted that real networks are scale-free, with degree distributions that follows a power-law. The models that describe these types of networks incorporate important characteristics such as growth and

Content from this work may be used under the terms of the Creative Commons Attribution 3.0 licence. Any further distribution of this work must maintain attribution to the author(s) and the title of the work, journal citation and DOI.
Published under licence by IOP Publishing Ltd

preferential attachment [5, 6] in contrast to random networks on which all the vertices have equal probability of connecting to each other.

Among the tools that can be used to characterize the properties of networks are Cellular Automata (CA). Originally studied as models of growth and self-reproducing processes, CAs have become a relevant tool for the study of spatio-temporal pattern formation [7]. Besides being used as modeling tools, CAs are also employed in the study of complexity. They can be defined as discrete dynamical systems composed by spatially distributed units over tessellations of the Euclidean space, which are governed by deterministic rules. Regular tessellations are the most common topologies used for modeling CAs, however, real-world topologies are built upon irregular tessellations. Related works have shown that CAs can be combined with networks in order to analyze topological properties of complex systems under the perspective of pattern formation. Previous works on small-world networks were performed by Watts and Tomassini [8, 9]. Both authors discuss the global computing capacities of CAs as the density classification problem in small-world topologies. Marr & Hütt studied the dynamics of evolving networks through the use of CAs [10, 11]. Their results indicate a strong association between entropy measures obtained from the spatio-temporal patterns and the degree distribution of a network.

In this study, we propose a method that performs the classification of theoretical network models based on spatio-temporal patterns evolved by a cellular automaton. The methodology consists in the modeling of a binary totalistic CA over the topology of the network and in the definition of the CA dynamics based on the density of living neighbors (cells that are in the *on* sate). Then, quantitative measurements are obtained through the evolved patterns. Specifically, we used the distribution of the Shannon entropy to compose the feature vector of each network. The dataset used in the experiments performed in this paper is composed by four different types of networks: random, small-world, scale-free and geographical. We used two parameters in the evaluation of the classification accuracy: the initial number of living neighbors, σ, and the threshold parameter, μ, related to the density of living neighbors. It was obtained a maximum accuracy of 95.56% which demonstrates the feasibility of the proposed method.

2. Proposed Methodology

A cellular automaton can be described by the quintuple $\mathcal{C} = < \mathcal{T}, \mathbb{S}, s_0, \mathcal{N}, \phi >$, where, $\mathcal{T}$ is the tessellation of the automaton, $\mathbb{S}$ is the set of states, s_0 is the initial configuration of the states in $t = 0$, $\mathcal{N}$ is the neighborhood function, and, finally, ϕ is the transition function that governs the dynamics of the automaton. The first step of the proposed method consists in the modeling of the CA over the network topology. Therefore, the network becomes the tessellation $\mathcal{T}$ of the automaton and each network node corresponds to one CA cell. We assumed a binary state space $\mathbb{S}$ and s_0 is controlled by the parameter σ, which is a threshold that defines the percentage of initial alive nodes.

Since the network is an irregular tessellation, the number of neighbors of each CA cell is restricted to the neighborhood of each network node. Therefore, function $\mathcal{N}$ maps each node to its degree, which is defined as: $k_i = \sum_{j=1}^{N} A_{ij}$, where, N is the total number of nodes and A is the adjacency matrix. $A_{ij} = 1$, if i is connected to j and $A_{ij} = 0$, otherwise. Finally, the transition function of the referred network CA is given as a function of the neighborhood density, ρ_i, defined as $\rho_i = \frac{1}{k_i} \sum_{j=1}^{N} A_{ij} s_j(t)$, where $s_j(t)$ is the state of node j in time t. Once we have a binary state space, $s_j(t)$ can assume 0 or 1. Therefore, ρ_i represents the rate of alive neighbors of node i. The transition function ϕ incorporates the dynamic of the nodes according to the following equation [10]

$$x_i(t+1) = \begin{cases} x_i(t), & \rho_i \leq \mu \\ 1 - x_i(t), & \rho_i > \mu, \end{cases} \tag{1}$$

where μ is a threshold for ρ_i under which the node maintains its state in the next time step, otherwise, the node inverts its state. Fig. 1-a) illustrates the evolution of a given network with $\mu = 0.5$ using Eq. 1. The first newtork ($t = 0$) represents the initial configuration of states for each node where black cells represent the "dead" nodes, while the white cells, the "alive" nodes, *i.e.*, $s_i = 0$ and $s_i = 1$, respectively. Moreover, the subsequent figures show the configuration of the state for each node during three time steps.

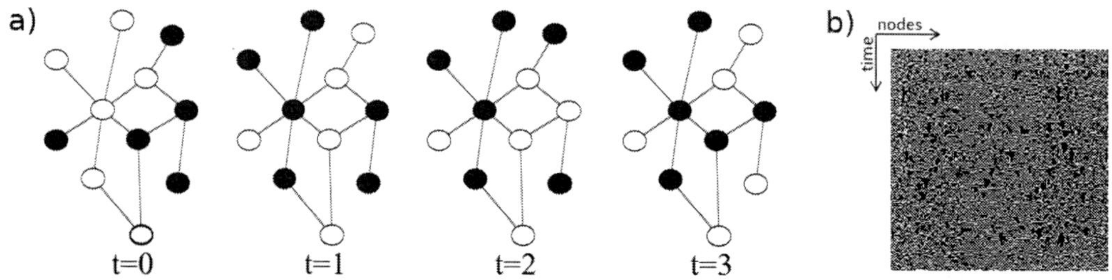

Figure 1. Evolution dynamics. a) Illustration of the configurations of the states of each node for a sample network using Eq. 1 as transition function. b) Space-time diagram for a small-world network with 200 nodes. The CA was evolved by 200 time steps.

The evolved pattern obtained by applying the transition function can be represented as a spatio-temporal diagram on which each column represents a network node and each row represents the states of all nodes at each time step. Fig. 1-b) illustrates an example for a small-world network. The nodes were ordered by k_i, therefore, the left-most nodes are the ones with the smallest values of k_i, and, the right-most, corresponds to the ones with the highest values of k_i. Further, we evaluated the evolution of each node as a time series containing zeros and ones. Then, the Shannon entropy was calculated for each node according to the following equation: $S_i = -(p_i^0 \log_2 p_i^0 + p_i^1 \log_2 p_i^1)$, where p_i^0 is the probability of finding zeros in the time series and p_i^1 is the probability of ones. The Shannon entropy quantifies how heterogeneous is the series and is normalized between $[0, 1]$. Finally, given the values of S_i for each node, we calculated the distribution of the Shannon entropy, Φ_S, for the whole network. Therefore, the vector Φ_S represents the network descriptor that characterize a network model.

3. Results and Discussion

In previous works, Marr&Hütt [12, 10] have shown that for different intervals of μ parameter, different spatio-temporal patterns can be obtained for a specific network topology. For low values of μ, oscillating patterns tend to prevail since the majority of the nodes will change their states at each time step. In contrast, for higher values of μ, the majority of the nodes will not change their states, and, consequently, stable patterns will prevail. Finally, for intermediary values of μ, chaotic and complex behavior tend to emerge. In this paper, we analyzed four different theoretical network models. Fig. 2-a) shows the variation of the average Shannon entropy (calculated for each network model by iterating over the Shannon entropy of each network node), for different values of μ. We can observe that the transition from oscillating patterns (high Shannon entropy) to stable patters (low Shannon entropy) occurs given distinct values of μ for each network model.

Based on this result, we evaluated the accuracy of network model classification as a function of the threshold parameter μ as well as the distribution of the alive and dead populations given by σ. We created a dataset composed of synthetic networks generated according to the following theoretical models: 1) random, with connection probability of $p = \langle k \rangle / n$; 2) small-world, with rewiring probability of $p = 0.1$; 3) scale-free, with power exponent $\gamma = 1$, and, 4) geographical,

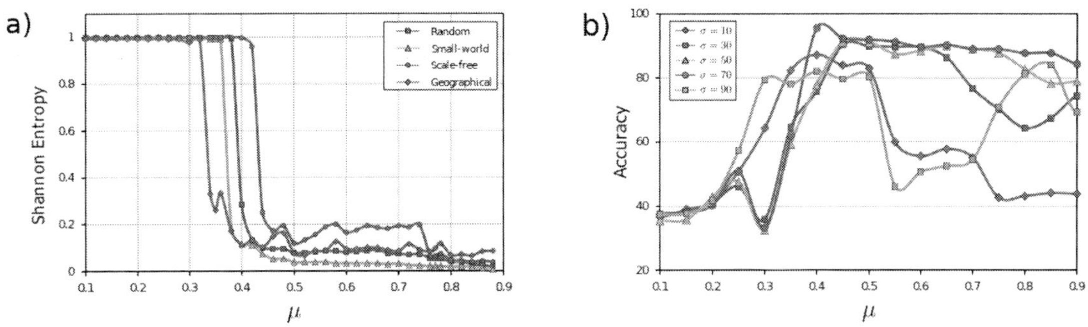

Figure 2. a) Shannon Entropy as a function of μ parameter for the four network models. b) Accuracy values (%) as a function of μ parameter for distinct initial conditions (σ parameter.

on which the Cartesian coordinates of the nodes are predefined. This dataset contains 100 networks of each model with mean degree $\langle k \rangle = 8$. Each network model represents a class in this experiment. By applying the proposed methodology, we obtained the Shannon entropy distribution, for each network, which were used as the network descriptor Φ_S. We have employed 10-fold cross-validation scheme and SVM (*Support Vector Machines*) classifier. In this context, accuracy values represent the number of correct classified instances. Fig. 2-b) presents the results for five different values of σ: 10, 30, 50, 70 and 90. Each curve of this figure represents a σ value. The x-axis corresponds to μ parameter and the accuracy obtained is shown on the y-axis.

We can see from Fig. 2-b) that the highest values of accuracy were obtained for intermediary values of μ and the influence of σ on accuracy becomes more apparent from $\mu = 0.4$. The highest accuracy obtained in the classifications of the correct network models was 95.56% for $\mu = 0.4$.

4. Conclusion

We have presented a method that performs network model classification based on the spatio-temporal pattern of a one parameter totalistic cellular automaton. This method achieved very good accuracy results and can be an alternative for network characterization.

Acknowledgements

G.H.B. Miranda and O.M. Bruno are grateful for the support from São Paulo Research Foundation (FAPESP), with grant 2015/05899-7. G.H.B. Miranda and J.Machicao are grateful for the support of Coordination for the Improvement of Higher Education Personnel (CAPES).

References

[1] Strogatz S H 2001 *Nature* **410** 268–276
[2] Albert R and Barabási A L 2002 *Reviews of modern physics* **74** 47
[3] Boccaletti S, Latora V, Moreno Y, Chavez M and Hwang D U 2006 *Physics reports* **424** 175–308
[4] Fortunato S 2010 *Physics Reports* **486** 75–174
[5] Barabási A L and Albert R 1999 *science* **286** 509–512
[6] Dorogovtsev S N and Mendes J F 2002 *Advances in physics* **51** 1079–1187
[7] Wolfram S 1984 *Nature* **311** 419–424
[8] Watts D J 1999 *Small worlds: the dynamics of networks between order and randomness* (Princeton university press)
[9] Tomassini M, Giacobini M and Darabos C 2005 *Complex Systems* **15** 261–284
[10] Marr C and Hütt M T 2009 *Physics Letters A* **373** 546–549
[11] Marr C and Hütt M T 2012 *Entropy* **14** 993–1010
[12] Marr C and Hütt M T 2005 *Physica A: Statistical Mechanics and its Applications* **354** 641–662

Molecular Dynamics Study of Polyethylene under Extreme Confinement

G Kritikos[1], A Sgouros[2], G G Vogiatzis[2] and D N Theodorou[2]

[1] School of Chemical Engineering, Aristotle University of Thessaloniki, GR-54124, Thessaloniki, Greece.

[2] School of Chemical Engineering, National Technical University of Athens, Zografou Campus, GR-15780 Athens, Greece.

E-mail: kritikgio@gmail.com

Abstract. We present results concerning the dynamics and the structure of adsorbed layers of molten polyethylene (PE) between two graphite surfaces. The molecular weight of the monodisperse PE chains reaches the entanglement regime. We study three cases of interwall distances, equal to two, three and four times the unperturbed radius of gyration (R_g) of PE chains. The confined system is equilibrated by use of efficient Monte Carlo (MC) algorithms. Conducting molecular dynamics (MD) simulations, we reveal the distribution of relaxation times as a function of distance from the graphite walls at the temperature of 450 K. From the atomic-level stresses we calculate a realistic estimate of the adhesion tension, which is not affected significantly by the width of the pore. Although the distance between the two walls is comparable to the width of the adsorbed layer, we do not record the formation of 'glassy bridges' under the studied conditions. The diffusion of polymer chains in the middle layer is not inhibited by the existence of the two adsorbed layers. Extreme confinement conditions imposed by the long range wall potentials bring about an increase in both the adsorption and desorption rates of chains. The presented results seem to cohere with a reduction in the calorimetric (heat capacity step) glass transition temperature (T_g).

1. Introduction

Upon cooling polymer melts, before the liquid to solid transition is reached, the viscosity and relaxation times increase dramatically. In some cases, extreme confinement conditions, in which the interwall distance (h_w) is comparable to the chain dimensions, have been related to a non-Einstein-like decrease in viscosity [1]. Moreover it has been shown [2–6] that attractive confining walls introduce one more mechanism (in addition to the caging effect and intramolecular conformational barriers) for time scale separation of glass forming polymers. This mechanism is the slow desorption kinetics of monomers from the surface, leading to slow layer exchange dynamics on the length scale of (at least) the radius of gyration of the chains. Several experimental works [4,7] have supported the idea that the confined/bounded layer is immobilized above the major T_g, in such a way that finally it does not participate in the glass transition.

In most atomistic simulation studies (MD, MC etc.) [2,5,6] the size of h_w has been considered much larger than the R_g. Nevertheless, in polymer nanocomposite science much attention has been focused on the case where layers of polymer adsorbed on neighboring particles overlap. In this situation an undefined mechanism seems to affect the mechanical properties of nanomaterials. The phenomenon is usually explained by the formation of 'glassy bridges' between the nanoparticles [8,9]. Moreover, it is not clear how these enhanced mechanical properties of polymer nanocomposites are related to

Content from this work may be used under the terms of the Creative Commons Attribution 3.0 licence. Any further distribution of this work must maintain attribution to the author(s) and the title of the work, journal citation and DOI.
Published under licence by IOP Publishing Ltd

polymer diffusion in the interparticle layer [1,9]. Our study focuses on the characteristics of the overlapping opposite adsorbed layers at a temperature far above T_g. Also, for the most confined system we have studied two cases of higher pressure. We have used combined MD and MC simulations in order to overcome the significant barriers of sampling under extreme confinement conditions.

2. Simulation details

The confined polymer systems were studied at the temperature of 450 K and pressure of 0.1 MPa. They consisted of 79 polyethylene (PE) chains of 200 carbons each. We have investigated three cases of confinement between the two opposite graphite surfaces located perpendicular to the z-axis (capped systems). A case of h_w equal to 100.0 Å (labeled as cap100), a case of h_w equal to 80.3 Å (cap80) and a case of h_w equal to 50.5 (cap50). In the case where one of the two solid surfaces of the cap80 system was replaced by vacuum, the supported system was labeled as sup80. The choice of the volume of the simulation box took into account the excluded volume of the graphite. In order to do so we have assumed that the density profile of the capped system near one of the substrates should be superposable with the density profile of the respective supported system [10]. In this way the middle layer has the density and the pressure of the bulk. The initial configurations of the confined polyethylene have been obtained by a MC builder capable of constructing polymeric chains of arbitrary geometry in heavily constrained environments [11], based on the quasi-Metropolis scheme of Theodorou and Suter [12]. Then, a MC equilibration of the initial configurations was carried out by use of connectivity altering moves, such as concerted rotation and double bridging [13,14]. All MD simulations have been conducted using the large-scale atomic/molecular massively parallel simulator (LAMMPS) [15].

We have used a united atom model for the description of the PE chains. Each methyl (CH_3) and methylene (CH_2) group along the chain backbone is treated as one interaction site [5,16]. Nonbonded interactions were described by the TraPPE force field [17] and bonded interactions (bond stretching, bond angle bending and torsional potential) were implemented following the work of Nath et al. [18]. The potential energy of interaction between a polymer atom and a semi-infinite crystalline graphite phase was calculated by the implementation of an efficient and accurate Fourier summation method designed by Steele [19]. According to the methodology introduced by Mansfield and Theodorou [20], the 10-4 sum was performed explicitly over the first 63 underlying layers of graphite atoms, while the contribution of the remaining layers was accounted for by a 9-3 potential. The Lennard-Jones interaction parameters between a polymer united atom and a graphite carbon atom were estimated according to the Lorentz-Berthelot rule.

3. Results and discussion

The wall potential used efficiently incorporates both the long range field of the graphite and the roughness of its basal plane. In figure 1 we present results concerning the density profile of the capped system compared with the respective density profile of a supported film. Although an accurate estimation of the polymer concentration in the graphite pore would strictly require a grand canonical approach, we can verify that the density fluctuations close to the wall are in agreement with the ones of a supported film of commensurate thickness [5]. In the same figure we distinguish three layers. Layer I is defined by the first density minimum at 6.25 Å from the wall. In this region the wall potential severely restricts displacement along the z-axis. A chain having at least one atom inside layer I is considered adsorbed [5]. Layer II, extends from 6.25 to 15.25 Å, where the presence of the wall field is weaker and so are the density fluctuations. Over the middle layer III, beyond 15.25 Å, density fluctuations almost vanish. The root mean squared radius of gyration R_g of the studied bulk PE was calculated as 25.8 Å. On the other hand, the extent of the adsorbed layer exceeds this distance, reaching up to 70 Å from each wall [4] according to the volume fraction profile (not shown). The bond order parameter for all capped systems (figure 2) confirms a middle layer of no order. The alignment parallel to the surface at the distance of 3.25 Å is caused by the trains and the loops [21], while the first maximum at 6.25 Å is interpreted as an alignment vertical to the wall at the distance of the first density minimum [4,22].

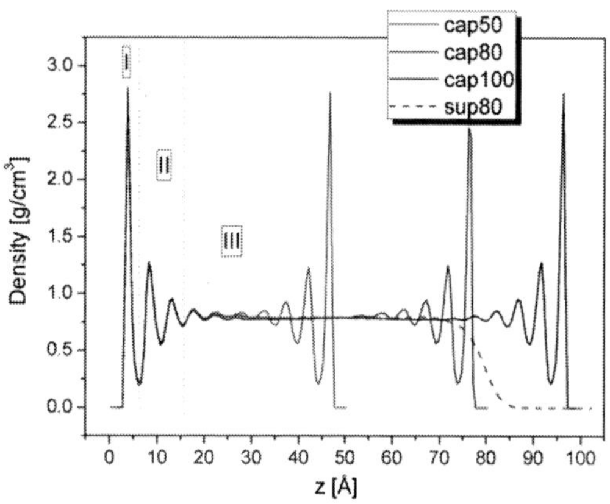

Figure 1. Density profiles of the three capped and one supported thin films.

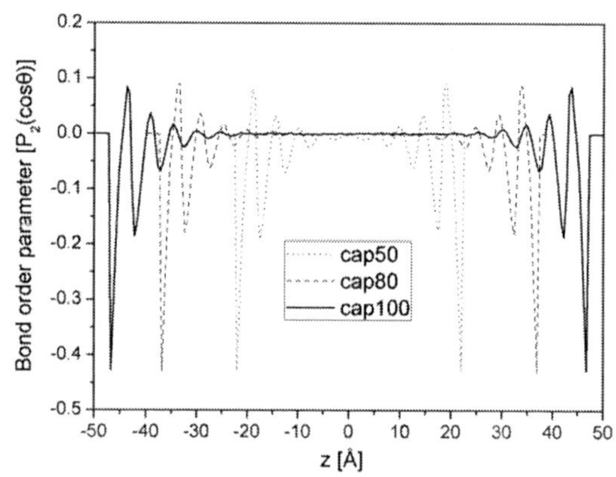

Figure 2. Bond order parameter of the three capped systems

An explanation of the observed structure in the adsorbed layer is possible if we measure the gradient in mobility in layers parallel to the wall. The results (figure 3) show that the displacement in the *xy*-plane (parallel to the graphite surfaces) is enhanced at distances where a significant suppression of the displacement along the *z*-axis (perpendicular to graphite surfaces) is present [5,10]. It is supported the idea that the small fraction of adsorbed chains which joins the decoupled layers I and II with different mobility in the *xy*-plane is stretched [22]. It is worth mentioning that a notable maximum peak in the volume fraction of chain free ends [23] is observed at a distance of 7 Å, inside the depletion layer of the density profile. The resulting density gradient seems to guide the first step from the non-adsorbed (free) state to the adsorbed state.

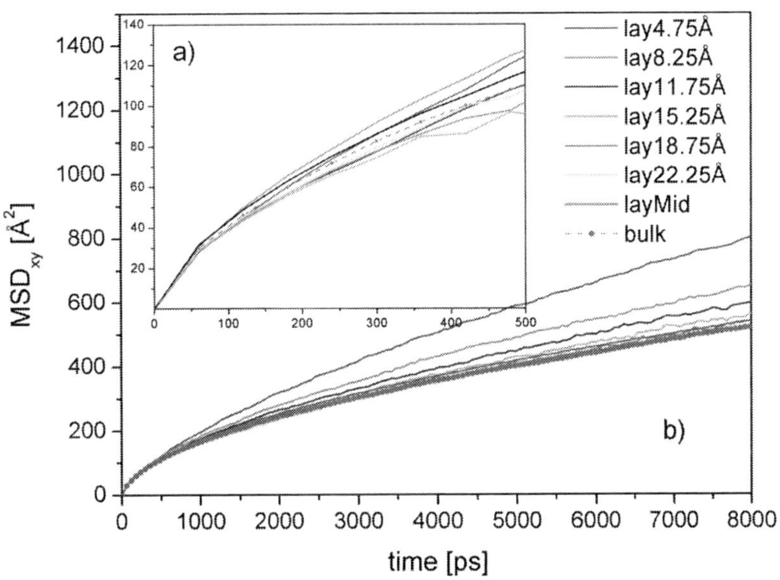

Figure 3. Mean square displacement (MSD) of the atoms based only on the x and y components of the displacement, parallel to the walls, for the cap80 system. The results are presented for six equal size layers (of 3.5 Å), parallel to the walls (region from 3 to 24 Å in z-axis) and the remaining middle layer. In the inset to the figure (a) we present results for short times while in the main figure (b) we present results for long times.

In the analysis presented in figure 3, an atom contributes to the displacement in a particular layer only if it is the same with the one the atom occupied at the time origin. At short times (lower than 1 ns, figure 3a) we have made the assumption that the atom should not escape the layer at intermediate times, while at longer times (figure 3b) in order to improve the statistics it is assumed that the displacement in the z-axis cannot exceed the width of the layer. The results indicate that although at short times, due to the wall roughness, the displacement in the second layer at 8.25 Å form the wall is faster, at longer times (several nanoseconds) it is the first layer at 4.75 Å that depicts the highest displacement in the xy-plane [5,10].

Chains adsorbed on both walls (bridges) are detected only in the case of the most confined system cap50. In the center of the cap50 system, the volume fraction of the observed bridges is 0.05, while the volume fraction of free chains is 0.47. Upon increasing the pressure in order to reach bulk densities of 0.81 and 0.84 g/cm^3, we do not observe a significant increase in the volume fraction of bridges. Efficient sampling at these high pressures was enabled by use of the MC algorithm. In our computational experiment, the used force field for PE at a temperature of 450 K and pressures of 0.1, 40 and 85 MPa did not confirm the formation of any kind of 'glassy bridges'.

The stress profile especially in the systems cap100 and cap80 confirms the existence of a middle layer with an isotropic pressure equal to 0.1 MPa. Moreover, the surface tension of a respective free film (both walls have been replaced by vacuum) of PE with 200 carbons was found equal to 23 ± 2 mN/m. The calculated adhesion tension for capped PE films of 200 and 260 carbons was equal to 90 ± 3 mN/m and 85 ± 2 mN/m respectively. We estimated a decrease of the adhesion tension as a function of the molecular weight [24], while under the studied conditions the adhesion tension was not affected by the interwall distance.

In figure 4 we depict the bond auto-correlation function (ACF). The results are presented for the layers defined in figure 1 and also for the loops and the tails of adsorbed chains. We observe that the increased decorrelation times are mainly attributed to layer I (trains). Also, tails decorrelate faster than loops. The same picture is supported by the torsional auto-correlation functions. In figure 5 we plot the decorrelation times of the torsion ACFs (fitted to the mKWW function [6]) as a function of the distance from the surface for the cap80 and cap50 systems. It is observed that the longest decorrelation time on the wall for the cap80 system is higher than the respective value for the cap50 system. In order to interpret this significant outcome, we have calculated the average time a chain lies adsorbed in each of the three capped systems. The residence times in the adsorbed state are 1.15 ± 0.05 ns, 1.21 ± 0.08 ns and 1.31 ± 0.09 ns for the cap50, cap80 and cap100 systems, respectively. The lower residence time in the cap50 system is justified by the high adsorption and desorption rates introduced by the bridges. It seems that, although the graphite wall establishes an adsorption/desorption process, under conditions of extreme confinement ($h_w < 3R_g$) the field of the opposite wall can cause an increase in the desorption rate [25].

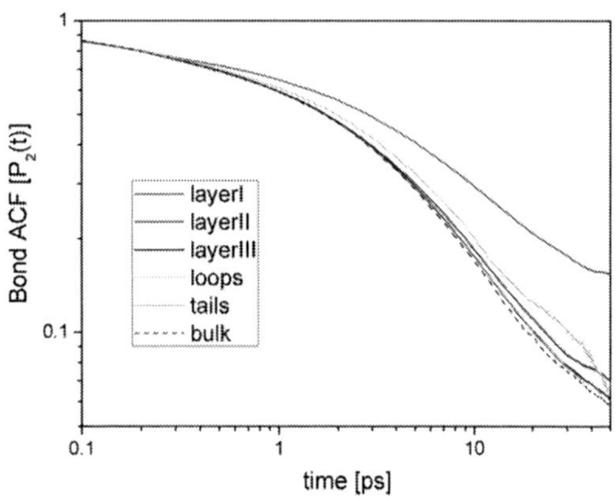

Figure 4. Bond autocorrelation functions (ACFs) for layers defined in figure 1 and also the loop and tail parts of the adsorbed chains for the cap80 system.

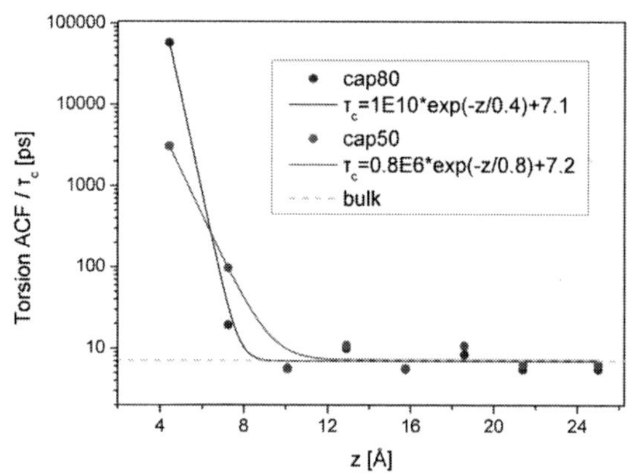

Figure 5. Comparison of the relaxation times estimated by the mKWW function for the cap80 and cap50 systems.

Finally using MD simulation trajectories of over 600 ns, we have calculated the diffusion coefficients of the centers of mass. The results are (MSD=1.5[MSD_x + MSD_y]): 3.5×10^{-7} cm^2s^{-1}, 4.0×10^{-7} cm^2s^{-1} and 3.7×10^{-7} cm^2s^{-1} for the cap50, cap80 and cap100 systems, respectively. In all cases the diffusion coefficients were higher than the one in the bulk (3.2×10^{-7} cm^2s^{-1}) [5,16].

4. Conclusions

We have presented results from atomistic simulations of PE under extreme confinement conditions, far above T_g. The initial configurations were equilibrated at all length scales by use of an efficient MC algorithm and trajectories of over 600 ns were produced with MD simulations. Even in the case of high pressure we do not record the formation of 'glassy bridges' under the studied conditions. The decoupling between layers of different mobility creates a density gradient where the concentration of chain free ends is increased. Moreover, the introduction of an opposite graphite wall (capped systems) seems to increase the desorption rate of chains. Our results agree with previous investigations [5] on thin supported PE films that indicate acceleration in dynamics compared to the bulk.

References

[1] Tuteja A, Duxbury P M and Mackay M E 2007 *Macromolecules* **40** 9427–34
[2] Yelash L, Virnau P, Binder K and Paul W 2010 *Phys. Rev. E - Stat. Nonlinear, Soft Matter Phys.* **82** 050801
[3] Theodorou D N, Vogiatzis G G and Kritikos G 2014 *Macromolecules* **47** 6964–81
[4] Kritikos G 2014 *Polym. (United Kingdom)* **55** 4658–70
[5] Daoulas K C, Harmandaris V A and Mavrantzas V G 2005 *Macromolecules* **38** 5796–809
[6] Pandey Y N, Brayton A, Burkhart C, Papakonstantopoulos G J and Doxastakis M 2014 *J. Chem. Phys.* **140** 1–13
[7] Wurm A, Ismail M, Kretzschmar B, Pospiech D and Schick C 2010 *Macromolecules* **43** 1480–7
[8] Merabia S, Sotta P and Long D R 2008 *Macromolecules* **41** 8252–66
[9] Dequidt A, Long D R, Sotta P and Sanséau O 2012 *Eur. Phys. J. E* **35** art. no. 61
[10] Kritikos G, Vergadou N and Economou I G 2016 *J. Phys. Chem. C* **120** 1013–24
[11] Vogiatzis G G and Theodorou D N 2014 *Macromolecules* **47** 387–404
[12] Theodorou D N and Suter U W 1985 *Macromolecules* **18** 1467–78
[13] Karayiannis N C, Mavrantzas V G and Theodorou D N 2002 *Phys. Rev. Lett.* **88** 105503
[14] Dodd L R, Boone T D and Theodorou D N 1993 *Mol. Phys.* **78** 961–96
[15] Plimpton S 1995 *Comput. Mater. Sci.* **4** 361–4
[16] Harmandaris V A, Mavrantzas V G and Theodorou D N 1998 *Macromolecules* **31** 7934–43
[17] Martin M G and Siepmann J I 1998 *J. Phys. Chem. B* **102** 2569–77
[18] Nath S K, Escobedo F A and Pablo J J de 1998 *J. Chem. Phys.* **108** 9905–11
[19] Steele W A 1973 *Surf. Sci.* **36** 317–52
[20] Mansfield K F and Theodorou D N 1991 *Macromolecules* **24** 4295–309
[21] Fleer 1980 *J. Phys. Chem.* **84** 178–90
[22] Kritikos G and Terzis A F 2013 *Eur. Polym. J.* **49** 613–29
[23] de Gennes P G 2000 *Eur. Phys. J. E* **2** 201–5
[24] Janczuk B 1985 *J. Colloid Interface Sci.* **108** 541–8
[25] Borah B J, Jobic H and Yashonath S 2010 *J. Chem. Phys.* **132** 144507

A Theoretical Approach to Engineering a New Enzyme

Greg Anderson[1], Raghu N. Behera[2], Ravi Gomatam[1]

[1]Bhaktivedanta Institute, Juhu Road, Juhu, Mumbai - 400049, India

[2]Birla Institute of Technology and Science Pilani, K.K. Birla Goa Campus, Zuarinagar, Goa - 403726, India

ganderson@bvinst.edu

Abstract. Density function theory, a subfield of quantum mechanics (QM), in combination with molecular mechanics (MM) has opened the way to engineer new artificial enzymes. Herein, we report theoretical calculations done using QM/MM to examine whether the regioselectivity and rate of chlorination of the enzyme chloroperoxidase can be improved by replacing the vanadium of this enzyme with niobium through dialysis. Our calculations show that a niobium substituted chloroperoxidase will be able to enter the initial steps of the catalytic cycle for chlorination. Although the protonation state of the niobium substituted enzyme is calculated to be different from than that of the natural vanadium substituted enzyme, our calculations show that the catalytic cycle can still proceed forward. Using natural bond orbitals, we analyse the electronic differences between the niobium substituted enzyme and the natural enzyme. We conclude by briefly examining how good of a model QM/MM provides for understanding the mechanism of catalysis of chloroperoxidase.

1. Introduction

Density functional theory (DFT), a subfield of quantum mechanics (QM), is successful in calculating the ground state energy of large molecules, but it cannot be applied to macromolecules like enzymes which have thousands of atoms [1]. On the other hand, molecular mechanics (MM), which applies Newtonian mechanics to molecules, has been applied to model the conformational states of enzymes; however, it does poorly in modeling chemical reactions in which covalent bonds are broken or formed. For modeling chemical reactions, DFT is needed. The combination of the DFT and MM (i.e., "QM/MM") has been applied to study a number of enzymes [2]. In QM/MM studies, DFT is used for modeling some portion of the active site, and MM is used to model the remainder of the enzyme.

Herein we utilize QM/MM to study vanadium chloroperoxidase (VCPO), which catalyzes chlorination of electrophilic substrates and compare our results with the literature [3-4]. QM/MM has also been applied to determine the ^{51}V chemical shifts of VCPO as well as the peroxo form of the enzyme [5]. DFT studies [6] have been carried on just the vanadium cofactor and methyl amine and guanidine to model the active site; however, QM/MM studies showed that long range electrostatic forces have very important effects on the protonation state of the vanadium co-factor [3].

VCPO catalyzes oxidation of chloride in the presence of cellular hydrogen peroxide:

Content from this work may be used under the terms of the Creative Commons Attribution 3.0 licence. Any further distribution of this work must maintain attribution to the author(s) and the title of the work, journal citation and DOI.
Published under licence by IOP Publishing Ltd

$$H_2O_2 + H^+ + Cl^- \rightarrow H_2O + HOCl$$
$$R\text{-}H + HOCl \rightarrow R\text{-}Cl + H_2O$$

In this process, the vanadium co-factor remains in the same oxidation state (i.e., +5) throughout the catalytic cycle, which is outlined in figure 1.

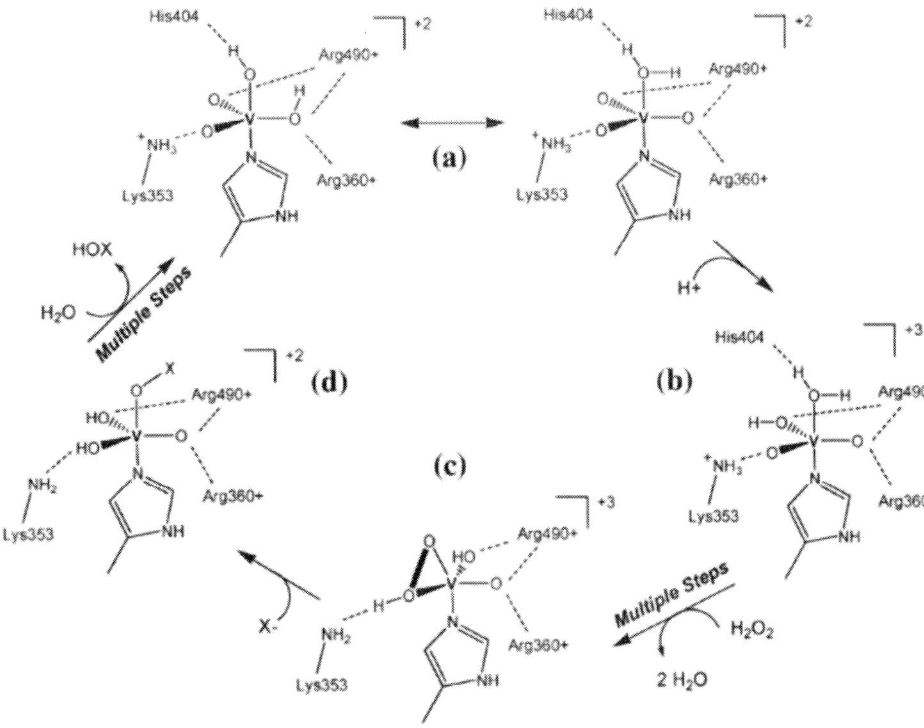

Figure 1. Catalytic cycle of VCPO (from [3])

An X-Ray crystal structure at a resolution of 2.15 Å reveals the active site as depicted in figure 2 [7]. This resolution is such that one cannot ascertain the protonation state of the VO_4 co-factor. The VO_4 co-factor has a trigonal bipyramidal structure and is held in the enzyme by hydrogen bonds from Lys353, Ser402, His404, Arg360 and Arg496 and possibly by a weak bond to His496. One can easily remove the vanadium cofactor by dialysis and reincorporate it [8]. There is no reason in principle why some other metal could not be substituted in the reincorporation step.

Our long range goal is to determine whether we can improve the rate and regioselectivity of VCPO by altering the transition metal in VCPO. We thus began this study by determining whether replacement of vanadium with niobium would even lead to a viable enzyme that could enter the catalytic cycle of VCPO.

2. Theory

The QM/MM approach to modelling of enzymes involves dividing the enzyme into a quantum mechanical part and molecular mechanical part. Thus, the overall energy of the enzyme is calculated as follows [9]:

$$E^{total} = E^{MM}_{MMpart} + E^{QM(MM)}_{QMpart} + E_{QM/MM} \quad (1)$$

The first term, E_{MMpart}^{MM}, is the MM energy of MM part of the system and is calculated with the OPLS-AA (Optimized Potentials for Liquid Simulations) force field [10].

The second term, $E_{QMpart}^{QM(MM)}$, is calculated by DFT and includes electrostatic interactions between the QM nuclei and the partial charges on the MM atoms. In DFT, the energy of the system is a functional of the probability density and can be factored into contributions from a kinetic energy functional, $T[\rho(r)]$, a potential energy functional, $J[\rho(r)]$, arising from classical Columbic interactions as well as an exchange correlation functional, $E_{XC}[\rho(r)]$:

$$E[\rho(r)] = T[\rho(r)] + J[\rho(r)] + E_{XC}[\rho(r)] \tag{2}$$

To find the energy, one assumes that "fictitious" orbitals, χ_i, can be assigned to represent each electron, 1, 2, 3, . . . *i*, . . . *N* where *N* is the total number of electrons in the system. The first two terms are calculated self-consistently in a way parallel to Hartree-Fock theory.

The remaining exchange-correlation energy functional is approximated by the adiabatic connection method. For this study, we used the hybrid B3LYP functional which involves parameterization as follows:

$$E_{XC} = (1-a)E_X^{LSDA} + aE_{XC}^{\lambda=0} + bE_X^{B88} + cE_C^{LYP} + (1-c)E_C^{LSDA} \tag{4}$$

The three parameters (*a*,*b*,*c*) were selected to fit a set of reference data (ie, the G2 data base) consisting of atomization energies, ionization energies, proton affinities and some total energies. DFT with B3LYP gives the energies within 1 kcal/mol for a surprisingly wide variety of systems [1]. Each of the terms of equation 4 is a complicated expression. For example, the first term is the following:

$$E_X^{LSDA} = \frac{A}{2}\left\{\begin{array}{l} \ln\dfrac{r_s}{r_s + b\sqrt{r_s} + c} + \dfrac{2b}{\sqrt{4c-b^2}}\tan^{-1}\left(\dfrac{\sqrt{4c-b^2}}{2\sqrt{r_s}+b}\right) - \\ \dfrac{bx_0}{x_0^2 + bx_0 + c}\left\{\begin{array}{l} \ln\left[\dfrac{\left(\sqrt{r_s}-x_0\right)^2}{r_s + b\sqrt{r_s} + c}\right] \\ + \dfrac{2(b+2x_0)}{\sqrt{4c-b^2}}\tan^{-1}\left(\dfrac{\sqrt{4c-b^2}}{2\sqrt{r_s}+b}\right) \end{array}\right\} \end{array}\right\} \tag{5}$$

The probability density is now represented by r_s instead of ρ and a different set of empirical constants (A, x_0, b and c) are needed.

The third term of equation (1), $E_{QM/MM}$, includes terms from van der Waals contributions which are calculated by MM. Additionally, one must treat the covalent bonds which are cut by the boundary between the QM and MM regions. We used the frozen orbital approach of Friesner and co-workers [2] in which the coefficients of the basis functions for the orbitals on the boundaries are set to zero; thus, the orbitals on the boundaries do not enter the self-consistent field calculation. Instead, molecular mechanics type parameters are introduced that provide an accurate energy function.

3. Methods

QM/MM calculations were done using the software, QSite [11]. The X-ray crystal structure (1VNI) of VCPO was downloaded from the Protein Data Bank (www.rcsb.org). QSite provides a preprocessing module that selectively removes waters from outside of the active site, i.e., which are more than 5 Å from vanadium. Because Impact, the software component of QSite which does MM, cannot process metal bonds, the preprocessing module makes the bond order zero between the vanadium and oxygens and between V and N of His496. Manually, we adjusted the charge on vanadium to +5 and the charge on the oxygens to -1 or -2 depending on whether the oxygen is protonated. This module also optimizes the hydrogen bond network and performs a restrained minimization with OPLS-AA force field to RMSD<0.3. For the QM calculations, we used DFT with B3LYP and a basis set of LACVP using electron core potentials for the metal and 6-31G** for the rest of the atoms in the QM region.. The QM region consisted of the side chains of Lys353, His496 with a VO_4 or NbO_4 cofactor, Asp292, and the full residues of Arg360, Arg490, His404, Gly403, and Ser402. Arg360 is adjacent to a Pro361 residue, and QSite does not allow backbone cuts in the proline ring. Therefore following the method of the literature [3], we converted Pro361 to alanine and included this full alanine residue in the QM part of the calculation. MM constraints consisted of the protein backbone and any atoms more than 20 Å from the metal. QM constraints were the amino acid backbone of Arg360, Arg490, His404, Gly403, Ser402 and His496 ring as well as the hydrogens of Lys353 facing away from the cofactor.

4. Results and Discussion

In order to assess the viability of a niobium substituted chloroperoxidase enzyme entering the catalytic cycle, we must first ask a large number of questions such as (a) what is the resting state of the niobium substituted peroxidase? (b) what is the lowest energy protonation state of each of the proposed intermediates (c) are the proposed steps still favorable upon niobium substitution? After doing QSite calculations, Kravitz et al. [3] concluded that the resting state of VCPO is an equilibrium between **1-V** and **2-V** based on the energy of the optimized structures (see figure 2). In our calculations, we find that **2-V** is lower in energy than **1-V** by 7 kcal/mole. The optimized structure, **2-V**, in comparison to **1-V** and **3-V** has bond lengths which are close to the bond lengths of X-ray crystal structure. In particular, for **2-V**, the V-O bond lengths for equatorial oxygens are all 1.69 Å, and the X-Ray data reveals 1.60 Å for V-O10 and V-O11 and 1.61 Å for V-O9. One possible explanation for this difference in results from those Kravitz et al. is the following. At the time of their research in 2005, QSite could only perform calculations on 8000 atoms, and VCPO is over 9000 atoms. Therefore, they did their calculations on a truncated version of the enzyme after justifying their truncation by examining the electrostatics of the enzyme. But even Kravitz et al. [3] noticed that long range electrostatic forces play a significant role in the relative energies of various protonation states of VCPO, and hence it seems possible that their analysis of electrostatic effects was incomplete, and their truncation affected the energies that they calculated. The other QM/MM study of VCPO neglected to even do calculations on **2-V** [4].

On substitution with niobium, we find a significant shift in energies of the various protonation states of the cofactor. Now, structure **3-Nb** becomes lower in energy than either **1-V** or **2-V**. Since **2-V** is lower in energy than **3-V**, it is puzzling why **3-Nb** should be lower in energy than **2-Nb**.

We undertook a study of the natural bond orbitals (NBO) in the optimized structures. Figure 3 summarizes the results of this study regarding the bond orders between the metal and oxygens. Namely, the co-factor of VCPO, (**2-V**), is actually VO_3. Furthermore, these results suggest the cofactor is not actually bonded to His496. Raugei et al. [4] in their QM/MM study of VCPO obtained the similar results on **1-V** and **3-V** by calculating Mayer's bond orders. Since the distance between V-N (2.15 Å) is less than the sum of the van der Waals radii of V (2.05 Å) and N (1.55 Å), it is possible that there are secondary bonding interactions [12].

Table I. Relative energies (in kcal/mol) of **1-3** shown in figure 2. The first column shows the results of Kravitz et al. [3]. The remaining two columns show the results of the present study where the metal (M) is either V or Nb.

	M=V	M=V	M=Nb
1	0.0	6.7	5.0
2	0.3	0.0	4.0
3	6.0	14.1	0.0

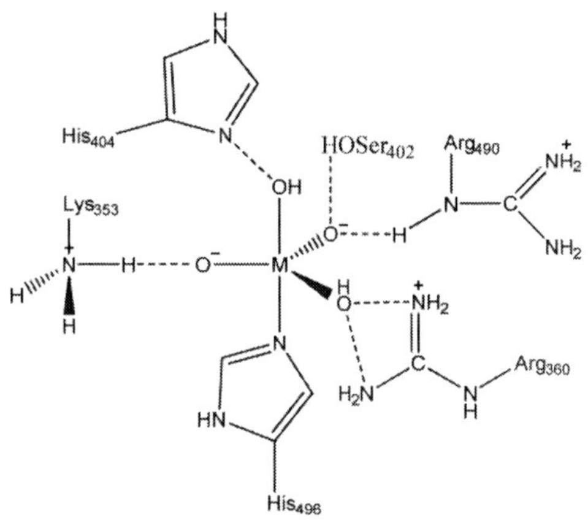

1-V: -2458921.2
1-Nb: -2449527.6

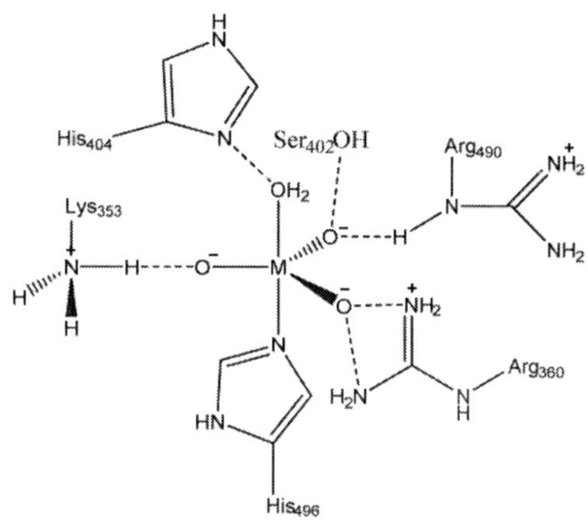

2-V: -2458927.9
2-Nb: -2449528.6

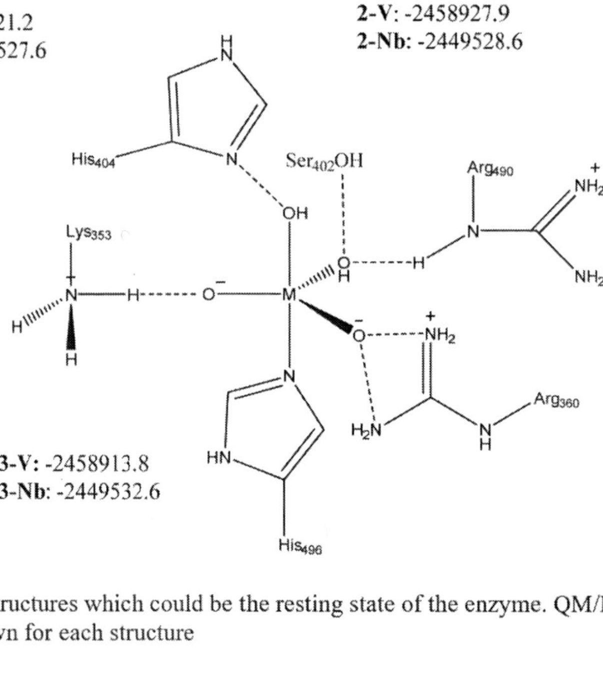

3-V: -2458913.8
3-Nb: -2449532.6

Figure 2. Various possible structures which could be the resting state of the enzyme. QM/MM energies in kcal/mol are shown for each structure

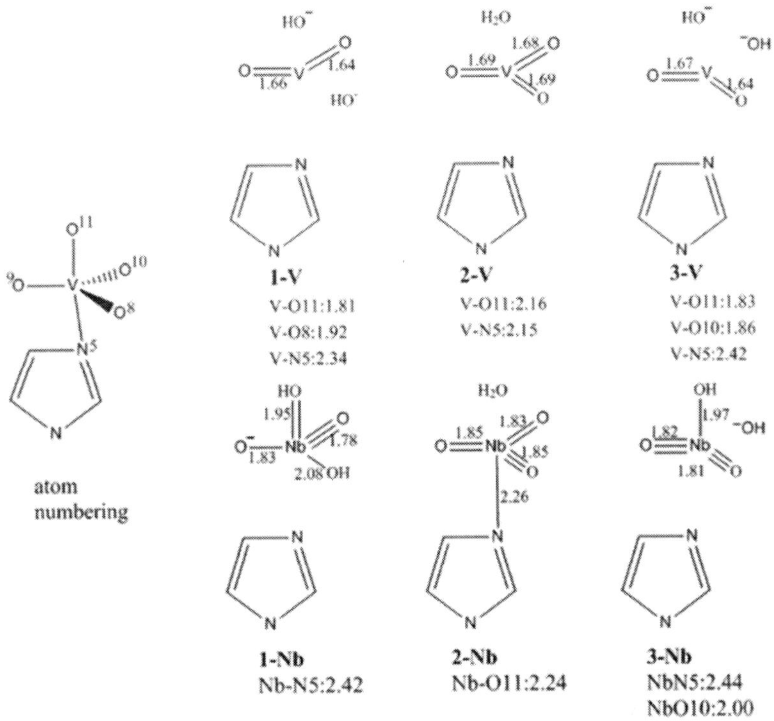

Figure 3. Bond orders based on a NBO study of optimized structures **1-3**. Bond lengths were obtained from structures optimized by QM/MM (i.e. QSite).

It is **2-V** which enters the catalytic cycle of the natural enzyme, VCPO. The question is whether **3-Nb** can enter the analogous catalytic cycle. In other words, is protonation of **3-Nb** to yield **4-Nb** (figure 4) favorable? Calculation of the free energy of **3-Nb** and **4-Nb** shows that protonation is a favorable reaction by 258 kcal/mol.

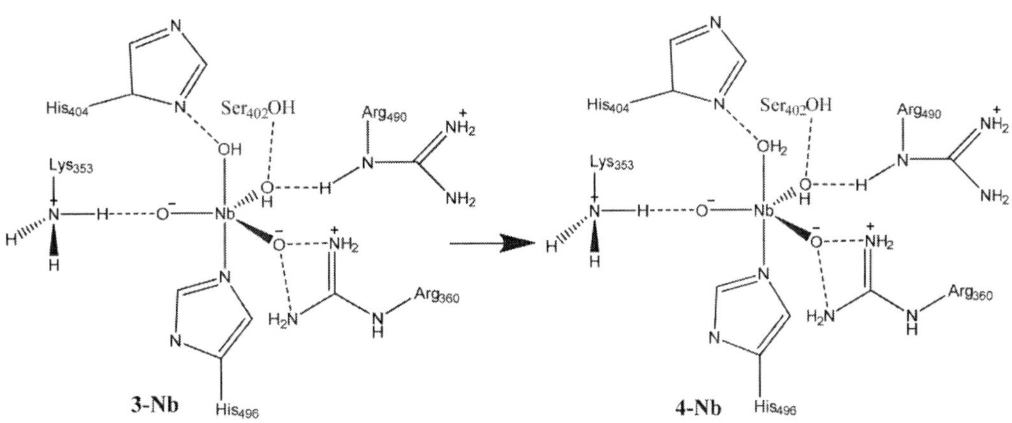

Figure 4. First step of the catalytic cycle with a niobium substituted enzyme

5. Conclusions

We have calculated the energies of various protonation states of VCPO and found **2-V** to be the most stable. If the vanadium of VCPO were replaced with niobium, we calculated the resting state of the enzyme will be a different protonation state (i.e. **3-Nb**) than in the natural enzyme. It is still thermodynamically very favourable for **3-Nb** to enter a catalytic cycle analogous to the one VCPO uses to chlorinate substrate. In future work, we will examine whether the niobium substituted enzyme will improve the rate of chlorination. This work will certainly have application in industrial processes wherein chlorination plays an important role.

In this conclusion, we would also like to take a step back and consider the kind of modeling that the density functional theory affords. For example, Cramer [13] in the introduction of his textbook on computational chemistry promises to present no equation without an effort to "provide an intuitive explanation for its form and the various terms within it"; however, in presenting the equations of DFT such as (5), he has to admit an "utter violation" of this promise since no physical meaning can be assigned to any term in (5). Because of this problem in DFT modeling, we should continue the search for other complementary ways of modeling matter [14].

6. References

[1] Koch W and Holthausen M C 2001 *A Chemist's Guide to Density Functional Theory* (Weinheim, Germany: Wiley-VCH Verlag GmbH)
[2] Friesner R A and Guallar V 2005 *Annu. Rev. Physicist. Chem.* **56** 389.
[3] Kravitz J Y, Pecoraro V L, Carlson H A 2005 *J. Chem. Theory Comput.*, **1** 1265
[4] Raugei S and Carloni P 2006 *J. Phys. Chem.* B **110** 3747
[5] Geethalakshmi K R, Waller M P, Thiel W, Bühl M 2009 *J. of Phys. Chem.* B **113** 4456
[6] Zampella G., Fantucci P, Pecoraro V L, and De Gioia L 2006 *Inorganic Chemistry* **45** 7133
[7] Macedo-Ribeiro S, Hemrika W, Renirie R,Wever R, Messerschmidt A 1999 *J. Biol. Inorg. Chem.* **4** 209
[8] Tracey A S, Willsky G R, Takeuchi G R 2007 *Vanadium: Chemistry, Biochemistry, Pharmacology and Practical Applications* (New York: CRC Press) p 162
[9] Lonsdale R, Harvey J N, Mulholland A J 2012 *Chem. Soc. Rev.* **41** 3025
[10] Jorgensen W L, Maxwell D S, Tirado-Rives J 1996 *J. Am. Chem. Soc.* **118** 11225
[11] QSite, version 6.7, Schrodinger, LLC, New York, NY, 2015
[12] Behera, R N and Panda A 2012 *Computational and Theoretical Chemistry* **999** 215
[13] Cramer C 2002 *Essentials of Computational Chemistry* (New York: Wiley)
[14] Gomatam R 2009 Quantum Theory, the Chinese Room Argument and the Symbol Grounding Problem, *Lecture Notes in Computer Science*, ed P. Bruza, et al., vol 5494, (Berlin: Springer) p 174; Gomatam R 1999 Quantum Theory and Observation Problem, *Journal of Consciousness Studies* **6** 173; Gomatam R 2014 Toward Placing the Concept of 'Chemical Element' on a New Quantum Footing, Annual Meeting of the International Society for the Philosophy of Chemistry, July 7-9, 2014, London

Flexible star polymer chain adsorption by a flat surface: a molecular dynamics simulation

Siham Zenak[1], Aicha Guenachi[2], Sid Ahmed Sabeur[3]

Laboratoire d'Etude Physique des Matériaux, Faculté de Physique, Université d'Oran des Sciences et de la Technologie, BP 1505, El M'naouer Oran 31000, Algeria

E-mail: s.siham111@gmail.com, guenachi.aicha@gmail.com, sidsabeur@gmail.com

Abstract. In this work, we have studied the adsorption of a single flexible star polymer chain by a flat surface. The molecular dynamics simulation has been validated for the case of a free star polymer chain using the diffusion experiment. The scaling laws found are in agreement with the Flory theory predictions and the Rouse model. For the adsorption, preliminary results were obtained for chains of different sizes N=31 to N=199 and different functionalities (f=3,4,6,8,10). For the case of semi-flexible star polymer chains, further investigation is needed to locate the critical point of adsorption when varying the potential interaction strength between the chain and the surface.

1. Introduction

The aim of this work is to study the adsorption of a star polymer chain by a flat surface for different chain lengths and different functionnalities[1] (number of arms f), and to determine how can the functionality of the chain affect the adsoption phenomena. First, we study the static and the dynamic properties of a free star polymer chain in dilute solution under good solvant conditions to get the scaling law $\tau = \tau_0 N^{2\nu+1}$ predicted by Flory[2], by measuring the gyration radius $R_g(t)$ which describes the three dimensional structure of the chain and the mean square displacement of the center of mass of the chain $g_3(t)$ which indicates that the chain is diffusing correctly in the simulation. In the second part of the work, we study the behavior of a star polymer chain attached to an attractive surface. A sufficient surface potential strength causes an adsorption transition, where the chain changes from a three dimensional structure to a bidimensional one and an expansion of the chain parallel to the surface. This behavior is well described by the parallel and perpendicular components of the gyration radius during the simulation. We will also try to determine the critical point of adsorption by measuring the number of adsorbed monomers for different chain lengths and different functionnalities. The work is presented as follows: In section 2 , we briefly present the model and the simulation method used in this study, the results are given and discussed in section 3.We summarize and conclude in section 4.

2. Model and simulation method

The star polymer chain is described by a coarse grained bead-spring model[3]. The interaction between two neighboring monomers is represented by the Finitely Extensible Nonlinear Elastic potential:

$$U_{FENE}(r) = \begin{cases} -\frac{1}{2}kR_{max}^2 \ln\left(1 - \frac{r^2}{R_{max}^2}\right) & r < R_{max} \\ \infty & r \geq R_{max} \end{cases} \tag{1}$$

Where $R_{max} = 1.5\sigma$ is the maximal distance between two bonded monomers and $k = 30\epsilon$ is the spring constant.

The interaction of nonbonded monomers is represented by the potential of Lennard-Jones, truncated and shifted at the minimum of the attractive part $r = 2^{1/6}\sigma$:

Content from this work may be used under the terms of the Creative Commons Attribution 3.0 licence. Any further distribution of this work must maintain attribution to the author(s) and the title of the work, journal citation and DOI.
Published under licence by IOP Publishing Ltd

$$U_{LJ}(r) = 4\epsilon\left[\left(\frac{\sigma}{r}\right)^{12} - \left(\frac{\sigma}{r}\right)^{6}\right] \quad (2)$$

Where $\epsilon = 1$ is the energy scale.

The interaction between the monomers and the surface is also represented by Lennard-Jones potential in the z direction:

$$U_{wall}(z) = 4\epsilon_{wall}\left[\left(\frac{\sigma}{z}\right)^{12} - \left(\frac{\sigma}{z}\right)^{6}\right] \quad (3)$$

Brownian dynamics[4] is used to simulate the behavior a star polymer chain under good solvant conditions, by integrating the equation of motion:

$$m\frac{d^2\vec{r}(t)}{dt^2} = -\nabla U(\vec{r}) - m\Gamma\frac{d\vec{r}(t)}{dt} + \vec{\eta}(t) \quad (4)$$

Here $-m\Gamma\frac{d\vec{r}}{dt}$ is the friction force and $\vec{\eta}(t)$ is a random force which corresponds to the interactions with the particles of the solvant. We have used the velocity-verlet algorithm[5] with a time step $\delta t = 0.002$ to integrate the equations of motion for a temperature $K_BT\epsilon = 1$ and a friction parameter $\Gamma = 1$. All the results are obtained in reduced units during the simulation.

3. Results

3.1. Free star polymer chain

We study the diffusion of a free star polymer chain with different lengths ($N = 31$, $N = 61$,$N = 82$, $N = 100$, $N = 200$). The properties that are measured are the mean squared displacement of the center of mass (MSQD)[6], defined as:

$$g_3(t) = \langle(\vec{r}_{cm}(t) - \vec{r}_{cm}(0))^2\rangle \quad (5)$$

and the gyration radius defined as:

$$\langle R_g^2\rangle = \frac{1}{N}\sum_{i=1}^{N}\langle(\vec{r}_i - \vec{r}_{cm})^2\rangle \quad (6)$$

Where $r_{cm}(0)$ is the center of mass at the begining of the simulation ($t = 0$), and $r_{cm}(t)$ is the center of mass at time t.

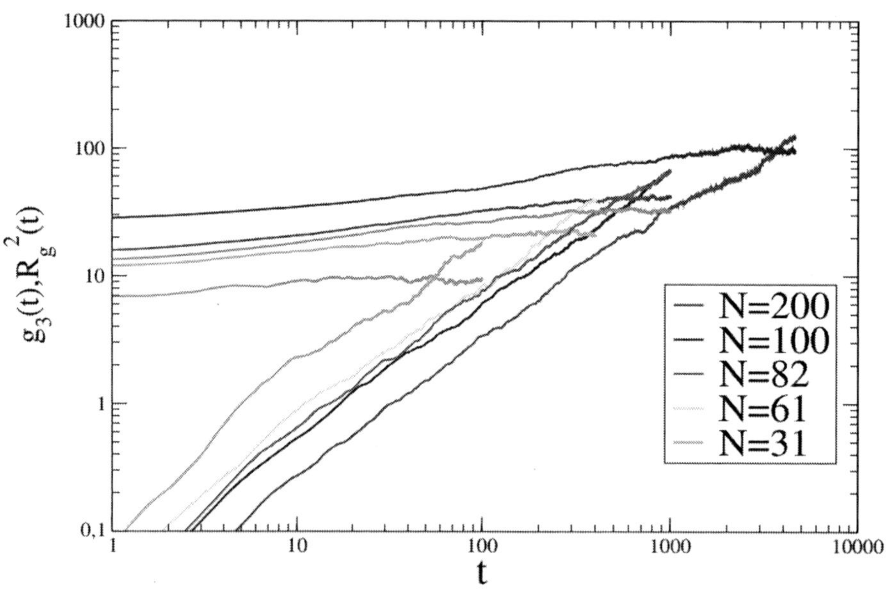

Figure 1: MSQD and gyration radius as a function of time in log-log scale

In figure 1, we observe two phases, the ballistic phase, where the chain has a viscoelastic behavior ($t < \tau$), and a diffusif phase where the chain simply diffuse in the solvant ($t > \tau$)

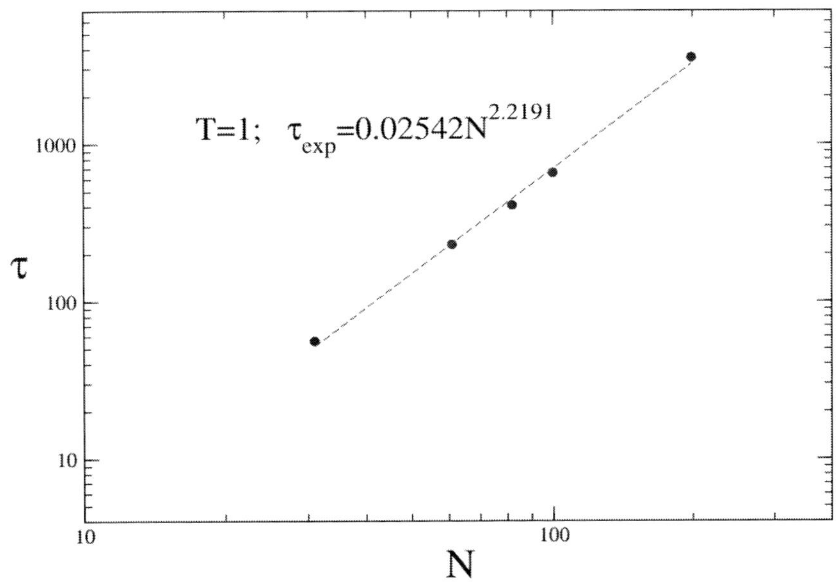

Figure 2: Relaxation time τ as a function of star polymer chain length N

By making a fit of type $y = AX^B$, we obtain the scale law:

$$\tau_{exp} = 0.02542 \times N^{2.2191} \tag{7}$$

From figure 2, we can see from the fit that the Flory formula $\tau = \tau_0 N^{2\nu+1}$ is satisfied where $\nu = \frac{3}{5}$. Since the scale law is confirmed, our molecular dynamics code is reliable to study the adsorption phenomena.

3.2. Star polymer chain attached to an attractive surface

The adsorption of star polymer chain is studied by attaching the end monomer of the last arm of the chain to the surface, the process of adsorption occurs when the strength of the attractive surface potential is sufficiently big, the star polymer makes a transition from a three dimensional structure to a two dimensional one. This transition can be observed by measuring the ratio $\frac{R_{g\perp}}{R_{g\parallel}}$ where $R_{g\perp}$ and $R_{g\parallel}$ are respectively the perpendicular and the parallel components of the gyration radius:

$$\langle R_{g\perp}^2 \rangle = \frac{1}{N} \sum_{i=1}^{N} \sum_{j>i}^{N-1} \langle (z_i - x_j)^2 \rangle \tag{8}$$

and

$$\langle R_{g\parallel}^2 \rangle = \frac{1}{N} \sum_{i=1}^{N} \sum_{j>i}^{N-1} \langle (x_i - x_j)^2 + (y_i - y_j)^2 \rangle \tag{9}$$

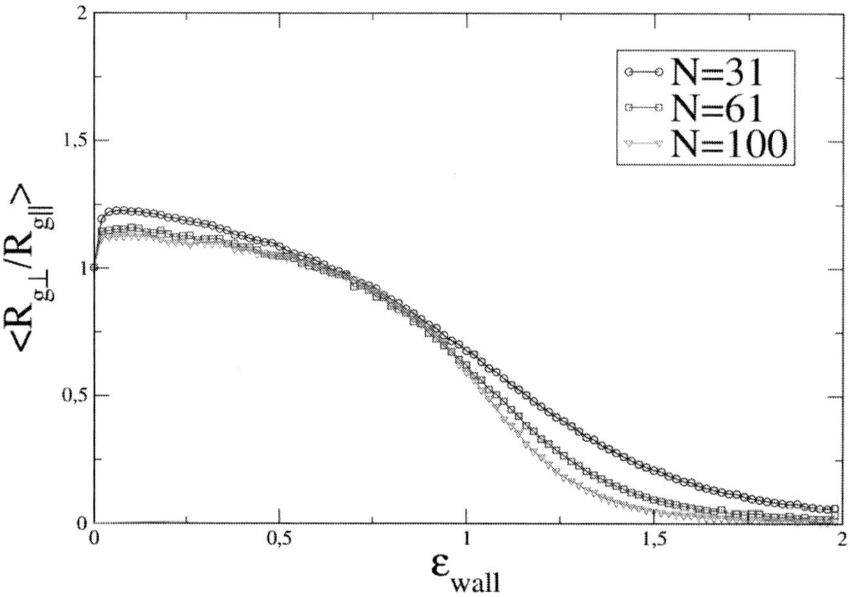

Figure 3: The ratio of perpendicular and parallel gyration radius parts as a function of ϵ_{wall} for different lengths N for a three arms star polymer chain

Figure 3 shows that for low potential stength of the surface ($\epsilon_{wall} \ll 1$), the chain behaves like a free polymer chain, and for high values of ϵ_{wall}, the $R_{g\parallel}$ increases, which means that the chain is spreading onto the surface. The transition phase appears for a critical ϵ_{wall} value, this corresponds to the area where the curves for $N = 31$, $N = 61$, $N = 100$ overlaps. The critical point of adsorption is lower than 1 and approximately ($\epsilon_{wall} \simeq 0.94 k_b T$). For $\epsilon_{wall} \leq 0.5$, the ratio is higher than 1, the chain is in the distortion phase. The chain is in the adsorption phase for $0.5 < \epsilon_{wall} \leq 1$. $R_{g\parallel}$ dominates for $\epsilon_{wall} > 1$ and this can be interpreted as a post-adsorption relaxation phase .

To estimate the precision in the simulation, we have made 20 samples for each run.The results with error bars are presented in figure 4 for the case of a star polymer with three arms ($f = 3$) and a chain length $N = 61$

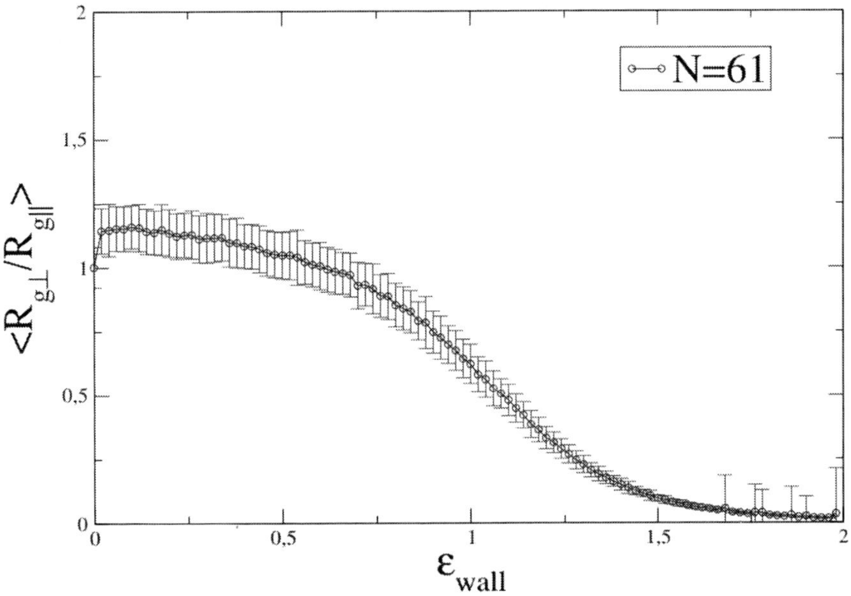

Figure 4: The ratio of perpendicular and parallel gyration radius parts as a function of ϵ_{wall} for $N = 61$ for a three arms star polymer chain. Error bars are obtained by calculating the standard deviation over 20 samples.

We have also calculated the rate of adsorption Θ which is the number of the monomers adsobed n_c divided by the total number of monomers N.

$$\Theta = \frac{n_c}{N} \tag{10}$$

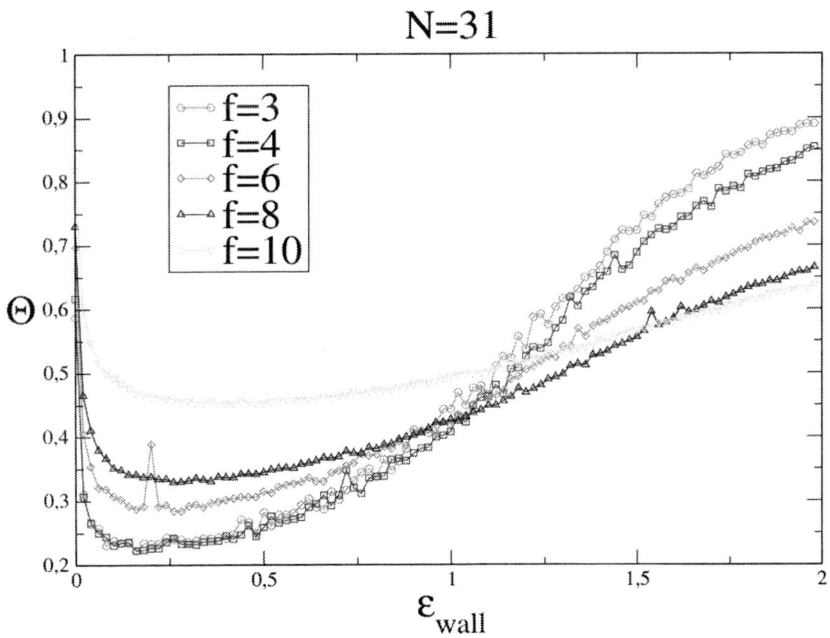

Figure 5: The rate of adsorption for a star polymer with $N = 31$ monomers and different functionalities (number of arms) as a function of ϵ_{wall}

From figure 5, we found that the functionality is a major factor which affect the number of adsorbed monomers. For $f = 3$, 88% of monomers are adsorbed at the end of the simulation, and only 63% of monomers are ; < 1.3

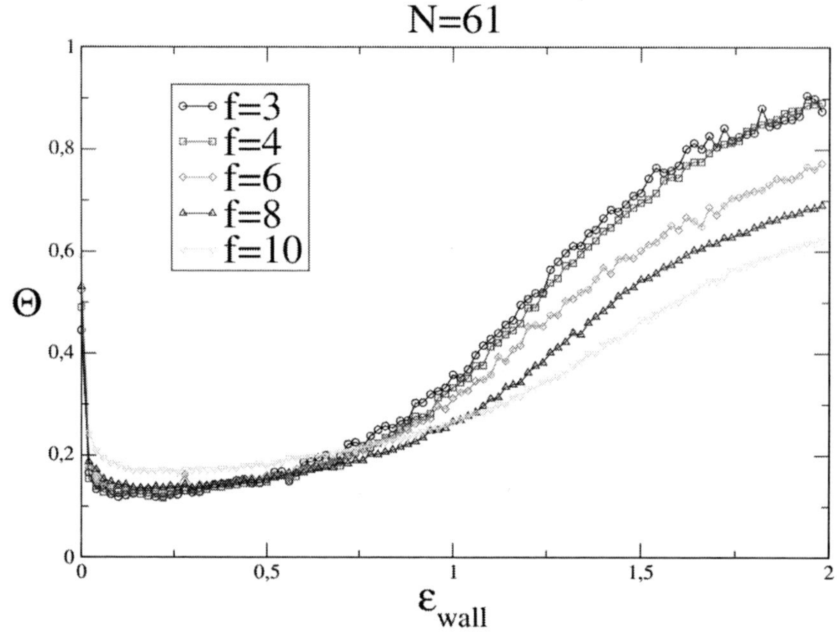

Figure 6: The rate of adsorption for a star polymer with 61 monomers and different functionalities (number of arms) in function of ϵ_{wall}

Figure 6 shows the same experiment for a chain length $N = 61$. For $f = 3$ and $f = 4$, 90% of the monomers are adsorbed, and only 62% of the monomers are adsorbed for $f = 10$, the critical adsorption point is located in $0.5 < \epsilon_{wall} < 0.8$.

4. Conclusion and perspectives

In this work, we have studied the behavior of a flexible star polymer chain attached to a flat surface. The critical point of adsorption has been approximately located by varying the strength of the interaction potential between the star polymer chain and the surface for different chain lengths and different functionalities. In a future work, it is intersting to estimate more exactly the critical point of adsorption by calculating the variation of energy of the chain as a function of the potential strength between the polymer chain and the surface and also look forward the behavior of semiflexible star polymer chains in contact with a surface and see if the critical point of adsoption changes by varying the degree of stiffness of the chain.

5. Acknowledgments

The authors thank Professor Andrey Michelv for fruitful discussions and insights. The computation was made possible on IBN BAJA supercomputer in the HPC center at USTOMB.

6. References

[1] A Chremos, PJ Camp, E Glynos, V Koutsos, Soft Matter 6(7), 1483-1493 (2010)
[2] P. G. De Gennes, Scaling Concepts in Polymer Physics, Cornell Univ. Press (1979)
[3] K. Kremer, G. Grest, J. Chem. Phys. 92(8),5057-5086 (1990)
[4] M. Doi, Introduction to polymer physics, Oxford Univ. Press (1996)
[5] L. Verlet, Phys. Rev. 159(1), 98-103 (1967)
[6] K. Binder, A. Milchev, Journal of Computer-Aided Materials Design,9, 33-74 (2002)

HuLiS, a program to teach mesomerism and more

Yannick Carissan, Nicolas Goudard, Denis Hagebaum-Reignier and Stéphane Humbel

Aix Marseille Université, Centrale Marseille, CNRS, iSm2 UMR 7313, 13397, Marseille, France

E-mail: yannick.carissan@univ-amu.fr

Abstract. The HuLiS program is presented as a tool to decompose a delocalized wave function as a linear combination of localized electronic structures. The principles of the energy based HL-CI and the overlap based schemes HL-P are developed. The results obtained for a set of 10 relevant organic molecules are shown to compare very well with high level quantum chemistry calculations. A trust factor τ is introduced and its use is shown on the allyl radical case where symmetry must be taken into account.

1. Introduction

The HuLiS program[1] deals with topological methods applied to chemistry. [1] It is a tool, which originally was devoted to teaching, but recent features make it very useful for research especially in the understanding of Valence Bond view of the chemical bond.[2, 3, 4]

The most commonly used topological method in chemistry was develop by Heinrich Hückel[5] in the late 1950's when computer time was not at reach for academic use. It deals with π electrons only and gives insights into mesomerism and aromaticity. The reader unfamiliar with this method should refer to ref. [6]. Nowadays, it is a great way to introduce computational chemistry as it requires simple math and gives access to meaningful results. As we use the initial formulation of the Hückel method, we restrict ourselves to the π part of the wave function. Furthermore, this method implies the following:

- one $2p_{z_A}$ orbital per atom A is used of energy α_A;
- orbitals of adjacent atoms are orthogonal;
- the off diagonal term of the Hamiltonian matrix for two adjacent atoms A and B has a negative value β_{AB};
- the molecular orbitals are eigenfunctions of the Hamiltonian matrix and their energies are their eigenvalues.

The values of α_A and β_{AB} are tabulated elsewhere.[7] It is sufficient here to remind that the original method was built for carbon atoms, thus $\alpha_C = \alpha$ and $\beta_{CC} = \beta$ both values being lower than zero. The Hückel method leads to delocalized molecular orbitals.

In the other hand, the Valence Bond approach of the chemical bond defines the bond as a local property: as in a Lewis picture, the bond is seen as an electron pair between two atoms, which leads to three structures, one neutral (one electron on each atom) and two ionic (two

[1] http://www.hulis.free.fr

Content from this work may be used under the terms of the Creative Commons Attribution 3.0 licence. Any further distribution of this work must maintain attribution to the author(s) and the title of the work, journal citation and DOI.
Published under licence by IOP Publishing Ltd

electrons on one atom and zero on the other and the other way around). This local view of the bond is a very natural way for chemists to rationalize properties and reactivity.

The HuLiS programs takes the best of both worlds: the delocalized wave function is written as a weighted combination of localized structures, which are meaningful. This is achieved by two methods presented in this contribution.

2. The Hückel derived VB methods

In HuLiS, the Hückel delocalized wave function Ψ of energy E is computed for the system under study. In a Valence Bond vision of the chemical bond, it is then written as a linear combination of localized electronic structures ψ_i of energies ϵ_i:

$$\Psi = \sum_i^N c_i \psi_i. \tag{1}$$

Two different schemes were developed to determine the c_i values. One is based on the energy of the delocalized structure whereas the other one is based on the overlap between the delocalized and the localized structures. Both use simple linear algebra and are described in the following section. As an illustration, calculations were performed on the set of molecules given in Figure ??, which results are summarized in Table ??.

2.1. The energy-based scheme: Hückel Lewis Configuration Interaction (HL-CI)

The HL-CI scheme aims at building a linear combination of localized structures

$$\tilde{\Psi} = \sum_i^N c_i \psi_i \tag{2}$$

which energy $\tilde{E}$ approaches the energy of the Hückel solution E. This schemes was built as an analogy of the Hückel method in a localized structures basis (when the Hückel method is built in an atomic orbital basis). For this analogy to be followed, one sets up the overlap between two different localized structures to zero.

$$S_{HLCI} = \begin{pmatrix} 1 & 0 & \cdots & 0 \\ 0 & 1 & \cdots & 0 \\ \vdots & \vdots & \ddots & \vdots \\ 0 & 0 & \cdots & 1 \end{pmatrix} \tag{3}$$

An effective Hamiltonian is written in the basis of the localized structures :

$$\hat{H}_{HLCI} = \begin{pmatrix} \epsilon_1 & B & \cdots & B \\ B & \epsilon_2 & \cdots & B \\ \vdots & \vdots & \ddots & \vdots \\ B & B & \cdots & \epsilon_N \end{pmatrix} \tag{4}$$

The diagonal terms are the energies of each localized structure and, by analogy with the Hückel method, the interaction term is set to a negative value B. It is very important to understand that the unlike in the Hückel method, it is not possible here to define adjacent localized structures. As a consequence, one cannot determine which off diagonal terms are zero, thus all off diagonal terms are set to B. A unique negative value of $B = B_0$ can be computed, which ensures that the eigenfunction it is associated with is such that $\tilde{E} = E$.[8] By construction, the eigenfunction associated to B_0 has all c_i coefficients positive. This scheme gives c_i values which are very close to advanced quantum chemistry methods such as Natural Resonance Theory (NRT) based on the Natural Bond Orbitals approach.[8, 9, 10, 11] However, it lacks of physics and does not account for out of phase interaction between localized structures if imposed by symmetry as all c_i are positive by construction.[12] This point is adressed in section 3.

2.2. The space-based scheme: Hückel Lewis Projection (HL-P)

To overcome the lack of physics of the HL-CI scheme the HL-P space based scheme was developed. In this scheme, energy is not used as a criterion, only overlap between structures is used. One uses the overlap between the localized structures and the Hückel solution:

$$S_\Psi = \begin{pmatrix} \langle\Psi|\Psi_1\rangle \\ \langle\Psi|\Psi_2\rangle \\ \vdots \\ \langle\Psi|\Psi_N\rangle \end{pmatrix} \tag{5}$$

and the overlap among the localized structures:

$$S = \begin{pmatrix} 1 & \langle\Psi_1|\Psi_2\rangle & \langle\Psi_1|\Psi_3\rangle & \cdots & \langle\Psi_1|\Psi_N\rangle \\ \langle\Psi_2|\Psi_1\rangle & 1 & \langle\Psi_2|\Psi_3\rangle & \cdots & \langle\Psi_2|\Psi_N\rangle \\ \langle\Psi_3|\Psi_1\rangle & \langle\Psi_3|\Psi_2\rangle & 1 & \cdots & \langle\Psi_3|\Psi_N\rangle \\ \vdots & \vdots & \vdots & \ddots & \vdots \\ \langle\Psi_N|\Psi_1\rangle & \langle\Psi_N|\Psi_2\rangle & \langle\Psi_N|\Psi_3\rangle & \cdots & 1 \end{pmatrix}. \tag{6}$$

The c_i coefficients are obtained by solving the following system:

$$S_\Psi = SC \tag{7}$$

with C the vector of all c_i coefficients.[12] In this scheme, a measure of the match between Ψ and $\tilde{\Psi}$ can be computed as the overlap between the two functions. It can be shown that this overlap that we call τ is the sum of the Coulson Chirgwin weights defined as:[12, 13]

$$w_i = c_i \sum_j^N c_j S_{ij}. \tag{8}$$

3. Comparison of the schemes

In order to compare the schemes among each other and to a more advanced quantum chemistry method, one computes the weights of each localized structure in $\tilde{\Psi}$. For the set of molecules presented in Figure 1 the weights and τ are given in Table 1. Both schemes give very similar results and compare very well to the *ab-initio* NRT method. A problem occurs with HL-CI when the symmetry of the total wave function imposes a c_i to be negative. As discussed in 2.1, HL-CI gives by construction only positive values for the c_i coefficients. In the case of allyl radical, the ground state wave function is of A_2 symmetry. In this case, it is the out of phase combination of the two localized structures, which is the appropriate VB solution (Figure 2). The calculation of τ for the HL-CI solution shows this failure (it drops to zero). Thus, despite the fact that τ was introduced in the HL-P framework, it is important to compute it for the HL-CI results.

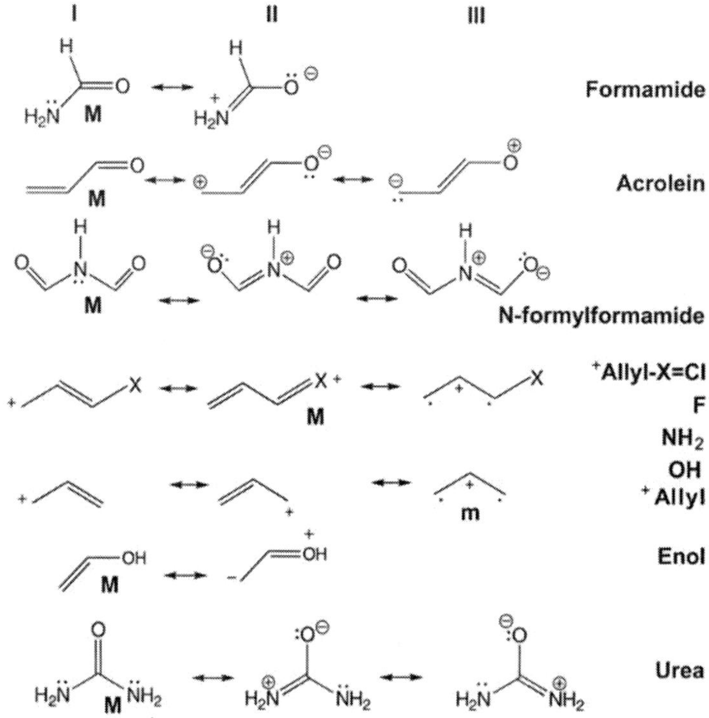

Figure 1. Structures used in the wave function decomposition for benchmark. Weights of structures are presented in Table 1. Reprinted with permission from [12]. Copyright 2008 American Chemical Society.

4. Conclusion

In this contribution, the HuLiS program was briefly presented. This program allows to have a double reading of a π wave function obtained with the Hückel method. The delocalized wave function can be accurately described as a linear combination of few meaningful localized structures. A criterion of the accuracy, τ can be computed and characterizes the set of localized structures used as well as the physical meaning of their linear combination. The two methods implemented show a good agreement with high level quantum chemistry calculation at a fraction of its computational cost.

References

[1] Cyrański M K, Ciesielski A, Krygowski T M and Stępień D K 2012 *Pure and Applied Chemistry* **84**
[2] Tiwary A S and K Mukherjee A 2008 *Molecular Physics* **106** 2271–2276
[3] Tiwary A S and Mukherjee A K 2009 *Journal of Molecular Structure: THEOCHEM* **909** 57–65
[4] Mboyi C D, Duhayon C, Canac Y and Chauvin R 2014 *Tetrahedron* **70** 4957–4968
[5] Hückel E 1957 *Zeitschrift für Elektrochemie, Berichte der Bunsengesellschaft für physikalische Chemie* **61** 866–890
[6] Levine I N 2013 7th ed (Boston: Pearson) ISBN 978-0-321-80345-0
[7] Van-Catledge F 1980 *Journal of Organic Chemistry* **45** 4801–4802
[8] Hagebaum-Reignier D, Girardi R, Carissan Y and Humbel S 2007 *Journal of Molecular Structure: THEOCHEM* **817** 99–109
[9] Glendening E D and Weinhold F 1998 *J. Comput. Chem.* **19** 593–609

Table 1. Weights obtained with HL-P as compared to NRT and HL-CI [8]. Structures used are presented in Figure 1. Reprinted with permission from [12]. Copyright 2008 American Chemical Society.

Molecule	NRT	HL-CI	HL-P	τ(%)
Formamide	67 / 33	66 / 34	65 / 35	90.9
N-formylformamide	45 / 27.5 / 27.5	46 / 27 / 27	43 / 29 / 29	83.6
Acrolein	87 / 13	79 / 21	85 / 15	85.3
	84 / 11 / 5	78 / 17 / 5	84 / 15 / 1	85.6
$^+$Allyl NH_2	23 / 77	29 / 71	25 / 75	84.5
	21 / 75 / 4	24 / 70 / 6	23 / 69 / 8	92.3
$^+$Allyl OH	30 / 70	39 / 61	39 / 61	84.5
	23 / 65 / 12	35 / 58 / 7	34 / 54 / 11	95.1
$^+$Allyl F	37 / 63	44 / 56	45 / 55	84.7
	29 / 56 / 15	40 / 52 / 8	39 / 49 / 12	96.4
$^+$Allyl Cl	37 / 63	38 / 62	36 / 64	84.0
	29 / 58 / 13	33 / 60 / 7	32 / 58 / 11	93.9
$^+$Allyl H	50 / 50	50 / 50	50 / 50	84.9
	39 / 39 / 22	46 / 46 / 8	44 / 44 / 13	97.4
Enol	86 / 14	93 / 7	93 / 7	96.3
Urea	60 / 20 / 20	46 / 27 / 27	51 / 24 / 24	82.6

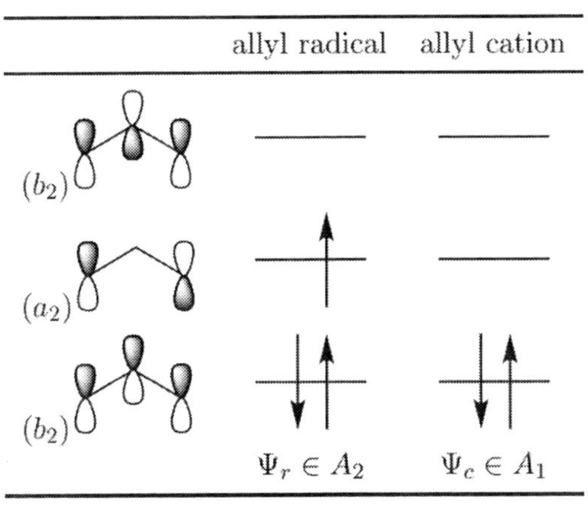

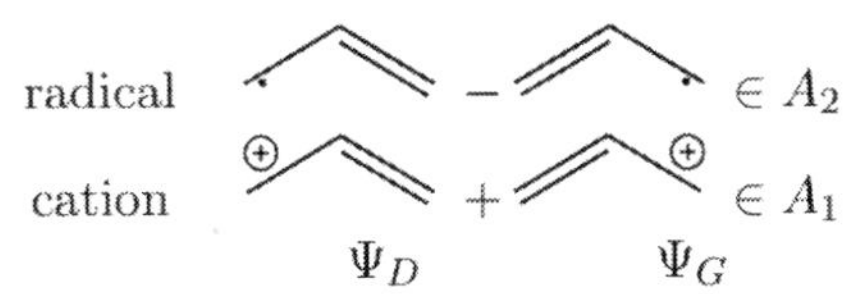

Method	allyl radical c_D	allyl radical c_G	allyl radical τ	allyl cation c_D	allyl cation c_G	allyl cation τ
HL-CI	$\frac{1}{\sqrt{2}}$	$\frac{1}{\sqrt{2}}$	0%	$\frac{1}{\sqrt{2}}$	$\frac{1}{\sqrt{2}}$	92%
HL-P	$\frac{1}{\sqrt{2}}$	$-\frac{1}{\sqrt{2}}$	92%	$\frac{1}{\sqrt{2}}$	$\frac{1}{\sqrt{2}}$	92%

Figure 2. Molecular orbital (left) and Valence Bond (right) visions of the allyl radical and cation wave function. The table shows that the minus sign before c_G obtained by HL-P describes the appropriate physics ($\tau = 92\%$) whereas HL-CI fails to do so (see text) because the A_2 symmetry imposes an out of phase interaction between the localized structures. Such an issue does not occur in the cation as its wave function is of A_1 symmetry.

[10] Glendening E D, Badenhoop J K and Weinhold F 1998 *Journal of Computational Chemistry* **19** 628–646
[11] Glendening E D, Landis C R and Weinhold F 2013 *Journal of Computational Chemistry* **34** 1429–1437
[12] Carissan Y, Hagebaum-Reignier D, Goudard N and Humbel S 2008 *J. Phys. Chem. A* **112** 13256–13262
[13] Chirgwin B H and Coulson C A 1950 *Proceedings of the Royal Society A: Mathematical, Physical and Engineering Sciences* **201** 196–209

The extended propagation equation of optical impulses in silica fibers

Y. Z. Pchelkina[1,2] , **I. V. Alimenkov**[1]

[1] Samara State Aerospace University (SSAU), 34, Moskovskoye shosse Street, Samara, 443086, Russia

[2] 148-32, Unyh Pionerov Prospect Street, Samara, 443105, Russia

E-mail: musina@yandex.ru

Abstract. We considered the new equation of distribution of optical impulses in silica fibers. We also found its exact solitonic solution and analyzed extreme cases.

Introduction

The field of the optical impulse which is distributed in a single-mode optical fiber supporting the linear polarization condition has the form [1], [3], [5].

$$\boldsymbol{E}(\boldsymbol{r},t) = \boldsymbol{e}_x F(x,y) A(z,t) \exp\{i(\beta_o z - \omega_o t)\} \tag{1}$$

were $F(x,y)$ – Gaussian function, which has the form: $\exp\{-(x^2+y^2)/w^2\}$ with radius of mode w, $A(z,t)$ - complex impulse envelope, ω_o – carrier frequency, $\beta_o = \omega_o n(\omega_o)/c$ – central wave number.

Materials and Methods

In case of Kerr-type nonlinearity for envelope of optical impulse there is an equation

$$i(\partial A/\partial z + \beta_1 \partial A/\partial t) + (1/2\beta_o)(\partial^2 A/\partial z^2) - (\beta_o/2)(\partial^2 A/\partial t^2) + \gamma|A|^2 A = 0 \tag{2}$$

here $\beta_1 = 1/v_g$ - inverse quantity of group velocity, β_2 - variance of group velocity, γ – index of Kerr-type nonlinearity. The value β_2 is negative value in zone of silica transparency. Thus, equation (2) is elliptic equation. If we discard one of two derivatives in equation (2), then we get parabolic equation, which is reduced to form of Schrödinger equation.

Usually [1-6] for silica fibers they use equation, which is received by rejection of second derivative with respect to coordinate. It is erroneous as it will be shown. Ordinarily, rejection of second derivative with respect to coordinate occurs at the initial stage of solution of wave equation in spectral representation [1], [3], [5], [6]. At more detailed approach we can get equation (2), in which we can to estimate importance degree of second derivatives. Given that in area of the minimum losses for silica: $\lambda \approx 1{,}55$ µm; $\beta_2 = -20$ (ps)^2/km; $n = 1{,}45$ and $z = ct_1/n$ we can notice that coefficients of second derivatives with respect to coordinate and with respect to time t are differ for more than two orders. Thus, for silica waveguide it is better to use equation

$$i(\partial A/\partial z + \beta_1 \partial A/\partial t) + (1/2\beta_o)(\partial^2 A/\partial z^2) + \gamma|A|^2 A = 0 \tag{3}$$

but not equation

Content from this work may be used under the terms of the Creative Commons Attribution 3.0 licence. Any further distribution of this work must maintain attribution to the author(s) and the title of the work, journal citation and DOI.
Published under licence by IOP Publishing Ltd

$$i(\partial A/\partial z + \beta_1 \partial A/\partial t) - (\beta_2/2)(\partial^2 A/\partial t^2) + \gamma|A|^2 A = 0 \tag{4}$$

Fundamental soliton of equation (4) in laboratory reference system is given by

$$A = \left(\sqrt{2b/\gamma}\exp\{ibz\}\right)/\left(ch\left(\sqrt{2b}(z - z_o - v_g t)/v_g\sqrt{|\beta_2|}\right)\right) \tag{5}$$

Here b - free argument and correction of central wave number, z_o - impulses coordinate at the initial time. Formula (5) is unsatisfactory; in view of velocity distribution of impulse is given by delta function. However, velocity of optical soliton [7], [8] is constant in the mean. The soliton solution of equation (3) in laboratory reference system is given by

$$A = \left((2b(1+b/2\beta_o))/\gamma ch^2\left(s\sqrt{2\beta_o b(1+b/2\beta_o)}\right)\right)^{1/2}\cdot\exp\{ibz\} \tag{6}$$

were

$$s(z,t) = z - z_o - vt, \tag{7}$$

$$v = v_g(1+b/\beta_o) \tag{8}$$

is free of shortcoming because of arbitrary of parameter b and after all, we will show a similar solution of equation (2):

$$A = \left((2b(1+b/2\beta_o))/\gamma ch^2\left(s\sqrt{2\beta_o b(1+b/2\beta_o)/(1-\beta_o\beta_2 v^2)}\right)\right)^{1/2}\cdot\exp\{ibz\} \tag{9}$$

The dimensionless quantity $|\beta_o\beta_2 v^2| \le 0{,}005$. And being that it is under the radical sign, and then equation (9) coincides with the equation (6) with high degree of probability. This argues in favor of the equation (3) for silica fibers.

Let us show that any soliton solution of equation (2) is some function of hyperbolic cosine, regardless of the type of nonlinearity. Let us to reduce this equation (2) for complex envelope of optical impulse to form of equation for optical intensity $I = I(s(z,t))$ by substitution

$$A(z,t) = \sqrt{2I}\exp\{ibz\} \tag{10}$$

were b – free argument and correction of central wave number β_o. This parameter is always expressed in terms of the peak intensity or the peak value of intensity.

Substitution of (10) into (2) gives us (after separation of real part from imaginary part of obtained solution) these two equations for optical intensity of envelope:

$$(1+b/\beta_o)(\partial I/\partial z) + (1/v_g)(\partial I/\partial t) = 0 \tag{11}$$

$$(1/\beta_o)\left[2I(\partial^2 I/\partial z^2) - (\partial I/\partial z)^2\right] - \beta_2\left[2I(\partial^2 I/\partial t^2) - (\partial I/\partial t)^2\right] = \left[b + (b^2/2\beta_o) - \Delta\beta(I)\right]8I^2. \tag{12}$$

Thus, we obtain two equations for one unknown function. Equation (12) is linear and homogeneous equation. This allows us to write its general solution, which is an arbitrary differentiable function $I = I(s(z,t))$, were

$$s(z,t) = z - z_o - vt, \tag{13}$$

$$v = v_g(1+b/\beta_o). \tag{14}$$

Here we take into account that $\beta_1 = 1/v_g$. In other words equation (11) defines an argument of the unknown function. In that case the form of the function remains arbitrary. Thus, the required function is a progressive wave with unknown wave profile and constant velocity (14). The wave profile describe by equation (12), which can be reduced as ordinary differential equation

$$\left[(1-\beta_o\beta_2 v^2)/\beta_o\right]\cdot\left[2I(d^2 I/ds^2) - (dI/ds)^2\right] = \left[b + (b^2/2\beta_o) - \Delta\beta(I)\right]\cdot 8I^2 \tag{15}$$

For ease of calculation let us proceed in (15) to the dimensionless independent variable τ:

$$\tau = s\cdot\left(b(b+2\beta_o)/(1-\beta_o\beta_2 v^2)\right)^{1/2} \tag{16}$$

As a result, equation (11) is of the form

$$2I\ddot{I} - \dot{I}^2 = 4I^2 - 8\gamma I^2\Delta\beta(I), \tag{17}$$

were

$$\gamma = \beta_o / (b(b + 2\beta_o)) \tag{18}$$

The derivative with respect to τ is designated as a point. As is easy to check that equation (17) is equivalent to the normal system of two Hamiltonian first-order equations:

$$\dot{I} = \partial H / \partial P_I \tag{19}$$

$$\dot{P}_I = -\partial H / \partial I \tag{20}$$

with Hamiltonian function

$$H = (P_I^2 - 1)I + 2\gamma B(I) \tag{21}$$

were $B(I)$ – antiderivative for function $\Delta\beta(I)$, namely

$$B(I) = \int \Delta\beta(I) dI \tag{22}$$

It is well known that the Hamiltonian system (19), (20) also describes one-dimensional point particle with coordinate I. Therefore, equation (17) is also the Euler-Lagrange equation for this one-dimensional mechanical particles. In classical dynamics of particles dimensionless variable τ, defined by equation (16), is used as time. Thus, the Hamiltonian system (19), (20), obtained for the intensity envelope of the optical impulse, has a mechanical sense, if the I is understood not as a field function, but as a coordinate of the point. Since the Hamilton function (21) does not depend explicitly on time, and then the mechanical energy of this point is stored.

The solutions $I(\tau)$, which asymptotically tends to zero at infinity, exist (if they certainly exist) with zero mechanical energy of point. This means that the envelope of the optical impulse in this case is a localized function. Such localized solutions of the equation (17) are of interest first of all.

The solution of Hamiltonian equations (19), (20), and therefore equation (17) is easily obtained using the theory of canonical transformations. As follows from Jacobi-Poincare theorem, if there exists a twice-differentiable function, $S(I, p, \tau)$ such that $\left|\partial^2 S / \partial I \partial p\right| \neq 0$, then transformation $(I, P_I) \leftrightarrow (q, p)$ is canonical equation and generated by function P_I with new Hamilton function $\overline{H}$:

$$P_I = \partial S / \partial I; \quad q = \partial S / \partial p \tag{23}$$

$$\overline{H}(q, p, \tau) = H(I(q, p, \tau), P_I(q, p, \tau), \tau) + (\partial S / \partial \tau)(I(q, p, \tau), p, \tau). \tag{24}$$

In the theory of canonical transformations for two-dimensional phase space, we can take as a basis the four types of generating functions, which depend on one new and one old variables. For further purposes, the most appropriate is a function that depends on the old impulse and the new coordinate: $F = R_2(P_I, q, \tau)$. Then the explicit form of the transformation can be found from equations

$$I = \partial F / \partial P_I; \quad p = \partial F / \partial q, \tag{25}$$

and new Hamiltonian function

$$\overline{H}(q, p, \tau) = H(I(q, p, \tau), P_I(q, p, \tau), \tau) - (\partial F / \partial \tau)(P_I(q, p, \tau), q, \tau). \tag{26}$$

Let us consider that the 1-st component in Hamiltonian is quiescent Hamiltonian and the 2-nd is disturbance, due to nonlinear response of the medium. Let's move from dynamic system $\{I, P_I, H\}$ to system $\{q, p, \overline{H}\}$, with interaction by using the generating function $F(P_I, q, \tau)$, such that in new variables the new Hamilton function would be defined by the disturbance only. This can be done by using the generating function

$$F = qArthP_I - q\tau \tag{27}$$

which is the complete integral of the unperturbed Hamilton-Jacobi equation: $\partial F / \partial \tau = (P_I^2 - 1)\partial F / \partial P_I$. From equations (21) follows the explicit form of the transformations:

$$I = qch^2(p + \tau); \quad P_I = th(p + \tau), \tag{28}$$

and from equation (22) follows the explicit form of the new Hamilton function:

$$\overline{H} = 2\gamma B(qch^2(p + \tau)), \tag{29}$$

new Hamilton equations

$$\dot{q} = \partial \overline{H} / \partial p; \quad \dot{p} = -\partial \overline{H} / \partial q \tag{30}$$

determine the dynamics of the new canonical variables q and p:

$$\dot{q} = 4\gamma q ch(p+\tau) sh(p+\tau)\Delta\beta(qch^2(p+\tau)), \quad \dot{p} = -2\gamma ch^2(p+\tau)\Delta\beta(qch^2(p+\tau)). \tag{31}$$

As the understanding of the interaction Hamiltonian (29) depends on time in hand $(p+\tau)$ the first integral of the system can be easily guessed

$$2\gamma B(qch^2(p+\tau)) - q = E = 0 \tag{32}$$

Truly, taking derivative with respect to time of this expression and taking into account (29) and (30) we will get the identity. Mechanical energy E is taken as zero (it is not the energy of the optical impulse, it is just an arbitrary constant, which in the mechanical interpretation makes sense of the mechanical energy of one-dimensional particles). To simplify the solution of the dynamic problem in the understanding of the interaction divide the first equation (31) for the second. The result is

$$\dot{q} / q = -2\dot{p} th(p+\tau) \tag{33}$$

The form of the last equation does not depend on the type of nonlinear response function, and is determined only by the canonical transformation (28), which, in turn, is determined by the unperturbed Hamiltonian $H_o = (P_I^2 - 1)I$. This unperturbed Hamiltonian is defined in the end of the linear part of equation (2). If one first integral of the system will be known, will reduce the system order by one, therefore, to one equation. As such equation it is necessary to take equation (33). As a second equation to solve the problem, you should take the first integral (32), from which it follows that q is some function of the hyperbolic cosine of the argument $(p+\tau)$. Substitution of this function into (33) gives the time dependence $p(\tau)$, and that completes the solution to the problem. So, for example, for Kerr nonlinearity $B(I) = \mu I^2$: $p = -2\tau$, $q = 1/2\gamma\mu ch^4(\tau)$ and it is an exact solution.

Conclusion

Thus, it is shown that the problem of finding localized solutions of the extended nonlinear equation the propagation of optical pulses in fibers (2) for silica fibers is reduced to a simpler equation of parabolic type (3). It differs significantly from generally accepted in the literature on fiber optics equation (4). Solutions of all three equations are in the laboratory frame of reference.

Besides, the problem of finding localized solutions of equation (2) is reduced by using the theory of Hamiltonian systems to two equations (32) and (33). The first of equations is algebraic.

Acknowledgements

This work was supported by the Ministry of education and science of the Russian Federation in the framework of the implementation of the Program of increasing the competitiveness of SSAU among the world's leading scientific and educational centers for 2013-2020 years.

References

[1] Agrawal, G.P. Nonlinear Fiber Optics. 2nd ed. – Academic Press - 1995. – 616p.
[2] Kivshar, Y.S. Optical solitons. From Fibers to Photonic Crystals / Y.S. Kivshar, G.P. Agrawal – Academic Press – 2003. – 540p.
[3] Agrawal, G.P. Applications of Nonlinear Fiber Optics (Optics and Photonics Series) 2nd ed. - Academic Press - 2008. – 528p.
[4] Shiva Kumar. Impact of Nonlinearities on Fiber Optic Communications - Springer - 2011. – 555p.
[5] Agrawal, G.P. Nonlinear Fiber Optics. – Academic Press. – 2013. – 629p.
[6] Guang S. He. Nonlinear Optics and Photonics - Oxford University Press – 2014. - 688p.
[7] Zaharov, V.E. Optical solitons and quasisolitons. – JETP, 86(5) – 1998 - pp. 1035-1046.
[8] Zaharov, V.E. Solitons and collapses. - Physics-Uspekhi, 55 (6)- 2012. - pp. 535-556.

Mathematical modeling of the instability of viscous fluid films

L.A. Prokudina

South-Ural State University, Chelyabinsk, Russia

E-mail: prokudina-la@mail.ru

Abstract. Nonlinear mathematical model of free surface fluid film is presents. Increment, frequency, phase velocity for thin layers of viscous liquids at low Reynolds numbers are calculated. The instability region is found. Optimal flow regimes of films of water and alcohol, corresponding to the maximum values of increment, are calculated.

1. Introduction

The relevance and practical importance of studying the movements of thin layers of a viscous liquid (fluid film) associated with their wide implementation in many devices of thermal power, chemical industry (fracking film, a column with plane-parallel attachment, absorbers, etc.) [1-11]. The combination of the small thickness of the fluid film and large contact areas, as well as the existence, under certain conditions, the circulating fluid flows allows significantly intensify the chemical, thermal, and diffusion processes. In many industrial processes of gas-liquid mass transfer and heat flow occur under conditions of instability of the interfacial surface. Instability is the cause of numerous physical phenomena, for example, rupture of the film in plastic film and packed devices.

2. Mathematical model of non-isothermal liquid film

The mathematical model of flow of fluid films under the action of gravity on the solid impermeable surface are Navier - Stokes equations, the continuity of the boundary conditions has the form [12]:

$$\begin{cases} \dfrac{\partial u_+}{\partial t_+} + u_+ \dfrac{\partial u_+}{\partial x_+} + v_+ \dfrac{\partial u_+}{\partial y_+} = -\dfrac{\partial P_+}{\partial x_+} + \mathrm{Fr}_x + \dfrac{1}{\mathrm{Re}}\left(\dfrac{\partial^2 u_+}{\partial x_+^2} + \dfrac{\partial^2 u_+}{\partial y_+^2}\right), \\ \dfrac{\partial v_+}{\partial t_+} + u_+ \dfrac{\partial v_+}{\partial x_+} + v_+ \dfrac{\partial v_+}{\partial y_+} = -\dfrac{\partial P_+}{\partial y_+} + \mathrm{Fr}_y + \dfrac{1}{\mathrm{Re}}\left(\dfrac{\partial^2 v_+}{\partial x_+^2} + \dfrac{\partial^2 v_+}{\partial y_+^2}\right), \\ \dfrac{\partial u_+}{\partial x_+} + \dfrac{\partial v_+}{\partial y_+} = 0, \end{cases} \tag{1}$$

$$y = 0 : u = 0, v = V_0; \tag{2}$$

$$y = \delta : \frac{1}{\mathrm{Re}}\left[2\frac{\partial u_+}{\partial x_+}\cdot\frac{\partial \delta_+}{\partial x_+} - 2\frac{\partial v_+}{\partial y_+}\cdot\frac{\partial \delta_+}{\partial x_+} - \left(\frac{\partial v_+}{\partial x_+} + \frac{\partial u_+}{\partial y_+}\right)\right] + M\frac{\partial \delta_+}{\partial x_+} + N\frac{\partial^2 u_+}{\partial x_+^2} = 0; \tag{3}$$

Content from this work may be used under the terms of the Creative Commons Attribution 3.0 licence. Any further distribution of this work must maintain attribution to the author(s) and the title of the work, journal citation and DOI.
Published under licence by IOP Publishing Ltd

$$P_+ = \frac{2}{\text{Re}}\left[\frac{\partial v_+}{\partial y_+} - \frac{\partial \delta_+}{\partial x_+}\left(\frac{\partial u_+}{\partial y_+} + \frac{\partial v_+}{\partial x_+}\right)\right] - \sigma_+ \cdot \frac{\partial^2 \delta_+}{\partial x_+^2} + P_0; \tag{4}$$

$$\frac{\partial \delta_+}{\partial t_+} = v_+ - u_+ \frac{\partial \delta_+}{\partial x_+}, \tag{5}$$

where u and v is the projection of the velocity on the coordinate axes Ox and Oy, respectively; Fr_x and Fr_y – the projection of the Froude number; M – the Marangoni parameter; σ – the surface tension coefficient, N is the parameter of the surface viscosity. Index "+" marks the dimensionless quantities.

In the framework of the presented model (1-5) was found the velocity distribution of the fluid film at small Reynolds numbers ($\text{Re} \le 1$) and was derived the equation for the free surface fluid film [13,14]

$$\frac{\partial \psi}{\partial t} = b_0 \frac{\partial^2 \psi}{\partial x \partial t} + b_1 \frac{\partial \psi}{\partial x} + b_2 \frac{\partial^2 \psi}{\partial x^2} + b_3 \frac{\partial^3 \psi}{\partial x^3} + b_4 \frac{\partial^4 \psi}{\partial x^4} + b_5 \psi \frac{\partial \psi}{\partial x} + b_6 \psi \frac{\partial^2 \psi}{\partial x^2} + \tag{6}$$

$$+ b_7 \left(\frac{\partial \psi}{\partial x}\right)^2 + b_8 \frac{\partial \psi}{\partial x} \cdot \frac{\partial^2 \psi}{\partial x^2} + b_9 \psi \left(\frac{\partial \psi}{\partial x}\right)^2,$$

where $\psi(x,t)$ is the deviation of the free surface of a fluid film from the thickness of the unperturbed layer. The coefficients of equation (6) have the form:

$$b_0 = \frac{5}{24}\text{Re}^2 F_x, \qquad b_1 = -\text{Re}\cdot \text{Fr}_x - \text{Re}\cdot \tau_x,\ b_2 = -\frac{\text{Re}}{3}\cdot \text{Fr}_y - \frac{\text{Re}}{2}\cdot M + \frac{3}{40}\text{Re}^3 \text{Fr}_x \left(\tau_x + \text{Fr}_x\right),$$

$$b_3 = -\frac{\text{Re}^2}{2}\cdot \text{Fr}_x N, \qquad b_4 = -\frac{\text{Re}}{3}\cdot \sigma,\quad b_5 = -2\cdot \text{Re}\cdot \text{Fr}_x,$$

$$b_6 = -\text{Re}\left(\text{Fr}_y + M\right) + \frac{3}{4}\text{Re}^3 \text{Fr}_x \left(\frac{\tau_x}{2} + \frac{3}{5}\text{Fr}_x\right), \qquad b_7 = b_6,$$

$$b_8 = b_3,\ b_9 = -\text{Re}\left(2\cdot \text{Fr}_y + M\right) + \frac{3}{2}\text{Re}^3 \text{Fr}_x \left(\tau_x + \frac{3}{2}\text{Fr}_x\right).$$

Dispersion equation was derived:

$$\left(\omega_r + i\omega_i\right)\left(b_0 k_x - i\right) + b_4 k_x^4 - b_2 k_x^2 + i\left(-b_3 k_x^3 + b_1 k_x\right) = 0. \tag{7}$$

Equation (7) allows us to calculate the wave characteristics of fluid films for small Reynolds numbers ($\text{Re} \le 1$) when viscosity forces exceed inertial forces.
Frequency

$$\omega_r = \frac{Y - X \cdot Z}{1 + Z^2}, \tag{8}$$

Increment

$$\omega_i = X + \omega_r \cdot Z . \tag{9}$$

The phase velocity

$$C_r = \frac{\omega_r}{k_x} . \tag{10}$$

The group velocity

$$\frac{d\omega_r}{dk_x} = \frac{Y' - Z \cdot X' - X \cdot Z'}{1 + Z^2} - \frac{2Z}{1 + Z^2}\omega_r , \tag{11}$$

where $X = a_1 k_x^4 - a_6 k_x^2$, $Y = a_4 k_x{}^3 - a_{11} k_x$, $Z = a_7 k_x$,

$X' = 4a_1 k_x^3 - 2a_6 k_x$, $Y' = 3a_4 k_x{}^2 - a_{11}$, $Z' = a_7$,

$a_1 = -\frac{1}{3}\mathrm{Re}\sigma$; $a_4 = -\frac{1}{2}\mathrm{Re}^2\mathrm{Fr}_x\mathrm{N}$; $a_6 = -\frac{1}{3}\mathrm{ReFr}_y + \frac{3}{40}\mathrm{Re}^3\mathrm{Fr}_x\left(\tau_x + \mathrm{Fr}_x\right)$;

$a_7 = \frac{5}{24}\mathrm{Re}^2\mathrm{Fr}_x$; $a_{11} = -\mathrm{Re}\left(\mathrm{Fr}_x + \tau_x\right)$.

3. A computational experiment

A computational experiment was spending for vertical fluid film of water and ethyl alcohol for calculation of the rate of growth of the perturbations, phase velocity for slow-moving fluid films (Re < 1). The regions of instability of fluid films were found. The fastest growing harmonics for a given range of Reynolds numbers were identified. Flow regimes of fluid film, corresponding to the maximum value of increment, are called optimal. The equation for finding the critical values of the wave number, corresponding to the optimal flow regime has the form:

$$3k_x^4 b_0^2\left(b_4 + b_0 b_3\right) + k_x^2[4\left(b_4 + b_0 b_3\right) - b_0^2\left(b_2 + b_0 b_1\right)] - 2\left(b_2 + b_0 b_1\right) = 0 .$$

The data for the calculation of the optimal regimes of two-phase flow are required at the design stage of film devices for increasing safety and reliability of structures.

The results of numerical experiments:

1. The flow of a fluid film under the action of gravity is unstable for all the studied range of Reynolds numbers. The criterion of instability $\omega_i > 0$. The value of the increment is represented on the Fig. 1, 3. The instability region for water is Fig.2 and ethyl alcohol Fig.4.

2. Dimensionless phase velocity varies in the range $2{,}99994 < c_r \le 3$ for water (Fig. 5) and $2{,}99984 < c_r \le 3$ for ethanol.

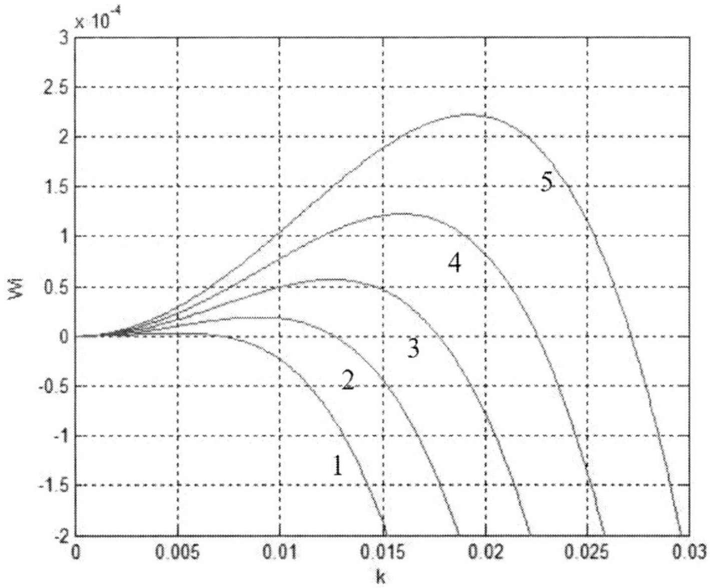

Figure.1. The dependence of the increment from the wave number for water: 1 – Re=0,2; 2- Re=0,4; 3 – 0,6; 4 – 0,8; 5 – 1

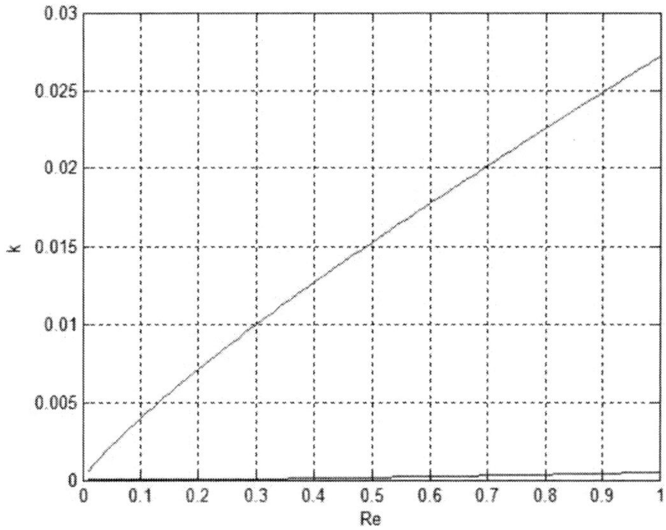

Figure.2. Instability region for water

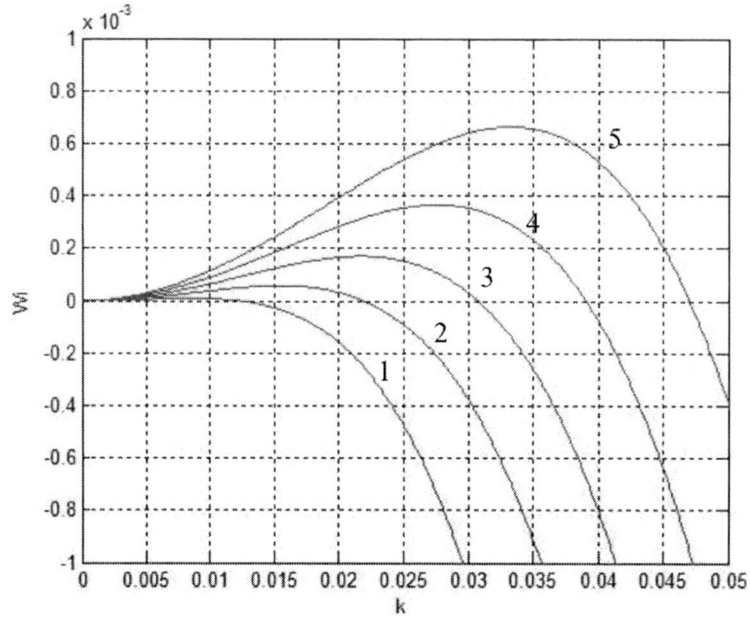

Figure.3. Dependence of the increment from the wave number for alcohol:
1 – Re=0.2; 2 – Re=0.4; 3 – Re=0.6; 4 – Re=0.8; 5 – Re=1

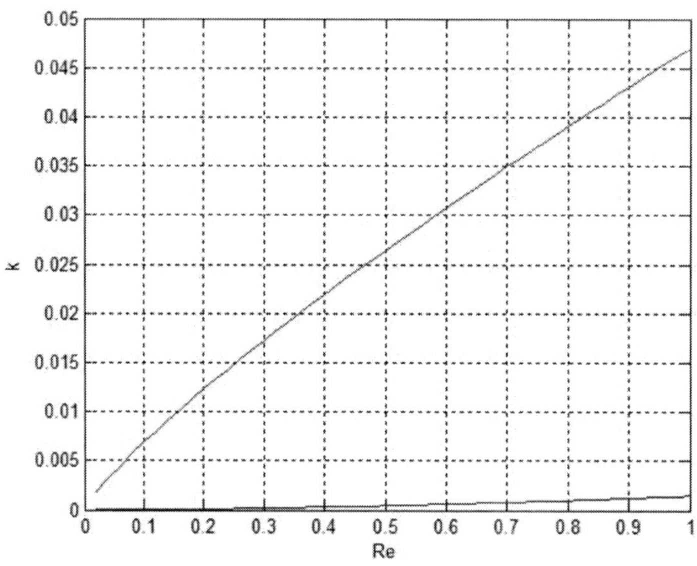

Figure.4. Region of instability for alcohol

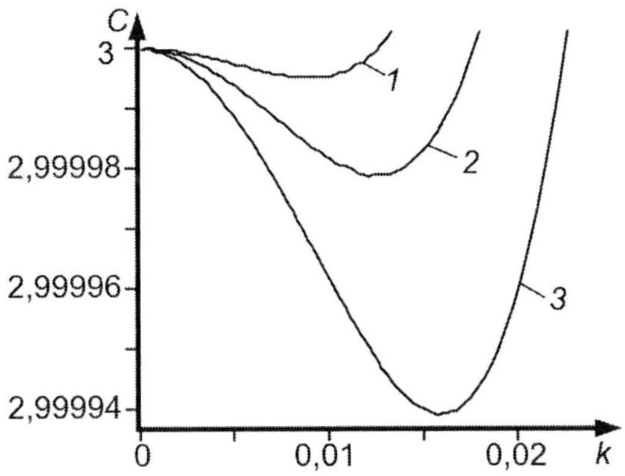

Figure.5. Phase velocity (water):
1 – Re=0,4; 2 – Re=0,6; 3 – Re=0,8

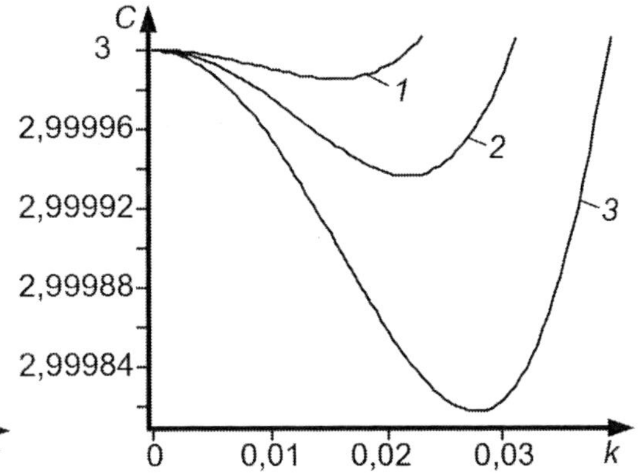

Figure.6. Phase velocity (alcohol):
1 – Re=0,4; 2 – Re=0,6; 3 – Re=0,8

3. With increasing of Reynolds number is observing the increase of the wave number corresponding to the maximum increment value $\omega_{i\max}$, which leads to the reduction of wavelength ($\lambda = \frac{2\pi}{k}$).

References

[1] Kapitsa P.L. 1948 Zh. Eksper. Teor. Fiz. **18** 3

[2] Kapitsa P.L. and Kapitsa S.P. 1949 *Wave flow of thin layers of viscous fluid*, Zh. Eksper. Teor. Fiz. **19** 105

[3] Yih C.S. 1963 *Stability of Liquid flow down an inclined plane*, Phys. Fluids. **6** 321

[4] Kholpanov L.P. 1965 *To the question of wave flow of thin liquid layers on a vertical plane*, Inzh._Fiz. Zh. **8** 488

[5] Massot C., Irani F. and Lightfoot E. N.1966 *Modified description of wave motion in a falling film*, AIChE J. **12** 445

[6] Krylov V.S., Vorotilin V.P. and Levich V.G. 1969 *The theory of wave motion of thin liquid films*, Theor. foundations of chemical. technology. **3**(4) 499 (in Russian)

[7] Krantz W.B. and Goren S.L. 1971 *Stability of thin liquid films flowing down a plane*, Ind. Eng. Chem. Fundam.**10** (1) 91

[8] Filippov A.G., Saltanov G.A. and Kukushkin A.N. 1988 *Fluid Flow and Heat and Mass Transfer in the Presence of Surfactants*, Energoizdat Moscow (in Russian)

[9] Kholpanov L.P. and Shkadov V.Ya. 1990 *Hydrodynamic and Heat and Mass Transfer with Free Surface*, Nauka, Moscow (in Russian)

[10] Gogonin I. and Shemagin I. 1993 *Heat exchange at film condensation and film boiling in the components of NPS*, Energoatomizdat (in Russian)

[11] Alekseenko S.V., Nakoryakov V.E. and Pokusaev B.G. 1994 *Wave Flow of Liquid Films*, Begell House, New York

[12] Prokudina L A and Vyatkin G P 1998 *Instability of a nonisothermal liquid film* Dokl. Phys **43**(10) 652

[13] Prokudina L A 2014 *Influence of Surface Tension inhomogeneity on the Wave Flow of a liquid Film* J. Eng. Phys. and Thermophys. **87** 165

[14] Prokudina L.A. 2014 *Nonlinear evolution of perturbations in a thin fluid layer during wave formation* J. of Experimental and Theoretical Physics **118** (3) 480
DOI: 10.1134/S1063776114020046

Soliton solutions of the (2+1)-dimensional complex modified Korteweg-de Vries and Maxwell-Bloch equations

K Yesmakhanova, G Bekova, G Shaikhova, R Myrzakulov

Eurasian International Center for Theoretical Physics, Department of Higher Mathematics and Department of General & Theoretical Physics, L.N. Gumilyov Eurasian National University, Astana 010008, Kazakhstan

E-mail: kryesmakhanova@gmail.com.

Abstract.

In this paper, we consider the (2+1)-dimensional complex modified Korteweg-de Vries and Maxwell-Bloch (cmKdVMB) equations. Lax pairs of cmKdVMB equations are presented. Using the Lax pair, we construct a Darboux transformation and namely one-fold transformations. The soliton solutions are obtained from the different "seeds" by using this Darboux transformation.

1. Introduction

The theory of nonlinear partial differential equations has attracted a lot of attention among researcher and is fundamentally linked to several basic developments in this area of soliton theory. It is well known that nonlinear equations such as the Korteweg-de Vries (KdV) equation, modified Korteweg-de Vries (mKdV) equation and the nonlinear Schrodinger (NLS) equation are the most typical and well-studied integrable evolution equations which describe nonlinear wave phenomena for a range of dispersive physical systems. In the study of nonlinear waves their solutions play an important role [1-12]. One of the generalizations of the mKdV equation is complex mKdV (cmKdV) equation which is one of the well-known and completely integrable equations in soliton theory. It possesses all the basic characters of integrable models. From a physical point of view cmKdV equation has been derived for, e.g. the fundamental evolution of nonlinear lattice, fluid dynamics, plasma physics, ultra-short pulses in nonlinear optics, nonlinear transmission lines and so on [13-14].

At the present time, the cmKdV equation is used in pair with Maxwell-Bloch system of equations, and so it called as complex modified Korteweg-de Vries and Maxwell-Bloch (cmKdVMB) equations. Moreover, this equation can be received by the reduction of the Hirota-Maxwell-Bloch (HMB) system of equations. In (1+1)-dimensions cmKdVMB equations were studied in [15-17] by the reduction of HMB system of equations.

In this work, we consider (2+1)-dimensional cmKdVMB equations which were suggested in [18]. Here we use the method of Darboux transformation. It has been proved to be an efficient way to find the different solutions from "seed" solutions this transformation allows to construct non-trivial analytical solutions of nonlinear partial differential equations [19-23].

Content from this work may be used under the terms of the Creative Commons Attribution 3.0 licence. Any further distribution of this work must maintain attribution to the author(s) and the title of the work, journal citation and DOI.
Published under licence by IOP Publishing Ltd

The paper is organized as follows. The (2+1)-dimensional cmKdVMB equations we present in Section 2. In Section 3, we construct the DT for the (2+1)-dimensional cmKdVMB equations. In Section 4, using the constructed one-fold DT, the one-soliton solutions of the (2+1)-dimensional cmKdVMB equations is given.

2. The (2+1)-dimensional complex modified Korteweg de Vries and Maxwell-Bloch equations

The complex modified Korteweg-de Vries and Maxwell-Bloch (cmKdVMB) equations reads as [18]

$$
\begin{aligned}
iq_t + iq_{xxy} - vq + i(wq)_x - 2ip &= 0, &(1)\\
v_x - 2i\delta(q^*_{xy}q - q^*q_{xy}) &= 0, &(2)\\
w_x - 2\delta(|q|^2)_y &= 0, &(3)\\
p_x - 2i\omega p - 2\eta q &= 0, &(4)\\
\eta_x + \delta(q^*p + p^*q) &= 0, &(5)
\end{aligned}
$$

where q, p are complex functions, v, w, η are real functions and ω, δ are real constants. This set of equations (1)-(5) is integrable by IST. The Lax pair corresponding to the cmKdVMB equations is given by [18], i.e.

$$
\begin{aligned}
\Psi_x &= A\Psi, &(6)\\
\Psi_t &= 4\lambda^2\Psi_y + B\Psi, &(7)
\end{aligned}
$$

where

$$
\begin{aligned}
A &= -i\lambda\sigma_3 + A_0, &(8)\\
B &= \lambda B_1 + B_0 + \frac{i}{\lambda+\omega}B_{-1}, &(9)
\end{aligned}
$$

with

$$
\begin{aligned}
B_1 &= iw\sigma_3 + 2i\sigma_3 A_{0y}, &(10)\\
A_0 &= \begin{pmatrix} 0 & q \\ -r & 0 \end{pmatrix}, &(11)\\
B_0 &= -\frac{i}{2}v\sigma_3 + \begin{pmatrix} 0 & -q_{xy} - wq \\ r_{xy} + wr & 0 \end{pmatrix}, &(12)\\
B_{-1} &= \begin{pmatrix} \eta & -p \\ -k & -\eta \end{pmatrix} &(13)
\end{aligned}
$$

and $r = \delta q^*, \quad k = \delta p^*$, where $\delta = \pm 1$. The compatibility condition of (6)-(7) is still

$$A_t - 4\lambda^2 A_y - B_x + [A, B] = 0. \tag{14}$$

3. DT for the (2+1)-dimensional cmKdVMB equations

In this section, we construct the DT for the (2+1)-dimensional cmKdVMB equations (1)-(5). In particular, we give in detail the one-fold DT and briefly the n-fold DT.

3.1. One-fold DT

Let Ψ and $\Psi^{[1]}$ are two solutions of the system (6)-(7) so that

$$\begin{aligned} \Psi_x^{[1]} &= A^{[1]}\Psi^{[1]}, & (15)\\ \Psi_t^{[1]} &= 4\lambda^2\Psi_y^{[1]} + B^{[1]}\Psi^{[1]}. & (16)\end{aligned}$$

We assume that these two solutions are related by the following transformation:

$$\Psi^{[1]} = T\Psi = (\lambda I - M)\Psi. \tag{17}$$

The matrix function T obeys the following equations

$$\begin{aligned} T_x + TA &= A^{[1]}T, & (18)\\ T_t + TB &= 4\lambda^2 T_y + B^{[1]}T. & (19)\end{aligned}$$

Then the relation between q, p, v, w, η and $q^{[1]}, p^{[1]}, v^{[1]}, w^{[1]}, \eta^{[1]}$ can be reduced from (18)-(19), which is the Darboux transformation of cmKdVMB equations. Comparing the coefficient of λ^i $(i = 0, 1, 2)$ of the sides the equation (18), we have

$$\begin{aligned} \lambda^0 &: M_x = A_0^{[1]}M - MA_0, & (20)\\ \lambda^1 &: A_0^{[1]} = A_0 + i[M, \sigma_3], & (21)\\ \lambda^2 &: iI\sigma_3 = i\sigma_3 I. & (22)\end{aligned}$$

The equation (21) gives

$$\begin{aligned} q^{[1]} &= q - 2im_{12}, & (23)\\ q^{*[1]} &= q^* - 2im_{21}, & (24)\end{aligned}$$

where

$$M = \begin{pmatrix} m_{11} & m_{12} \\ m_{21} & m_{22} \end{pmatrix}, \quad I = \begin{pmatrix} 1 & 0 \\ 0 & 1 \end{pmatrix}. \tag{25}$$

Hence we get $m_{21} = -m_{12}^*$ in our attractive interaction case that is if $\delta = +1$. After comparing the coefficient of λ^i $(i = 0, 1, 2)$ the equation (19) gives us the following relations

$$\begin{aligned} \lambda^0 &: -M_t = iB_{-1}^{[1]} - B_0^{[1]}M - iB_{-1} + MB_0, & (26)\\ \lambda^1 &: B_0^{[1]} = B_0 - MB_1 + B_1^{[1]}M, & (27)\\ \lambda^2 &: 4M_y = B_1^{[1]} - B_1, & (28)\\ (\lambda + \omega)^{-1} &: 0 = -i\omega B_{-1}^{[1]} - iB_{-1}^{[1]}M + i\omega B_{-1} + iMB_{-1}. & (29)\end{aligned}$$

Hence we get the DT

$$\begin{aligned} B_0^{[1]} &= B_0 - MB_1 + (B_1 + 4M_y)M, & (30)\\ B_1^{[1]} &= B_1 + 4M_y, & (31)\\ B_{-1}^{[1]} &= (M + \omega I)B_{-1}(M + \omega I)^{-1}. & (32)\end{aligned}$$

At the same, from the equation (30)-(31) we get

$$v^{[1]} = v + 4(m_{12}q_y^* + m_{12}^* q_y + 2im_{11}m_{11y} - 2im_{12}^* m_{12y}), \tag{33}$$

$$w^{[1]} = w - 4im_{11y} = w + 4im_{22y} \tag{34}$$

and we additionally have $m_{22} = m_{11}^*$. So the matix M has the form

$$M = \begin{pmatrix} m_{11} & m_{12} \\ -m_{12}^* & m_{11}^* \end{pmatrix}, \quad M^{-1} = \frac{1}{|m_{11}|^2 + |m_{12}|^2}\begin{pmatrix} m_{11}^* & -m_{12} \\ m_{12}^* & m_{11} \end{pmatrix}, \tag{35}$$

$$M + \omega I = \begin{pmatrix} m_{11} + \omega & m_{12} \\ -m_{12}^* & \omega + m_{11}^* \end{pmatrix}, \quad (M + \omega I)^{-1} = \frac{1}{\square}\begin{pmatrix} m_{11}^* + \omega & -m_{12} \\ m_{12}^* & \omega + m_{11} \end{pmatrix}, \tag{36}$$

where

$$\square = det(M + \omega I) = \omega^2 + \omega(m_{11} + m_{11}^*) + |m_{11}|^2 + |m_{12}|^2. \tag{37}$$

The equation (32) gives

$$\eta^{[1]} = \frac{(|\omega + m_{11}|^2 - |m_{12}|^2)\eta - pm_{12}^*(\omega + m_{11}) - p^* m_{12}(\omega + m_{11}^*)}{\square}, \tag{38}$$

$$p^{[1]} = \frac{p(\omega + m_{11})^2 - p^* m_{12}^2 + 2\eta m_{12}(\omega + m_{11})}{\square}, \tag{39}$$

$$p^{*[1]} = \frac{p^*(\omega + m_{11}^*)^2 - pm_{12}^{*2} + 2\eta m_{12}^*(\omega + m_{11}^*)}{\square}. \tag{40}$$

We now assume that

$$M = H\Lambda H^{-1}, \tag{41}$$

where

$$H = \begin{pmatrix} \psi_1(\lambda_1; t, x, y) & \psi_1(\lambda_2; t, x, y) \\ \psi_2(\lambda_1; t, x, y) & \psi_2(\lambda_2; t, x, y) \end{pmatrix} := \begin{pmatrix} \psi_{1,1} & \psi_{1,2} \\ \psi_{2,1} & \psi_{2,2} \end{pmatrix}. \tag{42}$$

Here

$$\Lambda = \begin{pmatrix} \lambda_1 & 0 \\ 0 & \lambda_2 \end{pmatrix} \tag{43}$$

and $det\ H \neq 0$, where λ_1 and λ_2 are complex constants. The matrix H obeys the system

$$H_x = -i\sigma_3 H\Lambda + A_0 H, \tag{44}$$

$$H_t = 4H_y^2 + B_1 H\Lambda + B_0 H + B_{-1} H\Sigma, \tag{45}$$

where

$$\Sigma = \begin{pmatrix} \frac{i}{\lambda_1 + \omega} & 0 \\ 0 & \frac{i}{\lambda_2 + \omega} \end{pmatrix}. \tag{46}$$

In order to satisfy the constraints of M and $B_{-1}^{[1]}$ as mentioned above, we first notes that

$$\Psi^+ = \Psi^{-1}, \quad A_0^+ = -A_0, \tag{47}$$

$$\lambda_2 = \lambda_1^*, \quad H = \begin{pmatrix} \psi_1(\lambda_1;t,x,y) & -\psi_2^*(\lambda_1;t,x,y) \\ \psi_2(\lambda_1;t,x,y) & \psi_1^*(\lambda_1;t,x,y) \end{pmatrix}, \tag{48}$$

$$H^{-1} = \frac{1}{\Delta}\begin{pmatrix} \psi_1^*(\lambda_1;t,x,y) & \psi_2^*(\lambda_1;t,x,y) \\ -\psi_2(\lambda_1;t,x,y) & \psi_1(\lambda_1;t,x,y) \end{pmatrix}, \tag{49}$$

where

$$\Delta = |\psi_1|^2 + |\psi_2|^2. \tag{50}$$

So the matrix M has the form

$$M = \frac{1}{\Delta}\begin{pmatrix} \lambda_1|\psi_1|^2 + \lambda_2|\psi_2|^2 & (\lambda_1 - \lambda_2)\psi_1\psi_2^* \\ (\lambda_1 - \lambda_2)\psi_1^*\psi_2 & \lambda_1|\psi_2|^2 + \lambda_2|\psi_1|^2) \end{pmatrix}. \tag{51}$$

Finally we can write the one-fold DT for the (2+1)-dimensional cmKdVMB equations as:

$$q^{[1]} = q - 2im_{12}, \tag{52}$$

$$v^{[1]} = v + 4(m_{12}q_y^* + m_{12}^*q_y + 2im_{11}m_{11y} - 2im_{12}^*m_{12y}), \tag{53}$$

$$w^{[1]} = w - 4im_{11y} = w + 4im_{22y}, \tag{54}$$

$$\eta^{[1]} = \frac{(|\omega + m_{11}|^2 - |m_{12}|^2)\eta - pm_{12}^*(\omega + m_{11}) - p^*m_{12}(\omega + m_{11}^*)}{\square}, \tag{55}$$

$$p^{[1]} = \frac{p(\omega + m_{11})^2 - p^*m_{12}^2 + 2\eta m_{12}(\omega + m_{11})}{\square}. \tag{56}$$

At last, we note that the expressions of m_{ij} can be rewritten in the determinant form as

$$m_{11} = \frac{\lambda_1|\psi_1|^2 + \lambda_2|\psi_2|^2}{\Delta} = \frac{\Delta_{11}}{\Delta}, \quad m_{12} = \frac{(\lambda_1 - \lambda_2)\psi_1\psi_2^*}{\Delta} = \frac{\Delta_{12}}{\Delta}, \tag{57}$$

where

$$\Delta_{11} = det\begin{pmatrix} \psi_1 & -\lambda_2\psi_2^* \\ \psi_2 & \lambda_1\psi_1^* \end{pmatrix}, \quad \Delta_{12} = -det\begin{pmatrix} \psi_1 & \lambda_1\psi_1 \\ \psi_2^* & \lambda_2\psi_2^* \end{pmatrix}. \tag{58}$$

4. Soliton solutions

Having the explicit form of the DT, we are ready to construct exact solutions of the (2+1)-dimensional cmKdVMB equations. As an example, let us present the one-soliton solution. To get the one-soliton solution we take the seed solution as

$$q = v = w = p = 0, \quad \eta = 1. \tag{59}$$

Then the corresponding associated linear system takes the form

$$\psi_{1x} = -i\lambda\psi_1, \tag{60}$$

$$\psi_{2x} = i\lambda\psi_2, \tag{61}$$

$$\psi_{1t} = 4\lambda^2\psi_{1y} + \frac{i}{\lambda + \omega}\psi_1, \tag{62}$$

$$\psi_{2t} = 4\lambda^2\psi_{2y} - \frac{i}{\lambda + \omega}\psi_2. \tag{63}$$

This system admits the following exact solutions

$$\psi_1 = e^{-i\lambda x + i\mu y + i(4\lambda^2\mu + \frac{1}{\lambda+\omega})t + \delta_1 + i\delta_2}, \tag{64}$$

$$\psi_2 = e^{i\lambda x - i\mu y - i(4\lambda^2\mu + \frac{1}{\lambda+\omega})t - \delta_1 - i\delta_2 + i\delta_0} \tag{65}$$

or

$$\psi_1 = e^{\theta_1 + i\chi_1}, \tag{66}$$

$$\psi_2 = e^{\theta_2 + i\chi_2}. \tag{67}$$

Here $\mu = \eta_1 + i\nu_1$, $\lambda = \alpha_1 + i\beta_1$, δ_i are real constants then

$$\theta_1 = \beta_1 x - \nu_1 y - (8\eta_1\alpha_1\beta_1 + 4\nu_1(\alpha_1^2 - \beta_1^2) - \frac{\beta_1}{(\alpha_1+\omega)^2 + \beta_1^2})t + \delta_1, \tag{68}$$

$$\chi_1 = -\alpha_1 x + \eta_1 y + (4\eta_1(\alpha_1^2 - \beta_1^2) - 8\nu_1\alpha_1\beta_1 + \frac{\alpha_1 + \omega}{(\alpha_1+\omega)^2 + \beta_1^2})t + \delta_2 \tag{69}$$

and $\theta_2 = -\theta_1$, $\chi_2 = -\chi_1 + \delta_0$. Then the one-soliton solution of the (2+1)-dimensional cmKdVMB equations (1)-(5) takes the form

$$q^{[1]} = 2\beta_1 e^{2i\chi_1 - i\delta_0}\,\mathrm{sech}[2\theta_1], \tag{70}$$

$$v^{[1]} = 16\beta_1(\alpha_1\nu_1 + \beta_1\eta_1)\,\mathrm{sech}^2[2\theta_1], \tag{71}$$

$$w^{[1]} = -8\beta_1\nu_1\,\mathrm{sech}^2[2\theta_1], \tag{72}$$

$$\eta^{[1]} = \frac{(\omega+\alpha_1)^2 + \beta_1^2(1 - 2\,\mathrm{sech}^2[2\theta_1])}{(\omega+\alpha_1)^2 + \beta_1^2}, \tag{73}$$

$$p^{[1]} = \frac{2i\beta_1 e^{2i\chi_1 - i\delta_0}(\omega + \alpha_1 + i\beta_1\tanh[2\theta_1])}{((\omega+\alpha_1)^2 + \beta_1^2)\cosh[2\theta_1]}. \tag{74}$$

Pictorial representations of one-soliton solutions (81)-(85) of the cmKdVMB equations is shown in figure 1 and figure 2

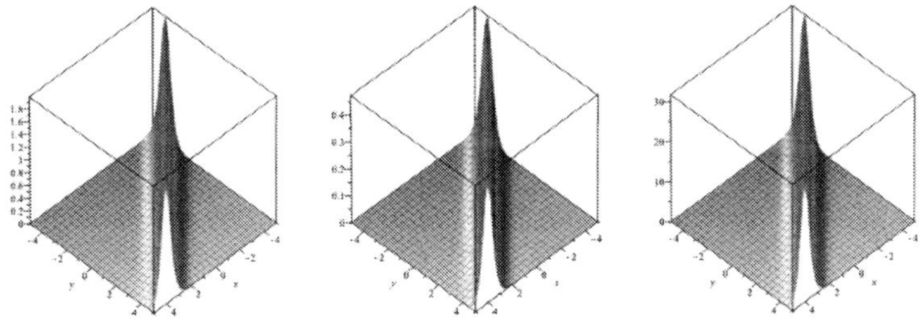

(a) (b) (c)

Figure 1: One-soliton solutions $a)||q^{[1]}||, b)||p^{[1]}||, c)v^{[1]}$ of the cmKdVMB equations when $\alpha_1 = 1, \beta_1 = 1, \omega = 1.5, \nu_1 = 1, \eta_1 = 1$

Acknowledgments
The work of K.Y., G.B., G.Sh. and M.R. is supported by the Ministry of Education and Science of Republic of Kazakhstan (the base part of the state task No.0888/GF4)

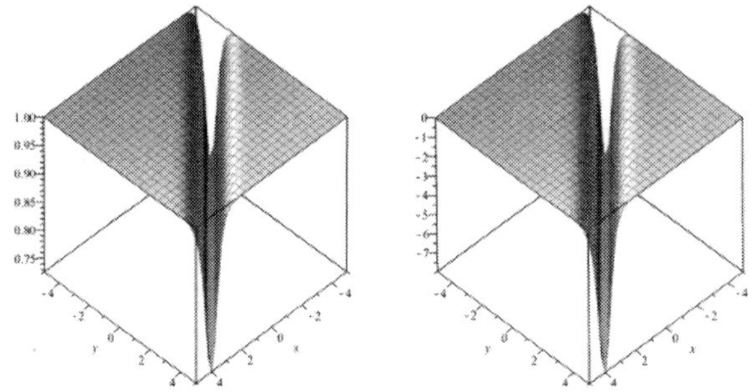

(a) (b)

Figure 2: One-soliton solutions $a)\eta^{[1]}, b)w^{[1]}$ of the cmKdVMB equations when $\alpha_1 = 1, \beta_1 = 1, \omega = 1.5, \nu_1 = 1, \eta_1 = 1$

References

[1] Agrawal G P 2012 *Nonlinear Fiber Optics* 5th edition (Academic Press, San Diego, CA)
[2] Zhunussova Zh Kh, Yesmakhanova K R, Tungushbaeva D I, Mamyrbekova G K, Nugmanova G N and Myrzakulov R 2015 *International Journal of Mathematical, Computational, Statistical, Natural and Physical Engineering* **9** 328-331
[3] Myrzakulov R, Mamyrbekova G K, Nugmanova G N, Yesmakhanova K R and Lakshmanan M 2014 *Physics Letters* A **378** 2118-2123
[4] Myrzakulov R, Vijayalakshmi S, Syzdykova R and Lakshmanan M 1998 *Journal of Mathematical Physics* **39** 2122-2139
[5] Myrzakulov R, Lakshmanan M, Vijayalakshmi S and Danlybaeva A 1998 *Journal of Mathematical Physics,* **39** 3765-3771
[6] Myrzakulov R, Danlybaeva A K and Nugmanova G N 1999 *Theoretical and Mathematical Physics* **118** 13 441-451
[7] Chen Chi and Zhou Zi-Xiang (2009) *Chinese Physics Letters,* **26** 8 080504
[8] Zhao-Wen Yan, Min-Ru Chen, Ke Wu and Wei-Zhong Zhao 2012 *Journal of the Physical Society of Japan* **81** 094006
[9] Yan Zhao-Wen, Chen Min-Ru, Wu Ke and Zhao Wei-Zhong 2012 *Communications in Theoretical Physics* **58** 463-468
[10] Esmakhanova K R, Nugmanova G N, Wei-Zhong Zhao and Ke Wu 2006 Integrable inhomogeneous Lakshmanan-Myrzakulov equation *Preprint* nlin/0604034
[11] Zhen-Huan Zhang, Ming Deng, Wei-Zhong Zhao and Ke Wu 2006 On the integrable inhomogeneous Myrzakulov-I equation *Preprint* nlin/0603069
[12] Myrzakulov R, Nugmanova G and Syzdykova R 1998 *Journal of Physics A: Mathematical & Theoretical* **31** 47 9535-9545
[13] Ablowitz M J and Clarkson P A 1991 *Solitons. Nonlinear Evolution Equations and Inverse Scattering"* (Cambridge University Press, Cambridge, NY)
[14] Jingsong He, Lihong Wang, Linjing Li, Porseizian K and Erdelyi R 2014 *Phys. Rev.* E **89** 062917
[15] Chuanzhong Li, Jingsong He and Porsezian K 2013 *Physical Review* E **87** 012913
[16] Chuanzhong Li and Jingsong He 2012 Darboux transformation and positons of the inhomogeneous Hirota and the Maxwell-Bloch equation *Preprint* nlin/1210.2501
[17] Jieming Yang, Chuanzhong Li, Tiantian Li and Zhaoneng Cheng 2013 *Chinese Physics Letters* **30** 104201
[18] Myrzakulov R, Mamyrbekova G K, Nugmanova G N and Lakshmanan M 2015 *Symmetry* **7** 3 1352-1375
[19] Matveev V B and Salle M A 1991 *Darboux transformations and solitons* (Springer-VerlagBerlin)
[20] Rogers C and Schief W K 2002 *Backlund and Darboux transformations: geometry and modernapplications in soliton theory* vol 30 (Cambridge University Press)
[21] Shaikhova G, Yesmakhanova K, Mamyrbekova G and Myrzakulov R 2014 Darboux transformation and solutions of the (2+1)-dimensional Schrodinger-Maxwell-Bloch equation *Preprint* nlin/1402.4669
[22] Yersultanova Z S, Zhassybayeva M, Mamyrbekova G, Nugmanova G and Myrzakulov R 2016 *Int. J. Geom. Methods Mod. Phys.* **13** 1550134
[23] Myrzakulov R, Vijayalakshmi S, Nugmanova G and Lakshmanan M 1997 *Physics Letters* A **233** 391-396

Journal of Physics: Conference Series **738** (2016) 012019 doi:10.1088/1742-6596/738/1/012019

Atomistic and Coarse-grained Simulations of Hexabenzocoronene Crystals

O G Ziogos[1], G Megariotis[1] and D N Theodorou[1]

[1] School of Chemical Engineering, National Technical University of Athens, Athens, Greece

E-mail: oziogos@mail.ntua.gr

Abstract. This study concerns atomistic and coarse-grained Molecular Dynamics simulations of pristine hexabenzocoronene (HBC) molecular crystals. HBC is a symmetric graphene flake of nanometric size that falls in the category of polyaromatic hydrocarbons, finding numerous applications in the field of organic electronics. The HBC molecule is simulated in its crystalline phase initially by means of an all-atom representation, where the molecules self-organize into well aligned molecular stacks, which in turn create a perfect monoclinic molecular crystal. The atomistic model reproduces fairly well the structural experimental properties and thus can be used as a reliable starting point for the development of a coarse-grained model following a bottom-up approach. The coarse-grained model is developed by applying Iterative Boltzmann Inversion, a systematic coarse-graining method which reproduces a set of target atomistic radial distribution functions and intramolecular distributions at the coarser level of description. This model allows the simulation of HBC crystals over longer time and length scales. The crystalline phase is analyzed in terms of the Saupe tensor and thermomechanical properties are probed at the atomistic level.

1. Introduction

One of the most influential and groundbreaking milestones of the past decade regarding materials science and engineering was undoubtedly the advent of graphene [1]. Graphene is a planar, two dimensional (2D) periodic arrangement of sp2 hybridized carbon atoms residing on the sites of a honeycomb lattice – practically an isolated graphite sheet – with extraordinary electronic [2], mechanical [3] and chemical properties [4]. Besides its pristine 2D geometric state, graphene can exist in other allotropic forms, such as one dimensional graphene nanoribbons [5] and aperiodic nanoflakes [6]. Graphene flakes of nanometric dimensions and variable geometries have been placed in the epicenter of organic electronic research as candidate material to be used as active media for organic light emitting diodes, photovoltaics, and field effect transistors [7]. In such devices, nanographene flakes self-organize into well structured supramolecular assemblies, with the polyaromatic cores in close proximity, thus enabling the interaction of their delocalized pi-pi orbitals. Electronic conductivity in the aforementioned molecular arrays is not based on coplanar to the graphene flake ballistic transport but on transplanar hopping transport mechanisms [8].

The current flagship of nanographenes is hexa-peri-hexabenzocoronene (HBC), either in its pristine state [9] or in its functionalized state, where a series of flexible side chains decorate the periphery of the molecule, altering profoundly the properties of the molecular crystals [10]. Pristine HBC is a

Content from this work may be used under the terms of the Creative Commons Attribution 3.0 licence. Any further distribution of this work must maintain attribution to the author(s) and the title of the work, journal citation and DOI.
Published under licence by IOP Publishing Ltd

planar molecule consisting of 42 carbon atoms and 18 peripheral hydrogen atoms in a perfectly hexagonal symmetric arrangement that crystallizes at monoclinic $P2_1/a$ molecular crystals [9].

In our studies, we employ empirical Molecular Dynamics (MD) simulations in order to equilibrate the molecular crystals and derive structural, thermodynamic and dynamical properties at equilibrium. It is a fact that atomistic MD calculations that consider explicitly all atoms in a molecular system and factor in all sorts of time consuming energy and force calculations, e.g. reciprocal-space electrostatic interactions, are typically limited to a few tenths of nanometers and a few thousands of nanoseconds, as far as the spatial and temporal domains are concerned. Bearing in mind these limitations and using as input information from detailed atomistic MD calculations, a bottom-up coarse-graining approach is adopted and utilized, allowing the examination of the systems under study at augmented spatial and temporal regimes, ranging up to the micrometric scale.

2. Computational Details

2.1. Initial Configurations

The fact that HBC crystallizes into a monoclinic lattice would normally force all calculation to be carried out by employing simulation boxes of general triclinic geometry. In order to facilitate coarse grained calculations, initial configurations with orthogonal simulation boxes are created, with a minuscule compromise regarding the crystallography of the system. To be more precise, HBC crystals are described by a monoclinic primitive unit cell with parameters a_0 = 18.431Å, b_0 = 5.119Å, c_0 = 12.929 Å and β_0 = 112.6° [9]. Upon careful geometric inspection, the molecular crystals can be also described by an alternative unit cell with parameters a = 34.06Å, b = 5.119Å, c = 38.79Å and β = 92.1°. In order to alter the unit cell vectors aiming to obtain an orthogonal cell, one would have to shift the end of vector $\boldsymbol{a}$ by approximately 1.248Å along the direction of $\boldsymbol{c}$. Defining four periodic molecular arrays along the direction of $\boldsymbol{c}$, the orthogonal unit cell is achieved by en masse translations of each array by 0.312Å, 0.624Å, 0.936Å and 1.248Å respectively. All the aforementioned geometric elements and unit cells are depicted in Figure 1.

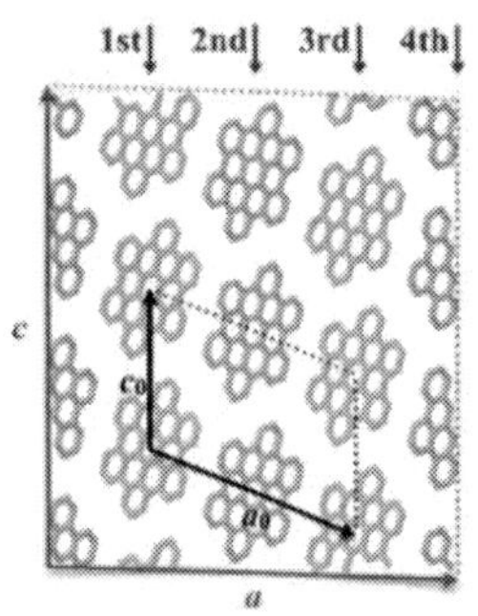

Figure 1. Top view of the $\boldsymbol{a}$, $\boldsymbol{c}$ unit cell for a HBC molecular crystal viewed along the direction of vector $\boldsymbol{b}$. Imbedded in this unic cell, the primitive $\boldsymbol{a_0}$, $\boldsymbol{c_0}$ unit cell is also drawn. The four short arrows over the unit cell highlight the four molecular arrays that are accordingly shifted along the $\boldsymbol{c}$ direction in order to transform the monoclinic unit cell to an orthogonal one. Peripheral hydrogen atoms are omitted for clarity.

Apart from the orthogonal initial configurations, monoclinic $P2_1/a$ molecular crystals are also created using the primitive unit cell in order to be used as input to MD calculations with triclinic simulation boxes.

2.2. Atomistic Molecular Dynamics Simulations

Empirical atomistic MD simulations are carried out in the isostress-isothermal statistical ensemble using the LAMMPS package [11]. The integration of the equation of motion is accomplished through the Martyna-Tobias-Klein numerical scheme [12] with a time step of 1fs, using a time constant of 100fs for temperate control and 2500fs for pressure control. Typical periodic boundary conditions are applied in each direction, allowing this way the examination of the systems in their bulk phase. For the quantification of the bonded interactions, the force field parameterization is taken from earlier work on substituted HBC molecules [13].

2.3. Coarse-grained Simulations

The coarse-grained representation of HBC molecule consists of six superatoms forming a regular hexagon, as seen in Figure 2.

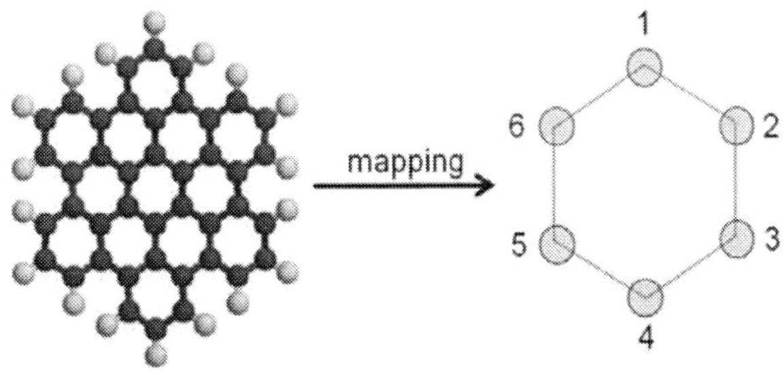

Figure 2. Atomistic and coarse-grained representations of HBC molecule.

In particular, all superatoms are of the same type (AR1) and are placed at the geometric center of the six outermost aromatic rings of the HBC polyaromatic core (see also Figure 1). As far as the interactions between the superatoms are concerned, they are categorized as intermolecular and intramolecular ones. The coarse-grained model is developed by applying the well-known Iterative Boltzmann Inversion (IBI) method which falls in the category of systematic coarse-graining techniques [14]. Due to the representation of HBC molecule by one superatom type, only one radial distribution function (RDF) should be reproduced at this level of description. All coarse-grained simulations were conducted with GROMACS package [15] in the canonical ensemble. Upon coarse-graining, the integration time step of the equations of motion increases from 1fs to 12fs, while the number of interaction sites per molecule is reduced from 60 to 6.

2.4. Structural characterization

The Saupe ordering tensor $\boldsymbol{Q}$, which is widely used in the study of liquid-crystalline phases, is employed for the analysis of the HBC crystalline phase at the atomistic and coarse-grained level. The tensor in question is defined by the following equation

$$\boldsymbol{Q} = \frac{1}{N}\sum_{i=1}^{N}\left\{\frac{3}{2}\hat{\boldsymbol{u}}_i\hat{\boldsymbol{u}}_i - \frac{1}{2}\mathbf{1}\right\} \tag{1}$$

where $\hat{\boldsymbol{u}}_i$ is a unit vector normal to the mesoscopic hexagonal ring (coarse-grained level) or the polyaromatic core (atomistic level). The sum in Equation 1 is taken over all molecules N of the system. For the special geometry of the crystal system under consideration, the eigevalues of $\boldsymbol{Q}$ tensor are computed analytically by the next set of equations [16]:

$$\begin{cases} \lambda_1 = -1/2 \\ \lambda_2 = \left(3\sin^2\theta - 1\right)/2 \\ \lambda_3 = \left(3\cos^2\theta - 1\right)/2 \end{cases} \tag{2}$$

where θ is the tilt angle which is formed by the normal unit vector and the columnar direction.

2.5. Mechanical characterization

The simulation of the molecular crystals using general triclinic simulation boxes prompts us with the capability of estimating the elastic moduli of the material using typical strain fluctuation methodologies [17]. The isothermal compliance tensor $\boldsymbol{S}^T$ is described by Equation 3:

$$S^T_{ijkl} = \frac{V_0}{k_B T}\left(\langle \eta_{ij}\eta_{kl}\rangle - \langle \eta_{ij}\rangle\langle \eta_{kl}\rangle\right) \tag{3}$$

where V_0 is the average system volume at zero stress conditions, $\boldsymbol{\eta}_{ij}$ the strain matrix element with respect to a zero stress average cell shape and the angular brackets correspond to time averages. The strain matrix is expressed as

$$\boldsymbol{\eta} = \frac{1}{2}\left(\tilde{\boldsymbol{h}}_0^{-1}\tilde{\boldsymbol{h}}\boldsymbol{h}\boldsymbol{h}_0^{-1} - \boldsymbol{1}\right) \tag{4}$$

where $\boldsymbol{h}_0$ is the reference cell shape matrix at zero stress and $\boldsymbol{h}$ the instantaneous cell shape matrix, with the tilde corresponding to the transpose matrix. The singular 9×9 compliance matrix calculated this way is transformed to a non-singular 6×6 matrix by applying the Voigt convention [18]. From this new matrix, the isothermal bulk modulus B and shear modulus G_R are derived in a straight-forward fashion via the following two equations [18,19]:

$$B = S_{11} + S_{22} + S_{33} + 2\left(S_{12} + S_{23} + S_{31}\right) \tag{5}$$

$$G_R = 15\times\left[4\left(S_{11}+S_{22}+S_{33}\right) - 4\left(S_{12}+S_{23}+S_{31}\right) + 3\left(S_{44}+S_{55}+S_{66}\right)\right]^{-1} \tag{6}$$

3. Results

The implementation of the IBI method entails the reproduction of a set of RDFs and intramolecular (or non-bonded) distributions at the coarse-grained level. As already detailed in the previous section, only one RDF (AR1-AR1) is considered for the HBC molecule. The fact that the system under consideration is found in the crystalline phase has as a consequence the existence of many peaks in the RDF AR1-AR1 which is a severe handicap for the full convergence of the IBI method. However, a partial convergence can be achieved as seen in Figure 3, in which the target and coarse-grained RDFs are depicted in black and red color respectively.

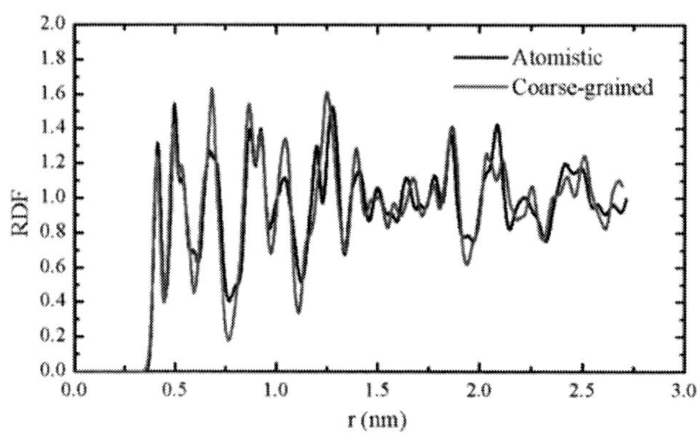

Figure 3. Comparison of atomistic and coarse-grained RDFs for the pair AR1-AR1.

The crystal phase predicted with the coarse-grained model is compared to the crystalline phase of the underlying atomistic model in terms of Q tensor eigevalues and θ tilt angle with the latter being experimentally observable. These properties are presented in Table 1, from which the consistency between the atomistic and coarse-grained model is evident. More specifically, in Table 1, the θ angles are compared for the two models while the eigenvalues from the simulations are compared with the corresponding analytical estimates computed by Equation 2.

Table 1. Saupe tensor eigenvalues and molecular tilt angle from detailed atomistic simulations, coarse-grained simulations and analytical theory.

	Simulation				Theory		
	λ_1	λ_2	λ_3	θ	λ_1	λ_2	λ_3
Atomistic model	-0.499	0.356	0.129	49.65°	-0.5	0.370	0.130
Coarse-grained model	-0.498	0.340	0.159	48.52°	-0.5	0.342	0.158

As far as the elastic response of HBC in its crystalline state is concerned, the molecular crystal exhibits a rather stiff behavior, since the calculated elastic moduli vary in the region of tenths of GPa. Since the molecules are in their pristine form without any peripheral flexible substituents, the rise in temperature – always below the melting point – is expected to have a rather small effect in the elastic moduli. This is indeed the case, as it is demonstrated in Figure 4, where the evolution of the bulk and shear moduli is depicted upon material heating.

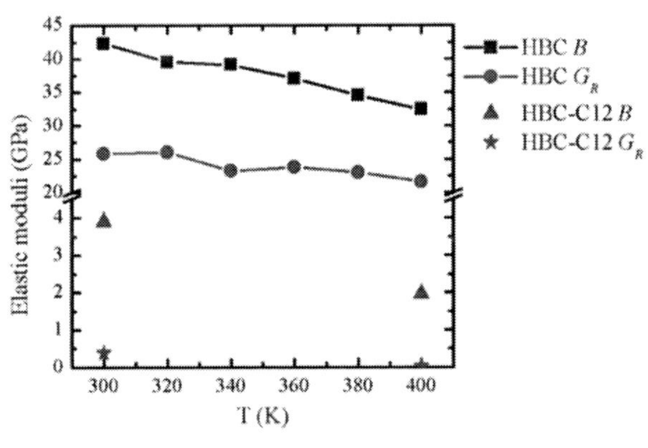

Figure 4. Temperature variation of bulk modulus B and shear modulus G_R of the HBC crystal. Values for the soluble HBC-C12 system are shown for comparison.

In order to emphasize the effect of the lack of peripheral side chains, the relevant elastic moduli for a grafted system [13] are also shown.

4. Conclusions

In this current work, we report the successful description of a molecular crystal of HBC molecules by means of detailed, fully atomistic and reduced, coarse-grained MD simulations. The two different levels of description are in fine agreement with each other, as well as with available experimental data. An analytical formulation for the eigenvalues of the Saupe orientation tensor for discotic materials is proposed and validated against simulation results. Finally, state-of-the-art atomistic simulations are applied in order to quantify the elastic response of this material. Judging by our findings, pristine HBC crystals exhibit a stiff behaviour, ranging between their soft soluble grafted counterparts and typical covalent crystals.

Acknowledgements
This research has been co-financed by the European Union (European Social Fund – ESF) and Greek national funds through the Operational Programme 'Education and Lifelong Learning' of the National Strategic Reference Framework (NSRF) – Research Funding Programme: THALIS. Investing in knowledge society through the European social fund [grant number MIS 379436] and by the IKY Fellowships of Excellence for Postgraduate Studies in Greece – Siemens Programme [award number 11126/13a]. Grigorios Megariotis and Doros N. Theodorou acknowledge generous funding by the Limmat Foundation. This work was also supported by computational resources granted from the Greek Research & Technology Network (GRNET) in the National HPC facility 'ARIS' under the project MultiCLC (ID pr001042).

References

[1] Geim A K and Novoselov K S 2007 *Nature Mater.* **6** 183
[2] Novoselov K S, Geim A K , Morozov S V, Jiang D, Katsnelson M I, Grigorieva I V, Dubonos S V and Firsov A A 2005 *Nature* **438** 197
Castro Neto A H, Guinea F, Peres N M R, Novoselov K S and Geim A K 2009 *Rev.Mod. Phys.* **81** 109
[3] Lee C, Wei X, Kysar J W and Hone J 2008 *Science* **321** 385
[4] Georgakilas V, Otyepka M, Bourlinos A B, Chandra V, Kim N, Kemp K C, Hobza P, Zboril R and Kim K S 2012 *Chem. Rev.* **112** 6156
[5] Li X, Wang X, Zhang L, Lee S and Dai H 2008 *Science* **319** 1229
[6] Müller S and Müllen K 2007 *Phil. Trans. R. Soc. A* **365** 1453
[7] Wu J, Pisula W and Müllen K 2007 *Chem. Rev.* **107** 718
[8] Feng X, Marcon V, Pisula W, Hansen M R, Kirkpatrick J, Grozema F, Andrienko D, Kremer K and Müllen K 2009 *Nature Mater.* **8** 421
Vehoff T, Baumeier B, Troisi A and Andrienko D 2010 *J. Am. Chem. Soc.* **132** 11702
[9] Goddard R, Haenel M W, Herndon W C, Krüger C and Zander M 1995 *J. Am. Chem. Soc.* **117** 30
[10] Elmahdy M M, Floudas G, Mondeshki M, Spiess H W, Dou X and Müllen K. 2008 *Phys. Rev. Lett.* **100** 107801
Hansen M R, Feng X, MacHo V, Müllen K, Spiess H W and Floudas G 2011 *Phys. Rev. Lett.* **107** 257801
[11] Plimpton S 1995 *J. Comput. Phys.* **117** 1
[12] Martyna G J, Tobias D J and Klein M L 1994 *J. Chem. Phys.* **101** 4177
[13] Ziogos O G and Theodorou D N 2015 *Mol. Phys.* **113** 2776
[14] Müller-Plathe F 2002 *Chem. Phys. Chem.* **3** 754
Megariotis G, Vyrkou A, Leygue A and Theodorou D N 2011 *Ind. Eng. Chem. Res.* **50** 546
[15] Hess B, Kutzner C, van der Spoel D and Lindahl E 2008 *J. Chem. Theory Comput.* **4** 435
[16] Megariotis G, Ziogos O G and Theodorou D N 2015 *AIP conf. Proc.* **1702** 19001
[17] Parrinello M and Rahman A 1982 *J. Chem. Phys.* **76**, 2662
Ray J R 1988 *Comput. Phys. Rep.* **8**, 109
[18] Nye F J 1969 *Physical properties of crystals* (London: Oxford University Press)
[19] Hill R 1952 *Proc. Phys. Soc. A* **65**, 349

Momentum distribution of occupation numbers in FeAs-based superconductors

V A Kashurnikov, A V Krasavin, Ya V Zhumagulov

National Research Nuclear University MEPhI (Moscow Engineering Physics Institute), 31 Kashirskoye shosse, 115409, Moscow, Russia

avkrasavin@gmail.com

Abstract. The momentum distribution of charge carriers in two-dimensional FeAs-clusters modeling iron-based superconductors was calculated within the limits of the two-orbital model. The calculation was implemented by use of quantum world-line Monte Carlo algorithm. Matsubara Green's function was calculated for clusters with sizes up to 10x10 FeAs-cells. Profiles of the distributions are presented for the entire Brillouin zone. The data indicate the presence of a jump near the Fermi level, which is the evidence of the Fermi-type distribution. The influence on the momentum distribution of the value of the interaction parameter is discussed.

1. Introduction

The form of the momentum distribution function of charge carriers, the presence or absence of the Fermi jump, can provide additional information for the study of electron correlations in high-temperature superconductors (HTSC) based on iron [1-4]. Clarification of the issue of the nature of a strongly correlated system, whether it is of Fermi type, or has Luttinger liquid features, will allow properly analyzing experimental and calculated data on various spectral characteristics of these compounds.

The description of physical properties of iron-based HTSC in the framework of the two-dimensional tight-binding model, such as the two- [5, 6], three- [7, 8], and five-orbital models [8, 9], which are typical generalized Hubbard models, virtually does not allow a correct analysis in the framework of various approximations, including the mean-field approximation. The complexity of these models leads to the use of exact quantum calculation methods to study iron-based HTSC.

Preliminary studies of electron correlations in large FeAs clusters developed earlier by the authors in a series of papers [10-14] with the use of the quantum world-line Monte Carlo algorithm (CTWL-algorithm) within the limits of the full two-orbital model, yielded a number of results correlated with other experimental and theoretical work.

The present work is devoted to the study of the momentum distribution of charge carriers and the analysis of possible Fermi nature of the distribution near the Fermi surface. Quantum algorithm developed in [10] allows the calculation of the Matsubara Green's function and the off-diagonal density matrix, which, in turn, allows the calculation of the momentum distribution and analyzing its dependence on the size of the system, the temperature and the interaction parameters.

We present data on the calculation of the charge carrier distribution function within the full two-orbital model for the clusters with sizes up to 10×10 FeAs cells. The profile of the distribution function for the entire Brillouin zone is obtained, and the jump of the distribution near the Fermi level is analyzed. There is a strong dependence on the value of the interaction parameters.

Content from this work may be used under the terms of the Creative Commons Attribution 3.0 licence. Any further distribution of this work must maintain attribution to the author(s) and the title of the work, journal citation and DOI.
Published under licence by IOP Publishing Ltd

2. Model

The minimal two-orbital model for FeAs-based HTSC takes into account the real crystal structure of these compounds, as well as anisotropy of physical properties. Since 3d-states of iron atoms provide the main contribution to the formation of the band structure near the Fermi level [5, 6], the Hamiltonian of the model is presented as follows:

$$H = H_{int} + H_{kin},$$

$$H_{int} = U\sum_{i,\alpha} n_{i,\alpha,\uparrow}n_{i,\alpha,\downarrow} + V\sum_{i} n_{i,x}n_{i,y} - \mu\sum_{i} n_i - J\sum_{i}\left(n_{i,x,\uparrow}n_{i,y,\uparrow} + n_{i,x,\downarrow}n_{i,y,\downarrow}\right)$$

$$-J\sum_{i}\left(a^{+}_{i,x,\downarrow}a_{i,x,\uparrow}a^{+}_{i,y,\uparrow}a_{i,y,\downarrow} + a^{+}_{i,x,\uparrow}a_{i,x,\downarrow}a^{+}_{i,y,\downarrow}a_{i,y,\uparrow} + a^{+}_{i,x,\uparrow}a_{i,y,\downarrow}a^{+}_{i,x,\downarrow}a_{i,y,\uparrow} + a^{+}_{i,y,\uparrow}a_{i,x,\downarrow}a^{+}_{i,y,\downarrow}a_{i,x,\uparrow}\right),$$

$$H_{kin} = -t_1\sum_{i,\sigma}\left(a^{+}_{i,x,\sigma}a_{i+x,x,\sigma} + a^{+}_{i,y,\sigma}a_{i+y,y,\sigma} + h.c.\right) - t_2\sum_{i,\sigma}\left(a^{+}_{i,y,\sigma}a_{i+x,y,\sigma} + a^{+}_{i,x,\sigma}a_{i+y,x,\sigma} + h.c.\right)$$

$$-t_3\sum_{i,\sigma}\left(a^{+}_{i,x,\sigma}a_{i+x+y,x,\sigma} + a^{+}_{i,x,\sigma}a_{i+x-y,x,\sigma} + a^{+}_{i,y,\sigma}a_{i+x+y,y,\sigma} + a^{+}_{i,y,\sigma}a_{i+x-y,y,\sigma} + h.c.\right)$$

$$-t_4\sum_{i,\sigma}\left(a^{+}_{i,x,\sigma}a_{i+x+y,y,\sigma} + a^{+}_{i,y,\sigma}a_{i+x+y,x,\sigma} + a^{+}_{i,x,\sigma}a_{i+x-y,y,\sigma} + a^{+}_{i,y,\sigma}a_{i+x-y,x,\sigma} + h.c.\right) \quad (1)$$

where operator $a^{+}_{i,x(y),\sigma}(a_{i,x(y),\sigma})$ creates (annihilates) an electron with spin σ on site $\boldsymbol{i}$ and orbital $x(y)$; t_i, $i = 1, \dots, 4$ are the hopping amplitudes between d_{xz} and d_{yz} orbitals defined as x and y, respectively; parameters U and V describe the Coulomb interaction within and between orbitals, respectively; J is the exchange integral; μ is the chemical potential.

As our research shows, namely the exchange term is important for the correct description of the correlation properties of the system and for implementation of a certain type of symmetry pairing of charge carriers. But at the same time it is this term which distinguishes the model (1) from the usual generalized Hubbard model and leads to difficulties in the implementation of quantum Monte Carlo algorithm. All the features of the numerical modeling, coding of basic states, and calculation of Green's and correlation functions in the framework of CTWL-algorithm are described in detail in [10, 11].

The set of realistic parameters of H_{kin} in (1) was taken from [10–12]:

$$t_1 = 0.058; t_2 = 0.22; t_3 = -0.0208; t_4 = -0.079. \quad (2)$$

The relationship between the parameters of H_{int} was determined as

$$V = 0.5U; J = 0.25U. \quad (3)$$

All energy quantities are measured in 0.2 eV units.

3. Matsubara Green's function and momentum distribution

The Matsubara Green's function

$$G_{ij\sigma\sigma'}(\tau) = -\langle T_\tau a_{i\alpha\sigma}(\tau)a^{+}_{j\beta\sigma'}(0)\rangle, \quad (4)$$

where i, j are the coordinates of Fe atoms; α, β are the orbitals; σ, σ' are the spin projections, was calculated for the analysis of the momentum distribution for clusters of sizes up to 10×10 FeAs-sells in the temperature range $\beta = 5 \div 20$, and the range of the interaction parameter $U = 2 \div 16$ while maintaining the relation (3).

Momentum distribution can be obtained from the data for the Matsubara Green's functions as follows:

$$n(\vec{k}) = -G(\vec{k}, \tau \to -0); \quad (5)$$

$$G(\vec{k}, \tau) = \sum_{ij} G_{ij\sigma\sigma}(\tau)e^{\vec{k}\vec{r}_{ij}}.$$

The distribution $n(\vec{k})$ was normalized according to average occupation numbers of orbitals calculated independently. The values of the chemical potential were chosen so that the calculation was carried out at half filling.

4. Results

It should be noted that the results for clusters of size 6×6 and greater are no longer depend on the size of the system, indicating the possibility of the applicability of the data obtained for the analysis of the properties of a macroscopic system.

Figure 1 (a) shows the profiles of the momentum distribution along the main crystallographic directions for 8×8 cluster, depending on the interaction parameter U. The "jump" in the occupation numbers is clearly visible. We define the Fermi level as the point of the maximum gradient of the distribution. The interaction significantly blurs the distribution profile, and there are no completely filled or empty areas. According to our results, the distribution weakly depends on temperature at $1/T \geq 5$ near the Fermi level. Figure 1 (b) also shows the profiles of the electron density of states along the main crystallographic directions, calculated from data on the Green's function by the method presented in [16].

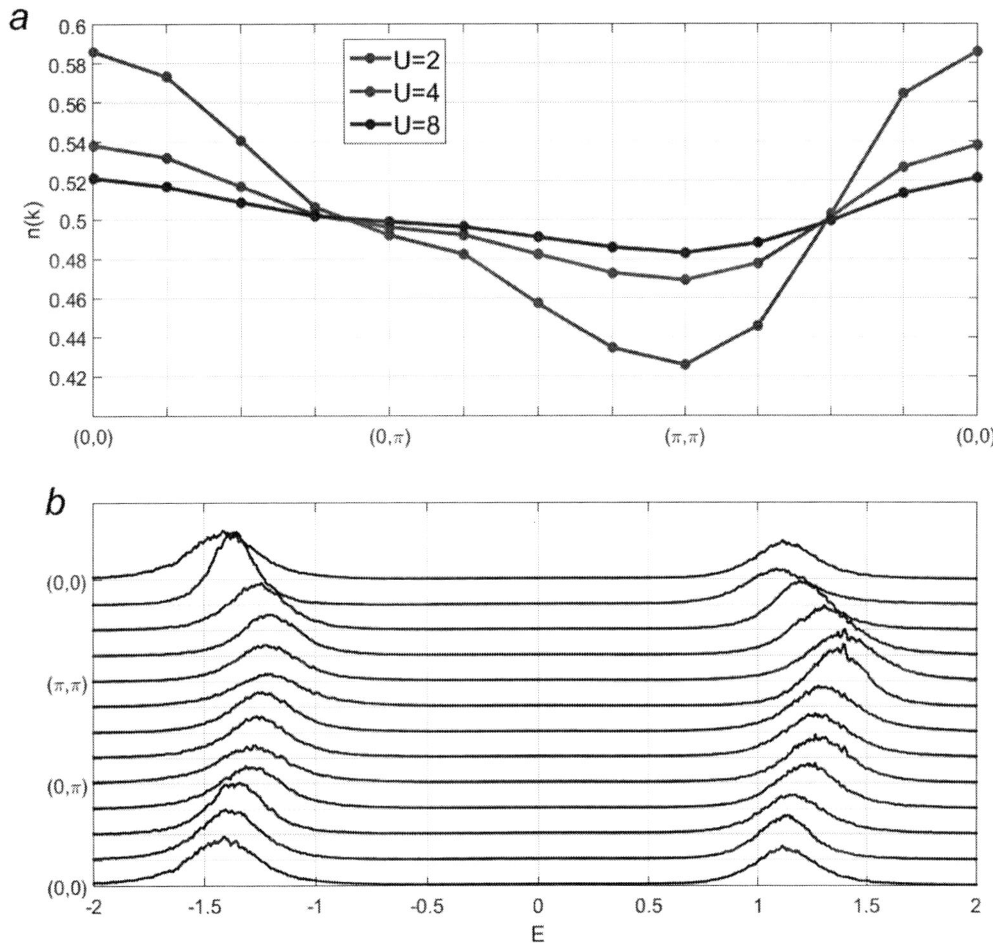

Figure 1. (a) Profile of the momentum distribution along the main crystallographic directions depending on the interaction parameter U. Temperature $1/T = 5$; cluster 8×8. (b) The density of electronic states for the same cluster; $U = 2$.

Figure 2 shows the distribution for the entire Brillouin zone. The Fermi surface corresponding to the half-filling is displayed in white. Note that the lines corresponding to different occupation numbers reflect the picture of the possible profiles of the Fermi surface for other fillings, assuming normal zone filling while maintaining the calculated distribution profile. The line of the maximum gradient corresponds to the half-filling. At $U = 8$ the dependence on the interaction is saturated, and for greater values of U the distribution profile does not change.

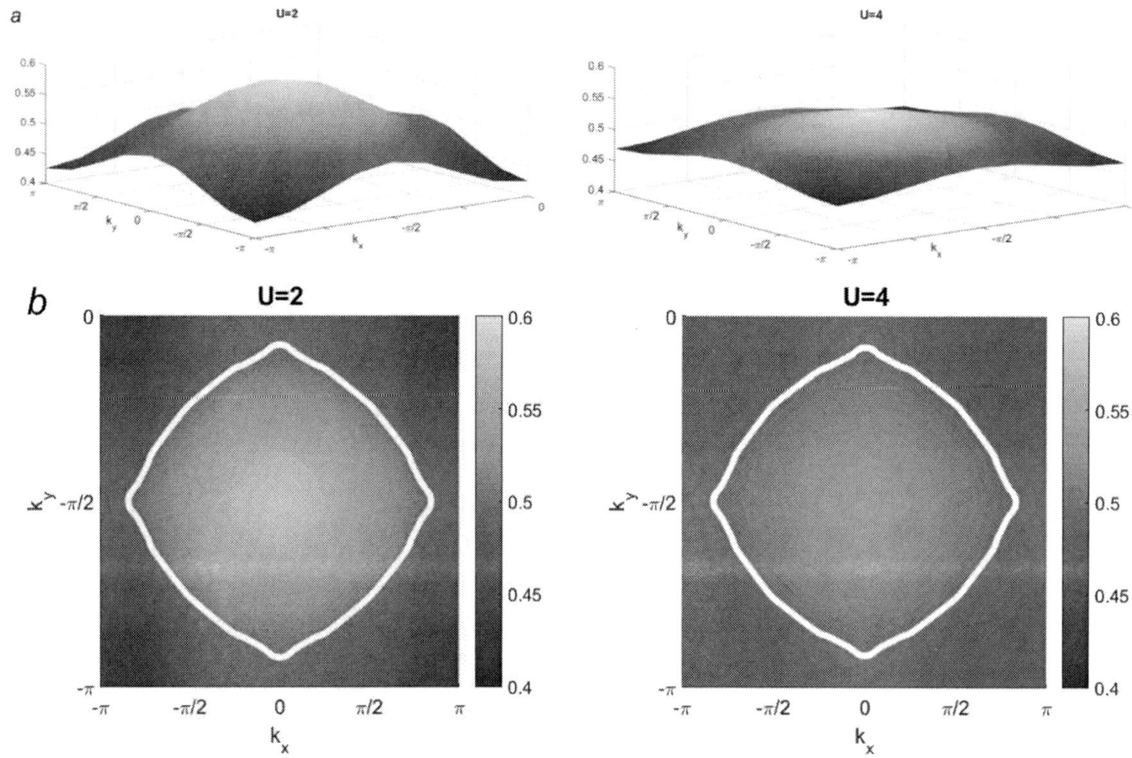

Figure 2. (a) The evolution of the momentum distribution for the entire Brillouin zone with the change the interaction parameter U. (b) The same in the projection on the xy plane. The Fermi surface is shown in white. Cluster 8×8; temperature $1/T = 5$.

5. Conclusions

Calculations of FeAs clusters of size up to 10×10 within the limits of the full two-orbital model were performed using generalized quantum Monte Carlo algorithm at half-filling and different values of temperature and the interaction parameter. The momentum distribution of charge carriers is restored from the Matsubara Green's function. The data for the distribution profile along the main crystallographic directions and for the entire Brillouin zone are presented. The jump of the distribution of occupation numbers is obtained, indicating the Fermi-liquid nature of the electronic ensemble. It is shown that this jump is sensitive to the interaction parameters.

Acknowledgements

The work was supported by Russian Found for Basic Research (projects # 14-08-00509 and # 15-02-02764).

References

[1] Y.Kamihara, T. Watanabe, M. Hirano, and H. Hosono, J. Am. Chem. Soc., **130** (2008) 3296.
[2] Yu. Isyumov, E. Kurmaev. High-Tc Superconductors Based on FeAs Compounds. Springer, 2010.
[3] E. Dagotto, Rev. Mod. Phys. **85** (2013) 849.
[4] I.A. Nekrasov, M.V. Sadovskii, JETP Lett. **99** (2014) 598.
[5] S. Raghu, Xiao-Liang Qi, Chao-Xing Liu et al., Phys. Rev. B **77** (2008) 220503(R).
[6] A. Moreo, M. Daghofer, J.A. Riera, and E. Dagotto, Phys. Rev. B **79** (2009) 134502.
[7] S.-L. Yu, J. Knang, and J.-X.Li, Phys. Rev. B, **79**, 064517 (2009).
[8] Q. Luo, G. Martins, D.-X.Yaoet al., Phys. Rev. B **82**, 104508 (2010).
[9] S. Graser, T.A. Maier, P.J. Hirschfeld, D.J. Scalapino, New J. Phys., **11**, 025016 (2009).
[10] V.A. Kashurnikov, A.V. Krasavin, JETP Lett. **97** (2013) 333.
[11] V. A. Kashurnikov and A. V. Krasavin, JETP Lett. **100**, 16 (2014)
[12] V.A. Kashurnikov, A.V. Krasavin, Phys. Lett. A **378** (2014) 2894.
[13] V.A. Kashurnikov, A.V. Krasavin, Physics Procedia **65** (2015) 21.
[14] V.A. Kashurnikov, A.V. Krasavin, Physics Procedia **71** (2015) 379.
[15] S. Liang, G. Alvarez, C. Sen, A. Moreo, and E. Dagotto, Phys. Rev. Lett. **109** (2012) 047001.
[16] V.A. Kashurnikov, A.V. Krasavin, Ya.V. Zhumagulov, JETP Lett., **103** (2016) 334.

Molecular modeling and simulation of atactic polystyrene/amorphous silica nanocomposites

I Mathioudakis[1], G G Vogiatzis[1], C Tzoumanekas[1] and D N Theodorou[1]
[1]Department of Materials Science and Engineering, School of Chemical Engineering, National Technical University of Athens, Athens, Greece

E-mail: imathiou@mail.ntua.gr

Abstract. The local structure, segmental dynamics, topological analysis of entanglement networks and mechanical properties of atactic polystyrene – amorphous silica nanocomposites are studied via molecular simulations using two interconnected levels of representation: (a) A coarse – grained level. Equilibration at all length scales at this level is achieved via connectivity – altering Monte Carlo simulations. (b) An atomistic level. Initial configurations for atomistic Molecular Dynamics (MD) simulations are obtained by reverse mapping well-equilibrated coarse-grained configurations. By analyzing atomistic MD trajectories, the polymer density profile is found to exhibit layering in the vicinity of the nanoparticle surface. The dynamics of polystyrene (in neat and filled melt systems) is characterized in terms of bond orientation. Well-equilibrated coarse-grained long-chain configurations are reduced to entanglement networks via topological analysis with the CReTA algorithm. Atomistic simulation results for the mechanical properties are compared to the experimental measurements and other computational works.

1. Introduction

The complexity of intermolecular interactions and confinement in polymer-nanoparticle systems leads to spatial variations in structure and dynamics at both the meso- and nanoscale, which are of the order of 100 nm – 1 μm and 1 – 100 nm, respectively. Nanomaterials fabricated by dispersing nanoparticles in polymer melts have the potential for unique optical, electrical, rheological, and permeability properties, as well as mechanical performance that far exceeds that of traditional composites [1, 2, 3, 4]. The attributes of a heterogeneous material depend on the properties of its component materials and on the characteristic length scale of their coexistence, as well as on the chemical and morphological details of the dispersion. Measurements of the rheological properties of polymer matrix nanocomposites in the melt have revealed unexpected phenomena of great practical importance for the molding technology of polymers. Of particular interest are the experiments of Mackay et al. [1, 2, 3], who showed that the viscosity of polymer melts with dispersed nanoparticles may be lower than that of their pure counterparts.

Kumar and Krishnamoorti [5] has presented in a review article that despite intense research activity on polymeric matrix nanocomposites, the exact mechanisms by which nanoparticles affect the properties in often unexpected ways still remain largely unknown. The importance of nanocomposites has been known for some time, e.g. in the use of carbon black for modifying the properties of elastomers for car tires. More recently, interest in these materials was stimulated by the discovery [6] that the incorporation of small amounts of nanoparticles in a polymer matrix can result in a

Content from this work may be used under the terms of the Creative Commons Attribution 3.0 licence. Any further distribution of this work must maintain attribution to the author(s) and the title of the work, journal citation and DOI.
Published under licence by IOP Publishing Ltd

simultaneous increase in both the elastic modulus (stiffness), and the energy required for their deformation to failure (toughness), contrary to what happens in the case of filler particles with macroscopic (>1mm) dimensions.

The dynamics of polystyrene nanocomposites has been studied at a microscopic level using neutron scattering techniques. Neutron scattering measurements offer the unique possibility of analyzing the spatial arrangement and dimensions of individual components and provide an excellent tool for assessing the microscopic dynamics of nanocomposites at various length scales. Using quasi – elastic neutron scattering (QENS) measurements, nanocomposites of polystyrene (PS), poly (methyl methacrylate) (PMMA) and tetramethylbisphenol – A polycarbonate (TMPC) with silica and fullerenes have been studied. Nanoparticles were found to bring about an increase in the glass transition temperature of these materials, as measured by differential scanning calorimetry (DSC) and dynamic mechanical spectroscopy (DMA). These authors suggested that the suppression of the local relaxation dynamics of the nanocomposite is consistent with an enhancement of cohesive interactions in the system, which may be the root of the increase in T_g for the nanocomposites. A very small increase of T_g for silica – polystyrene nanocomposites was observed by Kontou and Anthoulis [7].

Vogiatzis et al. [8] have investigated the structure of a polystyrene matrix filled with tightly cross – linked polystyrene nanoparticles, using Monte Carlo simulations of a coarse – grained model. Their work shows that the local structure of the polymer matrix in the vicinity of the nanoparticle is different in many ways from that of the corresponding bulk, both at the segment and the chain level. The local polymer density profile near the particle displays a maximum and the bonds develop considerable orientation parallel to the nanoparticle surface. Using atomistic MD simulations, Ndoro et al. [9, 10] studied the structure and dynamics around silica nanoparticles dispersed in a matrix of low molecular weight polystyrene. They observed that near a nanoparticle a polymer layer of increased density was formed, due to favorable interaction energy. Polymer bonds in this layer were highly oriented parallel to the surface of the particle, and segmental dynamics was slower relative to the bulk. The effect of the nanoparticle curvature and grafting density on the mean square displacement of free polystyrene chains and also on the mean relaxation time of various intramolecular vectors was investigated as a function of separation from the nanoparticle surface. Confinement, reduced surface curvature, and densification resulted in a reduction of the mean square displacement and an increase in the mean relaxation time of the C_a – H bond vector and chain end-to-end vector in the vicinity of the surface.

Brown et al. [11] have examined the structure and dynamics of a system containing an inorganic nanoparticle embedded in a polymer matrix. In their work, results are presented regarding the variation of the structure and dynamics with distance from the polymer – nanoparticle interface and as a function of pressure.

Li et al. [12] observed that highly entangled polymer chains were disentangled upon increasing the volume fraction of spherical non – attractive nanoparticles. The authors report a critical volume fraction controlling the crossover from polymer chain entanglements to nanoparticle – induced entanglements. While below this critical volume fraction the polymer chain relaxation accelerates upon filling, above this volume fraction the situation reverses.

Using a quasi-harmonic approximation, Lempesis et al. [13] have calculated and predicted mechanical properties of atactic PS in the elastic regime. The authors report a very good agreement of Young's modulus and Poisson's ratio with experimental measurements [27].

2. Experimental Methods

The simulation of the nanocomposites is based on a two-tier approach. Equilibrating the melt takes place at the coarse-grained level, where each polymer interaction site corresponds to a monomer unit and each nanoparticle is represented as a single sphere, using powerful connectivity - altering MC algorithms [16]. In particular, a polymer coarse-grained site stands for a PS diad (meso or racemo). PS coarse-grained sites are governed by bonded and non-bonded interactions derived from the atomistic force field via Iterative Boltzmann Inversion (IBI). Silica nanoparticles interact with coarse-grained sites through Hamaker integrated potentials derived directly from the atomistic LJ parameters [17].

Next we restore the atomistic detail through a process which is based on a combination of local MC moves, quasi-Metropolis reconstruction of atomistic sites and gradual minimization of the potential energy of the generated configuration. Specifically, reverse mapping of coarse-grained PS to united-atom structures is applied retaining implicit coarse-grained tacticity (1 coarse-grained site = 8 united atom carbons) [18].

Atomic-level interactions between polymer chains and nanoparticles play a central role in this work. Given the low polarity of PS segments, they were assumed to consist of excluded volume and London dispersion contributions, represented by Lennard-Jones (LJ) potentials. LJ parameters between polymer atoms and fullerene carbons were obtained using the Lorentz-Berthelot rules. For polymer-silica interactions the parameterization was based on previous work on zeolite/hydrocarbon systems which has yielded excellent predictions for sorption isotherms and diffusivities [19]. Any chemical bonding between nanoparticle surfaces and polymer was disregarded. Silica nanoparticles of prescribed diameter were generated by excising a spherical region from bulk amorphous silica simulated with MD and terminating dangling bonds at the surface as silanol groups, following ref [11].

Finally, to study the dynamics of the nanocomposites, we perform MD simulations in the atomistic representation using a united-atom PS forcefield [20] integrated into the LAMMPS [21] MD application.

The PS nanocomposite systems studied consist of silica particles molecules dispersed in a melt of atactic polystyrene. A single particle of diameter 3 nm, 6 nm or 12 nm, surrounded by polymeric melt composed of chains of molecular weight 20 to 200 kg/mol is simulated under nearly infinite dilution conditions. The coarse-grained MC simulations are performed in the canonical ensemble (NVT) at a temperature of 500 K, for 8 up to 20×10^{9} moves. Subsequent reverse mapping to the atomistic level takes 590×10^{6} MC moves. The MD simulation starts in the isothermal-isobaric (NPT) statistical ensemble at 500 K and 1 bar and continues in the canonical and microcanonical (NVE). The integration time step in the MD is set to 1 fs (10^{-15} s).

3. Results and Discussion

The estimated root – mean – square radius of gyration $\left\langle R_g^2 \right\rangle^{1/2}$ as a function of molar mass for neat and nanocomposite polystyrene systems is shown in Fig. 1. Very good agreement is observed for all the molar masses examined for neat polystyrene systems. As far as the nanocomposite polystyrene systems are concerned, the presence of the nanoparticle seems to affect slightly the root – mean – square radius of gyration.

A fusion of *trans* and *gauche* conformational states is apparently produced by the atomistic potential at $T = 500$ K. Vogiatzis and Theodorou [18] observed that torsion angle distributions in their reverse – mapped structures are extremely close to those obtained from a 80mer structure that was directly equilibrated by MD at the atomistic level, without intervention of any coarse – graining and reverse mapping. This observation implies that one can achieve well – equilibrated structures, down to the atomistic scale, using the reverse mapping procedure. In the case of *racemo* diads, the percentage of *gauche* states increases slightly in the presence of the silica nanoparticle. This effect may be attributed to the effort of the chains to engulf the silica nanoparticle. The overall content of trans conformations and the total percentage of $\overline{g}$ conformations in the reverse – mapped structures are in reasonable agreement with experimental NMR measurements [23].

Local structure around the silica nanoparticles has been quantified. At short distances from the surface of the nanoparticle, the local density is strongly perturbed. The radial distribution of polymer density around the nanoparticle is seen to exhibit two peaks before stabilizing at its bulk value. Profiles from the atomistic and from the coarse-grained simulation are in good agreement for the same particle diameter, underlining the consistency of the two approaches [24].

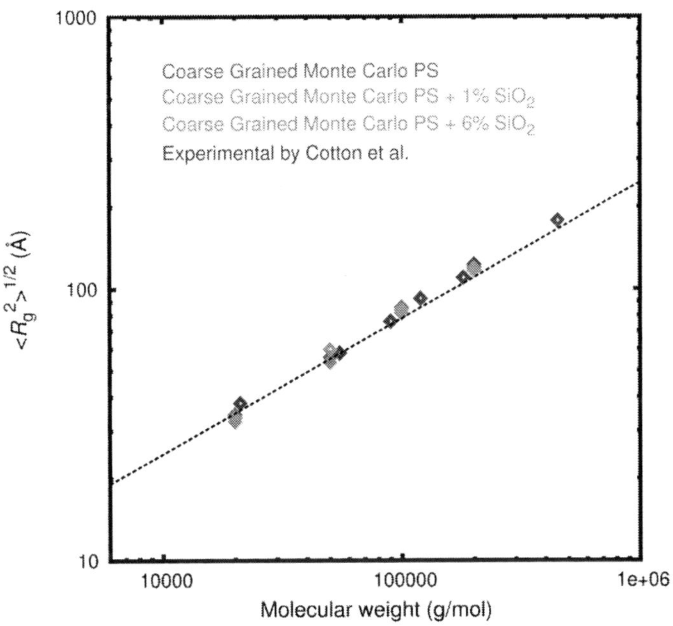

Figure 1. Average root mean squared radius of gyration of the coarse-grained chains of neat and nanocomposite polystyrene systems as a function of molar mass in the melt at 500 K (red, green and magenta rhomboid symbols). Neutron scattering measurements for high molar mass polystyrene are also included (blue rhomboid symbols) [22].

Based on the MD trajectories, segmental and local dynamics in the polymer have been characterized. Directly accessible are the orientational decorrelation functions of the phenyl stems. Characteristic times extracted from these decorrelation functions have been compared to dielectric spectroscopy (DS) measurements. In Fig. 2, a slight influence of the presence of the silica nanoparticle (dispersed at infinite dilution) is observed on global segmental dynamics. In this case, the first Legendre polynomial $P_1(t)$ is employed as a measure of the polymer segmental dynamics. This is a very common measure by calculating time-correlation functions of a vector, $\mathbf{v}_b$, representing a chemical bond. This measure of the segmental dynamics of a-PS can be compared with NMR data. It is quite common to fit the long – time behavior of orientational autocorrelation functions of this kind by a modified Kohlrausch – Williams – Watts (mKWW) function [25]. Here, orientational relaxation of a-PS C-CM phenyl ring – backbone vector is measured using $P_1(t)$ and characteristic times are calculated for bulk PS and PS/SiO_2 systems at volume fraction $\Phi_1 = 1\%$ and $\Phi_2 = 6\%$ respectively. We need to extract these characteristic times at different temperatures in order to calculate glass transition temperature T_g of bulk PS and PS/silica nanocomposites.

The study of local segmental dynamics, when a nanoparticle is present, requires a discretization of space, so as to quantify possible a nanoparticle effects. We choose to carry out an analysis based on a discretization of the simulation box using spherical shells, with the silica nanoparticle acting as the center of the shells. Shells surrounding the nanoparticle are of equal thickness. They were constructed to define different spatial regions as a way to investigate how far the silica surface influences the a-PS segmental dynamics. We clearly observe slower dynamical behavior of this vector near the surface of the silica nanoparticle. The local influence of the presence of the silica nanoparticle on the segmental dynamics of PS are shown in Fig. 3.

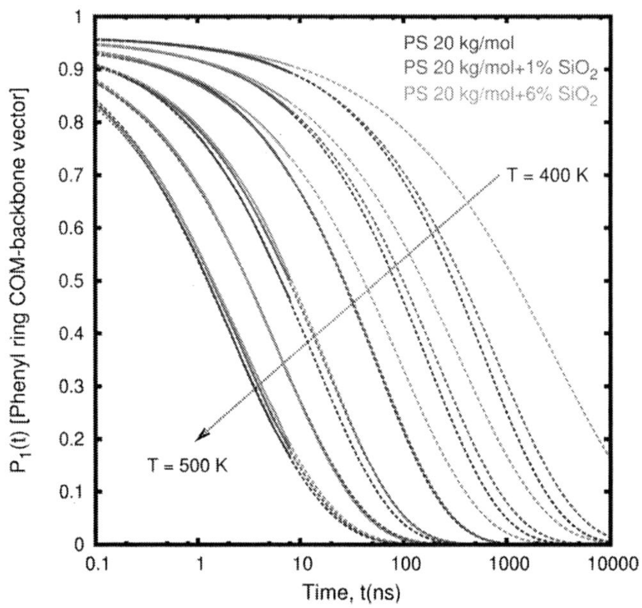

Figure 2. Orientational time autocorrelation functions of the phenyl stem $\mathbf{v}_{CH\text{-}CM}$ (solid lines), fitted to the mKWW equation (dashed lines).

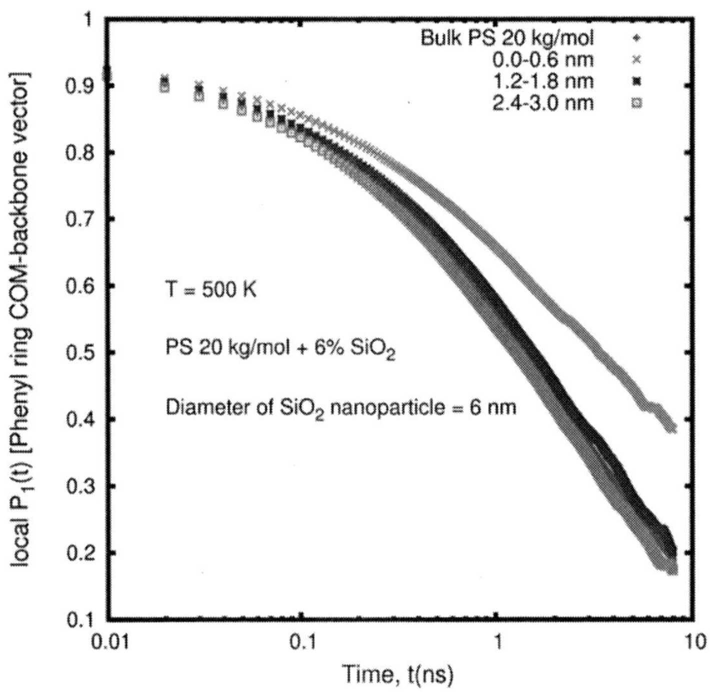

Figure 3. Local orientational relaxation of PS CH backbone – CM phenyl ring vector compared with distance from the surface of silica nanoparticle at T=500 K.

Topological analysis of coarse-grained neat and nanocomposite systems of PS has been performed by using the CReTA (contour reduction topological analysis) algorithm [26]. For a neat polystyrene system with 40 chains of 2000 diads each, N_e was estimated to be 126 diads, and since the average molar mass of a diad is 104.15 g mol^{-1}, the corresponding entanglement molar mass M_e is 13097 ± 200 g mol^{-1}. This value is in very good agreement with experimental estimates based on plateau modulus measurements [34]. For a nanocomposite polystyrene system with the same number of chains and molecular weight, the presence of silica nanoparticle with diameter 6 nm increases the N_e [24] from 126 diads, which is the value of neat polystyrene system, to 152 diads and the corresponding entanglement molar mass M_e accordingly.

After calculation of structural and dynamical properties, elastic constants such as the Young's modulus are studied using stress – strain computational experiments at room temperature on bulk PS and PS/SiO_2 systems at volume fraction Φ_1 = 1% and Φ_2 = 6% respectively. For bulk PS, Young's modulus was estimated to be E = 3.7 ± 0.2 GPa. This value is in a good agreement with experimental measurements [27] and other computational works [13]. Likewise, the room temperature Young's modulus for polystyrene – silica nanocomposite systems was calculated from the slope of a line fitted to the stress-strain simulation data at small deformations. The results were $E_{\Phi=1\%}$ = 4.9 ± 0.3 GPa and $E_{\Phi=6\%}$ = 5 ± 0.4 GPa, respectively. As we can see, an increase of the Young's modulus relative to bulk PS is brought about by the presence of the silica nanoparticles. Also, the bulk modulus k can be calculated using the strain fluctuations methodology [14, 15]. For bulk glassy PS at 300 K, the bulk modulus was estimated as k_{PS} = 3.56 ± 0.4 GPa and for the nanocomposite polystyrene systems as $k_{\Phi=1\%}$ = 3.75 ± 0.5 GPa and $k_{\Phi=6\%}$ = 3.94 ± 0.6 GPa. The result for bulk PS is in good agreement with the experimental value $k_{exp,PS}$ = 3 GPa [27].

4. Conclusion

In this work we have presented results on the local structure, segmental and local dynamics of polystyrene nanocomposites. Developing a hierarchical approach involving two interconnected levels, it was possible to prepare fully equilibrated configurations with high molecular weight chains. The dynamics of the nanocomposite was slower than in the pure melt at a temperature range above the glass transition temperature, in good agreement with available experimental observations. Local density around the nanoparticles was disturbed over a region of thickness of a few nanometers. Entanglements analysis based on the CReTA algorithm shows that the presence of silica nanoparticles increases entanglement molar mass M_e, either the silica nanoparticle is treated as a phantom particle or is allowed to interact with reduced chains. The presence of silica nanoparticles in a polystyrene matrix increases Young's modulus value of pure polystyrene. Methodologies for calculating the shear and bulk modulus of our systems through volume and strain fluctuations in the undeformed state have also been used. Our results are in very good agreement with available experimental measurements and other computational works.

References

[1] Mackay M E, Dao T T, Tuteja A, Ho D L, van Horn B, Kim H –C and Hawker C J 2003 *Nat. Mater.* **2** 762 – 66
[2] Tuteja A, Mackay M E, Hawker C J and van Horn B 2005 *Macromolecules* **38** 8000 – 11
[3] Tuteja A, Duxbury P M and Mackay M E 2007 *Macromolecules* **40** 9427 – 34
[4] Bockstaller M R, Mickiewicz R A and Thomas E L 2005 *Adv. Mater.* **17** 1331 – 49
[5] Kumar S K and Krishnamoorti R 2010 *R. Annu. Rev. Chem. Biomol. Eng.* **1** 37 – 58
[6] Schmidt D F, Clement F and Giannelis E P 2006 *Adv. Funct. Mater.* **16** 417 – 25
[7] Kontou E and Anthoulis G 2007 *J. Appl. Polym. Sci.* **105** 1723 – 31
[8] Vogiatzis G G, Voyiatzis E and Theodorou D N 2011 *Europ. Polym. J.* **47** 699 – 712

[9] Ndoro T V M, Voyiatzis E, Ghanbari A, Theodorou D N, Böhm M C and Müller – Plathe F 2011 *Macromolecules* **44** 2316 – 27
[10] Ndoro T V M, Böhm M C and Müller – Plathe F 2012 *Macromolecules* **45** 171 – 179
[11] Brown D, Marcadon V, Mélé P and Albérola N D 2008 *Macromolecules* **41** 1499 – 511
[12] Li Y, Kröger M and Liu W K 2012 *Phys. Rev. Lett.* **119** 118001
[13] Lempesis N, Vogiatzis G G, Boulougouris G C, van Breemen L C, Hütter M and Theodorou D N 2013 *Mol. Phys.* **111** 3430 – 3441
[14] Parinello M and Rahman A 1981 *J. Appl. Phys.* **52** 7182
[15] Parinello M and Rahman A 1982 *J. Appl. Phys.* **76** 2662
[16] Spyriouni T, Tzoumanekas C, Theodorou D N, Müller – Plathe F and Milano G 2007 *Macromolecules* **40** 3876 – 85
[17] Everaers R and Ejtehadi M R 2003 *Phys. Rev. E* **67** 041710
[18] Vogiatzis G G and Theodorou D N 2014 *Macromolecules* **47** 387 – 404
[19] Snurr R Q, Bell A T and Theodorou D N 1993 *J. Phys. Chem.* **97** 13742 – 52
[20] Lyulin A V and Michels M A J 2002 *Macromolecules* **35** 1463 – 72
[21] Plimpton S. 1995 *J. Comp. Phys.* **117** 1 – 19
[22] Cotton J P, Decker D, Benoit H, Farnoux B, Higgins J, Jannick G, Ober R, Picot C and des Cloizeaux J 1974 *Macromolecules* **7** 863 – 72
[23] Dunbar M, Novak B and Schmidt – Rohr K 1998 *Solid State Nucl. Magn. Reson.* **12** 119 – 37
[24] Mathioudakis I G, Vogiatzis G G, Tzoumanekas C and Theodorou D N 2016 *IEEE Trans. on Nanotech.* **15** 416 – 422
[25] Logotheti G E and Theodorou D N 2007 *Macromolecules* **40** 2235 – 45
[26] Tzoumanekas C and Theodorou D N 2006 *Macromolecules* **39** 4592 – 603
[27] Mark J E 1999 *Polymer Data Handbook* (Oxford University Press: New York)

Convex and concave successions of power-law decays in small-angle scattering

E M Anitas[1,2]

[1]Joint Institute for Nuclear Research, Dubna 141980, Moscow region, Russia

[2]Horia Hulubei National Institute of Physics and Nuclear Engineering, RO-077125 Bucharest, Romania

E-mail: anitas<at>theor.jinr.ru

Abstract. The small-angle scattering (SAS) structure factor from a new model of a 3D deterministic fractal in which the relative positions and the number of structural units vary with fractal iteration number is calculated. It is shown that, depending on the relative positions of scattering units inside the fractal, we can obtain various types of power-law successions, such as: convex/concave - when the absolute value of the scattering exponent of the first power-law decay is higher/smaller than that of the subsequent power-law decay, or any combination of them (i.e. convex-concave or concave-convex). The obtained results can explain experimental SAS (neutron or X-rays) data which are characterized by a succession of power-law decays of arbitrary length.

1. Introduction

Small-angle scattering is a well established technique [1, 2] for determination of structural properties of complex nano and micro systems such as: aggregation processes in polymeric solutions [3], morphology of crystalline amorphous block copolymers in solution [4], microstructure of elastomeric/composite membranes [5, 6], and is especially suited for those structures which are characterized by either exact or statistical self-similarity (deterministic and random fractals; [7–9]). Its main advantage is the possibility to distinguish between mass and surface fractals [10–14] through the value of the scattering exponent τ from the power-law dependence $I(q) \propto q^{-\tau}$, where $\tau = D_{\mathrm{m}}$ for mass fractals ($0 < D_{\mathrm{m}} < 3$), and $\tau = 6 - D_{\mathrm{s}}$ for surface fractals ($2 < D_{\mathrm{s}} < 3$). Here, D_{m} and D_{s} denote the mass, and respectively the surface fractal dimension, and q is the scattering vector magnitude.

For random fractals, besides the fractal dimension, one may obtain the overall size of the fractal from the Guinier region, and the size of the smallest structural 'subunit' composing the fractal, from the end of the fractal region. For deterministic fractals with a single scale it has been shown that additional structural information can be obtained from experimental SAS data, such as: the iteration number, the fractal scaling factor, or the total number of structural subunits composing the fractal [15, 16].

However, some experimental SAS data [17–21] show a succession of power-law decays with arbitrarily scattering exponents, and recently few theoretical models have been proposed both, for 'convex' / 'concave' data (where the absolute value of the scattering exponent of the first power-law regime is higher/smaller than that of the next power-law regime) [22, 23]. While for 'convex' data the model assumes a three-phase structure in which homogeneous structures with scattering length densities ρ_1 and ρ_2 are immersed in a homogeneous medium with density ρ_0, for the 'concave' data the basic assumption is a two-phase system in which the homogeneous structures of scattering length density ρ_1 are fractals whose scaling factor varies with iteration number.

Content from this work may be used under the terms of the Creative Commons Attribution 3.0 licence. Any further distribution of this work must maintain attribution to the author(s) and the title of the work, journal citation and DOI.
Published under licence by IOP Publishing Ltd

In this work we develop a new theoretical model based on deterministic mass fractals with a single scale, and we show that the behavior given by a succession of 'convex' / 'concave' SAS data, and more generally, by any combination of them, can be explained also by a two-phase system in which the relative positions and the number of structural subunits vary with fractal iteration number.

2. Models

In building our models, we consider a top-down approach in which a cube of edge l_0 with a ball of radius $r_0 = l_0/2$ in its center (initiator) is repeatedly divided in smaller pieces according to some prescribed rules (see below). We choose a Cartesian system of coordinates whose center coincide with the center of the initiator. The iteration rule consists in replacing the cube of edge l_0 with a given number $k_1 > 1$ of smaller cubes of edge length $l_1 = l_0/3$ (first iteration). Then, at m-th iteration the edge of the cube is $l_m = l_0/3^m$ and the radii of concentric balls are $r_m \equiv l_m/2 = r_0/3^m$. Here, we choose three different fractals (Fig. 1) with a significant difference in their fractal dimensions $D_{\rm m}$ so that also the differences in the slopes of scattering structure factor and the succession of 'convex'/'concave' power-law regimes to be easily recognized.

The so-obtained fractals consist of subunits with scattering length density ρ_u immersed in a solid matrix of scattering length density ρ_m. Thus, the scattering contrast will be given by $\Delta\rho = \rho_u - \rho_p$ and the scattering intensity becomes [1]:

$$I(q) = n\,|\Delta\rho|^2\, V^2 \langle |F(\boldsymbol{q})|^2 \rangle, \tag{1}$$

where n is the concentration of fractals, V is their volume, the angle brackets $\langle \cdots \rangle$ is the averaging over all possible orientations, and $F(\boldsymbol{q}) \equiv (1/V) \int_V e^{-i\boldsymbol{q}\cdot\boldsymbol{r}} \mathrm{d}\boldsymbol{r}$ is the normalized form factor.

We choose configurations in such a way that the scattering units are positioned symmetric with respect to the origin, and their number at m-th iteration is given by

$$k_m = \begin{cases} 8^m, & \mathit{for\ fractal\ I}, \\ 14^m, & \mathit{for\ fractal\ II}, \\ 20^m, & \mathit{for\ fractal\ III}. \end{cases} \tag{2}$$

Note that *fractal-I* is the well known triadic Cantor set, and *fractal-III* is the Menger sponge. Thus, the fractal dimensions are obtained in the limit of infinite number of iterations, and is given by [15]

$$D_{\rm m} = \lim_{m\to\infty} \frac{\log k_m}{\log(1/3)} = \begin{cases} 1.89, & \mathit{for\ fractal\ I}, \\ 2.40, & \mathit{for\ fractal\ II} \\ 2.71, & \mathit{for\ fractal\ III}. \end{cases} \tag{3}$$

According to the number of subunits given by Eq. (2), the corresponding generative function describing the relative positions of scattering subunits inside each fractal, can be written as:

$$G_m(\boldsymbol{q}) = \begin{cases} \Lambda_m(\boldsymbol{q}), & \mathit{for\ fractal\ I}, \\ \frac{1}{14}\left(8\Lambda_m(\boldsymbol{q}) + 2\Gamma_m(\boldsymbol{q})\right), & \mathit{for\ fractal\ II} \\ \frac{1}{20}\left(8\Lambda_m(\boldsymbol{q}) + 4\Gamma_m(\boldsymbol{q})\right), & \mathit{for\ fractal\ III}, \end{cases} \tag{4}$$

where $\Lambda_m(\boldsymbol{q}) \equiv \cos(l_m q_x)\cos(l_m q_y)\cos(l_m q_z)$, $C_m(\boldsymbol{q}) \equiv \cos(l_m q_x) + \cos(l_m q_y) + \cos(l_m q_z)$, and $\Gamma_m(\boldsymbol{q}) \equiv \cos(l_m q_x)\cos(l_m q_y) + \cos(l_m q_x)\cos(l_m q_z) + \cos(l_m q_y)\cos(l_m q_z)$ (Fig. 1).

Each of the individual fractal in Fig. 1 gives rise to a single power-law decay (with $\tau = D_{\rm m}$) in the scattering intensity. Thus, in order to observe *a succession* of power-law decays with different

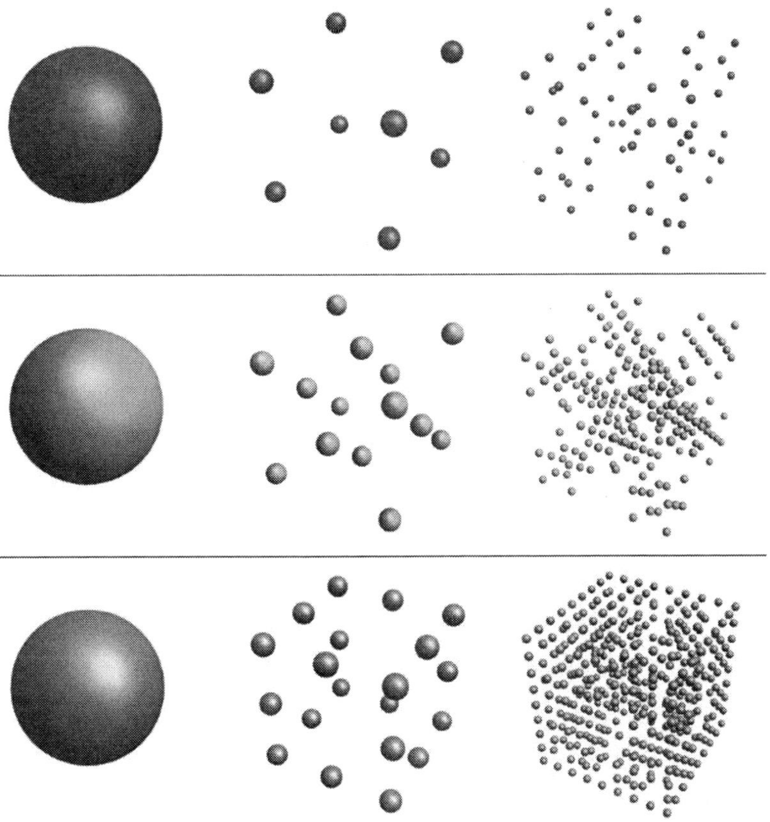

Figure 1. The initiator and first two iterations for models with $k_m = 4^m$ (upper fractal), $k_m = 8^m$ (middle fractal) and $k_m = 12^m$ (lower fractal), for arbitrarily iteration m.

exponents, we proceed as follows: for the first p iterations we consider the structure of a single fractal (from those defined in Eq. (2)), for the next q iterations we consider the structure of another fractal (also from those defined in Eq. (2)), and finally, for the next r iterations we consider the structure of the remaining fractal. Thus, depending on the particular fractal chosen for each p, q or r iterations, we shall expect any of the following type of successions of power-law decays: $q^{-1.89} \to q^{-2.40} \to q^{-2.71}$ ('concave' succession), $q^{-2.71} \to q^{-2.40} \to q^{-1.89}$ ('convex' succession), $q^{-2.40} \to q^{-2.71} \to q^{-1.89}$, or $q^{-2.40} \to q^{-1.89} \to q^{-2.71}$. For simplicity, in the following we choose $p = q = r = 2$ for each type of succession. However, the length of each individual fractal region can be varied by changing any of the values of p, q or r.

3. Scattering structure factor

In order to calculate the monodisperse scattering structure factor, we neglect the shape of fractal subunits, and thus, using Eq. (2) together with Eq. (4) we can write [15, 16]:

$$S_m^{\text{mono}}(q) = k_m \langle \prod_{i=1}^{m} |G_i(\boldsymbol{q})|^2 \rangle. \tag{5}$$

Then, the polydispersity involves a system of various sizes l, and the distribution function $D_N(l)$ is defined in such a way that $D_N(l)\mathrm{d}l$ represents the probability of finding a fractal with sizes between

$(l, \mathrm{d}l)$. Therefore, the polydisperse structure factor given by Eq. (5) becomes:

$$S_m^{\mathrm{poly}}(q) = k_m \int_0^{\infty} \langle \prod_{i=1}^{m} |G_i(\boldsymbol{q})|^2 \rangle D_N(l) \mathrm{d}l. \tag{6}$$

In the following, a log-normal distribution of mean length $\mu_0 \equiv \langle l \rangle_D$ and relative variance $\sigma_r = (\langle l^2 \rangle_D - l_0^2)^{1/2}/l_0$ is considered.

4. Results and discussions

In order to obtain all the possible combinations of power-law successions we consider the following types of fractals: $Type\ I$ - a fractal consisting of $fractal\ I$ for first two iterations, followed by $fractal\ II$ for 3-rd and 4-th iteration, and $fractal\ III$ for 5-th and 6-th iterations. In this way, the monodisperse structure factor is characterized by the following sequence of generalized power-law regimes [15] $q^{-1.89} \to q^{-2.40} \to q^{-2.71}$ (Fig. 2a - black curve). The succession of simple power-law regimes, as obtained in most of the experimental data, are recovered by taking into account the polydispersity, as discussed in section (3)/ The numerical results are presented in Fig. 2a - red curve, which clearly show the succession of simple regimes with decreasing absolute values of the power-law exponent, i.e. a 'concave' scattering curve.

The other possible combinations of power-law successions are obtained similar as for $fractal\ I$. For example, the $fractal\ II$ structure is built from the fractal consisting of $fractal\ II$ for first two iterations, followed by $fractal\ III$ at 3-rd and 4-th iterations, and then $fractal\ I$ at 5-th and 6-th iterations. The corresponding mono and polydisperse structure factor are shown in Fig. 2b (black, and respectively red curves), which clearly reveals a combination of 'concave' scattering corresponding to transition $q^{-2.40} \to q^{-2.71}$ followed by a 'convex' scattering corresponding to transition from $q^{-2.71} \to q^{-1.89}$. The results corresponding to SAS from $TypeIII$ and $TypeIV$ are shown in Fig. 2c and d, which illustrates the other possible transitions: $q^{-2.40} \to q^{-1.89} \to q^{-2.71}$, and respectively $q^{-2.71} \to q^{-2.40} \to q^{-1.89}$.

For all the considered $Types$, one can observe that beyond the last generalized/simple power law-decay we have $q > 1/(\beta_s^m l_0)$, and thus the structure factors shown in Fig. 2a-d tend to one. This implies that the asymptotic values are $1/k_m$ (see Eq. 2), which is in agreement with the numerical results (horizontal lines in Fig. 2a-d).

5. Conclusion

We have built a new theoretical model based on fractal structures which can explain experimental small-angle scattering curves characterized by a succession of power-law decays. Its main feature consists in changing the construction rule of the fractal after a finite number of iterations (here, at every second iteration), which give rise to a succession of power-law decays with different scattering exponents. We confirmed this by performing analytical calculations of mono and polydisperse structure factor. In addition, the obtained structure factor allows us to obtain the number of subunits in each fractal, from the asymptotic values at high q.

The proposed model is based on deterministic mass fractals and can be applied to 'convex'/'concave' data or any combination of them.

References

[1] Feigin L A and Svergun D I 1987 *Structure Analysis by Small-Angle X-Ray and Neutron Scattering* (NY: Plenum press)
[2] Glatter O and Kratky O 1982 *Small-angle X-ray Scattering* (London: Academic Press)
[3] Dumoulin M, Hamd W, Thune E, Rochas C and Guinebretiere R 2016 *J. Appl. Cryst.* **49** 1–9
[4] Radulescu A, Goerigk G and Richter D 2015 *J. Appl. Cryst.* **48** 1860–1869
[5] Anitas E M, Bica I, Erhan R V, Bunoiu M and Kuklin A I 2015 *Rom. Journ. Phys.* **60** 653–657
[6] Bica I, Anitas E M, Averis L M E and Bunoiu M 2015 *J. Ind. Eng. Chem.* **21** 1323–1327

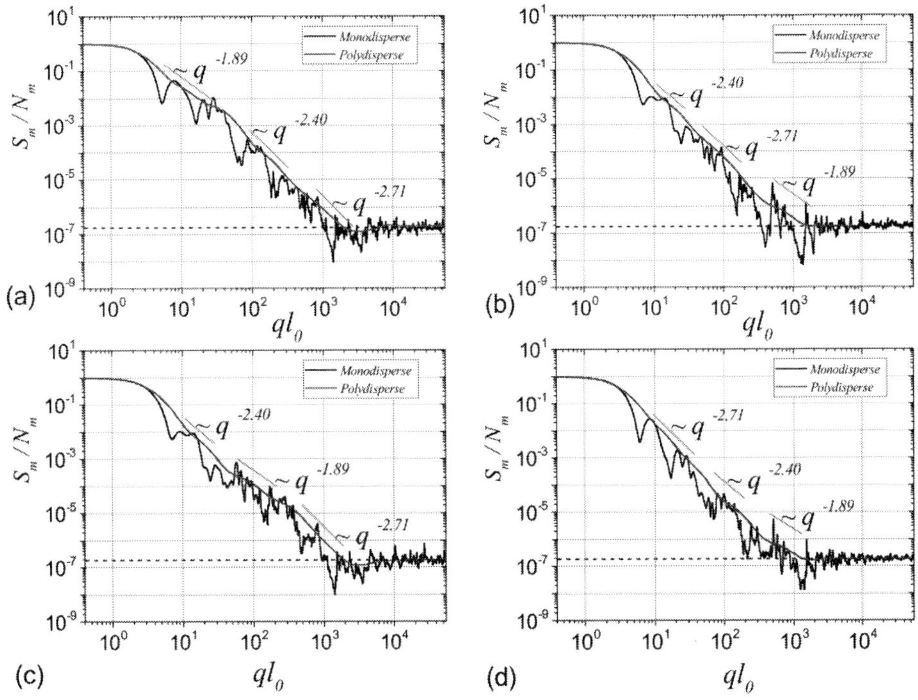

Figure 2. The fractal structure factor for first six iterations. a) *Type I*; b) *Type II*; c) *Type III*; and d) *Type IV* structures (see description in text for each *Type*).

[7] Gouyet J F 1996 *Physics and Fractal Structures* (Springer)
[8] Turcotte D L 1997 *Fractals and Chaos in Geology and Geophysics* (Cambridge University Press)
[9] Hamburger-Lidar D A 1996 *Phys. Rev. E* **54** 354
[10] Bale H D and Schmidt P W 1984 *Phys. Rev. Lett.* **53** 596
[11] Martin J E and Hurd A J 1987 *J. of Appl. Cryst.* **20** 61–78
[12] Teixeira J 1988 *J. Appl. Cryst.* **21** 781
[13] Schmidt P W 1991 *J. of Appl. Cryst.* **24** 414–435
[14] Beaucage G 1995 *J. Appl. Cryst.* **28** 717
[15] Cherny A Y, Anitas E M, Kuklin A I and Osipov V A 2011 *Phys. Rev. E* **84** 036203–1–036203–11
[16] Cherny A Y, Anitas E M, Osipov V A and Kuklin A I 2015 *Rom. Journ. Phys* **60** 658–663
[17] Balasoiu M, Craus M L, Kuklin A I, Plestil J, Haramus V, Islamov A H, Erhan R V *et al.* 2008 *Journal of Optoelectronics and Advanced Materials* **10** 2932–2935
[18] Zhao J, Shi D and Lian J 2009 *Carbon* **47** 2329
[19] Anitas E M, Balasoiu M, Bica I, Osipov V A and Kuklin A I 2009 *Optoelectronics and Advanced Materials - Rapid Communications* **3** 621–624
[20] Headen T F, Boek E S, Stellbrink J and M S U 2009 *Langmuir* **2** 422
[21] Golosova A A, Adelsberger J, Sepe A, Niedermeier M A, Lindner P, Funari S S, Jordan R *et al.* 2012 *J. Phys. Chem. C* **116** 15765
[22] Cherny A Y, Anitas E M, Osipov V A and Kuklin A I 2014 *J. Appl. Cryst.* **47** 198–206
[23] Anitas E M 2014 *Eur. Phys. J. B* **87** 139

Dyson equation for electromagnetic scattering of heterogeneous media with spatially disordered particles: properties of the effective medium

G Berginc[1]
THALES, *Elancourt*, FRANCE

E-mail: gerard.berginc@fr.thalesgroup.com

Abstract. In this paper, we consider the coherent component of the electromagnetic wave field inside random media. The subject of our interest concerns a random medium, consisting of a statistical ensemble of different scattering species and artificial material structures developed on base of dielectric or metallic resonant or non-resonant particles. The starting point of our theory is the multiple scattering theory, the averaged electric field satisfies a Dyson equation with a mass operator related to the effective dielectric permittivity of the homogenized structure. Quantum multiple scattering theory has been transposed into this electromagnetic case. We give a formal solution for the mass operator by introducing the T-matrix formalism. We show that the T-matrix satisfies a Lippman-Schwinger equation. Then, we introduce the Quasi-Crystalline Coherent Potential Approximation (QC-CPA), which takes into account the correlation between the particles with a pair-distribution function. The mass operator includes geometric effects, caused by resonant behavior due to the shape and size of particles, cluster effects because of correlations between particles. Significant modifications of particle scattering properties can be observed.

1.Introduction

The first intent of this paper is to discuss the relation between the Dyson equation for electromagnetic scattering of heterogeneous random media and the effective dielectric constant which characterizes the coherent part of an electromagnetic wave propagating inside a random medium containing metallic or dielectric particles. In this paper we show that we can obtain a general expression of the effective dielectric constant, which is the solution of a closed system of equations. In the high frequency limit the expression includes the vectorial case generalization of the result obtained by Keller, which has been derived, using a scalar theory. We obtain an important tensorial generalization of the notion of effective permittivity. The description of electromagnetic waves propagation in random media in terms of the properties of the constituents has been studied extensively in the past decades [1-19]. In most of published works, the basic idea is to calculate several statistical moments of the electromagnetic field to understand how the wave can interact with the the random medium [7, 9, 10, 11, 14, 15].
In the second section of this paper, we are concerned by the first moment which is the average electric field. Under some assumptions, it can be shown that the average electric field propagates as if the medium where homogeneous but with a renormalized permittivity, called effective permittivity. We introduce the multiple scattering formalism, the Dyson Equation and the Mass operator. We discuss the different hypotheses under which the effective medium theory is valid. The calculation of this parameter has a long history which dates back from the work of Clausius-Mossotti and Maxwell

[1] To whom any correspondence should be addressed.

Content from this work may be used under the terms of the Creative Commons Attribution 3.0 licence. Any further distribution of this work must maintain attribution to the author(s) and the title of the work, journal citation and DOI.
Published under licence by IOP Publishing Ltd

Garnett [19]. Most of the studies are concerned with the quasi-static limit where retardation effects are disregarded [17-24]. In order to take into account scattering effects, quantum multiple scattering theory has been transposed into the electromagnetic case [5, 7, 9, 10, 11, 14, 15], but as a rigorous analytic solution cannot be derived, several approximation schemes have been developed [5, 7, 10, 11, 14, 15, 25-36]. In section 3, we describe the Quasicrystalline Coherent Potential Approximation (QC-CPA) which takes into account the correlation between the particles [12, 14, 25-28]. In this section we examine the different steps to obtain the system of equations verified by the effective permittivity under the (QC-CPA) approach. In section 4, we add some new approximations to the (QC-CPA) approach which give us a tractable equation for the effective permittivity. The expression obtained contains the low frequency limit of the (QC-CPA) approach. At this limit, the (QC-CPA) equations can be written as a generalized Maxwell Garnett formula and are proven to be in good agreement with the experimental results [13, 14, 37-39]. Furthermore, the formula obtained contains also the approximate formula due to Keller, which has been derived in using scalar theory, but seems to be in good agreement with the experimental data for particles larger than a wavelength [40], these results are described in section 5. In section 6, we discuss the different results of this paper and the possible extension to magnetic response of the theory of Dyson equation for multiple scattering in random media with disordered high-index dielectric scatterers. This opens up the developing field of all-dielectric nanophotonics which allows to control both magnetic and electric response of structured matter.

2.Dyson equation and effective permittivity

In the following, we consider harmonic waves with $e^{-i\omega t}$ pulsation. We consider a disordered ensemble of $N \gg 1$ identical spheres of radius r_s with dielectric function $\varepsilon_s(\omega)$ within an infinite medium with dielectric function $\varepsilon_1(\omega)$. The field produced at $\boldsymbol{r}$ by a discrete source located at $\boldsymbol{r_0}$ is given by the dyadic Green function $\overline{\boldsymbol{G}}(\boldsymbol{r}, \boldsymbol{r}_0, \omega)$, which verifies the following propagation equation:

$$\nabla \times \nabla \times \overline{\boldsymbol{G}}(\boldsymbol{r}, \boldsymbol{r}_0, \omega) - \epsilon_V(\boldsymbol{r}, \omega)\, K_{vac}^2\, \overline{\boldsymbol{G}}(\boldsymbol{r}, \boldsymbol{r}_0, \omega) = \delta(\boldsymbol{r} - \boldsymbol{r}_0)\overline{\boldsymbol{I}} \tag{1}$$

where $K_{vac} = \omega/c$ with c the speed of light in vacuum and

$$\epsilon_V(\boldsymbol{r}, \omega) = \epsilon_1(\omega) + \sum_{j=1}^{N} [\epsilon_s(\omega) - \epsilon_1(\omega)]\, \Theta_s(\boldsymbol{r} - \boldsymbol{r}_j)$$

where $\boldsymbol{r}_1, \ldots, \boldsymbol{r}_N$ are the centers of the particles and Θ_s the spherical particle shape:

$$\Theta_s(\boldsymbol{r}) = \begin{cases} 1 & \text{if} \quad ||\boldsymbol{r}|| < r_s \\ 0 & \text{if} \quad ||\boldsymbol{r}|| > r_s \end{cases} \tag{2}$$

The equation (1) has a unique solution if we impose the radiation condition at infinity. The multiple scattering process by the particles is mathematically decomposed in introducing the Green function $\overline{\boldsymbol{G}}_1^{\infty}$, which describes the propagation within an homogenous medium with permittivity $\varepsilon_1(\omega)$, which verifies the following equation:

$$\nabla \times \nabla \times \overline{\boldsymbol{G}}_1^{\infty}(\boldsymbol{r}, \boldsymbol{r}_0, \omega) - \epsilon_1(\omega)\, K_{vac}^2\, \overline{\boldsymbol{G}}_1^{\infty}(\boldsymbol{r}, \boldsymbol{r}_0, \omega) = \delta(\boldsymbol{r} - \boldsymbol{r}_0)\overline{\boldsymbol{I}} \tag{3}$$

with the appropriate boundary conditions. In an infinite random medium, we have [14, 41, 42]:

$$\overline{\boldsymbol{G}}_1^{\infty}(\boldsymbol{r},\boldsymbol{r}_0,\omega) = \left[\overline{\boldsymbol{I}} + \frac{\nabla\nabla}{K_1^2}\right] \frac{e^{\mathrm{i}\,K_1\,||\boldsymbol{r}-\boldsymbol{r}_0||}}{4\pi||\boldsymbol{r}||} \tag{4}$$

where $K_1^2 = \epsilon_1(\omega)\,K_{vac}^2$.
Using this Green function, we decompose the Green function $\overline{\boldsymbol{G}}(\boldsymbol{r},\boldsymbol{r}_0,\omega)$ under the following form [5, 7, 9, 14, 15]:

$$\overline{\boldsymbol{G}} = \overline{\boldsymbol{G}}_1^{\infty} + \overline{\boldsymbol{G}}_1^{\infty} \cdot \overline{\boldsymbol{V}} \cdot \overline{\boldsymbol{G}} \tag{5}$$

where the operator notation is used:

$$[\overline{\boldsymbol{A}} \cdot \overline{\boldsymbol{B}}](\boldsymbol{r},\boldsymbol{r}_0) = \int \mathrm{d}^3\,\boldsymbol{r}_1\,\overline{\boldsymbol{A}}(\boldsymbol{r},\boldsymbol{r}_1) \cdot \overline{\boldsymbol{B}}(\boldsymbol{r}_1,\boldsymbol{r}_0) \tag{6}$$

The potential $\overline{\boldsymbol{V}}$ which describes the interaction between the wave and the N particles, is given by:

$$\overline{\boldsymbol{V}} = \sum_{i=1}^{N} \overline{\boldsymbol{v}}_{\boldsymbol{r}_i} \tag{7}$$

$$\overline{\boldsymbol{v}}_{\boldsymbol{r}_i}(\boldsymbol{r},\boldsymbol{r}_0,\omega) = (2\pi)^2\,\delta(\boldsymbol{r}-\boldsymbol{r}_0)\,\overline{\boldsymbol{v}}_{\boldsymbol{r}_i}(\boldsymbol{r},\omega) \tag{8}$$

$$\overline{\boldsymbol{v}}_{\boldsymbol{r}_i}(\boldsymbol{r},\omega) = [K_s^2 - K_1^2]\,\Theta_d(\boldsymbol{r}-\boldsymbol{r}_i)\overline{\boldsymbol{I}} \tag{9}$$

with $K_s^2 = \epsilon_s(\omega)\,K_{vac}^2$. It is useful to introduce the T matrix defined by [5, 6, 7, 19, 12, 14, 15]:

$$\overline{\boldsymbol{G}} = \overline{\boldsymbol{G}}_1^{\infty} + \overline{\boldsymbol{G}}_1^{\infty} \cdot \overline{\boldsymbol{T}} \cdot \overline{\boldsymbol{G}}_1^{\infty} \tag{10}$$

By iterating equation (5) and comparing it with the definition (10), we show that the T matrix verifies the following equation:

$$\overline{\boldsymbol{T}} = \overline{\boldsymbol{V}} + \overline{\boldsymbol{V}} \cdot \overline{\boldsymbol{G}}_1^{\infty} \cdot \overline{\boldsymbol{T}} \tag{11}$$

If we introduce the T matrix for each scatterer by:

$$\overline{\boldsymbol{t}}_{\boldsymbol{r}_i} = \overline{\boldsymbol{v}}_{\boldsymbol{r}_i} + \overline{\boldsymbol{v}}_{\boldsymbol{r}_i} \cdot \overline{\boldsymbol{G}}_1^{\infty} \cdot \overline{\boldsymbol{t}}_{\boldsymbol{r}_i} \tag{12}$$

we can decompose the T matrix for the system in a series of multiple scattering processes for the particles [6, 11, 13, 16]:

$$\overline{\boldsymbol{T}} = \sum_{i=1}^{N} \overline{\boldsymbol{t}}_{\boldsymbol{r}_i} + \sum_{i=1}^{N} \sum_{j=1,j\neq i}^{N} \overline{\boldsymbol{t}}_{\boldsymbol{r}_j} \cdot \overline{\boldsymbol{G}}_1^{\infty} \cdot \overline{\boldsymbol{t}}_{\boldsymbol{r}_i} + \cdots \tag{13}$$

This T matrix is useful to calculate the average field $<\overline{\boldsymbol{G}}>$ since we have:

$$<\overline{\boldsymbol{G}}>= \overline{\boldsymbol{G}}_1^{\infty} + \overline{\boldsymbol{G}}_1^{\infty} \cdot <\overline{\boldsymbol{T}}> \cdot \overline{\boldsymbol{G}}_1^{\infty} \tag{14}$$

The equivalent of the potential operator $\overline{\boldsymbol{V}}$ for the average Green function $<\overline{\boldsymbol{G}}>$ is the mass operator $\overline{\boldsymbol{\Sigma}}$ defined by the following equation:

$$<\overline{\boldsymbol{G}}>= \overline{\boldsymbol{G}}_1^{\infty} + \overline{\boldsymbol{G}}_1^{\infty} \cdot \overline{\boldsymbol{\Sigma}} \cdot <\overline{\boldsymbol{G}}> \tag{15}$$

As in equation (11), we have the following relationship between the average T matrix and the mass operator:

$$< \overline{\boldsymbol{T}} >= \overline{\boldsymbol{\Sigma}} + \overline{\boldsymbol{\Sigma}} \cdot \overline{\boldsymbol{G}}_1^{\infty} \cdot < \overline{\boldsymbol{T}} > \tag{16}$$

or:

$$\overline{\boldsymbol{\Sigma}} =< \overline{\boldsymbol{T}} > \cdot \left[\overline{\boldsymbol{I}} + \overline{\boldsymbol{G}}_1^{\infty} \cdot < \overline{\boldsymbol{T}} > \right]^{-1} \tag{17}$$

The mass operator corresponds to all irreducible diagrams in the Feynman representation [5, 7, 9, 11]. The equation (15) written in differential form is:

$$\begin{aligned} &\nabla \times \nabla \times < \overline{\boldsymbol{G}}(\boldsymbol{r}, \boldsymbol{r}_0, \omega) > -\epsilon_1(\omega)\, K_{vac}^2 < \overline{\boldsymbol{G}}(\boldsymbol{r}, \boldsymbol{r}_0, \omega) > \\ &- \int \mathrm{d}^3\boldsymbol{r}_1\, \overline{\boldsymbol{\Sigma}}(\boldsymbol{r}, \boldsymbol{r}_1, \omega) \cdot < \overline{\boldsymbol{G}}(\boldsymbol{r}_1, \boldsymbol{r}_0, \omega) >= \delta(\boldsymbol{r} - \boldsymbol{r}_0) \overline{\boldsymbol{I}}\,. \end{aligned} \tag{18}$$

For a statistical homogeneous medium we have:

$$\overline{\boldsymbol{\Sigma}}(\boldsymbol{r}, \boldsymbol{r}_1, \omega) = \overline{\boldsymbol{\Sigma}}(\boldsymbol{r} - \boldsymbol{r}_1, \omega) \tag{19}$$

$$< \overline{\boldsymbol{G}}(\boldsymbol{r}, \boldsymbol{r}_0, \omega) > =< \overline{\boldsymbol{G}}(\boldsymbol{r} - \boldsymbol{r}_0, \omega) > \tag{20}$$

Thus, we can use a Fourier transform:

$$\overline{\boldsymbol{\Sigma}}(\boldsymbol{k}, \omega) = \int \mathrm{d}^3\boldsymbol{r}\, \exp(-\mathrm{i}\boldsymbol{k} \cdot \boldsymbol{r})\, \overline{\boldsymbol{\Sigma}}(\boldsymbol{r}, \omega) \tag{21}$$

$$\overline{\boldsymbol{G}}(\boldsymbol{k}, \omega) = \int \mathrm{d}^3\boldsymbol{r}\, \exp(-\mathrm{i}\,\boldsymbol{k} \cdot \boldsymbol{r})\, \overline{\boldsymbol{G}}(\boldsymbol{r}, \omega) \tag{22}$$

and equation (18) becomes:

$$\left[||\boldsymbol{k}||^2\, (\overline{\boldsymbol{I}} - \hat{\boldsymbol{k}}\hat{\boldsymbol{k}}) - \epsilon_1(\omega)\, K_{vac}^2 \overline{\boldsymbol{I}} - \overline{\boldsymbol{\Sigma}}(\boldsymbol{k}, \omega) \right] \cdot < \overline{\boldsymbol{G}}(\boldsymbol{k}, \omega) >= \overline{\boldsymbol{I}} \tag{23}$$

For a statistical isotropic medium, we have:

$$\overline{\boldsymbol{\Sigma}}(\boldsymbol{k}, \omega) = \Sigma_{\perp}(||\boldsymbol{k}||, \omega)(\overline{\boldsymbol{I}} - \hat{\boldsymbol{k}}\hat{\boldsymbol{k}}) + \Sigma_{||}(||\boldsymbol{k}||, \omega)\, \hat{\boldsymbol{k}}\hat{\boldsymbol{k}} \tag{24}$$

with $\hat{\boldsymbol{k}} = \boldsymbol{k}/||\boldsymbol{k}||$ and then:

$$< \overline{\boldsymbol{G}}(\boldsymbol{k}, \omega) > = \left[||\boldsymbol{k}||^2\, (\overline{\boldsymbol{I}} - \hat{\boldsymbol{k}}\hat{\boldsymbol{k}}) - \epsilon_1(\omega)\, K_{vac}^2 \overline{\boldsymbol{I}} - \overline{\boldsymbol{\Sigma}}(\boldsymbol{k}, \omega) \right]^{-1} \tag{25}$$

$$\begin{aligned} &= \frac{\overline{\boldsymbol{I}} - \hat{\boldsymbol{k}}\hat{\boldsymbol{k}}}{||\boldsymbol{k}||^2 - (\epsilon_1(\omega) K_{vac}^2 + \Sigma_{\perp}(||\boldsymbol{k}||, \omega))} \\ &- \frac{\hat{\boldsymbol{k}}\hat{\boldsymbol{k}}}{\epsilon_1(\omega) K_{vac}^2 + \Sigma_{||}(||\boldsymbol{k}||, \omega)} \end{aligned} \tag{26}$$

In the following, we introduce two effective permittivity functions $\epsilon_e^{\perp}$ and $\epsilon_e^{||}$ defined by:

$$\epsilon_e^{\perp}(||\boldsymbol{k}||, \omega) K_{vac}^2 = \epsilon_1(\omega) K_{vac}^2 + \Sigma_{\perp}(||\boldsymbol{k}||, \omega) \tag{27}$$

$$\epsilon_e^{||}(||\boldsymbol{k}||, \omega) K_{vac}^2 = \epsilon_1(\omega) K_{vac}^2 + \Sigma_{||}(||\boldsymbol{k}||, \omega) \tag{28}$$

and (26) is written:

$$
\begin{aligned}
&< \overline{G}(\boldsymbol{k},\omega) > \\
&= \left[\overline{I} - \frac{\boldsymbol{k}\boldsymbol{k}}{\epsilon_e^{\perp}(||\boldsymbol{k}||,\omega)K_{vac}^2}\right]\frac{1}{||\boldsymbol{k}||^2 - \epsilon_e^{\perp}(||\boldsymbol{k}||,\omega)K_{vac}^2} \\
&+ \frac{\hat{\boldsymbol{k}}\hat{\boldsymbol{k}}}{\epsilon_e^{\perp}(||\boldsymbol{k}||,\omega)K_{vac}^2} - \frac{\hat{\boldsymbol{k}}\hat{\boldsymbol{k}}}{\epsilon_e^{\parallel}(||\boldsymbol{k}||,\omega)K_{vac}^2}.
\end{aligned} \tag{29}
$$

The Green function can be expressed in the space domain by the following formulation:

$$
\begin{aligned}
&< \overline{G}(\boldsymbol{r},\omega) > \\
&= \int \frac{\mathrm{d}^3\boldsymbol{k}}{(2\pi)^3}\left[\overline{I} + \frac{\nabla\nabla}{\epsilon_e^{\perp}(||\boldsymbol{k}||,\omega)K_{vac}^2}\right]\frac{\mathrm{e}^{\mathrm{i}\,\boldsymbol{k}\cdot\boldsymbol{r}}}{||\boldsymbol{k}||^2 - \epsilon_e^{\perp}(||\boldsymbol{k}||,\omega)K_{vac}^2} \\
&+ \int \frac{\mathrm{d}^3\boldsymbol{k}}{(2\pi)^3}\left[\frac{\mathrm{e}^{\mathrm{i}\,\boldsymbol{k}\cdot\boldsymbol{r}}}{\epsilon_e^{\perp}(||\boldsymbol{k}||,\omega)K_{vac}^2} - \frac{\mathrm{e}^{\mathrm{i}\,\boldsymbol{k}\cdot\boldsymbol{r}}}{\epsilon_e^{\parallel}(||\boldsymbol{k}||,\omega)K_{vac}^2}\right]\frac{\boldsymbol{k}\boldsymbol{k}}{||\boldsymbol{k}||^2}.
\end{aligned} \tag{30}
$$

After integration on the solid angle in equation (30) given by the expressions (A1) and (A2) in the appendix, we obtain the expression:

$$
\begin{aligned}
&< \overline{G}(\boldsymbol{r},\omega) > \\
&= \frac{1}{\mathrm{i}||\boldsymbol{r}||}\int_{-\infty}^{+\infty}\frac{\mathrm{d}K}{(2\pi)^2}\left[\overline{I} + \frac{\nabla\nabla}{\epsilon_e^{\perp}(K,\omega)K_{vac}^2}\right]\frac{K\,\mathrm{e}^{\mathrm{i}\,K\,||\boldsymbol{r}||}}{K^2 - \epsilon_e^{\perp}(K,\omega)K_{vac}^2} \\
&+ \frac{1}{\mathrm{i}||\boldsymbol{r}||}\nabla\nabla\int_{-\infty}^{+\infty}\frac{\mathrm{d}K}{(2\pi)^2}\left[\frac{\mathrm{e}^{\mathrm{i}\,K\,||\boldsymbol{r}||}}{\epsilon_e^{\perp}(K,\omega)K_{vac}^2} - \frac{\mathrm{e}^{\mathrm{i}\,K\,||\boldsymbol{r}||}}{\epsilon_e^{\parallel}(K,\omega)K_{vac}^2}\right]\frac{1}{K}
\end{aligned} \tag{31}
$$

where we have supposed that $\epsilon_e^{\perp}(||\boldsymbol{k}||,\omega) = \epsilon_e^{\perp}(-||\boldsymbol{k}||,\omega)$ and $\epsilon_e^{\parallel}(||\boldsymbol{k}||,\omega) = \epsilon_e^{\parallel}(-||\boldsymbol{k}||,\omega)$.

Using the residue theorem, we easily evaluate these integrals. However, we disregard the longitudinal excitation, which are solutions of $\epsilon_e^{\parallel}(K,\omega) = 0$ and $\epsilon_e^{\perp}(K,\omega) = 0$ since we only consider the propagation of the transversal electromagnetic field.
Furthermore, we see that the contribution of the pole $K = 0$ in the second term of equation (31) is null. In fact, the dyadic $\nabla\nabla$ operates on a constant as we have $\mathrm{e}^{\mathrm{i}\,K\,||\boldsymbol{r}||} = 1$ for this pole. Hence, we obtain the following expression for the Green function:

$$
< \overline{G}(\boldsymbol{r},\omega) >= \sum_{i=1}^{n}\left[\overline{I} + \frac{\nabla\nabla}{K_{e\,i}^2}\right]\frac{\mathrm{e}^{\mathrm{i}\,K_{e\,i}\,||\boldsymbol{r}||}}{4\pi||\boldsymbol{r}||} \tag{32}
$$

where $K_{e\,i}$ are the roots of $K_{e\,i}^2 = \epsilon_e^{\perp}(K_{e\,i},\omega)K_{vac}^2$, the roots verify the equation $\mathrm{Im}(K_{e\,i}) > 0$ imposed by the radiation condition at infinity. Sheng has called the roots $K_{e\,i}$ the quasi-modes of the random medium [9, 31]. If we only consider the root $K_e = K_{e\,j}$ which has the smallest imaginary part ($\mathrm{Im}(K_{e\,j}) = min_i[\mathrm{Im}(K_{e\,i})]$) and then the smallest exponential factor in equation (32), we define the effective permittivity by $\epsilon_e(\omega) = \epsilon_e^{\perp}(K_e,\omega)$. The average Green function is then equal to the Green function for an infinite homogenous medium with permittivity $\varepsilon_e(\omega)$:

$$
< \overline{G}(\boldsymbol{r},\omega) >= \overline{G}_e^{\infty}(\boldsymbol{r},\omega) \tag{33}
$$

where

$$\overline{G}_e^{\infty}(\boldsymbol{r},\omega) = \left[\overline{\boldsymbol{I}} + \frac{\nabla\nabla}{K_e^2}\right] \frac{e^{\mathrm{i}\,K_e\,||\boldsymbol{r}||}}{4\pi||\boldsymbol{r}||} \tag{34}$$

Thus, the effective medium approach is valid if we omit the longitudinal excitation in the medium and if the propagative mode with the smallest imaginary part is the main contribution in the development (32).

3.The coherent-potential and quasi-crystalline approximations

Previously, we have shown how the mass operator is related to the effective permittivity. To calculate the mass operator, we can use equations (13) and (17). However, we can improve this system of equations, thus we rewrite the Green function development (5) by replacing the Green function $\overline{G}_1^{\infty}$ with $\overline{G}_e^{\infty}$:

$$\overline{G} = \overline{G}_e^{\infty} + \overline{G}_e^{\infty} \cdot \overline{V}_e \cdot \overline{G} \tag{35}$$

where we have introduced a new potential $\overline{V}_e$:

$$\overline{V}_e = \sum_{i=1}^{N} \widetilde{v}_{e,r_i} \tag{36}$$

$$\widetilde{v}_{e,r_i}(\boldsymbol{r},\boldsymbol{r}_0,\omega) = (2\pi)^2\,\delta(\boldsymbol{r}-\boldsymbol{r}_0)\,\widetilde{v}_{e,r_i}(\boldsymbol{r},\omega) \tag{37}$$

$$\widetilde{v}_{e,r_i}(\boldsymbol{r},\omega) = [K_s^2 - K_e^2]\,\Theta_s(\boldsymbol{r}-\boldsymbol{r}_i)\overline{\boldsymbol{I}} + [K_1^2 - K_e^2]\overline{\boldsymbol{I}}\,. \tag{38}$$

As in the previous section, we introduce a T matrix:

$$\overline{G} = \overline{G}_e^{\infty} + \overline{G}_e^{\infty} \cdot \overline{T}_e \cdot \overline{G}_e^{\infty} \tag{39}$$

which can be decomposed as following:

$$\overline{T}_e = \sum_{i=1}^{N} \widetilde{t}_{e;r_i} + \sum_{i=1}^{N} \sum_{j=1, j\neq i}^{N} \widetilde{t}_{e;r_j} \cdot \overline{G}_e^{\infty} \cdot \widetilde{t}_{e;r_i} + \cdots \tag{40}$$

where we have defined a renormalized T matrix for the particles:

$$\widetilde{t}_{e,r_i} = \widetilde{v}_{e,r_i} + \widetilde{v}_{e,r_i} \cdot \overline{G}_e^{\infty} \cdot \widetilde{t}_{e,r_i} \tag{41}$$

We impose the following condition on the average field:

$$< \overline{G}(\boldsymbol{r},\omega) >= \overline{G}_e(\boldsymbol{r},\omega) \tag{42}$$

or,

$$< \overline{T}_e >= \overline{0} \tag{43}$$

considering equation (39). The condition (42) is the Coherent-Potential Approximation (CPA) [9, 11, 14, 25]. The expressions (43) and (40) form a closed system of equations for the unknown permittivity $\varepsilon_e(\omega)$. At the first order in density of particles, this system of equations gives:

$$\sum_{i=1}^{N} < \widetilde{t}_{e,\boldsymbol{r}_i} >= \mathbf{0} \tag{44}$$

In Fourier-space, the T matrix for one scatterer verifies the property:

$$\widetilde{t}_{e,\boldsymbol{r}_i}(\boldsymbol{k}|\boldsymbol{k}_0) = \mathrm{e}^{-\mathrm{i}(\boldsymbol{k}-\boldsymbol{k}_0)\cdot\boldsymbol{r}_i}\,\widetilde{t}_{e,o}(\boldsymbol{k}|\boldsymbol{k}_0) \tag{45}$$

where $\widetilde{t}_{e,o}$ is the T matrix for a particle located at the origin of coordinates. The average of the exponential term, introduced in equation (44) with the equation (45), gives for a statistical homogeneous medium:

$$< \sum_{i=1}^{N} \mathrm{e}^{-\mathrm{i}(\boldsymbol{k}-\boldsymbol{k}_0)\cdot\boldsymbol{r}_i} > = N \int \mathrm{d}^3\boldsymbol{r}\,\frac{1}{\mathcal{V}}\mathrm{e}^{-\mathrm{i}(\boldsymbol{k}-\boldsymbol{k}_0)\cdot\boldsymbol{r}} \tag{46}$$

$$= n\,(2\pi)^2\,\delta(\boldsymbol{k}-\boldsymbol{k}_0) \tag{47}$$

where we have defined the density of scatterers $n = N / \mathcal{V}$ with $\mathcal{V}$ the volume of the random medium. The condition (44) becomes:

$$\widetilde{t}_{e,o}(\boldsymbol{k}_0|\boldsymbol{k}_0) = \overline{\mathbf{0}} \tag{48}$$

This CPA condition has been used in several works [9, 31, 43]. It is worth mentioning that operator $\widetilde{t}_{e,\boldsymbol{r}_i}(\boldsymbol{k}|\boldsymbol{k}_0)$ is not the T matrix describing the scattering by a particle of permittivity $\varepsilon_s(\omega)$ surrounded by a medium of permittivity $\varepsilon_e(\omega)$. To describe this electromagnetic interaction, the operator (38) should have the following form:

$$\widetilde{v}_{e,\boldsymbol{r}_i}(\boldsymbol{r},\omega) = [K_s^2(\omega) - K_e^2(\omega)]\,\Theta_s(\boldsymbol{r}-\boldsymbol{r}_i)\overline{\boldsymbol{I}} \tag{49}$$

However, we see that $\widetilde{v}_{e,\boldsymbol{r}_i}(\boldsymbol{r},\omega)$ is different from the operator in equation (38), and especially we have $\widetilde{v}_{e;\boldsymbol{r}_i}(\boldsymbol{r},\omega) = [K_1^2 - K_e^2]\overline{\boldsymbol{I}}$ for $||\boldsymbol{r}-\boldsymbol{r}_i|| > r_s$, which does not correspond to definition (49), where $\widetilde{v}_{e;\boldsymbol{r}_i}(\boldsymbol{r},\omega) = 0$ when $\boldsymbol{r}$ is outside the particle.
Thus, $\widetilde{t}_{e,\boldsymbol{r}_i}$ is a non-local operator and cannot be obtained from the classical Mie theory [3, 25]. To overcome this difficulty, in some works [9, 43] $\widetilde{t}_{e,\boldsymbol{r}_i}$ is replaced by the scattering operator of a "structural unit". Nevertheless, this approach does not seem to have any theoretical justification. Hence, we prefer to use the more rigorous approach introduced in the scattering theory by disordered liquid metal [28] and adapted in the electromagnetic scattering case by Tsang *et al.* [11, 14].
In this approach, the non-local term $[K_1^2 - K_e^2]\overline{\boldsymbol{I}}$ is correctly taken into account by averaging equations (39) where we use the correct potential $\widetilde{v}_{e,\boldsymbol{r}_i}(\boldsymbol{r},\omega)$, defined by (38). A system of hierarchical equations is obtained where correlation functions between two or more particles are successively introduced.

The chain of equations is closed by using the Quasi-Crystalline Approximation (QCA), which disregards the fluctuation of the effective field, acting on a particle located at r_j, due to a deviation of a particle located at r_i from its average position [26].
This approximation describes the correlation between the particles, only with a two-point correlation function $g(\boldsymbol{r}_i, \boldsymbol{r}_j) = g(||\boldsymbol{r}_i - \boldsymbol{r}_j||)$. Under the QC-CPA scheme, we obtain the following expression for the mass operator [11, 14, 25, 28]:

$$\overline{\boldsymbol{\Sigma}}(\boldsymbol{k}_0, \omega) = n\,\overline{\boldsymbol{C}}_{e,o}(\boldsymbol{k}_0|\boldsymbol{k}_0) \tag{50}$$

$$\begin{aligned}\overline{\boldsymbol{C}}_{e,o}(\boldsymbol{k}|\boldsymbol{k}_0) &= \overline{\boldsymbol{t}}_{e,o}(\boldsymbol{k}|\boldsymbol{k}_0)\\ &+ n\int\frac{\mathrm{d}^3\boldsymbol{k}_1}{(2\pi)^3}\,h(\boldsymbol{k}-\boldsymbol{k}_1)\,\overline{\boldsymbol{t}}_{e,o}(\boldsymbol{k}|\boldsymbol{k}_1)\cdot\overline{\boldsymbol{G}}_e^{\infty}(\boldsymbol{k}_1)\cdot\overline{\boldsymbol{C}}_{e,o}(\boldsymbol{k}_1|\boldsymbol{k}_0)\end{aligned} \tag{51}$$

where

$$\overline{\boldsymbol{t}}_{e,o} = \overline{\boldsymbol{v}}_{e,o} + \overline{\boldsymbol{v}}_{e,o}\cdot\overline{\boldsymbol{G}}_e^{\infty}\cdot\overline{\boldsymbol{t}}_{e,o} \tag{52}$$

$$\overline{\boldsymbol{v}}_{e,o}(\boldsymbol{r}, \boldsymbol{r}_0) = (2\,\pi)\delta(\boldsymbol{r}-\boldsymbol{r}_0)\,\overline{\boldsymbol{v}}_{e,o}(\boldsymbol{r}) \tag{53}$$

$$\overline{\boldsymbol{v}}_{e,o}(\boldsymbol{r}) = [K_s^2 - K_1^2]\,\Theta_s(\boldsymbol{r})\overline{\boldsymbol{I}} \tag{54}$$

and

$$h(\boldsymbol{r}) = g(\boldsymbol{r}) - 1 \tag{55}$$

$$h(\boldsymbol{k}-\boldsymbol{k}_1) = \int \mathrm{d}^3\boldsymbol{r}\ \exp(-\mathrm{i}\,(\boldsymbol{k}-\boldsymbol{k}_1)\cdot\boldsymbol{r})\,h(\boldsymbol{r}) \tag{56}$$

$$\overline{\boldsymbol{G}}_e^{\infty}(\boldsymbol{k}) = \int \mathrm{d}^3\boldsymbol{r}\ \exp(-\mathrm{i}\boldsymbol{k}\cdot\boldsymbol{r})\,\overline{\boldsymbol{G}}_e^{\infty}(\boldsymbol{r}) \tag{57}$$

If we rewrite the potential (54) under the following form:

$$\overline{\boldsymbol{v}}_{e;o}(\boldsymbol{r}) = [\tilde{K}_s^2 - K_e^2]\,\Theta_s(\boldsymbol{r})\overline{\boldsymbol{I}} \tag{58}$$

where we have defined a new wave number $\tilde{K}_s^2 = K_s^2 - K_1^2 + K_e^2$, we see that the operator $\overline{\boldsymbol{t}}_{e,o}$ is the T matrix for a scatterer of permittivity $\tilde{\epsilon}_s = \epsilon_s - \epsilon_1 + \epsilon_e$ in a medium of permittivity ϵ_e. As described in the previous section, the effective propagation constant K_e is the root, which has the smallest imaginary part, of the equation:

$$K_e^2 = K_1^2 + \Sigma_\perp(K_e, \omega) \tag{59}$$

where the mass operator is decomposed under the form (24). Once the effective wave number K_e is obtained, the effective permittivity is given by:

$$\epsilon_e(\omega) = K_e^2/K_{vac}^2 \tag{60}$$

4.Some further approximations

Solving numerically the previous system of equations (50-60) is a difficult task. However, for the low frequency limit of this system of equation, an analytical solution can be obtained and has shown to be in good agreement with the experimental results [11, 14]. We have also to mention that the numerical resolution of the quasicrystalline approximation without the coherent potential approximation has been

developed [13, 38]. To reduce the numerical difficulties for this system of equations (50-60), we add two new approximations to the QC-CPA scheme:
➛ *A far-field approximation*, for an incident plane wave:

$$\boldsymbol{E}^i(\boldsymbol{r}) = \boldsymbol{E}^i(\boldsymbol{k}_0)\, e^{\mathrm{i}\,\boldsymbol{k}_0\cdot\boldsymbol{r}} \tag{61}$$

transverse to the propagation direction $\hat{\boldsymbol{k}}_0$:

$$\boldsymbol{E}^i(\boldsymbol{k}_0)\cdot\hat{\boldsymbol{k}}_0 = 0 \tag{62}$$

where $\boldsymbol{k}_0 = K_e\,\hat{\boldsymbol{k}}_0$ and $\hat{\boldsymbol{k}}_0\cdot\hat{\boldsymbol{k}}_0 = 1$, the scattered far-field, by a particle within a medium of permittivity $\varepsilon_e(\omega)$, is described by an operator $\overline{\boldsymbol{f}}(\hat{\boldsymbol{k}}|\hat{\boldsymbol{k}}_0)$. We obtain the following equation :

$$\boldsymbol{E}^s(\boldsymbol{r}) = \frac{e^{\mathrm{i}\,K_e||\boldsymbol{r}||}}{||\boldsymbol{r}||}\overline{\boldsymbol{f}}(\hat{\boldsymbol{k}}|\hat{\boldsymbol{k}}_0)\cdot\boldsymbol{E}^i(\hat{\boldsymbol{k}}_0) \tag{63}$$

which verifies transversality conditions:

$$\overline{\boldsymbol{f}}(\hat{\boldsymbol{k}}|\hat{\boldsymbol{k}}_0)\cdot\hat{\boldsymbol{k}}_0 = 0 \tag{64}$$

$$\hat{\boldsymbol{k}}\cdot\overline{\boldsymbol{f}}(\hat{\boldsymbol{k}}|\hat{\boldsymbol{k}}_0) = 0 \tag{65}$$

Moreover, the scattered field in the general case is expressed with the operator $\overline{\boldsymbol{t}}_{e,o}$ by:

$$\boldsymbol{E}^s(\boldsymbol{r}) = \int \mathrm{d}^3\boldsymbol{r}_1\,\mathrm{d}^3\boldsymbol{r}_2\,\overline{\boldsymbol{G}}_e^{\infty}(\boldsymbol{r},\boldsymbol{r}_1)\cdot\overline{\boldsymbol{t}}_{e,o}(\boldsymbol{r}_1|\boldsymbol{r}_2)\cdot\boldsymbol{E}^i(\boldsymbol{r}_2) \tag{66}$$

Using the properties of the Green function in equation (66), we obtained the scattered far-field in function of the operator $\overline{\boldsymbol{t}}_{e,o}(\boldsymbol{k}|\boldsymbol{k}_0)$, and comparing the result with equation (63), we can write:

$$\begin{aligned} 4\pi\,\overline{\boldsymbol{f}}(\hat{\boldsymbol{k}}|\hat{\boldsymbol{k}}_0) =& (\overline{\boldsymbol{I}} - \hat{\boldsymbol{k}}\hat{\boldsymbol{k}}) \\ &\cdot\overline{\boldsymbol{t}}_{e,o}(K_e\hat{\boldsymbol{k}}|K_e\hat{\boldsymbol{k}}_0)\cdot(\overline{\boldsymbol{I}} - \hat{\boldsymbol{k}}_0\hat{\boldsymbol{k}}_0) \end{aligned} \tag{67}$$

Our far-field approximation disregards the longitudinal component and the off-shell contribution in the operator $\overline{\boldsymbol{t}}_{e,o}$, and we write:

$$\overline{\boldsymbol{t}}(K_e\hat{\boldsymbol{k}}|K_e\hat{\boldsymbol{k}}_0) \simeq 4\pi\,\overline{\boldsymbol{f}}(\hat{\boldsymbol{k}}|\hat{\boldsymbol{k}}_0) \tag{68}$$

$$= 4\pi\,(\overline{\boldsymbol{I}} - \hat{\boldsymbol{k}}\hat{\boldsymbol{k}})\cdot\overline{\boldsymbol{f}}(\hat{\boldsymbol{k}}|\hat{\boldsymbol{k}}_0)\cdot(\overline{\boldsymbol{I}} - \hat{\boldsymbol{k}}_0\hat{\boldsymbol{k}}_0) \tag{69}$$

where the last equality comes from the properties (64-65).

➛ *A forward scattering approximation*: for scatterers whose dimensions are large compared to a wavelength, the magnitude of the scattered field is predominantly located in the forward direction (i.e. $|f(\hat{\boldsymbol{k}}_0|\hat{\boldsymbol{k}}_0)| \gg |f(-\hat{\boldsymbol{k}}_0|\hat{\boldsymbol{k}}_0)|$). With our forward approximation we only conserve the contribution of the amplitude of diffusion $f(\hat{\boldsymbol{k}}|\hat{\boldsymbol{k}}_0)$ in the direction of the incident wave $\hat{\boldsymbol{k}}_0$. Using the hypothesis (68), we can write:

$$\overline{\boldsymbol{t}}_{e,o}(K_e\hat{\boldsymbol{k}}|K_e\hat{\boldsymbol{k}}_0) = 4\pi\,\overline{\boldsymbol{f}}(\hat{\boldsymbol{k}}|\hat{\boldsymbol{k}}_0) \tag{70}$$

$$\simeq 4\pi\,\overline{\boldsymbol{f}}(\hat{\boldsymbol{k}}_0|\hat{\boldsymbol{k}}_0) \tag{71}$$

$$= 4\pi\,(\overline{\boldsymbol{I}} - \hat{\boldsymbol{k}}_0\hat{\boldsymbol{k}}_0) f(K_e,\omega) \tag{72}$$

where:

$$f(K_e,\omega) = \frac{\mathrm{i}}{K_e}\,S_1(0) = \frac{\mathrm{i}}{K_e}\,S_2(0) \tag{73}$$

with $S_1(0) = S_2(0)$ given by the Mie theory [3, 4, 44]. It is worth mentioning that the approximation (71) is also valid for small scatterers (Rayleigh scatterers).
In this case, the scattering amplitude $f(\hat{\boldsymbol{k}}|\hat{\boldsymbol{k}}_0)$ does not depend on the directions of the incident $\hat{\boldsymbol{k}}_0$ and scattered $\hat{\boldsymbol{k}}$ wave vectors, since we have:

$$\overline{\boldsymbol{t}}_{e,o}(\boldsymbol{k}|\boldsymbol{k}_0) = t_{e,o}(\omega)\overline{\boldsymbol{I}} \tag{74}$$

From equation (67), we show that:

$$\overline{\boldsymbol{f}}(\hat{\boldsymbol{k}}|\hat{\boldsymbol{k}}_0) = \overline{\boldsymbol{f}}(\hat{\boldsymbol{k}}_0|\hat{\boldsymbol{k}}_0) \tag{75}$$

and we also obtain the coefficient $f(K_e,\omega)$:

$$4\pi\, f(K_e,\omega) = t_{e,o}(\omega) \tag{76}$$

Furthermore, we see from equation (51), that for the order zero in density, we have:

$$\overline{\boldsymbol{C}}_{e,o}(\boldsymbol{k}|\boldsymbol{k}_0) = \overline{\boldsymbol{t}}_{e,o}(\boldsymbol{k}|\boldsymbol{k}_0) \tag{77}$$

and the forward approximation (71) can be applied to the operator $\overline{\boldsymbol{C}}_{e,o}(\boldsymbol{k}|\boldsymbol{k}_0)$ in the limit of low density. We will suppose that the forward approximation is valid for the operator whatever the order in density and we write:

$$\overline{\boldsymbol{C}}_{e,o}(\boldsymbol{k}|\boldsymbol{k}_0) \simeq \overline{\boldsymbol{C}}_{e,o}(\boldsymbol{k}_0|\boldsymbol{k}_0) \tag{78}$$

$$\simeq (\overline{\boldsymbol{I}} - \hat{\boldsymbol{k}}_0\hat{\boldsymbol{k}}_0)\, C^{\perp}_{e,o}(||\boldsymbol{k}_0||,\omega) \tag{79}$$

With this hypothesis, only the path of type 1 in figure 1 is considered. This approximation also implies that the operator $\overline{\boldsymbol{C}}_{e,o}(\boldsymbol{k}|\boldsymbol{k}_0)$ is transverse to the propagation direction $\hat{\boldsymbol{k}}_0$.
From the previous hypothesis and the QC-CPA equations (51), we obtain an expression of $\overline{\boldsymbol{C}}_{e,o}(\boldsymbol{k}|\boldsymbol{k}_0)$

$$\begin{aligned}\overline{\boldsymbol{C}}_{e,o}(\boldsymbol{k}_0|\boldsymbol{k}_0) &= 4\pi\, f(K_e,\omega)\overline{\boldsymbol{I}}_{\perp}(\hat{\boldsymbol{k}}_0)\\ &+4\pi\, n\, f(K_e,\omega)\,\overline{\boldsymbol{m}}(\boldsymbol{k}_0)\cdot\overline{\boldsymbol{C}}_{e,o}(\boldsymbol{k}_0|\boldsymbol{k}_0)\end{aligned} \tag{80}$$

where we have introduced the notation:

$$\overline{\boldsymbol{I}}_\perp(\hat{\boldsymbol{k}}_0) = (\overline{\boldsymbol{I}} - \hat{\boldsymbol{k}}_0\hat{\boldsymbol{k}}_0) \tag{81}$$

$$\overline{\boldsymbol{m}}(\boldsymbol{k}_0) = \int \frac{\mathrm{d}^3\boldsymbol{k}_1}{(2\pi)^3}\, h(\boldsymbol{k}_0 - \boldsymbol{k}_1) \cdot \overline{\boldsymbol{I}}_\perp(\hat{\boldsymbol{k}}_0) \cdot \overline{\boldsymbol{G}}_e^\infty(\boldsymbol{k}_1) \cdot \overline{\boldsymbol{I}}_\perp(\hat{\boldsymbol{k}}_0) \tag{82}$$

Then, we have:

$$\overline{\boldsymbol{C}}_{e,o}(\boldsymbol{k}_0|\boldsymbol{k}_0) = \left[\overline{\boldsymbol{I}}_\perp(\hat{\boldsymbol{k}}_0) - 4\pi n f(K_e,\omega)\overline{\boldsymbol{m}}(\boldsymbol{k}_0)\right]^{-1} \cdot \overline{\boldsymbol{I}}_\perp(\hat{\boldsymbol{k}}_0)\, 4\pi\, f(K_e,\omega)\,. \tag{83}$$

Using the classical properties of the Fourier transform, we obtain:

$$\overline{\boldsymbol{m}}(\boldsymbol{k}_0) = \overline{\boldsymbol{I}}_\perp(\hat{\boldsymbol{k}}_0) \cdot \int \mathrm{d}^3\boldsymbol{r}\, \mathrm{e}^{-\mathrm{i}\boldsymbol{k}_0\cdot\boldsymbol{r}}\, h(\boldsymbol{r})\, \overline{\boldsymbol{G}}_e^\infty(\boldsymbol{r}) \cdot \overline{\boldsymbol{I}}_\perp(\hat{\boldsymbol{k}}_0) \tag{84}$$

where we have used the translation invariance of the Green function: $\overline{\boldsymbol{G}}_e^\infty(\boldsymbol{r} - \boldsymbol{r}_0) = \overline{\boldsymbol{G}}_e^\infty(\boldsymbol{r}, \boldsymbol{r}_0)$. The Dyadic Green function has a singularity which can be extracted in introducing the principal value of the Green function [13, 14, 41, 45]:

$$\overline{\boldsymbol{G}}_e^\infty(\boldsymbol{r}) = P.V.\overline{\boldsymbol{G}}_e(\boldsymbol{r}) - \frac{1}{3K_e^2}\,\delta(\boldsymbol{r})\overline{\boldsymbol{I}} \tag{85}$$

where the principal value is defined by:

$$P.V. \int \mathrm{d}^3\boldsymbol{r}_0\, \overline{\boldsymbol{G}}_e^\infty(r - r_0) \cdot \overline{\phi}(r_0) = lim_{a\to 0} \int_{S_a(\boldsymbol{r})} \mathrm{d}^3\boldsymbol{r}_0\, \overline{\boldsymbol{G}}_e^\infty(r - r_0) \cdot \overline{\phi}(\boldsymbol{r}_0)\,, \tag{86}$$

with $\overline{\phi}(\boldsymbol{r}_0)$ a test function and $S_a(\boldsymbol{r})$ a spherical volume of radius a centered at $\boldsymbol{r}$. This principal value can be easily calculated, and we obtain [11, 14]:

$$P.V.G_e(\boldsymbol{r}) = \frac{e^{\mathrm{i}K_e||\boldsymbol{r}||}}{4\pi||\boldsymbol{r}||}\left[\left(1 - \frac{1}{\mathrm{i}K_e||\boldsymbol{r}||} - \frac{1}{K_e^2||\boldsymbol{r}||^2}\right)\overline{\boldsymbol{I}} - \left(1 - \frac{3}{\mathrm{i}K_e||\boldsymbol{r}||} - \frac{3}{K_e^2||\boldsymbol{r}||^2}\right)\hat{\boldsymbol{r}}\hat{\boldsymbol{r}}\right]. \tag{87}$$

Using polar coordinate in the integral (84):

$$\int \mathrm{d}^3\boldsymbol{r}\ . = \int_0^{+\infty} r^2\mathrm{d}r \int_{4\pi} \mathrm{d}^2\hat{\boldsymbol{r}} \tag{88}$$

and the integral on solid angles given in Appendix, we obtain the following result:

$$\overline{\boldsymbol{m}}(\boldsymbol{k}_0) = \left[-\frac{h(0)}{3K_e^2} + m(K_e)\right]\overline{\boldsymbol{I}}_\perp(\hat{\boldsymbol{k}}_0) \tag{89}$$

with

$$m(K_e) = \int_0^{+\infty} r\,\mathrm{d}r\, p(K_e\, r)\,[g(r) - 1]\mathrm{e}^{\mathrm{i}K_e r} \tag{90}$$

$$p(x) = \frac{\sin x}{x} - \left(\frac{\sin x}{x^3} - \frac{\cos x}{x^2}\right), \\ - \left(\frac{1}{\mathrm{i}x} + \frac{1}{x^2}\right)\left(\frac{\sin x}{x} - 3\left[\frac{\sin x}{x^3} - \frac{\cos x}{x^2}\right]\right) \tag{91}$$

where we have assumed $||\boldsymbol{k}_0|| = K_e$ because the mass operator (50) is evaluated with this value in the equation (59) to obtain the effective permittivity. As the particles cannot overlap, we have $g(0) = 0$ and then $h(0) = -1$. With equations (59, 24, 50, 83, 89), we derive an expression for the effective wave number K_e :

$$K_e^2 = K_1^2 + \frac{4\pi\, n\, f(K_e, \omega)}{1 - 4\pi\, n\, f(K_e, \omega)\left(\frac{1}{3K_e^2} + m(K_e)\right)} \tag{92}$$

where the scalar $m(K_e)$ is defined by equations (90,91).
The scalar $f(K_e, \omega)$ is the forward scattering amplitude: $\overline{\boldsymbol{f}}(\hat{\boldsymbol{k}}_0|\hat{\boldsymbol{k}}_0) = (\overline{\boldsymbol{I}} - \hat{\boldsymbol{k}}_0\hat{\boldsymbol{k}}_0) f(K_e, \omega)$ for a particle of permittivity $\tilde{\epsilon}_s = \epsilon_s - \epsilon_1 + \epsilon_e$ within a medium of permittivity ε_e . The relationship between the effective wave number K_e and the effective permittivity ε_e is given by:

$$\epsilon_e(\omega) = K_e^2 / K_{vac}^2 \tag{93}$$

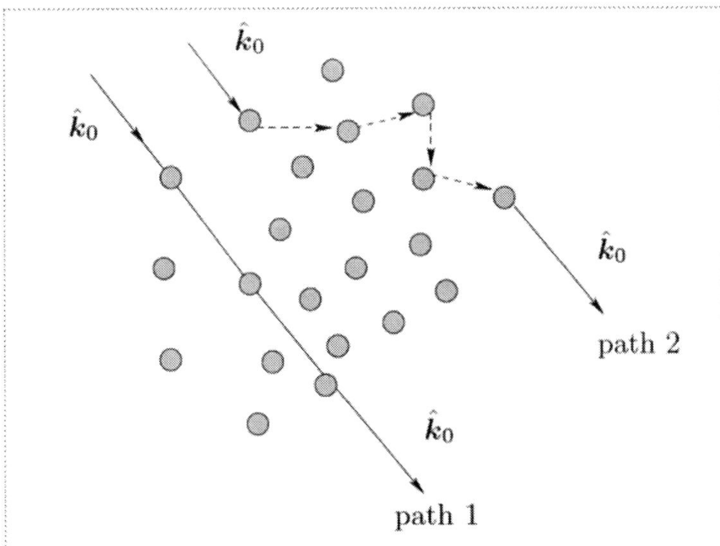

Figure 1. Two different paths which contribute to the mass operator $\overline{\boldsymbol{\Sigma}}(||\boldsymbol{k}_0||, \omega) = n\,\overline{\boldsymbol{C}}_{e,o}(\boldsymbol{k}_0|\boldsymbol{k}_0)$ in the (QC-CPA) approach. Only the path of kind 1 is taken into account in our forward scattering approximation (79).

5.High frequency and low frequency limits

In section 3, we have derived a general system of equation to obtain the effective permittivity. With some approximations we obtain a system which can be solved more easily. We have derived [46] a general formulation for an effective permittivity for a layer of an inhomogeneous random medium with different types of particles and bounded with randomly rough surfaces. We obtained an analytical expression for Rayleigh-type scatterers for the low frequency limit. This formula is important because it takes into account a broader model of nanoscale particles if we introduce the regularization parameter [46] of the principal value of the Green function. This parameter is linked to resonances of the considered scatterer. We are now going to show that the relation (92) that we have obtained contains also the Keller formula [40] in the high frequency limit. This formula has been shown to be in good agreement with experimental results for particles larger than a wavelength. The Keller formula can be obtained in considering the QC-CPA approach in the scalar case [11, 14]. The equations are formerly identical to equation (50-57) where the dyadic Green function

$$\overline{\boldsymbol{G}}_e^{\infty}(\boldsymbol{r},\boldsymbol{r}_0,\omega)=\left[\overline{\boldsymbol{I}}+\frac{\nabla\nabla}{K_e^2}\right]\frac{e^{\mathrm{i}\,K_e\,||\boldsymbol{r}-\boldsymbol{r}_0||}}{4\pi||\boldsymbol{r}||} \tag{94}$$

has to be replaced by the scalar Green function:

$$G_e^{\infty}(\boldsymbol{r},\boldsymbol{r}_0,\omega)=\frac{e^{\mathrm{i}\,K_e\,||\boldsymbol{r}-\boldsymbol{r}_0||}}{4\pi||\boldsymbol{r}||} \tag{95}$$

Using equation (A1), the first iteration of the scalar version of equation (51) gives:

$$\begin{aligned}K_e^2 &= K_1^2+4\pi\,n\,f(K_e,\omega)\\ &+(4\pi)^2\,n^2\,f^2(K_e,\omega)\int_0^{+\infty}\mathrm{d}r\,\frac{\sin K_e\,r}{K_e}[g(r)-1]\mathrm{e}^{\mathrm{i}K_e\,r}+\dots\end{aligned} \tag{96}$$

As was shown by Waterman et al [27], this development is valid if the following condition is verified:

$$(4\pi)^2\,n\,|f(K_e,\omega)|^2/K_e\ll 1 \tag{97}$$

In the geometric limit, the scattering cross section σ_s for a single particle is in good approximation given by $\sigma_s\simeq 2\,\pi\,r_s^2$.
As the cross-section is connected to the scattering amplitude by the relation $\sigma_s=\frac{8\pi}{3}|f(K_e,\omega)|^2$ and as the maximum density is $n=1/v_s$, we see that for particles larger than a wavelength the condition (97) is satisfied:

$$(4\pi)^2\,n\,|f(K_e,\omega)|^2/K_e\simeq 1/K_e\,r_s\ll 1 \tag{98}$$

The equation (96) has been derived by Keller [40] and this expression is in good agreement with experiments for large particles compared to the wavelength. If we now use a Taylor development in equation (92), we obtain:

$$\begin{aligned}K_e^2 &= K_1^2+(4\pi)^2\,n\,f(K_e,\omega)\\ &+(4\pi)^2\,n^2\,f^2(K_e,\omega)\left[\frac{1}{3\,K_e^2}+m(K_e)\right]+\dots\end{aligned} \tag{99}$$

This development is valid if the condition (97) is satisfied. In the geometrical limits, we can approximate the function $p(K_e\,r)$ in definition (90). For $K_e\,r\gg 1$ we have from equation (91):

$$p(x)\simeq\frac{sinx}{x},\quad x\gg 1 \tag{100}$$

The relation (99) becomes:

$$K_e^2 = K_1^2 + 4\pi n f(K_e,\omega) + (4\pi)^2 n^2 f^2(K_e,\omega)\left[\frac{1}{3K_e^2} + \int_0^{+\infty} dr \frac{\sin K_e r}{K_e}[g(r)-1]e^{iK_e r}\right] \tag{101}$$

We see that equation (101) differs from the equation (96), only by the factor $1/3K_e^2$, due to the singularity of the vectorial Green function and this factor cannot be derived from the scalar theory developed by Keller. We also notice that to solve numerically our new equation (92), we can utilize the same procedure that is used for the original Keller formula (96) with the Muller theory. Therefore we have derived a numerical tractable approximation to the (QC-CPA) scheme.

6.Discussion

The intent of this paper has been to establish a new formula for the effective dielectric constant which characterizes the coherent part of an electromagnetic wave propagating in a random medium containing disordered particles. The particles can be metallic, dielectric, spherical or non-spherical. We have discussed the relation between the Dyson equation and the effective permittivity of the medium. With this formulation, we have introduced the mass operator. The mass operator or self-energy operator includes geometric effects, caused by resonant behavior due to the shape and size of particles, cluster effects because of correlations between particles. The problem is solved if we obtain an expression of the mass operator. For that, the relation between the mass operator and the scattering operator is expressed. We give a formal solution for the scattering operator by introducing the T-operator formalism. We show that the T-operator, or the scattering operator satisfies a Lippman-Schwinger equation. This operator includes all the multiple scattering. Then we have introduced the Quasi-Crystalline Coherent Potential Approximation (QC-CPA), which takes into account the correlation between the particles with a pair-distribution function. This function describes the correlation between two distinct particles. To simplify the numerical calculation of the effective permittivity under the (QC-CPA) approach, we have added far-field and forward scattering approximations to (QC-CPA) scheme. In the low frequency limit, our equation is identical with the usual result obtained under the (QC-CPA) scheme and we have generalized the formulation for Rayleigh scatterers by introducing near plasmon resonances. For particles whose dimensions are of the order of wavelength or in the high frequency limit the expression includes the generalization, in the vectorial case, of the result obtained by Keller. We obtained an important tensorial generalization of the notion of effective permittivity. This technique of Dyson equation in wave multiple scattering by spatially disordered discrete medium leads to a dielectric permittivity tensor. This expression of the permittivity developed in this paper has been extended to random media with different types of particles and bounded by rough surfaces [46]. This extended formulation of the effective permittivity is then used to calculate the coherent fields and incoherent intensities scattered from a three-dimensional disordered medium with discrete scatterers and randomly rough interfaces.
The authors [47] show the physical nature of diamagnetic property in the effective magnetic permeability at coherent electromagnetic wave multiple scattering by statistical ensemble of independent perfectly reflected non–magnetic small spherical particles. This observation is made with the aid of the technique of Dyson equation developed in this paper for a random medium with disordered particles.

Appendix

$$\int_{4\pi} d^2\hat{\boldsymbol{r}}\, e^{-i\boldsymbol{k}_0\cdot\hat{\boldsymbol{r}}||\boldsymbol{r}||} = 4\pi\frac{\sin||\boldsymbol{k}_0||\,||\boldsymbol{r}||}{||\boldsymbol{k}_0||\,||\boldsymbol{r}||} \tag{A1}$$

$$
\begin{aligned}
&\int_{4\pi} \mathrm{d}^2\hat{\boldsymbol{r}}\, e^{-\mathrm{i}\,\boldsymbol{k}_0\cdot\hat{\boldsymbol{r}}||\boldsymbol{r}||}\hat{\boldsymbol{r}}\hat{\boldsymbol{r}} = 4\pi\left[\frac{\sin||\boldsymbol{k}_0||\,||\boldsymbol{r}||}{||\boldsymbol{k}_0||^3\,||\boldsymbol{r}||^3}\right.\\
&\left.-\frac{\cos||\boldsymbol{k}_0||\,||\boldsymbol{r}||}{||\boldsymbol{k}_0||^2\,||\boldsymbol{r}||^2}\right](\overline{\boldsymbol{I}}-\hat{\boldsymbol{k}}_0\hat{\boldsymbol{k}}_0)+4\pi\left[\frac{\sin||\boldsymbol{k}_0||\,||\boldsymbol{r}||}{||\boldsymbol{k}_0||\,||\boldsymbol{r}||}\right.\\
&\left.+2\frac{\cos||\boldsymbol{k}_0||\,||\boldsymbol{r}||}{||\boldsymbol{k}_0||^2\,||\boldsymbol{r}||^2}-2\frac{\sin||\boldsymbol{k}_0||\,||\boldsymbol{r}||}{||\boldsymbol{k}_0||^3\,||\boldsymbol{r}||^3}\right]\hat{\boldsymbol{k}}_0\hat{\boldsymbol{k}}_0\,,
\end{aligned}
\tag{A2}
$$

References

[1] Chandrasekhar S 1960 *Radiative Transfer* (New York: Dover Publications)

[2] van de Hulst H C 1980 *Multiple Light Scattering* vol 1 and 2 (New York: Academic Press)

[3] Bohren C and Huffman D 1983 *Absorption and Scattering of Light by Small Particles* (New York: Wiley-Interscience)

[4] Ishimaru A *Wave Propagation and Scattering in Random Media* vol 2 (New York: Academic Press)

[5] Frish U 1968 Wave propagation in random media *Probabilistic Methods in Applied Mathematics* vol 1 ed A T Baruch-Reid (New York: Academic Press) pp 75-198

[6] Lagendijk L and van Tiggelen B A 1996 *Physics Reports* **270** 143

[7] Apresyan L A and Kravtsov Y A 1996 *Radiation Transfer: Statistical and Wave Aspects* (Amsterdam: Gordon and Breach)wareman

[8] Barabanenkov Y N Kravtsov Y A Ozrin V D and Saichev A I 1991 *Progress in Optics* **XXIX** 65

[9] Sheng P *Introduction to Wave Scattering, Localization, and Mesoscopic Phenomena* 1995 (New York: Academic Press)

[10] Sheng P ed *Scattering and Localization of Classical Waves in Random Media* 1990 (Singapore: World Scientific)

[11] Tsang L Kong J A and Shin R 1985 *Theory of Microwave Remote Sensing* (New York: Wiley-Interscience)

[12] Tsang L Kong J A and Ding K H 2000 *Scattering of Electromagnetic Waves: Theories and Applications* vol 1 (New York: Wiley-Interscience)

[13] Tsang L Kong J A Ding K H and Ao C O 2001 *Scattering of Electromagnetic Waves: Numerical Simulations* vol 2 (New York: Wiley-Interscience)

[14] Tsang L and Kong J A 2001 *Scattering of Electromagnetic Waves: Advanced Topics* vol 3 (New York: Wiley-Interscience)

[15] Berginc G and Maradudin A A 2013 Scattering properties of random structures in thin films *Optical Thin Films and Coatings: From materials to applications* ed A Piegari and F Flory (Oxford: Woodhead Publishing Limited) pp 177-289

[16] Bellman R ed 1964 Stochastic Processes in Mathematical Physics and Engineering *Proceedings of Symposium in Applied Mathematics of the American Mathematical Society* vol 16 (Rhode Island: AMS)

[17] Scaife B K P 1989 Principles of Dielectric *Monographs on the Physics and Chemistry of materials* vol 45 (Oxford: Clarendon Press)

[18] Choy T C 1999 Effective Medium Theory: Principles and Applications *International Series of Monographs on Physics* vol 102 (Oxford: Clarendon Press)

[19] Sihvola A 1999 Electromagnetic Mixing Formulas and Applications *Electromagnetic Waves Series* vol 47 (London: IEE).

[20] Cummings K D Garland J C and Tanner D B 1984 *Phys. Rev. B* **30** 4170

[21] Felderhof B U 1989 *Phys. Rev. B* **39** 5669
[22] Barrera R G Monsivais G and W. L. Mochan 1988 *Phys. Rev. B* **38** 5371
[23] Agarwal G S and Inguva R 1984 *Phys. Rev. B* **30** 6108
[24] Lamb W Wood D M and Ashcroft N W 1980 *Phys. Rev. B* **21** 2248
[25] Tsang T and Kong J A 1980 *J. Appl. Phys.* **51** 3465
[26] Lax M 1952 *Phys. Rev.* **85** 621
[27] Waterman C and Truell R 1961 *J. Math. Phys.* **2** 512
[28] Korringa J and Mills R L 1972 *Phys. Rev. B* **5** 1654
[29] Davis V A and Schwartz L 1995 *Phys. Rev. B* **31** 5155
[30] Golden K M et al 1998 *IEEE Trans. Geosci. Remote Sensing* **36** 1655
[31] Jing X Sheng P and Zhou M 1992 *Phys. Rev. A* **46** 6513
[32] Bush K 1996 *Phys. Rev. B* **54** 893
[33] Walden C J 1995 *Phys. Rev. E* **52** 3115
[34] Walden C J 1998 *Phys. Rev. E* **57** 2377
[35] Stoyanov A J Howell B F Fischer E C Uberall H and Chouffani K 1999 *J. Appl. Phys.* **86** 3110
[36] Stoyanov A J Fischer E C and Uberall H 2001 *J. Appl. Phys.* **89** 4486
[37] Mandt C E Kuga Y Tsang L and Ishimaru A 1992 *Waves in Random Media* **2** 225
[38] Zurk L M Tsang L Ding K H and Winebrenner D P 1995 *J. Opt. Soc. Am. A* **12** 1772
[39] West R D Gibbs D Tsang L and Fung A K 1994 *J. Opt. Soc. Am. A* **11** 1854
[40] Keller J B 1964 Stochastic Equations and Wave Propagation in Random Media *Processes in Mathematical Physics and Engineering Proc. of Symposia in Applied Mathematics* vol 16 (Providence: AMS) pp. 145–170.
[41] van Bladel J 1991 *Singular Electromagnetics Fields and Sources* (Oxford: Clarendon Press)
[42] Tai C T 1994 *Dyadic Green Functions in Electromagnetic Theory* (New York: IEEE)
[43] Soukoulis C M and Datta S 1994 *Phys. Rev. B* **49** 3800
[44] Kerker M 1969 *The Scattering of Light* (New York: Academic Press)
[45] Hanson G W and Yakovlev A B 2002 *Operator Theory for Electromagnetics: an Introduction* (New York: Springer)
[46] Berginc G 2013 *Quantum Electronics* **43** 1055
[47] Barabanenkov Y N and Barabanenkov M Y 2013 Rigorous Dyson equation and quasi-separable T-scattering operator technique for study of magnetic response from ordered and disordered non-magnetic particles' ensembles at electromagnetic wave multiple scattering *Aspects in Optical Metrology IV: Proc. of SPIE* Vol 8789 ed B Bodermann K Frenner and R M Silver (Bellingham: SPIE) 87890O

Journal of Physics: Conference Series **738** (2016) 012024 doi:10.1088/1742-6596/738/1/012024

Criteria for Analyzing the Quality Estimate shake using two different impellers, "Rushton Turbine" and "PBT 45"

E C M Oliveira[1], M O M Silva[1], J R Nunhez[2] e L C L J Ribeiro[1]
[1] Faculdade de Tecnologia, Universidade Estadual de Campinas – Limeira, Brasil
[2] Faculdade de Engenharia Química, Universidade Estadual de Campinas – Campinas, Brasil.

E-mail: evacarolline.mo@gmail.com, mayarmaia@gmail.com, nunhez@feq.unicamp.br, lubi@ft.unicamp.br.

Abstract. The Computational Fluid Dynamics (CFD) is used to numerically simulate the behavior of fluid flow, and all the laws that rule the study of fluids, both the energy and mass transfer, chemical reactions, hydraulic behavior, and other applications. Among these innumerable applications, the local behavior of a mixing tank can be estimated using the CFD techniques. In order to follow-up studies in the literature giving them increased credibility, this project aimed to determine whether both the scheduling Potency/Volume and Torque/Volume maintain the necessary level of agitation in the mix tank to extrapolate the volume 10 times and 100 times, using an axial impeller type blades inclined at 45 degrees (PBT 45), and a radial impeller (Rushton turbine) with the use of the CFD. With the results, was analyzed if a criterion is superior to the other in terms of maintaining the level of agitation in scheduling. The case study made it possible to achieve the main objective, deepen studies in Computational Fluid Dynamics. Therefore, managed to the largest domain in tools offered by CFX-ANSYS 14.0 software, allowing future work to be conducted.

1. Introduction

Much of the production processes in industries, requires the stirring of the products at any stage. Agitation systems vary depending on the geometry of the tank and the type and number of impellers. In addition to geometry, other important variables are the properties of the fluid and the rotation speed of impeller [3]. These variables are used to calculate the power (P) and torque (Tq) responsible for the operational cost of the process and fixed costs of the tank, respectively. Such factors must be taken into account to provide the best conditions for the process considered and the greatest possible operational savings. In the case of agitation, the torque is the relationship between the power and the speed of rotation of the impellers.

The hardest part in the design of a impeller is the estimate of the size of the engine, what influences sharply in price from a stirring [3], usually this estimate is less when using the torque/volume criterion, however, this does not indicate that the level of agitation is maintained. Agitating systems are estimated from the data of number of power, used typically for turbulent flow, which can be understood as a "coefficient of drag" of impeller. From the use of computational fluid dynamics, this paper proposed studying the scale-up in mixing tanks from P/V criteria and Tq/V. The project allows estimating the power consumed by a mixing tank extrapolated from a tank pilot, analyzing if both escalation methods maintain the level of unrest as the initial tank, checking if a criterion is superior to

Content from this work may be used under the terms of the Creative Commons Attribution 3.0 licence. Any further distribution of this work must maintain attribution to the author(s) and the title of the work, journal citation and DOI.
Published under licence by IOP Publishing Ltd

another in terms of maintaining the level of unrest in the staggering. It permits the decision-making on the operating costs of a process to be carried out with greater reliability.

2. Methodological procedures

It was employed as pilot tank mixing tank with the impeller type PBT 45 [6], corresponding to a square tank of 0.24 m, with an approximate volume of 10.86 liters. The computational modeling was performed to extrapolate the tank 10 times and 100 times. With the same characteristics, was made a tank with the impeller type Rushton turbine, repeating the same extrapolations. Scale-up was carried out in the initial model. From the criterion P/V and Tq/V scaling has been increased by 10 times, using two levels of agitation for the tests, P/V = 0.2 kW/m ^{3} and P/V = 2.0 kW/m ^{3}. In sequence, increased the scale in 100 times, using the same two levels of agitation for the tests doing it to both impeller, PBT 45 and Rushton turbine. Adding to the final twenty simulations with different rotation speeds.

Fluid dynamic behavior of the fluid in turbulent regime mixing tanks was analyzed by developing a three-dimensional model. The numerical results have been obtained through the use of the software CFX-14.0 the ANSYS, was discredited by a mesh that aims at obtaining the variables, such as: speed, temperature, viscosity, and others, at specific points within the entire tank [1]. Here the interest was at inside of the tanks. To simulate the tank was with two domains: rotating and stationary. The stationary domain comprises the geometry of the tank, while the Rotary field is constituted by the geometry of the region around the impeller. It is observed that the geometry of the tank has the space corresponding to the rotation of the impeller domain.

Tetrahedral volumes formats were used and prisms for the creation of the meshes. It was considered the minimum size of the element, the rate of glineth and refinement located mainly in regions near the surfaces of the walls of the tank and of the impeller where if you expect high-speed gradients or turbulent kinetic energy. In figures 1, 2 and 3 it can be observed the meshes created. According to [5], for the great performance of turbulence models one of the key factors is the exact solution of the boundary layer velocity profile. That way, according to [1], one of the methods to control the size of the mesh in the boundary layer is the specification of the dimensionless distance y+ wall and the Reynolds number.

In the definition of the simulation in CFX-Pre, the characteristics about the setup were defined, where was the coupling of different domains. Physical and chemical data were provided, defined the mathematical models employed, as well as what the boundary conditions of the simulation. On Computational solution step with CFX-Solver Manager, during the simulation accompanied by graphics error waste and predefined monitoring points, the answers were generated in files simulation. In the last step, the post processing through the analyses do the results in CFD-Post, if the behavior of the fluid dynamics of the problem at any stage of the solution. Maximum values were rated first and of y+ in the impeller and on the wall of the tank, and then raised the speed profiles at different points in the tank and then analyze and generate the results.

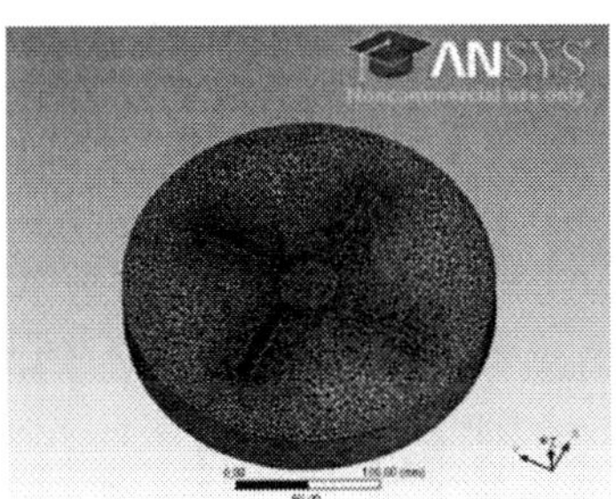

Figure 1. Mesh Impeller with straight blades inclined at 45 degrees (PBT 45).

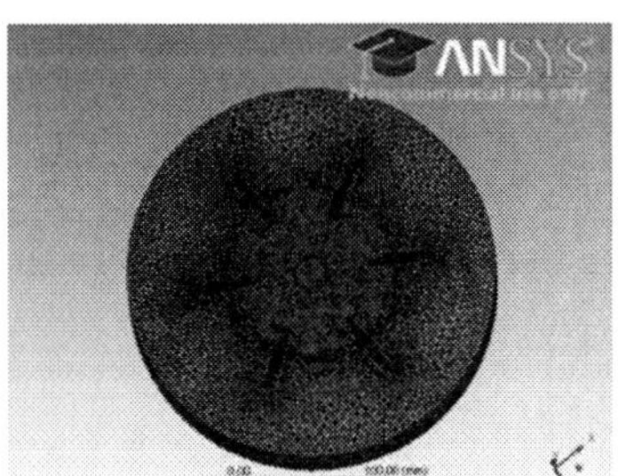

Figure 2. Mesh radial impeller (Rushton turbine).

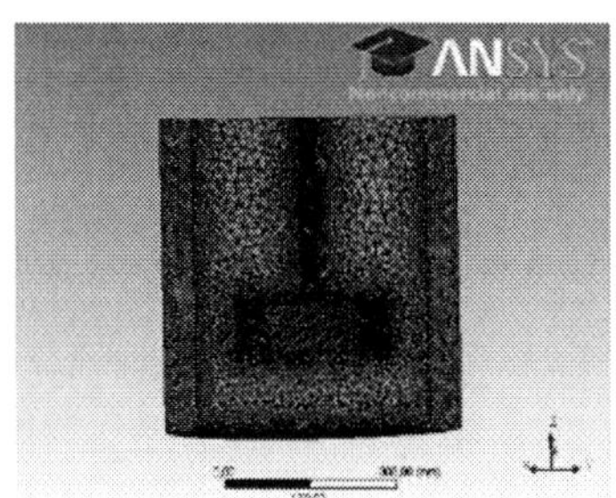

Figure 3. Mesh tank.

3. Results and discussion analysis

The first analysis refers to the validation result in CFD pilot tank with the experimental results. Experimental data was obtained from the University of Alberta -Canada, carried out by researchers Suzanne Kresta and Márcio Bezerra Machado [4]. Figure 4 shows the comparison of the experimental points with the results in CFD with the SST turbulence model used. This graph in Figure 4 demonstrates that the CFD model was able to capture the variations in speed with enough precision, proving the model validation in terms of profiling speed. The dimensionless speed is the relationship of local speed and speed at the end of the impeller.

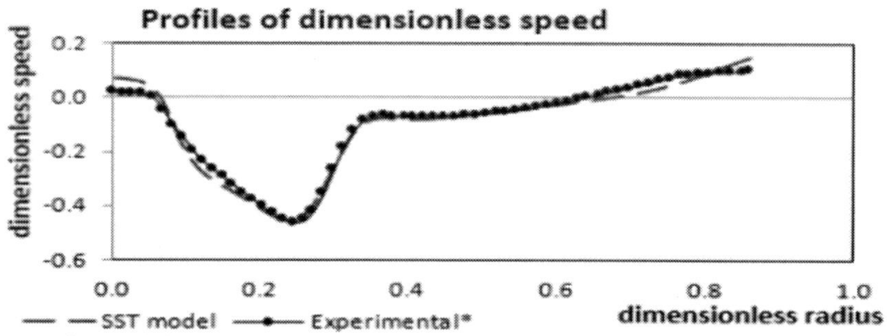

Figure 4. Comparison of the experimental model with the SST turbulence model using the tank pilot [6].

As the tanks grossed in 10 and 100 times the same computational model used and respected the same parameters of the tank pilot, it is considered that the model in 10 times and 100 extrapolated times also represented with details the speed profiles. The graphics were generated from the P/V criterion and the criterion Tq/V. Considering the comparisons to the two levels of agitation, P/V=0.2 kW/m^3 and P/V=2.0 kW/m^3, which are widely used in the chemical industry. In the next sequence analysis was to compare the dimensionless speed profiles in several lines in the tanks, both to the tank pilot, extrapolated in 10 times and the extrapolated in 100 times, ensuring the proportionality of the positions of the lines. In each line, compared the graphics of the tank pilot speed profiles with the extrapolated in 10 and 100 times. If the criterion for extrapolation holds the same conditions of the tank pilot, the three profiles (pilot, 10 and 100 times) must match. These processes were performed also for the two impellers studied, type PBT 45 and Rushton turbine. Figure 5 and 6 show the lines where the results were compared for each of the respective impellers. It is observed that the impeller type Rushton has a line (line 9), the main interest was just to show that at the end of the impeller the speed is greater, closer to the value 1, which was to be expected.

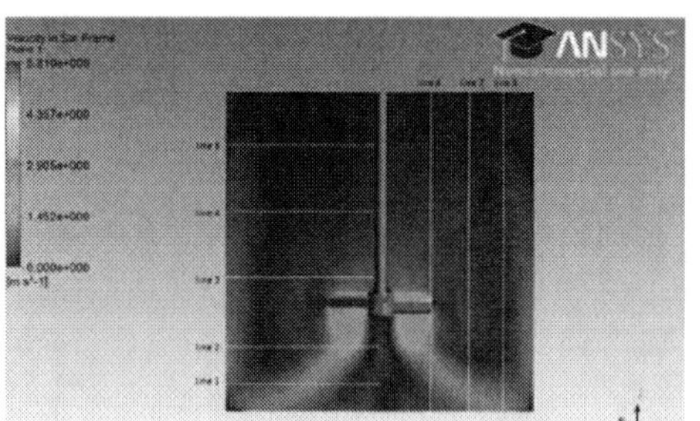

Figure 5. Comparison lines to the tank with the impeller straight blades inclined at 45 degrees (PBT 45).

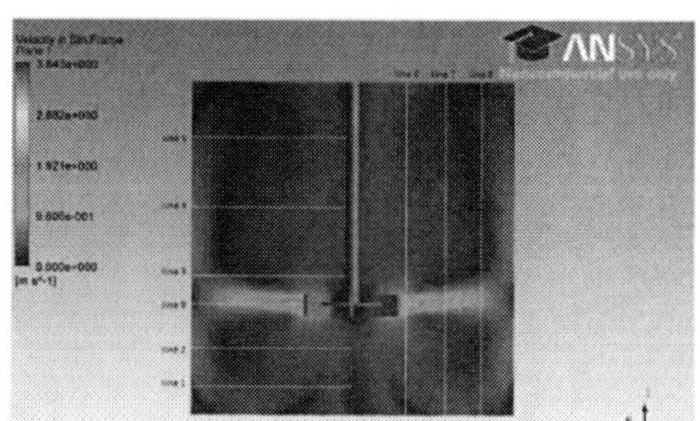

Figure 6. Comparison lines to the tank with the impeller type Rushton turbine.

In all graphs generated compared to each line the variation of dimensionless speed in relation to the dimensionless radius. Generally the speed profiles in the three cases were similar escalation. However, there is a coincidence between the profiles; this is was expected, because you know that is impossible in scale-up keep geometric and kinematic similarity at the same time [2]. Note that the speeds for each tank, despite showing peaks displaced, have similar magnitudes. Therefore, it can be said that the agitation level indicates a similar condition in terms of intensity of agitation. It is observed that both criteria P/V and Tq/V enabled extrapolated very similar conditions. The graphs of figures 7 and 8, for example, prove it.

After comparing both criteria of extrapolation, P/V and Tq/V, notice some variations in the profiles of dimensionless speed between methods, mainly the lines went more of impeller, and near the impeller speeds are greatest. However, in General, both the P/V criterion as Tq/V, kept on extrapolation of dimensionless similar speed profiles.

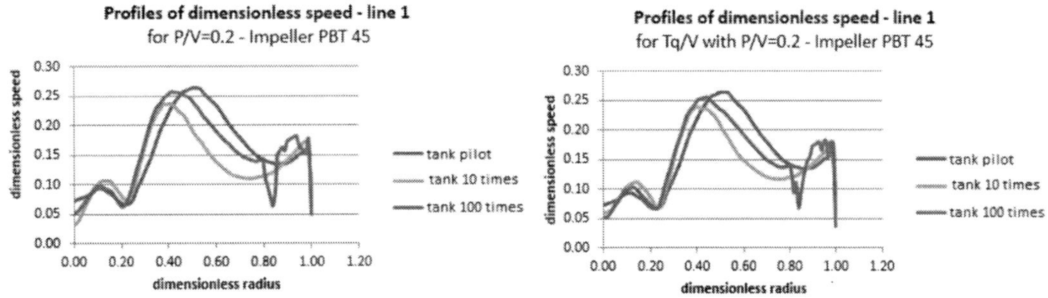

Figure 7. Comparison of dimensionless speed profiles in 1 line (Figure 5) for pilot tanks, extrapolated in 10 times and extrapolated in 100 times, with the impeller straight blades inclined at 45 degrees (PBT 45) and level of agitation for the P/V=0.2 kW/m^3.

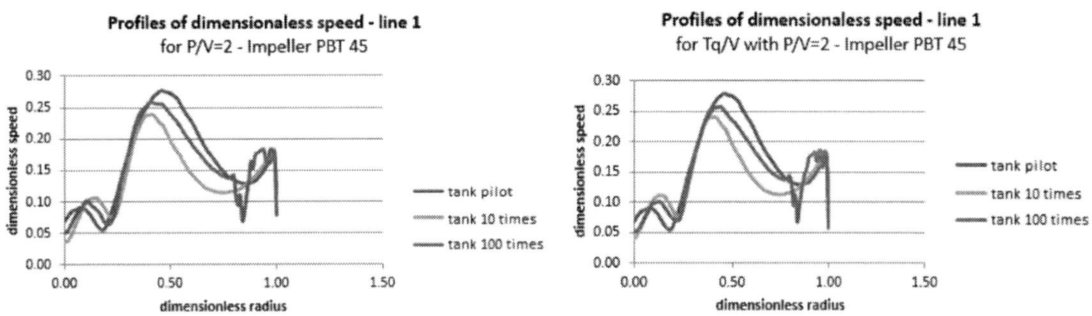

Figure 8. Comparison of dimensionless speed profiles in 1 line (Figure 5) for pilot tanks, extrapolated in 10 times and extrapolated in 100 times, with the impeller straight blades inclined at 45 degrees (PBT 45) and level of agitation for the P/V=2.0 kW/m^3.

4. Conclusion

The analysis done for the tank with the impeller type PBT 45 and tank with the impeller type Rushton turbine is similar, since the conclusions concerning comparisons of two extrapolation criteria studied, P/V and Tq/V, are the same for both types of impellers. The criteria of extrapolation managed to maintain in a similar manner to the level of agitation when extrapolated.

It is concluded that the criterion Tq/V is the most appropriate, since it leads to impellers projects with a lower cost installation, as well as maintenance.

Acknowledgements
The authors gratefully acknowledge the support provided by FAPESP process number 2012/12298-1 and 2012/12296-9 and the PRP/UNICAMP.

References
[1] ANSYS, Inc. User Manual. ANSYS CFX 14.0, 2011.
[2] HASSAN, Y. A. Multiphase Flow Measurements Using Particule Image Velocimetry. The Handbook of Fluid Dynamics, (ed.) Johnson, R. W., Press- Springer, Boca Raton FL, 1998.
[3] JOAQUIM JUNIOR, C. F.; CEKINSKI, E.; NUNHEZ, J. R.; URENHA, L. C. Agitação e mistura na indústria. LTC. Rio de Janeiro, 2007.
[4] MACHADO, M. B.; NUNHEZ,J. R.; NOBES, D.; KRESTA, S. M. Impeller characterization and selection: Balancing efficient hydrodynamics with process mixing requirements. AIChE Journal, Volume 58, ed 8, 2012, p. 2573-2588.
[5] SPOGIS, N. Metodologia para determinação de curvas de potência e fluxos característicos para impelidores axiais, radiais e tangenciais utilizando a fluidodinâmica computacional. Campinas, SP, 2002.
[6] VERGEL, J. L. G. Estudo da influência da malha computacional, modelos de turbulência e aspectos numéricos da modelagem CFD em impelidores PBT usando malhas não-estruturadas. Campinas, SP, 2013.

Modelling of the size effect's influence on the critical temperature of the nanowires' phase transition

Afremov Leonid, Petrov Aleksandr

Department of Theoretical and Nuclear Physics, The School of Natural Sciences, Far Eastern Federal University, 8, Sukhanova St., 690950, Vladivostok, Russian Federation

e-mail: afremov.ll@dvfu.ru, aleksandr.al.petrov@gmail.com

Abstract. Influence of the size effect on the magnetic phase transition's temperature and magnetic moment of the nanowires was modelled based on "average spin" method. As it was shown that Curie temperature was increased with nanowires' sizes increase. The calculated value of the critical exponent of the spin-spin correlations $\nu = 0.737$ is close to experimental value.

1. Introduction

Magnetic phase transition's temperature varies by noticeable amount for nanoparticles (NPs) [1-2], nanowires [3] and ultrathin films [4-5] compared with the bulk material. This is due to the nanostructure's size is commensurable number with the spin-spin correlation size. In this case nanodivides influence on the physical properties of the nanostructures so what is called size-effects to affect.

There are many experimental and theoretical studies about nanostructures' size-dependent magnetic properties of the phase transitions. So experimental studies of the size-dependent Curie temperature of EuS NPs were shown in [1, 2]. Measurements were shown decreasing of the phase transition temperature with decreasing NPs size based on magnetic isotherms method. The authors of [3] came to this conclusion used scaling correlation of Curie temperature's dependence on the nanowire diameter:

$$\varepsilon = \frac{T_c(d \to \infty) - T_c(d)}{T_c(d \to \infty)} = \left(\frac{\xi_0}{d}\right)^{1/\nu}, \quad (1)$$

and measured spin-spin correlations' critical exponent $\nu = 0.8 \pm 0.1$ and correlation's length $\xi_0 = 1.7 \pm 0.5\, nm$ at $T = 0\, K$. It should be noted ultrathin films' experimental studies [4] wherein the correlation (1) was used to calculate experimental data for ν.

Theoretical studies of the size-effect's influence on the phase transition's critical temperature based on d-f model and using Creen's function technique [6], Monte-Carlo method [7-11] and average spin method [12-13] are in agreement with experimental data.

The aim of this work is to study the influence of the nanowire's size on the magnetic phase transition temperature. The study was based on "average spin" method developed earlier [12, 14-15] and it main principles are given below.

Content from this work may be used under the terms of the Creative Commons Attribution 3.0 licence. Any further distribution of this work must maintain attribution to the author(s) and the title of the work, journal citation and DOI.
Published under licence by IOP Publishing Ltd

2. The model

Here we can consider two types of the nanowires: continuous and hollow in which the atoms are distributed over the side surface. Size-effect is determined only by the length of the hollow nanowire in the approximation of the nearest neighbours. We can consider the model of the hollow nanowire:

- Nanowire has the shape of a regular right prism with the height La where a is a lattice constant, L are the whole numbers;
- the atoms of the ferromagnetic are distributed uniformly over sites of simple square lattice with constant a and located over the side surface with magnetic moments m_{0k}.
- fields of interaction h between atom's spin magnetic moments are distributed randomly and there is direct exchange interaction between nearest neighbours only;
- atom's spin magnetic moments are oriented along axis oz perpendicular to the base of the prism (in the approximation of the Ising model).

According to [12, 14-15] the distribution function of random fields of exchange interaction h with $n-th$ atom is determined as:

$$W_n(h) = \iint \delta\left(h - \sum_{\{k\}_n} \varphi_k(\boldsymbol{m}_k, \boldsymbol{r}_k)\right) \prod_{\{k\}_n} \Phi(\boldsymbol{m}_k, \boldsymbol{r}_k) d\boldsymbol{m}_k d\boldsymbol{r}_k \,, \qquad (2)$$

where $\{k\}_n$-the nearest neighbors set with number n, $\varphi_k = \varphi_k(\boldsymbol{m}_k, \boldsymbol{r}_k)$-field at the origin of the coordinates created by atoms with magnetic moments $\boldsymbol{m}_k$ located at the coordinates $\boldsymbol{r}_k$, $\Phi(\boldsymbol{m}_k, \boldsymbol{r}_k)$-the distribution function for the magnetic moments and coordinates. In the case of identical atoms with magnetic moment m_{ko} distributed over crystalline lattice sites it is:

$$\Phi(\boldsymbol{m}_k, \boldsymbol{r}_k) = \delta(\boldsymbol{r}_k - \boldsymbol{r}_{k0})[\alpha_k \delta(\Theta_k) + \beta_k \delta(\Theta_k - \pi)]$$
$$[(1-p)\delta(m_k) + p\delta(m_k - m_{k0})]), \qquad (3)$$

where $\boldsymbol{r}_{k0}$ –the coordinates of lattice sites, α_k-the relative probability of the spin orientation along axis oz $(\Theta_k = 0)$; $\beta_k = 1 - \alpha_k$- the relative probability of the spin orientation against axis oz $(\Theta_k = \pi)$; Θ_k-the angle between $\boldsymbol{m}_k$ and axis oz.

There is the characteristic function written with equation (2) and (3):

$$A_n(\rho) = \prod_{\{k\}_n} ((1-p) + p[\alpha_k \exp\{i\rho\varphi_k(m_{k0}, r_{k0})\} +$$
$$+\beta_k \exp\{-i\rho\varphi_k(m_{k0}, r_{k0})\}]), \qquad (4)$$

and then we can write the distribution function $W_n(h) = \int A_n(\rho) exp\{i\rho h\} d\rho$. According to "average spin" method we replace instantaneous values α_k and β_k in equation (4) to their assembly average $\langle\alpha_k\rangle = \int \alpha_k W_n(h) dh$ and $\langle\beta_k\rangle = \int \beta_k W_n(h) dh$. Such replacement allow us to write self-consistent equations connecting magnetic moment μ_n with neighbor's magnetic moments defined by $W_n(h)$:

$$\mu_n = \langle\alpha_n\rangle - \langle\beta_n\rangle = \int_{-\infty}^{\infty} \tanh\left(\frac{m_{no}h}{k_B T}\right) W_n(h) dh. \qquad (5)$$

A system of equations can be represented as follow to determine the Curie temperature using the approach [13]:

$$x_n = \sum_{\{l\}} \int Tanh\left(\frac{h}{J\,t}\right) \frac{\partial W_n}{\partial \mu_l} x_l dh, \qquad (6)$$

where $x_n = \{\partial\mu_n/\partial\mu_1\}_{T\to T_C}$, the summation is over the nearest neighbours in (6), $t = k_B T/m_o J$ – the relative temperature, J- the exchange interaction constant.

Equation systems (5) and (6) allow us to study the dependence of the average magnetic moment on the temperature and Curie temperature dependence on the nanowire's length.

3. Results

The temperature dependence of the atomic average magnetic moment m is represented on the figure 1 for nanowire with length L calculated with (6) in the approximation that the exchange interaction constants are the same between atoms.

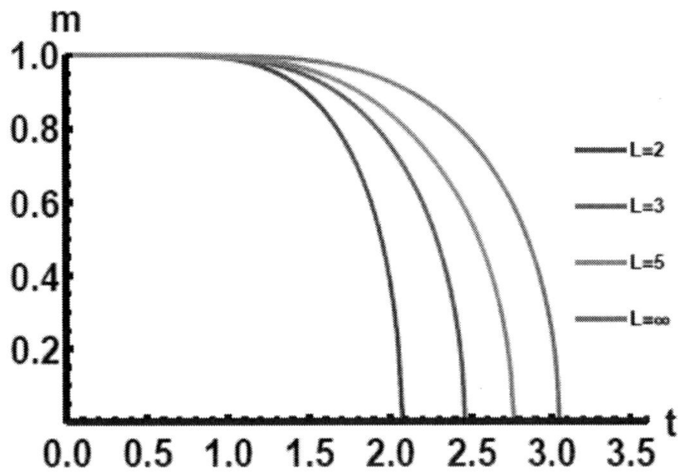

Fig 1. The dependence of the average magnetic moment m on the temperature t for nanowires with different lengths L measured in monolayers.

As could be expected the phase transition temperature T_c increases with L increasing as in ultrathin films [12, 13]. The above-mentioned property is associated with the nearest neighbours' number increasing with nanowire's length increasing.

The dependence of the relative change of Curie temperature $\varepsilon(L)$ on the nanowire's length (in monolayers) is represented in figure 2 and followed by (1) where $\xi_0 = 0.87\ ML$ and $\lambda = 1{,}36$ are the correlation legth at $T = 0$ and inverse value of the spin-spin correlations' critical exponent $\nu = 1/\lambda = 0.735$ respectively.

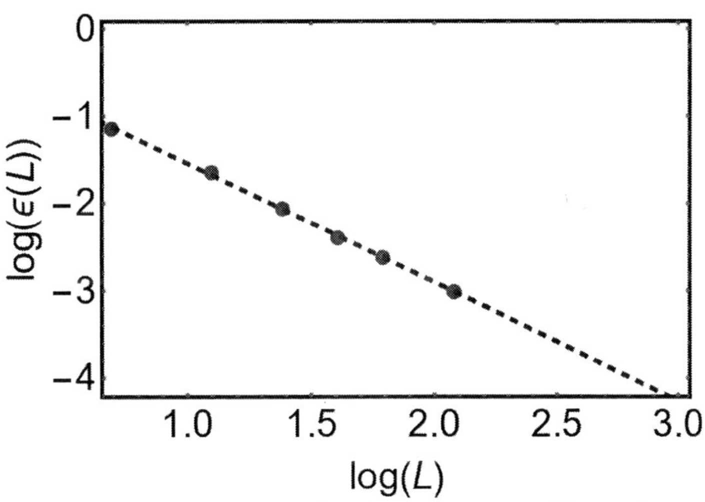

Fig 2. Dependence of the relative change of Curie temperature $\varepsilon(L)$ on the nanowire's length (in monolayers) in logarithmic scale.

4. Conclusion

Reducing of the nanowire's size leads to increased surface role resulting in a decrease in the number of nearest neighbors as a reduction of the thickness of ultrathin films [4, 5], and hence the phase transition temperature to drop. The critical exponent of the spin-spin correlations $\nu = 0.735$ is in agreement with experimental data.

5. Acknowledgements

The work was supported by Ministry of Education and Science, project № 559.2014.

References

[1] Regulacio M D, Bussman K, Lewis B, Stoll S L 2006 *Journal of American Chemical Society* **128** 11173.
[2] Regulacio M D, Srotoswini K, Zuniga E, Wang G, Dollahon N R, Yee G T, Stoll S L 2008 *Chem.Mat.* **20(10)** 3368.
[3] Lopez-Ruiz R, Magen C, Luis F and Bartolome J 2012 *Journal of Applied Physics* **112** 073906.
[4] Li Y, Baberschke K 1992 *Physical Review Letters* **68(8)** 1208-1211
[5] Bramfeld T S, Willis R F 2008 *Journal of Applied Physics* **103** 07C718
[6] Wesselinowa J M, Apostolov A T 2012 *Journal of Magnetism and Magnetic Materials* **324** 23
[7] Laosiritaworn Y, Poulter J, Staunton J B 2004 *Phys. Rev. B* **70** 104413
[8] Nefedev K V, Belokon V I, Kapitan V Y, Dyachenko O I 2014 *Journal of Physics: Conference Series* **490 (1)** 012163.
[9] Nefedev K V, Kapitan V Y 2013 *Advanced Materials Research* **718** 69-73.
[10] Nefedev K V, Kapitan V Y 2014 *Journal of nano- and electronic physics* 6**(3)** 03005.
[11] Belokon V I, Nefedev K V, Kapitan V Y, Dyachenko O I 2013 *Advanced Materials Research* 774-**776** 523-527.
[12] Afremov L L, Kirienko Y V, Petrov A A 2014 *Bulletin of the Russian Academy of Sciences* **78** 104
[13] Afremov L L, Petrov A A 2014 *Solid State Phenomena* **215** 227
[14] Belokon V I, Nefedev K V 2001 *Journal of Experimental and Theoretical Physics* **120(17)** 156
[15] Belokon V I, Semkin S V 1992 *J. Exp. Theor. P.* **100** 1254-1258

Importance of pressure reducing valves (PRVs) in water supply networks.

R. O. S. Signoreti[1], R. Z. Camargo[1], L. M. Canno[1], M. S. G. Pires[1], L. C. L. J. Ribeiro[1]

[1]Unicamp – State University of Campinas – College of Technology
R. Paschoal Marmo 1888, Centro, CEP 13484-332, Jd. Nova Itália, Limeira, SP, Brazil.

E-mail: rodrigo.signoreti@gmail.com, rafael_kolosso@hotmail.com, laura@ft.unicamp.br, marta@ft.unicamp.br, lubienska.ft@gmail.com

Abstract. Challenged with the high rate of leakage from water supply systems, these managers are committed to identify control mechanisms. In order to standardize and control the pressure Pressure Reducing Valves (VRP) are installed in the supply network, shown to be more effective and provide a faster return for the actual loss control measures. It is known that the control pressure is while controlling the occurrence of leakage. Usually the network is sectored in areas defined by pressure levels according to its topography, once inserted the VRP in the same system will limit the downstream pressure. This work aims to show the importance of VRP as loss reduction for tool.

1. Introduction

Recognizing the importance of water as a limited natural resource and considering the prospect of growth of urban populations, it is necessary an investigation of the distribution of water supply systems for addressing up common basic infrastructure to big centers around the world [1]. Basically the goal of a design of water distribution networks was not only to provide water to consumers, but in recent decades, in addition to providing water to consumers in satisfactory condition, researchers have concentrated efforts to reduce network costs, losses, among others, which can often represent amounts of million.

For a good operation of a water supply system, it is essential that put together the concern for water availability in quantity and its quality for human consumption. However, many water supply networks have presented operational deficiencies, causing losses. Given the scarcity of water that Brazil has faced in recent years, water losses in the supply network have become major problems, and should be treated and managed with preventive measures.

In particular the reduction of physical or actual losses allows reduce operating costs and maintenance. These also optimize the use of existing facilities providing an increase in treated water supply without expanding the production system [2]. So, among the actions recommended as best operational practices to optimize the operation of the distribution systems there is the pressure of managing the network. This management pressures have been reflected in the aspect of reducing excessive or unnecessary pressures, resulting in the installation of pressure reducing valves (PRVs). This paper discusses the use of PRVs as loss reduction tool.

Content from this work may be used under the terms of the Creative Commons Attribution 3.0 licence. Any further distribution of this work must maintain attribution to the author(s) and the title of the work, journal citation and DOI.
Published under licence by IOP Publishing Ltd

2. The water supply system and losses

An efficient urban water supply system must be operated in the daily horizon with the lowest cost of electricity and maintenance providing each network node consumer demands with desired pressure. Furthermore, the delivery system must be reliable in enabling the water supply under abnormal conditions caused by faults in its elements [3].

All the networks parts are local sections subject to losses, mainly in distribution. These losses can occur by deterioration in infrastructure and/or by the user for processing by unlawful means [4]. In this context, studies are conducted to improve operational practices to better performance of the systems such as pressure management in reducing network losses [5].

In a water supply system, the variation of the internal pressure level is linked to human consumption: nighttime with low consumption and daytime use with high [15]. It is in this period of lower consumption, night and early morning hours, that occurs leaks, when the static pressure is high and the low dynamic and reservoirs tend to reach their maximum level, which further raises pressures occurring disruptions in the system [17].

The pressure is the force that the water must have to go through the different elements of the distribution network and reach the consumer. When high, increases the frequency of breaks, reducing the lifetime of the pipes and the flow of leaks is also much higher [16].

By definition, we can say that losses supply systems is nothing more than the volume of water at the input of the system minus the authorized consumption, and thus, losses are grouped in physical losses and either (not consumed) and non-physical losses and either apparent (consumed, but not billed). These losses can determine the increase in running cost taking in to account if it is in capture, adduction, treatment and final distribution, so that there is an increase in the cost of operating revenue [6]. Through Figure 1, it can be seen that water losses are, mostly associated with the actual losses in the systems.

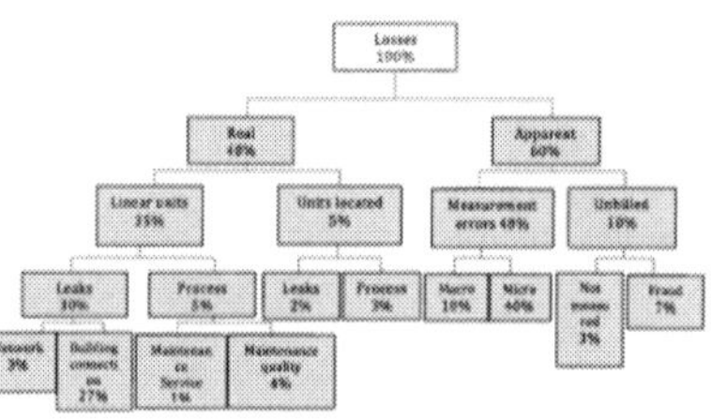

Figure 1. Hypothesis distribution losses in the water supply system [7].

The water losses are measured using established performance indicators characterizing as a strategic indicator, given its strong ties with many organizational processes (Main, Support and, Management) [8]. Many are the metrics that characterize the loss of water: one to measure in% other in liters / active connection.day, others in% or liters / connection.day but separating real loss, apparent loss, or encompassing both [9].

It is believed that by controlling the pressure it is possible to reduce the amount of losses from leakage, the occurrence of internal damage to the premises and fuel consumption related to the high-pressure network, providing the end user a service with more stabilized pressures [4]. Thus, the Pressure Reducing Valves (PRVs) are recommended to reduce excessive pressures, adapting them as far as possible. They make possible an improvement in losses combat management, allowing not only the management of pressures as well as the knowledge of the flow through the system and monitoring control points of the area enabling more accurate corrective actions [10].

3. The importance of PRVs

The main feature in a system that has PRVs is that its controlled area is bounded by closing border records ensuring not only the control pressures as well as knowledge of how to flow monitored and enabling continuous improvement management combat losses [10]. The PRVs are installed at specific points to thereby provide a reduced pressure downstream avoiding excessive pressures that can damage the system in certain parts of the network and ensuring an adequate level of service [5].

Therefore, a water supply network that controls the pressure automatically controls the occurrence of leaks, since these are a function dependent on the pressure occurring in the system. Thus, PRVs as dissipative structures are used in hydraulic systems as a way to standardize and control the pressures giving rise to localized head loss through dissipation of hydraulic energy by lowering the pressure values downstream.

PRVs can be controlled mechanically or electronically, working not only to a single pressure value but, for various pressure levels defined as a function of variations in water demand, allowing a more efficient service levels management and better hydraulic performance of the system [11].

This type of device has been installed on a large scale in various water systems. These valves ensure the minimum and maximum pressures, ie, static and dynamic pressures for the end consumer following the fixed limits. According to NBR No. 12,218 / 1994 the maximum static pressure in the distribution pipes should be 500 kPa and, the minimum dynamic pressure of 100 kPa [12]. Therefore, this valve is recommended in supply systems and reservoirs reducing the pressure and thus, maintaining an active leakage control with an economically viable cost-effective.

While experts recommend its use some sectors within the companies of operating systems suspect that the pressure reduction caused by PRVs may be causing the reduction in revenues of these companies with consequent reduction in the overall consumption of water connections in areas controlled with the device [5].

There are different VRPs and the most common are: controlled by spring (Figure 2), controlled by piston (Figure 3) and, controlled by diaphragm (Figure 4). The main function is to limit downstream pressure when it exceeds a certain value.

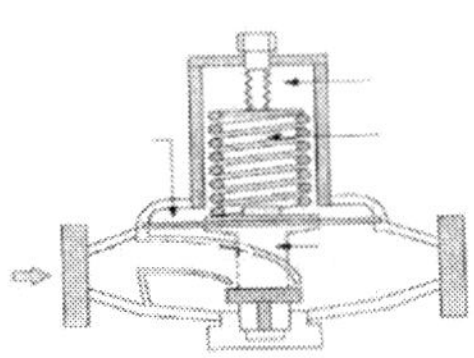

Figure 2. Pressure Reducing Valves - Controlled by spring [11].

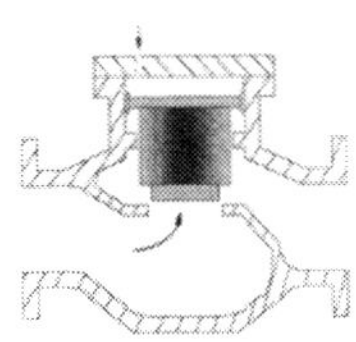

Figure 3. Pressure Reducing Valves - Controlled by piston [11].

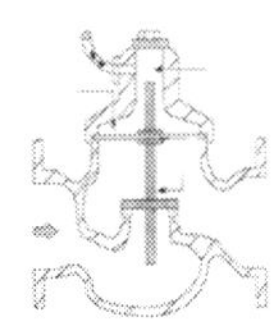

Figure 4. Pressure Reducing Valves - Controlled by diaphragm [11].

There are three operating states for PRVs: active, passive open and passive closed. The active state occurs when the valve downstream pressure is too high closing the valve, which increases the localized load loss and reduces the value of the downstream pressure to the desired value of the PRVs. The passive open state occurs when the upstream pressure is insufficient and less than that presented in the PRVs configuration. In these conditions the valve opens fully keeping the upstream and downstream the same pressure and introducing only the pressure drop in the open valve. The passive closed state is the total closure of the PRVs in the flow reversal conditions, i.e. when for some reason the downstream pressure results higher than the upstream pressure, in this case the valve behaves like a check valve [7].

The problem of inclusion of valves, for practical purposes, can be addressed as a pseudo device simulating the effects as incremental roughness in the pipe [13].

4. Conclusion

It can be seen that leaks produce negative effects both operationally and economically for companies. Systems with leaks need more energy pumping to compensate pressure losses and increase the risk of intrusion of contaminants compromising the water distribution quality [14]. These lead to high capital investments for system rehabilitation or combat the deterioration of structures. Currently, companies operate networks that were designed under conventional criteria of satisfaction and represent a large portion of aged components.

Thus losses are becoming increasingly larger these days when taking into account the progressive reduction of the sources of water and energy supply. Loss reduction policies structured on successful experiences are needed. After valves analysis we can notice they are now one of the fundamental elements for regulation and operation of pressurized hydraulic systems. Without them much of the facilities could hardly operate in the current conditions, which would require investments and higher operating costs.

The identification and use of different PRVs types of behaviors and influence on hydraulic systems (either for supply, distribution and irrigation) for the control pressures and reduction of leakage can be a strategy for loss reduction program. Therefore, optimizations of a supply system using PRVs (oriented by control pressures) can result in effectively reduce in loss in the system thereby improve the service. However, it is clear that the use of PRVs is extremely effective for controlling water losses and reduction of leakage but, should be understood as an accessory for specific situations and their indiscriminate use is not a recommended practice.

Acknowledgment
The authors would like to thank the institutions where they conduct their research for supporting this project and also the financial support provided by CNPq and PRP/UNICAMP.

References
[1] CHEUNG, P. B. Analysis of rehabilitation of water distribution networks for supply via multi-objective genetic algorithms. USP. São Cralos. 2004.
[2] PNCDA, DTA – A2 Losses indicators in Water Supply Systems. Brasília 2003.
[3] RIGHETTO, A.M. System Optimal Operation Water Distribution Urban. Serea. 2002.
[4] MORAIS, D.C.; CAVALCANTE, C.A.V.; ALMEIDA, A.T.. Prioritization loss control areas of water distribution networks. Operations Research,, v.30, n.1, p.15-32, Recife, jan/abril 2010.
[5] GALVÃO, J.R.B.. Relationship Pressure x consumption in areas controlled by pressure reducing valves (PRVs) - Case study: water distribution network in the metropolitan region of São Paulo. São Paulo: 2007, 247p. Dissertação (Mestre em Engenharia Hidráulica) – Escola Politécnica da Universidade de São Paulo, 2007.
[6] DALMO, F. C.; VIANA, A. N. C.; MAUAD, F. F. Use electronic Geofone in research and leak detection: Case Study. XVII Simpósio Brasileiro de Recursos Hídricos. São Paulo. 2007.
[7] ARAUJO, L. S. Loss Control in the sustainable management of water supply systems. Universidade Técnica de Lisbon, Lisbon, 2005.
[8] GONÇALVES, J. E. L.. Companies are large collections processes. São Paulo. 2.000. 14 p.
[9] FUNASA. Loss Reduction in Water Supply Systems. Brasília. 2014.
[10] PEREIRA, F.F.. Reducing valves pressure application in the reduction of real losses in water distribution networks in the East Business Unit - SABESP. São Paulo: 2009, 76p. Trabalho de Conclusão de Curso (Graduação em Engenharia Civil) – Universidade Anhembi Morumbi, 2009.
[11] RAMOS, H. COVAS, D, ARAÚJO, L.. Pressure reducing valves and energy production. 7th Congress of the water. Lisbon. 2004.
[12] GONÇALVES, E.; LIMA, C.L.. Control pressure and pressure regulating valves operation. Practical Guides - operation techniques in water supply systems, vol. 4. Sec. Nac. de San. Ambiental, 2007.
[13] ARAUJO, L.; RAMOS, H.; COELHO, S. Pressure control for leakage minimization in water distribution systems management. Water Resources Management, Springer. V.20, p133-149, 2006.
[14] ARANDIA PÉREZ, A. M. E. Study of great localicação Pressure Reducing Valves in Water Distribution Systems considering multiple objectives. Campo Grande. 2008.
[15] CARDOSO, A.P.P.M. Reduction of real losses in water supply system through advanced control pressure - applying a subsystem of Porto. 2013, 144p.
[16] COSTA, A.C.A. Reduction of real losses in public drinking water system for the management of pressure. 2014.
[17] GONÇALVES, R.O. Analysis of the physical losses in Santa's water supply system Cruz do Sul, RS.- University of Santa Cruz do Sul, RS. 2013 65p.

Molecular dynamics of halogenated graphene - hexagonal boron nitride nanoribbons

G. A. Nemnes[1,2], Camelia Visan[1], D. V. Anghel[1] and A. Manolescu[3]

[1]Horia Hulubei National Institute for Physics and Nuclear Engineering, 077126, Magurele-Ilfov, Romania

[2]University of Bucharest, Faculty of Physics, MDEO Research Center, 077125 Magurele-Ilfov, Romania

[3]School of Science and Engineering, Reykjavik University, Menntavegur 1, IS-101 Reykjavik, Iceland

E-mail: alexandru.nemnes@nipne.ro

Abstract. The hybrid graphene - hexagonal boron nitride (G-hBN) systems offer new routes in the design of nanoscale electronic devices. Using *ab initio* density functional theory calculations we investigate the dynamics of zig-zag nanoribbons a few interatomic distances wide. Several structures are analyzed, namely pristine graphene, hBN and G-hBN systems. By passivating the nanoribbon edges with hydrogen and different halogen atoms, one may tune the electronic and mechanical properties, like the band gap energies and the natural frequencies of vibration.

1. Introduction

Motivated by recent developments in achieving highly defined patterns in hybrid graphene - hexagonal boron nitride (G-hBN) materials [1, 2] we investigate the structural, electrical and vibrational properties of halogenated G-hBN nanoribbons. Having a lattice mismatch of less than 2%, graphene and hBN form an ideal mixture of two dimensional materials with rather different conductive properties: while graphene has a high electrical and thermal conductance, hBN has insulating properties.

By embedding hBN in graphene one opens the possibility of the field effect control over the active region. During the past few years several devices using G-hBN materials have been proposed, such as field effect transistors [3,4], spin filters [5–7], core-shell nanoflakes [8], tunneling double barrier structures [9] and thermoelectric devices [10–14].

We analyze here the electronic and vibrational properties of graphene, hBN and mixed G-hBN nanoribbons with the focus on the changes induced by passivation with hydrogen and different halogen atoms of different mass and electronegativity. Atomic-scale control has already been achieved and well-defined graphene nanoribbons have been produced [15]. Our computational methodology combines molecular dynamics (MD) with density functional theory (DFT) at the atomic level. The results indicate the possibility of tuning of the bandgap energy and the natural vibration frequency by changing the edge passivating atoms.

Content from this work may be used under the terms of the Creative Commons Attribution 3.0 licence. Any further distribution of this work must maintain attribution to the author(s) and the title of the work, journal citation and DOI.
Published under licence by IOP Publishing Ltd

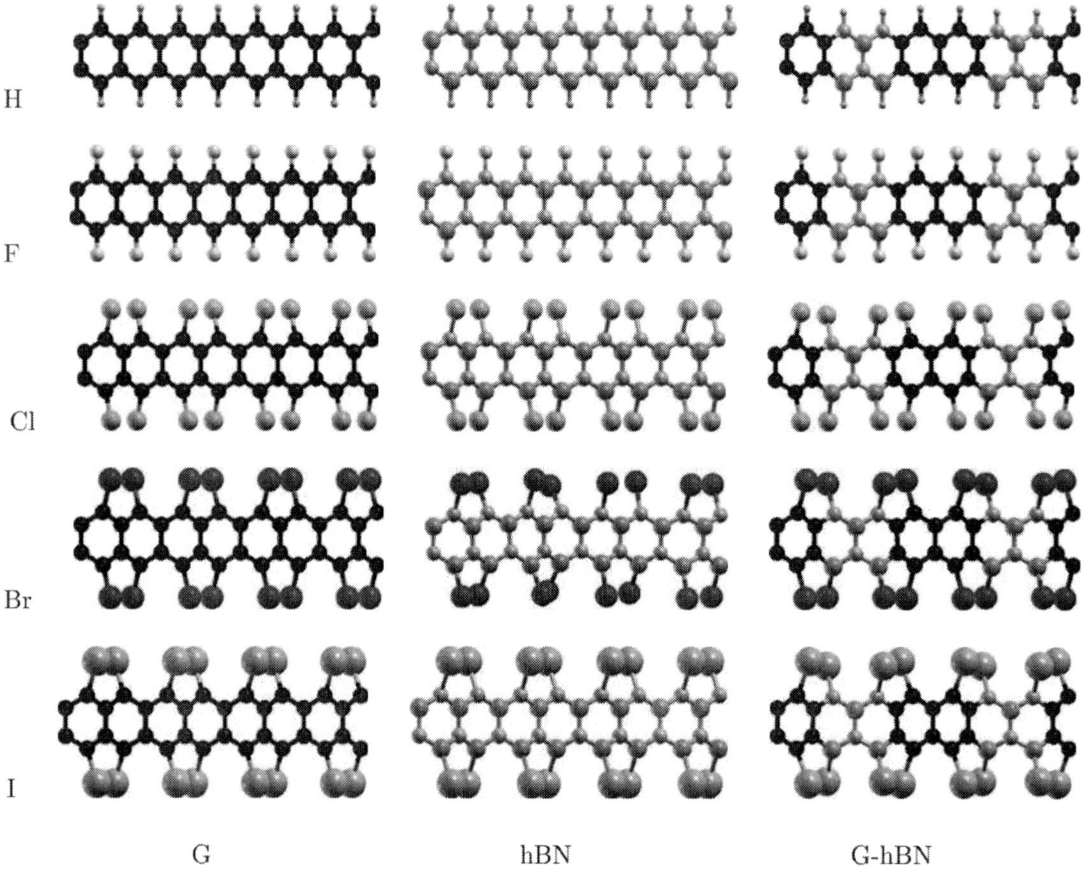

Figure 1. The structures: G, hBN and G-hBN nanoribbons, with the edges passivated by hydrogen (grey) and halogen atoms X = F (light blue), Cl (green), Br (red), I (grey). Boron and nitrogen atoms are depicted in pink and light blue, respectively.

2. Structures and Methods

We consider three types of structures: graphene, hBN, and G-hBN nanoribbons with passivated edges, as indicated in Fig. 1. As the passivating agent we take the hydrogen or one of the halogen atoms X = F, Cl, Br, I. The nanoribbon is oriented along the x direction, with y direction in-plane and z direction perpendicular to the nanoribbon plane.

The DFT calculations are performed using the SIESTA package [16]. We employ the local density approximation (LDA) in the parametrisation proposed by Ceperley and Alder [17]. The real space grid is fixed by the mesh cutoff energy of 200 Ry and a Monkhorst-Pack grid of $1 \times 1 \times 5$ was used. The systems are first relaxed until the residual forces are less than 0.01 eV/Å. Subsequently, a bent nanoribbon is obtained by linearly displacing the atoms off-equilibrium, in the direction perpendicular to the nanoribbon plane, with the maximum offset $\Delta = 1$Å at the middle of the nanoribbon. Starting from this initial configuration, we perform MD runs of the system coupled to a Nose thermostat, which is kept at a constant temperature of 300 K. The simulation time is 10 ps, with a MD time steps of 1 fs. The simulation time is sufficient to allow the system to perform several oscillations, such that the initial configuration should not affect significantly the results.

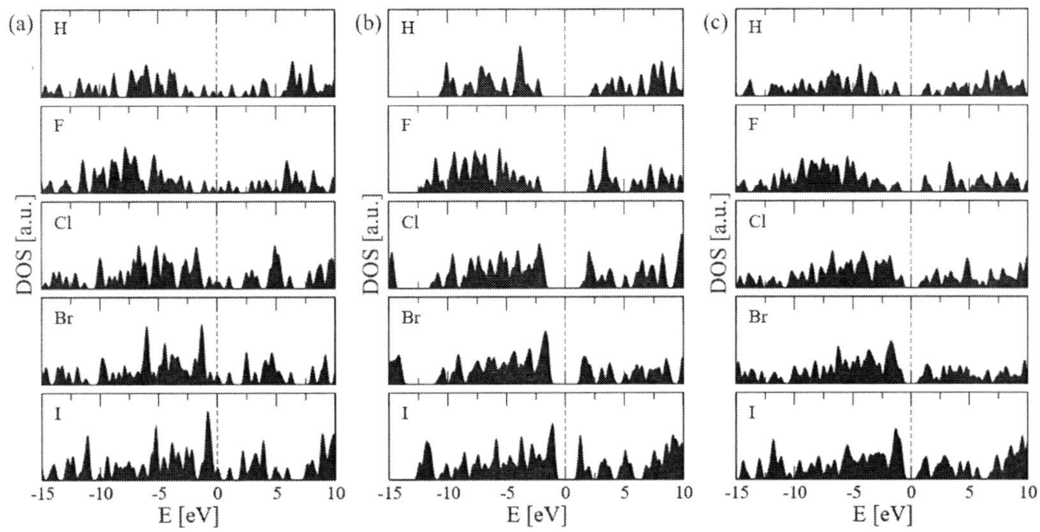

Figure 2. The density of states of graphene (a), hBN (b) and G-hBN (c) nanoribbons, for different types of passivation. The dashed vertical lines mark the Fermi energy.

3. Results and discussion

The structures are first relaxed until the pre-set equilibrium condition is reached. There is a significant structural difference between the systems passivated with H and F, compared to the other three. The H/F passivated nanoribbons retain planar configurations while in the other cases the heavier halogen atoms (Cl, Br, I) are shifted out of the nanoribbon plane. The minimum energy configurations are obtained for alternating passivating atoms with respect to the nanoribbon plane. There are two possibilities of alternating the atoms on the two edges: in-phase and out-of-phase. We find that the in-phase structures, in which the passivating atoms (Cl, Br, I) at about the same x coordinate are on the same side of the nanoribbon plane, have lower total energies. For this reason we shall analyze only such structures in the rest of the paper. The structural data for equilibrium configurations, represented by C-H/X, B-H/X and N-H/X bonding lengths, is listed in Table 1. After relaxation, the lengths of the nanoribbon segments are about 20 Å for all structures.

Table 1. Structural data – typical bonding lengths [Å].

	H	F	Cl	Br	I
C	1.11	1.32	1.71	1.88	2.07
B	1.02	1.31	1.75	1.89	2.13
N	1.22	1.37	1.71	1.91	2.06

Analyzing the density of states (DOS) of the considered structures, we observe a systematic behavior of the band gap energy. First, the graphene nanoribbons indicate metallic behavior,

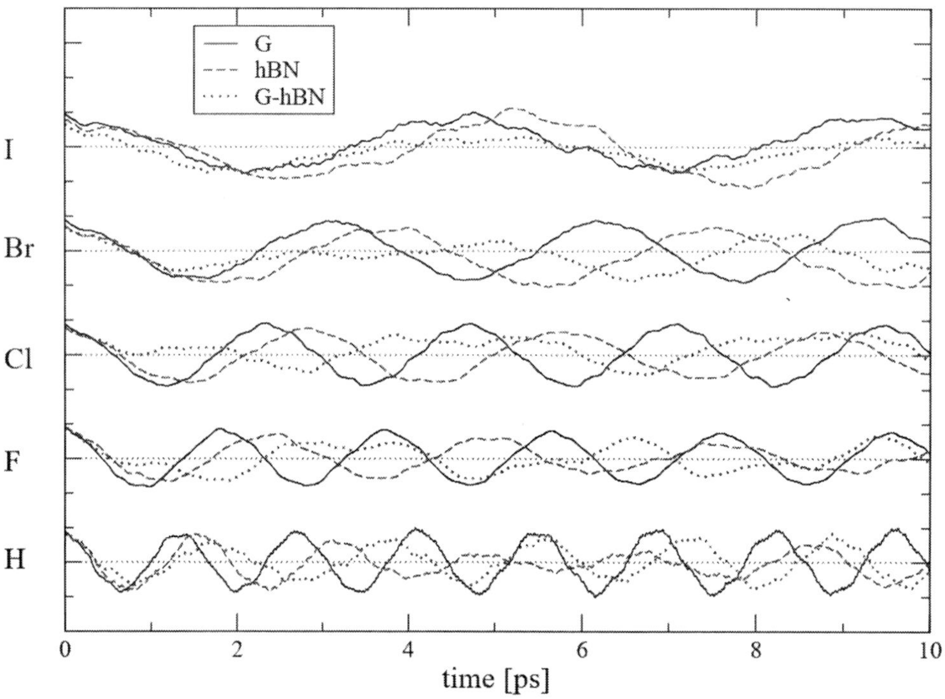

Figure 3. Average out-of-plane (z) displacement vs. time of the nanoribbon atoms: G (solid lines), hBN (dashed lines) and G-hBN (dotted lines). The spacing between two consecutive dotted lines is 1.5 Å.

independent on the passivation type. The DOS just below the Fermi energy typically increases with the halogen atomic number. Second, the hBN nanoribbons present a wide band gap of about 4 eV for H-passivation. However, the gap decreases monotonically with increasing mass of the halogen atom, reaching a minimum around 1.6 eV, corresponding to iodine passivation. Third, the mixed G-hBN systems have correspondingly smaller gaps than the one observed in hBN systems, with values ranging from around 2 eV (H-passivation) to about 1.1 eV (I-passivation). Moreover, taking into account that H- and F-passivated systems are planar, while the equilibrium configurations of the other three have out-of-plane halogen atoms, one may also see a resemblance in the DOS spectra within the two groups of systems, (H,F) and (Cl,Br,I), especially around the band gaps.

Next, the dynamical properties are investigated. After the equilibration, the nanoribbons are stretched out of equilibrium by displacing the atoms in the z direction by a distance proportional to their distance to the nearest end. In this way, the maximum displacement of 1 Å takes place in the middle, whereas the atoms at the extremities of the nanoribbon in the x direction are not moved. After stretching, the constraints are removed and the nanoribbon starts to oscillate. The natural frequencies of graphene nanoribbons range from 0.74 THz (H-passivation) down to 0.22 THz (I-passivation). A similar trend is found for hBN and G-hBN structures. Comparing hBN with graphene systems with the same passivation, the oscillating period increases systematically, with frequencies in the range 0.19 - 0.63 THz, which is consistent with other studies where it was found that h-BN sheet is a less stiff material as compared to graphene [18]. The less homogeneous G-hBN system enables more complex oscillating modes.

The atomic motion is further analyzed using the averaged squared displacement $\langle (x - x_0)^2 \rangle$

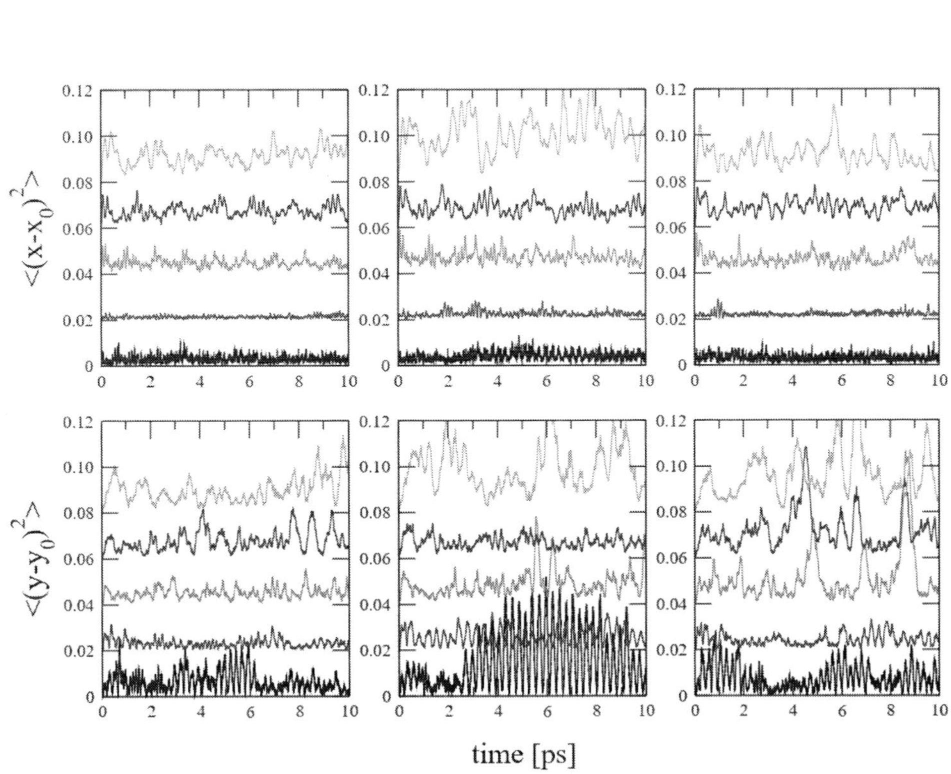

Figure 4. Averages of squared displacements [Å^2] along x- and y coordinates, with respect to equilibrium positions, for G, hBN, and G-hBN from left to right. The data sets represent H, F, Cl, Br and I passivated systems from bottom to top in each graph, each graph being shifted vertically by 0.02 Å^2 for clarity.

and $\langle (y - y_0)^2 \rangle$ positions with respect to the equilibrium positions, x_0 and y_0. The average is taken over all atoms at each time step. Figure 4 shows that all three systems (G, hBN, G-hBN) have roughly the same behavior with respect to changing the passivating atoms: the flourinated systems indicate the overall smallest average displacements, which can be explained by the planar, more stable configuration. The hydrogenated samples are also planar, however the much smaller mass of H introduces larger fluctuations. The other three halogenated structures (X=Cl,Br,I) are non-planar and they introduce larger fluctuations, which may potentially impact the transport properties and noise.

4. Conclusion

We investigated by *ab initio* DFT calculations the electronic and vibrational properties of graphene, hBN and G-hBN nanoribbons, with different types of passivation. By changing the halogen atoms, the physical properties may be tuned: the band gaps of hBN may be decreased by using halogen atoms with higher atomic number; the oscillating period is found to be in the low THz regime and it changes systematically with the mass of the border atoms; the higher stiffness of graphene, compared to hBN is evidenced by higher oscillation frequency. The type of the passivation proves to be essential for adjusting both electronic and vibrational properties and this may provide a route for optimizing mixed graphene - boron nitride nanostructures.

Acknowledgments

This work was supported by the National Authority for Scientific Research and Innovation (ANCSI) under grant PN16420202.

References

[1] Li Song, Lijie Ci, Hao Lu et al., Nanolett. 10, 3209 (2010).
[2] Yongji Gong, Gang Shi, Zhuhua Zhang et al., Nature Communications 5, 3193 (2014).
[3] L. Britnell, R. V. Gorbachev, A. K. Geim, L. A. Ponomarenko, A. Mishchenko, M. T. Greenaway, T. M. Fromhold, K. S. Novoselov, L. Eaves, Nature Commun. 4, 1794 (2013)
[4] G. Fiori, A. Betti, S. Bruzzone and G. Iannaccone, ACS Nano **6**, 2642 (2012)
[5] G. A. Nemnes, J. Nanomater. **2012**, 748639 (2012).
[6] G. A. Nemnes and S. Antohe, Mater. Sci. Eng. B **178**, 1347 (2013).
[7] Camelia Visan, Rom. Rep. Phys. **66**, 983 (2014).
[8] A. A. Nila, G. A. Nemnes, A. Manolescu, Rom. J. Phys. 60, 696 (2015).
[9] V. H. Nguyen, F. Mazzamuto, A. Bournel and P. Dollfus, J. Phys. D: Appl. Phys. **45**, 325104 (2012)
[10] K. Yang, Y. Chen, R. DAgosta, Y. Xie, J. Zhong, and A. Rubio, Phys. Rev. B **86**, 045425 (2012)
[11] Camelia Visan, J. Electron. Mater. **43**, 3470 (2014)
[12] Y. S. Liu, W. Q. Zhou, J. F. Feng, X. F. Wang, Chem. Phys. Lett. **625**, 14 (2015)
[13] S. I. Vishkayi, M. B. Tagani, H. R. Soleimani, J. Phys. D: Appl. Phys. **48**, 235304 (2015)
[14] G. A. Nemnes and Camelia Visan EPJ Web of Conferences **108**, 02045 (2016)
[15] J. van der Lit, M. P. Boneschanscher, D. Vanmaekelbergh, M. Ijäs, A. Uppstu, M. Ervasti, A. Harju, P. Liljeroth and I. Swart, Nat. Commun. 4, 2023 (2013)
[16] J. M. Soler, E. Artacho, J. D. Gale, A. Garca, J. Junquera, P. Ordejon, and D. Sanchez-Portal, J. Phys. Cond. Mater. **14**, 2745 (2002)
[17] D.M. Ceperley, B.J. Alder, Phys. Rev. Lett. **45**, 566 (1980)
[18] S. K. Singh, M. Neek-Amal, S. Costamagna, and F. M. Peeters, Phys. Rev. B **87**, 184106 (2013)

Two-body quantum propagation in arbitrary potentials

Federico Grasselli[1,2], Andrea Bertoni[2], and Guido Goldoni[1,2]
[1]Dipartimento di Scienze Fisiche, Informatiche e Matematiche, Università degli Studi di Modena e Reggio Emilia, Via Campi 213/a, Modena, Italy
[2]S3, Consiglio Nazionale delle Ricerche, Istituto Nanoscienze, Via Campi 213/a, Modena, Italy
E-mail: `federico.grasselli@unimore.it`

Abstract. We have implemented a unitary, numerically exact, Fourier split step method, based on a proper Suzuki-Trotter factorization of the quantum evolution operator, to propagate a two-body complex in arbitrary external potential landscapes taking into account exactly the internal structure. We have simulated spatially indirect Wannier-Mott excitons – optically excited electron-hole pairs with the two charges confined to different layers of a semiconductor heterostructure with prototypical 1D and 2D potentials emphasizing the effects of the internal dynamics and the insufficiency of mean-field methods in this context.

1. Introduction

When a composite object, say two attracting particles, is dynamically propagated, its internal structure cannot be *a priori* neglected, and energy might be transferred between the relative motion and the centre-of-mass (c.m.) degrees of freedom (DOFs) by an external one-body potential acting on the each particle separately. In the quantum regime, electron-hole correlations or, equivalently, c.m.–relative motion correlations must be carefully included even if the c.m. kinetic energy is not sufficient to excite any internal transition. In particular, mean-field approximations, neglecting dynamical correlations by construction, produce propagations which in general differ substantially from exact calculations.[1]

In this paper we discuss the exact unitary propagation in selected 1D and 2D configurations of a spatially indirect Wannier-Mott exciton (IX) in a GaAs-based semiconductor double-quantum-well (QW) heterostructure, i.e., a photogenerated electron-hole (e–h) pair in which the two charges are kept separated to different layers.

2. Theoretical model and quantum evolution algorithm

In the absence of external potential landscapes, the two-particle Hamiltonian reads

$$H_0 = -\frac{\hbar^2}{2M}\nabla^2_{\mathbf{R}} - \frac{\hbar^2}{2m}\nabla^2_{\mathbf{r}} + U_{\rm int}(\mathbf{r}) \tag{1}$$

where $\mathbf{R} \equiv (m_1\mathbf{r}_1 + m_2\mathbf{r}_2)/M$, and $\mathbf{r} \equiv \mathbf{r}_1 - \mathbf{r}_2$ are the c.m. and relative coordinates, respectively, $M \equiv m_1 + m_2$ and $m \equiv (m_1^{-1} + m_2^{-1})^{-1}$ being the total and reduced masses of the system, respectively. $U_{\rm int}(\mathbf{r})$ is the interaction attractive potential of the two particles,[1] giving a set of

[1] It is in principle not necessary the stricter dependence of $U_{\rm int}$ on the *modulus* of $\mathbf{r}$ in our propagation method.

Content from this work may be used under the terms of the Creative Commons Attribution 3.0 licence. Any further distribution of this work must maintain attribution to the author(s) and the title of the work, journal citation and DOI.
Published under licence by IOP Publishing Ltd

bound states $\phi_n(\mathbf{r})$, with eigenenergies ϵ_n, for the two-body internal dynamics. The c.m. and relative coordinate representation, which is separable in the absence of an external potential, is particularly convenient. First, a single-particle external potential, acting on the two particles at positions $\mathbf{r}_1, \mathbf{r}_2$, can be written as $U_{\text{ext}} = U_1(\mathbf{r}_1) + U_2(\mathbf{r}_2) = U_{\text{ext}}(\mathbf{R}, \mathbf{r})$, and in general breaks the separability of the exciton Hamiltonian: when $U_{\text{ext}} \neq 0$, we must deal with a general $(2 \times D)$-DOF problem, where D is the dimensionality. If, however, U_{ext} is not strong enough to dissociate the electron-hole pair, the relative coordinate $\mathbf{r}$ remains in a finite region about $\mathbf{r} = 0$. Therefore, for numerical propagation, the c.m. and relative coordinates representation is particularly convenient compared to formally equivalent $\mathbf{r}_1, \mathbf{r}_2$ representation.

In order to obtain the coherent quantum dynamics of the total wave function of the two-body complex, $\Psi = \Psi(\mathbf{R}, \mathbf{r}; t)$, we have to solve the time-dependent Schrödinger equation (TDSE). This can be done numerically by repeated application of the evolution operator, $\mathcal{U}(t, t + \Delta_t)$, between times t and $t + \Delta_t$, as

$$\Psi(\mathbf{R}, \mathbf{r}; t + \Delta_t) = \mathcal{U}(t + \Delta_t; t)\Psi(\mathbf{R}, \mathbf{r}; t). \tag{2}$$

We adopt the following Suzuki-Trotter factorization for $\mathcal{U}$:[2]

$$\mathcal{U}(t + \Delta_t; t) = e^{-\frac{i}{\hbar}U_{\text{tot}}\frac{\Delta_t}{2}} e^{-\frac{i}{\hbar}T\Delta_t} e^{-\frac{i}{\hbar}U_{\text{tot}}\frac{\Delta_t}{2}} + O(\Delta_t^3). \tag{3}$$

Here $T = \mathbf{P}^2/(2M) + \mathbf{p}^2/(2m)$ is the total kinetic energy operator, where $\mathbf{P}$ and $\mathbf{p}$ are c.m. and relative momenta, respectively, and $U_{\text{tot}} = U_{\text{int}}(\mathbf{r}) + U_{\text{ext}}(\mathbf{R}, \mathbf{r})$ is the total potential energy operator. The evolution operator is thus the product of terms, each diagonal either in direct or in reciprocal space (see Appendix in Ref. [3] for details). Therefore, by switching from position to momentum representation, and *vice versa*, through Fourier transformation, the application of $\mathcal{U}$ on Ψ reduces to simple array-by-array multiplication. This, and the use of fast Fourier transform algorithms, results in higher computational efficiency compared to other TDSE solvers, like those based on finite difference discretized Hamiltonian (e.g. the Crank-Nicolson method). This method is called the Fourier split step approach.[4, 5]. It is unitary, numerically exact as $\Delta_t \to 0$, and straightforwardly applicable to time dependent potentials. Also, it can be parallelized for massively parallel architectures.

Below we discuss the exact propagation of an indirect IX under different external potentials in 1D,[3] and 2D.[1] In all cases, we chose the initial state of the simulation as the product $\Psi(\mathbf{R}, \mathbf{r}; t = 0)$ of a c.m. Gaussian of width σ, centred at $\mathbf{R}_0$, and propagating with central wave vector $\mathbf{K}_0$, $\chi_G(\mathbf{R}) \sim \exp(i\mathbf{K}_0 \cdot \mathbf{R}) \exp[-(\mathbf{R} - \mathbf{R}_0)^2/4\sigma^2]$ and the ground state of the relative motion Hamiltonian, $\phi_0(r)$. Note, however, that no separability is assumed during the propagation.

3. 1D potential ramp

Below we consider specifically the cases of a 1D potential ramp, $U_{\text{ext}} = U_e(r_e) + U_h(r_h)$, where $U_{e(h)}(r_{e(h)}) = \mp e z_{e(h)}[(V_a - V_g) + (V_b - V_a)r_{e(h)}/(sL)]$, generally used to accelerate the IXs along the QW plane in devices [see Fig. 1(a)–inset].[6] This is somehow a special case, in that the c.m. and relative coordinates are separable, and $U_{\text{ext}} = U_{\text{bias}} + U^{\text{acc}}_{\text{c.m.}}(R) + U^{\text{acc}}_{\text{rel}}(r)$, where $U_{\text{bias}} = ed(V_a - V_g)$ is a constant bias energy, which does not affect the dynamics, and $U^{\text{acc}}_{\text{c.m.}}(R) = ed[(V_b - U_a)/(sL)]R$ is the c.m. accelerating potential. In a classical picture, $U^{\text{acc}}_{\text{c.m.}}(R)$ accelerates the electron-hole dipole according to $M\ddot{R} = -\partial_R U^{\text{acc}}_{\text{c.m.}}(R) = -ed[(V_b - V_a)/(sL)]$. Therefore, this case may serve as a test-bed of our calculations, since the classical and quantum dynamics of the c.m., given by $\partial^2_{tt}[\langle R\rangle(t)]$, should coincide according to Ehrenfest's theorem. The correspondence is shown in Fig. 1(a).

The potential $U^{\text{acc}}_{\text{rel}}(r) = -e(z_e + d m_e/M)[(V_b - V_a)/(sL)]\, r$ induces excitations of the internal degree of freedom and oscillations around $r = 0$, which can be quantitatively analyzed in terms of c.m.-averaged projections on the ϕ_n's, $a_n(t) = \int \left| \int \phi_n^*(r)\Psi(R, r; t)dr \right|^2 dR$. Figure 1(b) shows the transfer of energy between the ground and first excited state, with minor contributions from other states. Pictorially, this corresponds to oscillations of the internal wave packet around $r = 0$. For external potentials which are sufficiently strong with respect to the Coulomb binding energy of the e–h pair, direct, or tunneling-mediated dissociation is also possible [see Fig. 1(c)].[2]

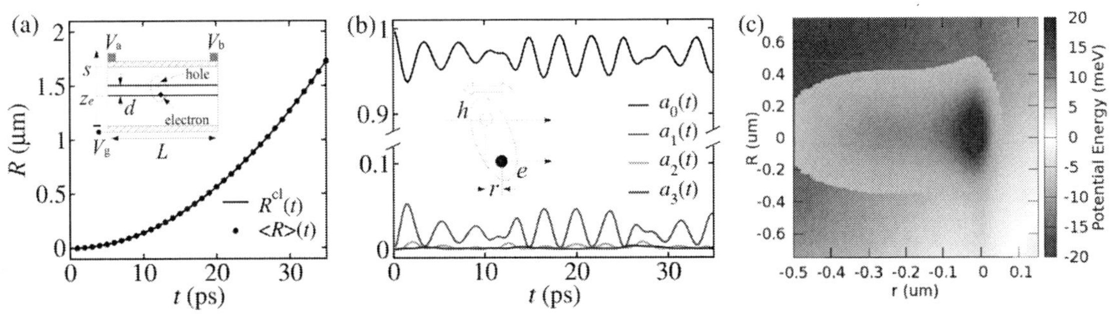

Figure 1. Propagation in a 1D potential ramp. (a) c.m. accelerated motion: classical (line) vs. Ehrenfest's (dots); inset: sketch of the physical system. (b) eigenstate population, showing the internal oscillating dynamics; σ = 150 nm. d = 20 nm, $V_a - V_b$ = 0.5 V. (c) snapshot of $|\Psi(R, r; t = 3\,\text{ps})|^2$ (green) together with total potential energy, $U_{\text{int}} + U_{\text{ext}}$, $U_{\text{int}}(r) = -1/(\varepsilon\sqrt{r^2 + d^2})$: example of dissociation in strong $V_a - V_b = 1.5$ V.

4. 2D potentials

We extend to 2D quantum wells the previous 1D model, and include non separable $U_{\text{ext}}(\mathbf{R}, \mathbf{r})$. Moreover, in order to have a more quantitative insight about the purely c.m. and internal dynamical correlations, and their effect on the IX propagation, we compare exact calculations with the so-called rigid exciton (RIX) model, in which the c.m. moves in the effective potential $U_{\text{eff}}(\mathbf{R}) = \int d\mathbf{r} U_{\text{ext}}(\mathbf{R}, \mathbf{r})|\phi_0(r)|^2$, where U_{ext} is averaged on the internal motion ground state $\phi_0(r)$, in which the IX is supposed to be frozen. In these rigid-particle calculations, only the c.m. wave function $\chi(\mathbf{R}; t)$ is propagated under $U_{\text{eff}}(\mathbf{R})$. On the contrary, in exact simulations, where *no specific form* on $\Psi(\mathbf{R}, \mathbf{r}; t)$ is assumed, the problem is a full 4-DOF-dynamics. Here, a very large number of grid points is needed ($\sim 2^{32}$ in typical 4-DOF-simulation), so it is essential to exploit massively parallel architectures.

To analyze results in 2D simulations we define the IX transmission $T = \int_{\mathbf{T}} d\mathbf{R} \rho_{\text{c.m.}}(\mathbf{R}; t \to \infty)$, i.e. the integral, over the region $\mathbf{T}$ beyond the external potential [see inset in Fig.2(a)], of the c.m. marginal probability $\rho_{\text{c.m.}}(\mathbf{R}; t) \equiv \int d\mathbf{r}|\Psi(\mathbf{R}, \mathbf{r}; t)|^2$ at asymptotic times. Note that in the RIX model $\rho_{\text{c.m}}(\mathbf{R})$ simply coincides with $|\chi(\mathbf{R}; t)|^2$.

We start considering the transmission of the IX wave packet in the scattering by a potential $U_{\text{ext}} = U_{e,0}[\theta(y_e)\theta(L_y - y_e)]$, with $U_{e,0} < 0$, i.e., an energy well for the electron alone, while the hole propagation is free, apart from Coulomb interaction [see Fig. 2(a)–inset]. This potential cannot be separated into the sum of c.m. and relative motion terms. The RIX model captures the full calculation results only when $U_{e,0}$ is shallow, i.e. well below the first internal excitation energy, $(\epsilon_1 - \epsilon_0) \approx 2.35$ meV.

As a second paradigmatic example, we simulate the IX scattering against a slit opened in a potential consisting of a barrier for the hole of height $U_{h,0}$ = +6 meV and an electron well of depth $U_{e,0}$ = −3 meV. As shown in Fig. 2(b–c), the shape of the exact diffraction patterns in transmission/reflection is not well reproduced by the RIX method. Moreover, the exact

[2] We use GaAs well in-plane effective masses $m_e = 0.067\,m_0$, $m_h = 0.111\,m_0$, and dielectric constant $\varepsilon = 12.9$.

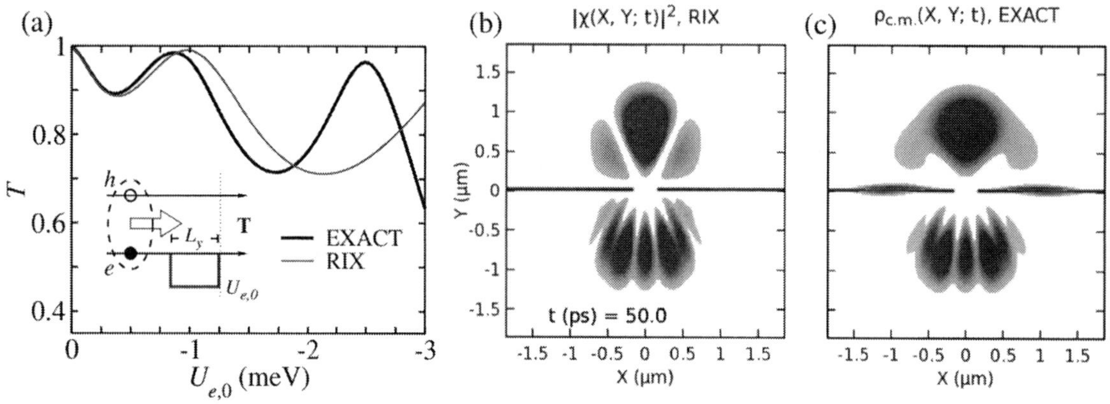

Figure 2. Results from 2D calculations, with initial c.m. kinetic energy of 0.4 meV. (a) Transmission T vs. electron well depth, $U_{e,0}$: (black) exact 4 DOF propagation; (red) RIX approximation. (a)–inset: sketch of the scattering; $\sigma = 80$ nm. $L_y = 40$ nm. (b-c) Scattering through a 240 nm-wide slit: snapshot at $t = 50$ ps in RIX and exact models (see text for details); $\sigma = 160$ nm. $L_y = 20$ nm.

calculation exposes that a significant part of the IX wave packet moves along the edges of the external potential, a behaviour which can be explained as the Coulomb bound state of the electron, captured by its well, and the hole, confined outside its potential barrier. Being a purely two-particle dynamical process, this behaviour is not captured by RIX simulations.

5. Conclusions

We showed that both c.m. and internal quantum dynamics of a bound system must be taken into account in a wide variety of collisional phenomena. The rigid-particle approximation cannot account for most of pure two-body behaviours (excitations and dissociation of the bound state, diffraction patterns, etc.), which arise in real-time-and-space exact quantum propagations. Even more refined mean-field approximations allowing internal dynamics, but keeping the total wave function factorized, neglect – by construction – the correlation between c.m. and internal DOFs, and prove to fail in reproducing correct results.[1] Two-body phenomena quantitatively dictate transmission/reflection coefficients, and might be of fundamental importance in quantum device engineering.

6. Acknowledgments

We acknowledge the CINECA award under Iscra C project IsC33-FUQUDIX for computing time on parallel architectures.

References

[1] F Grasselli, A Bertoni, and G Goldoni 2016 *Phys. Rev. B* **93** 195310
[2] M Suzuki 1993 *Proceedings of the Japan Academy, Series B***69** 161
[3] F Grasselli, A Bertoni, and G Goldoni 2015 *J. Chem. Phys* **142** 034701
[4] C Leforestier, R Bisseling, C Cerjan, M Feit, R Friesner, A Guldberg, A Hammerich, G Jolicard, W Karrlein, H-D Meyer, N Lipkin, O Roncero, and R Kosloff 1991 *J. Comput. Phys.* **94** 59
[5] A Castro, M A L Marques, and A Rubio 2004 *J. Chem. Phys* **121** 3425
[6] A Gärtner, A W Holleitner, J P Kotthaus, D Schuh 2006 *Appl. Phys. Lett.* **89** 052108

Modified stochastic variational approach to non-Hermitian quantum systems

Daniel Kraft and Willibald Plessas
Theoretical Physics, Institute of Physics, University of Graz, Universitätsplatz 5, A-8010 Graz, Austria

E-mail: plessas@uni-graz.at

Abstract. The stochastic variational method has proven to be a very efficient and accurate tool to calculate especially bound states of quantum-mechanical few-body systems. It relies on the Rayleigh-Ritz variational principle for minimizing real eigenenergies of Hermitian Hamiltonians. From molecular to atomic, nuclear, and particle physics there is actually a great demand of describing also resonant states to a high degree of reliance. This is especially true with regard to hadron resonances, which have to be treated in a relativistic framework. So far standard methods of dealing with quantum chromodynamics have not yet succeeded in describing hadron resonances in a realistic manner. Resonant states can be handled by non-Hermitian quantum Hamiltonians. These states correspond to poles in the lower half of the unphysical sheet of the complex energy plane and are therefore intimately connected with complex eigenvalues. Consequently the Rayleigh-Ritz variational principle cannot be employed in the usual manner. We have studied alternative selection principles for the choice of test functions to treat resonances along the stochastic variational method. We have found that a stationarity principle for the complex energy eigenvalues provides a viable method for selecting test functions for resonant states in a constructive manner. We discuss several variants thereof and exemplify their practical efficiencies.

1. Introduction

Usually the theory of quantum systems is formulated in a Hilbert space and is based on Hermitian operators for observables [1]. Most calculational methods for solving quantum-mechanical problems are adapted to such a framework. This is particularly true for variational techniques treating bound states. They are based on the fundamental Rayleigh-Ritz theorem for minimizing the energy expectation values by exploiting an ever larger subspace of the full Hilbert space. In particular, the stochastic variational method (SVM), as it is elaborated and outlined, for instance, in [2], is very flexible and well-suited for a wide range of quantum systems, especially because it allows to handle non-linear variational parameters occurring in a given problem (see also [3]).

Resonant quantum systems (with a finite number of exit/decay channels) elude such a conventional treatment. However, they occur in a wide range of different scales from molecules and atoms to nuclear and particle physics. Quantum-mechanical resonances fall outside the discrete and continuous (real) spectra of Hermitian Hamiltonians. They correspond to complex eigenvalues, with the real parts representing the *resonance energies* and the imaginary parts the *decay widths*. The resonance wave functions are not a-priori square-integrable, what makes them inapplicable for directly calculating observables.

Content from this work may be used under the terms of the Creative Commons Attribution 3.0 licence. Any further distribution of this work must maintain attribution to the author(s) and the title of the work, journal citation and DOI.
Published under licence by IOP Publishing Ltd

As a consequence, conventional variational techniques as characterized above cannot be applied, as they are adapted to eigenfunctions in square-integrable Hilbert spaces $L^2(\mathbb{R}^n)$. A common way out is to apply suitable similarity transformations (e.g., complex scaling), what changes the problem of locating the original resonance positions into finding eigenvalues of a transformed operator on a Hilbert space. This operator is, however, non-Hermitian. Basic strategies for dealing with non-Hermitian quantum-mechanical operators (Hamiltonians) are discussed in [4].

Unfortunately, after such (similarity) transformations are performed, standard variational methods are not directly applicable, since a minimization principle of eigenenergies, such as the Rayleigh-Ritz principle, is not adequate for complex eigenvalues. Rather an alternative selection principle for test functions has to be advocated. Instead of a Rayleigh-Ritz-type minimization a stationarity principle of complex eigenvalues can be designed, which can also deal with non-Hermitian Hamiltonians. Here we show that the SVM can be extended using a generalized selection criterion for test functions based on a stationarity principle. In this way it is made capable of treating also quantum-mechanical resonances in an efficient manner.

The findings presented in this paper are essentially based on results obtained in the master thesis [5], where also the mathematical foundation of the approach is given together with further details and additional examples.

2. Short characterization of resonances

In the conventional setting of quantum mechanics one considers a Hamiltonian

$$\hat{H} = -\frac{1}{2}\Delta + \gamma\hat{V}\,, \tag{1}$$

given on a square-integrable Hilbert space. It consists of the kinetic energy operator, here suitable normalized corresponding to the Laplacian Δ, and a potential operator $\hat{V}$ with a strength γ. The time evolution is produced by the operator $\hat{U}(t) = e^{-i\hat{H}t}$ (see, e.g., standard text books like [1], [6] or [7]). $\hat{U}(t)$ exists as a unitary operator, if $\hat{H}$ is time-independent and self-adjoint. The quantum-mechanical probability density is preserved and the energy eigenvalues are real. The bound states from the discrete spectrum of a given system (here, for simplicity assumed to be one-dimensional) follow from the eigenvalue equation

$$\hat{H}\psi(x) = E\psi(x) \tag{2}$$

and are represented by poles of the resolvent $\hat{G}(z) = (z - \hat{H})^{-1}$, with $z \in \mathbb{C}^1$, on the negative real axis of the physical sheet of the complex energy plane, whereas the scattering states lie on the branch cut at positive real energies (cf., e.g., [6]).

Resonances of the system, which can, for instance, be transmuted from bound states by reducing the potential strength γ, correspond to poles of the resolvent on the lower half-plane of the unphysical energy sheet and can be observed by sudden changes in the energy dependence of the (ℓ-th partial wave) scattering cross section. This resonance behaviour is usually parametrized by the Breit-Wigner formula

$$\sigma_\ell(E) = \frac{\left(\frac{\Gamma}{2}\right)^2}{(E - E_r)^2 + \left(\frac{\Gamma}{2}\right)^2}\,, \tag{3}$$

where the pole position is given by $\mathcal{E} = E_r - i\frac{\Gamma}{2}$, with E_r the so-called resonance energy and Γ the resonance decay width.

The poles of the resolvent or equivalently the scattering operator $S_\ell(z)$ are related to zeros of the Jost function $\mathscr{f}_\ell$, which represent the coefficients of the asymptotic parts of the regular solution $\phi_{\ell k}$ of the Schrödinger equation (2) for $r \to \infty$

$$\phi_{\ell k}(r) \to \frac{i}{2}\left[\mathcal{f}_\ell(k)\hat{h}_\ell^-(kr) - \mathcal{f}_\ell^*(k)\hat{h}_\ell^+(kr)\right] , \tag{4}$$

where $\hat{h}^\pm$ are the Riccati-Hankel functions. If we require outgoing boundary conditions and $\mathcal{f}_\ell(k)$=0 for a certain complex momentum k, i.e. a resonance pole at $k = k_r - ik_i$, where $k_r, k_i \in \mathbb{R}^1$ and $k_i > 0$, we get

$$\phi_{\ell k}(r) \to Ce^{ikr} = Ce^{ik_r r} \cdot e^{k_i r} \to \infty \tag{5}$$

as $r \to \infty$. This means $\phi_{\ell k} \notin L^2(\mathbb{R}^n)$ and therefore unsuitable for the calculation of observables.

3. Non-Hermitian Hamiltonians and c-product

3.1. Complex Hamiltonians

Asymptotically rising resonance wave functions (residues at complex poles, sometimes also termed Gamow functions) $\psi(x)$ can be regularized by similarity transformations (see [4]):

$$\psi_S = \hat{S}\psi, \quad \hat{H}_S = \hat{S}\hat{H}\hat{S}^{-1}. \tag{6}$$

The most important property of such transformations is that the eigenvalue spectrum of the original Hamiltonian remains invariant with the respective eigenstates:

$$\hat{H}\psi = \lambda\psi\,, \qquad \hat{H}_S\psi_S = \lambda\psi_S, \qquad \lambda \in \mathbb{C}^1\,. \tag{7}$$

Most frequently used is the complex-scaling method, which consists in rotating the space variable by a certain angle θ

$$(\hat{S}_\theta\psi)(x) = \psi_\theta(x) = \tilde{\psi}\left(e^{i\theta}x\right), \tag{8}$$

where $\tilde{\psi}$ is understood as the analytic continuation of ψ onto a suitable domain of rotated coordinates. This transformation makes ψ_θ asymptotically well-behaved for a desired range of momenta k

$$(\hat{S}_\theta\psi)(x \to \infty) \sim \exp\left(ie^{i\theta}kx\right) = e^{ik'x}, \tag{9}$$

where $k' = e^{i\theta}k$ and thus $\operatorname{Im} k' > 0$. Evidently, the complex-scaling transformation turns the Hamiltonian complex

$$\hat{H}_\theta = -\frac{1}{2}e^{-2i\theta}\Delta + \gamma\hat{V}\left(e^{i\theta}x\right) , \tag{10}$$

where, of course, analyticity of the potential $\hat{V}$ is required in the required domain.

Complex Hamiltonians may arise in a number of further cases related to resonances. For instance, a coupled-channels theory involving resonance decay channels produces a complex optical potential after a Feshbach reduction (see, e.g., [8]. For energies above the resonance threshold the corresponding reduced Hamiltonian will thus acquire an imaginary part and lead to complex eigenvalues relating to a resonance at $\mathcal{E} = E_r - i\frac{\Gamma}{2}$.

Standard variational methods adhering to the Rayleigh-Ritz principle cannot be applied to solving the eigenvalue problem of a complex Hamiltonian like in Eq. (10) or similar. A possible way out is a stationarity principle for finding the right position of the resonance poles. A prerequisite for formulating such a stationarity principle is the definition of an alternative to the usual scalar product on a Hilbert space.

3.2. c-product

The introduction of the c-product $(\cdot|\cdot)$ provides a substitution for the common scalar product $\langle\cdot|\cdot\rangle$ in case of non-Hermitian Hamiltonians. It is defined through the mapping $L^2(\mathbb{R}^n)\times L^2(\mathbb{R}^n)\to\mathbb{C}^1$ by

$$(\phi|\psi) = \int_{\mathbb{R}^n} \phi(x)\psi(x)\, dx. \tag{11}$$

With the c-product many of the usual properties of the inner product are carried over to the non-Hermitian case. Suppose that ϕ_S and ψ_S are similarity-transformed eigenstates of $\hat{H}_S$, then in particular

- $|(\phi|\psi)| \leq ||\phi||\,||\psi||, \ \ \forall\ \phi_S, \psi_S \in L^2(\mathbb{R}^n)$, i.e. the Cauchy-Schwartz inequality holds, and
- $\left(\phi_S|\hat{H}_S\psi_S\right) = \left(\hat{H}_S\phi_S|\psi_S\right) := \left(\phi_S|\hat{H}_S|\psi_S\right)$, i.e. $\hat{H}_S$ is "Hermitian" with respect to the c-product.

Similarly, the orthogonality property for eigenstates corresponding to different eigenvalues is guaranteed by the c-product.

4. Stationarity principle and the SVM

We are now prepared to formulate the stationarity principle for the selection of basis functions in a variational approach to non-Hermitian problems and to apply it especially in the SVM.

4.1. Stationarity principle

Given the similarity-transformed eigenvalue problem as in Eq. (7), we consider the Rayleigh quotient

$$R(\phi_S) = \frac{\left(\phi_S|\hat{H}_S|\phi_S\right)}{(\phi_S|\phi_S)}\,. \tag{12}$$

Then for some arbitrarily varied state $\phi_S = \psi_S + \epsilon\delta\psi_S$, with $\epsilon \to 0$, one gets

$$R(\phi_S) = \lambda + O(\epsilon^2)\,, \tag{13}$$

i.e. ψ_S is a critical point of R, where the critical value is the exact eigenvalue of $\hat{H}_S$.

In the context of a variational calculation searching for the optimal solution of a non-Hermitian eigenvalue problem, we consider $\psi_S(\boldsymbol{\alpha})$ as a function of a set of (non-linear) parameters $\boldsymbol{\alpha} = (\alpha_1, \alpha_2, ...\alpha_r) \in \mathbb{C}^r$. If for a particular set $\boldsymbol{\alpha}^0$ the eigenvalue equation is satisfied and $(\psi_S(\boldsymbol{\alpha}^0)|\psi_S(\boldsymbol{\alpha}^0)) \neq 0$, then

$$\frac{\partial R(\boldsymbol{\alpha}^0)}{\partial \alpha^i} = 0\,, \quad \forall \alpha_i\,. \tag{14}$$

This means that we have to look for a stationary energy eigenvalue in order to find the optimal (non-linear) variational parameters.

The linear variational parameters a_k of an expansion into test functions χ_k

$$\psi_S = \sum_{k=1}^{\infty} a_k\chi_k \approx \sum_{k=1}^{N} a_k\chi_k \tag{15}$$

can be determined as usual by the solution of the following system of equations:

$$\sum_{l=1}^{N} \left(\chi_k|\hat{H}_S|\chi_l\right) a_l = \lambda \sum_{l=1}^{N} (\chi_k|\chi_l)\, a_l\,. \tag{16}$$

4.2. Modified SVM

The stationarity principle outlined above can directly be applied for the random selection of basis functions in the SVM for non-Hermitian problems. Several variants are still possible. We have checked the performance of different selection criteria for the case of the Rittby potential (see Fig. 1)

$$V(x) = \left(\frac{1}{2}x^2 - J\right) e^{-\beta x^2} \tag{17}$$

with parameters $J = 0.8$ and $\beta = 0.1$, such that exactly one bound state is supported, shown by the (red) level on the l.h.s. of Fig. 1. The exact solutions for its resonances are known from the literature [9] or can be generated by alternative methods, such as the finite-difference discretisation after complex scaling as in [5].

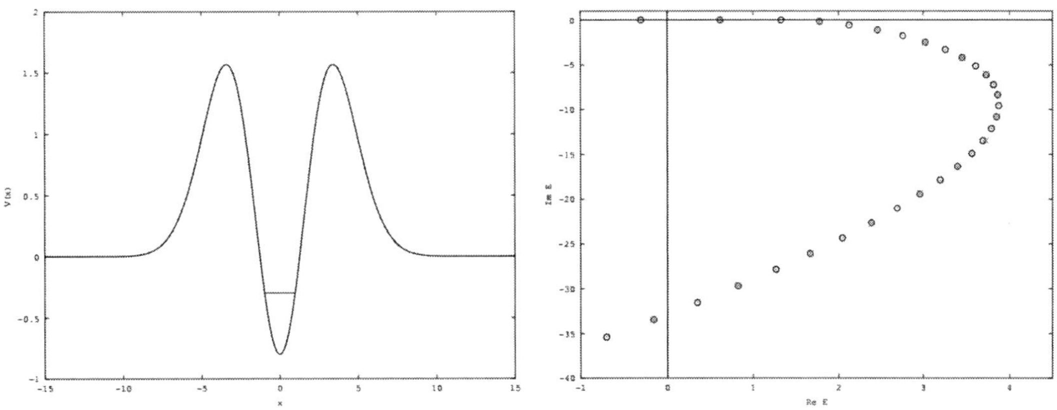

Figure 1. Spatial dependence of the Rittby potential (left) and its eigenvalues, as calculated with the finite-difference method (right). The (green) crosses are the results from [9] available for odd-n eigenvalues.

Application of the stationarity principle to the SVM with random selection of basis functions yields the results shown in Fig. 2. We have also tried additional criteria for the selection of test functions in order to explore speeding up the convergence. Among them were notably (for details see [5])

- the gradient method, i.e. the selection of test functions based on the minimization of the norm of the multi-dimensional gradient in the space of non-linear parameters $\nabla R^T = \left(\frac{\partial R}{\partial \alpha_1} \ldots \frac{\partial R}{\partial \alpha_r}\right)$, and
- a weight-based update strategy together with the orthonormalisation of the basis (ONB), where the least-relevant test functions are always omitted before generating new ones.

The rates of convergence for the three selection criteria described above are shown in Fig. 3. Again we exemplify them for the particular case of the n=3 resonance of the Rittby potential, whose exact position is at $\mathcal{E}_3 = 1.7846 - i0.1738$. As a measure we use the absolute values $|\mathcal{E}_3 - \mathcal{E}_3^{\text{approx}}|$, where $\mathcal{E}_3, \mathcal{E}_3^{\text{approx}} \in \mathbb{C}^1$, the latter being the approximate eigenvalue in each step.

It appears that improvements over a pure random selection of basis functions can be achieved by the more refined selection criteria outlined above. Especially the weight-based method with ONB immediately improves the approximation already with moderate basis sizes. The best accuracy of about 10^{-7}, which is probably sufficient for most applications, is reached for $N \approx 70$ test functions. For even bigger bases this precision is lost again, presumably due to the fact

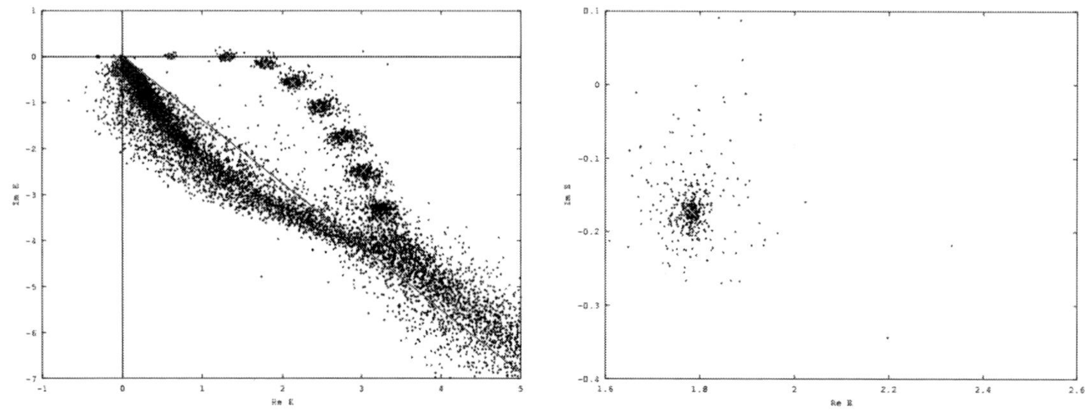

Figure 2. Left: Approximate eigenvalues (black dots) from stochastically selecting $N = 40$ random basis functions and varying only the linear parameters, according to the stationarity principle for the Rittby potential (17); 500 runs are combined. The (green) crosses give the exact resonance positions. Right: Zoomed distribution around the n=3 resonance.

that in this case the eigenvalue problem becomes badly conditioned. The gradient method offers itself as a straightforward and efficient tool for generating a reliable basis. It outperforms the random stochastic selection starting from a basis size of $N \approx 40$ and reaches its highest accuracy of $\leq 10^{-7}$ beyond $N \approx 80$.

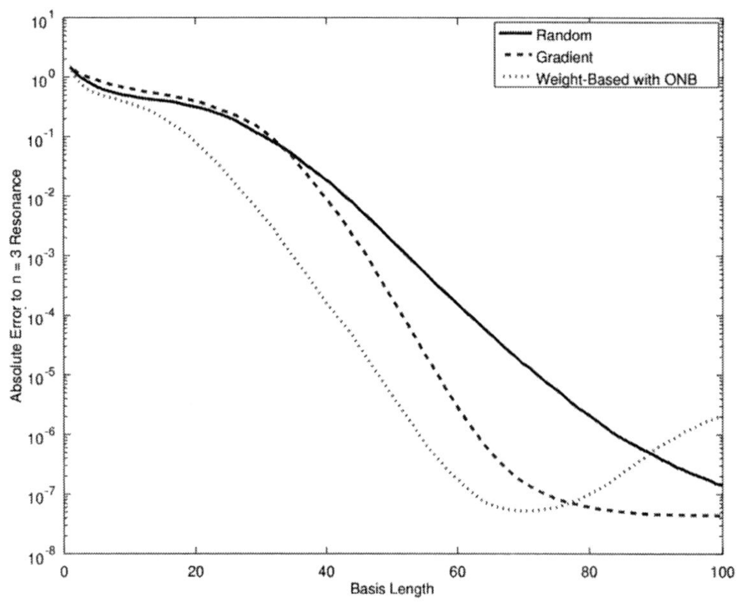

Figure 3. Convergence of the absolute error in the resonance position in case of the n=3 resonance of the Rittby potential (17) for different selection criteria of the basis (see the text): Random (solid/black), gradient (dashed/blue) and weight-based with orthonormalisation of the basis (dotted/red).

5. Conclusions

In this paper our main goal has been especially the adaptation of the SVM for the treatment of quantum-mechanical resonances associated with complex eigenvalues (i.e. poles in the lower half-plane of the unphysical sheet of the complex energy plane). They are governed by non-Hermitian Hamiltonians, which arise, for instance, from the use of complex optical potentials or after similarity transformations, such as the Zel'dovich or complex-scaling transformations. Since the Rayleigh-Ritz variational principle is not applicable in such cases, we advocated a stationarity principle for the energy expectation value. It has proven suitable for optimizing the linear and non-linear variational parameters within the SVM approach.

Along with the generalisation of the variational principle, we also studied refined criteria for selecting good candidate basis functions in the SVM. In addition to a purely stochastic selection of basis functions in the search for the non-linear variational parameters, we examined a gradient-based search as well as a weight-based method, replacing basis functions with small expansion coefficients by more relevant ones. Thereby we arrived at two particular variants of the SVM that can be applied to non-Hermitian problems in an efficient manner. They turned out to perform significantly better than a purely random basis selection or even a straight-forward finite-difference discretisation of the Hamiltonian.

References

[1] Prugovečki E 1981 *Quantum Mechanics in Hilbert Space* 2nd ed. (New York: Academic Press)
[2] Suzuki Y and Varga K 1998 *Stochastic Variational Approach to Quantum-Mechanical Few-Body Problems* (Berlin: Springer)
[3] Varga K 2016 *Quantum simulations of interaction of electromagnetic fields and nanostructures* (Invited Talk at IC-MSQUARE 2016, Athens)
[4] Moiseyev N 2011 *Non-Hermitian Quantum Mechanics* (Cambridge: Cambridge University Press)
[5] Kraft D 2013 *Stochastic Variational Approaches to Non-Hermitian Quantum-Mechanical Problems* (Graz: Graz University Master Thesis)
[6] Taylor J R 1972 *Scattering Theory: The Quantum Theory on Nonrelativistic Collisions* (New York: John Wiley & Sons)
[7] Sakurai J J 1994 *Modern Quantum Mechanics* revised ed. (Reading: Addison-Wesley)
[8] Schmidt R A, Canton L, Plessas W and Schweiger W 2016 *Towards a realistic description of hadron resonances* (Contribution to IC-MSQUARE 2016, Athens)
[9] Korsch H J, Laurent H and Möhlenkamp R 1982 *Phys. Rev.* A **26** 1802

Towards Noncommutative Topological Quantum Field Theory: New invariants for 3-manifolds

Zois, I.P.

PPC, Research Centre, 9, Leontariou Street, Pallini, Attika, Greece

i.zois@dei.com.gr and i.zois@exeter.oxon.org

Abstract. We present some ideas for a possible Noncommutative Topological Quantum Field Theory (NCTQFT for short) and Noncommutative Floer Homology (NCFH for short). Our motivation is two-fold and it comes both from physics and mathematics: On the one hand we argue that NCTQFT is the correct mathematical framework for a quantum field theory of all known interactions in nature (including gravity). On the other hand we hope that a possible NCFH will apply to practically every 3-manifold (and not only to homology 3-spheres as ordinary Floer Homology currently does). The two motivations are closely related since, at least in the commutative case, Floer Homology Groups constitute the space of quantum observables of (3+1)-dim Topological Quantum Field Theory. Towards this goal we define some new invariants for 3-manifolds using the space of taut codim-1 foliations modulo coarse isotopy along with various techniques from noncommutative geometry.

1. NCG as the underlying geometry for physical theories

There is accumulating evidence in various space-time dimensions that NCTQFT should provide a framework general enough to incorporate all 4 known interactions in nature.

Case I (4 dimensions): From the influential article due to G. 't Hooft on dimensional reduction in quantum gravity [1] where the holography principle was introduced, we know that quantum gravity should be a topological quantum field theory.

The argument is quite elegant: Given a (pseudo) Riemannian 4-manifold (M,g) we adopt the standard approach that the dynamical variable of the theory is the metric g and quantum field theory corresponds to computing the partition function. Thus one has to perform the path integral

$$Z(M) \sim \int Dg \exp(iS_M) \quad (1)$$

where S_M denotes the Einstein -Hilbert action on M

$$S_M = (1/2\kappa)\int_M R \quad (2)$$

with R the scalar curvature and in (1) the path integral is meant to integrate over all metrics g on the manifold M. If one were able to perform this path integral over all metrics, then the result should depend only on the topological data of the initial Riemannian manifold M. The first experimental evidence for holography (dimensional reduction in quantum gravity) was provided by the GEO 600 experiment. This is an interferometric gravitational wave detector outside Hannover in Germany. C.J. Hogan in a series of articles [2] gave an explanation of the mysterious (300- 1400)Hz "noise" as arising from non-locality mechanisms due to holography.

Concerning the other 3 interactions (strong, weak and electromagnetic) we know from Connes' et al. work [3] that they can be incorporated in a noncommutative (nc) space arising as the Cartesian

Content from this work may be used under the terms of the Creative Commons Attribution 3.0 licence. Any further distribution of this work must maintain attribution to the author(s) and the title of the work, journal citation and DOI.
Published under licence by IOP Publishing Ltd

product of a 4-dim spin Riemannian manifold times a discrete space of metric dim 0 and KO-dim 6 mod 8. The nc space has a corresponding algebra of the form

$$\mathbf{C} + \mathbf{H} + M_3(\mathbf{C})$$

where **C** denotes complex numbers, **H** denotes the quaternions and $M_3(\mathbf{C})$ denotes 3X3 complex matrices.

[Aside Note 1: The total algebra becomes $M_2(\mathbf{H}) + M_4(\mathbf{C})$ if we incorporate the higher dimensional analogue of the Heisenberg commutation relations and the phenomenologically more accurate Pati-Salam gauge symmetry, see [15]].

Hence if we combine the two, gravity plus gauge theories of Yang-Mills type (for the electroweak and strong interactions) we believe that NCTQFT should provide the correct framework to describe all 4 known interactions in nature.

Case II (higher dimensions, string/M-Theory): We know that nc spaces arise both as extra (nc) toroidal compactifications of matrix models (Connes-Douglas-Schwarcz [4]) and from open strings when a gauge B-field is turned on (Seiberg-Witten [5]). In both cases, assuming space-time has either 4 dimensions or higher, gauge fields present are the origins of noncommutativity of the underlying geometry.

2. Invariants for 3-manifolds emerging from flat connections

According to the Atiyah definition of topological quantum field theory [6], one starts with a manifold with boundary and the correlation functions of the theory take values in some vector spaces associated to the boundary manifold. In the simplest case where the 4-manifold has no boundary, the correlators are just numerical (topological) invariants of the (bulk) manifold. Let us focus on dimension 4 which is the macroscopic space-time dimension and carries greater geometric interest. Obviously in this case the boundary of a 4-manifold say M, will be a 3-manifold, say N, namely $N = \partial M$. We fix SU(2) as our working Lie group. Given a 3-manifold N (assumed closed, oriented and connected) with fundamental group $\pi_1(N)$, the set

$$R(N) := \mathrm{Hom}\,(\pi_1(N), SU(2)) \,/\, \mathrm{ad}(SU(2))$$

consisting of equivalence classes of representations of the fundamental group $\pi_1(N)$ to SU(2) modulo conjugation *tends to be discrete*. For example if N is a (rational) homology 3-sphere, then R(N) has finite cardinality and the trivial representation is isolated.

There is a 1:1 correspondence between the elements of the set R(N) above and elements of the set

$$A(N) := \{\text{flat } SU(2)\text{-connections on } N\} \,/\, (\text{gauge equivalence}).$$

The bijection is the holonomy of the flat connection. The crucial observation is this: Although R(N) depends on the homotopy type of N, we can get topological invariants of N (namely invariants under homeomorphism) if we use the moduli space A(N). Depending on how one "decorates" or manipulates the (gauge classes of) flat SU(2)-connections in the set A(N) above, one can obtain the following:

a. The low energy limit of the Jones-Witten invariant for N (see [7]).

For any flat connection A (plus a Riemannian metric on N), one can form the twisted de Rham differential

$$d_A = \partial + A \wedge \quad (3)$$

twisted by the flat connection A; this is the exterior covariant derivative with respect to the connection A; it is a differential, namely its square equals zero due to the flatness of the connection. Then one can form the twisted Laplacian Δ_A in the standard way

$$\Delta_A = d_A d_A^* + d_A^* d_A \quad (4)$$

(using the metric to define first the Hodge star operator " * " and then the adjoint operator d_A* in the usual way)

$$d_A^* = (-1)^{kn+n+1} * d_A * \quad (5)$$

(acting on k-forms on an n-manifold) and then compute the *Ray-Singer torsion* T(N,A) of the flat connection A as follows:

$$\log[T(N,A)] = (1/2)\,\Sigma_{i=0}^{3}(-1)^i\, i\, \zeta'(\Delta_{i,A}) \quad (6)$$

where $\Delta_{i,A}$ denotes the twisted Laplacian acting on i-forms and

$$\zeta'(\Delta_{i,A})(0) = (-d/ds)\,\zeta(\Delta_{i,A})|_{s=0} = \log D(\Delta_{i,A}) \quad (7)$$

is the ζ-function regularised determinant of the twisted Laplace operator. Finally we add all Ray-Singer torsions (for all gauge classes of flat connections) and we get a topological invariant for the manifold N. Thus the finiteness of the set A(N) is crucial for the convergence (finiteness in fact) of the torsion sum.

[Aside Note 2: The Ray-Singer torsion is defined using the twisted Laplacian which involves the choice of a Riemannian metric; Under certain conditions-vanishing of twisted de Rham cohomology groups- it can be proved that the torsion is nonetheless independent of the metric].

Recall that the ζ-function of the Laplace operator $\zeta(\Delta_i)$ is by definition (for complex s) equal to

$$\zeta(\Delta_i) = \Sigma\lambda_n^{-s}$$

where we sum over all non-negative eigenvalues λ_n.

b. The Casson invariant.

Take again G = SU(2) and pick a Hegaard splitting on N. Then, assuming that the set R(N) is regular (namely that the 1st twisted de Rham cohomology groups of N vanish for all flat connections), each (gauge class of) flat connection in the set A(N) acquires an orientation and then we take the difference between the number of positively minus the number of negatively oriented flat connections. Although the orientation depends on the Hegaard splitting, the difference behaves like an index and it is independent of the Hegaard splitting of N. The Casson invariant is the above difference of the number of positively minus the number of negatively oriented gauge classes of flat SU(2)-connections. Again the finiteness of both sets R(N) and A(N) is crucial for the definition of the Casson invariant.

c. Floer homology

Take again G =S U(2) and consider the Cherns-Simmons 3-form as a Lagrangian density on N. The corresponding action functional

$$S = (k/4\pi)\int_N \mathrm{tr}[A \wedge dA + (2/3)\, A \wedge A \wedge A] \quad (8)$$

defines a Morse function on the space of irreducible connections. Its critical points are the flat connections.

Then each element of A(N) aquires a ``label" which is the Morse index of the critical point; in ordinary finite dimensional Morse theory this is equal to the number of negative eigenvalues of the Hessian. But the Hessian of the Chern-Simons function is unbounded below and we get infinity as Morse index for every critical point. So naive immitation of ordinary finite dimensional Morse theory techniques do not work.

Floer in [8] observed the following crucial fact: if we pick a Riemannian metric on N, then considering the non compact 4-manifold R X N along with its corresponding Riemannian metric, a continuous 1-parameter family of connections A_t on N corresponds to a unique connection **A** on R X N; then, choosing the axial gauge (0th component of the connection vanishes), the gradient flow equation for the Chern-Simons function S on N corresponds to the instanton equation on the non compact 4-manifold R X N.

$$\partial_t A_t = *F_{A_t} \Leftrightarrow F_A^+ = 0 \quad (9)$$

Then consider the linearised instanton equation

$$d_A a = 0,$$

where a is a small perturbation. This operator is not elliptic; we perturb it to

$$D_A = - d_A{}^* + d_A{}^+$$

to make it elliptic. Then the finite integer Morse index for each critical point comes as the relative (with respect to the trivial flat connection) Fredholm index of the perturbed elliptic operator D_A. In this way the moduli space A(N) aquires a **Z**/8 grading and then we follow ideas from ordinary finite dimensional Morse theory: We define the Floer-Morse complex using as generators the critical points and the ``differential" is essentially defined by the flow lines of the critical points. Taking the cohomology in the usual way we get the Floer homology groups of N. The Euler characteristic of the Floer-Morse complex equals twice the Casson invariant (see [9]). Again the finiteness of the set A(N) and R(N) is crucial in this construction.

All the above constructions depend crucially on the fact that the set A(N) (or equivalently the set R(N)) is finite. Unfortunately this is true for a rather small class of 3-manifolds (eg homology 3-spheres). It would be a big generalisation if we could find another moduli space which is finite for a larger class of 3-manifolds and if possibly for all 3-manifolds. The key observation is that in fact such a set exists: *It is the space of taut codim-1 foliations modulo coarse isotopy*. As proved by Gabai in [10], this set is finite for all 3-manifolds (closed, oriented and connected). The use of foliations has an additional advantage: it brings noncommutative geometric techniques on stage.

3. New invariants for 3-manifolds emerging from foliations

A codim-1 foliation F on a 3-manifold N is given by an codim-1 (hence dim 2) integrable subbundle F of the tangent bundle TN of N. Locally this can be defined by a 1-form ω satisfying

$$\omega \wedge d\omega = 0 \quad (10)$$

A foliation is called *topologically taut* if there exists a circle S^1 intersecting transversally all leaves. It is called *geometrically taut* if there exists a Riemannian metric for which all leaves are minimal surfaces (they have mean curvature zero). It is called *homologically taut* if there exists a closed 2-form β which is positive along the leaves. *These 3 different definitions of tautness are equivalent.* Two foliations are called *coarse isotopic* if up to isotopy of each one of them their oriented tangent planes differ pointwise by angles less than π. For any 3-manifold N (assumed closed, oriented and connected) we define the set G(N) as follows:

$$G(N) = \{\text{taut codim-1 foliations on N}\} \,/\, (\text{coarse isotopy})$$

and let us denote by g(N) the cardinality of the set G(N). Then Gabai in [10] proved that g(N) is finite for all 3-manifolds (closed, oriented and connected) and it is a topological invariant of the 3-manifold N. The crucial fact is that although coarse isotopy depends on the choice of a Riemannian metric (since it involves the notion of angle), the number g(N) does not.

Armed with Gabai's result, there is the *possibility to define new invariants* and constructions on N by using different "labels" on elements of the set G(N).

a. The Godbillon-Vey (GV for short) invariant of the manifold N

Each codim-1 foliation F on the 3-manifold N has a 3-dim characteristic class which is the *Godbillon-Vey class of the foliation*. This is defined as follows: The integrability condition (10) is equivalent to $d\omega = \omega\wedge\theta$ for another 1-form θ. Then $\theta\wedge d\theta$ is a 3-dim real cohomology class of N. We add all the GV classes for all elements in G(N). The result will be another 3-dim real cohomology class of N and then we can evaluate it on the fundamental homology class of N and thus get a real number as a result. This is the *GV invariant of the manifold N* (not the foliation).

b. The sum of the Ray-Singer torsions for foliations

A codim-1 foliation F on a 3-manifold N is given by an codim-1 (hence dim 2) integrable subbundle F of the tangent bundle TN of N. Integrability means that the Lie bracket of vector fields is closed in F. Equiavlently, this means that the horizontal (or tangential or leafwise) exterior derivative d_F which is the usual exterior derivative restricted to take derivatives along the horizontal (or tangential or leafwise) directions only is a differential, namely $d_F^2 = 0$. We can use a Riemannian metric and define the Hodge star operator along with the adjoint horizontal differential $d_F{}^*$ and finally we can define the tangential Laplacian in the usual way (see section 2 above)

$$\Delta_F = d_F d_F{}^* + d_F{}^* d_F. \quad (11)$$

\Then we can define the Ray-Singer torsion for the foliation F in an analogous way as we did for flat connections (see equation 6 above)

$$\log[T(N,F)] = (1/2)\, \Sigma_{i=0}{}^{3}\, (-1)^i\, i\, \zeta'(\Delta_{i,F}) \quad (12)$$

with the same definition for the ζ-function regularized determinant (equation 7). Finally we can add all Ray-Singer torsions for each foliation coarse isotopy class in G(N) and thus get an invariant for the 3- manifold N.

c. Sum of Tangential Chern-Simons 3-forms

Again F is a codim-1 foliation on the 3-manifold N. Since the tangential exterior derivative d_F of the foliation F is a differential, we can form the tangential de Rham complex $(\Omega_F{}^{\bullet}(N), d_F)$, where $\Omega_F{}^{\bullet}$ denotes sections of the exterior bundle $\Lambda^{\bullet}(F^*)$. Taking the cohomology in the usual way we get the so-called *tangential cohomology groups* $H_F{}^{\bullet}(N)$. One can build the tangential version of the usual Chern-Weil theory of characteristic classes and finally construct the tangential Chern character from K-theory to tangential cohomology as a ring map

$$ch_F : K^0(N) \rightarrow +_n H_F{}^{2n}(N;C)$$

From the above constructions it follows that tangential Chern classes have their corresponding tangential Chern-Simmons forms (see [11]). One can take the tangential Chern-Simons 3-forms for elements in G(N), add them up and integrate over the fundamental 3-dim homology class of N. This will give another numerical invariant for N.

d. Transverse fundamental cyclic cocycles of foliations

In general, given a codim-q foliation F on an n-manifold N, there is the noncommutative approach to the study of foliations. This amounts to defining the holonomy groupoid Gr(F,N) of the foliation and then complete it to a C*-algebra, this is the corresponding foliation C*-algebra denoted C(F,N). This construction is complicated and can be found in detail in Connes' book on Noncommutative Geometry [12]. Moreover, to each foliation F as above on N (with the mild assumption that its normal bundle Q: = TN/F is orientable) one can construct its *transverse fundamental cyclic cocycle* (see Zois [13]). This is a q-dim class (where q is the codimension of the foliation F) in the cyclic cohomology $HC^q(C^{\infty}(F,N))$.

[Aside *Note 3:* Actually in the construction of the tfcc it is more convenient to use the algebra of smooth complex functions $C^{\infty}(F,N)$ which is a dense subalgebra of C(F,N) because the cohomology of C(F,N) is rather poor in many cases].

In our case at hand, taut foliations by definition (see topological tautness above) have a closed transversal which is an S^1. If a foliation admits a closed transversal, then its corresponding C*-algebra simplifies drastically and it is reduced to just the algebra of continuous functions on the transversal.

[Aside *Note 4:* This is actually true in the case where the transverse circle intersects every leaf only once. We make this simplifying assumption here].

Hence in our case where we consider taut codim-1 foliations on a 3-manifold N, the foliation C*-algebra will just always be $C(S^1)$, ie the algebra of continuous complex valued functions on S^1 vanishing at infinity. Consequently for each taut foliation coarse isotopy class in the moduli space G(N), its corresponding transverse fundamental cyclic cocycle defines an element in $HC^1(C^\infty(S^1))$, the first cyclic cohomology group of the algebra $C^\infty(S^1)$. One can add all the corresponding transverse fundamental cyclic cocycles for elements on G(N) and get again as a result an element in $HC^1(C^\infty(S^1))$. Thus we have to figure out what $HC^1(C^\infty(S^1))$ is.
There is a theorem of Connes (see [12]) according to which (for compact M) there is a canonical isomorphism

$$HC^k(C^\infty(M)) \approx Z^{dR}{}_k(M) + H^{dR}{}_{k-2}(M) + H^{dR}{}_{k-4}(M) + \ldots,$$

where $Z^{dR}{}_k(M)$ denotes the set of closed de Rham k-currents, and $H^{dR}{}_j(M)$ denotes the ith de Rham homology group.

Hence in our case at hand we have $HC^1(C^\infty(S^1)) \approx Z^{dR}{}_1(S^1)$.

Since the degree is the same as the dimension, one can only integrate over S^1 itself and $Z^{dR}{}_1(S^1)$ must be 1-dimensional, generated by the fundamental homology class $[S^1]$.

If the transverse circle does not intersect every leaf only once, then the foliation algebra will be more complicated. Nonetheless it will still be a simple C*-algebra though.

These new invariants need further study. In the generic case, where the assumption in Aside Note 4 above does not hold, what can go wrong is that the tfcc may be invariant under an equivalence relation which is more narrow than coarse homotopy and in fact each coarse homotopy class contains an infinite number of distinct tfcc. In this case convergence issues arise and one must be able to somehow control the infinite sum. (This is currently under investigation).

Looking further ahead, the possibility to define a noncommutative version of Floer Homology (see [14]) for all 3-manifolds using NCG tools and techniques starts from a Lagrangian density whose Euler-Lagrange equations yield taut codim-1 foliations as solutions. This is known as the "inverse problem" in the calculus of variations. Hopefully we shall be able to report on this elsewhere.

References

[1] 't Hooft, G.: Dimensional Reduction in Quantum Gravity, gr-qc/9310026.

[2] Hogan, C. J.: Measurement of Quantum Fluctuations in Geometry, Phys. Rev. D, Vol 77 (2008), page 104031.

[3] Chamseddine, A.H., Connes, A., Marcolli, M.: Gravity and the Standard Model with Neutrino Mixing, Adv. Theor. Math. Phys. 11 (2007), 991-1089.

[4] Connes, A., Douglas, M.R., Schwarcz, A.:Noncommutative Geometry and Matrix Theory: Compactifications on Tori, JHEP 9802:003 (1998).

[5] Seiberg, N., Witten, E.:String Theory and Noncommutative Geometry, JHEP 9909:032 (1999).

[6] Atiyah, M.F.: Topological Quantum Field Theories, IHES Publ. Math. 68 (1989).

[7] Witten, E.: Quantum Field Theory and the Jones Polynomial, Commun.Math.Phys. 121 (1989).

[8] Floer, A.: An instanton invariant for 3-manifolds, Commun. Math. Phys. (1989).

[9] Donaldson, S.K.: Floer Homology Groups in Yang-Mills Theory, Cambridge University Press 2002.
[10] Gabai, D.: Essential Laminations and Kneser Normal Form, Jour. Diff. Geom. Vol 53 No 3 (1999).
[11] Moore, C.C. and Schochet, C.: Global Analysis on Foliated Spaces, Cambridge University Press (2005).
[12] Connes, A.: Noncommutative Geometry, Academic Press (1994).
[13] Zois, I.P.: A new invariant for σ-models, Commun. Math. Phys. 209.3 (2000).
[14] Zois, I.P.: Noncommutative Topological Quantum Field Theory, Noncommutative Floer Homology, Noncommutative Hodge Theory, arXiv:1401.4080.
[15] Chamseddine, A.H., Connes, A., van Suijlekom, W.D.: Grand Unification in the Spectral Pati-Salam Model, arXiv:1507.08161

Dilatonic Brans-Dicke Anisotropic Collapsing Fluid Sphere And de Broglie Quantum Wave Motion

HOSSEIN GHAFFARNEJAD

Faculty of Physics, Semnan University, Semnan, IRAN, Zip Code: 35131-19111

E-mail: hghafarnejad@yahoo.com; hghafarnejad@profs.semnan.ac.ir

Abstract. Two dimensional (2D) analogue of vacuum sector of the Brans Dicke (BD) gravity [1] is studied to obtain dynamics of anisotropic spherically symmetric perfect fluid. Our obtained static solutions behave as dark matter with state equation $\gamma = \frac{p(\rho)}{\varrho} = -0.25$ but in non-static regimes behave as regular perfect fluid with barotropic index $\gamma > 0$. Positivity property of total mass of the fluid causes that the BD parameter to be $\omega > \frac{2}{3}$ and/or $\omega < -1$. Locations of the event and apparent horizons of the collapsing fluid are obtained in its static regime. In case $\omega > 0$ the apparent horizon is covered by event horizon where the cosmic censorship hypothesis is still valid. According to the model [1], we obtain de Broglie pilot wave of our metric solution which describes particles ensemble which become distinguishable via different values of ω. Incident current density of particles ensemble on the horizons is evaluated which describe the 'Hawking radiation'. The de Brogle -Bohm quantum potential effect is calculated also on the event (apparent) horizon which is independent (dependent) to values of ω.

1. Introduction

Around a collapsing star, gravity takes an ultra-strong intensity. Hence corresponding matter density and space-time curvature diverge to infinity. It is predicted from singularity theorems in general relativity [2] that end of stellar collapse leads to a visible naked or invisible covered singularity. Property of visibility of singularities are determined by the causal structure of collapsing process, satisfying the gravitational field equations. Internal dynamics of the collapse can be determine type of singularities. They will be naked and hence visible from view of external observer if collapse process delays the formation of the event horizon [3,4,5]. Penrose represented 'cosmic censorship hypotheses' [6], which explains properties of the final singularity of a gravitational collapse. In this conjecture, space-time singularities created from gravitational collapse should be usually covered by the event horizon hyper-surface. Hence it will be invisible from view of outside observers of collapsed object namely black hole [7]. However there is still a doubt for reality of censorship hypotheses and has been recognized as one of the most open problems in classical general theory of relativity. The censorship hypotheses may to be still valid in quantum gravity approach (see [8] and references therein). Black holes and naked singularities have different properties observationally. They have different characteristics in the gravitational lensing [9,10] and also many exact solutions of Einstein's gravitational field equations are obtained which admit naked singularities [11-18]. In the present paper we use vacuum sector of the BD gravity [19] in 2D analogue [1] and obtain non-static solutions of linearized field equations. Stress tensor of our solutions treats as anisotropic spherically symmetric perfect fluid. Equation of state takes dark and visible regular matter regimes during

Content from this work may be used under the terms of the Creative Commons Attribution 3.0 licence. Any further distribution of this work must maintain attribution to the author(s) and the title of the work, journal citation and DOI.
Published under licence by IOP Publishing Ltd

the collapse process under particular conditions on different values of the parameters of the system. Apparent and event horizons of the collapsing object are evaluated where the apparent horizon become inside of the event horizon for $\omega > 0$ at static regime. According to results of the work [1] we obtain de Broglie-Bohm quantum wave functional of the collapsing fluid sphere. Probability amplitude of particles ensemble and their radial velocities are calculated from incident current density (ICD) of particles ensemble. Mass of the collapsing fluid must be takes some positive real values numerically reaching to conditions $\omega < -1; \omega > \frac{2}{3}$. The work is organized as follows: In section 2 we call dilatonic BD gravity in 2D analogue [1] and obtain linearized metric solution. Corresponding stress tensor is obtained to be treat as an anisotropic spherically symmetric BD perfect fluid. Location of the horizons are determined as spherically symmetric plan waves. Using some suitable conditions we obtained time independent horizons location and mass function. In section 3 we use results of the work [1] to obtain de Brogle wave functional of the model and then evaluate ICD of particles ensemble moving on the horizons. de Brogle-Bohm quantum potential effect is also considered as corrections on our results. Section 4 denotes to summery and conclusion.

2. Dilatonic Brans-Dicke collapsing sphere

One can obtain effective parts of the BD gravity [19] defined on spherically symmetric curved space times $ds^2 = g_{ab}dx^a dx^b + \psi^2(x^a)\,(d\theta^2 + \sin^2\theta d\varphi^2)$ in 2D analogue [1] as $I[\phi, \psi, g_{ab}] = \frac{1}{2}\int d^2x\sqrt{g}\{\phi + \frac{1}{2}\phi\psi^2 R + \phi g^{ab}\partial_a\psi\partial_b\psi + 2\psi g^{ab}\partial_a\psi\partial_b\phi - \frac{\omega\psi^2}{2\phi}g^{ab}\partial_a\phi\partial_b\phi\}$ where ϕ and ω are BD field and parameter respectively. Experimentally ω is obtained to be as ≥ 40000 [20,21]. ψ^2 is 2-sphere conformal factor and R is Ricci scalar of 2D metric g_{ab} defined by 2D coordinates system x^a with $a \equiv 1, 2$. We solved linearized dynamical field equations and obtained metric solution as $Gds^2 \approx -(1 + \varepsilon x_0 e^{a\tau+b\rho})d\tau^2 + 2(1 + \varepsilon y_0 e^{a\tau+b\rho})d\tau d\rho + (1 + \varepsilon z_0 e^{a\tau+b\rho})d\rho^2$ where $|\varepsilon| << 1$ is order parameter of the perturbation series expansion, $t = \sqrt{G}\tau$ and $r = \sqrt{G}\rho$ (details of calculations are given in ref. [23]). Also we obtained that $\psi \approx \sqrt{eG}(1 + \varepsilon u_0 e^{a\tau+b\rho})$ and $G\phi \approx 1 + \varepsilon v_0 e^{a\tau+b\rho}$ in which a, b are integral constants and for moving waves with light velocity we must be set $a = \pm b$. Other parameters defined by $\omega, u_0, v_0, y_0, x_0$ and γ are obtained exactly against a, b in ref. [23]. In the next section we show ε is related to the corresponding quantum potential of de Brogle wave evaluated on the event horizon hypersurface. In case of static regime where $a = 0$ we will have $\gamma = -0.25$ describing a dark matter dominant of the gravitational system and $b = -\sqrt{\frac{6\omega-4}{\omega+1}}, \omega > \frac{2}{3}$. With these conditions our metric solutions become asymptotically flat at large scales $\rho >> 1$ for which $\frac{u_0}{v_0} = \frac{3\omega-2}{5}, \frac{x_0}{v_0} = \frac{y_0}{v_0} = \frac{8(3\omega-2)}{5}, \frac{z_0}{v_0} = \frac{2(2-3\omega)(2\omega-3)}{5(\omega+1)}$. In case of non static regime of the metric solution we must be set $(a, b) \neq 0$. In the latter case one can obtain moving position of event horizon by solving $g_{tt}(\tau, \rho) = 0$ as $a\tau + b\rho = -\ln|\varepsilon x_0|$ which in static regime become $\rho_{EH} = \sqrt{\frac{\omega+1}{6\omega-4}}\ln\left|\frac{4\varepsilon v_0}{5}(6\omega - 4)\right|$. Surface area of moving apparent horizon is $4\pi\psi^2(\tau, \rho)$. Its position is obtained by solving $g^{ab}\partial_a\psi\partial_b\psi = 0$ as $a\tau + b\rho = \ln\left|\frac{2(b^2+2ab-a^2)}{\varepsilon[(a^2+b^2)z_0+2(b^2+ab-a^2)y_0-(a-b)^2x_0]}\right|$ which is $\rho_{AH} = \sqrt{\frac{\omega+1}{6\omega-4}}\ln\left|\frac{4\varepsilon v_0}{5}\frac{(3\omega-2)(\omega+5)}{(\omega+1)}\right|$ in static regime. In the general relativistic limits $\omega \to +\infty$ we obtain $\rho_{EA} = 0.94$. One can determine mass function $m(r)$ up to the radius r by calculating $m(r) = \int_0^r 4\pi r^2 \varrho(r)dr$ which by setting $a = 0$, $r = \sqrt{G}\rho$, and $b = -\sqrt{\frac{6\omega-4}{\omega+1}}$, we will have $m^*(\rho) = \frac{m(\rho)}{8\pi\sqrt{G}v_0\varepsilon} = \sqrt{\frac{\omega+1}{6\omega-4}}\left\{1 - \frac{1}{2}\left[\left(\frac{6\omega-4}{\omega+1}\right)^2 - \left(\frac{6\omega-4}{\omega+1}\right) + 2\right]e^{-\sqrt{\frac{6\omega-4}{\omega+1}}\rho}\right\}$ (see ref. [23]). Positivity and reality of the mass quantity of the fluid restrict us to choose $\omega > \frac{2}{3}$ or $\omega < -1$.

3. de Broglie quantum gravity perspective

According to results of the work [1] (see also [23]) we can obtain phase $\bar{D}$ and amplitude $\bar{R}$ of de Broglie pilot wave $\Psi(\psi,\phi) = \sqrt{\bar{R}}e^{i\bar{D}}$ of the spherically symmetric BD action given in the previous section as $\phi\psi^2 = \left(\frac{3+2\omega}{2+\omega}\right)\frac{\bar{R}}{4}$, and $\psi = \sqrt{G}\bar{R}^{\left(\frac{1+\omega}{3+2\omega}\right)}e^{i\bar{D}\frac{\sqrt{4+2\omega}}{3+2\omega}}$. Inserting the solutions given at the previous section the parts of amplitude $\bar{R}$ and the phase $\bar{D}$ read $\bar{R}(\tau,\rho) = \left(\frac{8+2\omega}{3+2\omega}\right)\exp\{1+\varepsilon(u_0+v_0)e^{a\tau+b\rho}\}$ and $\bar{D}(\tau,\rho) = \frac{1}{\sqrt{(4+2\omega)(3+2\omega)}}\Big\{\frac{1}{2}-(1+\omega)\ln\left(\frac{8+4\omega}{3+2\omega}\right)+$ $\varepsilon[(2+\omega)u_0-(1+\omega)v_0]e^{a\tau+b\rho}\Big\}$ which by setting $a=0$ reduces to their static forme (standing waves) as $\bar{R}(\rho) = \left(\frac{8+2\omega}{3+2\omega}\right)\exp\{1+\varepsilon(3v_0/5)(\omega+1)e^{-\sqrt{\frac{6\omega-4}{\omega+1}}\rho}\}$ and $\bar{D}(\rho) = \frac{1}{\sqrt{(4+2\omega)(3+2\omega)}}\Big\{\frac{1}{2}-$ $(1+\omega)\ln\left(\frac{8+4\omega}{3+2\omega}\right)+\varepsilon(v_0/5)(3\omega^2-\omega-9)e^{-\sqrt{\frac{6\omega-4}{\omega+1}}\rho}\Big\}$. We calculate now ICD of particles ensemble containing the Plank mass $m_p = \frac{1}{\sqrt{G}}$ (in units $\hbar = c = 1$) via $J_{incident} = \frac{i\hbar}{2m_p}\{\Psi\partial_\rho\Psi^* - \Psi^*\partial_\rho\Psi\}$ as $J_{incident}(\rho) = \left(\frac{9-\omega-3\omega^2}{3+2\omega}\right)\sqrt{\frac{(4+\omega)(3\omega-2)}{(\omega+1)(\omega+2)}}\left(\frac{3\sqrt{2}v_0}{5}\right)\exp\left\{\frac{1}{2}-\sqrt{\frac{6\omega-4}{1+\omega}}\rho+\varepsilon\left(\frac{3v_0}{10}\right)(1+\omega)e^{-\sqrt{\frac{6\omega-4}{1+\omega}}\rho}\right\}$. Its value become $J_{incident}(\rho_{EH}) = \left(\frac{9-\omega-3\omega^2}{3+2\omega}\right)\sqrt{\frac{(4+\omega)}{(\omega+1)(\omega+2)(3\omega-2)}}\left(\frac{3\sqrt{2}}{8\varepsilon}\right)e^{\frac{5}{8}\left(\frac{3\omega-1}{3\omega-2}\right)}\Big\}$ on the event horizon hypersurface and $J_{incident}(\rho_{AH}) = \left(\frac{9-\omega-3\omega^2}{3+2\omega}\right)\sqrt{\frac{(4+\omega)(1+\omega)}{(3\omega-2)(\omega+2)}}\left(\frac{3\sqrt{2}}{4\varepsilon}\right)\exp\left\{\frac{15\omega^2+58\omega-37}{8(3\omega-2)(\omega+5)}\right\}$ on the apparent horizon hypersurface respectively. Quantum potential of the de Broglie wave is defined by $Q = -\frac{\nabla_\mu\nabla^\mu\sqrt{\bar{D}}}{\sqrt{\bar{D}}}$ which for our metric solutions in static regime become $Q(\rho) = -\varepsilon^2\left(\frac{3v_0}{5}\right)(3\omega-2)e^{-\sqrt{\frac{6\omega-4}{\omega+1}}\rho}$. Its value become $Q(\rho_{EH}) = -\frac{3}{8}\varepsilon$ on the event horizon hypersurface and $Q(\rho_{AH}) = -\frac{3}{4}\left(\frac{1+\omega}{5+\omega}\right)\varepsilon$ on the apparent horizon hypersurface respectively. The former result shows relation between quantum potential and perturbation order parameter ε. Namely one can evolve physical meaning of ε which is supported by quantum potential counterpart evaluated on the event horizon as independent of ω. Obviously our results and inferences must be corrected by regarding backreaction corrections of de Broglie-Bohm quantum potential effects on moving particle ensembles (see section 3 in ref. [24]) which is not considered here. It is suitable to define relative normalized ICD of particle ensemble as $\Delta(\omega) = \frac{J_{incident}(\rho_{EH})-J_{incident}(\rho_{AH})}{J_{incident}(\rho_{AH})} = \frac{\exp\{\frac{3\omega}{2(3\omega-2)(5+\omega)}\}}{2(1+\omega)}$ and relative normalized quantum potential as $\delta(\omega) = \frac{Q(\rho_{EH})-Q(\rho_{AH})}{Q(\rho_{AH})} = \frac{3-\omega}{2(1+\omega)}$. We plotted diagrams of Δ and δ against ω with solid and dash lines respectively in figure 1. We see potential barrier for $-1<\omega<3$ and potential well for $\omega<-1$ and/or $\omega>3$ where one can evolve quantum tunneling of particle ensemble! In other words diagram shows that for $-5<\omega<-4.7$ and $-1.6<\omega<-0.8$ we have $\Delta<0$ which means that the particles ensemble moves into the event horizon.

4. Concluding remarks

Two dimensional analogue of vacuum sector of the BD gravity model is used to solve its dynamical equations. Our classical solutions describe anisotropic perfect fluid collapsing sphere behaves as dark matter gravitational source in its static regime and as regular visible matter in its dynamical non-static regime. Total mass of the fluid is calculated against the BD parameter for which we must be choose $\omega \leq -1$ and/or $\omega > \frac{2}{3}$. In general relativistic approach $\omega \to +\infty$ ratio of total mass of the fluid per the Plank mass leads to $M^* = \frac{1}{\sqrt{6}} << 1$. It is eliminated with $\omega = -1$ and takes infinite value with $\omega = \frac{2}{3}$. In short, end of the collapsing fluid reaches to a black

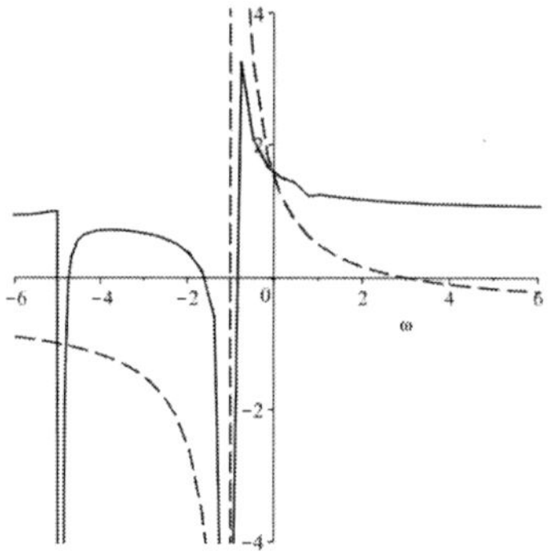

Figure 1. Diagram of Δ (the normalized relative ICD of particles ensemble) and δ (normalized relative quantum potential) is plotted against ω with solid and dash lines respectively.

hole structure and covered the causal singularity satisfying the cosmic censorship hypothesis. de Broglie-Bohm quantum pilot wave of the system is also obtained and corresponding ICD of particles ensemble is evaluated on the event and the apparent horizons. They describe statistical approach of thermal 'Hawking radiation'. Relative ICD of particles ensemble moving on the horizons take maximum value where the quantum potential diverges to an infinite value at $\omega = -1$.

References

[1] Ghaffarnejad H 2014 *Int. J. Theor. Phys.* **53** 2616 *Preprint* gr-qc/1309.6157
[2] Wald R M 1984 *General Relativity*, (The University of Chicago Press, Chicago).
[3] Joshi P S 2007 *Gravitational Collapse and Space-time Singularities* (Cambridge University Press).
[4] Singh T P 1999 *J. Astrophys. Astr.* **20** 221
[5] Ziaie A H, Atazadeh Kh and Tavakoli Y 2010 *Class. Quantum Grav.* **27** 075016; Erratum-ibid.27, 209801 *Preprint* gr-qc/1003.1725.
[6] Penrose R 1969 *Rev. Nuovo Cimento* **1** 252
[7] Ori A and Piran T 1990 *Phys. Rev. D* **42** 1068
[8] Ghaffarnejad H, Neyad H and Mojahedi M A 2013 *Astrophys. Space Sci* **346** 497 *preprint* physics.gen-ph/1305.6914
[9] Virbhadra K S and Keeton C R 2008 *Phys. Rev. D* **77** 124014 *preprint* gr-qc/0710.2333.
[10] Virbhadra K S and Ellis G F R 2002 *Phys. Rev. D* **65** 103004
[11] Mena F C Nolan B C and Tavakol R 2004 *Phys. Rev. D* **70** 084030 *preprint* gr-qc/0405041.
[12] Martin-Garcia J M and Gundlach C 2003 *Phys. Rev. D* **68** 024011
[13] Patil K D 2003 *Phys. Rev. D* **67** 024017
[14] Joshi P S 1995 *Phys. Rev. D* **51** 6778
[15] Lemos J 1992 *Phys. Rev. Lett.* **68** 1447
[16] Joshi P S and Dwivedi I H 1992 *Commun. Math.Phys.* **146** 333
[17] Ghosh S G and Dadhich N 2003 *Gen. Relativ. Gravit* **35** 359
[18] Szekeres P and Lyer V 1993 *Phys. Rev. D* **47** 4362
[19] Brans C and Dicke R 1961 *Phys. Rev.* **124** 925
[20] Will C M 1993 *Theory And Experiment In Gravitational Physics* (Cambridge University press) *Preprint* gr-qc/9811036.
[21] Will C M 2006 *Living Rev. Rel.* **9**; http://WWW.livingreviews.org/lrr-2006-3.
[22] Culetu H 2011 *Preprint* gr-qc/1101.2980
[23] Ghaffarnejad H 2014 *Preprint* gr-qc/1407.7367
[24] Ghafarnejad H 2006 *Astrophys. and Space Sci.* **301** 145

A Langevin Approach to a Classical Brownian Oscillator in an Electromagnetic Field

J. S. Espinoza Ortiz[1], F. C. Bauke[2], R. E. Lagos[3]
[1]Departamento de Física, IFQC, Universidade Federal de Goiás, Catalão, Goiás 75704/020, Brasil.
[2]Instituto de Física *Gleb Wataghin* , UNICAMP ,Campinas ,São Paulo 13083/970, Brasil.
[3]Departamento de Física, IGCE, UNESP, Rio Claro, São Paulo 13500/970, Brasil.

E-mail: jsespino@physics.org

Abstract. We consider a charged Brownian particle bounded by an harmonic potential, embedded in a Markovian heat bath and driven from equilibrium by external electric and magnetic fields. We develop a quaternionic-like (or Pauli spinor-like) representation, hitherto exploited in classical Lorentz related dynamics. Within this formalism, in a very straight forward and elegant fashion, we compute the exact solution for the resulting generalized Langevin equation, for the case of a constant magnetic field. For the case the source electromagnetic fields satisfy Maxwell's equations, yielding spinor-like Mathieu equations, we compute the solutions within the JWKB approximation. With the solutions at hand we further compute spatial, velocities and crossed time correlations. In particular we study the (kinetically defined) nonequilbrium temperature. Therefore, we can display the system's time evolution towards equilibrium or towards non equilibrium (steady or not) states.

1. Introduction

The ubiquitous Brownian motion remains an outstanding paradigm in modern physics [1, 2, 3, 4, 5, 6, 7, 8] (and references therein). The theoretical framework consisting of Kramers equation (a Fokker-Planck equation in phase space), Smoluchowski equation (asymptotic or overdamped contraction of the latter) and the associated stochastic Langevin equation, have been widely applied to diverse problems, such as: Brownian motion in potential wells, chemical reactions rate theory, nuclear dynamics, stochastic resonance, surface diffusion, general stochastic processes and evolution of nonequilibrium systems, in both classical and quantum contexts. More recent applications include thermodynamics of small systems, molecular motors, chemical and biological nanostructures in the burgeoning field of nanotechnology, mesoscopic motors power output and efficiency. It took approximately sixty years to report exact solutions for the Brownian motion of a charged particle in uniform and static electric and/or magnetic fields [6, 7, 8, 9]. Here we outline the extension of our previous work [5, 6, 7, 8] and consider a charged Brownian oscillator under electromagnetic fields.

In section 2 we write down the Langevin equation for a charged particle under an harmonic potential and external electric and magnetic fields (as reported in [6, 7] homogeneous forces can be included straightforwardly into our formalism). In section 3 we compute the exact solution of Langevin's equation when considering a time independent magnetic field, via a novel method (spinors and Pauli matrices), compute spatial, velocities and crossed time correlations; and outline some applications of our result, in particular we study the (kinetically defined)

Content from this work may be used under the terms of the Creative Commons Attribution 3.0 licence. Any further distribution of this work must maintain attribution to the author(s) and the title of the work, journal citation and DOI.
Published under licence by IOP Publishing Ltd

non equilibrium temperature. In section 4 we present the approximate (JWKB) solution of Langevin's equation under an electromagnetic field satisfying Maxwell equations. Finally in section 5 we present further applications and extensions of our work.

2. Langevin Equation for a Charged Brownian Oscillator under Electromagnetic Fields

Lets us consider a charged (carrier) Brownian oscillating particle, in contact with a reservoir at temperature T_{R}, under the influence of electromagnetic fields. The Langevin equation [5, 6, 7, 8, 9, 10, 11] is given by

$$m\frac{d^2\mathbf{x}}{dt^2} = -m\lambda\frac{d\mathbf{x}}{dt} - m\omega_0^2\mathbf{x} + q\left(\mathbf{E} + \frac{1}{c}\frac{d\mathbf{x}}{dt} \times \mathbf{B}\right) + \mathbf{F}^{\mathrm{r}}(t) \tag{1}$$

where $\mathbf{x}$ is the carrier's position (the carrier's velocity is $\mathbf{v} = d\mathbf{x}/dt$), m the mass, q the charge, $m\lambda$ Stokes' force constant ($\lambda^{-1} = \tau_{\mathrm{c}}$, collision time), ω_0 Hooke's frequency, c light velocity (in CGS units) and **E**&**B** the Electric and Magnetic external fields, respectively. $F^{\mathrm{r}}(t)$ is the random Brownian force, white Markovian noise with statistics (mean and correlations) given by [6, 7, 8, 9, 10, 11]

$$\langle \mathbf{F}^{\mathrm{r}}(t)\rangle = \mathbf{0} \quad , \quad \langle F_i^{\mathrm{r}}(t)F_j^{\mathrm{r}}(t')\rangle = 2\Gamma\delta_{ij}\delta\left(t-t'\right) \tag{2}$$

where the noise strength is given by $\Gamma = m\lambda k_{\mathrm{B}}T_{\mathrm{R}}$ (Fluctuation Dissipation Theorem [1, 2, 3]).

We consider homogeneous fields (constant or solely time dependent), choose $\mathbf{B} = B\hat{\mathbf{z}}$ and with the convenient definitions $\mathbf{f} = q\mathbf{E}/m$, $\mathbf{f}^{\mathrm{r}} = \mathrm{F}^{\mathrm{r}}/m$ and cyclotron frequency $\omega_c = qB/mc$, the Brownian Lorentz Langevin oscillator equations are cast as

$$\begin{aligned} \ddot{x} &= -\lambda\dot{x} + \omega_c\dot{y} - \omega_0^2 x + f_x + f_x^{\mathrm{r}}\,, \\ \ddot{y} &= -\lambda\dot{y} - \omega_c\dot{x} - \omega_0^2 y + f_y + f_y^{\mathrm{r}}\,, \\ \ddot{z} &= -\lambda\dot{z} - \omega_0^2 z + f_z + f_z^{\mathrm{r}}\,. \end{aligned} \tag{3}$$

Since the z (non magnetic) component is uncoupled from the xy (magnetic) motion hereafter we ignore the former, and when needed we call in the exact solution [1, 2, 3].

Define sigma real matrices (2×2), a commutative quaternion subalgebra. A is a "σ-matrix" if the real matrix A can be cast as

$$A = \begin{pmatrix} a_1 & -a_2 \\ a_2 & a_1 \end{pmatrix} = a_1\begin{pmatrix} 1 & 0 \\ 0 & 1 \end{pmatrix} + a_2\begin{pmatrix} 0 & -1 \\ 1 & 0 \end{pmatrix} = a_1 + a_2\sigma\,, \qquad \sigma^2 = -1\,.$$

As in quaternions we define the conjugate $\widetilde{A}$, the norm $\|A\|$ and the argument $\arg(A)$ associated to $A = a_1 + a_2\sigma$, respectively by $\widetilde{A} = a_1 - a_2\sigma$, $\|A\|^2 = A\widetilde{A} = \det A = a_1^2 + a_2^2$, $\arg(A) = a_2/a_1$. It follows $A^{-1} = \widetilde{A}/\|A\|^2$. Furthermore since $\sigma = \imath\sigma_y$ (the y component of Pauli's σ matrix) [10] we obtain the (de Moivre like) relations

$$\exp(A) = \exp(a_1)(\cos a_2 + \sigma\sin a_2) \quad , \quad \exp(iA) = \exp(ia_1)(\cosh a_2 + i\sigma\sinh a_2)\,,$$

and from the last equations, straightforward expressions for several functions such as $\cos A$, $\sin A$, $\ln A$, A^p, are obtained. In particular: $\cos A = \cos a_1\cosh a_2 - \sigma\sin a_1\sinh a_2$, $\sin A = \sin a_1\cosh a_2 + \sigma\cos a_1\sinh a_2$, $\ln A = \|A\| + \sigma\arctan\arg A$, and

$$\sqrt{A} = \sqrt{\frac{\|A\| + a_1}{2}} + sgn(a_2)\sigma\sqrt{\frac{\|A\| - a_1}{2}} \tag{4}$$

for $a_1 > 0$ (for $a_1 < 0$, we choose the convention $\sqrt{A} = i\sqrt{-A}$).

Now, define the "spinors"

$$q = \begin{pmatrix} x \\ y \end{pmatrix}, \qquad \phi = \begin{pmatrix} f_x + f_x^{\mathrm{r}} \\ f_y + f_y^{\mathrm{r}} \end{pmatrix} = f + f^{\mathrm{r}} \tag{5}$$

and as usual, the adjoint spinor $q^\dagger = (x, y)$. With $\Lambda_{\mathrm{c}} = \lambda + \omega_{\mathrm{c}}\sigma$, equation (3) is cast in the form

$$\frac{d^2q}{dt^2} + \Lambda_{\mathrm{c}}\frac{dq}{dt} + \omega_0^2 q = \phi\,. \tag{6}$$

Now, we proceed to solve this system of two second order linear inhomogeneous coupled equations, for two cases, namely: constant magnetic field (section 3) and plane waves fields, solutions of Maxwell equations (section 4).

3. Exact Solution for constant Magnetic Field

For a constant magnetic field, equation (6) has an exact solution, and was derived in a related, more general context [11, 12]. Here we apply a novel method of solution hinging on the quaternion scheme presented in the previous section. We believe this method to be more straightforward since in the quaternion commutative sub algebra, our resulting equations can be considered as representing a bona fide one dimensional problem. At the very end of the calculations, yielding observable quantities, only then we explicitly decompose any σ-matrices (say A) into its scalar components $A = a_1 + a_2\sigma$.

Disregarding the homogeneous solution (transient), the exact solution of equation (6) for constant magnetic field, is given by [13],

$$q(t) = \int_0^t d\tau G(t-\tau)\left(f(\tau) + f^{\mathrm{r}}(\tau)\right) \tag{7}$$

with the Green function

$$G(t) = \frac{1}{\Omega_{\mathrm{c}}}\sin(\Omega_{\mathrm{c}}t)\exp\left(-\frac{1}{2}\Lambda_{\mathrm{c}}t\right) = g_1(t) + \sigma g_2(t) \tag{8}$$

where $\Omega_{\mathrm{c}} = \sqrt{\omega_0^2 - \Lambda_{\mathrm{c}}^2/4}$. We can readily compute the (spinor) velocity $v(t) = dq(t)/dt$. Now, with a minimum of algebra we can compute all quadratic correlations $C_{ab}(t,\tau) = \langle a(t)b(\tau)\rangle$ with the rules given by equation (2), where a and b are any combination of the spinor components of position $q(t)$ and velocity $v(t)$. The $z(t)$ and $v_z(t)$ components are easily computed from either spinor component by just setting $\omega_{\mathrm{c}} = 0$. All expressions are readily integrated into elementary functions. In particular the case of time independent electric field hereafter considered in this section. Some interesting correlations are

$$K(t) = \frac{1}{2}m\langle v^\dagger(t)v(t) + v_z^2(t)\rangle = \frac{3}{2}k_B T(t)\,, \tag{9}$$

$$V(t) = \frac{1}{2}m\omega_0^2\langle q^\dagger(t)q(t) + z^2(t)\rangle\,, \tag{10}$$

yielding a kinetic time dependent definition of temperature [6, 7, 8] and average potential energy, respectively. For zero external fields both equal time correlations asymptotically converge towards the classical equipartition theorem value $3k_{\mathrm{B}}T_{\mathrm{R}}/2$. Interesting preliminary results involve the transition from a normal carrier profile to a hot carrier profile [8] as a mode softening process occurs $(\omega_0 \to 0)$.

4. Approximate (JWKB) solution for a Maxwell Electromagnetic Field

When the external electromagnetic fields are not constant some approximations are necessary in order to produce tractable and relevant results (for example in sun activated photovoltaic devices [14]). In particular for electromagnetic plane monochromatic (ω) waves propagating in the $\widehat{\mathbf{x}}$ direction and satisfying Maxwell equations [15], we have $\mathbf{B} = \widehat{\mathbf{z}}B\cos\omega\, t$ and $\mathbf{E} = \widehat{\mathbf{y}}B\cos\omega\, t$. After taking real parts (as usual in classical physics), the Brownian Maxwell Langevin equation is cast as:

$$\frac{d^2q}{dt^2} + \Lambda(t)\frac{dq}{dt} + \omega_0^2 q = f + f^{\mathrm{r}}\,, \tag{11}$$

again, as before, we decouple the planar xy motion from the perpendicular z motion. In this case we have $f_x = 0$ and $f_y = c\omega_c \cos\omega t$, and $\Lambda(t) = \lambda + \sigma\omega_c \cos\omega t$. Due to the time dependent $\Lambda(t)$ dissipative factor, the exact solution involves spinorial Mathieu functions. Nevertheless we can write the JWKB solutions [16], certainty a good approximation for small field $w_c \ll \lambda$,

$$q(t) = \int_0^t d\tau\, G(t,\tau)\,(f(\tau) + f^{\mathrm{r}}(\tau))\,, \tag{12}$$

$$G(t,\tau) = \frac{1}{\sqrt{\Omega(t)\,\Omega(\tau)}}\sin\left(\int_\tau^t d\theta\Omega(\theta)\right)\exp\left(-\frac{1}{2}\int_\tau^t d\theta\Lambda(\theta)\right)\,, \tag{13}$$

where $\Omega(t) = \sqrt{\omega_0^2 - \Lambda^2(t)/4}$. Notice, in this case that the Green function given in equation (11) is not dependent in the time difference $t-\tau$ as in the exact solution in section 3 for the constant magnetic field case, and stands as the JWKB approximation for damped spinor Mathieu functions. Correlation functions in this case cannot be reduced to elementary functions as in the previous case.

5. Conclusions and Future work

In this short communication, we outlined and presented a novel method to compute solutions for the Langevin equation for a charged Brownian oscillator under external fields, in particular electromagnetic fields. Two cases were considered, namely: for the constant magnetic field, the exact solution was presented; and for both time dependent electric and magnetic fields satisfying Maxwell equations, the approximate JWKB solution was obtained. We outlined too, how to compute quadratic (two time) correlation functions, related to relevant physical quantities such as nonequilibrium temperature. These correlation functions will allow us to study the evolution of this nonequilbrium Brownian process, the oscillator's dynamics and as mentioned above the mode softening transition. For the case of constant magnetic field work is in progress, and results will be presented in a subsequent long paper. As for the JWKB solution, work is at the preliminary stage. This work is relevant in different fields besides physics, including biology and economics. It contributes to further understand small (nano) systems, may they be mechanical, chemical or biological.

Acknowledgments

The authors would like to thank the support of the Goiás Research Foundation - FAPEG.

References

[1] S. Chandrasekhar (1943) Stochastic problems in physics and astronomy, Rev. Mod. Phys. **15**(1) 2-87.

[2] V. Balakrishnan, *Elements of Nonequilibrium Statistical Mechanics* (Ane Books Pvt. Ltd. 2014).

[3] W. T. Coffey, Yu. P. Kalmykov and J. T Waldron, *The Langevin Equation with applications to Stochastic problems in Physics, Chemistry and Electrical Engineering* (World Scientific, 2004) 2^{nd}ed.

[4] L. A. Barreiro, J. R. Campanha and R. E. Lagos The thermohydrodynamical picture of Brownian motion via a generalized Smoluchowsky equation, Physica **A 283** (2000) 160-165; arXiv:cond-mat/9910405 v3.

[5] L. A. Barreiro, J. R. Campanha and R. E. Lagos (2002) The thermohydrodynamical picture of a charged Brownian particle, Rev. Mex. Fís. **48,** Suppl. **3** 13-15; arXiv:cond-mat/0012187 v1.
[6] T. P. Simões and R. E. Lagos (2005) Kramers equation for a charged Brownian particle: The exact solution, Physica **A** **355** 274-282; arXiv:cond-mat/0409434 v2.
[7] R. E. Lagos and T. P. Simões (2011) Charged Brownian particles: Kramers and Smoluchowski equations and the hydrothermodynamical picture, Physica **A** **390** 1591-1601; arXiv:cond-mat/1305.0487 v2.
[8] F. C. Bauke and R. E. Lagos (2014) Hot Brownian carriers in the Langevin picture: Application to a simple model for the Gunn effect in GaAs , Physica **A** **393** 235-243; arXiv:cond-mat/1304.5474 v2.
[9] J. I. Jiménez-Aquino and M. Romero-Bastida (2007) Fokker-Planck-Kramers equations of a heavy ion in presence of external fields, Phys. Rev. **E** **76** 021106/9pp.
[10] J. I. Jiménez-Aquino (2011) Work-fluctuation theorem for a charged harmonic oscillator, J. Phys. **A**: Math. Theor. **44** 295002/16pp.
[11] J. I. Jiménez-Aquino and R. M. Velasco (2013) Power-fluctuation theorem for a Brownian oscillator in a thermal bath, J. Phys. **A**: Math. Theor. **46** 32500/15pp.
[12] W. M. Fisher and S. C. Rand (2010) Light-induced dynamics in the Lorentz oscillator model with magnetic forces, Phys. Rev. **A** **82** 013802/6pp.
[13] R. E. Lagos, H. Suhl and T. Tiedje (1983) Theory of electronhole kinetics in amorphous semiconductors under illumination: Application to solar cells , J. Appl. Phys. **54** 3951-7.
[14] E. Butkov, *Mathematical Physics* (Addison-Wesley Publishing Co. 1968).
[15] E. Merzbacher, *Quantum Mechanics* (John Wiley&Sons Inc. 3rd ed. 1998).
[16] P. M. Morse and H. Feshbach, *Methods of Theoretical Physics* (McGraw-Hill Book Co. Inc. 1953).

Generalizing spin and pseudospin symmetries for relativistic spin 1/2 fermions

P. Alberto[1], M. Malheiro[2], T. Frederico[2], A. de Castro[3]

[1] Physics Department and CFisUC, University of Coimbra, P-3004-516 Coimbra, Portugal
[2]Instituto Tecnológico de Aeronáutica, DCTA, 12228-900 São José dos Campos, São Paulo, Brazil
[3]Departamento de Física e Química, Universidade Estadual Paulista, 12516-410 Guaratinguetá, São Paulo, Brazil

E-mail: pedro.alberto@uc.pt

Abstract. We propose a generalization of pseudospin and spin symmetries, the SU(2) symmetries of Dirac equation with scalar and vector mean-field potentials originally found independently in the 70's by Smith and Tassie, and Bell and Ruegg. As relativistic symmetries, they have been extensively researched and applied to several physical systems for the last 18 years. The main feature of these symmetries is the suppression of the spin-orbit coupling either in the upper or lower components of the Dirac spinor, thereby turning the respective second-order equations into Schrödinger-like equations, i.e, without a matrix structure. In this paper we use the original formalism of Bell and Ruegg to derive general requirements for the Lorentz structures of potentials in order to have these SU(2) symmetries in the Dirac equation, again allowing for the suppression of the matrix structure of the second-order equation of either the upper or lower components of the Dirac spinor. Furthermore, we derive equivalent conditions for spin and pseudospin symmetries with 2- and 1-dimensional potentials and list some possible candidates for 3, 2, and 1 dimensions. We suggest applications for physical systems in three and two dimensions, namely electrons in graphene.

1. Introduction

Pseudospin symmetry has been a hot topic in nuclear physics since the late 60's, when it was introduced to explain the near degeneracy of some single-particle levels near the Fermi surface. Most of the formulations were then non-relativistic but in 1997 Ginocchio was able to relate it with a symmetry of the Dirac equation with scalar S and vector V mean-field potentials such that $V = -S + C$ where C is a constant [1]. There is also a related symmetry, the spin symmetry, that has been used explain the suppression of spin-orbit splittings in states of mesons with a heavy and a light quark. Actually, both symmetries had been described in the 70's independently by Smith and Tassie [2] and by Bell and Ruegg [3] as SU(2) symmetries of the Dirac equation with scalar (coupling with mass) and vector potentials (coupling with energy). There several reviews of this subject, two extensive ones, [4] and [5], and some more focused as [6] give a good overall account of the many results produced in these last 18 years, especially in understanding the nature of the symmetries in view of their application in several physical systems.

One remarkable feature of these symmetries is the suppression of either the spin-orbit or the so-called pseudospin-orbit coupling that are present in the second-order equations for the upper and lower Dirac spinor components, respectively. Since those terms arise from the coupling

Content from this work may be used under the terms of the Creative Commons Attribution 3.0 licence. Any further distribution of this work must maintain attribution to the author(s) and the title of the work, journal citation and DOI.
Published under licence by IOP Publishing Ltd

of those spinor components in their first-order Dirac equations, they have a non-trivial, i.e., different from identity, matrix structure. Therefore, their suppression amounts to have the upper (spin symmetry) or lower (pseudospin symmetry) spinors satisfying second-order equations of Shrödinger type, i.e, with no matrix structure. This means that one can suppress spin-orbit couplings even in a relativistic fermion system. This may be surprising at first, since spin-orbit coupling is known to be a relativistic correction to non-relativistic quantum mechanics with only vector potentials [7], and actually leads to conservation of orbital angular momentum in relativistic fermion systems with several kinds of potentials [8, 9, 10].

In this paper we show how to obtain other pairs of potentials in the Dirac equation that produce this same suppression, under spin or pseudospin symmetry conditions. We extend this analysis to 2- and 1-dimensional potentials. A detailed account of the procedure can be found in [11]. We also suggest possible physical systems to which the new symmetries may be applied to.

2. General Spin and pseudospin symmetries in the Dirac equation

2.1. Generators of the spin symmetry

The time–independent Dirac equation for a spin 1/2 particle with energy E, under the action of an external hermitian V potential (which may include a mass term) with a general Lorentz structure reads

$$H\psi = (\boldsymbol{\alpha} \cdot \hat{\boldsymbol{p}} + V)\psi = E\psi \ . \tag{1}$$

where α_i $(i = 1, 2, 3)$ are the 4×4 matrices related to the usual Dirac gamma matrices γ^μ, $\mu = 0, 1, 2, 3$, by $\alpha_i = \gamma^0\gamma^i$. Units $\hbar = c = 1$ are used.

In order to find the requirements under which the potential V meets the conditions for spin or pseudospin symmetries of the Dirac equation (1), we follow closely the procedure of Bell and Ruegg [3].

We consider operators $P_\pm = (I \pm O)/2$, with I being the identity matrix in spinor space, and a matrix is this space such that $O^2 = I$. These have the projector properties

$$P_\pm P_\pm = P_\pm \tag{2}$$
$$P_\pm P_\mp = 0 \ . \tag{3}$$

We also require that the anti-commutator $\{\alpha_i, O\}$ $i = 1, 2, 3$ is zero, such that

$$P_\pm \alpha_i = \alpha_i P_\mp \quad i = 1, 2, 3 \ . \tag{4}$$

In order that one may have spin or pseudospin-like symmetries, the potential V has to have the general form $V = V_O O + V_v I$, where V_O and V_v are functions of the coordinates, such that it can be written as

$$V = V_+ P_+ + V_- P_- \qquad \text{with} \qquad V_\pm = V_v \pm V_O \ . \tag{5}$$

We apply now the projectors $P_\pm$ to the Dirac equation (1) to get the two coupled equations

$$\boldsymbol{\alpha} \cdot \hat{\boldsymbol{p}}\, \psi_- + V_+ \psi_+ = E\psi_+ \tag{6}$$
$$\boldsymbol{\alpha} \cdot \hat{\boldsymbol{p}}\, \psi_+ + V_- \psi_- = E\psi_- \ , \tag{7}$$

where $\psi_\pm = P_\pm \psi$.

If we now set one of the potentials $V_\pm$, for example V_-, to a constant C_-, i.e., $V_v = V_O + C_-$, we may apply $\boldsymbol{\alpha}\cdot\hat{\boldsymbol{p}}$ to equation (7), and, by using the general property $\boldsymbol{\alpha}\cdot\boldsymbol{A}\,\boldsymbol{\alpha}\cdot\boldsymbol{B} = \boldsymbol{A}\cdot\boldsymbol{B} + \mathrm{i}(\boldsymbol{A}\times\boldsymbol{B})\cdot\boldsymbol{\Sigma}$ where $\boldsymbol{\Sigma} = \boldsymbol{\alpha}\times\boldsymbol{\alpha}/(2\mathrm{i}) = \gamma^5\boldsymbol{\alpha}$ is the spin matrix in four-dimensional spinor space, we get

$$\hat{\boldsymbol{p}}^2\,\psi_+ = (E - C_-)\boldsymbol{\alpha}\cdot\hat{\boldsymbol{p}}\,\psi_- = (E - C_-)(E - V_+)\,\psi_+ \tag{8}$$

using also eq. (6). This is a Schrödinger-type equation for ψ_+ with no matrix structure. Therefore, it is invariant under the infinitesimal spin transformation

$$\psi_+ \to \psi_+ + \delta\psi_+ = \psi_+ + \frac{\boldsymbol{\epsilon}\cdot\boldsymbol{\Sigma}}{2\mathrm{i}}\psi_+ \tag{9}$$

The corresponding transformation for ψ_- is

$$\delta\psi_- = \frac{\boldsymbol{\epsilon}}{2\mathrm{i}}\cdot\frac{\boldsymbol{\alpha}\cdot\hat{\boldsymbol{p}}\,\boldsymbol{\Sigma}\,\boldsymbol{\alpha}\cdot\hat{\boldsymbol{p}}}{\hat{\boldsymbol{p}}^2}\,\psi_- \,. \tag{10}$$

For the transformation of the full spinor $\psi = \psi_+ + \psi_-$, and defining $\boldsymbol{s} = \boldsymbol{\alpha}\cdot\hat{\boldsymbol{p}}\,\boldsymbol{\Sigma}\,\boldsymbol{\alpha}\cdot\hat{\boldsymbol{p}}/\hat{\boldsymbol{p}}^2$, we get

$$\delta\psi = \delta\psi_+ + \delta\psi_- = \frac{\boldsymbol{\epsilon}}{2\mathrm{i}}\cdot(\boldsymbol{\Sigma}\psi_+ + \boldsymbol{s}\psi_-) = \frac{\boldsymbol{\epsilon}}{2\mathrm{i}}\cdot(\boldsymbol{\Sigma}\,P_+ + \boldsymbol{s}\,P_-)\psi \,, \tag{11}$$

from which we can write the generator of this transformation as

$$\boldsymbol{S}_- = \boldsymbol{\Sigma}\,P_+ + \boldsymbol{s}\,P_- \,. \tag{12}$$

One can obtain the second-order equation for ψ_- from eqs. (6) and (7). It reads

$$\hat{\boldsymbol{p}}^2\,\psi_- + \frac{1}{E - V_+}(\nabla V_+ \times \hat{\boldsymbol{p}}\cdot\boldsymbol{\Sigma} - \mathrm{i}\,\nabla V_+\cdot\hat{\boldsymbol{p}})\psi_- = (E - C_-)(E - V_+)\,\psi_- \,. \tag{13}$$

If the potential V_+ is radial, in the second term in the left-hand side of the equation we can identify a spin-orbit coupling term and the Darwin term [6].

One can show that these generators satisfy a SU(2) algebra, i.e.,

$$[(S_-)_i, (S_-)_j] = 2\mathrm{i}\,\varepsilon_{ijk}(S_-)_k \tag{14}$$

and that they commute with the Hamiltonian $H_- = \boldsymbol{\alpha}\cdot\hat{\boldsymbol{p}} + V_+P_+ + C_-P_-$, provided one has $\{\boldsymbol{\alpha}, O\} = 0$ or $[P_\pm, \boldsymbol{\Sigma}] = 0$, which is actually a consequence of (4) [11].

2.2. Generators of the pseudospin symmetry

Of course, we could as well have set instead V_+ in (5) to a constant C_+. In that case, the roles of $\psi_\pm$ would be reversed and one would have another symmetry whose generator would be

$$\boldsymbol{S}_+ = \boldsymbol{\Sigma}\,P_- + \boldsymbol{s}\,P_+ \,, \tag{15}$$

which would commute with the Hamiltonian

$$H_+ = \boldsymbol{\alpha}\cdot\hat{\boldsymbol{p}} + V_-P_- + C_+P_+ \,. \tag{16}$$

Similarly as the case for spin symmetry, one can show that these generators commute with H_+ and also satisfy a SU(2) algebra.

The second-order equations for the upper and lower spinors would then be

$$\begin{aligned}\hat{\boldsymbol{p}}^2\,\psi_+ &+ \frac{1}{E - V_-}(\nabla V_- \times \hat{\boldsymbol{p}}\cdot\boldsymbol{\Sigma} - \mathrm{i}\,\nabla V_-\cdot\hat{\boldsymbol{p}})\psi_+ = (E - C_+)(E - V_-)\,\psi_+\\ \hat{\boldsymbol{p}}^2\,\psi_- &= (E - C_+)(E - V_-)\,\psi_- \,.\end{aligned}$$

This case, $V_v = -V_O + C_+$, is usually known as the pseudospin symmetry case of the Dirac Hamiltonian, since we have a normal spin transformation in the lower component of the Dirac spinor which has an inverse parity relative to the upper component and to the whole spinor.

3. Potentials allowing for general spin or pseudospin symmetries in 3, 2 and 1-dimensional space

From the previous section, in order to have one of these two SU(2) symmetries, the matrix O must satisfy the following relations:

(i) $O^2 = I$

(ii) $\{\alpha_i, O\} = 0 \qquad i = 1, 2, 3.$

As explained before, the condition $[O, \Sigma_i] = 0$ is also satisfied, as a consequence of condition (ii). These requirements are satisfied by the Hermitian matrices $\beta = \gamma^0$ and $\mathrm{i}\gamma^0\gamma^5$. The case of $O = \gamma^0$ leads to the well-known spin and pseudospin symmetries described in the Introduction.

If one weakens the second requirement, one can also consider the linear combination $O = \boldsymbol{\lambda}\cdot\boldsymbol{\alpha}$, such that $\boldsymbol{\lambda}$ is a constant unit vector ($\boldsymbol{\lambda}\cdot\boldsymbol{\lambda} = 1$). Indeed, if one requires that the anti-commutator $\{\boldsymbol{\alpha}\cdot\hat{\boldsymbol{p}}, O\}$ applied to ψ is zero, then $\boldsymbol{\lambda}\cdot\hat{\boldsymbol{p}}\,\psi = 0$. The effect, as before, is to fulfill the property $P_\pm\boldsymbol{\alpha}\cdot\hat{\boldsymbol{p}}\,\psi = \boldsymbol{\alpha}\cdot\hat{\boldsymbol{p}}P_\mp\,\psi$. Then the third condition can be satisfied in a weak way, considering transformations with infinitesimal parameters $\boldsymbol{\epsilon}$ such that

$$[\boldsymbol{\lambda}\cdot\boldsymbol{\alpha}, \boldsymbol{\epsilon}\cdot\boldsymbol{\Sigma}] = 2\mathrm{i}\,(\boldsymbol{\lambda}\times\boldsymbol{\epsilon})\cdot\boldsymbol{\alpha} = 0\ , \tag{17}$$

meaning that the vectors $\boldsymbol{\lambda}$ and $\boldsymbol{\epsilon}$ must be parallel. For instance, if $\boldsymbol{\lambda} = \hat{e}_z$, i.e., $O \equiv \alpha_3 = \gamma^0\gamma^3$, then one should have $\hat{p}_3\,\psi = 0$ and $\boldsymbol{\epsilon} = \epsilon\hat{e}_z$. In this case the symmetry generator would be the matrix Σ_3, generator of the two-dimensional rotation group in four-component spinor space, which is a realization of the unitary group $U(1)$. Our problem would be 2-dimensional, i.e., the spinor (and potentials), would depend only on coordinates x, y.

Another possibility for O would be the linear combination of the space components of the tensor operator in spinor space $\gamma^0\sigma^{0i} = \mathrm{i}\beta\alpha_i$, i.e., $O = \mathrm{i}\beta\boldsymbol{\lambda}\cdot\boldsymbol{\alpha}$. The first requirement would be met again by setting $\boldsymbol{\lambda}\cdot\boldsymbol{\lambda} = 1$. The second requirements and the commutator condition $[O, \Sigma_i] = 0$ would be met by setting, respectively,

$$\{\beta\boldsymbol{\lambda}\cdot\boldsymbol{\alpha}, \boldsymbol{\alpha}\cdot\hat{\boldsymbol{p}}\}\,\psi = \beta[\boldsymbol{\lambda}\cdot\boldsymbol{\alpha}, \boldsymbol{\alpha}\cdot\hat{\boldsymbol{p}}]\,\psi = 2\mathrm{i}\,\beta\boldsymbol{\lambda}\times\hat{\boldsymbol{p}}\cdot\boldsymbol{\Sigma}\,\psi = 0\ , \tag{18}$$

$$[\beta\boldsymbol{\lambda}\cdot\boldsymbol{\alpha}, \boldsymbol{\epsilon}\cdot\boldsymbol{\Sigma}] = \beta[\boldsymbol{\lambda}\cdot\boldsymbol{\alpha}, \boldsymbol{\epsilon}\cdot\boldsymbol{\Sigma}] = 2\mathrm{i}\,\beta\boldsymbol{\lambda}\times\boldsymbol{\epsilon}\cdot\boldsymbol{\alpha} = 0\ . \tag{19}$$

The first condition would be satisfied if $\boldsymbol{\lambda}\times\hat{\boldsymbol{p}}\,\psi = 0$ and the second one if $\boldsymbol{\epsilon}$ is parallel to $\boldsymbol{\lambda}$. If one chooses again $\boldsymbol{\lambda} = \hat{e}_z$, this would give rise to a 1-dimensional potential, depending only on z.

4. Discussion and conclusions

We have derived the general conditions under which a general potential plus a vector potential give rise to spin and pseudospin-like symmetries in the Dirac equation, i.e., lead to a Schrödinger-like equation for the upper or lower component of the Dirac spinor. In three-dimensional space, we showed that there are two potentials which satisfy those conditions: a scalar potential, giving rise to the usual the spin and pseudospin symmetries found independently by Smith and Tassie [2], and Bell and Ruegg [3] , and a pseudoscalar potential. In this last case the Dirac Hamiltonian would read

$$H = \boldsymbol{\alpha}\cdot\hat{\boldsymbol{p}} + \mathrm{i}\beta\gamma^5 V_{ps} + V_v\ , \tag{20}$$

with $V_{ps} = \pm V_v \mp C_\mp$. In physical terms, this would correspond to a system of massless fermions interacting with mean-field pseudoscalar and vector potentials which have the same magnitude up to a constant. One physical system in which this symmetry would be slightly broken would describe a fermion, say, a baryon, with a relatively small (effective) mass, interacting with a pion and ω meson.

On other hand, if we constrain the fermion motion to 2- and 1-dimensional space, there are additional potentials for which these symmetries can be realized. The respective spin or pseudospin symmetric Hamiltonians would look like

$$H_2 = \alpha_x \hat{p}_x + \alpha_y \hat{p}_y + \alpha_z V_z + V_{2v} \qquad (21)$$

$$H_1 = \alpha_z \hat{p}_z + \mathrm{i}\beta\alpha_z V_t + V_{1v} \ , \qquad (22)$$

with 2- and 1- dimensional mean-field potentials such that $V_z(x,y) = \pm V_{2v}(x,y)$ and $V_t(z) = \pm V_{1v}(z)$. $V_z(x,y)$ is a (space) vector potential, whereas $V_t(z)$ is a tensor potential.

Equation (21) describes a 2-dimensional massless fermion system with a energy coupling potential and a vector (γ^3) potential. There is actually a physical system that this Hamiltonian can be applied to: electrons in graphene, the so called Dirac electrons. These effective particles can be described by a massless 3+1 Dirac equation within the framework of interacting quantum field theories (see e.g. [12, 13]). One example is the continuum spectrum of the Dirac electron interacting with two dimensional potentials embedded in a 3+1 space [14]. Again, in that theory one has in general also potentials with Lorentz structure other than vector, and, in this case, the third component of a four-vector potential ($\gamma^0\gamma^3 = \alpha_z$) (note that the Lorentz character of $V_O O$ is given by its form in the covariant form of the Dirac equation, i.e., $\gamma^0 V_O O$). This opens the possibility of looking into what would be the effect of these symmetries on the continuum [14] and discrete spectrum of the Dirac electrons, as well as its breaking due the other potentials or to the fact that the condition $V_{2v} \pm V_z = C_\pm$ is not fulfilled. In this way, the Dirac electrons in graphene could be a tool to study the consequences of the generalized spin and pseudospin symmetries in a controllable form. We leave for a future work such detailed investigation.

Acknowledgments
PA would like to thank the Universidade Estadual Paulista, Guaratinguetá Campus, for supporting his stays in its Physics and Chemistry Department. MM and TF would like to thank FAPESP for support under the thematic project 2015/26258-4 and CNPq for partial support.

References

[1] J. N. Ginocchio, Phys. Rev. Lett. **78**, 436 (1997).
[2] G. B. Smith and L. J. Tassie, Ann of Phys **65**, 352 (1971).
[3] J. S. Bell and H. Ruegg, Nucl. Phys. **B98**, 151 (1975); Errata, Nucl. Phys. **B104**, 546 (1976).
[4] J. N. Ginocchio, Phys. Rep. **414** 165 (2005).
[5] H. Liang, J. Meng,, S.-G. Zhou, Phys. Rep. **570** 1 (2015).
[6] P. Alberto, A. Castro, M. Fiolhais, R. Lisboa and M. Malheiro, J. of Phys., Conference Series **490** (2014) 012069.
[7] P. Alberto, M. Fiolhais, and M. Oliveira, Eur. J. Phys. **19**, 553 (1998).
[8] R. Lisboa, M. Malheiro, A. S. de Castro, P. Alberto, and M. Fiolhais, Phys. Rev. C **69**, 024319 (2004); A. S. de Castro, P. Alberto, R. Lisboa and M. Malheiro, Phys. Rev. C **73**, 054309 (2006).
[9] P. Alberto, A. S. de Castro and M. Malheiro Phys. Rev. C **75**, 047303 (2007).
[10] A. S. de Castro and P. Alberto, Phys. Rev. A **86**, 032122 (2012).
[11] P. Alberto, M. Malheiro, T. Frederico, A. de Castro Phys. Rev. A **92**, 062137 (2015).
[12] R. Jackiw and S.-Y. Pi, Phys. Rev. Lett. **98**, 266402 (2007).
[13] O. Oliveira, C. E. Cordeiro, A. Delfino, W. de Paula, T. Frederico, Phys. Rev. B **83**, 155419 (2011).
[14] C. Popovici, O. Oliveira, W. de Paula, and T. Frederico, Phys. Rev. B **85**, 235424 (2012).

Statistical classification techniques in high energy physics (SDDT algorithm)

Petr Bouř, Václav Kůs and Jiří Franc

Department of Mathematics, Faculty of Nuclear Sciences and Physical Engineering, Czech Technical University in Prague, Trojanova 13, 12000 Praha 2, Czech Republic

E-mail: petr.bour@fjfi.cvut.cz, vaclav.kus@fjfi.cvut.cz, jiri.franc@fjfi.cvut.cz

Abstract. We present our proposal of the supervised binary divergence decision tree with nested separation method based on the generalized linear models. A key insight we provide is the clustering driven only by a few selected physical variables. The proper selection consists of the variables achieving the maximal divergence measure between two different classes. Further, we apply our method to Monte Carlo simulations of physics processes corresponding to a data sample of top quark-antiquark pair candidate events in the lepton+jets decay channel. The data sample is produced in $p\bar{p}$ collisions at $\sqrt{s} = 1.96$ TeV. It corresponds to an integrated luminosity of 9.7 fb^{-1} recorded with the D0 detector during Run II of the Fermilab Tevatron Collider. The efficiency of our algorithm achieves 90% AUC in separating signal from background. We also briefly deal with the modification of statistical tests applicable to weighted data sets in order to test homogeneity of the Monte Carlo simulations and measured data. The justification of these modified tests is proposed through the divergence tests.

1. Introduction and Data Sample

Experiments running on particle accelerators produce a vast amount of raw data containing not only the crucial signal but also a large number of observations originating from another physical processes (backgrounds). In this paper, we are concerned with signal separation from massively contaminated data measured with the D0 detector during Run II of the Fermilab Tevatron Collider [1]. We are interested in top-antitop quark pair production after a collision of accelerated protons and antiprotons with one lepton (electron or muon) and at least two jets in the final state. This study is based upon the full Tevatron data sample recorded by the D0 detector at $\sqrt{s} = 1.96$ TeV and, after applying data quality requirements, corresponding to an integrated luminosity of 9.7 fb^{-1} [2]. The selection in the lepton+jets decay channel is similar to the one described in [3].

Let us denote the particular class of observations by S (signal) and the class of different decay processes by B (backgrounds). In order to train our classification algorithm, we are provided with the Monte Carlo (MC) simulations coming from different MC event generators, but with known classification to S or B. The dimension of MC is limited to $m = 46$ selected physical variables. After validation of homogeneity between data and MC, we perform supervised classification. However, there is a weighting applied to MC we have to cope with.

Content from this work may be used under the terms of the Creative Commons Attribution 3.0 licence. Any further distribution of this work must maintain attribution to the author(s) and the title of the work, journal citation and DOI.
Published under licence by IOP Publishing Ltd

2. Divergence Measures and Variable Selection

Due to the nature of our separation techniques we use the specific divergence measures as a tool for wise variable selection in every step of our signal separation. Since some variables are strongly correlated with each other and they differ in signal separability, we select only those variables of the highest distinction between probability distributions of S and B. For this purpose, we define *normalized Rényi distance* of order α for distributions P, Q with probability densities p, q as $R_\alpha(P,Q) = \frac{1}{\alpha(1-\alpha)} \int p^\alpha q^{(1-\alpha)}\, \mathrm{d}\mu$, where $\alpha \in \mathbb{R}$, $\alpha \neq 0$, $\alpha \neq 1$, and μ is a dominating measure on $\mathbb{R}$. Moreover, by [4], we have $R_1(P,Q) = \lim_{\alpha\to 1} R_\alpha(P,Q)$, and $R_0(P,Q) = \lim_{\alpha\to 0} R_\alpha(P,Q)$, to be the reversed Kullback-Leibler divergence $KL(Q,P)$, and $KL(P,Q)$, respectively. It is easily seen that it holds $R_\alpha(P,Q) = R_{1-\alpha}(Q,P)$.

To evaluate the $R_\alpha(P,Q)$ divergence for weighted MC, we consider the modified weighted empirical distribution function (EDF) based on an analogy with EDF for weighted inputs. Let $\boldsymbol{X} = (X_1, \ldots, X_n)$ be iid random variables distributed by cumulative distribution function (CDF) $F(x)$ and let $(w_1, \ldots, w_n)$ be respective weights, where $W = \sum_{i=1}^n w_i$. We define the weighted empirical distribution function (WEDF) to be $F_n^W(x) = (1/W)\sum_{i=1}^n w_i I_{(-\infty,x]}(X_i)$, where $I_A(X)$ is the indicator of the set A. In the case of $w_i = 1$ for all $i \in \widehat{n}$, that is the unweighted recorded data, the definition of WEDF coincides with the usual EDF. The continuous version of WEDF serve for finding of the weighted histograms with equiprobable binnings, see [5] in detail. This continuous estimate of EDF is essential for robust equiprobable binning because the value of the normalized Rényi distance is independent on the bin width and also on the number of bins, provided we keep the number of observations in a bin at a reasonable level. Equiprobable binning with corresponding histograms of S and B classes for selected variables in MC can be seen in figure 1. Notice the considerable similarity of both distributions.

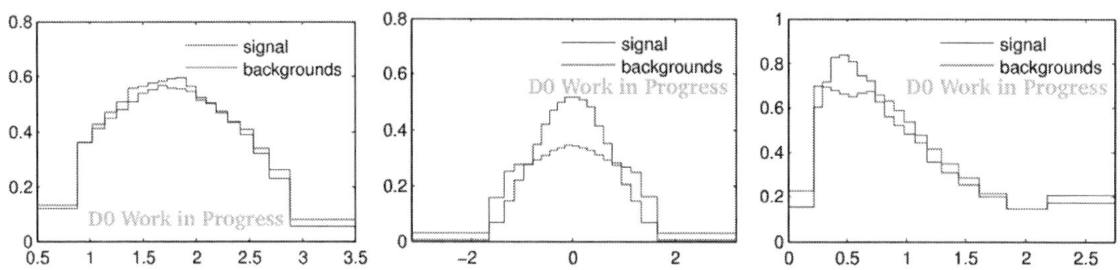

Figure 1. Weighted histograms with equiprobable binnings for variables $\Delta\phi\left(j^1, j^2\right)$, η^{j_1} and K_t^{minp} in MC channel Electron 2 Jets.

During our classification algorithm we need to select the variables potentially useful for separation of S and B. Therefore, we are looking for those variables with the highest value of divergence $R_\alpha(P,Q)$ between S and B. The question naturally arises whether we shall use the distribution of S or B in the first argument of $R_\alpha(P,Q)$. Which value of the parameter α should be chosen is also of some interest. Fortunately, there is a very satisfactory and consistent solution to this problem, due to the specific symmetry mentioned above. Observe the dependence of the symmetry with its divergence values on parameter α in figure 2. Thus we choose $\alpha = \frac{1}{2}$, which provides us with the unique fixed point for comparison of all variables since $R_{1/2}(S,B) = R_{1/2}(B,S)$.

3. Supervised Divergence Decision Tree

The binary supervised divergence decision tree (SDDT) was originally proposed and tested in [5]. This universal classification method is based upon certain localization of separation

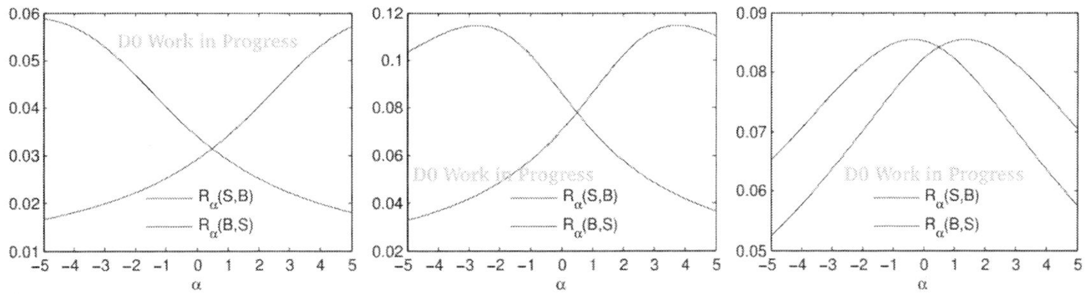

Figure 2. Dependence of normalized Rényi distance on the parameter α for distributions of S and B for variables $\Delta\phi\left(j^1, j^2\right)$, η^{j_1} and K_t^{minp} in MC channel Electron 2 Jets.

among nodes of the tree. We shall briefly describe the final algorithm. Suppose we are given weighted simulation MC $\subset \mathbb{R}^m$, where m stands for the number of MC variables. For every observation $x_i \in$ MC there is a weight $w_i > 0$. Denote by $X, Y \subset$ MC the training set and test set, respectively. Recall the definition of the signal class S and the backgrounds class B from section 1. Let $U = \{\{U_X\}, \{U_Y\}\}$ be a current node for the algorithm, where $U_X \subset X, U_Y \subset Y$. After the separation of the node U, we denote by $U_0, U_1 \subset U$ the sets of observations classified as B, S, respectively. For the simplicity, we write $U^{(k)}$ for the current node, where k is an unique binary code reflecting the path in the tree. Let $M^{(k)}$ be the separation model produced in the node $U^{(k)}$. For every $x_i \in U^{(k)}$ we obtain the probability $p_i^{(k)}$ of signal classification by the model $M^{(k)}$. Next, we insert X, Y into the root of the tree and declare the root as a current node $U = \{\{U_X\}, \{U_Y\}\} = \{\{X\}, \{Y\}\}$. From now on, the SDDT learning algorithm proceeds as follows:

(i) For appropriately chosen $\gamma > 0$, if $\sum_{x_i \in U_X \cap S} w_i < \gamma$ or $\sum_{x_i \in U_X \cap B} w_i < \gamma$ then we declare the current node U as a pure final leaf.

(ii) For appropriately chosen $d \in \widehat{m}$ and for every combination of d-dimensional selections from m variables, under use of histogram with equiprobable binning, we compute ϕ-divergence $D_\phi^{(j)}(S, B)$ in the current node U, where $j \in \{1, \ldots, \binom{m}{d}\}$. We used here the normalized Rényi distance of order α.

(iii) From all of the $\binom{m}{d}$ combinations we select the j^*-th combination such that the ϕ-divergence $D_\phi^{(j^*)}(S, B)$ is maximal in U, i.e. $j^* = \arg\max\{D_\phi^{(j)}(S, B) : j = 1, \ldots, \binom{m}{d}\}$.

(iv) Only for d selected variables in j^*-th combination we are training the nested classification based on generalized linear models ([6]) on the training set U_X. Further, we optimize and test the produced model $M^{(k)}$ on the test set U_Y.

(v) In the case where overfitting is detected, we repeat step iv with the different setting of parameters in the classification tool. Otherwise, we record the model $M^{(k)}$ and the probability $p_i^{(k)}$ for every $x_i \in U$.

(vi) We split the node U with the model $M^{(k)}$ per d selected variables in j^*-th combination into two new nodes U_0, U_1.

Following the algorithm we arrive at probability vector $D_i = (p_i^{(k_1)}, \ldots, p_i^{(k_{l_i})})$ for each $x_i \in \{X \cup Y\}$, where l_i is the number of levels in the tree for x_i. The resulting discriminant of probabilities of signal classification $D = (p_1, \ldots, p_N)$ can be obtained in several manners indicated in [5], including for example processing the discriminants D_i by artificial neural networks or even by the secondary application of SDDT to discriminants D_i. For the results in section 4 we take into account the arithmetic mean of the discriminant D_i.

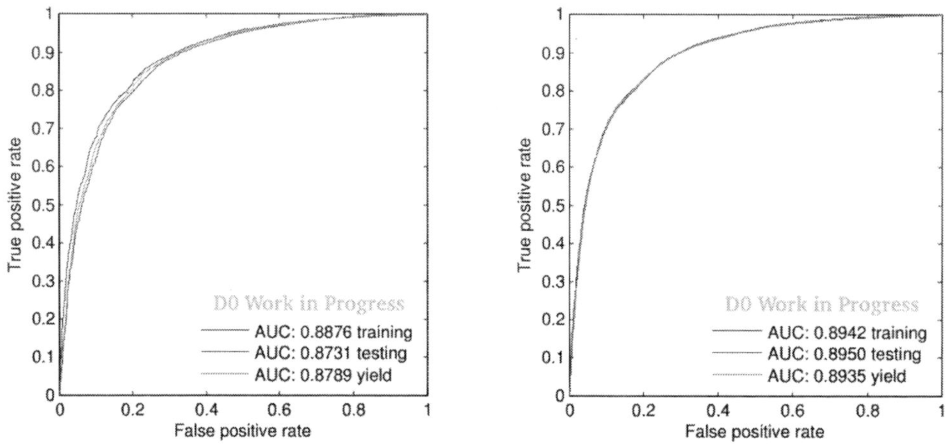

Figure 3. ROC for SDDT signal separation in MC Electron 4+ Jets and Muon 4+ Jets channels.

Table 1. AUC for SDDT signal separation in MC Electron 4+ Jets and Muon 4+ Jets channels.

	Electron 4+ Jets			Muon 4+ Jets		
d	training	test	yield	training	test	yield
2	0.8840	0.8694	0.8764	0.8895	0.8900	0.8884
3	0.8863	0.8710	0.8776	0.8908	0.8909	0.8901
4	0.8876	0.8731	0.8789	0.8942	0.8950	0.8935

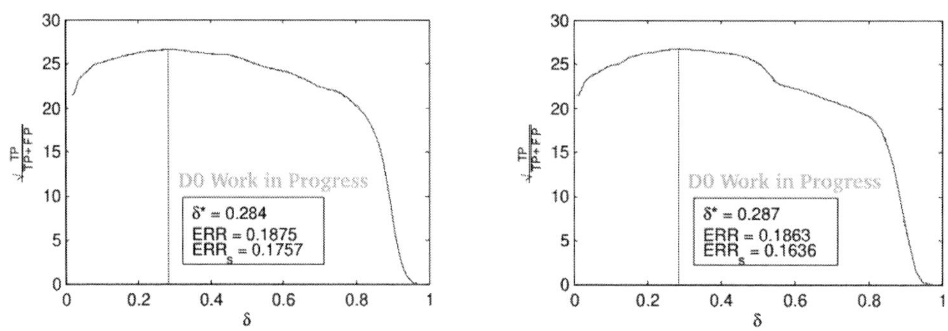

Figure 4. Optimal cut δ^* in $D = (p_1, \ldots, p_N)$ for final SDDT signal separation in MC channel Electron 4+ Jets for $d = 2$ and $d = 3$.

4. Study of signal separation in MC $t\bar{t}$ lepton+jets decay channel

Now let us describe the SDDT separation of signal S from MC D0. Prior to MC classification, we tested the performance of SDDT in order to reflect the dependence on the choice of parameters γ, d and α. The success of SDDT separation is measured via the Area Under Curve (AUC) of the Receiver Operating Characteristic (ROC), as observed in figure 3. Generally, AUC for SDDT separation grows with higher dimension d and remains without notable overfitting. As an example, see table 1, where AUC results are quite comparable with AUC obtained in [7]. We close this section by showing the example of the optimal cut δ^* such that the observations x_i are classified as signal S iff $p_i \geq \delta^*$. For this cutting point we can see the overall error of classification ERR and the error ERR_s within signals S in figure 4. The universality of the originally proposed SDDT method consists in almost arbitrary choice of nested separation technique. Among others, there is a great potential in optimization of the whole SDDT procedure through the appropriate choice of SDDT parameters γ, d, α. However, the profound research still needs to be carried out.

5. Addendum: Weighted Tests of Homogeneity

Prior to supervised learning of any classification method, it is vital to guarantee homogeneity of measured data and MC distributions. In order to avoid the robustness problems of a particular parametric family (for the normal distribution case see [8]), we pursued nonparametric homogeneity testing based upon EDFs of two data sets $\boldsymbol{X} = (X_1, \ldots, X_{n_1})$, $\boldsymbol{Y} = (Y_1, \ldots, Y_{n_2})$, with respective distribution functions F, G. By the homogeneity hypothesis, as our null hypothesis H_0, we understand $H_0 : F = G \quad \text{vs} \quad H_1 : F \neq G$ at significance level $\alpha \in (0, 1)$. The nature of homogeneity testing prompted us to look for the p-value, i.e., the lowest significance level for which we reject the hypothesis H_0. We considered two homogeneity tests based on EDFs of our data sets: two sample Kolmogorov-Smirnov test [9], and two sample Anderson-Darling test [10]. However, for weighted data samples we were forced to replace EDFs with their respective WEDFs. These modified tests for the weighted data samples do not have to obey the asymptotic statistical properties of the original procedures. Let us stress that the p-value obtained using the modified weighted tests can not be considered a regular approximate p-value without subsequent detailed research. However, we carried out the numerical verification of the modified approaches in [5] and found there is no significant difference observed in p-values related to regular unweighted tests.

Acknowledgments

This work was supported by the MEYS CZ grants LG15047, LM2015068, SGS15/214/OHK4/3T/14, and GA16-09848S.

References

[1] D0 Collaboration Abazov V M et al. 2006 The upgraded D0 detector *Nucl.Instrum.Methods Sect.A* **565**, 463
[2] Casey B C K et al. 2013 The D0 Run IIb Luminosity Measurement *Nucl. Instrum. Methods Sect. A* **698**, 208
[3] D0 Collaboration Abazov V M et al. 2014 Measurement of differential $t\bar{t}$ production cross sections in $p\bar{p}$ collisions *Phys. Rev. D* **90**, 092006
[4] Vajda I 2006 On divergences and informations in statistics and information theory *IEEE Trans. Inf. Theory* **52** 4394–412
[5] Bouř P 2015 *Testing of separation approaches with attributes selections for physical data (FNAL/CERN)* (Prague: SPMS 2015 proceedings, FNSPE CTU in Prague)
[6] McCulloch C E, Searle S R and Neuhaus J M 2008 *Generalized, Linear, and Mixed Models* (New York: Wiley)
[7] Stepanek M, Franc J and Kus V 2015 Modification of Gaussian mixture models for data classification in high energy physics *Journal of Physics: Conference Series* **574** 012150
[8] Frydlova I, Vajda I and Kus V 2012 Modified power divergence estimators in normal model- simulation and comparative study, *Kybernetika* **48**(4) 795–808
[9] Smirnov N J 1944 Approximate laws of distribution of random variables from empirical data *Usp. Mat. Nauk* **10** 179–206
[10] Engmann S and Cousineau D 2011 Comparing distributions: the two-sample Anderson-Darling test as an alternative to the Kolmogorov-Smirnoff test *J. Appl. Quant. Methods* **6** 1–17

Modeling a Kolmogorov-Type Magnetic Field in the Galaxy and its Effect on an Extragalactic Isotropic Flux of Ultra High Energy Cosmic Rays

Pantea Davoudifar

Research Institute for Astronomy and Astrophysics of Maragha (RIAAM)- Maragha, IRAN, P.O.Box: 441-55134

E-mail: dfpantea@riaam.ac.ir

Abstract. A model of turbulent galactic magnetic fields was developed in which, the type of turbulence were considered to be Kolmogorov. We tested the effect of this model on an isotropically distributed flux of ultra high energy cosmic ray in the extragalactic space. To do this, a giant Galactic halo (radius of $\sim 100Mpc$) was considered. Regular and random components of the Galactic Magnetic Fields were considered to have the mean observed relevant values and also satisfy a Kolmogorov field type. The deviation from isotropy then were calculated considering the propagation of ultra high energy protons in such a magnetic field and results were discussed to show how isotropic is the flux of ultra high energy cosmic rays in the extragalactic space. It is seen that considering an isotropic flux of ultra high energy cosmic rays in the intergalactic space for different choices of galactic magnetic field is not consistence with the distribution of observed ultra high energy events.

1. Introduction

The strength of Galactic and Extragalactic Magnetic Fields (GMF and EGMF) are possible to be estimated considering different observational techniques. In fact, most of our knowledge about magnetic fields in the outer space is obtained through studying different interactions between matter, waves and magnetic fields within the stellar atmospheres and in Interstellar (IS) and Intergalactic (IG) spaces. Extensive works were done to map [1] the magnetic fields of galaxies and also modeling their fields theoretically or numerically.

Where, physical processes such as Zeeman effect [2] are mostly used to study MFs within the galaxies, on the stellar atmosphere and in more dense regions of the Galaxy; Faradey Rotation (FR) [3] is widely used to evaluate mean values of MF strengths along the line of sight and is applicable both in the IS and IG spaces with low matter densities. In 1951, it was suggested that the gas in Orion nebula was turbulent with a Kolmogorov regime [4], but this suggestion later was challenged by other theories which essentially questioned the existence of a pervasive turbulence in the ISM [5]. Up to 1980s, many physical processes and observations were discussed considering a uniform ISM (for a good review see [6]) and the turbulence was more usually accepted at small scales. In 1980s experimental evidences were in the favor of large scale turbulence and compression from interstellar turbulence was considered as one of the main mechanisms of cloud-formation [7].

Content from this work may be used under the terms of the Creative Commons Attribution 3.0 licence. Any further distribution of this work must maintain attribution to the author(s) and the title of the work, journal citation and DOI.
Published under licence by IOP Publishing Ltd

In a special case, MF models are important at large distances when an Ultra High Energy Charged Particle (UHECP) travels to the Earth from its source. But, considering a top-down model to discuss about the origin of Ultra High Energy Cosmic Rays (UHECR), no special direction is possible to be considered for their sources. To examine this assumption, we studied a case of isotropically distributed flux of UHECR in IG space. We considered our Galaxy to be a point in the universe, as always it is assumed in cosmology. In a top-down model, an observer on this point measures the same flux of UHECR in all directions and also measure the same flux when moving on a sphere which we consider to be our Galaxy. The main question is that: how isotropic is the observed flux of UHECR inside this sphere?

2. Model of Galactic Magnetic Field

We have used a previously developed model [8, 9, 10], considering a Kolmogorov turbulent regime. The Galaxy considered to be consisted of a giant halo with a radius of 100 Mpc and the thickness of disk were considered to be 0.3 kpc with the radius of 30 kpc. For the GMF in disk, we used the model of Rand and Kulkarini [11] (Model A). Out of the disk, the regular component were considered to have a bipolar structure. To compare with other results we considered the regular component of GMF from model of Harari, Mollerach and Roulet [12] (Model B). The same Galactic dimensions were applied for model B.

We considered cells of orders of $100pc$ in halo and $50pc$ in Galactic disk [13] to define the turbulent Kolmogorov regime. For both cases the random components were considered to have the strength of $3\mu G$ in Halo and $1\mu G$ in Galactic disk.

3. Isotropy v.s. Anisotropy

To study the effect of GMF on the anisotropy of the observed UHE events, we assumed a sphere with the radios of $100Mpc$ (i.e our Galaxy). We studied the deviation from isotropy using:

$$\delta(\alpha,\delta)_i = \frac{N(\alpha,\delta)_i - \langle N\rangle(\alpha,\delta)_i}{\langle N\rangle(\alpha,\delta)_i}. \tag{1}$$

in which, $(\alpha,\delta)_i$ stands for the Galactic right ascension and declination of each bin on a sphere of radios of 100 Mpc. An initially isotropic flux of UHCR were considered to enter this sphere. We considered 100 protons enter along the radius of this sphere (i.e Galaxy) on each bin. The number of bins were considered 180×360. So the value of $\langle N\rangle(\alpha,\delta)_i$ were considered to be 100 for each bin. The new $(\alpha,\delta)_i$ of each event then were calculated considering the propagation of each UHECR in the Galaxy and obtaining the value of its deflection angle after traveling 100 Mpc. To do this we did 6480000 run and an output of 6480000 of new $(\alpha,\delta)_i$ were produced.

At this stage we counted the number of events in each Galactic bin (i.e. $(\alpha,\delta)_i$) and the values of $N(\alpha,\delta)_i$ were obtained.

Now considering equation 1 it is possible to calculate the deviation from isotropy for each bin of the sky.

4. The Results

We have shown the results for model A in figure 1.

The effect of a bipolar GMF is seen in figure 1. Considering model B the situation is changed. As in model B, GMF has not a bipolar structure out of the disk, the difference of north to south is not obvious like model A (figure 2).

The interesting result is that, considering an extragalactic origin of CR particles, an initially homogeneous and isotropic flux wont be observed isotropic inside the Galaxy. In fact, degrees of clustering maybe observed due to the turbulent behavior of the fields. And this may distorted our studies about the origin of UHECR.

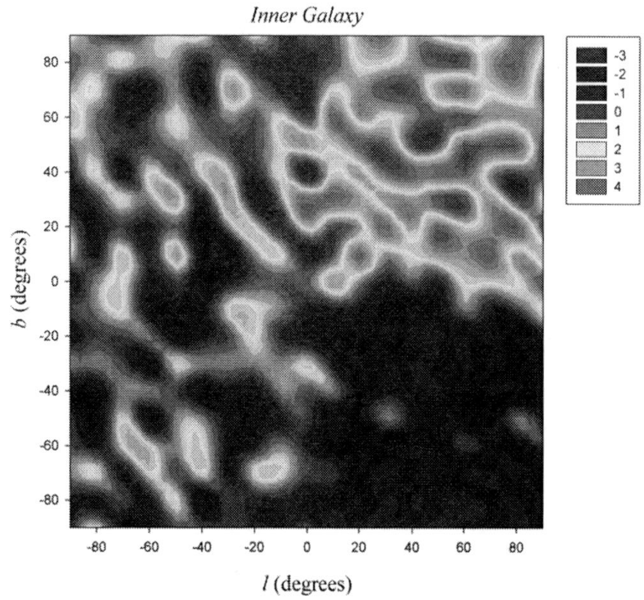

Figure 1. Deviation from Isotropy for EG protons with energies ranged in 10^{18-19}. The region of Inner Galaxy is mapped in which the deviation is multiplied by 100. Model A of GMF was used.

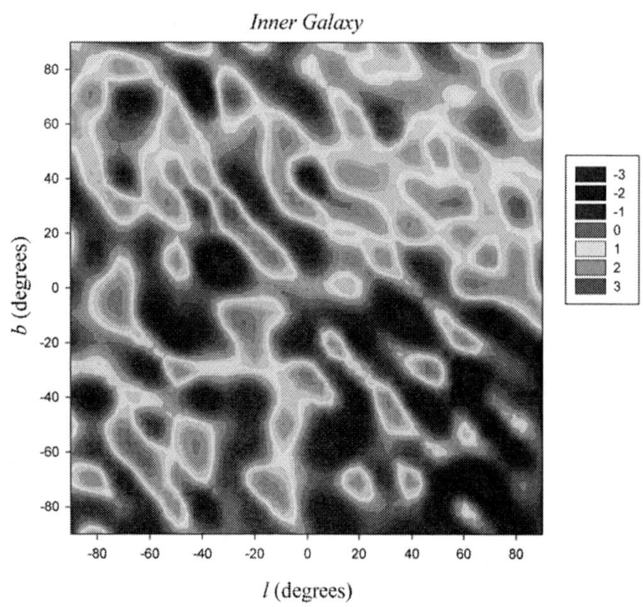

Figure 2. Deviation from Isotropy for EG protons with energies ranged in 10^{18-19}. The region of Inner Galaxy is mapped in which the deviation is multiplied by 100. Model B of GMF was used.

Other simulations shows that at higher energies the deviation from isotropy is smaller. Whilst for protons with energies in the range of $10^{18-19}eV$ the assumption of an isotropic extragalactic flux does not seem possible, for protons with energies of $10^{19-20}eV$ this assumption depends on the Galactic magnetic field model, and for particles with energies of $10^{20-21}eV$ the assumption seems possible.

Anyhow, it does not mean that for the highest energies a top-down model is verified but at least one may consider the observed flux is not affected but the Galactic magnetic field.

4.1. Acknowledgments
This work has been supported financially by Research Institute for Astronomy and Astrophysics of Maragha (RIAAM) under research project No. $1/4165 - 23$.

References
[1] http://www.esa.int/spaceinimages/Images/2014/05/, Milky Way's magnetic fingerprint
[2] Ruzmaikin A.A., Shukurov A.M., Sokoloff D.D, 1988, "Magnetic Fields of Galaxies", Kluwer Academic Publishers
[3] Frick P., Sokoloff D.D., Stepanov R., Beck R., 2009, Monthly Notices of Royal Astronomical Society Letters, 401, L24-L28
[4] Von Hoerner S., 1951, Zeitschrift fr Astrophysik, 30, 17-64
[5] Oort J.H., Spitzer L.Jr., 1955, Astrophysical Journal, 121, 6-23
[6] Elmegreen B.G., Scalo J., 2004, Anuual Review of Astronomy and Astrophysics, 42, 211-273
[7] Elmegreen B.G., 1991, The Physics of Star Formation and Early Stellar Evolution, ed. Lada C.J., Kylafis N.D., Dordrecht:Kluwer, 3560.
[8] Davoudifar P., Fatemi S.J., Clay R., Whelan B., 2011, Journal of Sciences, Islamic Republic of Iran, 22, 75-84
[9] Davoudifar P., Rowshan Tabari K., 2014, Journal of Physics: Conference Series, 490, 012029
[10] Davoudifar P., Rowshan Tabari K., 2015, Journal of Physics: Conference Series, 633, 012055
[11] Rand R.J., Kulkarini S.R., 1989, Astrophysical Journal, 343, 760-772
[12] Harari D., Mollerach S., Roulet S., 1999, Journal of High Energy Physics, 08, 022
[13] Lazarian A., Gouveia Dal Pino E.M., 2015, Magnetic Fields in Diffuse Media, Springer

Breakdown of the equivalence between gravitational mass and energy for a composite quantum body

Andrei G. Lebed

Department of Physics, University of Arizona, 1118 E. 4-th Street, Tucson, AZ 85721, USA and L.D. Landau Institute for Theoretical Physics, RAS, 2 Kosygina Street, Moscow, Russia

E-mail: `lebed@physics.arizona.edu`

Abstract. The simplest quantum composite body, a hydrogen atom, is considered in the presence of a weak external gravitational field. We define an operator for the passive gravitational mass of the atom in the post-Newtonian approximation of the general relativity and show that it does not commute with its energy operator. Nevertheless, the equivalence between the expectation values of the mass and energy is shown to survive at a macroscopic level for stationary quantum states. Breakdown of the equivalence between passive gravitational mass and energy at a microscopic level for stationary quantum states can be experimentally detected by studying unusual electromagnetic radiation, emitted by the atoms, supported by and moving in the Earth's gravitational field with constant velocity, using spacecraft or satellite

1. Introduction

The notion of gravitational mass for a composite body is known to be non-trivial in general relativity and related to several paradoxes. The role of the classical virial theorem in establishing of the equivalence between averaged over time gravitational mass and energy is discussed in detail in Refs.[1,2] for different types of classical composite bodies. In particular, for electrostatically bound two bodies, it is shown that the gravitational field is coupled to a combination $3K + 2U$, where K is the kinetic energy and U the potential energy. Since the classical virial theorem states that the following time average is equal to zero, $\langle 2K + U \rangle_t = 0$, we can conclude that averaged over time gravitational mass is proportional to the total amount of energy, E [1,2]:

$$\langle m^g \rangle_t = \langle 3K + 2U \rangle_t / c^2 = \langle K + U \rangle_t / c^2 = E/c^2. \quad (1)$$

2. Goal

The main goal of our paper is to study the quantum problem concerning the passive gravitational mass of a composite body. As the simplest example, we consider a hydrogen atom in the Earth's gravitational field, where we take into account only kinetic and Coulomb potential energies of an electron in a curved spacetime. We claim three main results in the paper (see also Refs. [3,4]). Our first result is that the equivalence between passive gravitational mass and energy in the absence of gravitational field survives, at the macroscopic level, in the quantum case. More strictly speaking, we show that the expectation value of the mass is equal to E/c^2 for stationary quantum states due to the quantum virial theorem [5]. Our second result is a breakdown of the equivalence between passive gravitational mass and energy, at the microscopic

Content from this work may be used under the terms of the Creative Commons Attribution 3.0 licence. Any further distribution of this work must maintain attribution to the author(s) and the title of the work, journal citation and DOI.
Published under licence by IOP Publishing Ltd

level, for stationary quantum states due to the fact that passive gravitational mass operator does not commute with energy operator. As a result, there exist a non-zero probability that a measurement of passive gravitational mass can yield a value, which is different from E/c^2, given by the Einstein's equation. Our third result is a suggestion of a realistic experiment to detect this inequivalence by measurement of electromagnetic radiation, emitted by a macroscopic ensemble of hydrogen atoms, supported by and moving in the Earth's gravitational field, using spacecraft or satellite.

3. Gravitational mass in classical physics

Here, we derive the Lagrangian and Hamiltonian of a hydrogen atom in the Earth's gravitational field, taking into account couplings of kinetic and potential Coulomb energies of an electron with a weak gravitational field. Note that we keep only terms of the order of $1/c^2$ and disregard all tidal effects. Therefore, we can write the interval, using a weak field approximation [6]:

$$ds^2 = -(1 + 2\phi/c^2)(cdt)^2 + (1 - 2\phi/c^2)(dx^2 + dy^2 + dz^2),\ \phi = -GM/R, \tag{2}$$

where G is the gravitational constant, c is the velocity of light, M is the Earth mass, R is the distance between center of the Earth and center of mass of a hydrogen atom (i.e., a proton). Then, in the local proper spacetime coordinates, which in the first approximation in ϕ/c^2 are

$$x' = (1 - \phi/c^2)x,\ y' = (1 - \phi/c^2)y, z' = (1 - \phi/c^2)z,\ t' = (1 + \phi/c^2)t, \tag{3}$$

the classical Lagrangian and action of an electron have the following standard forms:

$$L' = -m_e c^2 + \frac{1}{2} m_e (\mathbf{v}')^2 + \frac{e^2}{r'}\ ,\quad S' = \int L' dt', \tag{4}$$

where m_e is the bare electron mass, e and $\mathbf{v}'$ are the electron charge and velocity, respectively; r' is a distance between electron and proton. It is possible to show that the Lagrangian (4) can be rewritten in coordinates (x, y, z, t) as

$$L = -m_e c^2 + \frac{1}{2} m_e \mathbf{v}^2 + \frac{e^2}{r} - m_e \phi - \left(3 m_e \frac{\mathbf{v}^2}{2} - 2\frac{e^2}{r}\right)\frac{\phi}{c^2}, \tag{5}$$

which corresponds to the following Hamiltonian:

$$H = m_e c^2 + \frac{\mathbf{p}^2}{2m_e} - \frac{e^2}{r} + m_e \phi + \left(3\frac{\mathbf{p}^2}{2m_e} - 2\frac{e^2}{r}\right)\frac{\phi}{c^2}. \tag{6}$$

4. Gravitational mass in quantum physics

The Hamiltonian (6) can be quantized by substituting a momentum operator, $\hat{\mathbf{p}} = -i\hbar\partial/\partial\mathbf{r}$, instead of canonical momentum, $\mathbf{p}$:

$$\hat{H} = m_e c^2 + \frac{\hat{\mathbf{p}}^2}{2m_e} - \frac{e^2}{r} + \hat{m}_e^g \phi\ , \tag{7}$$

where we introduce passive gravitational mass operator of an electron to be proportional to its weight operator in a weak gravitational field (2),

$$\hat{m}_e^g = m_e + \left(\frac{\hat{\mathbf{p}}^2}{2m_e} - \frac{e^2}{r}\right)\frac{1}{c^2} + \left(2\frac{\hat{\mathbf{p}}^2}{2m_e} - \frac{e^2}{r}\right)\frac{1}{c^2}\ . \tag{8}$$

Note that the Hamiltonian (7),(8) contains the virial contribution (i.e., the last term) to the mass operator and, therefore, does not commute with electron energy operator, taken in the absence of the field. Below, we discuss some consequences of Eqs.(7),(8). Suppose that we have a macroscopic ensemble of hydrogen atoms with each of them being in a ground state with energy E_1. Then the expectation value of the electron mass operator (8) per atom is

$$< \hat{m}_e^g >= m_e + \frac{E_1}{c^2} + \left\langle 2\frac{\hat{\mathbf{p}}^2}{2m_e} - \frac{e^2}{r} \right\rangle \frac{1}{c^2} = m_e + \frac{E_1}{c^2}, \tag{9}$$

where the third term in Eq.(9) is zero in accordance with the quantum virial theorem [5]. Therefore, we conclude that the equivalence between passive gravitational mass and energy survives at a macroscopic level for stationary quantum states. Let us discuss how Eqs.(7),(8) break the equivalence between passive gravitational mass and energy at a microscopic level. Here, we illustrate the above mentioned inequivalence, using the following thought experiment. Suppose that at $t = 0$ we create a ground state wave function of a hydrogen atom, corresponding to the absence of gravitational field,

$$\Psi_1(r,t) = \Psi_1(r)\exp(-iE_1 t/\hbar) \ . \tag{10}$$

In gravitational field (2), wave function (10) is not anymore a ground state. It is possible to show that a general solution of the Schrödinger equation, corresponding to the Hamiltonian (7),(8), can be written as

$$\Psi(r,t) = (1-\phi/c^2)^{3/2} \sum_{n=1}^{\infty} a_n \Psi_n[(1-\phi/c^2)r] \exp[-im_e c^2(1+\phi/c^2)t/\hbar] \exp[-iE_n(1+\phi/c^2)t/\hbar]. \tag{11}$$

We wish to draw attention to the fact that wave function (11) is a series of eigenfunctions of passive gravitational mass operator (8). $\Psi_n(r)$ is a normalized wave function of an electron in a hydrogen atom in the absence of gravitational field, corresponding to energy E_n.

In accordance with quantum mechanics, probability that, at $t > 0$, an electron occupies excited state with energy $m_e c^2(1+\phi/c^2) + E_n(1+\phi/c^2)$ is

$$P_n = |a_n|^2, \ a_n = \int \Psi_1^*(r)\Psi_n[(1-\phi/c^2)r]d^3\mathbf{r} = -(\phi/c^2)\int \Psi_1^*(r) r \Psi_n'(r) d^3\mathbf{r},$$
$$\int \Psi_1^*(r) r \Psi_n'(r) d^3\mathbf{r} = V_{n,1}/(\hbar\omega_{n,1}), \ \hbar\omega_{n,1} = E_n - E_1, \ n \neq 1,$$
$$V_{n,1} = \int \Psi_1^*(r)\hat{V}(\mathbf{r})\Psi_n(r)d^3\mathbf{r}, \ \hat{V}(\mathbf{r}) = 2\frac{\hat{\mathbf{p}}^2}{2m_e} - \frac{e^2}{r}. \tag{12}$$

Let us discuss Eq.(12). We stress that it directly demonstrates that there is a finite probability,

$$P_n = |a_n|^2 = (\phi/c^2)^2 \ [V_{n,1}/(E_n - E_1)]^2 \ , \ n \neq 1, \tag{13}$$

that, at $t > 0$, an electron occupies n-th ($n \neq 1$) energy level, which breaks the expected Einstein's equation, $m_e^g = m_e + E_1/c^2$. In fact, this means that quantum measurement of passive gravitational mass in a quantum state with a definite energy (10) gives the following quantized values:

$$m_e^g(n) = m_e + E_n/c^2 \ . \tag{14}$$

5. Suggested experiment

Here, we describe a realistic experiment: We consider a hydrogen atom being in its ground state at $t = 0$, located at distance R' from the center of the Earth and having the wave function:

$$\tilde{\Psi}_1(r,t) = (1-\phi'/c^2)^{3/2}\Psi_1[(1-\phi'/c^2)r]\exp[-im_e c^2(1+\phi'/c^2)t/\hbar]\exp[-iE_1(1+\phi'/c^2)t/\hbar]. \quad (15)$$

The atom is supported by in the Earth's gravitational field and moving from the Earth at constant velocity, $u \ll \alpha c$ (where α is the fine structure constant), by spacecraft or satellite. It is possible to show that the electron wave function and the time dependent perturbation for the Hamiltonian (7)-(8), in the coordinate system related to center of mass of the atom, can be expressed as

$$\tilde{\Psi}(r,t) = (1-\phi'/c^2)^{3/2}\sum_{n=1}^{\infty}\tilde{a}_n(t)\Psi_n[(1-\phi'/c^2)r]\exp[-im_e c^2(1+\phi'/c^2)t/\hbar]$$
$$\exp[-iE_n(1+\phi'/c^2)t/\hbar], \quad (16)$$

$$\hat{U}(\mathbf{r},t) = \frac{\phi(R'+ut)-\phi(R')}{c^2}\left(3\frac{\hat{\mathbf{p}}^2}{2m_e} - 2\frac{e^2}{r}\right). \quad (17)$$

We wish to draw attention that in a spacecraft (satellite), which moves with constant velocity, gravitational force, which acts on each individual hydrogen atom, is compensated by some non-gravitational forces. This causes very small changes of a hydrogen atom energy levels and is not important for our calculations. Therefore, the atoms do not feel directly gravitational acceleration, $\mathbf{g}$, but feel, instead, gravitational potential, $\phi(R") = \phi(R'+ut)$.

It is important that, if excited levels of a hydrogen atom were strictly stationary, then probability to find the passive gravitational mass to be quantized with $n \neq 1$ (14) would be

$$\tilde{P}_n = (V_{n,1}/\hbar\omega_{n,1})^2[\phi(R')/c^2]^2 \;, n \neq 1,\; |\phi(R'+ut)| \ll |\phi(R')|, u \ll \omega_{n,1}R, \quad (18)$$

which coincides with Eq. (13). In reality, the excited levels spontaneously decay with time and, therefore, it is possible to observe the quantization law (14) indirectly by measuring electromagnetic radiation from a macroscopic ensemble of the atoms. In this case, Eq.(18) gives a probability that a hydrogen atom emits a photon with frequency $\omega_{n,1} = (E_n - E_1)/\hbar$ during the time of the experiment. To estimate the probabilities (18) we use the following numerical values of the Earth's mass, $M \simeq 6 \times 10^{24} kg$, and its radius, $R_0 \simeq 6.36 \times 10^6 m$. It is important that, although the probabilities (18) are small, the number of photons, N, emitted by macroscopic ensemble of the atoms, can be large. For instance, for 1000 moles of hydrogen atoms, N is estimated as

$$N(n \to 1) = 2.95 \times 10^8\; [V_{n,1}/(E_n - E_1)]^2, \quad N(2 \to 1) = 0.9 \times 10^8, \quad (19)$$

which can be experimentally detected. [Here, $N(n \to 1)$ stands for a number of photons, emitted with frequency $\omega_{n,1} = (E_n - E_1)/\hbar$.]

Acknowledgments

We are thankful to N.N. Bagmet (Lebed), V.A. Belinski, Steven Carlip, Li-Zhi Fang, Douglas Singleton, and V.E. Zakharov for useful discussions. This work was supported by the NSF under Grant DMR-1104512.

References

[1] Nordtvedt K 1994 *Class. Quantum Grav.* **11**, A119
[2] Carlip S 1998 *Am. J. Phys.* **66**, 409
[3] Lebed A G 2011 *Preprint* arXiv:1111.5365v1
[4] Lebed A G 2012 *Preprint* arXiv:1205.3134v1
[5] Park D 2005 *Introduction to the Quantum Theory, 3-rd Edition* (Dover Publications, New York)
[6] Misner C W, Thorne K S and Wheeler J A 1973 *Gravitation* (W.H. Freeman and Co, New York)

Langevin model for a Brownian system with directed motion

Francisco Ambía and Humberto Híjar

Grupo de Sistemas Inteligentes, Facultad de Ingeniería, Universidad La Salle, Benjamín Franklin 47, 06140, Ciudad de México, México

E-mail: humberto.hijar@lasallistas.org.mx

Abstract. We propose a model for an active Brownian system that exhibits one-dimensional directed motion. This system consists of two Brownian spherical particles that interact through an elastic potential and have time-dependent radii. We suggest an algorithm by which the sizes of the particles can be varied, such that the center of mass of the system is able to move at an average constant speed in one direction. The dynamics of the system is studied theoretically using a Langevin model, as well as from Brownian Dynamics simulations.

1. Introduction

Understanding how a microscopic physical system is able to perform directed motion when it is immersed in an isotropic thermal bath, is a problem that has received considerable interest through the last decades [1, 2]. This interest has been increased by the observations of the mechanisms that allow molecular motors to perform specialized tasks in the cells of living organisms [3]. It is now known that such molecular machines use the spontaneous fluctuations occurring in their surroundings together with ratchet mechanisms, in order to break the spatial symmetry of their dynamics [4]. The understanding of such mechanisms could lead in the future to the fabrication of nanomachines aimed at functions such as targeted drug delivery, stirring and pumping in microfluidic devices, and separation of biological macromolecules [5]. Due to the significant impact that these applications would have, developing simplified models for the Brownian motors that retain the basic physical phenomena and can be mathematically treated, could be very helpful to understand the principles that govern their dynamics.

Directed Brownian motion can be achieved by employing a ratchet potential, i.e., a periodic potential with no reflection symmetry that is able to produce a net current of particles from unbiased thermal fluctuations [1]. In a more general context, a system is said to constitute a Brownian motor when [2]: the spatio-temporal periodicity critically affects the rectification mechanism, the averages of all acting forces and gradients vanish, random forces play a principal role, and it is kept in a nonequilibrium state by a rupture of the detailed balance symmetry. One-dimensional rectified Brownian motion appears in molecular transport through cell membranes, single-file diffusion, and noise-assisted transport promoted by oscillating barriers [6, 7, 8].

Here, we introduce a model for a one-dimensional Brownian motor which, to the best of our knowledge, has not been studied previously from the present point of view. Our main purpose is to study the dynamics of the proposed system analytically, by means of a Langevin model, and numerically by means of Brownian Dynamics (BD) simulations.

Content from this work may be used under the terms of the Creative Commons Attribution 3.0 licence. Any further distribution of this work must maintain attribution to the author(s) and the title of the work, journal citation and DOI.
Published under licence by IOP Publishing Ltd

2. The model

Consider a system of two spherical particles restricted to move in one dimension. Let x_1 and x_2 denote the positions of these particles, and R_1 and R_2 their respective radii. It will be assumed that they interact through an elastic potential, $U = k(x_2 - x_1 - l)^2/2$, where k is the restoring coefficient and l is the equilibrium distance between the particles. The system will be immersed in a fluid with viscosity coefficient η. For mathematical simplicity, we will consider the mass of both particles to be equal to a constant m. R_1 and R_2 will be assumed to be sufficiently small such that the dynamics of the system can be well described by the over damped approximation. Using the center of mass position, $x_{\text{c.m.}} = (x_1 + x_2)/2$, and the relative coordinate, $x_{\text{r}} = x_2 - x_1$, the dynamics of the system can be described in terms of the following system of Langevin equations

$$\frac{dx_{\text{c.m.}}}{dt} = -\frac{1}{2}\left(\frac{\omega^2}{\beta_2} - \frac{\omega^2}{\beta_1}\right)(x_{\text{r}} - l) + \frac{1}{2}\left(\frac{1}{\beta_1}\mathcal{A}_1 + \frac{1}{\beta_2}\mathcal{A}_2\right), \tag{1}$$

$$\frac{dx_{\text{r}}}{dt} = -\left(\frac{\omega^2}{\beta_1} + \frac{\omega^2}{\beta_2}\right)(x_{\text{r}} - l) + \frac{1}{\beta_2}\mathcal{A}_2 - \frac{1}{\beta_1}\mathcal{A}_1, \tag{2}$$

where: $\omega = \sqrt{k/m}$; $\beta_i = aR_i/m$, with $a = 6\pi\eta$; and $\mathcal{A}_i$ is the stochastic force per unit mass acting on the ith particle, for $i = 1, 2$.

As it is usual, stochastic forces will be considered as Markov-Gaussian processes with zero mean and fluctuation-dissipation theorem

$$\langle \mathcal{A}_i(t')\mathcal{A}_j(t)\rangle = \frac{2k_B T\beta_i}{m}\delta(t' - t)\delta_{ij}, \tag{3}$$

where k_B is the Boltzmann constant, T is the temperature of the fluid, and no summation over repeated indices is implied.

If R_1 and R_2 are constants the formal solution of Eqs. (1) and (2) reads as

$$x_{\text{c.m.}} = x^o_{\text{c.m.}} + \tilde{\eta}\left(1 - e^{-\alpha t}\right)(x^o_{\text{r}} - l) + \int_0^t d\xi \left[\psi^{(1)}_{\text{c.m.}}(t - \xi)\mathcal{A}_1(\xi) + \psi^{(2)}_{\text{c.m.}}(t - \xi)\mathcal{A}_2(\xi)\right], \tag{4}$$

$$x_{\text{r}} = l + (x^o_{\text{r}} - l)e^{-\alpha t} + \int_0^t d\xi \left[\psi^{(1)}_{\text{r}}(t - \xi)\mathcal{A}_1(\xi) + \psi^{(2)}_{\text{r}}(t - \xi)\mathcal{A}_2(\xi)\right]. \tag{5}$$

In Eqs. (4) and (5), $x^o_{\text{c.m.}}$ and x^o_{r} are the initial values of the $x_{\text{c.m.}}$ and x_{r}, respectively; $\alpha = \omega^2\left(\beta_1^{-1} + \beta_2^{-1}\right)$, is the characteristic inverse time for the damped dynamics of the system; and $\tilde{\eta} = (R_2 - R_1)/2(R_1 + R_2)$. The auxiliary functions $\psi^{(i)}_{\text{c.m.}}$, $\psi^{(i)}_{\text{r}}$, have the following definitions for $i = 1, 2$,

$$\psi^{(i)}_{\text{c.m.}}(t) = \frac{m}{a}\left[\frac{e^{-\alpha t}}{2R_i} + \frac{1 - e^{-\alpha t}}{R_1 + R_2}\right], \quad \psi^{(i)}_{\text{r}}(t) = (-1)^i \frac{m}{a}\frac{e^{-\alpha t}}{R_i}. \tag{6}$$

3. Algorithm for directed motion

According to Eqs. (4) and (5), in the limit $t \gg \alpha^{-1}$, the average value of $x_{\text{r}} - l$ will vanish, while that of $x_{\text{c.m.}}$ will move by a total amount $\tilde{\eta}(x^o_{\text{r}} - l)$. Physically, this is a consequence of the fact that the bigger particle experiences a larger frictional force and acts as a kind of anchor for the coupled system, while the smaller is able to move due to the elastic interaction and drives the center of mass. As long as $x_{\text{r}} \neq l$, $x_{\text{c.m.}}$ will tend to move, on the average, in the direction determined by the sign of the product $\tilde{\eta}(x_{\text{r}} - l)$. Now, the perturbation brought about by the stochastic forces will drive x_{r} away from l persistently, and if $\tilde{\eta}$ could change its sign, due to a change in the values of R_1 and R_2, then the product $\tilde{\eta}(x_r - l)$ might be forced to keep its sign constant, thus giving directed motion to the coupled system.

In the present work, we will explore the consequences of varying R_1 and R_2 between two defined values R_{max} and R_{min} ($R_{max} > R_{min}$), according to the following prescription

$$\begin{array}{llll} R_1 = R_{\max} & \text{and} & R_2 = R_{\min} & \text{if } x_r < l, \\ R_1 = R_{\min} & \text{and} & R_2 = R_{\max} & \text{if } x_r > l. \end{array} \tag{7}$$

Notice that when R_1 and R_2 are allowed to change in time according to the protocol Eq. (7), then Eqs. (4) and (5) are valid only for a short time interval, τ_1, that lasts until the configuration of the system is updated. After that, the system evolves following similar equations but with modified functions such that $\tilde{\eta} \to -\tilde{\eta}$, $\psi^{(1)}_{\text{c.m.}} \to \psi^{(2)}_{\text{c.m.}}$, $\psi^{(2)}_{\text{c.m.}} \to \psi^{(1)}_{\text{c.m.}}$, $\psi^{(1)}_{\text{r}} \to -\psi^{(2)}_{\text{r}}$, and $\psi^{(2)}_{\text{r}} \to -\psi^{(1)}_{\text{r}}$. It can be shown from Eqs. (3) and (5), and from the previous symmetry properties, that the probability for observing an elongation of the system of magnitude $\tilde{x}_r$, at time t, given that it was $\tilde{x}^o_r$, at time zero, is the same as the one obtained for a system with fixed configuration.

On the other hand, in order to describe the statistical properties of $x_{\text{c.m.}}$, we will perform the following simplifying assumptions. First, that the stochastic processes of $x_{\text{c.m.}}$ during the time intervals of fixed configuration, τ_i, are statistically independent. Second, that the distribution of times τ_i has a small dispersion, in such a way that all configuration changes take place at regular time intervals with duration $\bar{\tau}$. Third, that observations of the system are performed after allowing the variable x_r to achieve its stationary distribution. Then, it can be shown that the probability for observing the center of mass at position $x^{(n)}_{\text{c.m.}}$, after n changes in configuration, given that its initial value was $x^{(0)}_{\text{c.m.}}$, is

$$W\left(x^{(n)}_{\text{c.m.}}, n \,|\, x^{(0)}_{\text{c.m.}}, 0\right) = \frac{1}{\sqrt{2\pi\sigma^2_n(\bar{\tau})}} \exp\left\{-\left[x^{(n)}_{\text{c.m.}} - \left(x^{(0)}_{\text{c.m.}} - \xi_n(\bar{\tau})\right)\right]^2 / 2\sigma^2_n(\bar{\tau})\right\}, \tag{8}$$

where the functions $\sigma^2_n(\bar{\tau})$ and $\xi_n(\bar{\tau})$ are defined as $\sigma^2_n(\bar{\tau}) = n\left[\sigma^2_{\text{c.m.}}(\bar{\tau}) + (1 - e^{-\alpha\bar{\tau}})^2 \sigma^2_r\right]$, and $\xi_n(\bar{\tau}) = n(1 - e^{-\alpha\bar{\tau}})\,|\tilde{\eta}|\,|\bar{x_r}|$ respectively.

In these definitions $\sigma^2_{\text{c.m.}}(\bar{\tau})$ represents the standard deviation for the distribution of $x_{\text{c.m.}}$ at time $\bar{\tau}$ which, for brevity, is not written explicitly; $\sigma^2_r = k_B T/k$, is the stationary value of the corresponding standard deviation of x_r; and $|\bar{x_r}| = \sigma_r\sqrt{2/\pi}$.

It can be shown that in the limit $\alpha\bar{\tau} \ll 1$, which can be achieved for large values of the k, we have $\sigma^2_n(\bar{\tau}) = n\bar{\tau}k_B T(R_{\min} + R_{\max})/2aR_{\min}R_{\max}$, and $\xi_n(\bar{\tau}) = vn\bar{\tau}$, where we have identified an average velocity for the Brownian motor as

$$v = \frac{1}{6\sqrt{2}\pi^{3/2}} \frac{R_{\max} - R_{\min}}{R_{\max}R_{\min}} \frac{\sqrt{kk_B T}}{\eta}, \tag{9}$$

which as it could be expected increases with the radius difference, the thermal energy of the bath, and the strength of the elastic interaction, but decreases with the viscosity of the medium.

4. Comparison with Brownian Dynamics Simulations

We implemented BD simulations that allowed us to solve Eqs. (1) and (2), based on a Runge-Kutta-Maruyama scheme. In our implementation the units of length, mass and energy where fixed by the values $R_{\min} = 1$, $m = 1$, and $k_B T = 1$, respectively. We performed numerical experiments where the values of the restitution coefficient and the viscosity were fixed at $k = 5\,k_B T/R^2_{\min}$, and $\eta = 5\sqrt{mk_B T}/R^2_{\min}$, respectively. Figure 1 *a*, shows the trajectories of five Brownian motors with different configurations characterized by the values of the percental radius difference, $\Delta = (R_{\max} - R_{\min})/R_{\min} = 0.05, 0.10, 0.15, 0.20, 0.25$. It can be observed that the system exhibits, indeed, directed motion even for small values of Δ. The straight lines shown in Fig. 1 *a*, represent the corresponding motion at the constant speed given by Eq. (9). In Fig. 1 *a*, time is given in simulations units $(R_{\min}\sqrt{m/k_BT})$.

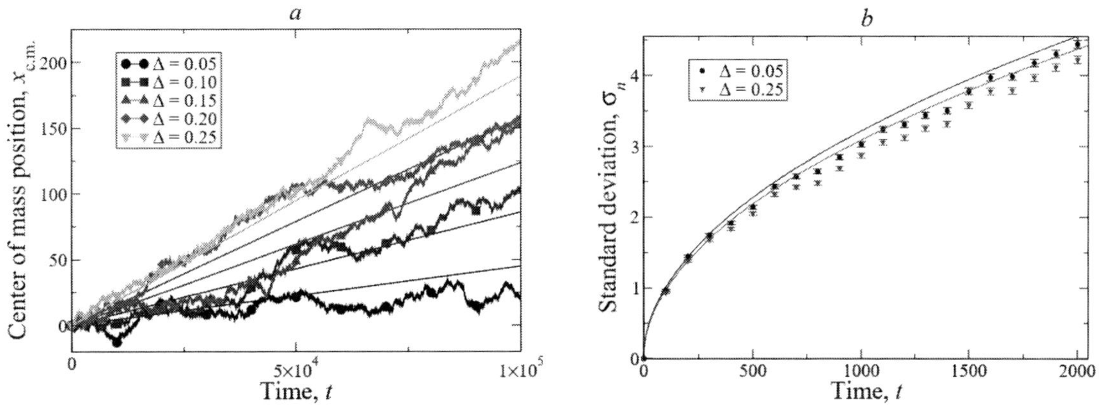

Figure 1. Comparison of the theoretical and numerical results. *a* Typical trajectories of Brownian motors with different configurations. Noise curves correspond with experimental results (symbols were included to guide the eye). *b* Standard deviation for the distribution function of the trajectory of two ensembles of Brownian motors with different configuration. Symbols represent numerical results and the continuous curves the analytical prediction.

We also conducted experiments where ensembles of 12500 independent Brownian motors were simulated. The standard deviation of the distribution of $x^{(n)}_{\mathrm{c.m.}}$ for such ensembles was numerically calculated. Figure 1 *b* shows the results of such experiments for two different types of motors with configurations $\Delta = 0.05$ and 0.25. A comparison is performed with the standard deviation expected from the analytical model. A very good agreement of the theoretical and numerical approaches can be observed for small simulation times, while a deviation between them can be noticed for large values of t. This can be understood due to the simplifying assumptions used to obtain Eq. (8).

We notice that our model is similar to the so called *pushmepullyou* microswimmer introduced by Avron et al. in Ref. [9]. The main difference from that model and ours is that in Ref. [9] thermal forces were neglected while, here, they are the main mechanism that promotes the motion of the system. Extensions of the present work to include the effects of hydrodynamic fields around the Brownian motor similar to those considered in Refs. [9] and [10] are in progress.

References

[1] Hänggi P and Bartussek R 1996 *Brownian rectifiers: how to convert Brownian motion into directed transport*, in *Nonlinear Physics of Complex Systems* (Berlin: Springer) p 294
[2] Hänggi P, Marchesoni F and Nori F 2005 *Ann. Phys. (Leipzig)* **14** 51
[3] Svoboda K, Schmidt C F, Schnapp B J and Block S M 1993 *Nature* **365** 721
[4] Hoffmann P M 2012 *Life's Ratchet* New York (Basic Books)
[5] Leibler S 1994 *Nature* **370** 412
[6] Bleil S, Reimann P and Bechinger C 2007 *Phys. Rev. E* **75** 031117
[7] Reichhardt C and Nori F 1999 *Phys. Rev. Lett.* **82** 414
[8] Speer D, Eichhorn R and Reimann P 2009 *Phys. Rev. Lett.* **102** 124101
[9] Avron J E, Kenneth O and Oaknin D H 2005 *New J. Phys.* **7** 234
[10] H. Híjar 2015 *Phys. Rev. E* **91** 022139

Scalar field cosmology via non-local integrals of motion

N Dimakis
Instituto de Ciencias Fisicas y Matematicas,
Universidad Austral de Chile, Valdivia, Chile

E-mail: nsdimakis@gmail.com

Abstract. In re-parametrization invariant systems, such as mini-superspace Lagrangians, the existence of constraints can lead to the emergence of additional non-local integrals of motion defined in phase space. In the case of a FLRW flat/non-flat space-time minimally coupled to an arbitrary scalar field, we manage to use such conserved quantities to completely integrate the system of equations of motion. This is achieved without constraining the potential in any way. Thus, obtaining the most general solution that encompasses all possible cosmological scenarios which can be based on the existence of a scalar field.

1. Introduction and general theory

Symmetries are widely used in the study of cosmological systems, especially within the mini-superspace approximation [1]-[8]. The study of Noether symmetries of both regular [9] and singular Lagrangians [10], [11] that are used in this process, provides the means to constrain quantities involved in these systems so as to yield them integrable ([12], [13]). In this work we are going to exploit certain non-local symmetries that appear in singular systems to fully integrate a scalar field with an arbitrary potential, minimally coupled to gravity under the ansatz of an FLRW space-time (spatially flat/non-flat). A work where an integration of this type, without imposing conditions on the potential, has been previously done in homogeneous cosmology for Bianchi types I and V [14].

It was proven in [10], that for singular systems described by Lagrangians of the form

$$\bar{L} = \frac{1}{2N(t)}\bar{G}_{\mu\nu}(q)\dot{q}^{\mu}(t)\dot{q}^{\nu}(t) - N(t)\,U(q) \tag{1.1}$$

all conformal Killing fields of $\bar{G}_{\mu\nu}$ can be used to write down integrals of motion. To observe that, let us for simplicity perform a re-parametrization of the form $N \to n = N\,V$. Then, it can be seen that for the equivalent system

$$L = \frac{1}{2n}G_{\mu\nu}(q)\dot{q}^{\mu}\dot{q}^{\nu} - n, \tag{1.2}$$

Content from this work may be used under the terms of the Creative Commons Attribution 3.0 licence. Any further distribution of this work must maintain attribution to the author(s) and the title of the work, journal citation and DOI.
Published under licence by IOP Publishing Ltd

where $G_{\mu\nu} = U\bar{G}_{\mu\nu}$, if there exist vector fields $\xi^\alpha(q)$ defined on the configuration space for which the relation $\pounds_\xi G_{\mu\nu} = \omega(q) G_{\mu\nu}$ holds, then the quantity

$$Q = \xi^\alpha p_\alpha + \int n(t)\omega(q(t))dt, \tag{1.3}$$

with $p_\alpha = \frac{\partial L}{\partial q^\alpha}$, defines integrals of motion on the phase space due to the existence of the Hamiltonian constraint

$$\mathcal{H} = \frac{1}{2} G^{\mu\nu} p_\mu p_\nu + 1 \approx 0 \tag{1.4}$$

It can be easily verified that, due to ξ being a conformal Killing vector with conformal factor ω, Q is a constant of motion

$$\frac{dQ}{dt} = \frac{\partial Q}{\partial t} + \{Q, \mathcal{H}\} = \omega \mathcal{H} \approx 0.$$

For a general description of constrained systems and their properties, see ([15],[16])

In the special case were $\omega = 0$, i.e. the ξ's are Killing vectors of the scaled mini-supermetric $G_{\mu\nu}$, relation (1.3) leads to autonomous integrals of motion that do not exhibit any explicit time dependence. On the other hand, when $\omega \neq 0$, non local integrals of motion emerge that are actually rheonomic, due to the explicit time dependence owed to the integral in (1.3).

Our starting point, for the derivation of the mini-superspace Lagrangian, is the action

$$S = \int d^4x \left(\sqrt{-g}R + \epsilon\, \phi_{,\mu}\phi^{,\mu} + 2V(\phi)\right) \tag{1.5}$$

where R is the Ricci scalar and g the determinant of the spacetime metric $g_{\mu\nu}$. We also allow for the existence of a phantom field by introducing a pure sign constant $\epsilon = \pm 1$. As is known, variation with respect to $g^{\mu\nu}$ yields the equations

$$R_{\mu\nu} - \frac{1}{2} R g_{\mu\nu} = T_{\mu\nu} \tag{1.6}$$

with $T_{\mu\nu} = \epsilon\, \phi_{,\mu}\phi_{,\nu} - \frac{1}{2}\left(\epsilon\, \phi^{,\kappa}\phi_{,\kappa} - 2V(\phi)\right) g_{\mu\nu}$, being the energy-momentum tensor for the matter part. Variation of (1.5) with respect to the scalar field leads to the Klein-Gordon equation

$$\epsilon\, \Box\phi - V'(\phi) = 0, \tag{1.7}$$

where the prime denotes differentiation over ϕ.

It can be easily shown that, under the ansatz

$$ds^2 = -N(t)^2 dt^2 + a(t)^2 \left(\frac{1}{1-kr^2} dr^2 + r^2 d\theta^2 + d\varphi^2\right), \tag{1.8}$$

the demand for the field to be spatially homogeneous $\phi = \phi(t)$, and a re-parametrization of the lapse function

$$N = \frac{n}{2a\left(a^2 V(\phi) - 3k\right)}. \tag{1.9}$$

the set of equations (1.6) and (1.7) is equivalent to the Euler-Lagrange equations derived by a Lagrangian of the form (1.2) with $q^\mu = (a, \phi)$ and the mini-supermetric being

$$G_{\mu\nu} = 4a^2\left(a^2 V(\phi) - 3k\right) \begin{pmatrix} -6 & 0 \\ 0 & \epsilon\, a^2. \end{pmatrix} \tag{1.10}$$

Any of the conformal Killing fields of $G_{\mu\nu}$ can now be used to define integrals of motion for the corresponding system.

2. The spatially flat case

In the $k = 0$ case, the mini-supermetric (1.10) exhibits a homothetic vector that is independent of the scalar field potential $V(\phi)$, namely $\xi = \frac{a}{6}\frac{\partial}{\partial a}$, that satisfies $\pounds_\xi G_{\mu\nu} = G_{\mu\nu}$. This results in the existence of a conserved quantity in phase space that is written as

$$Q = \frac{a}{6}p_a + \int n(t)dt = \frac{a}{6}\frac{\partial L}{\partial \dot{a}} + \int n(t)dt = -\frac{4\,a^5\dot{a}V(\phi)}{n} + \int n(t)dt = \kappa \tag{2.1}$$

with κ being a constant.

We proceed by fixing the gauge through the choice $\phi(t) = t$. At the same time we express the scaled lapse function $n(t)$ with the help of a new non-constant function $h(t)$ and we reparameterize the potential - that now can be considered as a function of time $V(t)$ - with respect to a new (again non-constant) function $A(t)$:

$$n(t) = \dot{h}(t), \qquad V(t) = \frac{(h(t)-\kappa)\,\dot{h}(t)}{4\dot{A}(t)}. \tag{2.2}$$

As a result, (2.1) reduces to a local expression that can be easily integrated to give

$$a(t) = \pm 6^{1/6}\left(A(t) + c_1\right)^{1/6}. \tag{2.3}$$

Substitution of this solution into the Euler-Lagrange equation for $a(t)$ leads to a first order differential equation for $A(t)$ that is solved by

$$A(t) = \frac{3\,m^4}{2}\exp\left(-\int \frac{\dot{h} \pm \left(\dot{h}^2 + \epsilon\,(\kappa - h)^2\right)^{1/2}}{\kappa - h}dt\right) - c_1, \tag{2.4}$$

with m being a non zero constant. In these solutions function $h(t)$ remains free, reflecting the arbitrariness of the potential $V(t)$.

It can be seen by (2.3) and (2.4), that the constant c_1 is not important for the solution, so we might as well consider it to be zero. The same is true for κ, since one need only define a new parametrization as $h(t) = \kappa + \exp\left(\frac{\omega}{2} - 3\int \frac{\epsilon}{\dot{\omega}}dt\right)$. It can be verified that with this re-parametrization of the lapse function $N(t)$ as given by (1.9), together with the scale factor $a(t)$ and the potential the line element can be eventually written (for $\dot{\omega} > 0$, if $\dot{\omega} < 0$ one needs only implement another parametrization and bring it in the same final form):

$$ds^2 = -m^4\dot{\omega}^2 e^{6\int(\epsilon/\dot{\omega})dt}dt^2 + e^{\omega/3}\left(dr^2 + r^2d\theta^2 + r^2\sin^2\theta d\varphi^2\right), \tag{2.5}$$

while the respective scalar field potential for each $\omega(t)$ is $V(t) = \frac{(\dot{\omega}^2 - 6\,\epsilon)e^{-6\int(\epsilon/\dot{\omega})dt}}{12m^4\dot{\omega}^2}$.

Solution (2.5) can be further simplified by performing a change in the time variable from t to ω. Since, ω is an arbitrary function of t, we choose to invert the relation $\omega(t)$ by the use of an arbitrary function $F(\omega)$ defined as follows:

$$\phi(t) = t = \int\sqrt{\epsilon\frac{F'(\omega)}{6}}d\omega = \phi(\omega), \tag{2.6}$$

the prime denoting differentiation with respect to the argument ω. The line element (2.5) - under a slight redefinition of the $F(\omega)$ function ($F(\omega) \to F(\omega) - \log m^4$) that does not alter (2.6) - can be written

$$ds^2 = -e^{F(\omega)}d\omega^2 + e^{\omega/3}\left(dr^2 + r^2 d\theta^2 + r^2\sin^2\theta d\varphi^2\right) \tag{2.7}$$

while the potential transforms to

$$V(\omega) = \frac{1}{12}e^{-F(\omega)}\left(1 - F'(\omega)\right). \tag{2.8}$$

The set (2.6) - (2.8) satisfies the Einstein plus Klein-Gordon equations in the new time variable ω, with $F(\omega)$ remaining of course an arbitrary function due to the fact that we have not adopted a particular form for the potential.

3. The $k \neq 0$ case

In this case there exists no homothetic vector of the mini-superspace metric $G_{\mu\nu}$. However, we shall make use of the conformal vector $\xi = \frac{\partial}{\partial\phi}$, with corresponding factor $\frac{a^2V'(\phi)}{a^2V(\phi)-3k}$. Subsequently, the following non-local integral of motion can be defined

$$\begin{aligned} Q = p_\phi + \int\frac{a(t)^2n(t)V'(\phi(t))}{a(t)^2V(\phi(t)) - 3k}dt = \frac{\partial L}{\partial\dot\phi} + \int\frac{a(t)^2n(t)V'(\phi(t))}{a(t)^2V(\phi(t)) - 3k}dt \\ = \frac{4\,\epsilon\, a^4\dot\phi\left(a^2V(\phi) - 3k\right)}{n} + \int\frac{a(t)^2n(t)V'(\phi(t))}{a(t)^2V(\phi(t)) - 3k}dt. \end{aligned} \tag{3.1}$$

It can be straightforwardly checked that, equation $Q = \kappa$ is the first integral of the Klein-Gordon equation (1.7) (or equivalently, of the Euler-Lagrange equation with respect to ϕ).

Again, we choose to adopt the gauge $\phi(t) = t$, and allow for the following reparameterizations

$$n(t) = \frac{2\dot h\left(a^2V - 3k\right)}{a^2\dot V}, \qquad V(t) = \int\frac{\dot w}{a^6}dt \tag{3.2}$$

with $w(t)$ and $h(t)$ being non constant functions of time. Now, the corresponding equation (2.1) reduces to a local expression that can be integrated to yield

$$h(t) = \frac{1}{2}\left(\kappa \pm \sqrt{4\,c_1 + \kappa^2 - 8\,\epsilon\, w}\right). \tag{3.3}$$

By choosing to perform another re-parametrization $w(t) = \frac{a\dot v}{\dot a} + \frac{1}{8\,\epsilon}\left(4\,c_1 + \kappa^2\right) - 6v$, in which we introduce a new function of time $v(t)$, together with a substitution of (3.2) and (3.3) into the quadratic constraint $\frac{\partial L}{\partial n} = 0$, we get a differential equation for $v(t)$, which is solved by

$$v(t) = \exp\left(\int\left(\frac{6\dot a}{a} - \frac{\epsilon\, a}{\dot a}\right)dt\right)\left(\int\frac{ka^5\exp\left(-\int\left(\frac{6\dot a}{a} - \frac{\epsilon\, a}{\dot a}\right)dt\right)}{2\dot a}dt + c_2\right). \tag{3.4}$$

where c_2 is an integration constant. If we set $a(t) = e^{\omega/6}$ and adopt a suitable time change

$$\phi(t) = t = \int\left[\frac{1}{6\,\epsilon}\left(\frac{S''(\omega)}{S'(\omega)} + \frac{1}{3}\right)\right]^{1/2}d\omega = \phi(\omega) \tag{3.5}$$

where $S(\omega) = \exp\left(12\,k\int e^{F(\omega)-\omega/3}d\omega\right) - \frac{6c_2}{k}$, the result can be significantly simplified. The ensuing line element is

$$ds^2 = -e^{F(\omega)}d\omega^2 + e^{\omega/3}\left(\frac{1}{1-kr^2}dr^2 + r^2 d\theta^2 + r^2\sin\theta d\varphi^2\right), \tag{3.6}$$

while the corresponding potential becomes

$$V(\omega) = \frac{1}{12}e^{-F(\omega)}\left(1 - F'(\omega)\right) + 2\,k\,e^{-\omega/3}. \tag{3.7}$$

As a result, the set of relations (3.5) - (3.7) satisfies Einstein's equation plus scalar field for the case $k \neq 0$, with $F(\omega)$ remaining again arbitrary. It can be seen that if one considers $k = 0$, then this solution becomes exactly the one given in the previous section. However, it has to be noted that in the process of deriving the above relations the assumption $k \neq 0$ has been taken into account.

4. Discussion

In this paper we derived the general solution of a general scalar field minimally coupled to gravity for a FLRW space-time, without having to impose any conditions over the potential. This result is not only important in the context of general relativity but also may be used for the derivation of solutions in other theories of gravity, which under certain transformations may be mapped to a minimally coupled scalar field action (such as $f(R)$ or scalar tensor theories of gravity).

A key element for the derivation of the result has been the re-parametrization invariance of the singular action (1.2). The choice of gauge in which the scalar field is time allowed us to "merge" and parameterize the potential with other functions of time. Finally, the complete integration of the system was made possible with the use of one of the infinite non-local symmetries that can be defined in phase space with the help of conformal Killing vectors of the two dimensional mini-supermetric.

Acknowledgments

The author acknowledges financial support by FONDECYT postdoctoral grant no. 3150016.

References

[1] Vakili B and Khazaie F 2012 *Clas. Quant. Grav.* **738** 035015
[2] Capozziello S, De Laurentis M, Odintsov S D 2012 *Eur. Phys. J.* **C72** 2068
[3] Terzis P A, Dimakis N and Christodoulakis T 2014 *Phys. Rev. D* **90** 123543
[4] Vakili B 2014 *Phys. Lett. B* **738** 488-492
[5] Zampeli A 2015 *J. Phys. Conf. Ser.* **633** 012019
[6] Zampeli A, Pailas T, Terzis P A and Christodoulakis T 2015 *Preprint gr-qc/1511.08382*
[7] Paliathanasis A and Capozziello S 2016 *Preprint gr-qc/1602.08914*
[8] Belinchón J A, Harko T and Mak M K 2016 *Astrophys. Space Sci.* **361** 52
[9] Tsamparlis M and Paliathanasis A 2011 *Gen. Relativ. Grav.* **43** 1861
[10] Christodoulakis T, Dimakis N and Terzis P A 2014 *J. Phys. A: Math. Theor.* **47** 095292
[11] Terzis P A, Dimakis N, Christodoulakis T, Paliathanasis A and Tsamparlis M 2016 *J. Geom. Phys.* 101 52-64
[12] Paliathanasis A, Tsamparlis M and Basilakos S 2011 *Phys. Rev. D* **84** 123514
[13] Dimakis N, Christodoulakis T, Terzis P A 2014 *J. Geom. Phys.* **77** 97-112
[14] Christodoulakis T, Grammenos T, Helias C, Kevrekidis P G and Spanou A 2006 *J. Math. Phys.* **47** 042505
[15] Dirac P A M 1964 *Lectures on Quantum Mechanics* (Yeshiva University, Academic Press, New York)
[16] Sundermeyer K 1982 *Constrained Dynamics* (Springer-Verlag)

On the formulation of a Bäcklund Wahlquist-Estabrook transformation for a supersymmetric Korteweg-de Vries equation

Alvaro Restuccia[1], **Adrián Sotomayor**[2]
[1]Physics Department, Antofagasta University, Chile
[2]Department of Mathematics , Antofagasta University, Chile
E-mail: alvaro.restuccia@uantof.cl, adrian.sotomayor@uantof.cl

Abstract. We present a local Bäcklund Wahlquist-Estabrook (WE) transformation for a supersymmetric Korteweg-de Vries (KdV) equation. As in the scalar case, such type of transformation generates infinite hierarchies of solutions and also implicitly gives the associated (local) conserved quantities. A nice property is that every of such hierarchies admits a nonlinear superposition principle, starting for an initial solution, including as a particular case the multisolitonic solutions of the system. We discuss the symmetries of the system and we present in an explicit way its local conserved quantities with the help of the associated Gardner transformation.

1. Introduction

The Bäcklund transformation in the context of KdV equation was introduced by Wahlquist and Estabrook [1] soon after Hirota introduced his approach to generate multisolitonic solutions for the same equation [2]. The Bäcklund (WE) transformation for KdV equation consists of a system of partial differential equations for two auxiliary functions w and w'. If w and w' solve the Bäcklund system then w and w' imply solutions of the proper KdV equation. In particular we can start with any solution w' of the Bäcklund system and solving for w we automatically get another solution of the KdV equation. Starting with the solution $w' = 0$, for example, the Bäcklund system allows to obtain in an iterative way multisolitonic solutions of the KdV equation. For this particular case the Bäcklund transformation generates and iterative procedure equivalent to the Hirota approach. The Bäcklund (WE) transformation may be thought as a nonlinear transformation from the space of solutions of the KdV equation into itself. Consequently it contains information on the symmetries of the KdV equation which are not a priori manifest. An important aspect is that one can obtain from the Bäcklund transformation information on the conserved quantities associated to these symmetries. In fact, we can obtain the sequence of infinite conserved local conserved quantities of the KdV equation from the Bäcklund transformation. The procedure to obtain them is to relate the difference $w - w'$ to a solution of the Gardner equation [3, 4, 5]. It can be shown that w_x and w'_x are then solutions of the KdV equation obtained from the Gardner transformation. The existence of Bäcklund transformations is thus an important property of some integrable systems, which allows us to disentangle the hidden symmetries of them. The extension of Bäcklund transformations for

Content from this work may be used under the terms of the Creative Commons Attribution 3.0 licence. Any further distribution of this work must maintain attribution to the author(s) and the title of the work, journal citation and DOI.
Published under licence by IOP Publishing Ltd

supersymmetric KdV equations [6, 7, 8, 9, 10, 11] is still an open problem in the subject. In [12] the Hirota approach has been extended for the supersymmetric KdV system analyzed by Mathieu in [6], arising from the KP hierarchy. So far an extension of it in the sense of Bäcklund (WE) has not been obtained.

In this work we introduce a Bäcklund transformation in the sense of (WE) for a supersymmetric KdV system. This system is invariant under Galileo transformations and hence it differs from the one in [6]. As we show in the next section the system reduces for a particular Grassmann algebra (with only one generator) to the integrable system considered in [13, 14, 15, 16, 17, 18]. The invariance under Galileo transformations is a fundamental property if we are seeking an application at low energy of the integrable system. The analysis of the supersymmetric model we consider in this work was not discussed in [6] because the equation for the even part of the superfield is exactly the KdV equation. However, its coupling to the odd part of the superfield is non-trivial and it has an interesting Poisson structure which was analyzed for a particular reduction in [18]. In addition, the supersymmetric model in [6] when expanded in terms of the basis of the Grassmann algebra yields exactly the KdV equation for the component of the superfield associated to the unitary element of the basis [19]. Consequently, in both supersymmetric systems, at some level, the KdV equation appears for some even component of the superfield.

Hence there are good reasons to analyze the supersymmetric model, invariant under Galileo transformations, we are going to consider in this work.

2. A supersymmetric Wahlquist-Estabrook (WE) transformation for the KdV equation

The supersymmetric extension of the Korteweg-de Vries equation we are going to consider is given by

$$\Phi_t + \Phi_{xxx} + (D\Phi)\Phi_x = 0. \tag{1}$$

In the context of equation (1) Φ is an odd superfield given by $\Phi(x,\theta) = \xi(x) + \theta u(x)$ and D is the covariant derivative given by $D = \frac{\partial}{\partial\theta} + \theta\frac{\partial}{\partial x}$. This means that Φ can be viewed as a function of the real variables x and t (we do not write t explicitly) with values on a given Grassmann algebra Λ with one preferred generator θ. It is a $N = 1$ supersymmetric extension for KdV equation. In that setting it is sufficient to take u and ξ to be even and odd respectively for every value of x.

Equation (1) is invariant under the supersymmetric transformation $\delta\Phi = \eta Q\Phi$ (where η is an arbitrary odd parameter and Q is given by $Q = \frac{\partial}{\partial\theta} - \theta\frac{\partial}{\partial x})$ which in components reads as $\delta u(x) = \eta\xi'(x)$ and $\delta\xi(x) = \eta u(x)$.

We notice that (1) is equivalent to the equation given in [6] for the particular value of $a = 0$, the only one for which it is invariant under Galileo transformations and hence which may have a physical interpretation at low energies. The Galilean transformation for which (1) remains invariant is given by

$$\begin{aligned} &x \to \tilde{x} = x + ct, \quad \tilde{\theta} = \theta, \quad (c \in \mathbb{R}), \\ &t \to \tilde{t} = t, \\ &\Phi \to \tilde{\Phi} = \Phi + c\theta. \end{aligned}$$

We have

$$\begin{aligned} \frac{\partial\Phi}{\partial t} &= \frac{\partial\tilde{\Phi}}{\partial\tilde{t}} + c\frac{\partial\tilde{\Phi}}{\partial\tilde{x}}, \\ D\Phi &= \tilde{D}\tilde{\Phi} - c. \end{aligned}$$

Journal of Physics: Conference Series **738** (2016) 012039 doi:10.1088/1742-6596/738/1/012039

Consequently, equation (1) remains invariant under the given Galileo transformation, in distinction to the supersymmetric KdV equation discussed in [6] which does not have this symmetry.

In this section we present a Bäcklund transformation for (1). We do so by introducing an auxiliary supersymmetric equation in terms of a Casimir potential:

$$\Omega_t + \Omega_{xxx} + \frac{1}{2}\Omega_x^2 = 0. \tag{2}$$

In equation (2) Ω is an even superfield potential for Φ (or an even casimir potential for Φ), that is $D\Omega = \Phi$ where Ω is given by $\Omega(x, \theta) = w(x) + \theta\zeta(x)$. We have that $D\Omega = \Phi$ implies that $\zeta = \xi$ and $w_x = u$. Equation (2) is equivalent to (1) after a redefinition of Ω.

Equation (1) in components reads as

$$\begin{aligned} u_t + u_{xxx} + uu_x &= 0 \\ \xi_t + \xi_{xxx} + u\xi_x &= 0. \end{aligned} \tag{3}$$

We also note that if we take the first derivative with respect to x in the second equation of (3) and redefine $\xi_x = \kappa$ we get the system

$$\begin{aligned} u_t + u_{xxx} + uu_x &= 0 \\ \kappa_t + \kappa_{xxx} + (u\kappa)_x &= 0. \end{aligned} \tag{4}$$

If we consider a Grassmann algebra with only one generator e we have

$$\begin{aligned} u &= u_1 \cdot 1 \\ \xi &= \xi_1 e \end{aligned} \tag{5}$$

for arbitrary real functions u_1, ξ_1 which we assume to be rapidly decreasing. In this case the system (3) is simply given by

$$\begin{aligned} (u_1)_t + (u_1)_{xxx} + u_1(u_1)_x &= 0 \\ (\xi_1)_t + (\xi_1)_{xxx} + u_1(\xi_1)_x &= 0 \end{aligned} \tag{6}$$

which is exactly the system in [17, 18] for $\lambda = 0$, see also [14, 15].

In what follows we propose, following Wahlquist and Estabrook, a method to generate solutions for (2) and hence we can directly generate solutions for (1). To do this we propose a (WE) transformation for (2), the system is given by

$$\begin{aligned} \Omega_x + \Omega'_x &= \eta - \tfrac{1}{12}(\Omega - \Omega')^2 \\ \Omega_t + \Omega'_t &= \tfrac{1}{12}\left[(\Omega - \Omega')^2\right]_{xx} - \tfrac{1}{2}\Omega_x^2 - \tfrac{1}{2}(\Omega'_x)^2. \end{aligned} \tag{7}$$

To prove that the above system works like a (WE) transformation for (2) we first sum of the second derivative of first equation of system (7) with the second equation of the same system, obtaining directly

$$Q(\Omega) + Q(\Omega') = 0, \tag{8}$$

where

$$Q(\Omega) = \Omega_t + \Omega_{xxx} + \frac{1}{2}\Omega_x^2. \tag{9}$$

If we can prove also that $Q(\Omega) - Q(\Omega') = 0$ we automatically get that Ω and Ω' are solutions for (2) and hence $D\Omega = \Phi$ is solution for (1). To do this we differentiate the first equation of (7) with respect to t and we then subtract to it the first derivative of the second equation of (7) with respect to x. We then obtain (using the compatibility condition $\Omega_{xt} = \Omega_{tx}$, and the same for Ω') that

$$-\frac{1}{6}(\Omega-\Omega')(\Omega-\Omega')_t = \frac{1}{6}\left[(\Omega-\Omega')(\Omega-\Omega')_x\right]_{xx} - \Omega_x\Omega_{xx} - \Omega'_x\Omega'_{xx},$$

which implies

$$\frac{1}{6}\left[(\Omega-\Omega')(\Omega-\Omega')_t + (\Omega-\Omega')(\Omega-\Omega')_{xxx} + (\Omega-\Omega')_{xx}(\Omega-\Omega')_x\right] + \\ +\frac{1}{3}(\Omega-\Omega')_x(\Omega-\Omega')_{xx} - \Omega_x\Omega_{xx} - \Omega'_x\Omega'_{xx} = 0.$$

But we have that

$$\begin{aligned} &\frac{1}{2}(\Omega-\Omega')_{xx}(\Omega-\Omega')_x - \Omega_x\Omega_{xx} - \Omega'_x\Omega'_{xx} = \\ =\ &\frac{1}{2}\left[\Omega_{xx}\Omega_x + \Omega'_{xx}\Omega'_x - \Omega_{xx}\Omega'_x - \Omega'_{xx}\Omega_x\right] - \Omega_{xx}\Omega'_x - \Omega'_x\Omega'_{xx} = \\ =\ &-\frac{1}{2}\left(\Omega_{xx}\Omega_x + \Omega'_{xx}\Omega'_x + \Omega_{xx}\Omega'_x + \Omega'_{xx}\Omega_x\right) = \\ =\ &-\frac{1}{2}\left(\Omega_{xx}+\Omega'_{xx}\right)\left(\Omega_x+\Omega'_x\right) = \\ =\ &\frac{1}{12}(\Omega-\Omega')(\Omega-\Omega')_x(\Omega_x+\Omega'_x) = \frac{1}{12}(\Omega-\Omega')\left(\Omega_x^2 - {\Omega'_x}^2\right). \end{aligned}$$

We get then that

$$\frac{1}{6}(\Omega-\Omega')\left[Q(\Omega) - Q(\Omega')\right] = 0. \tag{10}$$

If $\Omega - \Omega'$ has in its expansion in terms of the Grassmann algebra generators a non zero coefficient relative to the identity generator, we get that $Q(\Omega) - Q(\Omega') = 0$ and from this and (8) we conclude that $Q(\Omega) = Q(\Omega') = 0$, or equivalently , Ω and Ω' satisfying (7) are solutions of (2).

We have thus proved that (7) is a (WE) transformation for equation (2).

3. Relating the Bäcklund transformation to the Gardner system

Let us analyze the two solutions Ω and Ω' of (7). The corresponding $\Phi = D\Omega$ and $\Phi' = D\Omega'$ are solutions of the supersymmetric KdV system (1). The integrability condition of (7) yields, assuming that the components of the identity on the Grassmann expansion is different from zero,

$$Q(\Omega) - Q(\Omega') = 0.$$

If we evaluate explicitly the left hand member of this equation and use the first equation in (7) we obtain

$$\Omega - \Omega' = (\Omega-\Omega')_t + (\Omega-\Omega')_{xxx} + \frac{1}{2}\left[\eta - \frac{1}{12}(\Omega-\Omega')^2\right](\Omega_x - \Omega'_x). \tag{11}$$

We now define the superfield χ as

$$\chi = \frac{1}{2\epsilon}(\Omega-\Omega') + \frac{3}{\epsilon^2},$$

where ϵ is an even parameter related to η by $\eta = \frac{3}{\epsilon^2}$.

The superfield χ satisfies then

$$\chi_t + \chi_{xxx} + \chi\chi_x - \frac{1}{6}\epsilon^2\chi^2\chi_x = 0. \tag{12}$$

It is a parametric Gardner-type equation. Moreover, using the first of equations (7) we get

$$\Omega_x + \Omega'_x = 2\left(\chi - \frac{1}{6}\epsilon^2\chi^2\right). \tag{13}$$

Combining (11) and (13) we obtain

$$\Omega_x = \chi + \epsilon\chi_x - \frac{1}{6}\epsilon^2\chi^2, \tag{14}$$

$$\Omega'_x = \chi - \epsilon\chi_x - \frac{1}{6}\epsilon^2\chi^2. \tag{15}$$

(14) and (15) define the Gardner transformation. Concluding, given Ω and Ω' a solution of the system (7), χ is a solution of (12) and Ω, Ω' are related to χ by the Gardner transformation. (14) and (15) are related by changing $\epsilon \to -\epsilon$. The converse is also valid. Given a solution of the Gardner equation (12), the Gardner transformation (14) and (15) define Ω and Ω' which solve (7) and from them we obtain $\Phi = D\Omega, \Phi' = D\Omega'$ both solutions of the supersymmetric KdV equation (1).

From (14) and (15) we can expand χ as a formal series in ϵ. Since $\int_{-\infty}^{+\infty}\chi dx$ is a conserved quantity all the coefficients of the expansion in ϵ are conserved quantities. We then obtain an infinite sequence of conserved quantities. The first few of them are

$$\int_{-\infty}^{+\infty} u dx \quad , \quad \int_{-\infty}^{+\infty} u\xi_x dx,$$

$$\int_{-\infty}^{+\infty} \left(u^2\xi_x - 2u_x\xi_{xx}\right) dx \quad , \quad \int_{-\infty}^{+\infty} u^2 dx,$$

$$\int_{-\infty}^{+\infty} \left(\frac{1}{3}u^3 - (u_x)^2\right) dx.$$

4. Solutions for the supersymmetric system

We finally obtain solutions for the system (7) starting with the trivial solution $\Omega' = 0$. In this case the first equation of (7) remains $\Omega_x = \eta - \frac{1}{12}\Omega^2$. This equation, written in components is

$$w_x = \eta_1 - \frac{1}{12}w^2$$
$$\xi_x = \eta_2 - \frac{1}{6}w\xi,$$

where $\eta = \eta_1 + \theta\eta_2$. For example, taking the Grassmann algebra with two generators e_1 and e_2 the previous system in components reads:

$$w_{1x} = \eta_1^1 - \frac{1}{12}w_1^2$$
$$w_{12x} = \eta_1^{12} - \frac{1}{6}w_1 w_{12}$$
$$\xi_1 = \eta_2^1 - \frac{1}{6}w_1\xi_1$$
$$\xi_2 = \eta_2^2 - \frac{1}{6}w_1\xi_2,$$

where $w = w_1 \cdot 1 + w_{12}e_1e_2$,$\eta_1 = \eta_1^1 \cdot 1 + \eta_1^{12}e_1e_2$,$\xi = \xi_1e_1 + \xi_2e_2$ and $\eta_2 = \eta_2^1e_1 + \eta_2^2e_2$.

We obtain the solutions

$$w_1(\tau) = \sqrt{12\eta_1^1}\left[\frac{C_1e^{\frac{\tau}{r}} - 1}{1 + C_1e^{\frac{\tau}{r}}}\right] \tag{16}$$

(where C_1 is an arbitrary real constant, $r = \sqrt{\frac{3}{\eta_1^1}}$) and

$$w_{12}(\tau) = \frac{e^{\frac{\tau}{r}}}{\left(C_1e^{\frac{\tau}{r}} + 1\right)^2}\left[\eta_1^{12}\left(rC_1^2e^{\frac{\tau}{r}} + 2C_1\tau - re^{-\frac{\tau}{r}}\right) + C_{12}\right] \tag{17}$$

(where C_{12} is another arbitrary real constant), with similar expressions for ξ_1 and ξ_2.

To obtain the explicit dependance of t we use second equation of (7). After some calculations we have

$$\Omega_t = -\frac{1}{3}\eta\left(\eta - \frac{1}{12}\Omega^2\right) \tag{18}$$

which implies

$$\Omega_t + \frac{1}{3}\eta\Omega_x = 0 \tag{19}$$

and thus

$$\Omega = \Omega\left(x - \frac{1}{3}\eta t\right). \tag{20}$$

This means, with respect to (16) and (17), that $\tau = x - \frac{1}{3}\eta t$.

We notice that the above solutions are wave solutions only in the case we have $\eta_2 = 0$.

We notice also that (16) implies the very well known solutions for KdV equation. That is, if C_1 is positive we have the one soliton solutions and if C_1 is negative we have (the first) singular solutions for KdV equation.

The above solutions can be iterated as in the scalar case using the Bäcklund Walhquist-Estabrook transformation to generate infinite hierarchies of solutions.

5. Conclusions

We obtained a Bäcklund Walhquist-Estabrook transformation for a supersymmetric KdV system. We found the explicit relation between the (WE) transformation and the associated Gardner map. We showed that the given system is Galilean invariant and in that sense it may have interesting physical approximation at low energies, in a similar way as the very well known KdV equation. We obtained solutions for the system with the help of the Bäcklund Walhquist-Estabrook transformation and discussed its dependance in terms of the independent variables x and t.

Acknowledgments

A. R. and A. S. are partially supported by Project Fondecyt 1161192, Chile.

References

[1] H. D. Wahlquist and F. B. Estabrook, Phys. Rev. Lett. 31, number 23, 1386 (1973).
[2] R. Hirota, Phys. Rev. Lett. 27, number 18, 1192-1194 (1971).
[3] J. Satsuma, Prog. Theor. Phys. Vol. 52, 1396 (1974).
[4] H. H. Cheng, Bäcklund Transformations, Lecture Notes in Mathematics No. 515, Springer-Verlag, New York, 241-252(1976).

[5] P. J. Olver, "On the construction of deformations of integrable systems", University of Minnesota, Preprint (1981).
[6] P. Mathieu, J. Math. Phys. 29, 2499 (1988).
[7] P. Labelle and P. Mathieu, J. Math. Phys. 32, 923 (1991).
[8] S. Belucci, E. Ivanov and S. Krivonos, J. Math. Phys. 34, 3087 (1993).
[9] F. Delduc and E. Ivanov, Phys. Lett. B 309, 312-319 (1993).
[10] F. Delduc, E. Ivanov and S. Krivonos, J. Math. Phys. 37, 1356 (1996).
[11] Z. Popowicz, Phys. Lett. B 459, 150-158 (1999).
[12] Q. P. Liu and Xing-Biao Hu, J. Phys. A: Math. Gen. 38, 6371-6378 (2005).
[13] A. K. Karasu, J. Math. Phys. 38 (7), 3616-3622 (1997).
[14] S. Yu. Sakovich, J. Nonlin. Math. Phys. 6, N 3 255-262 (1999).
[15] P. Casati and G. Ortenzi, J. Geom. and Phys. 56, 418-449 (2006).
[16] L. C. Vega, A. Restuccia and A. Sotomayor, J. Phys. Conf. Series 490, 012024 (2014).
[17] L. C. Vega, A. Restuccia and A. Sotomayor, arXiv:1407.7743 v3 [math-ph].
[18] A. Restuccia and A. Sotomayor, Open Phys. 14:95-105 (2016).
[19] S. Andrea, A. Restuccia and A. Sotomayor, J. Math. Phys. 42, 2625-2634 (2001).

Random non-Hermitian tight-binding models

G. Marinello[†] and M. P. Pato[‡]

Instituto de Física, Rua do Matão 1371, Universidade de São Paulo, 05508-090, São Paulo, S.P., Brazil

E-mail: [†] gabriel.marinello.santos@usp.br

E-mail: [‡] mpato@if.usp.br

Abstract. For a one dimensional system tight binding models are described by sparse tridiagonal matrices which describe interactions between nearest neighbors. In this report, we construct open and closed random tight-binding models based in the tridiagonal matrices of the so-called β-ensembles of random matrix theory.

1. Introduction

Tigh-binding models appear in many applications in condensed matter physics [1, 2, 3]. For one dimensional system, they are described by tridiagonal matrices in which diagonal elements represent the sites while the off-diagonal elements describe the interactions between nearest neighbors of relevance to the problem at hand [4]. In the early 2000s and in another context, tridiagonal matrix models were used to provide an extension of certain classical cases of random matrix theory, which became known as the β-ensembles [3, 5, 6]. This ensemble became the basis of recent efforts to provide models which verify the pseudo-Hermitian condition [7], satisfied by PT-symmetric systems [8]. In the present work, we show that the pseudo-Hermitian β-ensemble also constitutes a basis of a random non-Hermitian tight-binding model.

A matrix H is pseudo-Hermitian if it is connected to its adjoint $H^\dagger$ by the relation

$$H^\dagger = \eta H \eta^{-1}, \tag{1}$$

which means that it shares with its adjoint the same set of eigenvalues [7]. As a consequence, eigenvalues of a pseudo-Hermitian are necessarily real or complex conjugate. In particular, real matrices are pseudo-Hermitian and a matrix η, in this case, must exist. On the other hand, if the matrix $\eta^{1/2}$ such that $\eta^{1/2}\eta^{1/2} = \eta$ and its inverse exist, and are Hermitian, then the matrix

$$K = \eta^{1/2} H \eta^{-1/2} = \eta^{-1/2} \eta H \eta^{-1} \eta^{1/2} = K^\dagger \tag{2}$$

is Hermitian and shares with H the same set of eigenvalues. In this case, all the eigenvalues of H are real.

Eqs. (1) and (2) shows that a non-Hermitian matrix can have the same eigenvalues of a Hermitian one. Considering in the opposite direction, Eq. (1) can be used to construct a pseudo-Hermitian matrix out of a Hermitian one. The procedure is simple in the case of tridiagonal matrices aa shown in Ref. [9] where a pseudo-Hermitian β-ensemble was constructed. That procedure is here extended by constructing a family of matrices isospectral to the matrices of

Content from this work may be used under the terms of the Creative Commons Attribution 3.0 licence. Any further distribution of this work must maintain attribution to the author(s) and the title of the work, journal citation and DOI.
Published under licence by IOP Publishing Ltd

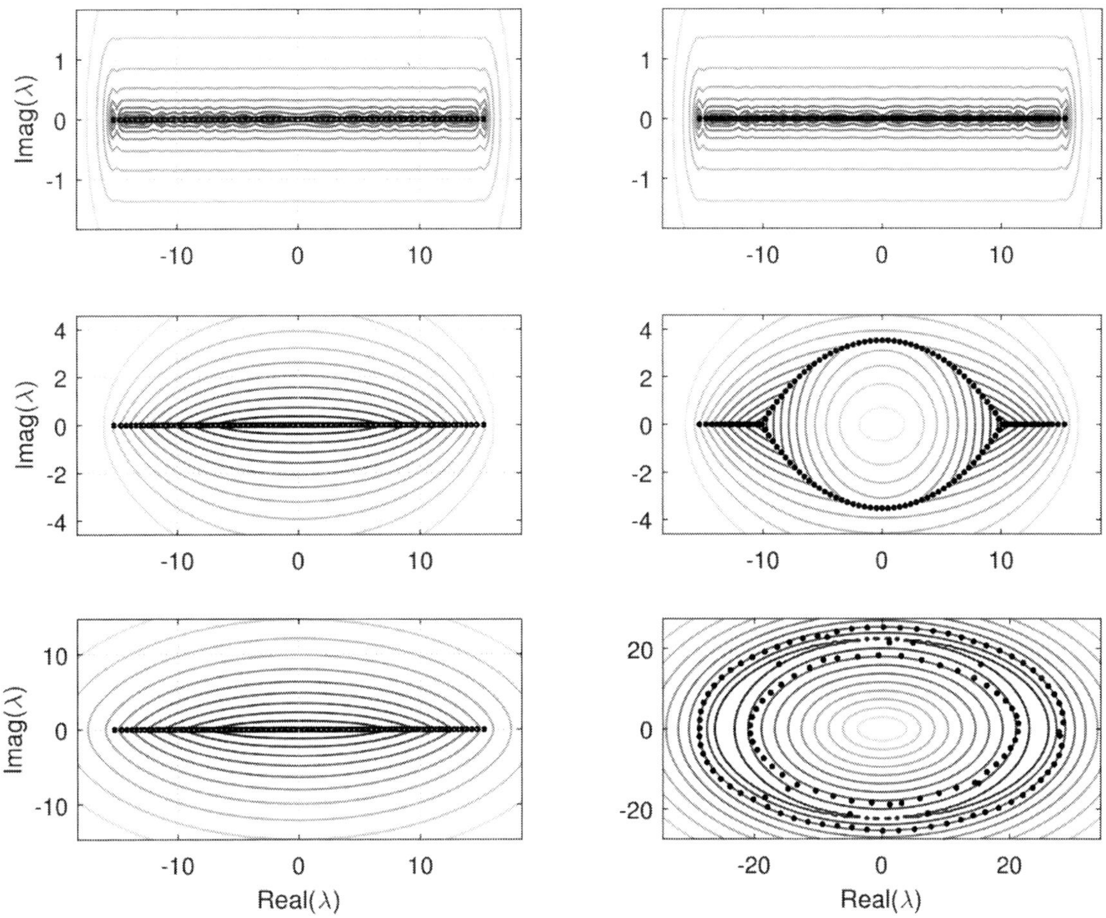

Figure 1. Pseudospectra for the average open chains (left) and closed chains (right) of size $N = 128$ and $\beta = 1$ with $\alpha = 0.0, 0.25, 1.00$ from top to bottom.

the β-ensemble. This family defines an open chain with random hopping in a consistent way the chain is closed and defines a random tight-binding model.

2. The model

Let us start by considering the real matrix

$$H(\alpha) = \begin{pmatrix} a_1 & d_1^{1+\alpha} & & & & d_N^{1-\alpha} \\ d_1^{1-\alpha} & a_2 & d_2^{\alpha+1} & & & \\ & \cdot & \cdot & \cdot & & \\ & & \cdot & \cdot & \cdot & \\ & & & \cdot & \cdot & \cdot \\ & & & d_{N-2}^{1-\alpha} & a_{N-1} & d_{N-1}^{1+\alpha} \\ d_N^{1+\alpha} & & & & d_{N-1}^{1-\alpha} & a_N \end{pmatrix} \tag{3}$$

where α is a real parameter greater or equal to zero. With the presence of elements at the corners the matrix describe a closed chain and, open one, if $d_N = 0$.

2.1. The Open Chain

Starting with the open case, if $\alpha > 0$, the matrix $H(\alpha)$ is non-Hermitian but its eigenvalues are real having the same values of the Hermitian matrix $H(\alpha = 0)$. In other words, the eigenvalues are not affected by the parameter α. This can be proved by constructing the diagonal matrix [10]

$$\eta(\alpha) = \text{diag}\left(1, d_1^{2\alpha}, d_1^{2\alpha} d_2^{2\alpha}, \ldots, \prod_{k=1}^{N-1} d_i^{2\alpha}\right) \tag{4}$$

which possesses a square root $\eta^{1/2}$ such that we can write

$$H(\alpha) = \eta(\alpha)^{1/2} H(0) \eta(\alpha)^{-1/2}. \tag{5}$$

As $H(0)$ is Hermitian $H(\alpha)$ contitutes an isospectral family of matrices. Another way of proving the isospectrality is by considering the recursion relation

$$P_n(x) = [a_n - x] P_{n-1}(x) - d_{n-1}^2 P_{n-2}(x) \tag{6}$$

satisfied by its characteristic polynomial in which the parameter α just does not appear.

If now the diagonal elements are taken from the normal distribution $N(0,1)$ and the off-diagonal ones from the χ_ν distribution with $\nu = \beta k$ degrees of freedom, for the k-th row, such that

$$a_k = \{X \mid f(X) = N(0,1), k = 1, ..., N\} \tag{7}$$

$$d_k = \left\{\frac{X}{\sqrt{2}} \;\middle|\; f(X) = \chi_{\beta k}, k = 1, ..., N-1\right\} \tag{8}$$

$$d_N = \begin{cases} 0, & \text{open chain} \\ \left\{\frac{X}{\sqrt{2}} \;\middle|\; f(X) = \chi_{\beta N}\right\}, & \text{closed chain} \end{cases} \tag{9}$$

then $H(0)$ is a matrix of the so-called β ensemble whose the joint density distribution of the eigenvalues generalizes the RMT eigenvalues from $\beta = 1, 2, 4$ to any real positive value [6]. For a fixed matrix size and sufficiently small values of β the diagonal is dominant and the eigenvalues become uncorrelated. In the other extreme, that is large values of β the subdiagonal elements dominate and become localized such that fluctuations are suppressed.

It also becomes convenient to define an average matrix for this ensemble

$$\langle H(\alpha)\rangle = \eta(\alpha)^{1/2} \langle H(0)\rangle \eta(\alpha)^{-1/2} \tag{10}$$

That may be accomplished by averaging each element independently in (7)-(9), to the mean of the respective distribution. Numerical evidence shows that the spectral behavior of sample matrices displays relatively small fluctuations around that of the average matrix.

For the open chain, that is $d_N = 0$, the eigenvectors of $H(\alpha)$ are obtained from those of $H(0)$ as

$$|\Psi(\alpha)\rangle = \eta^{1/2}(\alpha) |\Psi(0)\rangle \tag{11}$$

where $\eta(\alpha)$ is the similarity transformation of the pseudo-Hermitian condition (1) for a given parameter α.

Therefore, using Appendix A, we can show that besides being isospectral also the localization of the eigenvectors is not affected by the parameter α. The effect of the parameter is to make non-normal the non-Hermitian matrices of the family. This non-normality aspect of the matrices is revealed by investigating the pseudospectra of their eigenvalues. The eigenvectors of the pseudo-Hermitian open chain $H(\alpha)$ are obtained by a gauge transformation of the β-ensembles ones, a transformation which is not expected to produce change in their localization.

In the left column of Fig. 1, we present a comparison of the behavior of the pseudospectra (cf. Appendix B) of open chains of size $n = 64$ and $\beta = 1$ for the parameters $\alpha = 0.0, 0.25$ and 1.00 from top to bottom. The pattern shown in this figure of the pseudospectra is typical of non-normal matrices, and as a consequence indicates that these matrices should display instabilities under small perturbations. It was shown in Ref. [9] that the minimal perturbation for the transition to occur becomes arbitrarily small as the size of the matrix increases.

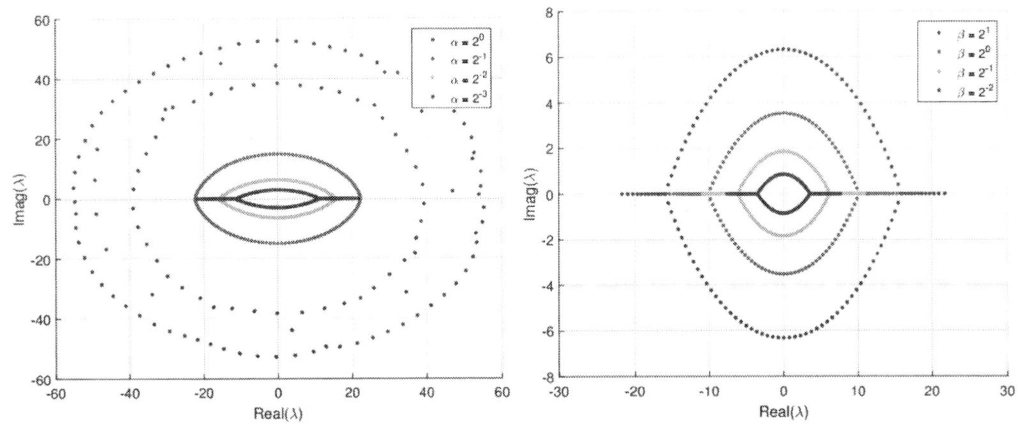

Figure 2. Spectra of the closed chain case for varying α of $N = 128$ and $\beta = 2.0$ average matrices (left) and for varying β of $N = 128$ and $\alpha = 0.25$ average matrices (right).

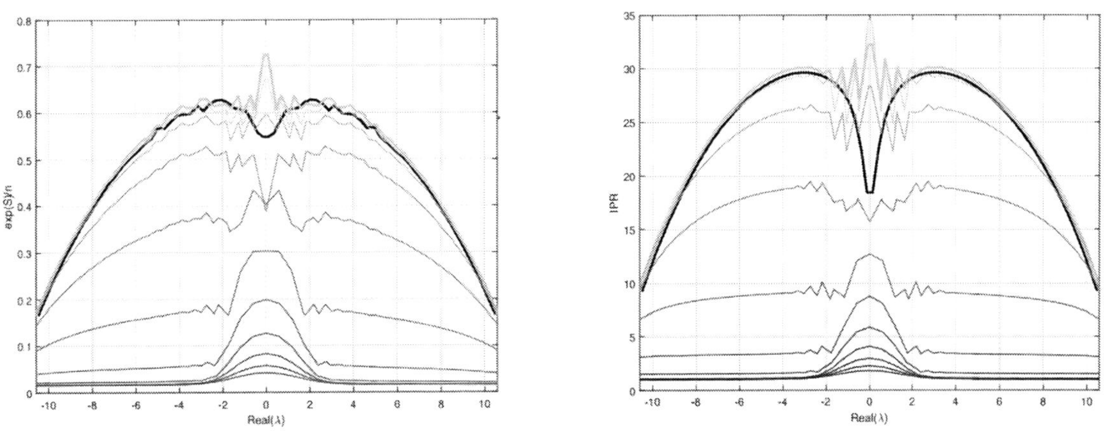

Figure 3. Entropy per lattice point (left) and Inverse Participation Ratio (right) for the open chain (black line), Hermitian closed chain (gray line) and increasing parameter α from 10^{-2} to 10^{-1} (top: red lines, increasing α corresponds to lower lines; bottom: blue lines, increasing α corresponds to lower lines).

2.2. *Closed Chain*

The analysis using Appendix A does not apply to the closed chain. The right column of Fig. 1 shows the pseudospectra for the average matrices of closed chains for varying α and Fig. 2 presents the eigenvalues of closed chain matrices varying parameters α (left) and β (right).

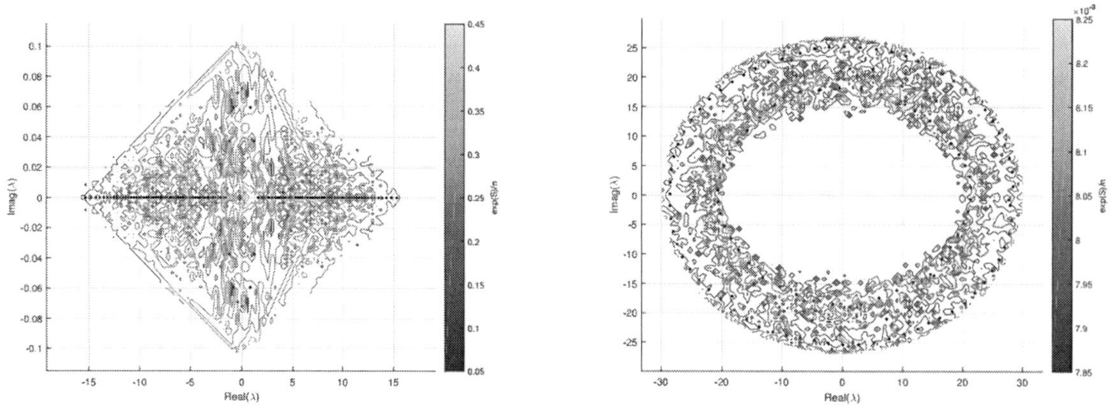

Figure 4. Entropy map for 5000 closed chain matrices of size $n = 128$ and $\beta = 1.0$, with $\alpha = 0.01$ (left) and $\alpha = 1.0$ (right). The black dots denote the eigenvalues of the average matrix.

Although the change in β does not change the characteristic shape of the spectra, the change in α causes dramatic change in the position of the eigenvalues in the complex plane as well as in the pseudospectrum. It is noteworthy that the structure seen in the bottom-right corner of Fig. 1, showing two concentric ellipses, differs from any of the behaviors commonly observed in the literature [4, 9, 11].

This suggests the possibility of changes in the properties also of the eigenvectors. This is illustrated in Figs. 3 and 4. In Fig. 3 the localization as measured through entropy and IPR are presented. There is a noticeable decrease in the fraction of occupied states as the parameter α increases. In Fig. 4, we can see a mapping for two particular values of α are presented. In the first case, $\alpha = 0.01$, the entropies display a low degree of localization, such that the eigenstates occupy a large fraction of the basis states for most of the matrices' eigenvalues. For the latter case, $\alpha = 1.00$, the states have become highly localized in the complex plane's vicinity to the average matrix's eigenvalues, having fractional occupancy of the basis states.

3. Conclusion

We have constructed a closed chain ensemble of random matrices, derived from the β-ensemble of random matrix theory. Numerical evidence was presented to illustrate the behavior of said matrices. In the tight-binding models of Refs. [4, 11], a spectra of oval shape with trailing real segments was seen. This was again seen in Refs. [9, 16]. The most noteworthy difference the present model has in regard to those previouly present in the literature is the presence, of a second elliptical structure in the spectrum, corroborated by the pseudospectral evidence, which appears spontaneously as the parameter α is increased.

Appendix A. The eigenvector delocalization

In order to study the localization of the eigenvectors, let us consider the Shannon entropy of a given state $\Psi(\alpha)$

$$S_{\eta(\alpha);N}[\Psi(\alpha)] = -\sum_{i=1}^{N} p_i[\Psi(\alpha)] \log\left(p_i[\Psi(\alpha)]\right) \tag{A.1}$$

of an eigenvector normalized in the inner product $\langle\phi|\,\psi\rangle_{\eta(\alpha)} \equiv \langle\phi|\,\eta(\alpha)\,|\psi\rangle$ [12], in which the probabilities of occupation of a given state $|i\rangle$ are given by

$$p_i = \langle\Psi(\alpha)|\,\eta(\alpha)\,|i\rangle\,\langle i|\,\Psi(\alpha)\rangle\,. \tag{A.2}$$

By taking exponential $N_c = \exp[S_{\eta(\alpha);N}(\Psi(\alpha))]$ we have a measure of the occupation of number of eigenstates of the basis. Assuming the scaling $N_c \sim N^D$, if $N_c \to 1$ when $N \to \infty$ the state is localized. For $0 < D < 1$ the eigenvector spreads over a fixed fraction of the states.

As long as the matrix $\eta^{1/2}|\eta = \eta^{1/2}\eta^{1/2}$ may be define, the eigenvectors $\Psi(\alpha)$ of matrices of the pseudo-Hermitian ensemble are expressed in terms of those $\Psi(0)$ of the Hermitian β-ensemble as

$$|\Psi(\alpha)\rangle = \eta(\alpha)^{1/2}\,|\Psi(0)\rangle\,. \tag{A.3}$$

For the case of the open chain, this relation implies that eigenvector localization independent of α as can easily be shown from Eqs. (A.2) and (A.3).

Additionally, the Inverse Participation Ratio

$$\mathrm{IPR} = \frac{1}{\sum\limits_{k=1}^{N} p_i^2} \tag{A.4}$$

is another way to measure the localization [13] of normalized vectors.

Appendix B. Pseudospectra

The pseudospectrum of a linear operator has been shown to be useful in analyzing operators for which there is no eigenvector basis (see *e.g.* [14, 15] and references therein). There are multiple equivalent ways by which the pseudospectrum of a linear operator can be defined. Here we use instead another definition that uses the resolvent of the matrix A, where the set z satisfies

$$|(z - A)|^{-1} < \frac{1}{\epsilon}, \tag{B.1}$$

with ϵ being positive and very small. The $||.||_2$ pseudospectrum may be related to the smallest singular value of the operator, that is for an operator H, the pseudospectrum is given by

$$\Sigma_\epsilon(H) = \{w \in \mathbb{C} : \sigma_{\min}(wI - H) \le \epsilon\}\,. \tag{B.2}$$

For a normal matrix A, that is one for which $AA^\dagger = A^\dagger A$, the pseudospectra yields circles in the complex plane centered in the eigenvalues. On the other hand, for non-normal matrices, the pseudospectra present behaviors not regularly distributed close to the eigenvalues.

Acknowledgements

The authors would like to acknowledge the support of Brazilian funding agencies CNPq and FAPESP.

References

[1] A. Altland and B. Simons. Cambridge University Press, 2006. C. M. Bender. In: Rep. Prog. Phys. 70.6 (2007), pp. 9471018.
[2] J. Feinberg and A. Zee, Nuc. Phys. B **522**, 599 (1999).
[3] P. J. Forrester. Princeton University Press, 2010. Walter A. Harrison. Dover Publications, 1989.
[4] N. Hatano and D. R. Nelson, Phys. Rev. Lett. **77**, 570 (1997).
[5] T. Baker and P. Forrester, Commun. Math. Phys. **188**, 175 (1997).
[6] I. Dumitriu and A. Edelman, J. Math. Phys., **43** 5830 (2002).
[7] A. Mostafazadeh, J. Math. Phys. **43**, 205, *ibid.* **43**, 2814, *ibid.* **43** 3944 (2002).
[8] C. M. Bender and S. Boettcher, Phys. Rev. Lett. **80**, 5243 (1998); C. M. Bender, S. Boettcher, and P. N. Meisinger, J. Math. Phys. **40**, 2201 (1999).
[9] G. Marinello and M.P. Pato. In: Physica A: Statistical Mechanics and its Applications 444 (2016), pp. 10491061.

[10] O. Bohigas and M. P. Pato, AIP Advances **3**, 032130 (2013).
[11] N. Hatano and D. R. Nelson, Phys. Rev. B **56**, 8651 (1997).
[12] The notation $\langle\phi|\,\psi\rangle = \sum_{i=1}^{N}\phi_i^*\psi_i$ denotes the usual inner product in $\mathbb{C}^N$ space. The fact that generally for a pseudo-Hermitian matrix, the right eigenvectors $|\psi\rangle$ are not pairwise orthogonal requires us to normalize with respect to the left eigenvectors $\langle\phi|$. Instead they are connected as $|\phi\rangle = \eta\,|\psi\rangle$, which can be readily shown using (1).
[13] A. Amir, N. Hatano and D. R. Nelson, arXiv:1512.05478v2 (2015).
[14] L. N. Trefethen and M. Embree, *Spectra and Pseudospectra: The Behavior of Nonnormal Matrices and Operators*, Princeton University Press (2005).
[15] R. Henry and D. Krejčiřík, J. Spectr. Theory, to appear.
[16] G. Marinello and M. P. Pato. In: F. Bagarello, R. Passante and C. Trapani (Eds.), Non-Hermitian Hamiltonians in Quantum Physics, Springer (2016).

Journal of Physics: Conference Series **738** (2016) 012041 doi:10.1088/1742-6596/738/1/012041

Wormholes with asymptotic Lifshitz scaling in Hořava gravity

Jorge Bellorín[1 2], **Alvaro Restuccia**[1 2], **Adrián Sotomayor**[3]
[1] Departamento de Física, Universidad Simón Bolívar, Caracas, Venezuela
[2] Departamento de Física, Universidad de Antofagasta, Antofagasta, Chile
[3] Departamento de Matemáticas, Universidad de Antofagasta, Antofagasta, Chile

E-mail: jbellorin@usb.ve, alvaro.restuccia@uantof.cl, adrian.sotomayor@uantof.cl

Abstract. We study static spherically symmetric solutions of the nonprojectable Hořava theory with and without cosmological constant. The solutions we find are two-side wormholes and (single-side) naked singularities. Interestingly, in the case of negative cosmological constant we find that in the exterior side the wormhole acquires an asymptotic scaling between space and time equal to the scaling of the Lifshitz solution, which was previously found to be a vacuum solution of the same theory. This result leads us to pose the question whether in the case of negative cosmological constant the asymptotic anisotropic Lifshitz scaling is a generic feature of the vacuum field equations rather than the asymptotic AdS-like scaling.

1. Introduction

Hořava theory has been proposed as an UV completion of general relativity with the property of being power-counting renormalizable. The foliation-preserving-diffeomorphisms gauge symmetry enables the introduction of higher order spatial derivatives (higher spatial curvature) in the potential without need of introducing higher order time derivatives. This mechanism, in principle, could avoid adding ghosts while rendering the theory renormalizable (careful analysis of the propagating modes is needed in any case).

In this work we study the problem of finding solutions, under a combination of analytical and numerical approaches, of the large-distance effective action of the nonprojectable Hořava theory [1, 2]. We impose three conditions on the set of solutions: staticity, spherical symmetry and vanishing of the shift vector. Unlike General Relativity (GR), in Hořava theory spherical symmetry does not imply staticity on the vacuum solutions. Moreover the asymptotic behavior of the solutions does not need to be the same dictated by GR. Thus, we impose staticity and spherically symmetry independently. Since the cosmological constant represents the lowest-order term in the Lagrangian, it has important consequences on the form of the solutions of the large-distance effective action. We study the cases of negative and vanishing cosmological constant, which can be englobed into a unique mathematical approach, as we shall see.

In [3] solutions of the Einstein-aether theory without cosmological constant under conditions analog to the ones we impose here were found. This theory is dual to the large-distance effective action of the Hořava theory if the aether vector is hypersurface orthogonal. Those authors found a wormhole-like geometry. In [4] the same problem was focused directly on the Hořava theory. In [5] a symmetric-wormhole ansatz was studied in Hořava theory.

Content from this work may be used under the terms of the Creative Commons Attribution 3.0 licence. Any further distribution of this work must maintain attribution to the author(s) and the title of the work, journal citation and DOI.
Published under licence by IOP Publishing Ltd

In this work we shall perform further developments including a negative cosmological constant. Part of our analysis is based on the introduction of a new radial coordinate that helps to clarify the geometry of the solutions. We shall obtain two kinds of solutions: a wormhole-like geometry formed by two sides joined by a throat and a single-side solution with a naked singularity at the origin. We shall pay special attention to the asymptotic behavior of the solutions. We shall see that in the case of the wormhole the asymptotia of its two sides is different, the wormhole is not symmetric. When the cosmological constant vanishes one of the sides is asymptotically flat whereas the other one is asymptotically singular. The asymptotically singular side can be regarded as a kind of internal space. When the cosmological constant is negative, there is again an internal side with singular asymptotia and an asymptotically regular side. It comes out the question on whether the regular side follows the AdS asymptotia or other kind of geometry instead.

2. Finding the solutions

The theory is written in ADM variables. The most general effective action for large distances with cosmological constant is [1, 2]

$$S = \int dt d^3x \sqrt{g} N (K_{ij}K^{ij} - \lambda K^2 - 2\Lambda + \xi R + \alpha a_i a^i) \,, \tag{1}$$

where $K_{ij} = \frac{1}{2N}(\dot{g}_{ij} - 2\nabla_{(i}N_{j)})$, K is its trace, R is the spatial Ricci scalar, $a_i = \partial_i \ln N$ and λ, Λ, ξ and α are independent coupling constants. Under the conditions we are going to use we may set $\xi = 1$ without loss of generality. When the value of the coupling constant λ is out of the kinetic conformal point, $\lambda \neq 1/3$, the theory propagates three degrees of freedom: the two tensorial modes that are also present in GR plus one extra scalar mode [2, 6, 7]. Sometimes this is believed to be the general situation and several studies are devoted to the physics of the extra mode. However, when $\lambda = 1/3$, which is the kinetic conformal point, the theory acquires two extra second-class constraints that eliminate the extra mode completely for all ranges of energy [8]. Another exception that is relevant for large-distance dynamics is the model studied in [9].

Here we present a summary of the approach for solving the field equations under the symmetry conditions we are interested in. The details can be seen in [10]. The static spherically symmetric spacetime metric with vanishing shift vector, $N_i = 0$, can be written in time×spherical coordinates as

$$ds^2 = -N(r)^2 dt^2 + \frac{dr^2}{f(r)} + r^2 d\Omega^2_{(2)} \,. \tag{2}$$

From the action (1) we derive the field equations and then evaluate them on this ansatz. We get that the system is well posed in the sense the number of independent equations matches with the number of unknowns ($N(r)$ and $f(r)$). From the resulting field equations the following quadratic equation can be obtained

$$\left(\frac{\sqrt{f}}{\rho} + \frac{r\sqrt{f}N'}{\rho N}\right)^2 - \left(\frac{\beta r\sqrt{f}N'}{\rho N}\right)^2 = 1 \,, \tag{3}$$

where the prime denotes differentiation with respect to r, $\rho \equiv 1 - \Lambda r^2$ and $\beta \equiv \sqrt{1 - \alpha/2}$. We find that for any $\chi \in (-\infty, +\infty)$ the solution to this quadratic equation can be written as (modulo coordinate transformations)

$$\frac{\sqrt{f}}{\rho} + \frac{r\sqrt{f}N'}{\rho N} = \cosh\chi \,, \qquad \frac{\beta r\sqrt{f}N'}{\rho N} = \sinh\chi \,. \tag{4}$$

A further combination of the field equations with this solution, regarding r, N and f as functions of χ, yields the following differential equation for $r(\chi)$

$$\frac{dr}{d\chi} = \left(\frac{1}{\beta} - \frac{\cosh\chi}{\sinh\chi}\right)\frac{r\rho^2}{H}, \tag{5}$$

where H is a factor that depends on χ and $r(\chi)$. The solution $r(\chi)$ of (5) can be interpreted as a coordinate transformation from χ to r. By doing further combinations a differential equation for $N(\chi)$ can be obtained and the radial component $g_{\chi\chi}$ can be algebraically determined in terms of $r(\chi)$. Thus, the search for solutions can be closely casted as a problem of solving first-order differential equations in χ, including the function $r(\chi)$.

When $\Lambda = 0$ the H factor becomes equal to one. In this case (5) can be explicitly solved for $r(\chi)$ [11]. We shall return to this case later on.

Next we perform the numerical integration of Eq.(5) when $\Lambda < 0$ providing several initial data $r(\chi_0)$ for $\chi_0 \in (-\infty, +\infty)$. We find that there are three classes of curves. In Fig. 1 we show the plot of a representative curve for each class.

In all of the three classes the solutions stop at some finite values of χ where the numerical integration diverges ($dr/d\chi$ diverges). We have classified these points into two kinds of divergence points according to whether $r(\chi)$ remains finite or not there. The point at which also $r(\chi)$ diverges, denoted by χ_s, is fixed for all solutions at the value $\tanh\chi_s = 1/2\beta$. It corresponds to the spatial infinity. We can show analytically that in the χ coordinate the spatial infinity is reached only at $\chi = \chi_s$ and at $\chi = +\infty$.

The other points of divergence at which $r(\chi)$ remains finite vary their location on each solution. It turns out that there is always a pair of different-class solutions that tend to the same divergence point of this kind (from above and below of the corresponding value of r). Thus, we can join such two solutions to form a single extended solution. In Fig. 1 we show how this joining works in the red and blue curves which are different solutions in χ and stop at the same joining point. We have checked numerically that the joining is continuous in the metric components and their derivatives as well as in the square-curvature scalar (Kretschmann scalar). Therefore, the divergence points where $r(\chi)$ remains finite are nothing but a failure of the χ coordinate and are the junction points of the extended solution.

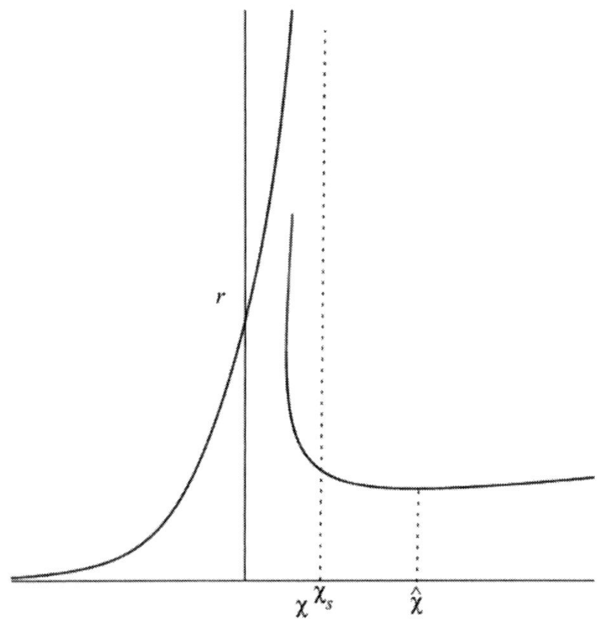

Figure 1. Each colored curve represents one of the three classes of numerical solutions in χ. χ_s is the divergence point that correspond to the spatial infinity. $\hat{\chi}$ is the minimum that one class of solutions has.

Thus, we finally get that there are two kinds of vacuum solutions that satisfy the conditions we have imposed. The most prominent geometric feature of the joined solution is that it has a minimum value of $r(\chi)$ located at $\tanh\hat{\chi} = \beta$, and it is valid at both sides of the minimum. This means that the solution has a throat at $\hat{\chi}$. The solution is regular at the throat and, unlike r, the coordinate χ is useful to parameterize the space around the throat in a single chart. The two sides of the throat extend themselves from the throat to spatial infinity. One of the sides is completely regular both in its inner points and at its infinite boundary. The other side is regular internally but exhibits an essential singularity at infinity. We regard this side as a kind of internal space. The wormhole is not symmetric between its two sides. We remark that the coordinate transformation from r to χ was purely spatial, thus there is no place for a horizon in this solution. We have that, except for the infinite boundary of the internal side, the solution is regular and free of horizons; any physical particle can pass through the throat from one side to the other one in both directions, except, again, if it reaches the singular boundary of the internal side. We have checked this regularity numerically with the metric components and their derivatives. We have also checked regularity of the solution at the throat, at the joining point and at the exterior asymptotic boundary by evaluating the square-curvature scalar. We have also checked that the square-curvature scalar diverges at the interior asymptotic boundary.

The other solution is composed of a single spatial side, extends itself from the origin to spatial infinity and at the origin exhibits a naked singularity. All these features have been checked with the metric components and the square-curvature scalar.

When the cosmological constant vanishes all the analysis we have presented in the coordinate χ is valid, with the advantage that in this case we can find the solution analytically. The solution of (5) is

$$r(\chi) = \frac{k e^{\chi/\beta}}{\sinh\chi}, \tag{6}$$

where k is an integration constant. The metric components can also be obtained analytically [11]. In the full range $\chi \in (-\infty, +\infty)$ these expressions comprise again two solutions that are analog to the ones of the $\Lambda < 0$ case: the wormhole with a throat (covered in $\chi \in (0, +\infty)$) and a single-side space with a naked singularity in the origin (covered in $\chi \in (-\infty, 0)$). The wormhole is very similar to the previous case. It has two nonsymmetric sides; the internal one is asymptotically singular whereas the exterior side is *asymptotically flat*. The solution with the naked singularity at the origin is also asymptotically flat.

3. Asymptotic Lifshitz scaling

Returning to the $\Lambda < 0$ case, now we present the interesting asymptotic behavior of the solutions. In [12] it was argued that Hořava theory, rather than a relativistic theory, provides the minimal holographic dual for Lifshitz-type field theories with anisotropic scaling between space and time. In that study those authors showed that the Lifshitz spacetime is a *vacuum* solution of the nonprojectable Hořava theory with negative cosmological constant. This spacetime can be written as

$$ds^2 = -\left(\frac{r}{\ell}\right)^{2z} dt^2 + \left(\frac{\ell}{r}\right)^2 dr^2 + \left(\frac{r}{\ell}\right)^2 dx^i dx^i \,. \tag{7}$$

The exponent z determines the degree of anisotropy between the time and the spatial direction r. In order to be a solution of the large-distance effective Hořava action its geometrical parameters z and ℓ must be related to the dynamical coupling constants Λ and α of the theory. The relations between constants those authors found can be written in the form (we set 3+1 spacetime dimensions whereas in [12] it is fixed $\ell = 1$)

$$z = \frac{1}{1-\alpha}, \qquad \ell^2 = -\frac{(2-\alpha)(3-2\alpha)}{2\Lambda(1-\alpha)^2}. \tag{8}$$

We stress that the Lifshitz spacetime is not spherically symmetric. In the coordinates of (7) the surfaces of t and r constant are not spheres. However, we find interesting to compare the Lifshitz spacetime with our solutions. To establish the connection we equate the bulk coordinate r of (7) with our original radial coordinate r. The asymptotic behavior of our solutions can be investigated analytically in the original system of field equations where the radius r is the independent variable. To achieve this we assume that if N and f diverge at $r \to \infty$, then they do so with some dominant powers of r. That is, we assume that as $r \to \infty$ these components diverge as $N = (r/\ell)^z$ and $f = (r/\ell)^b$, with $z, b, \ell > 0$. Under this assumption, the asymptotic analysis of the field equations derived from (1) (see [10] for details) shows that *necessarily* the form of the asymptotic solution in the time$\times$radial sector is dictated by the scaling of the Lifshitz solution; that is, necessarily $b = 2$ and z and ℓ are fixed in terms of α and Λ *exactly as in* (8). This is an analytical result. Now, in the case of the wormhole we have that only the exterior side is asymptotically regular. Therefore, in the exterior side there is a place for the asymptotic Lifshitz scaling (8). We have checked on the numerical solutions that effectively this is the case. The same holds for the asymptotic limit of the solution with the naked singularity at the origin.

Our results on the asymptotic behavior and the results of [12] lead us to ask whether this Lifshitz scaling, rather that the relativistic AdS scaling, is a generic feature of this family of vacuum solutions of the Hořava theory with a negative cosmological constant. That is, there could be general ansatz for geometries where a bulk coordinate r can be picked up. Then the generic asymptotic behavior could be a Lifshitz scaling in the form $g_{tt} \sim -r^{2z}$ and $g_{rr} \sim r^{-2}$, with z given in terms of α by (8).

4. Conclusions

The vacuum solutions of the large-distance effective action of complete nonprojectable Hořava gravity with cosmological constant $\Lambda \leq 0$ that satisfy the conditions of staticity, spherical symmetry and vanishing of the shift vector are two: a wormhole of two sides and a space with a naked singularity at the origin. We have shown that both the exterior side of the wormhole and the other solution acquire a kind of Lifshitz scaling asymptotically. Surprisingly, the dynamical exponent z (adopting the terminology of statistical physics) and even the coefficient of the "longitudinal" component of the metric coincide exactly with that of the Lishitz spacetime when this is casted as a vacuum solution of the theory. It would be interesting to elucidate whether in this theory this a general behavior for solutions that possess common structures in the set of isometries and noncompact directions. A related analysis on certain solutions of the Hořava theory with asymptotic Lifshitz scaling has been recently done in $2+1$ dimensions in [13]. It would be also interesting to study the case with positive cosmological constant, although likely other coordinate systems are more appropiate in this case.

Acknowledgments A. R. and A. S. are partially supported by Project Fondecyt 1161192, Chile.

References

[1] P. Hořava, *Quantum Gravity at a Lifshitz Point*, Phys. Rev. D **79** (2009) 084008, arXiv0901.3775 [hep-th].
[2] D. Blas, O. Pujolàs and S. Sibiryakov, *Consistent Extension Of Hořava Gravity*, Phys. Rev. Lett. **104** (2010) 181302, arXiv0909.3525 [hep-th].
[3] C. Eling and T. Jacobson, *Spherical solutions in Einstein-aether theory: Static aether and stars*, Class. Quant. Grav. **23** (2006) 5625 [Erratum-ibid. **27** (2010) 049801], grqc0603058 [gr-qc].
[4] E. Kiritsis, *Spherically symmetric solutions in modified Hořava-Lifshitz gravity*, Phys. Rev. D **81** (2010) 044009, arXiv0911.3164 [hep-th].
[5] M. Botta-Cantcheff, N. Grandi and M. Sturla, *Wormhole solutions to Horava gravity*, Phys. Rev. D **82** (2010) 124034, arXiv0906.0582 [hep-th].
[6] W. Donnelly and T. Jacobson, *Hamiltonian structure of Horava gravity* Phys. Rev. D **84** (2011) 104019, arXiv:1106.2131 [hep-th].
[7] J. Bellorín and A. Restuccia, *Consistency of the Hamiltonian formulation of the lowest-order effective action of the complete Horava theory* Phys. Rev. D **84** (2011) 104037, arXiv:1106.5766 [hep-th].
[8] J. Bellorín, A. Restuccia and A. Sotomayor, *A consistent Hořava gravity without extra modes and equivalent to general relativity at the linearized level* Phys. Rev. D **87** (2013) 084020 , arXiv1302.1357 [hep-th].

[9] J. Bellorín and A. Restuccia, *On the consistency of the Hořava Theory*, Int. J. Mod. Phys. D **21** (2012) 1250029, arXiv1004.0055 [hep-th].
[10] J. Bellorín, A. Restuccia and A. Sotomayor, *Solutions with throats in Hořava gravity with cosmological constant*, Int. J. Mod. Phys. D **25** (2016) 1650016, arXiv:1501.04568 [gr-qc].
[11] J. Bellorín, A. Restuccia and A. Sotomayor, *Wormholes and naked singularities in the complete Hořava theory*, Phys. Rev. D **90** (2014) 4, 044009, arXiv1404.2884 [gr-qc].
[12] T. Griffin, P. Hořava and C. M. Melby-Thompson, *Lifshitz Gravity for Lifshitz Holography*, Phys. Rev. Lett. **110** (2013) 081602, arXiv1211.4872 [hep-th].
[13] S. Basu, J. Bhattacharyya, D. Mattingly and M. Roberson, *Asymptotically Lifshitz spacetimes with universal horizons in* $(1+2)$ *dimensions*, arXiv:1601.03274 (2016) [hep-th].

Probing new CP violating observables in D meson decays

Yong-Feng Liu [1] **and Xian-Wei Kang** [2]
[1] Beijing Research Centre of Sustainable Energy and Building, Beijing University of Civil Engineering and Architecture, Beijing 100044, China
[2] Departamento de Física, Universidad de Murcia, E-30071 Murcia, Spain

E-mail: `Corresponding author: kxw198710@126.com (Xian-Wei Kang)`

Abstract. CP violation in the charm quark sector has not been examined very well as the case for strange and beauty ones. Some novel insights into the issue on the CP violation in D meson decay are discussed. Specifically, i) the T-violating observables in $D \to VV$ decays are constructed. Assuming CPT invariance T violation implies CP violation. This is a new idea and an alternative way for probing CP violation in D decays; ii) the decay of quantum correlated $D\overline{D}$ pair to vector mesons (denoted by V) is explored, which offers the new CP violating observables that have not been noticed before;

1. Introduction

CP violation in D decay is still a primary and hot topic in flavor physics. In the Standard Model (SM), this quantity is shown to be small, and in fact, up to now, has not been well determined in experiment. Any large signal of CP violation would be a smoking gun for new physics, see e.g. Refs. [1, 2]. The general consensus is that in SM the CP asymmetry is highly suppressed and out of the current experimental capability, however, in a New Physics (NP) model, CP violation can reach a relatively large quantity. In Ref. [2] the authors show the CP violation can yield order of 10^{-2}, which is comparable to the current experimental sensitivity. Or contrarily, the non-observation of these CP violating signal will at least put severe constraint on these NP models.

The decays $D \to PP$ (P is a pseudoscalar, typical examples are $D \to K\pi$, $\pi\pi$) have been extensively considered, however, we notice that the mode of D decaying to vector meson can provide rich information in angular distribution due to the polarization of spin-1 meson. In case of $D \to VV$ decay, we show the triple production (TP) term can occur which violates the time-reversal invariance. These signals can be accessed by a angular analysis in experiment. The quantum coherent double D decays, $\psi(3770) \to D\overline{D} \to (VV)(VV)$, is also quite interesting, since it provides new CP violating observables that have not been aware of before. We may denote $\psi(3770)$ resonance by ψ for simplicity. Below we will sketch this method. Replacing one VV pair by $K\pi$, one has $\psi(3770) \to D\overline{D} \to (VV)(K\pi)$, which is exploited to render a new information on the strong phase δ, required for the extraction of the angle γ in Cabibbo-Kobayashi-Maskawa (CKM) matrix [3, 4].

Content from this work may be used under the terms of the Creative Commons Attribution 3.0 licence. Any further distribution of this work must maintain attribution to the author(s) and the title of the work, journal citation and DOI.
Published under licence by IOP Publishing Ltd

2. Single D decays to VV

The mode of single D decaying to VV is used to explore the CP violation [5]. The vector meson can provide a polarization factor (denoted by ϵ) and thus the triple product like $\vec{k}\cdot(\epsilon_1\times\epsilon_2)$ can be composed. It is interesting to note that such triple product (TP) violates time-reversal (T) symmetry. Then if such structures can be isolated from the angular analysis, these would be measurable in experiment. By comparing a pair of CP-conjugate process, the term of correlation of triple product can be extracted in a way such that the clean T-violation can be probed [6]. For the process $D(p)\to V_1(k,\epsilon_1)V_2(q,\epsilon_2)$, where in the parenthesis the four-momenta and polarization is indicated for each vector meson, the amplitude can be written as [6],

$$\mathcal{M} = a\epsilon_1^*\cdot\epsilon_2^* + \frac{b}{m_1m_2}(p\cdot\epsilon_1^*)(p\cdot\epsilon_2^*) + i\,\frac{c}{m_1m_2}\epsilon^{\alpha\beta\gamma\delta}\epsilon_{1\alpha}^*\epsilon_{2\beta}^*k_\gamma p_\delta\,, \tag{1}$$

The coefficients a, b and c are general complex and can be parameterized as [6]

$$a=\sum_j a_je^{i\delta_{sj}}e^{i\phi_{sj}}\,,\quad b=\sum_j b_je^{i\delta_{dj}}e^{i\phi_{dj}}\,,\quad c=\sum_j c_je^{i\delta_{pj}}e^{i\phi_{pj}} \tag{2}$$

where j is only symbolic to show they can receive several pieces of amplitudes with their phases, δ is the phase coming from the strong interaction, and ϕ is the weak phase violating CP symmetry. Under CP transformation, δ does not change sign while ϕ does. Calculating the modulus squared, one obtains

$$\begin{aligned}|\mathcal{M}|^2 \;=\;& |a|^2|\epsilon_1^*\cdot\epsilon_2^*|^2+\frac{|b|^2}{m_1^2m_2^2}|(k\cdot\epsilon_2^*)(q\cdot\epsilon_1^*)|^2+\frac{|c|^2}{m_1^2m_2^2}|\epsilon^{\alpha\beta\gamma\delta}\epsilon_{1\alpha}^*\epsilon_{2\beta}^*k_\gamma p_\delta|^2\\ &+2\frac{Re(ab^*)}{m_1m_2}(\epsilon_1^*\cdot\epsilon_2^*)(k\cdot\epsilon_2^*)(q\cdot\epsilon_1^*)\\ &+2\frac{Im(ac^*)}{m_1m_2}(\epsilon_1^*\cdot\epsilon_2^*)\epsilon^{\alpha\beta\gamma\delta}\epsilon_{1\alpha}^*\epsilon_{2\beta}^*k_\gamma p_\delta+2\frac{Im(bc^*)}{m_1^2m_2^2}(k\cdot\epsilon_2^*)(q\cdot\epsilon_1^*)\epsilon^{\alpha\beta\gamma\delta}\epsilon_{1\alpha}^*\epsilon_{2\beta}^*k_\gamma p_\delta\,.\end{aligned} \tag{3}$$

Note that the term $\epsilon^{\alpha\beta\gamma\delta}\epsilon_{1\alpha}^*\epsilon_{2\beta}^*k_\gamma p_\delta$ in the rest frame of D meson takes just $\vec{p}\cdot(\epsilon_1^*\times\epsilon_2^*)$ which changes sign under the time-reversal transformation. With CPT invariance one can write the amplitude for $\bar{D}(p)\to\bar{V}_1(k,\epsilon_1)\bar{V}_2(q,\epsilon_2)$ as

$$\overline{\mathcal{M}} = \bar{a}\epsilon_1^*\cdot\epsilon_2^* + \frac{\bar{b}}{m_1m_2}(p\cdot\epsilon_1^*)(p\cdot\epsilon_2^*) - i\frac{\bar{c}}{m_1m_2}\epsilon^{\alpha\beta\gamma\delta}\epsilon_{1\alpha}^*\epsilon_{2\beta}^*k_\gamma p_\delta\,, \tag{4}$$

with

$$\bar{a}=\sum_j a_je^{i\delta_{sj}}e^{-i\phi_{sj}}\,,\quad \bar{b}=\sum_j b_je^{i\delta_{dj}}e^{-i\phi_{dj}}\,,\quad \bar{c}=\sum_j c_je^{i\delta_{pj}}e^{-i\phi_{pj}}\,. \tag{5}$$

And then

$$\begin{aligned}|\overline{\mathcal{M}}|^2 \;=\;& |\bar{a}|^2|\epsilon_1^*\cdot\epsilon_2^*|^2+\frac{|\bar{b}|^2}{m_1^2m_2^2}|(k\cdot\epsilon_2^*)(q\cdot\epsilon_1^*)|^2+\frac{|\bar{c}|^2}{m_1^2m_2^2}|\epsilon^{\alpha\beta\gamma\delta}\epsilon_{1\alpha}^*\epsilon_{2\beta}^*k_\gamma p_\delta|^2\\ &+2\frac{Re(\bar{a}\bar{b}^*)}{m_1m_2}(\epsilon_1^*\cdot\epsilon_2^*)(k\cdot\epsilon_2^*)(q\cdot\epsilon_1^*)\\ &-2\frac{Im(\bar{a}\bar{c}^*)}{m_1m_2}(\epsilon_1^*\cdot\epsilon_2^*)\epsilon^{\alpha\beta\gamma\delta}\epsilon_{1\alpha}^*\epsilon_{2\beta}^*k_\gamma p_\delta-2\frac{Im(\bar{b}\bar{c}^*)}{m_1^2m_2^2}(k\cdot\epsilon_2^*)(q\cdot\epsilon_1^*)\epsilon^{\alpha\beta\gamma\delta}\epsilon_{1\alpha}^*\epsilon_{2\beta}^*k_\gamma p_\delta\,.\end{aligned} \tag{6}$$

Comparing $|M|^2$ and $|\overline{M}|^2$, one finds

$$\mathcal{A}_T \propto Im(ac^*) = \sum_{i,j} a_i c_j \sin[(\phi_{si} - \phi_{pj}) + (\delta_{si} - \delta_{pj})] \tag{7}$$

and

$$\begin{aligned} \frac{1}{2}(\mathcal{A}_T + \bar{\mathcal{A}_T}) \propto \frac{1}{2}[Im(ac^*) - Im(\bar{a}\bar{c}^*)] &= \sum_{i,j} a_i c_j \sin(\phi_{si} - \phi_{pj}) \cos(\delta_{si} - \delta_{pj}) , \\ \frac{1}{2}(\mathcal{A}_T - \bar{\mathcal{A}_T}) \propto \frac{1}{2}[Im(ac^*) + Im(\bar{a}\bar{c}^*)] &= \sum_{i,j} a_i c_j \cos(\phi_{si} - \phi_{pj}) \sin(\delta_{si} - \delta_{pj}) . \end{aligned} \tag{8}$$

$\mathcal{A}_T$ itself is not a signal of T violation, since even $\phi = 0$ one can still get a non-zero A_T due to the contamination of the strong phase. However, $\mathcal{A}_T + \bar{\mathcal{A}_T}$ is a clean identification of the T-violation. The next step is to build the bridge between $\mathcal{A}_T$ and an experimentally accessable quantities.

Keeping in mind that a, b, c is related to the helicity amplitude, the angular distribution should contain these T-violating information. To that end, the helicity formalism developed by Jacob and Wick [7] is proved to be very useful. The two-body decay amplitude can be written as (see also Ref. [8])

$$\mathcal{A}(a \to 1 + 2) = \sqrt{\frac{2J+1}{4\pi}} D^{J*}_{M\lambda}(\phi_1, \theta_1, 0) H_{\lambda_1 \lambda_2}, \tag{9}$$

where J is the total angular momentum of particle a, M the spin projection along an arbitrarily chosen z-axis, (θ, ϕ) is the polar angle of $\vec{p}_1$ in the rest frame of a; λ's are the corresponding helicities, and H is the helicity amplitude. The Rotation matrix is defined as

$$D^j_{m'm}(\alpha, \beta, \gamma) = e^{-i\alpha m'} d^j_{m'm}(\beta) e^{-i\gamma m} \tag{10}$$

with the Wigner d function. In Eq. (9), we have chosen the choice $\gamma = 0$ [7]. For cascade two-body decay, the amplitude is just given the product of each reaction. Then the differential decay rate can be calculated, containing the helicity amplitudes $H_{1,1}$, $H_{0,0}$, $H_{-1,-1}$ corresponding to the helicity configurations for the two vector mesons. In a practical experimental analysis, the transversity basis can be also introduced,

$$A_{||} = \frac{1}{\sqrt{2}}(H_{1,1} + H_{-1,-1}), \quad A_0 = H_0, \quad A_{\perp} = \frac{1}{\sqrt{2}}(H_{1,1} + H_{-1,-1}) \tag{11}$$

which has a good CP transformation behavior with eigenvalues $+1$, $+1$ and -1 respectively. This point can be understood by [7, 8]

$$CP|JM, \lambda_1\lambda_2; 1, 2\rangle = \eta_1\eta_2(-)^{J-s_1-s_2}|JM, \lambda_1\lambda_2; \bar{1}, \bar{2}\rangle. \tag{12}$$

For the current case, $0^- \to 1^- + 1^-$, we have the total angular momentum $J = 0$, the spin of vector mesons $s_1 = s_2 = 1$, and the intrinsic parity $\eta_1 = \eta_2 = -1$, then $H_{\lambda_1,\lambda_2} = \bar{H}_{-\lambda_1,-\lambda_2}$. Here we write out the angular distribution for the process $D \to VV \to (PP)(PP)$,

$$\begin{aligned} \frac{d\Gamma}{d\cos\theta_1 d\cos\theta_2 d\phi} \propto\ & \frac{1}{2}\sin^2\theta_1 \sin^2\theta_2 \cos^2\phi |A_{||}|^2 + \frac{1}{2}\sin^2\theta_1 \sin^2\theta_2 \sin^2\phi |A_{\perp}|^2 \\ & + \cos^2\theta_1 \cos^2\theta_2 |A_0|^2 - \frac{1}{2}\sin^2\theta_1 \sin^2\theta_2 \sin 2\phi Im(A_{\perp}A^*_{||}) \\ & - \frac{\sqrt{2}}{4}\sin 2\theta_1 \sin 2\theta_2 \cos\phi Re(A_{||}A^*_0) + \frac{\sqrt{2}}{4}\sin 2\theta_1 \sin 2\theta_2 \sin\phi Im(A_{\perp}A^*_0) . \end{aligned} \tag{13}$$

The parts $(A_\perp A_0^*)$ and $(A_{||} A_\perp^*)$ are related to T-violating signal and can be measured in experiment. Such angular distribution formula with helicity amplitudes are also investigated in Refs. [9, 10]. We also plan to extend this framework to the interesting decay mode $B \to \pi\pi l \bar{\nu}_l$ [11].

3. Quantum correlated $D\overline{D}$ decay

3.1. $\psi(3770) \to D^0\overline{D}^0 \to (V_1V_2)(V_3V_4)$

The original idea is motivated by that CP violation occurs in the correlated $D^0\overline{D}^0$ decays to final state $f_a f_b$ with the same CP eigenvalue under the CP transformation [12]. This is simply because $f_a f_b$, that is produced from $\psi \to D^0\overline{D}^0$, is in the state of P-wave. As for e.g., the $(PP)(PP)$ or $(PP)(VP)$ decay product, the only observable is the decay rate or branching ratio, since the helicity is fixed in these cases, however, as mentioned before, for the $(VV)(VV)$ mode very rich angular information can be available due to the polarization of vector meson. We will describe our decay chain as

$$\psi \to D^0\overline{D}^0,\ D^0 \to V_1V_2,\ \overline{D}^0 \to V_3V_4, \tag{14}$$

where the vector meson decays to pseudoscalar pairs,

$$\begin{aligned} V_1 &\to M_1M_1', \quad V_2 \to M_2M_2', \\ V_3 &\to M_3M_3', \quad V_4 \to M_4M_4'. \end{aligned} \tag{15}$$

According to the idea that the final states must have different CP eigenstates, the transiversity combinations $(0,0)$, $(||,||)$, $(\perp,\perp)$, $(0,||)$, $(||,0)$, for the two pairs of vector mesons violate CP symmetry since A_0 and $A_{||}$ are CP-even while $A_\perp$ CP-odd. The next step is to pick out these pieces in the angular distribution and e.g. [13],

$$\int d\Gamma \frac{1}{128}\prod_{i=1}^{4}(5\cos^2\theta_i - 1) \sim |\mathcal{A}(D^0 \to V_1V_2)|^2|\mathcal{A}(D^0 \to V_3V_4)|^2|\rho^0_{V_1V_2} - \rho^0_{V_3V_4}|^2. \tag{16}$$

with the defination

$$\rho_f = \frac{\mathcal{A}(\overline{D}^0 \to f)}{\mathcal{A}(D^0 \to f)}. \tag{17}$$

For more details, we refer readers to the original paper [13] and more supplements in Ref. [14, 15]. Here we mainly concentrate at the physics behind. In experiment, the information of $|\rho^0_{V_1V_2} - \rho^0_{V_3V_4}|^2$ can be easily accessed in terms of the expression for branching ratio

$$BR(D^0\overline{D}^0 \to f_af_b) = 2\,BR(D^0 \to f_a)\,BR(D^0 \to f_b)\,|\rho_a - \rho_b|^2. \tag{18}$$

In above equation, all the small quantities due to the mixing have been neglected. One can parameterize $\rho_f = \eta_f(1+\delta_f)\exp(i\alpha_f)$ where δ_f represents CP- violation in decay, and α_f is the phase difference between D^0 and $\overline{D}^0$ to the same final state f. Then one has, neglecting δ_f,

$$|\rho_a - \rho_b|^2 = 4\sin^2\frac{\alpha_a - \alpha_b}{2}. \tag{19}$$

Up to now, there is no direct measurement for joint $D^0\overline{D}^0$ decay. Assuming non-observation of the CP-violating signals, taking the $(V_1V_2) = (\rho^0\rho^0)_0$ ("0" part), $(V_3V_4) = (\bar{K}^{*0}\rho^0)_{||}$ ("||" part) as example, one can put the constraint that the corresponding angle difference is smaller than 5 degrees around. The statistics knowledge on estimating the rare signal has been used, namely, taking the method proposed in Ref. [16], an upper limit 2.4 for the signal event is setted for non-observation of a signal in a true experiment with average background event number $b = 3$. This has been widely used in estimating the upper limit of rare signals, see e.g. for the $\Lambda - \overline{\Lambda}$ oscillation [17, 18]. In the estimate, we have considered the 20 fb^{-1} $\psi(3770)$ data. At last we stress that the two vector meson pairs should not be identical. More clearly, they are either different particles or in different transeversity combination (like $(0, ||)$ or $(||, 0)$, but can not be $(0, 0)$, $(||, ||)$ and $(\perp, \perp)$), otherwise, the branching ratio will be zero, as can be seen easily from Eq. (18).

3.2. $\psi(3770) \to D^0\overline{D}^0 \to (V_1V_2)(K\pi)$

This subsection is devoted to the better extraction of the strong phase δ occuring in the $D \to K\pi$ decay, that is needed to extract the CKM matrix angle γ in the Atwood-Dunietz-Soni (ADS) method [19]. In 1997, the authors ADS proposed a nice method that makes it possible for a clean extraction of angle γ, by combining the measurements of $B^- \to K^-D^0(D^0 \to f)$ and $B^- \to K^-\overline{D}^0(\overline{D}^0 \to f)$, where f is doubly-Cabibbo-suppressed mode of D^0 and thus a Cabibbo-favoured state of $\overline{D}^0$, but f is not necessary a CP-eigenstate. In fact, $K\pi$ is a typical example. In doing so, the knowledge on $D \to K\pi$ decay is an essential input for that procedure. One can introduce the strong phase δ by the ratio of amplitudes

$$r\, e^{i\delta} = \frac{\mathcal{A}(D^0 \to K^-\pi^+)}{\mathcal{A}(\overline{D}^0 \to K^-\pi^+)}. \tag{20}$$

Adapting the formula for the $4V$ modes to current case for $(V_1V_2)(K\pi)$ one will get the differential decay rate. It depends on the amplitudes of $D \to VV$ part, i.e., A_0, $A_{||}$, $A_\perp$ and also r and δ. The new feature is the emergence of the $|\sin\delta|$ term. Note that δ has been measured from the $D^0\overline{D}^0 \to (PP)(PP)$ mode, and the results show that δ is a quite small number since $\cos\delta \approx 1$ [20, 21]. Considering this fact, the sensitivity on $\sin\delta$ is much larger than the one for $\cos\delta$ and we expect to use this measurement to improve the errors on δ.

4. Conclusion

The processes $\psi(3770) \to D^0\overline{D}^0 \to (VV)(VV)$, $\psi(3770) \to D^0\overline{D}^0 \to (VV)(K\pi)$ and $D \to VV \to (PP)(PP)$ are explored, where P and V denote the pseudoscalar and vector mesons, respectively. The motivation is to fully exploit the polarization of vector meson and thus provides rich angular observables. Concerning the $4V$ mode, we show that the new CP violating observables can be constructed, and as for $(VV)(K\pi)$ we point out it can be used to extract the strong phase δ in the decay $D \to K\pi$. The appearance of $\sin\delta$ term makes it more welcome than $\cos\delta$ since δ is very small. Besides these quantum correlated $D^0\overline{D}^0$ decay we notice that $D \to VV$ is also a promising mode. Here the novelty is utilizing the T-violation. Assuming CPT invariance, T-violation is equivalent to a CP-violating signal. These proposals can be tested in BES or the future super tau-charm factory with much higher luminosity.

Acknowledgements

The author X.W.Kang would like to express his appreciation to J. Charles, S. Descotes-Genon and Hai-Bo Li. Some texts are based on the collaborative paper [5][13] with them. Concerning ordering of the reference, we have benefited much from J. A. Oller. We also acknowledge the email exchanges with Prof. G. Valencia and Prof. D. London. This proceeding is funded by the Importation and Development of High-Caliber Talents Project of Beijing Municipal Institutions (CIT&TCD20140311) and the Beijing Natural Science Fund (KZ201510016019).

References

[1] Delepine D, Faisel G and Ramirez C A 2013 *Phys. Rev.* D **87** 075017
[2] Grossman Y, Kagen A L and Nir Y 2007 *Phys. Rev.* D **75** 036008
[3] Cabibbo N 1963, *Phys. Rev. Lett.* **10**, 531 (1963)
[4] Kobayashi M and Maskawa T 1973, *Prog. Theor. Phys.* **49**, 652
[5] Kang X W and Li H B 2010, *Phys. Lett.* B **684**, 137 (*Preprint* arXiv:0912.3068 [hep-ph])
[6] Valencia G, *Phys. Rev.* D **39**, 3339
[7] Jacob M and Wick G C 1959, *Annals Phys.* **7**, 404; [Annals Phys. **281**, 774 (2000)].
[8] Richman J D 1984, *"An Experimenter's Guide to the Helicity Formalism"*, CALT-68-1148.
[9] Kang X W, Li H B, Lu G R and Zou B S 2010, *Chin. Phys.* C **34**, 1061 (*Preprint* arXiv:0911.2998 [hep-ph])
[10] Kang X W, Li H B, Lu G R and Datta A 2011, *Int. J. Mod. Phys.* A **26**, 2523 (*Preprint* arXiv:1003.5494 [hep-ph])
[11] Kang X W, Kubis B, Hanhart C and Meißner U G 2014, *Phys. Rev.* D **89**, 053015 (*Preprint* arXiv:1312.1193[hep-ph])
[12] Bigi I I Y and Sanda A I 1984, *Phys. Rev.* D **29**, 1393
[13] Charles J, Descotes-Genon S, Kang X W, Li H B and Lu G R 2010, *Phys. Rev.* D **81**, 054032 (*Preprint* arXiv:0912.0899 [hep-ph])
[14] Kang X W, Charles J, Descotes-Genon S and Li H B 2010, PoS ICHEP **2010**, 247 (*Preprint* [arXiv:1010.1687 [hep-ph]])
[15] Charles J, Descotes-Genon S, Kang X W and Li H B 2011, *Int. J. Mod. Phys. Conf. Ser.* **02**, 132
[16] Roe B P and Woodroofe M B 1999, *Phys. Rev.* D **60**, 053009 (*Preprint* physics/9812036)
[17] Kang X W, Li H B and Lu G R 2010, *Phys. Rev.* D **81**, 051901 (*Preprint* arXiv:0906.0230 [hep-ph])
[18] Kang X W, Li H B and Lu G R 2010, *"Probe* $\Lambda - \bar{\Lambda}$ *oscillation in* $J/\psi \to \Lambda\bar{\Lambda}$ *decay at BES-III,"* (*Preprint* arXiv:1008.2845 [hep-ph])
[19] Atwood D, Dunietz I and Soni A 1997, *Phys. Rev. Lett.* **78**, 3257 (*Preprint* hep-ph/9612433)
[20] Asner D M *et al* 2012 [CLEO Collaboration], *Phys. Rev.* D **86** 112001 (*Preprint* arXiv:1210.0939[hep-ex])
[21] Ablikim M *et al* 2014 [BESIII Collaboration], *Phys. Lett.* B **734**, 227 (*Preprint* arXiv:1404.4691[hep-ex])

Test baryon antibaryon oscillation in collider experiments

Yong-Feng Liu [1] **and Xian-Wei Kang** [2]

[1] Beijing Research Centre of Sustainable Energy and Building, Beijing University of Civil Engineering and Architecture, Beijing 100044, China

[2] Departamento de Física, Universidad de Murcia, E-30071 Murcia, Spain

E-mail: Corresponding author: kxw198710@126.com (Xian-Wei Kang)

Abstract. Searching for the New Physics (NP) phenomenon beyond Standard Model (SM) is still a main focus in particle physics. Here we propose to search for $\Lambda - \overline{\Lambda}$ oscillation in the decay $J/\psi \to \Lambda\bar{\Lambda}$ using BES detector. With one-year luminosity at BESIII, we can put a constraint that the $\Lambda - \overline{\Lambda}$ oscillation mass is smaller than 10^{-15} MeV at 90% confidence level, corresponding to the oscillation time of 10^{-6} second around, in case of non-observation of any signals. These measurements should provide very precious informations besides the neutron oscillation experiment. Also it would be the first-time access by experiment for $\Lambda - \overline{\Lambda}$ oscillation.

1. Introduction

Standard Model (SM) has becomes a fundamental theory in particle physics, but the efforts for searching for the New Physics (NP) phenomenon beyond SM never desists. The baryon-number-violating process that is prohibited by SM is one of the main directions, which is also related to the more basic question of understanding the baryon asymmetry in the universe, according to the Sakharov conditions [1]: Baryon number B violation; C symmetry and CP-symmetry violation; Interactions out of thermal equilibrium. We notice that the standard $D^0\overline{D}^0$ mixing technique can be extended to the Λ baryon sector, if the $\Lambda - \overline{\Lambda}$ oscillation indeed exists. Then we discuss its feasibility in collider experiments.

2. Formalism

We will fully follow the convention that is used in the neutral D and B meson mixing, see Ref. [2, 3, 4] for a more comprehensive details. However, note here we are discussing the $\Lambda - \overline{\Lambda}$ oscillation, which is a $\Delta B = 2$ process, and thus certainly a NP phenomenon. We sketch our main idea, and the technical details can be found in Refs. [5, 6].

For the $\Lambda - \overline{\Lambda}$ system, we can write a Schrödinger-like equation as

$$i\frac{\partial}{\partial t}\begin{pmatrix} \Lambda(t) \\ \overline{\Lambda}(t) \end{pmatrix} = \mathbf{M}\begin{pmatrix} \Lambda(t) \\ \overline{\Lambda}(t) \end{pmatrix}, \tag{1}$$

where $\mathbf{M}$ matrix plays the role of Hamiltonien and

$$\mathbf{M} = \begin{pmatrix} m_\Lambda - \Delta E_\Lambda & \delta m_{\Lambda\overline{\Lambda}} \\ \delta m_{\Lambda\overline{\Lambda}} & m_{\overline{\Lambda}} - \Delta E_{\overline{\Lambda}} \end{pmatrix}, \tag{2}$$

Content from this work may be used under the terms of the Creative Commons Attribution 3.0 licence. Any further distribution of this work must maintain attribution to the author(s) and the title of the work, journal citation and DOI.
Published under licence by IOP Publishing Ltd

where m is the mass of Λ (or $\overline{\Lambda}$); δm is the oscillation mass, or can be understood as a coupling between Λ and $\overline{\Lambda}$ indicating the strength of the oscillation. Note the identical off-diagonal element due to the CPT invariance. ΔE is an external energy, and will be zero in the case of absence of the external field. In case of a magnetic field, $\Delta E_\Lambda = -\vec{\mu}_\Lambda \cdot \vec{B}$ and $\Delta E_{\overline{\Lambda}} = -\vec{\mu}_{\overline{\Lambda}} \cdot \vec{B}$ with $\vec{\mu}$ the magnetic moment. The minus sign in $m - \Delta E$ in the diagonal element is puly a convention, and writing as + corresponds to exchange of the role of Λ and $\overline{\Lambda}$. Considering $\Lambda - \overline{\Lambda}$ oscillation, one can write the eigenstates as

$$
\begin{aligned}
|\Lambda_H\rangle &= \frac{1}{\sqrt{2}}(\sqrt{1+z}|\Lambda\rangle + \sqrt{1-z}|\overline{\Lambda}\rangle),\\
|\Lambda_L\rangle &= \frac{1}{\sqrt{2}}(\sqrt{1-z}|\Lambda\rangle - \sqrt{1+z}|\overline{\Lambda}\rangle),
\end{aligned} \tag{3}
$$

with $z \equiv \frac{2\Delta E}{\Delta m}$; $\Delta E \equiv |\Delta E_\Lambda| = |\Delta E_{\overline{\Lambda}}|$ and $\Delta m \equiv m_H - m_L = 2\sqrt{(\Delta E)^2 + \delta m^2_{\Lambda\overline{\Lambda}}}$, i.e., the mass difference between heavy and light Λ. It turns out that the probability of finding a $\overline{\Lambda}$ in a beam of Λ after time t is described by [7]

$$
\mathcal{P}(\overline{\Lambda}, t) = \frac{\delta m^2_{\Lambda\overline{\Lambda}}}{\delta m^2_{\Lambda\overline{\Lambda}} + (\Delta E)^2} \sin^2(\sqrt{\delta m^2_{\Lambda\overline{\Lambda}} + (\Delta E)^2} \cdot t). \tag{4}
$$

For free Λ in absence of external field, $\Delta E = 0$, and Eq. (4) becomes

$$
\begin{aligned}
\mathcal{P}(\overline{\Lambda}, t) &= \sin^2(\delta m_{\Lambda\overline{\Lambda}} \cdot t)\\
&\approx (\delta m_{\Lambda\overline{\Lambda}} \cdot t) = \left(\frac{t}{\tau_{\Lambda\overline{\Lambda}}}\right)^2,
\end{aligned} \tag{5}
$$

where we have defined the relation between oscillation mass and oscillation time. A derivation with more details can be found in Appendix A in Ref. [8]. The recent status of the study of the neutron anti-neutron oscillation is summarized in Ref. [9]. Hereafter, we consider the possible search of $\Lambda - \overline{\Lambda}$ oscillation in $J/\psi \to \Lambda\overline{\Lambda}$ coherent decay. Here "coherent" means the $\Lambda\overline{\Lambda}$ should occur in $C = -1$, with C the charge parity. We take the results from PDG for the $D^0 - \overline{D}^0$ mixing [2], i.e., the general expression for the time-dependent decay rate is

$$
\begin{aligned}
R(f_1, f_2; t) &\equiv \frac{d\Gamma(J/\psi \to \Lambda_{\text{phys}}\overline{\Lambda}_{\text{phys}} \to f_1 f_2)}{dt}\\
&= \frac{1}{4}\mathcal{N}e^{-\Gamma|t|}\Big[(|a_1|^2 + |a_2|^2)\cosh(y_\Lambda \Gamma t) + (|a_1|^2 - |a_2|^2)\cos(x_\Lambda \Gamma t)\\
&\quad + 2\mathcal{R}e(a_1 a_2^*)\sinh(y_\Lambda \Gamma t) + 2\mathcal{I}m(a_1 a_2^*)\sin(x_\Lambda \Gamma t)\Big],
\end{aligned} \tag{6}
$$

where $\mathcal{N}$ is a common normalization factor and will be cancelled in the following observables. In Eq. (6), terms proportional to $|a_1|^2$ are associated with decays that occur without any net oscillation, while terms proportional to $|a_2|^2$ indeed follow from a net oscillation. The other terms are their interference. The mixing parameters are defined as

$$
\begin{aligned}
m &\equiv \frac{m_H + m_L}{2}, \quad \Delta m \equiv m_H - m_L,\\
\Gamma &\equiv \frac{\Gamma_H + \Gamma_L}{2}, \quad \Delta\Gamma \equiv \Gamma_H - \Gamma_L,\\
x_\Lambda &\equiv \frac{\Delta m}{\Gamma}, \quad y_\Lambda \equiv \frac{\Delta\Gamma}{2\Gamma}.
\end{aligned} \tag{7}
$$

The time-integrated quantity $R(f_1, f_2)$ can then be expressed as

$$R(f_1, f_2) = \frac{1}{4}\mathcal{N}\left[(|a_1|^2 + |a_2|^2)\frac{1}{1-y_\Lambda^2} + (|a_1|^2 - |a_2|^2)\frac{1}{1+x_\Lambda^2}\right]. \tag{8}$$

Note that t is the time difference between Λ and $\overline{\Lambda}$ and should be integrated from $-\infty$ to ∞. The quantities a_1 and a_2 are

$$\begin{aligned} a_1 &\equiv A_{f_1}\overline{A}_{f_2} - \overline{A}_{f_1}A_{f_2} \\ a_2 &\equiv z(A_{f_1}\overline{A}_{f_2} + \overline{A}_{f_1}A_{f_2}) - \sqrt{1-z^2}(A_{f_1}A_{f_2} - \overline{A}_{f_1}\overline{A}_{f_2}) \end{aligned} \tag{9}$$

with $A_{f_i} \equiv \langle f_i|\mathcal{H}|\Lambda\rangle$, $\overline{A}_{f_i} \equiv \langle f_i|\mathcal{H}|\overline{\Lambda}\rangle$ $(i = 1, 2)$.

In what follows, we will discuss two separate cases: i) in the external field such that $z \to 1$ ii) in the absence of the external field, then $z = 0$.

- z=1
  At BES-III, the external magnetic field is about 1.0 T, and then $\Delta E \sim 2 \times 10^{-11}$ MeV, $z \to 1$. For $f_1 = f_2 = p\pi^-$

$$R(p\pi^-, p\pi^-; t) = \mathcal{N}e^{-\Gamma|t|}|A_{p\pi^-}|^2|\overline{A}_{p\pi^-}|^2\frac{x_\Lambda^2 + y_\Lambda^2}{2}(\Gamma t)^2, \tag{10}$$

  where we have neglected CP violation and expanded the terms up to order of x_Λ^2 or y_Λ^2. Similarly,

$$R(p\pi^-, \overline{p}\pi^+; t) = \mathcal{N}\frac{1}{2}e^{-\Gamma|t|}|A_{p\pi^-}|^2|\overline{A}_{\overline{p}\pi^+}|^2. \tag{11}$$

  where since this mode is a favourable one and we keep only the leading term. The ratio of Eq. (10) to Eq. (11) is

$$\mathcal{R}(t) = 2|\lambda_{p\pi^-}|^2\frac{x_\Lambda^2 + y_\Lambda^2}{2}(\Gamma t)^2, \quad \lambda_{p\pi^-} = \frac{\overline{A}_{p\pi^-}}{A_{p\pi^-}} \tag{12}$$

  and the time-integrated case reads

$$\mathcal{R} = 2|\lambda_{p\pi^-}|^2(x_\Lambda^2 + y_\Lambda^2). \tag{13}$$

- z=0
  If there is no external field, one obviously has $z = 0$. Then the above equation should be modified by

$$\begin{aligned} \mathcal{R}(t) &= \frac{1}{2}\frac{x_\Lambda^2 + y_\Lambda^2}{2}(\Gamma t)^2, & (14) \\ \mathcal{R} &= \frac{x_\Lambda^2 + y_\Lambda^2}{2}. & (15) \end{aligned}$$

3. Potential experimental searches

Now we discuss the experimental feasibility to measure the oscillation mass. With huge data sample, one can measure both the $\mathcal{R}$ and $|\lambda_{p\pi^-}|$. The time information can be also accessed, or in other words, both $\mathcal{R}$ and $\mathcal{R}(t)$ can be measured. From these quantities, the oscillation parameters will be determined reliably. As a first step, we can get an estimated value for $\delta m_{\Lambda\bar{\Lambda}}$ in the absence of an external field from Eq. (15). Assuming $y_\Lambda = 0$ one will get

$$\delta m_{\Lambda\overline{\Lambda}} = \frac{1}{\sqrt{2}}\sqrt{\mathcal{R}}\Gamma. \tag{16}$$

About 10×10^9 J/ψ samples can be collected per year at BEPC-II in Beijing [10, 11]. Provided that no signal events of $J/\psi \to \Lambda_H\Lambda_L \to (p\pi^-)(p\pi^-)$ or $(\bar{p}\pi^+)(\bar{p}\pi^+)$ are observed, one should bound the upper limit $\mathcal{R} \leq 4 \times 10^{-7}$ and correspondingly $\delta m_{\Lambda\overline{\Lambda}} \leq 10^{-15}$ MeV at the confidence level of 90%, inferring from the knowledge of interval estimation for very rare signal (see also Refs. [12, 13, 14, 15, 16]). At a super tau-charm factory [17], the luminosity will be increased to near 10^4 times compared to BESIII, and then the constraint for the oscillation is more severe, $\delta m_{\Lambda\overline{\Lambda}} \leq 10^{-17}$ MeV.

Note that $\Upsilon(4S)$ also has the same quantum numbers as J/ψ, i.e., $I^G(J^{PC}) = 0^-(1^{--})$, which can also decays to the coherent $\Lambda\overline{\Lambda}$ states. But considering the branching ratios and the data sample from Belle and BaBar dectors, BESIII is again the most promising candidate to detect it. The true measurement is desired to get more valuable information. At last we note that in Refs. [18, 19, 20, 21, 22] for the neutron-antineutron interaction, they are the Standard-Model calculation, and these neutron-antineutron oscillation effects are surely omitted. We also notice that the channel $B \to \Lambda\overline{\Lambda}l\bar{\nu}_l$ is interesting to prove the $\Lambda\overline{\Lambda}$ interaction as well as the weak interaction in B meson decay, looking form our experience in $B \to \pi\pi l\bar{\nu}_l$ decay [23, 24].

4. Conclusion

In conclusion, we propose a new way to probe the baryon-number violating signal by using the collider experiment. The expression for the oscillation mass has been derived. BESIII is capable of such measurements. For a first step forward, we estimate the oscillation mass in the absence of the external magnetic field. With one year's luminosity at BES-III, we can set an upper limit of $\delta m_{\Lambda\overline{\Lambda}} < 10^{-15}$ MeV at 90% confidence level, corresponding to about 10^{-6} s of $\Lambda - \overline{\Lambda}$ oscillation time. It will be the first search of $\Lambda - \overline{\Lambda}$ oscillation experimentally.

Acknowledge

The author X.W.Kang would like to express his appreciation for useful discussions from Hai-Bo Li. Some texts are based on their common publication [5]. This proceeding is funded by the Importation and Development of High-Caliber Talents Project of Beijing Municipal Institutions (CIT&TCD20140311) and the Beijing Natural Science Fund (KZ201510016019).

References

[1] Sakharov A D 1967, *JETP Lett.* **5** 24.
[2] Olive K A et al 2014. (Particle Data Group), *Chin. Phys.* C, **38**, 090001 and 2015 update
[3] Li H B and Yang M Z 2006, *Phys. Rev.* D **74**, 094016 (*Preprint* hep-ph/0610073)
[4] X. D. Cheng, K. L. He, H. B. Li, Y. F. Wang and M. Z. Yang 2007, *Phys. Rev.* D **75** 094019 (*Preprint* arXiv:0704.0120 [hep-ex])
[5] Kang X W, Li H B and Lu G R 2010, *Phys. Rev.* D **81**, 051901 (*Preprint* arXiv:0906.0230 [hep-ph])
[6] Kang X W, Li H B and Lu G R 2010, *"Probe $\Lambda - \bar{\Lambda}$ oscillation in $J/\psi \to \Lambda\bar{\Lambda}$ decay at BES-III,"* (*Preprint* arXiv:1008.2845 [hep-ph])
[7] Mohapatra R N and Marshak R E 1980, *Phys. Lett.* B **94** 183
[8] Bergevin M 2010, *"Search for Neutron anti-neutron oscillation at the SUDBURY Neutrino Observatory"*, PhD thesis, Univeristy of Guelph, Canada.
[9] Phillips D G et al 2016, *Phys. Rept.* **612**, 1 (*Preprint* arXiv:1410.1100 [hep-ex])
[10] D. M. Asner et al. 2009, *Int. J. Mod. Phys.* A **24**, S1 (2009) (*Preprint* arXiv:0809.1869 [hep-ex])
[11] Wang Y F 2007, *eConf* C **070805**, 38 (*Preprint* arXiv:0711.4199 [hep-ex])
[12] Charles J, Descotes-Genon S, Kang X W, Li H B and Lu G R 2010, *Phys. Rev.* D **81**, 054032 (*Preprint* arXiv:0912.0899 [hep-ph])

[13] Kang X W and Li H B 2010, *Phys. Lett.* B **684**, 137 (*Preprint* arXiv:0912.3068 [hep-ph])
[14] Kang X W, Li H B, Lu G R and Datta A 2011, *Int. J. Mod. Phys.* A **26**, 2523 (*Preprint* arXiv:1003.5494 [hep-ph])
[15] Kang X W, Charles J, Descotes-Genon S and Li H B 2010, PoS ICHEP **2010**, 247 (*Preprint* [arXiv:1010.1687 [hep-ph]])
[16] Charles J, Descotes-Genon S, Kang X W and Li H B 2011, *Int. J. Mod. Phys. Conf. Ser.* **02**, 132
[17] M. Bona et al. 2007 [SuperB Collaboration], Pisa, Italy: INFN (2007) 453 (*Preprint* arXiv:0709.0451 [hep-ex])
[18] Kang X W, Haidenbauer J and Meißner U G 2014, *JHEP* **1402**, 113 (*Preprint* arXiv:1311.1658 [hep-ph])
[19] Haidenbauer J, Kang X W and Meißner U G 2014, *Nucl. Phys. A* **929**, 102 (*Preprint* arXiv:1405.1628 [nucl-th])
[20] Kang X W, Haidenbauer J and Meißner U G 2015, *Phys. Rev. D* **91** (*Preprint* arXiv:1502.00880 [nucl-th])
[21] Haidenbauer J, Hanhart C, Kang X W and Meißner U G, *Phys. Rev. D* **92** (Preprint arXiv:1506.08120 [nucl-th])
[22] Kang X W 2015, PoS(CD15) 108 (*Preprint* arXiv:1508.06189 [hep-ph])
[23] Kang X W, Kubis B, Hanhart C and Meißner U G 2014, *Phys. Rev.* D **89**, 053015 (*Preprint* arXiv:1312.1193[hep-ph])
[24] Kang X W, Kubis B, Hanhart C and Meißner U G 2015, PoS (CD15) 053.

Entropy, Inequality and Statistical Complexity for the Modified Pöschl-Teller Potential in Non-Extensive Formalism

Derya Gemici-Deveci[1,2], Ekrem Aydıner[1]
[1]Department of Physics, İstanbul University, 34134, İstanbul, Turkey
[2]Department of Opticianry, İstanbul Kemerburgaz University, İstanbul, Turkey

E-mail: derya.deveci@kemerburgaz.edu.tr

E-mail: ekrem.aydiner@istanbul.edu.tr

Abstract. In this study we have obtained non-extensive entropies and have showed that these entropies satisfy the non-extensive uncertainty relation for the wave function corresponding to the modified Pöschl-Teller potential. We have also computed statistical complexity depending on entropic index q for the same wave function and we have showed that complexity decreases by increasing of q.

1. Introduction

According to the Heisenberg inequality any two non-commuting conjugate observables cannot be simultaneously measured with arbitrary precision [1, 2]. Considering two Hermitian operators $\widehat{x}$ and $\widehat{p_x}$ corresponding to the observables x and p_x in a finite N-dimensional Hilbert space, the uncertainty principle in the Robertson form [3] reads $\Delta\widehat{x}\Delta\widehat{p_x} \geq \frac{1}{2}|\langle\psi|[\widehat{x},\widehat{p_x}]|\psi\rangle|$ where $\Delta\widehat{x}$ and $\Delta\widehat{p_x}$ denote the standard deviation of the distributions. These deviations can be obtained from given wave functions. However, this form of the uncertainty relation faces several main obstacles in practice. First one is due to the state dependence of the right hand side of the above equation, since this prevents one from obtaining a fixed lower bound [4, 5, 6]. The second obstacle is coming from the using of the standard deviations on its left hand side, because the standard deviation relies heavily on the second term, and it cannot be used when the second term diverges. [7]. To overcome the mentioned difficulties, it was proposed to that entropic uncertainty relations are based on information entropies that are more appropriate uncertainty measures such as Shannon, Fisher, Reyni and Tsallis. For example, Bialynicki-Birula and Mycielski (BBM) inequality [8] is given in terms of Shannon information entropies as $S_q + S_p \geq N(1 + \ln\pi)$ where $S_q = -\int \mid \psi(q) \mid^2 \ln(\mid \psi(q) \mid^2 dq$ is the position space information entropy, $S_p = -\int \mid \psi(p) \mid^2 \ln(\mid \psi(p) \mid^2 dp$ is the momentum space information entropy, and N is dimension of the space. The entropic uncertainty relations derived in terms of the Shannon entropy remove the first aforementioned obstacle, namely, the dependence on the state when the uncertainty is assessed. However, the second obstacle is remaining of the divergent second moment. The resolution of both obstacles is possible by using the non-additive Tsallis information entropy as shown in Ref. [9].

Content from this work may be used under the terms of the Creative Commons Attribution 3.0 licence. Any further distribution of this work must maintain attribution to the author(s) and the title of the work, journal citation and DOI.
Published under licence by IOP Publishing Ltd

In this study we discuss non-extensive uncertainty inequality and complexity for the modified Pöschl-Teller (MPT) potential in non-extensive formalism. We have obtained Tsallis entropies for the MPT potential and have showed that these entropies satisfy non-extensive uncertainty inequality. We also discuss complexity for this potential.

2. Entropy, Inequality and Statistical Complexity

2.1. Non-extensive Entropy

A generalized non-extensive form of entropy was introduced in 1988 by Tsallis [10]

$$S_q = \frac{1}{q-1}\left(1 - \sum_{i=1} p_i^q\right) \qquad (\sum_{i=1}^{W} p_i = 1 \ , k > 0) \tag{1}$$

where q is a real parameter sometimes called simply entropic index. This expression is reduced to Boltzmann-Gibbs prescription in the limit of $q \to 1$. The expression also provides a link between the microscopic dynamics and macroscopic thermodynamics. The cases $q < 1$, $q > 1$ and $q = 1$ correspond to superadditivity, subadditivity and additivity, respectively [10]. In the literature there are some researches [11]-[12] about obtaining insight onto the mathematical and physical meaning of q parameter and its associated non-extensive formalism. Also there are some researches that calculated the value of the entropic index q which is the measure of complexity [13].

2.2. Entropic Inequality

The probability densities in X-space are $|\psi(r)|^2$ and that in the P-space (momentum) is $\left|\tilde{\psi}(k)\right|^2$, both of which are normalized to unity. The corresponding Tsallis entropies for continuous form are defined as [14, 15]

$$S_p(X) = \frac{1}{p-1}\left(1 - \int d^N r \, |\psi(r)|^{2p}\right) \qquad S_q(P) = \frac{1}{q-1}\left(1 - \int d^N k \left|\tilde{\psi}(k)\right|^{2q}\right) \tag{2}$$

with condition $1/p + 1/q = 2$. The inequaility for non-extensive entropies is given by (See: Ref. [15])

$$[1 + (1-q) S_q(P)]^{1/2q} \le \left(\frac{\pi}{q}\right)^{N/4q} \left(\frac{\pi}{p}\right)^{-N/4p} [1 + (1-p) S_p(X)]^{1/2p} \tag{3}$$

in the limit of p and q going to unity, Eq. 3 reduce to BBM inequality.

2.3. Statistical Complexity

A simple measure for statistical complexity was proposed by LMC (Lopez-Ruiz-Mancini-Calbet) [16]

$$C^{LMC} = S.D \tag{4}$$

where S is the information entropy and D is the disequilibrium can be taken as some kind of distance to an equiprobability. In non-extensive formalism [13], LMC complexity can be written as

$$C_q(\rho) = S_q(P) \, .D(\rho(r)) \tag{5}$$

where $S_q(P)$ is the Tsallis entropy and D is the disequilibrium given in a simple form $D = \int \rho^2(r) dr$ where ρ is the probability density [13, 17].

2.4. MPT Potential

The MPT is given by $V(r) = -V_0/\cosh^2(\alpha r)$ where α is a parameter and V_0 is the potential depth. An approximate solution of Schrödinger equation in the spherical coordinates for MPT potential is given by Agboola [18] as

$$\Psi_{nlm}(r,\Omega) = r^{\frac{N-1}{2}} R_{nl}(r) Y_l^m(\Omega) \tag{6}$$

$Y_l^m(\Omega)$ is the spherical part and R_{nl} is the radial part of the wave function. The radial part of the wave function is

$$R_{nl}(s) = C_{n_r} s^{2\nu}(1-s)^{\frac{\epsilon}{2}} P_{n_r}^{\nu,\epsilon}(1-2s) \tag{7}$$

where P_{n_r} is Jacobi polynomials [19]

$$P_n^{a,b}(s) = \frac{(-1)^n}{n!2^n(1-s)^a(1+s)^b}\frac{d^n}{ds^n}[((1-s)^{a+n})(1+s)^{b+n}] \quad \text{with} \quad s = \tanh^2(\alpha r) \tag{8}$$

and C_{n_r} is the normalization constant

$$C_{n_r}\sum_{k=0}^{n_r}\frac{(-n_r)_k(\epsilon+\nu+n_r+1)_k\left(4\nu+\frac{1}{2}\right)_k}{\left(\epsilon+4\nu+\frac{1}{2}\right)_k(\nu+1)_k k!}\times\sum_{j=0}^{n_r}\frac{(-n_r)_j(\epsilon+\nu+n_r+1)_j\left(4\nu+k+\frac{1}{2}\right)_j}{\left(\epsilon+4\nu+k+\frac{1}{2}\right)_j(\nu+1)_j j!} = \frac{\alpha}{B\left(4\nu+\frac{1}{2},\epsilon\right)} \tag{9}$$

where ν, α, ϵ are constants of wave function, $(...)_{k,j}$ are Poschammer functions and $n_r = 0, 1, 2, ...$ for the ground state and excited states.

3. Numerical Results

We deal with only radial part of wave function (6) and we set $x = r$. In order to see the form of radial wave function for MPT potential, the probability densities in position and momentum spaces (i.e., $|\psi(x)|^2$ and $\left|\tilde{\psi}(k)\right|^2$) are given in Figure 1 for different parameters.

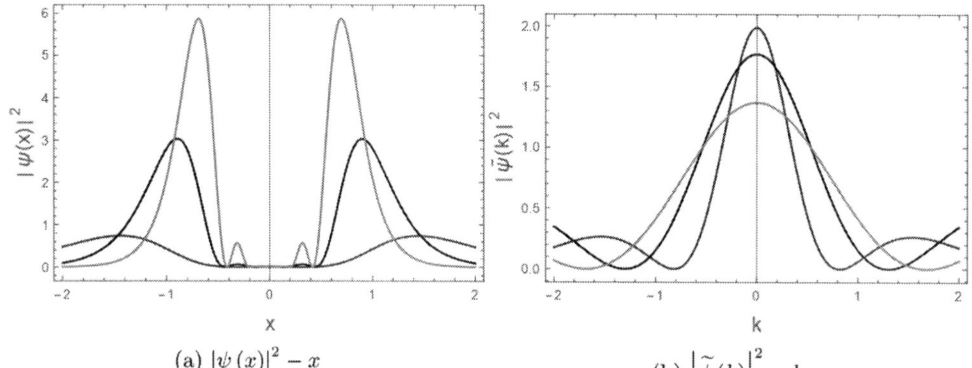

(a) $|\psi(x)|^2 - x$ (b) $\left|\tilde{\psi}(k)\right|^2 - k$

Figure 1: The probability density in position and momentum space (red: $n = 0$, $\alpha = 1$, blue: $n = 1$, $\alpha = 2$, green: $n = 2$, $\alpha = 3$).

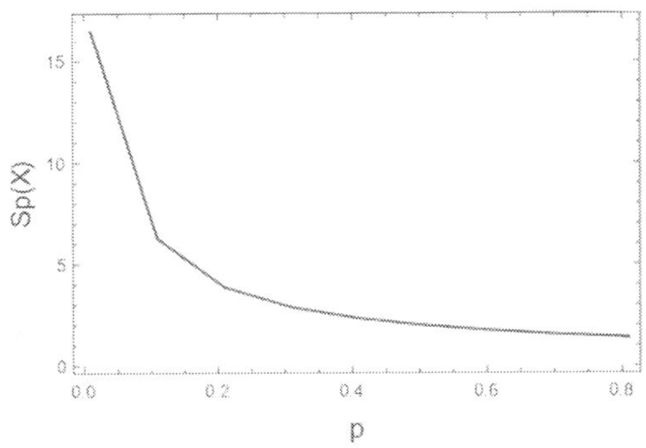

(a) $S_p(X) - p$

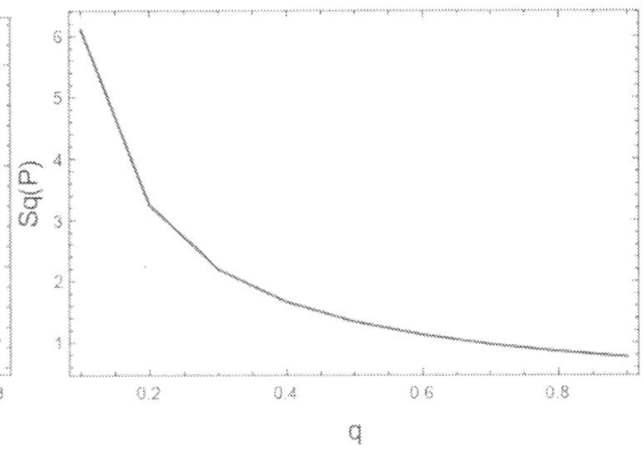

(b) $S_q(P) - q$

Figure 2: The non-extensive entropies in position and momentum space for $n = 0$.

As it can be seen from Figure 1 (a) that position-space probability density of the wave function has non-Gaussian form. This indicates that the wave function has also non-Gaussian form. Therefore to cope with main obstacles mentioned above we can prefer to use a suitable entropic uncertainty relation instead of Heisenberg one. In Figure 2, non-extensive entropies $S_p(X)$ and $S_q(P)$ for ground state $n = 0$ $(\alpha = \nu = \epsilon = 1)$ have been plotted versus p and q, respectively. As it can be seen clearly from Figure 2 (a) and (b) For small p and q values, non-extensive entropies take maximum values and they decrease by increasing p and q values. In the limit of $p, q \rightarrow 1$ Tsallis entropies reduce to Shannon entropies. For $p < 1$ and $q < 1$, these potentials satisfy non-extensive uncertainty inequality given Eq. 3. In Table 1, the results are given for different values of p and q with condition $1/p + 1/q = 2$ [15].

Table 1: The table for non-extensive inequality for the Modified Posch Teller Potential

p	q	$\frac{[1+(1-p)S_p(X)]^{1/2p}}{[1+(1-q)S_q(P)]^{1/2q}}$	$\geq$	$\left(\frac{\pi}{q}\right)^{-N/4q}\left(\frac{\pi}{p}\right)^{N/4p}$
3/4	3/2	1.42513	$\geq$	0.164791
5/6	5/4	1.24473	$\geq$	0.169184
8/7	6/7	1.7092	$\geq$	0.876745

Furthermore, we have obtained statistical complexity for radial wave function in Eq. (7) by using statistical complexity measure of C_q in Eq. (5). Non-extensive statistical complexity C_q for first excited state is given in Figure 3. As it can be seen in the Figure 3 that the non-extensive complexity decreases by increasing of the q values. For $q = 0$ complexity has maximum value, however, in the limit $q \rightarrow 1$, complexity goes to zero. One can see that Tsallis's q parameter plays important role on the complexity since equiprobable states in the system and information entropies change depending on the q values.

4. Conclusion

In this study we briefly introduce entropy, uncertainty inequality and statistical complexity in the non-extensive formalism. Later we have obtained the non-extensive entropies for the wave function corresponding to MPT potential and have showed that these entropies satisfy the

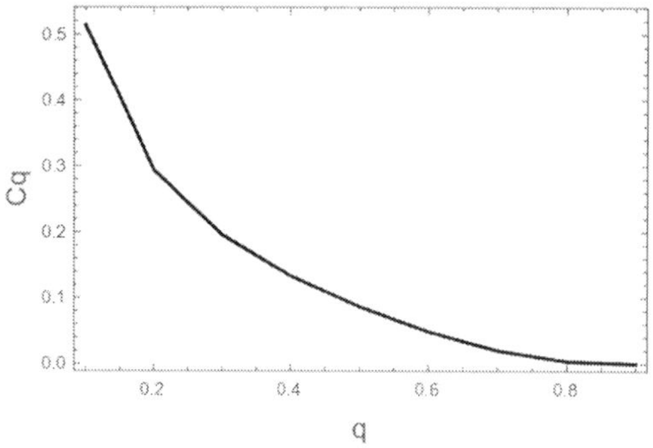

Figure 3: Complexity as a function of q for the first excited state ($n = 1$), ($0 < q < 1$)

non-extensive uncertainty relation. We have also computed statistical complexity depending on Tsallis's index q for the same wave function and we have showed that complexity decreases by increasing of q. We conclude that using of the non-extensive entropies and uncertainty for non-Gaussian wave function such as MPT wave function is appropriate to overcome main obstacles mentioned in the introduction.

Acknowledgments
Author D.G.D. thank to Seyit Deniz Han for his kind assistance with this work. This work was supported by Scientific Research Projects Coordination Unit of Istanbul University. Project numbers: BEK-2016-20733.

References
[1] Heisenberg, W., "Uber den anschaulichen Inhalt der quantentheoretischen Kinematik und Mechanik", Zeitschrift fur Physik, 43, 3, 172 (1927).
[2] Kennard, E. H., "Zur Quentenmechanik einfacher Bewegungstypen", Zeitschrift fr Physik, 44, 4, 326 (1927).
[3] Robertson, H. P., "The uncertainity principle", Phys. Rev. 34, 163 (1929).
[4] Maassen, H. and Uffink, J. B. M.,"Generalized entropic uncertainty relations" Phys. Rev. Lett. **60**, 1103 (1988).
[5] Deutsch, D., Phys. Rev. Lett."Uncertainty in Quantum Measurements" **50**, 631 (1983).
[6] Uffink, J. B. M. and Hilgevoord,"Uncertainty principle and uncertainty relations" J., Found. Phys. **15** 925 (1985).
[7] Lillo, F. and Mantegna, R. N., "Anomalous spreading of power-law quantum wave packets" Phys. Rev. Lett. **84**, 1061 (2000).
[8] Bialynicki-Birula, I., Mycielski, J., "Uncertainty relations for information entropy in wave mechanics", Comm. Math. Phys., 44, 2, 129 (1975).
[9] Abe, S., Martinez, S., Pennini, F. and Plastino, A.,"Nonextensive thermodynamic relations" Phys. Lett. A **295**, 74 (2002).
[10] Tsallis, C., "Possible generalization of Boltzmann-Gibbs statistics", Journal of Statistical Physics, 52, 479487 (1988).
[11] Tsallis, C., "I. Nonextensive Statistical Mechanics and Thermodynamics: Historical Background and Present Status", Springer, Nonextensive Statistical Mechanics and Its Applications, 560, 3-98 (2001).
[12] Zelenyi, L. M., Milovanov, A. V., "Fractal Properties of Sunspots", Soviet Astronomy Letters, 17, 425 (1991).
[13] Abe, S., Landsberg, P. T., Plastino, A. R., and Yamano T., "Nonadditive statistical measure of complexity and values of the entropic index q", arXiv:cond-mat/0402217v1 (2004).
[14] C. Tsallis, in: New Trends in Magnetism, Magnetic Materials and their Applications, eds. J.L.Morán-Lopez and J.M. Sánchez (Plenum Press, New York, p.451, 1994).
[15] Rajagopal, A. K., "The Sobolev inequality and the Tsallis entropic uncertainty relation", Physics Letters A, 205, 1, 32 (1995).
[16] Lopez-Ruiz, R., Mancini, H., Calbet, X., "A statistical measure of complexity", Phys. Lett. A, 209, 5-6, p. 321-326 (1995).

[17] Yamano, T., "A statistical complexity measure with nonextensive entropy and quasi-multiplicativity", J. Math. Phys. 45, 1974 (2004).

[18] Agboola, D., "Solutions to the Modified Pöschl-Teller Potential in D-dimensions", Chin. Phys. Lett., 27, 4, 040301 (2010).

[19] Askey, R. and James, A. W., "Some basic hypergeometric orthogonal polynomials that generalize Jacobi polynomials", American Mathematical Soc., 319 (1985).

Journal of Physics: Conference Series **738** (2016) 012045 doi:10.1088/1742-6596/738/1/012045

Towards a realistic description of hadron resonances

R A Schmidt[1], L Canton[2], W Schweiger[1] and W Plessas[1]

[1] Theoretical Physics, Institute of Physics, University of Graz, Universitätsplatz 5, A-8010 Graz, Austria

[2] Istituto Nazionale di Fisica Nucleare, Via F. Marzolo 8, I-35131 Padova, Italy

E-mail: `plessas@uni-graz.at`

Abstract. We report on our attempts of treating excited hadron states as true quantum resonances. Hitherto the spectroscopy of mesons, usually considered as quark-antiquark systems, and of baryons, usually considered as three-quark systems, has been treated through excitation spectra of bound states (namely, confined few-quark systems), corresponding to poles of the quantum-mechanical resolvent at real negative values in the complex energy plane. As a result the wave functions, i.e. the residua of the resolvent, have not exhibited the behaviour as required for hadron resonances with their multiple decay modes. This has led to disturbing shortcomings in the description of hadronic resonance phenomena. We have aimed at a more realistic description of hadron resonances within relativistic constituent-quark models taking into account explicitly meson-decay channels. The corresponding coupled-channels theory is based on a relativistically invariant mass operator capable of producing hadron ground states with real energies and hadron resonances with complex energies, the latter corresponding to poles in the lower half-plane of the unphysical sheet of the complex energy plane. So far we have demonstrated the feasibility of the coupled-channels approach to hadron resonances along model calculations producing indeed the desired properties. The corresponding spectral properties will be discussed in this contribution. More refined studies are under way towards constructing a coupled-channels relativistic constituent-quark model for meson and baryon resonances.

1. Introduction

The description of hadron resonances represents a big challenge in all current approaches to quantum chromodynamics (QCD). Particularly in the framework of constituent-quark models, hadronic resonances are usually treated as excited bound states rather than as resonant states with finite widths. Calculations of strong decays have thus shown short-comings generally producing too small decay widths [1, 2, 3, 4]. To remedy this situation we are investigating a coupled-channels (CC) approach taking into account explicit meson, especially pionic, degrees of freedom.

The CC approach has been tested before within a simple scalar toy model, leaving out all spin and flavor dependences. It turned out that the coupling to a mesonic channel shifts the ground-state mass down and generates the resonant state with a finite width, whose magnitude is dependent essentially on the coupling strength to the meson channel [5, 6]. Recently we have obtained results for the πNN and the $\pi N\Delta$ systems, including all spin and flavor degrees of freedom.

Our theory relies on a relativistically invariant CC mass operator in general coupling a hadron state to n further (decay) channels. We exemplify the approach for a single baryon B, specifically

Content from this work may be used under the terms of the Creative Commons Attribution 3.0 licence. Any further distribution of this work must maintain attribution to the author(s) and the title of the work, journal citation and DOI.
Published under licence by IOP Publishing Ltd

the N or the Δ, coupled to a one-pion channel by a relativistically invariant mass operator in matrix form:

$$\begin{pmatrix} M_{\tilde{B}} & K \\ K^{\dagger} & M_{\tilde{B}\pi} \end{pmatrix} \begin{pmatrix} |\psi_B\rangle \\ |\psi_{B\pi}\rangle \end{pmatrix} = m \begin{pmatrix} |\psi_B\rangle \\ |\psi_{B\pi}\rangle \end{pmatrix} . \tag{1}$$

Herein, $M_{\tilde{B}}$ and $M_{\tilde{B}\pi}$ are the invariant mass operators of the bare baryon state $\tilde{B}$ and the $\tilde{B}\pi$ channel, respectively. K contains the transition dynamics. This means that the bare baryon $\tilde{B}$ gets dressed by pion loops (see Fig. 1).

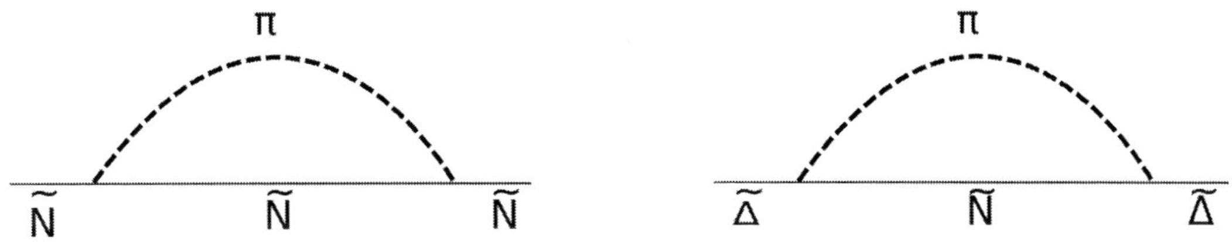

Figure 1. Pion-loop diagrams considered for the mass renormalization of bare baryons $\tilde{N}$ and $\tilde{\Delta}$.

After eliminating the $B\pi$ channel by the Feshbach method one ends up with the following eigenvalue problem for the dressed baryon ground or resonant state $|\psi_B >$:

$$\left[M_{\tilde{B}} - K\left(m - M_{\tilde{B}\pi} + i0\right)^{-1} K^{\dagger}\right] |\psi_B >= m|\psi_B > . \tag{2}$$

Here, evidently an optical potential occurs, which becomes complex above the $\tilde{B}\pi$ threshold. It should be noted that the mass eigenvalue m appears also in the optical-potential term on the left-hand side of Eq. (2). Beyond the resonance threshold it acquires an imaginary part leading to a finite decay width.

The transition dynamics contained in K is deduced from the following Lagrangian densities

$$\mathcal{L}_{\pi NN} = -\frac{f_{\pi NN}}{m_\pi} \bar{\Psi}\gamma_5\gamma^{\mu}\Psi\partial_{\mu}\Phi , \tag{3}$$

$$\mathcal{L}_{\pi N\Delta} = -\frac{f_{\pi N\Delta}}{m_\pi} \bar{\Psi}\Psi^{\mu}\partial_{\mu}\Phi + h.c. , \tag{4}$$

where Ψ and Ψ^{μ} represent the N and Δ fields, which are coupled in pseudovector form by the π field Φ with strengths $f_{\pi NN}$ and $f_{\pi N\Delta}$, respectively. This leads to transition matrix elements from the bare $\tilde{N}$ and $\tilde{\Delta}$ states to the channels including the explicit pions (with mass m_π) for the cases of πNN

$$< \tilde{N}|\mathcal{L}_{\pi\tilde{N}\tilde{N}}(0)|\tilde{N},\pi;\vec{k}''_{\pi} >= \sum \frac{i\, f_{\pi\tilde{N}\tilde{N}}}{m_\pi} \bar{u}(k_{\tilde{N}},\Sigma_{\tilde{N}})\gamma^{\mu}\gamma_5 u(k''_{\tilde{N}},\Sigma''_{\tilde{N}})(k''_{\pi})_{\mu} \tag{5}$$

and $\pi N\Delta$

$$< \tilde{\Delta}|\mathcal{L}_{\pi\tilde{N}\tilde{\Delta}}(0)|\tilde{N},\pi;\vec{k}''_{\pi} >= \sum \frac{i\, f_{\pi\tilde{N}\tilde{\Delta}}}{m_\pi} \bar{u}^{\mu}(k_{\tilde{\Delta}},\Sigma_{\tilde{\Delta}}) u(k''_{\tilde{N}},\Sigma_{\tilde{N}})(k''_{\pi})_{\mu} . \tag{6}$$

Here $u(k_N, \Sigma_N)$ are the spin-$\frac{1}{2}$ Dirac spinors of the N and $u^\mu(k_\Delta, \Sigma_\Delta)$ the spin-$\frac{3}{2}$ Rarita-Schwinger spinors of the Δ. In the rest frame of the baryon B the eigenvalue equation (2) finally turns into the following explicit form

$$\Bigg(m_{\tilde{B}} + \int \frac{d^3 k''_\pi}{(2\pi)^3} \frac{1}{2\omega''_\pi 2\omega''_{\tilde{N}} 2m_{\tilde{B}}} \mathcal{F}_{\pi\tilde{N}\tilde{B}}(|\vec{k}''_\pi|) < \tilde{B}|\mathcal{L}_{\pi\tilde{N}\tilde{B}}(0)|\tilde{N}, \pi; \vec{k}''_\pi > \\ \times \left(m - \sqrt{m^2_{\tilde{N}} + \vec{k}''^2_\pi} - \sqrt{m^2_\pi + \vec{k}''^2_\pi} + i0 \right)^{-1} \mathcal{F}^*_{\pi\tilde{N}\tilde{B}}(|\vec{k}''_\pi|) \\ \times < \tilde{N}, \pi; \vec{k}''_\pi|\mathcal{L}^\dagger_{\pi\tilde{N}\tilde{B}}(0)|\tilde{B} > \Bigg) < \tilde{B}|\psi_B >= m < \tilde{B}|\psi_B > \quad (7)$$

with B standing for N or Δ and all quantities with a tilde referring to bare particles. The wave functions of the baryon states $< \tilde{B}|\psi_B >$ are represented by free momentum eigenstates denoted by $< \tilde{B}|$ or equivalently by free velocity states $< \tilde{B}; \vec{v} = 0|$ (for pertinent details see Ref. [5]). The processes corresponding to the optical potential in Eq. (7) are graphically represented in Fig. 1.

In Eq. (7) we have inserted form factors $\mathcal{F}_{\pi\tilde{N}\tilde{B}}$ for the extended meson-baryon vertices. They are taken from three different models, namely, a relativistic constituent-quark model (RCQM) [7, 8, 9] as well as two phenomenological meson-nucleon models, namely, the one by Sato and Lee (SL) [10] and the one by Polinder and Rijken (PR) [11]. The corresponding parametrizations are all given in Ref. [9] according to the form

$$F(\vec{q}^2) = \frac{1}{1 + \left(\frac{\vec{q}}{\Lambda_1}\right)^2 + \left(\frac{\vec{q}}{\Lambda_2}\right)^4} . \quad (8)$$

The cut-off parameters occurring in Eq. (8) and the values of the coupling constants are summarized in Tab. 1. The functional dependences of the various vertex form factors are shown in Figs. 2 and 3. It should be noted that the RCQM vertex form factors are derived from a microscopic calculation on the quark level, while the SL and PR form factors are determined by a fit to meson-nucleon phenomenology.

Table 1. πNN and $\pi N\Delta$ coupling constants as well as cut-off parameters entering into Eq. (8) for the three different form-factor models used in the present work (cf. Ref. [9]).

		RCQM	SL	PR
	$\frac{f_N^2}{4\pi}$	0.0691	0.08	0.013
N	Λ_1	0.451	0.453	0.940
	Λ_2	0.931	0.641	1.102
	$\frac{f_\Delta^2}{4\pi}$	0.188	0.334	0.167
Δ	Λ_1	0.594	0.458	0.853
	Λ_2	0.998	0.648	1.014

By solving the eigenvalue equation (7) with the mass m calibrated to the physical nucleon mass $m_N = 939$ MeV we find the bare nucleon mass $m_{\tilde{N}}$ and thus the influence of the pion loop. Tab. 2 contains the results for the pion dressing of the nucleon ground state. It is seen that all three different form-factor models lead to very similar magnitudes for the mass differences $m_N - m_{\tilde{N}}$ of about 100 MeV.

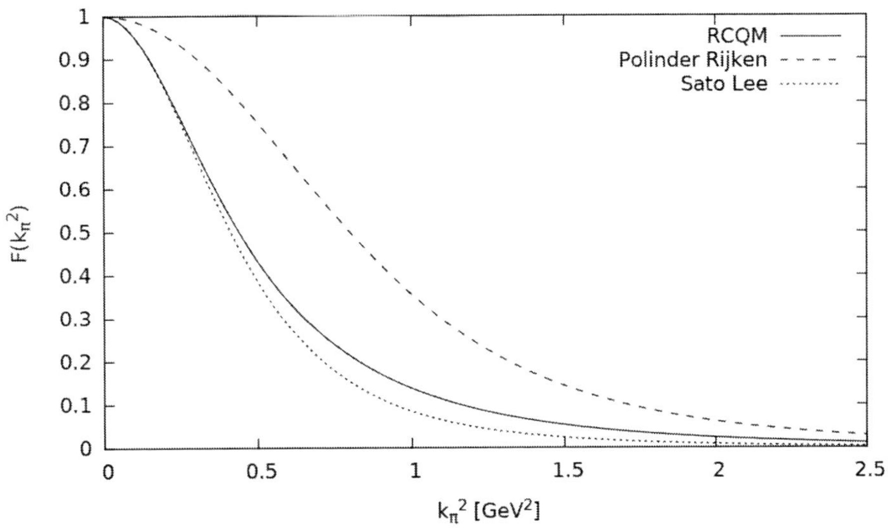

Figure 2. Momentum dependences of the three different form-factor models in case of the πNN system.

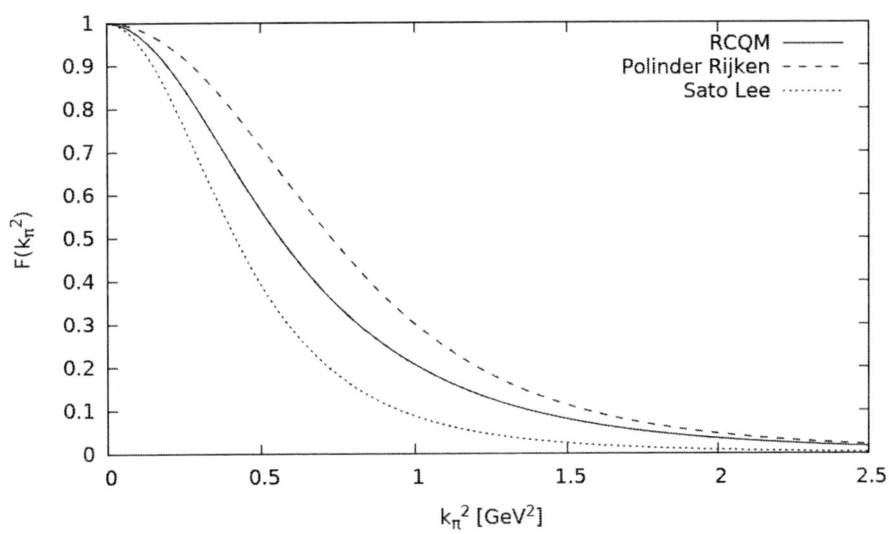

Figure 3. Momentum dependences of the three different form-factor models in case of the $\pi N\Delta$ system.

In the case of the Δ resonance we are interested in the mesonic effects on both the mass as well as the decay width. In the first instance, we employ a bare intermediate nucleon $\tilde{N}$ as is shown in the graph on the r.h.s. of Fig. 1. The corresponding results are given in Tab. 3. Again the pionic effects on the masses are quite similar for the three different form-factor models. The π-decay widths, however, show bigger variations. Still, they are all too small as compared to the phenomenological value.

A more realistic description of the $\Delta \to N\pi$ decay width is obtained by replacing the bare $\tilde{N}$

Table 2. One-loop effects on the nucleon mass m_N from coupling to the πNN channel ($m_{\tilde{N}}$ being the bare mass and m_N the dressed one).

	RCQM	SL	PR
$m_{\tilde{N}}$	1.067	1.031	1.051
$m_N - m_{\tilde{N}}$	-0.128	-0.092	-0.112

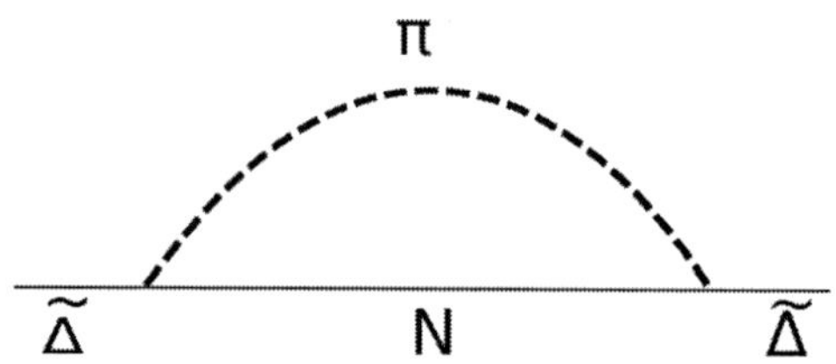

Figure 4. Pion-loop diagram for the $\pi N\Delta$ system with an intermediate physical nucleon with mass $m_N = 939$ MeV.

Table 3. Mesonic effects on the Δ mass $\mathrm{Re}\,(m_\Delta)$ and π-decay width Γ from coupling to the $\pi N\Delta$ channel, according to the loop diagram on the r.h.s. of Fig. 1. The bare nucleon masses $m_{\tilde{N}}$ are the same as in Tab. 2.

	RCQM	SL	PR
$m_{\tilde{N}}$	1.067	1.031	1.051
$m_{\tilde{\Delta}}$	1.300	1.295	1.336
$\mathrm{Re}\,(m_\Delta) - m_{\tilde{\Delta}}$	-0.068	-0.062	-0.104
$\Gamma = 2\,\mathrm{Im}\,(m_\Delta)$	0.0026	0.017	0.0048

Table 4. Mesonic effects on the Δ mass $\mathrm{Re}\,(m_\Delta)$ and π-decay width Γ from coupling to the $\pi N\Delta$ channel, according to the loop diagram in Fig. 4.

	RCQM	SL	PR
$m_{\tilde{N}}$	0.939	0.939	0.939
$m_{\tilde{\Delta}}$	1.318	1.306	1.358
$\mathrm{Re}\,(m_\Delta) - m_{\tilde{\Delta}}$	-0.086	-0.073	-0.125
$\Gamma = 2\,\mathrm{Im}\,(m_\Delta)$	0.042	0.069	0.039

with mass $m_{\tilde{N}}$ in the intermediate state by the physical nucleon N with mass $m_N = 939$ MeV, as depicted in Fig. 4.

The corresponding results are given in Tab. 4. It is immediately seen that the decay widths get much enhanced, while the effects on the masses are only slightly changed. We expect the larger phase space for the pionic decay to be responsible for the enhancement of the decay widths.

At this stage an open problem is left with regard to dressing the vertex form factors and the coupling strengths in our work. Corresponding studies have earlier been undertaken, e.g., by both Sato and Lee [10] as well as Polinder and Rijken [11]. We may expect a further improvement of our results by following a similar way. However, it constitutes a difficult task to realize such a framework consistently in our approach.

2. Summary

We have considered one-loop pionic effects on the N mass as well as on the resonance energy and the hadronic decay width of the Δ in a relativistically invariant CC approach. Thereby, pionic contributions on these observables are taken into account explicitly. The results show the desired behaviour: The dressed masses turn out to be smaller than the bare ones and in addition a decay width naturally comes about for the resonant Δ state. Further investigations are presently under way for the N^* resonances, and effects of additional mesonic channels are studied.

Acknowledgment

This work was supported by the Austrian Science Fund, FWF, through the Doctoral Program on *Hadrons in Vacuum, Nuclei, and Stars* (FWF DK W1203-N16).

References

[1] Melde T, Plessas W and Wagenbrunn R F 2005 *Phys. Rev.* C **72** 015207;
[2] Melde T, Plessas W and Sengl B 2007 *Phys. Rev.* C **76** 025204
[3] Sengl B, Melde T and Plessas W 2007 *Phys. Rev.* D **76** 054008
[4] Melde T, Plessas W and Sengl B 2008 *Phys. Rev.* D **77** 114002
[5] Kleinhappel R 2010 *Diploma Thesis*
[6] Kleinhappel R and Schweiger W 2010 in: *Dressing Hadrons* (Proceedings of the Mini-Workshop 2010, Bled, Slovenia) 33; arXiv:1010.3919
[7] Glozman L Y, Plessas W, Varga K and Wagenbrunn R F 1998 *Phys. Rev.* D **58** 094030
[8] Glozman L Y, Papp Z, Plessas W, Varga K and Wagenbrunn R F 1998 *Phys. Rev.* C **57** 3406
[9] Melde T, Canton L and Plessas W 2009 *Phys. Rev. Lett.* **102** 132002
[10] Sato T and Lee T S H 1996 *Phys. Rev.* C **54** 2660
[11] Polinder H and Rijken T A 2005 *Phys. Rev.* C **72** 065210;

Computation of charged current neutrino-Te reactions cross sections

V. Tsakstara [1], T.S. Kosmas[2], J. Sinatkas[3]
[1]University of Western Makedonia, GR-53100 Florina, Greece
[2]Theoretical Physics Section, University of Ioannina, GR 45110 Ioannina, Greece
[3]Information Engineering Department, TEI of Western Macedonia, GR-52100 Kastoria, Greece

E-mail: vtsaksta@cc.uoi.gr

Abstract. Neutrino-nucleus reactions, involving both neutral current (NC) and charged current (CC) interactions are important probes in modern neutrino physics searches. In the present work, we study the concrete CC reactions $^{130}Te(\nu_\ell, \ell^-)^{130}I$ and $^{130}Te(\tilde{\nu}_\ell, \ell^+)^{130}Sb$ which are of current experimental interest for the CUORE and COBRA experiments operating at Gran Sasso underground laboratory in Italy. The nuclear wave functions for the required initial and final nuclear states are derived by employing the proton-neutron (p-n) quasi-particle random phase approximation (QRPA) which has been previously tested in our neutral-current ν-nucleus studies for Te isotopes.

1. Introduction

The last few decades, charged current (CC) neutrino-nucleus scattering became a quite interesting research topic in astro-nuclear physics, particle physics and cosmology [1, 2]. Neutrinos are extremely sensitive probes for investigating the fundamental electroweak interactions and the neutrino properties (via their interactions with nucleons and nuclei) as well as the stellar evolution (several neutrino processes take place in core-collapse supernova, in the sun's interior and in other stellar systems) [3]. Thus, solar neutrinos, for example, provide significant information for the thermonuclear reactions (p-p and CNO-cycle processes) occurring inside the sun, while supernova neutrino reactions are important for modelling supernova explosions, etc. [3, 4, 5].

On the other hand, promising nuclear detectors have extensively employed as micro-laboratories in neutrino searches through their conventional NC and CC processes with neutrinos that involve vector and axial-vector interactions. Nuclear detector responses to neutrinos are, furthermore, useful tools for investigating exotic (non-standard) interactions at low and intermediate neutrino energies [6, 7, 8, 9]. In particular, many terrestrial experiments, being in operation or designed to operate the next years for detection of laboratory neutrinos (beta-beam neutrinos, spallation neutron source neutrinos, etc.) and astrophysical neutrinos (solar, supernova, geo-neutrinos, etc.), are good sources of information for the exotic properties of neutrinos (ν-oscillations, neutrino magnetic moments, etc.) [8, 9].

In the present work, we examine the role of ^{130}Te isotope as neutrino detector by studying its conventional charged-current neutrino-induced reactions at low and intermediate neutrino energies ($0 \leq \epsilon_i \leq 100$ MeV). This work is an extension of our previous studies [3, 4, 5] in

Content from this work may be used under the terms of the Creative Commons Attribution 3.0 licence. Any further distribution of this work must maintain attribution to the author(s) and the title of the work, journal citation and DOI.
Published under licence by IOP Publishing Ltd

which we have comprehensively evaluated the NC reactions of this isotope with neutrinos and anti-neutrinos. Tellurium isotopes are contents of the detectors of the COBRA and CUORE experiments at Gran Sasso.

2. Brief description of the formalism

Within the context of the current-current interaction theory, the standard model effective Hamiltonian for conventional charged changing reactions is written as [1]

$$\mathcal{H} = \frac{G\cos\theta_c}{\sqrt{2}} j_\mu(\mathbf{x}) J^\mu(\mathbf{x}), \tag{1}$$

where $G = 1.1664 \times 10^{-5}\ GeV^{-2}$ is the weak coupling constant and θ_c is the Cabibo angle. j_μ and J^μ stand for the leptonic and hadronic (nucleonic) currents, respectively. The leptonic current j^μ has the form (V-A theory)

$$j^\mu = \bar{u}_\ell(k_\ell)\gamma^\mu(1-\gamma_5)u_{\nu_\ell}(k_{\nu_\ell}), \tag{2}$$

where u_{ν_ℓ}, u_ℓ (with the normalization $\bar{u}u = 1$), are Dirac spinors for the neutrino and lepton having four-momenta $k_i = (\varepsilon_i, \mathbf{k_i})$, $i = \nu_\ell, \ell$, respectively. The corresponding Feynman diagrams describing the CC neutrino nucleus interactions is shown in Fig. 1.

The nucleonic current J_μ in Eq. (1) (neglecting, as usually, contributions of the second class currents) is given by

$$J_\mu = \bar{u}_p(p_p)\left[F_1(q^2)\gamma_\mu + F_2(q^2)\frac{i\sigma_{\mu\lambda}q^\lambda}{2M} + F_A(q^2)\gamma_\mu\gamma_5 + F_P(q^2)q_\mu\gamma_5\right]u_n(p_n), \tag{3}$$

(M denotes the nucleon mass). The functions of the four momentum transfer q^2 (with

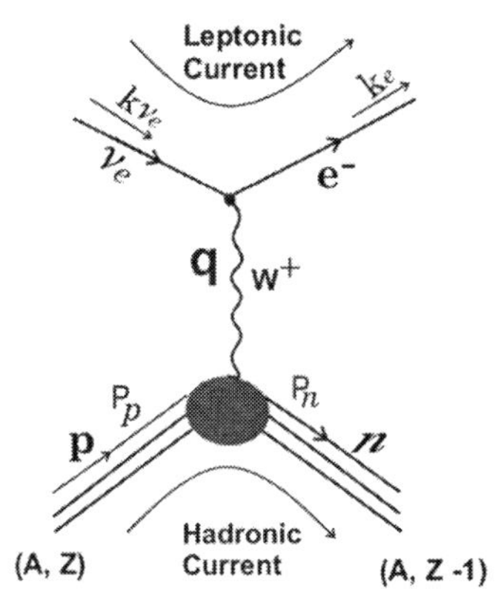

Figure 1. Feynman diagrams describing the neutrino nucleus interactions at nuclear level via the exchange of intermediate vector bosons ($W^\pm$) for charged changing currents.

$q = p_n - p_p$): F_1, F_2, F_A, and F_P, are the well known Dirac, Pauli, axial vector, and pseudo-scalar form factors, respectively. In the convection used in the present work, q^2, is written as

$$q^2 = q^\mu q_\mu \equiv q_0{}^2 - \mathbf{q}^2 = (\varepsilon_{\nu_\ell} - \varepsilon_\ell)^2 - (\mathbf{k}_{\nu_\ell} - \mathbf{k}_\ell)^2 \tag{4}$$

In the case of the neutrino reaction the total energies of the neutron (E_n) and proton (E_p) are written as

$$E_n = \sqrt{\mathbf{p}_n^2 + M_n^2}, \qquad E_p = \sqrt{(\mathbf{p}_n + \mathbf{q})^2 + M_p^2} \approx \sqrt{\mathbf{p}_n^2 + \mathbf{q}^2 + M_p^2}\,. \tag{5}$$

Table 1. Nuclear parameters needed for cross sections calculations of the CC neutrino reactions $^{130}Te(\nu_\ell, \ell^-)^{130}I$ and anti-neutrino-^{130}Te reaction $^{130}Te(\tilde{\nu}_\ell, \ell^+)^{130}Sb$.

Isotope	Z, N	Model Space	$Q - value(MeV)$	J_{gs}
^{130}Te	52, 78	1p-0f-2s-1d-0g-2p-1f-0h	-0.4173	0^+
^{130}Sb	51, 79	1p-0f-2s-1d-0g-2p-1f-0h	5.0671	8^-
^{130}I	53, 77	1p-0f-2s-1d-0g-2p-1f-0h	2.9440	5^+

where $M_n \approx M_p \approx M$ (the approximation in E_p holds if $\mathbf{q} \cdot \mathbf{p}_n \approx 0$). For the anti-neutrino reaction, the leptonic and hadronic currents are the complex conjugates of Eqs. (2) and (3).

The differential cross section for CC neutrino-nucleus scattering, after applying a multipole decomposition of the weak hadronic current (method of Donnelly-Walecka) [4], is written as

$$\frac{\mathrm{d}^2\sigma_{i\to f}}{\mathrm{d}\Omega \mathrm{d}\omega} = \frac{G^2}{\pi}\frac{|\vec{k}_\ell|\varepsilon_\ell}{(2J_i+1)}F(Z,\varepsilon_\ell)\left(\sum_{J=0}^{\infty}\sigma_{CL}^J + \sum_{J=1}^{\infty}\sigma_T^J\right). \tag{6}$$

where the summations in the right hand side contain the contributions σ_{CL}^J, for the Coulomb $\widehat{\mathcal{M}}_J$ and longitudinal $\widehat{\mathcal{L}}_J$, and σ_T^J, for the transverse electric $\widehat{\mathcal{T}}_J^{el}$ and magnetic $\widehat{\mathcal{T}}_J^{mag}$ multipole operators (see Ref. [10, 11] for definitions of these operators and other notations). They include both polar-vector and axial-vector weak interaction components. The function $F(Z, \varepsilon_\ell)$ (known as Fermi function) takes into account the Coulomb-final-state interaction between the final nucleus and the outgoing lepton (electron in the case of ν_e scattering). In Eq. (6) $i(f)$ denote the initial (final) state and Ω (ω) the solid angle and excitation energy of the daughter nucleus.

3. The QRPA for charged current reactions

For CC neutrino-nucleus reactions considered in the present work, low or intermediate energy neutrinos (or anti-neutrinos) are inelastically scattered from a target nucleus (A,Z). There are only incoherent channels open in these processes. The initial and final nuclear isotopes are assumed to be spherically symmetric. The neutrino-detector, ^{130}Te (initial nucleus), has a ground state with spin equal to $|J^\pi\rangle = |0^+\rangle$ (even-even nucleus). The ground state spin of the final nuclei, ^{130}I and ^{130}Sb, are listed in Table 1 together with the model space chosen and $Q-$values of the studied processes.

The required wave functions for the initial and final nuclear states for charged current neutrino-nucleus induced reactions, can be derived in the context of the proton-neutron (p-n) quasi-particle random phase approximation (QRPA) [3, 4, 5]. In this method, the mth excited state with total angular momentum J projection M and parity π, denoted by $|J_m^\pi M\rangle$, is created by acting on the QRPA ground state with the phonon operator

$$\hat{Q}_{J^\pi M}^{m\dagger} = \sum_{k,l}\left[X^m(kl,J)A^\dagger(kl,JM) + Y^m(kl,J)\tilde{A}(kl,JM)\right] \tag{7}$$

where X and Y are the forward and backward going amplitudes which can be determined from the common QRPA matrix equations (see, e.g., Ref. [4]).

The quasi-particle pair-creation and pair-annihilation operators, $A^\dagger$ and $\tilde{A}$, are defined as

$$A^\dagger(kl,JM) \equiv \left[a_k^\dagger a_l^\dagger\right]_M^J = \sum_{m_k,m_l}\langle j_k m_k j_l m_l|JM\rangle a_{km_k}^\dagger a_{lm_l}^\dagger$$

$$\tilde{A}(kl,JM) = (-1)^{J-M}A(kl,J-M). \tag{8}$$

In the latter definitions, the square brackets denote angular-momentum coupling while $a^\dagger$ and a denote quasi-particle creation and annihilation operators, respectively.

Table 2. Harmonic oscillator size parameter b and pairing interaction parameters for protons (g^p_{pair}), and neutrons (g^n_{pair}). The relevant theoretical values of proton/neutron energy gaps ($\Delta^{th}_p, \Delta^{th}_n$) that reproduce well the corresponding empirical gaps, $\Delta^{exp}_{p,n}$ for ^{130}Te, are also listed.

Isotope	Abundance (%)	b_{HO} (fm)	g^n_{pair} (MeV)	g^p_{pair} (MeV)	Δ^{exp}_p (MeV)	Δ^{th}_p (MeV)	Δ^{exp}_n (MeV)	Δ^{th}_n (MeV)
^{130}Te	33.8	2.257	1.037	0.843	1.017	1.214	1.0170	1.2069

4. Results and Discussion

For the present cross section calculations of CC (anti)neutrino-nucleus reactions, we have chosen a set of interesting neutrino detectors employed in current underground experiments, i.e. the COBRA and CUORE experiments operating at Gran Sasso, Italy. We focus on the ^{130}Te target isotope content of the CdTe and CdZnTe semiconductors, detector materials of COBRA experiment, and the TeO_2, detector material of the CUORE experiment.

We study the two charged-current (anti)neutrino-nucleus processes:

$$^{130}Te + \nu_\ell \longrightarrow ^{130} I + \ell^- , \qquad ^{130}Te + \tilde{\nu}_\ell \longrightarrow ^{130} Sb + \ell^+ \tag{9}$$

The first is the charged current reaction of ^{130}Te with neutrinos (producing a lepton ℓ^-), and the second with anti-neutrinos (producing an anti-lepton ℓ^+). The corresponding neutral current processes of both of them have previously been studied theoretically by using the method of Ref. [4].

At first, the strong pairing interaction between the nucleons was treated in the framework of the BCS theory. As usually, we adjusted the pairing force and obtain realistic quasi-particle energies separately for protons and neutrons as shown in Table 2. The fit was performed in such a way that the lowest quasi-particle energy to approach the phenomenological pairing gaps $\Delta_{n/p}$ (for neutrons and protons). The latter are also given from empirical expressions in terms of the proton/neutron separation energies $S_{n(p)}$ [4, 5].

For the derivation of the nuclear states of the final nuclei we used the proton-neutron quasi-particle RPA (p-n QRPA) based on the Bonn C-D two-nucleon strong interaction. The reliability of this QRPA method was checked through the reproducibility of the experimental spectra for the daughter isotopes ^{130}I and ^{130}Sb. The theoretical and experimental spectra were constructed and compared as shown in Ref. [2]. Results of the cross sections, obtained for realistic state-by-state calculations in inelastic neutrino-Te scattering processes will be published elsewhere.

References

[1] Kosmas T S and Oset E, 1996 *Phys. Rev.* **C 53** 1409.
[2] Sinatkas J, Tsakstara, V, Kosmas T S, 2015 *J. Phys.: Conf. Ser.* **633** 012142.
[3] Tsakstara V and Kosmas T S, 2011 *Phys. Rev.* **C 84** 064620.
[4] Tsakstara V and Kosmas T S, 2011 *Phys. Rev.* **C 83** 054612.
[5] Tsakstara V, Kosmas T S, 2012 *Phys. Rev.* **C 86** 044618.
[6] Kosmas T S, Vergados J D, 1988 *Phys. Lett.* **B 215** 460–464.
[7] Kosmas T S, Vergados J D, 1989 *Phys. Lett.* **B 217** 19–24
[8] Papoulias D K, Kosmas T S, 2014 *Phys. Lett.* **B 728** 482-488.
[9] Papoulias D K, Kosmas T S, 2015 *Phys. Lett.* **B 747** 454-459.
[10] Chasioti V C, Kosmas T S, 2009 *Nucl. Phys.* **A 829** 234-252.
[11] Kosmas T S, Vergados J D, 1992 *Nucl. Phys.* **A 536** 72-86.

The choice of reaction path during ternary diffusion process

Bartek Wierzba[1]

[1]Rzeszów Uniwersity of Technology, Faculty of Mechanical Engineering and Aeronautics, al. Powstańców Warszawy 12, 35-959 Rzeszów, Poland

E-mail: bwierzba@prz.edu.pl

Abstract. The complete understanding of the phenomenological process related to the evolution of the ternary multiphase system is still lacking. In this paper the ternary interdiffusion process in multiphase system is discussed by means of numerical modeling. The entropy production principle is proposed to chose the proper diffusion path during the process (e.g. internal oxidation). The bi-velocity method of the three-component multi-phase system is presented. The simulations show the local entropy production curve that determines the thickness of the two-phase zone.

1. Introduction

Many materials are multicomponent alloys and consist of several grains and phases. Their macroscopic properties and reliability highly depend on the microstructure and its evolution in time. It has been shown extensively that the phase field technique is capable of dealing in a phenomenological and also quantitative way with the complex processes and morphologies involved in microstructure evolution [1, 2, 3, 4, 5, 6, 7].

As it is well known, the second law of thermodynamics indicates only the general direction of the evolution of the system, the tendency of a closed system's entropy to reach the maximum, the tendency of free energy to be minimum at fixed volume and temperature, and the tendency of Gibbs's potential to reach minimum at fixed temperature and pressure [8, 9]. In this paper we will try to answer the fundamental question: how is the choice made, if more than one diffusion path to relaxation is possible?

The interplay between inter-diffusion in the multi-phase system involves thermodynamics as well as kinetic considerations [10]. It follows that the knowledge of the thermodynamics in the ternary system preferably in the form of a ternary phase diagram is a necessity [11]. On the ternary A–B–C system with a miscibility gap, the diffusion effects can be graphically represented at the concentration triangle as a diffusion path. The diffusion path can go across single and two-phase regions in the ternary diagram. When the diffusion path crosses the tie lines, the two-phase zone can grow. The points of the intersection of the diffusion path with the phase boundaries in the phase diagram serve as a basis to identify the sequential single and two-phase layers present in the diffusion zone.

In this paper the bi-velocity model is proposed to calculate the time evolution of the reactive diffusion in multi-phase system. The bi-velocity model allow to calculate the local entropy production

[1] To whom any correspondence should be addressed.

Content from this work may be used under the terms of the Creative Commons Attribution 3.0 licence. Any further distribution of this work must maintain attribution to the author(s) and the title of the work, journal citation and DOI.
Published under licence by IOP Publishing Ltd

[12] and Kirkendall effect [13] during the diffusion process. The local entropy production is the missing element in proper choice of diffusion path. It will be presented, that the maxims on the local entropy production curve corresponds to the intersection points between the diffusion path and phase boundaries.

2. Model.

To calculate the ternary diffusion in multiphase system the bi-velocity method [14, 15, 16] will be used. It is applicable in multicomponent - multiphase systems but we will further consider the three-component system in which m layers, separated by the interfaces of finite thicknesses, can grow. The main equation describing the concentration (in this case the average concentration) is mass conservation law for each component and for each phase.

$$\frac{\partial \overline{c}_i}{\partial t}+\frac{\partial}{\partial x}\left(\overline{c}_i\overline{\upsilon}_i\right)=0, \qquad i=1,2,3 \tag{1}$$

where: $\overline{c}_i$ and $\overline{\upsilon}_i$ denote an average concentration and velocity of the i-th component, respectively. The schematic picture representing the diffusion between two phases α and β with two phase zone formation $\alpha+\beta$ is presented on figure 1. The straight lines represent the boundaries between the phase and two phase zone. The dotted lines shows the conodes (tie lines), note that on one conode the chemical potentials are equal. The conodes are evenly distributed over the boundaries. The average concentration is defined inside two phase region. The average concentration can be estimated from the composition triangle:

$$\overline{c}_i=f_\alpha\, c_{i,\alpha}+\left(1-f_\alpha\right)c_{i,\beta}, \quad i=1,2,3. \tag{2}$$

where f_α denote the volume fraction in a two-phase region between α and β phase, $c_{i,j}$ denote the concentration of the i-th component in j-th phase. In the two-phase system the following relation holds:

$$f_\alpha+f_\beta=1, \quad 0\le f_\alpha\le 1. \tag{3}$$

Let: $f_\alpha=f$ and $f_\beta=1-f$, then f equals 1 in α phase region ($\overline{c}_i=c_{i,\alpha}$) and 0 within β phase region ($\overline{c}_i=c_{i,\beta}$). In $\alpha+\beta$ phase zone: $0<f<1$.

Following the Darken, the average flux $\overline{c}_i\overline{\upsilon}_i$ is divided into two parts, i.e. the diffusion, $\overline{c}_i\overline{\upsilon}_i^{\,d}$, and drift, $\overline{c}_i\overline{\upsilon}^{drift}$:

$$\overline{c}_i\overline{\upsilon}_i=\overline{c}_i\overline{\upsilon}_i^{\,d}+\overline{c}_i\overline{\upsilon}^{drift}, \qquad i=1,2,3, \tag{4}$$

The diffusion flux is expressed by 2-nd fick law:

$$\overline{c}_i\overline{\upsilon}_i^{\,d}=-\overline{D}_i\frac{\partial \overline{c}_i}{\partial x}, \qquad i=1,2,3 \tag{5}$$

Finally, the drift velocity can be defined by volume continuity equation [14] as:

$$\overline{\upsilon}^{drift}=\sum_{i=1}^{3}\overline{D}_i\frac{\partial \overline{N}_i}{\partial x}, \qquad i=1,2,3 \tag{6}$$

where $\overline{N}_i=\overline{c}_i/\sum_{i=1}^{3}\overline{c}_i$.

The average entropy production rate in case of isothermal process (i.e. the temperature gradient, pressure are negligible) is define as [17]:

$$A^s=R\sum\nolimits_i\frac{\left(\overline{c}_i\,\overline{\upsilon}_i^{\,d}\right)^2}{\overline{c}_i\,\overline{D}_i}, \tag{7}$$

where R denote the gas constant and the average diffusion coefficient [18] equal

$$\bar{D}_i = D_{i,\alpha}\frac{\partial c_{i,\alpha}}{\partial \bar{c}_i} + D_{i,\beta}\frac{\partial c_{i,\beta}}{\partial \bar{c}_i}, \qquad i=1,2,3 \tag{8}$$

The final form of the proposed model describe the following set of the equations (average mass conservation for each component and entropy density equation):

$$\frac{\partial \bar{c}_i}{\partial t} + \frac{\partial}{\partial x}\left(-\left(D_{i,\alpha}\frac{\partial c_{i,\alpha}}{\partial \bar{c}_i} + D_{i,\beta}\frac{\partial c_{i,\beta}}{\partial \bar{c}_i}\right)\frac{\partial \bar{c}_i}{\partial x} + \bar{c}_i\sum_{i=k}^{3}\left(D_{k,\alpha}\frac{\partial c_{k,\alpha}}{\partial \bar{c}_k} + D_{i,\beta}\frac{\partial c_{k,\beta}}{\partial \bar{c}_k}\right)\frac{\partial \bar{N}_k}{\partial x}\right) = 0, \qquad i=1,2,3 \tag{9}$$

and

$$A^s = -R\sum_{i=1}^{3}\frac{\left(D_{i,\alpha}\frac{\partial c_{i,\alpha}}{\partial \bar{c}_i}\frac{\partial \bar{c}_i}{\partial x} + D_{i,\beta}\frac{\partial c_{i,\beta}}{\partial \bar{c}_i}\frac{\partial \bar{c}_i}{\partial x}\right)^2}{\bar{c}_i\left(D_{i,\alpha}\frac{\partial c_{i,\alpha}}{\partial \bar{c}_i} + D_{i,\beta}\frac{\partial c_{i,\beta}}{\partial \bar{c}_i}\right)} \tag{10}$$

Finally the two-phase zone in bi-velocity phase field method can be approximated by the gradients of concentrations in α and β phase:

$$\frac{\partial \bar{c}_i}{\partial t} + \frac{\partial}{\partial x}\left(-D_{i,\alpha}\frac{\partial c_{i,\alpha}}{\partial x} - D_{i,\beta}\frac{\partial c_{i,\beta}}{\partial x} + \bar{N}_i\sum_{i=k}^{3}\left(D_{k,\alpha}\frac{\partial c_{k,\alpha}}{\partial x} + D_{i,\beta}\frac{\partial c_{k,\beta}}{\partial x_k}\right)\right) = 0, \qquad i=1,2,3 \tag{11}$$

and

$$A^s = -R\sum_{i=1}^{3}\frac{\left(D_{i,\alpha}\frac{\partial c_{i,\alpha}}{\partial x} + D_{i,\beta}\frac{\partial c_{i,\beta}}{\partial x}\right)^2}{\bar{c}_i\left(D_{i,\alpha}\frac{\partial c_{i,\alpha}}{\partial \bar{c}_i} + D_{i,\beta}\frac{\partial c_{i,\beta}}{\partial \bar{c}_i}\right)} \tag{12}$$

The component mass conservations law Eq. (11) can be rearranged to:

$$\frac{\partial \bar{c}_i}{\partial t} + \frac{\partial}{\partial x}\left(-D_{i,\alpha}\frac{\partial c_{i,\alpha}}{\partial x} + \bar{N}_i\sum_{i=k}^{3}D_{k,\alpha}\frac{\partial c_{k,\alpha}}{\partial x} - D_{i,\beta}\frac{\partial c_{i,\beta}}{\partial x} + \bar{N}_i\sum_{i=k}^{3}D_{i,\beta}\frac{\partial c_{k,\beta}}{\partial x_k}\right) = 0, \qquad i=1,2,3 \tag{13}$$

Then using the definition describing the average concentration, Eq. (2):

$$\frac{\partial\left(f_\alpha c_{i,\alpha} + (1-f_\alpha)c_{i,\beta}\right)}{\partial t} + \frac{\partial}{\partial x}\left(-D_{i,\alpha}\frac{\partial c_{i,\alpha}}{\partial x} + \left(f_\alpha N_{i,\alpha} + (1-f_\alpha)N_{i,\beta}\right)\sum_{i=k}^{3}D_{k,\alpha}\frac{\partial c_{k,\alpha}}{\partial x}\right) +$$
$$\frac{\partial}{\partial x}\left(-D_{i,\beta}\frac{\partial c_{i,\beta}}{\partial x} + \left(f_\alpha N_{i,\alpha} + (1-f_\alpha)N_{i,\beta}\right)\sum_{i=k}^{3}D_{i,\beta}\frac{\partial c_{k,\beta}}{\partial x_k}\right) = 0, \qquad i=1,2,3 \tag{14}$$

By simple algebraic rearranging above expression can be rewritten as follow:

$$\begin{aligned}&\left(c_{i,\alpha}-c_{i,\beta}\right)\frac{\partial f_\alpha}{\partial t}+f_\alpha\frac{\partial c_{i,\alpha}}{\partial t}+\left(1-f_\alpha\right)\frac{\partial c_{i,\beta}}{\partial t}\\ &-D_{i,\alpha}\frac{\partial^2 c_{i,\alpha}}{\partial x^2}+\frac{\partial}{\partial x}\left(f_\alpha N_{i,\alpha}\sum_{i=k}^{3}D_{k,\alpha}\frac{\partial c_{k,\alpha}}{\partial x}\right)+\frac{\partial}{\partial x}\left(\left(1-f_\alpha\right)N_{i,\beta}\sum_{i=k}^{3}D_{k,\alpha}\frac{\partial c_{k,\alpha}}{\partial x}\right)-\\ &-D_{i,\beta}\frac{\partial^2 c_{i,\beta}}{\partial x^2}+\frac{\partial}{\partial x}\left(f_\alpha N_{i,\alpha}\sum_{i=k}^{3}D_{i,\beta}\frac{\partial c_{k,\beta}}{\partial x_k}\right)+\frac{\partial}{\partial x}\left(\left(1-f_\alpha\right)N_{i,\beta}\sum_{i=k}^{3}D_{i,\beta}\frac{\partial c_{k,\beta}}{\partial x_k}\right)=0,\qquad i=1,2,3\end{aligned}\tag{15}$$

Finally, assuming that the mass conservation law holds for each phase separately the following expressions are valid:

$$\frac{\partial c_{i,\alpha}}{\partial t}+\frac{\partial}{\partial x}\left(-D_{i,\alpha}\frac{\partial c_{i,\alpha}}{\partial x}+N_{i,\alpha}\sum_{k=1}^{3}D_{k,\alpha}\frac{\partial c_{k,\alpha}}{\partial x}\right)=0,\qquad i=1,2,3\tag{16}$$

and

$$\frac{\partial c_{i,\beta}}{\partial t}+\frac{\partial}{\partial x}\left(-D_{i,\beta}\frac{\partial c_{i,\beta}}{\partial x}+N_{i,\beta}\sum_{k=1}^{3}D_{k,\beta}\frac{\partial c_{k,\beta}}{\partial x}\right)=0,\qquad i=1,2,3\tag{17}$$

thus, Eqs. (15) can be rewritten in form:

$$\begin{aligned}0=&\left(c_{i,\alpha}-c_{i,\beta}\right)\frac{\partial f_\alpha}{\partial t}+\left(N_{i,\alpha}\sum_{i=k}^{3}D_{k,\alpha}\frac{\partial c_{k,\alpha}}{\partial x}-D_{i,\alpha}\frac{\partial c_{i,\alpha}}{\partial x}-N_{i,\beta}\sum_{i=k}^{3}D_{i,\beta}\frac{\partial c_{k,\beta}}{\partial x_k}+D_{i,\beta}\frac{\partial c_{i,\beta}}{\partial x}\right)\frac{\partial f_\alpha}{\partial x}\\ &+\frac{\partial}{\partial x}\left(\left(1-f_\alpha\right)\left[-D_{i,\alpha}\frac{\partial c_{i,\alpha}}{\partial x}+N_{i,\beta}\sum_{i=k}^{3}D_{k,\alpha}\frac{\partial c_{k,\alpha}}{\partial x}\right]-f_\alpha\left[-D_{i,\beta}\frac{\partial c_{i,\beta}}{\partial x}+N_{i,\alpha}\sum_{i=k}^{3}D_{i,\beta}\frac{\partial c_{k,\beta}}{\partial x_k}\right]\right),\qquad i=1,2,3\end{aligned}\tag{18}$$

Equation (18) shows that the two-phase zone can be described by the derivative of the phase fraction in time. Note, that the two-phase zone formation near the α phase boundary ($f_\alpha \approx 1$) depend mainly on the gradient of concentration of components from β phase, thus two-phase zone will grow:

$$0=-D_{i,\beta}\frac{\partial c_{i,\beta}}{\partial x}+N_{i,\alpha}\sum_{i=k}^{3}D_{i,\beta}\frac{\partial c_{k,\beta}}{\partial x_k}\tag{19}$$

analogous situation can be shown near the β phase boundary.

2.1. The calculations performed at each time step:

To solve above equations the following steps should be accomplished:

- Solve diffusion flux equations (for each component), Eq. (5), volume continuity equation, Eq. (6) and the equation for local entropy production, Eq. (7). Calculate diffusion velocities, $\bar{\upsilon}_i^d$, $i=1,2,3$, drift velocity, $\bar{\upsilon}^{drift}$ and local entropy production rate, A^s. The interphase boundaries are identified with the two maxims of the local entropy-production rate where the driving force for the irreversible mass transport is highest.
- Solve the conservation laws, Eqs. (1). Calculate concentrations, $\bar{c}_i$, $i=1,2,3$ for advancement $t_k \to t_{k+1}$. Determine the conode which cross the $\bar{c}_i$, $i=1,2,3$ in the two-phase region. The conodes are evenly distributed over the boundaries;
- Compute the intersection points of the conodes with the boundaries. These intersections determine the concentrations of the components in each phase, $c_{i,\alpha}$, $c_{i,\beta}$ for $i=1,2,3$;
- Calculate the volume fractions of each phase, f_α and f_β, Eq. (3);

2.2. The initial steps:

- Determine the boundaries of the two phase region and represent them by the polynomials dependent on the mole fractions of two from the three components;
- define the conodes (tie lines), *k*, which are evenly distributed along the boundaries

The remaining initial data necessary in the calculations include: terminal composition of the diffusion couple (representing the terminal points of the diffusion path), diffusion coefficients and molar volumes of the components in all phases, processing time and temperature (results section).

2.3. Numerical method [19]

The numerical treatment of the bi-velocity method comes down to solving a set of ordinary differential equations $dy/dt = f(y,t)$, where $f: R^n \to R^n$, $y = y(t) \in R^n$. To solve the ODEs e.g. the explicit Euler method can be used. In this method the passage from time t_k to $t_{k+1} = t_k + \Delta t$ was carried out by one-step evaluation. The advancement $t_k \to t_{k+1}$ was executed in which $y_{k+1} = y_k + \Delta t \cdot f(y_k, t_k)$ with $f(y,t)$ evaluated for $t = t_k$ and $y = y_k$ [19].

In the present calculations, the method of lines to solve numerically the ODEs system resulting from the space discretization of the bi-velocity set of equations was used [20]. The uniform grid, contained 100 mesh points, was used and the concentrations and pressure were defined at points x_k. The applied discretization led to the system of ordinary differential equations (ODEs) in time variable. For example, the equations for diffusion flux (Eq. (5)) of all three components were transposed by finite difference approximation in one dimensional space, as follows:

$$^{k}\bar{J}_i^d(t) := \bar{J}_i^d(x_k,t) = -\bar{D}_i(x_k,t)\,\frac{\partial \bar{c}_i}{\partial x}(x_k,t) \approx -\bar{D}_i(x_k,t)\frac{^{k+1}\bar{c}_i - {}^{k-1}\bar{c}_i}{x_{k+1} - x_{k-1}}, \quad i = 1,2,3 \tag{20}$$

For boundary nodes the one-sided uniform finite differences were used:

$$^{0}\bar{J}_i^d(t) \approx -\bar{D}_i(x_0,t)\frac{^{1}\bar{c}_i - {}^{0}\bar{c}_i}{x_1 - x_0}, \quad i = 1,2,3 \text{ and} \tag{21}$$

$$^{n}\bar{J}_i^d(t) \approx -\bar{D}_i(x_n,t)\frac{^{n}\bar{c}_i - {}^{n-1}\bar{c}_i}{x_n - x_{n-1}}, \quad i = 1,2,3 \tag{22}$$

where: $^{k}\bar{c}_i = \bar{c}_i(x_k,t)$.

The space derivatives in the above equations were approximated by two point (first derivative) and three point (second derivative) uniform finite differences:

$$\frac{\partial \Phi}{\partial x}(x_k,t) \approx \frac{^{k+1}\Phi - {}^{k-1}\Phi}{x_{k+1} - x_{k-1}} \tag{23}$$

and

$$\frac{\partial^2 \Phi}{\partial x^2}(x_k,t) \approx \frac{^{k+1}\Phi(x_k - x_{k-1}) + {}^{k-1}\Phi(x_{k+1} - x_k) - {}^{k}\Phi(x_{k+1} - x_{k-1})}{0.5(x_{k+1} - x_{k-1})(x_{k+1} - x_k)(x_k - x_{k-1})} \tag{24}$$

Both were calculated from Taylor expansion.

The obtained discretization formulae (Eqs. (20)-(24)) are sufficient to implement the computations, but to solve the ODEs a time integrator is needed. The time integrator, used in the present computations, was based on Runge-Kutta method. Exactly, the adaptive step size Runge-Kutta-Fehlberg modification was used. The six evaluations of the functions from the fifth-order Runge-Kutta algorithm were used to make the another combinations implemented in the fourth-order Runge-Kutta method. A difference between these two estimates served as an estimate of the truncation error. Hence, the step size was adjusted [21].

2.4. Computer implementations.

In particular, the phase boundaries, which are margins for the single-phase and multi-phase regions, must be known, fig. 1.

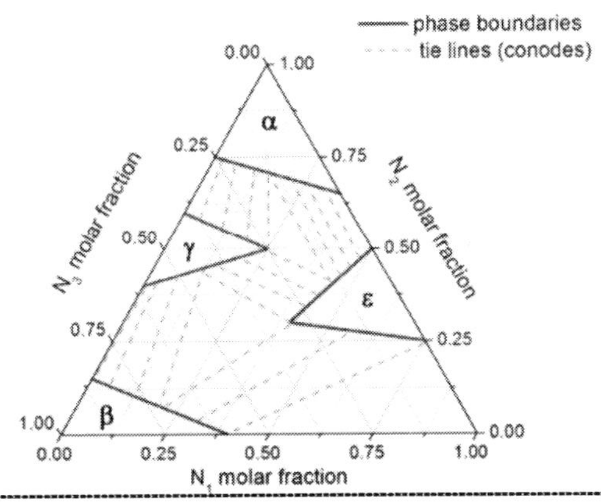

Figure 1. The schematic view of the concentration triangle. The straight lines denotes the boundaries between phases (α, β, γ and ε). The dashed lines represents the tie lines connecting the phases.

They provide the boundary conditions for the solution of equations governing diffusion. During calculations (in each time step) the concentration profiles for the three components are calculated. Hence diffusion path and other properties can be determined.

3. Results.

In order to calculate the diffusion between α and β phase three possible diffusion path was assumed and calculated: $\alpha \to \gamma \to \beta$, $\alpha \to \varepsilon \to \beta$. The kinetic data used in simulations are presented in Table 1.

Table 1. The diffusion coefficients, $D_{i,j}\left[cm^2s^{-1}\right]$, and initial concentrations, at. %, used in simulations of reactive diffusion in two-phase system.

Diffusion coefficient	Component 1	Component 2	Component 3
$D_{i,\alpha}$	1.0 10^{-8}	5.0 10^{-8}	3.0 10^{-8}
$D_{i,\gamma}$	1.0 10^{-9}	5.0 10^{-9}	3.0 10^{-9}
$D_{i,\beta}$	1.0 10^{-8}	5.0 10^{-8}	3.0 10^{-8}
$D_{i,\varepsilon}$	1.0 10^{-7}	5.0 10^{-7}	3.0 10^{-7}
Phase	Terminal composition, at. %		
α	13	78	9
γ	7	50	43
β	20	3	77
ε	53	35	12

3.1. Diffusion path $\alpha \rightarrow \gamma \rightarrow \beta$.

On figures 2 and 3 the time evolution of the concentration profile and corresponding evolution of the local entropy production is shown. The presented data describes the changes of the average composition of the system parallel to the mass transport direction, as it results from the concentration profiles. It is seen that the diffusion path passes the $\alpha+\gamma$ and $\gamma+\beta$ two-phase regions. The thickness of the two-phase zones can be determined from the local average entropy production, Fig. 3. The maxims on the local entropy production curve are correlated with the points where the diffusion path cross the interphase boundaries. It is evident that the entropy produced at the heat and mass diffusion does not depend on the frame of reference and it can serve additional parameter characterizing the formation of the diffusion zone in two-phase system.

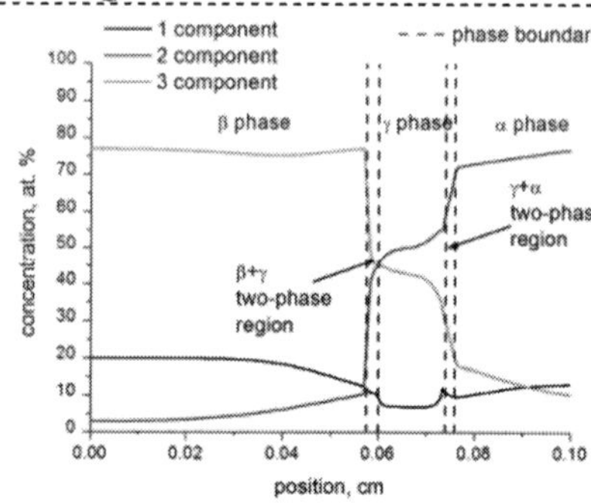

Figure 2. The concentration profile of the arbitrary three-phase system between $\alpha \rightarrow \gamma \rightarrow \beta$ phases after 1h of reactive diffusion process. The vertical lines denotes the phase boundaries.

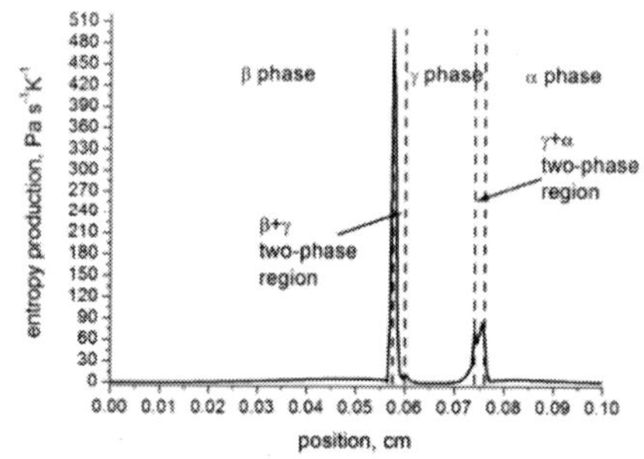

Figure 3. The local entropy production profile of the arbitrary three-phase system between $\alpha \rightarrow \gamma \rightarrow \beta$ phases after 1h of reactive diffusion process. The vertical lines denotes the phase boundaries.

3.2. diffusion path is $\alpha \rightarrow \varepsilon \rightarrow \beta$

On figures 4 and 5 the time evolution of the concentration profile and corresponding evolution of the local entropy production is shown. The presented data describes the changes of the average composition of the system parallel to the mass transport direction, as it results from the concentration profiles. It is seen that the diffusion path passes the $\alpha+\varepsilon$ and $\varepsilon+\beta$ two-phase regions. The diffusion coefficients in ε phase are grater than in γ phase, thus the thickness of the ε phase is larger.

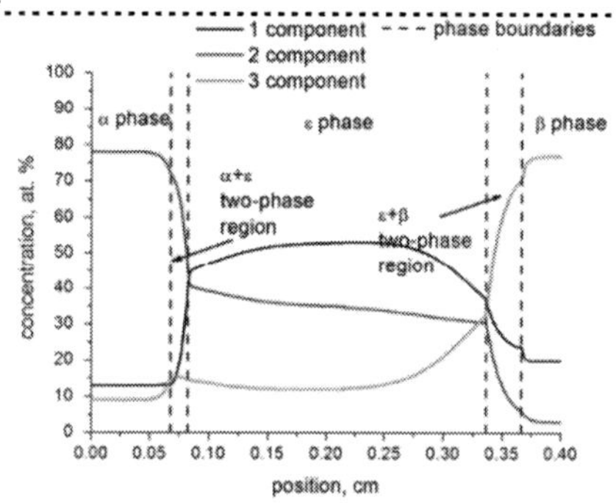

Figure 4. The concentration profile of the arbitrary three-phase system between $\alpha \rightarrow \varepsilon \rightarrow \beta$ phases after 1h of reactive diffusion process. The vertical lines denotes the phase boundaries.

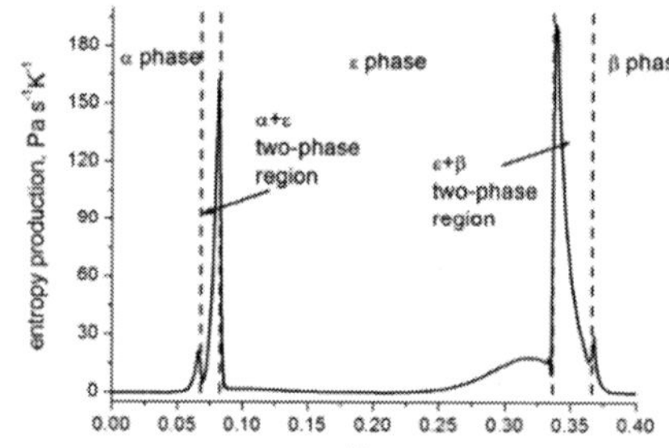

Figure 5. The local entropy production profile of the arbitrary three-phase system between $\alpha \rightarrow \varepsilon \rightarrow \beta$ phases after 1h of reactive diffusion process. The vertical lines denotes the phase boundaries.

4. Conclusions.

The bi-velocity concept of calculating the evolution of the concentration profile and entropy production was presented. It was shown that the entropy production curve can be used to estimate the thickness of the formed two-phase zones (the maxims on the curve correspond to the boundaries of the phases). The method can be further generalized to calculate the kinetic of the diffusion reaction processes in multiphase systems, e.g. internal oxidation or aluminization processes.

Acknowledgements. This work has been supported by the National Science Centre (NCN) in Poland, decision number 2014/15/B/ST8/00120.

References.

[1] Moleans N 2011 Acta. Mat. **59** 1077
[2] Chen L.Q. 2002 Annu Rev Mater Res **32** 113
[3] Boettinger W., Warren J., Beckermann C., Karma A. 2002 Annu Rev Mater Res **23** 163
[4] Singer I. Loginova, Singer H. 2008 Rep Prog Phys **71** 106501
[5] Emmerich H. 2008 Advances in Physics **57** 1
[6] Moelans N., Blanpain B., Wollants P. 2008 Computer Coupling of Phase Diagrams and Thermochemistry (Calphad) **32** 268.
[7] Steinbach I. 2009 Modelling Simul. Mater. Sci. Eng. **17** 073001
[8] de Groot S.R., Mazur, P. 1962 Non-equilibrium Thermodynamics, North-Holland, Amsterdam
[9] Gusak A. M. 2010 Diffusion-controlled Solid State Reactions WILEY-VCH Verlag Germany
[10] Wierzba B., Skibiński W. 2013 Physica A **392** 4316
[11] Dalvi A. D., Coates D. E. 1972 Oxid. of Met. **5** 113
[12] Wierzba B., 2012 Physica A **391** 56
[13] Wierzba B., 2013 Physica A **392** 2860
[14] Danielewski M., Wierzba B., 2010 Acta Mater. **58** 6717
[15] Wierzba B., Danielewski M. 2011 Phil. Mag. **91** 3228
[16] Wierzba B., Danielewski M. 2011 Physica A **390** 2325
[17] Wierzba B. 2012 Physica A **391** 56
[18] Wierzba B., Tkacz-Śmiech K. 2013 Physica A **392** 1100
[19] Wierzba B., Tkacz-Smiech K., Nowotnik A. 2013 CVD **19** 267
[20] Schiesser W.E. 1991 The Numerical Method of Lines, Academic Press, San Diego
[21] Press W. H., Flannery B. P., Teukolsky S. A., Vetterling W. T., Numerical Recipes in C: The Art of Scientific Computing, Cambridge University Press 1992.

Euclidean Closed Linear Transformations of Complex Spacetime and generally of Complex Spaces of dimension four endowed with the Same or Different Metric

Spyridon Vossos[1, 3] and Elias Vossos[2]

[1]Chemist, [2]Physicist, School of Science, National and Kapodistrian University of Athens, Panepistimiopolis, GR15784 Zografos, Athens, Hellas

E-mail: spyrolia@hotmail.com

Keywords: 5th Euclidean postulate, complex space, Einstein equations, Euclidean complex relativistic mechanics, Euclidean metric, general relativity, human senses, isometry, linear spacetime transformation, Lorentz matrix, Lorentz metric, Lorentz transformation, Minkowski space, Schwarzschild metric, spacetime, special relativity, real space, universal speed, velocity, Vossos matrix, Vossos transformation.

PACS: 02.10.Ud, 02.40.Dr, 03.30.+p, 03.50.De, 03.65.-w, 04.20.-q

Abstract. Relativity Theory and the corresponding Relativistic Quantum Mechanics are the fundamental theories of physics. Special Relativity (SR) relates the frames of Relativistic Inertial observers (RIOs), through Linear Spacetime Transformation (LSTT) of linear spacetime. Classic Special Relativity uses real spacetime endowed with Lorentz metric and the frames of two RIOs with parallel spatial axes are always related through Lorentz Boost (LB). This cancels the transitive attribute in parallelism, when three RIOs are related, because LB is not closed transformation, causing Thomas Rotation. In this presentation, we consider closed LSTT of Complex Spacetime, so there is no necessity for spatial axes rotation and all the frames are chosen having parallel spatial axes. The solution is expressed by a 4x4 matrix (Λ) containing components of the complex velocity of one Observer wrt another and two functions depended by the metric of Spacetime. Demanding isometric transformation, it emerges a class of metrics that are in accordance with the closed LSTT and the transformation matrix contains one parameter ω depended by the metric of Spacetime. In case that we relate RIOs with steady metric, it emerges one steady number (ω_I) depended by the metric of Spacetime of the specific SR. If ω_I is an imaginary number, the elements of the Λ are complex numbers, so the corresponding spacetime is necessarily complex and there exists real Universal Speed (U_I). The specific value $\omega_I=\pm i$ gives Vossos transformation (VT) endowed with Lorentz metric (for $g_{ii}=1$) of complex spacetime and invariant spacetime interval (or equivalently invariant speed of light in vacuum), which produce the theory of Euclidean Complex Relativistic Mechanics (ECRMs). If ω_I is a real number ($\omega_I \neq 0$) the elements of the Λ are real numbers, so the corresponding spacetime is real, but there exist imaginary U_I. The specific value $\omega_I=0$ gives Galileo Transformation (GT) with the invariant time, in which any other closed LSTT is reduced, if one RIO has small velocity wrt another RIO. Thus, we have infinite number of closed LSTTs, each one with the corresponding SR theory. In case that we relate accelerated

[3]Spyridon Vossos, 39 Epidavrou Str., GR-15233 Halandri, Athens, Hellas, e-mail: spyrolia@hotmail.com.

Content from this work may be used under the terms of the Creative Commons Attribution 3.0 licence. Any further distribution of this work must maintain attribution to the author(s) and the title of the work, journal citation and DOI.
Published under licence by IOP Publishing Ltd

observers with variable metric of spacetime, we have the case of General Relativity (GR). For being that clear, we produce a generalized Schwarzschild metric, which is in accordance with any SR based on this closed complex LSTT and Einstein equations. The application of this kind of transformations to the SR and GR is obvious. But, the results may be applied to any linear space of dimension four endowed with steady or variable metric, whose elements (four-vectors) have spatial part (vector) with Euclidean metric.

1. Introduction

Relativity Theory (RT) is the foundation stone of modern Physics which combined with Quantum Mechanics (QMs), leads to the Relativistic Quantum Mechanics (RQMs). Moreover, there exist many different approaches of RT, each one with the corresponding RQMs. For instance, *Galileo Transformation* (GT) endowed with the corresponding metric, produces *Newtonian Physics* (NP), which is associated with the classic QMs producing *Schrödinger Equation.* Thus, many low velocities phenomena, like the atomic spectra without fine structure, were explained. On the other hand, *Lorentz Transformation* (LT) endowed with the corresponding metric, produces Classic Special Relativity (CSR), which is associated with the Classic RQMs, producing *Klein-Gordon Equation.* Thus, many high velocities phenomena and the fine structure of atomic spectra were explained [1,2].

In this presentation, we prove that there exist two types of complex Linear Spacetime Transformation (LSTTs) with common solution the GT, which can be applied not only to the Special Relativity (SR), but also to General Relativity (GR), because the production of the corresponding matrices has become without adapting one specific metric. In addition, any complex *Cartesian Coordinates* (CCs) of the theory may be turned to the corresponding real CCs, in order to can be realized through human senses. Moreover, we produce a *generalized Schwarzschild metric,* which is in accordance with any SR based on this closed complex LSTT and Einstein equations. This new modeling of study gives us the capability studying at the same time *Einstein* RT, NP, or any other Theory of Physics that is in accordance with closed Linear Spacetime Transformations. This is achieved because the coefficients of spacetime metric are contained in the transformation matrix. Besides NP is obtained, not only by the low velocity limit, but also by the zero limit of the space coefficient of spacetime metric.

2. The Matrix (Λ) and the Metric (g) of the Closed Linear Transformation

In a 3D complex 'space' endowed with *Euclidean metric*, there exists a frame Oxyz having real CCs. Another real independent variable ('time') and the aforementioned coordinates produce a real four-vector. There exist two types of closed linear 'spacetime' transformation to this real four-vector: **one** with the 'time' being depended by the 'spatial' position where the 'event' happens and **another** with the 'time' being independent by the position. The **first type** has real or imaginary Invariant 'Speed' (U), in contrast to the **second type** which has only infinite U. Moreover, demanding the transformation having **isometry** [3], the **first type** transformation matrix is totally calculated and contains except for the 'velocity' of the frame O$'x'y'z'$, a parameter ω, with

$$\omega^2 = g_{ii}/g_{00} \quad (1)$$

where g_{00} and g_{ii} are the metric's coefficients of time and space respectively. The **second type** is turn to GT**.** Taking the limit $\omega \to 0$ to the **first type,** may emerges GT. So, in case of isometry, the **second type is embedded to the first type** transformation. Besides the **first type** is divided to two cases: **one case** that the 'time' and the 'space' have coefficients of metric of 'spacetime' with different sign, where the transformation leads to complex 3D 'space' with real U and **the other case** that 'time' and 'space' have coefficients of metric with the same sign, where the transformation leads to real 3D 'space' with imaginary U. **Time remains real**, in any case. Below, we present the typical matrix ($\Lambda_{\tau\upsilon\pi}$), the general matrix ($\Lambda_{(\beta)}$), the covariant matrix of spacetime metric (g), the universal speed (U) and the domain of spacetime (C^4) that corresponds to the transformation of a contravariant four-vector:

$$dX' = \Lambda dX \quad (2)$$

$$(A_{\tau\upsilon\pi}, A_{(\beta)}, g, U, \mathrm{C}^4)$$

$$\lambda = \omega\left|\vec{\beta}\right| b = ?$$

$\lambda \neq 0$

$$\Lambda_{\tau\upsilon\pi} = b\begin{bmatrix} 1 & \omega^2\beta & 0 & 0 \\ -\beta & 1 & 0 & 0 \\ 0 & 0 & 1 & \omega\beta \\ 0 & 0 & -\omega\beta & 1 \end{bmatrix},$$

$$\Lambda_{(\beta)} = b\begin{bmatrix} 1 & \omega^2\beta^T \\ -\beta & \mathrm{I}+\omega\,\mathrm{A}_{(\beta)} \end{bmatrix}, U \in \{\Re, I\}.$$

isometry

$$\Lambda_{\tau\upsilon\pi(\omega,\beta)} = \gamma_{(\mathrm{i}\,\omega\beta)}\begin{bmatrix} 1 & \omega^2\beta & 0 & 0 \\ -\beta & 1 & 0 & 0 \\ 0 & 0 & 1 & \omega\beta \\ 0 & 0 & -\omega\beta & 1 \end{bmatrix},$$

$$\Lambda_{(\beta)} = \gamma_{(\mathrm{i}\,\omega\vec{\beta})}\begin{bmatrix} 1 & \omega^2\beta^T \\ -\beta & \mathrm{I}+\omega\,\mathrm{A}_{(\beta)} \end{bmatrix},$$

$$g = g_{ii}\begin{bmatrix} \frac{1}{\omega^2} & 0 & 0 & 0 \\ 0 & 1 & 0 & 0 \\ 0 & 0 & 1 & 0 \\ 0 & 0 & 0 & 1 \end{bmatrix}.$$

$\omega = ?$

$\omega = \zeta\mathrm{i}$

$$\Lambda_{\tau\upsilon\pi(\mathrm{i}\zeta,\beta)} = \gamma_{(\zeta\vec{\beta})}\begin{bmatrix} 1 & -\zeta^2\beta & 0 & 0 \\ -\beta & 1 & 0 & 0 \\ 0 & 0 & 1 & \mathrm{i}\zeta\beta \\ 0 & 0 & -\mathrm{i}\zeta\beta & 1 \end{bmatrix},$$

$$\Lambda_{(\mathrm{i}\zeta,\vec{\beta})} = \gamma_{(\zeta\vec{\beta})}\begin{bmatrix} 1 & -\zeta^2\beta^T \\ -\beta & \mathrm{I}+\mathrm{i}\zeta\,\mathrm{A}_{(\beta)} \end{bmatrix},$$

$$g = g_{ii}\begin{bmatrix} -\frac{1}{\zeta^2} & 0 & 0 & 0 \\ 0 & 1 & 0 & 0 \\ 0 & 0 & 1 & 0 \\ 0 & 0 & 0 & 1 \end{bmatrix}, U = \frac{1}{|\zeta|}\mathrm{c},\ X \in \mathrm{RC}^3.$$

$\omega = \zeta \in \Re$

$$\Lambda_{\tau\upsilon\pi(\zeta,\beta)} = \gamma_{(\mathrm{i}\zeta\beta)}\begin{bmatrix} 1 & \zeta^2\beta & 0 & 0 \\ -\beta & 1 & 0 & 0 \\ 0 & 0 & 1 & \zeta\beta \\ 0 & 0 & -\zeta\beta & 1 \end{bmatrix},$$

$$\Lambda_{(\zeta,\vec{\beta})} = \gamma_{(\mathrm{i}\zeta\vec{\beta})}\begin{bmatrix} 1 & \zeta^2\beta^T \\ -\beta & \mathrm{I}+\zeta\,\mathrm{A}_{(\beta)} \end{bmatrix},$$

$$g = g_{ii}\begin{bmatrix} \frac{1}{\zeta^2} & 0 & 0 & 0 \\ 0 & 1 & 0 & 0 \\ 0 & 0 & 1 & 0 \\ 0 & 0 & 0 & 1 \end{bmatrix}, U \in I,\ X \in \mathrm{R}^4.$$

$\lambda = 0$

$$\Lambda_{\tau\upsilon\pi} = \begin{bmatrix} b & 0 & 0 & 0 \\ -h\beta & h & 0 & 0 \\ 0 & 0 & h & 0 \\ 0 & 0 & 0 & h \end{bmatrix},$$

$$\Lambda_{(\beta)} = \begin{bmatrix} b & \mathrm{O} \\ -h\beta & h\mathrm{I} \end{bmatrix}, U = +\infty.$$

isometry

$$\Lambda_{\Gamma\tau\upsilon\pi(\beta)} = \begin{bmatrix} 1 & 0 & 0 & 0 \\ -\beta & 1 & 0 & 0 \\ 0 & 0 & 1 & 0 \\ 0 & 0 & 0 & 1 \end{bmatrix},$$

$$\Lambda_{\Gamma(\vec{\beta})} = \begin{bmatrix} 1 & \mathrm{O} \\ -\beta & \mathrm{I} \end{bmatrix},$$

$$g_\Gamma = \begin{bmatrix} g_{00} & 0 & 0 & 0 \\ 0 & 0 & 0 & 0 \\ 0 & 0 & 0 & 0 \\ 0 & 0 & 0 & 0 \end{bmatrix}, U = +\infty, \mathrm{R}^4.$$

where

$$\beta = \begin{bmatrix} \beta_x \\ \beta_y \\ \beta_z \end{bmatrix},\ \delta = \begin{bmatrix} \delta_x \\ \delta_y \\ \delta_z \end{bmatrix},\ \mathrm{A}_{(\beta)} = \begin{bmatrix} 0 & \beta_z & -\beta_y \\ -\beta_z & 0 & \beta_x \\ \beta_y & -\beta_x & 0 \end{bmatrix} \tag{3}$$

We observe that

$$\mathrm{A}_{(\beta)}\,\delta = \left[-\vec{\beta}\times\vec{\delta}\right] = \left[\vec{\delta}\times\vec{\beta}\right] \tag{4}$$

Using vectors, the transformation becomes

$$\mathrm{c}\,\mathrm{d}t' = \gamma_{(\mathrm{i}\,\omega\beta)}(\mathrm{c}\,\mathrm{d}t + \omega^2\vec{\beta}\cdot\mathrm{d}\vec{x}),\ \mathrm{d}\vec{x}' = \gamma_{(\mathrm{i}\,\omega\beta)}\left(\mathrm{d}\vec{x} - \vec{\beta}\,\mathrm{c}\,\mathrm{d}t - \omega\vec{\beta}\times\mathrm{d}\vec{x}\right) \tag{5}$$

The norm of the position four-vector for observers **with the same ω** is the corresponding **invariant quantity**:

$$\mathrm{d}S^2 = \mathrm{d}X^T g\,\mathrm{d}X = g_{00}\,\mathrm{c}^2\mathrm{d}t^2 + g_{ii}\,\mathrm{d}\vec{x}^2 = g_{ii}\left[\left(\frac{g_{00}}{g_{ii}}\right)\mathrm{c}^2 t^2 + (1)\,\mathrm{d}\vec{x}^2\right] = g_{ii}\left[\left(\frac{1}{\omega^2}\right)\mathrm{c}^2\mathrm{d}t^2 + (1)\,\mathrm{d}\vec{x}^2\right] \tag{6}$$

Researching for a possible **Invariant Speed** (U) for observers **with the same ω**, it emerges

$$\omega^2 = -\frac{\mathrm{c}^2}{U^2} \tag{7}$$

In case of SR, the matrices form a group of elements

$$d = (\Lambda_{(\omega_1,\vec{\beta})},\, b) \tag{8}$$

with operation

$$d_1 * d_2 = (\Lambda_{(\omega_1,\vec{\beta}_2)}\,\Lambda_{(\omega_1,\vec{\beta}_1)},\, \Lambda_{(\omega_1,\vec{\beta}_2)}\,b_1 + b_2) \tag{9}$$

where b_1^{μ} is μ-coordinate measured by O′, if O measures $x^{\nu} = 0$ and b_2^{μ} is μ- coordinate measured by O′′, if O′ measures $x_{\mathrm{M}}^{\prime\nu} = 0$.

The **Universal Speed** (U_1) for RIOs, is

$$U_1^2 = -\frac{\mathrm{c}^2}{\omega_1^2} \tag{10}$$

3. Time – Proper Time

Let have a particle P, moving with velocity $\vec{\upsilon}_P$ wrt observer O ($\vec{\upsilon}'_P$ wrt observer O′) in a spacetime. The *generalized definition of Proper Time* (PT) is

$$\mathrm{d}\tau^2 = \frac{\mathrm{d}S'^2}{g_{00}\,\mathrm{c}^2} \tag{11}$$

Using (6) and (1), we have

$$\mathrm{d}\tau^2 = \frac{g_{ii}}{g_{00}\,\mathrm{c}^2}\left(\frac{1}{\omega^2}\mathrm{c}^2\mathrm{d}t'^2 + \mathrm{d}\vec{x}'^2\right) = \frac{\omega^2}{\mathrm{c}^2}\left(\frac{1}{\omega^2}\mathrm{c}^2\mathrm{d}t'^2 + \mathrm{d}\vec{x}'^2\right) = \mathrm{d}t'^2 + \frac{\omega^2}{\mathrm{c}^2}\mathrm{d}\vec{x}'^2 = \mathrm{d}t'^2\left(1 + \frac{\omega^2}{\mathrm{c}^2}\vec{\upsilon}_P'^2\right) \tag{12}$$

Thus, the relation between the time and the proper time is

$$\frac{\mathrm{d}t'}{\mathrm{d}\tau} = \gamma'_{(\mathrm{i}\,\omega\vec{\beta}_P)} \tag{13}$$

For GT with $\omega \to 0$, it is $\gamma'_{(i\omega\vec{\beta}_P)} = 1$. So d$\tau$=d$t'$=d$t$.

In case of RIOs, (13) becomes

$$\frac{\mathrm{d}t'}{\mathrm{d}\tau} = \gamma'_{(\mathrm{i}\,\omega_1\vec{\beta}_P)} \tag{14}$$

So, for *Vossos Transformation* (VT) with $\omega_I=\pm i$, it emerges $\gamma'_{(i\omega_I\vec{\beta}_P)}=\gamma'_{(\vec{\beta}_P)}$. Thus, we have the same result as CSR [4].
In any case using PT, we can define four-velocity, four-momentum etc building the whole structure of SR and GR.

4. *Generalized Schwarzschild metric* in accordance with closed complex Linear Spacetime Transformation

For being clear the meaning of this theory, we apply it to the unique solution of *Einstein equations* in vacuum with spherical symmetry that is *Schwarzschild metric,* according to *Birkoff's theorem* [5]. Thus, it emerges a *generalized Schwarzschild metric* that is in accordance with any acceptable SR having real, imaginary or infinite universal speed. We define the relativistic potential Φ around a center of gravity as

$$\Phi=\frac{U_I^{\,2}}{2}\ln\left(1-\frac{r_{SI}}{r}\right) \tag{15}$$

where r is the distance between the center of the gravity and the spatial position of the event and r_{SI} is the *generalized Schwarzschild radius*

$$r_{SI}=\frac{2GM}{U_I^{\,2}} \tag{16}$$

The definition of *Schwarzschild radius* is

$$r_S=\frac{2GM}{c^2} \tag{17}$$

Thus (16) combined with (17) and (10) gives

$$r_{SI}=-\omega_I^{\,2}r_S \tag{18}$$

and (15) becomes

$$\Phi=\frac{-c^2}{2\omega_I^{\,2}}\ln\left(1+\frac{\omega_I^{\,2}r_S}{r}\right)=-\frac{c^2}{2}\frac{r_S}{r}+\ldots=-\frac{GM}{r}+\ldots \tag{19}$$

We observe that if $\omega_I=\pm i$ (VT), it emerges the original *Schwarzschild potential.* Moreover, if $\omega_I\to 0$ (GT), we compute

$$\lim_{\omega_I\to 0}\Phi=\frac{-c^2}{2}\lim_{\omega_I\to 0}\left[\frac{1}{\omega_I^{\,2}}\ln\left(1+\frac{\omega_I^{\,2}r_S}{r}\right)\right]=\frac{-c^2}{4}\lim_{\omega_I\to 0}\left[\frac{1}{\omega_I}\frac{\frac{2\omega_I r_S}{r}}{1+\frac{\omega_I^{\,2}r_S}{r}}\right]=-\frac{c^2}{2}\frac{r_S}{r}=-\frac{GM}{r} \tag{20}$$

In figure 1, we show the parametric plot of the relativistic potential Φ wrt r/r_s, for different values of ω_I. We observe that for $\omega_I\to\pm\infty$, the relativistic potential becomes zero–function.
Now, we examine the case that $g_{00I}\leq 0$ and $g_{iiI}\geq 0$, it is $g_{rrI}\geq 0$, too. Besides $\omega=\zeta i$, with $\zeta\in\Re$.
The *generalized Schwarzschild metric* is

$$dS^2=dX^{T}g\,dX=\begin{bmatrix}cdt & dr & d\theta & d\phi\end{bmatrix}\cdot\begin{bmatrix}-(-g_{00I})e^{A} & 0 & 0 & 0\\ 0 & g_{rrI}e^{B} & 0 & 0\\ 0 & 0 & r^2 & 0\\ 0 & 0 & 0 & r^2\sin^2\theta\end{bmatrix}\cdot\begin{bmatrix}cdt\\ dr\\ d\theta\\ d\phi\end{bmatrix} \tag{21}$$

where

$$A=\frac{2}{U_I^{\,2}}\Phi \tag{22}$$

This combined with *Einstein equations* in vacuum gives

$$\mathrm{d}S^2 = g_{00\mathrm{I}}\left(1-\frac{r_{\mathrm{SI}}}{r}\right)\mathrm{c}^2\mathrm{d}t^2 - \frac{1}{g_{00\mathrm{I}}\left(1-\frac{r_{\mathrm{SI}}}{r}\right)}\mathrm{d}r^2 + r^2\,\mathrm{d}\theta^2 + r^2\sin^2\theta\,\mathrm{d}\phi^2 \tag{23}$$

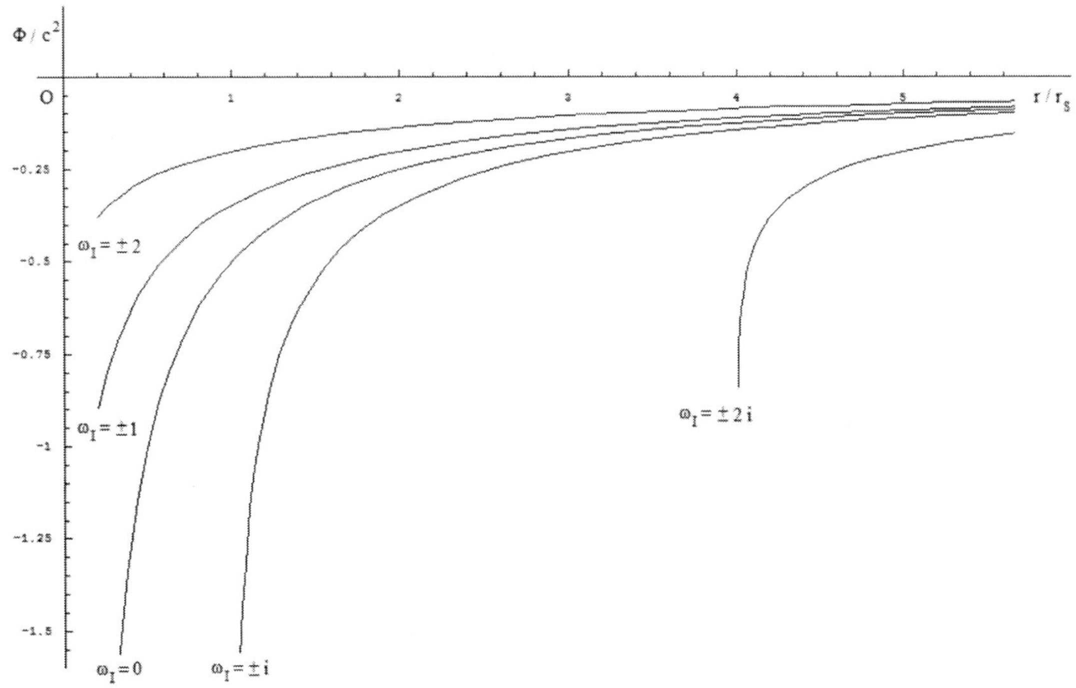

Figure 1. Parametric plot of the relativistic potential Φ wrt r/r_s, for different values of ω_I.

The *isotropic form* of the *generalized Schwarzschild metric* is

$$\mathrm{d}S^2 = g_{00}\,\mathrm{c}^2\mathrm{d}t^2 + g_{ii}\left(\mathrm{d}\tilde{r}^2 + \tilde{r}^2\,\mathrm{d}\theta^2 + \tilde{r}^2\sin^2\theta\,\mathrm{d}\phi^2\right) \tag{24}$$

with

$$\mathrm{d}\tilde{x}^2 = \mathrm{d}\tilde{r}^2 + \tilde{r}^2\,\mathrm{d}\theta^2 + \tilde{r}^2\sin^2\phi\,\mathrm{d}\phi^2 \tag{25}$$

$$\tilde{x} = \tilde{r}\sin\theta\cos\phi\,,\ \tilde{y} = \tilde{r}\sin\theta\sin\phi\,,\ \tilde{z} = \tilde{r}\cos\theta \tag{26}$$

$$\tilde{r} = \frac{|\zeta_{\mathrm{I}}| r_{\mathrm{S}}}{4}\left(\frac{2r}{|\zeta_{\mathrm{I}}|^2 r_{\mathrm{S}}}\left(1+\sqrt{1-|\zeta_{\mathrm{I}}|^2\frac{r_{\mathrm{S}}}{r}}\right)-1\right)^{\frac{1}{\sqrt{-g_{00\mathrm{I}}}}} \tag{27}$$

$$r = \frac{|\zeta_{\mathrm{I}}|^2 r_{\mathrm{S}}}{4}\left[1+\left(\frac{|\zeta_{\mathrm{I}}| r_{\mathrm{S}}}{4\tilde{r}}\right)^{\sqrt{-g_{00\mathrm{I}}}}\right]^2\left(\frac{4\tilde{r}}{|\zeta_{\mathrm{I}}| r_{\mathrm{S}}}\right)^{\sqrt{-g_{00\mathrm{I}}}} \tag{28}$$

$$g_{00} = g_{00\mathrm{I}}\left(\frac{1-\left(\frac{|\zeta_{\mathrm{I}}| r_{\mathrm{S}}}{4\tilde{r}}\right)^{\sqrt{-g_{00\mathrm{I}}}}}{1+\left(\frac{|\zeta_{\mathrm{I}}| r_{\mathrm{S}}}{4\tilde{r}}\right)^{\sqrt{-g_{00\mathrm{I}}}}}\right)^2 \tag{29}$$

$$\tilde{g}_{ii} = \frac{|\zeta_1|^4 r_S^2}{16\tilde{r}^2}\left(1+\left(\frac{|\zeta_1| r_S}{4\tilde{r}}\right)^{\sqrt{-g_{00\mathrm{I}}}}\right)^4 \left(\frac{4\tilde{r}}{|\zeta_1| r_S}\right)^{2\sqrt{-g_{00\mathrm{I}}}} \tag{30}$$

$$|\zeta| = \frac{4^{\sqrt{-g_{00\mathrm{I}}}-1}|\zeta_1|^{2-\sqrt{-g_{00\mathrm{I}}}}\left(1+\left(\frac{|\zeta_1| r_S}{4\tilde{r}}\right)^{\sqrt{-g_{00\mathrm{I}}}}\right)^3 \left(\frac{\tilde{r}}{r_S}\right)^{\sqrt{-g_{00\mathrm{I}}}-1}}{\sqrt{-g_{00\mathrm{I}}}\left|1-\left(\frac{|\zeta_1| r_S}{4\tilde{r}}\right)^{\sqrt{-g_{00\mathrm{I}}}}\right|} \tag{31}$$

Besides

$$g_{ii\mathrm{I}} = \begin{cases} 0 \ , & -1 \prec g_{00\mathrm{I}} \le 0 \\ |\zeta_1|^2 \ , & g_{00\mathrm{I}} = -1 \\ +\infty \ , & g_{00\mathrm{I}} \prec -1 \end{cases} \tag{32}$$

The universal speed is

$$U = \frac{1}{|\zeta|}\mathrm{c} \tag{33}$$

so, the relative change of U wrt $U_{(+\infty)}$ is

$$\Delta U_r = \frac{U_{(r)} - U_{(+\infty)}}{U_{(+\infty)}} = \frac{|\zeta_1|}{|\zeta|} - 1 \tag{34}$$

The transformation inside the infinitesimal area around the surface of a sphere with center the same as the center of the gravity and radius r is

$$\begin{bmatrix} \mathrm{c}\,\mathrm{d}t' \\ \mathrm{d}\tilde{x}' \\ \mathrm{d}\tilde{y}' \\ \mathrm{d}\tilde{z}' \end{bmatrix} = \gamma_{(\zeta\vec{\beta})} \begin{bmatrix} 1 & -\zeta^2\beta_x & \zeta^2\beta_y & \zeta^2\beta_z \\ -\beta_x & 1 & \zeta\beta_z & -\zeta\beta_y \\ -\beta_y & -\zeta\beta_z & 1 & \zeta\beta_x \\ -\beta_z & \zeta\beta_y & -\zeta\beta_x & 1 \end{bmatrix} \cdot \begin{bmatrix} \mathrm{c}\,\mathrm{d}t \\ \mathrm{d}\tilde{x} \\ \mathrm{d}\tilde{y} \\ \mathrm{d}\tilde{z} \end{bmatrix} \tag{35}$$

In case of *Lorentz metric* ($g_{00\mathrm{I}} = -1$ and $g_{ii\mathrm{I}} = 1$), it is $\boldsymbol{\zeta_1 = \pm 1}$ and we have the SR which corresponds to VT with the metric of spacetime

$$\tilde{g}_{00} = -\left(\frac{1-\frac{r_S}{4\tilde{r}}}{1+\frac{r_S}{4\tilde{r}}}\right)^2 \ , \ \tilde{g}_{ii} = \left(1+\frac{r_S}{4\tilde{r}}\right)^4 \tag{36}$$

so, it emerges the original *isotropic form* of *Schwarzschild metric* [6] with

$$|\zeta| = \frac{\left(1+\frac{r_S}{4\tilde{r}}\right)^3}{\left|1-\frac{r_S}{4\tilde{r}}\right|} \tag{37}$$

and

$$\tilde{r} = \frac{r_S}{4}\left(\frac{2r}{r_S}\left(1+\sqrt{1-\frac{r_S}{r}}\right)-1\right) \tag{38}$$

5. Experimental

Appling the above results of the original *isotropic form* of *Schwarzschild metric*:
(i) to the **Earth** with mass M=5.9742×10^{24} kg, on the sea level with r=6378140 m and taking into account that G=6.67428(67)×10^{-11}m^{3}kg^{-1}s^{-2} and c=299792458 ms^{-1} (exact) [7], it emerges r_S=8.873578523×10^{-3}m≈0.88cm. Thus r_s/r=1.391248628×10^{-9}, $\tilde{r}$ =6378139.995 m, ζ=±1.000000001, U=299792457.6 m s^{-1} and ΔU_r=-1.36×10^{-9} and
(ii) to the **Sun**, with mass is M=1.9891×10^{30} kg, on the surface with radius is r=6.9599×10^8 m, r_S=2954.443279 m, we have r_s/r=4.244950759×10^{-6} m, $\tilde{r}$ =6.959885228×10^8 m, ζ=±1.000004245, U=299791185.4 m s^{-1} and ΔU_r=-4.244912×10^{-6}.

References

[1] Kleppner D 1999 A short history of atomic physics in the twentieth century *Rev. Mod. Phys.* **71** (2) S78
http://www.cstam.org.cn/Upfiles/200732678933.pdf
[2] Thomas L H 1926 The Motion of the Spinning Electron *Nature (London)* **117** 514
[3] Tsamparlis M 2010 *Special relativity: An introduction with 200 problems and solutions* (Berlin Heidelberg: Springer-Verlag) p 10, eq 1.15, p 240
[ISBN: 978-3-642-03836-5, e-ISBN: 978-3-642-03837-2]
[4] Vossos S and Vossos E 2015 Euclidean Complex Relativistic Mechanics: A New Special Relativity Theory *J. Phys.: Conf. Ser.* **633** 012027
http://dx.doi.org/10.1088/1742-6596/633/1/012027
[5] Rindler W 2006 *Relativity: Special, General and Cosmological* (New York: Oxford University Press) p 228-231
[ISBN: 978-0-19-856732-5]
[6] Rindler W 2006 *Relativity: Special, General and Cosmological* (New York: Oxford University Press) p 237, eq 11.26
[ISBN: 978-0-19-856732-5]
[7] Lide D R ed 2009 *CRC Handbook of Chemistry and Physics* 89th Edition (Internet Version) (FL: CRC Press/Taylor and Francis, Boca Raton) pp 1-1,14-2

5th International Conference on Mathematical Modeling in Physical Sciences (IC-MSquare 2016) IOP Publishing
Journal of Physics: Conference Series **738** (2016) 012049 doi:10.1088/1742-6596/738/1/012049

Smallest number of incident directions for topological derivative imaging: a numerical study

Won-Kwang Park

Department of Mathematics, Kookmin University, Seoul, 136-702, Korea

E-mail: parkwk@kookmin.ac.kr

Abstract. Various studies have confirmed that only a small number of incident field directions are needed to apply the topological derivative in inverse scattering problem. In this paper, we consider a topological derivative-based technique with a small number of such directions for imaging thin, curve-like dielectric inhomogeneities embedded in a homogeneous domain, and explore a suitable condition via various numerical simulation results.

1. Introduction

Identifying the shape of unknown crack-like thin defects in homogeneous domains is an interesting and remarkable inverse scattering problem. For this purpose, various iterative and non-iterative algorithms have been developed. From various studies [1, 2, 3], it has emerged that iterative-based methods require the derivation of a complex Fréchet derivative, *a priori* information of the unknown target, considerable computational resources, and a good initial guess. In contrast, non-iterative techniques have no such requirements, although a large number of incident fields with various directions and corresponding scattered field data are essential [4, 5, 6, 7].

The topological derivative is a non-iterative imaging technique for which, unlike various other non-iterative techniques, good results can be achieved even with a small number of incident field directions [8, 9]. However, the fewest such directions that give acceptable results are not known. Motivated by this, we aim to identify the smallest number through numerical simulations.

This paper is organized as follows. In Section 2, we introduce the direct scattering problem and the topological derivative. In Section 3, we present various numerical results and explore the smallest number of incident directions. Our conclusions are given in Section 4.

2. Direct scattering problem and topological derivative

Let $\Omega \subset \mathbb{R}^2$ be a homogeneous domain with a smooth boundary $\partial\Omega$. In this paper, we assume that a homogeneous thin dielectric inhomogeneity Γ of small thickness $2h$ is completely hidden in Ω. For this, we set Γ to be of the form:

$$\Gamma = \{\mathbf{x} + \eta\mathbf{n}(\mathbf{x}) : \mathbf{x} \in \sigma, -h \leq \eta \leq h\},$$

where σ is a simple smooth curve that describes the supporting curve of Γ and $\mathbf{n}(\mathbf{x})$ is the unit normal to σ at $\mathbf{x}$. In this paper, $\omega = 2\pi/\lambda$ is a fixed angular frequency, where λ is a given wavelength such that $h \ll \lambda$, and all materials are characterized by their dielectric permittivity

Content from this work may be used under the terms of the Creative Commons Attribution 3.0 licence. Any further distribution of this work must maintain attribution to the author(s) and the title of the work, journal citation and DOI.
Published under licence by IOP Publishing Ltd

at ω. For simplicity, let ε_0 and $\varepsilon_\star$ be the permittivities of Ω and Γ, respectively. We can then define the piecewise constant permittivity $\varepsilon(\mathbf{x})$ as

$$\varepsilon(\mathbf{x}) := \begin{cases} \varepsilon_0 & \text{for} \quad \mathbf{x} \in \Omega \backslash \overline{\Gamma} \\ \varepsilon_\star & \text{for} \quad \mathbf{x} \in \Gamma \end{cases}$$

Let $u^{(n)}(\mathbf{x};\omega)$ be the time-harmonic total field that satisfies the Helmholtz equation

$$\begin{cases} \Delta u^{(n)}(\mathbf{x};\omega) + \omega^2 \varepsilon(\mathbf{x}) u^{(n)}(\mathbf{x};\omega) = 0 & \text{for} \quad \mathbf{x} \in \Omega, \\ \nabla u^{(n)}(\mathbf{x};\omega) \cdot \boldsymbol{\nu}(\mathbf{x}) = \nabla e^{i\omega \boldsymbol{\theta}_n \cdot \mathbf{x}} \cdot \boldsymbol{\nu}(\mathbf{x}) & \text{for} \quad \mathbf{x} \in \partial\Omega, \end{cases} \tag{1}$$

with transmission conditions on the boundary of Γ. Here, $\boldsymbol{\nu}(\mathbf{x})$ denotes the unit normal vector at $\mathbf{x} \in \partial\Omega$, and $\boldsymbol{\theta}_n$ denotes a vector on the unit circle $\mathbb{S}^1$ such that

$$\boldsymbol{\theta}_n = [\cos(\theta_n), \sin(\theta_n)]^T = \left[\cos\left(\frac{2\pi(n-1)}{N-1}\right), \sin\left(\frac{2\pi(n-1)}{N-1}\right)\right]^T.$$

In the same manner, let $u_{\mathrm{B}}^{(n)}(\mathbf{x};\omega) = e^{i\omega\boldsymbol{\theta}_n \cdot \mathbf{x}}$ be the solution of (1) without Γ.

To introduce the topological derivative, let us consider the following energy functional:

$$\mathcal{E}(\Omega;\omega) := \frac{1}{2}\sum_{n=1}^{N} \int_{\partial\Omega} |u^{(n)}(\mathbf{x};\omega) - u_{\mathrm{B}}^{(n)}(\mathbf{x};\omega)|^2 dS(\mathbf{x}). \tag{2}$$

The topological derivative measures the influence of creating a small inclusion at a certain point inside Ω. In this paper, we create a dielectric inhomogeneity, say Σ, of small diameter r at $\mathbf{z} \in \Omega \backslash \partial\Omega$, and denote the corresponding domain as $\Omega|\Sigma$. Then, based on [9], the topological derivative $d_T\mathcal{E}(\mathbf{z};\omega)$ satisfies the following asymptotic formula:

$$\mathcal{E}(\Omega|\Sigma;\omega) = \mathcal{E}(\Omega;\omega) + \varphi(r;\omega) d_T\mathcal{E}(\mathbf{z};\omega) + o(\varphi(r;\omega)),$$

and can be represented as

$$d_T\mathcal{E}(\mathbf{z};\omega) = \mathrm{Re}\sum_{n=1}^{N}\left(u_{\mathrm{A}}^{(n)}(\mathbf{z};\omega)\overline{u_{\mathrm{B}}^{(n)}(\mathbf{z};\omega)}\right),$$

where $u_{\mathrm{A}}^{(n)}(\mathbf{x};\omega)$ satisfies the following adjoint problem:

$$\begin{cases} \Delta u_{\mathrm{A}}^{(n)}(\mathbf{x};\omega) + \omega^2 u_{\mathrm{A}}^{(n)}(\mathbf{x};\omega) = 0 & \text{for} \quad \mathbf{x} \in \Omega, \\ \nabla u_{\mathrm{A}}^{(n)}(\mathbf{x};\omega) \cdot \boldsymbol{\nu}(\mathbf{x}) = u^{(n)}(\mathbf{x};\omega) - u_{\mathrm{B}}^{(n)}(\mathbf{x};\omega) & \text{for} \quad \mathbf{x} \in \partial\Omega. \end{cases} \tag{3}$$

The points at which $d_T\mathcal{E}(\mathbf{z};\omega)$ attains its maximum positive value are expected to be on σ. Hence, the shape of Γ can be identified via the map of $d_T\mathcal{E}(\mathbf{z};\omega)$ [10]. The shape of Γ is obtained in only one iteration. Hence, this is faster than the traditional iterative method.

Journal of Physics: Conference Series **738** (2016) 012049 doi:10.1088/1742-6596/738/1/012049

3. Simulation results and smallest number of incident directions

In this section, the results of numerical simulations with small N are presented. Before starting, it is worth mentioning that the application of multi-frequencies guarantees better imaging performance than the application of a single frequency (see [8] for a detailed discussion). Hence, we consider the following normalized multi-frequency topological derivative:

$$\mathcal{I}_{\mathrm{TD}}(\mathbf{z}; K) := \frac{1}{K}\sum_{k=1}^{K} \frac{d_T\mathcal{E}(\mathbf{z};\omega_f)}{\max[d_T\mathcal{E}(\mathbf{z};\omega_f)]}, \quad k = 1, 2, \cdots, K. \tag{4}$$

To perform the numerical simulations, the homogeneous domain Ω is chosen as a unit circle centered at the origin, and two σ_j specifying the thin inclusions Γ_j are chosen as

$$\sigma_1 = \left\{[s - 0.2, -0.5s^2 + 0.5]^T : -0.5 \le s \le 0.5\right\}$$
$$\sigma_2 = \left\{[s + 0.2, s^3 + s^2 - 0.6]^T : -0.5 \le s \le 0.5\right\}.$$

The thickness h of every Γ_j is set to 0.02, ε_0 is chosen as 1, and $K = 10$ different frequencies with $\omega_1 = 2\pi/0.7$ and $\omega_{10} = 2\pi/0.3$ are applied. Let ε_j denote the permittivity of Γ_j, $j = 1, 2$, and set $\varepsilon_j = 5$. In all of our results, a white Gaussian noise with 15 dB signal-to-noise ratio (SNR) has been added to the unperturbed data $u^{(n)}(\mathbf{x};\omega_k)$.

Figure 1 exhibits the maps of $\mathcal{I}_{\mathrm{TD}}(\mathbf{z}; 10)$ with $N = 1, 2, \cdots, 6$ when the inhomogeneity is Γ_1. Based on the results, we can clearly observe that, when N is a small odd number, it is very hard to identify the shape of Γ_1. Note that it is still impossible to recognize the shape of Γ_1 when $N = 2$, but very good results can be obtained when N is an even number greater than 4. A similar phenomenon can be seen in Figure 2 when the inhomogeneity is Γ_2.

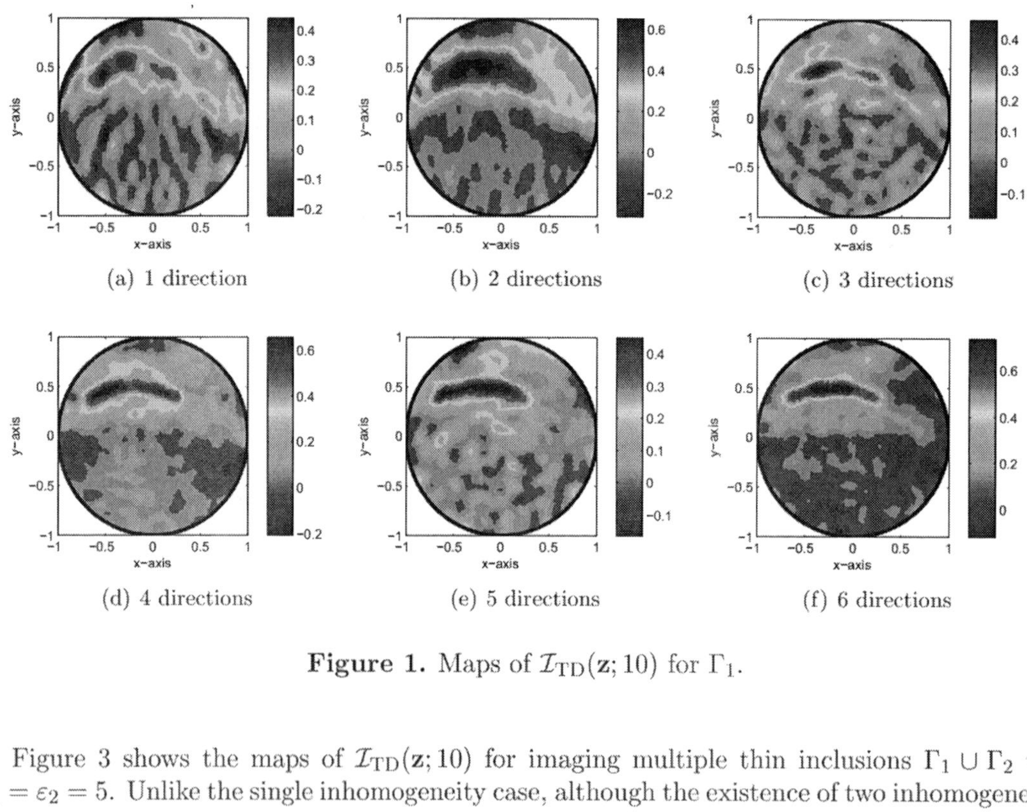

(a) 1 direction (b) 2 directions (c) 3 directions

(d) 4 directions (e) 5 directions (f) 6 directions

Figure 1. Maps of $\mathcal{I}_{\mathrm{TD}}(\mathbf{z}; 10)$ for Γ_1.

Figure 3 shows the maps of $\mathcal{I}_{\mathrm{TD}}(\mathbf{z}; 10)$ for imaging multiple thin inclusions $\Gamma_1 \cup \Gamma_2$ with $\varepsilon_1 = \varepsilon_2 = 5$. Unlike the single inhomogeneity case, although the existence of two inhomogeneities

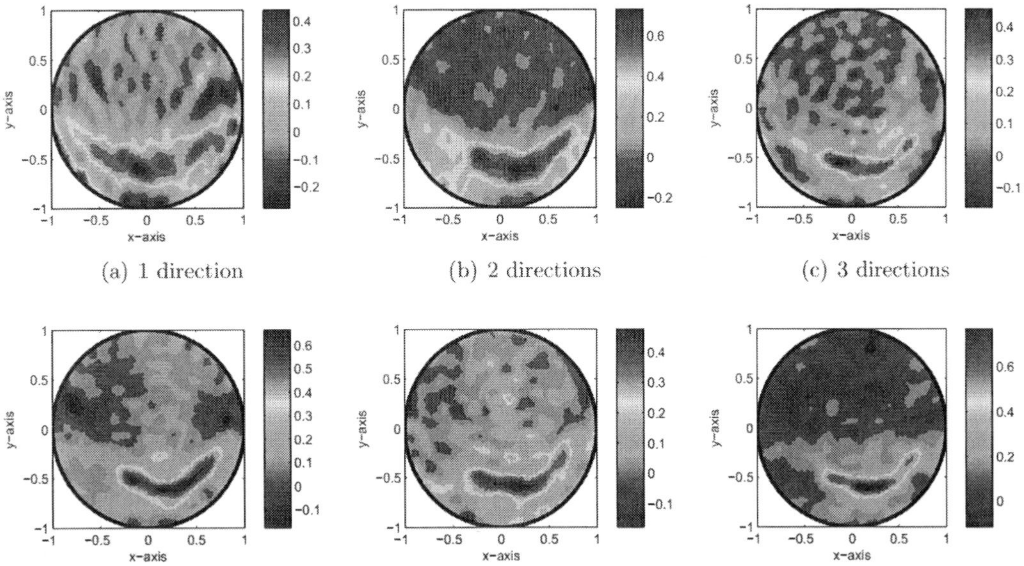

(a) 1 direction (b) 2 directions (c) 3 directions

(d) 4 directions (e) 5 directions (f) 6 directions

Figure 2. Maps of $\mathcal{I}_{\mathrm{TD}}(\mathbf{z}; 10)$ for Γ_2.

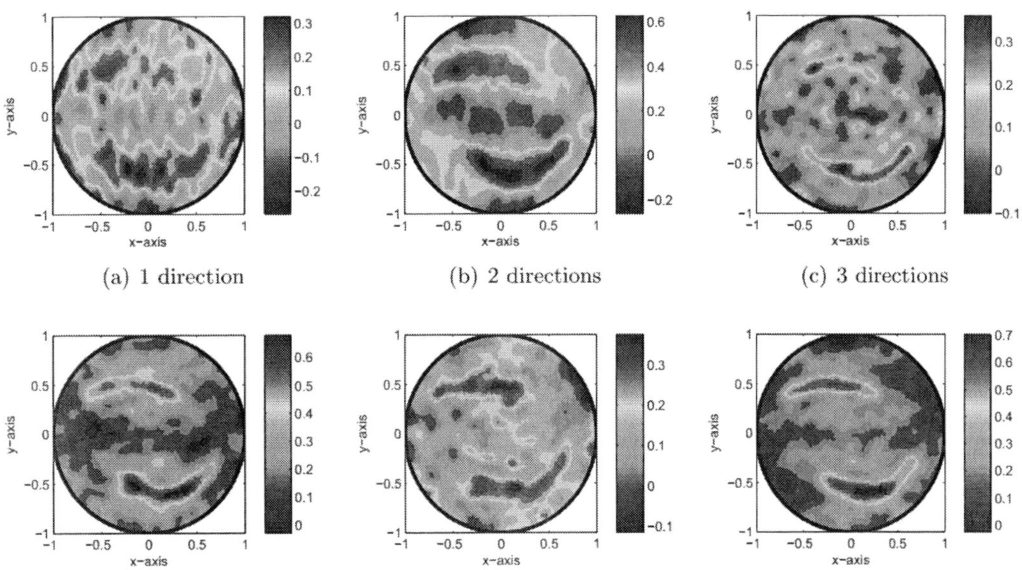

(a) 1 direction (b) 2 directions (c) 3 directions

(d) 4 directions (e) 5 directions (f) 6 directions

Figure 3. Maps of $\mathcal{I}_{\mathrm{TD}}(\mathbf{z}; 10)$ for $\Gamma_1 \cup \Gamma_2$ with same permittivity.

can be recognized, it is hard to identify their shape. However, when N is an even number greater than 4, an outline of the inhomogeneities is retrieved.

For the final example, let us consider the imaging of $\Gamma_1 \cup \Gamma_2$ under the same configuration as the previous example except that $\varepsilon_1 = 10$ and $\varepsilon_2 = 5$. It is interesting to observe that, although the shape of Γ_2 cannot be retrieved, the shape of Γ_1 retrieved by our method is close to the true one. Similar to the previous example, we can observe that the shape of Γ_2 can be obtained when N is an even number greater than 4.

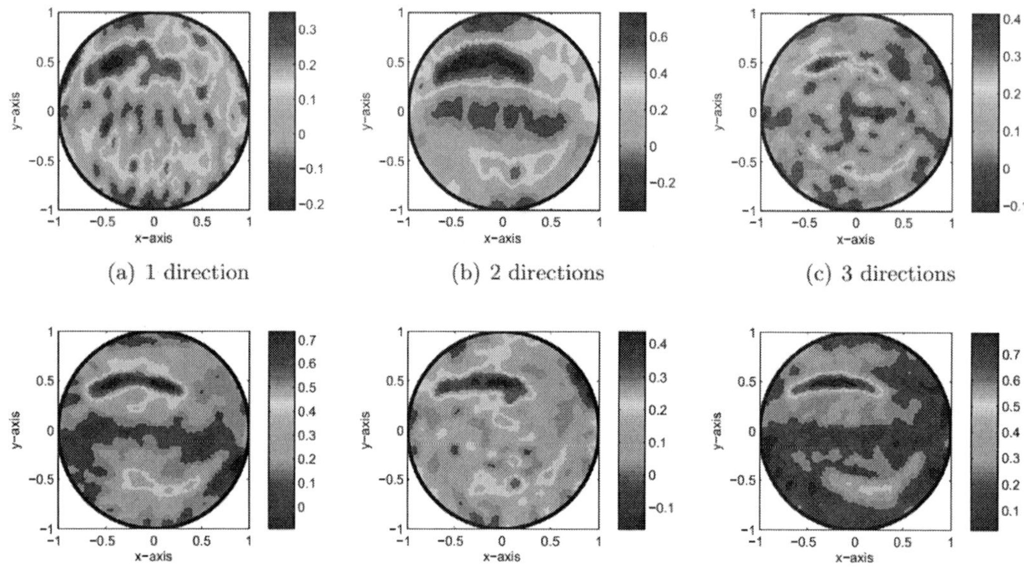

(a) 1 direction (b) 2 directions (c) 3 directions

(d) 4 directions (e) 5 directions (f) 6 directions

Figure 4. Maps of $\mathcal{I}_{\mathrm{TD}}(\mathbf{z}; 10)$ for $\Gamma_1 \cup \Gamma_2$ with different permittivities.

4. Concluding remark

We considered the applicability of the topological derivative with a small number of fields of incident directions for the one-step iterative imaging of thin dielectric inhomogeneities in a homogeneous domain. From various results for single and multiple inhomogeneities, we can conclude that, unlike many non-iterative techniques, the topological derivative requires a small number of incident directions, and this must be an even number greater than 4. As this finding is somewhat heuristic, we intend to develop the underlying mathematical theory in future work.

Acknowledgments

This research was supported by the Basic Science Research Program through the National Research Foundation of Korea (NRF) funded by the Ministry of Education (No. NRF-2014R1A1A2055225) and the research program of Kookmin University in Korea.

References

[1] Carpio A and Rapun M L 2008 *Inverse Problems* **24** 045014
[2] Dorn O and Lesselier D 2006 *Inverse Problems* **22** R67–R131
[3] Kress R 1995 *Math. Meth. Appl. Sci.* **18** 267–293
[4] Ammari H, Iakovleva E and Lesselier D 2005 *Multiscale Model. Simul.* **3** 597–628
[5] Ammari H, Iakovleva E and Moskow S 2003 *SIAM J. Math. Anal.* **34** 882–900
[6] Park W K 2015 *SIAM J. Appl. Math.* **75** 209–228
[7] Kirsch A and Ritter S 2000 *Inverse Problems* **16** 89–105
[8] Park W K 2013 *J. Math. Anal. Appl.* **404** 501–518
[9] Park W K 2012 *J. Comput. Phys.* **231** 1426–1439
[10] Ammari H, Kang H, Lee H and Park W K 2010 *SIAM J. Sci. Comput.* **32** 894–922

Anand constitutive model of lead-free solder joints in 3D IC device

Liang Zhang[1,2], Zhi-quan Liu[2], Yu-tong Ji[3]

[1] School of Mechanical and Electrical Engineering, Jiangsu Normal University, Xuzhou 221116, China
[2] Institute of Metal Research, Chinese Academy of Sciences, Shenyang110016, China
[3] School of Information and Art, Jiangsu Vocational Institute of Architectural Technology, Xuzhou 221116, China

Email:zhangliang@jsnu.edu.cn

Abstract. Anand constitutive relation of SnAgCu and SnAgCu-nano Al solders were studied under uniaxial tension, and the constitutive model was used in the finite element simulation to analyze the stress-strain response of lead-free solder joints in 3D IC devices. The results showed that the nine parameters of the Anand model can be determined from separated constitutive relations and experimental results. Based on Anand model, the finite element method was selected to calculate the stress-strain response of lead-free solder joints, it was found that in the 3D IC device the maximum stress-strain concentrated in the concern solder joints, the stress-strain of SnAgCu-nano Al solder joints was lower than that of SnAgCu solder joints, which represented that the addition of nano Al particles can enhance the reliability of lead-free solder joints in 3D IC devices.

1. Introduction

Since the implementation of the Restriction of Hazardous Substances (RoHS) directives nearly a decade ago, there have been new developments and several improvements in lead-free solders to replace the conventional SnPb solders for using in the electronics industry[1,2]. Among serial lead-free solders investigated, SnAg/SnAgCu base solders with excellent properties were proposed as the most promising substitute[3,4]. However, the low creep resistance[5] and brittle bulk IMC[6] maybe two important drawback of these solder in service for high-density devices.

For 3D IC structure, three-dimensional (3D) stacking of integrated circuits (ICs) possesses unique advantages, which include the enhancement of device density per volume, utilization of short vertical interconnection for improved electrical performance, and the capability of integrating multiple functions into a single package[7]. In Ni/SnAg/Ni 3D IC application, it is found that the optimal weight percentage of Ag is between 2.4 and 3.5 wt.% to inhibit the formation of microvoids [8]. Moreover, the Sn, SnAgCu also were reported as interconnection materials in 3D IC, which can determine the reliability of electronic devices in service. Due to large difference in the coefficient of thermal expansion (CTE)[9] of the different constituents in the 3D assembly, thermal stress were created in solder joints during thermal process. Therefore, the reliability of solder joints in 3D IC should be studied further in service.

In this paper, SnAgCu and SnAgCu bearing 0.1% Al nanoparticles were selected to be used in 3D IC, the Anand constitutive model of these solders was obtained by tensile tests, and the model was

Content from this work may be used under the terms of the Creative Commons Attribution 3.0 licence. Any further distribution of this work must maintain attribution to the author(s) and the title of the work, journal citation and DOI.
Published under licence by IOP Publishing Ltd

incorporated into finite element simulation to analyze the stress-strain response of solder joints in service, which can be utilized to evaluated the reliability of 3D IC device.

2. Anand model

2.1. Experimental procedure

SnAgCu solders bearing Al nano-particles were prepared via mechanically incorporating 0.1% of about 20-25nm Al nanoparticles into Sn3.8Ag0.7Cu solder paste for about 15 minutes to promote uniform particle distribution. Then the paste with no-clean flux was printed on Cu substrates, then the samples were heated in an industrial reflow oven, the molten alloys were chill cast ingots in a ceramics mold. Then they were solidified by nominally air-cooling. All solder specimens were heat treated at 125℃ for an hour to stabilize the microstructure of the SnAgCu and SnAgCu-nanoAl solder alloys. The dog-shaped bulk solder specimens for the uniaxial tensile test were prepared, by machining from extruded SnAgCu and SnAgCu-nanoAl solder bar. The solder specimens have a total length of 65mm, a gage length of 15mm, and a diameter of 3mm.

A series of uniaxial tensile tests of the SnAgCu and SnAgCu-nanoAl solders were performed with a number of different strain rates and temperature conditions. The tests were conducted at tensile strain rates ranging from $1.0 \times 10^{-4}\ s^{-1}$ to $1.0 \times 10^{-2}\ s^{-1}$, and at temperatures of -55℃, -25 ℃,25 ℃, 75 ℃, and 125 ℃. True strain control mode was employed and true strain rate was maintained constant during the tests until specimen failure occurred.

2.2. Anand equations

Anand and Brown have established the constitutive equations for large, isotropic, viscoplastic deformations but small elastic deformations, and the equation is the single-scalar internal variable model [10,11]. Within the frame work of this model, there appears a scalar valued function for the equivalent plastic strain rate and an evolution equation for the internal variable[12]. Anand model incorporates viscoplastic, time-dependent plastic phenomenon where the development of plastic strain is dependent on the rate of loading, viscoplasticity is defined by unifying plastic and creep deformations[13]. In this constitutive model, two basic features can be found obviously[14,15]. Firstly, this model need no explicit yield condition and no loading/unloading criterion, the plastic strain is assumed to take place at all nonzero stress values, although at low stresses the rate of plastic flow may be immeasurably small; secondly, this model employs a single scalar as an internal variable to represent the isotropic resistance to plastic flow offered by the internal state of the material, the single scalar variable s represents an averaged isotropic resistance to macroscopic plastic flow offered by the underlying isotropic strengthening mechanisms.

The internal variable *s* in the model characterizes isotropic hardening mechanism such as dislocation density, solid solution hardening, and subgrain and grain size effects, etc., and the relationship between deformation resistance and the equivalent stress can be represented as[16],

$$\sigma^* = cs \tag{1}$$

where c is a function of strain rate and temperature as expressed in the following equation, s is the deformation resistance with the dimensions of stress, σ^* is the equivalent stress for the steady plastic flow. Moreover, c is defined as follows:

$$c = \frac{1}{\xi}\sinh^{-1}\left[\left(\frac{\dot{\varepsilon}_p}{A}\exp\left(Q/RT\right)\right)^{m}\right]$$

(2)

With the combination of equations (1) and (2), the flow equation of the Anand model was constructed to exactly accommodate the strain rate dependence on the stress at constant structure [17]:

$$\dot{\varepsilon}_p = A\exp\left(-\frac{Q}{RT}\right)\left[\sinh\left(\xi\frac{\sigma}{s}\right)\right]^{1/m} \tag{3}$$

where $\dot{\varepsilon}_p$ is the inelastic strain rate, A is pre-exponential factor, Q is the activation energy, R is universal gas constant, T is the temperature, ξ stands for the materials constant, m is the strain rate sensitivity.

Moreover, the evolution equation for the internal variable s is assumed to be the form as

$$\dot{s} = h(\sigma, s, T)\dot{\varepsilon}_p \tag{4}$$

where

$$h = h_0\left|1-\frac{s}{s^*}\right|^a sign\left(1-\frac{s}{s^*}\right) \quad a \ge 1 \tag{5}$$

Therefore, the evolution equation for the internal variable s is derived with the combination of equations (4) and (5) as

$$\dot{s} = \left\{h_0\left|1-\frac{s}{s^*}\right|^a sign\left(1-\frac{s}{s^*}\right)\right\}\dot{\varepsilon}_p \tag{6}$$

where

$$s^* = \hat{s}\left[\frac{\dot{\varepsilon}_p}{A}\exp(Q/RT)\right]^n \tag{7}$$

where h_0 represents the hardening/softening constant, n stands for the strain rate sensitivity for the saturation value of deformation resistance, a is the strain rate sensitivity of hardening/softening, s^* describes the saturation value of s associated with a set of given temperature and strain rate, and $\hat{s}$ is a coefficient.

Anand model includes two kinds of equations, the first equation deals with the relationship between saturation stress and strain rate under certain temperature, while the second equation deals with relationship between strain and stress under certain strain rate and temperature[18]. From equations (1), (2) and (7), together with $\sigma^* = cs^*$,the equations are as follows:

$$\sigma^* = \frac{c\hat{s}}{\xi}\left[\frac{\dot{\varepsilon}_p}{A}\exp(Q/RT)\right]^n \sinh^{-1}\left[\left(\frac{\dot{\varepsilon}_p}{A}\exp(Q/RT)\right)^m\right] \tag{8}$$

Equation (8) includes the saturation stress, temperature and strain rate.

From equations (1) and (6), with isothermal, and $s^* > s$ conditions, we can obtain:

$$\frac{d\sigma}{d\varepsilon_p} = ch_0\left|1-\frac{\sigma}{\sigma^*}\right|^a sign\left(1-\frac{\sigma}{\sigma^*}\right),\ a \ge 1 \tag{9}$$

To estimate the remaining parameters, all of the stress-strain data from the monotonic tests is sued to fit the following stress-strain relationship[19]:

$$\sigma = \sigma^* - \left[(\sigma^* - \sigma_0)^{(1-a)} + (a-1)\left\{(ch_0)(\sigma^*)^{-a}\right\}\varepsilon_p\right]^{1/(1-a)},\ a \ne 1 \tag{10}$$

where $\sigma_0 = cs_0$ and s_0 is the initial value of s.

2.3. Parameters analysis

For the Anand constitutive model, nine parameters need to been tested respectively. These materials constants A , Q/R , $\hat{s}$, h_0 , ξ , m , n , a and s_0 can be determined by the following standard procedures[20]:

(1) Obtained the saturation stresses from uniaxial tensile tests with constant rates and temperatures.

(2) Determined the value of A ,Q , $\hat{s}/\xi$,h_0 ,ξ ,m and n in equation (8) from the data acquired in step (1) by a non-linear least square fit. The nonlinear least squares fit to the experimental data was obtained using the Levenburg-Marquardt Algorithm which was coupled to a driver subroutine that exercised the constitutive model and generated uniaxial tension for any prescribed set of material parameters.

(3) Determined $\hat{s}$ and ξ from the values obtained in step(2). The parameter ξ was selected that the constant c in equation(2) was less than unity, and $\hat{s}$ was then determined from the combined term $\hat{s}/\xi$.

(4) The combined constants ch_0 ,cs_0 and a in equation(10) were determined from the constant strain rate data by a least square fit. With the value of c gotten in step(3), h_0 and s_0 were conclude.

The materials parameters of Anand model for the SnAgCu and SnAgCu-nanoAl lead-free solders determined by this procedure are listed in Table 1.

Table 1 Material parameters of Anand model for lead free solders

Solder	A/s^{-1}	$Q/R/K^{-1}$	$\hat{s}/MPa$	h_0/MPa	ξ	m	n	a	s_0/MPa
SnAgCu-nanoAl	22100	8015	59	4132	4.1	0.15	0.0154	2.0	27
SnAgCu	24300	8710	62.3	3541.2	5.8	0.183	0.019	1.9	39.5

3. Finite element simulation

The quarter model of 3D IC structure was established because of the geometric symmetry of the assembly, which can reduce the computational procedure. The SOLID45 linear element is used for meshing all the materials except solder joints, the VISCO107 linear element (8-nodes element), which describes the viscoplastic material behavior, is utilized to mesh solder joints in finite element software. The model of 3D IC structure under consideration consists of lead-free solder joints (SnAgCu-nanoAl and SnAgCu), Cu pillars and two Chips, which is shown in Fig. 1. And selective mesh refinement is used to concentrate highly refined element in the solder joints where is most likely to fail, sparse elements for other materials. Moreover, zero displacement constraints of vertical direction of the cross-area were applied to the cross-sections of the 1/4 model, namely all nodes on the symmetric surface (X=0, Z=0) were fixed in the corresponding directions (X, Z), and the node at the origin (X=Y=Z=0) was fixed in any directions.

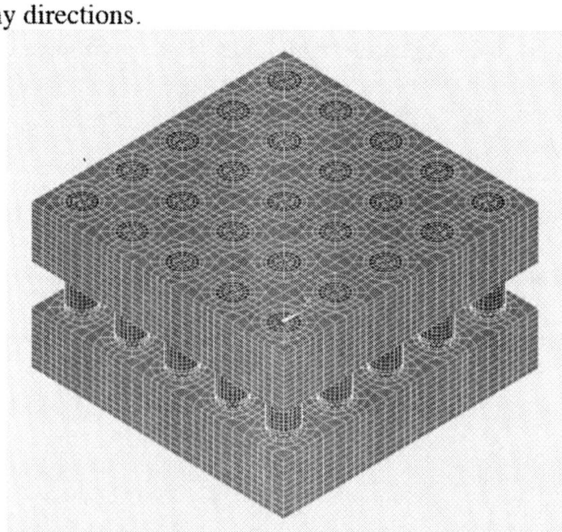

Fig.1 Quarter model of 3D IC structure

In the finite element simulation, accelerated thermal cycling loading (218K-398K) was selected to evaluate the reliability of lead-free solder joints in 3D IC, and the thermal cyclic loading was selected to be imposed on the 3D IC structure. The temperatures ranging from 218K to 398K was carried out for the testing, and the reference temperature (initial temperature) is 298K , dwell time at all peak temperature is 15min, the rates of descend and ascend temperature are 12 K/min. Moreover, it is assumed that in the finite element simulation no stress occurs at the initial temperature. For the thermal cycle loading, at all nodes in the 3D IC structure were loaded a uniform temperature distribution.

Because the thermal mismatch among the chips, Cu, solder joints is linearly dependent on the DNP (distance to neutral point), the highest von Mises stress can be found in the corner solder joints. Fig.2 show the von Mises stress distribution of the solder joints in the array, it is demonstrated that the von Mises stress increases obviously with the increase of DNP, the corner solder joint is the critical solder joints, and the largest von Mises stress was observed on the corner solder joint near the chip.

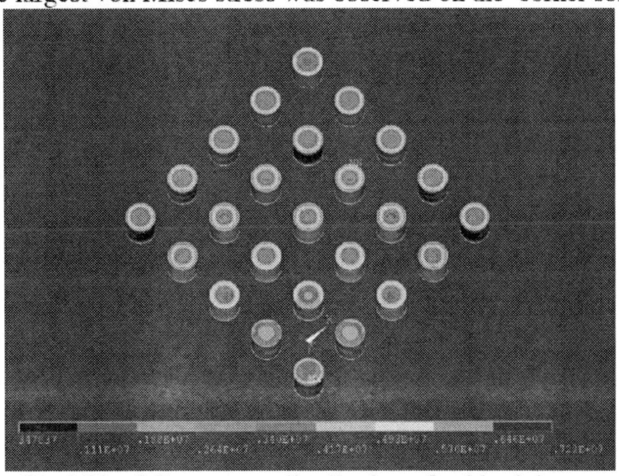

Fig.2 Von Mises stress distribution of solder joints

Due to the CTE (coefficient of thermal expansion) of the materials in the 3D IC, the stress-strain can be happened in the solder joints, so with the finite element simulation results, the fatigue life of SnAgCu/ SnAgCu-nanoAl solder joints in 3D IC structure can be calculated. The modified Coffin–Manson equation is still the most convenient tool for estimating the thermal fatigue life of solder joint. Engelmaier developed the model and utilized to solder joints fatigue prediction, the Engelmaier modified fatigue equation can be written as

$$N_f = \frac{1}{2}\left(\frac{\Delta\gamma}{2\varepsilon'_f}\right)^{1/c} \tag{11}$$

where N_f is the fatigue life, $\Delta\gamma$ is the inelastic shear strain range per cycle calculated using finite element simulation , ε' is the fatigue ductility coefficient which is taken as $2\varepsilon'_f = 0.514$, c is the fatigue ductility exponent which is approximately -0.5708 for SnAgCu based solder joints. It should be noted that, in the present study, the modified Coffin–Manson equation is applied using the data from finite element elements with the inelastic highest shear strain.

Table 2 Fatigue lives of SnAgCu-nanoAl and SnAgCu solder joints

Solder joints	Inelastic shear strain	Fatigue life
SnAgCu-nanoAl	0.0065	1057.34
SnAgCu	0.0052	1563.12

Table 2 shows the fatigue life of SnAgCu and SnAgCu-nanoAl solder joints calculated using the modified Coffin–Manson equation based on finite element simulation. The fatigue life of and SnAgCu-nanoAl solder joints is higher than that of SnAgCu solder joints, reflecting the increase of the fatigue resistance in the solder joint to some extent, the addition of nano-Al particles can increase the fatigue life by 47.8% compared with the SnAgCu solder joints during thermal cycling loading. The phenomena can be attributed to the reduction of the fatigue damages and the change of microcrack propagation sits during the fracture process, which depends on the refinement of intermetallic compounds due to the Al nanoparticles addition[21].

4. Conclusions

The Anand model of SnAgCu and SnAgCu-nano Al solders were studied under uniaxial tension, the model was incorporated into finite element simulation in 3D IC structure. It was found that in the 3D IC device the maximum stress-strain concentrated in the concern solder joints, the stress-strain of SnAgCu-nano Al solder joints was lower than that of SnAgCu solder joints, the fatigue life which represented that the addition of nano Al particles can enhance the reliability of lead-free solder joints in 3D IC devices.

Acknowledge

This study was funded by the Natural Science Foundation of China(51475220), the Qing Lan Project, the China Postdoctoral Science Foundation funded project(2016M591464) and the State Foundation of Laboratory of Advanced Brazing Filler Metals & Technology (Zhengzhou Research Institute of Mechanical Engineering) (SKLABFMT-2015-03).

References

[1] Kanlayasiri K, Sukpimai K. 2016.Journal of Alloys and Compounds, **668**, 169.
[2] Zhang L, Tu K N. 2014. Materials Science & Engineering: R: Reports, **82**,1.
[3] Ji F, X S B, Zhang L, Gao L L, Sheng Z, Dai W. 2011. Chinese Journal of Mechanical Engineering, **24**,428.
[4] Sun L, Zhang L, Zhong S J, Ma J, Bao L. 2015. Journal of Materials Science: Materials in Electronics, **26**, 9164.
[5] Zhang L, Fan X Y. Guo Y H, He C W. 2014. Electronic Materials Letters, **10**,645.
[6] Yang L, ZhangY C, Dai J, Jing Y F, Ge J G, Zhang N.2015. Materials and Design, **67**,209.
[7] Shen Y L, Johnson R W. 2013. Microelectronics Reliability, **53**,79.
[8] Yu J J, Yang C A, Lin Y F, Hsueh C H, Kao C R. 2015. Journal of Alloys and Compounds, **629**,16..
[9] Amalua E H, Ekerea N N, Zarmaia M T, Takyib G. 2015. Finite Elements in Analysis and Design, **107**,13.
[10] Anand L.1985. International Journal of Plasticity, **1**,213.
[11] Brown S B, Kim K H, Anand L.1989. International Journal of Plasticity, **5**, 95.
[12] Wu R, McCluskey F P. 2008 .Thermal and Thermomechanical Phenomena in Electronic Systems, Orlando, FL, May, 683-686.
[13] Ye H, Lin M H, Bassaran C. 2002. Finite Elements in Analysis and Design, **38**, 601.
[14] Hu X D, Ju D Y.2006. Transactions of Nonferrous Metals Society of China, **16**, 586.
[15] Chen X F, Liang L H, Liu Y, Wang Q.2009. Chinese Journal of Applied Mechanics, **26**,248.
[16] Chen X, Chen G, Sakane M. 2004.The Ninth Intersocirty Conference on Thermal and Thermomechanical Phenomena in Electronic Systems, June1-4, 447.
[17] Cheng Z N, Wang G Z, Chen L, Wilde J, Becker K. 2000. Soldering & Surface Mount Technology, **12**, 31.
[18] Pang J H L, Low P T H, Xiong B S. 2004.The Ninth Intersociety Conference on Thermal and Thermomechanical Phenomena in Electronic Systems, June1-4,131.
[19] Bahate D, Chan G, Subbarayan G. 2008. IEEE Transactions on Components and Packaging

Technologies, **31**, 622.
[20] Bai N, Chen X, Gao H. 2009. Materials and Design, **30**,122.
[21] Zhang L, Han J g, Guo Y h, Sun L. 2015. Journal of Materials Science: Materials in Electronics, **26**, 3615.

Monte Carlo entropic sampling applied to Ising-like model for 2D and 3D systems

C M Jureschi[1,2], J Linares[3,*], P R Dahoo[4], Y Alayli[1]

[1] LISV, UVSQ, Université Paris Saclay, 78140 Vélizy, France.
[2] Faculty of Electrical Engineering and Computer Science & MANSiD, USV, Suceava, 720229, Romania.
[3] GEMaC, UVSQ, Université Paris Saclay, 78035 Versailles Cedex, France.
[4] LATMOS, UVSQ, Université Paris Saclay, F-78280, Guyancourt.

E-mail: jorge.linares@uvsq.fr

Abstract. In this paper we present the Monte Carlo entropic sampling (MCES) applied to an Ising-like model for 2D and 3D system in order to show the interaction influence of the edge molecules of the system with their local environment. We show that, as for the 1D and the 2D spin crossover (SCO) systems, the origin of multi steps transition in 3D SCO is the effect of the edge interaction molecules with its local environment together with short and long range interactions. Another important result worth noting is the co-existence of step transitions with hysteresis and without hysteresis. By increasing the value of the edge interaction, L, the transition is shifted to the lower temperatures: it means that the role of edge interaction is equivalent to an applied negative pressure because the edge interaction favours the HS state while the applied pressure favours the LS state. We also analyse, in this contribution, the role of the short- and long-range interaction, J respectively G, with respect to the environment interaction, L.

1. Introduction

Simulation of physical processes is a challenge for scientists trying to understand how the physical processes that occur inside materials drive their properties. Thus, in physics besides analytical methods which can solve exactly either finite or infinite systems, there are also other numerical methods which can solve exactly or with an approximation only finite system. Monte Carlo methods are the most utilized ones by researches in order to model probabilistic or stochastic systems whose analytical solutions are too complicated or impossible to be obtained. Concerning SCO materials [1-6], the Monte Carlo methods are used in combination with certain models such as: Ising-like model, atom-phonon coupling (APC) model or mechanical-elastic model. All these models are used to simulate the evolution of SCO materials when they are subjected to external perturbations such as: thermal or/and pressure variation, light irradiation, applied external magnetic or electrical fields. The importance granted to these materials is due to their potential application in displays, nanoelectronics, data storage or as temperature or/and pressure sensors. These materials which are characterized by two stable states, a diamagnetic low spin (LS) state and a paramagnetic high spin (HS) state, present the great advantage that they can be controlled at the atomic level.

Content from this work may be used under the terms of the Creative Commons Attribution 3.0 licence. Any further distribution of this work must maintain attribution to the author(s) and the title of the work, journal citation and DOI.
Published under licence by IOP Publishing Ltd

Recent investigations using Monte Carlo methods applied to an Ising-like model [7-9] have shown that the behaviour of SCO materials is influenced by the network architecture of the system, the system's size or the size of particles [10]. Also, the role of edge effects for 1D, 2D and 3D SCO network systems has been reported by considering that the edge molecules are fixed in HS states. Some important results have been reported in the recent years for 1D and 2D SCO network system when the edge molecule are active.

In this work the role of edge molecules as active molecules that interact with their local environment are analyzed using MCES method applied to an Ising-like model.

2. The model

To represent the two spin states of molecules, Wajnflasz et al. [7] introduced a spin operator, σ, which can take the value -1 when the molecule is in the LS state and the value +1 when the molecule is in the HS state. The degeneracy ratio of the two spin states is given by $g = g_{HS} / g_{LS}$. In this context, by taking into account the interaction of edge molecules with their local environment and by considering the edge molecules as active molecules, Chiruta *et al.* [11] added a new parameter, *L*, to the Ising type Hamiltonian. For a system of *N* molecules with N_s edge molecules the Hamiltonian is given by:

$$H = \frac{\Delta - k_B T \ln g}{2} \sum_{i=1}^{N_t} \sigma_i - J \sum_{<i,j>} \sigma_i \sigma_j - G < \sigma > \sum_{i=1}^{N_t} \sigma_i - L \sum_{i=1}^{N_s} \sigma_i \, , \tag{1}$$

where Δ is the gap energy, k_B is Boltzmann constant, J is the short range interaction and G is the long range interaction.

Defining the dimensionless macroscopic variable, *m*, *s* and *c* as:

$$m = \sum_{i=1}^{N} \sigma_i, \quad s = \sum_{<i,j>} \sigma_i \sigma_j \quad and \quad c = \sum_{k=1}^{N_s} \sigma_k^{'} \, , \tag{2}$$

the Hamiltonian from equation (1) can be written as:

$$H = \left(\frac{\Delta - k_B T \ln g}{2} - G < \sigma > \right) m - Js - Lc \, , \tag{3}$$

where

$$h_f = -\left(\frac{\Delta - k_B T \ln g}{2} - G < \sigma > \right). \tag{4}$$

According to the Hamiltonian from equation (3), the average magnetization can be written as follows:

$$< \sigma > = \frac{\sum\limits_{i=1,NL} \frac{m_i}{NL} d(m_i s_i c_i) \exp\left(-\frac{1}{k_B T} (-h_f m_i - J s_i - L c_i) \right)}{\sum\limits_{i=1,NL} d(m_i s_i c_i) \exp\left(-\frac{1}{k_B T} (-h_f m_i - J s_i - L c_i) \right)} \, , \tag{5}$$

where *NL* is the number of distinct configuration of states *<m,s,c>* and *d(m,s,c)* is the number of configuration for a given set of values.

3. Monte Carlo Entropic Sampling

We used the Monte Carlo entropic sampling (MCES) method described by Shteto *et al.* [12,13] to generate a table that contains the dimensionless macroscopic variables, *m*, *s*, *c* , and their density, *d(m,s,c)* that is the degeneracy of the macrostate *(m,s,c)*. The principle of MCES is that to obtain a desired distribution P, it is necessary to introduce an appropriate distribution as a bias in the detailed balance equation expressed as:

$$P_i W(i \to j) = P_j W(j \to i), \tag{6}$$

and the biasing probability was chosen as the inverse of the desired restricted density of states.

$$P_i \propto \frac{1}{d(m_i, s_i, c_i)}, \tag{7}$$

In this way, configurations with weakly degenerate macrostates are favored and those with highly degenerate states are damped. In this case, the detailed balance equation (6) can be written as:

$$\frac{W(i \to j)}{W(j \to i)} = \frac{P_j}{P_i} = \frac{d(m_i, s_i, c_i)}{d(m_j, s_j, c_j)}. \tag{8}$$

Because in the first Monte Carlo step the density of the state $d(m,s,c)$ is unknown we put all $d(m,s,c)$ equal to 1. So, after iteration k the density will be $d_k(m,s,c)$. Then, using $d_k(m,s,c)$ as a bias, a MC sampling is ran; it is termed a 'Monte Carlo stage' and it yields a histogram of the frequency of the macrostates: $H_k(m,s,c)$:

$$H_k(m,s,c) \propto d(m,s,c) \frac{1}{d_k(m,s,c)}. \tag{9}$$

The resulting restricted density of states is obtained after applying a correction for the bias:

$$d_{k+1}(m,s,c) \propto d_k(m,s,c)\, H_k(m,s,c). \quad (10$$

From the table of the $d(m,s,c)$ thus built from the (m, s, c) states, the partition function can be calculated using the following expression:

$$Z = \sum_{(m,s,c)} d(m,s,c) \exp(-\beta(-hm - Js - Lc)), \tag{11}$$

from which all the thermodynamic properties of the system can be analytically derived.

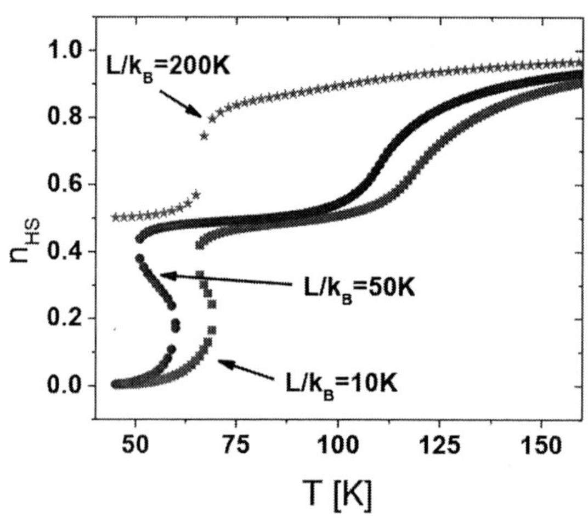

Figure 1. The thermal behaviour for a 3D SCO system for different values of parameter L. The computational parameters are: Δ/k_B = 600 K, J/k_B = -40 K, G/k_B = 200 K, ln(g) = 6, N = 125 (5 x 5 x 5).

In figure 1 is presented the influence of edge effects. As we can see, by increasing the parameter value of the edge molecules interaction with their local environment, the transition is shifted to the low temperature and if it is strong enough then the transition is incomplete and without hysteresis.
The role of the system's shape was analyzed in the last years [14]. It is well known that thermal behaviors of systems with the same number of molecules but with different shapes differs from one system to another. As was reported by Chiruta et al [14], a chain system consisting of 1600 molecules shows less cooperativity than in a ladder system or a square system containing the same number of molecules. So, the transition for a chain system is more gradual, even without hysteresis, compared to a ladder or a square system where the transition is abrupt and with a large hysteresis width. In figure 2 is reported the thermal behaviors for a system with different shapes but with edge effect. As we can see in figure 2 the behavior is strong influenced by the edge effect. By decreasing the system's cooperativity the transition is shifted to lower temperatures and occurs with hysteresis.

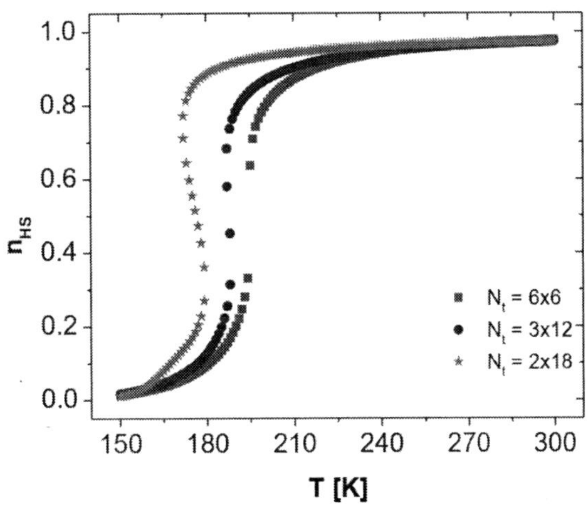

Figure 2. The thermal behaviour for different shapes of a 2D system containing 36 molecules. The computational parameters are: Δ/k_B = 1300 K, J/k_B = 15 K, G/k_B = 172.7 K, L/k_B = 120K, ln(g) = 6.01, N = 36.

4. Conclusions
In this work we studied the thermal behaviour for an Ising-like system under submitted to edge effects. In this context we used MCES method applied to an Ising-like model (where the edge effect is represented by the L parameter) to show the influence of edge effect in 2D and 3D SCO systems. It is important to mention that the edge interaction favors the HS state and that by increasing the value of parameter L or by increasing the ratio between the number of edge and inner molecules, the transition is shifted to the lower temperature.

Acknowledgments
CHAIR Materials Simulation and Engineering, UVSQ, Université Paris Saclay is gratefully acknowledged.

References
[1] Gütlich P 2013 *European Journal of Inorganic Chemistry* **2013** 581-91
[2] Harding D J, Harding P and Phonsri W 2016 *Coordination Chemistry Reviews* **313** 38-61
[3] Jureschi C M, Linares J, Boulmaali A, Dahoo P, Rotaru A and Garcia Y 2016 *Sensors* **16** 187
[4] V. Mishra, R. Mukherjee, J. Linares, C. Balde, C. Desplanches, J.F. Létard, E. Collet, L. Toupet, M. Castro, F. Varret, *Inorg. Chemistry* 2008, 47, 7577-7587

[5] C. Enachescu, R. Tanasa, A. Stancu, F. Varret, J. Linares, E. Codjovi, *Phys. Rev. B* 72, 054413 (2005)
[6] S. Klokishner, J. Linares, F. Varret *Chemical Physics* 255 (2000) 317-323
[7] Wajnflasz J and Pick R 1971 *J. Phys. Colloques* **32** C1-91-C1-2
[8] Bousseksou A, Nasser J, Linares J, Boukheddaden K and Varret F 1992 *Journal De Physique I* **2** 1381-403
[9] Boukheddaden K, Linares J, Codjovi E, Varret F, Niel V and Real J A 2003 *Journal of Applied Physics* **93** 7103-7105
[10] Muraoka A, Boukheddaden K, Linares J and Varret F 2011 *Physical Review B* **84** http://dx.doi.org/10.1103/PhysRevB.84.054119
[11] Chiruta D, Jureschi C-M, Linares J, Dahoo P, Garcia Y and Rotaru A 2015 *The European Physical Journal B* **88** 1-5
[12] Shteto I, Linares J and Varret F 1997 *Physical Review E* **56** 5128-37
[13] Linares J, Enachescu C, Boukheddaden K and Varret F 2003 *Polyhedron* **22** 2453-2456
[14] Chiruta D, Jureschi C-M, Linares J, Garcia Y and Rotaru A 2014 *Journal of Applied Physics* **115** - doi: 10.1063/1.4864035

Simulations of Edge Effect in 1D Spin Crossover Compounds by Atom-Phonon Coupling Model

J Linares[1,*], D Chiruta[1], C M Jureschi[2,3], Y Alayli[2], C O Turcu[3] and P R Dahoo[4]

[1] GEMaC, UVSQ, Université Paris Saclay, 78035 Versailles Cedex, France.
[2] LISV, UVSQ, Université Paris Saclay, 78140 Vélizy, France.
[3] Faculty of Electrical Engineering and Computer Science & MANSiD, USV, Suceava, 720229, Romania.
[4] LATMOS /IPSL, UVSQ, Université Paris-Saclay, UPMC, CNRS, France

E-mail: jorge.linares@uvsq.fr

Abstract. We used the atom-phonon coupling model to explain and illustrate the behaviour of a linear nano-chain of molecules. The analysis of the system's behaviour was performed using Free Energy method, and by applying Monte Carlo Metropolis (MCM) method which take into account the phonon contribution. In particular we tested both the MCM algorithm and the dynamic-matrix method and we expose how the thermal behaviour of a 1D spin crossover system varies as a function of different factors. Furthermore we blocked the edge atoms of the chain in its high spin state to study the effect on the system's behaviour.

1. Introduction

Spin crossover (SCO) is a typical example of molecular bistability, in which a diamagnetic low-spin (LS) and a paramagnetic high spin (HS) states are reversibly switchable by the application of different physical causes such as a magnetic field, pressure, light, temperature or electrical field [1-8]. Although the interest in SCO materials is essentially directed towards future multi-sensing and nano-electronics devices, the ST transition still remains one of the main goals for the physics community.

The ST transition behaviour which depends not only on external factors but also on internal factors in solid state materials, is known to be characterized by different responses such as a gradual transition, an abrupt transition, a multi step-transition with or without hysteresis or an incomplete transition under pressure. To explain the competition between the LS and HS states and moreover, to give a clear image of the spin transition phenomenon, various models such as the Ising-like model [9-12], the atom-phonon coupling model (APC) [13,14] or the mechano-elastic model [15,16], have been developed over the years.

Considering spring-like atoms characterized by three elastic constants function of the atoms' state, Nasser proposed in 2001 [13] the APC model to give a clear and wide view of the ST process in the SCO compounds. In this model three constants are defined such as: $\lambda = C_{LL}$ - the atoms are in the low spin state, $\mu = C_{LH}$ - one atom is in the low spin state and the other atom is in the high spin state, $\nu = C_{HH}$ - the atoms are in the high spin state.

2. The model

Since a detailed description of the model can be found in this paper [14], only some aspects are illustrated in this section and only some equations which are prerequisite to describe this work are

Content from this work may be used under the terms of the Creative Commons Attribution 3.0 licence. Any further distribution of this work must maintain attribution to the author(s) and the title of the work, journal citation and DOI.
Published under licence by IOP Publishing Ltd

recalled. To study the behaviour of SCO systems in the framework of the APC model, a circular chain of N atoms interconnected by a spring is considered. The elastic constants which have already been presented in the introduction are assumed to verify the inequalities: $\lambda > \mu > \nu$. To present the electronic state of the atoms, each electronic state of the atoms is associated to a fictitious spin operator σ which can take two eigenvalues +1 or -1 as a function of the atom state (HS or LS). The Hamiltonian is given as a sum of two terms, the spin Hamiltonian and phonon Hamiltonian:

$$H = H_{spin} + H_{phonon} \tag{1}$$

with:

$$H_{spin} = \sum \frac{\Delta}{2}\sigma_i \text{ and } H_{phonon} = \sum g^{nhs} B(f_i) \tag{2}$$

In equation 2, the spin Hamiltonian depends on Δ, the difference in energy between the two electronic levels (LS) and (HS) and the phonon Hamiltonian depends on the degeneracy g^{nhs} and the Boltzmann factor $B(f_i)$.

The Hamiltonian given by Equation 1 can be solved in two different cases as discussed in sections 2.1 and 2.2.

2.1. Dynamic Matrix method and Monte Carlo Metropolis algorithm

According to the dynamic matrix method described in [17], the partition function can be written as:

$$Z_{ph}^{i} = g^{nhs_i} e^{-\frac{\beta\Delta}{2}m_i} \prod_{i=1,N} \left[2\sinh\left(\frac{\hbar\omega_i}{2k_B T} \right) \right]^{-1} \tag{3}$$

Using dynamic matrix we obtain the frequencies and the partition function, Z, for each configuration. It is important to note that some configurations that have the same frequencies can be grouped. To overcame this aspect we define "*nbre(nhs)*" the parameter which is the sum of the configuration with the same "*nhs*" value and with the same eigen values (frequencies).

$$Z_{ph} = \sum_{i=1,N} \left\{ nbre(nhs) Z_{ph}^{i} \right\} \tag{4}$$

Using MCM algorithm described in [18] the switching probability between the old state and the new state, *P*, is given by the next equation:

$$P_{state} = \frac{g^{nhs_{new}} e^{-\frac{\beta\Delta}{2}m_{new}} \prod_{i=1,N} \left[2\sinh\left(\frac{\hbar\omega_{new}}{2k_B T} \right) \right]^{-1}}{g^{nhs_{old}} e^{-\frac{\beta\Delta}{2}m_{old}} \prod_{i=1,N} \left[2\sinh\left(\frac{\hbar\omega_{old}}{2k_B T} \right) \right]^{-1}} \tag{5}$$

2.2. Free energy

In the free energy method the magnetization M is calculated as follows:

$$M = \frac{\sum m_i \deg \xi_i}{\sum \deg \xi_i} \frac{1}{N} \tag{6}$$

where *deg* is the number of configurations with the same eigen value.

The high spin fraction n_{HS} giving the ratio between the number of atoms in the high spin state and the total number of atoms is defined as:

$$n_{HS} = (1 + M)/2 \tag{7}$$

Using the following formula to calculate the free energy, F, we are able to study the stability of the system's state:

$$F(n_{HS}) = -T\ln\left(\sum_{i=1}^{N}\xi_i\right) \tag{8}$$

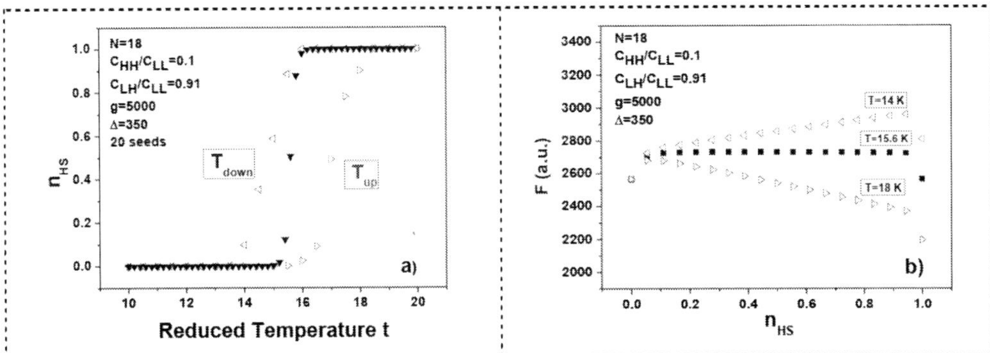

Figure 1. a) the simulated HS fraction, n_{HS}, as a function of the temperature and b) the free energy for three different temperature for an **A** type system.

Figure 1 a) and b) show the evolution of the HS fraction, n_{HS}, as a function of temperature and the free energy for three different temperatures. In figure 1 a) besides the red curve, when the temperature is increasing and the blue curve when the temperature is decreasing (obtained by transfer matrix method and Monte Carlo Metropolis algorithm), the results obtained by the numerical calculation previously described are plotted in black.

3. The role of edge atoms

To explain the role of edge atoms two types of systems were considered: an **A** type system in which the molecules are either HS and (or) LS and a **B** type system in which we add and fix at the beginning and at the end, one atom in the HS state to an **A** type system. In this case the system **A** can be **HS......HS-LS......LS** while the system **B** is of the form **HS+LS......HS-LS......LS+HS**. The evolution of HS fraction, n_{HS}, for an **A** type system is reported in figure 1 a) with the free energy in figure 1 b). In figures 2 a) and b) are reported the HS fraction and the free energy for three different temperatures for a **B** type system. The comparison of the results reported in figure 1 a) with figure 2 a) shows that by adding and fixing the edge atoms in HS state, the transition temperature is shifted to a lower temperature and the width of hysteresis loop is decreased. Moreover the transition is more abrupt.

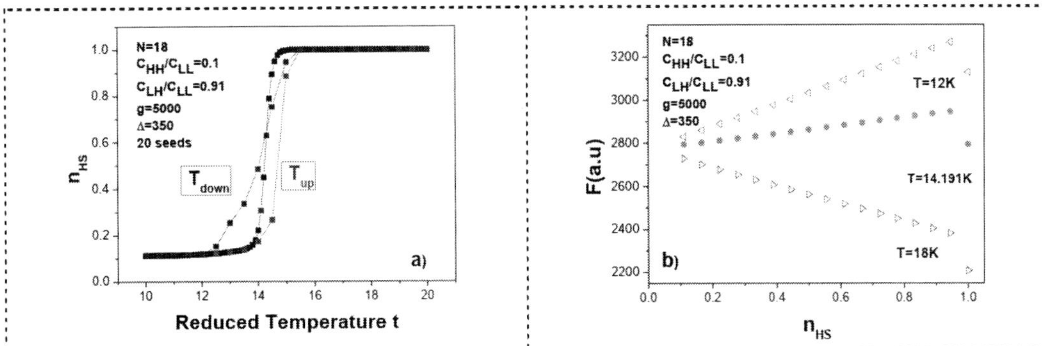

Figure 2. a) the simulated HS fraction, n_{HS}, as a function of the temperature and b) the free energy for three different temperatures for a **B** type system.

4. Conclusions

In this work the thermal behaviour of a 1D spin crossover system under the action of edge atoms (HS), in the framework of the atom-phonon coupling model is studied, using Monte Carlo Metropolis algorithm combined with transfer matrix method and the Free energy method. By taking into account the phonons contributions, the hysteretic behaviour of a 1D system is reproduced.

Acknowledgments

The authors would like to thank Professor J.A. Nasser for useful discussions. CHAIR Materials Simulation and Engineering, UVSQ, Université Paris Saclay is gratefully acknowledged.

References

[1] Bousseksou A, Molnar G, Salmon L and Nicolazzi W 2011 *Chemical Society Reviews* **40** 3313-35
[2] Gütlich P 2013 *European Journal of Inorganic Chemistry* 581-91
[3] Gutlich P, Ksenofontov V and Gaspar A 2005 *Coord Chem Rev* **249** 1811 - 29
[4] Rotaru A, Dugay J, Tan R P, Gural'skiy I A, Salmon L, Demont P, Carrey J, Molnar G, Respaud M and Bousseksou A 2013 *Advanced Materials* **25** 1745-9
[5] Jureschi C-M, Rusu I, Codjovi E, Linares J, Garcia Y and Rotaru A 2014 *Physica B: Condensed Matter* **449** 47-51
[6] Jureschi C M, Linares J, Rotaru A, Ritti M H, Parlier M, Dîrtu M M, Wolff M and Garcia Y 2015 *Sensors (Basel, Switzerland)* **15** 2388-98
[7] Mishra V, Mukherjee R, Linares J, Balde C, Desplanches C, Létard J F, Collet E, Toupet L, Castro M and Varret F 2008 *Inorg Chemistry* **47** 7577-7587
[8] Klokishner S, Linares J and Varret F 2000 *Chemical Physics* **255** 317-323
[9] Wajnflasz J and Pick R 1971 *J. Phys. Colloques* **32** C1-91-C1-2
[10] Bousseksou A, Nasser J, Linares J, Boukheddaden K and Varret F 1992 *Journal De Physique I* **2** 1381-403
[11] Boukheddaden K, Linares J, Codjovi E, Varret F, Niel V and Real J A 2003 *Journal of Applied Physics* **93** 7103-7105
[12] Shteto I, Linares J and Varret F 1997 *Physical Review E* **56** 5128-5137
[13] Nasser J A 2001 *European Physical Journal B* **21** 3-10
[14] Nasser J A, Boukheddaden K and Linares J 2004 *European Physical Journal B* **39** 219-27
[15] Enachescu C, Nishino M, Miyashita S, Stoleriu L and Stancu A 2012 *Physical Review B* **86** 054114
[16] Chakraborty P, Enachescu C and Hauser A 2013 *European Journal of Inorganic Chemistry* 770-80
[17] Chiruta D, Jureschi C M, Linares J, Nasser J and Rotaru A 2015 *Physica B: Condensed Matter* **476** 61-70
[18] Rotaru A, Carmona A, Combaud F, Linares J, Stancu A and Nasser J 2009 *Polyhedron* **28** 1684-7

Logistic systems with linear feedback

Leonid Son, Dmitry Shulgin, Olga Ogluzdina

Ural Federal University, 19 Myra st., Ekaterinburg, Russia

ldson@yandex.ru

Abstract. A wide variety of systems may be described by specific dependence, which is known as logistic curve, or S-curve, between the internal characteristic and the external parameter. Linear feedback between these two values may be suggested for a wide set of systems also. In present paper, we suggest a bifurcation behavior for systems with both features, and discuss it for two cases, which are the Ising magnet in external field, and the development of manufacturing enterprise.

1. Introduction

Many systems, which are described in terms of external variable x and internal parameter y, may be characterized by specific dependence $y(x)$ known as logistic function, or S-function [1]:

$$y = S(x) = \frac{M\exp\{\alpha(x-C)\}+m}{\exp\{\alpha(x-C)\}+1}. \tag{1}$$

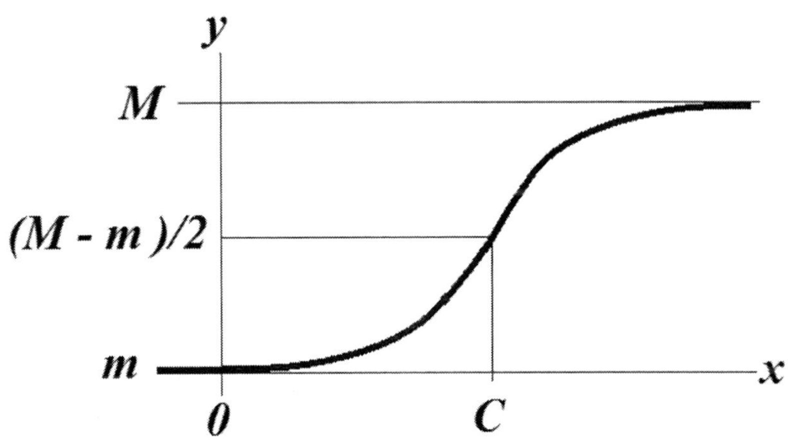

Fig.1. S-curve and its parameters.

Content from this work may be used under the terms of the Creative Commons Attribution 3.0 licence. Any further distribution of this work must maintain attribution to the author(s) and the title of the work, journal citation and DOI.
Published under licence by IOP Publishing Ltd

This function arises for a system, which internal parameter y varies with x from its minimal value m at $x \to -\infty$ to the maximal M at $x \to +\infty$. The "velocity" of this varying is proportional to the deviation of y from its extreme values:

$$\frac{dy}{dx} = \alpha(y-m)(M-y), \tag{2}$$

where α is the coefficient of the proportionality. Function (1) is the solution of differential equation (2), and C is the integration constant, which corresponds to the choice of the x origin. The graph of the function (1) is presented in fig.1. S-curve describes population growth [2,3], technological transformations [4], dependence of the concentration on chemical potential [5], etc. In the present article, we discuss systems, in which linear feedback between the applied external variable and the internal parameter may be suggested. In the next section, we describe common features of such feedback, and in subsequent sections, we concern some cases from physics and economics.

2. Linear feedback and bifurcation.

Linear feedback implies that external effect described by the variable x is formed in accordance with the system behavior characterized by the parameter y:

$$x = ay + b. \tag{3}$$

Thus, the system has to correspond to those (x, y) that arise from the solution of equation

$$\frac{1}{a}x - \frac{b}{a} = \frac{M\exp\{\alpha(x-C)\}+m}{\exp\{\alpha(x-C)\}+1} \tag{4}$$

Depending on the a,b values, this equation may have 1,2 or 3 solutions, and that cases are presented in fig.2, where the graphical solution of eq.(4) is presented.

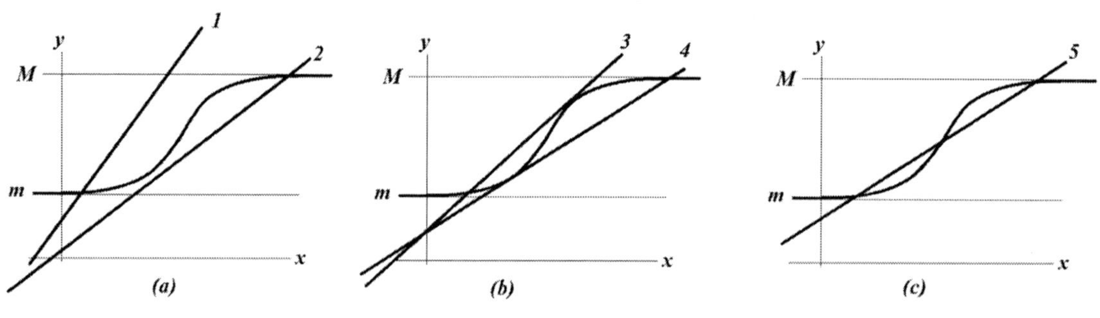

Fig.2. (a)-single solution of eq.(4), (b) – two solutions, (c) – three solutions.

Thus, varying a and b, one can produce a transformation of the system from the position near $y \cong m$ (fig.2a, line 1) to the $y \cong M$ (fig.2a, line 2). The sequence of the left – side (straight line) positions of eq.(4) is 1 – 3 – 5 – 4 – 2. The stability of a certain solution may by estimated in terms of iteration procedure, which reproduces the system behavior: initial external impact x_0 results in system response $S(x_0)$, then the next – step value may be presented as $x_1 = aS(x_0) + b$, etc. Thus, one gets the iteration procedure

$$x_{n+1} = aS(x_n) + b \ . \tag{5}$$

From the other hand, procedure (5) reproduces Seidel method for eq.(4) numerical solution. With respect to (5), fig1(a) demonstrates two single stable points, fig.1(c) – two stable points, separated by the unstable one. Fig.1(b) shows the appearance of a couple of stable and unstable points, i.e. to the bifurcation that corresponds to the cusp catastrophe [6]. The cusp point in the (a,b) plane has the coordinates

$$a_c = \frac{4}{\alpha(M-m)}, \qquad b_c = C - \frac{2(M+m)}{\alpha(M-m)}. \tag{6}$$

3. Case 1: Ising model in the mean – field approximation.

The Ising model describes magnet in following terms (see, for example, [7]). Consider certain space lattice, in sites of which unit spins are placed. With respect to the applied external magnetic field h, each spin on site r may have only two projections $s(r) = \pm 1$ on its direction. Neighboring spins interact with each other, so that the energy of the system is

$$H = -\frac{1}{2}\sum_{r,r'} Js(r)s(r') - h\sum_{r} s(r). \tag{7}$$

In the first term, summation goes over all pairs of nearest neighbors r, r'. Here, J is so called exchange integral, which is the gain in energy for neighboring spins to have equal orientations. For ferromagnets, J is positive. The second term is the energy of spins interaction with external field. To proceed with the thermodynamics of the system, one has to calculate the partition function

$$Z(T,h) = \sum_{\{s(r)\}} \exp\left\{-\frac{H}{T}\right\}, \tag{8}$$

where the summation goes over all possible configurations $\{s(r)\}$. Exact calculation in 3-dimensional space is still unsolved mathematical task. The most useful approximation is the mean – field approximation, which looks as follows. Consider single site surrounded by neighbors with mean spin projection value σ. For this site, the energy is

$$H(s) = -(\frac{\nu J}{2}\sigma + h)s, \qquad s = \pm 1, \tag{9}$$

where ν is the number of nearest neighbors. Instead of (8), one gets single site statistics with only two possible configurations. The partition, and the mean value $< s >$ can be easily calculated:

$$Z = \exp\left\{-(\frac{\nu J}{2T}\sigma + \frac{h}{T})\right\} + \exp\left\{+(\frac{\nu J}{2T}\sigma + \frac{h}{T})\right\}, \tag{10}$$

$$< s > = \frac{1}{Z}\left[\exp\left\{+(\frac{\nu J}{2T}\sigma + \frac{h}{T})\right\} - \exp\left\{-(\frac{\nu J}{2T}\sigma + \frac{h}{T})\right\}\right] \tag{11}$$

Obvious self-consistency condition $< s > = \sigma$ leads to the equation

$$\frac{T}{\nu J}x - \frac{2h}{\nu J} = \frac{\exp(x)-1}{\exp(x)+1}\,, \tag{12}$$

where

$$x = \frac{\nu J}{T}\sigma + \frac{2h}{T}\quad. \tag{13}$$

Equation (12) coincides with (4) at following parameters

$$M = +1,\ m = -1,\ a = \frac{\nu J}{T},\ b = \frac{2h}{T},\ \ C = 0\,. \tag{14}$$

Thus, one can consider the Ising magnet as an example of logistic system with linear feedback. The cusp catastrophe corresponds to the hysteresis loop. Relations (6) give for the the cusp point

$$h_c = 0,\ T_c = \frac{J\nu}{2}\,, \tag{15}$$

which is exactly the Curie point of second order phase transition.

4. Case 2: Manufacturing enterprise.

In this section, we discuss an enterprise in terms of its sales proceeds y and working capital (current assets) x. For an arbitrary case, the dependence $y(x)$ may be characterized by

$$\frac{dy}{dx} = c_0 + c_1 y - c_2 y^2\,. \tag{16}$$

This dependence implies that at small y, the rate of its growth with respect to x is a linear function $c_0 + c_1 y$ with positive c_0, c_1. At larger y, the quadratic term $-c_2 y^2$, $c_2 > 0$ eliminates the rate of growth. Equation (16) suggests a general shape of dependence of the rate of y growth on its absolute value. From the other hand, (16) coincides with (2) after following notations

$$\alpha = c_2,\ M = \frac{c_1}{2c_2} + \sqrt{\left(\frac{c_1}{2c_2}\right)^2 + \frac{c_0}{c_2}}\ ,\ \ m = \frac{c_1}{2c_2} - \sqrt{\left(\frac{c_1}{2c_2}\right)^2 + \frac{c_0}{c_2}} \tag{17}$$

Thus, S-function (1) describes general dependence of sales proceeds on current assets. The integration constant C should be chosen as to fulfill obvious condition $y(0) = 0$. It always may be done, because in (17) $m < 0$. From the other hand, any enterprise may be described in terms of the resources theory introduced by Barney [8,9]. According to the theory, every enterprise resource contributes into several features, which characterize the enterprise with respect to its Competitive Advantage. The features are V (valuable, i.e. can bring sales proceeds), R (rare, i.e. has low direct competition), I (imperfectly imitable, i.e. satisfies the need that cannot be satisfied by other products, or has low indirect competition), and O (organization, which characterizes technological complexity). These features form so called VRIO criteria for a manufacturing enterprise. The VRIO criteria correlate with the $y(x)$ S-dependence rather well (see fig.3). For the manufacturing cycle, current assets arise from the sales products of the previous cycle, and from the external funding. Thus, one gets the iteration procedure (5), where $0 < a < 1$ is the dimensionless coefficient of refinancing, and b is the external funding.

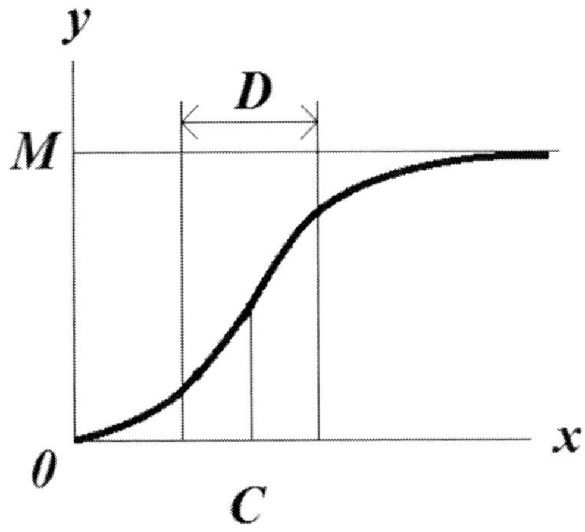

Fig.3. The dependence of sales proceeds on current assets according to (16,17). The maximal sales proceeds M corresponds to the V-feature, C shows the minimal assets that are necessary to get visible proceeds (O-feature), and the curve width D characterizes direct and indirect competition (R, I features). Thus, the VRIO criteria fit the $y(x)$ curve well.

Consider the case $b = 0$. Possible positions of the feedback line with respect to the S-curve are shown in fig.4. Line 1 corresponds to the small a coefficient, and there is only single stable zero point solution. Line 3 corresponds to the large coefficient of refinancing, but one has to remember that there is an upper limit $0 < a < 1$. The most probable situation corresponds to the line 2, which deals with the cusp catastrophe. If the initial cycle funding $x_0 < x_0^c$, then the iteration procedure (5) leads to the zero solution, i.e. the enterprise fails. When $x_0 > x_0^c$, the process gets into the attraction domain of the upper solution, and the enterprise becomes successful. Fig.4. allows one to formulate "startup rule": the current assets should rise after the very first production cycle.

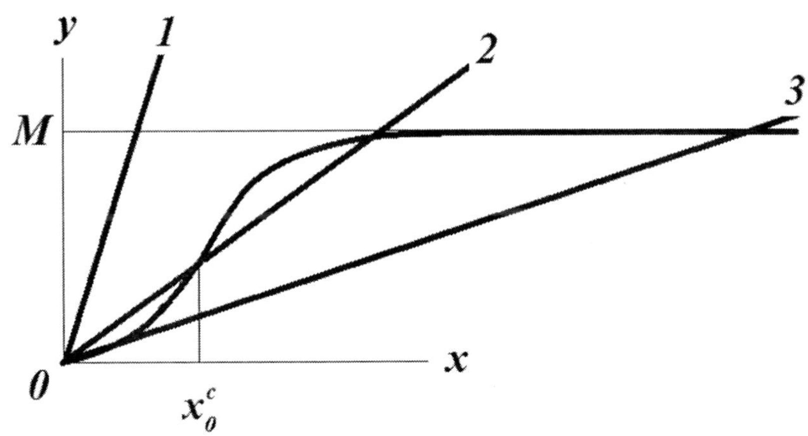

Fig.4. Manufacturing enterprise as S-system with linear feedback. Lines 1,2,3 correspond to the enlarging coefficient of refinancing.

5. Conclusion

We considered two cases from very different areas of science. The cases demonstrate wide universality of simple mathematical construction, which combines S-dependence of the internal variable on the external one, and the linear feedback. The combination results in the cusp catastrophe, and may be suggested for a wide variety of nonlinear systems.

6. Acknowledgements

The work was supported by Russian Ministry of Science and Education (Government job 2014/392 projects 2391 and 1177), and by Goverment of the Russian Federation, act 211 contract № 02.A03.21.0006.

References

[1] Verhulst, Pierre-Francois 1838 *Notice sur la loi que la population poursuit dans son accroissement* Correspondance mathématique et physique **10** 113–121
[2] Kingsland S E 1995 *Modeling nature: episodes in the history of population ecology* Chicago: University of Chicago Press ISBN 0-226-43728-0.
[3] Yukalov V I, Yukalova E P, Sornette D 2009 *Punctuated evolution due to delayed carrying capacity* Physica D: Nonlinear Phenomena **238** (17) 1752
[4] Ayres R U 1989 *Technological Transformations and Long Waves* International Institute for Applied Systems Analysis Lazenburg, Austria. ISBN 8-7046-0083-6
[5] Son LD, Sidorov VE, Popel PS, Shulgin DB 2015 *Nonergodicity in binary alloys* J. Phys: Conference series **633** 1 12023-12027
[6] Arnold V I 1992 *Catastrophe Theory* 3rd ed. Berlin: Springer-Verlag
[7] Baxter R 1982 *Exactly Solved Models in Statistical Mechanics* London: Academic Press
[8] Barney J 1991 *Firm Resources and Sustained Competitive Advantage* Journal of Management **17** 1 99-120
[9] Barney J 1995 *Looking inside for Competitive Advantage* The Academy of Management Executive **9** 4 49-61

Formalization of an environmental model using formal concept analysis - FCA

Rubén D. Bourdon-García[1,2] **and Javier D. Burgos-Salcedo**[3]

[1] Technology Development Center - CDTEC, Saint Matthew Foundation for the High Education, Transv. 17 25 - 25, 111411 Bogotá D.C., Colombia

[2] GERMINA Research Group, Universidad Nacional de Colombia, Av. 30 45 - 03, 453 - 217, 111321 Bogotá D.C., Colombia

[3] CIINAS Corporation, Av. 32 13 - 52, Torre 1, Apart. 1707, 110311 Bogotá D.C., Colombia

E-mail: cdtec@fus.edu.co, rdbourdong@unal.edu.co, jdburgoss@corporacionciinas.org

Abstract. Nowadays, there is a huge necessity to generate novel strategies for social-ecological systems analyses for resolving global sustainability problems. This paper has as main purpose the application of the formal concept analysis to formalize the theory of Augusto Ángel Maya, who without a doubt, was one of the most important environmental philosophers in South America; Ángel Maya proposed and established that Ecosystem-Culture relations, instead Human-Nature ones, are determinants in our understanding and management of natural resources. Based on this, a concept lattice, formal concepts, subconcept-superconcept relations, partially ordered sets, supremum and infimum of the lattice and implications between attributes (Duquenne-Guigues base), were determined for the ecosystem-culture relations.

1. Introduction

This paper has as its foundation the model of environmental interpretation of Augusto Ángel Maya, a very important philosophical conception for the understanding of the current environmental problems and possible solutions. Augusto Ángel Maya's model presents two dimensions related with environmental problems, the cultural system and the ecosystem, as follows:

First of all, the two orders are recognized as independent. The environmental problem consists of both, the ecosystem and the cultural dimensions as they have their own laws of operation. If mankind had to adapt to play a role in the ecosystem, there would not be any environmental problems... The environmental problem arises since the human species does not occupy any niche within the ecosystem. The fact that the human species has achieved a certain independence ... is the result of the evolutionary process... The relationship between ecosystem and culture does not happen only through technology, but it also involves the way men relate with one another [1].

The ecosystem [1] is one of the main dimensions of Ángel's model and it consists of the functional components such as energy flow, trophic levels, biogeochemical cycles, ecological niche, ecological balance and resilience and areas of life on earth (biomes) that explain multiple forms of the existing ecosystem and the succession in it. In addition, besides the ecosystem, there is also a cultural dimension or a cultural system [1], where its components are, the cosmovision of nature or symbolic world, population, technological paradigm and ultimately, both economic

Content from this work may be used under the terms of the Creative Commons Attribution 3.0 licence. Any further distribution of this work must maintain attribution to the author(s) and the title of the work, journal citation and DOI.
Published under licence by IOP Publishing Ltd

and political social relations, as productive and reproductive, established between different populations and within themselves.

Augusto Ángel identifies at least three phases of relations between the ecosystem and the cultural system by appealing to their theoretical differences, which in practice are closely linked. A first **A** relation is recognized which goes from the ecosystem to cultural systems. Culture has to adjust to the external environment and it is known that a culture of arid land is different from one built in the middle of the rainforest. The **B** relation goes from the cultural system to ecosystems. It is called the ratio of impact. Every culture, in its formation process, transforms the ecosystem. The third relation, or ratio **C**, is given the name nemesis. A culture beyond the limits of the external environment must change or disappear due to the pressure of its impacts. Therefore, relation **C** is the nature's answer against non-adaptive cultures. Ángel aims to synthesize the patterns of Ecosystem-Culture relations more than to exhaustively describe the environmental components of the two dimensions. Our goals were to update and to formalize the Augusto Ángel Maya's environmental model. In figure 1 there is a representation of Augusto Ángel Maya's Model (AAM's Model).

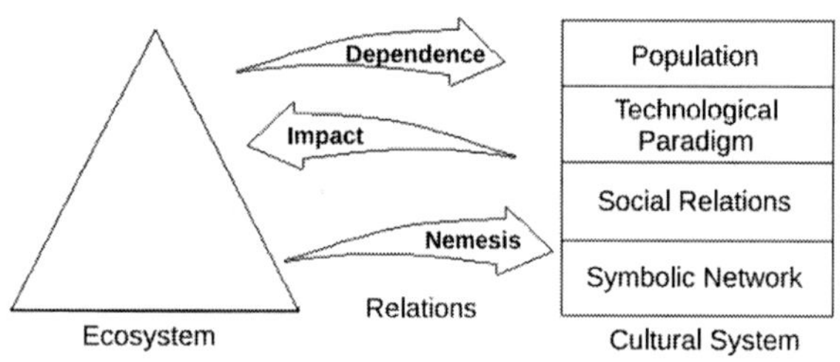

Figure 1. Augusto Ángel Maya's Model of Ecosystem-Culture relations. Based on the image included in [1]. It doesn't any have restriction of use.

2. Theory and Calculations

2.1. Augusto Ángel Maya's updated model

The ecosystem, culture and its components were updated by us, while retaining part of the definitions and connotations already submitted by Ángel. From the general systems theory and some principles of complexity, we classified the ecosystem and cultural system as evolutionary complex systems. The AAM's model was updated, refined and expanded in its environmental dimensions and timescale, to cover the entire biosphere as much as possible (life systems of planet earth), so that it could be formalized as a model of operating networks. In the Ecosystem, the biogeochemical cycles were separated into two parts: circulation through the ecosystem or biochemical circulation (nutrient cycling) and geochemical circulation (environment's fluids, natural resources and mineral deposits); this is due to the fact that biochemical and geochemical circulations have different circulation times through a biogeochemical cycle or full replacement (e.g. nitrogen has a replacement time in the atmosphere and sediments of a scale of 10^7 years, which is very different from the replacement time on the soil which is 2000 years old [2]) and its main reservoir in the world (for example, Nitrogen has 4×10^{21} g in its main reserve in the atmosphere, Sulfur has 2×10^{22} g in its main reserve in the lithosphere and Carbon has $7,7 \times 10^{19}$ g in sediments and rocks [2]) is found outside the ecosystem.

The components of the updated ecosystem with [2, 3, 4, 5, 6, 7, 8] are: Energy flow, food webs, nutrient cycles (biochemical circulation), fundamental and realized niche, morphostasis and morphogenesis; the environment's fluids constituted by the atmosphere and water bodies (geochemical circulation); natural resources and mineral deposits, both organic and inorganic (geochemical circulation); and the living areas or biomes and aquatic ecosystems. On the other hand, cultural system updated components are: The community formed by human populations and domesticated species, as well as animals and plants, their needs and their satisfaction [9]; the symbolic network or human symbols represented on the art, the letters and oral tradition; the technological paradigm composed of science, technology and technical, traditional and ancestral knowledge; social networks are produced by both, transverse and vertical relations [10]; and the sociocultural spatiality or culture's development space [11, 12].

2.2. Formal concept analysis and representation
The study of relations, between environmental dimensions (ecosystem and culture) as a network, is addressed from the formal concept analysis (FCA), since its versatility doesn't need a minimal number of concepts (nodes), nor does it lose information. It is even possible to recover additional information [13]. The FCA theory was developed by Wille in 1982 and the central notion is the *formal context* [14], which is defined as a structure set $\mathbb{K} := (A, B, I)$ where A is a set of *objects*, B is a set of *atttibutes* and I is a binary relation between A y B, i.e. $I \subseteq A \times B$; aIb or $(a, b) \in I$ indicates that object a has attribute b. The *formal context* can be represented by a cross table where rows are headed by the *objects*' names and its columns headed by the *attibutes*' names. A cross (X) in row a and column b means that, in the formal context, it holds aIb.

The following derivation operators are defined for arbitrary X $\subseteq$ A y Y $\subseteq$ B:

$$X \to X^I := \{b \in B \mid aIb \text{ for all } a \in X\}$$
$$Y \to Y^I := \{a \in A \mid aIb \text{ for all } b \in Y\}$$

A *formal concept* of the *formal context* (A, B, I) is defined as a pair (A_1, B_1) with $A_1 \subseteq A$, $B_1 \subseteq B$, $A = B^I$ y $B = A^I$, respectively. Set A_1 is called the *extent* and set B_1 is called the *intent* of the formal concept (A_1, B_1).

Once the components are determined to formalize the updated model, set A of objects and set B of attributes are defined starting from the components of the ecosystem and the cultural system presented above. For instance, the biogeochemical cycles were divided into nutrient cycles, in the ecosystem, and out of the ecosystem by the environment's fluids, natural resources and mineral deposits. Each one is presented in [15]. By using the properties of the substances' cycles in the ecosystem (nutrient cycles), the concepts and attributes of table 1 were defined.

Table 1. Nutrient cycles object and its attributes.

OBJECT	ABBREVIATION (OBJECT, attribute)	MEANING
Nutrient cycles	CIRCBIO	
	recicsust	Recycling of the elements through the food chain under different escales
	medivida	Processing cycles chemicals mediated a vital process
	sustavida	Combination of elements that make up all living organisms
	procevida	Chemical processes that form all living organisms
	pericorto	With short periods and means of circulation and recirculation of substances (hours, days, months, years, decades or hundreds of years)

For example, attribute *recicsust* belongs to CIRCBIO which belongs to the ECOSISTEM. Therefore, attribute *recicsust* belongs to these objects for inclusion and it was indicated with a cross (X) on the formal context [15] for the intersection between attribute *recicsust* and

CIRCBIO and ECOSISTEM objects. For others attributes, the same procedure was done. By using all A objects (36) and B attributes (176), the formal context of the environment (AAM's updated model) was obtained [15].

3. Results and Discussion

Starting from the formal context, the conceptual lattice was built and for this, the ConExp software [16] was used. An excerpt of the conceptual lattice (PART II) is presented in figure 2 where nodes represent the environment's formal concepts (A_i, B_i).

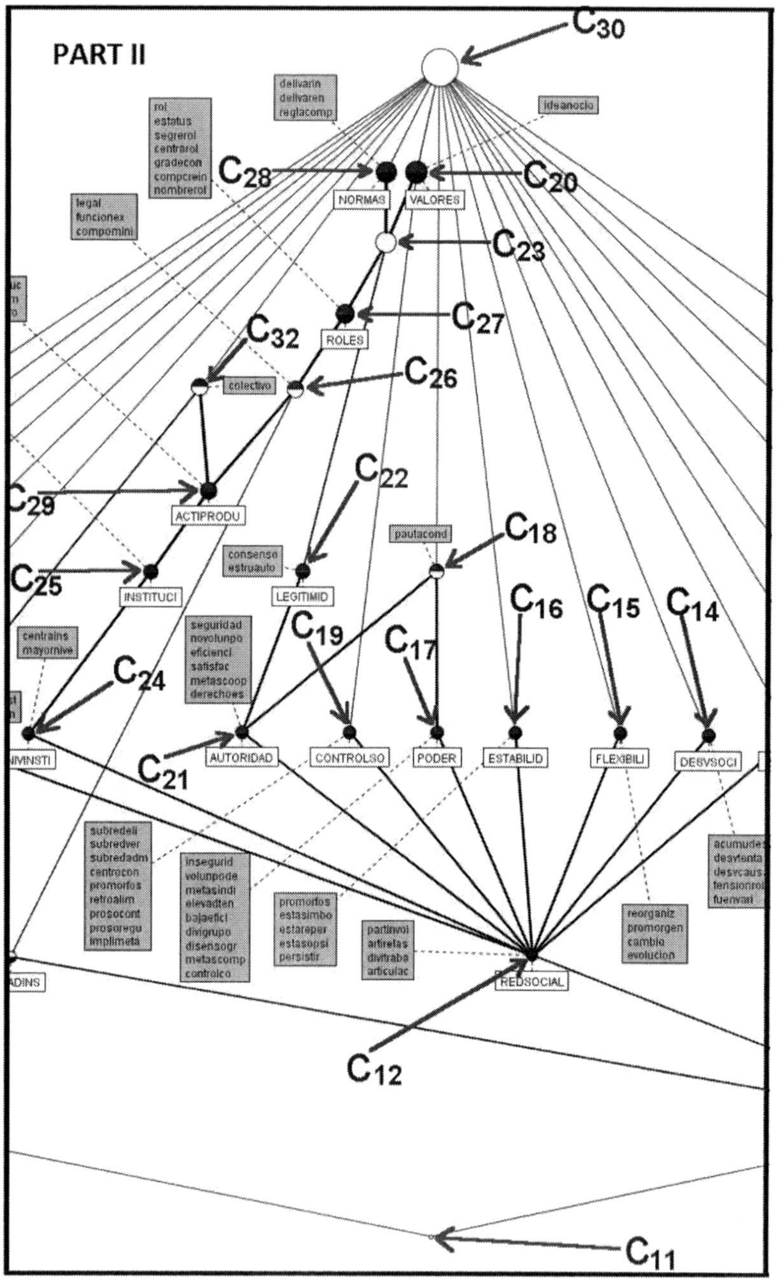

Figure 2. Sociocultural Compartment (PART II) of the Conceptual Lattice of the Environment.

For the lattice, the objects can be identified due to the white boxes below each node and the attributes due to the gray boxes associated with each other. In the whole conceptual lattice of environment [15], one has separated its components (Ecosystem and Cultural System), concepts, objects, attributes and relations, but both components connected by common concepts or attributes. However, the Formal Concept Analysis allows to separate the environmental components and to review each one independently.

Now to get a *formal concept*, from the lattice, *...each concept [formal] is represented by a little circle [in the lattice] so that its extension (intension) consists of all the objects (attributes) whose names can be reached by a descending (ascending) path from the circle* [14].

The total number of concepts found was 44 (C_0 - C_{43}) and for relations, it was 70, but the formal context has over 200 binary relations [15], in fact, the lattice representation condenses the relations because it avoids redundancy on the binary relations. The height of the lattice is 10 because there are 10 different hierarchical levels of concepts or a maximum number of subconcept-superconcept relations in a partially ordered set of the lattice. The hierarchy of concepts tells us about the connections between concepts and its levels or importance. 30 partially ordered sets of the environment were obtained, the one with the lowest number of concepts was $C_{11} < C_{10} < C_8 < C_{30}$ (or $C_{11} < C_{10} < C_9 < C_{30}$) and the one with the largest number of concepts was $C_{11} < C_{10} < C_{12} < C_{24} < C_{25} < C_{29} < C_{26} < C_{27} < C_{23} < C_{20} < C_{30}$ (or $C_{11} < C_{10} < C_{12} < C_{24} < C_{25} < C_{29} < C_{26} < C_{27} < C_{23} < C_{28} < C_{30}$).

In the conceptual lattice of the environment, the infimum of the lattice (Inf**MA**) is concept C_{11} and it contains the empty set in its extension and all attributes of the context in its intension, it is, of course, the smallest element of the lattice and it corresponds to the formal concept of the environment, because it is the only one that contains all the attributes of context. In addition, the lattice has supreme (Sup**MA**), which is the C_{30} concept and contains the set of all objects in the environment in its extension and the empty set in its intension. It is, of course, the largest element of the lattice as there is no common attribute in the context. The only element that shares intention in the lattice is the one with the empty set.

Finally, the implications between attributes or Duquenne-Guigues base were determined. An implication between attributes $P \Rightarrow C$ for the (A, B, I) context, consists of two P and C subsets of B attributes set of (A, B, I) context, where the P set is called the premise and C is called the conclusion of that involvement. The $P \Rightarrow C$ implication for the (A, B, I) context is valid (kept in) as long as what's next is true:

For every **a** by A: If all attributes of the P premise apply to the object, then all attributes of the C conclusion also apply to **a**.

In total, 297 implications between attributes for the context of the environment were found. The implications have the following format:

$\# <$ Number of objects $>$ Premise $\Rightarrow$ Conclusion ; $\#$ it simply refers to the number of involvement in the list; the Number of objects means, how many objects of the context, the implication is valid for; Premise and Conclusion are usually lists of names of attributes (or intentions) that take place in the premise and in the conclusion as well.

It was necessary to reorganize the implications that apply to the same attributes and to determine the general condition that may be necessary, necessary and sufficient, etc. The reorganization of the implications between attributes is exemplified by implications 34 - 38 from the Duquenne-Guigues base of the environmental formal context [15].

$34 < 2 >$ lenguaes $\Rightarrow$ tradioral tradiarti prerrohum cosmovisi; $35 < 2 >$ tradioral $\Rightarrow$ lenguaes tradiarti prerrohum cosmovisi; $36 < 2 >$ tradiarti $\Rightarrow$ lenguaes tradioral prerrohum cosmovisi; $37 < 2 >$ prerrohum $\Rightarrow$ lenguaes tradioral tradiarti cosmovisi; $38 < 2 >$ cosmovisi $\Rightarrow$ lenguaes tradioral tradiarti prerrohum. 34- 38 implications are accomplished for the following objects: CULTURA (cultural system) and SIMBOLOS (symbolic network). All these attributes are located in the premise and conclusion. This means that every attribute is a necessary and sufficient condition for the rest of the attributes and this is a necessary and sufficient condition for all attributes. The two attributes can be interpreted as follows:

lenguaes $\Rightarrow$ tradioral requires, tradioral $\Rightarrow$ lenguaes; then, lenguaes $\Leftrightarrow$ tradioral

This necessary and sufficient condition is met for all attributes of symbols or symbolic network and can be interpreted as follows: The complex structures of written language (lenguaes), the oral (tradioral) and artistic traditions (tradiarti) must be submitted simultaneously, since they are a prerogative of human species (prerrohum) and they reflect their cosmovision of nature (cosmovisi).

Next, the most important implications between attributes for the conceptual lattice of the environment are discussed. In [15] there are all implications between attributes.

63 - 68 implications are valid for the following objects: CULTURA, REDSOCIAL (social networks), NIVINSTI (institutional network) and INSTITUCI (institutions). They are interpreted as follows: on the institutional network, the existence of productive activities determines the existence of organized groups of people in institutional units with specific functions, with purposes, goals and specific mechanisms for developing social function and also that they present explicit rules. *This means that without productive activities, civilizations can not exist.* Before we can even consider colonizing another planet (like Mars), we need to establish productive activities.

106 - 111 implications are valid for the following objects: CULTURA, REDSOCIAL and AUTORIDAD (authority). These conditions are interpreted as follows: if the character pattern of social behavior and the legitimacy of authority take place, the imposed duty ocurrs either within a social dimension, cooperative and collective goals, high satisfaction in people and social security or high group efficiency and it is possible for power not to be manifested in people.

120 - 122 implications are valid for the following objects: CULTURA, REDSOCIAL, NIVINSTI (institutional level), ACTIPRODU (productive activities), INSTITUCI (institutional networking) and UNIDADINS (institutional unit). The interpretation is that if there are social roles, there will be a minimum institutional component with explicit and legalized functions. It can be synthesized that in the sociocultural system, performances of roles are critical to shaping the legal minimum institutional units.

206 - 209 implications apply to the following objects: CULTURA and REDSOCIAL which are interpreted as follows: If there are components of social networks, vertical and transverse articulation of human beings through social relations, there is social division of labor and participation of individuals too, starting from the formal context of the environment used.

4. Conclusions

To make Ángel's environmental interpretation model quantitatively operative, it was necessary to include more realism, that is, to secure the greatest possible degree of correspondence between the environmental concepts of the model and mathematical statements. This was the primary task undertaken in this paper by means of formal concept analysis and network approach that culminated in the design of a formal network model, which enabled clarification in the relations or implications between components of environment.

It was not intended to exhaustively cover all aspects of the environment in terms of interactions, but to present a path of work for the elucidation of relations and interactions. The current environmental researchers (and their trends), despite the criticism, do not present proposals to approach interactions, but sometimes admit that the focus of interactions is a better proposal with real processes.

With the use of FCA, it was possible to get novel hypotheses and relations about the environment from its formal context, in order to be verified with the real world.

The FCA allowed clear out the connections or relations between the environmental components in a graphic sense with the conceptual lattice and it condensed the binary relations of the formal context. Therefore, the FCA is recommended to do the first formal representation of any phenomenon because it allows to clear out the phenomenon in objects, attributes and concepts related.

The constraints and relations between attributes, objects and concepts were determined for 191 implications. These 191 implications or association rules were interpreted into text, so that they could become accurate. Full results can be found at Bourdon's dissertation [15].

References

[1] Ángel A 2013 *The Challenge of Life. Ecosystem and Culture. An Introduction to the Study of the Environment (in spanish)* (Bogotá D.C.: Online publication, www.augustoangelmaya.com)
[2] Odum E and Barrett G 2005 *Fundamentals of Ecology* (Pacific Grove, CA: Thomson BROOKS/COLE)
[3] Curtis H, Barnes S, Schnek A and Massarini A 2015 *Invitation to Biology (in spanish)* (Buenos Aires: Editorial Médica Panamericana)
[4] Krebs C 2009 *Ecology: The Experimental Analysis of Distribution and Abundance* (San Francisco: Benjamin Cummings)
[5] Solomon E, Berg L and Martín A 2011 *Biology* (Belmont: BROOKS/COLE Cengage Learning)
[6] Starr C, Taggart R, Evers C and Starr L 2013 *Biology: The Unity and Diversity of Life* (Belmont: BROOKS/COLE Cengage Learning)
[7] Schlesinger W and Bernhardt E 2013 *Biogeochemistry. An Analysis of Global Change* (Oxford: Academic Press)
[8] Stiling P 2002 *Ecology. Theories and Applications* (New Jersey: Prentice Hall)
[9] Max-Neef M, Elizalde A and Hopenhayn M 1998 *Human Scale Development. Concepts, Applications y some Reflections (in spanish)* (Barcelona: Nordan-Comunidad e Icaria Editorial S.A.)
[10] Buckley W 1993 *Sociology and Modern Systems Theory (in spanish)* (Buenos Aires: Amorrortu Editores)
[11] Delgado O 2003 *Debates on Space in Contemporary Geography (in spanish)* (Bogotá D.C.: Unibiblos)
[12] Dollfus O 1990 *The Geographical Space (in spanish)* (Barcelona: Oikos-tau S.A.)
[13] Mens K and Tourwé T 2005 Delving source code with formal concept analysis *Computer Languages, Systems & Structures* **31** 183-97
[14] Wille R 1999 *Formal Concept Analysis. Mathematical Foundations* ed B Ganter *et al* (Berlin: Springer) pp 1-33
[15] Bourdon R D 2009 *Formalization of Augusto Ángel Maya's Environmental Model in Networks Representation (in spanish)* [Thesis] Bogotá D.C.: Universidad Nacional de Colombia
[16] Yevtushenko S 2000 *Proc. Nat. Conf. on Artificial Intelligence (Russia)* Vol 7 (Russia: KII-2000) p 127

Understanding dynamics of large-scale atmospheric vortices with moist-convective shallow water model

M. Rostami and V. Zeitlin
LMD-ENS, 24 Rue Lhomond, 75005 Paris, France
E-mail: zeitlin@lmd.ens.fr

Abstract. Atmospheric jets and vortices which, together with inertia-gravity waves, constitute the principal dynamical entities of large-scale atmospheric motions, are well described in the framework of one- or multi-layer rotating shallow water models, which are obtained by vertically averaging of full "primitive" equations. There is a simple and physically consistent way to include moist convection in these models by adding a relaxational parameterization of precipitation and coupling precipitation with convective fluxes with the help of moist enthalpy conservation. We recall the construction of moist-convective rotating shallow water model (mcRSW) model and give an example of application to upper-layer atmospheric vortices.

1. Introduction

We study instabilities of idealised atmospheric upper-layer vortices of the type of cut-off lows frequently happening in the atmosphere, and the role of humidity, condensation and related moist convection upon their saturation. We use the simplest, albeit retaining all essential properties, moist-convective shallow shallow-water model and compare behavior of the vortices in "dry" and "moist-precipitating" environments, with the moisture being a passive tracer (M) and having a condensation sink (MP) which creates a moist-convective vertical flux in the latter. Adding evaporation source gives moist-precipitating-evaporating (MPE) configuration. The approach follows a similar study of atmospheric jets [1]. Condensation, having a switch character, is an essentially nonlinear phenomenon, and hence the techniques of linear stability analysis are inapplicable in moist-precipitating configurations. So we perform linear stability analysis of "dry" vortices, and then use the obtained unstable modes to initialize numerical simulations of both "dry" and moist-precipitating (and evaporating) saturation of the instability. We use pseudospectral collocation method in polar coordinates [2] for linear stability analysis, and well-balanced high-resolution finite-volume numerical scheme, adapted for mcRSW in [3], for nonlinear simulations.

2. The model and the vortex configuration

Equations of the two-layer mcRSW introduced in [3] read:

$$\begin{aligned}&\frac{D\boldsymbol{v}_1}{Dt}+f\hat{z}\times\boldsymbol{v}_1=-g\nabla_H(h_1+h_2),\ \frac{D\boldsymbol{v}_2}{Dt}+f\hat{z}\times\boldsymbol{v}_2=-g\nabla_H(h_1+sh_2)+\frac{\boldsymbol{v}_1-\boldsymbol{v}_2}{h_2}\beta P_1\\&\frac{Dh_1}{Dt}+\nabla.(h_1\boldsymbol{v}_1)=-\beta P_1,\ \frac{Dh_2}{Dt}+\nabla.(h_2\boldsymbol{v}_2)=+\beta P_1,\ \partial_t Q_1+\nabla.(Q_1\boldsymbol{v}_1)=-P_1+E.\end{aligned}\tag{1}$$

Content from this work may be used under the terms of the Creative Commons Attribution 3.0 licence. Any further distribution of this work must maintain attribution to the author(s) and the title of the work, journal citation and DOI.
Published under licence by IOP Publishing Ltd

D/Dt is material derivative, $\boldsymbol{v}_i$ - horizontal velocity in layer $i = 1, 2$, f is Coriolis parameter, h_i - layer thicknesses, θ_i - potential temperatures, $s = \frac{\theta_2}{\theta_1} > 1$ - stratification. The model is obtained from the primitive equations by vertical averaging between the material surfaces in isobaric pseudo-height coordinates and adding additional convective flux due to the latent heat release. This convective flux is linked to the water-vapor condensation in the lower humid layer P_1 via moist enthalpy conservation, and gives a sink (source) in the mass conservation equation in the lower (upper) layer. Due to the convective mass exchange this flux leads to the Rayleigh drag in the upper-layer momentum equation. Condensation provides a sink in the bulk moisture content Q_1 in the lower layer, while evaporation E gives a source. Condensation is parameterised as follows: $P_1 = \frac{Q_1 - Q^s}{\tau} H(Q_1 - Q^s)$, where $H(.)$ is step function, and Q^s is a saturation value. A parameterisation $E = \delta.|\vec{\mathbf{v}}|$ crudely corresponds to a vortex system evolving over the ocean.

Writing the "dry" two-layer system in polar coordinates it is easy to see that flow configurations in cyclo-geostrophic equilibrium, i.e. the equilibrium between centrifugal, Coriolis and pressure forces:

$$(\frac{V_1}{r} + f)V_1 = g\partial_r (H_1 + H_2), \quad (\frac{V_2}{r} + f)V_2 = g\partial_r (H_1 + sH_2) \tag{2}$$

are exact solutions corresponding to axisymmetric vortices. We are interested in isolated vortices satisfying (2) and having zero circulation at infinity, and work with alpha-Gaussian vortices with the following azimuthal velocity distribution, in non-dimensional variables:

$$V^*(r^*) = \pm\epsilon r^* \overline{2}^{\,\alpha} e^{\overline{\;2\;}^{\,(-r^{*\alpha}+1)}}, \quad \alpha \geq 1. \tag{3}$$

Here the positive sign corresponds to cyclones and the negative one to anticyclones. Taking the profile (3) for V_1, supposing $H_1 + sH_2 = 0$, and taking $H_1 + H_2$ as the primitive of the l.h.s. of the first equation (2) calculated with (3), allows to find unambiguously H_1 and H_2 for a vortex with non-zero velocity uniquely in the upper layer.

3. Linear stability problem

Linearising the "dry" version ($P_1 = 0$) of (1) about the above-described vortex, and looking for solutions of the linearised equation in the form of azimuthal harmonics $\propto e^{i(l\theta - \omega t)}$, where θ is the polar angle, leads to a linear eigenproblem for the frequencies ω. Eigenvalues with non-zero imaginary part correspond to instabilities. The resulting eigenproblem is discretised in radial direction following [2] and solved by the pseudo-spectral collocation method. Fig. 1 shows a typical most unstable mode resulting from the linear stability analysis. This mode corresponds to the classical baroclinic instability, and is used to initialise nonlinear simulations.

4. Nonlinear saturation of the instability. Comparison of dry and moist scenarios

Nonlinear saturation of the instability was studied by direct numerical simulations initialised with the main vortex configuration with superimposed most unstable mode of weak amplitude (several per cent). A useful diagnostics of the outputs of such simulations is provided by potential vorticity (PV) $q_i = \frac{\zeta_i + f}{h_i}$, $i = 1, 2$, which is a Lagrangian invariant of the dry system layer-wise, or PV anomaly (PVA) $q_i - f/H_i$, where H_i are non-perturbed thicknesses of the layers. Here $\zeta_i == \partial_x v_i - \partial_y u_i$ $(i = 1, 2)$ is relative vorticity. We present in Fig. 2, 3 the evolution of the PVA, respectively in upper and lower layers, during the saturation of the baroclinic instability of the upper-layer cyclone. As we see, the initial upper-layer vortex is streched due to the perturbation, elongates and forms secondary vortices of opposite sign, while their lower-layer counterparts are also being formed. The vortex finally splits in two secondary dipoles which run away in opposite directions. This scenario of "dipolar splitting" is known in literature [4],

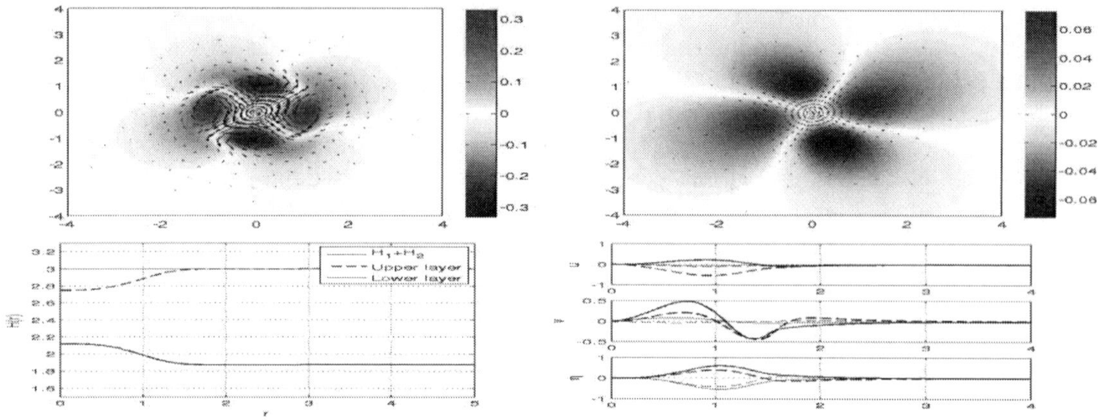

Figure 1. Most unstable mode. Top: pressure and velocity in upper (left) and lower (right) layers. Bottom: H_1 and H_2 (left), radial structure of the mode in upper (thick) and lower (thin) layers. Dashed (Solid): imaginary (real) real parts. ($\alpha = 4$, $l = 2, \epsilon = 0.08$, $s = 1.37$).

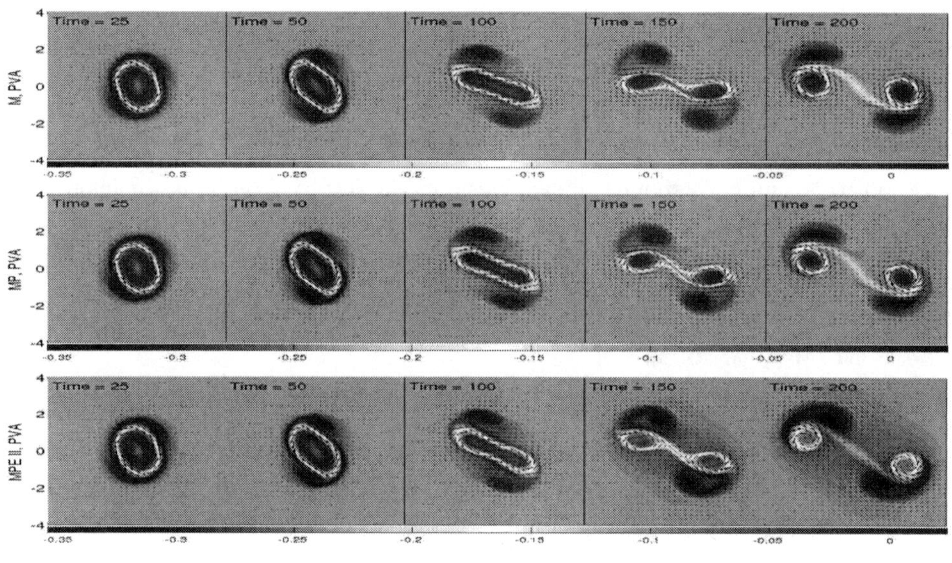

Figure 2. Evolution of PVA in the upper layer during the evolution of the upper-layer cyclone

and thus serves as a benchmark of the "dry" scenario (M). Yet, our simulations clearly show that precipitation, especially in conjunction with evaporation substantially enhances the cyclonic partners of the dipoles, especially in the lower layer, and thus modifies the scenario. The changes are spectacular in the precipitating and evaporating environment.

An important characteristic of the saturation of the baroclinic instability is related inertia-gravity wave emission. The sources of inertia-gravity waves in the atmosphere are not well quantified, in spite of the important dynamical role they are playing, e.g. in the transport of energy and momentum. The wave activity is diagnosed with the help of wind divergence field. It has a peak during initial stretching of the vortex, but no enhancement related to precipitation

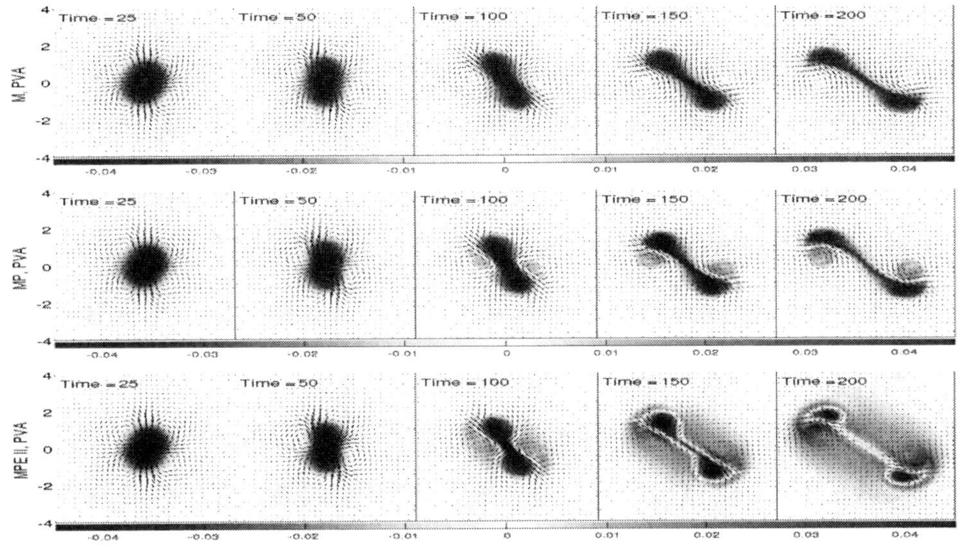

Figure 3. Same as in Fig. 2, but for the lower layer.

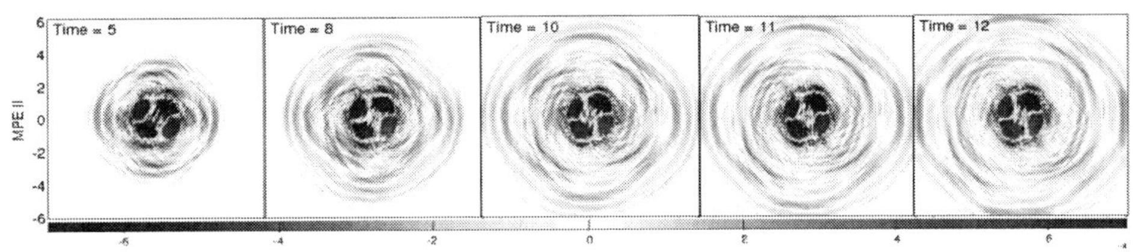

Figure 4. Wave emission in the MPE II environment ($\alpha = 4, \epsilon = 0.08, s = 1.37, H_0 = 3, H_2/H_1 = 0.6, \gamma = 0.05$. Propagation of a packet of inertia-gravity waves is clearly visible.

events. The wave-field corresponding to the maximum of wave emission is presented in Fig. 4

5. Summary and conclusions

Thus, we performed a linear stability analysis of shielded upper-layer atmospheric vortices and used the results of this analysis to initialise fully nonlinear numerical simulations of the saturation of their dominant instability in "dry", moist-precipitating and moist-precipitating and evaporating scenarios. We have found that moist effects significantly change the saturation scenario, leading to spectacular enhancement of secondary lower-layer vortices.

References

[1] Lambaerts, J., Lapeyre, G., and Zeitlin, V., 2012, *J. Atmos. Sci.*, **69**, 1405-26.
[2] Lahaye, N and Zeitlin, V., 2015, *J. Fluid Mech.*, **762**, 5-34
[3] Lambaerts, J., Lapeyre, G., Zeitlin, V. and Bouchut, F., 2011, *Phys. Fluids*, **23**, 046603.
[4] Baey, J.-M. and Carton, X. 2002, *J. Fluid Mech.*, **460**, 15175.

A model for seed dispersion and vegetation growth

Jaqueline Maria da Silva[1], **Maurício Vieira Kritz**[2]

[1] Universidade Federal dos Vales do Jequitinhonha e Mucuri - UFVJM. Campus Mucuri. Rua do Cruzeiro, 01. Jardim São Paulo. Teófilo Otoni-MG. Brazil.
[2] Laboratório Nacional de Computação Científica - LNCC. Rua Getúlio Vargas, 333. Quitandinha. Petrópolis-RJ. Brazil.

E-mail: jaqueline.silva@ufvjm.edu.br

E-mail: kritz@lncc.br

Abstract. The study of processes associated with vegetation grow is very important to understand the dynamics of flooded ecosystems and their sustainable management. We present a cell-centered individual-based probabilistic model for the dynamics of tree-populations, that is further tailored towards the environmental conditions present in the Amazon floodplains.

1. Introduction

The study of processes associated with vegetation grow is very important to understand the dynamics of flooded ecosystems and their sustainable management [5]. In the Amazon floodplains, the seed dispersion process is strongly influenced by the annual flood. Seed dispersion and germination occur in the beginnings of a plant's life cycle and variations in them strongly affect the distribution, structure and dynamics of tree populations [4].

Although difficult to collect in the Amazon floodplains, due to environmental conditions, information about the production and deployment of seeds by trees, as well as their germination, is available. Existing models for vegetation dynamics usually consider environmental factors as age, height, seed distribution, mortality, shadowing and the flood in an integrated manner [2, 1].

We propose a computational, non-deterministic but probabilistic, model to study the impact of seed dispersion and germination on the behavior of tree-populations. This model is data-meagre, that is, it requires the collection of very little data to estimate its parameters, that are mostly probabilities and rates [3] .

This model is for generic populations of trees. Annual flood effects will be considered implicitly, through the data and parameters used, that originate from field research in flooded areas and by considering in the algorithm some details of the tree-growth and the seed-dispersion processes that are characteristic of the flooded areas. Our model also considers a simplified, minimally necessary, age structure and light incidence. Despite being probabilistic, the model behaves similarly to usual population models with finite carrying capacity.

2. The Model

The model for the dynamical behavior of a tree population highlighted here is computational, having no mathematical counterpart, non-deterministic and parametrized by probabilities and rates estimated directly from field data. It is a hybridization of cellular automata and life-cycle

Content from this work may be used under the terms of the Creative Commons Attribution 3.0 licence. Any further distribution of this work must maintain attribution to the author(s) and the title of the work, journal citation and DOI.
Published under licence by IOP Publishing Ltd

models. Besides, it contemplates seed dispersion and the influence of light as the main source of interaction among individuals. Since interactions are represented by means of probabilities as well as growth and mortality rates, the introduction of other species of vegetation in the model is straightforward. It is enough to consider the dependencies of parameters of each population on the density of other species individuals.

To represent the spatial distribution of seeds and the growth of germinated seeds, our model makes use of neighborhood-regions containing mature trees. In them, mature-trees influence the germination of seeds and the growth of young trees. The probability of a seed settling at any point in the first neighborhood varies only with the distance from the mature tree. We make the hypothesis that every year each mature tree produces a fixed number of seeds that are distributed in accordance with this probability density. Seed-germination and the growth process depend on solar light availability. In particular, since the canopy of mature-trees contain sufficiently dense leaves and twigs structure, the light permeating the canopy will vary differently along the day depending on how close a point on the ground is and influence the growth of surrounding trees underneath. Indeed, for most species, the closer a young-tree is from a mature-tree, smaller will be its chances to survive. In some cases it may occur the death of these individuals by shading.

We assume that the tree populations are constrained to a contiguous two - dimensional region C that is subdivided into rectangular non-intercepting cells. This is an individual based cellular model. In counterpoint, the temporal behavior of plant populations, grounded on individual's lifecycles is handled through compartmental models, parametrized by birth and mortality rates. The compartments are not uniform, represent the minimum lifecycle stadiums associated with different maturation time intervals that are needed. Special characteristics of the Amazon Forest flooded areas determine these stadiums and data collected there provide the model parameters.

2.1. Seed Distribution

The probability density function, $\mathcal{D}$, of a seed settling with medium distribution σ at a point P distancing r from a mature-tree is:

$$\mathcal{D}(r,\sigma) = \frac{1}{\pi\sigma^2} e^{(-\frac{r}{\sigma})^2} \tag{1}$$

The probability $\mathcal{P}(r,\sigma)$ of a seed to fall in a neighborhood dA is:

$$\mathcal{P}(r,\sigma) = \mathcal{D}(r,\sigma) \times dA \tag{2}$$

2.2. Light Incidence

In the model, light competition is represented by a mortality probability. We make the hypothesis that young-trees or mature-trees mortalities by suffocation μ_s is associated with local shadowing d_{ls}:

$$\mu_s = \mu_s(d_{ls}) \tag{3}$$

3. Numerical Results

Several tests were executed varying the values of the diameter of the settling-neighbourhood σ. The initial condition was kept fixed and has nine trees on the left of the region $\mathcal{C}$, that sit on different neighbourhoods. Figure 1 and 2 show the dynamic evolution of the population for two values of σ, for the first 200 years.

It is possible to observe that the seed dispersion parameters strongly influence the vegetation dynamical behaviour. For $\sigma = 4$, the vegetation is already in balance after 150 years occupying 100% of the region $\mathcal{C}$ while for $\sigma = 2$ the trees occupy only around 80% of $\mathcal{C}$ after 200 years.

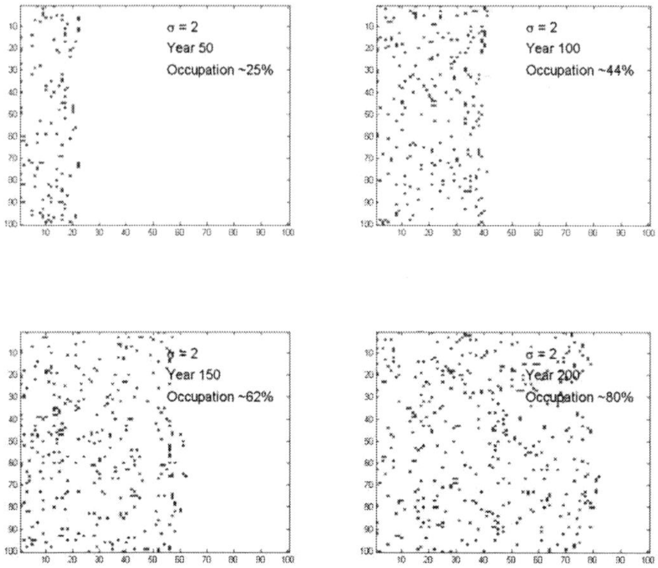

Figure 1. Spatial distribution of mature-trees in region C for $\sigma = 2$ and $t = 50$, 100, 150, and 200 yrs.

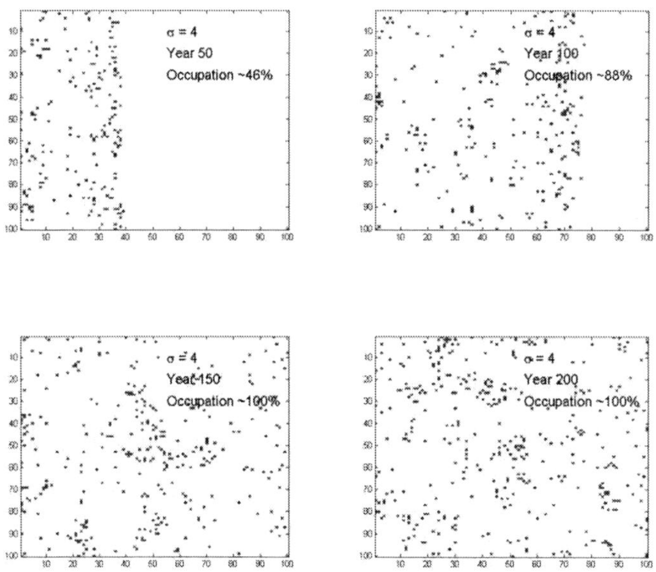

Figure 2. Spatial distribution of mature-trees in region C for $\sigma = 4$ and $t = 50$, 100, 150, and 200 yrs.

4. Conclusions

Although non-deterministic and probability-ruled, the model introduces is statistically sound showing reasonable freedom around a secular tendency that mimics deterministic population models of finite carrying capacity [3].

Figure 3 shows the evolution of the total population along time for $\sigma = 2, 3, 4$ and 5. It is possible to observe that for $\sigma = 5$, the dynamical behavior reaches it steady state before 100 years.

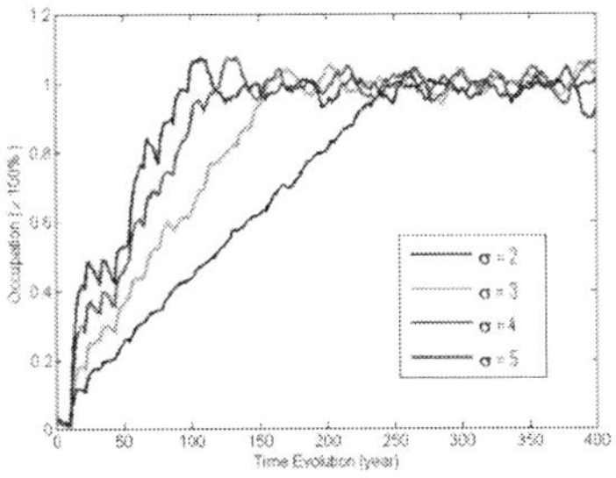

Figure 3. Time evolution of the total number of trees in the population, for the first 400 yrs and σ = 2, 3, 4 e 5.

On the other hand, for $\sigma = 2$, it reaches the steady state only after 200 years. This occurs because for $\sigma = 2$ the settling probability rules concentrate most seeds under the canopies of mature trees. The bigger the value of σ, the faster the population dynamics reaches the steady state, because initially most seeds settle outside the shading zone of mature trees. As the number of mature trees grow, it becomes more difficult for young trees to survive. Bigger the values of σ represent better flood influence in the seeds dispersion. In fact, the flood transports seeds to places distant from its mother-tree.

References

[1] H. Balzter, P. W. Braun and W. Kohler, Cellular automata models for vegetation dynamics, Ecological Modelling, Volume 107, Pages 113-125, (1998).

[2] H. Bossel and H. Krieger, Simulation model of natural tropical forest dynamics, Ecological Modelling, Volume 59, Pages 37-71, (1991).

[3] J. M. Silva, Modelos para a Dinâmica da Vegetação em Áreas Alagáveis Amazônicas, Ph.D. Thesis, Laboratório Nacional de Computação Científica - LNCC, Petrópolis-RJ. Outubro de 2011.

[4] W. J. Junk, Nielsen, L. B. and Nielsen, I. C. and Baslev, H., Flood tolerance and tree distribution in Central Amazonian Floodplains, (Chap.) Tropical Forests: Botanical dynamics, speciation and diversity, London, Academic Press, (1989).

[5] M. V. Kritz, J. M. da Silva and C. Mazza, Modelos e Sustentabilidade nas Paisagens Amazônicas Alagáveis, Editora SBMAC - Sociedade Brasileira de Matemática Aplicada e Computacional, (2012).

Simulation of magnetic coatings on textile fibers

T Blachowicz[1] and A Ehrmann[2]
[1] Institute of Physics – Center for Science and Education, Silesian University of Technology, 44-100 Gliwice, Poland
[2] Faculty of Engineering and Mathematics, Bielefeld University of Applied Sciences, 33619 Bielefeld, Germany

E-mail: tomasz.blachowicz@polsl.pl

Abstract. While the properties of conductive fibres and coatings on textiles can easily be measured and calculated, magnetic coatings of fibres, yarns and fabrics still lack descriptions of their physical properties. Since magnetic textiles can be used for a variety of applications, from magnetic filters to invisible water-marks to magnetic coils and sensors, simulations would be supportive to understand and utilize their properties. The article gives an overview of different coatings on textile fibres, varying the magnetic materials as well as the fibre composition, giving rise to the interactions between neighbouring coated fibres. In this way, it is possible to understand the strong shape anisotropy which must be taken into account when the magnetic properties of textiles are to be tailored. Additionally, the differences between several possible magnetic coating materials become visible. This study can help adjusting the magnetic properties of textile fabrics to a desired application.

1. Introduction

In the area of smart textiles, conductive properties of fibres, yarns, and fabrics are being investigated intensively in simulation and experiment. Theoretical examinations of magnetic textiles, however, are scarce. Only few articles report on modelling magnetic cores of textile coils [1] or magnetic microfibers [2,3].

This is especially unsatisfactory since several applications can be imagined which are based on the magnetic properties of textiles, necessitating a deeper insight into the behaviour of intrinsically magnetic fibres or magnetic coatings on fibres and textile fabrics.

Magnetic coatings can, e.g., be used to change other properties of a finishing layer. Magnetic nanoparticles with hydrophobic properties can become superhydrophobic when coated in an external magnetic field. For this experiment, Fe_3O_4 (magnetite) nanoparticles were coated with a fluoroalkyl silica layer, resulting in significantly increased contact angles for field coating, which was attributed to the formation of nano- and microstructured surface features [4]. Alternatively, maghemite or magnetite particles were introduced into a polysiloxane matrix to produce similar switchable magnetic properties [5].

Applying a magnetic field after the coating process can change the hydrophobic properties, such as contact angle and roll-off angle, to enable cleaning of hydrophobic surfaces [6,7]. This effect was attributed to an increase in the surface roughness.

Magnetic fields can also be used to change other properties of textiles containing magnetic particles. Magneto-rheological fluids can, e.g., be integrated in spacer fabrics to produce composites which an external magnetic field can harden [8]. The optical properties of a textile can be modified by

Content from this work may be used under the terms of the Creative Commons Attribution 3.0 licence. Any further distribution of this work must maintain attribution to the author(s) and the title of the work, journal citation and DOI.
Published under licence by IOP Publishing Ltd

an external magnetic field if structural colours are formed by magnetic nano-structures, e.g. for utilization as pixels in a colour-changing pattern [9].

Measuring magnetic properties of textiles can be used, e.g., for magnetic ink in order to overcome the low efficiency of artificial eyes [10] or for brand protection [11].

Several other applications exist in the area of smart textiles, such as magnetic coils for sensors and actuators [12,13], shielding of static magnetic fields [14], magnetic stimulus in shape memory polymers [15,16], textile whiteboards [17], textile antennae with artificial magnetic conductor planes [18,19], making medical implants visible by included superparamagnetic nanoparticles, [20], etc.

Due to the lack of simulations in this area and the large number of possible applications, this article depicts the influences of the magnetic materials chosen for fibre coating as well as the interdependence between neighbouring fibres with magnetic coating.

2. Methods

Round geometries, such as fibre coatings, can be modelled better by Magpar which is based on the finite element method [21] than by OOMMF [22] or similar programs which use finite differences. Using finite tetrahedral elements with dimensions of maximum 3 nm, dynamical integration of the Landau-Lifshitz-Gilbert equation of motion was used to simulate hysteresis loops.

For the comparison of different magnetic coating materials, iron (Fe) and cobalt (Co) layers on fibres with lengths 10 μm and diameters 1 μm were modelled. Single fibres as well as fibre bundles (cf. Fig. 1) were simulated. For Fe and Co, the exchange constants were chosen as $A_{Fe} = 2.0\cdot10^{-11}$ J/m and $A_{Co} = 1.3\cdot10^{-11}$ J/m, the magnetic polarization at saturation $J_{Fe} = 2.1$ T and $J_{Co} = 1.76$ T, and the Gilbert damping constant $\alpha = 0.01$.

The simulation was performed sweeping from zero external field up to +900 kA/m (with the external field in the sample plane), and back, up to -900 kA/m and to + 900 kA/m again to close the hysteresis loop, using a field sweeping speed of 10 kA/(m ns).

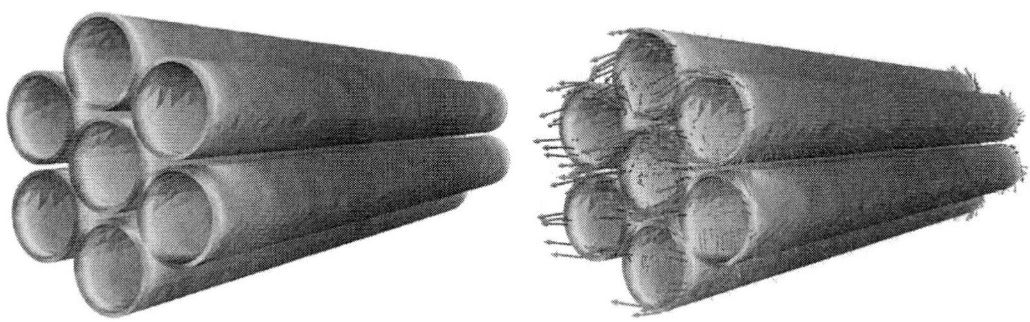

Figure 1. Geometry of simulated fibre bundle modelled with Co coating. Demagnetizing fields are imaged by material colours (left panel) and as the vector fields (right panel).

3. Results

The results of Fe single fibre and fibre bundle coatings are depicted in Fig. 2. In both cases, the maximum magnetization which is reached during the hysteresis loop is significantly decreased for angles unequal 0° (i.e. parallel to the fibre / fibre bundle), showing the strong shape anisotropy which causes the magnetization to stay in the fibre plane. For 0° orientation, the hysteresis loops for a single fibre and a whole fibre bundle are very similar; for the other angles, however, coercive fields and loop shapes differ. In all cases, the coercivities are relatively small, showing values of maximum 25 mT.

It should be mentioned, however, that modelling magnetic coatings on fibres of the dimensions used here show relatively smooth hysteresis loops with irreversible parts (i.e. an open loop) even in case of 90°, which has been shown to be significantly different in smaller fibres with identical aspect ratio of 1:10 [3].

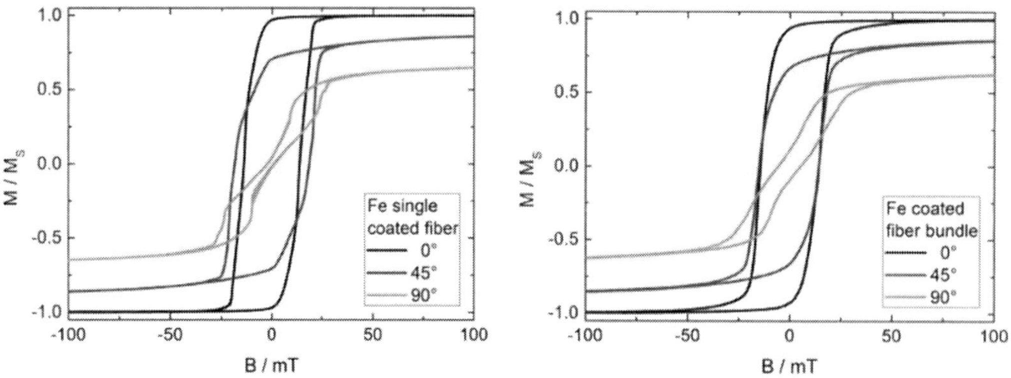

Figure 2. Hysteresis loops, simulated for a single Fe coated fibre (left panel) or an Fe coated fibre bundle (right panel) of fibre lengths 10 μm and diameters 1 μm.

Fig. 3 depicts the identical geometries, modelled for a Co coating on single fibres and fibre bundles. Obviously, the hysteresis loop is in all cases much broader than for an Fe coating. Additionally, the loop shape is more rectangular, indicating coherent rotation of all magnetic moments in the samples, as opposed to domain wall processes and other partial magnetization reversal processes.

In both the single Co coated fibre and the fibre bundle, the coercive field for a field orientation of 45° is strongly reduced. This effect, although not visible for the Fe coated fibres, is typical for orientations deviating from the easy axis. Especially in nano-systems, however, completely different angle dependencies may occur.

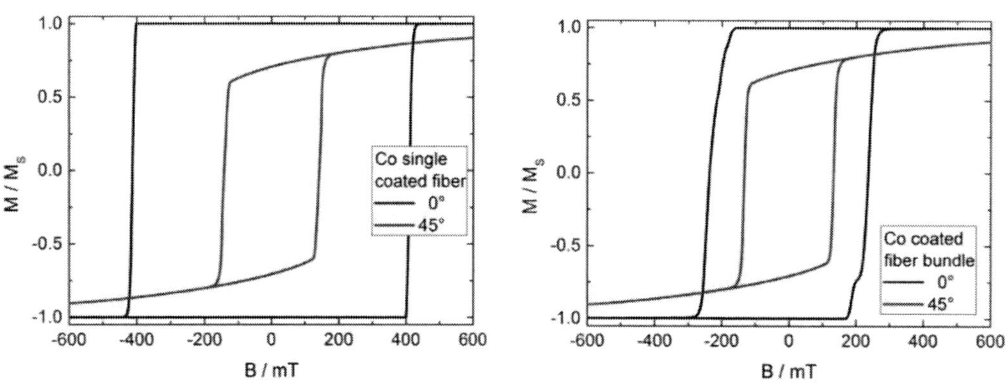

Figure 3. Hysteresis loops, simulated for a single Co fibre (left panel) or a Co fibre bundle (right panel) of fibre lengths 10 μm and diameters 1 μm.

It should be mentioned that the simulated geometries, while being relatively small from a textile point of view, are already quite large for micromagnetic simulations. Table 1 gives an overview of the simulation durations of the results depicted here, using a cluster with 16 cores. This shows one of the largest problems in modelling magnetic coatings on textile fabrics, yarns and fibres: While the thin coating layers necessitate micromagnetic simulations to gain reliable results, the relatively large coated areas cause large models with respectively long computation times.

Table 1. Simulation durations for the results shown in this article.

Material \ geometry	Single fibre	Fibre bundle
Fe	4-8 days	24-28 days
Co	3-9 days	15-31 days

4. Conclusion

In conclusion, we have shown the influence of material and orientation on the magnetic properties of fibre coatings, with coercivities being more than one order of magnitude larger for Co than for Fe and a strong shape anisotropy reducing out-of-plane magnetization. Especially for Co coatings, significant differences occur between single fibre and fibre bundle simulations which have to be taken into account in preparation of real magnetic textile samples.

Acknowledgements
This work was partially supported by the internal project BK-219/RIF/2015 of the Silesian University of Technology.

References

[1] Zieba J, Frydrysiak M 2012 *Smartex Res. J.* **1** 102-110
[2] Wiak S, Firych-Nowacka A, Smólka K 2010 *COMPEL Int. J. Comput. Math. Electr. Electron. Eng.* **29** 1159-1171
[3] Ehrmann A and Blachowicz T 2015 *J. Inst. Eng. India Ser. E* **96** 145-150
[4] Fang J, Wang H X, Xue Y H, Wag X A, Lin T 2010 *ACS Applied Materials & Interfaces* **2** 1449-1455
[5] Schollmeyer E 2007 *Tekstil* **56** 75-80
[6] Verplanck N, Coffinier Y, Thomy V, Boukherroub R 2007 *Nanoscale Research Letters* **2** 577-596
[7] Ho T, Ghochaghi N, Tepper G 2013 *J. Appl. Polymer Science* **130** 2352-2358
[8] Mistik S I, Shah T, Hadimani R L, Siores E 2012 *Journal of Intelligent Material Systems and Structures* **23** 1277-1283
[9] Kim J, Choi S E, Lee H, Kwon S 2012 *Advanced Materials* **25** 1415-1419
[10] You F 2014 Patent CN203444492-U
[11] Benesovsky P 2008 4th ITC & CD
[12] Rubacha M, Zieba J 2006 *Fibres & Textiles in Eastern Europe* **14** 49-53
[13] Rubacha M, Zieba J 2007 *Fibres & Textiles in Eastern Europe* **15** 101-104
[14] Weber M O, Akter F, Ehrmann A 2013 *Industria Textila* **64** 184-187
[15] Meng H, Li G Q 2013 *Polymer* **54** 2199-2221
[16] Kauffmann-Weiss S, Scheerbaum N, Liu J, Klauss H, Schultz L, Mäder E, Häßler R, Heinrich G, and Gutfleisch O 2012 *Adv. Engineering Materials* **14** 20-27
[17] Etre K, Stein R, Dunn E, Commander J 2014 *Journal of the American Institute for Conservation* **53** 211-218
[18] Kamardin K, Rahim M K A, Samsuri N A, Jalil M E, Majid H A 2015 *Microwave and Optical Technology Letters* **57** 197-200
[19] Yan S, Soh P J, Vandenbosch G A E 2014 *IEEE Transactions on Antennas and Propagation* **62** 6487-6490
[20] Mertens M E, Koch S, Schuster P, Wehner J, Wu Z J, Gremse F, Schulz V, Rongen L, Wolf F, Frese J, Gesche V N, van Zandvoort M, Mela P, Jockenhoevel S, Kiessling F, Lammers T 2015 *Biomaterials* **39** 155-163
[21] Scholz W, Fidler J, Schrefl T, Suess D, Dittrich R, Forster H and Tsiantos V 2003 *Comp. Mat. Sci.* **28** 366
[22] Donahue M J, Porter D G 1999 OOMMF User's Guide, Version 1.0. Interagency Report NISTIR 6376, National Institute of Standards and Technology, Gaithersburg, MD

Stability of magnetic nano-structures with respect to shape modifications

T Blachowicz[1] and A Ehrmann[2]
[1] Institute of Physics – Center for Science and Education, Silesian University of Technology, 44-100 Gliwice, Poland
[2] Faculty of Engineering and Mathematics, Bielefeld University of Applied Sciences, 33619 Bielefeld, Germany

E-mail: tomasz.blachowicz@polsl.pl

Abstract. Magnetic nano-structures can be used for various applications. Examinations of nano-structured systems often aim at decreasing pattern sizes due to their possible utilization in data storage media, in order to enhance the possible information density in a given area. This scaling process, however, is limited by the resolution of the lithography process which is used to produce the nano-particles. Thus the influence of shape modifications on the magnetic properties is important to be examined, especially the correlation between small form changes and magnetization reversal processes or coercive fields. In a recent project, square nano-particles from permalloy were simulated using Magpar. Simulations were performed for the ideal geometric shape used in the lithography process, for the realistic shape of the produced nano-particles as obtained by SEM, and for intermediate steps between these extreme shapes. This study allows for estimation of the reliability of magnetic properties of nano-structures with respect to shape modifications in the lithography process.

1. Introduction

Magnetic nano-structures are currently being investigated intensively due to their possible utilization in magneto-electronic devices. In such nano-structured particles, the shape anisotropy often dominates over magneto-crystalline and magneto-elastic anisotropies. This allows for creating magnetic nano-particles with desired magnetic properties or even novel magnetic states.

Especially ring-shaped particles are often examined due to the strongly reduced stray fields, allowing for data storage applications [1,2]. In such ring-shaped nano-structures, so-called flux-closed vortex states occur in which the core regions of usual vortex states [3-5] are excluded.

Such open geometries may be formed as round or elliptical rings [6-8], square rings [9], triangles [10] etc. Among these, square rings have been shown to be of special importance since they can lead to stable intermediate states at remanence which turns them into useful candidates for quaternary memory cells, enabling storage of 2 bits per particle and thus doubling the usual amount of data per magnetic unit [11,12]. Former investigations have shown a severe dependence of the magnetic properties of such systems on the particle shapes [13,14].

This article thus depicts the dependence of a micromagnetic simulation on the shape of a square nano-ring. It focusses on comparing simulations for the ideal shape – as used for the lithographic process – and for the real particle's shape as well as shapes between these extrema.

Content from this work may be used under the terms of the Creative Commons Attribution 3.0 licence. Any further distribution of this work must maintain attribution to the author(s) and the title of the work, journal citation and DOI.
Published under licence by IOP Publishing Ltd

Journal of Physics: Conference Series **738** (2016) 012058 doi:10.1088/1742-6596/738/1/012058

2. Methods

The micromagnetic simulation program Magpar is used to model the round edges in the best possible way [15]. Being based on the finite element method, Magpar approximates spherical systems better than programs using finite differences, such as OOMMF [16].

In Magpar, simulations are performed by dynamically integrating the Landau-Lifshitz-Gilbert equation. Finite tetrahedral elements with dimensions of maximum 3 nm were modelled; demagnetizing fields were taken into account exactly by reducing the element sizes along the edges. For the simulation of the permalloy (Py) particles, the exchange constant was chosen as $A = 1.3 \cdot 10^{-11}$ J/m, the magnetic polarization at saturation $J = 1$ T, and the Gilbert damping constant $\alpha = 0.01$.

The simulation was carried out starting at zero external field, sweeping to + 200 kA/m (with the external field in the sample plane), reversed to -200 kA/m and back to + 200 kA/m again to close the hysteresis loop. The field sweeping speed was 10 kA/(m ns), comparable to typical values in MRAM applications [17].

Experimentally, samples were prepared using e-beam lithography on a silicon(100) wafer with a double layer resist and different exposure doses to create walls of different thickness. The samples were metalized by a 15 nm permalloy layer, followed by a 1-2 nm titan cap layer to avoid oxidation of the Py. Afterwards, the resist was stripped.

The experimental sample shape is depicted in Fig. 1. The sample width was set to 400 nm, the wall diameter to 100 nm and the height to 10 nm. The corners were cut diagonally.

The magnetic properties of the samples were examined using MOKE (magneto-optical Kerr effect) with a diode bridge. Due to the laser spot diameter of approximately 30 μm, each measurement result is superposed of the signals of several thousand nano-structures.

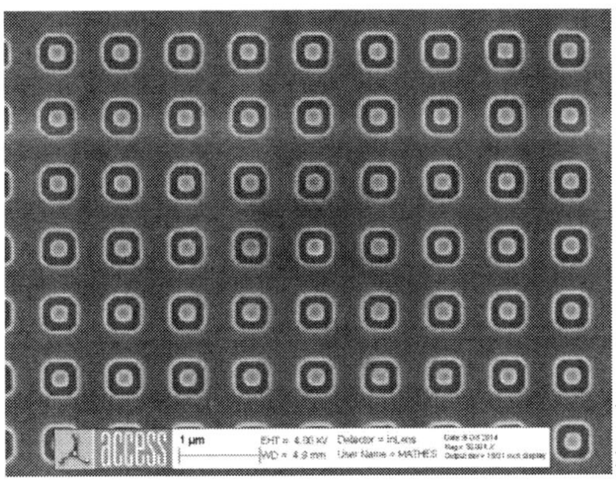

Figure 1. SEM picture of lithographically produced magnetic nano-particles.

3. Results

Fig. 2 depicts the shapes simulated for this examination.

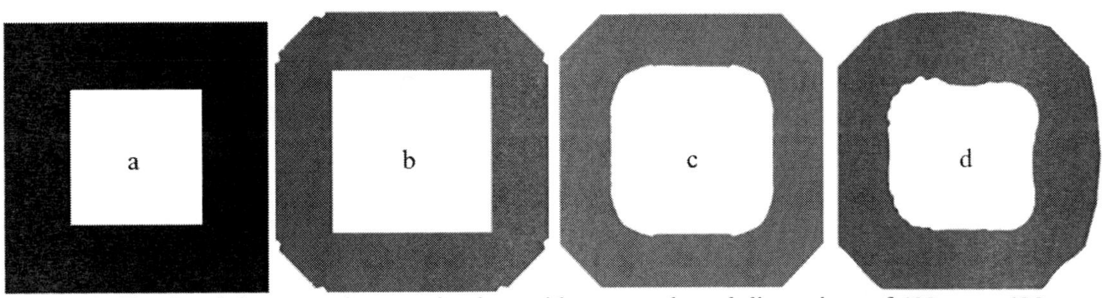

Figure 2. Simulated shapes under examination, with average lateral dimensions of 400 nm x 400 nm x 10 nm.

Starting from an ideal square ring (a), the corners were flattened (b), the inner corner rounded (c), and finally the shape of a real particle obtained by SEM (scanning electron microscopy) was modelled.

In Fig. 3, simulations of the coercive fields of samples (a)-(d) are shown. Obviously, the symmetry of the particle shape is reflected in the symmetry of the angle-dependent coercive fields. While the fourfold symmetry of sample (a) results in a fourfold symmetry of the simulated coercive field, no such symmetry is observed in the simulated coercive field of particle (d). The other two samples with rounded corners and – in case of sample (c) – slightly reduced symmetry show differences to both extremal particles and also have some features in common with them. Particle (b) still shows a fourfold symmetry and quantitatively similar coercive fields as sample (a), with significant differences around hard and soft axes (45° and 0°). Sample (c) resembles the coercive fields of particle (a), with clear symmetry breaking areas. Finally, sample (d) shows strong deviations from the original symmetry.

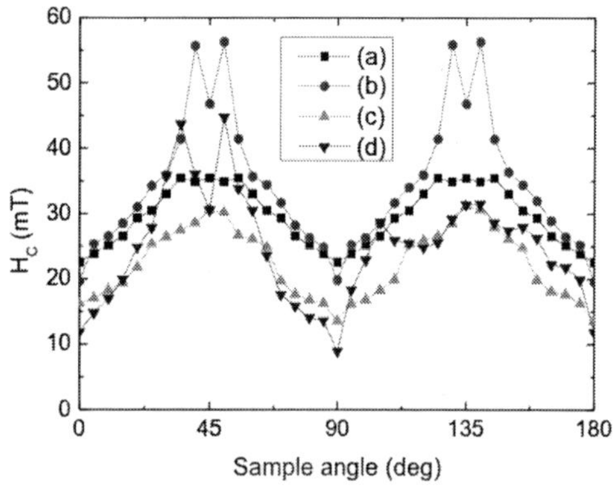

Figure 3. Simulations of the coercive fields of the samples depicted in Fig. 2.

These simulations show the strong influence of shape modifications on the angle-dependent coercive fields – and thus the modifications which can be expected due to undesired deviations from the ideal shape in the lithography process. To test the effect of such shape deviations experimentally, lithographically produced samples (c*) and (d*) – which were used as real bases for the particle models (c) and (d) – were investigated using MOKE. Fig. 4 shows the angle-dependent results of these measurements.

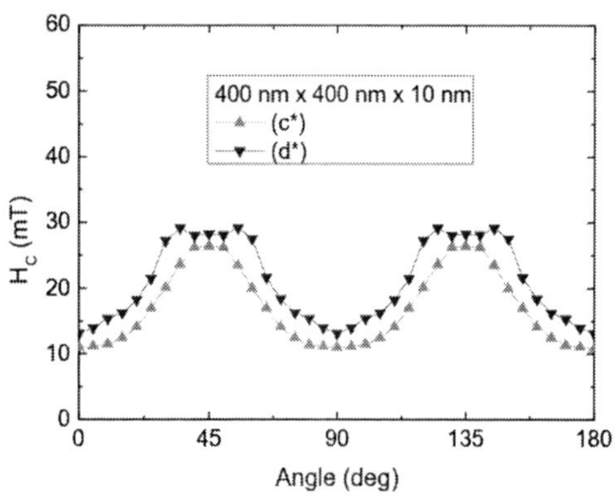

Figure 4. Measurements of real samples with different shapes.

Experimental and theoretical results are similar, with the experimental values being more symmetric and even than the simulated ones. This finding can easily be explained by the averaging processes over several particles which take place in the experiment.

The peaks arising in the simulation of sample (d) near 45° are strongly reduced in the measurement which can also be attributed to an averaging effect in the experiment.

Quantitatively, theoretical and experimental results are also similar.

These results show that micromagnetic simulations can support the identification of shape influences on the magnetic properties of nano-particles, especially since no averaging effects blur the results, opposite to the experiment. On the other hand, averaging over a large number of nano-particles in the experiment helps understanding principle dependencies of the particle shape on the coercive fields, since small deviations from the desired shape are averaged out.

4. Conclusion

To conclude, we have shown the significant influence of nano-particle shapes on their magnetic properties which can be modelled using micromagnetic simulations.

These principal findings are reflected in the experimental results, however, in a less pronounced way due to averaging over several single nano-particles in MOKE measurements. In this way, both approaches – experimentally and theoretically – can support and complement each other.

Acknowledgements

This work was partially supported by the internal project BK-219/RIF/2015 of the Silesian University of Technology.

References

[1] Zhang W and Haas S 2010 *Phys. Rev. B* **81** 064433
[2] Zhu F Q, Fan D L, Zhu X C, Zhu J G, Cammarata R C and Chien C L 2004 *Adv. Mater.* **16** 2155
[3] Eltschka M, Wötzel M, Rhensius J, Krzyk S, Nowak U, Kläui M, Kasama T, Dunin-Borkowski R E, Heyderman L J, van Driel H J and Duine R A 2010 *Phys. Rev. Lett.* **105** 056601
[4] Kim J-S, Boulle O, Verstoep S, Heyne L, Rhensius J, Kläui M, Heyderman L J, Kronast F, Mattheis R, Ulysse C and Faini G 2010 *Phys. Rev. B* **82** 104427
[5] Redondo C, Sierra B, Moralejo S and Castano F 2010 *J. Magn. Magn. Mater.* **322** 1969
[6] Subramani A, Geerpuram D, Domanowski A, Baskaran V and Metlushko V 2004 *Physica C* **404** 241
[7] Wang J, Adeyeye A O and Singh N 2005 *Appl. Phys. Lett.* **87** 262508
[8] Gao X S, Adeyeye A O, Goolaup S, Singh N, Jung W, Castaño F J and Ross C A 2007 *J. Appl. Phys.* **101** 09F505
[9] Vavassori P, Grimsditch M, Novosad V, Metlushko V and Ilic B 2003 *Phys. Rev. B* **67** 134429
[10] Thevenard L, Zeng H T, Petit D and Cowburn R P 2010 *J. Magn. Magn. Mater.* **322** 2152
[11] Blachowicz T and Ehrmann A 2011 *J. Appl. Phys.* **110** 073911
[12] Tillmanns A, Oertker S, Beschoten B, Güntherodt G, Leighton C, Schuller I K and Nogués J 2006 *Appl. Phys. Lett.* **89** 202512
[13] Blachowicz T, Ehrmann A, Steblinski P and Palka J 2013 *J. Appl. Phys.***113** 013901
[14] Ehrmann A, Blachowicz T, Komraus S, Nees M-K, Jakobs P-J, Leiste H, Mathes M and Schaarschmidt M 2015 *J. Appl. Phys.* **117** 173903
[15] Scholz W, Fidler J, Schrefl T, Suess D, Dittrich R, Forster H and Tsiantos V 2003 *Comp. Mat. Sci.* **28** 366
[16] Donahue M J, Porter D G 1999 OOMMF User's Guide, Version 1.0. Interagency Report NISTIR 6376, National Institute of Standards and Technology, Gaithersburg, MD
[17] Tehrani S, Engel B, Slaughter J M, Chen E, DeHerrera M, Durlam M, Naji P, Whig R, Janesky J and Calder J. 2000 *IEEE Trans. Magn.* **36** 2752

Modeling of the optical properties of a two-dimensional system of small conductive particles.

Kondikov A. A.[1,2] , **Tonkaev P. A.**[1,2], **Chaldyshev V. V.**[1,2], **Vartanyan T.A.**[3]
[1] Peter the Great Saint-Petersburg Polytechnic University, St.Petersburg, Russia
[2] Ioffe Institute, St.Petersburg, Russia
[3] ITMO University, St.Petersburg, Russia

E-mail: kondikov@bk.ru

Abstract. Software was developed for quick numerical calculations and graphic display of the absorption, reflection and transmittance spectra of two-dimensional systems of small conductive particles. It allowed us to make instant comparison of calculation results and experimental data. A lattice model was used to simulate nearly distributed particles, and the coherent-potential approximation was applied to obtain a solution to the problem of interacting particles. The Delphi programming environment was used.

1. Introduction

Surface plasmons give rise to several unusual optical phenomena, such as changing of optical response of materials at different scales [1], detection of chemical and biological species [2]. Strong interaction between light and surface plasmons in nanoparticles provides for new opportunities of practical usage, such as transportation of energy [3] and spectroscopy [4]. Optical properties of metal nanoparticles can be tuned by changing their shape, size, and environment, creating promising research fields like plasmon-based photonics. Especially interesting for applications are metal nanoparticle arrays and their combinations with semiconductors or insulators [5]. According to calculations [2], sensors created by using such hybrid materials will show extraordinary sensitivity to biological molecules. Theory predicts that those materials can be used to create high-speed switches [6]. In light of the above, there is a need in reliable instruments for instant calculations and modelling of optical properties of such hybrid structures that will allow quick comparison of calculation results with experimental data.

In this work we present a program for calculation of the optical properties of the layer of particles arranged in a plane. Such arrangement is common for different research samples and devices. We compare our calculations with experimental optical reflection spectra for a system of Ag particles on the GaAs surface.

2. Experiment

The samples were semi-insulating GaAs substrates with thin Ag films sputtered on them. The Ag film thickness was 50 Å. In order to obtain Ag nanoparticles the samples were roasted at the temperature of 200°C. Particles thus obtained were randomly dispersed over the GaAs surface. The characteristic particle size was 35-40 nm (characteristic radius 17-20 nm).

Optical reflection spectra were measured at room temperature for *s*- and *p*-polarized light at different angles of the light incidence. The light source was an LS-1 tungsten bulb connected to an

Content from this work may be used under the terms of the Creative Commons Attribution 3.0 licence. Any further distribution of this work must maintain attribution to the author(s) and the title of the work, journal citation and DOI.
Published under licence by IOP Publishing Ltd

optical fiber. The optical reflection spectra were recorded using an Ocean Optics HR4000 spectrometer and were processed by Spectrasuite software.

In addition to the samples of interest we also measured reflection spectra from a bare GaAs substrate and from a sapphire plate. The latter was used as a reference for normalization since the optical properties of sapphire are well documented and easily described by the Fresnel's law. The reflection of the bare GaAs substrate was subtracted from the optical reflection spectra of the samples in order to reveal the contribution of Ag particles to the total reflection.

An experimental spectrum of light reflection from GaAs substrate with Ag nanoparticles is shown at Fig. 1. The spectrum was recorded at 25° of incidence of *p*-polarized light.

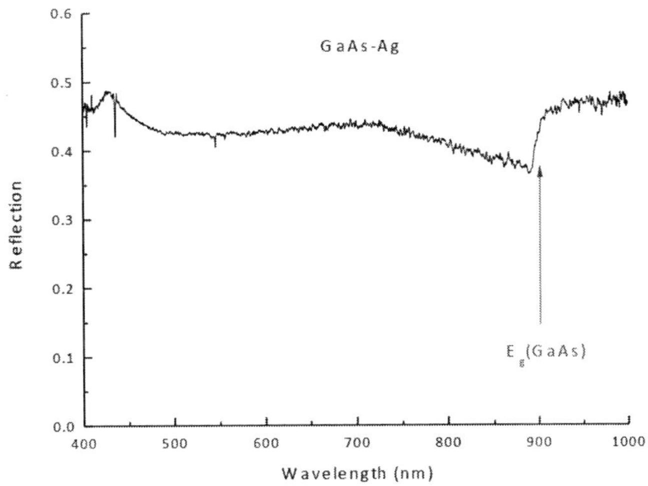

Figure 1. Experimental reflection spectrum recorded for *p*-polarized light at 25° of incidence. Green arrow shows the location of energy gap of GaAs.

Fig. 2 (a,b) shows experimental reflection spectra recorded with *p*-polarized light at the angles of incidence taken 25°, 35°, 40° and 45°, from which the reflection by bare GaAs substrate was subtracted. Nevertheless, there is a "step" at 900 nm caused by abrupt changes in optical properties of GaAs near its energy gap (1.4 eV). As the major contribution by Ag nanoparticle array one can observe the maximum at 700 nm and shoulder at about 530 nm on all reflection spectra. The amplitude of the shoulder is increasing with the increasing angle of incidence, whereas the amplitude of maximum at 700 nm is decreasing. Maximum of reflection is about 10% in absolute units. The half width at half maximum (HWHM) of the peak at 700 nm is estimated as 140 nm.

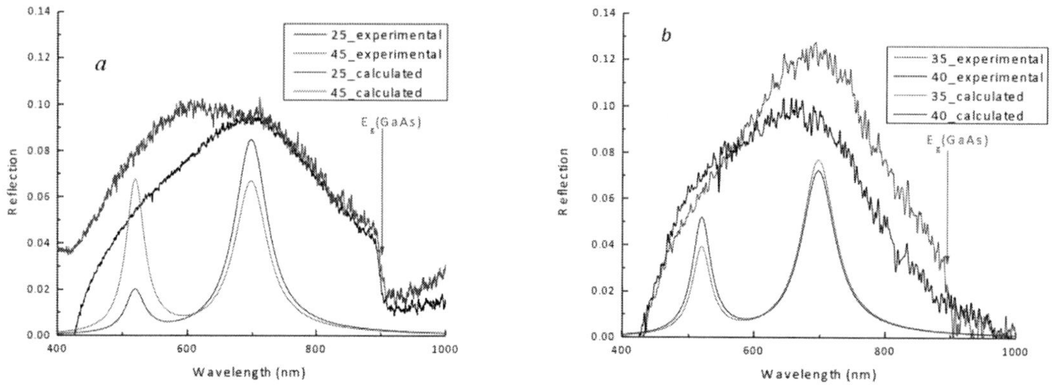

Figure 2. Experimental (noisy curves) and calculated (smooth curves) spectra of optical reflection obtained for angles 25°, 45° (*a*) and 35°, 40° (*b*) of incidence of *p*-polarized light. Green arrows show the location of the GaAs energy gap.

3. Calculations

The system of Ag nanoparticles was considered as a two-dimensional square periodic lattice with period a. The particles were assumed to be spherical with identical radius R.
The bulk plasma frequency is given by the following formula [7]

$$\omega_p = \sqrt{\frac{4\pi n_e e^2}{m_{eff}}}, \quad (1)$$

where e is the electron charge, n_e is the electron concentration and m_{eff} is the effective mass of the electron. The dielectric function of plasma then will be

$$\varepsilon(\omega) = \varepsilon_0 - \frac{\omega_p^{\ 2}}{\omega(\omega + \frac{i}{\tau})}, \quad (2)$$

where τ is the damping constant. In the pure Drude model ε_0 is set to unity. On the other hand, to account for the interband transitions in real metals different values of ε_0 are used.

The polarizability of a single sphere is [8]

$$\alpha(\omega) = R^3 \frac{\varepsilon(\omega) - \varepsilon_c}{\varepsilon(\omega) + 2\varepsilon_c}, \quad (3)$$

where ε_c is the dielectric function of the environment.
Substituting (2) into equation (3) gives

$$\alpha_A(\omega) = R^3 \left(1 + \frac{3\omega b(\omega + \frac{i}{\tau})}{\Omega^2 - \omega(1+2b)(\omega + \frac{i}{\tau})} \right), \quad (4)$$

where $\Omega = \omega_p / \sqrt{3\varepsilon_0}$ and $b = \frac{\varepsilon_c}{\varepsilon_0}$. It is more convenient for calculation to use $\gamma = \hbar / \tau$ instead of τ.

In addition to the incident electromagnetic wave there is a secondary field induced by the ensemble of nearby particles. It makes a contribution to the total polarizability. Taking into account the secondary electromagnetic field the diagonal components of the tensor of polirazability are [9]

$$\alpha_{\parallel}^{0}=R^{3}\left(\frac{1}{(1+R^{3}U_{0})}+\frac{3\omega b(\omega+\frac{i}{\tau})}{((1+R^{3}U_{0})(\Omega_{\parallel}^{2}-\omega(1+2b)(\omega+\frac{i}{\tau})))}\right) \quad (5)$$

$$\alpha_{\perp}^{0}=R^{3}\left(\frac{1}{(1+R^{3}U_{0})}+\frac{3\omega b(\omega+\frac{i}{\tau})}{((1+R^{3}U_{0})(\Omega_{\perp}^{2}-\omega(1+2b)(\omega+\frac{i}{\tau})))}\right) \quad (6)$$

$$\Omega_{\parallel}^{2}=\Omega^{2}(1-\frac{1}{2}R^{3}U_{0}) \quad (7)$$

$$\Omega_{\perp}^{2}=\Omega^{2}(1+R^{3}U_{0}). \quad (8)$$

Here the contribution of induced secondary field is taken into account in approximation of a regular square lattice of identical nanoparticles. In this approximation the lattice sum gives

$$U_{0}=9.03a^{-3}, \quad (9)$$

where a is the lattice constant.

In accord with (5) and (6) the parallel and perpendicular components of the polarizability are substantially different. They have different resonant frequencies, $\Omega_{\perp}$ and $\Omega_{\parallel}$. Hence, there should present two peaks in the reflectance and transmittance spectra.

The total reflectance of the system for light incident at an arbitrary angle θ is [8]

$$r=\frac{-A(\cos^{2}\theta\alpha_{\parallel}^{0}-\sin^{2}\theta\alpha_{\perp}^{0})-AB\alpha_{\parallel}^{0}\alpha_{\perp}^{0}e^{2i\theta}}{1+B(\alpha_{\perp}^{0}-\alpha_{\parallel}^{0})-A(\cos^{2}\theta\alpha_{\parallel}^{0}+\sin^{2}\theta\alpha_{\perp}^{0})-AB\alpha_{\perp}^{0}\alpha_{\parallel}^{0}e^{2i\theta}}, \quad (10)$$

$$A=\frac{2\pi\omega i}{ca^{2}\cos\theta}, \quad (11)$$

$$B=\frac{2\pi\omega}{ca^{2}\sin\theta}. \quad (12)$$

Using this theory we developed a program, which makes quick numerical calculations and graphic display of the absorption, reflection and transmittance spectra. It provides an opportunity to save and load experimental spectra for easy comparison. The interface of the program is shown in Fig. 3.

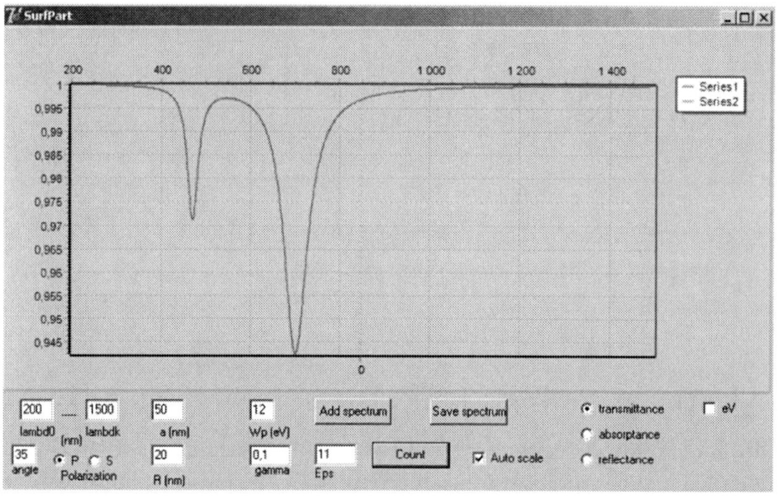

Figure 3. Program interface showing calculation of the transmission spectrum for silver nanoparticles on the surface of GaAs. The lattice constant is a = 50 nm. The particle radius is R = 20 nm. The angle of *p*-polarized light incidence is 35°. The transmittance is calculated for the wavelength range from 200 nm to 1500 nm. The material of particles has plasmonic frequency ω_p = 12 eV with damping parameter γ = 0.1 eV. The ambient dielectric function is 11.

4. Discussion and conclusions

In order to compare the calculations with experiment we should specify the parameters of the system. Based on the sample images provided by scanning electron microscope (SEM) we considered the characteristic particle size *R* to be in a range from 14 to 22 nm and the effective lattice constant of about $a = 50\pm5$ nm. The dielectric functions for Ag and GaAs are well documented [9]. We took ω_p = 12 eV an $\gamma = \hbar/\tau = 0.1$ eV as common parameters of Ag and $\varepsilon_c = 12$ as a simple representation of the GaAs substrate.

The calculated optical reflection spectra are shown in Fig.2 along with experimental ones. In this calculation the particle radius is R = 18 nm, which provides the best agreement of the two calculated reflection peaks with the features in the experimental spectra. The comparison of calculations with the result of experiment shows that the dominant contribution to the reflection at relatively low angles of the light incidence (25-35°) comes from in-plane component of total polarizability. In case of larger angles of the light incidence the experimentally observed major peak shifts to shorter wavelength and a shoulder at around 530 nm becomes pronounced. The calculation shows that this transformation originates from the resonance related to the out-of-plane component of total polarizability. Our simple model shows a reasonable similarity with the experimental results. It predicts the positions of peaks in reflection spectra and their transformation at different angles of the light incidence.

However, the width of calculated peaks is different from the experimental ones. The homogeneous broadening in our model is HWHM = 25 nm with common parameters of Ag particles, whereas the experimental HWHM = 140 nm. This discrepancy may be explained by the fact that in a real sample, particles have different sizes and forms. It is well known that these parameters strongly impact the resonance frequency [10]. Our calculations show that the observed HWHM about 140 nm corresponds to variation of the particle radius from 10 to 25 nm. This dispersion is well consistent to our SEM observations, where a spread of particle radii can be estimated as 15-21 nm. Also our model does not consider the nonlinearity of dielectric function of the substrate. As well, its imaginary part is not included in the calculation. This can result in some broadening of the optical reflection spectra.

However, our estimate shows that the inhomogeneous broadening gives major contribution to the observed reflection spectra.

Thus, the program was developed for quick numerical calculations and graphic display of optical absorption, reflection and transmittance spectra of a layer of plasmonic nanoparticles formed on a dielectric substrate. In overall the calculated spectra are reasonably similar to the experimental ones. The program will be improved in the future by adding the possibility to make calculations by using loaded complex dielectric function of the environment.

5. Acknowledgements

Authors are grateful to P.V. Gladskyh, I.A. Gladskyh and N.A. Toropov for sample preparation. The work was supported by the Russian Foundation for Basic Research, grant No 16-02-00932 A.

References

[1] Kreibig U, Vollmer M 1995 Optical Properties of Metal Clusters *Springer* **25** 532

[2] Hatef A, Sadeghi S M, Boulais E, Meunier M 2013 Quantum dot–metallic nanorod sensors via exciton–plasmon interaction *Nanotechnology* **24** 015502

[3] Ozbay E 2006 Plasmonics: merging photonics and electronics at nanoscale dimensions *Science* **331** 189-193

[4] Haynes C L, McFarland A D, Van Duyne R P 2005 Surface-Enhanced Raman Spectroscopy *Anal. Chem.* 77 338-346

[5] Hua-Jun Chen, Ka-Di Zhu 2015 Surface Plasmon Enhanced Sensitive Detection for Possible Signature of Majorana Fermions via a Hybrid Semiconductor Quantum Dot-Metal Nanoparticle System *Scientific Reports* **5** 13518

[6] Wen-Xing Yang, Ai-Xi Chen, Ziwen Huang, Ray-Kuang Lee 2015 Ultrafast optical switching in quantum dot-metallic nanoparticle hybrid systems *Opt. Express* **23** 13032-13040

[7] Born M, Wolf E 1999 Principles of Optics *London: Cambridge University Press*

[8] Persson B N J, Liebsch A 1983 Optical properties of two-dimensional systems of randomly distributed particles *Physical Review B* **28** 4247

[9] Nolte D D 1994 Optical scattering and absorption by metal nanoclusters in GaAs *Journal of Applied Physics* **76** 3740

[10] Bohren C F, Huffman D R 1983 Absorption and Scattering of Light by Small Particles *New York: John Wiley & Sons*

Research of resonant light reflection by a periodic system of GaAs/AlGaAs quantum wells.

Tonkaev P.A. [1] [2]
Kondikov A.A. [1] [2]
Chaldyshev V. V. [1] [2]

[1]Peter the Great Saint-Petersburg Polytechnic University, St.Petersburg, Russia
[2]Ioffe Institute, St.Petersburg, Russia

E-mail: tonkaev-pavel@mail.ru

Abstract. Measurements of the optical reflection spectra from periodic structures with two quantum wells in the elementary cell have been done. The dependencies of the light reflection on the angle of the light incidence, polarization and temperature were studied. An analysis of the experimental data showed that the pattern with 60 cells is a good Bragg reflector with reflectivity more than 90% in the maximum of the spectral band.

1. Introduction

Structure with dielectric function that periodically varies in space is very interesting from both fundamental research and applied physics standpoint. These structures are called photonic crystals. The periodicity can be created in one, two or three dimensions. One-dimensional photonic crystals are often referred to as distributed Bragg reflectors. They found many applications in such devices as vertically emitting lasers, band filters, etc [1].

Resonant Bragg structures (RBS) are a type of photonic crystals [2,3]. In these structures a periodic change of the dielectric function is caused by exciton states in quantum wells. Optical properties of resonant Bragg structures have been discussed in several theoretical works [3,4,5]. In such structures exciton-polariton resonance mode arises when the Bragg reflection frequency coincides with the exciton frequency in the quantum wells. The oscillator strength of this mode is proportional to the number of quantum wells, so it is super-radiant.

Effect of external electric field on optical reflection from RBS is especially interesting for application in active photonic devices. The energies and oscillator strength of the quantum well excitons depend on the voltage applied. As a result the resonant optical response of the RBS can be detuned or modified [6,7].

Creating RBS is a quite difficult task. A system with a large number of quantum wells is required to obtain a strong super-radiant mode. The creation of such structures is complicated due to significant growth time and large thickness. The structures must be uniform with low homogeneous and inhomogeneous broadening of the excitonic states in quantum wells. The best growth technology currently available for such purpose is molecular-beam epitaxy of GaAs with AlGaAs barriers. There are several studies dedicated to RBS in this system [6-11].

Content from this work may be used under the terms of the Creative Commons Attribution 3.0 licence. Any further distribution of this work must maintain attribution to the author(s) and the title of the work, journal citation and DOI.
Published under licence by IOP Publishing Ltd

In this work we study optical reflection from GaAs/AlGaAs RBS as a function of temperature, polarization and angle of the light incidence. We also studied changes in the optical reflection when an external bias is applied to the RBS.

2. Samples and experimental details

The structures were grown by molecular beam epitaxy on conductive 2-inch GaAs substrates with (001) orientation. The design of structure is shown in Figure 1. Structure consists of 60 periodically repeating elementary supercells. Each supercell contains two GaAs quantum wells separated by a thin AlAs barrier. The thicknesses of the GaAs quantum wells and AlAs barriers are 7.8 nm and 4.8 nm, espectively. A thick (97.3 nm) AlGaAs barriers with he aluminum content of 20% separate the pairs of uantum wells. The top of the structure is covered with nti-reflective conductive ITO layer.

The optical reflection spectra with *s* and *p* olarizations for various angles of the light incidence vere measured in the temperature range from 10K to 70K. An Ocean Optics LS-1 lamp was used as light ource. A Glan prism was used as a light polarizer. The ight was transmitted from the source to the sample and rom sample to the detector by fiber optic cables. An cean Optics HR4000 spectrometer was used to egister the reflected optical signal, which allowed us o measure in the range from 190 to 1100 nm. The pectra were recorded from unbiased sample, as well as vith applied voltage of +4V and -4V. The electric urrent did not exceed 3 μA.

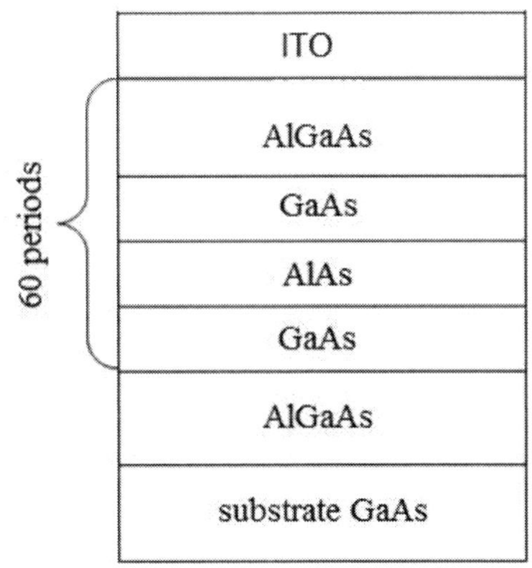

Figure 1. Structure of sample.

3. Results of experiment

Figure 2 shows experimental optical reflection spectra recorded with s-polarized light for the incident angles of 23°, 45° and 67° at the temperature of 10K. Characteristic features of spectra are marked with arrows. Marker 1 indicates the band gap of the AlGaAs barrier, which is 1.78 eV at 10K and does not depend on the angle of light incidence. Marker 2 indicates the exciton resonant energy in the quantum wells, which is 1.58 eV and also do not depend on the angle of light incidence. Arrows 3 in the spectra point out to the positions of Bragg peaks. The position of these peaks is associated with periodic changes of the refractive index in the structure. With increasing the angle of incidence the Bragg peak shifts to higher energies in accord to the Bragg's Law. The amplitude of Bragg peak is 25%, 54% and 70% at the angles of 23°, 45° and 67°, respectively. Small smooth peaks over a wide range of the spectra are Bragg satellites. ITO layer provides low non-resonant reflection in the spectral range of interest at small angles of the light incidence.

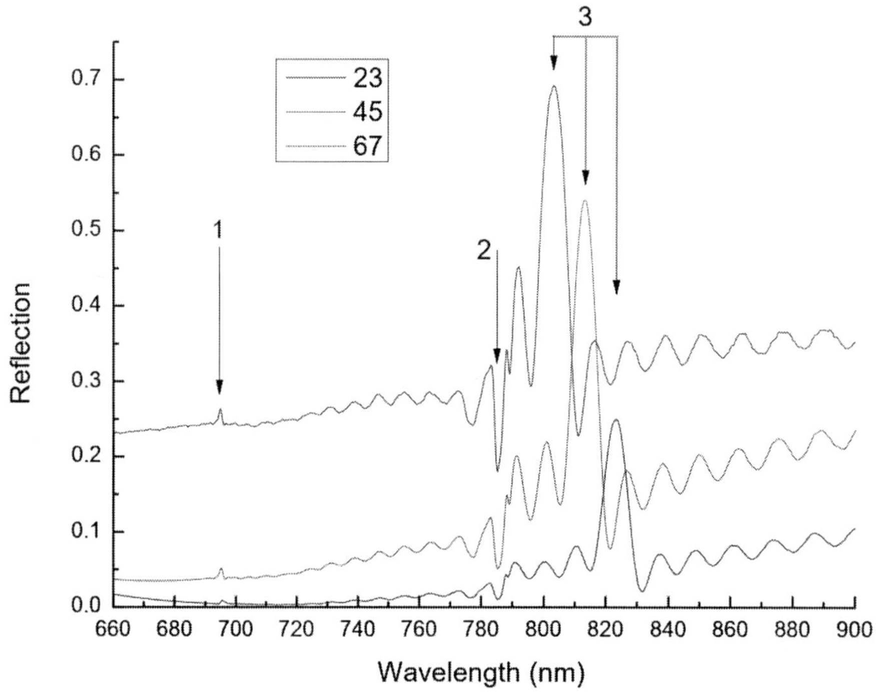

Figure 2. Optical reflection spectra recorded at 10K for s-polarized light incident at angles of 23°, 45° and 67°. Arrows show the characteristic features the spectrum.

Figure 3 shows optical reflection spectra recorded at different temperatures, namely 10K, 130K, 200K and 270K. The light has *p*-polarization and its angle of incidence is 23°. The peak positions of quantum well exciton are signed by arrows. The labelled exciton energies are derived from our theoretical calculation. It is obvious that the quantum well exciton peak shifts towards lower energy with increasing temperature, which is a result of shrinking of the band gap. At low temperature (10 and 130 K) the arrows clearly indicate features in the experimental reflection spectra. When the temperature is elevated to 200 and father to 270K, these excitonic features become hardly visible.

The major reflection peak at any temperature originates from Bragg reflection. The amplitudes of Bragg peaks are 25%, 29%, 49%, and 92% for temperatures of 10K, 130K, 200K and 270K, respectively. The width of this peak is 16.6±0.6 meV and it apparently does not depend on temperature.

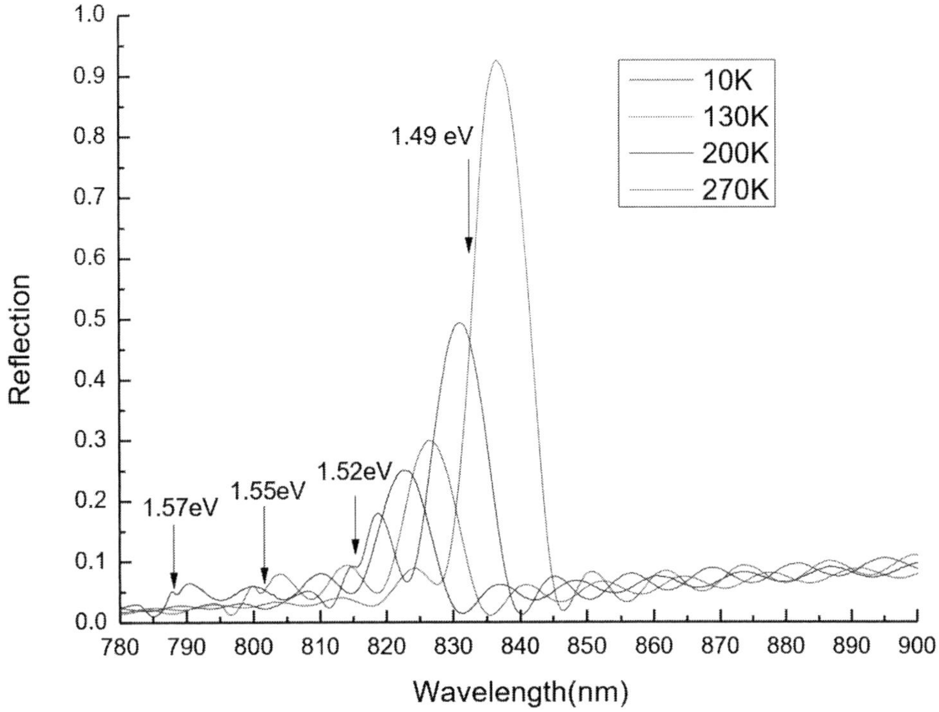

Figure 3. Optical reflection spectra from GaAs/AlGaAs RBS at 10K, 130K, 200K and 270K. The light is *p*-polarized and the angle of incidence is 23°.

The optical reflection spectra were also measured with a DC voltage applied between the substrate and ITO cap layer. In order to precisely reveal the impact of the external electric field we calculated differential spectra by subtracting the spectra without bias from biased ones recorder under the same conditions. Next, we calculated the ratio of these differential spectra to the spectrum recorded unbiased. Figure 4 shows thus obtained normalized differential spectra $\Delta R/R$, which corresponds to the temperature of 170K and light incidence angle of 23°. Noise (~1%) is visible in the spectrum. Arrow 1 matches with the energy gap in the AlGaAs barrier, which is 1.74 eV at 170K. Arrow 2 indicates a sharp feature with amplitude of ~ 9% for voltage of -4V and ~4% for +4V. This feature occurs at the energy of the quantum well excitons, which is 1.537 eV. The full width of the peak is 7.2 meV. Third arrow corresponds to the change in the Bragg reflection. It is centered at 1.496 eV. The apparent width is in between 13 and 23 meV.

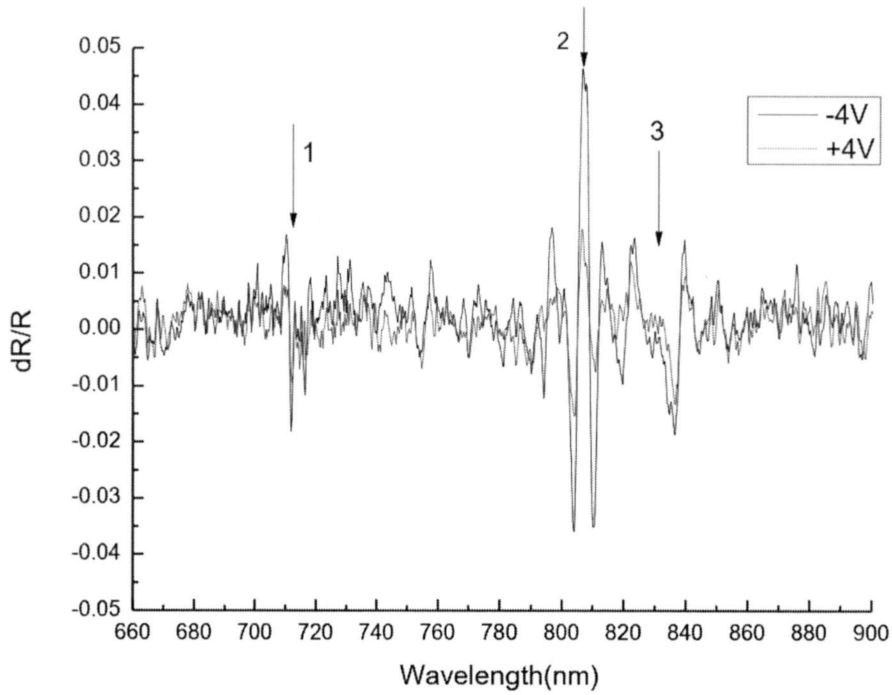

Figure 4. Dependence of a relative change in the reflection with a voltage of -4V and +4V. Temperature is 170K. Angle of incidence is 23°. Arrows indicate barrier band gap (1), quantum-well excitons (2) and Bragg band (3) features.

4. Discussion and conclusions

In this work we studied optical reflection from RBS structure consisting of 60 pairs of GaAs quantum wells separated by AlGaAs barriers. The condition of the formation of the collective exciton-polariton mode in RBS is the coincidence of the Bragg and exciton resonances. There are two ways to achieve the double resonance conditions. One of them is changing the angle of the light incidence at a constant temperature. In this case, the position of main Bragg peak is shifted. The other one is changing the temperature with a fixed angle of the light incidence, which results in a shift of the resonant frequency of the quantum well excitons.

The first way is illustrated in Figure 2 where the spectra were recorded at a constant temperature for various angles of the light incidence. In all the represented spectra the exciton and Bragg resonances are detuned from each other, so that the resonance condition is not reached. It is evident, however, that the magnitudes of both resonances increase when the Bragg and exciton peak approach each other. It is a result of interaction of the exciton-polariton and lattice-polariton modes.

In order to achieve the double resonance conditions we used the second way, i.e. variation of the temperature. One can see in Figure 3 that both exciton and Bragg peaks shift toward longer wavelength when temperature is elevated. Since the exciton shifts stronger, the two resonances can be gradually tuned to each other. Similarly to tuning by the incident angle, the tuning by temperature results in a growth of the Bragg peak amplitude as it approaches the exciton resonance. However, the exciton peak disappears from the spectrum above certain temperature. In order to follow its position we performed quantum-mechanical calculations taking into account temperature dependence of the band gap of GaAs quantum wells and AlGaAs barriers. The calculated energies of excitonic transitions are shown in the Figure 3. The arrows mark corresponding wavelengths. One can see the features under the arrows in the low-temperature experimental spectra. The calculations show that at a temperature of 270 the exciton and the Bragg peak are just slightly detuned. As a result, the amplitude of the optical reflection becomes as high as 92% with the band width as large as 16 meV. This result is well consistent with previous observations [12].

External electric field affects the optical reflection as shown in Figure 4. The first sharp peak at 1.74 eV in the spectrum indicates the band gap of the AlGaAs barrier. It is evident in Figure 2 that the optical reflection as low as 0.8% at this wavelength. Taking into account relatively small variation $\Delta R/R \sim 3\%$ we can consider this modulation of the optical reflection to be not significant. The second sharp peak at 1.537 eV in Figure 4 corresponds to electro-reflection by excitons in quantum wells. The signal is substantially asymmetric in respect to positive and negative bias. It seems to be a result of the existence of some built-in electric field. Amplitude of the relative modulation of excitonic reflectivity is 9%, when the absolute optical reflection is 9% at this wavelength. So, the electro-modulation of the optical reflection is enhanced for the quantum well excitons compared to the bulk excitons in AlGaAS barriers.

A weak electric field should not affect the Bragg reflection, which originates from some contrast in the refraction indices of GaAs and AlGaAs. In the Figure 4 one can see that the signal is almost absent near the middle of the Bragg band indicated by arrow 3. The signal appears at the edges of the band, being opposite in sign. Such signal can be produced by a small shift of the Bragg band, which may become possible due to interplay of the lattice-polaritons and exciton-polaritons [11].

Thus, we studied optical properties of GaAs/AlGaAs RBS with 60 periods of supercell containing two quantum wells. Our study showed that a periodic system of the quantum well excitons can be arranged as a good Bragg reflector with reflectivity of 92% and bandwidth of 16 meV near room temperature. A change in the optical reflection with an applied electric field was observed for both the excitonic resonance and Bragg band.

Acknowledgment
Authors are grateful to E.V. Nikitina and A.V. Babichev for sample preparations. The research by P.A. Tonkaev was partially supported by the Russian Foundation for Basic Research, grant #№ 14-29-07243 ofim.

References

[1] Hjelme D R, Bjerkan L, Neegard S, Rambech J S, and Aarsnes J V 1997 Application of Bragg grating sensors in the characterization of scaled marine vehicle models *Appl. Opt.* **36**, 328-336
[2] Ivchenko E L, Poddubnyi A N 2006 Resonant Three-Dimensional Photonic Crystals *Physics of the Solid State* **48** 581–588
[3] Ivchenko E L, Nesvizhskii A I, Jorda S 1994 Bragg reflection of light from quantum-well structures *Physics of the Solid State* **36** 1156-1161
[4] Voronov M M, Ivchenko E L, Poddubnyĭ A N, Chaldyshev V V 2006 Special frequencies in the optical reflectance spectra of resonant Bragg structures *Physics of the Solid Stat* **48** 1814–1819
[5] Voronov M M, Ivchenko E L, Kosobukin V A, Poddubnyĭ A N 2007 Specific features in reflectance and absorbance spectra of one-dimensional resonant photonic crystals *Physics of the Solid State* **49** 1792–1802
[6] Chaldyshev V V,. Kundelev E V, Yuechao Chen, Poddubny A N, Vasil`ev A P, Zhiheng Liu 2012 Resonant optical reflection by periodic systems of the GaAs/AlGaAs quantum well excitons, *Phys. Status Solidi C* **9** 1557-1559
[7] Chaldyshev V V, Sholohov D E, Vasil'ev A P 2010 An $(AlGaAs/GaAs/AlGaAs)_{60}$ resonant Bragg structure based on the second quantum-confinement level of heavy-hole excitons in quantum wells *Semiconductors* **44** 1222–1226
[8] Chaldyshev V V, Shkolnik A S, Evtikhiev V P and Holden T 2006 Optical reflection and contactless electroreflection from GaAlAs layers with periodically arranged GaAs quantum wells

Semiconductors **40** 1710
[9] ChaldyshevV V, ShkolnikA S, Evtikhiev V P and Holden T 2007 Modulation optical spectroscopy of excitons in structures with GaAs multiple quantum wells separated by tunneling-nontransparent barriers *Semiconductors* **41** 1455
[10] Chaldyshev V V, Shkolnik A S, Evtikhiev V P and Holden T 2008 Optical spectroscopy of a semi-insulating GaAs/AlGaAs multiple quantum well system near double excitons-polariton and Bragg resonance. *J.Mater.Sci.: Materials in Electronics* **19** 699.
[11] Goldberg D, Deych L I, Lisyansky A A, Shi Z, Menon V M, Tokranov V, Yakimov M and Oktyabrsky S 2009 *Nature Photonics* **3** 662
[12] Chaldyshev V V, Kundelev E V, Nikitina E V, Egorov A Yu, Gorbatsevich A A 2012 Resonance Reflection of Light by a Periodic System of Excitons in GaAs/AlGaAs Quantum Wells *Semiconductors* **46** 1016–1019

Relation Between the Cell Volume and the Cell Cycle Dynamics in Mammalian cell

A.C.G. Magno[1], I.L. Oliveira[1] and J.V.S. Hauck[2]

[1] Federal University of Juiz de fora, Juiz de Fora-Minas Gerais, Brasil
[2] Federal Institute of Rio de Janeiro, Engenheiro Paulo de Frontin- Rio de Janeiro, Brasil

E-mail: `alessandra.cristina.gomes.mgn@gmail.com`

Abstract. The main goal of this work is to add and analyze an equation that represents the volume in a dynamical model of the mammalian cell cycle proposed by Gérard and Goldbeter (2011) [1]. The cell division occurs when the $cyclinB/Cdk1$ complex is totally degraded (Tyson and Novak, 2011)[2] and it reaches a minimum value. At this point, the cell is divided into two newborn daughter cells and each one will contain the half of the cytoplasmic content of the mother cell. The equations of our base model are only valid if the cell volume, where the reactions occur, is constant. Whether the cell volume is not constant, that is, the rate of change of its volume with respect to time is explicitly taken into account in the mathematical model, then the equations of the original model are no longer valid. Therefore, every equations were modified from the mass conservation principle for considering a volume that changes with time. Through this approach, the cell volume affects all model variables. Two different dynamic simulation methods were accomplished: deterministic and stochastic. In the stochastic simulation, the volume affects every model's parameters which have molar unit, whereas in the deterministic one, it is incorporated into the differential equations. In deterministic simulation, the biochemical species may be in concentration units, while in stochastic simulation such species must be converted to number of molecules which are directly proportional to the cell volume. In an effort to understand the influence of the new equation a stability analysis was performed. This elucidates how the growth factor impacts the stability of the model's limit cycles. In conclusion, a more precise model, in comparison to the base model, was created for the cell cycle as it now takes into consideration the cell volume variation

1. Introduction

There is an increasing interest in modeling the cell cycle, specially the mammalian cell cycle. This may be explained as some diseases, like the cancer, or even the aging process are directly related to the cell cycle. Thus, understanding the cell cycle and how it is affected by those problems is essential for proposing new treatments. On that account, several models for the cell cycle were proposed. There are, for instance, models from a simple yeast cell to a complex human cell. There are mainly two methods of simulation, the ordinary deterministic and the stochastic. The deterministic method simulates the average concentration, which is in a large system precise and simple. On the other hand, a stochastic method simulates each reaction individually, following the probability of each reaction occurs. This is computationally more expensive, but can capture the stochastic noise that in small systems, like a cell, may lead to a different result in relation to the average.

Content from this work may be used under the terms of the Creative Commons Attribution 3.0 licence. Any further distribution of this work must maintain attribution to the author(s) and the title of the work, journal citation and DOI.
Published under licence by IOP Publishing Ltd

This work proposes an improvement for the model of the mammalian cell cycle proposed by [1, 3], to explicitly take in account the cell volume variation. This new model is then analyzed in order to verify how the volume variation affects the dynamics of the differential equations. Finally, the model is simulated using a deterministic and a stochastic method.

2. Proposed Model

This work develops an extension for the model proposed by [1]. The original model is composed by a set of differential equations (1-6). However, those equations do not contain any information related to the cell volume. In an effort to improve the model's completeness, information about how the cell volume changes along the cell cycle will be added.

$$\frac{dM_d}{dt}=v_{sd}\cdot\left(\frac{GF}{K_{gf}+GF}\right)-V_{dd}\cdot\left(\frac{M_d}{K_{dd}+M_d}\right) \quad (1)$$

$$\frac{dE2F}{dt}=V_{1e2f}\cdot\left(\frac{E2F_{tot}-E2F}{K_{1e2f}+E2F_{tot}-E2F}\right)\cdot(M_d+M_e) -V_{2e2f}\cdot\left(\frac{E2F}{K_{2e2f}+E2F}\right).M_a \quad (2)$$

$$\frac{dM_e}{dt}=v_{se}.E2F-V_{de}\cdot M_a\cdot\left(\frac{M_e}{K_{de}+M_e}\right) \quad (3)$$

$$\frac{dM_a}{dt}=v_{sa}.E2F-V_{da}.Cdc20.\left(\frac{M_a}{K_{da}+M_a}\right) \quad (4)$$

$$\frac{dM_b}{dt}=v_{sb}\cdot M_a-V_{db}.Cdc20.\left(\frac{M_b}{K_{db}+M_b}\right) \quad (5)$$

$$\frac{dCdc20}{dt}=V_{1cdc20}\cdot M_b\cdot\left(\frac{Cdc20_{tot}-Cdc20}{K_{1cdc20}+Cdc20_{tot}-Cdc20}\right) -V_{2cdc20}\cdot\left(\frac{Cdc20}{K_{2cdc20}+Cdc20}\right) \quad (6)$$

As we are working with concentrations ($\frac{\mu M}{L}$), the Equations (1-6) are only valid when the cell volume is constant. So, when the cell volume is not constant, these equations are no longer valid. However, these equations were developed from the mass conservation principle, therefore, it is the mass that is conserved not the concentration. Let N_e be the equivalent of M_e in moles $[M]$ and V the cell volume $[L]$. Now, the Equation (3) can be rewrite as in Equation (7), having $M_e \cdot V = N_e$.

As the cell volume is not constant, the product rule must be applied on the left side of the Equation (7). The equation is then divided by V and the original term $\frac{dM_e}{dt}$ is isolated on the left side, resulting in the Equation (8).

It is reasonable to suppose that a cell growth is proportional to its volume, following an exponential growth during an active cell cycle [4]. Thus, the cell volume may be represented by the Equation (9).

Using this expression (Eq. 9) for $\frac{dV}{dt}$ in the equation (8) results in a simpler equation (10). This equation now takes into account the cell volume variation, and the term $-\mu M_e$ is called dilution factor due the cell growth. The same procedure can be applied to each equation of the original model, resulting in a new set o differential equation composed by the transformed equations in addition to the volume differential equation (9). The cell division is done when the $cyclinB/Cdk1$ complex is totally degraded, like the work presented by [2]. At this moment the volume V is cut by half. In the stochastic version, the amount of molecules of each chemical is also reduced to its half.

$$\frac{d(VM_e)}{dt}=V\left[v_{se}.E2F-V_{de}\cdot M_a\cdot\left(\frac{M_e}{K_{de}+M_e}\right)\right] \quad (7)$$

$$\frac{dV}{dt}=\mu V \quad (9)$$

$$\frac{dM_e}{dt}=v_{se}.E2F-V_{de}\cdot M_a\cdot\left(\frac{M_e}{K_{de}+M_e}\right)-\frac{M_e}{V}\frac{dV}{dt} \quad (8)$$

$$\frac{dM_e}{dt}=v_{se}.E2F-V_{de}\cdot M_a\cdot\left(\frac{M_e}{K_{de}+M_e}\right)-\mu M_e \quad (10)$$

2.1. Stochastic

As cells are small, the stochasticity may impact the behavior of the system [5]. Thence, a stochastic version of this model is also proposed. To do so, each variable and parameter having a concentration unit ($\frac{\mu M}{L}$) must be converted to number of molecules. This number is directly proportional to the cell volume. This conversion is done by multiplying the value in concentration unit by a factor of $\Omega = V_c \cdot N_A \cdot 10^{-6}$, where V_c is the cell volume in liters (L), N_A is the Avogadro constant and the 10^{-6} factor is to convert the parameters unit from $\frac{\mu M}{L}$ to $\frac{M}{L}$ [6].

In order to model the system as a stochastic process, each differential equation has to be separated in two [4]. The first one represents the activation a new molecule, while the second represents the deactivation of a molecule. The stochastic model is represented in the Table (1). Additionally, we implement the direct method, proposed by [6], for simulating the stochastic model.

Table 1: Stochastic model.

Reaction Id	Reaction	Propensity of the reaction
1	$\longrightarrow$ Md	$v_{sd}.\Omega.\left(\frac{GF}{K_{gf}.\Omega+GF}\right)$
2	Md $\longrightarrow$	$V_{dd}.\Omega.\left(\frac{Md}{K_{dd}.\Omega+Md}\right)$
3	$\longrightarrow$ E2F	$V_{1e2f}.\left(\frac{E2F_{tot}.\Omega-E2F}{K_{1e2f}.\Omega+E2F_{tot}.\Omega-E2F}\right).(Md+Me)$
4	E2F $\longrightarrow$	$V_{2e2f}.\left(\frac{E2F}{K_{2e2f}.\Omega+E2F}\right).Ma$
5	$\longrightarrow$ Me	$v_{se}.E2F$
6	Me $\longrightarrow$	$V_{de}.Ma.\left(\frac{Me}{K_{de}.\Omega+Me}\right)$
7	$\longrightarrow$ Ma	$v_{sa}.E2F$
8	Ma $\longrightarrow$	$V_{da}.Cdc20.\left(\frac{Ma}{K_{da}.\Omega+Ma}\right)$
9	$\longrightarrow$ Mb	$v_{sb}.Ma$
10	Mb $\longrightarrow$	$V_{db}.Cdc20.\left(\frac{Mb}{K_{db}.\Omega+Mb}\right)$
11	$\longrightarrow$ Cdc20	$V_{1cdc20}.Mb.\left(\frac{Cdc20_{tot}.\Omega-Cdc20}{K_{1cdc20}.\Omega+Cdc20_{tot}.\Omega-Cdc20}\right)$
12	Cdc20 $\longrightarrow$	$V_{2cdc20}.\Omega.\left(\frac{Cdc20}{K_{2cdc20}.\Omega+Cdc20}\right)$

This model is easily adapted to a model with a non constant volume. As we are working with number of molecules, the equations are already valid for a variable volume. However, the conversion factor Ω will no longer be constant. Even though this seems a small change, it is not, because Ω affects the value of several parameters.

3. Results

This section will present an analysis of the the proposed method, along with the results found when the model is both stochastically and deterministically simulated. In an effort to stimulate the reproducibility, the values used for each parameter of the model can be found in [7].

3.1. Stability analysis

In order to present a stable oscillatory behavior, an ODE system must have at least one limit cycle [8]. The original model, proposed in [1], has a limit cycle. Thus, the model proposed here must preserve this limit cycle. As our model introduces an extra parameter, an analysis of the impact of this new parameter is presented.

The first thing one must do to perform a stability analysis is to find the stable points. These points can be found by solving a nonlinear system which is obtained when each derivate is made equal to zero[9]. Although the system obtained has several solutions, when the solutions are limited to the positive Real numbers there are only two solutions P_1 and P_2 (Tab. 2). However, the second point P_2 is also not valid because the $E2F$ value exceeds the maximum value allowed of 3.0 defined by the parameter $E2F_{tot}$. So, the following analysis will only focus in the valid point P_1.

Table 2: Equilibrium points

	$Cdc20$	$E2f$	Ma	Mb	Me	Md
P_1	0.1738902240	0.2448834860	0.2901560559	0.3365279086	0.08954976057	0.165495
P_2	1.786431803	3.010457122	2.715508604	1.834578112	0.1783832744	0.165495

In an effort to further analyze the equilibrium point P_1, a linearized version of the system was used in its the neighborhood. In this simplified version the eigenvalues were computed for inferring the stability of the point. These values are found in [7]. As there are complex eigenvalues, we can conclude that the system presents an oscillatory behavior in the neighborhood of the point P_1[9].

For validating our model, we must verify the impact of the parameter μ over the equilibrium point. To perform this analysis, the matcont[10] extension of the well-known MATLAB software was used. The result shows that the system behavior is preserved until the critical point $\mu = 0.11$ (Fig. 1). At this point, a Hopf bifurcation occurs, this means that a limit cycle is created or vanished at this point [11]. Hopefully, this values is much bigger than the actual value computed for being used $\mu = 0.0385$.

The Figures (Fig. 2) and (Fig. 3) show the phase plane $M_a \times M_b$ for several values of μ. However, the first one shows distinct cycles on the plane, while the second is a 3D graph composed by phase plane $M_a \times M_b$ and a height given by μ. These graphs evince that as μ approaches 0.11, the limit cycle shrinks until it reaches the critical point 0.11 where it vanishes. In fact, this evinces that the parameter μ influences the stability of the limit cycle, but it also evinces that it only changes the system's behavior when its value is much greater than the value we estimate (0.0385 against 0.11).

3.2. Deterministic Simulation

The Figure 4 shows the results of simulating the original model proposed by [1], while the Figure 5 shows the results of our new model. Even though our model presents lower concentrations, it is clear that they have a similar behavior. These lower concentrations are expected as we introduce a factor of dilution. One may also notice that the length of the cycle is slighted shorter, to be precise the cycle reduces from $23,573$ to $21,816$ hours, in our model. However, this is not sufficient to invalidate our model.

3.3. Stochastic Simulation

In an effort to produce reliable results, the results are presented in terms of concentrations. This allows a direct comparison between the stochastic version and the deterministic version. In order to obtain the concentration C, the following operation was done $C = \frac{N_m}{\Omega}$, where N_m is the number of molecules.

The graph illustrated in Figure 6 shows the results for our new model. The volume here varies between $0.334pL$ to $0.7pL$ ($0.334pL$ is the average volume of human granulocytes [12]), following the Differential Equation 7. This result is close to deterministic version, indicating that our stochastic model is also correct.

4. Conclusion and Future Works

In this work a new mathematical model for the mammalian cell cycle was developed. This new model takes into account not only the concentrations of some key elements, but also explicitly models the cell volume variation. Finally, this model also has a key moment where the cell division explicitly occurs, cutting by a half the cell volume.

Another important contribution of this work is the analysis of the impact of the new parameter μ over the stability of the model. This analysis concludes that only high values of μ affects the model's stability. As values of this magnitude are not viable from a physiological point of view, we may conclude that parameter impact, over the stability, is very small.

As a future work, we can include the influence of a chemotherapeutic on the cell cycle model. Among several chemotherapeutic, an interesting options is the cisplatin family, which inhibits the DNA synthesis. Therefore, it can be included in the model affecting the phase S.

Another future can be adapting the model for representing the cell cycle of a cancer cell. This may enhance the comprehension of which parts of the cell cycle are affected by each type cancer, guiding the development of efficient drugs.

5. Figures

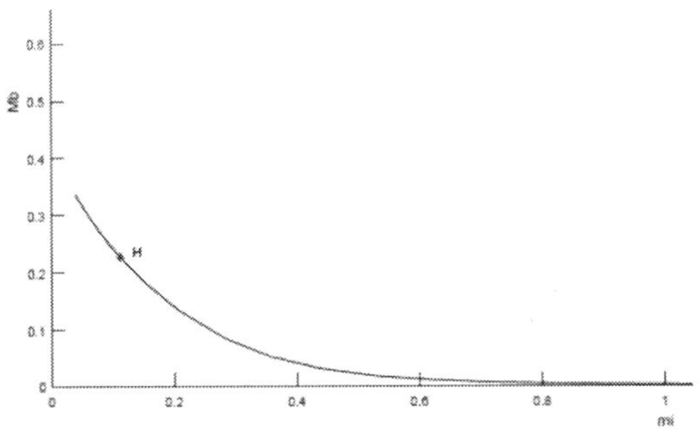

Figure 1: Diagram of bifurcation. Equilibrium value of Mb in relation the μ value. The point H represents the Hopf bifurcation ($\mu = 0.11$).

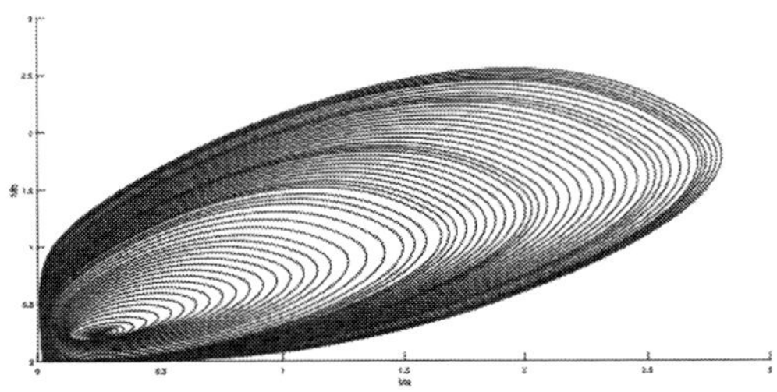

Figure 2: Phase plane $M_a \times M_b$. Each curve represents a distinct value of μ

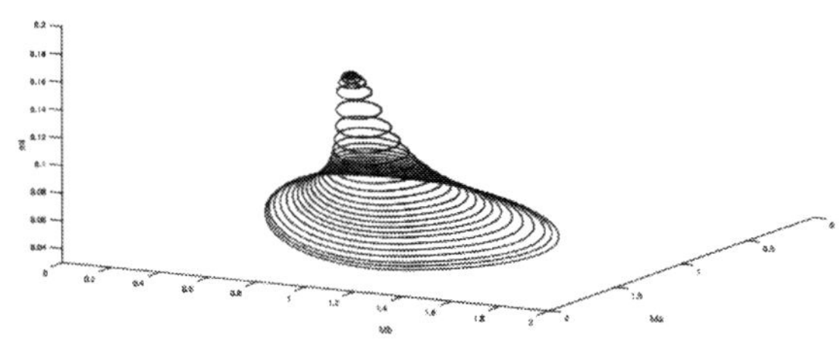

Figure 3: Graph of several phase planes $M_a \times M_b$ with μ being the height.

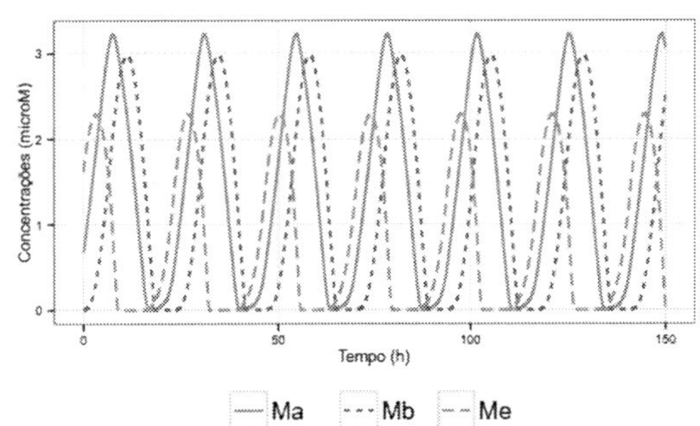

Figure 4: Complexes of $cyclinA/Cdk2$ (M_a), $cyclinB/Cdk1$ (M_b) and $cyclinE/Cdk2$ (M_e). Deterministic simulation of the cell cycle with constant volume[1].

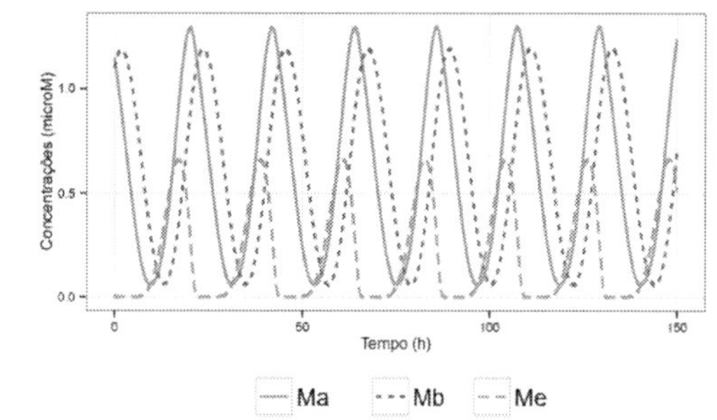

Figure 5: Complexes of $cyclinA/Cdk2$ (M_a), $cyclinB/Cdk1$ (M_b) and $cyclinE/Cdk2$ (M_e). Deterministic simulation of the model proposed here.

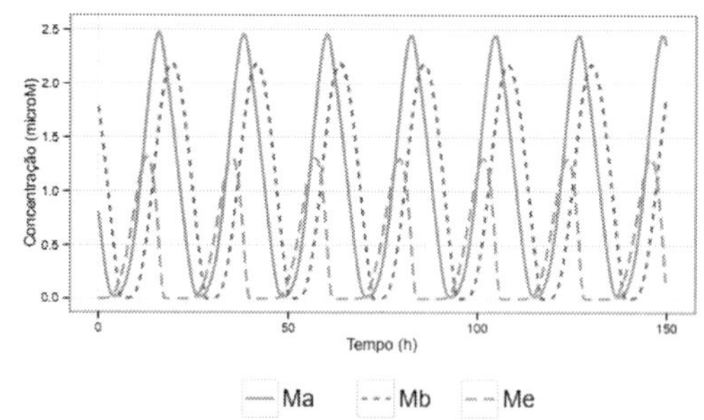

Figure 6: Complexes of $cyclinA/Cdk2$, $cyclinB/Cdk1$ and $cyclinE/Cdk2$. Mean concentrations obtained by the average of 500 stochastic simulations of the model with variable volume.

6. Acknowledgements
The financial support by IFRJ is greatly acknowledged.

References
[1] Gérard C and Goldbeter A 2011 *Interface Focus* rsfs20100008
[2] Tyson J J and Novak B 2001 *Journal of Theoretical Biology* **210** 249–263
[3] Gérard C, Gonze D and Goldbeter A 2012 *FEBS Journal* **279** 3411–3431
[4] Oliveira I L 2008
[5] Elowitz M B, Levine A J, Siggia E D and Swain P S 2002 *Science* **297** 1183–1186
[6] Gillespie D T 1977 *The journal of physical chemistry* **81** 2340–2361
[7] Gomes Magno A C 2016 *Relação Entre o Volume da Célula e Dinamica do Ciclo Celular em Mamíferos* Master's thesis Universidade Federal de Juiz de Fora, Instituto de Ciências Exatas, Departamento de Ciência da Computção, Programa de Pós-Graduação em Ciência da Computção
[8] Hurewicz W 1964 *Lectures on ordinary differential equations* (Courier Corporation)
[9] ZILL D G and CULLEN M R 2001 Equações diferenciais. vol. 2. trad. alfredo alves de farias
[10] Dhooge A, Govaerts W and Kuznetsov Y A 2003 *ACM Transactions on Mathematical Software (TOMS)* **29** 141–164
[11] de Souza G H 2015 *Bifurcação de Hopf em sistemas de equações diferenciais com simetria* Master's thesis Universidade Estadual de Maringá, Centro de Ciências Exatas, Departamento de Matemática, Programa de Pós-Graduação em Matemática
[12] Nibbering P H, Zomerdijk T P, Corsèl-Van Tilburg A J and Van Furth R 1990 *Journal of immunological methods* **129** 143–145

5th International Conference on Mathematical Modeling in Physical Sciences (IC-MSquare 2016) IOP Publishing
Journal of Physics: Conference Series **738** (2016) 012062 doi:10.1088/1742-6596/738/1/012062

Determining the Critical Point of a Sigmoidal Curve via its Fourier Transform

Ayse Humeyra Bilge and Yunus Ozdemir
Faculty of Engineering and Natural Sciences, Kadir Has University, Istanbul, Turkey,
Department of Mathematics, Anadolu University, Eskisehir, Turkey.
E-mail: ayse.bilge@khas.edu.tr, yunuso@anadolu.edu.tr

Abstract. A sigmoidal curve $y(t)$ is a monotone increasing curve such that all derivatives vanish at infinity. Let t_n be the point where the nth derivative of $y(t)$ reaches its global extremum. In the previous work on sol-gel transition modelled by the Susceptible-Infected-Recovered (SIR) system, we observed that the sequence $\{t_n\}$ seemed to converge to a point that agrees qualitatively with the location of the gel point [2]. In the present work we outline a proof that for sigmoidal curves satisfying fairly general assumptions on their Fourier transform, the sequence $\{t_n\}$ is convergent and we call it "the critical point of the sigmoidal curve". In the context of phase transitions, the limit point is interpreted as a junction point of two different regimes where all derivatives undergo their highest rate of change.

1. Introduction

Let $y(t)$ be a monotone increasing function with horizontal asymptotes y_1 and y_2 as $t \to \pm\infty$ and with $\lim_{t\to\pm\infty} y^{(n)}(t) = 0$ for all $n \geq 1$, i.e, a sigmoidal curve. Let t_n be the point where the nth derivative of $y(t)$ reaches its global extremum. If the sequence $\{t_n\}$ converges, its limit is called the "critical point" of the sigmoidal curve $y(t)$.

The present work is motivated by an observation on the behaviour of the derivatives of sigmoidal curves representing gelation phenomena, modelled by the Susceptible-Infected-Removed (SIR) system of differential equations [2]. We have seen that the points t_n where higher order derivatives reach their absolute extreme values seemed to accumulate at a point in between the zeros of the second and first derivatives. The location of this accumulation point agreed qualitatively with the so-called "gel point". Later on, we computed higher order derivatives for a number of different sigmoidal functions and we observed the same type of accumulation behaviour interpreted as the existence of a critical point in the sense defined above. For odd sigmoidal curves, we observed that the critical point was always located at $t = 0$, i.e, at the zero of the first derivative, as expected, but we were unable to prove this even for the simplest functions. After trying numerous techniques we could explain the existence of the critical point in terms of the Fourier transform of the first derivative of the sigmoidal curve. In this paper, we outline the steps leading to the proof of the existence of the critical point. Complete proofs that are quite technical and lengthy will be presented elsewhere.

Content from this work may be used under the terms of the Creative Commons Attribution 3.0 licence. Any further distribution of this work must maintain attribution to the author(s) and the title of the work, journal citation and DOI.
Published under licence by IOP Publishing Ltd

2. Preliminaries

We briefly outline basic properties of Fourier transform. For simplicity assume that $f(t)$ is in L^1. Then its Fourier transform $\mathcal{F}(f) = F$ and the inverse transform $\mathcal{F}^{-1}(F) = f$ are defined by

$$F(\omega) = \frac{1}{\sqrt{2\pi}} \int_{-\infty}^{\infty} f(t)\, e^{-i\omega t}\, dt, \qquad f(t) = \frac{1}{\sqrt{2\pi}} \int_{-\infty}^{\infty} F(\omega)\, e^{i\omega t}\, d\omega$$

respectively. The effect of differentiation in the time domain is multiplication by $i\omega$ in the frequency domain, i.e, $\mathcal{F}(f^{(n)}(t)) = (i\omega)^n F(\omega)$. Multiplication and convolution in the time and frequency domains are related by $\mathcal{F}(f(t)\, g(t)) = \frac{1}{\sqrt{2\pi}} F(\omega) * G(\omega)$. The "modulation" of low frequency signal in the time domain is the multiplication of this signal by a sinusoidal function of fixed (usually high) angular frequency ω_0. In the frequency domain, the Fourier transform of the low frequency function is convolved with the Fourier transform of the sinusoid. The Fourier transform of a pure sinusoid is represented by Dirac δ functions occurring at $\pm\omega_0$ and convolution carries the spectrum of the low frequency signal to the frequencies $\pm\omega_0$. It follows that multiplication in the time domain by a complex exponential results in a shift in the frequency domain. Similarly, multiplication by a linear phase factor in the frequency domain leads to a shift in the time domain, as given below:

$$\mathcal{F}(f(t)e^{i\omega_0 t}) = F(\omega - \omega_0), \qquad \mathcal{F}^{-1}(e^{-i\alpha\omega} F(\omega)) = f(t - \alpha).$$

(See [3, 4] for more details.)

3. The existence and non-existence of the critical point

We will use the generalized logistic growth family of curves and its limiting function the Gompertz curve to illustrate the existence and the nonexistence of the critical point and its location. The generalized logistic growth curve with horizontal asymptotes at -1 and 1 is given by

$$y(t) = -1 + 2\left[1 + ke^{-\beta t}\right]^{-1/\nu},$$

where $k > 0$, $\beta > 0$ and $\nu > 0$. The parameter k can be adjusted by a time shift, β corresponds to a scaling of time and ν is the key parameter that determines the shape of the growth. For $\nu = 1$, $k = 1$ and $\beta = 2$ we obtain the standard logistic growth $y(t) = \tanh(t)$. The Gompertz curve obtained as the limit of the generalized logistic family for $k = 1/n$, $\nu = 1/n$, as $n \to \infty$ is the function

$$y(t) = -1 + 2e^{-e^{-\beta t}}.$$

We present the behavior of the derivatives of these functions in Figure 1. The critical point of the standard logistic curve is located at $t = 0$; the choice $k = 1$ ensures that the critical point of the generalized logistic curve is also is located at the same point. The Gompertz curve has no critical point, because the points t_n move to negative infinity.

One ingredient for the existence of a critical point is the fact that higher derivatives of the sigmoidal function behave as localized humps modulated by sinusoids of increasingly high frequencies and these wave packets accumulate near a certain point, as shown below in Figure 2 (see [1] for the Fourier transform of the first derivative of the generalized logistic curve).

The wave packet behavior in the time domain corresponds to the band-pass property of its Fourier transform. That is, if higher derivatives resemble more and more closely sinusoids of increasing frequencies, their Fourier transforms become more and more localized humps located at higher and higher frequencies. The wave packet behavior is not sufficient to ensure the existence of the critical point, as in the case of the Gompertz functions these wave packets move to negative infinity. These wave packets should be agglomerate as the order of differentiation

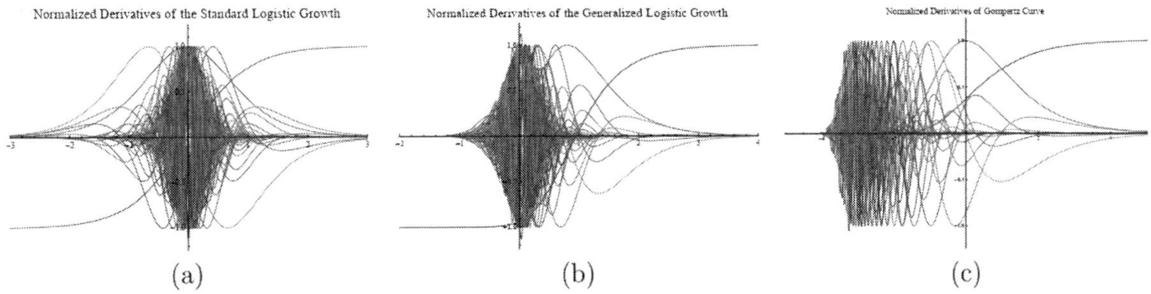

Figure 1. (a) Normalized derivatives of the standard logistic growth; (b) Normalized derivatives of the generalized logistic growth ($\beta = 1, k = 1, \nu = 1/5$) up to order 30. The behavior of the generalized logistic is more or less the same except that the zeros of even derivatives are not fixed. (c) The Gompertz function is the limit of the generalized logistic family, the critical point move to negative infinity.

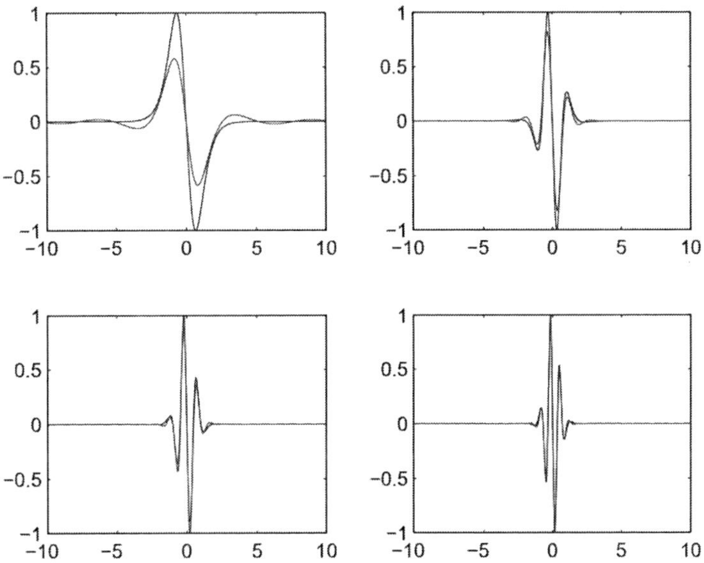

Figure 2. Comparison of the derivatives $f^{(n)}(t)$ of the standard logistic function with sinusoids of frequency ω_n.

increases. The property that ensures this is the "asymptotically constant phase" condition, given in Proposition 1 below.

The main result is the proposition below, whose proof will be published elsewhere. We claim that if $f(t)$ satisfies asymptotically constant phase and band-pass hypotheses, to be specified below, then the critical point is located at $t = 0$. Note that the asymptotically constant phase condition is trivially satisfied when $f(t)$ is even.

Proposition 1 *Let $f(t)$ be the first derivative of a sigmoidal curve $y(t)$ and $f^{(n)}(t)$ be its nth derivative. If*

i the Fourier transform of $f(t)$ has the form $F(\omega) = |F(\omega)|e^{-i\alpha\omega}e^{i\psi(\omega)}$ where α is a constant *and $\psi(\omega)$ has horizontal asymptotes,*

ii for $\omega > 0$, $\omega^n|F(\omega)|$ has a single maximum at ω_n and ω_n's are unbounded,

iii the spectrum is localized in the sense that there are constants ω_a and ω_b (depending on n), such that

$$\lim_{n\to\infty}\int_{|\omega|<\omega_a}\omega^n|F(\omega)|\,d\omega = \lim_{n\to\infty}\int_{|\omega|>\omega_b}\omega^n|F(\omega)|\,d\omega = 0,$$

then the sigmoidal curve $y(t)$ has a critical point located at $t = \alpha$.

The first condition is the key "asymptotically constant phase" assumption. If $F(\omega)$ is as in (i), then an appropriate shift in time eliminates this phase factor and the phase of $F(\omega)$ becomes asymptotically constant. The requirement of the existence of a single maximum in (ii) is technical; what we need is ensure that, as n goes to infinity, the the Fourier spectrum of the nth derivatives be shifted to the region where the phase is nearly constant. Finally (iii) is again a technical assumption to ensure that the spectrum of the nth derivative is localized. The proof consists of expressing $|f^n(t)|$ using the Fourier inversion formula and proving that for large n, it is less than $|f^n(0)|$. Intermediate steps include the determination of the location of the spectrum of the nth derivative, from the equality $\omega/n = F(\omega)/F'(\omega)$. The behavior of the solutions is shown in Figure 3, for the standard logistic growth.

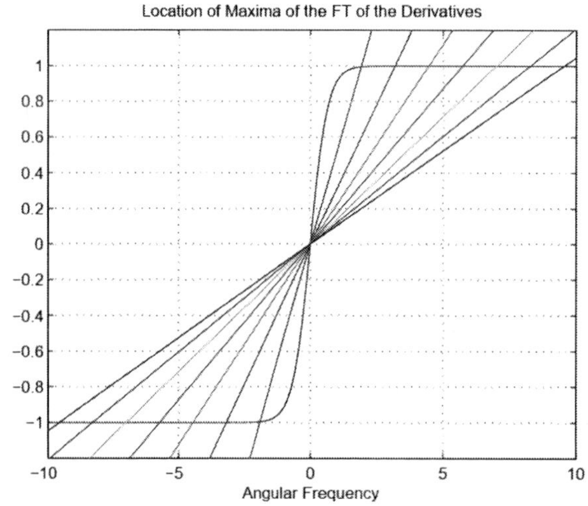

Figure 3. Graphical solution of the equation $\omega/n = F(\omega)/F'(\omega)$ for the standard logistic growth.

Finally we note that the asymptotically constant phase assumption leads to an intrinsic definition of even and odd components of a function $f(t)$, by choosing the origin of the time axis in such a way that the Fourier transform has asymptotically constant phase.

[1] Bilge, A.H., Ozdemir, Y. *The Fourier Transform of the First Derivative of the Generalized Logistic Growth Curve.* (2015), arXiv:1502.07182 [math.CA].
[2] Bilge, A.H., Pekcan, O. *A mathematical Description of the Critical Point in Phase Transitions.* Int. J. Mod. Phys. C **24** (2013).
[3] Duoandikoetxea, J. *Fourier analysis, Graduate Studies in Mathematics*; **29**, AMS, Providence, RI, 2001.
[4] Papoulis, A. *The Fourier Integral and its Applications*; McGraw-Hill Co., New York, 1962.

Mesoscopic Simulations of Crosslinked Polymer Networks

Grigorios Megariotis[a,1], Georgios G. Vogiatzis[1], Ludwig Schneider[2], Marcus Müller[2], Doros N. Theodorou[1]
[1]School of Chemical Engineering, National Technical University of Athens, Athens, Greece
[2]Institut für Theoretische Physik, Georg-August Universität, Göttingen, Germany

E-mail: gregm@mail.ntua.gr

Abstract. A new methodology and the corresponding C++ code for mesoscopic simulations of elastomers are presented. The test system, crosslinked cis-1,4-polyisoprene, is simulated with a Brownian Dynamics/kinetic Monte Carlo algorithm as a dense liquid of soft, coarse-grained beads, each representing 5-10 Kuhn segments. From the thermodynamic point of view, the system is described by a Helmholtz free-energy containing contributions from entropic springs between successive beads along a chain, slip-springs representing entanglements between beads on different chains, and non-bonded interactions. The methodology is employed for the calculation of the stress relaxation function from simulations of several microseconds at equilibrium, as well as for the prediction of stress-strain curves of crosslinked polymer networks under deformation.

1. Introduction

Atomistic and mesoscopic simulations are widely employed for the study of polymer systems, since they provide insights that are complementary to the information derived from experiments. Although the full-atom or united-atom representations provide an accurate description of polymers, their long relaxation time constitutes a severe obstacle to such approaches, and thus the development of mesoscopic (or coarse-grained) models is needed to cover longer time and length scales. One of the main characteristics of polymer melts and polymer networks is the entanglement effect, arising due to the uncrossability of polymer chains, which gives rise to complicated topological constraints [1]-[4]. The tube model, which considers a single chain in a mean field, is one of the most significant models for the description of entangled polymers [5], [6]. The model in question has been refined by introducing additional relaxation mechanisms such as contour length fluctuation and constraint release [7]-[9]. Also, slip-link and slip-spring models [10]-[22] have been introduced for the description of entanglement effects. In slip-link models a number of discrete points restrict the lateral motion of the chains, whereas the chain can slide through them [10]-[12]. Likhtman introduced a single-chain slip-link model [15]. Alternatively the lateral motion of the chain is restricted by slip-springs connecting pairs of beads [21], with slip-springs being created or destroyed at chain ends. The additional attraction in melts or solutions that is induced by slip-springs can be exactly compensated by a pairwise repulsive potential. Both concepts, of slip-links and slip-springs, have been widely employed in mesoscopic multi-chain simulations for the computation of viscoelastic properties of polymer systems [16]-[22].

[a] To whom any correspondence should be addressed.

Content from this work may be used under the terms of the Creative Commons Attribution 3.0 licence. Any further distribution of this work must maintain attribution to the author(s) and the title of the work, journal citation and DOI.
Published under licence by IOP Publishing Ltd

Polymer networks are created by interconnection of polymer chains, with the connection points being named crosslinks. If the crosslinked polymer chains are highly flexible and mobile, the material exhibits very high deformability and practically complete recovery [23]. Materials exhibiting such properties are known as elastomers (or rubber-like materials), whereas their elastic behaviour is called rubber-like elasticity. Its interpretation may be achieved by a number of molecular theories [23], [24]. The concept of slip-links has also been successfully introduced in mesoscopic multi-chain simulations of crosslinked polymer networks [23]-[25]. Atomistic simulations are also possible; however, the time and length scales are rather limited in this approach [3], [28]-[31].

2. Models and Methods

A full description of the methodology and the corresponding C++ code will be provided in two upcoming studies; the first study refers to polymer melts [22], while the second one is focused on crosslinked polymer networks [32]. The polymer chains are considered as sequences of beads, which are connected via (entropic) springs, with each coarse-grained bead encompassing a large number of Kuhn segments (e.g 5-10). In the context of our methodology, a polymer network consists of three kinds of beads, i.e., end-points, internal nodal points and crosslinks. The temporal evolution of our polymer network configurations is generated by Brownian dynamics (BD) simulation in the high friction limit, while the entanglement effect is introduced through slip-springs connecting neighbouring polymer chains. A slip-spring is a spring whose ends change as the simulation time elapses. These springs may hop between neighbouring beads along the polymer chains and may also be created and destroyed at the end-points, with the total number of slips-springs remaining constant. All the aforementioned mechanisms are governed by a kinetic Monte Carlo scheme, with only one adjustable parameter (v_{diff}), obeying the condition of microscopic reversibility [22]. This adjustable parameter can be regarded as a rate constant for a slip-spring to overcome the free-energy barrier of its hopping along the polymer chain. Initial configurations of monodisperse polymer melts are obtained by a field-theory inspired Monte Carlo simulation of a soft, coarse-grained model, wherein chains are represented as freely jointed sequences of Kuhn segments subject to a coarse-grained Helfand Hamiltonian [33] which restrains the density from departing from its mean value anywhere in the system [34], [35]. Crosslinked polymer networks are created by applying crosslinking in the coarse-grained representation of the polymer melts. In particular, two algorithms for crosslinking are available in our C++ code, i.e., end-grafting and random-crosslinking. In the former procedure, all end-points of the polymer chains are connected with internal nodal points of other neighbouring polymer chains. When an internal nodal point is connected with an end-point, a trifunctional crosslink is created and the end is no longer available for interconnection with other polymer chains. The resultant polymer network is free of dangling chains, since there are no end-points. A graphical representation of end-grafting is provided in figure 1a. As far as the random-crosslinking is concerned, a number of prescribed points (N_c) are randomly dispersed in the simulation box. These points serve as the crosslinks for the polymer network and are always tetrafunctional, i.e., four strands emanate from each. In more detail, the procedure follows the scheme:

- A number of pairs of internal nodal points belonging to two neighboring chains are selected by a sorted list. The sorting criterion is the distance from the considered crosslink. Among these pairs the closest pair is chosen.
- The aforementioned pair is deleted, along with the four strands emanating from it.
- The two polymer chains are connected at the considered crosslink by creating four new strands.

This random crosslinking procedure is schematically presented in the figure 1b. In the context of our methodology, the system is fully described from a thermodynamic point of view by a Helmholtz free-energy containing three explicit terms:

$$A = A_{\text{entropy springs}} + A_{\text{slip springs}} + A_{\text{nb}} \quad (1)$$

The first term is the contribution from the entropic springs, which connect successive beads along polymer chains, whereas the second term is included to account for the entanglement effect. In particular, the following equations are employed:

$$A_{\text{entropy springs}} = \sum_{ij} \frac{3}{2} k_B T \frac{R_{ij}^2}{n_{ij} b^2} \tag{2}$$

$$A_{\text{slip springs}} = \sum_{kl} \frac{3k_B T}{2b_{\text{sl}}^2} (\mathbf{R}_k - \mathbf{R}_l)^2 \tag{3}$$

The sum in eq. 2 is over all strands (or subchains) of the polymer melt or crosslinked polymer network whereas b and n_{ij} stand for the Kuhn length and the number of Kuhn segments in the strand, respectively. As far as the second contribution to the Helmholtz free-energy is concerned, it is computed as a sum over all slip-springs connecting two beads belonging to different neighbouring polymer chains. There is an additional parameter b_{sl} controlling the stiffness of the slip-spring, which can be set equal to the entanglement tube diameter of the considered polymer system. To deal with non-bonded interactions in the polymer network representation, an additional free energy contribution is included:

$$A_{\text{nb}} = \int d^3r \, a_{\text{vol}}[\rho(\mathbf{r})] = \sum_{\text{cells}} V_{\text{cell}} \, a_{\text{vol}}(\rho_{\text{cell}}) \tag{4}$$

In Eq. (4), $\rho(\mathbf{r})$ is the local density at a position $\mathbf{r}$ and $a_{\text{vol}}(\rho)$ is a free energy density, which is obtained from the local density using an equation of state (EOS). In this study, we make use of the Sanchez-Lacombe EOS [36]. In the above equation, the volume integral is approximated as a sum over all cells of an orthogonal grid introduced in the primary simulation box. $V_{\text{cell}} = L_x^{\text{grid}} L_y^{\text{grid}} L_z^{\text{grid}}$ is the volume of the rectangular parallelepiped defining the cell.

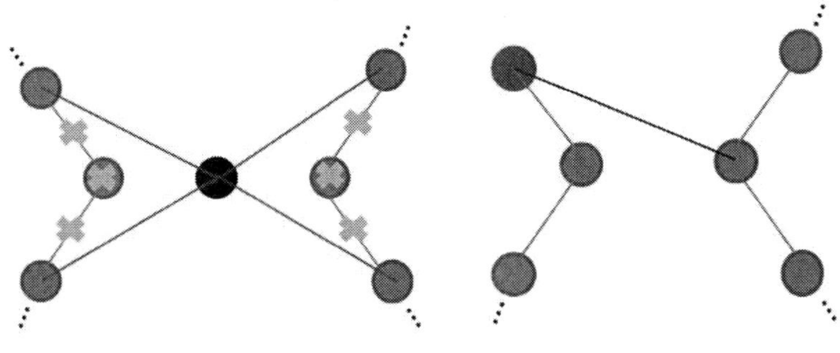

(a) (b)

Figure 1. (a) Schematic representation of random crosslinking with all crosslinks being tetrafunctional. (b) End-grafting: all end-points are connected with internal beads of neighbouring polymer chains. The resultant polymer network is free of end-points and all crosslinks are trifunctional by construction.

3. Results

A polymer network of cis-PI is created by random-crosslinking which is applied on a monodisperse polymer melt containing 172 linear polymer chains. A chain consists of 53 beads with each bead corresponding to 10.21 Kuhn segments. This polymer network contains 300 crosslinks, with their initial spatial distribution being visualized in figure 2a, while the whole polymer network is depicted in figure 2b. The entanglement effect is introduced by 496 slip-springs and the adjustable parameter (v_{diff}) is chosen equal to 10.0s^{-1}. Furthermore, an end-grafted polymer network was created from a cis-PI polymer melt containing 178 polymer chains. Each chain consists of 513 Kuhn segments. All BD simulations are conducted at constant temperature (T=400K) and volume with the integration time step

(Δt) being equal to 50ps. The undeformed polymer network is simulated in a cubic simulation box of edge length L=30nm ($L_x = L_y = L_z$) and a cubic grid of 5^3 cells of the same size ($N_{\mathbf{cell/dim}} = 5$) is introduced for the computation of non-bonded interactions. Our analysis begins with the computation of some conformational properties. First, we want to check whether the unperturbed chain conformation of the initial configuration is retained during the BD simulations. To this end, the strand length distribution and the end-to-end distance distribution of the originally linear chains, which formed the network, are presented and compared with the corresponding theoretical predictions given by the following equations, respectively:

$$P(\mathbf{r}) = \left(\frac{3}{2\pi n_{\text{Kuhn/strand}} b^2}\right)^{3/2} \exp\left(-\frac{3}{2}\frac{r^2}{n_{\text{Kuhn/strand}} b^2}\right) \tag{5}$$
$$P(r) = 4\pi r^2 P(\mathbf{r})$$

$$P^{\text{end-to-end}}(\mathbf{r}) = \left(\frac{3}{2\pi n_{\text{Kuhn/strand}} n_{\text{strands/chain}} b^2}\right)^{3/2} \exp\left(-\frac{3}{2}\frac{r^2}{n_{\text{Kuhn/strand}} n_{\text{strands/chain}} b^2}\right) \tag{6}$$
$$P^{\text{end-to-end}}(r) = 4\pi r^2 P^{\text{end-to-end}}(\mathbf{r})$$

In the above sets of equations, b symbolizes the Kuhn length of the simulated polymer while $n_{\text{Kuhn/strand}}$, $n_{\text{strands/chain}}$ stand for the number of Kuhn segments per strand and the number of strands per polymer chain, respectively. Figures 3a and 3b show the strand-length and end-to-end-distance distributions, as computed from the simulation, along with the theoretical predictions of eqs 5 and 6. The excellent agreement demonstrates that the slip-spring-induced attraction is negligible for this choice of parameters. The non-bonded scheme adopted in the context of our methodology allows us to predict the isothermal compressibility of the polymer network. From the simulation point of view, the property in question is computed by the following fluctuation equation:

$$\kappa_T = \frac{V\left\langle\left(\delta n_{\text{Kuhns/cell}}\right)^2\right\rangle}{k_B T \left\langle n_{\text{Kuhns/cell}}\right\rangle^2} \tag{7}$$

The averages $\langle ... \rangle$ are taken over all cells of the grid, which has been introduced in the primary simulation box due to the special nature of the non-bonded interactions. The averages in the numerator and denominator of eq 7 are the variance and the mean value of the number of Kuhn segments in each cell, respectively. The value calculated from eq 7 equals 1.278×10^{-4} bar^{-1} and is in very good agreement with the estimate from the Sanchez-Lacombe EOS (1.270×10^{-4}), which characterizes the long-wavelength behaviour. It is reminded that this is the underlying EOS of the non-bonded interactions (see also section 2). Regarding the computation of linear viscoelastic properties, the stress relaxation function, as defined in eq 8, is of great importance:

$$G(t) = \frac{V}{3k_B T}\left[\left\langle\tau^{xy}(t_0+t)\tau^{xy}(t_0)\right\rangle + \left\langle\tau^{xz}(t_0+t)\tau^{xz}(t_0)\right\rangle + \left\langle\tau^{yz}(t_0+t)\tau^{yz}(t_0)\right\rangle\right] \tag{8}$$

V is the volume of the simulation box and the brackets indicate autocorrelation functions concerning the off-diagonal elements of the stress tensor $\boldsymbol{\tau}$. In the case of polymer melts, $G(t)$ is employed for the computation of storage and loss moduli and the zero-shear-rate viscosity [18], [19], [22], [37]. The stress relaxation function of a crosslinked polymer network approaches a constant value at long times, and this value corresponds to the shear modulus of the rubbery material. A more accurate way of determining the shear modulus is through recording the stress-strain curve of a virtual elongation experiment at rather small deformations. The different rheological behavior between a polymer melt and a crosslinked polymer network is presented in figure 4a, in which the corresponding stress relaxation functions are provided. The black curve labelled *Rouse* in the corresponding legend of the figure refers to the stress relaxation function of the polymer melt, which was used for the end-grafting. Due to the absence of entanglements in our soft, coarse-grained model, the dynamics in a melt is well described by the well-known Rouse model [38]. In addition, the shear modulus ($G =$ (0.31±0.03)bar) estimated by a virtual elongation experiment is also included. The stress relaxation function for the polymer network created by random-crosslinking is provided in figure 4b. The bigger noise, compared with that of the corresponding curve in figure 4a, is due to the presence of dangling chains, which is an inherent characteristic of this kind of crosslinking.

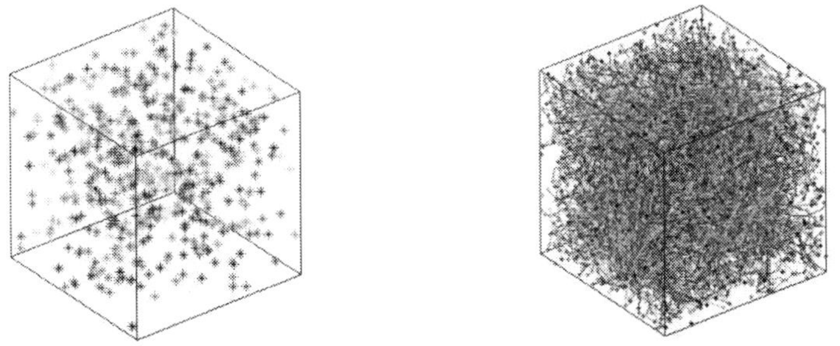

(a) (b)

Figure 2. (a) Spatial distribution of 300 crosslinks in the simulation box after the implementation of random crosslinking. (b) Visualization of a polymer network created by random-crosslinking.

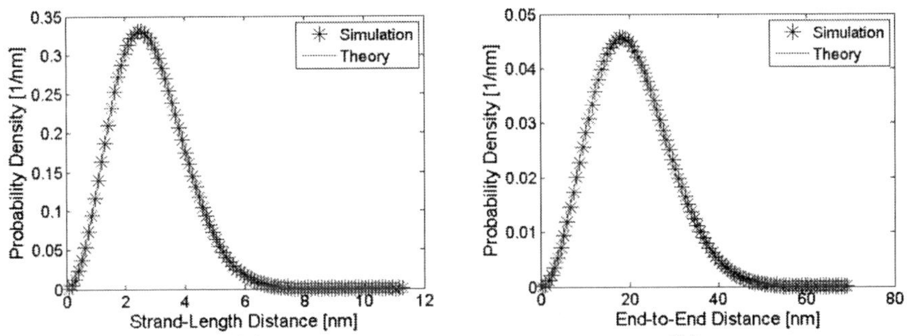

(a) (b)

Figure 3. (a) Strand length distribution and (b) end-to-end distance distribution of the polymer chains comprising the polymer network created by random-crosslinking.

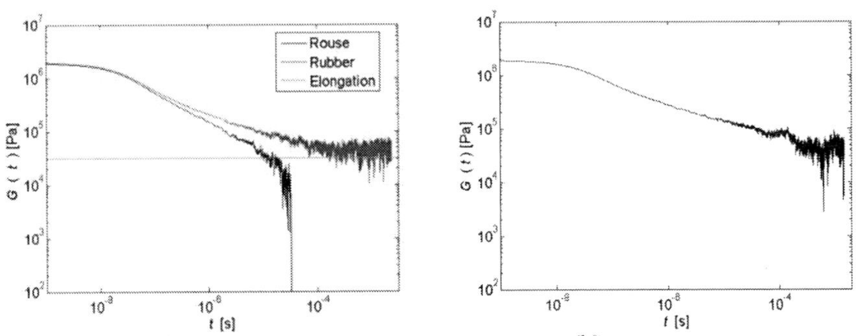

(a) (b)

Figure 4. (a) Stress relaxation functions of a polymer network created by end-grafting (*Rubber*) and of a polymer melt without slip-springs (*Rouse*). (b) Stress relaxation function of a polymer network created by random-crosslinking.

4. Conclusions
A new methodology is presented in this study concerning the simulation of crosslinked polymer networks at the mesoscale. Such polymer systems are described from thermodynamic point of view by a Helmholtz free-energy including three terms: a term for entropic elasticity, another term for the entanglement effect, and a term for non-bonded interactions. The isothermal compressibility of the polymer network is predicted from equilibrium density fluctuations and is in very good agreement with the prediction estimated by the Sanchez-Lacombe EOS. Also, two conformational properties were investigated to establish that strands and chains are unperturbed by the presence of slip-springs and non-bonded interactions. Viscoelastic properties can be calculated through the computation of the stress relaxation function of undeformed polymer networks and through virtual elongation experiments.

5. Acknowledgments
The major part of this work was funded by *Volkswagen Foundation* in the context of the project *Mesoscopic Simulations of Viscoelastic Properties of Networks*. In addition, the work was funded by the European Union through the projects *COMPNANOCOMP* and *COMPNANOCOMP-DPI* under grant number 295355 and by the *Limmat Foundation*. Grigorios Megariotis and Doros N. Theodorou also thank the *Limmat Foundation* for giving them the opportunity to continue this research on polymer networks. Georgios G. Vogiatzis thanks the *Alexander S. Onassis Public Benefit Foundation* for a doctoral scholarship.

6. References
[1] Doi M and Edwards S F 1986 *The Theory of Polymer Dynamics* (Clarendon Press: Oxford U.K.)
[2] Tzoumanekas C and Theodorou D N 2006 *Macromolecules* **39** 4592
[3] Anogiannakis S F, Tzoumanekas C and Theodorou D N 2012 *Macromolecules* **45** 9475
[4] McLeish T C B 2002 *Advances in Physics* **51** 1379
[5] Edwards S F 1967 *Proc. Soc. London* **92** 9
[6] de Gennes P G 1971 *J. Chem. Phys.* **55** 572
[7] Doi M 1983 *J. Polym. Sci.: Polym. Phys.* **21** 667
[8] Milner S and McLeish, T C B 1998 *Phys. Rev. Lett.* **81** 725
[9] Likhtman A E and McLeish T C B 2002 *Macromolecules* **35** 6332
[10] Doi M and Edwards S F 1978 *J. Chem. Soc. Faraday Trans.* **74** 1802
[11] Ball R, Doi M, Edwards S and Warner M 1982 *Polymer* **22** 1010
[12] Edwards S and Vilgis T 1986 *Polymer* **27** 483
[13] Schieber, J D, Neergaard J and Gupta S 2003 *J. Rheol.* **47** 213
[14] Nair D M and Schieber J D 2006 *Macromolecules 39* 3386
[15] Likhtman A E 2005 *Macromolecules* **38** 6128
[16] Masubuchi Y, Takimoto J, Koyama K, Ianniruberto G, Marrucci G and Greko F 2001 *J. Chem. Phys.* **115** 4387
[17] Uneyama T and Masubuchi Y 2011 *J. Chem. Phys.* **135** 184904
[18] Ramírez-Hernández A, Müller M and de Pablo J J 2013 *Soft Matter* **9** 2030
[19] Ramírez-Hernández A, Detcheverry F A, Peters B L, Chappa V C, Schweizer K S, Müller M and de Pablo J J 2013 *Macromolecules* **46** 6287
[20] Uneyama T and Masubuchi Y 2012 *J. Chem. Phys.* **137** 154902
[21] Chappa V C, Morse D C, Zippelius A and Müller M 2012 *Phys. Rev. Lett.* **109** 148302

[22] Vogiatzis G G, et al, *in preparation*
[23] Mark J E 2007 *Rubberlike Elasticity: A Molecular Primer* (Cambridge University Press)
[24] Treloar L R G 1975 *The Physics of Rubber Elasticity* (Oxford Classics Series, 2005)
[25] Oberdisse J, Ianniruberto G, Greco F and Marrucci G 2002 *Europhys. Lett.* **58** 530
[26] Oberdisse J, Ianniruberto G, Greco F and Marrucci G 2006 *Rheol Acta* **46** 95
[27] Masubuchi Y, Uneyama T, Watanable H, Ianniruberto G, Greco F and Marrucci G 2010 *J. Chem. Phys.* **132** 134902
[28] Rottach D R, Curro J G, Budzien J, Grest G S, Svaneborg C and Everaers R 2007 *Macromolecules* **40** 131
[29] Yang S and Qu J 2012 *Polymer* **53** 4806
[30] Morozinis A K, Tzoumanekas C, Anogiannakis S D and Theodorou D N 2013 *Polymer Science-Series C* **55** 212
[31] Varshney V, Patnaik S S, Roy A K and Farmer B L 2008 *Macromolecules* **41** 6837
[32] Megariotis G, et al. *in preparation*
[33] Helfand E and Tagami Y 1972 *J. Chem. Phys.* **56** 3592
[34] Vogiatzis G G, Voyiatzis E and Theodorou D N 2011 *Eur. Polym. J.* **47** 699
[35] Vogiatzis G G and Theodorou, D N 2013 *Macromolecules* **46** 4670
[36] Sanchez I C and Lacombe R H 1976 *J. Phys. Chem.* **80** 2352
[37] Vogiatzis G G 2015 *Multiscale Simulations of Polymer-Matrix Nanocomposites, PhD Thesis* (NTUA, School of Chemical Engineering)
[38] Rouse P E 1953 *J. Chem. Phys.* **21** 1272

n-vicinities method for three dimensional Ising Model

Boris Kryzhanovsky and Leonid Litinskii

Scientific Research Institute for System Analysis of Russian Academy of Science, 117218 Nakhimovskii prospect Str. 36-1, Moscow, Russian Federation

E-mail: kryzhanov@mail.ru, litin@mail.ru

Abstract. The n-vicinities method for approximate calculations of the partition function of a spin system was proposed previously. The equation of state was obtained in the most general form. In the present publication these results are adapted to the Ising model on the D-dimensional cubic lattice. The state equation is solved for an arbitrary dimension D and the behavior of the free energy is analyzed. For large values of D ($D > 2$) the obtained results are in good agreement with the ones obtained by means of computer simulations. For small values of D, there are noticeable discrepancies with the exact results.

1. Introduction

In the papers [1]-[3] we develop the n-vicinities method for approximate calculation of the partition function. The method can be described as follows. Let $N >> 1$ be the dimension of the problem, and $\mathbf{s}_0 \in \mathbf{R}^N$ be the initial configuration whose n-vicinity Ω_n is the set of all the configurations that differ from $\mathbf{s}_0$ by the values of n coordinates ($n = 0,1,...,N$). The distribution of the energies of the states from Ω_n fits the Gaussian density reasonable well [3]. The mean energy E_n and the variance σ_n^2 can be accurately expressed in terms of the parameters of the connection matrix [1], [2]. Then in the asymptotic limit summing over the states from Ω_n in the expression for the partition function can be replaced by integration of $\exp(-\beta E)$ over the Gaussian measure, and summing over the all n-vicinities by integration between 0 and 1 over the parameter $x = n / N$. As a result the partition function takes the form of the integral $Z_N \approx \int dx \int dE \exp[-N \cdot F(x,E)]$, which is calculated with the aid of the saddle-point method. The function of two variables $F(x,E)$ depends on the following parameters: the inverse temperature β, the external magnetic field (if it is present), the trace of the connection matrix squared and some other numerical parameters. That is $F(x,E) \equiv F(x,E,\beta,H...)$. In the most general form the expression for $F(x,E)$ was obtained in [2], [3]. In the present publication we adapt this expression for the Ising model on the D-dimension hypercube, derive the equation of state, plot and examine the graphs of the free energy. Because of the size of the publication here is only the summary of our calculations. The details will be presented in [4].

2. Basic Expressions

Content from this work may be used under the terms of the Creative Commons Attribution 3.0 licence. Any further distribution of this work must maintain attribution to the author(s) and the title of the work, journal citation and DOI.
Published under licence by IOP Publishing Ltd

Let $\mathbf{T}=\left(T_{ij}\right)_1^N$ be the connection matrix corresponding to the D-dimensional Ising model, $\mathbf{s}_0=(1,1,...,1)\in\mathbf{R}^N$ is the ground state of the spin system, Ω_x is the x-vicinity of the ground state, where $x=n/N\in[0,1]$. Let us calculate the energy per one spin: if $\mathbf{s}\in\Omega_x$, and H is the value of the homogeneous magnetic field, then

$$E(\mathbf{s},H)=-\frac{1}{N}\sum_{ij}T_{ij}s_is_j-\frac{1}{N}H\sum_i s_i=E(\mathbf{s})-H(1-2x).$$

By E_0 we denote the energy of the ground state: $E_0=E(\mathbf{s}_0)=-\frac{1}{N}\sum_{ij}T_{ij}=-2D$. Then the asymptotic expression for the energy averaged over Ω_x is $E_x=E_0(1-2x)^2$, and the variance is equal to $\sigma_x^2=\frac{32\sum_{ij}T_{ij}^2}{N^2}x^2(1-x)^2$. In [1]-[3] it was shown that the expression for the partition function can be reduced to the form: $Z_N\sim\int_0^1dx\int_{E_0}^{|E_0|}\exp[-N\cdot F(x,E)]dE$, where $N>>1$,

$$F(x,E)=L(x)+\beta[E-H\cdot(1-2x)]+\frac{1}{2}\left(\frac{E-E_x}{\sigma_x}\right)^2,$$

and $L(x)=x\ln x+(1-x)\ln(1-x)$.

To find the global minimum of the function $F(x,E)$ it is necessary to solve the set of equations:

$$\frac{\partial F}{\partial E}=\beta+\frac{E-E_x}{\sigma_x^2}=0,\quad \frac{\partial F}{\partial x}=\ln\frac{x}{1-x}+\frac{E-E_x}{\sigma_x}\frac{\partial}{\partial x}\left(\frac{E-E_x}{\sigma_x}\right)+2\beta H=0.$$

Solving the first equation for β and substituting the result into the second one, we obtain the minimization problem for the function of one variable $f(x)=L(x)+\beta E_x-\frac{\beta^2\sigma_x^2}{2}-\beta H(1-2x)$ restricted to the inequality $\beta+\frac{E_0-E_x}{\sigma_x^2}\le 0$. The restriction means that among all the minimums we examine only those for which $\frac{\partial F}{\partial E}=0$. We use the other normalization and more convenient notations:

$$b=2\beta\frac{\sum_{ij}T_{ij}^2}{\sum_{ij}T_{ij}},\quad \gamma=\frac{\left(\sum_{ij}T_{ij}\right)^2}{N\sum_{ij}T_{ij}^2},\quad h=\frac{NH}{\sum_{ij}T_{ij}}.$$

Then the final form of the problem is as follows: for each value of the "renormalized" inverse temperature b we have to find the global minimum of the function

$$f(x)=L(x)-\frac{\gamma b}{2}\left[(1-2x)^2+8b(x(1-x))^2+h(1-2x)\right] \tag{1}$$

at the interval $(0,x_b)$, where

$$x_b=\begin{cases}\frac{1}{2}, & \text{when } b\le 1\\ \frac{1-\sqrt{1-1/b}}{2}, & \text{when } b\ge 1\end{cases}. \tag{2}$$

When b exceeds 1, the right-hand boundary of the interval where we are looking for the global minimum becomes depending on b.

The free energy is equal to $f(b)=\min_{x\in(0,x_b)}f(x)$.

Note. For the D-dimensional Ising model with the connection J between the nearest neighbors we have: $\sum_{ij} T_{ij} = 2DNJ$, $\sum_{ij} T_{ij}^2 = 2DNJ^2$, $b = 2\beta J$, $\gamma = 2D$, $h = H/(2DJ)$. Here the characteristic γ is equal to the number of the nearest neighbors of each spin. However, generally the interactions along the different directions of the lattice can be different. In this case γ is no longer an integer. For example, for the two-dimensional Ising model with different interaction constants along the horizontal (J) and vertical directions (K) we have $\gamma = 2\left(1 + \frac{2}{K/J + J/K}\right)$. Then γ can takes any value from the interval $2 < \gamma < 4$. Similarly, for the three-dimensional Ising model with different interaction constants (J, K, L), γ can takes any value from the interval $2 < \gamma < 6$. In the general case, γ is the effective coordination number that describes the interaction of the spin with the nearest vicinity.

3. Analysis of the state equation when H=0

Let us set $H = 0$. The case of the nonzero magnetic field will be described in [4]. We seek the global minimum of the function $f(x)$ (1) with $h = 0$. For that we have to solve the state equation

$$\frac{\partial f(x)}{\partial x} = \ln\frac{x}{1-x} + 2\gamma b(1-2x)\left[1 - 4bx(1-x)\right] = 0. \tag{3}$$

It is evident that $x_0 = 0.5$ is the solution of this equation for any values of b and γ. x_0 is called the trivial solution. For very small values of b ($b \sim 0$) the function $f(x)$ decreases monotonically from $f(0) = -\gamma b/2$ up to $f(x_0) = -\ln 2 - \gamma b^2/4$ (see figure 1a). Consequently, for small b the trivial solution x_0 is the only solution of equation (3), and the free energy takes the form

$$f(b) = \min_x f(x) = -\ln 2 - \frac{\gamma b^2}{4}, \quad b \sim 0. \tag{4}$$

The interval $\gamma > 16/3$. It can be shown [4] that for γ from this interval the point x_0 remains the only solution of equation (3) until b is less than the critical value

$$b_c = \frac{1 - \sqrt{1 - 4/\gamma}}{2}. \tag{5}$$

When $b < b_c$, the free energy has the form (4). Once b exceeds the value b_c, x_0 becomes the maximum point of the function $f(x)$. To the left from x_0 appears another minimum point $x_1(b)$: $x_1(b) < x_0$ (see figure 1b). In this case, to calculate the free energy one has to substitute $x_1(b)$ in equation (1). When $b = b_c$ the phase transition of the second kind takes place in the system.

A further increase of the parameter b is accompanied by the deepening of the minimum and steady shifting of the minimum point towards 0. When b becomes larger than 1, the right boundary of the interval becomes the variable that is equal to x_b (see equation (2)). When $b \to \infty$ the interval $(0, x_b)$ contracts to the origin of coordinates and the global minimum $x_1(b)$ tends to zero. As it has to be, when the temperature tends to zero, the spin system tends to its ground state $\mathbf{s}_0 = (1, 1, ..., 1)$.

Table 1. Theoretical and experimental values of b_c for D-dimensional Ising models

	3D	4D	5D	6D	7D
Our theory	0.2113	0.1464	0.1127	0.0918	0.0774
Computer simulations	0.2216	0.1489	0.1139	0.0923	0.0777

Let us substitute the values $\gamma = 6, 8, 10, 12, 14$ into equation (5) and calculate the critical values of the inverse temperature for the 3D-, 4D-, 5D-, 6D- and 7D- Ising models, respectively. The obtained numbers are shown in the upper row of the table 1. In the lower row we present the values of b_c that are the results of computer simulations (see [5]-[7]). We see a good agreement between our theoretical estimates and the results of computer simulations.

Thus, equation (5) provides us with a reasonable estimate of the critical temperature for the Ising model on the D-dimension hypercube when $D > 8/3$. We would like to remind that the parameter γ has not to be even integer-valued numbers (see the note in the end of the previous section).

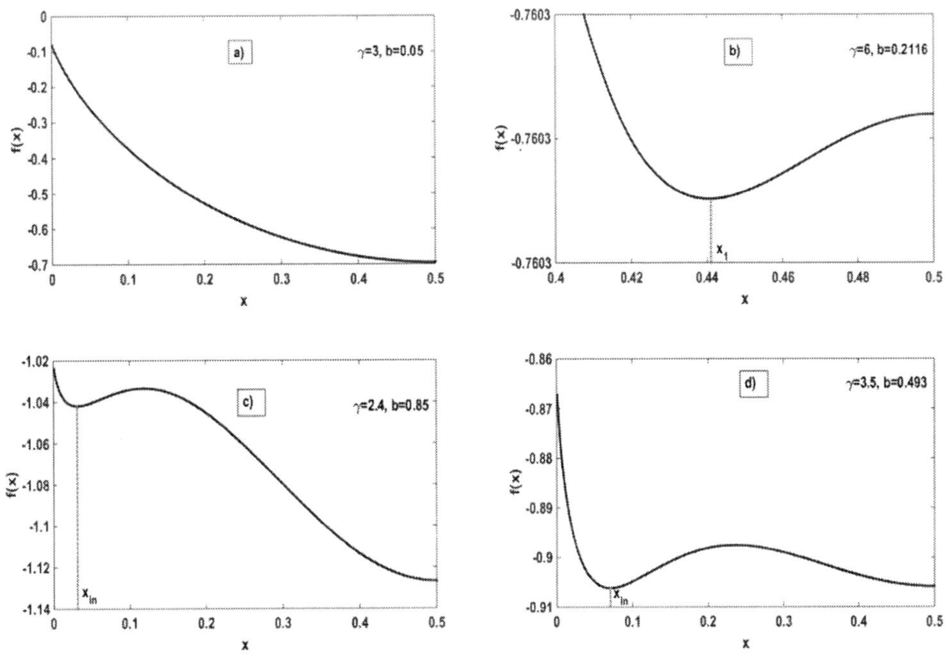

Figure 1. The behaviour of the function $f(x)$ for different values of γ and b (see the body of the text).

When $\gamma < 16/3$, with increase of the parameter b the behavior of the function $f(x)$ changes. Namely, now it certainly has a minimum in the inner point $x_{in}(b)$ of the interval $(0, x_0)$. The depth of this minimum has to be compared with $f(x_0)$. Different situations are possible depending on whether γ is less or larger than 2.75.

The interval $\gamma < 2.75$ (figure 1c). It turns out that in this case the local minimum in the point x_{in} is never deeper than the minimum in the point x_0 [4]. Then as long as $b \leq 1$ the global minimum of the function $f(x)$ is in the point x_0. When the value of b becomes larger than 1 and the right boundary x_b becomes depending on b, and the global minimum of the function $f(x)$ is in the point x_b:

$$f(b) = \begin{cases} -\ln 2 - \dfrac{\gamma b^2}{4}, & b \leq 1 \\ L(x_b) - \dfrac{\gamma}{4}(2b-1), & b \geq 1 \end{cases} . \tag{6}$$

During the further increase of $b\,(b \to \infty)$ the global minimum is either in the point x_b or it jumps to the point of the current minimum $x_{in}(b)$. The last happens if for a value of b the minimum $f(x_b)$

becomes equal to $f(x_{in})$. In any case, when b grows indefinitely, the minimum point tends to 0 and the system tends to the ground state.

Note, when $b=1$ there is a discontinuous of the first derivative of the free energy (6): $f_b'|_{b=1-0}=-\gamma/2$ and $f_b'|_{b=1+0}=(1-\gamma)/2$, which demonstrates the presence of the phase transition of the first kind. This is true for any $\gamma<2.75$, in particular for $\gamma=2$ that corresponds to the 1D Ising model. In the same time, the exact solution for the one-dimension Ising model does not show phase transitions for finite temperatures [8]. We have to admit that in the region $\gamma<2.75$ our method provides questionable results.

The interval $2.75<\gamma<16/3$ (figure 1d). In this case, for a value of b the inner local minimum $x_{in}(b)$ becomes necessarily equal to the minimum in the point x_0 [4]. It takes place when $b=b_j$ and by $x_j=x_{in}(b_j)$ we denote the corresponding value of x. For $b=b_j$ the global minimum jumps from the point x_0 to the point x_j. The magnetization of the system changes in a step-wise way from $m_0=0$ to $m_j=1-2x_j$, indicating the phase transition of the first kind. With further increase of b ($b\to\infty$) the minimum point $x_{in}(b)$ tends to 0 steadily and the system tends to its ground state.

The value $\gamma=4$ that corresponds to the 2D Ising mode belongs to the interval in question: $\gamma\in[2.75,16/3]$. From our approach it follows that in the 2D Ising system the phase transition of the first kind occurs when $b_c\approx 0.3912$. In the same time, the exact solution of Onsager shows the phase transition of the second kind for the noticeably larger value $b_c\approx 0.4407$ [8]. When $2.75<\gamma<16/3$ the results of our method are also questionable.

4. Discussion and conclusions

Let us illustrate the above statements by some computer calculations. In figure 2 for different values of γ the dependences of the magnetization $m=1-2x$ on the renormalized inverse temperature b are shown. At first when b is not too large, the global minimum is in the point $x_0=0.5$, and the magnetization of the state is equal to zero. When the parameter b reaches the critical value the global minimum leaves the point x_0 and from this moment the magnetization is a nonzero one. For $\gamma=6$ it happens when $b=b_c$ and after that the magnetization varies smoothly, without discontinuities. For other values of γ it happens when $b=b_j$ and at that the global minimum jumps from the point x_0 into the point x_j (see above). For $\gamma<16/3$ the jump of the magnetization $m=1-2x_j$ is clearly seen in figure 2.

The value of the mean energy $<E>=\sum_{\mathbf{s}}E(\mathbf{s})\exp(-\beta\cdot E(\mathbf{s}))$ behaves itself in the same way. The cumbersome calculations of this characteristic will be also presented in [4]. In Fig.3 for different values of γ the dependences of the renormalized mean energy $E=<E>/E_0$ as functions of b are shown. At first E is linear in b; when b becomes larger than the critical value, the behavior of E changes. When $\gamma=6$, the function $E(b)$ is a continuous function that decreases up to -1. When $\gamma\in[2.75,16/3]$ the value of E has a jump in the critical point. After that the system tends to the ground state. The jump manifests the presence of the gap in the distribution of the energies of states. However, we have no arguments why there is the energy gap when $\gamma<16/3$, and it is absent when $\gamma=6$.

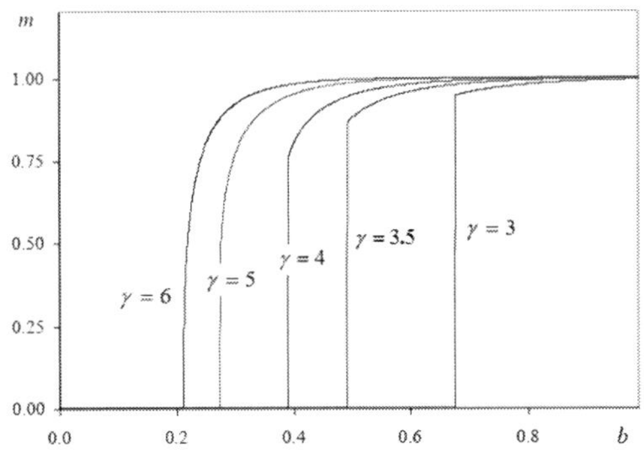

Figure 2. Magnetization m as function of b for different values of γ .

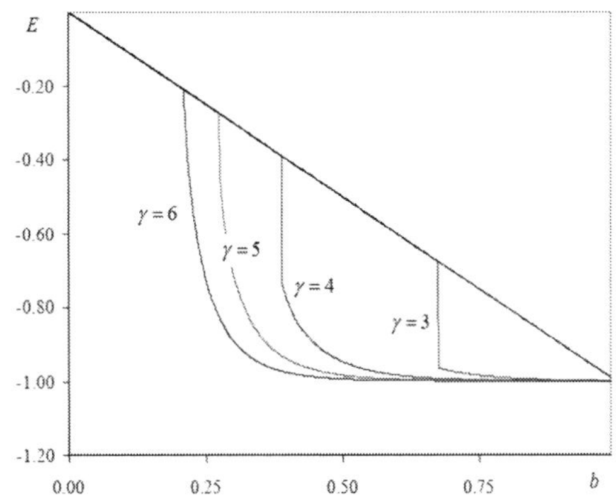

Figure 3. Renormalized mean energy E as function of b for different values of γ .

The essential fault of our approach is the presence of the jump of the magnetization in the 2D Ising model. This error is the result of distortions due to the approximation of the true distribution of energies by the Gaussian density. Without getting into details, let us note that analyzing the equation of state we see how a small distortion of the curve's shape allows to eliminate the unnecessary jump of the magnetization. We hope that this can be done by an improvement of our model. For example, a more accurate approximation can be used in place of the Gaussian density. This is the subject of our further analysis. Even in the present form when we ignore the jumps of the magnetization and are interested in the behavior of the free energy only, the results obtained with the aid of our approach are reasonable enough.

In figure 4 for the 2D Ising model we show the dependence of the inverse critical temperature β_c on the ratio of the interaction constants along the two axes: $J = 1$ and K varies from 0.2 to 1. We see that our results are in sufficiently good agreement with the exact solution.

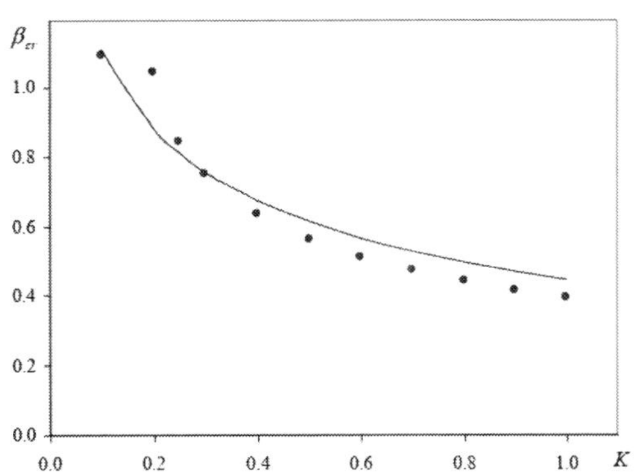

Figure 4. 2D Ising model ($\gamma = 4$). Dependence of the critical temperature on K: solid line is the Onsager theory, circles is our results.

In figure 5 the graphs of the free energy for different values of K are shown. The solid line is the free energy $f_{ex}(\beta)$ calculated according the Onsager theory, the circles are the results of our approach $f_{app}(\beta)$. It is clearly seen that the difference between our results and the exact theory are noticeable only in the region near the critical temperature β_c. To make this statement more evident, in figure 6 we present the graphs of the relative error $\left(f_{app}(\beta)-f_{ex}(\beta)\right)/f_{ex}(\beta)$. Its maximal value corresponds to the critical temperature β_c and it is of the order of 1%.

Let us summarize. For cubic lattices of high dimensions ($\gamma > 16/3$) our theory is in good agreement with the well-known results. There are no exact results for the high dimensional Ising models ($D > 2$). For these dimensions our approach, as well as the computer simulations, predicts the phase transition of the second kind. The experimental values of the critical temperatures for a rather long sequence of the values of γ coincide sufficiently accurately with the ones obtained with the aid of our formula (5) – see table 1.

When $\gamma < 16/3$ our theory is not sufficiently accurate. The most important fault is the presence of the phase transition of the first kind in the cases where there are no phase transitions at all ($\gamma = 2$), or the phase transition has to be of the second kind ($\gamma = 4$). The aim of our further analysis is to correct this defect.

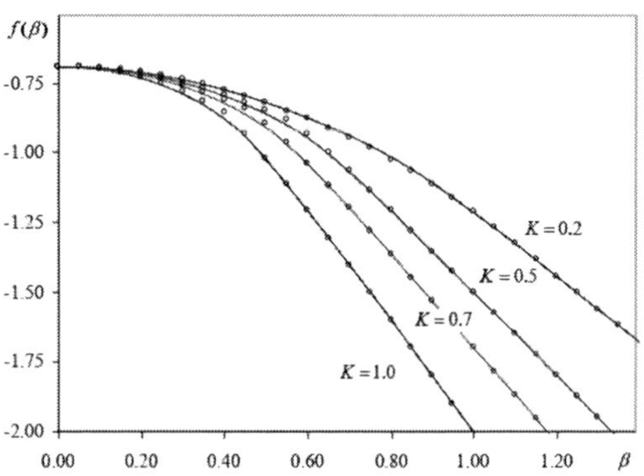

Figure 5. 2D Ising model ($\gamma = 4$). Free energy $f(\beta)$ for $K = 0.2,\ 0.5,\ 0.7,\ 1.0$: solid lines are the Onsager theory, circles are our results.

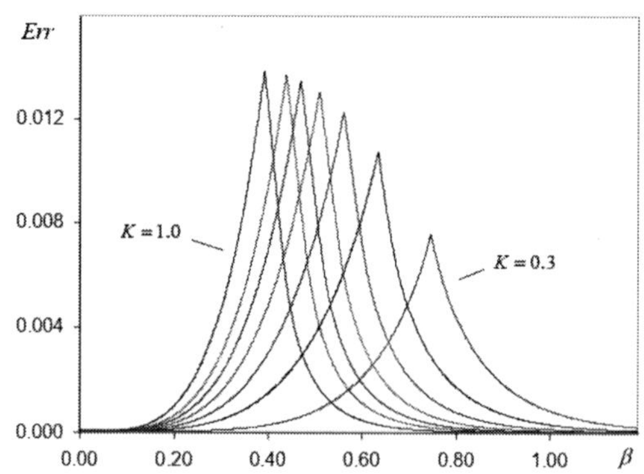

Figure 6. 2D Ising model ($\gamma = 4$). Relative errors as functions of β for $K = 1, 0.8, 0.7, 0.6, 0.5, 0.4, 0.3$ (from left to right).

The work was supported by the financial support of RBRF grants №15-07-04861 and 16-01-00626.

References

[1] Kryzhanovskiy B, Litinskii L 2014 http://iopscience.iop.org/1742-6596/574/1/012017/pdf/1742-6596_574_1_012017.pdf
[2] Kryzhanovsky B V, Litinskii L B 2014 *Doklady Mathematics* **90** 1
[3] Kryzhanovsky B, Litinskii L 2015 *Opt. Mem. & Neu. Net. (Inf. Optics)* **24** 165
[4] Kryzhanovsky B, Litinskii L 2016 (in preparation)
[5] Blote H W J, Shchur L N, Talapov A L 1999 *Int. J. Mod. Phys.* C **10** 1137
[6] Lundow P H, Markstrom K 2004 The critical behaviour of the Ising model on the 4-dimensional lattice Preprint arXiv:1202.3031v1
[7] Lundow P H, Markstrom K 2015 *Nucl. Phys.* **B 895** 30
[8] Baxter R J 1982 *Exactly solved models in statistical mechanics* (London: Academic Press)

Dispersive models describing mosquitoes' population dynamics

W M S Yamashita[1], L T Takahashi[1] and G Chapiro[1]
[1]Universidade Federal de Juiz de Fora, 36036-900, Juiz de Fora-MG, Brazil
E-mail: grigori@ice.ufjf.br

Abstract. The global incidences of dengue and, more recently, zica virus have increased the interest in studying and understanding the mosquito population dynamics. Understanding this dynamics is important for public health in countries where climatic and environmental conditions are favorable for the propagation of these diseases. This work is based on the study of nonlinear mathematical models dealing with the life cycle of the dengue mosquito using partial differential equations. We investigate the existence of traveling wave solutions using semi-analytical method combining dynamical systems techniques and numerical integration. Obtained solutions are validated through numerical simulations using finite difference schemes.

1. Introduction

There is a renewed interest in controlling the proliferation of *Aedes aegypti* mosquito, which is the transmitter of Dengue virus, and more recently Chikungynya and Zika viruses, [1]. In order to minimize both social and economic costs, [2] analyze the Dengue vector control problem in a multi-objective optimization approach using a dynamic mathematical model representing the mosquitoes' population. There are several modeling approaches appearing in the study of biological invasions and spread of diseases. In [3, 4], the authors study spacial population dynamics of *Aedes aegypti* using partial differential equations (PDEs) describes the life cycle of the mosquito *Aedes aegypti*. In [5] was studied nonlinear modification of this model considering different diffusion (dispersion) and advection (wind) effects. We follow the same approach here using techniques presented in [6] applied to generalized non-linear problem.

The paper is organized as follows. Section 2 presents the mathematical model. Section 3 discusses the existence of the solution for this model in a form of traveling wave. In Section 4 we apply the algorithm and validate the results comparing them to the direct numerical simulation obtained by using Finite Difference Scheme. The final conclusions are presented in Section 5.

2. Mathematical model

We consider one-dimensional model from [5] considering $p = 0$ and $q = 1$. To simplify the vital biological dynamics of mosquito, the model considers only two subpopulations: (1) eggs, larvae and pupae compose aquatic immobile phase ($A(x,t)$) and (2) female mosquitoes compose winged mobile phase ($M(x,t)$). The dimensional form of the model is

$$\begin{cases} \bar{M}_t &= \bar{D}\,\bar{M}_{xx} - 2\bar{\nu}\bar{M}/\bar{k}_1\bar{M}_x + \bar{\gamma}\bar{A}(1-\bar{M}/\bar{k}_1) - \bar{\mu}_1\bar{M}, \\ \bar{A}_t &= \bar{r}(1-\bar{A}/\bar{k}_2)\bar{M} - (\bar{\mu}_2+\bar{\gamma})\bar{A}, \end{cases} \tag{1}$$

Content from this work may be used under the terms of the Creative Commons Attribution 3.0 licence. Any further distribution of this work must maintain attribution to the author(s) and the title of the work, journal citation and DOI.
Published under licence by IOP Publishing Ltd

where $\bar{M}$ is the winged phase population, $\bar{A}$ is the aquatic phase population, $\bar{D}$ is the diffusion term, $\bar{\nu}$ is the advection term, $\bar{\mu}_i$ is the mortality rate ($i = 1$ winged phase; $i = 2$ aquatic phase), $\bar{k}_i$ is the carrying capacity of each phase, $\bar{\gamma}$ is the specific rate of maturation of the aquatic phase into winged phase and $\bar{r}$ is the oviposition rate of the female mosquitoes. Following [3], the model can be written in dimensionless form:

$$\left\{ \begin{array}{rcl} M_t &=& M_{xx} - 2\nu M M_x + \gamma A(1-M)/k - \mu_1 M, \\ A_t &=& k(1-A)M - (\mu_2+\gamma)A, \end{array} \right. \tag{2}$$

where the partial derivatives in M are denoted as $\partial M/\partial x_j = M_{x_j}$, same for A.

3. Traveling wave solution

In this section we investigate whether System (2) possesses a traveling wave solution, following the same steps of [6]. We change coordinates from (x,t) to (ξ,t), where $\xi = x - ct$ is called as traveling variable with constant propagation speed c. Following [7] we look for stationary solution in variable (ξ) as $M(x,t) = m(\xi)$ and $A(x,t) = a(\xi)$, where $m(\xi)$ and $a(\xi)$ correspond to wave profiles of winged and aquatic mosquitoes population densities, respectively. Performing traveling coordinates substitution, we rewrite System (2) as a system of ODEs:

$$\left\{ \begin{array}{rcl} m'(\xi) &=& h(\xi) \\ h'(\xi) &=& (2\nu m(\xi) - c)h(\xi) + (\mu_1 + \gamma a(\xi)/k)m(\xi) - \gamma a(\xi)/k \\ a'(\xi) &=& k(a(\xi)-1)m(\xi)/c + (\mu_2+\gamma)a(\xi)/c, \end{array} \right. \tag{3}$$

where prime indicates derivative in ξ. Solving the system $(m', h', a') = (0,0,0)$, we find two stationary solutions (also called equilibria) of System (3)

$$A = (0,0,0),\ B = (m^*, 0, a^*),\ \text{where } a^* = \frac{k(\gamma - \mu_1\mu_2 - \gamma\mu_1)}{\gamma(k+\mu_2+\gamma)} \quad \text{and } m^* = \frac{\gamma - \mu_1(\mu_2+\gamma)}{\gamma + k\mu_1}. \tag{4}$$

Here A represents the lack of both populations and B represents the maximum population in both phases.

The main goal consists in studying the existence of the traveling wave solution of System (3) in the form of heteroclinic orbit connecting equilibria A to B (or B to A) in the following sense. The propagating profile corresponds to the solution connecting the region with no mosquitoes upstream the wave ($\xi \to -\infty$) to the region with mosquitoes downstream the wave ($\xi \to \infty$):

$$\lim_{\xi\to-\infty}(m(\xi), h(\xi), a(\xi)) = (0,0,0) \quad \text{and} \quad \lim_{\xi\to+\infty}(m(\xi), h(\xi), a(\xi)) = (m^*, 0, a^*), \tag{5}$$

where $h \to 0$ because m converges to constant state. The other case is analogous. In both cases constants m^* and a^* are given in Eq. (4). The matrices of derivatives of the flux in System (3) at equilibria A and B are

$$J(A) = \begin{bmatrix} 0 & 1 & 0 \\ \mu_1 & -c & -\dfrac{\gamma}{k} \\ -\dfrac{k}{c} & 0 & \dfrac{\mu_2+\gamma}{c} \end{bmatrix}, \quad J(B) = \begin{bmatrix} 0 & 1 & 0 \\ \mu_1 + \dfrac{\gamma}{k}a^* & 2\nu(m^*) - c & \dfrac{\gamma}{k}(m^*-1) \\ \dfrac{k}{c}(a^*-1) & 0 & \dfrac{km^*+\mu_2+\gamma}{c} \end{bmatrix}. \tag{6}$$

4. Numerical results

In this section we study numerically PDE System (1) and compare is with ODE System (3) using the algorithm de [6]. Both studies use same parameter values found in the literature. Dimensional parameter values are, [3], $\bar{D} = 1.25 \times 10^{-2}$, $\bar{\nu} = 5.0 \times 10^{-2}$, $\bar{\gamma} = 0.2$, $\bar{r} = 30$, $\bar{k}_1 = 25$, $\bar{k}_2 = 100$, $\bar{\mu}_1 = 4.0 \times 10^{-2}$ $\bar{\mu}_2 = 1.0 \times 10^{-2}$, where space ($\bar{x}$) is measured in km and time ($\bar{t}$) in days. Dimensionless parameter values are, [3], $\nu = 8.164 \times 10^{-2}$, $\gamma = 6.66 \times 10^{-3}$, $k = 2.5 \times 10^{-1}$, $\mu_1 = 1.33 \times 10^{-3}$, $\mu_2 = 3.33 \times 10^{-4}$.

4.1. Direct numerical simulation

System (1) is solved numerically using Crank-Nicolson finite difference scheme. We set initial data for the variables M and A at time $t = 0$ as $M_0 = 10$ and $A_0 = 50$. Dirichlet boundary conditions are considered for both left and right sides. Applying numerical scheme we obtain the solution of System (1) for times $t = 0, 3, 6, 9, 15$ (days), see Fig. 1. This figure shows evidence that the solution of System (1) contains two traveling waves propagating forward and backward. From the biological point of view these results indicate the invasion phenomena in the population dynamics of *Aedes aegypti* with one profile propagating along the flow (from left to right with speed $C_{al} = 0.2776\, km/day$ and dimensionless speed $c_{al} = 0.4534$) and another profile propagating against the flow (from right to left with speed $C_{ag} = -0.2775\, km/day$ and dimensionless speed $c_{ag} = -0.4531$). This observation motivates the rigorous search for traveling wave solutions.

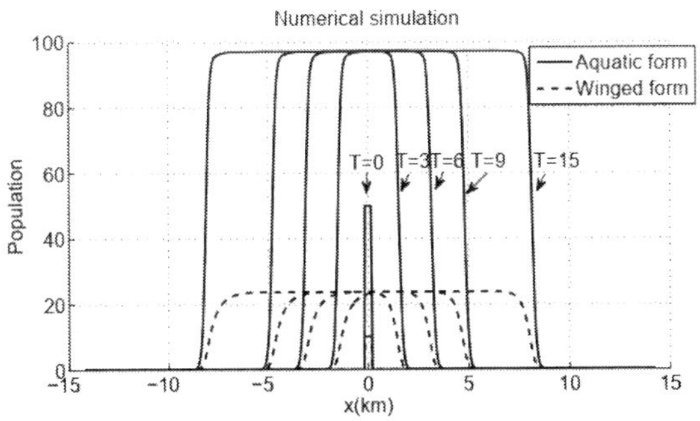

Figure 1. Numerical solution of System (1) for times $t = 0, 3, 6, 9, 15$ days, using the dimensional parameter values.

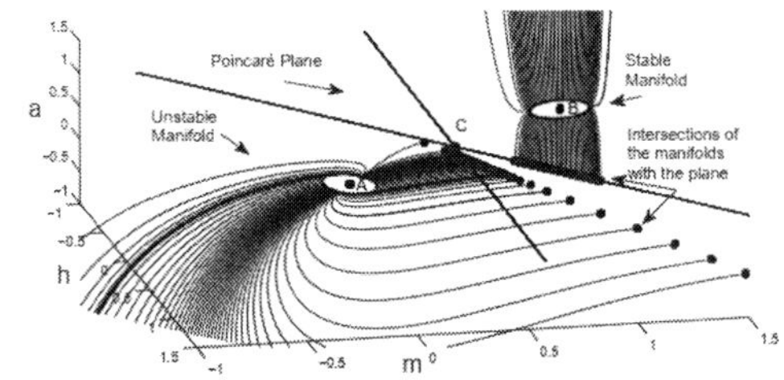

Figure 2. Numerical integration starts at small circles contained in E_A^u and E_B^s centered at equilibria A and B, respectively. Poincaré plane π is indicated by two straight lines; it intersects unstable manifold W_A^u and stable manifold W_B^s in points indicated by stars ($*$).

4.2. Semi-analytical method: invasion against the flow

In this section we follow the steps of the semi-analytical method described in [6] in order to verify the existence of the traveling wave solution of System (3), i.e., the heteroclinic orbit connecting equilibrium A to equilibrium B (boundary conditions given by Eq. (5)). The algorithm for the orbit connecting B to A is analogous. We consider dimensionless parameter values given in Section 4 (introduction) with $c_{ag} = -0.4531$ obtained in the numerical simulation.

(i) Equilibria solutions of System (3) are $A = (0, 0, 0)$ and $B = (0.951, 0, 0.971)$ with corresponding matrices of derivatives given by $J(A)$ and $J(B)$, Eq. (6).

(ii)
- $J(A)$ possesses two eigenvalues with positive real part: $\lambda_1 = 0.1999$ and $\lambda_2 = 0.4743$, with corresponding eigenvectors w_1 and w_2. These eigenvectors define the unstable subspace E_A^u which approaches locally the unstable manifold W_A^u;
- $J(B)$ possesses two eigenvalues with negative real part: $\lambda_3 = -0.0467$ and $\lambda_4 = -0.5403$, with corresponding eigenvectors w_3 and w_4. These eigenvectors define the stable subspace E_B^s which approaches locally the stable manifold W_B^s.

(iii) Using point $C = (A + B)/2$ and the vector $N = \overrightarrow{CB}$, we defined the Poincaré plane π. The plane π intersects both stable and unstable manifolds transversally, see Fig. 2, because $\zeta(C) \cdot N \neq 0$, where $\zeta(m, h, a) = (m', h', a')$ is given by Eq. (3).

(iv) Numerical integration starts at circles contained in E_A^u and E_B^s centered at equilibria A and B, respectively. The intersection sets between integration curves and π are denoted by

P_A^u and P_B^s. The lines P_A^u and P_B^s intersect on the Poincaré plane at point $P_A^u \cap P_B^s = P$, see Fig. 3. This ensures the existence of the heteroclinic orbit connecting equilibrium A to equilibrium B.

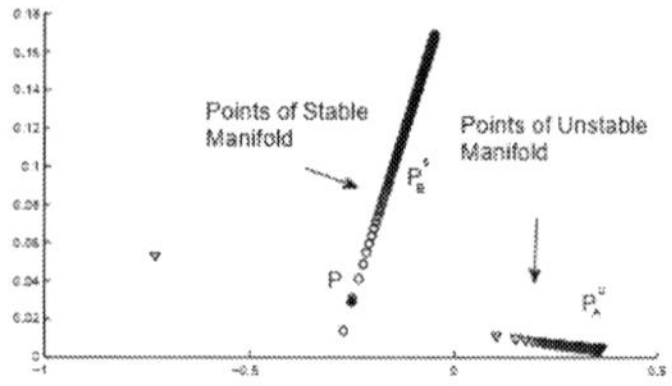

Figure 3. Poincaré plane π from Fig. 2 indicating the intersection sets P_A^u ($P_A^u \approx W_A^u \cap \pi$) and P_B^s ($P_B^s \approx W_B^s \cap \pi$). Point P represent the intersection between both manifolds.

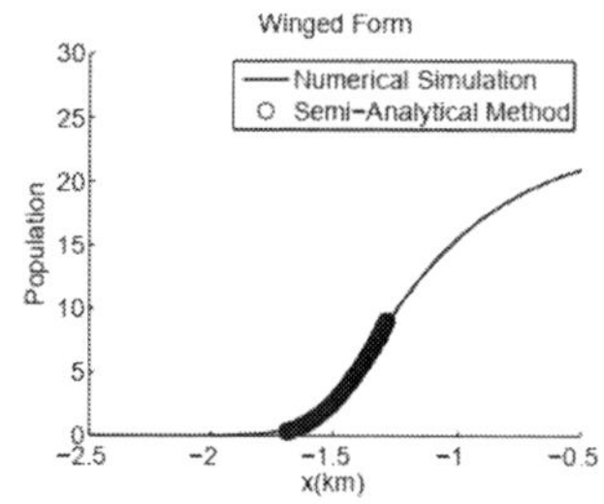

Figure 4. Against the flow invasion profiles for the winged phase obtained by using semi-analytical method and direct numerical simulation.

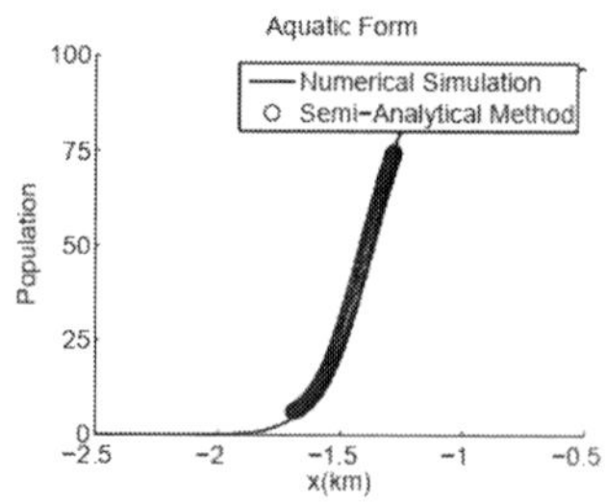

Figure 5. Against the flow invasion profiles for the aquatic phase obtained by using semi-analytical method and direct numerical simulation.

(v) In order to validate the proposed semi-analytical method we compare the solution profile obtained by the proposed method with the one from direct numerical simulation for the winged phase and aquatic phase, see Figs. 4 and 5. The small numerical difference between profiles observed on Figs. 4 and 5 are due to diffusion effect of the numerical scheme and the radius of the small circles used for starting the integration.

5. Conclusions

In this work we complete the results presented in [5] showing that both waves appearing in the numerical simulation are traveling waves. Our semi-analytical method verified the existence of traveling wave profiles appearing in the system describing population dynamics of *Aedes aegypti* mosquitoes. The results were validated through numerical simulations.

Acknowledgments

This work was supported, in part, by CNPq and CAPES.

References

[1] Carneiro L and Travassos L 2016 *Microbes and Infect.*
[2] Dias W O, Wanner E F and Cardoso R T 2015 *Math. Biosci.* **269** 37–47
[3] Takahashi L, Maidana N, Ferreira Jr W, Pulino P and Yang H 2005 *B. Math. Biol.* **67** 509–528
[4] Maidana N and Yang H 2008 *Math. Biosci.* **215** 64–77
[5] Freire I and Torrisi M 2013 *Nonlinear Anal.-Real* **14** 1300 – 1307
[6] Yamashita W, Takahashi L and Chapiro G 2016 *Manuscript submitted*
[7] Volpert A, Volpert V and Volpert V 2000 *Traveling Wave Solutions of Parabolic Systems* (*Translations of mathematical monographs* vol 140) (American Mathematical Society) ISBN 9780821811436

Multi-point quasi-rational approximants for the modified Bessel function $I_1(x)$

P. Martin[1], J. Olivares[2], L. Cortés-Vega[2] and A. Sotomayor[2]

[1] Physics Department, Antofagasta University, Chile
[2] Department of Mathematics , Antofagasta University, Chile

E-mail: pablo.martin@uantof.cl, jorge.olivares@uantof.cl, luis.cortes@uantof.cl, adrian.sotomayor@uantof.cl

Abstract. Approximants for the modified Bessel function $I_1(x)$ has been found using the multi-point quasi-rational technique. The approximations here determined has good accuracy for any positive value of the variable, and it seems to be adequate for most of the works where this function are used. Furthermore, the approximants are simple to calculate numerically in a direct way or using any usual MAPLE or MATLAB software.

1. Introduction

The modified Bessel function is a very important tool in several areas of Physics and Mathematics [1, 2, 3, 4]. The power series is an entire function, that is, it has an infinite radius of convergence. However precise calculations for regular values of x requires sometimes a lot of terms from the power series (several hundreds for high vales of x), which limits their use. The asymptotic expansions of these functions are also well-known [1, 2, 3, 4, 5], but its use is restricted to very high values of the independent variable x, besides the asymptotic expansions are semi-convergent series (convergent only in the Poincaré sense), which is also a problem, because to obtain the best accuracy different numbers of terms have to be taken depending of the values of x. There are some polynomials approximations [3], but there are good only in small range of the x-variable.

On the other way recently has been published a technique to obtain precise approximants to functions and eigenvalues [6, 7, 8, 9, 10, 11] using simultaneously power series and asymptotic expansions and building a bridge between both expansions using rational approximants, as in Padé method, but adding also some non-rational auxiliary functions.

The way to obtain the approximant with the present technique will be presented in section 2. The last section will be left for the conclusion.

2. Theoretical treatment

The power series of the modified Bessel function $I_1(x)$ is

$$I_1(x) = \frac{x}{2}\sum_{k=0}^{\infty}\frac{1}{k!(k+1)!}\left(\frac{x^2}{4}\right)^k. \tag{1}$$

Content from this work may be used under the terms of the Creative Commons Attribution 3.0 licence. Any further distribution of this work must maintain attribution to the author(s) and the title of the work, journal citation and DOI.
Published under licence by IOP Publishing Ltd

The asymptotic expansion for this function is also well known

$$I(x) \sim \frac{e^x}{\sqrt{2\pi x}}\left[1 - \frac{3}{8x} - \frac{15}{128x^2} - \cdots\right] \tag{2}$$

A bridge between both expansions has to be built by using the MPQA (multi-point quasi-rational approximant) technique. This requires to build a bridge using both powers simultaneously and using simple functions combined with rational ones.

It is clear from the asymptotic expansion that there is a fractional power at the infinite, but a simple auxiliary function as $\left(\frac{1}{\sqrt{x}}\right)$ can not be used because this kind of functions has the good behavior at the infinite, but at the same time appears a singularity as ramification at $x = 0$. However this singularity at $x = 0$ does not correspond to $I_1(x)$, and in this will be put at the approximant not convenient singularity.

To avoid this problem we should use an auxiliary function as $\left(\frac{1}{x+\lambda}\right)$ with $\lambda > 0$. In this way the inconvenient singularity will be in the negative axis, that is, outside of the region of approximation, which in our case will be $x \geq 0$ or right half complex plane. There is also a new reason to discuss this auxiliary function, since the power series for $I_1(x)$ has only odd powers of x, then in order to obtain a more efficient and adequate auxiliary function this has to be chosen as $\left(\frac{x}{(1+\lambda x^2)^{\frac{3}{4}}}\right)$. The simplest of these approximants will be with $\lambda = \frac{1}{4}$. Similar analysis to the previous one can me performed for the exponential behavior at the infinite and in this way the simplest approximant found for the function $I_1(x)$ is written as

$$\widetilde{I}_1(x) = x\left(e^x + e^{-x}\right)\left(1 + \frac{x^2}{4}\right)^{-3/4} \cdot \frac{\left(\frac{1}{2} + \frac{x^2}{32(\sqrt{\pi}-1)}\right)}{\left(2 + \frac{3\sqrt{\pi}\,x^2}{8(\sqrt{\pi}-1)}\right)}, \tag{3}$$

where the coefficients of the polynomial numerator and denominator has been chosen in such a way that the two first power series of the functions $I_1(x)$ and $\widetilde{I}_1(x)$ are incidents as well as the leading term of the asymptotic expansion.

The relative error from $0 < x < 100$ is also shown in figure 1.

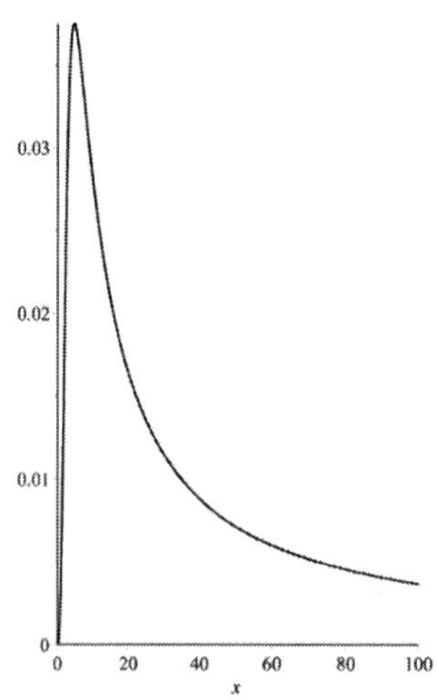

Figure 1. In this figure, the relative error of the appoximant is shown for $0 < x < 100$.

Since the function $I_1(x)$ increases exponentially largest value of x are not usual but the relative error is also small.

The maximum relative error is for values of x nearby 5. This is the usual behavior of the error using the multi-point quasi-rational approximant technique (MPQA) since the approximants are very precise for small and large values of the variable, and therefore the largest errors are for intermediate values. In any case the errors are always obtained using a small even for approximants low number of parameters. Including more parameters the largest relative errors decreases quickly, sometimes including one or two additional parameters the largest error decreases in magnitude order.

3. Conclusion

A precise approximant for the modified Bessel function of order one has been determined using the MPQA technique. The approximant is very simple and it is good for every positive value of the variable x. Using only a few parameters a good accuracy is obtained. This shows the goodness of this way to obtain approaches to spacial functions and in particular modified Bessel functions. Using this technique the largest error is usually for intermediate values, that is, no small and no large. Here we have determined an approximant with small number of parameters and the largest error was about 3 per cent nearby $x \sim 5$. Approximants with larger accuracy can be obtained using one or two parameters more, which are left for future works.

Acknowledgments

L. Cortés–Vega and A. Sotomayor are partially supported by Project FONDECYT 1161192, Chile.

References

[1] G. N. Watson, "A treatise on the theory of Bessel functions", 2nd. ed., Cambrigde, England: Cambrigde University Press (1966).
[2] J. D. Jackson, "Clasical Electrodynamics", 2nd. ed., John Wiley and Sons, Inc., Ch.3 (1975).
[3] M. Abramovitz and I. A. (Eds) Stegun, "Handbook of Mathematical Functions", 9th printing, Dover, New York 358-364 (1972).
[4] G. Arfken, "Mathematical Methods for Pysicists", 3rd. ed., Academic Press, Orlando Fl., USA 573-596 (1985).
[5] F. W. J. Olver, "Asymptotic and Special Functions", Academic Press, New York (1974).
[6] P. Martin, E. Castro, J. L. Paz and A. De Freitas, "Multipoint quasi-rational approximants in Quantum Chemistry", Chapter 3 of New Developments in Quantum Chemistry", Transworld Research Network, Kerala, India 55-78 (2009).
[7] P. Martin, E. Castro and J. L. Paz, Rev. Mex. Fis. 58, 301 (2012).
[8] L. Ladera and P. Martin, J. Comput. Phys. 73, 481 (1987).
[9] P. Martin and L. Guerrero, J. Math. Phys. 26, 705 (1985).
[10] P. Martin and L. Guerrero, J. Comp. Phys. 85, 487 (1989).
[11] P. Martin, Rodriguez-Nuñez and J. L. Marquez, Phys. Rev. B 45, 8359 (1992).

Quantum Metrology: Towards an alternative definition for the meter

Ioannis P. ZOIS

Testing, Research & Standards Centre, Public Power Corporation, 9, Leontariou Street, GR- 153 51, Pallini, Athens, Greece

i.zois@dei.com.gr and i.zois@exeter.oxon.org and izois@ppcr.gr

Abstract. The motivation for this work came from an attempt to give an alternative definition for the meter, the SI unit for measuring length. As a starting point towards this goal we present the underlying theory behind our approach which uses ideas from quantum field theory and non-commutative geometry, in particular the notion of an odd K-cycle which is based on the Dirac operator (and its inverse, the Dirac propagator). Using physics terminology, the key point in our strategy is this: instead of measuring ordinary length in space-time we measure the "algebraic (or spectral) length" in the space of quantum states of some fermion acted upon by the Dirac propagator.

1. Introduction

The SI unit for measuring lengths is the meter. This is defined as the length of the path traveled by light in vacuum during a time interval of 1/299,792,458 second (see [3], [25], [26], [17]). It is clear that this definition is based on A. Einstein's (Special and General) Theory of Relativity (see [6], [10], [23], [27], [28], [29]). A basic ingredient of the theory of relativity as much as physics is concerned is the fact that the speed of light is constant, namely it is the same for all inertial observers (this follows from numerous experiments, starting with the famous Michelson - Morley experiment in 1887). In fact the fundamental difference between the relativity principle of Einstein and that of Galileo is the exchange of the invariance of simultaneity with the invariance of the speed of light.

The mathematics used in relativity theory is (pseudo) - Riemannian Geometry (see [9], [15], [21]). The mathematical structure studied in Riemannian geometry is the Riemannian manifold (or Riemannian space): A Riemannian manifold is a pair *(M, g)* where *M* is a real smooth manifold whose tangent space at each point is equipped with a Riemannian metric *g*, that is a positive definite symmetric bilinear form (inner product) which varies smoothly on M. If g is not positive definite then one has a pseudo-Riemannian metric.

A metric enables one to measure lengths, angles, areas, volumes, to calculate curvatures, gradients of functions, divergences of vector fields etc. The space-time we live in is an example of a pseudo-Riemannian manifold of dimension 4. Since by definition a manifold locally looks like the Euclidean space $\mathbf{R}^n$ and since the dimension of space-time is 4 (at least macroscopically), then our space-time locally looks like $\mathbf{R}^4$. An important geometric difference between the Special (abbreviated to SR) and

Content from this work may be used under the terms of the Creative Commons Attribution 3.0 licence. Any further distribution of this work must maintain attribution to the author(s) and the title of the work, journal citation and DOI.
Published under licence by IOP Publishing Ltd

the General (abbreviated to GR) Theory of Relativity is the fact that in the former space-time is flat, namely both locally and globally is modeled on $\mathbf{R}^4$ whereas in the later space-time is a pseudo-Riemannian manifold with some nonzero curvature and hence it is only locally modeled on $\mathbf{R}^4$.

During the last 20 years or so an important break through occurred in geometry, this is the discovery of noncommutative geometry (see [4], [5], [31], [32], [33]). This new geometry enormously generalizes Riemannian geometry and it has deep connections to quantum physics (quantum field theory in particular). These new (as far as metrology is concerned) ingredients (noncommutative geometry and quantum field theory) constitute the basis of our work: Our idea is we believe quite clear, we imitate what happens in *cartography:* Instead of measuring the distance between say London and Paris, we use a map with its scale S. We measure the distance on the map DM and then we can calculate the real distance RD by the simple formula

$$RD = DM \ X \ S.$$

In the same fashion, instead of measuring the real spacetime distance (distance on a Riemannian manifold), we use an equivalent "algebraic (or spectral) distance" on a mathematical structure called "odd K-cycle of the corresponding C*-algebra" of the Riemannian manifold. This plays the role of the map with scale. A key ingredient of this structure is the Dirac operator used in quantum physics to describe fermions (relativistic Dirac equation). Then the proposition proved serves as a "translator" converting distances (and data) from the real spacetime (Riemannian manifold) to this fictitious space (odd K-cycle). It's the analogue of the simple equation used in cartography.

There is another motivation for our approach though: During the *24th General Conference of Standards and Measures on 21st October 2011 in Serves France,* there was an unanimous vote for the redefinition of several of the fundamental units, some of them based on the Planck's constant h. It is common knowledge nowadays that Planck's constant is the fundamental constant in quantum physics, it captures the so called particle -wave duality, the fundamental principle of quantum physics. Mathematically speaking Planck's constant measures the noncommutativity between position and momentum as expressed in Heisenberg's uncertainty principle, thus the connection with noncommutative geometry. One could say that this is the official introduction of Quantum Physics in metrology. Our approach relies heavily on quantum field theory.

The purpose of this article is to set the mathematical (geometric) basis for a possible application of this new geometry and quantum physics to metrology (hence the term quantum metrology) and especially to the definition of the fundamental units. We hope this will be the beginning of a fruitful interaction between mathematical physics and metrology.

2. The underlying geometry of the current definition of the meter

2.1 Minkowski geometry

In SR space-time is modeled on the vector space (but also trivially a manifold) $\mathbf{R}^4$ equipped with the so-called Minkwski (pseudo) metric. The fact that one needs the Minkowski and not the usual Euclidean metric in 4 dimensions can be readily seen for example from the wave equation that the electromagnetic field satisfies in vacuum:

$$\Box A_\mu = 0, \qquad (1)$$

where $\Box$ is the d'Alembert operator:

$$\Box = (1/c^2)dt^2 - \nabla^2 \quad (2)$$

and A_μ is the 4-potential. d'Alembert operator is nothing other than the Laplace operator in $\mathbf{R}^4$ using the Minkowski metric. It is well-known that one of the theoretic motivations for the discovery of SR is the fact that Maxwell's equations are not compatible with Galileo's relativity principle.

Choosing a basis, symmetric bilinear forms can be represented by symmetric matrices , in our case by a real 4 X 4 matrix. If we pick a η-Cartesian basis (an inertial reference frame as it is known in physics literature), Minkowski metric can be represented in its canonical form by the matrix

$$\eta = \text{diag}\,(1, -1, -1, -1) \quad (3)$$

where we adopt the signature (+ - - -). Hence the inner product between two vectors with coordinates (relative to the chosen basis) $X = (X^0, X^1, X^2, X^3)$ and $Y = (Y^0, Y^1, Y^2, Y^3)$ will be given by the following expression

$$\eta\,(X,Y) = X^{\circ}Y^{\circ} - X^1Y^1 - X^2Y^2 - X^3Y^3 \quad (4).$$

Given a metric it is clear that one can define the length of a vector $X = (X^0, X^1, X^2, X^3)$ (in physics terminology since these are elements of $\mathbf{R}^4$ they are called 4-vectors), which will be equal to

$$|X| = \eta\,(X,X) = X^0X^0 - X^1X^1 - X^2X^2 - X^3X^3 = (X^0)^2 - (X^1)^2 - (X^2)^2 - (X^3)^2 \quad (5)$$

One can immediately define angles, for example the cosine of an angle say θ, between two vectors X and Y will be given by the relation

$$\cos\theta = \eta\,(X,Y) / (|X||Y|) \quad (6)$$

In physics one uses the (infinitesimal) line element (proper-time)

$$d\tau^2 = (cdt)^2 - dx^2 - dy^2 - dz^2 \; (7)$$

Since the Minkowski metric is not positive definite it is apparent that expressions like (4), (5) and (7) can take positive, negative and zero values. In physics one uses the terms time-like, space-like and null respectively to characterize 4-vectors. For reasons which are out of the scope of this article, in SR we accept that only time-like vectors have physical meaning (namely only time-like vectors can be used to describe physical quantities) and that null 4-vectors describe photons (particles traveling with the speed of light). Thus if one wants to describe the path (position vector) of a photon, equation (7) will give

$$0 = (cdt)^2 - dx^2 - dy^2 - dz^2$$

hence

$$(cdt)^2 = dx^2 + dy^2 + dz^2$$

thus

$$(cdt)^2 = ds^2 \; (8)$$

Equation (8) gives the current definition of the meter in SI stated at the start of section 1 (this is the infinitesimal version of the equation, the finite version is obtained easily by integration).

2.2 Riemannian geometry

In the case of Riemannian geometry things are generalized as follows: To begin with, instead of the space $\mathbf{R}^4$ one has a real smooth manifold, we briefly recall its definition: A real smooth manifold of dimension (in general) n is a Hausdorff topological space which is locally homeomorphic to the Euclidean space $\mathbf{R}^n$, that is for every point of the manifold there exists a neighborhood (an open set containing it) which is homeomorphic to an open ball of dimension n (homeomorphic means that there exists a map which is 1-1, onto, continuous with a continuous inverse). These maps which locally model the manifold onto the space $\mathbf{R}^n$ are called coordinate charts. It is clear that coordinate charts may intersect and any point on the manifold can be arranged to belong to different charts. In smooth manifolds the extra condition is that the maps which represent these coordinate transformations on intersections should be smooth (since these coordinate transformations are essentially maps from $\mathbf{R}^n$ to $\mathbf{R}^n$, the term "smooth" means the usual thing we know from vector calculus, that is infinitely times differentiable). This definition enables one to carry over the usual vector calculus onto manifolds which constitute an important generalisation of spaces like $\mathbf{R}^n$. The classic example of a smooth real manifold of dimension two is the 2-sphere S^2.

The smoothness condition guarantees that the manifold has a well-defined tangent space at each point (this is a simple higher dimensional generalisation of the tangent vector of a smooth curve). The tangent space is nothing other than a vector space $\mathbf{R}^n$. Hence a Riemannian metric is a positive definite symmetric bilinear form on the tangent space which varies smoothly from point to point. Consequently our discussion in the previous section concerning metrics (symmetric bilinear forms) on vector spaces $\mathbf{R}^n$ can be immediately applied to this more general setting. Hence a Riemannian manifold is a pair (M,g) where M is a smooth real manifold and g is a Riemannian metric. It is clear that in SR the space-time manifold can be identified with its tangent space whereas in GR this is not necessarily true (compare $\mathbf{R}^2$ with the 2-sphere S^2, $\mathbf{R}^2$ is the tangent space on every point of the 2-sphere).

If one chooses a local coordinate system on a Riemannian manifold (the analogue of a basis in a vector space), then the metric can be represented by a symmetric n X n matrix yet the entries of this matrix will no longer be numbers but functions of the local coordinates (since the metric varies from point to point). More concretely, according to tensor algebra, the metric is a covariant tensor (field) of order 2. Then the line element on a Riemannian manifold (M, g) of dimension n, by choosing local coordinates denoted x^μ (where $\mu,\nu = 1,2,..,n$), will be given by the following formula

$$ds^2 = g_{\mu\nu}\, dx^\mu dx^\nu \quad (9)$$

where we have made use of the Einstein convention (repeated indices are summed). It is a fundamental physical principle that GR locally reduces to SR (this fact makes the notion of manifold appropriate to describe space-time itself, at least classically and at long scales). Furthermore photons (namely light) according to GR follow geodesics, in Riemannian geometry geodesics are the analogue of straight lines in Euclidean geometry, these are the paths - curves of minimal (more accurately extreme) length. Thus the geodesic distance d(P,Q) between two points P, Q on some Riemannian manifold is given by the length of the geodesic which connects the two points using equation (9). Essentially it is the infimum of the length of all possible paths connecting the two points P and Q where the length of each path is calculated using equation (9).

3. Noncommutative geometry

Noncommutative geometry is perhaps the most important development in geometry during the last 20 years or so. It was discovered by Alain Connes and Daniel Quillen.

The starting point of noncommutative geometry (abbreviated to NCG in the sequel) is Gelfand's theorem which we briefly explain: One the one hand let's consider the category whose objects are compact Hausdorff topological spaces and whose morphisms (arrows) are homeomorphisms among them. On the other hand let's consider the category whose objects are unital commutative C*-algebras and whose morphisms (arrows) are star preserving homomorphisms. Gelfand's theorem then states that these two categories are equivalent and the equivalences are given by the (mutually inverse) functors C(-) and Spec(-). The former assigns to each compact Hausdorff topological space X the commutative unital C*-algebra C(X) consisting of continuous complex valued functions defined on X. The later assigns to every commutative unital C*-algebra A the compact Hausdorff topological space Spec(A) which is the spectrum of the algebra A (the spectrum of an algebra consists of its characters, namely non-zero star preserving homomorphisms from the algebra to the field of the complex numbers **C**). These functors satisfy the following conditions

$$\mathrm{Spec} \circ C = 1$$

and

$$C \circ \mathrm{Spec} = 1,$$

namely C(Spec (A)) = A and Spec (C(X)) = X, where 1 is the identity functor and " ∘ " denotes the composition of functors. If the topological space is not compact but only locally compact, then the corresponding algebra has no unit (this difficulty can be overcome by adjoining a unit which increases complexity).

Gelfand's theorem implies that the topological information of a space is encoded in the corresponding algebra of continuous complex functions and in fact the topology can be recovered from the algebra. In particular this holds true for the topology of a Riemannian manifold. The natural question then is if there is a way to recover the metric data as well. It turns out that this can indeed be done using the notion of an (odd) K-cycle (see [22]). The definition is the following:

Definition 1: An (odd) K-cycle on an algebra A is a quadruple (A, π, H, D) where A is an algebra (commutative or not), H is a separable Hilbert space, π is a unitary representation of the algebra A on the Hilbert space H and D is an unbounded self-adjoint operator acting on the Hilbert space H satisfying the following properties:

1. The commutator [D,a] = Da - aD (where a is an element of the algebra A) is a bounded operator for any element a of the algebra A.
2. The operator $(1+D^2)^{-1}$ is compact.

(If the algebra A is non-unital then we slightly modify the above definition by demanding that the operator $[a(1+D^2)]^{-1}$ is compact for any element a of the algebra A).

The way one can recover the metric data using the notion of a K-cycle is deferred to section 5 where it will be presented as a Lemma.

K-cycles are the basic ingredients of K-Homology (see [31], [16]). This theory is the dual theory to the topological K-Theory due to Atiyah and Grothendieck (see [1], [11] and [13]) which has been proved a very useful tool in topology, geometry but also in modern theoretical physics (superstring theory, M- Theory, the theory of branes etc).

4. The Dirac operator

The Dirac operator will play a central role in our discussion and hence it deserves a brief review. Elementary particles belong to two big families, the bosons and the fermions. According to the spin statistics theorem in theoretical physics (see [24], we are not interested here in anyons, braid statistics etc), the former have integer spin and the obey the Bose-Einstein statistics whereas the later have half-integer spin, the obey the Dirac-Fermi statistics and the Pauli exclusive principle. Bosons are described by the Klein-Gordon equation (a second order partial differential equation of hyperbolic type) which is the relativistic version of the Schrodinger equation used in quantum mechanics (essentially this is the Laplace operator in 4-dimensional Minkowski space with signature (+ - - -)):

$$(\square + \mu^2)\Psi = [(1/c^2)dt^2 - (\partial_x{}^2 + \partial_y{}^2 + \partial_z{}^2) + \mu^2]\,\Psi = 0 \quad (10)$$

where $(\partial_x{}^2 + \partial_y{}^2 + \partial_z{}^2)$ is the Laplace operator in 3 spatial (Euclidean) dimensions (using Cartesian coordinates) and $\mu = 2\pi mc/h$ (m denotes the rest mass, c denotes the speed of light and h denotes Planck's constant).

Fermions again in 4-dimensional Minkowski space are described by Dirac's equation (this is written in the original Dirac's form):

$$(\beta mc^2 + \sum a_k P_k c)\,\Psi = [(i\,h)/(2\pi)]\,\partial_t\Psi \quad (11)$$

The index k takes the values 1,2,3 (these are the three spatial coordinates), m denotes the rest mass of the fermion, c denotes the speed of light and P denotes the momentum in the quantum mechanical sense, namely P_k corresponds to the partial derivative with respect to the corresponding spatial coordinate

$$P_k \leftrightarrow -ih\partial_k.$$

The "mysterious" quantities β and a_k are in fact complex 4X4 matrices satisfying the relations

$$(a_k)^2 = \beta^2 = I_4 \qquad (12)$$

where I_4 denotes the identity 4X4 matrix and

$$a_i a_j + a_j a_i = 0 = a_i\beta + \beta a_i \qquad (13)$$

(all indices i,j,k take values 1,2,3). One can rewrite Dirac's equation in covariant relativistic form as follows (see [7], [8], [20], [14] and [30]):

$$[(-ih/2\pi)\gamma^\mu\partial_\mu + mc]\,\Psi = 0 \quad (14)$$

where $\mu = 0,1,2,3$ and γ^μ denote the famous Dirac matrices (these are complex 4X4 matrices) which constitute a basis for the 4-dimensional Clifford algebra (in particular this is called the Dirac algebra) and they satisfy the relations

$$\{\gamma^\mu, \gamma^\nu\} = 2\,\eta^{\mu\nu} \qquad (15)$$

The symbol {,} denotes the anti-commutator (namely $\{\alpha, \beta\} = \alpha\beta + \beta\alpha$) and $\eta^{\mu\nu}$ denotes the (contravariant) Minkowski metric we saw in section 2. The following relation holds

$$\eta_{\mu\kappa}\,\eta^{\kappa\nu} = \delta_\mu{}^\nu \qquad (16)$$

where $\delta_\mu{}^\nu$ denotes the Kronecker δ.

In order to relate the covariant form with Dirac's original form of the equation we set $\beta = \gamma^0$ and $\gamma^k = \gamma^0 a^k$ (indices are lowered and raised using the metric):

$$\gamma^0 = \begin{pmatrix} I_2 & 0 \\ 0 & -I_2 \end{pmatrix},$$

$$\gamma^1 = \begin{pmatrix} 0 & \sigma_x \\ -\sigma_x & 0 \end{pmatrix},$$

$$\gamma^2 = \begin{pmatrix} 0 & \sigma_y \\ -\sigma_y & 0 \end{pmatrix} \text{ and}$$

$$\gamma^3 = \begin{pmatrix} 0 & \sigma_Z \\ -\sigma_Z & 0 \end{pmatrix}$$

where again I_2 denotes the identity 2X2 matrix and σ_χ, σ_Y, σ_z denote the Pauli 2X2 complex matrices.

The quantity $\gamma^\mu \partial_\mu$ is called the *Dirac operator.* The Dirac equation gave a precise description of the electron (relativistic particle of spin ½), it explained accurately the emission spectrum of the hydrogen atom (especially the hyperfinite structure) and it predicted the existence of antimatter. From the mathematical viewpoint the Dirac equation can be explained as an eigenvalue equation where the rest mass is proportional to the eigenvalue of the momentum operator. The Dirac operator can be interpreted as the square root of the Laplace operator. Let us also note here that all mass particles known today (quarks, leptons and neutrinos) are fermions and hence they are described by the Dirac equation.

The inverse of the Dirac operator is called the *Dirac propagator* in physics (or 2 point function). This is nothing other than the Green function for the (linear) Dirac operator. In position space this is given by the following formula

$$S_F(x-y) = \int \frac{d^4p}{(2\pi)^4} e^{-ip(x-y)} \frac{\gamma^\mu p^\mu + m}{p^2 - m^2 + i\varepsilon} = \left(\frac{\gamma^\mu (x-y)_\mu}{|x-y|^5} + \frac{m}{|x-y|^3}\right) J_1(m|x-y|) \quad (17)$$

where $J_1(m\,|x-y|)$ is the Bessel function of 1st kind. By a Fourier transform, in momentum space the Dirac propagator is given by the equation

$$SF(p) = \frac{\gamma^\mu p_\mu + m}{p^2 - m^2 + i\varepsilon} \quad (18)$$

The physical meaning of the propagator is that following Feynman's path integral approach, it is the basic ingredient in the expression that calculates the probability amplitude for a particle (in our case for a fermion) to travel from one place to another in a given time (position space) or to travel with certain energy and momentum (momentum space).

If we have an arbitrary Riemannian manifold (assumed closed for simplicity) instead of 4-dimensional Minkowski space, the above construction can be generalized (in a not so straightforward way). This was done by Atiyah - Hirzebruch - Patodi - Saphiro - Singer (see [19], [12] and [2]) as follows: Given a Riemannian manifold say M of dimension n, the Riemannian metric reduces the structure group GL(n) (invertible n X n real matrices) of the tangent bundle TM of M to the special orthogonal group SO(n) (this is the group of orthogonal n X n real matrices with unit determinant). In simple words this is a mathematical expression of the fact that given a metric one can define lengths and angles and hence one can define orthonormal bases, namely bases whose vectors are mutually vertical with length 1. Next one defines the spin bundle S over M which is the 2:1 lift of the bundle of orthonormal frames. The spin bundle S is a principal bundle with structure group Spin(n), which is the double cover of the group SO(n). If the above lift is possible, then one says that M is a spin manifold. A manifold is spin if the second Stiefel-Whitney class of its tangent bundle vanishes (the manifold is orientable if its 1st Stiefel-Whitney class vanishes).

In order to describe the Dirac operator for Riemannian manifolds one needs to define Clifford modules. More concretely, let (V,Q) be a vector space (over a commutative field k) equipped with a quadratic form Q. The Clifford algebra Cl(V,Q) is an associative algebra with unit which is defined as the quotient of the free tensor algebra $FT(V) = \Sigma\otimes^r V$ by the ideal I(V) of FT(V) generated by the elements of the form $v \otimes v + Q(v)1$ where v belongs in V and $\otimes$ denotes tensor product, namely:

$$Cl(V,Q) = \Sigma \otimes^r V / I(V) \quad (19).$$

Let us recall from the theory of Lie groups that for $n \geq 3$ there exists the universal covering homomorphism ξ_0 : Spin(n) → SO(n) with kernel {1,-1} (this is the meaning of the statement that the Lie group Spin(n) is the double cover of the Lie group SO(n)). Let us assume that $n \geq 3$ and let E be an oriented Riemann vector bundle of dimension n over some manifold M. A spin structure on E is a principal fibre bundle P(Spin(n)),E) with structure Lie group Spin(n) along with a double cover map

$$\xi: P(Spin(n)),E) \rightarrow P(SO(n),E) \quad (20)$$

where P(SO(n),E) denotes the bundle of oriented orthonormal frames of the vector bundle E that satisfy the relation $\xi(pg) = \xi(p)\xi_0(g)$ for every element p in P(Spin(n)),E) and g an element of the Lie group Spin(n).

A spin manifold is an oriented Riemannian manifold M with a spin structure on its tangent bundle TM (the dimension of the tangent bundle equals the dimension of the manifold).

Now every orthogonal transformation of $\mathbf{R}^n$ induces an orthogonal transformation on the Clifford algebra $Cl(\mathbf{R}^n)$ which maps the tensor algebra to itself and preserves the ideal. This induced map on the Clifford algebra is a homomorphism (it preserves multiplication) and hence one obtains a representation

$$cl(p(n)): SO(n) \rightarrow Aut\,(Cl(\mathbf{R}^n)) \quad (21).$$

If E is an oriented Riemann vector bundle as above over some manifold M, then the Clifford bundle Cl(E) of E is defined as

$$Cl(E) = P(SO(n),E) \times_{cl(p(n))} Cl(\mathbf{R}^n)$$

where

$$cl(p(n)): SO(n) \rightarrow Aut\,(Cl(\mathbf{R}^n))$$

is the representation coming from the induced transformations on the orthogonal transformations on $\mathbf{R}^n$. Since the vector bundle E is Riemann that means that one has an inner product defined on its fibres and Cl(E) is the associated bundle of Clifford algebras. Alternatively, Cl(E) could be defined as the quotient of the free tensor bundle $\Sigma\otimes^r E$ of E by the ideal bundle I(E), where I(E) is the bundle over M whose fibres are the two-sided ideals of the free tensor algebra of the fibre of E generated by the elements of the form

$$v \otimes v + \|v\|^2$$

where v is an element of the fibre of E. Hence one has (in analogy with equation (19) for vector spaces) that

$$Cl(E) = \Sigma \otimes^r E / I(E).$$

It is clear that Cl(E) is a Clifford algebra bundle over M and the fibre-wise multiplication on Cl(E) gives an algebra structure on the space of sections Γ (Cl(E)) of Cl(E).

Let M be a Riemannian manifold of dimension n with Clifford bundle Cl(M) (namely Cl(M) is the Clifford bundle of the tangent bundle TM of M, i.e. Cl(M) = Cl(TM)) and let S denote an arbitrary bundle of left modules over Cl(M) (that is a bundle over M whose fibre S_x over a point x of M is a left module over the Clifford algebra $Cl(M)_x$, S is called the spinor bundle). Assume further that S is Riemann with a Riemann connection (connection compatible with the metric on the manifold with vanishing torsion). Under these assumptions one can define in a natural way a first order differential operator D acting on sections of S which is called a *Dirac (type) operator* as follows

$$D\sigma = \Sigma\, e_j \cdot \nabla e_j\, \sigma \quad (22)$$

where the $e_i,\dots,e_n$ give an orthonormal basis of T_xM, σ is a section of S, ∇ denotes the covariant derivative on S defined by the connection and "$\cdot$" denotes the Clifford multiplication. The square D^2 of the Dirac operator is called *the (corresponding) Dirac Laplacian.*

5. The basic Lemma

The central idea of this article is given in the following Lemma:

Fundamental Lemma:

"One can recover the Riemannian metric from the fundamental K-cycle defined by the Dirac operator on a spin manifold M; In particular one has that

$$d(p,q) = \sup \{| f(p) - f(q) | ; f \in C(M) \text{ with } \| [D,f] \| \leq 1\} \quad (23)$$

where p and q are points in M and d(p, q) denotes their geodesic distance ".

Proof:

Let (M,g) be a closed n-dim Riemannian spin manifold. The fundamental K- homology class (A, π, H, D) of M (an odd K-cycle, see definition in section 3) is defined as follows: As algebra A we take the algebra C(M) of complex continuous functions on M (we could take smooth functions as well). Clearly C(M) is a vector space and it can be turned into an algebra using point-wise multiplication:

$$(fg)(x) = f(x)g(x) \qquad (24)$$

where f, g are elements of C(M) and x is a point of M (note that in the LHS of (24) we have multiplication in C(M) whereas in the RHS of (24) we have the usual multiplication of complex numbers). In fact C(M) is a unital C*-algebra where the unit is given by the constant function 1, star "*" operation is just complex conjugation and the norm ||.|| is given by

$$\|f\| = \max \{ |f(x)| \, \forall x \in M\}. \quad (25)$$

It is easy to check that C(M) is complete with respect to the metric induced by the above norm (well, in fact this is part of Gelfand's theorem).

As Hilbert space H we take the space $H = L^2(M,S)$, namely the space of square integrable sections of the (irreducible) spinor bundle S of M, its dimension is equal to $2^{(n/2)}$ .The scalar product in $H = L^2(M,S)$ is the usual one of the measure $d\mu(g)$ associated with the Riemannian metric g, i.e.

$$(\psi,\varphi) = \int d\mu(g)\psi^*(x)\varphi(x) \quad (26)$$

where ψ, φ are elements of the Hilbert space H, "*" denotes complex conjugation and the scalar product in the spinor space is the natural one in $\mathbf{R}^k$ where $k = 2^{(n/2)}$.

As unitary representation π we take the action of C(M) as multiplicative operators on H: If f is an element of C(M) and ξ an element of H, then $(f\xi)(x) = f(x)\,\xi(x)$.

Finally, as self-adjoint operator D we take the Dirac operator D of the spin manifold M associated with the Levi-Civita connection $\omega = dx^\mu\,\omega_\mu$ of the metric g. It is worth describing the Dirac operator in greater detail using local coordinates: Let (e_a, a = 1,2,...,n) be an orthonormal basis of vector fields related to the natural basis (∂_μ, $\mu = 1,2,...,n$) via the n- beins $e_a{}^\mu$ so that the components of the metric g are related by

$$g^{\mu\nu} = e_a{}^\mu e_b{}^\nu n^{ab} \quad (27)$$

where η^{ab} denotes the flat metric (the Greek indices μ,ν are the "curved" indices and the Latin indices a,b are the "flat" indices, all take values 1,2,..,n = dimM, curved indices are lowered and raised using the Riemann metric g whereas flat indices are lowered and raised using the flat metric η). The coefficients ($\omega_{\mu a}{}^b$) of the Levi-Civita connection (metric and torsion free) of the metric g are defined via

$$\nabla^\mu e_a = \omega_{\mu a}{}^b e_b \quad (28)$$

Moreover let Cl(M) denote the Clifford bundle over M whose fibre over x in M is the (complexified) Clifford algebra $Cl(T_xM)$ and let us denote by $\Gamma(M, Cl(M))$ the module of its sections. One gets an algebra homomorphism

$$\gamma : \Gamma(M, Cl(M)) \to B(H)$$

defined by

$$\gamma(dx^\mu) = \gamma^\mu(x) = \gamma^a e_a{}^\mu \quad (29)$$

The curved gamma matrices satisfy a similar relation with equation (15):

$$\{\gamma^\mu(x), \gamma^\nu(x)\} = 2 g^{\mu\nu} \quad (30)$$

The lift ∇^s of the Levi-Civita connection to the spinor bundle will be

$$\nabla^s{}_\mu = \partial_\mu + \omega_\mu{}^s = \partial_\mu + (1/2)\, \omega_\mu{}^{ab} \gamma^a \gamma^b \quad (31)$$

Then the Dirac operator defined by

$$D = \gamma \circ \nabla \quad (32)$$

can be written in local coordinates as

$$D = \gamma(dx^\mu)\, \nabla^s{}_\mu = \gamma^\mu(x)(\partial_\mu + \omega_\mu{}^s) = \gamma^a e^\mu{}_a (\partial_\mu + \omega_\mu{}^s) \quad (33)$$

It is straightforward to check that the conditions of the definition of an (odd) K-cycle are satisfied and thus (C(M), L^2 (M,S), · , D) is an (odd) K-cycle (this is an odd class in the K- Homology of M, namely an element of the 1st K-homology group $K_1(M)$ of M).

In order to recover the geodesic distance one works as follows: From the action of C(M) as multiplicative operators on H = L^2 (M,S) one finds that

$$[D, f]\, \psi = (\gamma^\mu \partial_\mu f)\psi$$

where ψ is in the Hilbert space and f is in C(M). Moreover the commutator [D, f] is itself a multiplicative operator as well, namely

$$[D, f] = (\gamma^\mu \partial_\mu f) = \gamma(df)$$

As a consequence its norm is

$$\| [D,f] \| = \sup |(\gamma^\mu \partial_\mu f)\, (\gamma^\nu \partial_\nu f)^*|^{1/2} = \sup |\, \gamma^{\mu\nu}(\partial_\mu f)(\partial_\nu f)^*|^{1/2} \quad (34)$$

Now the RHS of equation (34) is just the Lipschitz norm of f (see [4]) which is given by

$$\|f\|_{Lip} = sup_{x \neq y} \frac{|f(x)-f(y)|}{d_\gamma(x,y)} \quad (35)$$

where $d_\gamma(x,y)$ is the usual geodesic distance on M given by the well-known formula

$$d_\gamma(x,y) = \inf_\gamma \{\text{length of paths } \gamma \text{ from x to y}\}.$$

Hence one has that

$$\|[D,f]\| = sup_{x\neq y}\frac{|f(x)-f(y)|}{d_\gamma(x,y)}.$$

Now the condition $\| [D,f] \| \leq 1$ in the Lemma (equation (23)) automatically gives that

$$d(p,q) \leq d_\gamma(p,q) \quad (36)$$

To invert the inequality sign we fix a point q and we consider the function $f_{\gamma,q}(x) = d_\gamma(x,q)$. Then $\| [D, f_{\gamma,q} \| \leq 1$ and substituting this into (23) we get

$$d(p,q) \geq |f_{\gamma,q}(p)-f_{\gamma,q}(q)|= d_\gamma(p,q).$$

This final equation along with (36) prove the result.

Remarks:

1. As a simple example consider M = **R** and D = d/dx. Then the condition $\| [D,f] \| \leq 1$ reads $\sup |df/dx| \leq 1$ and the supremum is saturated by the function f(x) = x + cosy which gives the usual distance.
2. Let us recall for brevity that the Lipschitz norm of some function f in C(M), where M is some closed Riemannian spin manifold, is given by the following formula: $\|f\|_{Lip} = \|[d+d^*, \pi_1(f)]^2\|_{op} = (1/2)\ \|[[\Delta, \pi_2(f)], \pi_2(f)]\|_{op} = \|[D, \pi(f)]\|^2{}_{op}$ where the index "op" stands for the operator norm, d is the exterior derivative, d* is its (Hodge) adjoint (using the Riemannian metric), $\Delta = dd^* + d^*d$ denotes the Laplace operator, D denotes the Dirac operator and π_1, π_2 and π denote the representations of the algebra C(M) onto the Hilbert spaces L^2 (M, Λ), L^2(M) and L^2 (M, S) of square integrable forms, functions and spinors respectively (where functions act multiplicatively in all cases).
3. A slightly different method to recover the geodesic distance between two points on the Riemannian spin manifold M would be to make use of the corresponding characters: Let x,y be to points on M; then from Gelfand's theorem these points correspond to two characters χ_x and χ_y respectively (namely χ_x and χ_y belong to the spectrum Spec(C(M)) of C(M)) defined as follows: $\chi_x(f) = f(x)$ and $\chi_y(f) = f(y)$ for every f in C(M). Then the geodesic distance d(x,y) between the two points x,y of M for the Riemannian metric g of M is given by the relation
$$d(x,y) = \sup \{ | \chi_x (f) - \chi_y (f) | ; f \in C(M), \| [D,f] \| \leq 1 \}$$ where the norm $\| [D,f] \|$ is the operator norm in the Hilbert space H, namely
$$\| [D,f] \| = \min \{k \geq 0 : \| [D,f]\, v \| \leq k \|v\| \text{ for every element v of H}\}.$$ An equivalent definition for the operator norm which has similarities with the original proof is that
$$\| [D,f] \| = \sup \{\| [D,f]\, v\| / \|v\|, \text{ v in H with } v \neq 0 \}.$$
4. The Dirac propagator (equations 17, 18) written in 4-dim Minkowski space in section 4 gives an opportunity to measure distances using the argument (x-y) which is the space distance separating two points. The fundamental Lemma proved in section 5 essentially declares that

$$ds = D^{-1},$$

namely length can be measured in the space of the corresponding quantum states by using the Dirac propagator and this is applied to curved manifolds as well as to noncommutative spaces which are not necessarily manifolds. In relativistic quantum field theory the propagator is the Green function of the corresponding wave operator of a particle and according to the Feynman path integral approach it is the basic ingredient in the expression that calculates the probability amplitude for a particle (in our case for a fermion, see [20] for example) to travel from one place to another in a given time (position space) or to travel with certain energy and momentum (momentum space).

The advantages of this approach for measuring distance and length are the following: This formalism is generally covariant and hence it might be proved useful in the future to measure time apart from spatial distance. GR is a generally covariant theory yet this is not currently the case for quantum field theory: Since there is no quantum theory of gravity yet there is a clear distinction between space and time coordinates (this is due to the existence of the arrow of time, the third thermodynamic law etc). In addition since this formalism involves the Dirac operator it enables one to use any fermion (although electrons are perhaps the most practically convenient fermions). It may be the case that spacetime is not a manifold after all but some noncommutative space; this formalism could also be applied in this case to make use of the full force of noncommutative geometry (see [18], [32] and [33]). Finally our approach has its roots in modern quantum physics (quantum field theory) which has taught us that the true invariant structure of nature (at least at small scales) is the spectra of the momentum and energy operators.

References

[1] Atiyah, M. F.: "*K-Theory*", 1989, Advanced Book Classics (2nd Edition), Addison Wesley.

[2] Atiyah, M. F.: *"Collected Works"*, Vol 1,2,3,4,5,6, 2005, Oxford University Press.

[3] Barrow, J. D.: *"Constants of Nature, Theories of Everything"*, 1991 Clarendon Press Oxford.

[4] Connes, A.: *"Noncommutative Geometry"*, 1994, Academic Press.

[5] Connes, A. and Marcolli, M.: *"Noncommutative Geometry, Quantum Fields and Motives"*, 2007 American Mathematical Society.

[6] Dirac, P. A. M.: *"General Theory of Relativity"*, 1996, Princeton.

[7] Dirac, P. A. M.: *"The Quantum Theory of the Electron"*, Proc. R. Soc. A117, No 778, p.p. 610-624, 1928, London UK.

[8] Dirac, P. A. M.: *"Principles of Quantum Mechanics"*, 1982, 4the Edition, Clarendon Press Oxford.

[9] Do Carmo, M. P.: "*Riemannian Geometry*", 1992, Birkhauser, Boston.

[10] Einstein, A.: *"Relativity: The Special and General Theory"*, 2000, Bartleby N.York.

[11] Friendlander, E. and Grayson, D.: "*Handbook of K-Theory*", 2005 Springer.

[12] Friedrich, T.: *"Dirac Operators in Riemannian Geometry"*, 2000, American Mathematical Society.

[13] Grothendieck, A.: *"Elements de Geometrie Algebrique"*, (rediges avec la collaboration de J. Dieudonne), 1971 Grundlehren der Mathematischen Wissenschaften 166, 2nd Edition, Springer.

[14] Itzykson, C. and Zuber, J.-B.: "*Quantum Field Theory*", 2006, Dover.

[15] Jost, J.: "*Riemannian Geometry and Geometric Analysis*", 2008, Springer.

[16] Karoubi, M.: "*Homologie Cyclique et K-Theorie*", 1987, Asterisque.

[17] 70th PBT Symposium: "*Fundamental Constants in Physics and Metrology*", Metrologia 22.3, 1986.

[18] Kellendonk, J. and Zois, I.P.: " *Rotation Numbers, Boundary Forces and Gap labeling*", J. Phys. A: Math. Gen. 38 (2005) 3937-3946.

[19] Lawson, H.B. and Michelsohn, M.- L.: "*Spin Geometry*", 1989, Princeton.

[20] Peskin, M.E. and Schroeder, M.V.: "*Quantum Field Theory*", 1995, Addison - Wesley.

[21] Petersen, P.: "*Riemannian Geometry*", 2006, Springer.

[22] Quillen, D.G., Segal, G.B., Tsou, S.T.: "*The Interface of Mathematics and Particle Physics*", 1990, Clarendon Press, Oxford.

[23] Ridler, W.: "*Relativity: Special, General and Cosmological*", 2001, Oxford University Press.

[24] Streater, R.F., Wightman, A.S.: "*PCT, Spin - Statistics and all that*", 2000 Princeton University Press.

[25] Taylor, B.N.: "*Constants, Fundamental*", 1991 Encyclopaedia of Physics, 2nd Edition BCH Publishers.

[26] Thompson, A. and Taylor, B.N.: "*Guide for the Use of the International System of Units (SI)*", NIST Special Publication 811, 2008 Edition.

[27] 't Hooft, G.: "*Introduction to General Relativity*", 2001, Rinton N.Jersey.

[28] Wald, R.M.: "*General Relativity*", 1984, University of Chicago Press.

[29] Weinberg, S.: "*Gravitation and Cosmology: Principles and Applications of the General Theory of Relativity*", 1972, Wiley.

[30] Weinberg, S.: "*The Quantum Theory of Fields*", 1995, Vol 1,2,3, Cambridge University Press.

[31] Zois, I. P.: "*18 Lectures on K-Theory*", 2010 CERN arXiv: 1008.1346.

[32] Zois, I.P.: "*A new invariant for σ models*", Commun. Math. Phys. Vol 209, p.p. 757 - 783, 2009, Springer.

[33] Zois, I P.: "*Black Hole Entropy, Topological Entropy and The Baum-Connes Conjecture in K-Homology*", J.Phys.A35:3015,2002

Monte Carlo - Metropolis Investigations of Shape and Matrix Effects in 2D and 3D Spin-Crossover Nanoparticles

Salim Guerroudj[1,2], Rafael Caballero[3], Francisco De Zela[3], Catalin Jureschi[2,4], Jorge Linares[5*], Kamel Boukheddaden[5*]

[1] Département de Sciences Physique, UVSQ, Université Paris Saclay, F-78035 Versailles Cedex, France.
[2] LISV, Université de Versailles Saint-Quentin-en-Yvelines, Université Paris Saclay, Velizy 78140, France
[3] Departamento de Ciencias, Sección Física, , Pontificia Universidad Católica del Perú, Apartado 1761 - Lima, Peru
[4] Faculty of Electrical Engineering and Computer Science & Research Center MANSiD, Stefan cel Mare University, Suceava 720229, Romania; catalin.jureschi@gmail.com
[5] GEMaC, Université de Versailles Saint-Quentin-en-Yvelines, CNRS-UVSQ (UMR 8635), Université Paris Saclay, Versailles Cedex 78035, France

E-mail: jorge.linares@uvsq.fr, Kamel.bouheddaden@uvsq.fr

Abstract. The Ising like model, taking into account short- , long-range interaction as well as surface effects is used to investigate size and shape effects on the thermal behaviour of 2D and 3D spin crossover (SCO) nanoparticles embedded in a matrix. We analyze the role of the parameter t, representing the ratio between the number of surface and volume molecules, on the unusual thermal hysteresis behaviour (appearance of the hysteresis and a re-entrance phase transition) at small scales.

1. Introduction

There has been a growing interest, in recent years, in the design of SCO nanoparticles [1-6] with controlled shape and environment, due to their serious potential applications as sensors, switchable and reversible memories or nano-actuators. The Fe-II based SCO solids are able to switch thermally (but also under several other external stimuli such us: light, pressure, magnetic field etc.) between two spin states, namely the paramagnetic high spin (HS) and diamagnetic low-spin (LS) states, characterized by the associated respective spin values S=2 and S=0. The experimental studies confirmed the role of the surface effects leading to unusual and non-trivial size dependence of their bistable character. In particular in some cases, the SCO nanoparticles may show a cooperative switching accompanied with a thermal hysteresis at small sizes while they exhibit gradual conversions (no phase transitions) at bigger sizes, which goes

Content from this work may be used under the terms of the Creative Commons Attribution 3.0 licence. Any further distribution of this work must maintain attribution to the author(s) and the title of the work, journal citation and DOI.
Published under licence by IOP Publishing Ltd

against common sense of usual thermodynamics. Even thought, in some special cases, re-entrant phase transitions are obtained with size. The physical mechanisms at the origin of these intriguing behaviours remain however still misunderstood and so their modelling becomes mandatory. In this contribution, we investigate the above cited effects on SCO materials, by means of an Ising-like Hamiltonian [7-11] including short and long-range interactions as well as the interaction of the edge molecules of the system with their local environment. The Hamiltonian is solved in the frame of Monte Carlo Metropolis (MCM) procedure for 2D ($N_x \times N_y$) rectangular-shaped and 3D parallelepiped ($N_x \times N_y \times N_z$) SCO nanoparticles. The thermal dependence of the high-spin fraction is then derived for different particle sizes and the results are discussed with the help of analytical predictions allowing to evaluate the size dependence of the transition temperature and the conditions of occurrence of the re-entrant phase transition. In practice, the latter emerged from the Monte Carlo (MC) simulations below some critical nanoparticle size and for well identified conditions on the model parameter values.

2. The model

Based on the previous theoretical investigations of Linares and co-workers [5, 11] accounting for the matrix effect in a spin-crossover system including short and long range interactions, the Hamiltonian of the SCO nanoparticle writes:

$$H = \frac{\Delta - k_B T \ln g}{2} \sum_{i=1}^{N_t} \sigma_i - J \sum_{<i,j>} \sigma_i \sigma_j - G < \sigma > \sum_{i=1}^{N_t} \sigma_i - L \sum_{i=1}^{N_s} \sigma_i \, , \tag{1}$$

In (1), Δ is the energy difference E(HS) — E(LS) between the spin states of isolated molecules, G is the long-range part of the interaction, which is identified as due to lattice phonons and volume change accompanying the spin transition, J is a coupling parameter accounting for short-range ferroelastic interactions between spin states, here limited to first-neighbouring molecules. The last energetic contribution L (>0), limited to surface atoms (their number is equal to N_s), describes the additional "negative" ligand-field felt by the surface atoms.

The HS and LS states are associated with the +1 and -1 values of the fictitious spin operator, σ, with their respective different degeneracies $g+$ and $g-$, the ratio of which, is denoted, $g = \frac{g+}{g-}$. These degeneracies account for both internal (spin, orbit and intermolecular vibrations) and external (lattice phonons) degrees of freedom.

In order to simplify the expression of the Hamiltonian (1), we introduce dimensionless macroscopic variables, m, s and c, given by:

$$m = \sum_{i=1}^{N_t} \sigma_i \, , \tag{2}$$

$$s = \sum_{<i,j>} \sigma_i \sigma_j \tag{3}$$

$$and \; c = \sum_{i=1}^{N_s} \sigma_i \tag{4}$$

which are respectively associated with the total fictitious magnetization, two spin correlations and, the surface fictitious magnetization. Within these parameters, the total Hamiltonian (1) re-writes as:

$$H = \left(\frac{\Delta - k_B T \ln g}{2} - G < \sigma > \right) m - Js - Lc \, , \tag{5}$$

The associated bulk and surface order parameters of the system, denoted here m_V and m_s are then simply defined as:

$$m_V = \langle \sigma_V \rangle = \frac{\sum_{i=1}^{N_t} \sigma_i}{N_t}, \tag{6}$$

$$\text{and } m_s = \langle \sigma_s \rangle = \frac{\sum_{i=1}^{N_s} \sigma_i}{N_t} \tag{7}$$

For a 2D square-shaped lattice, the quantity, $N_t = N_x^2$, is the total number of atoms, which can be expressed as the sum of the number of atoms in the bulk (core) $N_c = (N_x - 2)^2$ and those in the edge $N_s = 4(N_x - 1)$. Table 1 summarizes the values of the parameters N_t, N_c and N_s for a square lattice and for a rectangular one.

Table 1. Number of atoms for different lattice.

	Total	**Edges**	**Core**
Square Lattice	N_x^2	$4(N_x-1)$	$(N_x-2)^2$
Rectangular Lattice	N_xN_y	$2(N_x+N_y-2)$	$(N_x-2)(N_y-2)$

For the sake of precision, two methods have been used depending on the sample's size and the geometry dimension. For small (resp. large) samples, namely for $N_t < 144$ (resp. $N_t > 144$) atoms, we solved the Hamiltonian (1) exactly by using the entropic sampling method (resp. MC simulations) already introduced in the previous papers [5, 11].

3. Results and discussions

First, we consider a 2D SCO system, in which we look for the size-dependence of the re-entrance phenomenon and in a second stage; we study the shape effect, on the thermal properties of a SCO nanoparticle, at fixed number of molecules.

3.1. 2D Size Effects

For small samples, we solved exactly Hamiltonian (1) in the canonical approach using the entropic sampling method, in order to investigate the size-dependence of the thermal hysteresis of the high-spin fraction, n_{HS}, defined as $n_{HS} = \frac{1+m}{2}$ where m is the net fictitious magnetization per site. The results of the simulations are summarized in the figures 1 and 2, corresponding to the respective lattice sizes 3 x 3 and 5 x 5. Used parameters were: $\Delta/k_B = 840$ K, $G/k_B = 115$ K, $L/k_B = 120$ K, $J/k_B = 15$ K and $\ln(g) = 6.9$. If we take into account that a typical distance between metal centers is of about 9 Å [12] a lattice size of 3 x 3 is equivalent to a square nanoparticle of 2.7 nm x 2.7 nm

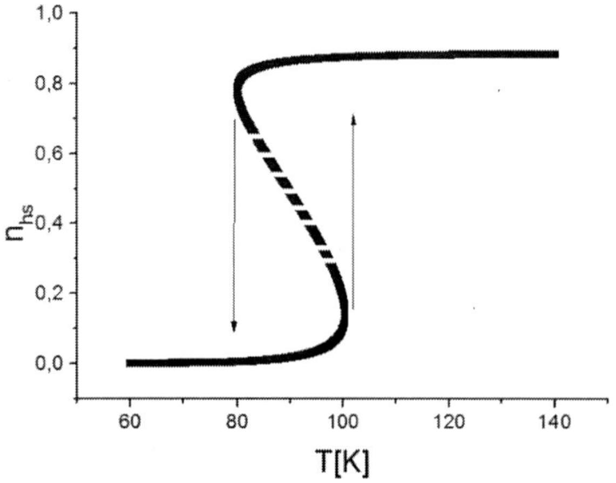

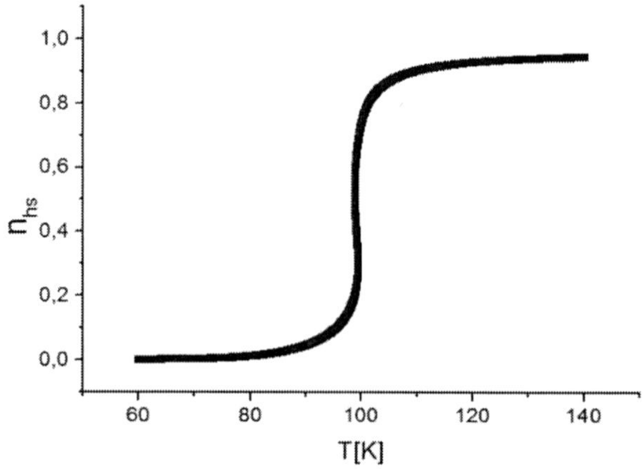

Figure 1. Exact thermal dependence of the HS fraction for a SCO nanoparticle of 9 (3 x 3) molecules, equivalent to ~3.25 nm^2 size.

Figure 2. Exact thermal dependence of the HS fraction for a SCO nanoparticle of 25 (5 x 5) molecules, equivalent to ~ 13 nm^2 size.

As we can see in Figs. 1 and 2, the width of the thermal hysteresis decreases as the nanoparticle size increases, as a result of surface active molecules. Concomitantly, the transition temperature is lowered from ~100 K to 90 K when the size of the nanoparticle decreases from 5×5 to 3×3, respectively. The last temperature is indeed very close to the expected transition temperature of the surface atoms, evaluated to be $\frac{\Delta - 2L}{k_B \ln g} \approx 87K$, which in excellent agreement with the simulation results. In the case of 3×3 molecules, 90% of the molecules are situated at the surface (here on the edges of the lattice), and therefore the surface effect exceeds that of the core, leading to the first-order transition of Fig. 1. In this case, the SCO transition takes place at the surface of the nanoparticle and drives that of the core. In the second case of Fig. 2, corresponding to a nanoparticle of 5×5 size, the thermal hysteresis vanishes, and the system exhibits a gradual spin transition between the LS and the HS states. In this case, more than 1/3 of the SCO molecules are located in the core of the nanoparticle, so their contribution is significant and competes with that of the edge atoms.

The exact resolution based on the entropic sampling method has serious limitations when increasing the lattice's size, as for most of the exact calculations at dimensions higher that one. Therefore, Monte Carlo (MC) simulations are used to investigate bigger samples in which we kept the same parameter values as previous exact calculations. Here, we use a MC resolution, based on Metropolis transition rates. The temperature is changed with 0.5 K step, and at each temperature 1000 MC steps are used to reach the thermal equilibrium, followed by 100 MC steps, along which we perform all the statistics. The results of the simulations are displayed in Figs. 3 and 4 for the respective lattice sizes 20×20 and 25×25. Thus a gradual spin transition is obtained until we reach a threshold size at which the thermal hysteresis reappears, indicating the presence of a re-entrance phenomenon.

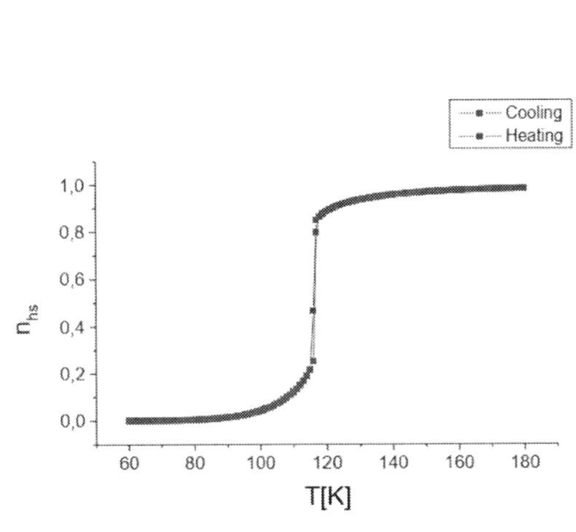

Figure 3: Simulated HS fraction for a SCO nanoparticle of 400 (20x20) molecules, equivalent to a square platelet of ~292 nm² size.

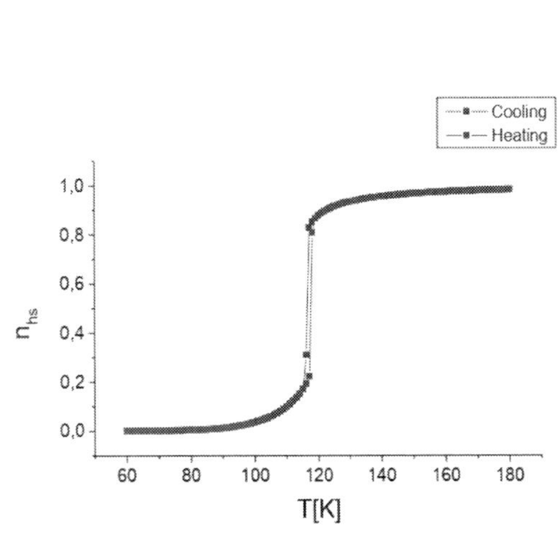

Figure 4: Simulated HS fraction for a SCO nanoparticle of 625 (25x25) molecules, equivalent to square platelet of ~ 466 nm² size.

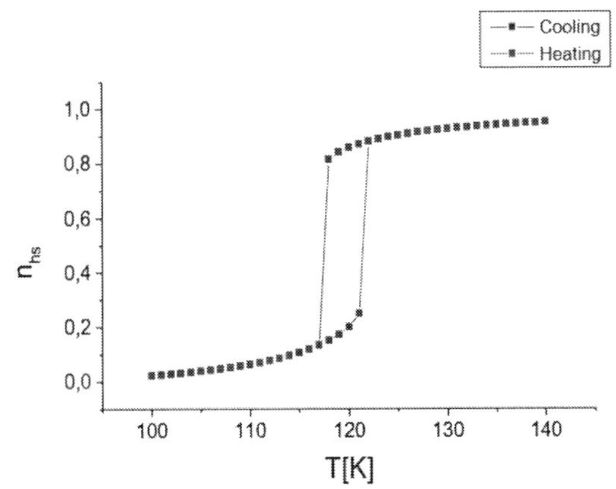

Figure 5: Simulated HS fraction for a SCO nanoparticle of 2500 (50x50) molecules configuration, equivalent to a square platelet of ~1945 nm² size.

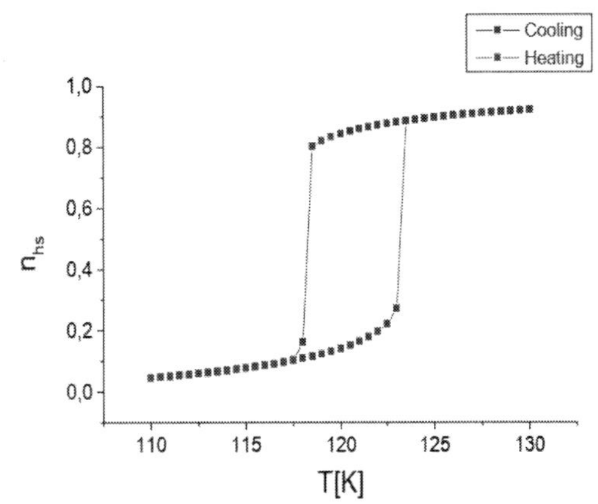

Figure 6: Simulated HS fraction for a SCO nanoparticle of 14400 (120x120) molecules configuration, equivalent to square platelet of ~ 11470 nm² sizes.

As shown in Fig. 3, for the case of 400 molecules does not lead to a thermal hysteresis for the present temperature kinetics used in the simulations: 1000 MC steps are used to reach the thermal equilibrium, followed by 100 MC steps. Here the ratio between the number of surface and bulk atoms is equal to ~11%, which remains quite small to drive a spin transition at the surface. On the other hand the transition temperature at the surface is located around 90 K, while that of the infinite system (in which the surface effects are negligible) is given by $\frac{\Delta}{k_B \ln g} \approx 122K$, a results which matches very well the results of Figs. 5 and 6. The case of 20×20 size is interesting, because it leads to realize the existence of a threshold size in the existence of a bistability for the bulk atoms. This is due to short range contribution whose effect depends on the size [13]. The thermal hysteresis appears clearly at bigger size as shown in figures 5 and 6, the transition from LS to HS of the surface and bulk atoms takes place at two well separated temperatures (90 K and 120 K), with a negligible contribution of the of the edge molecules due to their weak number compared to that of bulk atoms.

3.2. 2D Shape Effects

This section is devoted to the analysis of the shape effect in SCO nanoparticles. Indeed, the ratio surface/volume atom in a nanoparticle strongly depends on the shape of the nano-object. Here, we consider a nanoparticle containing a fixed number (let's say 36) of SCO molecules and we analyse all possibilities of shape constructions for square- and rectangular-shaped lattice configurations. We associate to each of them the parameter t, which represents the ratio between surface and volume numbers of molecules. Thus, for a square 6 x 6, we obtain $t = 0.56$, while for rectangular-shaped SCO nanoparticles of size, 9 x 4, 12 x 3 and 18 x 2, the respective t values are 0.61, 0.72 and 1. As shown in Fig 7a, the increase of t value drives the appearance of a first-order transition accompanied with a hysteresis loop, the width of which increases with t. Interestingly, we found that the transition temperature (that is approximately the centre of the hysteresis) follows a decreasing linear plot (see Fig. 7b) with the ratio t, which demonstrates that surface atoms are at the origin of this behaviour.

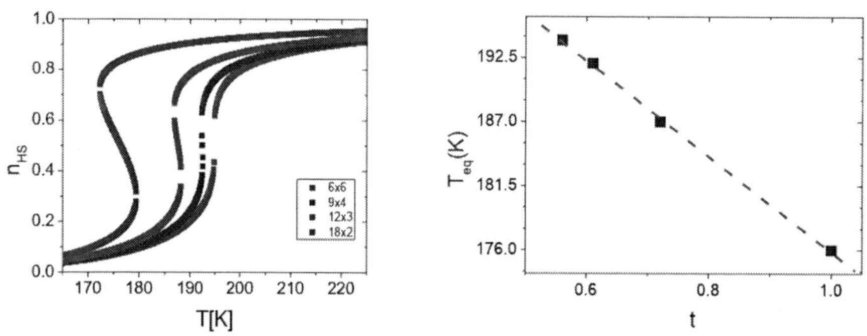

Figure 7. (a) The simulated thermal behaviour of the total high-spin fraction of a SCO nanoparticle of 36 atoms for different shapes: square 6x6 (with a surface/volume ratio t=0.56) and rectangles: 9 x 4 (t=0.61), 12 x 3 (t=0.72), 18 x 2 (t=1, all the molecules are located at the surface). (b) A linear plot showing the perfect correlation between the transition temperature and the surface/volume ratio, t. The computational parameters are: Δ/k_B = 1300 K, G/ k_B = 172.7 K, J/ k_B = 15 K, L/ k_B= 120 K, ln(g) = 6.01

3.3. Size Effects at 3D

A similar study as that of 2D is performed here for the 3D case. The computations are realized using MCM simulations on cubic nanoparticles containing 64 (4^3), 125 (5^3), 512 (8^3), 1728 (12^3), 36^3 and 48^3 particles for the same set of the model parameters. A bistability is observed for a nanoparticle of 4^3=64 molecules (mainly due to surface effects), which vanishes at higher dimensions, almost from 6^3 until 11^3 particles, before to re-appear again (due to bulk effects) from the size 12^3.

Conclusion

In this communication, we studied the effect of the size and shape of the SCO nanoparticles on their thermal bistability features. The studies were performed exactly using the entropic sampling method for the small nanoparticles and extended to bigger samples using MC simulations. For both 2D and 3D configurations, we confirmed the existence of a re-entrant phenomenon in the bistability characteristics of the nanoparticles as function of their size, in excellent agreement with recent experimental findings [1]. Furthermore, we demonstrated that the control of the shape of the SCO nanoparticles allows the control of the number of SCO atoms at the surface, thus affecting considerably their thermal properties, and so it may pave the way to the enhancement of their switching properties at small scales.

References.

[1] Peng H, Tricard S, Félix G, Molnár G, Nicolazzi W, Salmon L and Bousseksou A 2014 *Angewandte Chemie* **126** 11074-8
[2] Galan-Mascaros J R, Coronado E, Forment-Aliaga A, Monrabal-Capilla M, Pinilla-Cienfuegos E and Ceolin M 2010 *Inorganic Chemistry* **49** 5706-14
[3] Catala L, Volatron F, Brinzei D and Mallah T 2009 *Inorganic Chemistry* **48** 3360-70
[4] Forestier T, Kaiba A, Pechev S, Denux D, Guionneau P, Etrillard C, Daro N, Freysz E and Letard J-F 2009 *Chemistry-a European Journal* **15** 6122-30
[5] Linares J, Jureschi C and Boukheddaden K 2016 *Magnetochemistry* **2** 24
[6] Rotaru A, Varret F, Gindulescu A, Linares J, Stancu A, Létard J, Forestier T and Etrillard C 2011 *The European Physical Journal B - Condensed Matter and Complex Systems* **84** 439-49
[7] Wajnflasz J and Pick R 1971 *J. Phys. Colloques* **32** C1-91-C1-2
[8] Bousseksou A, Nasser J, Linares J, Boukheddaden K and Varret F 1992 *Journal De Physique I* **2** 1381-403
[9] Linares J, Spiering H and Varret F 1999 *European Physical Journal B* **10** 271-5
[10] Boukheddaden K, Linares J, Codjovi E, Varret F, Niel V and Real J A 2003 *Journal of Applied Physics* **93** 7103-5
[11] Linares J, Jureschi C-M, Boulmaali A and Boukheddaden K 2016 *Physica B: Condensed Matter* **486** 164-8
[12] Pillet S, Hubsch J and Lecomte C 2004 *European Physical Journal B* **38** 541-52
[13] Muraoka A, Boukheddaden K, Linares J and Varret F 2011 *Physical Review B* **84**

Modeling of Wind Effects on Stratified Flows in Open Channels: A Model for the Istanbul Strait (Bosphorus)

A.H.Bilge(1), **G.Kirkil**(1), **S.Burak**(2), **M. Incegul**(1)
(1)Faculty of Engineering and Natural Sciences, Kadir Has University, (2) Institute of Marine Sciences and Management, Istanbul University
E-mail: ayse.bilge@khas.edu.tr

Abstract. Stratified flows in open channels arise as a result of density or surface level differences. If the channel is connected to a basin at one or both ends, strong winds originating from the basin cause the "wind setup" effect that increases the water level at the entrance of the channel. On the other hand, along the channel, persistent winds in the upper layer flow direction lead to an increase of the drift velocity and to a decrease in upper layer flow depth. The Istanbul Strait (Bosphorus) connecting the Black and the Marmara Seas, is characterized by a stratified flow caused by the surface level and salinity difference between these basins, consisting of a southward upper layer flow and a northward lower layer flow. Along the strait, there are three hydraulic control points; the north sill, a midway contraction reach and the south sill. Under wind effects, the northern and southern entrances of the strait behave as an estuary whereas the midway reach to the south of the contraction acts as as an open channel. In winter, when the sea level difference is relatively low, the wind setup due to southerly winds may cause a blockage and even reversal of the upper layer flow. On the other hand in spring when there is excessive river discharge, northerly winds increase the influx of Black Sea waters into the strait and may lead to a blockage of the lower layer. We claim that strong northerly winds may cause a decrease of the upper layer depth beyond the contraction and we propose a simple model for its estimation in terms of the wind and water flow speeds.

1. Introduction

Sea basins connected by straits have been the subject of many research papers, e.g. [5] [6] [16] [2]. Salinity, temperature and/or surface level differences between the basins lead to stratified currents, such as in the Gibraltar Strait connecting the Atlantic Ocean and the Mediterranean, the Istanbul Strait (Bosphorus) connecting the Black Sea and the Marmara, the Kerch Strait connecting the Azov Sea to the Black Sea [9] . In these stratified flows, the Atlantic Ocean, the Black Sea and the Azov Sea act as low salinity basins. Perturbations to the main flow regime can arise as a result of tides, sea level differences or winds.

The exchange flow regime of the Istanbul Strait has important effects on the delicate ecological balances in the Black and the Marmara Seas. The effects of strong winds are one of the major causes of perturbations. The well known effects are the blockage and even reversal of the upper layer under the effect of southerly winds and the blockage of the lower layer under the effect of northerly winds [1], [3], [7], [8], [10],[11].

Content from this work may be used under the terms of the Creative Commons Attribution 3.0 licence. Any further distribution of this work must maintain attribution to the author(s) and the title of the work, journal citation and DOI.
Published under licence by IOP Publishing Ltd

The Istanbul Strait (Bosphorus) is a 31 km long narrow channel in the north-south direction joining the Black Sea and the Marmara Sea basins. It has a width of 3.329 km and 2.826 km at the north and south entrances respectively. The Black Sea's water level is about 30 cm higher than the Marmara Sea, due to high influx from rivers into the Black Sea, causing a net southward flux throughout the year. The more saline (36-38 psu) Mediterranean water flows at the bottom layer from south to north while the less saline (18-22 psu) Black Sea water flows at the top layer in the reverse direction. The south and north exits of the strait are marked by sills at depths 33 and 61 meters, respectively, forming hydrological control sections. There is a third control section about 10 km north of the southern entrance, caused by a "contraction".

In this work we point out a possible effect of northerly winds on the flow of the upper layer. It is well known that strong northerly winds cause a wind set up at the north entrance of the strait, that increases the sea level difference and the upper layer flux. As a result, the northern section of the strait is filled completely with Black Sea waters and this leads to a complete blockage of the lower layer current at the Contraction zone. As one moves to south beyond the Contraction, the flow speed of the upper layer increases considerably and there is more or less horizontal interface layer up to the south entrance. Motivated by a result in the literature on wind effects measured at the Ishikari river estuary [15], we suggest that strong northerly winds may cause a decrease in the thickness of the upper layer current along the section of the strait between the Contraction and the south entrance. Experimental work that supports this argument is given in [12]

2. The Model

Wind effects on open channel flows are studied within the framework of an experimental single layer flow setup [12], [14]. In this work, the authors have found if the wind blows for a sufficiently long period of time, a new flow regime is set up and the flow depth decreases, by momentum transfer between the wind and the water.

We propose a simple model for the decrease in the fluid level due to wind induced drift under the assumption of constant flux. For this computation we integrate the vertical drift velocity profile up to the unknown, the new lower level equilibrium height h and determine h from the constant flux assumption. To our knowledge there is no discussion of the wind induced open channel drift in the Bosphorus.

We combine experimental results of [12], [13] and [8] to choose a realistic drift velocity profile. In [13], the authors analyze wind driven and density currents in the Petrozavodsk Bay of Onega Lake, between the Baltic Sea and the White Sea. This paper reports measurements of the drift current (Figure 1a in [13]) due to wind effects. The wind induced flow is confined to the upper 1/6th of the 23 meter depth. It drops from about 50 cm/sec on the surface to zero sharply. The lower level density current in the reverse direction is more or less uniform and it has values around 6-8 cm/sec. Finally, in [8] we have detailed velocity profiles under various conditions at the south entrance.

Flow velocities are taken from [6], where the authors report current and salinity measurements along the Bosphorus, during early September in 1994 (days 250 to 260), over a period quite atmospheric conditions. The surface flow speed is 0.2 at the north entrance and it increases to 0.6-0.8 around the Contraction.

Vertical profile of streamwise velocity for open channel flows can approximately be represented by a power law [4]:

$$u(z) = U_{max} \left(z/h\right)^{1/N}, \quad 0 \le z \le h,$$

where h is the channel depth, z is the vertical distance from the bottom of the channel, $u(z)$ is the flow velocity and U_{max} is its maximum. The power law parameter N usually changes between 4 and 12, depending on the boundary friction. The most commonly used value is

$N = 7$. If the open channel flow is stratified with densities ρ_1 and ρ_2 ($\rho_1 > \rho_2$) and the flows are in opposite direction (as in the case of Istanbul Strait); velocity profiles of lower and upper layers each follow a power law with a smooth transition in between them (Figure 1).

When the wind blows on the free surface, in the direction of the flow, with a velocity V, an increased shear stress cause an increase in the flow speed of via a momentum transfer mechanism resulting in a new velocity profile $u'(z)$. This increase in the flow velocity induced by the wind dies out exponentially with depth. Nevertheless, if the discharge is constant, this leads to a reduction in the flow height to h'. Assuming constant cross-section and constant discharge, the flow heights h and h' are related by

$$q = \int_0^h u(z)\, dz \;=\; \int_0^{h'} u'(z)\, dz. \tag{1}$$

This is the effect that has been reported in an experimental setup in [12].

We study a problem where two layers in an open channel stratified flow flowing in opposite directions (Figure 1). The flow velocity at the interface layer is 0. With no wind effect on the free surface, the lower layer has a height of h_1 and maximum velocity of $U_{max,1}$ to the right and upper layer has a height of h_2 and maximum velocity of $U_{max,2}$ to the left. We assume that the flux of the upper layer is constant, the total fluid height is constant and and the interface layer acts as a movable boundary. When wind blows on the free surface, in the direction of the upper layer flow, with a velocity V, the height of the upper layer decreases to h_2', as discussed above. Since the total height is assumed to be constant the lower level height increases to h_1'. these values being related by

$$h_1 + h_2 = h_1' + h_2'.$$

We assume furthermore that the effect of wind on the open channel flow is limited to the upper layer only. The lower layer velocity profile is slightly adjusted to match with the new velocity profile in the upper layer. Then, Eqn(1) is still valid for a double layer flow.

The assumption of constant total height is valid in the reach at the south of the Contraction; the increases on the water levels are confined to the entrances of the Istanbul Strait. The validity of the constant flux assumption is supported by the existence of the Contraction. The mechanism leading to the increase of the height of the lower level may be either reduced velocity or increased flux, but the cause is practically irrelevant for the purposes of our analysis.

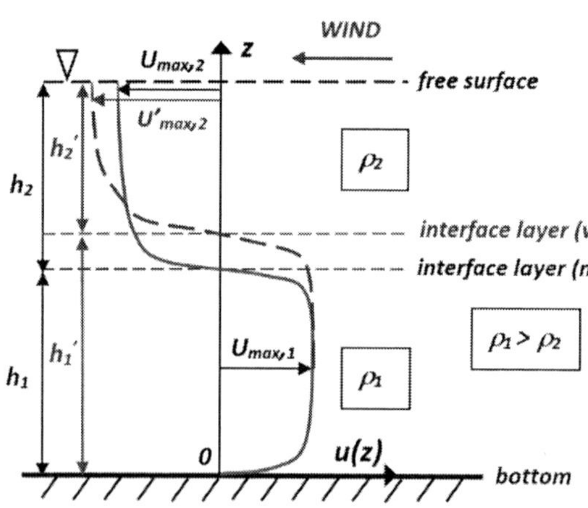

Figure 1. Bi-directional open channel flow (solid curve) and the wind induced velocity profile (dashed curve).

3. Simulation of Wind Effects

We consider a bi-directional flow in a channel with constant cross-section. Let h_1 and h_2 be the heights of the lower and upper levels respectively and let z denote the vertical distance from the channel bottom. The fluid velocities of these layers are respectively $u_1(z)$ (in the negative direction) and $u_2(z)$ (in the positive direction) with boundary conditions,

$$u_1(0) = 0, \quad \frac{du_1}{dz}(0) = \infty, \quad \frac{du_2}{dz}(h) = 0.$$

A general model that would fit these boundary conditions are

$$\text{For} \quad 0 \leq z \leq h_1 : \quad u_1(z) = -Az^{\alpha}[h_1 - z]^{\beta},$$

$$\text{For} \quad h_1 \leq z \leq h_1 + h_2 : \quad u_2(z) = B[z - h_1]^{\gamma}\left[h_1 + h_2(1 + \delta/\gamma) - z\right]^{\delta},$$

where A and B are positive and the exponents α, β, γ, δ are in the range $(0, 1)$. The smoothing of the first derivative at the interface can be achieved by spline interpolations at the boundaries of mixing layers say $h_1 - a$ and $h_1 + b$, by matching velocities and their derivatives, but this will be omitted in the present work.

In our simulation, we choose the following normalized values for fluid layers.

$$h = h_1 + h_2 = 1, \quad h_1 = 5/8h, \quad h_2 = 3/8h, \quad A = -1/2, \quad B = 1, \quad \alpha = \beta = \gamma = \delta = 1/7.$$

With these values, the velocity profile of the upper layer is practically constant up to a reasonable depth and in the calculation of wind effects on the upper layer height, we assume that $u_2(z)$ is approximately equal to U_{max}. As mentioned above, the wind effect is considered to die out as a Gaussian pulse, hence we propose a wind induced velocity profile in the form

$$v(z) = V_{max} e^{-k^2\left(\frac{h-z}{h_2}\right)^2}.$$

Let h_1' and h_2' be the layer heights under wind effects. The constant discharge equation is expressed as

$$\int_{h-h_2}^{h} u_2(z)\, dz = \int_{h-h_2'}^{h} u_2(z)\, dz + \int_{h-h_2'}^{h} v(z)\, dz.$$

Approximating $u_2(z)$ by U_{max}, we get

$$U_{max}(h_2 - h_2') = V_{max} \int_{h-h_2'}^{h} e^{-k^2\left(\frac{h-z}{h_2}\right)^2} dz.$$

After a change of variable $s = (h - z)/h_2$, we obtain

$$U_{max}(h_2 - h_2') = V_{max} \frac{h_2}{k} \int_0^{k\, h_2'/h_2} e^{-s^2}\, dz.$$

Since

$$\int_0^x e^{-t^2} dt = \frac{\sqrt{\pi}}{2} Erf(x),$$

after dividing both sides by h_2, we obtain the relation

$$U_{max}\left(1 - \frac{h_2'}{h_2}\right) = V_{max} \frac{\sqrt{\pi}}{2k} Erf(\left(\frac{k\, h_2'}{h_2}\right).$$

This gives a relation between the ratios h_2'/h_2 and V_{max}/U_{max}. We evaluate V_{max}/U_{max} numerically in terms of h_2'/h_2 in the range $(0.5, 1)$ and plot their relation for $k = 2, 3, 4, 5$. The results are presented in Figure 2a,b where we show the steady state (no-wind) velocity profile together with wind induced velocities (Figure 2a) and the reduction in the upper layer height h_2'/h_2 as a function of the relative wind speed V_{max}/U_{max} (Figure 2b) for the values of $k = 2, \ldots 5$.

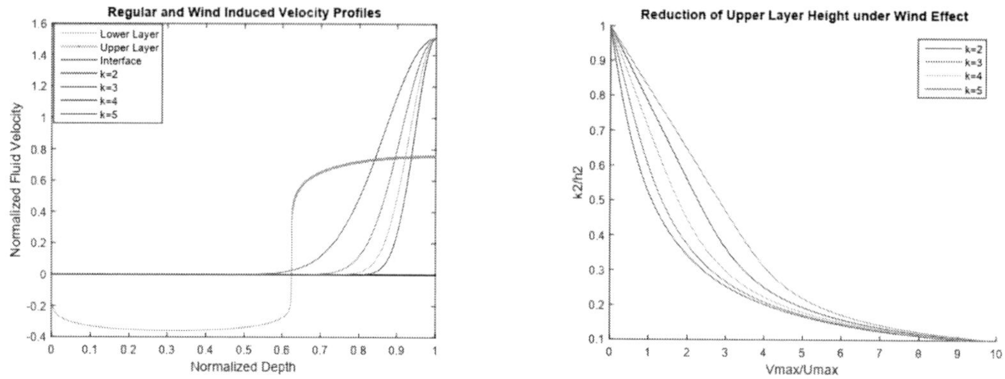

Figure 2 (a) Bi-directional velocity profile $u(z)$ and wind induced Gaussian velocity $v(z) = V_{max}e^{-k^2\left(\frac{h-z}{h_2}\right)^2}$ for $k = 2,\ldots,5$ (b) k_2/h_2 as a function of the relative wind speed V_{max}/U_{max}, where $U_{max} = u(h)$, k_2 and h_2 are respectively upper layer heights with and without wind effects.

References

[1] H. Altiok, A.Aslan, S. vezN. Kiratli, S. Tas, A.E. Muftuoglu, K.I. Sur, E. Okus, Influence of the extreme conditions on the water quality and material exchange flux in the strait of Istanbul, Journal of Marine Systems, 139, (2014) 204-215.

[2] A.Andersen, F. Jakobsen, B. Alpar, The water level in the Bosphorus Straight and its dependence on atmospheric forcing, Deutsche Hydrographische Zeitschrift , German Journal of Hydrography, vol. 49, (1997) N. 4.

[3] S. Burak, Evaluation of Pollution Abatement Policies in the Marmara Sea with Water Quality Monitoring, Asian Journal of Chemistry Vol. 20, No. 5 (2008), 4117-4128.

[4] H. Chanson, The Hydraulics of Open Channel Flow: An Introduction, Elsevier, (2004).

[5] Farmer, D.M., Armi, L., 1986. Maximal two-layer exchange over a sill and through the combination of a sill and contraction with barotropic flow. J. Fluid Mech. 164, 5376.

[6] M.C. Gregg, E. Ozsoy and M.A. Latif, Quasi-steady exhange flow in the Bosphorus, Geophysical Research Letters, Vol. 26, No 1, pp. 83-86, (1999)

[7] M.C. Greg, E. Ozsoy, Flow, water mass changes and hydrolics in the Bosphorus, Journal of Geophysical Research, vol 107, No. C3, 10.1029/2000JC000485, (2002).

[8] I. Guler, Y. Yuksel, A.C. Yalciner, E. Cevik, C. Ingerslev, Measurement and evaluation of the hydrodynamics of secondary currents in and near a strait connecting large water bodies- A field study, Ocean Engineering, vol 33, (2006) 1718-1748.

[9] V. A. Ivanov , L. V. Cherkesov, T. Ya. Shulga Analysis of the influence of water exchange through the Kerch Strait on the surge processes and currents in the Azov Sea, Physical Oceanography December 2010, Volume 20, Issue 4, pp 239-250 Translated from Morskoi Gidrofizicheskii Zhurnal, No. 4, pp. 314, JulyAugust, 2010.) Strait. J. Geophys. Res. 116, C08038. http://dx.doi.org/10.1029/2010JC006861.

[10] Jarosz, E.W., Teague, J., Book, J.W., Beiktepe, ., 2011b. Observed volume fluxes in the Bosphorus Strait. Geophys. Res. Lett. 38, L21608. http://dx.doi.org/10.1029/ 2011GL049557.

[11] T. Oguz, E. Ozsoy, M. Latif, H. Sur, U. Unluata, 1990: Modelling of hydraulically controlled exchange flow in the Bosphorus Strait. J. Phys. Oceanogr., 20, 7, 945-965.

[12] E.J. Plate and C.R. Goodwin, The influence of wind on open channel flow, Coastal Engineering Report

[13] B.I. Samolyubov and I.N. Ivanova, The evolution of velocity profiles and turbulant viscosity in a system of urrents with wind induced and density flows, Physics of the Earth, Atmosphere and Hydrosphere, Fizika, 2014, No 5. pp. 60-66

[14] J. Wu, An estimation of wind effects on dispersion in wide channels, Water Resources Research, Vol. 3, no.5, 1097-1104, (1969).

[15] K. Yokoo, S. Yoshida, C.P. Caulfield, P.F. Linden, I. Ito, Effect on wind on highly stratified flow, in WIT Transaction on the Built Environment, Coastal Engineering VII, Vol 78, Coastal Engineering, Paper DOI 10.2495/CE050031.

[16] H. Yuce, 1996. Mediterranean water in the Strait of Istanbul (Bosphorus) and the Black Sea exit. Estuar. Coast. Shelf Sci. 43, 597616

Analysis of the Solar magnetograms evolution using a spectral gap of critical nets

I. S. Knyazeva[1,2], N. G. Makarenko[1], I. Makarenko[3], A. Vdovina[3] and F. A. Urtiev[1]

[1] Pulkovo observatory, 65 Pulkovskoye ch., St.Petersburg, 196140, Russia
[2] St. Petersburg State University, 7/9 Universitetskaya nab., St. Petersburg, 199034, Russia
[3] School of Mathematics & Statistics, Newcastle University, Hershel Newcastle upon Tyne, NE1 7RU, UK

E-mail: iknyazeva@gmail.com, ng-makar@mail.ru, irina.makarenko@ncl.ac.uk, alina.vdovina@ncl.ac.uk, urtiev@gmail.com

Abstract. Analyzing the dynamics of the photospheric magnetic field of the Sun is one of the most important problems in Solar Physics. Different estimates of the complexity of magnetograms of the Sun Active Regions (AR) are used to predict the time and the strength of the solar flares, but the quality of the forecasts are still insufficient. A magnetogram is a highly variable discrete image with a very large number of local extrema. We use an idea of extraction of stable critical points within a framework of the scale-space theory. Two sequential convolutions of the image with the same Gaussian kernel and calculating the difference between the produced images allow to get a stable estimation of the Laplacian of the image. A critical graph is constructed using maxima and minima of the Laplacian. Dynamics of critical graphs can be used for diagnostics of dynamical regimes of ARs. The so-called spectral gap is proposed to be used as a numerical descriptor. This is the difference between the two largest eigenvalues of the discrete Laplacian of the graph constructed on critical networks. We investigated several ARs and found that there was a sudden increase in the spectral gap values one or two days before the flares.

1. Introduction

An early forecast of solar eruptive events remains an important objective of extreme space weather studies. Most of the failures of spacecrafts are caused by impact of the solar wind particles in the near-Earth space. Possible physical mechanisms in the magnetic field of active regions are known, each of which, or a combination of which, in certain circumstances, can induce a large solar flare [1]. However, these conditions are still uncertain. There are a large number of the so-called solar flare precursors, which preceded or accompanied the solar flares in the past. But it is admitted that a specific precursor does not necessary precede a solar flare and the appearance of the precursor does not necessarily cause the solar flare. There are different machine learning systems, based on different precursors, for example [2, 3].

Traditional precursors are based on the estimation of complexity of the observed patterns in the magnetic field of active regions. As a measure of complexity some morphological characteristics are commonly used, namely: the area, the number of spots in the active region, the maximum gradient, the length and complexity of a neutral line, the number of singular points of the field, the current helicity and others (see references in [4]). As a rule, these

Content from this work may be used under the terms of the Creative Commons Attribution 3.0 licence. Any further distribution of this work must maintain attribution to the author(s) and the title of the work, journal citation and DOI.
Published under licence by IOP Publishing Ltd

characteristics have been measured on the line-of-sight magnetograms of solar active regions, which includes only one component of the magnetic field. It is important to mention that vector magnetograms with three components of the magnetic fields are now becoming available [5].

Existing models of solar flares do not contradict the theoretical conception of the physical mechanisms leading to a flare, but they are not a direct consequence. Statistical properties of the observed sequence of events make the situation even more complicated. Reappearance of a flare at the same AR is described by the laws of the heavy-tailed Pareto type [6]. This means that the probability of occurrence of the large events is exponentially small. Therefore, searches of new approaches to describe different dynamic regimes in the dynamics of fields before the emergent events is an actual and still unsolved problem. The main idea of this project is to find a simple and stable descriptor of the magnetogram, which could be easily tracked during the time. In our previous works [4, 7] we computed different topological characterstics of the active regions, and we now continue to explore some of them. To get a descriptor let's use some interesting points, such as local extrema of solar magnetic fields, that do not change under transformations. We picked up the idea of constructing such points from [8]. The authors considered only such critical points that are stable under the consecutive smoothing.

In the process of the AR evolution a number and values of local extrema are changing. Therefore, the structure of a critical network is also changing. So, the dynamics of critical graphs can be used to diagnose dynamical regimes of ARs. As a numerical descriptor we propose to use the so-called spectral gap. It is the difference between the two largest eigenvalues of the discrete Laplacian of the graph constructed on critical networks. We tracked of changes in the spectral gap for three different active regions and found that there is a sharply increase of its values from 20 to 40 hours before the flare.

2. Method description

In our study, we have used a sequence of HMI magnetograms made with the Helioseismic and Magnetic Imager (HMI) on-board Solar Dynamics Observatory (SDO) (hmi.stanford.edu). A magnetogram of Solar magnetic fields is a matrix of data varying between −5000 and 5000, but most of the values are in the range of −500 to 500. Changes between the pixels not smooth, which makes the extraction of stable patterns from magnetograms very difficult.

Due to high variability of a magnetogram it is characterized by a very large number of local extrema. We use the idea of extraction of stable critical points within a framework of the scale-space theory, proposed in [8]. The authors use not all extrema of the image, because there are too many such extrema in the case of high variability of contrasts, but only the so-called beta-resistant features. The idea of the beta-stability is based on the properties of the Laplacian of the image calculated by the design, which is called scale-space. Scale-space was introduced in a number of papers [9, 10] and subsequently became widespread in image analysis. In the framework of this approach an image is represented as a set of one-parameter smooth images, depending on the size of the smoothing window used for the suppression of small-scale features. Smoothing is performed using a Gaussian kernel

$$g_t(x) = \frac{1}{\sqrt{2\pi t}}\exp(-x^2/2t), \qquad I_t(x) \equiv f * g_t = \int I(y)g_t(y-x)dy, \tag{1}$$

where t is a scale parameter.

On the other hand, the scale-space description can be presented as a solution of the heat equation with the Neumann boundary conditions, where temperature plays the role of the width of the smoothing kernel. In addition, it can be shown that the Laplacian of the image is quite accurately approximated by the difference between the image smoothed t times with a Gaussian

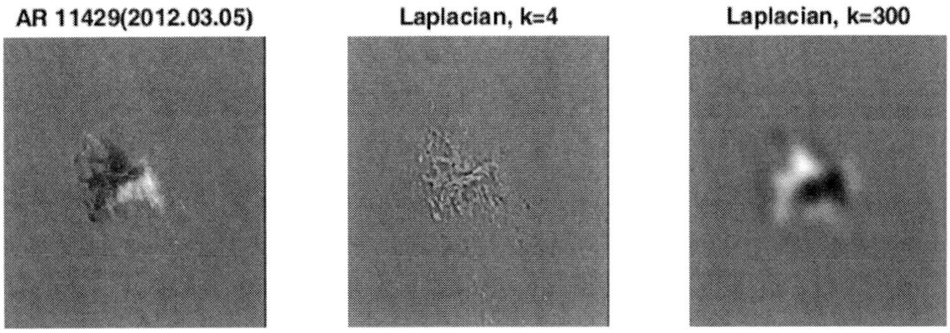

Figure 1. AR 11429 and two Laplacians with a different level of smoothing

kernel and the image smoothed $t+1$ times with the same Gaussian kernel:

$$\delta_t L = \frac{1}{2} \bigtriangledown^2 L, \quad L(x,y,0) = I(x,y) \qquad \Rightarrow \qquad L_t = \bigtriangledown^2 I_t(x) \approx I_{t+1} - I_t. \tag{2}$$

2.1. Beta-stable Laplacian

We can convolve the input image I with the Gaussian kernel G_σ a large number of times to give a scale space representation $\{I_k\}$:

$$\underbrace{I_k = G_\sigma * G_\sigma * G_\sigma}_{k} * I,$$

where σ is the width of the smoothing kernel and k is the index of the scale.

For each fixed scale the Laplacian of the image divides the whole image into convex areas of brightness (with the positive Laplacian) and concave areas of brightness (with the negative Laplacian). It turns out that when we zoom in, the number of such regions is stabilized. When the number of areas with positive and negative Laplacians does not vary, we say about its stability at this scale. A scale at which the number of regions, mentioned above, does not change any more is called a sustainable scale. Thus, we have to determine the stable singular points of the image. The maxima and minima of a sustainable Laplacian are the stable singular points of the image.

2.2. Nets construction

After the set of critical points is defined, we construct a graph in accordance with the following rules: the nodes of the graph are the critical points; a local minimum is connected to a local maximum, if there is an upward path between them, that is the pixels values must increase along this path. Example of two possible ascending path computed by simple matrix is presented at Fig. 2 This graph is called a critical network. The authors [8] showed that the critical net is invariant under affine transformations and various distortions of images, and the graph has a minimum number of self-intersections, that is, almost-planar.

2.3. The laplacian spectrum of graph

Full information about spectral graph theory could be found in the books [11, 12]. Recall, that $G(V, E)$ be a graph with a vertex set V of cardinality n and an edge set E of cardinality m. We will assume that it is undirected and finite. Let d_j be the degree of a vertex j. Let A be

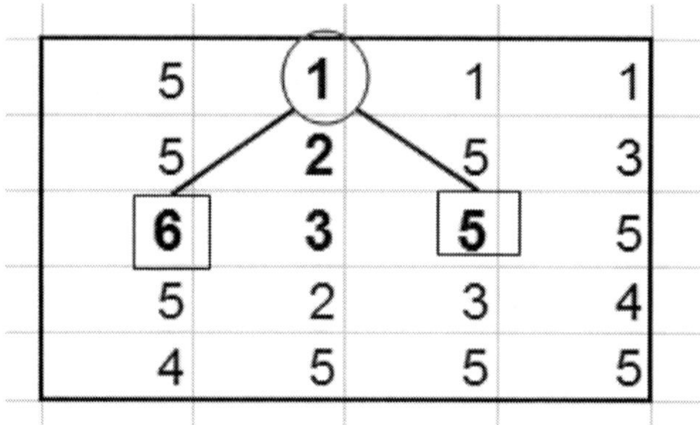

Figure 2. AR 11429 and two Laplacians with a different level of smoothing

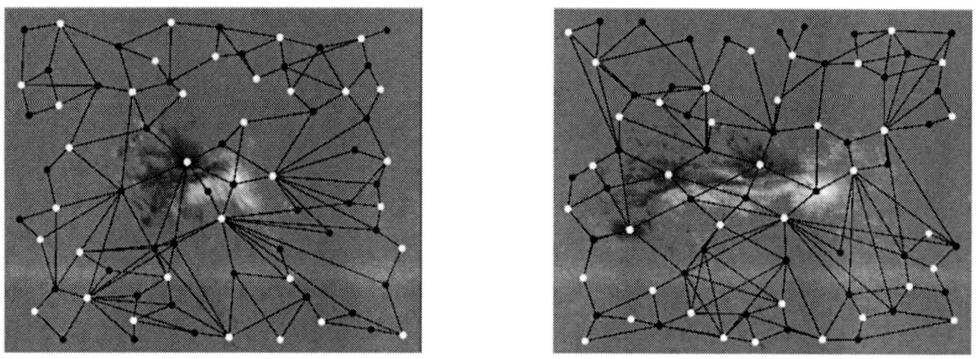

Figure 3. Critical nets for AR 11429 in two different time moments

the $n \times n$ $\{0, 1\}$ adjacency matrix, such that $A_{ij} = 1$ if and only if $ij \in E(G)$. Let D be the $n \times n$ diagonal matrix with $D_{jj} = d_j$. Then $L = D - A$ will be a combinatorial Laplacian matrix associated with the graph.

For this Laplacian matrix L of the graph (not to be confused with the Laplacian of the image!), the eigenvalues λ of the graph can be found by solving the equation

$$Lx = \lambda x.$$

As this matrix is symmetric along its diagonal, then, by definition, the eigenvalues are real. The set of eigenvalues of the Laplacian matrix is known as the spectrum of the graph, $\lambda_1 \geq \lambda_2 \geq \ldots \geq \lambda_n$. Any given Laplacian matrix has a unique set of eigenvalues. The spectrum of a graph is also known to be invariant, as the eigenvalues do not depend on the labeling of vertices. The first largest eigenvalue λ_1 of the graph is a kind a "average degree" of graph, the second λ_2 eigenvalue is a charactersitc of algebraic connectivity of the graph, which depends from the number of vertexes and the way of connection. The spectral gap Υ is defined as the difference between two largest eigenvalues of the spectrum,

$$\Upsilon = \lambda_1 - \lambda_2.$$

If the number of edges of the graph is close to the maximum, when every vertex is connected to every other vertex, and the graph is said to be dense. A graph in which the number of edges is close to the minimal number of edges is called a sparse graph. If we talk about evolution of complex network, increase in spectral gap means decreasing of algebraic connectivity, so network become more sparce, or growth of the critical points.

3. Results

Three Active Regions of the Sun have been examined, namely, AR 11158, AR 11429, and AR 11520. This choice was motivated by the criteria of complexity of magnetic fields of these active regions and the solar flare activity. The activity of the regions is measured as the amount of flares produced. The Solar flares are classified as A, B, C, M and X according to peak flux of X-rays near the earth measured on the GOES spacecraft. For prediction only the strongest M and X flares are usually considered, because they are the most dangerous for near-space systems. The location on the solar disk, where a flare released, is also important, as the most dangerous flares happen at the center of the solar disk. All three observed regions, listed above, produced X and M-Class flares at the center of the disk. We have already used these regions in our previous works, where dynamics of topological features have been investigated [4].

Each active region is represented by a set of fragments of the line-of-sight magnetogram having the size of 600×600 pixels. We took a time step between the fragments as 1 hour. An active region passes through the solar disk within $5-6$ days. So, the whole set of images analyzed for each region consists about 130 fragments. For each fragment we computed the beta-stable Lapalcian of the image, then constructed the critical nets using the beta-stable Lapalcian, and, finally, estimated the spectral gap of the Laplacian matrix associated with the critical net.

As a result for each active region we received a time series describing the spectral gap evolution. Fig. 3-5 shows the evolution of the spectral gap for the active regions AR 11158, AR 11429, and AR 11520.

By the abscissa axis, time t in dates is shown, by the ordinate axis the dimensionless quantity of the spectral gap is shown. The flare strength is indicated by the stems with the point in all figures. We translated the class of a flare into the numerical quantity using the standard procedure: the M-flares and X-flares were multiplied by 10 and 100, correspondingly.

As we see, in all three cases there is the sharp rise in the values of the spectral gap withing $24-48$ hours before the flare. From the physical point of view this means that either the degree of vertexes decreases or the number of vertexes increases, or both. In term of critical nets this means that the number of beta-stable extrema raises significantly, what corresponds to the increase in topological complexity. This result agrees with the hypothesis that before the solar flares a new emerging flux arises in the magnetic field and forms some new stable structures.

Let's consider the dynamics of each ARs.

AR 11429. The group appeared at the eastern limb on 3 March 2012 as a simple compact group, but next day it evolved to the large complex group (see Fig. 2). $X5.4$ flare released on 7 March at 00:24 UT, which was the second largest flare in magnitude since 2010. From March 9 the activity in the region began to decay slowly. This behaviour is accurately traced on a graph of the spectral gap in Fig 3. We see the significant rise in the gap during the first three day, with sharp increase by the 25 hours before the $X5.4$ flare occured, and another increase after the flare, which was followed by a slow decay.

AR 11520. The group 11520 appeared on the solar disk on 8 July 2012. Just after the appearance of the group, the magnetic class FHC has been assigned to it, which corresponds to a large, complex group with a high probability of flares. Strong activity and powerful flares were expected from this group, but only one $X1.4$ flare was on 12 July just at the center of disk. After that the activity started to decrease. Two M-Class flares released near the West limb of the Sun. There is no increase at the beginning of the graph in the dynamics of the spectral gap (see Fig. 4). This corresponds to the situation that the group appeared was already very complex and developed, although there is a jump withing two days before the flare, followed by a reduction in the spectral gap. It could be considered as a predictor of the $X1.2$ flare which released on 12 July. The similar increase was on 14 July, before the flare $M1.7$ on 17 July.

AR 11158. AR 11158 appeared on 12 February 2011 not from the eastern limb, but just near the center of the disk in the form of a compact group. After that it developed rapidly and a day later it reached the next class of complexity δ and the $M6.6$ flare was produced on 13 February. Two day later the $X2.2$ flare released and further activity of this region started to decrease. Again, on the plot of the spectral gap (Fig. 5), we see a rather big rise started two days before the X flare, and then the decrease in the spectral gap just before the flare, as well as several peaks at the second part of evolution graph coinciding with the weaker flares.

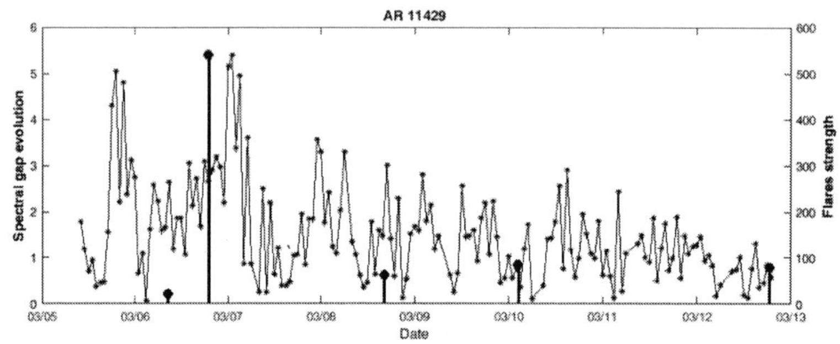

Figure 4. Evolution of the spectral gap for AR 11429.

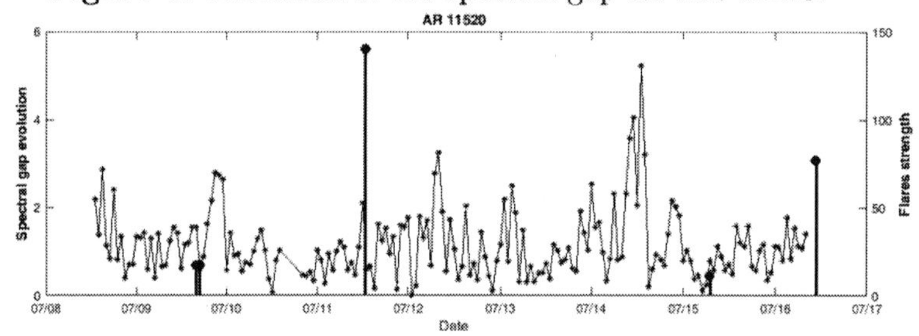

Figure 5. Evolution of the spectral gap for AR 11520.

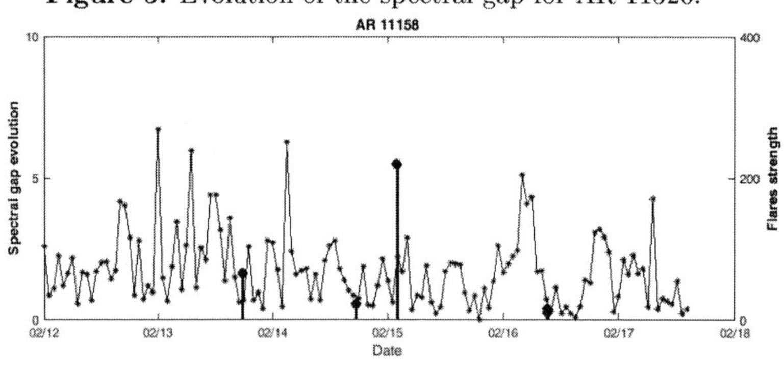

Figure 6. Evolution of the spectral gap for AR 11158.

4. Discussion

The aim of our research is developing effective descriptors of magnetic fields in the Solar active regions. The spectral gap can be considered as one of them. Combination of different descriptors might be used for predictive system construction. There are different machine learning systems by the present date, which have already been used for the solar magnetograms processing and the solar flare predictions. But there are an unlimited scope to make the predictions more accurate. The area of mathematics, connected with computational topology and complex network analysis, has developed significantly recently due to a huge number of practical problems and development of scientific computing. The powerful mathematical approaches for analyzing the distinguishing features of magnetograms can significantly improve the accuracy of the predictions of the solar flares. Our next step will be testing of some topological features. We developed this technique in our previous work for the features of critical nets in machine learning systems of solar flare prediction.

Acknowledgments

This work was partially supported by a grant from the Russian Science Foundation, RScF Project no. 15-29-01280 "Geometrical and topological methods of analysis of high-resolution images"), a grant from the Russian Science Foundation, RScF Project no. 16-32-00251 "Investigation of dynamical regimes of magnetic field in Solar active regions", and the Leverhulme Trust Grant "Observations and models of turbulent flows: a topological approach" (2015-2019), RPG-2014-427.

References

[1] Longcope D W 2005 *Living Rev. Solar Phys* **2**
[2] Ahmed O W, Qahwaji R, Colak T, Higgins P A, Gallagher P T and Bloomfield D S 2013 *Solar Physics* **283** 157–175
[3] Bobra M G and Couvidat S 2015 *The Astrophysical Journal* **798** 135
[4] Knyazeva I and Makarenko N 2014 *Astronomy Reports* **58** 145–155
[5] Scherrer P H, Schou J, Bush R, Kosovichev A, Bogart R, Hoeksema J, Liu Y, Duvall Jr T, Zhao J, Schrijver C *et al.* 2012 *Solar Physics* **275** 207–227
[6] Boffetta G, Carbone V, Giuliani P, Veltri P and Vulpiani A 1999 *Physical review letters* **83** 4662
[7] Knyazeva I, Makarenko N and Livshits M 2011 *Astronomy reports* **55** 463–471
[8] Gu S, Zheng Y and Tomasi C 2010 *Computer Vision–ECCV 2010* (Springer) pp 663–676
[9] Witkin A P 1984 *Acoustics, Speech, and Signal Processing, IEEE International Conference on ICASSP'84.* vol 9 (IEEE) pp 150–153
[10] Lindeberg T 2013 *Scale-space theory in computer vision* vol 256 (Springer Science & Business Media)
[11] Chung F R 1997 *Spectral graph theory* vol 92 (American Mathematical Soc.)
[12] Newman M 2010 *Networks: an introduction* (OUP Oxford)

Development and simulation of microfluidic Wheatstone bridge for high-precision sensor

N D Shipulya[1], S A Konakov[1], V VKrzhizhanovskaya[1,2,3]

[1] St. Petersburg Polytechnic University, 195251, St. Petersburg, Russia
[2] University of Amsterdam, 1098 XH Amsterdam, The Netherlands
[3] ITMO University, 197101, St. Petersburg, Russia

E-mail: Natalya.Shipulya27@mail.ru, konakov.st@gmail.com

Abstract. In this work we present the results of analytical modeling and 3D computer simulation of microfluidic Wheatstone bridge, which is used for high-accuracy measurements and precision instruments. We propose and simulate a new method of a bridge balancing process by changing the microchannel geometry. This process is based on the "etching in microchannel" technology we developed earlier (doi:10.1088/1742-6596/681/1/012035). Our method ensures a precise control of the flow rate and flow direction in the bridge microchannel. The advantage of our approach is the ability to work without any control valves and other active electronic systems, which are usually used for bridge balancing. The geometrical configuration of microchannels was selected based on the analytical estimations. A detailed 3D numerical model was based on Navier-Stokes equations for a laminar fluid flow at low Reynolds numbers. We investigated the behavior of the Wheatstone bridge under different process conditions; found a relation between the channel resistance and flow rate through the bridge; and calculated the pressure drop across the system under different total flow rates and viscosities. Finally, we describe a high-precision microfluidic pressure sensor that employs the Wheatstone bridge and discuss other applications in complex precision microfluidic systems.

1. Introduction

Lab-on-a-chip microfluidic systems are modern instruments for advanced material processing in biology, analytical chemistry and synthesis [1-3]. There are many simple or complex microfluidic systems for controlled gas or liquid flow through microchannel [4, 5]. Some non-trivial problems of high-precision measurements require new approaches and microfluidic devices.

In our work we investigate a microfluidic Wheatstone bridge (MWB). The system is a functional copy of a traditional electric Wheatstone bridge, which has a wide application in precision measurements. A microfluidic Wheatstone bridge was used for measuring viscosity and elasticity of blood [6]. It is a promising area of application, because MWB accurately assessed blood pulsation and flow in blood vessels, including arteries and capillaries, without any damage to the red blood cells. In [7] MWB was used for rapid sampling and confinement of suspended particles in aqueous solutions. MWB was also applied for measuring liquid electrokinetic properties and flow control based on biomolecular separation processes [8].

Usually microfluidic valves and other complex electronic systems with feedback loop are used to balance the microfluidic Wheatstone bridge [6-8]. This approach has disadvantages, because it involves additional parts in microfluidic system and reduces durability and measurement accuracy. To solve this problem, we suggest to develop a "perfectly balanced" MWB. A common microfabrication

Content from this work may be used under the terms of the Creative Commons Attribution 3.0 licence. Any further distribution of this work must maintain attribution to the author(s) and the title of the work, journal citation and DOI.
Published under licence by IOP Publishing Ltd

technology does not allow to produce a precisely balanced MWB, so we need a new technological approach. In this work we describe a novel method to make a "permanently balanced" microfluidic Wheatstone bridge, and present the results of the MWB investigation under different conditions.

2. Microfluidic Wheatstone bridge

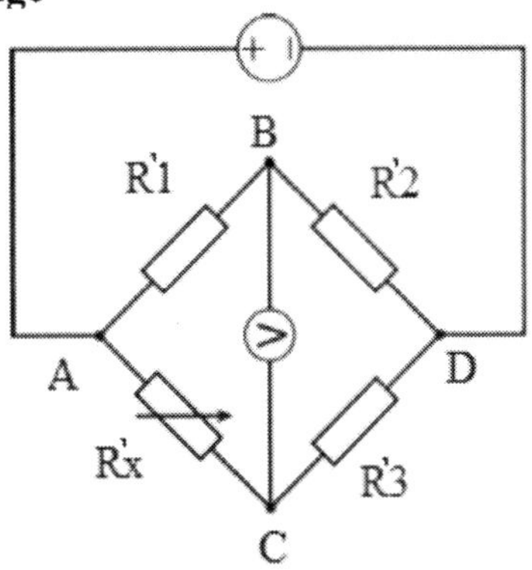

Figure 1. Electric circuit of the Wheatstone bridge.

A classical Wheatstone bridge [9] is an electrical circuit shown in Figure 1. This scheme is used for measuring resistance R_2' by balancing the bridge with a varying resistance R_x'. If the bridge is balanced then the voltage between points C and B is zero. In this case equation (1) describes the branch resistance ratios:

$$\frac{R_1'}{R_x'} = \frac{R_2'}{R_3'} \tag{1}$$

where R' is an electrical resistance.

Based on the analogy of electric circuit to incompressible Newtonian fluid flow, fluid flow is analogous to electric current, pressure drop to voltage, and hydraulic resistance to electrical resistance. This way we can design a microfluidic Wheatstone bridge shown in Figure 2a.

A prototype of the MWB have been produced in our lab (see Figure 2b). The microchannels shown in Figure 2a are etched in the silicon wafer. The microfluidic structure was covered by a glass wafer, and then interconnectors for liquid supply were mounted (not shown in Figure 2b).

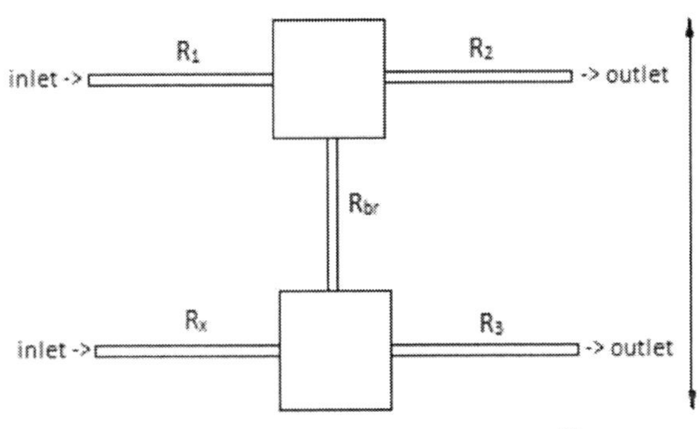

a)

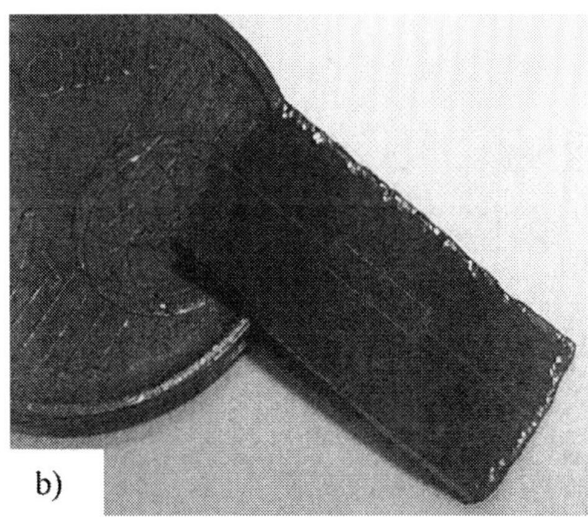

b)

Figure 2a. Scheme of the microfluidic Wheatstone bridge.

Figure 2b. Silicon-glass prototype of the microfluidic Wheatstone bridge.

For analytical estimations we used a definition of the microchannel hydraulic resistance [10]:

$$R = \frac{\Delta P}{Q} \quad (2)$$

where R is hydraulic resistance, ΔP is pressure difference between the inlet and outlet, Q is flow rate.

Hydraulic resistance for microchannel with rectangular cross-section was calculated from approximation [11]:

$$R = \frac{12\cdot\eta\cdot L}{1-0.63\cdot\left(\frac{h}{w}\right)}\cdot\frac{1}{h^3\cdot w} \quad (3)$$

where η is dynamic viscosity, L is length of the microchannel, h is height, w is width of the channel.

Based on equations (1-3) it is possible to design a microfluidic Wheatstone bridge. Experimental production of the MWB prototype has shown that it is difficult to make a balanced MWB because of the technology defects that slightly modify the microchannel resistance.

To solve this problem, we propose a new method to produce a balanced MWB by fine tuning the microchannel geometry. Our method is based on etching in microchannel technology, described in our previous work for silicon on glass wafer [12]. The main idea of the process has three steps: The first step is etching a microfluidic structure in a glass wafer. The second step is bonding glass with silicon wafer. And the third step is supplying silicon etchant into the channel. The solution etches the silicon and changes the geometry of the channel; the hydraulic resistance changes, too. This technology can be applied for many materials and devices. This idea of fine-tuning the microchannel resistivity can be applied to make a permanently balanced MWB without any additional valves and control systems.

To investigate the behavior of the MWB and the process of its balancing, we consider the MWB configuration shown in Figure 2a. Geometrical parameters are listed in Table 1. The bridge was calculated for Newtonian fluids with density range from 600 to 1200 kg/m^3, dynamic viscosity range 1.2-1.9·10^{-5} Pa·s and pressure difference between the inlet and outlet ΔP_{MWB} varied from 1000 to 15000 Pa. Under these conditions the highest Reynolds number is 1.9, which means a laminar flow in the microchannel.

Table 1. Parameters of the microfluidic Wheatstone bridge.

Channel (see Figure 2a)	Length, L (mm)	Height, h (μm)	Width, w (μm)	Resistance, $R\times10^{12}$ (Pa/m^3·s)
Channel R1	1.75	30	40	32.80
Channel R2	2.00	30	50	25.44
Channel R3	2.00	30	45	30.31
Channel RX	1.75	30	30-42, increment 1	29.97-62.26
Bridge Channel R_{Br}	0.87	30	150	2.62

3. Computational Fluid Dynamics model

To investigate the microfluidic Wheatstone bridge behavior, we used a computational fluid dynamics approach in 3D geometrical model. We solve the mass conservation law (4) and Navier-Stokes equations (5) for a steady-state laminar flow of incompressible fluid:

$$div(\boldsymbol{V}) = 0, \quad (4)$$

$$\frac{1}{\rho}\mathbf{grad}\,(\mathrm{p}) = \nu\,\Delta(\boldsymbol{V}) \quad (5)$$

where ρ is density, p is pressure, V is velocity vector, ν is kinematic viscosity.

Boundary conditions for flow velocity on all surfaces of microchannels are:

$$\boldsymbol{V}|_{wall} = 0 \quad (6)$$

Pressure at inlet is fixed at 100000 Pa. Pressure at outlet was varied in the range from 90000 Pa to 99000 Pa in a series of numerical experiments.

Equations (4, 5) were solved numerically by the finite volume method with conjugate gradient algorithm with a preconditioner.

We considered two series of numerical experiments: In the first series, we simulated a balancing process of MWB. For this purpose width of Channel "X" (R_X) was varied from 30 to 42 μm under the fixed pressure difference in MWB. The condition where velocity and pressure drop in the bridge channel (R_{Br}) are zero we will call a balance point (and the MWB is perfectly balanced). In the second series, we change the fluid viscosity for balanced and imbalanced MWB.

4. Results and discussion

Figure 3 presents velocity distribution in the MWB. To measure velocity in the bridge channel (R_{Br}), we take point A in the middle (see Figure 4). Between points B and C we measure the pressure drop (ΔP_{Br}) in the bridge channel.

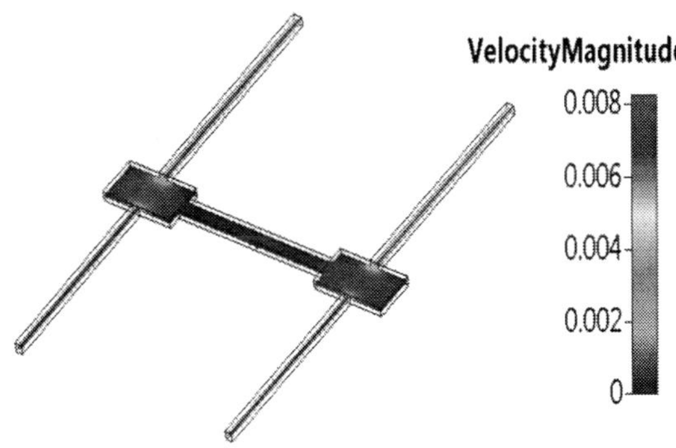

Figure 3. Velocity distribution in the MWB.

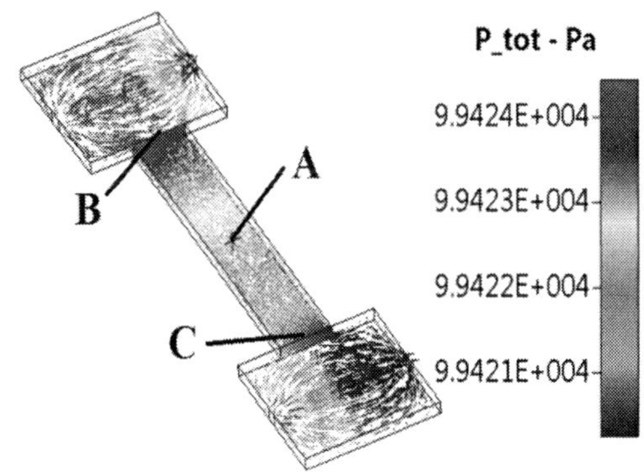

Figure 4. Velocity in the bridge channel

Influence of the width of Channel R_X on pressure drop and velocity in the bridge microchannel is shown in Figure 5. From these data we can find the balance point of MWB. In our case it is 36.39 micron of channel R_X width. It is important to note that the point does not change its position when we change the pressure difference in MWB (or total flow rate in it). Velocity magnitude for imbalanced MWB and P_{MWB} = 5000 Pa has the values consistent with the data given in [7].

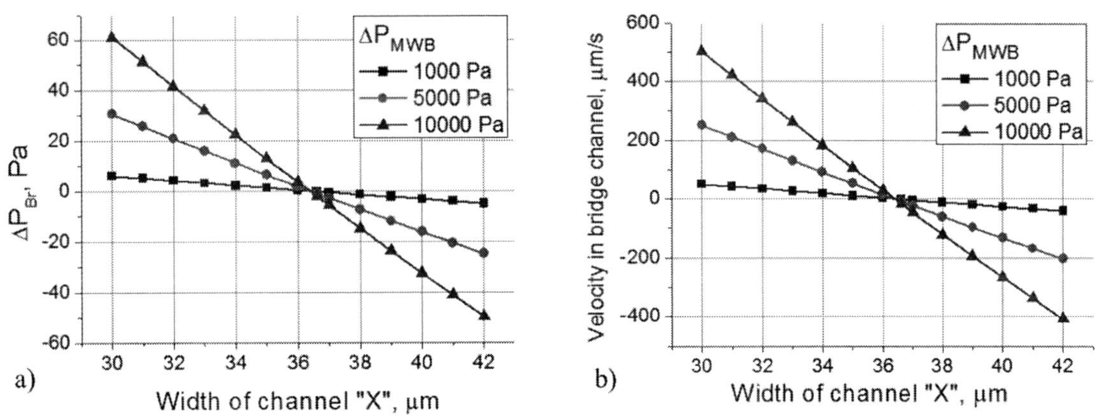

Figure 5. Pressure drop (a) and velocity (b) in the bridge microchannel for different width of the channel R_X and pressure in the MWB. Viscosity is 1.58910^{-5}Pa·s

In the second series of numerical experiments we investigated velocity in the bridge microchannel for different fluid viscosities for two cases of the MWB geometry: near the balance configuration and far from it. The results are presented in Figure 6. The dependences show a similar trend, but have a significant difference in values for balanced and imbalanced bridge: velocity magnitude in the imbalanced bridge (Figure 6b) is 200 times higher than that in the balanced bridge (Figure 6a). Based on this data we conclude that for the perfectly balanced bridge, velocity in the bridge microchannel should be zero and independent from viscosity and temperature. Bridge near the balance point has a high stability from viscosity variation. In addition, it means a temperature stability because liquids usually have a strong dependence of its viscosity from temperature

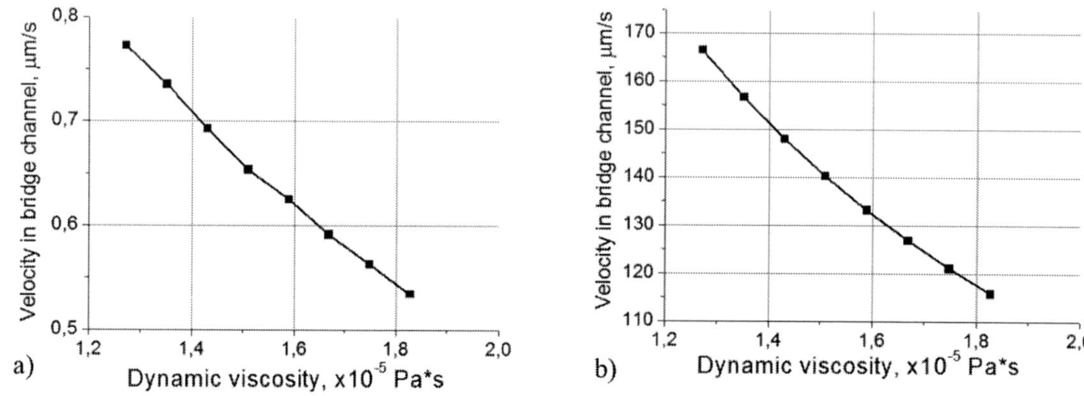

Figure 6. Velocity in the bridge microchannel for different viscosity of fluid and ΔPMWB =5000 Pa. a)Width of "X" channel is 36.4 µm (near balance configuration); b) Width of "X" channel is 40 µm.

One of the most interesting properties of the MWB is the ability to reduce pressure drop in the bridge (ΔP_{Br}) compared to the pressure drop in the system (ΔP_{MWB}) by several orders of magnitude. Let us define the pressure reduction coefficient K as:

$$K=\Delta P_{MWB}/\Delta P_{Br} \quad (7)$$

where ΔP_{MWB} is pressure difference between inlet and outlet of the microfluidic Wheatstone bridge, ΔP_{Br} is pressure difference between two ends of the bridge microchannel.

Figure 7 illustrates the dependence of ΔP_{Br} on ΔP_{MWB} for different width of channel "X". We note its linear trend and conclude that the angle of the slope strongly depends on balancing the MWB. Figure 8 shows a significant growth of pressure reduction coefficient (K) as the width of microchannel "X" is reducing and approaching the balance configuration (width of channel "X" is near 36.39µm). This property of pressure reduction can be applied in high-precision microfluidic devices.

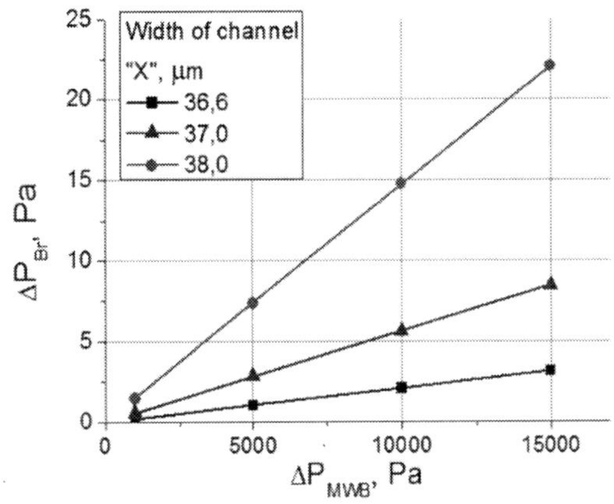

Figure 7 ΔP_{Br} as a function of ΔP_{MWB} for different width of channel "X".

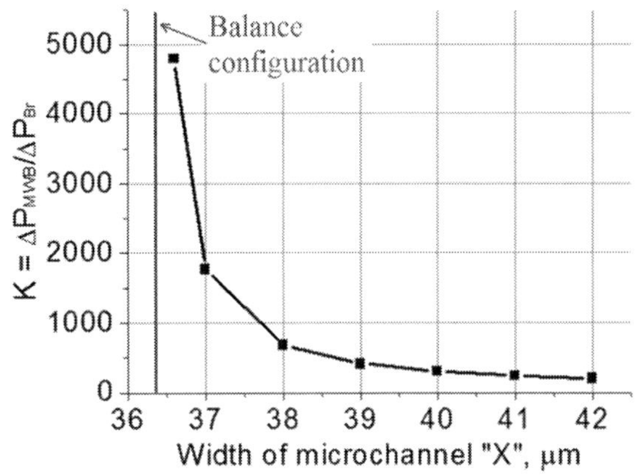

Figure 8 Pressure reduction coefficients for different width of the microchannel "X".

5. Application of microfluidic Wheatstone bridge to high-precision sensors

Based on the studied properties of the microfluidic Wheatstone bridge, we propose a design of high–precision microfluidic pressure sensor. The architecture of the system is shown in Figure 9.

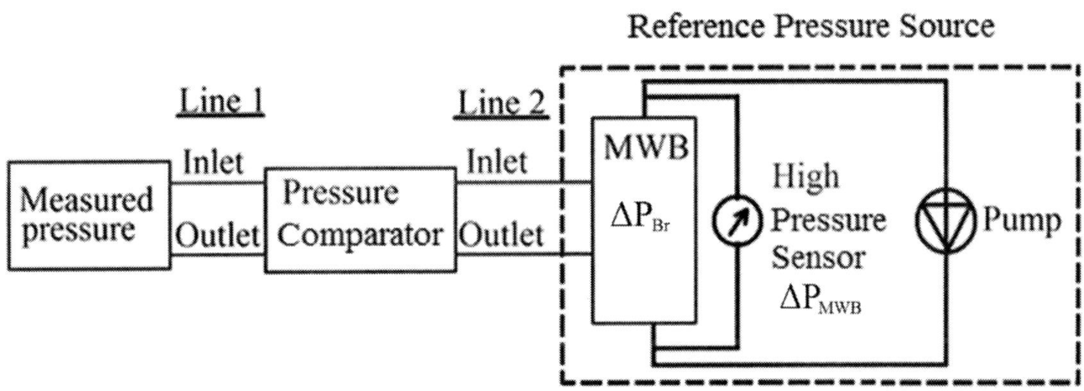

Figure 9. Architecture of the high accuracy microfluidic pressure sensor based on the MWB.

The device has two main parts: the Reference Pressure Source and the pressure Comparator. The Comparator has two measurement lines. Line1 connects to the place where pressure difference should be measured. Line2 connects to the Reference Pressure Source. Pressure difference in Line2 can be varied in some range. The Comparator alternately measures pressure difference in Line 1 and Line 2 and signals when the parameters are equal.

The design of the Reference Pressure Source includes circulation circuit, microfluidic pump microfluidic Wheatstone bridge and High Pressure Sensor. The pump drives a fluid flow in the circuit by providing a pressure drop in the MWB (ΔP_{MWB}).Value of the pressure drop is measured by the High Pressure Sensor. ΔP_{Br} is "K" times lower than ΔPMWB (see equation 7). This pressure drop (ΔP_{Br}) is supplied to Line 2 of Pressure Comparator.

Varying the flow in the circulation circuit, we change a pressure difference in Line 2. When the Pressure Comparator detects an equality of the pressure drops in L1 and L2, we note a value ΔP_{MWB} from High Pressure Sensor. Measured pressure difference in Line 1 (ΔP_{L1}) is calculated by equation 8

$$\Delta P_{L1}= \Delta P_{L2}=\Delta P_{Br}=\Delta P_{MWB}/K \quad (8)$$

where ΔP_{L1} is measured pressure drop (Line 1), ΔP_{L2} is pressure in Line 1 (equal to pressure difference in bridge channel ΔP_{Br}), ΔP_{MWB} is pressure drop in MWB, which is measured by high pressure sensor, K is pressure reduction coefficient.

Based on equation (8) we conclude that error of measurements by High-Pressure Sensor is reduced in K times. This properties of our microfluidic pressure sensor is very important, because it allows to measure a low pressure difference with high accuracy. For example, if ΔP_{MWB} =100 000 Pa, pressure reduction coefficient K = 5000 and error of high pressure sensor is 0.5% (typical value for High-Pressure Sensor) or ε=500 Pa, then the error of low pressure measurement is ε=0.1 Pa. This value corresponds to the best devices for high-precision measurements. It is possible to improve the metrological characteristics of described systems by increasing the pressure reduction coefficient. Such a microfluidic pressure sensor can be applied for measuring a pressure drop in ventilation systems, clean rooms and filters. Using additional devices, it is possible to measure also velocity, flow rate, or fluid viscosity with very high accuracy.

6. Conclusions and future work

In this work we investigated the microfluidic Wheatstone bridge. Starting from the classical electrical Wheatstone bridge, we design a microfluidic structure with similar operation principles. In addition, we propose to use a technology of etching in microchannel for balancing the MWB, improving the stability of the device and simplifying it. Analytical models of hydraulic resistance were used to design the geometry of the MWB. Applying Navier-Stokes equations for a laminar flow of viscous incompressible liquid, we created a 3D computational model of the microfluidic Wheatstone bridge and carried out two series of numerical experiments. We considered a pressure reduction coefficient in MWB as a function of the bridge balance. It was shown that the microfluidic structure near the balance point has a very high pressure reduction coefficient (K=5000) and its value can be changed during the production process by etching in microchannel technology. Based on the properties of the MWB, we developed the principles of high-precision microfluidic pressure sensor, which has perfect metrological characteristics and can be applied in many areas.

Acknowledgment. This work was supported by the Government of Russian Federation under "5-100-2020" Programme.

References

[1] Sackmann E K, Fulton A L, Beebe D J 2014 *Nature*, *507*(7491), 181-189.
[2] Liu D, et al. *Advanced Materials* 2015, 27.14: 2269-2269.
[3] Nightingale A M, deMello J C, 2013 *Advanced Materials*, *25*(13), 1806-1806.
[4] Mark D, Haeberle S, Roth G, von Stetten F, Zengerle R,.2010 *Chemical Society Reviews*, *39*(3), 1153-1182.
[5] Castillo-Leon J, Svendsen W E 2014 *Lab-on-a-Chip Devices and Micro-Total Analysis Systems: A Practical Guide*. (Springer).
[6] Kang Y J, Lee S-J 2013 *Biomicrofluidics*, *7*(5), 054122.
[7] Tanyeri M, Ranka M, Sittipolkul N and Schroeder C M 2011 *Lab on a Chip*, *11*(24), 4181-4186.
[8] Plecis A and Chen Y 2008*Analytical chemistry*, *80*(10), 3736-3742.
[9] Morris A S, Langari R 2012 *Measurement and instrumentation: theory and application*. (Academic Press).
[10] Tabeling P 2005 *Introduction to Microfluidics* (Great Britain: Oxford University Press)
[11] Bruus H 2007 *Theoretical Microfluidics* (Denmark: Oxford University Press)
[12] Konakov S A, Krzhizhanovskaya V V, 2016 *Journal of Physics: Conference Series,***681**(1), 012035

Simulating Relaxation Channels of CO_2 in Clathrate Nano-cages

P. R. Dahoo[1,2], R. Puig[2], A. Lakhlifi[3], C. Meis[2,4] and J. D. Gale[2,5]

[1] LATMOS /IPSL, UVSQ Université Paris-Saclay, UPMC Univ. Paris 06, CNRS, F-78280, Guyancourt, France.

[2] CHAIR Materials Simulation and Engineering, UVSQ, Université Paris Saclay, 78035 Versailles Cedex, France.

[3] Institut UTINAM-UMR 6213 CNRS- Université de Franche-Comté, Observatoire de Besançon, 25010 Besançon Cedex, France.

[4] INSTN- CEA, Université Paris Saclay – 91191 Gif-sur-Yvette, France.

[5] Nanochemistry Research Institute/Curtin Institute for Computation, Department of Chemistry, Curtin University, PO Box U1987, Perth, WA 6845, Australia.

E-mail: pierre-richard.dahoo@uvsq.fr

Abstract. The energy levels of CO_2 in the small (*s*) and large (*l*) nano-cages of cubic sI clathrates are calculated in the Born-Oppenheimer approximation using pairwise atom-atom interaction potentials. In the *s* cage, the centre of mass of CO_2 oscillates with small amplitudes, asymmetrically about the cage center with harmonic frequencies around 100 cm^{-1}. In the *l* cage, oscillations are anharmonic with large amplitude motions in a plane parallel to the hexagonal faces of the cage and the corresponding frequencies are calculated to be 55 cm^{-1} and 30 cm^{-1}. Librational harmonic frequencies are calculated at 101.7 cm^{-1} and 56.0 cm^{-1} in the *s* cage and at 27.9 cm^{-1} and 46.4 cm^{-1} in the *l* cage. Results show that the coupling between the CO_2 molecule and the nano-cage is quite different for the low frequency translational, rotational or librational modes and the high frequency vibrational modes, which consequently leads to different relaxation channels.

1. Introduction

A clathrate is a crystalline solid with a cage-like structure of water molecules surrounding atoms or small molecules. As a compact assembly of nano-cages, clathrates are found as three different kinds of Archimedes polyhedra characterized by a cubic structure-I (sI), a cubic structure-II (sII) or a hexagonal structure-H (sH) [1]. The unit cell of sI consists of 46 water molecules with two types of cages; a small cage (*s*) that is a polyhedron composed of 12 pentagonal faces, and a large cage (*l*) that is a polyhedron consisting of 12 pentagonal faces and two opposite hexagonal faces [2]. Clathrate hydrates are the focus of many astrophysical investigations, as well as a wide range of applications. These materials are considered to be responsible for the anomalous desorption behaviour of gaseous parent molecules in comets [3] by providing a dissociation pressure curve intermediate between the pure compound sublimation equilibrium and the water ice phase. This curve can be calculated by the application of the van der Waals-Platteeuw model, leading to a simple van't Hoff's law [4-5]. Application of this model allows the prediction of the N_2/CO and Ar/CO ratios of trapped N_2 and Ar in

Content from this work may be used under the terms of the Creative Commons Attribution 3.0 licence. Any further distribution of this work must maintain attribution to the author(s) and the title of the work, journal citation and DOI.
Published under licence by IOP Publishing Ltd

clathrates that match those measured in the coma of the Jupiter family Comet 67P/Churymov Gerasimenko [6]. Because of the consequences for global warming and climate change if clathrates release trapped carbon dioxide or methane, research work in this field is of particular importance, including the study of the interactions between the trapped molecule and the host water framework. One aim of such research is to determine how effective the absorption of infrared radiation by trapped CO_2 can be in the warming of clathrates and the subsequent release of CO_2. Depending on the competition between radiative and non-radiative relaxation, only a limited temperature rise of the clathrate may result because of stimulated emission (S.E.) of radiation, as observed for CO_2 trapped in low temperature rare gas matrices [7] and which should not lead to the break down of the nano-cages. The time and temperature dependence of the S.E. therefore gives some insight into the vibrational dynamics of clathrate-trapped CO_2.

2. The Site inclusion model and interaction potential energy

A theoretical model based on an extension of the site inclusion model used to study C_3, CO_2 and N_2O molecules [8-10] trapped in rare gas matrices is used to simulate the spectra of the linear CO_2 molecule trapped in clathrate sI matrix using 12-6 Lennard-Jones (LJ) interatomic potentials and effective charges that are included in order to account for the electrostatic interactions between atoms of CO_2 and H_2O. The model assumes the validity of the Born-Oppenheimer approximation, to separate the vibrational degrees of freedom (high frequency motions) from the orientational and translational degrees of freedom (low frequency motions) of the system formed by the molecule and the nano-cage. The total potential energy of the trapped molecule inside the nano-cage is written as:

$$V = \sum_j V_{Mj}(\mathbf{r}_{0j}) + \sum_{jj',j\neq j'} V_{jj'}(\mathbf{r}_{jj'}) \tag{1}$$

where the term V_{Mj} is the molecule-cage interaction and $\mathbf{r}_{0j}$ is the distance vector between the center of mass (c.m.) of CO_2 and the jth atom of the cage (*s* or *l*). The second term $V_{jj'}$ represents the interaction potential between two atoms of the cage: *i*) a jth atom of one H_2O molecule located at position r_j and *ii*) a j'^{th} atom of another H_2O molecule of the cage located at position $r_{j'}$ with respect to an absolute frame tied to the cage.

The interaction potential energy V_{Mj} between the CO_2 molecule and an atom of the cage is the sum of 12-6 LJ potentials representing the repulsion-dispersion contributions and the electrostatic interaction from effective charges q_i (i=1,2,3 for CO_2) and q_j on the atoms of the H_2O molecules of the *s* or *l* cage, such that:

$$V_{Mj}(\mathbf{r}_{0j}) = \sum_{i=1}^{3} \left[4\varepsilon_{ij} \left\{ \left(\frac{\sigma_{ij}}{r_{ij}} \right)^{12} - \left(\frac{\sigma_{ij}}{r_{ij}} \right)^{6} \right\} + \frac{1}{4\pi\varepsilon_0} \frac{q_i q_j e^2}{r_{ij}} \right] \tag{2}$$

where ε_{ij} and σ_{ij} are the mixed LJ potential parameters from the Lorentz-Berthelot combination rules $\varepsilon_{ij} = \sqrt{\varepsilon_{ii}\varepsilon_{jj}}$ and $\sigma_{ij} = \frac{\sigma_{ii} + \sigma_{jj}}{2}$. The distance vector, $\mathbf{r}_{ij}$ between *i* and *j* atoms can be expressed in the molecular frame (G, **x**, **y**, **z**) or the crystal frame (O, **X**, **Y**, **Z**) as given in reference [5]. The orientational degrees of freedom, $\mathbf{\Omega} = (\theta,\varphi)$, of the molecule and the position vector of any atom, *j*, are defined in the absolute crystal frame (O, **X**, **Y**, **Z**).

In the rigid crystal approximation, the potential energy hypersurface of the molecule in its cage, *n* (*n* = *s* or *l*), can then be written as:

$$V_M^{(n)}(\xi^0,\Omega,\{Q\}) = V_M^{(n)e}(\xi^0) + V_M^{(n)}(\xi^0,\{Q\}) + V_M^{(n)}(\xi^0,\Omega) + \Delta V_M^{(n)}(\xi^0,\Omega,\{Q\}) \tag{3}$$

where $V_M^{(n)e}(\xi^0)$ is the minimum of the potential energy corresponding to the equilibrium configuration; $V_M^{(n)}((\xi^0),\{Q\})$ is the vibrational dependence in terms of normal coordinates

$\{Q\}$; $V_M^{(n)}((\xi^0),\Omega)$ is the orientational dependence in terms of Euler angles ($\mathbf{\Omega} = (\theta,\varphi,\chi)$, χ=0 for a linear molecule); $\Delta V_M^{(n)}((\xi^0),\Omega,\{Q\})$ characterizes the translation-orientation-vibration dynamic coupling term, which can induce the relaxation of vibrational modes onto the translational and orientational modes; and (ξ^0) represents the displacement vectors of the c.m. of H_2O molecules of the cage and of the trapped CO_2, determined by the Green's functions of the perfect crystal.

From the potential energies calculated for the low frequency motions, the Hamiltonian can be solved for the eigen-values of the allowed translation and orientation in each cage (*s* or *l*);

$$H_M^{(n)} = H_{vib}^{(n)} + T_{rot}^{(n)} + T_{trans}^{(n)} + V_M^{(n)}(\xi^0,\Omega,\{Q\}) \tag{4}$$

where the first three terms $H_{vib}^{(n)} + T_{rot}^{(n)} + T_{trans}^{(n)}$ correspond to the free CO_2 molecular Hamiltonian.

3. Results and discussion

Calculations were performed for the CO_2 molecule trapped in a clathrate matrix containing 4x4x4 unit cells (up to 2,944 molecules of H_2O). The calculated potential energy hypersurfaces are plotted in Figure 1 for CO_2 a) in cage *s* and b) in cage *l* as a function of its orientation. The equilibrium configuration for cage *s* corresponds to minimum energy value -275 meV (-2,200 cm^{-1}). The molecule is at the cage centre, perpendicular to two pentagonal faces and with an allowed translation motion around the centre of the cage. For the cage *l*, the minimum energy value is - 372 meV (-2,976 cm^{-1}). The molecule is displaced from the cage centre by about 0.5 Å and can have a translational motion of large amplitude in a direction parallel to the two hexagonal faces.

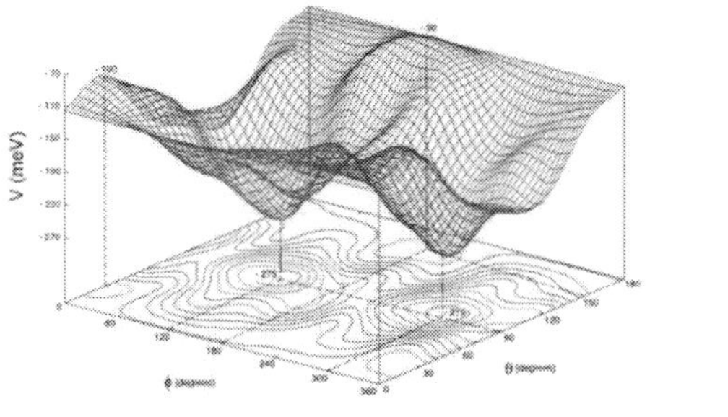

Figure 1 a). Potential energy of CO_2 as a function of orientation in cage *s*.

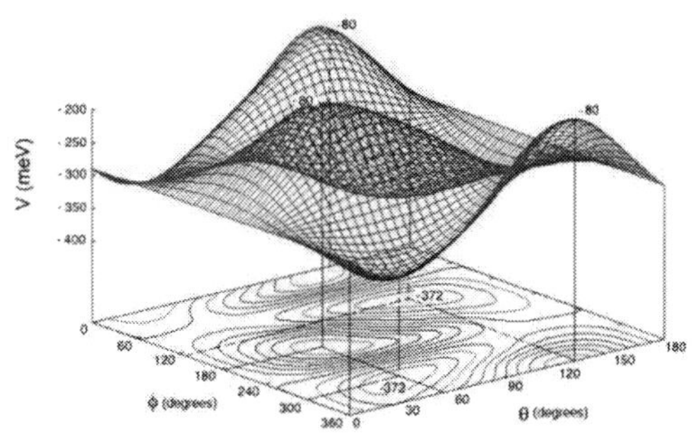

Figure 1 b). Potential energy of CO_2 as a function of orientation in cage *l*.

The contour plots of the orientational potential energy surfaces for CO_2 trapped in the clathrate sI structure type are calculated at its c.m. equilibrium translational positions in cages *s* and *l*. The two contour plots show a certain symmetry. In the small cage, the rotational symmetry of 180 deg about the direction defined by θ = 90 deg and φ = 180 deg corresponds to the symmetry of the molecule i.e.: $V_M(\varphi,\theta) = V_M(\varphi+\pi, \pi-\theta)$. Analysis of these contour plots shows that the orientational motions of the molecule are strongly hindered in the two cavities. They correspond to two-dimensional librational motions around θ^e = 90 deg and φ^e = 115 deg in cage *s*, and around θ^e = 130 deg and φ^e = 110 deg in cage *l*.

In each cage, the rotational kinetic Hamiltonian, T_{rot}, and the orientational potential energy, $V_M(\varphi,\theta)$, of the molecule are expanded as a power series with respect to new variables ($u_\theta = \cos(\theta)$ and $u_\varphi = \varphi-\varphi^e$ for cage *s* and $u_\theta = \sin(\theta-\theta^e)$ and $u_\varphi = \varphi-\varphi^e$ for cage *l*), which allows the labeling of the energy states with appropriate librational quantum numbers (*j,m*). As the harmonic terms are much greater than the anharmonic ones, the energy levels can first be determined in the harmonic approximation and then their calculation refined by applying standard perturbation techniques. The eigen-solutions of the orientational Hamiltonian are expressed as:

$$E_{jm} = \hbar\omega_\theta(j + \tfrac{1}{2}) + \hbar\omega_\varphi(m + \tfrac{1}{2})$$
$$|jm\rangle = N_j N_m \exp(-(u_\theta^2 + u_\varphi^2)/2) H_j(u_\theta) H_m(u_\varphi) \tag{5}$$

where ω, N and H are the librational harmonic frequencies, the normalization factors and the Hermite polynomials, connected to the two librational motions. The quantum numbers j and m are independent integers. The calculated harmonic frequencies are *i)* ω_θ= 101.7 cm^{-1} and ω_φ = 56.9 cm^{-1} in cage *s* and *ii)* ω_θ= 27.9 cm^{-1} and ω_φ= 46.4 cm^{-1} in cage *l*. The calculated orientational levels up to 200 cm^{-1}) show that there are only few well separated levels in cage *s*, while in cage *l* several intermixed levels are present. This can be explained by less hindered librational motions in the cage *l* when compared to motions in the cage *s*.

For high vibrational frequency modes, calculations show that the direct electrostatic interaction between the charges does not reproduce the shifts determined from observed transitions but that the coupling is much weaker in contrast to the coupling with low frequency modes. Decreasing the value of the effective charges leads to calculated vibrational shifts that compare better with available experimental shifts which are less than a few percent of the vibrational transition frequencies.

These results can be compared to the calculations performed on CO_2 trapped in low temperature rare gas matrix. The librational energies compare with those calculated for CO_2 in Ar [9] and in which laser induced fluorescence techniques [7] lead to the observation of amplified vibrational stimulated emission rather than a thermal effect resulting from energy transfer through non-radiative processes. In matrices, O_3 relaxes through binary collisions rather than by a multiphonon process [11] because the coupling of the vibrational degrees of freedom and the matrix is negligible. In comparison with the results reported in this work, the coupling scheme is consistent with a relaxation process by binary collisions. The high frequency vibrational degrees of freedom are almost isolated in the nano-cages because of a very weak coupling scheme which is unfavourable for relaxation by a multiphonon process through an energy gap law [11]. Then vibrationally excited CO_2 is expected to relax radiatively rather than by non-radiative processes. This analysis is also consistent with the results reported in reference [6] regarding molecules trapped in clathrates of Comet 67P/Churymov Gerasimenko, which remain isolated in the nano-cages until the comet passes by a radiative source like the sun. The various trapped species are then freed allowing for possible chemical reactions to occur as speculated from a recent laboratory experimental work [12].

4. Conclusions

In this work, the low frequency energy states of CO_2 trapped in the *s* and *l* nano-cages of cubic type I clathrate are numerically calculated and analyzed from a theoretical model. The results show that the molecule is rather isolated in its cage and that non-radiative relaxation processes are expected to be slower than radiative ones just as for CO_2 isolated in rare gas argon matrix.

Acknowledgments

The authors would like to thank CHAIR Materials Simulation and Engineering, UVSQ, Université Paris Saclay for funding.

References

[1] Sloan E. D., 2004 *Am. Mineral.* **89**, 1155.
[2] Suga H., Matsuo T. and Yamamuro O., 1992 *Pure & Appl. Chem.* **64**, 17.
[3] Delsemme A. H. and Swings P., 1952 *Ann. d. Astrophys.* **15**, 1.
[4] van der Waals J. H., Platteeuw J. C., 1959 *Adv. Chem. Phys.* **2**, 1.
[5] Lakhlifi A., Dahoo P. R., Picaud S. and Mousis O., 2015 *Chem. Phys.* **448**, 53.
[6] Mousis O., Lunine J. I., Luspay-Kuti A., Guillot T. *et al*, 2016 *The Astrophys. J. Lett.*, **819**, L33.
[7] Chabbi H., Dahoo P. R., Dubost H., Gauthier-Roy B., Vasserot A. M. and Abouaf-Marguin L., 2000 *Low Temp. Phys.* **26**, 9.
[8] Lakhlifi A., Dahoo P. R., Vala M., Szczepanski J. and Ekern S., 1997 *Chem. Phys.* **222**, 241.

[9] Lakhlifi A., Chabbi H., Dahoo P. R. and Teffo J. L., 2000 *Eur. Phys. J. D.* **12**, 435.
[10] Dahoo P. R., Lakhlifi A., Chabbi H. and Coanga J. M., 2006 *J. of Mol. Struc.* **786,** 157.
[11] Dahoo P. R., Jasmin D., Brosset P., Gauthier-Roy B. and Abouaf-Marguin L., 1998 *J. Chem. Phys.* **108**, 8541.
[12] Meinert C., Myrgorodska I., de Marcellus P., Buhse T., Nahon L., Hoffmann S. V., Le Sergeant d'Hendecourt L. and Meierhenrich U. J., 2016 *Science* **352**, 208.

Monte Carlo Method Applied to the ABV Model of an Interconnect Alloy

P R Dahoo[1,2*], J Linares[2,3], D Chiruta[2,3], C Chong[4], P Pougnet[5], C Meis[2,6] and A El Hami[7]

[1] LATMOS /IPSL, UVSQ Université Paris-Saclay, UPMC Univ. Paris 06, CNRS, F-78280, Guyancourt, France.
[2] CHAIR Materials Simulation and Engineering, UVSQ, Université Paris Saclay, 78035 Versailles Cedex, France.
[3] GEMaC, CNRS-UMR 8635, Université de Versailles St. Quentin en Yvelines, 78035 Versailles, France
[4] LISV, Université de Versailles St. Quentin en Yvelines, 78035 Versailles, France
[5] VALEO, BP 68532 Cergy, 95892 Cergy Pontoise Cedex, France.
[6] INSTN- CEA-Saclay – 91191 Gif-sur-Yvette Cedex, France.
[7] INSA de ROUEN Laboratoire de Mécanique, Pôle Technologique du Madrillet, BP 8-76801 Saint Etienne du Rouvray, France.

E-mail: pierre-richard.dahoo@uvsq.fr

Abstract. A Monte Carlo (MC) simulation of a 2D microscopic ABV (metal A, metal B and void V) Ising model of an interconnect alloy is performed by taking into account results of Finite Element methods (FEM) calculations on correlated void-thermal effects. The evolution of a homogeneous structure of a binary alloy containing a small percentage of voids is studied with temperature cycling. The diffusion of voids and segregation of A type or B type metals is a function of the relative interaction energy of the different pairs AA, BB, AB, AV and BV, the initial concentrations of A, B and V and local heating effect due to the presence of clusters of voids. Voids segregates in a matrix of A type, of B type or AB type and form large localized clusters or smaller delocalized ones of different shapes.

1. Introduction

This work focuses on the physics of failure which is at the heart of the degradation observed in the interconnecting alloys (ICA) used as solder joints in mechatronic devices. The presence of intermetallics, voids, delamination at adjacent boundaries in a power mechatronics module [1] reduces the reliability of these devices. The analysis of the root cause of the failures can either be performed from experimentation or from simulation. The purpose of this work is to investigate how the results of Monte Carlo's simulation in the frame of the Ising ABV model [2] may provide a solution to the design of reliable mechatronic devices. Simulation is a means to drive the designer towards the best possible reliable design solutions when the model is used within the reliability based design optimization (RBDO) method [3] which combines design and reliability optimization. RBDO which leads to more efficient design and a rigorous monitoring of the design is then able to predict and detect failure modes of the designed system. Different types of innovative materials to meet regulations

Content from this work may be used under the terms of the Creative Commons Attribution 3.0 licence. Any further distribution of this work must maintain attribution to the author(s) and the title of the work, journal citation and DOI.
Published under licence by IOP Publishing Ltd

about hazardous substances are assembled in a power mechatronics module. The active silicon based transistor chip that is interconnected to a copper busbar through lead alloy solder joints acts as a switch that controls the electronic current density within the chip. Although the copper busbar is cooled from the bottom by an aluminium cooling water case through an electrical insulating ceramic layer, the joule heating give rise to a thermal gradient in assemblies. As a result delamination and propagation of cracks enhanced by the differences in the coefficient of thermal expansion (CTE) of the materials in contact reduces the reliability of the device. Ageing of interconnections (solder, wire bonding ...) and bulk materials (copper busbar, silicon chip, insulating ceramic ...) occurs through repeated current density cycling. Figure 1 a) and b) shows the defects that are present in interconnect alloys before and after thermal shocks.

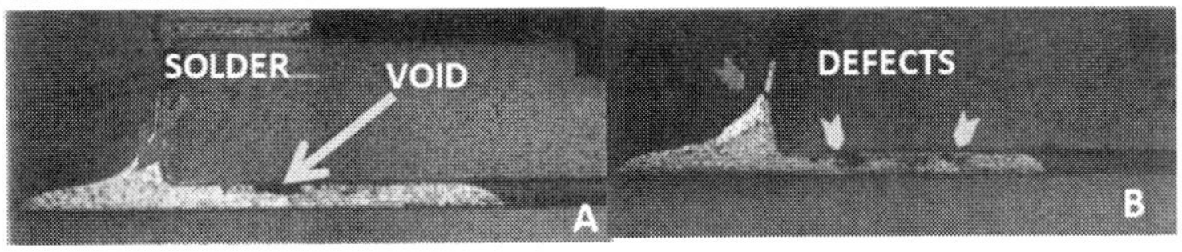

Figure 1: a) Cross-section after solder paste thermal reflow b) Cross-section after 1000 thermal shocks test -40°C/140 °C 1h/1h

2. Finite element methods to model the thermal effect of voids

In Figure 2, the x-ray tomography image of a component soldered on a FR4 substrate [4], shows the air bubbles on the thickness of the lead-free solder ICA. These defects can occupy between 10% and 60% of the surface and throughout the thickness of the solder.

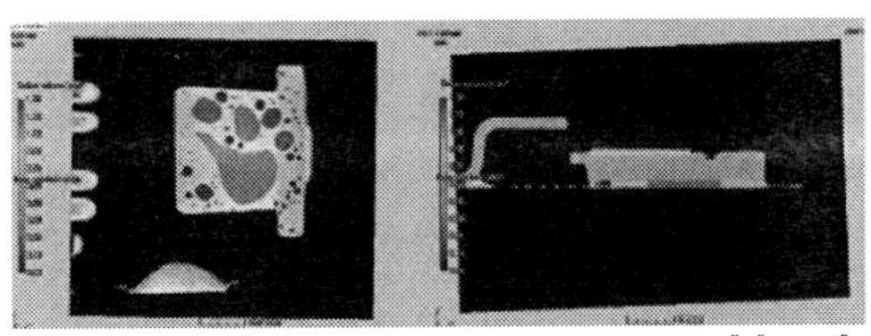

Figure 2: *x-ray tomography image of an electronic component soldered on a FR4 substrate, showing air bubbles on the thickness of the lead-free solder*

Thermal effects in the ICA in the presence porosities were modeled and simulated by FEM. Results are shown in figure 3. The main results are that the increase in the temperature of the silicon chip is a function of the spatial distribution and size of defects, with a maximum temperature localized above the larger defect. A study based on the shape of the defects had also shown similar results in the evolution and mapping of temperature. These defects behave as heat traps that cause local temperature rise [5]. If the defects are uniformly distributed the risk of lift off of the chip is more likely to occur.

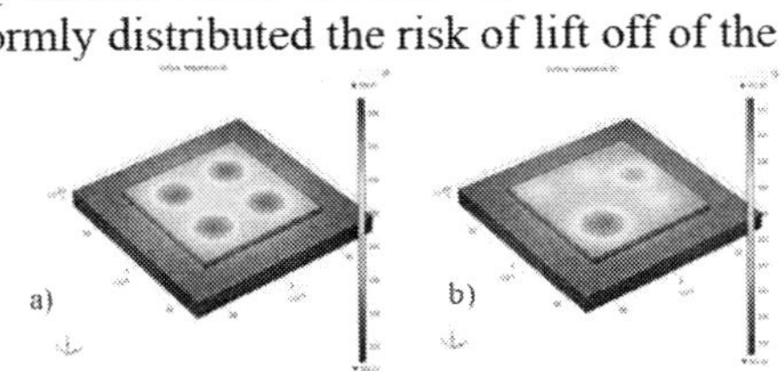

Figure 3: *Temperature mapping with circular defects distributed in the ICM: a) regularly (radius 2.2 mm, porosity 40%), b) at random (maximum radius 1.8 mm, porosity 40%)*

The position of the defect influences the maximum temperature reached (62°C). It is higher if a defect is near the center of the chip than close to the borders. In dynamic mode when the mechatronics module is subjected to power cycles, the difference in temperature will necessarily increase. Consequently, these heat trapping zones will induce thermo-mechanical stresses in the chip. Simulation results show that the maximum temperature depends on the degree of porosity. For the same degree of porosity it is localized on the defect of larger radius. One finds: T max (n = 1)> T max (n = 2)> T max (n = 3) where n is the number of defects. There is an increase of the maximum temperature with the size of the defects as the porosity ratio increases. The maximum deformation increases with the porosity up to 21% and beyond this threshold it decreases. RBDO method can be applied to determine the thermo-mechanical properties of the materials assembled in a mechatronic device. This method takes into account the uncertainties of the physical parameters in the context of ageing of interconnections of the devices during operation, and then optimize the finite element model used to determine the elastoplastic behavior of the material. The ABV model can also be implemented in the RBDO method to simulate the properties of innovative materials and identify the different schemes along which voids and cracks develop in the alloy and thus limit the uncertainties on the thermo-mechanical parameters.

3. The ABV model

The Ising model is a generic model that can be applied to different phase transition physics as can be observed for example in Magnetic Spin Cross-Over (SCO) compounds [2], Binary alloy, Grain boundary energy (Potts model). Based on the equivalence between the Ising-lattice gas and a binary alloy, the ABV model was developed by Yaldram and Binder [6-8], where A and B stand for two different metals and V for void or a vacancy. It has been widely used to study the configurational statistics of alloys on a microscopic basis using phenomenological atomic interaction parameters or values obtained from numerical simulation [9-10]. This model has been discussed by Tafa, Puri and Kumar [11] in connection to several other models BEG, FL, PS, PG or JM as referenced in [11] in terms of phase separation in ternary mixtures leading to characteristics features of the domain growth process.

In order to get an insight in the solutions to implement to reduce observed degradation leading to failures in the interconnexion of a mechatronic module, a Monte Carlo simulation in the Metropolis scheme is applied using the ABV model. Diffusion or aggregation effects are discussed in terms of initial conditions and interaction energies between the metal atoms. To study the temperature effect on the homogeneous structure of a binary alloy of constant composition containing voids, the simulation is performed on a 2D microscopic Ising model by using the following three states Hamiltonian which can be derived from the physical configuration [11]:

$$H = H_0 - J\sum_{\langle ij \rangle}^{NN} \sigma_i \sigma_j - K\sum_{\langle ij \rangle}^{NN} \sigma_i^2 \sigma_j^2 + U\sum_{\langle ij \rangle}^{NN} (\sigma_i^2 \sigma_j + \sigma_i \sigma_j^2) \quad (1)$$

where σ can be set to -1, 0, 1. In equation 1, the different parameters K, J and U are defined by the following equations:

$$\begin{aligned} J &= \frac{1}{4}(2\varepsilon_{AB} - \varepsilon_{AA} - \varepsilon_{BB}) \\ K &= \varepsilon_{AV} + \varepsilon_{BV} - \varepsilon_{VV} - \frac{1}{4}(\varepsilon_{AA} + \varepsilon_{BB} + 2\varepsilon_{AB}) \\ U &= \frac{1}{4}(\varepsilon_{AA} - \varepsilon_{BB}) + \frac{1}{2}(\varepsilon_{BV} - \varepsilon_{AV}) \end{aligned} \quad (2)$$

Where ε_{ik} ($i = A, B$ or V and $k = A, B$ or V) are the interaction energy parameters between the metal atoms A, B and void V. In order to model the observed diffusion of intermetallics in interconnexion materials, the interaction parameters ε_{AA}, ε_{BB} and ε_{AB} are adjusted and the time evolution of this material pseudo-spin configuration is calculated. There are no pairwise interaction terms involving voids but phase separation is mediated by vacancies. The physical processes that enable segregation are due to vacancy atom exchanges rather than exchanges between A and B. J determines the ordering of the system. In the absence of vacancies, the systems tends to mix if J < 0 and become homogeneous (complete disorder) if J > 0, then the system tends to segregate in A and B atoms. Parameters K/J and U/J account for the order or disorder that result from one of these two possible configurations. The results are discussed in terms of intensity of the interaction parameters, void percentage and local heating effect as modelled by FEM during cycling.

4. Results and discussion

Simulations were performed on a square containing 10000 atoms with an approximate atomic radius of 1.6 10^{-10} m on a surface area of 256 nm^2. Different cases were considered to determine the effect of the interaction energies on the segregation of A and B atoms as shown in figures 4 to 6. In figures 4 and 5, the initial conditions correspond to 20% of voids, with A and B in equal proportions (40%) on the square lattice and in figure 6, 80% of A, with B and V in equal proportions (10%) and the pair interaction are scaled with respect to $\varepsilon_{VV}=1$ and the temperature T.

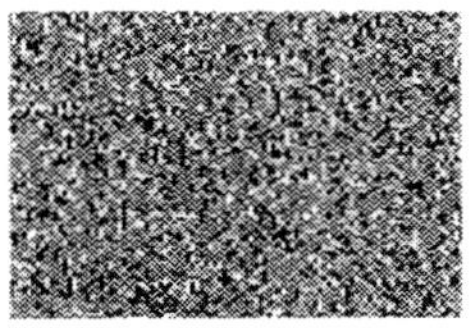

Figure 4 a). Start (1 Mc cycle), A: white, B: black, V: grey (A=B = 40%, V= 20%, $\varepsilon_{AA}=\varepsilon_{AB}=\varepsilon_{AV}=0.6$; $\varepsilon_{BB}=\varepsilon_{BV}=0.4$)

Figure 4b). Segregation (300 Mc cycles), A: white, B: black, V: grey (A=B = 40%, V= 20%, $\varepsilon_{AA}=\varepsilon_{AB}=\varepsilon_{AV}=0.6$; $\varepsilon_{BB}=\varepsilon_{BV}=0.4$)

In the three figures 4 to 6, voids segregate with AV interfaces only. In figure 4 B segregates slightly with AB interfaces only and segregation is more pronounced in figure 5.

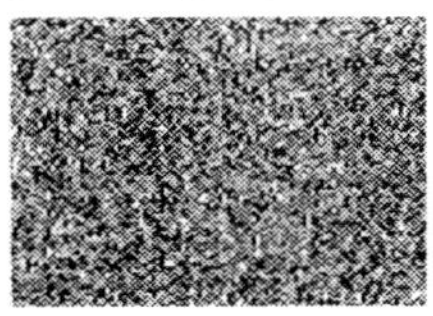

Figure 5 a). Start (1 Mc cycle), A: white, B: black, V: grey (A=B = 40%, V= 20%, $\varepsilon_{AA}=\varepsilon_{AB}=\varepsilon_{AV}=\varepsilon_{BB}=0.6$; $\varepsilon_{BV}=0.4$)

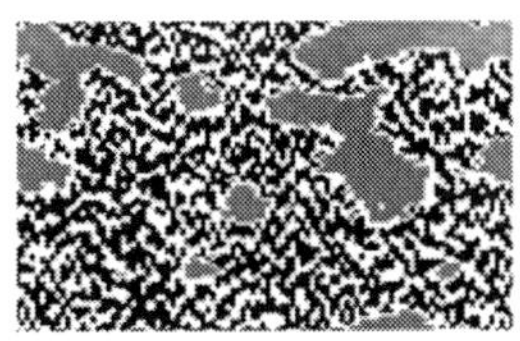

Figure 5 b). Segregation (300 Mc cycles), A: white, B: black, V: grey (A=B = 40%, V= 20%, $\varepsilon_{AA}=\varepsilon_{AB}=\varepsilon_{AV}=\varepsilon_{BB}=0.6$; $\varepsilon_{BV}=0.4$)

Figures 4 and 5 show no segregation of A while in figure 6, A segregates. In figure 4, voids segregates in an AB matrix and in figure 5, voids and B segregates in an A matrix.

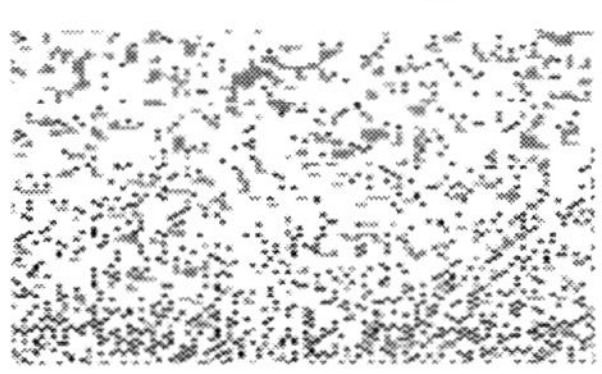

Figure 6 a). Start (1 Mc cycle), A: white, B: black, V: grey (A = 80%, B=V= 10%, $\varepsilon_{AA}=\varepsilon_{AB}=\varepsilon_{AV}=0.6$; $\varepsilon_{BB}=\varepsilon_{BV}=0.4$)

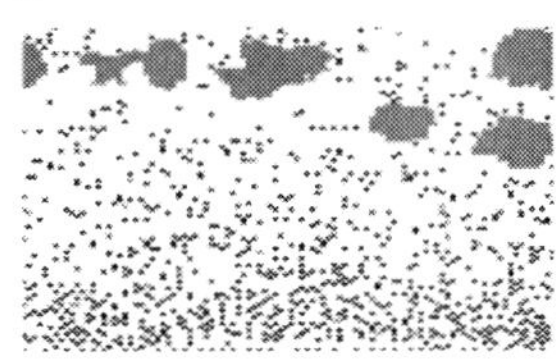

Figure 6 b). Segregation (300 Mc cycles), A: white, B: black, V: grey (A = 80%, B=V= 10%, $\varepsilon_{AA}=\varepsilon_{AB}=\varepsilon_{AV}=0.6$; $\varepsilon_{BB}=\varepsilon_{BV}=0.4$)

In figure 6, voids and A segregates in a B matrix.
Different cases in interaction energies have been simulated either by changing the relative values of the interaction energies or by reproducing static temperature cycling between 300K and 600 K.
When all energies are equal, then larger size clusters of vacancies are favoured. When $\varepsilon_{AA}=\varepsilon_{AB}=\varepsilon_{AV}=-0.6E_0<\varepsilon_{BB}=\varepsilon_{BV}=-0.4E_0$, then clusters of smaller sizes that spread in an AB matrix with A at the borders are simulated. When segregation by vacancies is enhanced through the reference energy E_0, vacancy clusters tend to increase in size and get localized. If $\varepsilon_{AV}=-0.6E_0 < \varepsilon_{BV}=-0.4E_0$, then voids drives A spatially with A at the borders of the vacancy clusters.
Cycling of temperature between 300K and 600K do not have a marked effect on the distribution of the vacancy clusters. But when local temperature rise are taken into account in the algorithm to mimic the trapping of heat by the vacancy clusters, then a 1D spreading effect of the clusters is obtained which may be interpreted as crack propagation. The results obtained show that the ABV model can be used in correlation to FEM methods to study the thermo-mechanical evolution of interconnect alloy or material during the operation of a power mechatronic device and reduce by numerical simulation the uncertainties on the coupling between thermal and mechanical parameters of materials assembled in such devices. Then RBDO method can be applied for reliability issues.

5. Conclusions

In this work, using a numerical method, the thermal effect in the presence of voids in an interconnect alloy is studied by FEM and the ABV model. Results of FEM are implemented in the ABV model.Whatever the interaction energies between the atoms of the alloy, the final configuration ends with segregation of voids. Voids segregates in a matrix of A type, of B type or AB type and form large localized clusters or smaller delocalized ones of different shapes.

Acknowledgments

The authors would like to thank CHAIR Materials Simulation and Engineering, UVSQ, Université Paris Saclay and the Memoire and Catrene-Euripides Thor's Projects for support.

References

[1] Dahoo P R, Alayli N, Girard A, Pougnet P, Tan K L and Morelle J M, 2010 *Mater. Res. Soc. Symp. Proc. Warrendale, PA,* **1195** 183
[2] Chiruta D, Linares J, Dahoo P R and Dimian M, 2012, *J. of Appl. Phy.s* **112** 074906
[3] El Hami A., Radi B., *Uncertainty and optimization in structural mechanics*, ISTE, London, John Wiley, New York, 2013
[4] Khettab M., *Studyof the influence of resin potting at the IML (Insulated Metal Leadframe), in the packaging of a current commutator mechatronic module, PhD Dissertation, UVSQ, Versailles, 2014*
[5] Adamidi E., Chong C., Khettab M., Alombert-Goget G., Dahoo P.R. and F. Maroteaux, *Thermal Impact of solder voids in the electronic packaging of power modules: Modelling and Simulation of thermal effect on IGBT*, EURIPÏDES FORUM, Gratz, Austria, (June 2012)
[6] Yaldram K and Binder K, 1991 *J. of Stat. Phys.* **62** 161
[7] Yaldram K and Binder K, 1991 *Acta Metallurgica Et Materialia* **39** 707
[8] Yaldram K and Binder K, 1991 *Zeitschrift Fur Physik B-Condensed Matter* **82** 405
[9] Gale J D, 1997, *J.C.S Faraday Transactions*, **93**, 629
[10] Soler J M, Artacho E, Gale J D, Garcia A, Junquera J, Ordejon and Sanchez-Portal D, 2002, *Journal of Physics: Condensed Matter*, **14**, 2745
[11] Tafa K, Puri S and Kumar D, 2001, *Phys. Rev E.*, **64**, 056139-1

Effect of anisotropy on the current-voltage characteristics of layered high-temperature superconductors with internal ferromagnetic defects

V.A. Kashurnikov, A.N. Maksimova, I.A. Rudnev, D.S. Odintsov

National Research Nuclear University MEPhI (Moscow Engineering Physics Institute)

115409, Kashirskoe shosse, 31, Moscow, Russia

e-mail: nastymaksimova@yandex.ru

Abstract. Dynamic of the three-dimensional vortex system in layered high-temperature superconductors with internal nanosized ferromagnetic defects has been studied by using Monte Carlo simulations. Magnetization processes in the array of ferromagnetic defects under the field of Abrikosov vortices have been taken into account. The current-voltage characteristics under applying of external dc magnetic field have been obtained. It has been shown that S-type nonlinearity of E(J) curve which have been obtained previously for two-dimensional vortex system also appears in this more realistic anisotropic three-dimensional case. The effect of vortex-vortex inter-layer coupling on the S-nonlinearity has been analyzed.

1. Introduction

The effect of nanosized particles as pinning centers on magnetic and transport properties of high-temperature superconductors has been widely analyzed till recent time. To improve the transport properties, the ferromagnetic nanoparticles as pinning centers can be used due to additional interaction energy between vortex and magnetic moment of the nanoparticle [1-3]. The superconductor-ferromagnet mutilayers are also fabricated to study the mutual influence of vortices and domain walls and to analyze the transport properties of such structures [4-7]. Spherical and ellipsoidal magnetic particles have been used as pinning centers in [1-2]. In the frame of London approximation the interaction energy between vortex and nanoparticle was obtained and the increase in critical current (estimated from the width of magnetization loop) was demonstrated. The authors of Ref. [3] considered the ensemble of ferromagnetic particles inside superconductor and the vortex lattice as a self-consistent system and numerically demonstrated some new effects, such as S-type nonlinearity on current-voltage characteristic, due to magnetization reversal of nanoparticles and the enhancement of critical current under certain conditions. Superconductor-ferromagnet multilayers are also widely studied. Formation of domain structure in ferromagnetic layer and interaction of vortex lattice with domain walls give additional opportunities for vortex pinning and tuning superconductive properties. The effect of magnetic-domain geometry on vortex pinning and, therefore, magnetic flux penetration and critical current was studied [4], the hysteretic vortex pinning in superconductor-ferromagnet multilayer was demonstrated [5]. The effect of domain wall width was also analyzed [6,7]. Thus, the study of ferromagnetic inclusions in the superconductor and superconductor-ferromagnet hybrid structures are subject of a sufficient practical interest. Since it seems difficult to fabricate

Content from this work may be used under the terms of the Creative Commons Attribution 3.0 licence. Any further distribution of this work must maintain attribution to the author(s) and the title of the work, journal citation and DOI.
Published under licence by IOP Publishing Ltd

superconductive samples with ferromagnetic inclusions of given parameters, the numerical simulations might be useful. A macroscopic magnetic inclusion can be represented by an ensemble of nanosized ferromagnetic particle with a magnetic moment or by effective potential profile for vortices. In this paper, we perform Monte Carlo simulations for three-dimensional layered high-temperature superconductor with nanosized ferromagnetic defects magnetized with self field of transport current in presence of external dc magnetic field.

2. Model and simulation results

We performed our simulations within three-dimensional model of layered high-temperature superconductor [8] by using Monte Carlo method. This model represents a vortex line as a stack of interacting planar vortices (pancakes). The Gibbs thermodynamic potential incorporates various types of in-pane interaction and inter-plane interaction. The details of simulation method for one superconducting layer have been previously reported in [9-11].

To provide artificial pinning centers, we used nanosized ferromagnetic particles as internal defects. The size of individual nanoparticle is much less than the average size of ferromagnetic domain so that such a particle can be described as a magnetic moment with a constant absolute value and variable direction. The direction of each moment depends on local magnetic field and so that is related to local vortex configuration near magnetic defect. For simulations, we describe superconductor and ferromagnetic particles as a self-consistent system and introduce an additional process – reorientation of magnetic moment – into Monte Carlo algorithm which was previously reported by us in [12,13].

We calculated current-voltage characteristics using method developed in [11] for three-dimensional $Bi_2Sr_2CaCu_2O_{8-\delta}$ slab at temperature range 1-10 K. Ferromagnetic nanoparticles have a size about several ξ, the magnetic moments of the particles are of order of 10^3 - 10^4 Bohr magnetons, which is typical for ferromagnetic materials. We suggest that the direction and absolute value of the magnetization of the particles change within the magnetization process depending on local field of Abrikosov vortices. The superconducting slab has a finite size L in x direction and periodic boundary conditions with period b in y direction and with period $N_L d$, where d=2.7 nm is a thickness of a superconducting layer, in z direction. In our simulations L=5 μm and b=3 μm that is much larger than characteristic distances at which vortices interact. We considered a three-dimensional superconducting slab which contains N_L~10 superconducting CuO_2 layers. In our algorithm, the production of vortex line containing less than N_L pancakes is allowed.

Let us consider a superconducting slab with internal ferromagnetic defects.

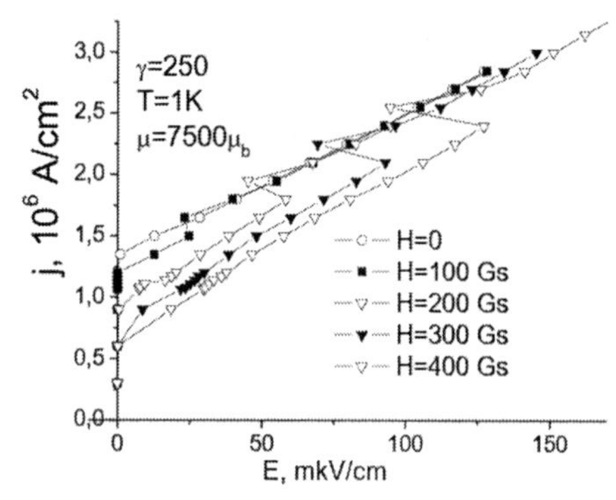

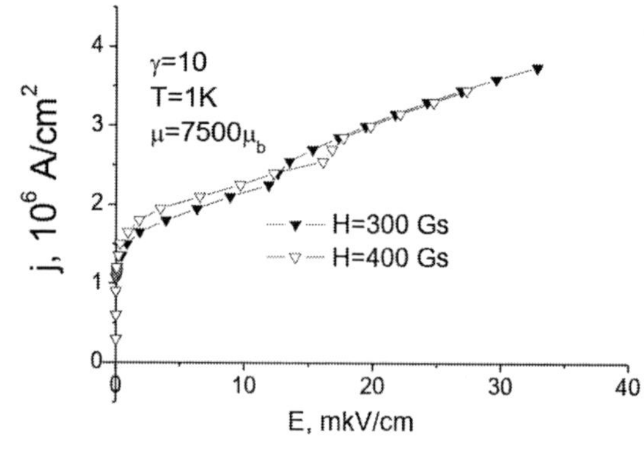

Fig.1. Current-voltage characteristics of anisotropic layered superconductor with internal ferromagnetic defects under applying external dc magnetic field. μ is magnetic moment if individual particle, μ_b is Bohr magneton, γ is anisotropy parameter.

Fig.1 represents the current-voltage characteristics in the presence of external dc magnetic field at different anisotropy parameters. One can easily see that in the case of high anisotropy (γ=250) the $j(E)$ curves at magnetic field H>200 Gs have S-type nonlinearity reported by us in [13]. However, in the case of low anisotropy the nonlinearity is less pronounced. To explain qualitatively this fact, we consider a periodic lattice of point ferromagnetic defects with variable period in the x direction (which is perpendicular to the direction in which vortices penetrate into the slab, fig.2(a)). The vortex which appears at the boundary interacts approximately with two nearest magnetic defects and two vortices pinned by them. The resulting interaction potentials at three different periods a in x direction (magnetic particles are fully magnetized) are plotted in the fig.2(b). It is easy to see that for a=220 nm and a=250 nm there is a potential barrier for the vortex and for a=300 nm the barrier is absent and vortices easily penetrates deep into the slab. Due to our quantitative explanation [13], the barrier is essential for S-type nonlinearity. j(E) curves in fig.1 were obtained for a=300 nm. The nonlinearity at γ=250 appears in this case is due to additional pinning of incomplete vortex lines on the boundary. In fig.3 the current-voltage characteristics at a=250 nm are shown.

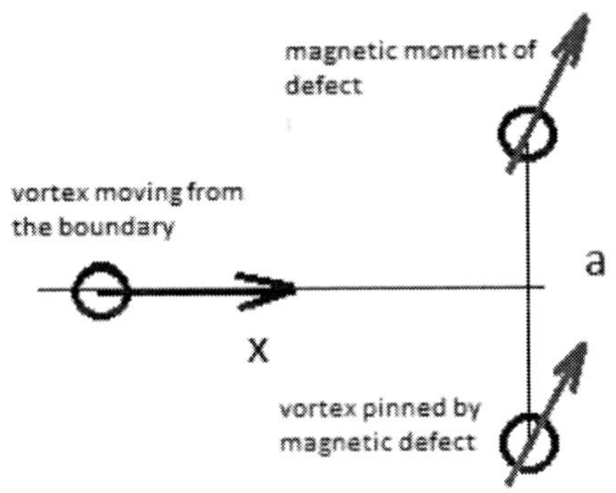

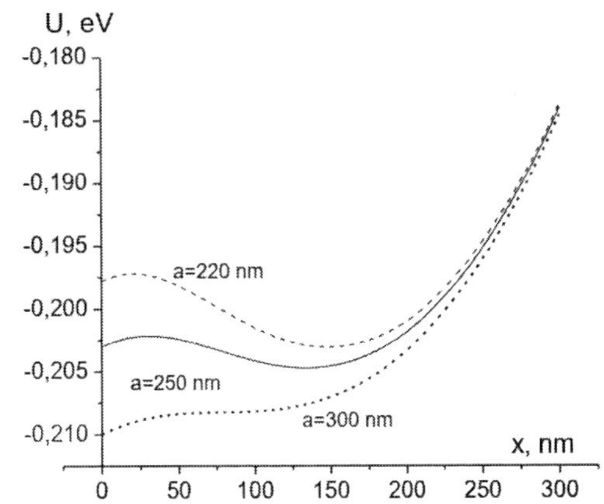

a) b)

Fig.2. The plot of interaction potential for vortex moving between two magnetic defects with pinned vortices. Vortex moves antiparallel to x-axis.

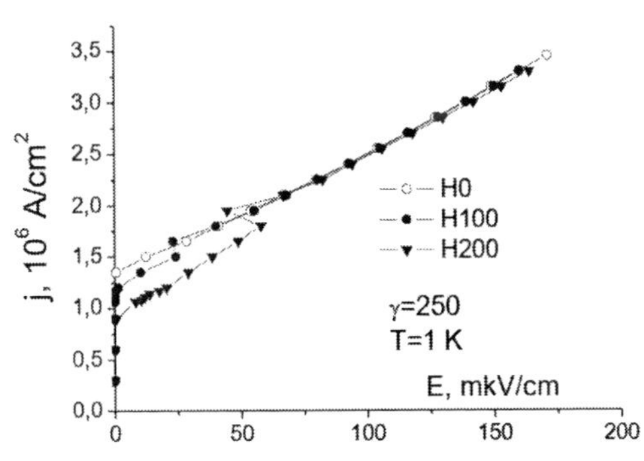

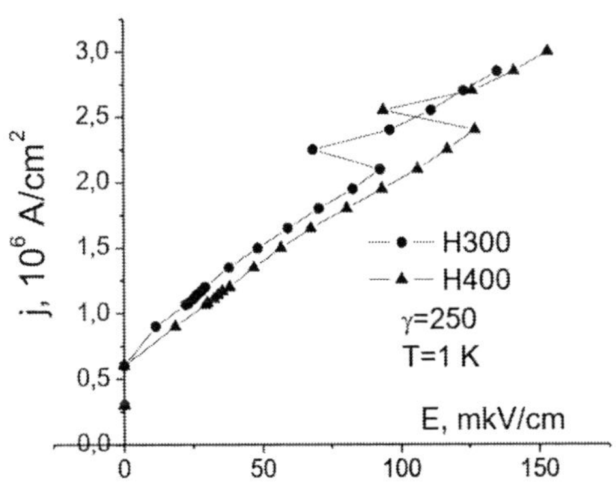

Fig.3. Current-voltage characteristics for superconductor with high anisotropy (a=250 nm).

3. Conclusion

In conclusion, a description of ferromagnetic nanoparticles with variable magnetization as internal pinning centers was introduced into three-dimensional model of layered high-temperature superconductor. The magnetization of such a superconductor with self-field of transport current under application of dc magnetic field was considered, the current-voltage characteristics were obtained. It was shown that the S-type nonlinearity obtained previously in two-dimensional model remains in more realistic three-dimensional case.

Acknowledgements

Research was done with the financial support of RFBR under grant № 15-02-02764.

References

[1] Snezhko A, Prozorov T, Prozorov R 2005, Phys. Rev. B 71 024527
[2] Chenguang Huang, Huadong Yong, Youhe Zhou 2014, Acta Mechanica Solida Sinica, Vol. 27, No. 1
[3] Kashurnikov V A, Maksimova A N, Rudnev I A 2014, Physics of the Solid State 56 (5), 894-911
[4] Marta Z. Cieplak, Adamus Z, Konczykowski M, Zhu L Y, Cheng X M, and Chien C L 2013, Phys. Rev. B 87, 014519
[5] Visani C, Metaxas P J, Collaudin A, Calvet B, Bernard R, Briatico J, Deranlot C, Bouzehouane K, and Villegas J E 2011, Phys. Rev. B 84, 054539
[6] Marta Z. Cieplak, Zhu L Y, Adamus Z, Konczykowski M, and Chien C L 2011, Phys. Rev. B 84, 020514(R)
[7] Zhu L Y, Marta Z. Cieplak, and Chien C L 2010, Phys. Rev. B 82, 060503(R)
[8] W.E. Lawrence, S. Doniach, in *Proceedings of LT 12, Kyoto*, 1970, edited by E.Kanda (Keigaku, Tokyo, 1971), 361
[9] Odintsov D S, Rudnev I A, Kashurnikov V A 2006, *JETP,* vol. 103, no. 1, pp. 66-76
[10] Odintsov D S, Rudnev I A, Kashurnikov V A 2007, *JETP*, vol. 105, pp. 253-255
[11] Rudnev I A, Odintsov D S, Kashurnikov V A 2008, *Phys. Lett. A*, vol. 372, pp. 3934-3936
[12] Kashurnikov V A, Maksimova A N, Rudnev I A 2014, *Physics of the Solid State*, vol. 56 (5), pp. 894-911
[13] Kashurnikov V A, Maksimova A N, Rudnev I A 2014, *JETP Letters*, vol. 100, no. 7, pp. 439–445

The analysis of the influence of the intellectual capital on the results of the commercial activity of financial institutions

Oleg Shkolik, Larisa Chirkova, Polina Chirkova
Ural Federal University, 620002, 19 Mira street, Ekaterinburg, Russia
E-mail: o.shkolik@mail.ru

Abstract. Developing (underdeveloped) countries are territories of slow economic growth (catch-up growth). Perspectives of their economic growth largely depend on developing and introducing financial and technological innovations in the sphere of the financial markets. The level and quality of those innovations should enable provision of faster growth of the financial sector of the national economy by rising stability and effectiveness of the financial institutions. Powerful and stable financial sector is the basic element for attracting investments and upsurge of liquidity in the economic system of a developing country that aims to have developed economy. Intellectual capital is the most important of the fundamental factors of production in the financial sphere. It is a catalytic element of the process of the economic development. From this position, the researchers' collective develops and presents a mathematical model which characterizes the connection between the intellectual capital and financial results of the commercial activity of financial institutions. The model is applied in the analysis of the activity of financial institutions that are part of the EEU.

1. Introduction

According to an approach firmly established in science and practice, hard assets and financial assets are reckoned among major production factors. Therefore, tangible assets are traditionally considered to be the most important resources of any business, since they form company's competitive advantages and create its added value.

We uphold an opposite position stating that at this particular historical moment intellectual capital matters the most for an efficient functioning of an economic system and increase of quality level of goods and services production.

The current epoch is a period of transition to a postindustrial society, characterized by the highest level of priority given to intellectual capital instead of hard assets or financial capital. Due to scientific and technical progress and a major change of a social consumption focus, a new production method was introduced into the global economic system. High dynamics of IT field and consequential high consumption of digital gadgets and products have gradually shifted emphasis towards intangible production. In the given circumstances intangible capital becomes a dominant among other production factors.

Consequently, issues of interpretation, use and commercialization of intellectual capital become of high importance.

Content from this work may be used under the terms of the Creative Commons Attribution 3.0 licence. Any further distribution of this work must maintain attribution to the author(s) and the title of the work, journal citation and DOI.
Published under licence by IOP Publishing Ltd

The main function of intellectual capital is to boost significantly company's growth of revenue basing on creation and implementation of required by the company systems of knowledge, objects and relations which, in turn, guarantee highly efficient business.

Usage of intellectual capital defines the pace and character of production technologies and products renewal which later becomes the main competitive advantage on the market. In fact, intellectual capital is a system of company's capital consistent intellectual advantages on the market. It is a catalytic element for the process of economic development.

2. Empirical basis and scientific hypotheses

A limited number of publications are devoted to research of intellectual capital as a dominant production factor. These publications widely observe theoretical and practical aspects of intellectual capital [1, 3]. These publications include research works by Bontis, Brennan, Doyle, Sveiby, Chen and other scientists. These papers study issues of intellectual capital structure, quantitative research and intellectual capital indicators, econometric evaluation of levels and methods of influence of intellectual capital on a company's total performance.

One of the main scientific restrictions of these studies is that they were conducted based on statistics of developed countries. As it follows, a very few number of applied scientific research is dedicated to research of backward countries' experiences. This situation is rooted to the fact that these countries are not only areas of slow economic development, but also they are characterized by delay in development of modern knowledge and business innovative activity.

The outcomes of research works support the thesis on a positive influence of intellectual capital on current business operations and resulting company's total performance. Nevertheless, these results also demonstrate impossibility to develop a unified methodology of intellectual capital measuring and its influence on company's performance evaluation procedures.

The results of these studies also highlight the fact that in a developed economy the level of intellectual capital influence on company's total performance is higher than the one in a developing economy. Yet it is the backward countries that require the most the research of all aspects of intellectual capital and development of its evaluation methods.

Moreover, an influence of intellectual capital on company's financial results is characterized by synergy effect of its structural elements.

In the context of modern economy quantitative methods of intellectual capital research define a degree of determinateness of company's financial results and its intellectual capital value.

Analysis of the conducted works' empirical base has demonstrated that there are methodological obstacles in the process of formulation of analytical model of intellectual capital. First off, there is no one opinion on the structure of intellectual capital. During its structural modulation the researchers include in it different constituent elements which differ by quantity as well as by quality. There are up to five components as a maximum. There are at least two components as a minimum. A three-components-model of intellectual capital came into the most widespread acceptance [1, 9]. One more methodological obstacle in intellectual capital modeling is an absence of a unified criterial approach to estimated figures which demonstrate intellectual capital's cost performance. In sum, a wide range of quantity, quality and cost indicators is used in research activities [2].

Our position tends to the concept of a three-component structure of intellectual capital. Specifically, we believe it is reasonable to distinguish the following structural elements of intellectual capital:

1. Human capital - company's ability to use the whole of its employees' intellectual capacities in order to get economic benefits. Employees' intellectual capacities include their inherent abilities based on knowledge, skills and expertise.

2. Relationship capital - company's ability to use resources which are related to company's external relationships aimed at gaining economic benefits. It includes such resources as

customers' loyalty, backward and distributor linkages, achieved business cooperation, goodwill etc.

3. Structural capital - company's ability to generate economic benefits by using independent results of employees' intellectual activity, where such results become company's property. This includes such self-standing items as patents, copyrights, trademarks and other non-material components considered by IFAC as intangible assets.

It should be noted, that application of the given structure is not unconditional and it should vary depending on branch, national and legal specialties.

3. Formulation of the problem and of the research methodology

The results of most studies in the field of intellectual capital are received basing on a major data extract. Therefore, existing evaluation models of intellectual capital and its influence on company's total performance have a wide application range, but they cannot be applied in specific (self-standing) economy sectors. A financial market can be seen as such specific economy sector. Disintegration of this sector is determined by specific production features. Firstly, no material product is produced at this market. Secondly, exactly intellectual production is the basis of production activity. In other words, in this economic field a non-material financial service is being provided.

The financial service is provided in form of investment consulting, investment management, that is via intellectual activity. Tangible assets fulfill an auxiliary function here; therefore intellectual capital is the most important production factor at this market. Results of other studies support the authors' opinion. Works by Peter Doyle confirm that in the structure of financial sector assets the main part is taken by structural capital (intangible assets) and relationship capital (brand, goodwill etc.) (table 1.) [5]. It should be stressed that in financial sector an amount of tangible assets is even less than in IT field (the field of absolute intellectual production).

Table 1. The assets structure of companies in different industries, %

	Tangible assets	Intangible assets	Relationship capital
Retail	70	15	15
Heavy industry	70	25	5
Municipal or urban engineering	70	30	0
Mechanical engineering	50	20	30
Pharmaceutics	40	50	10
Food industry & agriculture	40	5	55
Information technology	30	50	20
Financial markets	20	50	30

The financial market demands high level of technological effectiveness of operation processes carried out by financial and investment institutions. The technology of production process is provided by intellectual program products. In this context it is referred to products providing operational interaction of a financial service supplier and its consumer. Consequently intellectual aspects of the financial service production are implemented not only by means of direct intellectual activity, but also through the sale of intellectual program products.

4. Research subject, data being used, and construction of models

As research subject there were considered financial institutions and investment companies functioning at the financial market of the Eurasian Economic Union. It is defined by several factors. First of all, this international union unites developing (backward) countries. Second of all, in these countries the operation of the financial market is synchronized by economic and legal characteristics. Consequently, the statistical data of these countries is comparable and the models offered by us are applicable to this whole macroeconomic agglomeration. As statistical data there were taken official analytic reports and databases of regulatory and supervisory authorities of EEU member countries [4, 6, 7, 8].

The authors offer two models. The first model allows to define the value of an investment company's intellectual capital:

$$A_{IC} = HC + \frac{IA + m}{T_{IC}} + \frac{B + GW}{T_C} \tag{1}$$

Where A_{IC} - value of investment company's intellectual capital;

HC - value of human capital equal to total costs on personnel, including wages, monetary and physical stimulating payments (bonuses, rewards etc.);

IA - value of intangible assets shown in IFRS;

m - costs for modernization and maintenance of intangible assets paid in the current period (month, quarter, year) which enhance the increase of assets' accounting value;

T_{IC} - residual amount of time periods (months, quarters, years) of use of intangible assets' objects;

$B + GW$ - appraised value of objects comprising relationship capital (brand, goodwill etc.);

T_C - number of years of company's existence since foundation or rebranding.

The second model allows to evaluate the extent of intellectual capital's influence on investment company's total performance. This model is based on calculation of operating income received from the use of intellectual capital in investment activity.

$$OI_{IC} = V_B k_B + R_t k_t + R_d - A_{IC} \tag{2}$$

Where OI_{IC} - operating income generated by use of intellectual capital;

V_B - volume of financial turnover of broker operations with the use of intellectual program products;

k_B - investment company's commission rate on broker operations;

R_t - financial result on trust management investment operations;

k_t - investment company's commission rate on trust management investment operations;

R_d - financial result from investment operations with own resources.

Using these models, we have calculated a value of operating income per value of intellectual capital involved in production activity. In this case the operating income value is equal to the value of earnings before interest and taxes (EBIT) in terms of intellectual production:

$$EBIT_{IC} = (V_B k_B + R_t k_t + R_d) - \left(HC + \frac{IA + m}{T_I C} + \frac{B + GW}{T_C}\right) \tag{3}$$

Besides that, in the frames of this research a Basic earning power (BEP) ratio was calculated by the authors, which characterizes profitability of intellectual capital use:

$$BEP = \frac{EBIT_{IC}}{A_{IC}} \tag{4}$$

5. Empirical results

As a result of calculations it was found that an average value of operating income from intellectual capital use in investment companies operation activity within the past 5 years on EEU financial market ranged from 78,8% to 93,5% from an investment company's total performance. Summing up, one can state that intellectual capital makes for over three parts of investment companies' total performance. It was also found that the financial market is an economic branch where intellectual capital possesses one of the highest profitability levels. It is also proved by the fact that in the frames of this study the maximum value of the Basic earning power ratio exceeded 1000 units.

6. Conclusion

In summary, we have presented two models which allow not only to evaluate the level of intellectual capital influence on financial institutions total performance, but also to define investment companies' intellectual capital value. The authors have confirmed and proved the thesis stating that intellectual capital is a major production factor in modern economic surroundings. Moreover, the calculations have demonstrated that intellectual capital is the most important production factor in the field of investment activity on the financial market.

Acknowledgments

The work was supported by Government of the Russian Federation, Act 211 contract 02.A03.21.0006.

References

[1] Bontis N 1998 *Management Decision* 2 **36** 63-76
[2] Brennan N and Connell B 2000 *J. of Int. Capital* 3 **1** 206-240
[3] Chen M 2004 *J. of Nat. Taiwan U. of Science and Tech.* 1 **9** 12-35
[4] 2015 *Investments in the Kyrgyz Republic for 2010-2014* (NSC Kyr. Rep.) p 224
[5] Doyle P 2008 *Value-based Marketing* (John Wiley & Sons) p 382
[6] 2016 *Statistical Bulletin of the Bank of Russia* 1 **272** p 305
[7] 2015 *Statistical Bulletin of the National Bank of Kazakhstan* 1 **242** p 146
[8] 2016 *Statistical Bulletin of the National Bank of Kazakhstan* 1 **254** p 158
[9] Sveiby K 2001 *Meth. for Measuring Intangible Assets* (http://www.sveiby.com.au/BookContents.html)

Neutron monitor generated data distributions in quantum variational Monte Carlo

A S Kussainov[1,2] and N Pya[3]

[1] Physics and Technology Department, al-Farabi Kazakh National University, Almaty, Kazakhstan
[2] National Nanotechnology Laboratory of Open Type, al-Farabi Kazakh National University, Almaty, Kazakhstan
[3] Department of Mathematics, School of Science and Technology, Nazarbayev University, Astana, Kazakhstan

E-mail: arman.kussainov@gmail.com

Abstract. We have assessed the potential applications of the neutron monitor hardware as random number generator for normal and uniform distributions. The data tables from the acquisition channels with no extreme changes in the signal level were chosen as the retrospective model. The stochastic component was extracted by fitting the raw data with splines and then subtracting the fit. Scaling the extracted data to zero mean and variance of one is sufficient to obtain a stable standard normal random variate. Distributions under consideration pass all available normality tests. Inverse transform sampling is suggested to use as a source of the uniform random numbers. Variational Monte Carlo method for quantum harmonic oscillator was used to test the quality of our random numbers. If the data delivery rate is of importance and the conventional one minute resolution neutron count is insufficient, we could always settle for an efficient seed generator to feed into the faster algorithmic random number generator or create a buffer.

1. Introduction

Though hardly conceivable today, one may run into a situation when a random number could not be generated sequentially on a conventional computer. Specific statistical or computational tasks may lead to decision to use the random number look-up tables instead of pseudorandom software generators [1]. These tables have a long and successful history of applications. Tippett [2] is credited as the inventor of the random number table in 1927.

The sources for these tables are fairly different, and while ones came from the census data, the others are produced from the previously computed tables, like logarithmic, see Fisher and Yates [3]. Regardless of the order or location in the tables, these numbers should have the desired properties of the local or global randomness [4]. The mechanisms of sampling these numbers from the table are parallel topics of discussion.

One may also come across a situation when the iterative processes behind the regular pseudorandom software generator are not available at the very low level of data acquisition and processing, or inaccessible because of the rate, slow or fast, these data are sampled. In these cases we may want to take an advantage of the prerecorded data pattern, or equivalently, to create a buffer of the preregistered numbers [5], [6].

Content from this work may be used under the terms of the Creative Commons Attribution 3.0 licence. Any further distribution of this work must maintain attribution to the author(s) and the title of the work, journal citation and DOI.
Published under licence by IOP Publishing Ltd

This approach is still intrinsically related to hardware random number generators or so called true random number generator [7]. These are devices that generate random numbers from the real physical processes. It is assumed though that the data are being registered in a stable state of a system, away from any transient phenomena and completely unpredictable. The last condition should be a subject of an experimental test.

In our study, we have selected Markov chain Monte Carlo (MCMC) algorithm [11] as a fine example of the quantum Monte Carlo method to test the random number distributions generated from an experiment. MCMC algorithm is used to approximate the ground state of a quantum system using variational method by calculating an integral related to the average value of energy of a system.

MCMC algorithm proceeds by randomly moving about the space spanned by the integral. It is restricted in the choice of the next step by the certain set of rules called importance sampling. In basic case, we need two individual random number distributions to move a single walker around, where the walker is an object following the single chain of successive positions selected in accordance with the previously mentioned rules. The first one is the normal distribution driving our walker toward the region of maximum probability defined by the trial set of wavefunctions and another one is a uniform random number distribution used to sample the regions with low-density probability along the path.

We chose to use the neutron counts data from the ground–based neutron detectors. These are designed to measure the secondary products of collisions of the high-energy charged particles from the outer space with the nuclei of atmospheric gases [8].

The registered data have a rich spectral composition modulated by the Earth rotation, solar cycles and multiple other events. It also has an intensive random component we intend to use.

2. Methods

We have considered one and three dimensional quantum harmonic oscillators as a quantum system to model by the MCMC method [9].

One dimensional, time independent, Schrodinger equation is given by the following expression

$$-\frac{\hbar}{2m}\frac{d^2\varphi(x)}{dx^2} + \frac{1}{2}mk^2x^2\varphi(x) = E\varphi(x) \tag{1}$$

where k is the force constant, $\hbar$ is the reduced Planck constant and m is the mass of the oscillating particle. If we know the complete set of N eigenfunctions from the following superposition notation

$$\Phi(x) = \sum_{i=0}^{N-1} c_i\varphi_i(x) \tag{2}$$

then, assuming the orthonormality of the eigenfunctions, the average value of energy $\langle E\rangle$ of the oscillator could be calculated from the following expressions

$$\langle E\rangle = \frac{\int_{-\infty}^{+\infty} dx\Phi(x)^* H\Phi(x)}{\int_{-\infty}^{+\infty} dx\Phi(x)^*\Phi(x)} = E_0 + \frac{\sum_{i=0}^{N-1} |c_i|^2 (E_i - E_0)}{\sum_{i=0}^{N-1} |c_i|^2} \tag{3}$$

where H is the Hamiltonian given by the left part of equation (1) and E_0 is the exact value of the ground state energy.

Variational Monte Carlo method uses equation (3) as a starting point. Replacing $\Phi(x)$, the unknown superposition of the eigenfunctions, by the trial wavefunction $\Phi_{T,a}(x)$, we are starting to vary the parameter α. If we were right in our choice of the trial wave function, the calculated averaged energy, as a function of α, will have a minimum for the ground state value and we could reconstruct the corresponding eigenfunction.

One of the distinct features of this method is that we are not sampling the whole configuration space from negative to positive infinity indiscriminately but rather traversing it in the manner prescribed by the Metropolis–Hastings algorithm, which is a subcase of MCMC. The targeted wave function's tails, though stretching from negative to positive infinity, are vanishing to zero pretty fast in just a few steps away from the origin, and its overall shape is similar to the normal distribution. The Metropolis–Hastings algorithm samples the integral's space, guided by the weight function $\omega(x)$, as given in the equation (4)

$$\langle E\rangle = \int_{-\infty}^{+\infty} dx\omega(x)E_L(x), \text{ where } \omega(x) = \frac{|\Phi_{T,\alpha}(x)|^2}{\int_{-\infty}^{+\infty} dx' \, |\Phi_{T,\alpha}(x')|^2} \text{ and } E_L(x) = \frac{H\Phi_{T,\alpha}(x)}{\Phi_{T,\alpha(x)}} \quad (4)$$

where $\Phi_{T,a}(x)$ is the trial wavefunction of our choice.

That is we place the n walkers randomly and uniformly across the selected region, not far away from the origin, at coordinates $\{x_n\}_t$, where index t is used for the current set of coordinates. The next set of the walkers' coordinates $\{x_n\}_{t+1}$ is determined by the following rules for each n

$$\begin{gathered} x_{t+1} = x_t + \delta, \\ \text{if } \tfrac{\omega(x_{t+1})}{\omega(x_t)} \geq 1, \text{ the step is accepted,} \\ \text{if } \tfrac{\omega(x_{t+1})}{\omega(x_t)} < 1, \text{ accepted only if it is larger than a random number from } [0,1] \end{gathered} \quad (5)$$

here δ is the step's size which is determined based on the problems conditions and in general is subject to changes depending on the acceptance ratio. If this step is accepted the local energy $E_L(x)$ is calculated at this new walker's coordinates and contributes to the integral in equation (4). All the walkers are sampling the integral in equation (4) independently and may cross their paths.

If, for one dimensional case, the trial function is chosen to be $\Phi_{T,a}(x) = exp(-\alpha x^2)$, then the value of $\langle E\rangle$ could be calculated according to the formula

$$\langle E\rangle = \frac{1}{counts} \sum_{i=1}^{counts} E_L(x_i) \text{ ,where } E_L(x) = \alpha + x^2(0.5 - 2\alpha^2) \quad (6)$$

Here, the *counts* are the number of accepted steps across the all walkers' trajectories. Expression for E_L in the equation (6) was derived from the equation (4) and explicit notation for the trial wave function $\Phi_{T,a}(x)$. We also assume that $\hbar = m = k = 1$.

Three dimensional case is slightly different by the form of the Schrodinger equation we need to solve. Assuming the radial symmetry, it states that

$$-\frac{\hbar}{2m}\left[\frac{d^2\varphi(r)}{dr^2} + \frac{2}{r}\frac{d\varphi(x)}{dr}\right] + \frac{e^2}{r}\varphi(r) = E\varphi(r) \quad (7)$$

and

$$E_L(r) = -0.5\left[\alpha^2 - 2\alpha/r\right] - 1/r \quad (8)$$

In fact, the computational part of the problem may be directly constructed from the one dimensional case. All we need is to represent a single three dimensional step as a sequence of three one dimensional steps in each of the three dimensions.

3. Results

For our purposes, we have selected the 38000 minutes long data piece. It was simultaneously registered by the 6 channels of unit A of the neutron monitor, at the Tien-Shan mountain high-elevation research station. This monitor is located 3340 meters above the sea level. The data were registered for about 633 hours, starting at 00.00 UTC, January 23rd, 2016 [13].

As we said before, we need two individual random numbers distributions to move a single walker around. The first one is the normal distribution and another one is a uniform random number distribution used to sample the regions with low-density probability, see equations (4)-(5).

A piecewise cubic spline fit with breaks (knots) every 250 points implemented by the `splinefit` function in `Octave` package on `Debian`, has been used to detect characteristic variations in the intensities of the cosmic rays and to subtract them as a baseline from the raw data thus producing the Gaussian white noise, see Fig. 1 (a)-(c).

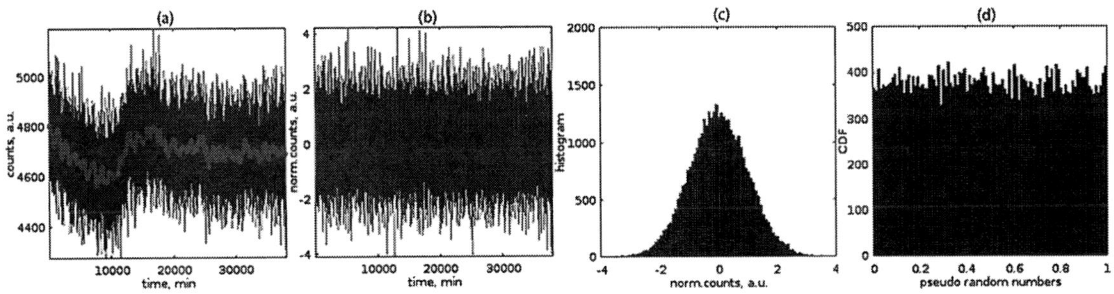

Figure 1. Original data (a) fitted with the `splinefit` function, extracted white noise (b), histogram (c) of its spectral composition and cumulative distribution function (d) to produce a uniform random distribution.

The standardization procedure, see equation (9), further transforms the filtered noise component to a new one with zero mean and unit variance, which is our final normal distribution, see Fig. 1 (b)-(c)

$$x' = (x - \bar{x})/s, \tag{9}$$

where s and $\bar{x}$ are the standard deviation and mean value of the extracted noise before standardization procedure.

It is known from the probability integral transform that data values that are generated from any given continuous distribution, normal in our case, can be transformed to random deviates with a uniform distribution on the interval [0,1] as $U = F(x)$, where $F(x)$ is the standard normal cumulative distribution function. We used this to produce a uniform random distribution plotted on Fig. 1 (d).

As a reference simulations we have used a pseudo–random generator provided by the `g++` compiler on `Debian` Linux OS. In this reference case, Box-Muller transform [12] was used to go from a uniform random number distribution to a normal random distribution. In order to compare these data with our experimentally generated random numbers distributions we need the number of walkers to be the same in both cases.

If the walkers are placed not far away from the origin, we need no more than 1000 steps, including thermalization stage, to reach and amply sample the area of maximum probability.

All channels are highly correlated in general. The sources of extracted noise in each individual channel may be different from just a local background radiation and other independent processes in the registration hardware and could correlate as well. Fitting the data with splines may be not good enough to get rid of the possible statistical imprints of the common registered events.

Nevertheless, we obtained a nearly perfect match, see Fig. 2, when we pulled the new values of the random normal and random uniform deviates from all available channels, in a successive order, one by one from each channel. We have got this match from as many as 17 walkers.

As one can see, the quality of experimental data, after few simple transformations, is reasonably good to bring the walker even from the remote location to the maximum probability point, see Fig. 2 (a). The data were good enough not to introduce a special procedure to treat the outliers in the raw data.

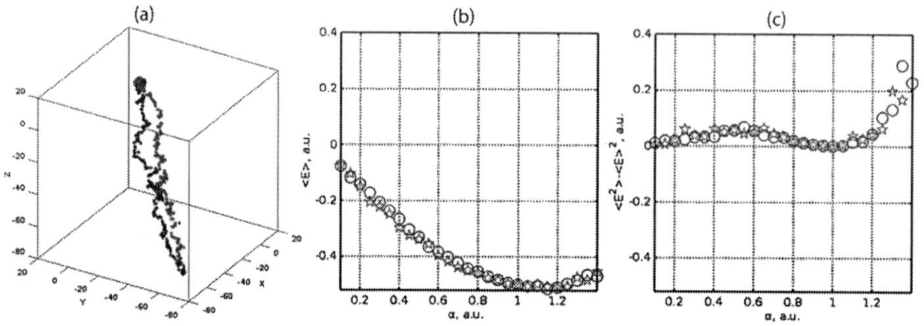

Figure 2. The 3D trajectories (a) of individual walkers driven to the maximum probability region by a regular pseudorandom numbers (red), and by a random numbers from our experimental data (blue).Computed average energy values (b) and computed values of the variance (c), where blue open circles stand for a regular pseudorandom numbers and red pentagrams stand for an experimental data.

The next two plots to the right from trajectories' plot, represent the average energy and its variance versus variation parameter $\alpha = 0$. As expected, we can clearly see the minimum at $\alpha = 0.5$ because the E_L loses its α dependence at this point, see equations (6) and (8). This value of α corresponds to the zero variance value on the right plot which pinpoints the ground state in our problem. The slight differences observed on the picture are related to the quality of the random numbers used in our simulations.

Supporting our MCMC simulations, the one-sample Kolmogorov-Smirnov test `kstest(x)` returns a positive test decision for the null hypothesis that our data indeed come from a standard normal distribution. The formal application of the full range of normality tests, as well as the tests of global and local randomness, has been left for a further study.

4. Conclusions

We have successfully used the neutron monitor data as a source for the normal and uniform random number generators to drive a MCMC method. The desired stochastic component was extracted by fitting the data with splines and subtracting this fit from the raw data. Further scaling of our data to zero mean and variance of one was sufficient to obtain a stable standard normal random variate. Cumulative distribution function was used as a source of the uniform random numbers.

We are able to obtained the exact values of the variational parameter α and local energy value E_L required to pinpoint the ground state of the three dimensional quantum harmonic oscillator. The negligible difference in the calculated values of the local energy for different α values as compared to a reference case may be attributed to the multiple factors related not only to the quality of the noise extraction but to a thorough analysis of this noise sources in different channels. For some problems, as ours, this quality is good enough or may be improved by additional treatment of the raw data. Distributions under consideration pass additional goodness of fit test for normality.

Acknowledgments
This study was supported by grant #3824/GF4 provided by the Science Committee at the Ministry of Science and Education of Republic of Kazakhstan to the principal investigator at the National Nanotechnology Open Laboratory. We also would like to thank Prof. Yelena White, from East Georgia State College, Swainsboro, GA, USA, for editing the manuscript.

References
[1] Mitchell R L and Stone C R 1977 *IEEE Transactions on Computers* **C-26** 1006-08
[2] Tippett L H C 1927 *Random sampling numbers. no. 15 Tracts for Computers* (Cambridge: Cambridge Univ.Press) reprinted in 1952
[3] Fisher R A, Yates F 1948 *Statistical Tables for Biological, Agricultural and Medical Research* 3rd ed. (London: Oliver&Boyd)
[4] Kendall M G, Smith B B 1938 *Journal of the Royal Statistical Society* **101** 147-66
[5] Kravtsov K S, Radchenko I V, Kulik S P, Molotkov S N 2015 *Journal of the Optical Society of America* B **32** 1743
[6] Self J, Macke M C 2015 *ArXiv e-prints* arXiv:1512.07303
[7] Dynes J F, Yuan Z L, Sharpe A W and Shields A J 2008 *Appl. Phys. Lett.* **93** 031109
[8] Simpson J A 2000 *Space Science Reviews* **93** 11-32
[9] Foulkes W M C 2001 *Rev. Mod. Phys.* **73** 33
[10] Dodge Y, Cox D, Commenges D 2003 *The Oxford Dictionary of Statistical Terms* (Oxford: The Oxford University Press)
[11] Metropolis N, Rosenbluth A W, Rosenbluth M N, Teller A H, Teller E 1953 *Journal of Chemical Physics* **21** 1087-92
[12] Box G E P, Muller M E A 1958 *The Annals of Mathematical Statistics* **29** 610-11
[13] Shepetov Yu V 2014 *The Tien-Shan Mountain Station's Database* (Moscow: Institute of Terrestrial Magnetism, Ionosphere and Radio Wave Propagation named after Nikolay Pushkov of the Russian Academy of Sciences) http://cr29.izmiran.ru/vardbaccess/frames-vari.html

Dynamical Analysis of Stock Market Instability by Cross-correlation Matrix

Tetsuya Takaishi
Hiroshima University of Economics, Hiroshima 731-0192, JAPAN
E-mail: tt-taka@hue.ac.jp

Abstract. We study stock market instability by using cross-correlations constructed from the return time series of 366 stocks traded on the Tokyo Stock Exchange from January 5, 1998 to December 30, 2013. To investigate the dynamical evolution of the cross-correlations, cross-correlation matrices are calculated with a rolling window of 400 days. To quantify the volatile market stages where the potential risk is high, we apply the principal components analysis and measure the cumulative risk fraction (CRF), which is the system variance associated with the first few principal components. From the CRF, we detected three volatile market stages corresponding to the bankruptcy of Lehman Brothers, the 2011 Tohoku Region Pacific Coast Earthquake, and the FRB QE3 reduction observation in the study period. We further apply the random matrix theory for the risk analysis and find that the first eigenvector is more equally de-localized when the market is volatile.

1. Introduction

The stock market is a complex system that undergoes unstable periods that result in financial crises in some cases. Measuring systemic risk is an important task to monitor current market status and possibly to avoid a future financial crisis. In financial crises, many stocks are interconnected and move collectively. The level of interconnectedness can be measured by cross-correlations between stocks, and there are a variety of works on cross-correlations that include the random matrix theory (RMT)[1, 2, 3, 4, 5] and the principal component analysis (PCA)[6, 7, 8, 9]. In this study, we calculate cross-correlations between stocks traded on the Tokyo Stock Exchange from January 5, 1998 to December 30, 2013 and apply the PCA and the RMT to analyze the dynamical properties of cross-correlations. In particular, we focus on the market instability and investigate when the market was volatile during the study period.

2. Cross-correlation matrix

We analyze the daily closing price data of stocks traded on the Tokyo Stock Exchange from January 5, 1998 to December 30, 2013, which corresponds to 3932 working days. We choose 366 stocks listed on the Topix 500 index.

Let $r_i(t)$ be a return for stock i ($i = 1, ..., N$) at time t ($t = 1, ..., T$) defined by the log-price difference as

$$r_i(t) = \ln p_i(t) - \ln p_i(t-1), \tag{1}$$

Content from this work may be used under the terms of the Creative Commons Attribution 3.0 licence. Any further distribution of this work must maintain attribution to the author(s) and the title of the work, journal citation and DOI.
Published under licence by IOP Publishing Ltd

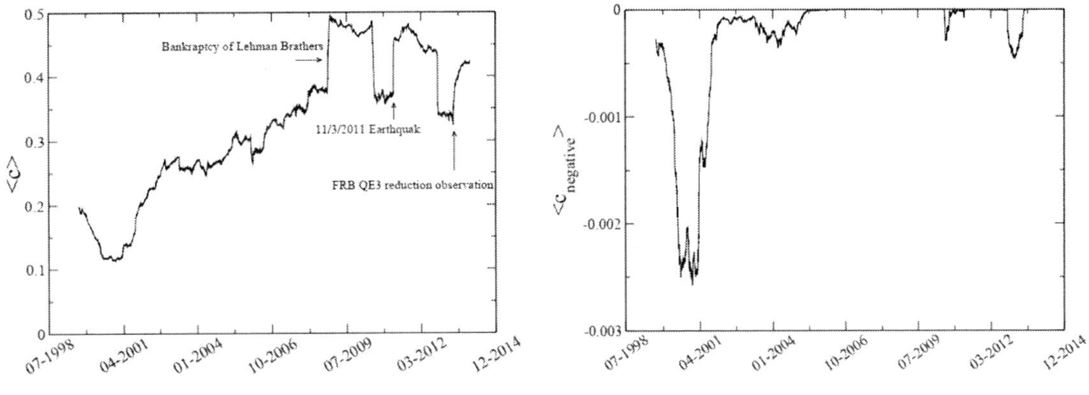

Figure 1. (Left) Average off-diagonal elements of the cross-correlation matrix. (Right) Average negative off-diagonal elements of the cross-correlation matrix. Each average is taken over 400 days in the rolling window.

where $p_i(t)$ is the price for stock i on day t. We also define the normalized return $m_i(t)$ as

$$m_i(t) = \frac{r_i(t) - \langle r_i \rangle}{\sigma_i}, \tag{2}$$

where $\langle ... \rangle$ indicates the time series average and σ_i is the standard deviation of r_i.

Using the normalized return $m_i(t)$, an equal-time cross-correlation matrix is calculated as $c_{ij} = \langle m_i m_j \rangle$, where an average, i.e. $\langle ... \rangle$, is taken over a period of the rolling window. In this study, we consider a rolling window of 400 working days, which roughly corresponds to two years. By definition, the elements of the cross-correlation matrix are restricted to $-1 \leq c_{ij} \leq 1$.

Fig.1(Left) shows the dynamical evolution of the average off-diagonal matrix element $\langle c \rangle$ given by

$$\langle c \rangle = \frac{2}{N(N-1)} \sum_{i>j}^{N} c_{ij}, \tag{3}$$

where $N = 366$. From the figure, we recognize that there exist three points where $\langle c \rangle$ increases abruptly. According to the historically observed events, these points correspond to the bankruptcy of Lehman Brothers, the Tohoku region pacific coast earthquake on 11/3/2011, and the FRB QE3 reduction observation as indicated in the figure. In Fig.1(Right), we also show the average of negative off-diagonal elements,

$$\langle c_{negative} \rangle = \frac{2}{N(N-1)} \sum_{i>j, c_{ij}<0}^{N} c_{ij}. \tag{4}$$

Notably, in the recent years, the contribution of negative off-diagonal elements to the cross-correlation matrix becomes less than that around 2000. In particular, at volatile stages, negative off-diagonal components disappear and most stocks are positively correlated.

3. Dynamical behavior of Cumulative Risk Fraction

In order to further investigate the dynamical properties of cross-correlation matrices, we apply the principal component analysis (PCA). Billio *et al.*[6] suggested to use the PCA to quantify the systemic risk and introduced the cumulative risk fraction (CRF) as a risk measure. The PCA has also been used to measure the systemic risk[7, 8, 9]. To construct the CRF, we first compute

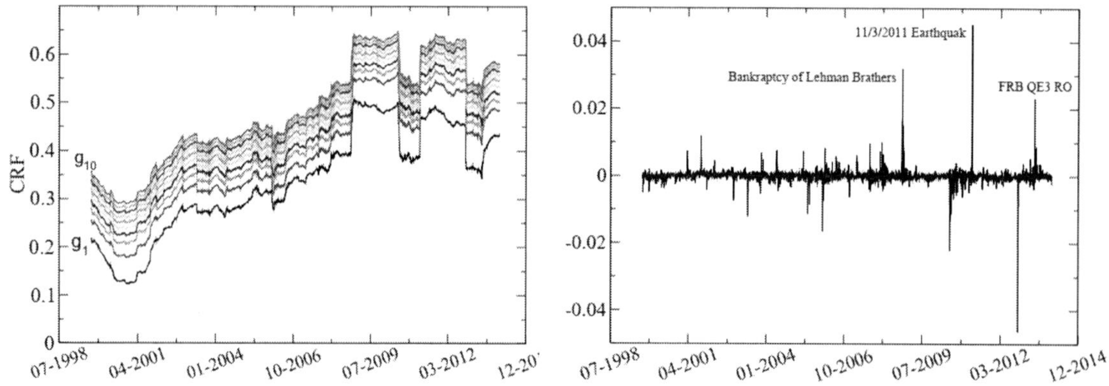

Figure 2. (Left) Time evolution of the CRF with 400-day rolling window and (Right) the change of g_1.

the eigenvalues of cross-correlation matrices, denoted as $\lambda_1, \lambda_2, \ldots, \lambda_N$, where all eigenvalues are sorted as $\lambda_1 > \lambda_2 > \ldots > \lambda_N$. Then, we calculate the CRF defined by[6]:

$$g_k = \frac{\omega_k}{\Omega}, \tag{5}$$

where Ω is the total variance of the system given by $\Omega = \sum_{i=1}^{N} \lambda_i$ and ω_k is the risk associated with the first k principal components given by $\omega_k = \sum_{i=1}^{k} \lambda_i$. The CRF quantifies the portion of the total variance explained by the first k principal components over the total variance[7]. Usually, the first few principal components explain most of the system variance. In the periods of financial crisis, many stocks are highly interconnected and their prices easily move together. In such periods, the CRF is expected to increase considerably because the system variance also increases.

Fig.2(Left) shows the time evolution of the CRF g_k for $k = 1, \ldots, 10$. We find that the structures of time evolution of g_k for $k = 1, \ldots, 10$ are very similar. This indicates that the first eigenvalue dominates in the CRF. We also find that the structure of the CRF resembles that of average off-diagonal elements of the cross-correlation matrix, and the CRF increases abruptly at the same points as observed in the CRF, that is, Fig.1(Left).

4. Changes of the Cumulative Risk Fraction

Zheng *et al.*[8] introduced the changes of the CRF to effectively quantify the points where the potential risk is high. The changes of the CRF is defined by

$$change_k(t) = g_k(t+1) - g_k(t). \tag{6}$$

The time evolution of $change_1(t)$ is presented in Fig.2(Right). We find that the change of the CRF shows pronounced positive spikes at the same three points where we observed the three abrupt increases in the CRF. Note that large negative spikes are artificially caused by the period of the rolling window.

5. Random Matrix Theory

Let $y_i(t)$ be an independent, identically distributed random variable with $i = 1, ..., N$ at time $t = 1, ..., T$. Then, we define the normalized variable:

$$w_i(t) = \frac{y_i(t) - \langle y_i \rangle}{\sigma_{y_i}}, \tag{7}$$

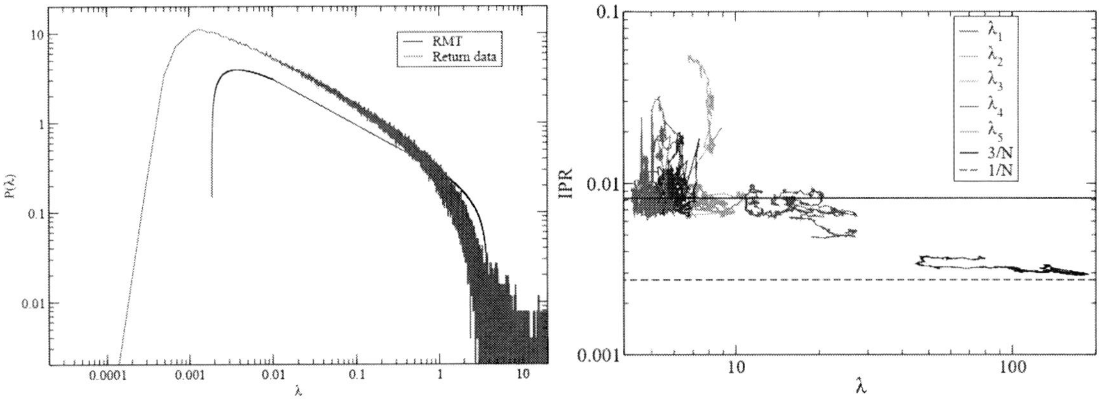

Figure 3. (Left) Eigenvalue distributions from the RMT and empirical return data. (Right) IPR versus eigenvalues for $\lambda_1, ..., \lambda_5$.

where σ_{y_i} is the standard deviation of y_i. The equal time cross-correlation between variables $y_i(t)$ is given by $W_{ij} = \langle w_i w_j \rangle$. The matrix W is called Wishart matrix. For $N \to \infty$ and $T \to \infty$ with $Q = T/N > 1$, an eigenvalue distribution of the matrix W is theoretically given by[10, 11]:

$$\rho(\lambda) = \frac{Q}{2\pi} \frac{\sqrt{(\lambda_+ - \lambda)(\lambda - \lambda_-)}}{\lambda}, \quad \lambda_\pm = 1 + \frac{1}{Q} \pm 2\sqrt{\frac{1}{Q}}. \tag{8}$$

Fig.3(Left) compares an eigenvalue distribution of the matrix W with that of the empirical cross-correlation matrix c, where $Q = 400/366 \approx 1.1$. The eigenvalue distribution of the empirical cross-correlation matrix differs considerably from the result from the RMT expectation. In particular, we find that for the empirical cross-correlation matrix, there exist many eigenvalues less than λ_- and larger than λ_+.

Another interesting entity, the inverse partition ratio that characterizes the eigenvectors, is defined by

$$IPR_k = \sum_{j=1,}^{N} (v_k^j)^4, \tag{9}$$

where v_k^j is the j-th component of the eigenvector for the k-th eigenvalue. In the RMT, the eigenvector components are de-localized and distributed as a Gaussian distribution. In such a case, the expectation of the IPR is $3/N$. On the other hand, when the eigenvector components are localized, for example, only one component has a non-zero value, the expectation of the IPR would be 1. There also exists another de-localized case in which all eigenvector components are equally de-localized, and in this case, the expectation of the IPR is $1/N$. Fig.3(Right) shows the IPR versus the eigenvalue $\lambda_k, k = 1, ..., 5$. The IPRs for $k = 2, ..., 5$ approach $3/N$, which is the RMT expectation. On the other hand, the IPR for $k = 1$, the largest eigenvalue, is near $1/N$, which means that the eigenvector components are equally de-localized. Fig.4 shows the time evolution of IPR_1 and the change of IPR_1, where the definition of the change of IPR_1 is the same as eq.(6). The IPR_1 seems to decrease and approach $1/N$ at the three points found in the cross-correlation and the CRF, although their signals are not very clear. This observation on the IPR_1 indicates that when the market is volatile, the largest eigenvalue components are more equally de-localized, which is different from the RMT expectation, and in such a case, the IPR approaches $1/N$.

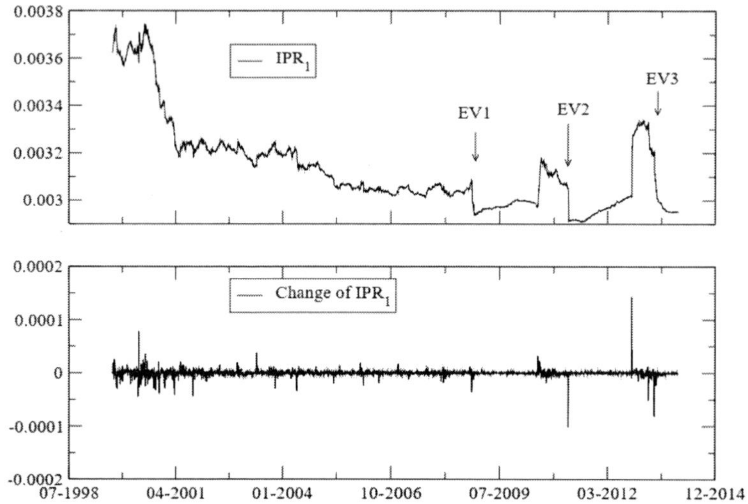

Figure 4. (Top) Time evolution of IPR and (Bottom) the change of IPR for λ_1. EV1,...,EV3 correspond to the three events: the bankruptcy of Lehman Brothers, the Tohoku Region Pacific Coast Earthquake, and the FRB QE3 reduction observation, respectively.

6. Conclusions

We have analyzed the cross-correlation matrices between 366 stocks traded on the Tokyo Stock Exchange from January 5, 1998 to December 30, 2013. We find that both the average off-diagonal elements of cross-correlation matrices and the cumulative risk fraction show abrupt increases at three points that correspond to three volatile stages of the Japanese stock market: the bankruptcy of Lehman Brothers, the Tohoku Region Pacific Coast Earthquake, and the FRB QE3 reduction observation. The change of the CRF also identifies these three points. From comparison with the random matrix theory, we find that the empirical cross-correlation matrix differs from the random matrix and, especially, the first eigenvector is more equally delocalized when the market is volatile. The cross-correlation matrices contain relevant information on the financial market status. By carefully analyzing the dynamical properties of the cross-correlations, we could monitor the risk that the financial markets confront.

Acknowledgement

Numerical calculations in this work were carried out at the Yukawa Institute Computer Facility and the facilities of the Institute of Statistical Mathematics. This work was supported by JSPS KAKENHI Grant Number 25330047.

References

[1] Plerou V, Gopikrishnan P, Rosenow B, Amaral L A N and Stanley H E 1999 *Phys. Rev. Lett.* **83** 1471-1474
[2] Laurent Laloux L, Cizeau P, Bouchaud J-P and Potters M 1999 *Phys. Rev. Lett.* **83** 1467
[3] Plerou V, Gopikrishnan P, Rosenow B, Amaral L A N and Stanley H E 2000 *Physica. A.* **287** 374-382
[4] Plerou V, Gopikrishnan P, Rosenow B, Amaral L A N, Guhr T and Stanley H E 2002 *Phys. Rev. E* **65** 066126
[5] Utsugi A, Ino K and Oshikawa M 2004 *Phys. Rev. E* **70** 026110
[6] Billio M, Getmansky M, Lo A W and Pelizzon L 2012 *J. Financ Econ.* **104** 535
[7] Kritzman M, Li Y Z, Page S and Rigobon R 2011 *J. Portf. Manag.* **37** 112
[8] Zheng Z, Podobnik B, Feng L and Li B 2012 *Sci. Rep.* **2** 888
[9] Ren F and Zhou W Z 2014 *PLoS ONE* **9(5)** e97711
[10] Edelman A 1998 *SIAMJ. Matrix Anal. Appl.* **9** 543
[11] Sengupta A M and Mitra P P 1999 *Phys. Rev. E* **60** 3389

Testing the shape of distributions of weather data

Ana L P Baccon[a] **and José T Lunardi**[a,b]

[a] Department of Mathematics and Statistics, State University of Ponta Grossa, Brazil
[b] School of Physics and Astronomy, University of Glasgow, UK

E-mail: `jttlunardi@uepg.br` (JTL)

Abstract. The characterization of the statistical distributions of observed weather data is of crucial importance both for the construction and for the validation of weather models, such as weather generators (WG's). An important class of WG's (e.g., the Richardson-type generators) reduce the time series of each variable to a time series of its residual elements, and the residuals are often assumed to be normally distributed. In this work we propose an approach to investigate if the *shape* assumed for the distribution of residuals is consistent or not with the observed data of a given site. Specifically, this procedure tests if the same distribution shape for the residuals noise is maintained along the time. The proposed approach is an adaptation to climate time series of a procedure first introduced to test the shapes of distributions of growth rates of business firms aggregated in large panels of short time series. We illustrate the procedure by applying it to the residuals time series of maximum temperature in a given location, and investigate the empirical consistency of two assumptions, namely i) the most common assumption that the distribution of the residuals is Gaussian and ii) that the residuals noise has a time invariant shape which coincides with the empirical distribution of all the residuals noise of the whole time series pooled together.

1. Introduction

The statistical characterization of the distributions of the relevant stochastic variables is of crucial importance in the construction and empirical validation of stochastic models, such as climate models. In particular, weather generators (WG's) play an important role in agrometeorological studies, such as in crop models, as well as in studies of hydrological changes, risk analysis, and as part of climate changes models [1, 2, 3, 4, 5, 6]. The main goal of a WG is to produce synthetic time series for weather variables that statistically resemble the actual ones from which the generator is calibrated. Several WG's were proposed in the literature, and several studies were performed to investigate their adequacy to describe weather variables in several specific sites (see, for instance, [3, 6, 7, 8, 9]).

An important class of WG's are the Richardson-type generators [1], which describes the time series of a weather variable as an autoregressive process for its residuals [10], which are obtained after the removal of the annual course of the mean and standard deviation of the observed values. Autocorrelations and cross-correlations among residuals of different variables are taken into account, and the stochastic terms associated with each variable is often assumed to be independent gaussian shocks (white noise). Besides, each variable is modeled conditionally to the precipitation status of each day (dry or wet). The parameters of the model, such as the distribution parameters for precipitation and for the lengths of dry and wet periods, as well as the cross and autocorrelations of the autoregressive process for the several variables must be

Content from this work may be used under the terms of the Creative Commons Attribution 3.0 licence. Any further distribution of this work must maintain attribution to the author(s) and the title of the work, journal citation and DOI.
Published under licence by IOP Publishing Ltd

estimated from the aggregated historical data. It is generally assumed in WG's that all years are statistically similar, and then several years of data (typically of the order of 20-30 years) for a given site are pooled together to the aim of parameter estimations (see [3] for an outline of Richardson-type generators and a description of the parameter estimation procedures).

An important issue in the process of validation of a WG is to investigate if the observed residuals are in fact distributed according to the assumptions of the underlying model. For instance, if the residuals are not normally distributed then a WG which rely on the assumption of normality may not be able to accurately describe extreme events in the corresponding variable. Statistical tests for normality consider in general data aggregated from several years [11]. However, aggregation of data within large periods may hide temporal changes in the residuals distribution, and the distribution of the pooled annual data (or data corresponding to a given month but with several years pooled together) may not represent the actual distribution of the residuals if its parameters, or the shape of this distribution, change with time.

In this work we are interested in testing the empirical consistency of the assumption that the shape of the residuals distribution of a weather variable is constant in time.[1] To this aim we adapt a procedure first introduced in reference [12], which was originally designed to test the assumption that (generally short) time series corresponding to the growth rates of business firms aggregated in a large panel of firms can be considered as being withdrawn from distributions which share the same shape, even if some of their parameters (namely the mean and variance) may vary among the firms. In Section 2 we present the basic ideas underlying the test procedure and its adaptation to climate time series. In Section 3 we illustrate the application of the procedure to a simple model for just one weather variable, namely the maximum temperature in a given site, conditioned on the precipitation status (wet or dry) of each day. In the last Section we discuss our results and present our conclusions.

2. The test procedure

In reference [12] a procedure was proposed to test the assumption that the (generally short) time series of growth rates of business firms aggregated in a large balanced panel of firms all share the same shape for the growth rate distribution. Here we present a brief outline of the basic ideas of such a procedure in the original context, and after we show how to adapt such a procedure to a climatte time series, which we are interest in in this work. For deeper details about the test and its power we refer the reader to [12].

2.1. Outline of the original test

Consider a large balanced panel (a matrix) formed by N rows and T columns. The i-th row contains T time records (annual growth rates) for the firm i ($i = 1, ..., N$). The null hypothesis is that all the firms in the panel share the same shape for their growth rate distributions, without necessarily sharing the same parameters. Thus, for example, a typical null could assert that all the time series in the panel were withdrawn from normal distributions, which may have different means and variances. The basic assumptions behind the null hypothesis are the following (below R_i denotes the random variable (r.v.) associated with the growth rates of the firm i).[2]

A1. the time records $\left\{r_i^j\right\}_{j=1}^{T}$ of firm i are T independent outcomes of the same r.v. R_i.

A2. different firms grow independently of each other.

A3. all the time series are stationary.

[1] More precisely, in the general case we must refer to the distribution of the residuals *noise* [see Eq. (3)]. When the residuals noise distribution is stable, as it is the case of Gaussian noise, the residuals and the residuals noise will belong to the same family of distributions and, therefore, will share the same *shape*.

[2] The notation used here is slightly different from that used in [12]

A4. each variable R_i may have an idiosyncratic mean μ_i and standard deviation σ_i, and it is assumed that *these are the only possible differences* between the R_i's. Therefore, according to this assumption the standardized r.v.

$$X_i = \frac{R_i - \mu_i}{\sigma_i} \tag{1}$$

will be *the same* for all i. Let us denote this variable by X and its distribution function by F_X. It is precisely this distribution which characterize the null about the common shape assumed to be shared by all the distributions of the firms (rows) in the panel.

Now, if one knows the actual values of all the idiosyncratic parameters μ_i and σ_i, then by simply pooling together all the standardized outcomes $\left\{x_i^j\right\}_{i=1,\dots,N}^{j=1,\dots,T}$, the resulting aggregated distribution would resemble the distribution of X. In order to test the null F_X one could simply apply a goodness-of-fit test to compare this distribution with that of the aggregated observed data standardized by rows. However, in general we do not know the actual idiosyncratic parameters μ_i and σ_i, and at the best these should be estimated from each row in the panel; when T is large enough such estimates will be accurate, and the lack of knowledge about the exact idiosyncratic parameters would not be an obstacle to do the standardization and perform the suggested test. However, as it was shown in [12], when T is not large (roughly $T \lesssim 50$) such estimates may be very rough, and *even if the null is correct the aggregate distribution of the outcomes standardized with respect to these roughly estimated parameters may deviate significantly from* F_X. In [12] it was also shown that

$$Z_i = \frac{R_i - \hat{\mu}_i}{\hat{\sigma}_i} = \frac{X_i - \bar{X}_i}{s_{X_i}} = \frac{X - \bar{X}}{s_X}, \tag{2}$$

where $\hat{\mu}_i$ and $\hat{\sigma}_i$ are the (generally innacurate) time series estimates for μ_i and σ_i, and $\bar{X}_i$, s_{X_i} are the (unknown) time series estimates for the mean and standard deviation of the standard r.v. (1), and similarly for $\bar{X}$ and s_X. From now on we will reserve the word "standardization" for standardization with respect to the actual parameters μ_i and σ_i, and will use the expression "z-transformation" for the standardization with respect to the corresponding time series estimates $\hat{\mu}_i$ and s_{X_i}.

Equation (2) is central to designing a test procedure able to deal with the issue of short time series in the panel. It states that under the null hypothesis the variables Z_i obtained by z-transforming the time series variables R_i *must be the same for all firms* and *is distributed identically to the variable* $\tilde{Z} \equiv \frac{X-\bar{X}}{s_X}$, where X is the r.v. corresponding to the null F_X. Therefore, what we should to do to test the null is simply *to compare (by using a goodness-of-fit test) the distribution of all the z-transformed growth rates pooled together with the modified null distribution* $F_{\tilde{Z}}$ (for details of how to obtain this distribution by a Monte Carlo simulation see [12]).

2.2. Adaptation to climate time series

For the sake of illustration of application of the above procedure to climate time series we consider a simple model consisting of a single variable V, whose daily time series is denoted by $\left\{V^j\right\}_{j=1}^{M}$, and M is assumed to be large (in the next Section we consider 21 years of data, which correspond to more than 7 thousands of daily records). To obtain a residual series the daily observations are normalized with respect to the mean and standard deviation of the corresponding day, which are modeled by a truncated Fourier series with an annual period. Two Fourier models are considered, depending on the wet or dry status of the day. After such a normalization is done, the resulting residuals Y^j are assumed to follow an AR(1) process [3, 10]

$$Y^j = \rho Y^{j-1} + \left(1 - \rho^2\right) \epsilon^j, \qquad j = 1, \cdots, M, \tag{3}$$

where the index j labels the day in the whole time series, ρ is the lag-1 autocorrelation, assumed to be constant within each calendar month, but which may differ among different calendar months (but not changing along the years), and ϵ^j are independent random shocks, whose distributions are often assumed in the literature to be a standard normal (white noise).

Our aim in this work is to adapt the procedure of the previous subsection to test the null hypothesis that a particular *shape* (e.g., Gaussian) of the distribution of the residuals noise ϵ^j is time invariant. The test will apply even if the mean μ^j and standard deviation σ^j may change in the time; however, in order to apply the test outlined above we will make the assumption that *within short time periods of length T ($T \lesssim 30$) the variations of the noise's mean and variance is negligible.* Now, by spliting the whole time series in subperiods of length T we can build a panel of N rows and T columns, where N is the number of subperiods.[3] Together with the null hypothesis we are interested in to test, it is straightforward to see that the set of assumptions made so far in our simple climate model (3) is compatible with the assumptions A1 – A4 of the previous subsection. Therefore, the test procedure outlined there can be straightforwardly applied to test the null hypothesis about the constancy of the shape of the residuals noise distribution in the model (3).

3. An illustrative example

To illustrate the application of the proposed procedure we considered a time series of daily data for the maximum temperature and precipitation for the locality of Chicago, USA (station "KMDW"), available in the software Wolfram Mathematica®. We considered data covering a period of 21 years, from January 1st, 1995 to December 31th, 2015, which resulted in a time series of length $M = 7,665$.

The time series of precipitations was used just to classify each day in the time series as wet or dry – we considered a day as wet if its accumulated precipitation was ≥ 0.1 mm; otherwise, it was considered as dry. After, data on maximum temperature for each calendar day was split into two groups (according to the dry or wet status), and for each group it was computed the mean and standard deviation. The annual course of the means and the standard deviations were approximated separately by a third order Fourier fitting, one fitting for each group (dry or wet). After eliminating the annual course of the mean and standard deviation (by using the values for each calendar day as given by the third order Fourier fitting), the normalized residues Y^j ($j = 1, \cdots, M$) were obtained.

To estimate the 1-lag autocorrelation ρ for each calendar month we proceeded as follows. We used the well-known Yule-Walker (biased) estimator for the autocorrelation associated to each month in the time series, resulting in a set of 21 (since we have 21 years of observations) biased estimates for each calendar month [13]. Then, we taken the average of these 21 values and corrected for the bias, obtaining an estimate for the calendar month m given by $\rho^{(m)} = \frac{n\bar{\rho}_1+1}{n-4}$, where n is the number of days of the calendar month m and $\bar{\rho}_1$ is the mean of the set of 21 biased Yule-Walker estimates for that month.[4] These estimates are shown in table 1.

Turning the estimated autocorrelations into (3) we computed the time series for the residual noise ϵ^j. When data for maximum temperature *or* precipitation were lacking for a day (this occurred only for 263 days in 21 years of data) we filled that day in the noise time series by

[3] $N = [M/T]$, where [.] denotes the integer part.

[4] Due to the lacking of data in some days, this average was computed with approximately 16 values for every month.

Table 1. Estimates for the monthly autocorrelations in the model (3).

	Jan	Feb	Mar	Apr	May	Jun	Jul	Aug	Sep	Oct	Nov	Dec
ρ	0.75	0.70	0.69	0.59	0.63	0.72	0.59	0.62	0.68	0.78	0.63	0.70

Table 2. p-Values of the KS test for two nulls and several subperiods lengths.

Length (days)	Gaussian	Empirical
7	0.024	0.838
14	0.008	0.937
21	0.030	0.871
28	0.002	0.594

randomly choosing a value from the remaining time series. Finally, we set the null hypothesis about the shape of the distributions of residual noises, split the whole time series in short subperiods of length T and applied the adapted procedure of Subsection 2.2. As for the goodness-of-fit test, which was the last stage of our procedure, we used the Kolmogorov-Smirnov (KS) test available in the software Wolfram Mathematica®.

3.1. Results

We considered two specific nulls. The first one consider that the invariant shape of the residuals noise distribution is Gaussian, which is the most often assumed shape in the literature. The second null we considered was that the invariant shape coincides with that empirically observed when we pool together all the observed residuals noise for the whole time series (this particular assumption implies that the parameters of this distribution does not change in the course of time as well). *En passant*, the last version of the widely used generator LARS-WG uses semi-empirical distributions for the noises and obtain better results in the simulation of maximum and minimum temperatures when compared with its previous versions which used Gaussian noises [4].

We tested the above two nulls by partitioning the time series in subperiods of lengths $T = 7, 14, 21$ and 28 days. The results are shown in table 2. Considering a 5% significance level we observe that the Gaussian null should be rejected for all the lengths considered. On the other hand, at the same significance level the "empirical" null should not be rejected in any of the cases. It is not shown in the table, but the empirical null is consistently accepted also for larger subperiods of lengths 90, 180, 365, 1825 (5 years) and 7665 (21 years, the length of the whole time series). Figures 1 and 2 illustrates the goodness of fit of the two considered nulls for subperiods of lengths 7 and 28 days.

4. Discussions and conclusion

We proposed an adaptation of the testing procedure introduced in [12] to investigate the null hypothesis that the *shape* of the distribution of the residuals noise of a climate time series is time invariant, even if its parameters are not. We illustrated the application of the adapted procedure to a time series of a single weather variable (the maximum temperature) in a given site, and showed that it was very effective in testing for two specific nulls, one stating that the

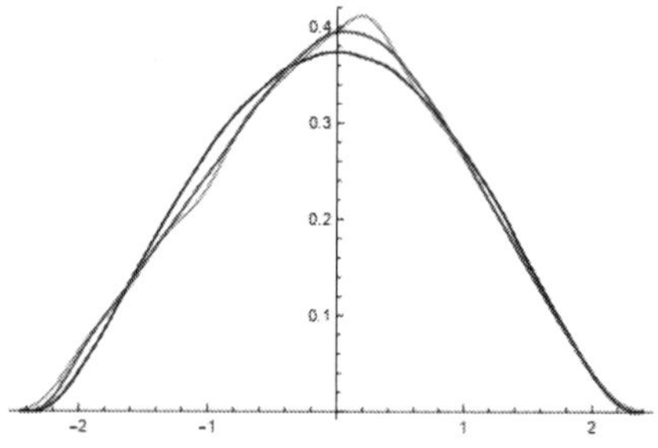

Figure 1. Histograms of the z-transformed empirical data (orange), and the corresponding histograms for the Gaussian (blue) and empirical (red) nulls, for subperiods of $T = 7$ days.

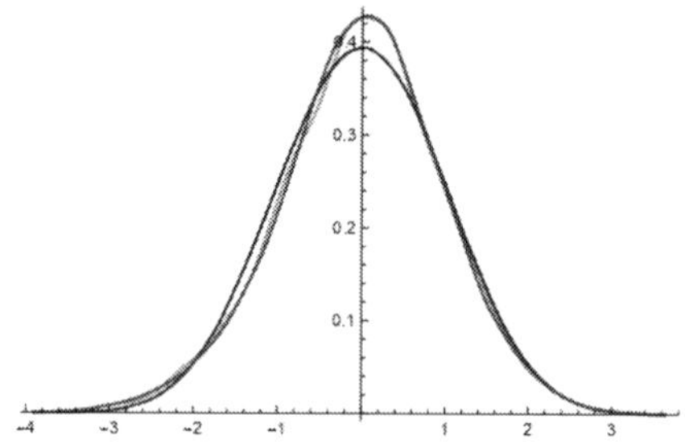

Figure 2. The same as figure 1, but with $T = 28$ days.

time invariant shape is Gaussian and another stating that the empirically observed distribution for the residual noises is time invariant (time invariant shape and parameters). From the results we concluded that the Gaussian assumptions for the residuals does not describe well the observed data in the context of this single variable model; instead, in this context the assumption that the residuals noise maintain a time invariant distribution performs better. Further investigations in the context of this simple model may consider analytical assumptions to describe the empirical distribution. As such empirical distributions often present a non vanishing skewness (see, for example, [4, 11]; it seems to be the case also in the situations illustrated in figures 1 and 2) we could guess that members of the class of asymmetric exponential power distributions [14], which generalizes the Gaussian by changing the tails behavior and by introducing skewness, could be good candidates for it. It is worth to emphasize that the fact that the empirical distribution of the noises was not reject by the test is by no means obvious, because the resulting empirical distributions obtained by aggregating all the time series noises could be just an effect of aggregating different variables with different distributions – our test provided evidence that this is not the case for the data considered here.

Finally, the proposed procedure can be straightforwardly extended to deal with multivariate stochastic models presenting auto and cross correlations among the variables, and thus it can be an additional and useful tool in the studies of validation of actual multivariate WG's for diverse sites. Such an extension, along with a wide study of validation of some WG's for diverse locations is presently under our investigation. The results will appear elsewhere.

Acknowledgments

JTL thanks CNPq/Brazil for partial financial support (Grant PDE 233512/2014-4).

References

[1] Richardson C W 1981 *Water Resources Research* **17** 182
[2] Kilsby C, Jones P, Burton A, Ford A, Fowler H, Harpham C, James P, Smith A and Wilby R 2007 *Environmental Modelling & Software* **22** 1705
[3] Fodor N, Dobi I, Mika J and Szeidl L 2010 *Meteorology and Atmospheric Physics* **107** 91
[4] Semenov M A and Stratonovitch P 2010 *Climate Research* **41** 1
[5] Iizumi T, Takayabu I, Dairaku K, Kusaka H, Nishimori M, Sakurai G, Ishizaki N N, Adachi S A and Semenov M A 2012 *Journal of Geophysical Research: Atmospheres* **117** D11114
[6] Chen J and Brissette F P 2014 *International Journal of Climatology* **34** 3089
[7] Semenov M A, Brooks R J, Barrow E M and Richardson C W 1998 *Climate Research* **10** 95
[8] Semenov M 2008 *Climate Research* **35** 203

[9] Fodor N, Dobi I, Mika J and Szeidl L 2013 *The Scientific World Journal* **2013** 571367
[10] Mearns L O, Katz R W and Schneider S H 1984 *Journal of Climate and Applied Meteorology* **23** 1601
[11] Harmel R D, Richardson C W, Hanson C L and Johnson G L 2002 *Journal of Applied Meteorology* **41** 744
[12] Lunardi J T, Miccichè S, Lillo F, Mantegna R N and Gallegati M 2014 *Journal of Economic Dynamics and Control* **39** 140
[13] Krone T, Albers C and Timmerman M 2015 *Quality & Quantity* 1
[14] Bottazzi G and Secchi A 2011 *Industrial and Corporate Change* **20** 991–1030

The Entropy Estimation of the Physics' Course Content on the Basis of Intradisciplinary Connections' Information Model

Gnitetskaya Tatyana

The School of Natural Sciences, Far Eastern Federal University, 8, Suhanova St. Vladivostok, 690950 Russian Fed.

gnitetskaya.tn@dvfu.ru

Abstract. In this paper the information model of intradisciplinary connections and semantic structures method are described. The information parameters, which we use in information model, are introduced. The question we would like to answer in this paper is – how to optimize the Physics Course' content. As an example, the differences between entropy values in the contents of physics lecture with one topic but different logics of explanations are showed.

1. Introduction

Research results in the field of microphysics obtained in a relatively short period of time caused a significant correction of a worldview proposed by physics. New ideas about the world gave birth to new General Physics Course content studied in universities. However, if we include in this course of physics all the achievements and provide detailed explanations, its volume will increase incredibly. At the same time, a new outlook can be formed only on the basis of a new concept of physics course. Thus, along with the problem of selection of new Physics Course content, the problem of optimization of information in this content is not less important. According to the author, this problem can be solved if we develop methods of reducing educational information entropy. This approach corresponds to modern psychology concepts on the presence of channel consciousness capacity, which depends on the entropy rate of information perceived by it [1Lomov]. Any interpretation of information and its synthesis is nothing but the invariant transformation of dozens semantic units and the establishment of connections between them. Therefore, the author proposes to simulate educational information and carry out an assessment of its entropy on the Basis of Intradisciplinary Connections' Information Model [2G-1]. This article describes a method of entropy evaluation on the example of the content of Physics Course content.

2. Intradisciplinary Connections' Information Model

Connection is a philosophical category, which allows setting a system's elements comparison criteria and perform the assessment of its integrity. One of the varieties of connections outlined in philosophy is a "substrate providing transfer of information..." [3-FS]. This interpretation of connection is the basis of the intradisciplinary connections' information model, which is based on a intradisciplinary connections' graph model developed by the author [4G-1]. The intradisciplinary connections' graph model is a directed graph tree built between the element of the structure where this notion (or the law, or other element of knowledge) is introduced and the element of the structure where this concept (or other element of knowledge) is used. In fact, the intradisciplinary connections' graph model, where

Content from this work may be used under the terms of the Creative Commons Attribution 3.0 licence. Any further distribution of this work must maintain attribution to the author(s) and the title of the work, journal citation and DOI.
Published under licence by IOP Publishing Ltd

graphs nodes are represented as semantic structures, updated by the calculation of informational characteristics, is considered by the author as an intradisciplinary connections' information model. Semantic structure is a hierarchical structure of semantic units (physical concepts) and connections between them (intradisciplinary connections). Semantic structure can be built upon the content of any logical information object - whether it's a paragraph, the contents of which is devoted to the description of physical phenomena stated in the title, or a lecture or a training module, the contents of which are also underlined by their titles. The lower level of generalization is occupied by the concepts that have been studied previously (in the previous paragraph, section, in other subject, or are introduced in this paragraph), the next level of generalization is occupied by the concepts that include the ones outlined below, and so on up to a graph node where the concept studied in this paragraph or information subject (the content of which is revealed by the structure) is located. Connections in the structure are intradisciplinary and downward-directed. Semantic structure is isomorphic to a semantic graph by a number of qualitative characteristics. Therefore, the information model introduces the concept of the semantic graph.
Entropy determining the amount of information contained in the homogeneous semantic graph within the classical information theory in the binary system is calculated as follows:

$$S(k,J) = J^{\frac{log_2((k-1)J+1)}{log_2 k}} \quad (1)$$

k here is a beam coherence rank (equal to the number of connections of the selected item with its subdominant).
$J = 1 + k + k^2 + \cdots k^{\varphi-1}$- the total number of semantic elements in the semantic graph with homogeneous coherence rank ordered by the level of the information field $\{V_\varphi\}$.
Levels counting in the model is downward-directed. The first term of the expression stands for the highest semantic unit located at the top of the concept graph model idea where $n_1 = 1$ is an addend, (k) is the number of elements of the second level, $n_2 = k$, since the exponent is $\varphi = 1$. Last number of the expression is defined by the number of elements at the model's lowest level equal to $n = k^{\varphi-1}$.
In a heterogeneous structure, coherence rank $\boldsymbol{k}$ is replaced by the average number of connections per one beam (4):

$$k_{\text{cp}} = \frac{J_c}{m} \quad (2)$$

where m_i is the number of beams with coherence rank k_i, J_c is the number of connections in the structure.
The amount of information contained in the model with normal completeness (when all finite elements belong to the same level) and the average coherence rate can be represented as a square of entropy:

$$U_0 = S^2(k_{\text{cp},} J) = J^2 \left(\frac{((k_{\text{cp}}-1)J+1)}{log_2 k_{\text{cp}}}\right)^2 \quad (3)$$

The degree of abstraction as the degree of structure organization can be obtained from:

$$\varphi(k,J) = \frac{log_2((k-1)J+1)}{log_2 k} \quad (4)$$

U_0 here stands only for an inner content of the concept model. In addition to the internal organization of the structure elements, the concept model includes an external configuration, which is in close relation with the model structure. The configuration of the semantic graph depends on the degree of branching. The configuration characteristic may be a coefficient of relative entropy $E = \frac{H'}{H}$,где $H = log_2 J_k$ – maximum entropy of finite elements of J_k graph.

$$H' = -\sum_{i=1}^{J_k} q_i \, log_2 q_i \quad (5)$$

H'– is relative entropy of the graph finite elements, q_i is probability of the model finite element. Probability distribution function can be expressed in terms of the reduced level of abstraction:

$$qA_{\alpha\beta}V_\beta = \frac{qB_{\beta-1}V_{\beta-1}}{\varphi(A_{\alpha\beta})\sum_{\alpha=1}^{k}\left(\varphi(A_{\alpha\beta})\right)^{-1}} \quad (6)$$

$q_{A_{\alpha\beta}}V_\beta$ here is a probability of the semantic unit $A_{\alpha\beta}$, located on V_β level of abstraction (see fig. 1), $q_{A_{\beta-1}}V_{\beta-1}$ – is a probability of the highest semantic unit $B_{\beta-1}$ belongs to $V_{\beta-1}$ graph level; $\boldsymbol{k}$ – the number of lower semantic units $A_{\alpha\beta}$, in a given connection beam; $\alpha = 1,2,\ldots k$,β is the stage level number.

3. Results

We will show how to reduce entropy of the content of lectures on physics with the help of the Intradisciplinary Connections' Information Model.

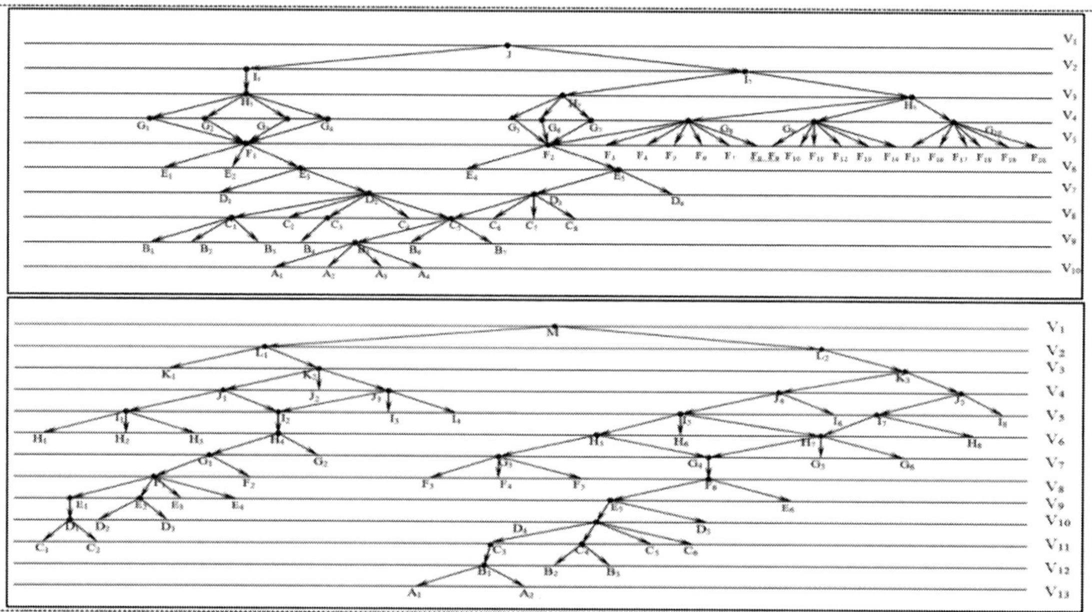

Fig.1. The "Dynamics of a material point" lecture semantic graphs.
The first option is at the top, the second is at the bottom of the figure

Fig. 1 shows two of the possible four options of lectures on physics on the topic "Dynamics of a material point". The lecture discusses the following aspects: Newton's Second Law. the concept of power, mass, the types of interaction, the relativistic equation of particle dynamics, the relativistic mass, the relativistic momentum, Newton's third law and the limits of its applicability. The sequence of these aspects may vary, as the filling with educational material. We considered four options of the lecture. The options differ by various locations of information blocks responsible for the formation of the concepts of interaction and mass in the structure of the lecture. The content of the lecture in the first option (see top of Figure 1) is consistent with the sequence of the above questions, where interaction and its types are considered after the introduction of Newton's Second law, where mass is introduced as the coefficient of proportionality between force and acceleration, and detailed presentation of the concept of mass follows only after that. The second option of the lecture is the following – at first, the types of interactions and mass are expounded in details, and then Newton's Second law is introduced. The semantic graph of the semantic structure of the second option is shown in the figure below. The third option begins as the second, but mass is introduced before momentum and the relativistic equation

of motion of a particle. In the fourth option, the types of interactions are introduced at the beginning of the lecture, and mass is expounding in details in the end after the introduction of Newton's laws and discussion of the limits their applicability.
Fig.1 shows that the graph corresponding to the second option of the lectures has a more complex configuration, it has a greater number of levels of generalization. The coefficient of relative entropy ***E*** is a quantitative characteristic of the semantic structure of the configuration.
The calculation of information characteristics of these methods of presentation is given in the table that shows that the first option of the lecture has the least entropy square value of the enclosed information and can be recommended for the presentation of educational material on the subject as the optimal.

Table. Information characteristic of four options of the "Dynamics of a material point" lecture presentation

Lecture presentation option No.	J	$U_0 Kbit^2$	E	H	H'	$S^2=U$ $Kbit^2$	$T_{min,}$ c
1	93	0,27	0,6	5,77	3,44	0,45	2,9
2	92	0,26	0,29	5,75	1,68	0,89	5,8
3	92	0,26	0,34	5,77	1,95	0,78	5,1
4	93	0,27	0,54	5,77	3,09	0,51	3,3

It is interesting that, knowing entropy of the semantic graph, consciousness capacity (16 bit/s [6]) and information retention time in random access memory (RAM) block of consciousness (10 seconds [7]), it is possible to carry out a rough estimate of the minimum time needed for perception of information concluded in the graph by consciousness. The table lists the minimum perception time for all of the cases examined. It can be seen that the first – optimal – option of the lecture has a less minimum time value than for other options. It draws a conclusion Thus, we can conclude about the greater availability of material with lower entropy.
Therefore, we can recommend evaluating entropy of educational material with an intention of It's optimization.

References

[1] Lomov B.F. Man and technique Engineering psychology essays.- M .:. Soviet radio. – p. 464.
[2] Philosophical Encyclopedic dictionary . M .: Soviet encyclopedia, 1983.- p. 840.
[3] GnitetskayaT. GraphModelofIntradisciplinaryConnectionsin Example of General Physics Course Journal of Physics: Conference Series 633 (2015) 012091 doi:10.1088/1742-6596/633/1/012091
[4] Gnitetskaya T.Information Model of Intradisciplinary Connections in the Context of a General Physics Course Proceedings 2015 International Conference on Simulation Modeling and Mathematical Statistics (SMMS, 2015), November 22-23 Chang Mai, Thailand, DEStech Publications, Inc. USA, 2015. – 519 pp. (P. 271-275), ISBN: 978-1-60595-112-6
[5] Shannon C.E. and. Weaver W, The Mathematical Theory of Communication Urbana, Illinois, The University of Illinois Press, (1949) 29-125.
[6] Archangelsky S.I. The educational process in high school, its legitimate bases and methods. - M.: Higher School, 1980.–p. 68.
[7] Klaus G. Kybernetic und Erkenntnistheorie, VEB – Deutschr Verlag der Wissenschaften, Berlin 1966, s. 28, 175

5th International Conference on Mathematical Modeling in Physical Sciences (IC-MSquare 2016) IOP Publishing
Journal of Physics: Conference Series **738** (2016) 012080 doi:10.1088/1742-6596/738/1/012080

Computation of masses and binding energies of some hadrons and bosons according to the rotating lepton model and the relativistic Newton equation

C.G. Vayenas[1,2,*], A.S. Fokas[3,4] & D. Grigoriou[1]

[1]LCEP, 1 Caratheodory St., University of Patras, Patras GR 26504, Greece
[2]Division of Natural Sciences, Academy of Athens, 28 Panepistimiou Ave., GR-10679 Athens, Greece
[3]Department of Applied Mathematics and Theoretical Physics, University of Cambridge, Cambridge, CB3 0WA, UK
[4]Department of Electrical Engineering, University of Southern California, Los Angeles, California, 90089-2560, USA

E-mail: cgvayenas@upatras.gr

Abstract. We compute analytically the masses, binding energies and hamiltonians of gravitationally bound Bohr-type states via the rotating relativistic lepton model which utilizes the de Broglie wavelength equation in conjunction with special relativity and Newton's relativistic gravitational law. The latter uses the inertial-gravitational masses, rather than the rest masses, of the rotating particles. The model also accounts for the electrostatic charge-induced dipole interactions between a central charged lepton, which is usually a positron, with the rotating relativistic lepton ring. We use three rotating relativistic neutrinos to model baryons, two rotating relativistic neutrinos to model mesons, and a rotating relativistic electron neutrino - positron (or electron) pair to model the $W^{\pm}$ bosons. It is found that gravitationally bound ground states comprising three relativistic neutrinos have masses in the baryon mass range ($\sim$ 0.9 to 1 GeV/c^2), while ground states comprising two neutrinos have masses in the meson mass range ($\sim$ 0.4 to 0.8 GeV/c^2). It is also found that the rest mass values of quarks are in good agreement with the heaviest neutrino mass value of 0.05 eV/c^2 and that the mass of $W^{\pm}$ bosons ($\sim$ 81 GeV/c^2) corresponds to the mass of a rotating gravitationally confined $e^{\pm} - \nu_e$ pair. A generalized expression is also derived for the gravitational potential energy of such relativistic Bohr-type structures.

1. Introduction

The potential importance of gravitational interactions between neutrinos was first discussed by Wheeler who actually introduced the term "neutrino geons" [1] long before it was shown that neutrinos have non-zero mass [2, 3].

In a recent book [4] and series of papers [5, 6, 7, 8, 9] we have shown that this is indeed the case for relativistic neutrinos with energies above 100 MeV [4, 5]. The gravitational attraction between such highly relativistic neutrinos becomes stronger than the Coulombic repulsion of two unit electric charges at the same distance [5].

This can be easily shown by just using gravitational rather than rest masses in Newton's universal gravitational law [4, 5]. The gravitational mass, m_g, of a particle with rest mass m_o

Content from this work may be used under the terms of the Creative Commons Attribution 3.0 licence. Any further distribution of this work must maintain attribution to the author(s) and the title of the work, journal citation and DOI.
Published under licence by IOP Publishing Ltd

is equal to its inertial mass, m_i, and the latter equals $\gamma^3 m_o$ where $\gamma(1 = (1 - \mathrm{v}^2/c^2)^{-1/2})$ is the Lorentz factor.

The equality

$$m_i = \gamma^3 m_o \tag{1}$$

was first shown by Einstein in his pioneering 1905 special relativistic paper [10, 11, 12] and this result was shown recently to hold for arbitrary particle motion, including circular motion [4, 5]. In view of equation (1), Newton's universal gravitational law becomes

$$F = \frac{G m_{1,o} m_{2,o} \gamma^6}{r^2}, \tag{2}$$

or

$$F = \frac{G m_{1,o} m_{2,o} \gamma_1^3 \gamma_2^3}{r^2}. \tag{3}$$

In the case of eq. (2), γ involves the velocity v between the two particles, while equation (3) refers to circular motion and the Lorentz factors γ_1 and γ_2 correspond to the velocities, v_1 and v_2, of the two particles with respect to the center of rotation [4, 5].

It has been found, surprisingly, that the use of equation (3) in conjunction with the de Broglie wavelength equation in Bohr-type models with gravity as the attractive force describes gravitationally confined fm-size structures with the masses and most of the properties of hadrons [4, 5]. Here we provide a comparative summary of the key results of these studies which have been recently extended to include mixed rotating $\nu_e - e^{\pm}$ structures. The latter have been shown to have the properties, such as masses, of $W^{\pm}$ and Z^o bosons [6, 7].

In the recent works [13, 14, 15, 16] it is argued that hadrons may be viewed as microscopic black holes, where gravitational collapse is prevented by the uncertainty principle. Furthermore, in [17, 18] Bohr-type models are used in order to establish a link between black holes and quantum gravity.

2. Mathematical models

2.1. Baryons

For the case of baryons the model comprises three rotating neutrinos or antineutrinos [4, 5] and the equation of motion for circular ground state orbits is [4, 5]

$$\gamma m_o \frac{\mathrm{v}^2}{r} = \frac{G m_o^2 \gamma^6}{\sqrt{3} r^2}, \tag{4}$$

coupled with the de Broglie wavelength equation

$$\lambda\!\!\!^{-} = \frac{\hbar}{\gamma m_o \mathrm{v}}. \tag{5}$$

Assuming $\lambda\!\!\!^{-} = r$ one obtains [4, 5]

$$\mathrm{v} \approx c \quad ; \gamma = 3^{1/12}(m_{Pl}/m_o)^{1/3} = 7.163 \times 10^9 \quad ; \quad r = \hbar/\gamma m_o c = 0.63 \times 10^{-15}\, m, \tag{6}$$

where $m_{Pl}(= (\hbar c/G)^{1/2})$ is the Planck mass. From (5) and (6) it follows that the mass, m_B, of the confined baryon-type state is

$$m_B = 3^{13/12}(m_{Pl} m_o^2)^{1/3}. \tag{7}$$

Exact agreement of eq. (7) with the neutron mass ($m_B = 939.565\ MeV/c^2$ [20, 21]) is obtained for $m_o = 0.04372$ [4, 5]. This is in excellent agreement with the current best experimental estimates of the heaviest neutrino mass of $0.051 \pm 0.01\ eV/c^2$ [2] and with the currently recommended value of $0.048\ eV/c^2$ [19].

2.2. Mesons

In this case the model comprises a neutrino-antineutrino pair and equation (4) becomes

$$\gamma m_o \frac{\mathrm{v}^2}{r} = \frac{G m_o^2 \gamma^6}{4r^2}. \tag{8}$$

Solving this equation in conjunction with the de Broglie equation (5), one obtains

$$\mathrm{v} \approx c \quad ; \quad \gamma = 2^{1/3}(m_{Pl}/m_o)^{1/3} = 8.234 \times 10^9 \quad ; \quad r = \hbar/\gamma m_o c = 0.548\ fm. \tag{9}$$

From (8) and (9) it follows that the mass, m_M, of the confined meson-type state is

$$m_M = 2^{4/3}(m_{Pl} m_o^2)^{1/3} = 2(2 m_{Pl} m_o^2)^{1/3}. \tag{10}$$

For $m_o = 0.04372\ eV/c^2$ [4, 5] one obtains $m_M = 720\ MeV/c^2$, which is in reasonable qualitative agreement with the mass of ρ mesons, i.e. 775 MeV/c^2 [20, 21].

Note that this problem has been already treated in reference [4] pp. 78-80. Due to an error of a factor of 2 in eq. 6.32, the first factor given for eq. (10) is $2^{7/6}$ instead of the current $2^{4/3}$.

It has been shown recently [6] that equation (10) can be generalized for a rotating pair of particles with arbitrary rest masses $m_{1,o}$ and $m_{2,o}$. In this case equation (10) takes the form

$$m_M = 2^{4/3}(m_{Pl} m_{1,o} m_{2,o})^{1/3} = 2(2 m_{Pl} m_{1,o} m_{2,o})^{1/3}. \tag{11}$$

It is worth noting that, as long as the centipetal force exerted on all the rotating particles is the same, the following equation is always found to be valid [6, 7]:

$$\gamma_1 m_{1,o} = \gamma_2 m_{2,o}, \tag{12}$$

which follows directly from the application of eq. (8) to each particle separately.

2.2.1. Bosons : Equation (11) has been also derived recently for the mass a rotating relativistic electron-neutrino pair, yielding [6]

$$m = 2m_W = 2^{1/3}(m_{Pl} m_e m_o)^{1/3} = 2(81.74)\ GeV/c^2 \tag{13}$$

Remarkably $m_W (= 81.74\ GeV/c^2)$ corresponds within 1.3% to the experimental mass of the $W^{\pm}$ boson (80.42 GeV/c^2). The suggestion that the $W^{\pm}$ boson appears to be a relativistic $\nu_e - e^{\pm}$ pair decaying to two fragments of equal mass $\gamma_e m_e = \gamma_\nu m_\nu$, is not too surprising if one recalls that the main decay products of the $W^{\pm}$ bosons are electrons, positrons and neutrinos [20, 21].

The same approach has been used recently [6] to compute the masses of the Z^o and of the H bosons. The former is modeled as a rotational $e^+ - e^- - \nu_e$ structure, and it yields

$$m = 2m_Z = 2^{1/2}(m_{Pl} m_e m_\nu)^{1/3} = 2(91.75\ GeV/c^2). \tag{14}$$

Remarkably the thus computed m_Z value corresponds within 1% to the experimental value of 91.19 GeV/c^2 [7].

In general, agreement with experimental values can be further improved by considering the Coulombic charge-charge, charge-induced dipole and induced dipole-induced dipole forces between positrons, electrons and electrically polarized neutrinos [6, 7].

3. Relativistic gravitational potential energy

For rotational structures with relativistic velocities comprising a number N of rotating particles the following analysis can be used for the computation of the relativistic gravitational potential energy of the particles.

We start from the generalized form of equation (4), namely

$$F_j = \gamma_j m_{o,j} \frac{\mathrm{v}_j^2}{r} = \frac{G m_{o,i} m_{o,j} \gamma_i^3 \gamma_j^3}{A_{ij} r^2} \tag{15}$$

where i and j are integers up to N, A_{ij} is a constant depending on N and on the particular structure geometry. For example for the three neutrino model it is $A = \sqrt{3}$, (eq. (4)) and for mesons it is $A = 4$, (eq. (8)).

The angular momentum of the particle, $m_{o,j}$, L_j, is given by

$$L_j = \gamma_j m_{o,j} \mathrm{v}_j r. \tag{16}$$

Recalling that L_j is conserved and assuming that $L_j = n_j \hbar$, where n_j is an integer we find

$$\frac{\gamma_j m_{o,j}}{n_j} = \frac{\gamma_i m_{o,i}}{n_i}. \tag{17}$$

Equations (15) and (17) together with the approximation $\mathrm{v} \approx c$, imply

$$\left(\frac{2A_{ij}}{r_{s,j}}\right) r = \gamma_j^5 \left(\frac{n_i}{n_j}\right)^3 \left(\frac{m_{o,j}}{m_{o,i}}\right)^2, \tag{18}$$

where $r_{s,j}$ is the Swarzschild radius of $m_{o,j}$. Replacing in the first of equations (15) γ_j via equation (18) we find

$$F = C_{ij} m_{o,j} c^2 \frac{1}{r^{4/5}} \quad ; \quad C_{ij} = \left(\frac{2A_{ij}}{r_{s,j}}\right)^{1/5} \left(\frac{n_j}{n_i}\right)^{3/5} \left(\frac{m_{o,i}}{m_{o,j}}\right)^{2/5} \tag{19}$$

Hence

$$V_G(r) - V_G(r_{min}) = \int_{r_{min}}^{r} F dr' = -5 C_{ij} m_o c^2 (r^{1/5} - r_{min}^{1/5}). \tag{20}$$

Since r_s and r_{min} are of the order of 10^{-58} to 10^{-63} m for $e^{\pm}$ and ν_e respectively, it follows that the last term in parenthesis is negligible and thus using (18), it follows that

$$V_G(r) = -5 \gamma_j m_{o,j} c^2. \tag{21}$$

This is an important result, already obtained in [4] for the three-neutrino problem and here shown to be valid for arbitrary elementary particle and geometry as long as $\gamma >> 1$, thus $\mathrm{v} \to c$. We note that it also holds for bosons, such as the $\mathrm{W}^{\pm}$ bosons [6] where it leads to

$$V_G(r) = -5 \gamma_e m_e c^2 = -5 m_Z c^2 \tag{22}$$

where m_Z is the boson mass. However in [6], where the Z^o boson mass m_Z was calculated within $\sim 1\%$ without any adjustable parameters, due to an error in eq. (38) the result reached for $V_G(r)$ in eq. (40) is different from (22) and incorrect, affecting also the ensuing numerical results of section 5 of [6]. Thus the correct equation (38) in ref. [6] is the above equation (22) of the present paper.

In conjunction with the fact that the total kinetic energy of a N particle structure is

$$T = \sum_i (\gamma_i - 1) m_{o,i} c^2 \tag{23}$$

equation (21) shows that the total Hamiltonian, $\mathcal{H}(= V_G + T + RE)$, where RE is the rest energy of the leptons, can be is negative only for $N < 5$.

This implies, that only up to five leptons can participate to the rotating lepton ring of these Bohr-type structures and that pentaquarks, if they exist, are extremely unstable.

4. Conclusions

The methodology based on the relativistic Newton equation coupled with the de Broglie wavelength equation allows for the computation of the masses of hadrons and bosons with very good ($\sim 1\%$) accuracy.

The gravitational potential energy $V_G(r)$ is the same for all rotating hadrons and bosons (eq. 21) which supports the conclusion that the strong force can be modeled efficiently as a relativistic gravity via the relativistic Newton equation.

References

[1] Dieter RB and Wheeler JA (1957) Interaction of Neutrinos and Gravitational Fields. Rev. Mod. Phys. 29(3):465-479.
[2] Mohapatra RN et al (2007) Theory of neutrinos: a white paper. Rep Prog Phys 70:1757-1867.
[3] Kajita T(2009) Neutrino oscillations: Discovery, current status, future directions. Int. J. Mod. Phys. A, 24:3437-3446.
[4] Vayenas CG, Souentie S (2012), Gravity, special relativity and the strong force: A Bohr-Einstein-de-Broglie model for the formation of hadrons. Springer, ISBN 978-1-4614-3935-6.
[5] Vayenas CG, Souentie S, Fokas A (2014) A Bohr-type model of a composite particle using gravity as the attractive force. arXiv:1306.5979v4 [physics.gen-ph]; Physica A 405:360-379.
[6] Vayenas CG, Fokas AS, Grigoriou D (2016) On the structure, masses and thermodynamics of the $W^{\pm}$ bosons. Physica A 450:37-48.
[7] Fokas AS, Vayenas CG (2016) On the structure, mass and thermodynamics of the Z^o bosons. Physica A, submitted.
[8] Vayenas CG (2014) Mathematical modeling of mass generation via confinement of relativistic particles. J. Phys.: Conf. Ser. 490:012084.
[9] Vayenas CG, Fokas AS, Grigoriou D (2015) Gravitational mass and Newton's universal gravitational law under relativistic conditions. J. Phys.: Conf. Ser. 633:012033.
[10] Einstein, A.: Zür Elektrodynamik bewegter Körper. Ann. der Physik., Bd. XVII, S. **17**, 891-921 (1905); English translation On the Electrodynamics of Moving Bodies (*http : //fourmilab.ch/etexts/einstein/specrel/www/*) by G.B. Jeffery and W. Perrett (1923).
[11] French AP (1968) Special relativity. W.W. Norton and Co., New York
[12] Freund J (2008) Special relativity for beginners. World Scientific Publishing, Singapore
[13] Das S & Vagenas EC (2009) Phenomenological Implications of the Generalized Uncertainty Principle. Can. J. Phys. 87:233-240.
[14] Vayenas CG and Grigoriou D (2015) Microscopic black hole stabilization via the uncertainty principle. J. Phys.: Conf. Ser. 574, 012059(1-8).
[15] Carr B, Modesto L, Prémont-Schwarz I (2011) Generalized Uncertainty Principle and Self-dual Black Holes. arXiv:1107.0708v1 [gr-qc].
[16] Carr BJ (2014) The Black Hole Uncertainty Principle Correspondence. arXiv:1402.1427v1 [gr-qc].
[17] Corda C (2013) Black hole quantum spectrum. Eur. Phys. J. C 73:2665-6.
[18] Corda C (2015) Bohr-like black holes. AIP Conf. Proc. 1648, 020004.
[19] Fritzsch H (2015) Neutrino masses and flavor mixing. Modern Physics Letters A, 30(16):1530012(1-6).
[20] Griffiths D (2008) Introduction to Elementary Particles. 2nd ed. Wiley-VCH Verlag GmbH & Co. KgaA, Weinheim.
[21] Tully CG (2011) Elementary Particle Physics in a Nutshell, ISBN13: 978-0-691-13116-0, Princeton University Press.

Classification of excessive domestic water consumption using Fuzzy Clustering Method

A. Zairi Zaidi[1] and Khairul A. Rasmani[2]

[1,2] Faculty of Computer and Mathematical Sciences, Universiti Teknologi MARA Seremban Campus, 70300 Seremban, Negeri Sembilan, Malaysia.

E-mail: ahmadzairi@gmail.com, khairulanwar@ns.uitm.edu.my

Abstract. Demand for clean and treated water is increasing all over the world. Therefore it is crucial to conserve water for better use and to avoid unnecessary, excessive consumption or wastage of this natural resource. Classification of excessive domestic water consumption is a difficult task due to the complexity in determining the amount of water usage per activity, especially as the data is known to vary between individuals. In this study, classification of excessive domestic water consumption is carried out using a well-known Fuzzy C-Means (FCM) clustering algorithm. Consumer data containing information on daily, weekly and monthly domestic water usage was employed for the purpose of classification. Using the same dataset, the result produced by the FCM clustering algorithm is compared with the result obtained from a statistical control chart. The finding of this study demonstrates the potential use of the FCM clustering algorithm for the classification of domestic consumer water consumption data.

1. Introduction

Water is an important natural resource to all living things. Every organism uses water for living needs and for other activities to fulfil their needs. This indicates that the greater the world population, the greater the demand for water. Nowadays, it can be seen that the increase in demand for water is not in accordance with the supply of water. There is insufficient water supply to meet the increasing demand in many places [1]. This shows that conservation of water is much needed in response to the water shortage. Therefore, efficient management of clean and treated water is necessary to avoid excessive consumption, which might lead to wastage of water for unnecessary purposes.

Classification of excessive domestic water consumption is crucial; the process involves the complexity that arises from uncertain information in the consumer data resulting from uncontrolled factors, such as the behaviour of the consumer in terms of socio-economic and socio-cultural aspects. Hence, it is expected that different individuals will have different behaviour regarding water usage. Besides that, the environmental factor is believed to make a significant contribution to the variation of per capita domestic water consumption for different localities. Therefore, it is difficult to determine excessive domestic water usage for each premise solely on the basis of per capita domestic consumption. Without a proper method to classify the data as 'excessive' or 'non-excessive' water usage, it is

[1] To whom any correspondence should be addressed.

Content from this work may be used under the terms of the Creative Commons Attribution 3.0 licence. Any further distribution of this work must maintain attribution to the author(s) and the title of the work, journal citation and DOI.
Published under licence by IOP Publishing Ltd

difficult to create a mathematical model using supervised learning methods. With the absence of class labels, there is great potential to employ an unsupervised learning technique for the study of the classification of excessive domestic water consumption. Unsupervised learning techniques are methods that can deal with unlabelled data, whereby the algorithm will carry out inferences and finally classify the data into homogenous classes.

The fuzzy clustering technique has been applied in many studies, such as the classification of coal seams [2], software quality prediction [3], classifying sets of oral cancer cell data [4] and market segmentation [5]. In this study, it is proposed that the classification of excessive domestic water consumption is conducted using the Fuzzy C-Means (FCM) clustering algorithm. Consumer data is used in this study instead of the duration and flow of water because the data collection process is easier and does not require a specific physical tool. FCM is an unsupervised learning algorithm that is used to group a data set into a number of clusters, whereby every data point has its membership degree of belongingness to every cluster [6].

2. Background

Appropriate clean water management, such as water distribution network management, the estimation and production of future water consumption and sophisticated water management systems are necessary to conserve water for current and future use. The complexity of water management and demand from consumers lead to the need for an effective water management system. Savic & Walters [7] suggested 'hydroinformatics' as an approach to water network management and maintenance. The researchers argued that the discipline of 'hydroinformatics' helps in the development and application of innovative technology for managing the water distribution network and they indicate three suitable methods that can be applied in the water industry: a geographical information system, artificial neural networks and a genetic algorithm. It has been claimed that the cost saving of using these methods ranges from 5%-50%. Makropoulos & Butler [8] discussed the nature of uncertainty in urban water management, which involves the determination of optimum locations for facilities in a particular area with respect to economic and environmental factors. It has been suggested that the type-1 and type-2 fuzzy inference system (FIS) can be useful tools to solve water management problems with uncertainties and can also be adopted for spatial decision support systems (SDSS) in urban water management.

Water consumption prediction and estimation plays a crucial part in treated water management. It helps in forecasting water consumption for better management of clean water in the future. Naizhuo & Yunsheng [9] conducted a study to estimate domestic water consumption in China and the United States at province/state level using night-time satellite imagery. Since a large population leads to high cost and difficulty in collecting data by traditional statistical methods, the study aims to investigate the potential of using night-time satellite imagery where the sum light derived from the imagery data is compared with the actual water consumption data. The result of the study shows that the proposed method has potential for the estimation of domestic water consumption at national or sub-national level in the future.

Altunkaynak, Özger, & Çakmakci, [10] proposed the use of a fuzzy model to predict future water consumption for Istanbul city. The Adaptive Neuro Fuzzy Inference System (ANFIS) was employed to estimate the membership function parameters and the Takagi-Segeno Fuzzy Inference System was used to generate the output. With overall relative error of less than 10%, the model can be considered as practically acceptable to predict water consumption in Istanbul city. Şen & Altunkaynak [11] investigated the use of a fuzzy approach in a different scenario, whereby a fuzzy model was established for predicting individual drinking water consumption. The researchers argued that many statistical approaches lead to loss of information or miscalculation since they involve crisp interval values, whereas statistical assumption made for regression analysis may not frequently be achieved in practical terms. Thus, a fuzzy model based on the Mamdani approach was proposed in the study, in which vague data such as human weight, weather, temperature and human physical activity were utilised to develop the model. The proposed model was considered acceptable for practical individual drinking water consumption rate predictions since the average relative error is less than 5%.

Corona-Nakamura et al. [12] proposed a model to identify automatically the output of domestic water consumption based on duration and flow of water. It has been argued by the researchers that a suitable model could be easily interpreted by humans and can include information provided by users or experts. In the study, the researchers use the Adaptive-Neuro Fuzzy Inference System (ANFIS) to classify types of water consumption activity in residential premises. The finding of the research shows that the model can be a good solution to the problem: it has a good recognition percentage and small errors. Moreover, Corona-Nakamura, Ruelas, Ojeda-Magana, & Finch [13] conducted a research study to develop an unsupervised learning model that can automatically identify the point of water consumption in a house. The unsupervised learning model is beneficial since it can adapt to particular cases automatically. The researchers employed the Fuzzy C-Means (FCM) technique for estimating the class centre and an Improved Gustafson-Kessel algorithm for clustering of the data set. It was claimed that the developed model has a high recognition percentage and can be the basis of a supervisory system for water conservation.

It can be observed from the literature that, although much research has been conducted related to water consumption prediction, little research has focussed on the study of domestic water consumption at residential premises by utilising consumer data. Even though the consumer data includes uncertain information, especially on the quantity of water used for each activity related to individual water consumption, the collection of such data is relatively easier as it will not involve any specific measuring tools. Additionally, it can be seen that there is already research that employs a fuzzy clustering method for the classification of water consumption [12, 13]. But, this research employed data that was more complicated to collect by consumers or water supply companies. Therefore, consumer data can be explored to find its potential use for research on the classification of water consumption using a fuzzy clustering method. This motivates the present study that focuses on employing the Fuzzy C-Means (FCM) clustering algorithm for the classification of domestic consumer water consumption data.

3. Fuzzy Clustering Method

Data clustering is the process of grouping data into the same group: the data are similar to each other in the same group and dissimilar from the data in other groups [14]. Fuzzy clustering is one of the many existing data clustering techniques. In hard clustering, data are grouped to exactly one cluster. On the other hand, in fuzzy clustering (or soft clustering), data can be grouped into more than several clusters where the data will have membership degree to each cluster in the interval [0, 1]; the larger the membership degree, the greater the association of the data to a particular cluster [15]. The Fuzzy C-Means (FCM) clustering algorithm [16] is the most commonly used algorithm in fuzzy clustering analysis [15]. This algorithm will group a data set into a known number of clusters. It aims to minimise iteratively the objective function $J(U, V)$, which is given by:

$$J(U,V)=\sum_{i=1}^{n}\sum_{j=1}^{c}\mu_{ij}^{m}\left\|x_i-v_j\right\|^2 \quad (1)$$

where n is the number of data points, c is the number of clusters, x_i is ith data point in the data set, v_j is the jth cluster centre, μ_{ij} is the value of membership degree of ith data point in the jth cluster, $m\in[1, \infty)$ is the fuzziness index which determines the level of clustering result fuzziness [17], U is the partition matrix, V is the cluster centre matrix and $\|x_i - v_j\|$ is the distance between a data point x_i and the cluster

centre v_j and it is based on Euclidean distance. This algorithm will iterate until there is no further minimisation of the objective function by updating membership degree u_{ij} and cluster centre v_{ij}.
There are several steps involved in the process of grouping data set using this FCM algorithm. Each iteration is labelled as *r*. The steps are explained as follows:

Step 1: Decide the number of clusters, *c*, where $2 \leq c < n$. In order to employ the FCM, the number of clusters must be known.

Step 2: Initialise partition matrix $U(r=0)$ randomly.

Step 3: Calculate the cluster centres matrix *V* using the equation.

$$v_j = \frac{\sum_{i=1}^{n} \mu_{ij}^{m} x_i}{\sum_{i=1}^{n} \mu_{ij}^{m}} \tag{2}$$

Step 4: Calculate the distance matrix *D* using the equation:

$$d_{ij} = \left\| x_i - v_j \right\|^2 \tag{3}$$

Step 5: Update the partition matrix, $U(r + 1)$

$$\mu_{ir} = \frac{1}{\sum_{r=1}^{c} \left(\frac{\left\| x_i - v_j \right\|}{\left\| x_i - v_r \right\|} \right)^{\frac{2}{m-1}}} \tag{4}$$

Step 6: If $\left\| U^{r+1} - U^{r} \right\| < \varepsilon$ where ε is a termination criterion in [0, 1] and *r* are iteration steps, stop the iteration or else repeat Step 3 to update the cluster centres and the partition matrix *U*.

4. Experimental Setup

The purpose of this research is to classify consumer water consumption data using the Fuzzy C-Means (FCM) clustering algorithm. A data set from the study by Mohd. Hanif et al. [18] has been used in this study. It consists of 439 observations, which represent 439 residential premises. Eleven variables are considered in this study. They are the number of households, the number of toilets, the number of times teeth are brushed, the number of times hands and face are washed, the number of times a bath/shower is taken, the number of times the toilet is flushed, the number of times clothes are washed by hand, the number of times clothes are washed in the washing machine, the number of times cooking takes place, the number of times the car is washed and the average amount of monthly water usage (in litres).

In this study, the FCM clustering algorithm available in MATLAB® software was used. The algorithm initially guesses the cluster centres for the data set [6] and iteratively updates the cluster centres and membership degree for each data point until the objective function (Equation 1) is minimised. When the iteration is stopped, the cluster centres will be located in the right position, whereby each data point has membership degree to each cluster. The higher the membership degree, the greater the association of a data point to a cluster.

The algorithm will also create a matrix of cluster centres showing the coordinates of cluster centres for each data point, a partition matrix containing the membership degree for each data point in each cluster and the values of objective function for each iteration. In this study, the number of clusters, c was set to two which are "likely" and "unlikely" in order to indicate the excessiveness of water consumption. The fuzzier, m was set by default to two since there is no experimentation and domain knowledge of the value [15].

5. Findings and Discussion

Using the MATLAB® software, the FCM algorithm took 91 iterations to optimise the objective function. The result shows that, after the 91st iteration, no further process is made by the algorithm. This is because all data points have been grouped to their clusters with their own cluster centres and membership degree. At this point, the data points are associated with each cluster by membership degree. Figure 1 shows the example of the graph of the average amount of monthly water usage against a predictive variable. It can be seen that there are two partitions in the data set with their cluster centres marked with two symbols, 'O' and '×'. Such analysis is very useful to show that the FCM algorithm can be used to separate the data into different classes, which can be labelled as 'likely' and 'unlikely' cases of excessive water usage. This analysis is also very useful in demonstrating how FCM can be used to analyse how each predictive variable contributed to the final classification outcomes.

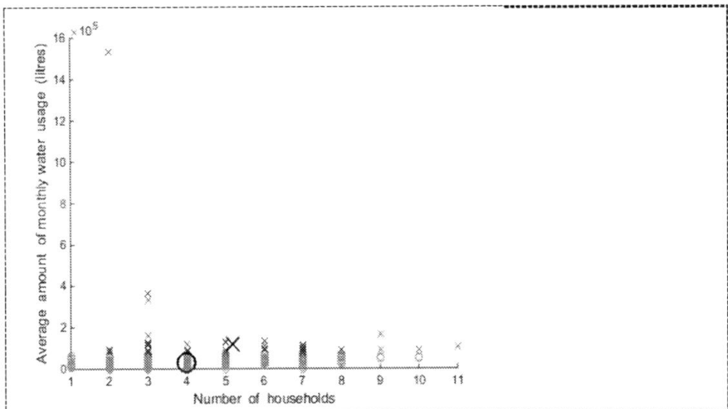

Figure 1. Example of clustering result for average monthly water usage against a predictive variable.

In order to validate the clustering outcome, the result produced by the FCM algorithm is compared with the result obtained from the study by Mohd Hanif et al. [18], where the Individual Hotelling T^2 control chart was used. Table I shows the clustering result using FCM and the result from the statistical control chart, which shows that the similarity percentage between the classification outcomes is 86.33%.

Table 1. Comparison of Result obtained using Hotelling T^2 Control Chart and the result produced by FCM clustering method

Method	Likely cases	Unlikely cases	Similarity Percentage
Individual Hotelling T^2 control chart	41	398	86.33%
Fuzzy C-Means clustering algorithm	47	392	

6. Conclusion

In this paper, it has been shown that the Fuzzy C-Means (FCM) clustering algorithm has been successfully employed to classify excessive domestic water consumption using consumer data. The result from this study was compared with other unsupervised learning method, which demonstrates the potential use of FCM for the classification of domestic consumer water consumption data. Note that, in this study, there is no validation data available. Hence, further research should look for the use of observed data collected from house visits to obtain class labels so that comparison and validation can be made with other supervised learning classification methods.

Acknowledgments

This research work is funded by the Ministry of Higher Education, Malaysia under the Fundamental Research Grant Scheme (FRGS) with reference number 600-RMI/FRGS TD 5/3 (1/2015). The authors also would like to thank the Institute of Research Management and Innovation (IRMI), Universiti Teknologi MARA, Malaysia.

References

[1] Arbués F, García-Valiñas M Á, and Martínez-Espiñeira R 2003 *The Journal of Socio-Economics* **32** 81-102
[2] Sahu H B, Mahapatra S S, and Panigrahi D C 2012 *Fuel Processing Technology* **104** 115-120
[3] Yuan X, Khoshgoftaar T M, Allen E B, and Ganesan K 2000 *Proc. of the 3rd IEEE Symp. on Application-Specific Systems and Software Engineering Technology* 85-90
[4] Wang X Y, Garibaldi J, and Ozen T 2003 *Proc. of the Int. Conf. on Australian and New Zealand Intelligent Information Systems* 233-238
[5] Tsuen-Ho H, Kao-Ming C, and Hei-Chun C 2000 *Proc. of the Ninth IEEE Int. Conf. on Fuzzy Systems* 621-626
[6] Ghosh S and Dubey S K 2013 *Int. J. of Advanced Computer Science and Applications* **4** 36-39
[7] Savic D A and Walters G A 1999 *Anti-Corrosion Methods and Materials* **46** 415-425
[8] Makropoulos C and Butler D 2004 *J. of Hydroinformatics* **6** 3-18
[9] Naizhuo Z and Yunsheng Z 2010 *Proc. of the 2010 4th Int. Conf. on Bioinformatics and Biomedical Engineering* 1-4
[10] Altunkaynak A, Özger M, and Çakmakci M 2005 *Water Resources Management* **19** 641-654
[11] Şen Z and Altunkaynak A 2009 *Expert Systems with Applications* **36** 11745-11752
[12] Corona-Nakamura M A, Ruelas R, Ojeda-Magana B, and Andina D 2008 *Proc. of the World Automation Congress* 1-9
[13] Corona-Nakamura M A, Ruelas R, Ojeda-Magana B, and Carr Finch D W 2009 *Proc. of the IEEE Int. Conf. on Systems, Man and Cybernetics* 3751-3756
[14] Hammouda K and Karray F 2000 *SYDE 625: Tools of intelligent systems design, course project, University of Waterloo Canada*
[15] Jipkate B R and Gohokar V V 2012 *Int. J. of Computational Engineering Research* **2** 737-739
[16] Bezdek J C, *Pattern Recognition with Fuzzy Objective Function Algorithms*: Kluwer Academic Publishers, 1981.
[17] Azar A T, El-Said S A, and Hassanien A E 2013 *Computer Methods and Programs in Biomedicine* **111** 1-16
[18] Mohd Hanif H, Rasmani K A, and Mohamed Ramli N 2013 *Proc. of the IJAS American-Canadian Int. Conf. on Academic Disciplines 2013*

Modelling of nonlinear filtering Poisson time series

Vladimir V Bochkarev and Inna A Belashova

Kazan Federal University, Russia

E-mail: inkin91-91@mail.ru

Abstract. In this article, algorithms of non-linear filtering of Poisson time series are tested using statistical modelling. The objective is to find a representation of a time series as a wavelet series with a small number of non-linear coefficients, which allows distinguishing statistically significant details. There are well-known efficient algorithms of non-linear wavelet filtering for the case when the values of a time series have a normal distribution. However, if the distribution is not normal, good results can be expected using the maximum likelihood estimations. The filtration is studied according to the criterion of maximum likelihood by the example of Poisson time series. For direct optimisation of the likelihood function, different stochastic (genetic algorithms, annealing method) and deterministic optimization algorithms are used. Testing of the algorithm using both simulated series and empirical data (series of rare words frequencies according to the Google Books Ngram data were used) showed that filtering based on the criterion of maximum likelihood has a great advantage over well-known algorithms for the case of Poisson series. Also, the most perspective methods of optimisation were selected for this problem.

1. Introduction

Widespread methods of time series filtering, especially wavelet thresholding, involve usage of data in which probability of fluctuations distribution is subject to the normal law [1]. Algorithms of wavelet thresholding are based on data dimension reduction strategy. It is assumed that to describe information contained in the series, it's enough to have several nonzero wavelet coefficients, which number is considerably smaller than the length of the series.

The minimum square error criterion is used for series with a normal distribution of variations. This results in a well-established strategy of data filtering. It implies that wavelet coefficients, which modulus values are below a certain threshold, are set to nil. If the distribution of variations differs significantly from the normal one, this strategy cannot produce good results. In this case, more general criterion of maximum likelihood is normally used to select non-zero coefficients, as well as their numerical values.

For data, which distribution differs from the normal one, the likelihood function depends on the parameters in a complicated way. Therefore, to find the parameters, we use numerical optimization methods such as genetic algorithms, simulated annealing method and deterministic algorithms (method confidence regions) [2,3].

In this article, we examined filtration of time series based on the criterion of maximum likelihood by the example of the series, which distribution is governed by Poisson's law. Poisson's law is introduced as a universal distribution for a number of independent rare events and is widely used for solving practical problems. The amount of network treatment, the number of device failures, the

Content from this work may be used under the terms of the Creative Commons Attribution 3.0 licence. Any further distribution of this work must maintain attribution to the author(s) and the title of the work, journal citation and DOI.
Published under licence by IOP Publishing Ltd

number of cases of rare diseases, the frequency of rare words and many other values are distributed by Poisson's law.

One of the interesting practical applications can be visualization and analysis of time series of word usage frequencies obtained using diachronic text corpora, for example, Google Books Ngram. The Google Books Ngram (https://books.google.com/ngrams) corpus, based on the largest electronic collection of texts Google Books, allows obtaining time series of word usage frequencies for 8 languages for the last five centuries. The project focuses primarily on the study of changes in the language and culture [5]. The online service Google Books Ngram Viewer is available on the project page. It allows visualization of time series of the selected words frequencies, using anti-aliasing by means of the moving average for better visual representation. Using the moving average can result in a distorted view of rapid variations of frequencies, jumps and bursts associated with certain historical events. Best results are obtained using nonlinear wavelet filters. However, to apply them to the series of rare words frequencies, they need to be adapted to the Poisson distribution.

This work has two objectives:

- To check how far significant benefit can be obtained from filtering of Poisson time series based on the criterion of maximum likelihood;
- To select the most appropriate algorithms to optimize the likelihood function.

To assess the quality of filtering, we use both simulated test Poisson series with known parameters and actual series of rare words frequencies. The quality of filtration for the series of empirical data can be evaluated on the basis of qualitative considerations that take into account the Poisson distribution properties.

2. Method

Likelihood function algorithm is used as a quantitative measure of approximation quality for a time series with the Poisson distribution [4]:

$$llh = const + \sum_{t} \{x_t \log \lambda_t - \lambda_t\}$$

Here λ_t is a mathematical expectation of the process at time t. The objective is to reduce dimension of the investigated time series by presenting the sequence λ_t as a wavelet series with a small number of items. It can be shown as follows:

$$\lambda_t \approx \sum_{(j,k)\in I} c_{jk}\psi_{jk}(t)$$

At that, the number of items in the formula (the content of set I) should be significantly smaller than the length of the series.

Two problems are to be solved: 1) to select the most suitable set of non-zero coefficients I; 2) to determine the optimal numerical values for these coefficients. The first task is more difficult than the second one because discrete optimisation should be performed with a great number of options. In this article, we tried to solve this problem using stochastic search algorithms suspenders (simulated annealing, genetic algorithms), as well as the greedy algorithm. The second problem can be easily solved using any known algorithm of non-linear optimisation.

Thus, the filtration algorithm includes the following steps:

- Selection of the initial set of non-zero coefficients;
- Defining a new set of non-zero coefficients;
- Varying the values of nonzero coefficients in order to minimize the quality function – the maximum likelihood function;
- Checking the termination condition and if it is not satisfied – going to the second step.

Let`s consider these steps in detail. Initially, the set of non-zero coefficients and the values of these coefficients can be selected in accordance with a known wavelet thresholding algorithms (in this paper, we used the fixed form threshold). The final result of applying such approach will certainly be no worse (for the value of the likelihood function) than using the standard method.

The set of non-zero coefficients is given as a binary vector, where 1 corresponds to non-zero coefficient and 0 corresponds to zero coefficient. In each of the above algorithms, the rules of varying this vector include saving the number of units (i.e., non-zero wavelet coefficients). When using the genetic algorithms, the vector turns to be a chromosome, to which mutation and crossover operations are applied at each step. After that, numerical values are specified for all the obtained coefficients, the log-likelihood function is calculated and the selection is performed. The desired final result is a set of non-zero coefficients at which the value of the likelihood function is maximum.

We tested the following filtering methods:

- Standard wavelet-thresholding;
- Method with optimization of the log-likelihood function by simulated annealing;
- Method with finding the optimal values of coefficients determined by the search (the dog leg algorithm was used). At that, the selection of non-zero coefficients was not changed.
- Method with optimization of the log-likelihood using genetic algorithms. At that, the genetic algorithms were used both for selection a set of non-zero coefficients and location their optimal values;
- Method which is a combination of the two previous methods; In this case, the selection of the set of nonzero coefficients is performed using genetic algorithms and the numerical values are calculated using the dog leg algorithm (hereinafter, this option is called a "combined method");
- Method with optimization of log-likelihood using "greedy" algorithm.

Let`s consider the last method. As the below examples show, the best results are obtained by using the genetic algorithms and the combined method. However, if the data size is great, it will take a lot of time to determine the best set of non-linear coefficients. In order to speed up the filtering, the "greedy" algorithm was used. It's realized in the following way: first, we assess how far the log-likelihood function is changed as a result of zeroing of one or another coefficient. Then, at each step, we set to nil the coefficient, the removal of which changes the objective function least of all and update the values of the coefficients. The algorithm is realized when a predetermined number of non-zero coefficients is left.

3. Results and discussion

A comparative testing of different filtering algorithms was performed using statistical modelling. Poisson series with time-dependent parameter $\lambda_t$were simulated. In most cases, filtering using the criterion of maximum likelihood shows significantly better results compared with the standard wavelet thresholding. If various optimization algorithms are compared, the best results are obtained if the combined algorithm is used (the set I is selected by the genetic algorithm, and the values of the coefficients are found by the determined search). Figure 1 shows an example of the filtering result for the test case «Noisy Bumps» from the Wavelet Toolbox of the Matlab package. It should be noted that different filtering methods are compared using the same number of non-zero coefficients. As can be seen from this figure, filtering methods based on the maximum likelihood criterion described both broad and narrow peaks doing the task better than the standard wavelet thresholding. Special attention should be paid to the intervals 120-160 and 600-700. There are two relatively narrow peaks in the initial series λ_t. These peaks are not seen as two peaks but as a single broad peak when the standard wavelet-thresholding is used. However, these peaks are distinguished as two peaks if the combined filtering method is used. This can be important for real practice. For quantitative comparison, Table 1 shows values of log-likelihood for the given example using different filtering methods.

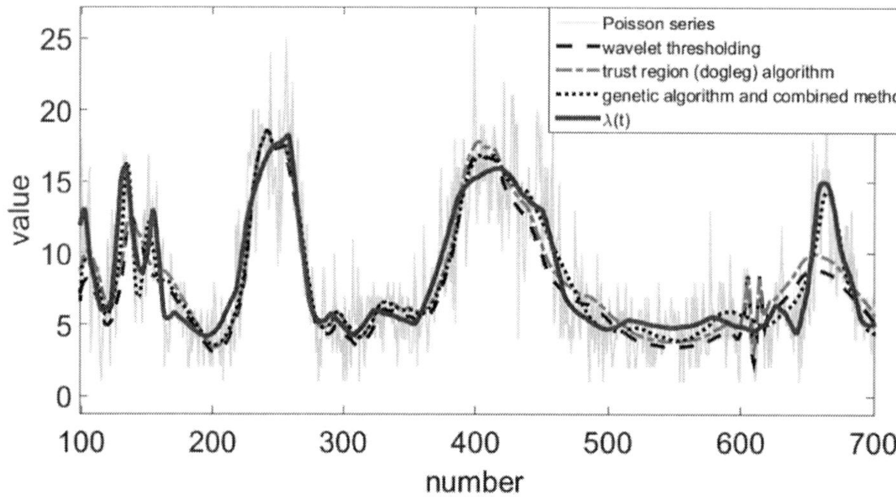

Figure 1 Comparison of the filtering results with different methods of the Poisson series (the example of "Noisy Bumps" from the Wavelet Toolbox package MATLAB)

Table 1 Comparison of the filtering results with different methods of the Poisson series (the example of "Noisy Bumps" from the Wavelet Toolbox package MATLAB)

Wavelet filtering	Genetic algorithms	Simulated annealing	Determined search	Combined method
$8.1178 \cdot 10^3$	$8.2146 \cdot 10^3$	$8.1178 \cdot 10^3$	$8.1643 \cdot 10^3$	$8.2147 \cdot 10^3$

If we take the largest and the smallest value of the log-likelihood function shown in the table and determine how many times one model is more probable than the other, it turns out to be a great value - 1.211 • 1042 times. Thus, using the likelihood function as a criterion for evaluation of filtering has significant advantage over the quadratic error functions for Poisson series. Good results of filtering the experimental data also confirm this.

As an example, Figure 2 shows the results of filtering of some frequencies for the English word «Karelian». It can be seen that the standard wavelet-thresholding cannot distinguish significant peaks associated with historical events. Even the biggest burst of 1939-1941 disappears (associated with the Soviet-Finnish war of 1939-1940 and 1941-1944). Attention should be paid to the unjustified extension of the transition interval in the second half of the 30s. Such disadvantages are not observed when filtering based on the criterion of maximum likelihood is used.

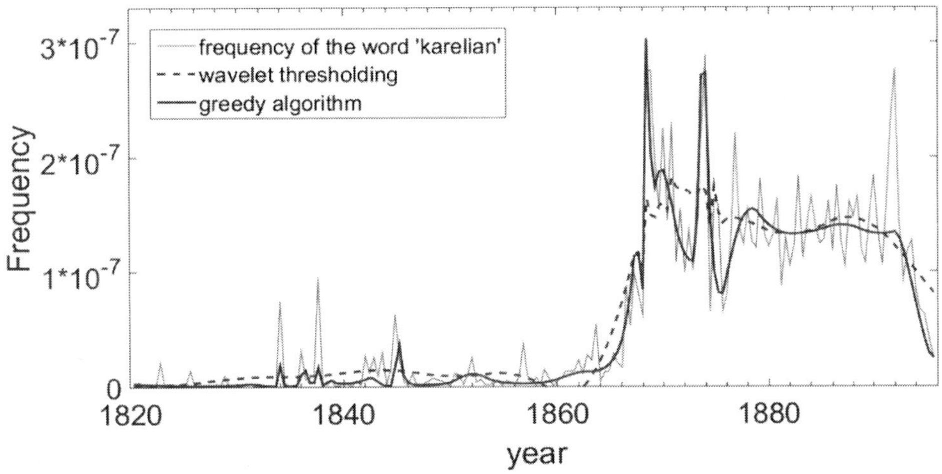

Figure 2 Comparison of the filtering results of the standard wavelet filtering and filtering based on the criterion of maximum likelihood (with optimization of the greedy algorithm). Frequency of use of the word «Karelian» is shown

Based on the qualitative assessment of the actual data filtering (frequency of use of rare words, the number of sunspots), it can be said that filtering based on the criterion of the maximum likelihood function (with optimization using a greedy algorithm) produces the best results, describing not only the trend, but also distinguishing the significant peaks.

4. Conclusion

The obtained results confirm that using the criterion of maximum likelihood has an advantage over minimum mean square error for the case of Poisson series filtering. Using genetic algorithms for selection of non-zero wavelet coefficients allows obtaining high-quality filtering, although it requires a significant amount of calculations. The use of the greedy algorithm can significantly reduce the calculation time with a slight deterioration in the quality.

5. Acknowledgments

This research was supported by the Russian Foundation for Basic Research (grant № 15-06-07402).

References

[1] S. Mallat A wavelet tour of signal processing. (Third Edition), Elsevier, 2009, 799 pp
[2] John R. Koza Genetic Programming: On the Programming of Computers by Means of Natural Selection. The MIT Press, 1992
[3] W. Sun and Y.-x. Yuan, Optimization theory and methods : nonlinear programming. New York: Springer, 2006
[4] Yu.S. Maslennikova, V.V. Bochkarev, D.S. Voloskov Modelling of word usage frequency dynamics using artificial neural network // 2014 J. Phys.: Conf. Ser. 490 012180 (http://iopscience.iop.org/1742-6596/490/1/012180)
[5] Michel, J.-B., et al.: Quantitative analysis of culture using millions of digitized books. Science, 331,176–182 (2010)

Vaccination Strategies: a comparative study in an epidemic scenario

D. B. Prates[1], C. L. T. F. Jardim[2], L. A. F. Ferreira[3], J. M. da Silva[4] and M. V. Kritz[5]

Universidade Federal dos Vales do Jequitinhonha e Mucuri, Minas Gerais - Brazil

E-mail: derekbomfim@hotmail.com[1], caiodt4@hotmail.com[2], leticiaalifig@hotmail.com[3], jaqueline.silva@ufvjm.edu.br[4] and kritz@lncc.br[5]

Abstract. Epidemics are an extremely important matter of study within the Mathematical Modeling area and can be widely found in the literature. Some epidemiological models use differential equations, which are very sensible to parameters, to represent and describe the diseases mathematically. For this work, a variation of the SIR model is discussed and applied to a certain epidemic scenario, wherein vaccination is introduced through two different strategies: constant vaccination and vaccination in pulses. Other epidemiological and population aspects are also considered, such as mortality/natality and infection rates. The analysis and results are performed through numerical solutions of the model and a special attention is given to the discussion generated by the paramenters variation.

1. Introduction

Mathematical models are essential tools that allow scientific and descriptive analysis, since their versatility admit the obtainment of several valuable results for a better comprehension of natural phenomena. Amongst a great number of matters studied in the mathematical modelling area, the epidemiology is one of big importance, since it may bring very useful results and is able to aid public health agencies to better deal with epidemics.

The SIR model (KERMACK, 1927), pioneer at the epidemiological modelling area, describes the spread of an infectious disease by proposing the division of a particular population in three individual categories: susceptible (S), infected (I) and recovered (R). A system of differential equations is used to run this model, which is fed with parameters representing population/epidemic intrinsic characteristics, such as the infection rate. More recent studies have developed mathematical models considering specific aspects for distinct epidemics, such as seasonal influence, cross-immunity and disease containment measures (STONE, 2000).

Attempts of epidemic control and containment in terms of vaccination programs are often seen and suggested. The governmental programs present, among their goals, the prevention of infant/older children diseases, a major health issue in any society. In 2012, the Global Vaccine Action Plan (GVAP) was approved in order to improve the access to vaccines around the world; thus, thousands of deaths can be avoided until 2020.

Several mathematical models allow the vaccination study in a certain population. Vaccination strategies are considered a measure that presents one of the best outcomes in preventing infectious diseases (ZHI-LONG HE, 2015). In (ANDERSON, 1995), an efficiency analysis is

Content from this work may be used under the terms of the Creative Commons Attribution 3.0 licence. Any further distribution of this work must maintain attribution to the author(s) and the title of the work, journal citation and DOI.
Published under licence by IOP Publishing Ltd

performed for the case when a certain parcel of the population is submitted to a vaccination campaign. The eradication condition of infectious diseases is achieved when the parcel of successfully vaccinated individuals is bigger than a specific critical value. The insertion of a vaccination campaign that could achieve this critical value of vaccinated individuals is often unreachable due to economical and time limitations.

In this work, it will be presented and analyzed a model that contains two vaccination strategies: a constant vaccination, wherein a certain parcel of the population that is born is immediately vaccinated; and vaccination in pulses, wherein a certain parcel of the population is vaccinated periodically. It is proposed the utilization of a function that illustrates the vaccination in pulses behavior. An implementation of the parameters variation study is performed. Finally, epidemiological scenarios wherein both strategies are alternately inserted in the same population and scenarios wherein only one of them is applied are numerically compared and discussed.

2. Model and Methods

The traditional SIR model was modified in order to be used with a time variant population considering the influence of the campaign of constant vaccination (STONE, 2000), where a certain parcel of the population ρ is vaccinated right after they are born, preventing its possibility of infection. Then, they are directly introduced in the group of recovered individuals. In order to use such model, it is necessary to describe some parameters that represent characteristics inherent of each disease. Thus, β, α and γ represents infection rate, recovery rate and mortality and natality rates, respectively. Considering *S(t)* the class of susceptible individuals, *I(t)* the class of infected individuals and *R(t)* the class of recovered individuals, the equations that describe the SIR model in this specific vaccination situation are (SABETI, 2011):

$$\begin{cases} \frac{dS}{dt} = (1-\rho_1)\gamma - (\beta I + \gamma)S \\ \frac{dI}{dt} = \beta IS - (\gamma + \alpha)I \\ \frac{dR}{dt} = \alpha I + (\rho_1 - R)\gamma \end{cases} \tag{1}$$

Some vaccination programs applied by health public agencies base themselves in periodic vaccination campaigns (AGUR, 1993), which happens to be an highly suggested method by the literature. In these cases, groups considered at risk are repeatedly vaccinated until the level of transmission and infection is reduced or stopped. So, defining ω as the period of the pulses and ϕ as their phase, the pulse modeling of this kind of vaccination is proposed by the insertion of a function ρ_1 given by:

$$\rho_1 = \left\{ \begin{array}{cc} 0 & if\cos(\omega t + \phi) \leq 0 \\ \rho(\cos(\omega t + \phi)) & if\cos(\omega t + \phi) > 0 \end{array} \right\} \tag{2}$$

In this analysis, the set of parameters used is $\beta = 1800$, $\gamma = 0.02$, $\alpha = 100$ (STONE, 2000) and $\phi = 0$.

3. Results and Discussions

The constant vaccination model has two equilibrium points: $P_1 = ((1-\rho), 0)$, the infection free equilibrium, and $P_2 = (\frac{\gamma+\alpha}{\beta}, \frac{\gamma(\beta-\gamma-\alpha-\beta\rho)}{\beta(\gamma+\alpha)})$, the epidemic equilibrium. It is shown in Figure 1a) that the infected individuals behavior of a population not submitted to a vaccination campaign presents more infection peaks and tends to an equilibrium point with a bigger number of infected people, in comparison with the behavior of the same population submitted to either of the vaccination strategies discussed here. It is also visible that, the bigger the parcel of the vaccinated population is, the bigger is the amplitude between infection peaks (and, therefore, the smaller is the number of these peaks in a same time interval).

Figure 1b) presents a phase plan of the susceptible and infected individuals. It is perceptible that the bigger the population parcel vaccinated is, the faster the epidemic stabilizes. When $\rho = 90\%$, for instance, the system tends fast towards the equilibrium point. This result is clearly explained by the Principle of the Mass Action, which basically says that the infection rate in an epidemic scenario is proportional to the number of infected and susceptible individuals currently present in that population.

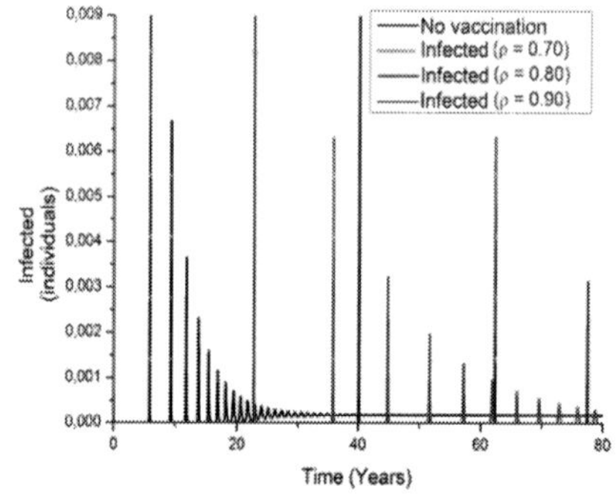

(a) Graphics of infected population versus time submitted to a constant vaccination, where ρ is assumed either 0% (no vaccination whatsoever), 70%, 80% or 90%.

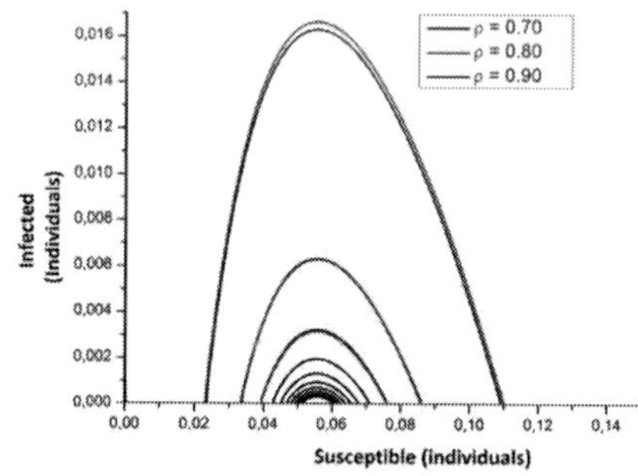

(b) Phase plan of infected population versus suscetible population, where ρ is assumed either 70%, 80% or 90%.

Figure 1. Constant vaccination results

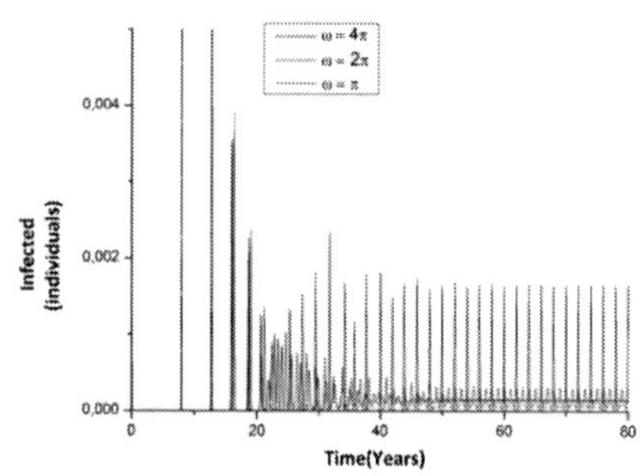

(a) Infected individuals versus time for pulse vaccination campaigns with different periods.

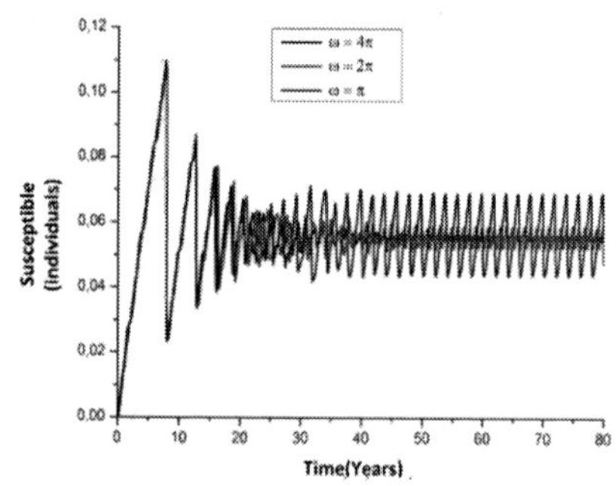

(b) Susceptible individuals versus time for pulse vaccination campaigns with different periods

Figure 2. Pulse vaccination results

Figure 2a) makes a comparison between three pulse vaccination strategies. The first strategy consists in vaccinating the population twice a year, whereas the second strategy vaccines the population once a year. Finally, the third strategy proposes a vaccination pulse happening in every two years. Such period change for the pulse is achieved by changing the value of ω, responsible for the frequency of the pulse function. Figure 2b) shows the susceptible population dynamics under the influence of these vaccination strategies. It is noticed that the bigger ω is, the more stable is their behavior.

In the Figure 3a) and 3b) the dynamics of the infected individuals are presented, in three situations: without the influence of the vaccination, with a constant vaccination technique and in a pulse vaccination scenario. It is seen that the population not submitted to a vaccination campaign tends to an equilibrium point with a higher number of infected individuals when compared to the vaccinated populations. The strategy of constant vaccination presents more infection peaks, but with smaller amplitudes: fewer susceptible individuals are infected in each infection peak.

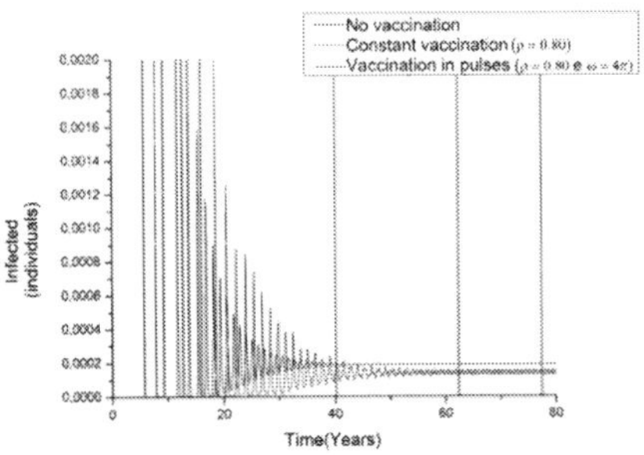

(a) Infected populations behavior through time for different types of vaccination campaigns

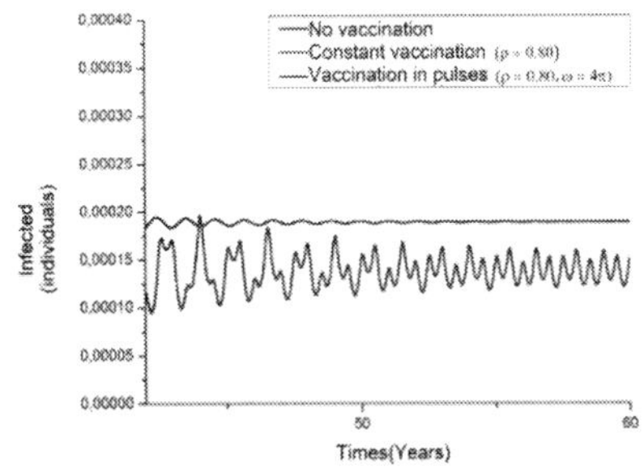

(b) Zoomed view window from Figure 3a)

Figure 3. Comparison between different vaccination strategies

4. Numerical comparison between vaccine insertion strategies

For this analysis, $\rho = 0.2$, $\omega = 4$ and $\phi = 0$ are chosen and a the differential equation system is computationally simulated in three cases: when just one of the vaccinations strategies is applied in the population and when both of them are alternately applied in the population. For the last case, a algorithm would solve each time step for both of the strategies and then use the one it found to work more efficiently, based on a specific efficiency parameter. In the case here described, this parameter is the number of infected people (the lesser infected after a infinitesimal time, more efficient the algorithm considers the strategy to be for that specific time).

Figure 4a) presents the behavior of the three classes in the simulation of a traditional SIR model (wherein no vaccination is applied). The analysis of Figure 4b) shows that the susceptible population in the alternated simulation tends to be smaller than the others. This is expected, since the vaccinated population is automatically moved from susceptible to removed class. Thus, the smaller number of susceptible individuals does not mean more infected individuals, but shows the effectiveness of the vaccination campaign.

Figure 4c) shows that the smaller number of infected individuals in the alternated simulation in fact translates how the strategy of using the algorithm for determining the alternated vaccination works. Finally, Figure 4d) shows that the alternated vaccination strategy makes the removed population tend to a number larger than those reached by the pulse vaccination and constant vaccination strategies alone. Again, this exemplifies the strategy effectiveness by vaccinating a larger number of vaccinated population.

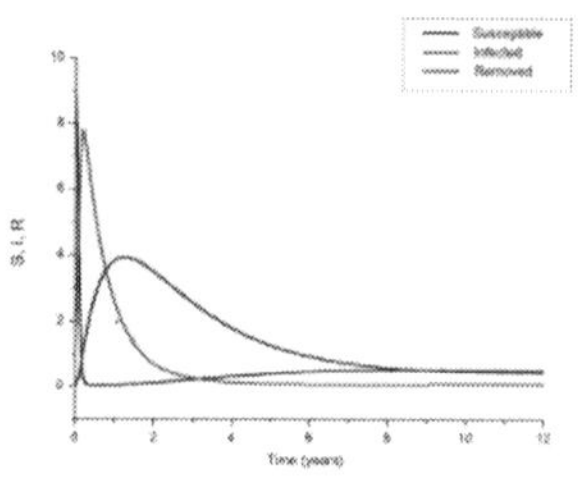

(a) Behavior of the three individual classes through time

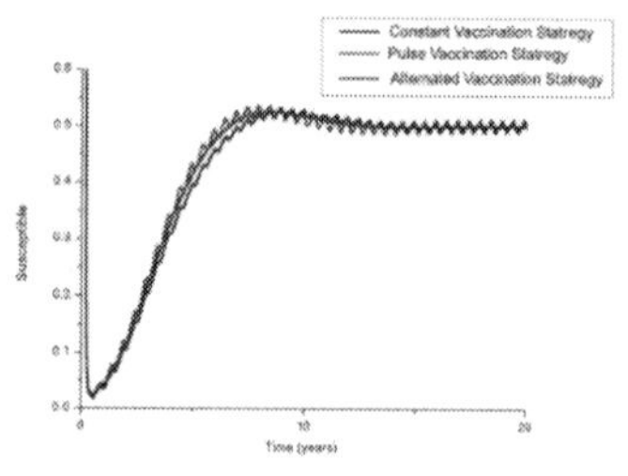

(b) Susceptible population behavior in different simulations

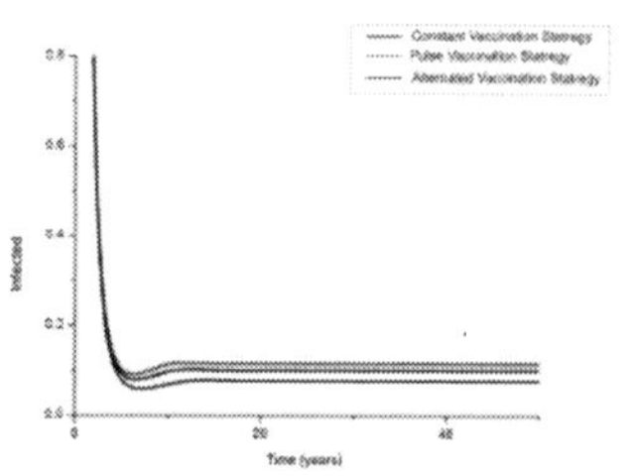

(c) Infected population behavior in different simulations

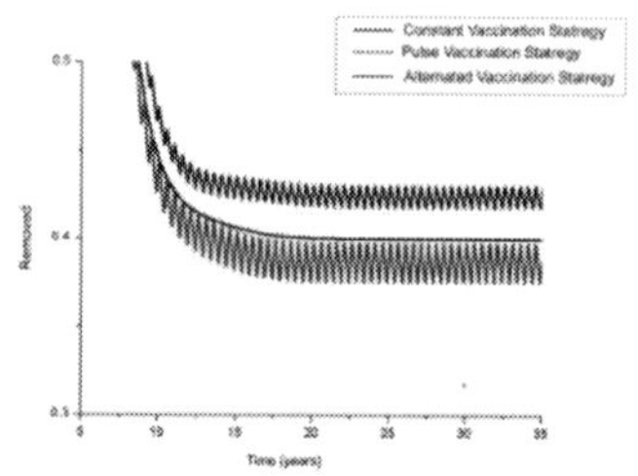

(d) Removed population behavior in different simulations

Figure 4. Simulation results for $\beta = 3.0, \gamma = 0.5, \alpha = 1.0, S_0 = 8.95$ e $I_0 = 1.05$.

5. Conclusion

The modified model has shown itself as an efficient tool for describing and foreseeing the epidemic behavior of a population under vaccination strategies. The proposal of the function ρ for the pulse vaccination strategy seems reasonable for governmental vaccination campaigns and worked well, accordingly to previous works in literature (STONE & SABETI, 2011). In the case of this work, the pulse vaccination has shown better and faster results and stability, which does not mean that will occur every time. However, the numerical analysis performed with an alternated vaccination strategy indicates that a better outcome could be reached if an algorithm like the one used for this work decided the best sequence of vaccinations to apply in each time interval. Since it is a perk of mathematical models its versatility and how they could be used to analyze different epidemics and sets of parameters, the methods and model here presented could be used in order to aid prevention studies and containment measures by health agencies, and even generate cost-benefit and time-sensitive campaigns.

References

[1] Stone, L., Shulgin, B. and Agur, Z., (2000) *Theoretical examination of the pulse vaccination policy in the SIR epidemic model* Mathematical and computer modelling, v. 31, n. 4, p. 207-215.

[2] Sabeti, M., Castilho, A. R. (2011). *Modelo Epidêmico Discreto SIR com estrutura etária e aplicação de vacinação em pulsos e constante*, Tese de Doutorado, UFPE.

[3] Agur, Z., Cojocaru, L., Mazor, G., Anderson, R. M., Danon, Y. L. (1993). *Pulse mass measles vaccination across age cohorts.* Proceedings of the National Academy of Sciences, v. 90, n. 24, p. 11698-11702.

[4] Barros, A. M. R. (2007) *Modelos Matemáticos de equações diferenciais ordinárias aplicados à epidemiologia* UFLA.

[5] Jardim, C. L. T. F., Prates, D. B., Silva, J. M., Ferreira, L. A. F., and Kritz, M. V. (2015). *Simulations of a epidemic model with parameters variation analysis for the dengue fever.* In Journal of Physics: Conference Series (Vol. 633, No. 1, p. 012008). IOP Publishing.

[6] Kermack, W.; McKendrick, A. (1927) *Contributions to the mathematical theory of diseases.* Proceedings of the Royal Society, **141** 1933.

[7] R. Anderson and R. May, (1995) *Infectious Disease of Humans, Dynamics and Control*, Oxford University Press, New York, NY, USA.

[8] Prates, Dérek B., Silva, Jaqueline M., Gomes, Jessica L. and Kritz, Maurcio V., *An epidemiological model with vaccination strategies*, AIP Conference Proceedings, 1738, 480035, 2016.

Flow study in channel with the use computational fluid dynamics (CFD)

W. D. Oliveira[1], M. S. G. Pires[1], L. M. Canno[1], L. C. L. J. Ribeiro[1]

[1]Unicamp – State University of Campinas – College of Technology
R. Paschoal Marmo 1888, Centro, CEP 13484-332, Jd. Nova Itália, Limeira, SP, Brazil.

E-mail: wesleydias96@gmail.com, marta@ft.unicamp.br, laura@ft.unicamp.br, lubienska.ft@gmail.com

Abstract. The Computational Fluid Dynamics (CFD) is a tool used to numerically simulate fluid flow behavior, and all the laws that govern the study of fluids is the mass transfer and energy, chemical reactions, hydraulic behaviors, among others applications. This tool mathematical equation solves the problem in a specific manner over a region of interest, with predetermined boundary conditions on this region. This work is to study the flow channel through the CFD technique.

1. Introduction

The flow is classified into free surface, or simply free if, whatever the transversal section, the liquid is always in contact with the atmosphere. This is the situation of the flow in rivers, streams or channels. Rain water galleries, tunnels, channels and, gutter work as free conduits collectors drains. As characteristics of this type of flow, it can be said that it necessarily occurs by gravity and that any disturbance in localized portions can lead to changes in the cross section of the chain in other parts [1]. Therefore, are considered channels all conduits leading water with a free surface, with open or closed section [2].

Considering a channel the movement is continuous it is necessary that no new entries or out of new liquid occurs, so the quantity of liquid remains constant thereby maintaining the flow in the channel. If the flow section will not change, and the nature of the walls remain the same, while maintaining the constant depth and speed, the motion is considered uniform and the channel will also be considered uniform.

The most common flow when the fluid is water, which has relatively low viscosity, are classified as turbulent. In the case of free turbulent flow, ie those that occur in channels usually be divided into fluvial and torrential regime. The river is characterized when the average speed in a particular section is less than a certain critical value. The torrential is characterized when the average speed in a particular section is greater than a certain critical value.

In fluid flow, the numerical model is established through the classical transport equations: conservation of momentum, mass and energy, combined with the turbulence models [3] [4] [5] [6]. To reproduce the flow, from a mathematical point of view, the resolution of the equations is highly complex, which leads to the need to use computers. A strongly recommended methodology to solve these models is the Computational Fluid Dynamics (CFD), which aims to numerically solve the equations to simulate fluid dynamics, and thus solve many practical problems.

Content from this work may be used under the terms of the Creative Commons Attribution 3.0 licence. Any further distribution of this work must maintain attribution to the author(s) and the title of the work, journal citation and DOI.
Published under licence by IOP Publishing Ltd

The computational model should be able to describe the physical behavior, mimicking the experimental system behavior; should be evaluated and compared to experimental, as the end result numerical system; must be capable of supporting theories or hypotheses, which explain the observed behavior; must be able to predict future behavior, that is, the effects produced by changes in system variables or mode of operation [7].

Currently, there are several commercial software using the computational fluid dynamics, such as PHOENICS CFX, STAR-CD, FLUENT, FLOW-3D and ANSYS CFX. In a simplified manner, these tools describe complex surfaces that are subdivided into numerous particles that form the mesh, in each of these particles are carried algebraic equations that are related to specific parameters of the problem. The resulting set of equations will be solved interactively, resulting in a complete description of the flow over the entire area. Therefore, the purpose of this paper is to study channels using CFD.

2. Computational Fluid Dynamics (CFD)

The Computational Fluid Dynamics (CFD) is characterized as numerical simulation of phenomena involving fluid flow. Applications of this technique of analysis and simulation of industrial processes started in 1995, a part of the process, equipment or an accessory, was examined individually in order to obtain information about the flow patterns that can support a performance increase study [8]. CFD technique can be used in: - aerodynamics and spacecraft vehicles; - Hydrofoil vessels; - Combustion engines and gas turbines; - Flow within the diffuser; - Cooling equipment, including micro-circuits; - Chemical process engineering: mixing and separation, polymer modeling; - Environmental engineering: distribution of pollutants and effluents; - Hydrology and oceanography: flows in rivers, estuaries and oceans; - Meteorology: weather; - Biomedical engineering: blood flow through arteries and veins; - Rotor pumps in order to optimize the geometry; - Heat exchangers, in order to maximize thermal exchanges for analysis of the effects of the use of baffles and vanes and alternative geometric configurations; - Cyclone of the catalytic cracking units oil in order to develop designs with high collection efficiency and low pressure drop; - Tanks reactors agitated as no ideal model reactor for the prediction of dead zones and circulation and the effects of baffles and stirrers in mixing conditions; among others [9].

The equations of the mathematical model used by CFD are subjected to initial and boundary conditions, and represent a particular problem. The analytical solution of these equations is possible only for very simple flows. To analyze real problems numerical methods are used. Traditional methods for the numerical solution of differential equations are the Finite Differences Methods, Finite Volumes and Finite Element. In commercial packages, preference is given to the Finite Volumes depending on their robustness and their conservative characteristics [10]. Through CFD simulations can predict data such as power consumption, flow pattern and solids concentration. It is therefore a powerful tool for predicting three-dimensional flow and distribution of pollutant concentration.

CFD technology has become a key part in the design and analysis of products and processes of many companies, due to their ability to predict the performance of equipment and processes even before they are produced or implemented. Due to the increased processing power of computers, the evolution of graphics and interactivity in the 3-D image manipulation tool has become less expensive, reduced the simulation time and hence the cost of use [11].

These models are useful in granting processes and frameworks of water bodies, produce estimates of the concentration of conservative and non-conservative pollutants or releases are continuous or instantaneous [12].

3. Results and Discussions

The system studied is composed of a channel portion constructed in reduced scale, corresponding to approximately 9 meters long with session 20x20 cm and two large reservoirs buried who assisted the circulation process of the fluid (Figure 1). In all studies that use such tools (computational fluid dynamics) experimental results are necessary because serves as comparative confirmation of results obtained through studies. We raised all the characteristics of the channel by sampling and following the pollutant plume (Figures 2 and 3). The channel with the collectors can be seen in Figure 4.

Figure 1. Channel stretch.

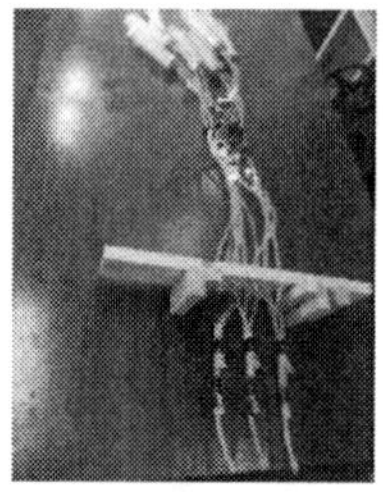

Figure 2. System Collector.

Figure 3. Channel with collectors.

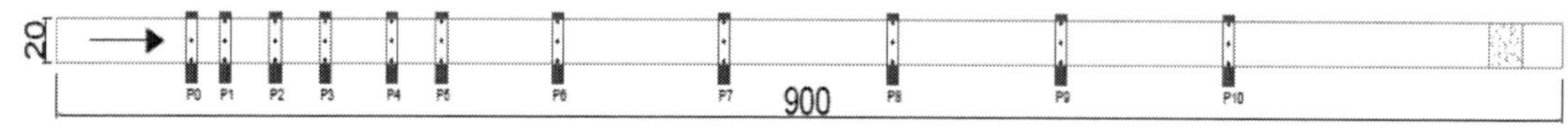

Figure 4. Profile channel with collection points.

With the data collected from surveys carried out in the channel (Figure 5), it was possible to conduct studies in the CFD. The first stage of studies in the tool is the construction of the channel geometry and the study area mesh, called a pre-processing stage. The mesh refinement is directly related to the accuracy of the results presented by the simulation and, it is a very laborious step (Figure 6). Then comes the simulation, the first processing stage, which defines the domain and the boundary conditions of the study area, the fluid properties and initial conditions for convergence processing (Figure 7). The second stage of processing is the resolution of the transport equations. And last post-processing, where is possible view the results. The values found experimentally and simulated are similar (Figure 8). With the modeled system it is possible predict future behavior, both in case of changes in system variables or in operation mode.

Característica	Pontos										
	P0	P1	P2	P3	P4	P5	P6	P7	P8	P9	P10
Área Molhada (m^2)	0,029	0,031	0,032	0,033	0,034	0,035	0,037	0,04	0,043	0,046	0,049
Velocidade (m/s)	0,448	0,419	0,406	0,394	0,382	0,371	0,351	0,325	0,302	0,283	0,265
Vazão (m^3/s)	0,013	0,013	0,013	0,013	0,013	0,013	0,013	0,013	0,013	0,013	0,013

Figure 5. Characterization of channel collection points.

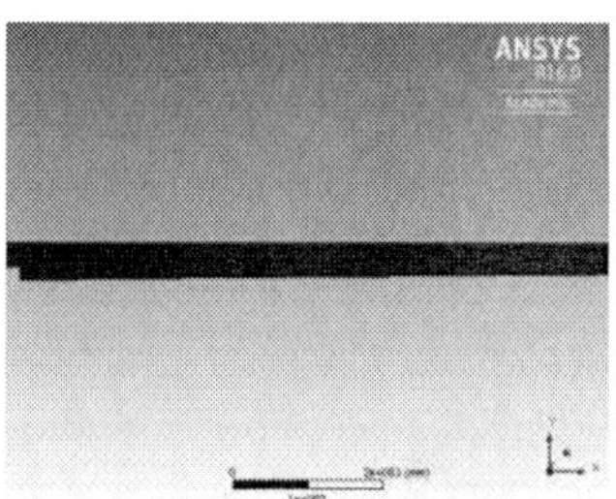

Figure 6. Mesh Definition.

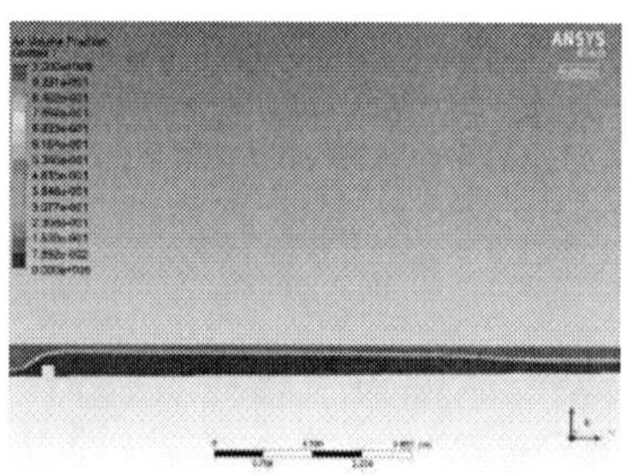

Figure 7. Volume setting to be simulated.

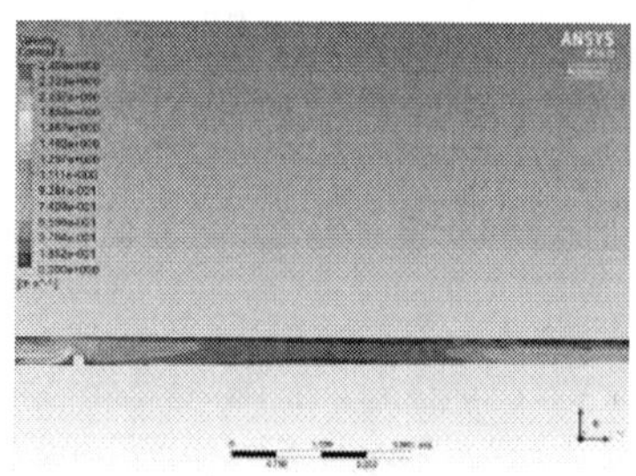

Figure 8. Simulation speed in the channel.

4. Conclusion

Even if the results provided by the computational technique are limited by the parameters used in the simulation, as in the experimental method, for the user is very easy to change parameters such as geometry, temperature and speed, as these are only "input data" for the simulator. In this sense, computational techniques are much closer to experimental than theoretical. So nowadays it is common to find the term "numerical experiments" in reference to the simulations of the same phenomena performed with different parameters [13]. The three strategies are compared to solve the fluid mechanics problems: experimental technique has more realistic advantages and present as disadvantages the required equipment, scale problems, simulation difficulty operating costs; Theory has the advantage of being more general, closed formula and as disadvantages restricted to simple geometries and physical processes generally restricted to linear problems; the numerical technique has the advantages the no restriction on linearity, geometries and complicated processes, temporal evolution of the process and disadvantages truncation errors, prescription of the appropriate boundary conditions, computational costs [14]. Despite all the flexibility that computational technique offers, it cannot solve many of the real problems involving flow fluid [15].

The channels through CFD studies are still scarce. Thus, it was proposed for this research some channel experimental studies in laboratory. It is possible to see that the model was able to describe the physical behavior, mimicking the experimental system behavior. CFD can be used to validate a component in the project stage and can be used to find ways to improve the characteristics of components already implemented. The simulation of computational fluid dynamics is practical and can change design feature with simple and fast modification in the Model geometry.

Acknowledgment

The authors would like to thank the institutions where they conduct their research for supporting this project and also the financial support provided by CNPq and PRP/UNICAMP.

References

[1] PORTO, R. M. Hidráulica Básica. 2.ed. São Carlos. EESC-USP. 2001.
[2] AZEVEDO NETTO, FERNANDEZ, M. F., Araújo, R., Ito, A. E. Manual de Hidráulica. 8.ed. 2002.
[3] LAUNDER, B. E.; SPALDING, D. B. Lectures in Mathematical Models of Turbulence. London: Academic Press, England, 1972.
[4] PATANKAR, S. V.; SPALDING, D. B. Calculation Procedure for Heat, Mass and Momentum Transfer in Three-Dimensional Parabolic Flows. Int. J. Heat Mass Transfer. v. 15, p. 1789. 1972.
[5] MALHOTRA, A.; BRANION, R. M. R.; HAUPTMANN, E. G. Modelling the Flow in a Hydrocyclone. Canadian J. Chem. Eng. v. 72, p. 953-960. 1994.
[6] CULLIVAN, J. C.; WILLIAMS, R. A.; CROSS, C. R. Understanding the Hydrocyclone Separator Through Computational Fluid Dynamics. Trans IChemE. v. 81, p. 455-466. 2003.
[7] FONTES, Carlos E.; SILVA, Luiz F. L. R.; LAGE, Paulo L. C.; RODRIGUES, Ricardo C. Introdução a Fluidodinâmica Computacional, 1. ed. Rio de Janeiro: Escola Piloto Virtual-UFRJ, 2005.
[8] OLINO, A. L. M. Otimização do impelidor KPC utilizando Fluidodinâmica Computacional (CFD). Campinas-SP: 2010. Dissertação (Mestrado) – Universidade Estadual de Campinas.
[9] SOARES, C. Desenvolvimento de uma metodologia para avaliação numérica e experimental do escoamento líquido/vapor em colunas de destilação. Campinas-SP: 2005. Tese (Doutorado) - Universidade Estadual de Campinas.
[10] MALISKA C.R. Transferência de Calor e Mecânica dos Fluidos Computacional. Florianópolis: LTC, 2004.
[11] LOMBARDI, J. C. Análise de distribuição de pressão em válvulas de diafragma poroso. 2005. 125 f. Tese (Doutorado em computação aplicada) – Instituto Nacional de Pesquisa Espaciais, São José dos Campos.
[12] DEVENS, J.A. Quantificação do coeficiente de dispersão longitudinal em pequenos cursos d'água naturais com o uso de traçador ambientalmente neutro. Dissertação de mestrado. Universidade federal de Ouro Preto - Ouro Preto, 2006.
[13] FLETCHER, W.J. and R.J. TREGONNING, 1992. Distribution and timing of spawning by the australian pilchard, Sardinops sagax neopilchardus off Albany, Western Australia. Aust. J. Mar. Freshwat. Res. 43:1437-1449.
[14] TANNEHILL, I.R. 1947. Drought and Its Causes and Effects. Princeton University Press, 597 pp.
[15] SIMÃO, J. A. C. Modelação em Hidráulica Fluvial e Ambiente. 2. Ed. COIMBRA. 2009.

Study of Dispersion Coefficient Channel

K. R. Akiyama[1], C. K. Bressan[1], M. S. G. Pires[1], L. M. Canno[1], L. C. L. J. Ribeiro[1]

[1]Unicamp – State University of Campinas – College of Technology
R. Paschoal Marmo 1888, Centro, CEP 13484-332, Jd. Nova Itália, Limeira, SP, Brazil.

E-mail: kahakiyama@hotmail.com, carolkley@hotmail.com, marta@ft.unicamp.br, laura@ft.unicamp.br, lubienska.ft@gmail.com

Abstract. The issue of water pollution has worsened in recent times due to releases, intentional or not, of pollutants in natural water bodies. This causes several studies about the distribution of pollutants are carried out. The water quality models have been developed and widely used today as a preventative tool, ie to try to predict what will be the concentration distribution of constituent along a body of water in spatial and temporal scale. To understand and use such models, it is necessary to know some concepts of hydraulic high on their application, including the longitudinal dispersion coefficient. This study aims to conduct a theoretical and experimental study of the channel dispersion coefficient, yielding more information about their direct determination in the literature.

1. Introduction

Environmental degradation has become a problem in the world. The deterioration of water quality pollution has become one of the biggest environmental problems in recent years. Water resources are being increasingly damaged, suffering from increased pollution loads that cause sediment transport and chemical and organic water contamination. This water pollution leads to special care, so the Brazilian environmental legislation is increasingly restrictive. Therefore, efforts can be made to mitigate the adverse impacts of urbanization and industrialization, it is urgent that the parameters that guide the pollutant transport equation are true to reality of pollutant dispersion to ensure strategy to manage resources regarding emissions pollutants into rivers.

One of parameters for analyzing the body fluid of a water quality is the scattering coefficient, which informs the capacity of the water body must carry and disperse the pollutants. The scattering coefficient is also called overall mass transfer coefficient. A way to study the scattering coefficient is conducting experiments using tracers. The tracer can monitor the plume of pollution and through collections determine the pollutant concentration curve. This paper discusses the theoretical and experimental study of the scattering coefficient in the laboratory channel.

2. Scattering coefficient

Regarding the growing concern over pollution of water resources, several studies about the spatial and temporal distribution of pollutant concentration are being carried out, as well the development of mathematical models that try to predict this distribution [1]. According to [5] the models require well-evaluated parameters to produce reliable prognostic. In rivers and streams, it is necessary to know the average flow velocity and the coefficient of longitudinal dispersion [2]. When a constituent is

Content from this work may be used under the terms of the Creative Commons Attribution 3.0 licence. Any further distribution of this work must maintain attribution to the author(s) and the title of the work, journal citation and DOI.
Published under licence by IOP Publishing Ltd

immersed in a fluid, it passes through two physical transport processes: advection and molecular diffusion. The molecular diffusion occurs due to the random movement of molecules from a region of higher concentration to a region of lower concentration. According to [3], it can be rendered by the following mathematical expression:

$$j_A = -\rho . D_{AB.} \nabla m_A \tag{1}$$

where: j is the diffusive mass flow of species A (kg / (sm2)); ρ is the specific mass of the mixture (kg / m3); D_AB is the molecular diffusion coefficient; and ∇m_A is the gradient of the mass fraction of species.

Advection, in turn, is due to the transport movement of fluid medium [4]. According to [2], the total rate of mass transport is the result of the addition of advective and diffusive portions, which is expressed by Equation 2 according to the Law of Fick considering uniform flow in steady:

$$\frac{\partial C}{\partial t} + U.\frac{\partial C}{\partial x} = D_L.\frac{\partial^2 C}{\partial x^2} \tag{2}$$

where C is the average concentration in the cross section located at a distance x from the release point (ML-3); U is the average flow velocity (LT-1); D_L is the longitudinal dispersion coefficient (L2T-1); and x and t are space variables (L) and temporal (T), respectively.

Group the variables associated to the mass transfer in the form of dimensionless numbers is common. According to [3] the dimensionless parameters are important because they allow the use of results obtained in a surface subjected to a set of convective conditions in geometrically similar surfaces subject to entirely different conditions. Such conditions may be, for example, the nature of the fluid, fluid velocity and the size of the surface. In the literature there are several dimensionless numbers, Reynolds number (Re) Schmidt number (Sc), Sherwood number (Sh), Stanton number (Stm). The Sherwood number, in particular, is a dimensionless parameter concentration on the surface and provides a measure of convective mass transfer, which occurs at the surface. It is also possible to work with the average value of Sherwood, which depends only on the number of Reynalds (Re) and Schmidt (Sc):

$$\overline{Sh} = f(Re, Sc) \tag{3}$$

The longitudinal dispersion is related to the ability of water bodies to disperse substances that come into contact with its waters, and the scattering coefficient indicates this capacity [1]. In the literature there are two ways to obtain the scattering coefficient: by direct methods and indirect methods.

Direct methods are one of the most reliable ways of obtaining the scattering coefficient. This is obtained from experiments in the field of stimulus-response tracers [2]. Among the direct present methods in the literature can be cited: the moment method, the spreading method, the graphics Krenkel methods and Chatwin, the peak concentration method, the method of the reference concentration and setting method [5]. According to [1], currently, the propagation method is the most accepted method to estimate the longitudinal dispersion coefficient.

Several authors have established mathematical expressions that enable the indirect and quickly determination of longitudinal dispersion coefficient [4]. The use of mathematical expressions affects the work to a limitation that should be considered, and some differences between the values obtained by equations with data obtained in the field by several authors can be reported.

3. Methodology

First was necessary defines a tracer to perform the test dispersion in the channel. Thus, the chosen tracer was sodium chloride (NaCl) due to its low cost, ease of measurement and lack of toxicity. Chosen tracer experiment was performed in order to obtain the calibration curve of Figure 1. This experiment aim to obtain conductivity for different amounts of salt in distilled water. The calibration curve was obtained by serial dilutions from a stock solution of 1000 mg / L, shown in Figure 2. The equipment used for reading the conductivity of the concentrations was Mars MB - 11 (Figure 3) calibrated with conductivity solution 146.7 S / cm.

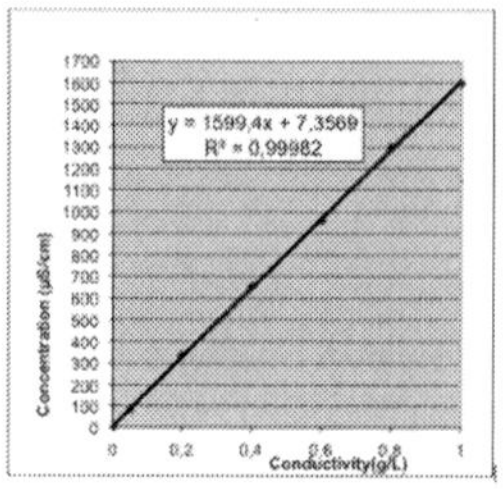

Figure 1. Calibration curve plotter.

Figure 2. The method of successive dilutions.

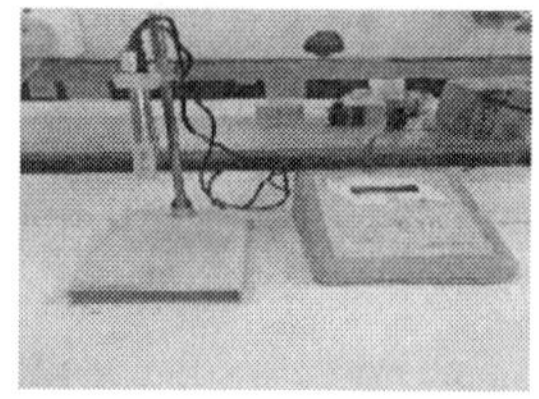

Figure 3. Conductivity Marte MB - 11.

The channel used is transparent with a length of 2.9 mm, a width of 0.10m, and a depth of 30cm (Figure 4). The concentration of tracer present in the storage vessel (Figure 5) was thrown into the channel at a flow rate of 5 l / s, with a distance of 5 cm from the top of the channel and a height of 20 cm. To collect the samples concentration was chosen three points along the channel called P1, P2 and P3 at distances of 55 cm, 56 cm and 275 cm respectively. A control point called P0 was also chosen. For each point, the sample was taken from three different depths, called top, middle and bottom to 20 cm, 13 cm and 6 cm, respectively. The collections were made by syringe connected to each depth, as shown in Figure 6. The collection points and channel data can be seen in Figure 7 and Figure 8.

Figure 4. Channel for Hydraulic experiments.

Figure 5. Storage tank mixing with tracer.

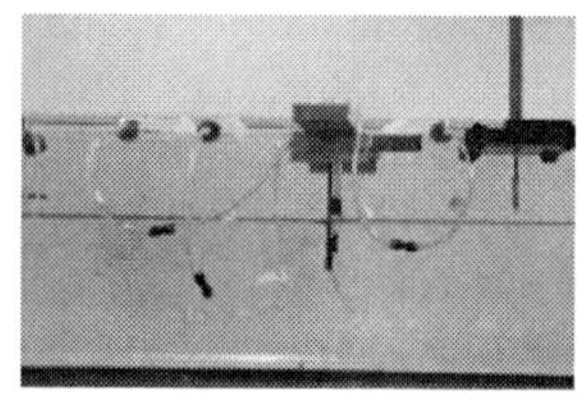

Figure 6. Collection Equipment [6].

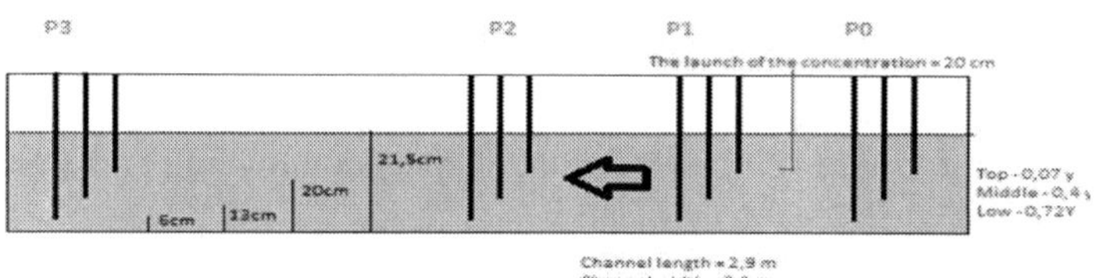

Figure 7. Collection points in the channel.

Characteristic	Point
Haight of Water Depth (cm)	21,5
Channel Base (cm)	10
Channel height (cm)	30
Channel Area (cm^2)	300
Wet Area (cm^2)	215
Wet Perimeter (cm)	53
Hydraulic radius (cm)	4,06

Characteristic	Point
Wet Area (m^2))	0,0215
Flow (m^3/s)	0,004973
Velocity (m/s)	0,231

Figure 8. Channel data.

4. Results and Discussions

We can observe in the results of sampling using the NaCl with the discharge side three sampling points at each point the concentrations obtained through the calibration curve tracer. It is observed the behavior of the dispersion of the tracer at time P1 a peak concentration that is justified because the point is closest to the dump so (Figure 9), the other two points - P2 and P3 - had their behavior with a lower concentration (Figure 10 e Figure 11). The value found for the dispersion coefficient, was 0.00211978.

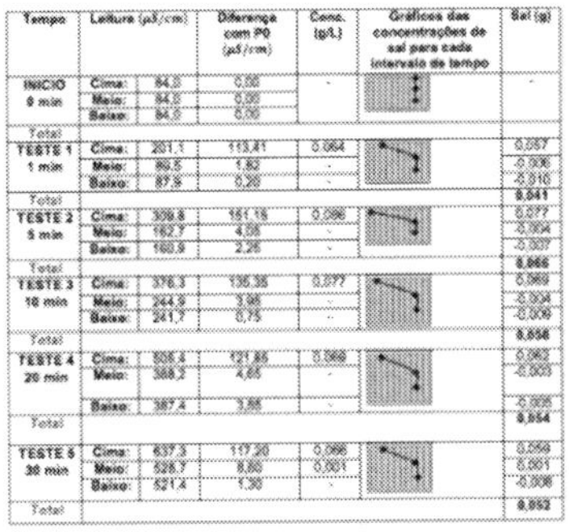

Tempo	Leitura (µS/cm)		Diferença com P0 (µS/cm)	Conc. (g/L)	Gráficos das concentrações de sal para cada intervalo de tempo	Sal (g)
INICIO 0 min	Cima:	84,0	0,00	-		-
	Meio:	84,0	0,00			
	Baixo:	84,0	0,00			
Total						
TESTE 1 1 min	Cima:	201,1	113,41	0,064		0,057
	Meio:	86,5	1,82	-		-0,006
	Baixo:	87,9	0,20	-		-0,010
Total						**0,041**
TESTE 2 5 min	Cima:	309,8	151,15	0,086		0,077
	Meio:	162,7	4,05	-		-0,004
	Baixo:	160,9	2,25	-		-0,007
Total						**0,066**
TESTE 3 10 min	Cima:	376,3	135,35	0,077		0,069
	Meio:	244,9	3,95	-		-0,004
	Baixo:	241,7	0,75	-		-0,009
Total						**0,056**
TESTE 4 20 min	Cima:	505,4	121,85	0,069		0,062
	Meio:	388,2	4,65	-		-0,003
	Baixo:	387,4	3,85	-		-0,005
Total						**0,054**
TESTE 5 30 min	Cima:	637,3	117,20	0,066		0,059
	Meio:	528,7	8,60	0,001		0,001
	Baixo:	521,4	1,30	-		-0,008
Total						**0,052**

Figure 9. Results of samples taken at the point P1 and graphs of concentrations for each collection interval at that point.

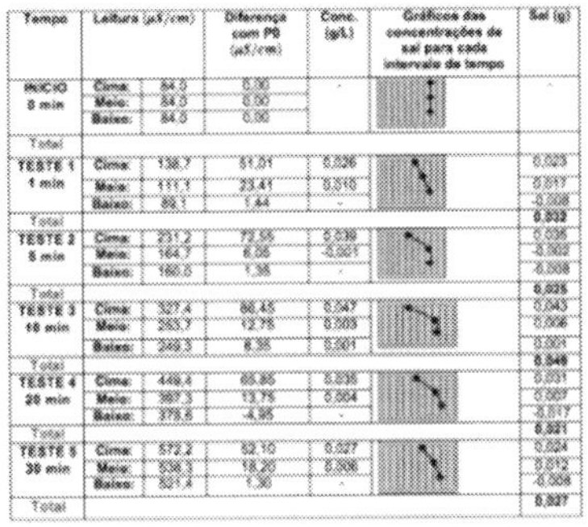

Tempo	Leitura (µS/cm)		Diferença com P0 (µS/cm)	Conc. (g/L)	Gráficos das concentrações de sal para cada intervalo de tempo	Sal (g)
INICIO 0 min	Cima:	84,0	0,00	-		-
	Meio:	84,0	0,00			
	Baixo:	84,0	0,00			
Total						
TESTE 1 1 min	Cima:	138,7	51,01	0,026		0,023
	Meio:	111,1	23,41	0,010		0,017
	Baixo:	89,1	1,44	-		-0,008
Total						**0,032**
TESTE 2 5 min	Cima:	231,2	72,55	0,039		0,035
	Meio:	164,7	6,05	-0,001		-0,002
	Baixo:	160,0	1,35	-		-0,008
Total						**0,025**
TESTE 3 10 min	Cima:	327,4	86,45	0,047		0,043
	Meio:	253,7	12,75	0,003		0,006
	Baixo:	249,3	8,35	0,001		0,001
Total						**0,049**
TESTE 4 20 min	Cima:	449,4	65,85	0,035		0,031
	Meio:	397,3	13,75	0,004		0,007
	Baixo:	378,6	-4,95	-		-0,017
Total						**0,021**
TESTE 5 30 min	Cima:	572,2	52,10	0,027		0,024
	Meio:	538,3	18,20	0,006		0,012
	Baixo:	521,4	1,30	-		-0,008
Total						**0,027**

Figure 10. Results of samples taken at the point P 2 and graphs of concentrations for each collection interval at that point.

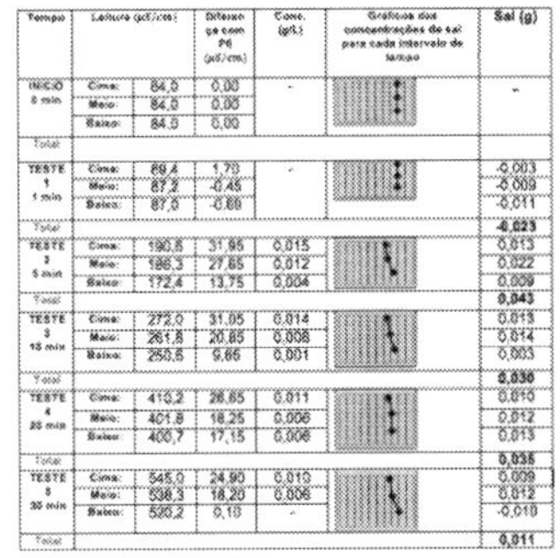

Tempo	Leitura (µS/cm)		Diferença com P0 (µS/cm)	Conc. (g/L)	Gráficos das concentrações de sal para cada intervalo de tempo	Sal (g)
INICIO 0 min	Cima:	84,0	0,00	-		-
	Meio:	84,0	0,00			
	Baixo:	84,0	0,00			
Total						
TESTE 1 1 min	Cima:	89,4	1,70	-		-0,003
	Meio:	87,2	-0,45			-0,009
	Baixo:	87,0	-0,60			-0,011
Total						**-0,023**
TESTE 2 5 min	Cima:	190,6	31,95	0,015		0,013
	Meio:	186,3	27,65	0,012		0,022
	Baixo:	172,4	13,75	0,004		0,009
Total						**0,043**
TESTE 3 10 min	Cima:	272,0	31,05	0,014		0,013
	Meio:	261,8	20,85	0,008		0,014
	Baixo:	250,6	9,65	0,001		0,003
Total						**0,030**
TESTE 4 20 min	Cima:	410,2	28,65	0,011		0,010
	Meio:	401,8	18,25	0,006		0,012
	Baixo:	400,7	17,15	0,006		0,013
Total						**0,035**
TESTE 5 30 min	Cima:	545,0	24,90	0,010		0,009
	Meio:	538,3	18,20	0,006		0,012
	Baixo:	520,2	0,10	-		-0,010
Total						**0,011**

Figure 11. Results of samples taken at the point P 3 and graphs of concentrations for each collection interval at that point.

5. Conclusion

The obtained calibration curve, which allows observe the readings on conductivity for different concentrations were good, since the coefficient of determination was very close to 1 (R2 = 0.998). With this, we concluded that the calculated concentrations in the dispersing test in the channel from the conductivity of the samples are very efficient. By the results obtained in the dispersion test in the channel was observed that the samples taken from P1 had a higher concentration with respect to samples P2 and P3. It was also observed concentrations in deeper layers of P2 and P3 when times changed, and were not detected in P1. An increase in the value of these concentrations was also observed. The graphics allow analyze that the concentrations in P1 were very high, whereas in P3 has almost uniform concentrations of salt in water. P2 presents an intermediate position compared with P1 and P3. In P3 the three depths were much closer values compared to the other two points.

The results showed that there was dispersion of the initial concentration of NaCl spatially and temporally, based on data found for the coefficient at each point.

Acknowledgment

The authors would like to thank the institutions where they conduct their research for supporting this project and also the financial support provided by CNPq and PRP/UNICAMP.

References

[1] BARBOSA JR, A.R.; SILVA, G.Q.; NEVES, B.V.B.; DEVENS, J.A. Direct methods of determination of longitudinal dispersion coefficient of natural watercourses. Minas-Ouro Preto, pag. 27-32, jan. mar. 2005.

[2] OLIVEIRA, V.V.D. Quantification and evaluation of longitudinal dispersion coefficient in watercourses. University of São Paulo - São Carlos, 2013.

[3] INCROPERA, F.P. e DEWITT, D.P. Heat and Mass Transfer Fundamentals. LTC: Books Scientific and Technical. 5ª edição, 2003.

[4] SOARES, P.A. Estimation and application of longitudinal dispersion coefficients and reaeration in natural water bodies. University regional of Blumenau-Blumenau, 2011.

[5] DEVENS, J.A. Quantification of longitudinal dispersion coefficient in small natural de'água courses with the use of environmentally neutral tracer. University federal of Ouro Preto - Ouro Preto, 2006.

[6] LOPES, M. S. Modeling of a hydraulic rectangular channel for the experimental study of a dispersion concervativo tracer. FT/UNICAMP. 2016.

Stochastic method for modeling of the rarefied gas transport coefficients

V Ya Rudyak[1] and **E V Lezhnev**[2]

[1] Doctor of Physics and Mathematics, Professor, Honored Worker of Higher School of Russia, member of the MAS HS, member of the American Nano Society, Head of the Department of Theoretical Mechanics NGASU (Sibstrin),
Novosibirsk State University of Civil Engineering,
113, Leningradskaya Str. Novosibirsk, 630008, Russia

E-mail: valery.rudyak@mail.ru

[2] Ph.D., assistant professor of higher mathematics NSTU
Novosibirsk State Technical University,
20, Av. K. Marksa, Novosibirsk, 630073, Russia

E-mail: lionlev@yandex.ru

Abstract. In this paper, we propose an algorithm for computation of the transport coefficients of rarefied gas, which is based on stochastic modeling of phase trajectories considered molecular system. The hard spheres potential is used. The number of operations is proportional to the number of used molecules. Naturally in this algorithm the conservation laws are performed. The efficiency of the algorithm is demonstrated by the calculation of the viscosity and diffusion coefficients of several noble gases (argon, neon, xenon, krypton). It was shown that the algorithm accuracy of the order of 1-2% can be obtained by using a relatively small number of molecules. The accuracy dependence on the number of used molecules, statistics (number of the used phase trajectories) and calculation time was analyzed.

1. Introduction

Transport processes in gases and liquids are due to the complex processes of energy, momentum, and mass transfer in physical space. Information on the corresponding transport coefficients is mainly obtained experimentally, although in rarefied gases, this can be done using the kinetic theory of gases [1, 2]. For dense gases and liquids, the transfer coefficients can be modeled by molecular dynamics simulations. However, to obtain sufficient calculation accuracy even in the simplest cases, it is necessary to use tens of thousands of molecules, which requires very long computation time [3].

For rarefied gases, the molecular dynamics method cannot be used to model transport coefficients because it is necessary to model very large number of particles. The using of the kinetic theory is also not a trivial task (one should be able to calculate the so-called Ω-integrals, and, to obtain sufficient accuracy, it is necessary to perform the rather long and cumbersome procedure of solving the corresponding integral equations, etc.). The procedure of modeling transport coefficients could be significantly accelerated by developing a method for simulating the dynamics of molecular systems. In

Content from this work may be used under the terms of the Creative Commons Attribution 3.0 licence. Any further distribution of this work must maintain attribution to the author(s) and the title of the work, journal citation and DOI.
Published under licence by IOP Publishing Ltd

this connection it should be noted that the molecular dynamics method does not give real phase trajectories of the considered molecular system. Then, using a sufficiently large number of simulated trajectories of the initial molecular system, one can expect to obtain satisfactory data on its average characteristics such as transport coefficients (note that the same is done in molecular dynamics simulations too). The purpose of this study is to develop a simulation method suitable for modeling transport processes in rarefied and moderately dense gases. In this work, the proposed algorithm was implemented for a system of molecules in the form of hard spheres.

2. Algorithm for simulating the dynamics of molecules

For the simulation of the dynamics of a molecular system to be reasonable, it is necessary to satisfy certain conditions. In particular, in the construction of any algorithm, the conservation laws must hold. The most time-consuming process in modeling the real dynamics of a molecular system is the implementation of the free movement of the molecules and their collisions. In molecular dynamics simulations of collisions, one has to perform N^2 operations (N is the number of molecules in the system). In the present algorithm, the collision process is assumed to occur randomly. Molecules of the simulated gas are placed in a simulation box which is a rectangular parallelepiped (or cube). Periodic boundary conditions are used to simulate the gas in the bulk. As a result, along with the main box, the evolution of all the copies surrounding it is taken into account. Initially, the molecules are distributed uniformly over the simulation volume in accordance with the predetermined number density. The velocities of the molecules are distributed according to the Maxwell distribution at a given temperature T,

$$f_M(\vec{v}_i) = \left(\frac{m}{2\pi kT}\right)^{3/2} \exp\left(-\frac{m|\mathbf{v}_i|^2}{2kT}\right), \tag{1}$$

but so that the total momentum of the system of molecules is equal to zero. Here m and $\mathbf{v}_i$ are the mass and velocity of the molecule i, respectively. The created system is not equilibrium in general case. To reach the equilibrium state it is carried out the preliminary calculation after that the velocity distribution of the modeling molecules is the Maxwell's one (1).

Simulation of the dynamics of a rarefied gas starts with compiling a list of molecules. At the initial time t, all the molecules are listed in a random order. Then, the time interval $\tau_1 = d/\mathrm{v}_{\max}$ is chosen, where $\mathrm{v}_{\max}$ is the maximum absolute value of the velocity of molecules of the system at the given time and d is the diameter of the molecule. Generation of the list for the time $(t+\tau_1)$ begins with a consideration of molecule 1. Since, in general, to calculate the transport coefficients of a rarefied gas, it is sufficient to know only the velocities of the molecules, the evolution in the physical space may not be considered (for dense gases, it should be considered). Therefore we can actually consider a spatially homogeneous system and investigate only the dynamics of change in the velocities of the molecules.

Thus, at time t, the molecules have velocities $(\mathbf{v}_1, \mathbf{v}_2, ..., \mathbf{v}_N)$. To determine whether molecule 1 collides with another molecule in the time interval τ_1, a random number uniformly distributed in the interval (0;1) is generated. If it is less than the average probability of collision $P_{\tau_1} = 4\tau_1 n d^2 \sqrt{\pi kT/m}$ (see [1]) in the time interval τ_1, then molecule j which collides with molecule 1 is randomly chosen from the remaining (N–1) molecules. Here $n = N/V$ is the numerical density of the gas and V is the volume of the simulation cell. As a result, the velocities of the colliding molecules change according to the laws of elastic collision

$$\mathbf{v}_1' = \mathbf{v}_1 + (\mathbf{v}_{1j} \cdot \mathbf{e})\mathbf{e}, \quad \mathbf{v}_j' = \mathbf{v}_j + (\mathbf{v}_{j1} \cdot \mathbf{e})\mathbf{e},$$

where $\mathbf{v}_{j1} = (\mathbf{v}_j - \mathbf{v}_1)$ is the vector of relative velocity and $\mathbf{e}$ is the unit vector from centre of molecule j to molecule 1.

If the generated number is greater than the average probability of collision, molecule 1 does not collide in the time interval τ_1 and its velocity does not change. If it collides, the velocities of molecules 1 and *j* in the initial list change. The remaining molecules are treated similarly. As a result, a new list of the velocities of the molecules $(\mathbf{v}_1', \mathbf{v}_2',..., \mathbf{v}_N')$ is generated.

After the generation of the list at the time $(t+\tau_1)$, the next time interval $\tau_2 = d/\mathrm{v}_{max}$ is chosen, where v_{max} is the maximum absolute value of the velocity of molecules of the system at the time $(t+\tau_1)$ and the procedure is repeated. This procedure is repeated within the given computation time period t_s, which is equal to $t_s = \tau_1 + \tau_2 + \tau_3 + ... + \tau_k$. The result of the computation is the full set of the velocity coordinates of all the molecules of the simulated system at successive times.

3. Modeling of transport coefficients

The described algorithm is applied to model the transport processes in rarefied gas. In particular, using this algorithm it may be calculated the transport coefficients. In this paper the viscosity and self-diffusion coefficients were calculated. The transport coefficients of the molecular system considered are calculated using fluctuation-dissipation theorems, which relate transport coefficients to the evolution of the corresponding correlation functions. In the literature, these relations are called the Green-Kubo formulas [4].

The self-diffusion coefficient *D* is the integral on autocorrelation velocity function (ACVF) χ of the modeled system

$$D = \frac{1}{3}\int_0^{t_p} \chi(0,\tau)d\tau , \tag{2}$$

where t_p is so named plateau time of the calculation of the integral (2) [3]. ACVF is calculated over all molecules velocities at

$$\chi(0,t) = \frac{1}{Nl}\sum_{i=1}^{N}\sum_{j=0}^{l-1}[\mathbf{v}_i(0)\cdot\mathbf{v}_i(j\Delta t)] . \tag{3}$$

Here $t = l\Delta t$, l is the number of the time steps where ACVF was calculated, Δt is the integration step, l is the number of the independent phase trajectories.

For the rarefied gas ACVF decays exponentially. The corresponding relaxation time is of the order of mean free path time, however the plateau values of the transport coefficients (the values after which they change little over time) are reached at times which are about an order of magnitude larger (25–30 of mean free path time). Comparison of the modeling and experimental data for a few inert gases (argon, neon, and krypton) shows their high accuracy (see Table 1). For example, in the simulation using an ensemble of 3200 molecules and one thousand independent phase trajectories, the accuracy of the self-diffusion coefficient was around 1–2%, which is within the experimental error.

Table 1. Comparison of simulation data of the self-diffusion coefficient D with the experimental data D_e [5].

	Ar	Kr	Ne	Xe
D, cm^2/c	0.158	0.081	0.456	0.049
D_e, cm^2/c	0.156	0.08	0.452	0.048
Δ,%	1.2	1.2	0.9	2.1

The viscosity coefficient is calculated by means of the following formula

$$\eta = \frac{1}{VkT}\int_0^{\tau_p}\left\langle \sigma_{xy}(0)\sigma_{xy}(t)\right\rangle dt\,,\ \sigma_{xy}(t) = \frac{1}{3l}\sum_{j=1}^{l}\sum_{i=1}^{N} m[v_x^i(j\Delta t)v_y^i(j\Delta t)]\,.$$

The correlation of the stress tensor σ_{xy} decays exponentially too. The comparison of the calculated and experimental data of the viscosity coefficient η with the experimental value η_e is presented in Table 2. Here the following effective diameters d were used, for Ar d= 0.366, for Kr d= 0.42, for Ne d= 0.26, for Xe d =0.494 nm [6]. The obtained accuracy of the determination of the viscosity coefficient is rather high.

Table 2. Comparison of simulation data of the viscosity coefficient with the experimental data D_e [5].

	Ar	Kr	Ne	Xe
$\eta \cdot 10^{-6}$, Па·с	22.51	25.6	31.9	23.4
$\eta_e \cdot 10^{-6}$, Па·с	22.7	25.5	31.7	23.3
Δ, %	0.8	0.4	0.6	0.4

4. Conclusions

Thus, it is shown that the proposed stochastic simulation algorithm is quite suitable for modeling transport processes in rarefied gases. It is much more cost-effective than the molecular dynamics method. The computation time is found to increase linearly with the number of molecules used.

The accuracy of the simulation is depended, of course, on the number of the molecules used. The high accuracy is realized beginning from determine number of particles. As a rule the number of needed particles should be of the order of a few thousands. It is connected with the appearance in the system so named repeated collisions (see [7, 8]). In rarefied gas such collisions are absent. We developed the special procedure to exclude these collisions. In this case the needed number of particles is reduced in times.

Another important factor that determines the accuracy of the simulation is the number of ensemble members over which the averaging is performed. Test calculations have shown that the accuracy of the calculations is $\Delta \sim 1/\sqrt{l}$, where l is the number of ensemble members over which the averaging is performed.

Acknowledgments

This work was supported by the Russian Science Foundation (grant No. 14-19-00312).

5. References

[1] Chapman S and Cowling T G 1990 *The Mathematical Theory of Non-Uniform Gases* (Cambridge: Cambridge University Press) p 457
[2] Ferziger J H and Kaper H G 1972 *Mathematical Theory of Transport Processes in Gases* (Amsterda-London: North-Holland Publishing Company) p 550
[3] Rudyak V Ya, Belkin A A, Ivanov D A and Egorov V V 2008 *High Temperature* **46** 30
[4] Zubarev D N1974 *Nonequilibrium Statistical Thermodynamics* (New York: Consultants Bureau) p 412
[5] Physical Values. Handbook 1991 Eds Grigoriev I S and Meilikhova E Z (Moscow: Energoatomizdat) p 1234
[6] Bird G A 1976 Molecular Gas Dynamics (Oxford: Clarendon Press) p 297
[7] Gimelshtein S F and Rudyak V Ya 1991 *Soviet Technical-Physics Letters* **17(19)** 74
[8] Rudyak V Ya 1991 *Fluid Dynamics* **26** p 909

Comparative Study of Shrinkage and Non-Shrinkage Model of Food Drying

N. Shahari[1], N. Jamil[1], KA. Rasmani[1]
[1]Faculty of Computer and Mathematical Sciences,
Universiti Teknologi MARA Negeri Sembilan, Malaysia

Email:norazni@ns.uitm.edu.my

Abstract. A single phase heat and mass model has always been used to represent the moisture and temperature distribution during the drying of food. Several effects of the drying process, such as physical and structural changes, have been considered in order to increase understanding of the movement of water and temperature. However, the comparison between the heat and mass equation with and without structural change (in terms of shrinkage), which can affect the accuracy of the prediction model, has been little investigated. In this paper, two mathematical models to describe the heat and mass transfer in food, with and without the assumption of structural change, were analysed. The equations were solved using the finite difference method. The converted coordinate system was introduced within the numerical computations for the shrinkage model. The result shows that the temperature with shrinkage predicts a higher temperature at a specific time compared to that of the non-shrinkage model. Furthermore, the predicted moisture content decreased faster at a specific time when the shrinkage effect was included in the model.

1. Introduction

The single phase heat and mass model, which is modelled as a homogeneous medium, has been widely used to describe the movement of water and heat during the drying of food [1-6]. In this case, moisture transport involves two dependent processes: the evaporation of moisture (in terms of liquid water) at the solid surface that needs heat from the drying air and the internal diffusion of liquid from the centre of the fruit to the surface. Overall, there are many advantages to using a single phase heat and mass model to capture the behaviour of heat and mass during drying for use in the development of enhanced food processing. Firstly, these models are generally fairly simple and contain few parameters. It is also relatively straightforward to include changes in shape and size as a result of water removal and internal collapse during the drying process. This may arise from a moisture gradient within the particle that induces microstructure stresses, leading to shrinkage [7].

Appearance analyses made by consumers on dried fruits relate to visual and physical appearance [8]. One of the physical characteristics is shrinkage: it is rarely possible to ignore shrinkage in the food material and this physical change must be taken into account in evaluating the profiles of moisture and temperature. Since these structural changes and the transport process are involved simultaneously and continuously during the drying process, several researchers have developed a shrinkage equation by transforming the space coordinate in the equation of heat and mass transport (for example see [6, 9, 10]). Other researchers have developed the shrinkage equation

Content from this work may be used under the terms of the Creative Commons Attribution 3.0 licence. Any further distribution of this work must maintain attribution to the author(s) and the title of the work, journal citation and DOI.
Published under licence by IOP Publishing Ltd

experimentally and then conducted the calculation at each time step in the equation of the heat and mass transport model [2, 11, 12]. In recent years there have been many studies of the heat and mass transfer model with the inclusion of shrinkage, such as the reviews by Katekawa and Silva [13] and Mayor and Sereno [7]. However, a comparison of shrinkage and non-shrinkage that could affect the accuracy of the distribution of the moisture and temperature profile has been little investigated.

The purpose of this present study is to make a comparative study by neglecting or considering the shrinkage effect subject to the convective drying process, based on a single phase continuum model of heat and mass transport [1, 14] With respect to the importance of obtaining a product with the desired quality, the effect of the drying air temperature to evaluate the moisture and temperature profile is also investigated.

2. Mathematical Model Formulation

A one-dimensional model is used: in this case, an infinite slab of finite thickness L, with moisture content $M(x,t)$ across the slab and unsteady temperature $T(x,t)$ is expressed by the well-known system of partial differential equations (PDEs) for moisture and energy transport [2]. The fruit sample is hygroscopic porous media, yet is assumed to be on a fictitious continuum. Weak internal evaporation and the transport of vapour within the dehydrated fruits towards the external food surface have not been considered. Non-dimension scaled variables are used in the formulation: a representative diffusion timescales, scaled moisture content, and scaled temperature are defined by

$$\tau=\frac{D_0 t}{L_0{}^2},\quad \overline{M}=\frac{M}{M_0},\quad \overline{T}=\frac{T-T_0}{T_{air}-T_0},\quad \xi=\frac{x}{L_0}\ and\ s(\tau)=\frac{L(t)}{L_0} \tag{1}$$

2.1. Continuum model (Fixed Volume Model)

Shahari and Hibberd [1] consider models with no shrinkage ($x = L_0$) associated with isothermal and non-isothermal diffusion and constant moisture diffusivity ($D = D_0$), with the governing of heat and mass transfer model as,

$$\frac{\partial \overline{M}(\xi,\tau)}{\partial \tau}=\frac{\partial^2 \overline{M}(\xi,\tau)}{\partial \xi^2},\quad \frac{\partial \overline{T}(\xi,\tau)}{\partial \tau}=Le\frac{\partial^2 \overline{T}(\xi,\tau)}{\partial \xi^2} \tag{2}$$

$$\text{at } \tau=0: \qquad \overline{M}=1 \text{ and } \overline{T}=0 \tag{3}$$

$$\text{at } \xi=0: \qquad \frac{\partial \overline{M}}{\partial \xi}=0, \quad \text{and} \quad \frac{\partial \overline{T}}{\partial \xi}=0, \tag{4}$$

$$\text{at } \xi=1: \qquad \frac{\partial \overline{M}}{\partial \xi}=-Bi_m(\overline{C}_{sur}-1) \text{ and } \frac{\partial \overline{T}}{\partial \xi}=-Bi(T_{sur}-1)+\overline{\lambda}\frac{\partial \overline{M}}{\partial \xi} \tag{5}$$

2.2. Continuum model with shrinkage effect

During drying, the shape and size of the food particles are constantly changing as a result of water removal and internal collapse. Shahari and Hibberd [6] consider a model with interface $L(t)$ decreasing with drying and, using the transformation $\xi=\frac{x}{L(t)}$, the surface interface corresponds to a fixed value $\xi=1$. Using the same non-dimension scale variable as above, the resulting system for moisture, temperature and shrinkage becomes

$$\frac{\partial \overline{M}}{\partial \tau}=\frac{1}{s^2(\tau)}\frac{\partial^2 \overline{M}}{\partial \xi^2}+\xi\frac{1}{s(\tau)}\frac{ds}{d\tau}\frac{\partial \overline{M}}{\partial \xi}, \tag{6}$$

$$\frac{\partial \overline{T}}{\partial \tau}=\frac{1}{s^2(\tau)}Le\frac{\partial^2 \overline{T}}{\partial \xi^2}+\xi\frac{1}{s(\tau)}\frac{ds}{d\tau}\frac{\partial \overline{T}}{\partial \xi}, \tag{7}$$

$$\frac{ds}{d\tau}=\frac{\rho_s}{\rho_w}\frac{M_0}{s(\tau)}\frac{\partial \overline{M}}{\partial \xi}. \tag{8}$$

Initial conditions associated with constant conditions

$$\tau=0:\ \overline{T}(\xi,0)=0,\ \overline{M}(\xi,0)=1,\ s(0)=1. \tag{9}$$

Taking symmetry boundary conditions in the mid-plane of the drying slice gives

$$\xi=0:\quad \frac{\partial \overline{M}}{\partial \xi}=0,\quad \text{and}\quad \frac{\partial \overline{T}}{\partial \xi}=0. \tag{10}$$

At the surface, moisture and temperature boundary conditions of the drying body in contact with drying air become

$$\xi=1:\frac{\partial \overline{M}}{\partial \xi}=-s(\tau)\,Bi_m\,C_{air}\,(\frac{C_{sur}}{C_{air}}-1), \tag{11}$$

$$\frac{\partial \overline{T}}{\partial \xi}=s(\tau)Bi(1-\overline{T}_{sur})+\lambda\frac{\partial \overline{M}}{\partial \xi}. \tag{12}$$

In the above, several dimensionless controlling parameters are defined by

$$Bi_m=\frac{h_m L_0}{DM_0},\quad Bi=\frac{hL_0}{k},\quad Le=\frac{\alpha}{D_0},\quad \lambda=\frac{H_v M_0 D_0 \rho_s}{k(T_{sur}-T_0)}.$$

Where M is the local moisture content (dry base), T is the local temperature, D_0 represents the effective diffusion coefficient, ρ_s is the density of dry solid, C_p heat capacity, k thermal conductivity, x is thickness and t the time. H_v is the heat of vaporization, T_{air} is air temperature and T_{sur} is the surface temperature of the fruit, h is the heat transfer coefficient, h_m is the mass transfer coefficient, T_0 is the initial temperature, M_0 is the initial moisture. C_{air} is the concentration of moisture in the air and C_{sur} is the concentration in the form of liquid water film at the food surface (example: see[15]) and $\alpha=\frac{k}{\rho_s C_p}$ is thermal diffusivity.

3. Numerical procedures

A numerical solution for equation (2) was obtained using the method of lines [16]. Discretizing the fixed integration region $0<\xi<L$ into N subinterval gives $N+1$ ordinary differential equation. This system was then solved numerically using MATLAB ODE45 solver, which is based on an order-5 Runge Kutta-Fehlberg method (RK45).

Equations (6) and (7) were solved together with (8), which involves a shrinkage equation obtained using the method of lines [16]. This is the case of the Stefan problem in which the interface condition is associated with melting at a moving interface. Discretising equations (6) and (7) with a central difference approximation gives,

$$\frac{d\overline{M}_j}{d\tau} = \frac{1}{s(\tau)^2}\left(\frac{\overline{M}_{j-1} - 2\overline{M}_j + \overline{M}_{j+1}}{(\Delta\xi)^2}\right) + \xi_j \frac{1}{s(\tau)}\frac{ds}{d\tau}\left(\frac{\overline{M}_{j+1} - \overline{M}_{j-1}}{2\Delta\xi}\right) \qquad j = 1,2,3\ldots N+1 \quad (13)$$

$$\frac{d\overline{T}_j}{d\tau} = \frac{1}{s(\tau)^2}\left(\frac{\overline{T}_{j-1} - 2\overline{T}_j + \overline{T}_{j+1}}{(\Delta\xi)^2}\right) + \xi_j \frac{1}{s(\tau)}\frac{ds}{d\tau}\left(\frac{\overline{T}_{j+1} - \overline{T}_{j-1}}{2\Delta\xi}\right) \qquad j = 1,2,3\ldots N+1 \quad (14)$$

A higher order discretization of the gradient term at the boundary is taken in line with the recommendation for the analogous Stefan problem by Furzeland (in Crank [17]). Discretization of the shrinkage condition in equation (8) is given by

$$\frac{ds}{d\tau} = \frac{\rho_s}{\rho_w}\frac{M_0}{s(\tau)}\left(\frac{3\overline{M}_N - 4\overline{M}_{N-1} + \overline{M}_{N-2}}{2\Delta\xi}\right) \quad (15)$$

For drying simulation, the input parameter is given in table 1 for generic drying conditions for fruit. Based on this input parameter, the following reference non-dimension parameter was obtained.

Table 1. Input parameter used in the simulation

Bi=0.3	Bi_{m-}=20	M_0=0.8 kg water/kg moist sample
λ =0.3	Le=5	
T_{air}-60^{0}C	H_v =2345KJ/kg	D_0=8x10^{-10}m^2/s
RH=20%	ρ_s=1080 kg/m^3	h =20-250 W/m^{2}K
L_0=5 mm	ρ_w=1000 kg/m^3	h_m=2x10^{-6}m/s
C_p=2.344KJ/kg K	α= k/ρ_sC$_p$= 4x10^{-9}m^2/s	k=0.475W/mK

4. Result and Discussion

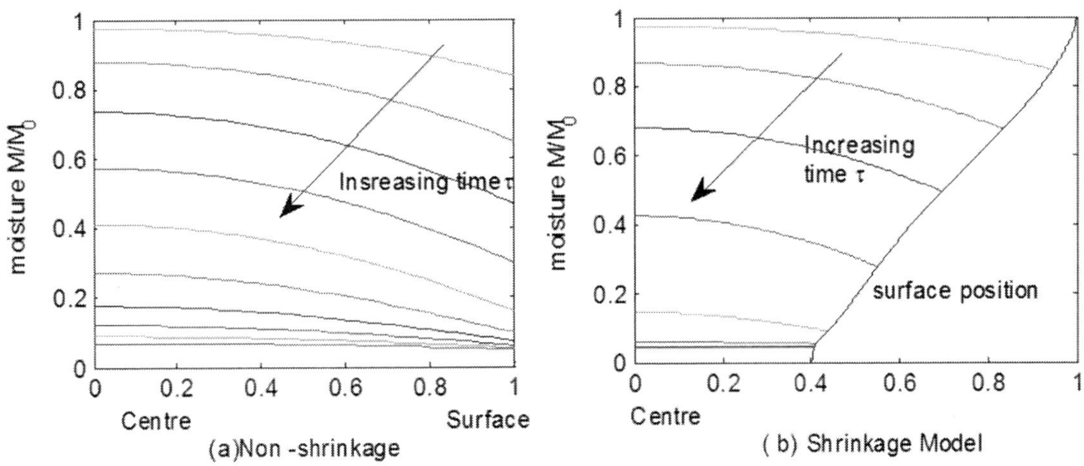

Figure 1: Profile of moisture through the sample with elapsed time tau=0-3 in step 0.25 for (a) Non-shrinkage model (b) shrinkage model

Figure 1 shows the moisture profile through the sample of the fruit with increasing time for the shrinkage and non-shrinkage models. Moisture decreased as time increased, but this process was a little slower for the non-shrinkage model compared to the shrinkage model. Figure 1(b) also includes the location of the free surface during shrinkage; the shrinkage of the product is shown by the change in the thickness of the fruit. The final value of shrinkage in this model can be determined readily as only solid mass is left at the end of drying, which in this case is 0.4. A similar pattern profile of shrinkage has been shown by other authors (for example see Chemkhi et al. [9], Crapiste et al. [18]).

However, Chemkhi et al [9] consider elastic deformation and Crapiste et al. [18] employ density values to account for shrinkage velocity.

Figure 2 shows a temperature profile in the food as a function of both position and time during drying for the non-shrinkage and the shrinkage model. The temperature at each location increases with the drying time, which is due to the higher drying temperature. It was found that the temperature profile rises rapidly in the early period of heating due to the difference between the air temperature and the food temperature. As the heating period progresses, the rise in temperature attains an almost uniform profile. At the end of drying, the temperature rises rapidly throughout the food as only small quantities of heat are needed for evaporation. The overall dimension varies with time depending on the amount of moisture loss for each section, which can be seen in figure 2(b). The temperature distribution increased much faster for the shrinkage model compared with the non-shrinkage model.

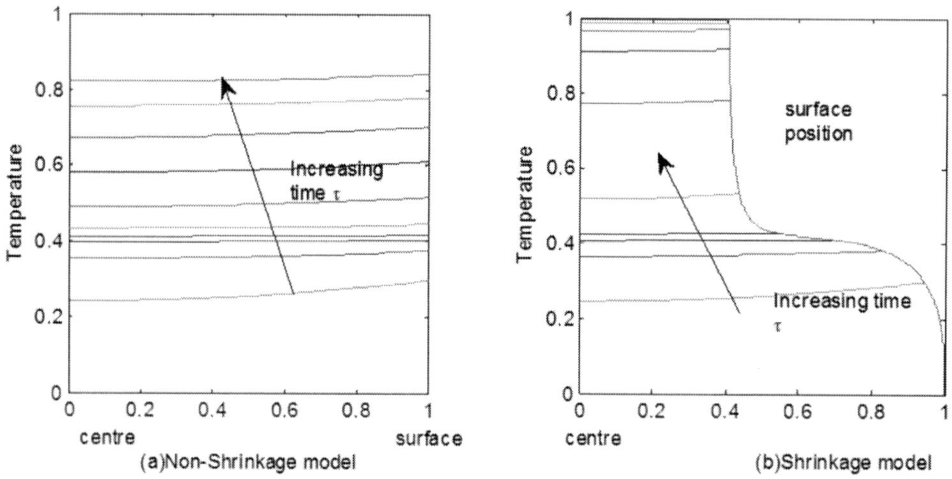

Figure 2: Temperature profile inside the fruit with increasing time tau=0-3 instep 0.25 for (a)Non-shrinkage model (b)shrinkage model

Figure 3 represents the comparison of moisture and temperature gradients at the centre of the fruits, with and without the shrinkage effect. It can be seen that the temperature with shrinkage predicts a higher temperature at a specific time, compared to that with non-shrinkage. However, the predicted moisture content decreased faster when the shrinkage effect was included in the model. As a result of the thickness of the sample and decreases due to shrinkage, the moisture has less distance to cover and hence it reaches the surface faster, with the result that the time taken to dry a food becomes shorter. Therefore, the model with shrinkage needs a shorter time for drying. When shrinkage occurs, there are volumetric and dimensional changes to the product. This occurs because the tissues in the food are unable to hold their structure when the space taken by the water is continuously removed. This suggests that there are significant differences in predicted moisture and temperature gradients.

Figure 4 shows the profile of moisture and temperature at the centre of the fruit when different air temperatures are used during drying. The distribution of moisture and temperature at the centre is different at a specific time during drying and the detection of this region is a very important aspect in terms of food safety. Locating the area of potential improvement will improve product quality while reducing product development and prototype cost.

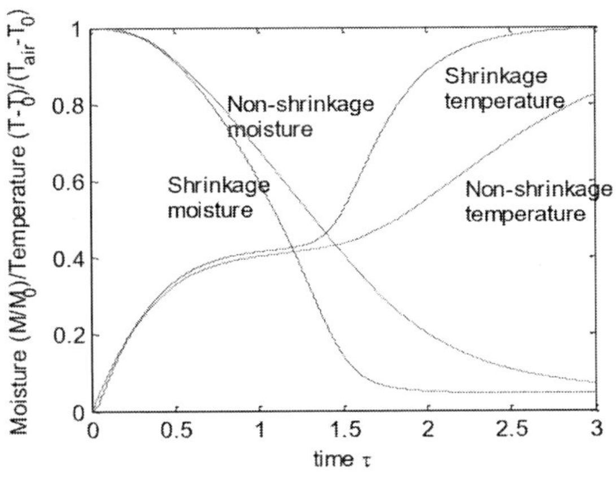

Figure 3: Comparison of moisture and temperature at the centre for shrinkage and non-shrinkage model, drying at 60°C and 20%RH

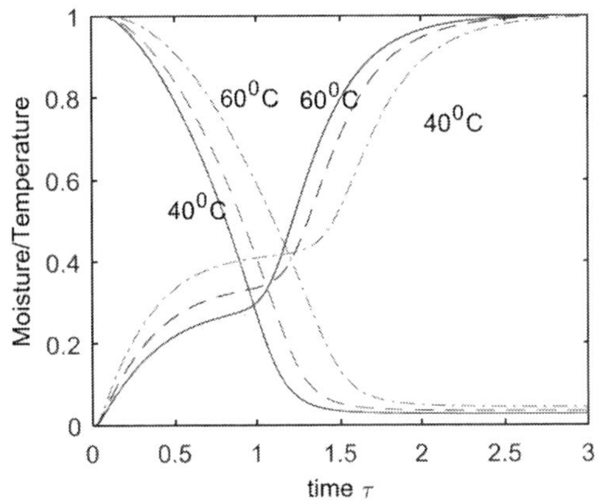

Figure 4: Comparison of moisture and temperature at the centre of the food with different air temperatures 40°C, 50°C and 60°C (for shrinkage model)

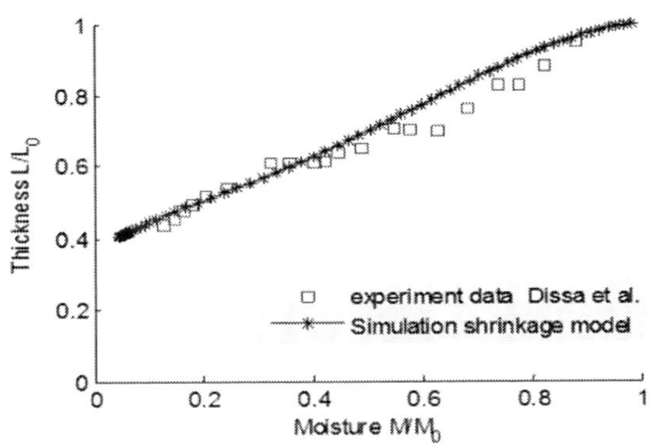

Figure 5: Comparison between experiment data [19] with simulation study of variation of shrinkage (V/V_0) with respect to the moisture content

Figure 5 shows the experiment measurement of shrinkage (V/V_0) taken from experiment drying of mango by Dissa et al. [19] at 60°C, which was compared with the simulation model of shrinkage. The thickness is assumed to decrease linearly with a reduction in moisture content. The trend deviates with higher moisture content but there is good agreement at lower moisture content: when moisture content is very low, the decrease in moisture is influenced by drying air and a very small decrease in thickness can be observed

5. Conclusion

This paper adopts a simple one dimensional model of heat and mass transfer that regards fruit as a homogenous structure, as discussed in [1, 6]. Using a numerical solution, the behavior of moisture and temperature subject to shrinkage was predicted for constant diffusivity. Shrinkage has been included in this study by including a velocity equation in the differential equation without adopting the separated procedure by means of regression analysis of the shrinkage data. Moisture content decreases more quickly in the shrinkage model, which concludes that the model with shrinkage effect needs a shorter time for drying, consistent with the experiment finding [20]. The result of the simulation study shows that the water in the food is removed, amounting to about 60% of the initial volume. This is consistent with the experiment finding by [21] which shows mango shrank by 83% of the initial volume, 77% of the initial volume of banana drying [22] and 62% of dried onion at 60°C [23].

Acknowledgments

This work was supported by Ministry of Higher Education, Malaysia under Grant No 600-RMI/RAGS 5/3 (147/2014) and Universiti Teknologi MARA, Malaysia.

References

[1] Shahari N and Hibberd S 2013 Analysis of Single Phase Moisture and Heat Model of Food Drying, *4th International Conference of Mathematical Models in Engineering and Computer Science,(Brosov, Romania)* 137
[2] Wang N and Brennan J 1995 *Journal of Food Engineering* **24** 47
[3] Villa-Corrales L, Flores-Prieto J, Xaman-Villasenor J and Garcia- Hernandez E 2010 *Journal of Food Engineering* **98** 198
[4] Janjai S, Lamlert N, Intawee P, Mahayothee B, Haewsungcharern M, Bala B and Muller J 2008 *Biosystem Engineering* **99** 523
[5] Barati E and Esfahani J A 2012 *International Journal of Thermal Sciences 56* **56** 86
[6] Shahari N and Hibberd S 2012 Mathematical Modelling of Shrinkage Effect during Drying of Food, *2012 IEEE Colloquium on Humanities, Sciences & Engineering Research (CHUSER 2012),(Kota Kinabalu, Sabah Malaysia)* 242
[7] Mayor L and Sereno A 2004 *Journal of Food Engineering* **61** 373
[8] Shahari N, Nursabrina M and Suhairah A Z 2015 The effects of drying on physical properties of bilimbi slices (Averrhoa bilimbi l.), *AIP Conference Proceedings,(Penang, Malaysia, 6-8 Nov 2013)* 070
[9] Chemkhi S, Zagrouba F and Bellagi F 2005 *Brazilian Journal of Chemical Engineering* **22** 153
[10] Hussain M M and Dincer I 2003 *Journal of Heat and Mass Transfer* **46** 4033
[11] Aprajeeta J, Gopirajah R and Anandharamakrishnan C 2015 *Journal of Food Engineering* **144** 119
[12] Ramallo L A and Mascheroni R H 2013 *Journal of Food Process Engineering* **36** 66
[13] Katekawa M E and Silva M A 2006 *Drying Technology* **24** 5
[14] Pavon-Melendez G, Hernandez J, Salgado M and Garcia M 2002 *Journal of Food Engineering* **51** 347
[15] Dunn G J, Wilson S K, Duffy B R, David S and Sefiane K 2008 *Colloids and Surfaces A : Physicochemical and Engineering Aspects* **323** 50
[16] Schiesser W E 1991 *The Numerical Method of Lines*. (San Diego: Academic Press)
[17] Crank J 1984 *Free and moving boundary problem*. (New York: Clarendon-Oxford Science Publication)
[18] Crapiste G H, Whitaker S and Rotstein E 1988 *Chemical Engineering Science* **43** 2919
[19] Dissa O A, Desimorieux H, Bathiebo J and Koulidiati J 2008 *Journal of Food Engineering* **88** 429
[20] Simal A, Femenia A, Garau M C and Rossello C 2005 *Journal of Food Engineering* **66** 323
[21] Yan Z, Sousa-Gallagher M J and Oliveira F A R 2008 *Journal of Food Engineering* **84** 340
[22] Hassini L, Azzouz S, Peczalski R and Belghith A 2007 *Journal of Food Engineering* **79** 47
[23] Abasi S, Mousavi S M, Mohebi M and Kiani S 2009 *Iranian Journal Of Chemical Engineering* **6** 57

Journal of Physics: Conference Series **738** (2016) 012088 doi:10.1088/1742-6596/738/1/012088

Differentiation on spaces of triangulations and optimized triangulations.

Jean-Pierre Magnot
Lycée Jeanne d'Arc
Avenue de Grande Bretagne
F-63000 Clermont-Ferrand

E-mail: jean-pierr.magnot@ac-clermont.fr

Abstract. We describe a smooth structure, called Frölicher space, on CW complexes and spaces of triangulations. This structure enables differential methods for e.g. minimization of functionnals. As an application, we exhibit how an optimized triangulation can be obtained in order to get an optimized H^1 approximation in a prescribed a class of triangulations.

Introduction

1. Frölicher structures on CW complexes

1.1. Frölicher spaces

Definition 1.1 • *A* ***Frölicher*** *space is a triple* $(X, \mathcal{F}, \mathcal{C})$ *such that*
- $\mathcal{C}$ *is a set of paths* $\mathbb{R} \to X$,
- *A function* $f : X \to \mathbb{R}$ *is in* $\mathcal{F}$ *if and only if for any* $c \in \mathcal{C}$, $f \circ c \in C^\infty(\mathbb{R}, \mathbb{R})$;
- *A path* $c : \mathbb{R} \to X$ *is in* $\mathcal{C}$ *(i.e. is a* ***contour****) if and only if for any* $f \in \mathcal{F}$, $f \circ c \in C^\infty(\mathbb{R}, \mathbb{R})$.

• *Let* $(X, \mathcal{F}, \mathcal{C})$ *et* $(X', \mathcal{F}', \mathcal{C}')$ *be two Frölicher spaces, a map* $f : X \to X'$ *is* ***differentiable*** *(=smooth) if and only if* $\mathcal{F}' \circ f \circ \mathcal{C} \in C^\infty(\mathbb{R}, \mathbb{R})$.

Any family of maps $\mathcal{F}_g$ from X to $\mathbb{R}$ generate a Frölicher structure $(X, \mathcal{F}, \mathcal{C})$, setting [6, 7]:
- $\mathcal{C}(\mathcal{F}_g) = \{c : \mathbb{R} \to X$ such that $\mathcal{F}_g \circ c \subset C^\infty(\mathbb{R}, \mathbb{R})\}$
- $\mathcal{F}(\mathcal{C}(\mathcal{F}_g)) = \{f : X \to \mathbb{R}$ such that $f \circ \mathcal{C} \subset C^\infty(\mathbb{R}, \mathbb{R})\}$.

A Frölicher space carries a natural topology, which is the pull-back topology of $\mathbb{R}$ via $\mathcal{F}$, see e.g. [1]. In the case of a finite dimensional differentiable manifold, the underlying topology of the Frölicher structure is the same as the manifold topology. In the infinite dimensional case, these two topologies differ very often.

The classical properties of a tangent space are not automatically checked in the case of Frölicher spaces following [3]. There are two possible tangent spaces:
-the **internal** tangent space, defined by the derivatives of smooth paths,
- the **external** tangent space, made of derivations on $\mathbb{R}-$valued smooth maps.

The internal tangent space is not necessarily a vector space, were as the external tangent space is. For example, consider the diffeological subspace of $\mathbb{R}^2$ made of the two lines $y = x$ and $y = -x$. They cross at the origin, the external tangent space is of dimension 2, the internal tangent space is made of two directions that cannot be combined by addition. This is why some authors sometimes call the internal tangent space by "tangent cone".

Content from this work may be used under the terms of the Creative Commons Attribution 3.0 licence. Any further distribution of this work must maintain attribution to the author(s) and the title of the work, journal citation and DOI.
Published under licence by IOP Publishing Ltd

Proposition 1.2 *Let X be a Frölicher space equipped with an equivalence relation $\mathcal{R}$. The quotient map $\pi : X \to X/\mathcal{R}$ defines by push-forward a diffeology on $X/\mathcal{R}$.*

We can even state the same results in the case of infinite products, in a very trivial way by taking the cartesian products of the contours. Let us now give the description of what happens for projective limits of Frölicher spaces.

1.2. Topological gluing of Frölicher spaces
Let us assume that X is a topological space, and that there is a collection $\{(X_i, \mathcal{F}_i, \mathcal{C}_i)\}_{i\in I}$ of Frölicher spaces, together with continuous maps $f_i : X_i \to X$. Then we can define a Frölicher structures on X setting

$$\mathcal{F}_{I,0} = \{f \in C^0(X,\mathbb{R}) | \forall i \in I, \quad f \circ f_i \circ \mathcal{C}_i \subset C^\infty(\mathbb{R},\mathbb{R})\},$$

wa define $\mathcal{C}_I$ the contours generated by the family $\mathcal{F}_{I,0}$, and $\mathcal{F}_I = \mathcal{F}(\mathcal{C}_I)$.

Example: Frölicher structure on CW complexes.
A CW-complex is a topological space built by induction:
- 1st step: gluing a family of intervals of the type $[0;1]$ along their broder, in order to have a graph. We get a $1-$CW complex.
-Assume that we have obtained a $n-$CW complex X_n We glue topologicaly some $(n+1)$ disks

$$D_{n+1} = \{x \in \mathbb{R}^{n+1} | ||x|| \leq 1\}$$

identifying the plots of the border S^n with some plots of X_n.

Since at each step we glue together some Frölicher space, the obtained CW complex is a Frölicher space.

1.3. Frölicher structure of a triangulation
Let M be a smooth manifold for dimension n. Let

$$\Delta_n = \{(x_0, ..., x_n) \in \mathbb{R}_+^{n+1} | x_0 + ... + x_n = 1\}$$

be the $n-$ simplex. a triangulation of M is a family $\sigma = (\sigma_i)_{i\in I}$ where $I \subset \mathbb{N}$ is a set of indexes, finite or infinite, each σ_i is a smooth map $\Delta_n \to M$, and such that:

(i) $\forall i \in I, \sigma_i$ is an embedding.
(ii) $\bigcup_{i\in I} \sigma_i(\Delta_n) = M$. (open covering)
(iii) $\forall (i,j) \in I^2$, $\sigma_i(\Delta_n) \cap \sigma_j(\Delta_n) \subset \sigma_i(\partial\Delta_n) \cap \sigma_j(\partial\Delta_n)$. (intersection along the borders)
(iv) $\forall (i,j) \in I^2$ such that $D_{i,j} = \sigma_i(\Delta_n) \cap \sigma_j(\Delta_n) \neq \emptyset$, for each $(n-1)$-face F of $D_{i,j}$, the "transition maps" $\sigma_j^{-1} \circ \sigma_i : \sigma_i^{-1}(F) \to \sigma_j^{-1}(F)$ are affine maps.

Under these (well-known) conditions, we can apply section 1.2, and equip the triangulated manifold (M, σ) with a Frölicher structure $(\mathcal{F}, \mathcal{C})$, generated by the smooth maps σ_i. The following result is obtained from the construction of $\mathcal{F}$ and $\mathcal{C}$:

Theorem 1.3 *The inclusion $(M, \mathcal{F}, \mathcal{C}) \to M$ is smooth.*

2. Smooth structures on the space of triangulations

2.1. The Frölicher structure

Now, let us fix the set of indexes I and fix a so-called **model triangulation** σ. We note by $\mathcal{T}_\sigma$ the set of triangulations σ' of M such that the domains $D_{i,j} = \sigma_i(\Delta_n) \cap \sigma_j(\Delta_n)$ and $D'_{i,j} = \sigma'_i(\Delta_n) \cap \sigma'_j(\Delta_n)$ are diffeomorphic (in the Frölicher category). Based on the properties described in [6], since $\mathcal{T}_\sigma \subset C^\infty(\Delta_n, M)^I$, we can equip $\mathcal{T}_\sigma$ with the trace Frölicher structure, in other words, the Frölicher structure on $\mathcal{T}_\sigma$ whose generating family of contours $\mathcal{C}$ are the contours in $C^\infty(\Delta_n, M)^I$ which lie in $\mathcal{T}_\sigma$. In practice, as we shall see below, we often consider a subspace of $\mathcal{T}_\sigma$.

2.2. Application: optimized approximations

We wish here to describe applications to optimization of H^1- approximation of functions and differential forms, in the sense of [9], which is commonly used in numerical analysis. M is assumed Riemannian. In the two examples, we consider the Sobolev space $H^1(M, \mathbb{R})$ or $H^1(\Omega^*(M, \mathbb{R}))$, both equipped with self-adjoint positive injective pseudo-differential operator A of order 2. With this setting, it is well-known that $(A.,.)_{L^2}$ is a H^1-scalar product, where $(.,.)_{L^2}$ is the L^2-scalar product. For example, we can assume that $A - Id$ is the Laplace-Beltrami operator on $H^1(M, \mathbb{R})$, or the Hodge Laplacian on $H^1(\Omega^*(M, \mathbb{R}))$. These approximations are the base of the finite elements method.

2.2.1. Approximation of functions in $H^1(M, \mathbb{R})$ For a fixed map $f \in C^0(M, \mathbb{R})$, the approximation $f_{\sigma'}$ of f with respect to the triangulation $\sigma' \in \mathcal{T}_\sigma$ is given by the following constraints:

- $f_{\sigma'} = f$ at the 0-vertex of the triangulation
- $\forall \sigma'_i \in \sigma', f_{\sigma'} \circ \sigma_i$ is an affine map $\Delta_n \to \mathbb{R}$.

Definition 2.1 *An optimized $\mathcal{T}_\sigma$ triangulation $\sigma_{op}(f)$ is a triangulation that minimize*

$$\Phi : \sigma' \mapsto (A(f_{\sigma'} - f), f_{\sigma'} - f)_{L^2}.$$

This is a smooth map, and the condition $D\Phi = 0$ is necessary at the minimizing point σ_{op}.

2.2.2. Approximation of differential forms in $H^1(\Omega^*(M, \mathbb{R}))$ Let $\alpha \in \Omega^k(M, \mathbb{R})$. Following [9], for any triangulation $\sigma' \in \mathcal{T}_\sigma$, there exists a discretization of α with respect to σ' which can be realized as a H^1-differential form. Then, one can get, the same way, an optimized triangulation σ_{op}.

3. Optimized triangulations in H^1 approximation

If one considers the whole space $\mathcal{T}_\sigma$, one quicly see that the minimization functionnal furnished only a reparametrization of the function f.

Let us consider the following example: let $f(x) = \frac{x+x^2}{2}$ defined on $[0;1]$, and let us fix the model triangulation as the trivial parametrization $[0;1] \to [0;1]$. An optimal triangulation $\sigma_{op} : [0;1] \to [0;1]$ is then a smooth map such that

$$\sigma_{op} = f(x).$$

This shows that, if we wish one minimize Φ on $\mathcal{T}_\sigma$,

- we have got here an optimized triangulation $\sigma_{op} = f(x)$, which is a toy example of what can happen when, for example, the first derivative of f is Lipschitz and when M is a compact domain;

- defining an optimal triangulation leads to choose a function σ_{op} in a too large class of functions.

In the finite elements method, finding a base triangulation is not the main feature, and one needs to get quickly a "not so bad" triangulation. This is why one usually considers, when M is a bounded domain of an Euclidian space, the space of affine triangulations:

$$AffT_\sigma = \left\{\sigma' \in T_\sigma | \forall i, \sigma'_i \text{ is affine } \right\}.$$

4. Conclusion and perspectives

Under the lights of this approach, which was not fully justified till the definition of the Frölicher structure of a triangulation in [8] and the remark that the space of triangulations could be equipped with a smooth structure in the present communication, we can sketch two promising directions:

- the systemetic review of existing procedures for choosing a triangulation, which can be very sophisticated, but often justified filling technical gaps by intuition. The present setting may furnish a technical tool to fully justify existing methods.
- the exploration of the variation of functionnal integrals in the space of connections, which are often defined via cylindrical integrals over H^1 approximations with respect to a sequence of triangulations.

[1] Batubenge, A.; Tshilombo, M.H.; Topologies on product and coproduct Frölicher spaces; *Demonstratio Math.* **47**, no4 (2014) 1012-1024
[2] Cherenack, P.; Ntumba, P.; Spaces with differentiable structure an application to cosmology *Demonstratio Math.* **34** no 1 (2001), 161-180
[3] Dugmore, D.; Ntumba, P.;On tangent cones of Frölicher spaces *Quaetiones mathematicae* **30** no1 (2007) 67-83
[4] Dugmore, D.; Ntumba, P.; Cofibrations in the category of Frölicher spaces: part I *Homotopy, homology and applications* **9**, no 2 (2007) 413-444
[5] Frölicher, A; Kriegl, A; *Linear spaces and differentiation theory* Wiley series in Pure and Applied Mathematics, Wiley Interscience (1988)
[6] Kriegl, A.; Michor, P.W.; *The convenient setting for global analysis* Math. surveys and monographs **53**, American Mathematical society, Providence, USA. (2000)
[7] Magnot, J-P.; *q-deformed Lax equations and their differential geometric background.* Lambert Academic Publishing (2015)
[8] Ntumba, P.; DW complexes and their underlying topogical spaces. *Quaestiones Mathematicae* **25**, no 1 , 119-134 (2009)
[9] Whitney, H.; *Geometric Integration Theory.* Princeton University Press, Princeton, NJ (1957)

Concurrence of bulk and surface order reconstruction to the relaxation of frustrated nematics

Antonino Amoddeo
Department of Civil, Energy, Environmental and Materials Engineering, Università "Mediterranea" di Reggio Calabria, Via Graziella 1, Feo di Vito, I-89122 Reggio Calabria, Italy

E-mail: antonino.amoddeo@unirc.it

Abstract. Applying appropriate electric pulses to a nematic liquid crystal confined between plates, the bulk order reconstruction can occur, a mechanism allowing the switching between topologically different nematic textures without any director rotation. Using a moving mesh finite element method we describe the order tensor dynamics for a nematic inside an asymmetric π-cell, putting in evidence as textural distortions induced by strong asymmetries can be relaxed via both bulk and surface order reconstruction, occurring close to a confining plate with different time duration.

1. Introduction

Nematic Liquid Crystals (NLCs) confined between flat glass plates, in presence of appropriate electric and mechanical stresses can experience textural modifications relaxing the induced distortion [1]. In a similar device, the π-cell, the plates are treated in order to impose an initial orientation to the adhered nematic molecules, commonly known as the pre-tilt angle. The electro-optical properties of nematics depend on their morphology, as they are rod-like molecules exhibiting orientational order and cylindrical symmetry. A unit vector, the director $\boldsymbol{n}$, accounts for the orientation of the NLC long axis, while the scalar order parameter S describes the degree of order along such direction, and their behaviour, if the applied distortions have length scales comparable to the biaxial coherence length ξ_b [2], becomes biaxial, as two distinct optical axes arise. Biaxial phenomena in NLCs have been extensively investigated, both experimentally and theoretically, see [3-6] and references therein, due to their role in the switching properties of electronic devices such as Liquid Crystal Displays (LCDs), which theoretical description requires the coupling of the $\boldsymbol{n}$ and S parameters within the frame of the Landau-de Gennes order tensor $\mathbf{Q}$ theory [7].

In fig. 1 we depict a π-cell, in which a thin film of nematic material is confined between the upper and lower glass plates, with pre-tilt angles θ_U and θ_L, respectively, generating a splayed texture [1], figure 1(a). The cell is said symmetric because $\theta_U = \theta_L$, otherwise it is asymmetric; both θ_U and θ_L are measured with respect to the internal face of each plate. An electric field applied perpendicularly to the boundary plates, cause the initial splay texture to be switched into the topological different π-bent one, figure 1(c), through the intermediate state of figure 1(b), where the biaxial order reconstruction take place in the cell centre: here, the two uniaxial textures depicted in figure 1(a) and 1(c) are connected through a wealth of transient biaxial states in the nanometric scale, allowing the local order of the nematic phase to change without any macroscopic director rotation [1].

Content from this work may be used under the terms of the Creative Commons Attribution 3.0 licence. Any further distribution of this work must maintain attribution to the author(s) and the title of the work, journal citation and DOI.
Published under licence by IOP Publishing Ltd

The modelling of the **Q**-tensor dynamics inside a π-cell containing a NLC results in a Partial Differential Equations (PDEs) system, solved using the Finite Element Method (FEM) implementing an adaptive grid numerical technique, the Moving Mesh Partial Differential Equation (MMPDE) [3-6]. Briefly, the domain discretization with an appropriate grid of points, typical of the FEM technique, is controlled according the monitor parameter $\nabla\mathbf{Q}$, while nodal connectivity and number of mesh points inside the domain are kept constant.

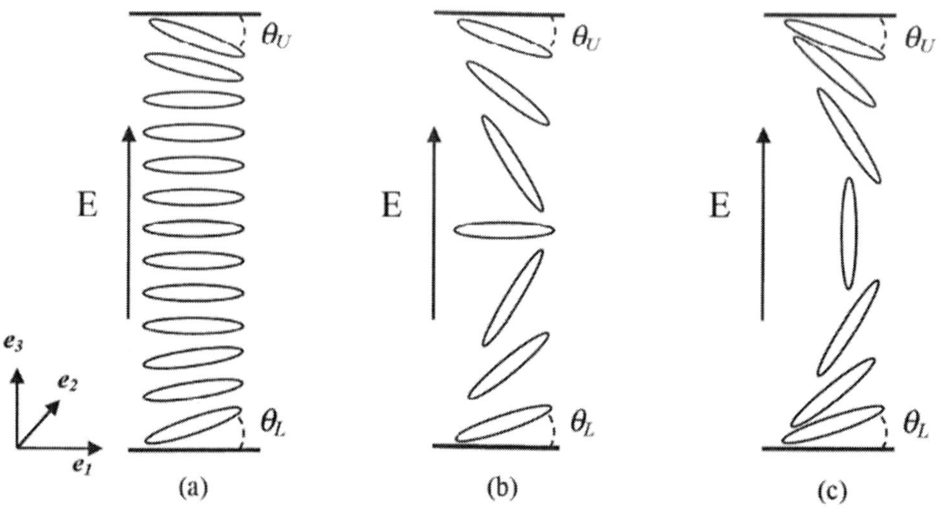

Figure 1. Outline of the π-cell with nematic molecules in different states: (a) horizontal alignment with a slight splay; (b) intermediate state, with a thin biaxial wall in the centre; (c) mostly vertical alignment (π-bent).

This allows the node points on which the PDE system is solved to be clustered into regions of high spatial variability where more detail is required, and it results an efficient and effective numerical tool appropriate to model the order dynamics of NLCs confined between plates, and the dynamics of complex systems more in general, as biological ones [8,9]. We present a MMPDE computation of the biaxial order dynamics for a 5CB NLC confined in an asymmetric π-cell, focusing on the switching mechanisms relaxing the nematic distortion. Throughout the simulations we neglect any material flow effect, as we are interested only determining the order evolution within the applied electric pulse duration, in which the order reconstruction transition is always confined [6,10].

2. Basic model

We assume that inside the π-cell, the free energy functional $\boldsymbol{F}$, because of the infinite anchoring energy, has contributions coming from the bulk only, which are distinguished in electric, elastic and thermotropic ones, respectively F_e, F_d, and F_t. The second rank order tensor $\mathbf{Q}$, expressed in the orthonormal basis of its eigenvectors $\{\boldsymbol{e}_1, \boldsymbol{e}_2, \boldsymbol{e}_3\}$ [7,11], describes both orientation and order of the NLC: the uniaxial nematic phase is characterized by the eigenvector $\boldsymbol{e}_i=\boldsymbol{n}$ associated to s_{max}; when all the three eigenvalues are different, the calamitic molecules are in the biaxial phase, on the contrary if they vanish the NLCs are in the isotropic phase. Neglecting ion effects and imposing the balance between $\boldsymbol{F}$ and the dissipation function D, the generalized Euler-Lagrange equations are obtained for each of the five independent parameters q_i corresponding to the degrees of freedom of a rod-like nematic molecule,

$$\frac{\partial D}{\partial \dot{q}_i} + \frac{\partial F_t}{\partial q_i} + \frac{\partial F_e}{\partial q_i} + \frac{\partial F_d}{\partial q_i} - \frac{\partial}{\partial x_j}\left(\frac{\partial F_d}{\partial q_{i,j}}\right) - \frac{\partial}{\partial x_j}\left(\frac{\partial F_e}{\partial q_{i,j}}\right) = 0 \quad i=1\ldots5 \tag{1}$$

where j denotes differentiation with respect to the spatial coordinates, and summation over repeated indices is assumed. The electric potential U, entering in $\boldsymbol{F}$ through F_e, is governed by

$$\nabla \cdot \boldsymbol{D} = \nabla \cdot \left(-\varepsilon_0 \boldsymbol{\varepsilon} \nabla U + \boldsymbol{P}_S\right) = 0 \tag{2}$$

in which $\boldsymbol{D}$ is the displacement field, ε_0 is the vacuum dielectric permeability, $\boldsymbol{\varepsilon}$ is the dielectric tensor and $\boldsymbol{P}_S$ is the spontaneous polarization vector. To solve the six coupled PDEs (1) and (2), due to the infinite anchoring energy of the nematic molecules we impose essential boundary conditions on the confining plates, obtaining the dynamical evolution of the system. The induced biaxial order is evaluated from [1]

$$\beta^2 = 1 - 6\frac{tr\left(\mathbf{Q}^3\right)^2}{tr\left(\mathbf{Q}^2\right)^3} \in [0,1] \tag{3}$$

where $\beta^2 = 0$ and $\beta^2 = 1$ mean, respectively, uniaxial and biaxial nematic textures.

In one spatial dimension, we define a physical domain $\Omega_p = [0,1]$ and a computational one $\Omega_c = [0,1]$; then we denote a PDEs solution over the physical domain $\Omega_p = [0,1]$ as $u(z,t)$, $z \in \Omega_p$, while if $\xi \in \Omega_c$ is a computational coordinate, we define a coordinate transformation from $\Omega_p \times (0,T]$ to $\Omega_c \times (0,T]$ as $\xi = \xi(z,t)$, and the reverse $z = z(\xi,t)$ holds. The integration domain is discretized with a fixed number of grid points, and the definition of the monitor function is based on a method proposed by Huang et al. [12], which ensures a good quality control of the meshes and final convergence of the FEM solution [13,14]

$$M(u(z,t)) = \int_0^1 \left(\left|\frac{\partial u(z,t)}{\partial z}\right|\right)^{1/2} \mathrm{d}z + \left(\left|\frac{\partial u(z,t)}{\partial z}\right|\right)^{1/2} \tag{4}$$

The equidistribution of (4) [13,15] in each subinterval of the integration domain gives the mesh equation

$$M(z(x,t))\frac{\partial}{\partial \xi} z(x,t) = \int_0^1 M(u(s,t))\,ds = C(t) \tag{5}$$

and since at a larger monitor function value corresponds a denser mesh, a mesh map control with improved resolution is obtained.

Full details about the used multipass algorithm, numerical procedure and settings can be found in [3-6], we remark only that to solve (1) and (2) we replace the unknown $u(z,t)$ in (4), with $tr(\mathbf{Q}^2)$, a quantity which is rapidly varying when the degree of order is not constant; the physical domain has been discretized with a mesh of 285 grid points using the physical parameters typical for 5CB nematic liquid crystal at ΔT=-1°C [16]. We modelled a one-dimensional asymmetric π-cell of thickness 1μm, to which a rectangular electric pulse is applied perpendicularly to the plates at t=0 s for a duration Δt = 0.25 ms, sampling the dynamical evolution of the $\mathbf{Q}$ tensor with a time step size δt= 0.1 μs.

3. Results and discussion

Recently we have shown [17], both experimentally and numerically, that applying an electric pulse to a 5CB NLC confined in an asymmetric π-cell, the arising distortion is relaxed through an increasing biaxial order close to the boundary plate where the anchoring angle is smaller, and to solve the PDEs system arising from the problem modelling we used the FEM based on a uniform discretization of the integration domain. In the present simulation, instead, we solve the PDEs system using the MMPDE technique to show as, enhancing the nematic frustration, a better resolution puts in evidence the relaxation mechanism observed close the boundary plates. In figure 2 we show the biaxiality maps obtained with $\theta_L = 19°$ and $\theta_U = -3°$, for amplitudes of the applied pulse of, respectively, 15 V/μm (a), 22 V/μm (b), 25 V/μm (c) and 40 V/μm (d), linearly mapped in a grey levels scale between the black (zero biaxiality) and the white (maximum biaxiality) colours. The vertical axis corresponds to the cell thickness, while in the horizontal one is represented the solution evolution in the 0 ms $\leq t \leq$ 0.1 ms interval, in logarithmic scale for plotting convenience. In each panel of the figure, the inset shows the magnification, not spatially scaled, of the biaxiality within 15 nm under the upper boundary plate. At t = 0 s, whatever the amplitude of the applied field is, close to the upper surface the nematic molecules

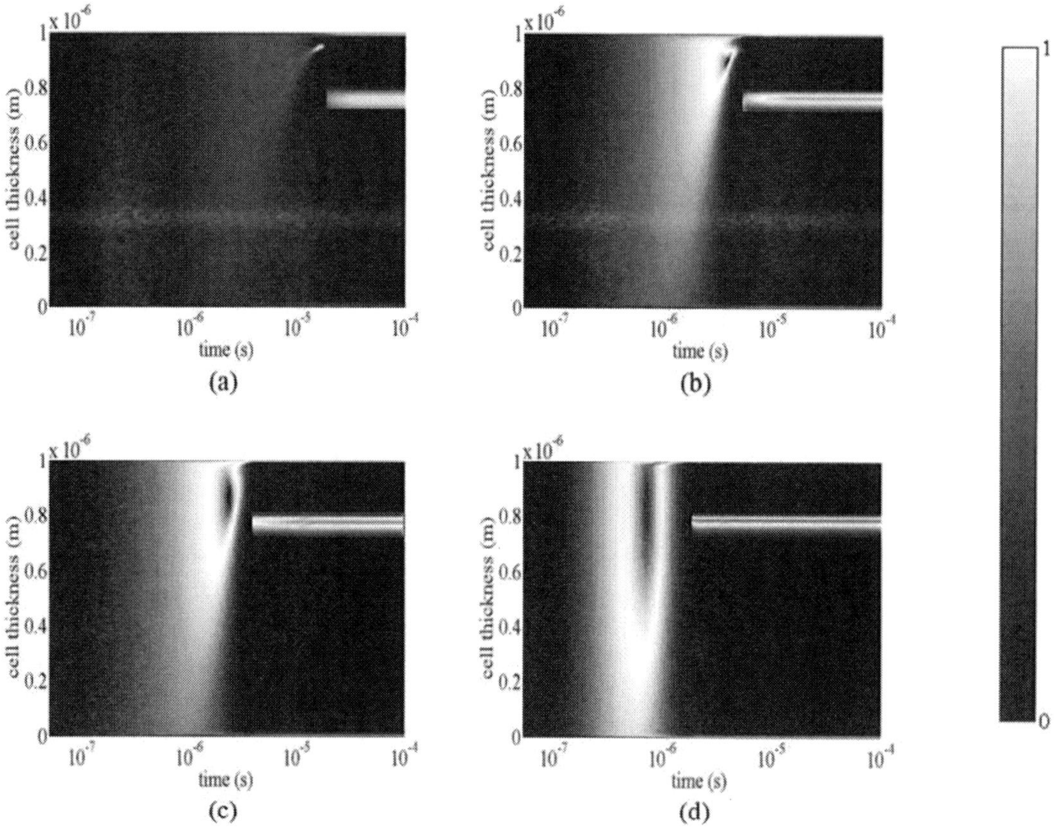

Figure 2. Surface contour plots of the biaxial order evolution for $\theta_L = 19°$, $\theta_U = -3°$, and electric pulse amplitudes of 15 V/μm (a), 22 V/μm (b), 25 V/μm (c) and 40 V/μm (d). The insets show the magnification of the biaxiality, not spatially scaled, within 15 nm under the upper boundary plate.

stay almost planar to the plate, while the bulk nematic texture has an asymmetric splayed configuration compatible with the prescribed boundary conditions. Applying a pulse of amplitude 15 V/μm, figure 2(a), for $t > 0$ s the director starts to align along the field direction, while a biaxial region of thickness about 20 nm, hence comparable with ξ_b, connects the quasi-planar nematic texture lying close to the upper surface, to the vertical nematic molecules located in the bulk: the nematic distortion tends towards the upper boundary plate where it relaxes by lowering the nematic order [3,17], and the starting uniaxial phase is locally replaced by growing biaxial domains. Just underneath the upper surface a bulk order reconstruction occurs around t = 10 µs, characterized by the ring-like biaxial region surrounding a planar uniaxial state [1]; for t > 10 µs the initial splayed texture is replaced by a bend one, and a uniaxial order is restored everywhere except for the biaxiality growing close to the boundary surfaces, due to the anchoring conditions imposed: near the upper surface, in particular, the competition between the molecules lying in a quasi planar configuration and the nematic texture aligned vertically along the field direction induces a strong distortion which is relaxed by a growing biaxial wall [17], see the inset of figure 2(a). Increasing the pulse amplitude, it is well known that the biaxial structures become faster [5], and at the same time the bulk order reconstruction spreads progressively along the cell thickness, losing its cylindrical symmetry [6]: in the present case, as the competition between the quasi planar texture and the vertical one becomes stronger close to the upper boundary plate, inside the surface biaxial wall an order reconstruction transition occurs, lasting for the whole pulse duration, see the inset of figures 2(b)-2(d), concurring together the bulk order reconstruction transition to the relaxation of the textural distortion [18].

4. Conclusions

Our MMPDE simulations are consistent with previous computations and suggest that, although with different temporal durations, surface and bulk order reconstruction transitions cooperate to relax the textural distortions induced in frustrated NLCs confined in a π-cell with strong asymmetries as well as strong electric distortions. The afforded problem involves characteristic lengths and times with large-scale differences, and although it is simplified and in one space dimension, it is a useful model representative of higher-dimensional real problems, confirming the validity of our numerical method also in simulating markedly distorted physical systems as in the present case.

References

[1] Barberi R, Ciuchi F, Durand G, Iovane M, Sikharulidze D, Sonnet A and Virga E G 2004 *Eur. Phys. J.* E **13** 61–71
[2] Biscari P and Cesana P 2007 *Contin. Mech. Thermodyn.* **19** 285–98
[3] Amoddeo A, Barberi R and Lombardo G 2012 *Phys. Rev.* E **85** 061705–10
[4] Amoddeo A, Barberi R and Lombardo G 2010 *Comput. Math. Appl.* **60** 2239–52
[5] Amoddeo A, Barberi R and Lombardo G 2011 *Liq. Cryst.* **38** 93–103
[6] Amoddeo A, Barberi R and Lombardo G 2013 *Liq. Cryst.* **40** 799–809
[7] de Gennes P G and Prost J 1993 *The Physics of Liquid Crystals* (Oxford: Clarendon Press)
[8] Amoddeo A 2015 *Comput. Math. Appl.* **69** 610–9
[9] Amoddeo A 2015 *Cogent Physics* **2** 1050080
[10] Brimicombe P D and Raynes E P 2005 *Liq. Cryst.* **32** 1273–83
[11] Virga E G 1994 *Variational Theories for Liquid Crystals* (London: Chapman and Hall)
[12] Huang W, Ren Y and Russell R D 1994 *SIAM J. Numer. Anal.* **31** 709–30
[13] Beckett G, Mackenzie J A, Ramage A and Sloan D M 2001 *J. Comput. Phys.* **167** 372–92
[14] Beckett G and Mackenzie J A 2000 *Appl. Numer. Math.* **35** 87–109
[15] de Boor C 1974 Good approximation by splines with variable knots II *Conf. on the Numerical Solution of Differential Equations (Dundee, UK, 1973) (Lecture Notes in Mathematics* vol 363*)* ed G A Watson (Berlin: Springer-Verlag) pp 12–20
[16] Ratna B R and Shashidhar R 1977 *Mol. Cryst. Liq. Cryst.* **42** 113–25
[17] Lombardo G, Amoddeo A, Hamdi R, Ayeb H and Barberi R 2012 *Eur. Phys. J.* E **35** 9711–16
[18] Amoddeo A 2015 *Journal of Physics: Conference Series* **574** 012102

Potential of Lattice Boltzmann Method to Determine the Ohmic Resistance in Porous Materials

Mayken Espinoza-Andaluz[1,2], Martin Andersson[1], Bengt Sundén[1]

[1] Department of Energy Sciences, Lund University, PO Box 118, 22100 Lund, Sweden
[2] Escuela Superior Politécnica del Litoral, ESPOL, PO Box 09-01-5863 Guayaquil, Ecuador

E-mail: mayken.espinoza_andaluz@energy.lth.se; masespin@espol.edu.ec

Abstract. The lattice Boltzmann method (LBM) is a suitable tool for solving transport phenomena that occur in gas- and liquid phases at different length scales, especially when complex geometries such as porous media are involved. However, investigations about applications of LBM in the solid electrical conducting material have not been carried out yet. Since in fuel cells (FCs) the multifunctional layers play an important role during the energy conversion process, and such layers consist of porous material, the ohmic resistance of porous materials represents a crucial characteristic to be studied to predict the internal ohmic losses. The purpose of this paper is to show the feasibility of LBM to determine the ohmic resistance of electrical conducting materials whose dimensions are modified considering the cross-sectional area and length. Characteristics, limitations and recommendations of LBM applied to solid electrical conducting materials calculating the ohmic resistance are presented considering the coupling of the methodology with the Ohm's Law. Additionally, the behavior of the ohmic resistance for a given porous material is presented.

1. Introduction

Predicting the internal ohmic losses involved in the energy conversion process is crucial in order to improve the microstructural architecture of the different multifunctional layers inside fuel cells (FCs). Most of these multifunctional layers are porous materials which represent a complex study of fluid and solid interactions.

Given the mentioned characteristics, several studies of the fluid behavior through porous media have been carried out, by the authors, using the lattice Boltzmann method (LBM) [1 – 2]. However, LBM has not been applied to analyze the similarities between the hydraulic and electrical behavior. Considering the mentioned analogy, LBM can result in a powerful tool to analyze certain electrical variables in solid conducting materials such as current density (flow of electrons) and ohmic resistance.

The purpose of this work is to prove the feasibility of applying the LBM in the solid conducting material instead of the fluid phase. The use of LBM in the solid conducting material gives us the opportunity to calculate the ohmic resistance in conductive materials and the possibility to predict the ohmic resistance for porous materials. Since LBM was originally a methodology for recovering hydraulic behavior from the Navier-Stoke equations, some considerations should be taken into account when it is applied to the flow of electrons. For this study the momentum and continuity equations are

Content from this work may be used under the terms of the Creative Commons Attribution 3.0 licence. Any further distribution of this work must maintain attribution to the author(s) and the title of the work, journal citation and DOI.
Published under licence by IOP Publishing Ltd

solved using LBM, which allow to determine the equivalent variables in the electrical system. The corresponding analogies are given in Table 1.

Table 1. Equivalences between hydraulic and electric variables with corresponding units.

	Hydraulic			Electric		
Type	**Variable**	**Symbol**	**Units**	**Variable**	**Symbol**	**Units**
Quantity	Volume	V	m^3	Charge	q	C
Potential	Pressure	p	N/m^2	Potential	φ	$N\,m/C$
Flux	Volumetric flow rate	ϕ_V	m^3/s	Current	I	C/s
Flux density	Velocity	v	m/s	Current density	j	$C/m^2\,s$

2. Methodology

The models are developed using the D3Q19 scheme, and the studied domains are shaped as 3D volumes with uniform cross-sectional areas along the length. The first stage of the investigation is to find the range of values of the equivalent current density, i.e., can be related with the velocity in hydraulic studies, in which LBM can present an equivalent ohmic behavior. The ohmic behavior implies: if the potential difference increases, also the current density increases. Additionally, the relationship between these two variables must be direct and linear. This measure is required since LBM has some limitations and would not give accurate results in higher velocities [3]. For this part, the domain is constructed as 5 *lu* side of the cross-sectional area and 15 *lu* length to assure excellent agreement of description in hydraulics [4]. The same domain is modified with the inclusion of non-conducting solid particles in the last part of this study.

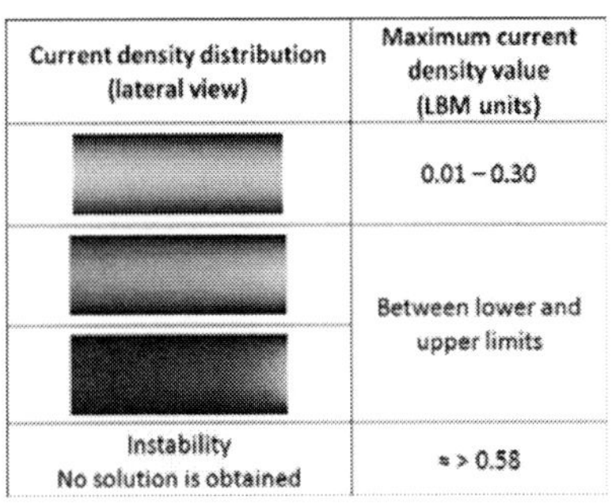

Figure 1. Lateral view of the current density distribution when potential difference is increasing. Higher current density values lead to instability.

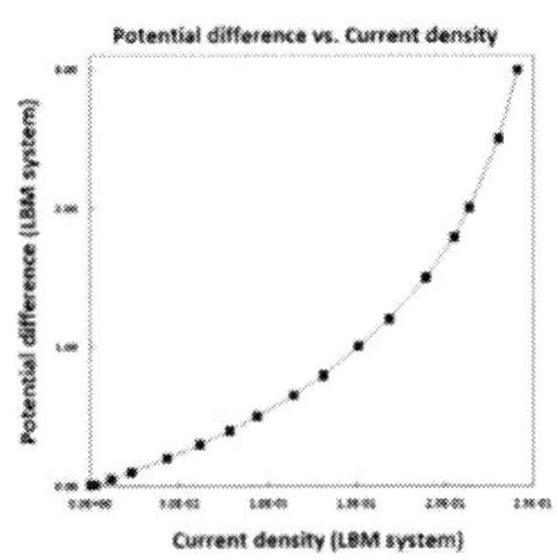

Figure 2. Potential difference vs. current density for the first domain analyzed. The ohmic behavior is obtained at lower current density values.

The first results show that there is a limitation on the maximum current density value obtained within the LBM implementation. Considering the material resistivity as a constant and that the flow of electrons is uniformly distributed, the current density values are not allowed to reach certain values to avoid instabilities.

Ohm's law establishes a relationship between the gradient of the applied potential and the current density:

$$j = -\sigma \nabla \varphi \tag{1}$$

where j is the current density (in A/m^2), φ is the potential function (in V), and σ the conductivity of the material (in $\Omega^{-1}m^{-1}$). Several values of potential differences are applied to domains with different sizes in order to demonstrate the validity of the equivalent Ohm's law. The results are presented in the next section.

3. Results and discussion

Considering equation (1) and knowing that the conductivity is the reciprocal value of the resistivity (ρ), the simplified version of the Ohm's law is expressed as I = V/R; where I is the electric current, V is the potential difference applied, and R is the ohmic resistance of the studied material. Figure 3 shows the results obtained when the dimensions of the conducting material are varied.

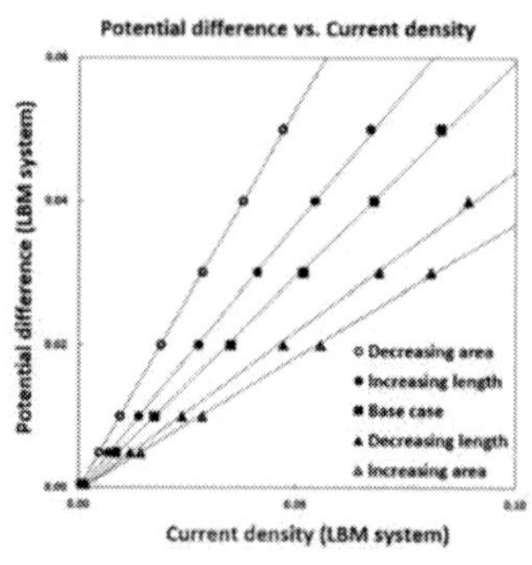

Figure 3. Ohmic region behavior for several domain sizes. Base case corresponds to a 20 *lu* side cube.

Ohm's law tells that the slope of the curves is proportional to the ohmic resistance materials. To corroborate the obtained results and to analyze the behavior of the ohmic resistance as a function of the material length and cross-sectional area, several measurements were effected.

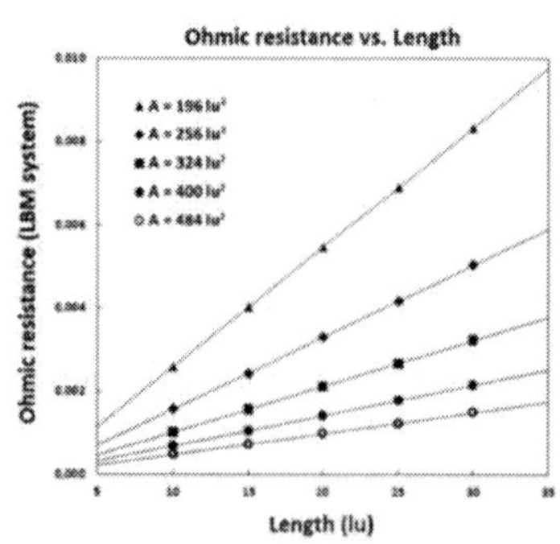

Figure 4. Obtained ohmic resistance values for a certain potential difference when the length varies.

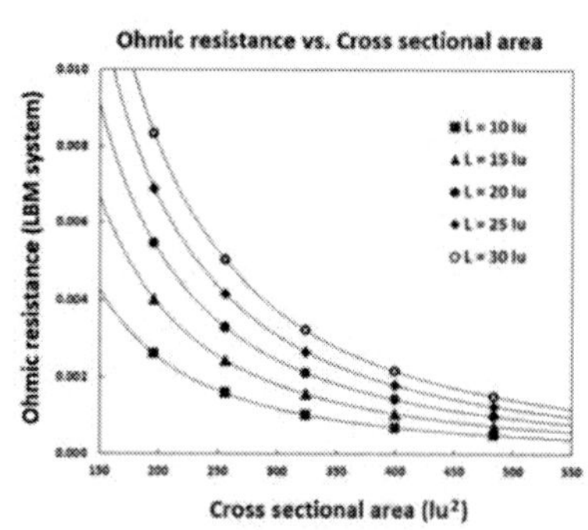

Figure 5. Obtained ohmic resistance values for a given potential difference when the cross-sectional area varies.

In physics, Ohm's law is expressed as E = ρ j; where E is the electric field (in V/m). If the electric field is assumed constant along the length of the conductor (L), the potential difference can be determined as V = E L. Using the mentioned relationships in equation (1) and solving for V/I, the ohmic resistance can be calculated as R = ρ (L/A). Results presented in Fig. 4 and Fig. 5 agree with the latest expression, i.e., the ohmic resistance increases in a linear manner when the length increases and decreases as an inverse function when cross-sectional area increases.

Although the behavior of the curves showed in Fig. 4 and Fig. 5 agree with the expected in ohmic materials, it is important to notice that there is an additional issue that should be addressed related to the material resistivity. Independently the size of the material and the potential difference applied, the resistivity is an intrinsic characteristic of the material, i.e., it must be constant. However, the values obtained during our simulations required some adjustments, which can be achieved relating the resistivity and the relaxation parameter in LBM, and also the boundary conditions can be revised.

Finally, some measurements of the incidence over the ohmic resistance when non-conducting particles are included in the material domain, i.e., porous material, are carried out. These non-conducting particles represents the void space volume. Figure 6 shows the variation of the ohmic resistance with a certain percentage of non-conducting particles. As expected, the ohmic resistance increases with the increasing percentage of such particles.

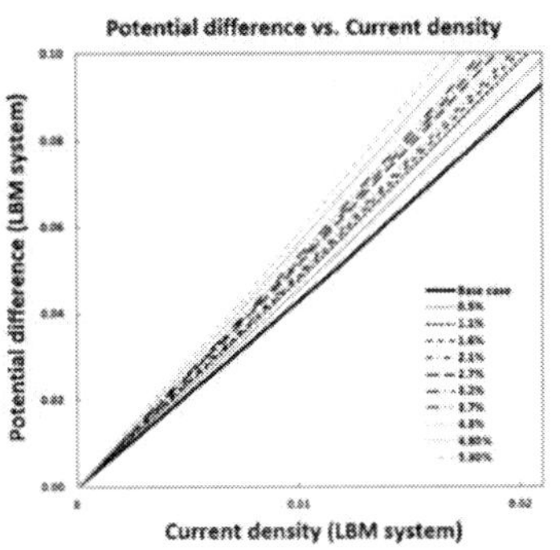

Figure 6. Base case corresponds to a volume domain of 5 *lu* side of cross-sectional and 15 *lu* length.

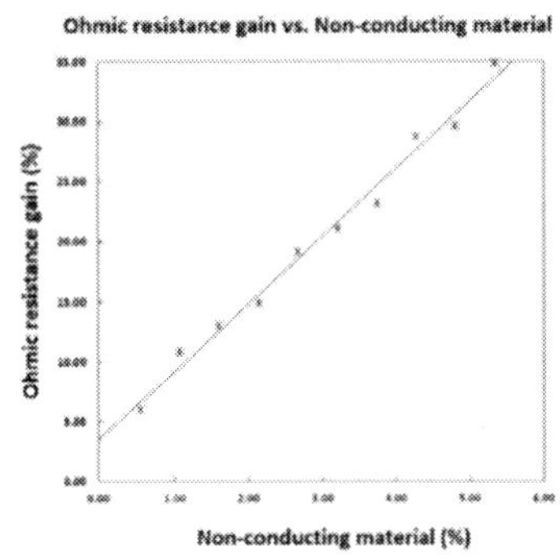

Figure 7. Trend of the ohmic resistance gain relative to the base case.

4. Conclusions and further studies

Although LBM has proven to be suitable to calculate the ohmic resistance for uniform shape conducting materials and porous materials, additional studies with different domain sizes and higher presence of non-conducting materials are required. Another issue to be addressed is related to the material resistivity; which can be associated to the relaxation parameter used in LBM. Additionally, the boundary conditions applied to the electric conductive surfaces can be analyzed.

References

[1] Espinoza, M., Sundén, B., and Andersson, M. (2014). Highlights of Fuel Cell Modeling From a Lattice Boltzmann Method Point of View. In ASME 2014 International Mechanical Engineering Congress and Exposition (pp. V06AT07A058-V06AT07A058). American Society of Mechanical Engineers.

[2] Espinoza, M., Andersson, M., Yuan, J., and Sundén, B. (2015) Compress effects on porosity, gas-phase tortuosity, and gas permeability in a simulated PEM gas diffusion layer. Int. J. Energy Res., **39**: 1528–1536. doi: 10.1002/er.3348.

[3] Mohamad, A. A. (2011). Lattice Boltzmann method: fundamentals and engineering applications with computer codes. Springer Science & Business Media.

[4] Sukop, M. C., and Thorne Jr, D. T. (2006). Lattice Boltzmann Modeling: An Introduction for Geoscientists and Engineers.

A mathematical method for the turbulent behavior of crowds using agent particles

Teruaki Ohnishi
Institute of Science and Technology for Society, Urayasu, Chiba 2790012, Japan

E-mail: ohnishi2015@yahoo.co.jp

Abstract Among the people moving as a group there appear social and psychological forces together with physical forces such as friction and resistance. With the definition that the *field of the crowd* is the region of those forces continuously extending with varying strength, and with the pre-requisite that the spatial distribution of the crowd, *i.e.*, the distribution of the field, varies according to the hydrodynamic rule by the Navier-Stokes equation, a methodology was proposed to describe the behavior of the crowd composed of many agent particles as the movement of a compressible, turbulent fluid. A numerical calculation was exemplified for the dynamic behavior and spatial distribution of crowds during movements when there appears a conflict between groups with different characters, imaging for instance the medieval battle of Breitenfeld.

1. Introduction

In our everyday life, we often find the movement of the crowd as if it strongly resembles in the motion of a fluid especially when many people come together to execute a common task [1]. Under some circumstances their movement may be restricted by some social and psychological factors such as norms, bonds, and items under tacit understandings within their own group. Individuals in the group thus make invisible mutual-interaction among them. The collectivity of the crowd becomes high in the case of a strong interaction which restricts the unlimited movement of the individuals. When the population density of the crowd exceeds a certain critical value, a sort of viscous character originating from physical and psychological frictions may appear. Moreover in such a dense crowd the exchange of information is limited only within a local region so that the movement of the crowd in a group may become incoherent. The movement of the crowd, therefore, may be similar to the turbulent motion of a compressible viscous fluid.

When the crowd disperses and the individuality of the people increases, various characters of individuals become to appear discretely so that we must consider the crowd as an agglomeration of independent individuals. When we describe the collective motion of an object consisted of individual particles, one of the useful methods is to adopt the agent particles (or simply, the agents hereafter) [2], and by doing so we have obtained a great deal of valuable knowledge about the behavior of pedestrians, the moving patterns of social insects, the attitude change or the opinion dynamics of individuals, etc [3-8]. In this paper we will also treat the crowd as an agglomeration of agents. The behavior of the crowd, however, will be simulated more realistically if we introduce a unified method by which we possibly represent the motion of the crowd from its dense state to a dilute state throughout the whole process of its behavior while the discrete nature of individual agents is held. Along this line we like to propose a mathematical method to realize the possibility as such by using the Navier-Stokes equation. For the feasibility study of this method, we then try to model the mass

Content from this work may be used under the terms of the Creative Commons Attribution 3.0 licence. Any further distribution of this work must maintain attribution to the author(s) and the title of the work, journal citation and DOI.
Published under licence by IOP Publishing Ltd

movement of troops with different characters as the fluid motion of the crowd by investigating their behavior during a conflict.

In the next section mathematical method are given and in section 3 several type of problems are considered to which we can apply this method. In section 4 the treatment of forces is described when we simulate the mass motion of the crowd during a conflict by our method, followed by an example calculation in section 5, supposing a medieval battle. Section 6 gives concluding remarks.

2. Turbulent equation for discrete particles and its numerical treatment

Individuals in the crowd interact with tacit, psychological forces which mutually impose social restrictions on each other, even if they do not physically contact, through the inspection of surrounding situations and their colleague's movement. Such situation can be reworded as that the spatial extension of the crowd just corresponds to the field where such fluid forces extend. When the individual in the crowd makes a relative motion to each other, it may give rise to the viscosity and the turbulence in the group, resulting in the distortion of the force field. Although the local strength of the field changes according to the change of the distribution of the crowd, the field continuously extends from the inner region of the crowd to the outward. We have been accustomed to the consideration that the field made by interacting particles is a sort of fluid, such as the fluid model of nucleus, the droplet model of nuclear fission, the field of quarks and gluons and so on. The fact that the granular materials containing inter grain water act as a fluid to a certain limit [9-11] is quite indicative from our viewpoint. If the flow of the crowd really resembles in the flow of the field, individuals in the crowd can be said to move according to the kinetic rule determined by the field. When we replace the crowd with representative agents, we can deduce the strength of the local field in terms of the local density of the agents whose flow is determined by the kinetic rule of the field. According to such a conception, introducing the agents as many number as they are proportional to the number of people in the real crowd, we estimate the movement of the crowd as the flow of the filed at the position of each agent by using the turbulent equation.

The behavior of a force field, *i.e.* the behavior of people or the agents in our case, is given by the following turbulent equation of compressible fluid and the equation of continuity.

$$\frac{\partial \mathbf{V}}{\partial t} = -(\mathbf{V}\cdot grad)\,\mathbf{V} + \frac{\eta}{\rho}\Delta\mathbf{V} + \frac{1}{\rho}(\varsigma + \frac{\eta}{3})grad\cdot div\,\mathbf{V} - \frac{1}{\rho}grad|\mathbf{P}| + \frac{\mathbf{F}}{\rho} \tag{1}$$

$$\frac{\partial \rho}{\partial t} = -div(\rho\mathbf{V}) \tag{2}$$

where **V** is the velocity of the crowd, ρ the density, η the coefficient of viscosity, ζ the volumetric viscosity, and **P** and **F** are respectively the pressure and the force acting on the crowd.

When we intend to simulate the collective motion of the crowd, one of the inherencies to be considered is the finiteness of moving velocity. Hence, although we obtain the fluid velocity of the agent from equation (1), it should be limited to a certain maximum value V_{max}. Moreover the viscosity arising from the crowd motion intuitionally seems to increase with the crowd density so that we set simply as $\eta/\rho \equiv H$ and $(\zeta+\eta/3)/\rho \equiv Z$, and treat the quantities H and Z as constants as a first approximation. Moreover since the fourth and fifth terms on the right hand side of equation (1) give forces acting on the unit mass of the crowd, those terms give values proportional to the force acting on an individual agent when we initially distribute the agents with the number just in proportion to the local density of the crowd. We therefore omit the factor ρ in the fourth and fifth terms of equation (1). Furthermore equation (2) is no more required so far as the number of agents is conserved.

The two-dimensional velocity of the agent m on the direction k (=x and y), $V_{m,k}$, is now given by

$$\frac{\partial V_{m,k}}{\partial t} = -V_{m,k}\cdot\frac{\partial V_{m,k}}{\partial k} - V_{m,\ell}\cdot\frac{\partial V_{m,k}}{\partial \ell} + H\cdot\sum_{q=k,\ell}\frac{\partial^2 V_{m,q}}{\partial q^2} + Z\cdot\frac{\partial}{\partial k}\sum_{q=k,\ell}\frac{\partial V_{m,q}}{\partial q} - \frac{\partial P_m}{\partial k} + F_{m,k} \tag{3}$$

where ℓ=y and x, and P_m and $F_{m,k}$ are respectively the pressure and the force acting on the agent m. When every agent is a member of one of the groups, we deduce the partial derivative of $V_{m,k}$ with respect to the coordinates k as

$$\frac{\partial V_{m,k}}{\partial k} \cong (\langle V_{m,k}\rangle_{+} - \langle V_{m,k}\rangle_{-})/(d_{k+} + d_{k-}) \tag{4}$$

Here $\langle V_{m,k}\rangle_{+}$ and $\langle V_{m,k}\rangle_{-}$ are the averages of the weighted sum of the agents belonging to the same group within the respective sides of +k and –k of the circle of a radius R_O as given by

$$\langle V_{m,k}\rangle_{\pm} = \pm\sum_{j}{}' V_{j,k} \exp(\theta_{mj}\mathbf{i}) \cdot \exp(-r_{mj}/\lambda) \tag{5}$$

where Σ'represents the sum without including the m itself, $\theta_{mj}=\arg(\boldsymbol{r}_{mj})$, $\boldsymbol{i}$ is the imaginary unit vector. Moreover the quantity $d_{k\pm}$ is the average of the distance on the $\pm k$ directions for all agents within the half circles on the $+k$ and $-k$ respective sides as given by

$$d_{k\pm} = \frac{1}{n_{k\pm}}\sum_{j}{}' X_j \exp(\theta_{mj}\mathbf{i}) \tag{6}$$

where $n_{k\pm}$ are the total number of agents within the respective half circles. The weight in equation (5) is assumed to exponentially decrease with the radial distance, and λ is the attenuation distance. The regions $\Sigma_{x\pm}$, $\Sigma_{y\pm}$ where the Σ' on the sides of $\pm x$ and $\pm y$ should be taken are depicted in Fig.1.

On the other hand the second partial derivative of $V_{m,k}$ is given by

$$\frac{\partial^2 V_{m,k}}{\partial k^2} \cong \left(\left\langle\frac{\partial V_{m,k}}{\partial k}\right\rangle_{+} - \left\langle\frac{\partial V_{m,k}}{\partial k}\right\rangle_{-}\right)/(d_{k+} + d_{k-}) \tag{7}$$

Here $\langle \partial V_{m,k}/\partial k\rangle_{\pm}$ can be derived by replacing $V_{j,k}$ in equation (5) by $\partial V_{j,k}/\partial k$. The other second derivatives are also given by the same manner. The values of partial derivatives of V, $\partial V_{m,\lambda}/\partial k$, $\partial^2 V_{m,\lambda}/\partial k^2$ and $\partial^2 V_{m,\lambda}/\partial k\,\partial \ell$, asymptotically approach null when the number density of agents decreases so that equation (3) approaches the usual equation of motion for a point mass. Discretizing equation (3) with respect to time, we have the velocity component $V^{t}_{m,k}$ of the agent m in the direction k at a time t as

$$V^{t}_{m,k} = \min\left(V^{t-\Delta t}_{m,k} + \left(\frac{\partial V_{m,k}}{\partial t}\right)^{t-\Delta t}\Delta t,\ V_{\max} V^{t-\Delta t}_{m,k} / \left|V^{t-\Delta t}_{m,k}\right|\right) \tag{8}$$

Here the superscript t indicates the value at a discrete time t, and the quantity Δt is a time-step. Using this quantity, we derive the position of m in the direction k at t, $X^{t}_{m,k}$, as

$$X^{t}_{m,k} = X^{t-\Delta t}_{m,k} + 0.5(V^{t}_{m,k} + V^{t-\Delta t}_{m,k})\Delta t + 0.25\left(\frac{\partial V^{t}_{m,k}}{\partial t} + \frac{\partial V^{t-\Delta t}_{m,k}}{\partial t}\right)(\Delta t)^2 \tag{9}$$

3. What type of problems can we simulate with this method ?

In the case of the movement of groups with different characters, conflicts will possibly occur if the space where one group invades is already possessed by the other group. In such a case the extinction of one of the groups, the homogenization of a group to the other, or the expulsion or the flight of the group to the outer region will occur as a result of the conflict. Some studies on such subjects have been done about the rise and fall of a community by using an agent model [16, 17], and the battle between medieval armies at Agincourt using a continuity theory [18].

With regard to the application of our model to the competing social phenomena, we can think of the following cases which have a common intention of motion of the individuals each of which movement is, however, seemingly random, such as (1) the behavior of the groups of wild animals which cause rivalries over their territory, (2) the behavioral pattern of people in the street demonstration and the adversarial people regulating their movement, (3) the behavior of soldiers in the troops during the battles in a close combat, especially before the medieval era, and so on. Since there

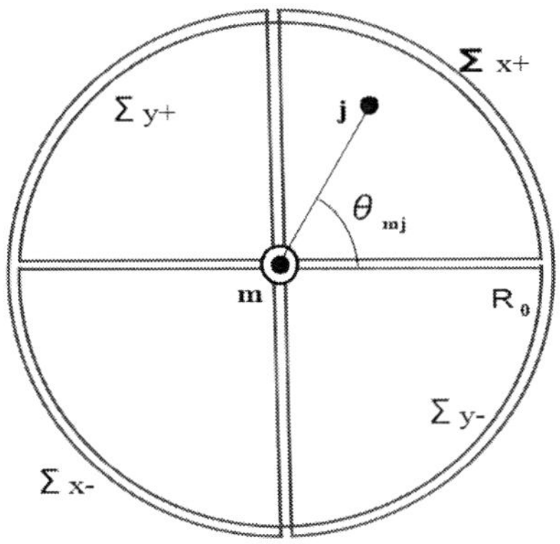

Figure 1. Regions to be performed for summation

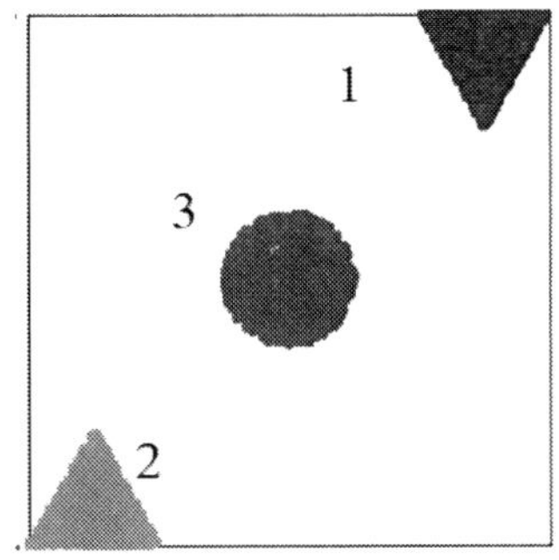

Figure 2. Initial distribution of agents.

exist quite few documents for those real competitions on the movement of individual people or animal, which describe the gradual shift of motion from a laminar flow with a high density of the crowd to a turbulent flow of a low density, direct comparison of the simulation with real situations is difficult to make. Notwithstanding the rareness of described documents, we imagine the situation of the battle of Breitenfeld in Germany, for instance, as follows [12, 13, 14].

In 1631, two armies were confronting each other at Breitenfeld, the one was the joint army of Sweden and Saxony headed by King Gustav, whose center were constituted from many battalions with cavalries and maskets on both wings. The other was the Imperial army led by Tilly whose center was formed from a number of Tercios also with cavalries and maskets on both wings. The Tercio is the combat unit composed of infantries with spears in its center with maskets on all sides around it.

The movement of those infantries was of a quite streamline flow at the beginning as well as the movement of cavalries. The center of Imperial battalion first forwarded to attack the enemy. Although its left wing was intercepted by the Swedish cavalries and maskets to result in running back in a scattered manner, the right wing of Imperial army made the Saxony army run backward. The Swedish Gustav made his army on the right wing move to the left, moreover made the cavalries which were standing by on the right wing carry the attack all the way to the weak central part of Imperial army finally to succeed in breaking it into pieces. Once it occurred the close battle between infantry and/or cavalries, the battle field became to be of turbulence and infantry fell into serious disorders to flee randomly, whose situation can also be imagined from the print of those days [15]

In the next section assumed forms are described regarding the forces acting between the agents to simulate such conflicts by our method.

4. Example formulation of forces

The pressure $\boldsymbol{P}$ in equation (1) is for the external pressure originating from the direct and indirect coercion and the psychological fear due to the enemy in the case of conflict. Such a pressure may be given as a function of the distance $|X_g\text{-}X|$ between the adversarial group and the agent itself as

$$\mathbf{P} = \xi_0 \sum_{\phi} (\mathbf{X} - \mathbf{X}_{\phi g}) \Big/ |(\mathbf{X} - \mathbf{X}_{\phi g})|^{\alpha} \tag{11}$$

where ξ_0 and α are constants, X is the position vector of the agent, $X_{\phi g}$ is the position corresponding to the gravitational center of the adversarial group ϕ. We simply take α=3 in the example calculation.

On the other hand the force F is consisted from:(1) the resistance f_1 against the isolated existence away from the agent's group, (2) the cohesive force f_2 originating from the mutual dependence and the solidarity within the same group, and from social norms restraining agents, (3) the repulsive force f_3 originating from the population pressure and the exclusion due to the conflict within the same group,

(4) the belligerent force f_4 against the distant enemy when each agent moves toward the gravitational center of adversarial group far from it, (5) the belligerent force f_5 against the nearest member of the enemy which approaches the agent within a critical range r_c, (6) the force f_6 to run away from the gravitational center of the enemy when the enemy is far from the group to which the agent belongs, and (7) the force f_7 to run away from the nearest member of the enemy when it approaches the agent within a range r_c.

The force f_1 is assumed, as the pressure P, to be proportional to some powers of distance $|X_g\text{-}X|$ as $f_1= \xi_1(X_g\text{-}X)|X_g\text{-}X|^{\beta}$, where ξ_1 and β are constants, and X_g is the vector for the gravitational center of the group, setting β=1, for instance, in what follows.

The forms of f_2~f_7 are all simply assumed to have an exponential dependence on the distance. At the position X of the agent m the forces f_2 and f_4, for instance, are respectively given by

$$\mathbf{f}_2 = \xi_2 \cdot \gamma_m \sum_{\kappa} \gamma_\kappa \varepsilon_{m\kappa} \exp(-\Psi_2 r_{m\kappa}) \tag{12}$$

$$\mathbf{f}_4 = \varepsilon_{m\Phi}\, \xi_4 \exp(-\Psi_{4\Phi} r_{m\Phi}) \tag{13}$$

Here γ_m is a coefficient, as will be described later, which relates to the strength of the force for the agent m as a group leader, ξ and Ψ with suffixes are all constants, $r_{m\kappa}$ is the distance between the agents m and κ, $\varepsilon = r/r$, the suffix κ indicates the other agents within the same group, ϕ represents the nearest agent in the other groups which approaches m, and Φ represents the gravitational center of the other group. To cope with the infinite approach of people within the crowd, a strong repulsive force is assumed between two agents when the distance r between them becomes less than r_a.

Generally speaking there exists more than one leader in a group whose members usually follow them. The leaders draw closer the group members as attractors, but they apt to become independent from the other leaders or antagonistic to each other, intending to acquire the power and supremacy. To model such a situation, we introduce 5 leaders (or attractor agents) in each group and invest in them the cohesive force f_2 with γ_m (□1) times stronger than the force acting between the ordinary agents. Moreover the force is assumed to become repulsive between the leaders; γ_m □- γ_m.

When the agents belonging to different groups closely approach each other, a conflict caused by the forces such as (12) and (13) may probably arise. In the case of the approach within the distance less than r_c, the agents belonging to the weaker group is assumed (i) to become extinct with a probability p_1, (ii) to be homogenized or to adapt themselves to the stronger group with p_2, and (iii) to safely exist as before with p_3, where $p_1+p_2+p_3$ =1. The resultant effects of the extinction and the change in characters of the agents are fed back to the movement of the system as a fluid.

5. Simulation of a conflict

The agents are assumed to move within a rectangular region of a side length L(=1.0), and to randomly distribute at an initial time in some limited regions 1, 2 and 3 where the different groups of agents 1, 2 and 3 with the number of these groups n_1=2000, n_2= n_3=1000 are respectively exist, as shown in figure 2..We also assume the offensiveness increases and the flight character decreases both in the order of the group 3, the group 2 and the group 1. We use here the following values of constants and parameters as a base case: R_0=0.0625, λ=0.0313, r_c=0.02, r_a=0.001, V_{max}=0.141, γ_m=900, H_1=H_2=2×10^{-4}, H_3=5×10^{-5}, Z_1=Z_2=2.5×10^{-4}, and Z_3=5.2×10^{-5}. The Reynolds number is $Re = V_{max}L/\eta \simeq V_{max}L/H_1 \simeq 10^3$, which corresponds to a developed turbulence.

The simulation is shown in figure 3. The values of constants regarding the forces appeared in the previous section are all taken common for these (a)~(d) in figure 3, which seem unessential for the comparison of qualitative results. The second column (b) in this figure shows the base case for a time sequence t =50, 75, 100, 125 and 200, where the time means the number of time step. The values of parameters used in this base case give the comparable strengths of viscosity and turbulence. The offensive agents 1 swiftly shift their positions changing their appearance to run down the weak, and the running-away agents 2 and 3 change their outward form to effectively flee from the enemy. The

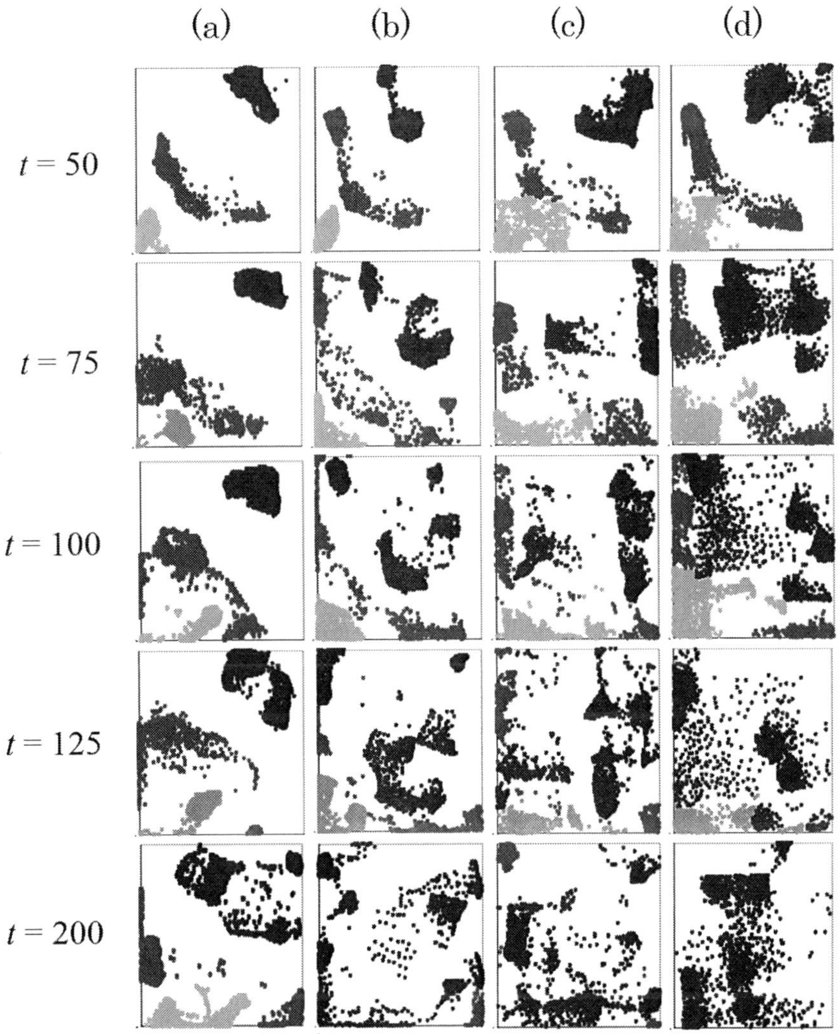

Figure3.:Time behavior of the group 1 (blue), 2 (light blue), and 3 (red) during the battle, where the group 2 and 3 being extinct by the group 1 and the group 3 by the group 2 both with a probability 1.0 when the close encounter takes place between them. Column (b) is the base case, whereas (a) is the case with the viscosity two times higher than (b), (c) is the non-viscous case and (d) is the case obtained with the ordinary equation of motion. The movement of those agents can be seen in animations [19].

offensive agents 1 form a wide wing during t =125～200 to enclose the weeks, showing the solidarity as a crowd [19].

The column (c) is for the case $\eta=\zeta=0$ but $(\mathbf{v}\bullet\mathrm{grad})\mathbf{v}\neq0$ in the Navier Stokes equation which reduces to the Euler equation. In the case of zero viscosity there does not exist any interaction and hence any social and psychological fields between agents. Although this causes the larger velocity of agents than the base case (b), it appears distinct edges of moving mass with a constant velocity because of the existence of a maximum velocity V_{max}. Furthermore the case (d) corresponds to the ordinary equation of motion with zero viscosity, eliminating the turbulent term, $(\mathbf{v}\bullet\mathrm{grad})\mathbf{v}=0$. The agents are not subject to any inner forces as physical and psychologically frictional forces, but only the outer forces. The dispersive nature of agents is strong so that their coherency is weak, showing a tendency of individual movement especially of the agents 1. Since the moving velocity becomes large, the clear edges of moving mass also appear in this case as in the case (c). The case (a), on the other hand, is for the viscosity two times stronger, $H_1=H_2=4\times10^{-4}$, $H_3=1\times10^{-4}$, $Z_1=Z_2=5\times10^{-4}$, and $Z_3=1.05\times10^{-4}$, than the base case (b). In this case the interactive force between agents becomes strong and the agents coagulate, indicating strong collectivity. Furthermore, since the viscosity is strong comparing with the turbulent nature, the moving velocity decreases to result in the motion of the agents as a group.

The viscosity thus acts as a factor to strengthen the coherency of the crowd. Such viscosity depends on the characteristics of the crowd as a condensed matter, namely on the strengths of physical friction and social and psychological interactions. Although there exist some examples of hydrodynamic analyses for the medieval battle of that type [1,18], it is also unclear whether the deformation of the crowd pattern as shown in figure 3 emerges in reality. For the clarification of those points and for making our model a quantitative tool we require the field observation regarding the collective motion of the crowd and close comparison between them and the simulations.

6. Conclusions

There are so many cases in social sciences where individual elements with different characters gather as a group to act for one common purpose such as the movement of races and refugees, the conflicts and the battles between races and tribes, and some types of demonstration. Our method in this paper can show a solution for such questions as how the crowd exerted by internal and external pressures diffuses in time and space, how its expansion changes in the case of the encounter with the other crowds with different characteristics. Such problems are interesting in that their academic discipline is just at around the midway of social sciences and natural sciences. It is also interesting to make this method more precise by using the behavioral data of the crowd observed in the real field.

References

[1] Hughes R L 2003 *Ann.Rev.Fluid Mech.* **35** 169-182
[2] Epstein J M and Axtell R 1996 *Growing artificial societies: Social science from the bottom up* (The Brookings Institution/MIT Press, Cambridge, MA)
[3] Helbing D, Schweitzer F and Molnar P 1997 *Phys.Rev.* **E56** 2527-2539
[4] AlGadhi S A H and Mahmassani H S 1991 *Transp Res. Record* No.1320 260-268
[5] Helbing D and Molnar P 1995 *Phys.Rev.* **E51** 4282-4286
[6] Piazza F 2010 *Phys.Rev.* **E82** 026111
[7] Schweitzer F and Holyst J A 2000 *Europ Phys.J.* **B15** 723-732
[8] Ohnishi T and Viquez A 2009 http://www.sts.or.jp/EngFile/TOAV.pdf
[9] Marston J O, Mansoor M M and Thoroddsen S T 2013 *Phys. Rev.* **E88** 010201
[10] Marston J O, Li.E Q and Thoroddsen S T 2012 *J.Fluid Mech.* **704** 5-36
[11] de Bruyn J R 2012 *J.Fluid Mech.* **704** 1-4
[12] Bennett M, Jorgensen C, Pavkovic M, Rice R S, Schneid F S and Scott C 2005 *Fighting techniques of the early modern world* (Amber Books Ltd., London)
[13] Battle of Breitenfeld(1631AD) http://www.juniorgeneral.org/30yw/30yw.html
[14] The battle of Breitenfeld http://www.youtube.com/watch?v=GGRajXx1h3s
[15] Schlacht bei Breitenfeld(1631) http://libris.kb.se/bib/544922
[16] Gumerman G I, Swedlund A C, Dean J S and Epstein J M 2003 *Artificial Life* **9**(4) 435-444
[17] Axelrod R 1997 *J.Conflict Resolution* **41**(#2) 203-226
[18] Clements R R and Hughes R L 2004 *Math. Comp. Simulation* **64** 259-269
[19] Ohnishi T http://ohnishi.ecnet.jp/RecentWorks/anime.html

Windows of Coherence: The Environment for organic substance

Yehuda Roth
Oranim Academic College of Education, Israel
E-mail: yudroth@gmail.com

Abstract. The following will introduce the A.I.E.S model, in which nonlinear maps determine the evolution of an inanimate substance to form an organic substance (Biopoiesis). We will demonstrate that when relating this model to a window of coherence that appears in the chaotic bifurcation diagram, a biological characteristics of warm-blood animate systems appears. Our model provides a mathematical platform in understanding Biopoiesis, that is, the process by which living organisms develop from inanimate matter.

I. Introduction

In the book entitled, *What is Life?*, Nobel laureate physicist E. Schrödinger states that "life feeds on negative entropy," namely, by the increase of order[1, 2]. Indeed, today life is still associated with order increasing (see examples in references [3],[4]).

At first glance, it looks as though this decrease in entropy violates the second law of thermodynamics, in which entropy always increases. However, since every living system is attached to an external environment, we can say that this reduction in entropy is always compensated by a growth of entropy in the attached thermal reservoir. As an example, we can observe an evolution of DNA molecules that have been analyzed numerically using a thermodynamic path integrals technique[5, 6]. Numeric calculations show a reduction in entropy on the account of a solvent environment.[7].

Nobel Prize-winning chemist Ilya Prigogine[8] noticed that some shapes, such as snowflakes (fractals in current terminology), possess natural order. Moreover, he described animate systems that exhibit behavior of the bifurcation type, namely, oscillation between different states. These early observations were primal clues for a generalized theory that identifies life with the evolution of nonlinear maps, such as the cellular automaton model [9, 10, 11].

Following other works, we will describe the ordering process by nonlinear maps. We will focus on a process in which inanimate systems, by reducing entropy, reproduce organic substance.

II. The nonlinear spanning map

Suppose we have a system evolving according to a nonlinear map. The system that is attached to a thermal bath is at a constant temperature T, such that, like warm-blooded animals, it is not necessarily in the reservoir temperature. Thus, our system can be out of the thermodynamical equilibrium. Yet, the constant temperature indicates that it is in a steady state.

Content from this work may be used under the terms of the Creative Commons Attribution 3.0 licence. Any further distribution of this work must maintain attribution to the author(s) and the title of the work, journal citation and DOI.
Published under licence by IOP Publishing Ltd

We start with a 1-D system which evolves with an iterating nonlinear equation[12]

$$x_{n+1} = f_R(x_n), \quad \forall n \;\; \alpha_n \in [0,1] \tag{1}$$

where all values of x are within the interval $[0,1]$, and R is a control parameter that determines the map type as either regular or chaotic. We assume that the R-parameter is related to temperature, such that as R rises, so does the temperature.

If, for example, we consider the logistic map $x_{n+1} = Rx_n(1-x_n)$, we can assume models like $R = 4\left(1-e^{-T/\tau}\right)$ or $R = 4e^{-T_0/T}$ where, τ and T_0 are constants characterizing the animate system. The 4-factor is a constraint determined by the logistic map model.

For a large variety of maps, increasing the strength parameter R (associated with heating the system from low to high temperatures) such that the map evolves from regular toward chaotic behaviors, follows similar steps[13].

The first step (a relatively small R) is characterized by maps that converge into a single stable point, calculated as

$$x_{n+1} = x_n \equiv x^{[1]} \Rightarrow x^{[1]} = f_R^{[1]}\left(x^{[1]}\right). \tag{2}$$

For a higher level of R, the map reaches two alternating values (the Andronov Hopf bifurcation)[14], as shown in fig. 1. Yet, it is possible to redefine a map that retrieves a single

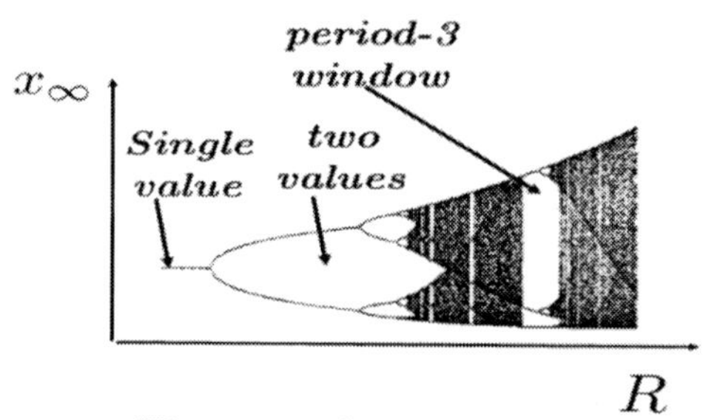

Figure 1. An example of a bifurcation diagram showing a window of coherence

stable value simply by using a double step iteration

$$x_{n+2} = f_R^{[2]}(x_n). \tag{3}$$

where $f_R^{[2]}(x_n) = f_R(f_R(x_n))$ is the map that iterates x_n twice.

A steady value is then obtained when

$$\begin{aligned} x_{n+2} = x_n \equiv x^{[2]} \Rightarrow x^{[2]} = f_R^{[2]}\left(x^{[2]}\right) \\ f_R(f_R) \equiv f_R^{[2]} \equiv f_R^{[1]} \circ f_R^{[1]}, \end{aligned} \tag{4}$$

where now the superscript [2] indicates that although the map $f_R^{[2]}$ retrieves a single final value, it is composed of two basic maps.

For a higher value R, the map terminates with four alternating values that give rise to the *four-rank map* $f_R^{[4]} = f_R^{[2]} \circ f_R^{[2]}$. Increasing R further, the same splitting continues until the map reaches the chaotic regime, in which it is impossible to redefine maps that stabilize at a finite single value.

In general, we define $f_R^{[m]}$—a *map of rank m* (m stands for "map"')—as

$$f_R^{[m]} = \underbrace{f_R^{[1]} \circ f_R^{[1]} \circ \dots}_{\leftarrow m \ times \rightarrow} = f_R^{[m/2]} \circ f_R^{[m/2]}. \tag{5}$$

Although we expect a chaotic behavior at high R, the bifurcation diagram shown in fig. 1 shows a much more complex behavior than just a simple division of period multiplications (1,2,4,...values), and chaos [15]. Amongst all the chaos, we find a stable period-3 orbit as shown in fig. 1. More windows of 5-period and 6-period are also found. Thus, in order to redefine the 1-rank map to retrieve a single stable value, we apply the appropriate rank map such as the 3-rank map $f_R^{[3]}(x_n)$.

III. The Animate state and the inanimate ensemble

Suppose we have a large dimensional Hilbert space spanned by a basis of states that are labeled by the states $|i\rangle$. Among this basis of states exists a unique single state, such as a DNA molecule or other organic substance that is associated with a single state $|\Omega\rangle$. This Hilbert space can be composed as a tensor-product of many spaces defined by different ingredients of the organic substance. The organic substance-state $|\Omega\rangle$ is defined to be unique as it defines the animate system, like a DNA molecule. All other states are related to an inanimate ensemble of states $|\aleph\rangle$, defined as $|\aleph\rangle = \sum\limits_{i \neq \Omega} a_i |i\rangle$, where the $\{a_i\}$-set selections determine the ensemble. To generalize our model further, we allow the coefficients a_i to be time dependent and complex. They can also be temperature dependent and random. Thus, except for the single organic substance state, $|\Omega\rangle$, all other complexities are introduced through the inanimate states ensemble.

Defining possible sets of the $\{a_i\}$ coefficients, it can be seen that $\forall \ \{a_i\} \quad \langle\Omega|\aleph\rangle = 0$, meaning that there is a pronounced distinction between the two concepts, being in the ordered organic substance state $|\Omega\rangle$ rules out the possibility of being part of an inanimate ensemble and vice-versa.

Using a Gröver-type state[16, 17], we present the A.I.E.S (Animate Inanimate Ensemble of States) model through a state ensemble $|n\rangle_{\aleph}$ such that

$$|n\rangle_{\aleph} = \alpha_n^{[p]} |\aleph\rangle + \beta_n^{[p]} |\Omega\rangle. \tag{6}$$

The coefficients $\alpha_n^{[p]}$ and $\beta_n^{[p]}$ generate the state-time evolution as determined by the nonlinear map, where the superscript $[p]$ is the p-rank map. The coefficient s's subscript at $|n\rangle_{\aleph}$ defines each state in the ensemble through the selection of $\{a_i\}$, namely, the $|\aleph\rangle$-state.

For each $|\aleph\rangle$, we expect the coefficients $\alpha_n^{[p]}$ and $\beta_n^{[p]}$ to behave as follows:

(i) Normalization condition
$\left|\alpha_n^{[p]}\right|^2 + \left|\beta_n^{[p]}\right|^2 = 1.$

(ii) Biopoiesis[18]
The process for which organic compounds, such as a DNA molecule, arise from inorganic matter through natural processes. For a p-rank map that converges into a single value, all states (most of them are inorganic), will line up to form the organic substance as represented by the state state $|\Omega\rangle$.

In order to fulfill these two conditions, we suggest a state of the form

$$\begin{aligned} |n\rangle_{\aleph} &= \sqrt{\Delta x_n^{[p]}}\, |\aleph\rangle + \sqrt{1 - \Delta x_n^{[p]}}\, |\Omega\rangle \\ \Delta x_n^{[p]} &\stackrel{\text{def}}{=} x_{n+p} - x_n. \end{aligned} \tag{7}$$

It is seen that for a map that converges into a single value, $x_{n+p} \rightarrow x_n$, we find that $\forall \aleph \lim_{n\to\infty} |n\rangle_{\aleph} = |\Omega\rangle$.

IV. windows of coherence-The environment for organic substance generation

Ergodicity can be related to chaotic maps[19]. The ergodic hypothesis is fundamental to the definition of the thermodynamic equilibrium. We can therefore say that in a high temperature range, our system tends to reach a thermodynamical equilibrium, namely to be in a chaotic state. However, observing bifurcation diagrams for a variety of maps, we find that even with a large R, there are narrow regimes ("windows"') for which the map exhibits a regular behavior (see Fig. 1).

Observing a p-rank map converging into a single value $x^{[p]}$, we notice that there are infinite possibilities for selecting initial conditions for the state $|n\rangle_{\aleph}$. However, under the appropriate p-rank map, it reduces to the ordered organic substance state, thereby reducing entropy. At first glance, it looks as though this behavior violates the second law of thermodynamics. However, since our analysis assumes the system to be attached to an external thermal bath, we can say that this reduction in entropy is always compensated by a growth of entropy in the attached thermal reservoir. In other words, a system based upon a window of coherence always emits heat that is absorbed by the environment.

One cannot ignore the similarity between this high temperature windows of coherence and warm-blooded animals systems. They both operate in a very narrow temperature spectrum ($\sim 35^oC - 42^oC$ in the human body) so as to increase order. Let us add by saying that all of these characteristics require an "internal engine" and other temperature regulation systems (such as sweating in the human body) to maintain this narrow temperature spectrum. When those regulating systems stop functioning, the system temperature lines up with the environment temperature, thereby terminating the organic substance reproducing process.

V. Discussion

We proposed the A.I.E.S model, in which an organic substance was presented by a state. After introducing the state $|n\rangle_{\aleph}$, which is the organic substance state in superposition with the inanimate ensemble of states, we introduced the state evolution as determined by a nonlinear map. In our model, the strength parameter R was assumed to be related to the temperature of the organic substance system. However, it possible to re transform the strength parameter R to depend on other parameters such as the Ph- level. We showed that in an environment induced by a window of coherence and by tracking the evolution of the nonlinear maps, the disordered inanimate ensemble reproduces an ordered organic substance, thereby reducing entropy and emitting heat. Such a system can survive only in a very narrow spectrum of temperatures in which the window of coherence is defined.

Our A.I.E.S model allows the existence of many particles states even at a spectrum of high temperatures. Thus, the A.I.E.S model can be implemented in other fields, such as quantum computers, operating at high temperatures.

References

[1] Schrödinger E., *What Is Life?* (Cambridge University Press 1992).
[2] Luke A.J. O'Ne *What is Life? The Next Fifty Years: Speculations on the Future of Biology*, (Cambridge University Press, 1997).
[3] Morowitz H. J., *Beginnings of Cellular Life: Metabolism Recapitulates Biogenesis*, (Yale University Press, 1992).
[4] http://www.panspermia.com/seconlaw.htm
[5] Joli M. , *Journal of Physics: Conference Series* **410**, (2013) 012038
[6] Zoli M., *Phys. Rev. E.*, **79**, 041027, (2009); M. Zoli, *Phys. Rev. E.*, **81**, 05191, (2010)
[7] Joli M., *J. Chem. Phys*, **141**, 174112 (2014)

[8] Prigogine I., *From Being To Becoming*, New York: W. H. Freeman and Company, (1980). p 123.
[9] Wolfram S., *Statistical Mechanics of Cellular Automata*. Reviews of Modern Physics **55 (3)**, (1983).
[10] Andrew I., *Cellular Automata: A Discrete Universe*, (World Scientific, 2001) ISBN 9789812381835.
[11] Andrew I., Halpern P., *Complex Systems*, **1** (1987)
[12] Roth Y. 2015, *J. Phys.: Conf. Ser* **574**, 012085 (2015)
[13] J. Awrejcewicz J. and Hagedorn P., *Nonlinearity, Bifurcation and Chaos - Theory and Applications*, (InTech, 2012).
[14] Marsden J. E., McCracken M., *The Hopf Bifurcation and its Applications*, (Springer-Verlag, 1976)
[15] Gilmore R., Lefrank M., *The Topology of Chaos*. (New York, USA. John Wiley Sons 2002).
[16] arXiv:quant-ph/0301079
[17] Roth Y., *SOP Transactions on Theoretical Physics*, **1**, Number 1, pp.23-26, (2014)
[18] Lal A. K., *Astrophysics and Space Science*, **317** 267-278 (2008)
[19] Eckmann J. P., Ruelle D., *Review of Modern Physics*, **57,3**, (1985)

Evaluation of connectedness between the University courses of Physics and Chemistry basing on the graph model of inter-subject links

Tatyana Gnitetskaya[1], **Elena Ivanova**[2]
[1-2]The School of Natural Sciences, Far Eastern Federal University,
8 Sukhanova St., Vladivostok, 690950, Russia

E-mail: gnitetskaya.tn@dvfu.ru

Abstract. An application of the graph model of inter-subject links to University courses of Physics and Chemistry is presented in this article. A part of inter-subject space with directions of inter-subject links from Physics to Chemistry in the group of physical concepts has been shown. The graph model of inter-subject links includes quantitative indicators. Its numerical values are given in the article. The degree of connectedness between the data of Physics and Chemistry courses is discussed for the courses considered. The effect of the courses placement within a curriculum on the value of their connectedness is shown. The placement of courses within a curriculum can provide the study of the courses at the same time or consecutive study, when one course precedes another.

1. Introduction

Currently, there is need for a new engineering profession. The specialists owning it perform the functions of non-technical nature. They are engineers-ecologists, engineers-managers in various fields, and others. Training of such specialists is distinguished by the presence of natural sciences in the curriculum, which must include Physics and Chemistry as obligatory disciplines.
At the same time, differentiation of training courses can lead to a distortion of the students' ideas about the surrounding World, the main feature of which is its internal unity. It is possible to establish interdisciplinary connection with the help of the graph model for interdisciplinary connection, proposed by T. Gnitetskaya [1]. Revealing of the same interdisciplinary relationship in training courses will allow to establish the picture of interconnectedness of these courses content and therefore may contribute to the formation of academic content of the courses in Physics and Chemistry, which would meet the requirements of integrity.

2 Graph Model of interdisciplinary connections

Quantitative evaluation of connectedness in Physics and Chemistry courses was performed using a graph model of interdisciplinary connections, as described in this work [1]. Graph model has allowed performing a comparative analysis of content connectedness of training courses in Physics and Chemistry used in the preparation of engineers-managers.

[1] To whom any correspondence should be addressed.

Content from this work may be used under the terms of the Creative Commons Attribution 3.0 licence. Any further distribution of this work must maintain attribution to the author(s) and the title of the work, journal citation and DOI.
Published under licence by IOP Publishing Ltd

The model assumes construction of interdisciplinary space for the courses under review. Description of its content and procedure of forming are shown below. We had to form two interdisciplinary spaces of Physics [2] and Chemistry [3] courses, which differ from each other by the time of studying the Physics course. In the first case (fig.1) the courses in Chemistry and Physics are studied simultaneously, in the second case the study of Physics course precedes Chemistry course (fig.2). Within the framework of the used graph model interdisciplinary connections are realized through the connection object, which is one of any element of knowledge (concepts, theories, laws, etc.) in Physics course. We examined a group of physical concepts. The connection has the direction from Physics to Chemistry and is installed between the whole Physics course (basic course) and the elements of the structure of Chemistry course (related course).
Interdisciplinary space of Physics and Chemistry courses represents a table in which structural elements of Chemistry course are distributed horizontally, and physical concepts - vertically. The cells at the intersection of rows and columns are marked if the physical concepts are used in this element of the Chemistry course' structure. For example, physical and chemical space is constructed as follows: a diamond icon is placed in the cell that corresponds to the structural element of Chemistry, which uses this physical concept. The cell is painted if this structural element of related Chemistry course coincides with the moment of formation of physical concept under review in basic Physics course. Thus, there are four types of cells - with diamond, shaded, shaded with diamond, and empty cells. For example, the concept of the mass is studied in Physics course prior to its use in the course of Chemistry - the first cell is painted. Diamonds are located in those cells (1, 2, 3, 6, 7, 9, 10, 11, 12, 13, 14, 15, 17, 20, 22), where the concept of mass is used in the course of Chemistry. Within the framework of the theory it can be calculated with the following quantitative characteristics of interdisciplinary connections.
Relative maximum length of interdisciplinary connection:

$$L^{(ex)}(EG^{V}_{\mu}) = \frac{i_{\max} - k_{\mu} + 1}{N} \qquad (1)$$

where i_{max} – is the number of structural elements, where the connection is terminated; $k\mu$ –is the root node or the elements' number, where a physical concept is used in Chemistry course for the first time; N - the number of the structural elements in Chemistry course.
For example, for the concept of the mass the relative maximum connection length is equal to 1.00, and for the concept of viscosity- 0.91. The relative strength of ILC

$$C^{(ex)}(EG^{V}_{\mu}) = \frac{F^{(ex)}(EG^{V}_{\mu})}{N} \qquad (2)$$

Where, $F^{(ex)}(EG^{V}_{\mu}) = \sum_i f_i^{(ex)}(EG^{V}_{\mu})$, where, $f_i^{(ex)}(EG^{V}_{\mu}) = \dfrac{(i-k_{\mu}+1) - R^{(ex)}_{k_{\mu}\, i}(EG^{V}_{\mu}))}{(i-k_{\mu}+1)}$

Here, R - the number of unmarked cells.

The relative strength of connection is established between the elements of the structure through any concept. For example, for the concept of mass the relative connection strength is equal to 0.61 and 0.10 for the concept of viscosity.
Knowledge of the values of these characteristics of interdisciplinary connections for each element of knowledge in a group of concepts [1] allows building physical concepts according to their significance in the course of Chemistry. Where the fundamental interdisciplinary core of Physics and Chemistry will include physical concepts, which values of relative maximum lengths of interdisciplinary connections are $L = 1$, and the values of the relative strengths of connection belong to the interval $1{,}00 \leq C \leq 0{,}73$ [4].

Journal of Physics: Conference Series **738** (2016) 012093 doi:10.1088/1742-6596/738/1/012093

Fundamental concepts are those that are used in each section of related course. In Table 2 such fundamental concepts include: atom, energy, temperature, molecule, and electron. Where R – is the number of unmarked cells in the table for a given concept. For example, for the concept of the mass unmarked cells are the cells 4, 5, 8, 16, 18, 19, 21.

3 Connectedness of courses in Physics and Chemistry

Calculation of the quantitative characteristics - length and strength of the connection - made it possible to calculate the connectedness of courses in Physics and Chemistry. The calculation results are summarized in Table 1.

Table 1. Connectedness of courses in Physics by authors A. Remizov, A. Potapenko and Chemistry byN. Glinka on a group of physical concepts in the direction of interdisciplinary connections from Physics to Chemistry

	Related course – Chemistry, N. Glinka Basic Course – Physics, A. Remizov, A. Potapenko	
	Connectedness of trainingcourses for a group of concepts, C	Full strength of connection, F
Physics course precedes Chemistry course	0,14	35,09
Physics course is studied simultaneously with the course of Chemistry	0,12	31,50

Typically, engineers' courses in Physics and Chemistry are given simultaneously from the second semester of the 1-st training course at the university. In this case, the connectedness of Physics and Chemistry courses for a group of physical concepts is 0.12.

Table 2. The space of interdisciplinary content of the basic course in Physics and related Chemistry course by authors [2, 3], in the case where courses in Physics and Chemistry are studied simultaneously

Element of chemistry course structure / Physical notions	Atomic-molecular theory	Periodic law of Mendeleev	Structure of the atom.	Chemical link .Struct. of molec.	The struct. of the solid, liquid	Regularities of chem. reactions	Water. Solutions	Electrolyte solutions	Red-Ox reactions. Electrochem.	Disperse systems. colloids	Hydrogen	Halogens	VI A group	V A group	IV A group	Properties of metals. Alloys	The 1A group of the PS	Complex compounds	The 2nd group of the PS	3rd group of the PS	IVB, VB, VIB, VIIB groups PS	VIII group	Relat. max. length L	Relat. max. streng th C
	1	2	3	4	5	6	7	8	9	10	11	12	13	14	15	16	17	18	19	20	21	22		
1 Atom	♦	♦	♦	♦	♦	♦	♦	♦	♦	♦	♦	♦	♦	♦	♦	♦	♦	♦	♦	♦	♦	♦	1,00	1,00
2 Energy	♦		♦	♦	♦	♦	♦	♦	♦	♦	♦	♦	♦	♦	♦	♦	♦	♦	♦	♦	♦	♦	1,00	1,00
3 Temperature	♦		♦	♦	♦	♦	♦	♦	♦	♦	♦	♦	♦	♦	♦	♦	♦	♦	♦	♦	♦	♦	1,00	0,88
4 Molecule	♦	♦	♦	♦	♦	♦	♦	♦	♦	♦	♦	♦	♦	♦	♦		♦	♦		♦		♦	1,00	0,85
5 Electron	♦	♦	♦	♦			♦		♦		♦	♦	♦	♦	♦	♦	♦	♦	♦	♦	♦	♦	1,00	0,75
6 Mass	♦	♦	♦			♦	♦		♦	♦	♦	♦	♦	♦	♦		♦			♦		♦	1,00	0,61
: :																							:	:
18 Spectrum			♦															♦				♦	0.91	0.13
																							:	:
254 Deuterons	♦																						0.05	0.05

Also, a fragment of the table shown in Table 2 shows that the number of physical concepts is large enough – 254. Frequently, students face the challenge of not understanding of some physical concepts since in the course of Chemistry they are used in the beginning of the study and in Physics course these concepts will be studied only at the end of the study. These are such concepts as α-rays, infrared rays and many others.

Table 3. The space of interdisciplinary content of the basic course in Physics and related Chemistry course authors [2, 3], in the case where the Physics course precedes the study of Chemistry course.

	Element of chemistry course structure / Physical notions	Atomic-molecular theory	Periodic law of Mendeleev	Structure of the atom.	Chemical link .Struct. of molec.	The struct. of the solid, liquid	Regularities of chem. reactions	Water. Solutions	Electrolyte solutions	Red-Ox reactions. Electrochem.	Disperse systems. colloids	Hydrogen	Halogens	VI A group	V A group	IV A group	Properties of metals. Alloys	The 1A group of the PS	Complex compounds	The 2nd group of the PS	3rd group of the PS	IVB, VB, VIB, VIIB groups PS	VIII group	Relat. max. length L	Relat. max. strength C
		1	2	3	4	5	6	7	8	9	10	11	12	13	14	15	16	17	18	19	20	21	22		
1	Atom	♦	♦	♦	♦	♦	♦	♦	♦	♦	♦	♦	♦	♦	♦	♦	♦	♦	♦	♦	♦	♦	♦	1.00	1.00
2	Energy	♦		♦	♦	♦	♦	♦	♦	♦	♦	♦	♦	♦	♦	♦	♦	♦	♦	♦	♦	♦	♦	1.00	0.91
3	Temperature	♦		♦	♦	♦	♦	♦	♦	♦	♦	♦	♦	♦	♦	♦	♦	♦	♦	♦	♦	♦	♦	1.00	0.91
4	Molecule	♦	♦	♦	♦	♦	♦	♦	♦	♦	♦	♦	♦	♦	♦	♦		♦	♦		♦		♦	1.00	0.90
5	Electron	♦	♦	♦	♦			♦		♦		♦	♦	♦	♦	♦	♦	♦	♦	♦	♦	♦	♦	1.00	0.73
6	Mass	♦	♦	♦			♦	♦		♦	♦	♦	♦	♦	♦	♦		♦			♦		♦	1.00	0.61
:	:																							:	:
40	Isotope	♦		♦				♦				♦								♦	♦			0,91	0,21
																								:	:
254	Prism	♦																						0.05	0.05

For comparison of the results the connectivity of courses in Physics and Chemistry was calculated, as well as a full-strength connection for the case when the study of the Physics course is preceded by the study of Chemistry course. The value of connectedness in this case will be 0.14, and the value of full connection strength increases up to 35.09.

Conclusions

Thus, in forming-up both the content of reduced natural science courses and the process of teaching these courses it is necessary to analyze their connectedness level and to conduct suitable adequate course content optimization. It is evident that increasing the value of connectedness of courses will increase the level of assimilation of knowledge by the students that in turn leads to the formation of complete ideas about the World. Therefore, we can recommend studying Physics course before learning Chemistry course while preparing students, future engineers-managers.

References

[1] Gnitetskaya T. Modeling of Interdisciplinary Connections in Science Courses Journal of Physics: Conference Series. Volume 490, Issue 1, 2014, DOI: 10.1088 / 1742-6596 / 490/1/012068 2nd International Conference on Mathematical Modeling in Physical Sciences in 2013, IC-MSQUARE 2013.

[2] Remizov A., A.Potapenko. The course of Physics, textbook for high schools / A.Remizov. - 2nd ed., Stereotype. - M .: Drofa, 2004 - 720 pages.

[3] Glinka N. General Chemistry: Textbook / N.Glinka. - Ed. 2013. 752 pages.

Journal of Physics: Conference Series **738** (2016) 012094 doi:10.1088/1742-6596/738/1/012094

Numerical model describing optimization of fibres winding process on open and closed frame

M Petrů[1], J Mlýnek[2], T Martinec[3]
[1]Deparment of Machinery Construction, Institute for Nanomaterials, Advanced Technology and Innovation, Technical University of Liberec Liberec, Studentska 2, 461 17, Czech Republic
[2]Department of Mathematics and Didactics of Mathematics, Faculty of Science, Humanities and Education, Technical University of Liberec Liberec, Studentska 2, 461 17, Czech Republic
[3]Department of Physical Measurements, Institute for Nanomaterials, Advanced Technology and Innovation, Technical University of Liberec Liberec, Studentska 2, 461 17, Czech Republic

E-mail: michal.petru@tul.cz

Abstract. This article discusses a numerical model describing optimization of fibres winding process on open and closed frame. The quality production of said type of composite frame depends primarily on the correct winding of fibers on a polyurethane core. It is especially needed to ensure the correct angles of the fibers winding on the polyurethane core and the homogeneity of individual winding layers. The article describes mathematical model for use an industrial robot in filament winding and how to calculate the trajectory of the robot. When winding fibers on the polyurethane core which is fastened to the robot-end-effector so that during the winding process goes through a fibre-processing head on the basis of the suitably determined robot-end-effector trajectory. We use the described numerical model and matrix calculus to enumerate the trajectory of the robot-end-effector to determine the desired passage of the frame through the fibre-processing head. The calculation of the trajectory was programmed in the Delphi development environment. Relations of the numerical model are important for use a real solving of the passage of a polyurethane core through fibre-processing head.

1. Introduction

Composites are generally lightweight, high strength and flexible, corrosion resistant and with a long lifespan. Applying composites is significant especially in the aerospace, aircraft and automotive industries. The use of industrial robots in composite production greatly reduces production cost, production time and minimizes scrap rate. The technology based on a winding carbon or glass roving by robot on a non-bearing polyurethane core is currently widespread in the manufacture of composites [1-5]. In this article we focus on the case where the core is formed of a polyurethane frame (which can be also closed) with a circular cross section. An industrial robot is used in the process of winding. This method provides full control over the placement, the laying direction, and the amount of fibres on the frame as well as the homogeneity of the fibre structure. The final composite is obtained after winding of the filament rovings on the frame and subsequent thermal curing of the frame. The objective of this

Content from this work may be used under the terms of the Creative Commons Attribution 3.0 licence. Any further distribution of this work must maintain attribution to the author(s) and the title of the work, journal citation and DOI.
Published under licence by IOP Publishing Ltd

article is to study and mathematically describe the trajectory of the robot-end-effector during the process winding of three layers of carbon filament rovings on a 3D formed frame with a circular cross section. The fibre-processing head is fixedly placed in the workspace of the robot and coordinates of its parts are specified in the basic coordinate system of the robot. The outer guide lines rotate around a common axis, the intermediate guide line is static. The frame is attached to the robot-end-effector. The passage of the frame through fibre-processing head is controlled by the movement of the robot-end-effector. When polyurethane frame passes through fibre-processing head the strands are successively wound on the surface of the frame at a targeted angle. First the outer rotating guide line ensures winding strands under 45° (relative to the axis of the head and the moving direction of the frame), subsequently the middle static guide winds the second layer of strands under 0° and the second outer rotation winds the final layer of strands under -45°. Our goal is for the frame to pass through the fibre-processing head perpendicular to the guide lines of the head as far as possible. Because the position of the fibre-processing head in the workspace of the robot is fixed and the middle guide line is static the required position of the longitudinal fibres on the surface of the wound frame can only be met by suitably turning the non-bearing frame. The principle of the winding solution is shown in Figure 1,2.

Figure 1. Attaching a closed frame to the robot-end-effector.

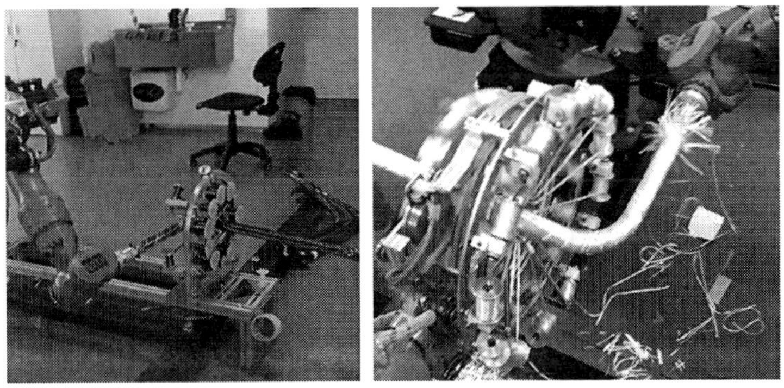

Figure 2. Fibre-processing head with one (left) and three guide lines on the frame.

2. Numerical Model for describe of robot trajectory of winding fibers on the frame

The numerical model for describe of robot trajectory of winding fibers on the frame is see in Figure 3. Within the described numerical model, we will consider the right-handed Euclidean coordinate system E_3 of the robot (*BCS*). Subsequently, we will consider the local right-handed Euclidean coordinate system of the robot-end-effector (*LCS*). To avoid any misunderstanding, we will label the points and vectors with coordinates in *BCS* with the subscript *BCS* and points and vectors with coordinates in *LCS* with the subscript *LCS*. Activities of the industrial robot are controlled using a robot control unit (in our case unit KR C4) and the library of instructions. For our purposes it is crucial to set the desired position of the robot-end-effector using the library of instructions. The position and orientation of the robot-end-effector is defined by using the *LCS*. The origin of the *LCS* is positioned in the robot-end-effector while at the same time the robot-end-effector is oriented in the direction of the positive part of the *z*-axis in the *LCS* with regard to the *BCS*. The actual position of the *LCS* with regard to the *BCS* is determined by the values listed in the "tool-center-point" (*TCP*). The tool centre point contains six values $TCP = (x, y, z, a, b, c)$. The first three parameters specify the coordinates of the origin of the *LCS* in regard to the *BCS*. The values *a*, *b* and *c* indicate the angle of the rotation of the *LCS* around the axis *z*, *y* and *x* with regard to the *BCS*. The non-bearing polyurethane frame can be described by the central axis *o* see Figure 4 and radius r_{TUBE} of the cross-section of the frame. The central axis of the frame *o* is entered in the *LCS* of the robot-end-effector using a discrete set of *N* points stored in the array $B: array[1..N, 1..3]$, where $B[i,1]$ indicates the *x*, $B[i,2]$ indicates the *y* and $B[i,3]$ indicates the *z* coordinate of the i-th point $B[i]_{LCS}$ lying on the axis o, Figure 4. At the same time, the array $vector1B: array[1..N,1..3]$ is entered, where $vector1B[i, 1]$, $vector1B[i, 2]$ and $vector1B[i, 3]$ indicate the coordinates of the unit vector $vector1B[i]_{LCS}$ tangent to the axis o at point $B[i]_{LCS}$. In addition, the array $vector2B: array[1..N,1..3]$ is defined, where $vector2B[i, 1]$, $vector2B[i, 2]$ and $vector2B[i, 3]$ indicate the coordinates of the unit vector $vector2B[i]_{LCS}$ which when passing the point $B[i]_{LCS}$ through the fibre-processing characterizes the necessary rotation of the frame about an axis head. All the time $vector1B[i]_{LCS} \perp vector2B[i]_{LCS}$ holds. We assume that the discrete set of points $B[i]_{LCS}$ lying on the axis o is specified sufficiently densely and defines with a sufficient accuracy the shape of the frame. We suppose the distance of two consecutive point $B[i]_{LCS}$ and $B[i+1]_{LCS}$ has a constant distance h. If the polyurethane frame is closed, then the first entered point on the axis o is identical to the last entered point ($B[1]_{LCS} \equiv B[N]_{LCS}$, $vector1B[1]_{LCS} \equiv vector1B[N]_{LCS}$ and $vector2B[1]_{LCS} \equiv vector2B[N]_{LCS}$). The coordinates of the individual components of the fibre-processing head are entered in the BCS coordinate system. The first outer rotating guide line is presented by the circle $k1$ with the centre $S1_{BCS} = [x_{S1}, y_{S1}, z_{S1}]_{BCS}$, the second outer rotating guide line is presented by the circle $k2$ with the centre $S2_{BCS} = [x_{S2}, y_{S2}, z_{S2}]_{BCS}$, both circles $k1$ and $k2$ have the same radius r_{CIRCLE} (we assume $r_{\mathrm{CIRCLE}} > r_{\mathrm{TUBE}}$). Points $S1$ and $S2$ lie on the axis s of the fibre-processing head, Figure 5.

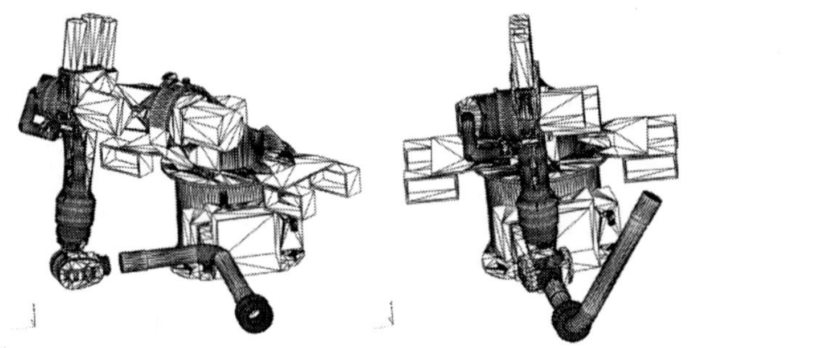

Figure 3. Numerical model for describe of robot trajectory of winding fibers on the frame.

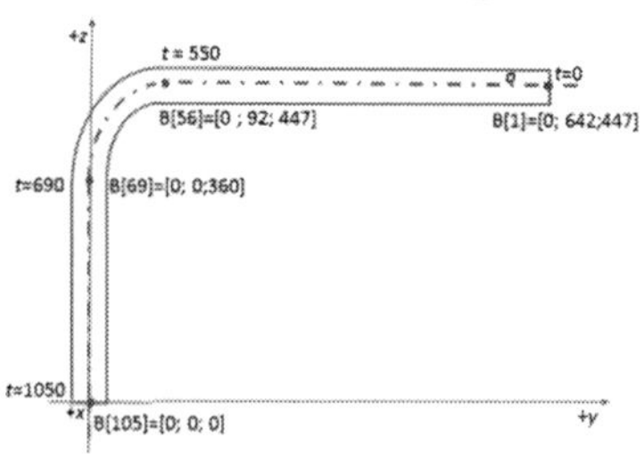

Figure 4. Geometry of a polyurethane frame.

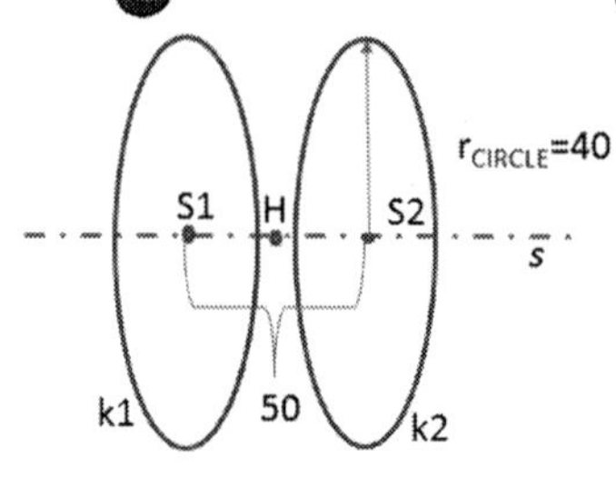

Figure 5. Points $S1$ and $S2$ lie on the axis s of the fibre-processing head.

The middle guide line is static and enables the placement of the fibres in a longitudinal direction is not necessary to consider for the model and the subsequent calculations. Point H_{BCS} is entered, this point lies in the middle of the abscissa $S1_{BCS}S2_{BCS}$. Subsequently, the unit vectors $vector1H_{BCS}$ (this vector indicates the direction of passage frame through the fibre-winding head) and vector $vector2H_{BCS}$ are entered, while the relation $vector1H_{BCS} \perp vector2H_{BCS}$ is valid. We in our test examples selected $vector1H_{BCS} = = (S1_{BCS} - S2_{BCS}) / \|S1_{BCS} - S2_{BCS}\|$, where $\|S1_{BCS} - S2_{BCS}\|$ is the length of the abscissa $S1_{BCS}S2_{BCS}$. The point H_{BCS} together with the vectors $vector1H_{BCS}$ and $vector2H_{BCS}$ allow us to calculate the passage polyurethane frame through the fibre-processing head while the frame possibly rotates around the tangent of its axis o in point $B[i]_{BCS}$.

1.1. Mathematical model of calculating the trajectory of the robot-end-effector

In the derivation procedure of calculation of trajectory, we used results listed in [6-8]. Note that the Denavit-Hartenberg method is often used to determine the trajectory of robot [9].

1.1.1. Calculation of the Passage of Frame Thorough the Fibre-processing Head

We calculate for $i=1,\dots N$ the TCP_i of the robot-end-effector so that at the same time $B[i]_{BCS} \equiv H_{BCS}$, $vector1B[i]_{BCS} \equiv vector1H_{BCS}$, $vector2B[i]_{BCS} \equiv vector2H_{BCS}$. After determining TCP_i all of the TCP_{i-1} parameters are continuously changed to the parameters of to the TCP_i. By this procedure we determine partial passage of the frame through the fibre-processing head. The initial *TCP* indicates the position and orientation of the robot-end-effector before the start of the passage of the frame through fibre-processing head.

1.1.2. Determining the Transformation Matrix

We assume that $TCP_{i-1} = (x_{i-1}, y_{i-1}, z_{i-1}, a_{i-1}, b_{i-1}, c_{i-1})$ is entered. With the aim of determining the transformation matrix T_i (which we will apply to *LCS*) we perform the following steps.

i. We determine the coordinates of the vector $vector1B[i]_{LCS}$ in the coordinate system *BCS* (under the assumption that the coordinate systems *BCS* and *LCS* have the same origin, this assumption is used when finding the required rotation matrix). The matrix Q_{i-1} of the rotation *LCS* towards *BCS* is in the form Equation (1), (see [2], pp. 31)

$$Q_{i-1} = Rot(z, a_{i-1})\, Rot(y, b_{i-1})\, Rot(x, c_{i-1}), \tag{1}$$

where $Rot(z,a_{i-1})$ is the orthogonal matrix of rotation of *LCS* around axis *z* at angle a_{i-1}, $Rot(y,b_{i-1})$ orthogonal matrix of rotation of *LCS* around axis *y* at angle b_{i-1} and $Rot(x,c_{i-1})$ orthogonal matrix of rotation of *LCS* around axis *x* at angle c_{i-1}.

Subsequently, we can express the coordinates of the vector $vector1B[i]_{LCS}$ in the *BCS* system in the form $vector1B[i]_{BCS} = Q_{i-1} vector1B[i]_{LCS}$.

ii. By using the scalar product of vectors $vector1H_{BCS}$ and $vector1B[i]_{BCS}$ we determine their deviation α.

iii. We determine the cross product $p_{BCS} = vector1H_{BCS} \times vector1B[i]_{BCS}$.

iv. Vector p_{BCS} is orthogonal to vectors $vector1H_{BCS}$ and $vector1B[i]_{BCS}$.

We normalize vector p_{BCS}, i.e. ensure its unit length $p_{BCS} = p_{BCS}/\|p_{BCS}\|$. We perform the rotation of vector $vector1B[i]_{BCS}$ by angle α around vector p_{BCS} (we consider the correct orientation of the angle α, i.e. rotate vector $vector1B[i]_{BCS}$ to vector $vector1H_{BCS}$). Then $vector1H_{BCS} = Rot(p_{BCS}, \alpha)\, vector1B[i]_{BCS}$. If we denote $p_{BCS} = (n_1, n_2, n_3, 0)^T$ then the matrix $Rot(p_{BCS}, \alpha)$ will have the form by (2), you can see [2].

$$Rot(p_{BCS}, \alpha) = \begin{bmatrix} c + n_1^2(1-c) & n_1n_2(1-c) - n_3s & n_1n_3(1-c) + n_2s & 0 \\ n_1n_2(1-c) + n_3s & c + n_2^2(1-c) & n_2n_3(1-c) - n_1s & 0 \\ n_1n_3(1-c) - n_2s & n_2n_3(1-c) + n_1s & c + n_3^2(1-c) & 0 \\ 0 & 0 & 0 & 1 \end{bmatrix} \tag{2}$$

where s and c indicate $s = \sin\alpha$, $c = \cos\alpha$. At the same time we determine the vector l_{BCS} (3).

$$l_{BCS} = Rot(p_{BCS}, \alpha) \cdot Q_{i-1} vector2B[i]_{LCS} \quad (3)$$

The cases where $\alpha = 0$ or $\alpha = \pi$ need to be solved separately.

v. By using the scalar product of the vectors we determine the deviation β of vectors $vector2H_{BCS}$ and l_{BCS}.

vi. The resulting matrix of rotation Q_i has the form by (4).

$$Q_i = Rot\left(vector1H_{BCS}, \beta\right) \cdot Rot\left(p_{BCS}, \alpha\right) \cdot Rot(z, a_{i-1}) \cdot Rot(y, b_{i-1}) \cdot Rot(x, c_{i-1}) \quad (4)$$

where the elements of the matrix $Rot\left(vector1H_{BCS}, \beta\right)$ are analogously defined as elements of the matrix $Rot\left(p_{BCS}, \alpha\right)$. In (4) we use the correctly determined angle orientation of angle β, i.e. we rotate the vector l_{BCS} (3) around the vector $vector1H_{BCS}$ to the vector $vector2H_{BCS}$.

vii. The translate vector is defined (5)

$$u(i)_{BCS} = H_{BCS} - Q_i B[i]_{LCS} - \left(x_{i-1}, y_{i-1}, z_{i-1}, 0\right)^T, \quad (5)$$

where matrix Q_i is determined by the relation (4). We remember, that $x_{i-1}, y_{i-1}, z_{i-1}$ are the first three parameters of TCP_{i-1}. The resulting transformation matrix T_i in the i-th step of the passing of the frame through the fibre-processing head is in the form (6).

$$T_i = Trans\left(x_{u(i)}, y_{u(i)}, z_{u(i)}\right) \cdot Q_i, \quad (6)$$

where the translate matrix $Trans\left(x_{u(i)}, y_{u(i)}, z_{u(i)}\right) = \begin{pmatrix} 1 & 0 & 0 & x_{u(i)} \\ 0 & 1 & 0 & y_{u(i)} \\ 0 & 0 & 1 & z_{u(i)} \\ 0 & 0 & 0 & 1 \end{pmatrix}$ and $x_{u(i)}, y_{u(i)}, z_{u(i)}$ are components of vector $u(i)_{BCS}$ in (5).

After performing of the LCS of the robot-end-effector the corresponding transformation matrix T_i determined by relation (7) will be valid $H_{BCS} \equiv B[i]_{BCS} = T_i \cdot B[i]_{LCS}$ and $vector1B[i]_{BCS} \equiv vector1H_{BCS}$, $vector2B[i]_{BCS} \equiv vector2H_{BCS}$.

1.1.3. Calculation of the Euler Angles

Any right-handed rotation Euclidean space E_3 around the given unit vector p by angle φ is determined by the orthogonal matrix $Q = Rot(p, \varphi)$, its elements are in the form (3) and $\det Q = 1$. The rotation matrices form the orthogonal group (5). Each rotation matrix can be written as a product of the matrices of rotation around the coordinate axes z, y and x, i.e. $Q = Rot(z, a) \cdot Rot(y, b) \cdot Rot(x, c)$ (see [2], pp. 31), where the matrices $Rot(z,a)$, $Rot(y,b)$ and $Rot(x,c)$ are in the form (2); a, b and c are the corresponding Euler angles. We note that the determination of Euler angles a, b and c is not unique (see [5]). By multiplying the left and right side from the left of the matrix $Rot\left(z, a_i\right)^{-1}$ because the matrix $Rot\left(z, a_i\right)$ is orthogonal $Rot\left(z, a_i\right)^{-1} = Rot\left(z, a_i\right)^T$) and modified in this way we can determine the rotation angles a_i, b_i and c_i (8). When calculating the angles of rotations we use the *ATAN2* function (part of the library of most programming languages), which calculates from the two input parameters arg_1 and arg_2 the value of the function arctangent for argument arg_1/arg_2. Moreover, the signs of both input parameters are used to determine the quadrant in which the resulting value function is located (it is valid that $-\pi < ATAN2(arg_1, arg_2) \leq \pi$ part). We write the matrix of rotation Q_i in the form by (7).

$$Q_i = \begin{pmatrix} q_{11}(i) & q_{12}(i) & q_{13}(i) & 0 \\ q_{21}(i) & q_{22}(i) & q_{23}(i) & 0 \\ q_{31}(i) & q_{32}(i) & q_{33}(i) & 0 \\ 0 & 0 & 0 & 1 \end{pmatrix}. \tag{7}$$

$$\left.\begin{aligned} a_i &= ATAN2(q_{21}(i), q_{11}(i)), \\ b_i &= ATAN2(-q_{31}(i), q_{11}(i)\cos a_i + q_{21}(i)\sin a_i), \\ c_i &= ATAN2(q_{13}(i)\sin a_i - q_{23}(i)\cos a_i,\ q_{22}(i)\cos a_i - q_{12}\sin a_i) \end{aligned}\right\} \tag{8}$$

Thus, we determined rotation angles a_i, b_i, c_i in (8) and the corresponding TCP_i defining the position and orientation of the *LCS* of the robot-end-effector in relation to the *BCS* of the robot. The TCP_i can be expressed in the form by (9).

$$TCP_i = (x_{u(i)}, y_{u(i)}, z_{u(i)}, a_i, b_i, c_i) \tag{9}$$

where the elements of the vector $u(i)_{BCS}$ are given by the relation (5).

3. Results and summary

We focus on the practical problem of the passage of the non-bearing polyurethane frame with a circular cross-section through the fibre-processing head. The central 2D axis o of the frame is composed of two interconnected perpendicular arms, Figure 4. The axis *o* is entered into the *LCS* of the robot-end-effector using a discrete set of points $B[i]_{LCS}$, $1 \leq i \leq N = 105$. The total length *d* of the axis *o* from the starting point $B[1]_{LCS}$ to the end point $B[N]_{LCS}$ of the axis is $d = 1050[\text{mm}]$. The continuous distance of the points $B[i]_{LCS}$ on the axis *o* from the starting point is denoted by the variable *l*. The distance between two consecutive points $B[i]_{LCS}$ and $B[i+1]_{LCS}$ is $h = 10[\text{mm}]$. Vectors $vector1B[i]_{LCS}$ and $vector2B[i]_{LCS}$ are specified, radius of the frame is $r_{\text{TUBE}} = 20[mm]$. The fibre-processing head is represented by the circles $k1$ and $k2$ with the centres $S1_{BCS} = [-1000, 1105, 990]$ and $S2_{BCS} = [-1000, 1035, 990]$ having the same radius $r_{\text{CIRCLE}} = 40[\text{mm}]$. The length of abscissa $S1_{BCS}S2_{BCS} = 70[\text{mm}]$. It is valid $H_{BCS} = (S1_{BCS} + S2_{BCS})/2$, $vector\,1H_{BCS} = (S1_{BCS} - S2_{BCS})/\|S1_{BCS} - S2_{BCS}\|$, $vector\,2H_{BCS} = (0, 0, 1)$. As stated in section II.C., the middle static line enabling the winding of the fibres to be placed in a longitudinal direction is not required for calculating the trajectory of the robot-end-effector. The frame needs to be rotated at a distance of 550 [mm] from the beginning of the axis *o* to the distance of 690 [mm] around a tangent to the frame of axis *o* during the passing of point $B[i]_{BCS}$ (for $56 \leq i \leq 69$) by point H_{BCS} when placing the longitudinal fibres relative to the axis *o* of the frame. This is a uniform right-handed rotary motion with an overall rotation angle of $\omega = \pi$. The rotation is performed during the passage of the bent portion of the frame through the fibre-processing head. The calculation of the trajectory of the robot-end-effector from numerical model referred to in the previous chapter was applied to the described problem. Figure 6 shows the results of numerical modelling for position of the robot-end-effector when passing polyurethane frame through winding head. In part a) shows the position of the robot-end-effector (first three parameters of *TCP*) and part b) the orientation of the robot-end-effector (last three parameters of *TCP*). The described numerical model with algorithm allows the accurate calculation and determination of the 3D trajectory of the robot-end-effector of the industry robot during the production of composite profile using the dry fibre winding technology on a open and closed non-bearing frame.

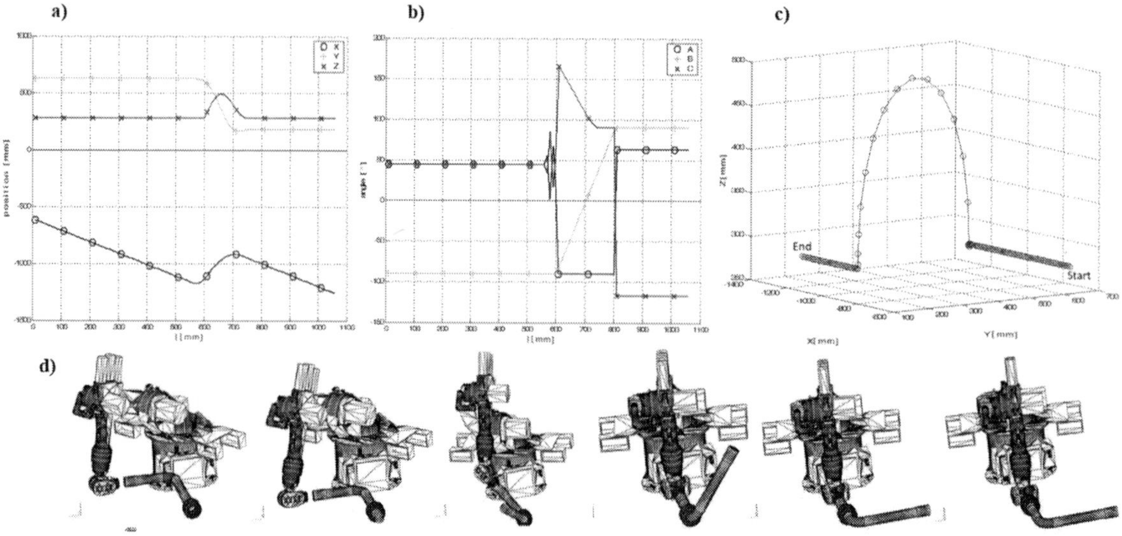

Figure 6. Results of numerical modeling of the course of the *TCP* during the passing of the frame through the fibre-processing head, a) parameter values of the first three parameters of *TCP*, b) parameter values of the last three parameters of *TCP*, c) trajectory of the robot-end-effector: start and end, d) time response of numerical model for optimal trajectory of fibres winding process.

Acknowledgments

The results of this project No. LO1201 were obtained through the financial support of the Ministry of Education, Youth and Sports, Czech Republic in the framework of the targeted support of the "National Programme for Sustainability I".

References

[1] Shirinzadech B, Alici G, Foog CW and Cassidy G 2004 Fabrication process of open surfaces by robotic fibre placement, *Robotics and Computer-Integrated Manufacturing*, vol 20, pp 17-28.
[2] Sciavicco L and Siciliano B 2004 *Modelling and Control of Robot Manipulators*, (London: Springer), 2004, p 378.
[3] Jazar R N 2007 *Theory and applied robotics,* (New York: Springer), p 883
[4] Sevcik L, Lepsik P, Petrů M, Mašín I and Martonka R 2014 Modern Methods of Construction Design, *Lecture Notes in Mechanical Engineering*, (New York: Springer), p 631
[5] Petrů M, Kovář R, Martinec T, Srb P, Lufinka A and Kulhavy P 2015 Analysis and study of vibrations of a clamping device used for winding carbon fibers into the core of a frame, *Proc. of the EAN 2015 - 53rd Conference on Experimental Stress Analysis (Cesky Krumlov)*, pp. 461-467.
[6] Martinec T, Mlýnek J, Petrů M 2015 Calculation of the robot trajectory for the optimum directional orientation of fibre placement in the manufacture of composite profile frames, *Robotics and Computer-Integrated Manufacturing*, vol 35, pp. 42-54.
[7] Salabough G G, *Computing Euler angles form a rotation matrix*, www.soi.city.ac.uk/~sbbh635/publications/euler.pdf.
[8] Tian L. and Collins C 2004 An effective robot trajectory planning method using a genetic algorithm, *Mechatronics*, vol. 14(5), pp. 455-470.
[9] Baločková L 2013 The method for solving kinematics of an industrial robot, *Applied Mechanics and materials*, vol 283, pp. 279–281.

On the generalization of statistical thermodynamic functions by a Riccati differential equation.

J J Peña, A Rubio-Ponce and J Morales

Universidad Autónoma Metropolitana - Azcapotzalco,
CBI - Area de Física Atómica Molecular Aplicada,
Av. San Pablo 180, 02200 México, D.F., MEXICO.

Corresponding author: jjpg@correo.azc.uam.mx

Abstract. In this work, we propose a non-linear differential equation of Riccati-type, where the standard partition function *Z(T)* is taken as its particular solution leading to their generalization $Z_g(T)$; from there, other related statistical thermodynamic functions are generalized. As an useful application of our proposal, other thermodynamic functions, namely, the internal energy, heat capacity, Helmholtz free energy and entropy, associated to the model of the ideal monatomic gas in *D*-dimensions are generalized. According to our results, thermodynamic properties derived from the standard partition functions by means of ordinary statistical mechanics are incomplete. In fact, although asymptotically with the increasing of temperature the generalized statistical thermodynamic functions reduce to the standard ones, these contain an extra term which is dominant at very low temperature indicating that standard findings should be corrected.
PACS: 05.30.-d, 03.65.Fd, 51.30.+i.
Keywords: Thermodynamic functions, Riccati equation.

1. Introduction

In statistical mechanics, almost all thermodynamic properties of a system can be obtained from the standard partition function *Z(T)* which is related with the standard internal energy of the system *U(T)* through the derivative of ln*Z(T)* [1]. This means that the existence of a partition function guarantees the knowledge of the thermodynamic properties of system [2]. This statement seems to be equivalent to that of quantum mechanics about solvable potentials, which establish that a potential is exactly solvable on condition of having an associated Witten superpotential [3,4]. In that case, the choice of the Witten superpotential as a particular solution of the involved Riccati equation leads us to identify the potential under study [5], to generalize the former potential as well as to find their isospectral partners [6]. Thus, because we are interested on the generalization of thermodynamic partition functions, we propose to follow a similar treatment to the one outlined above [7]. That is, the aim of this work is to propose a nonlinear differential equation of Riccati type where the thermodynamic variables *U(T)* and *Z(T)* are involved in such a way that all other thermodynamic properties of a system, that come from *Z(T)* or *U(T)*, can be generalized. In next section, the equations that generalize the partition function and the internal energy are presented along with the generalized equations that correspond to the entropy, specific heat and Helmholtz free energy. As an useful application of the proposed Riccati-type equation for statistical thermodynamics, in section III we consider the

Content from this work may be used under the terms of the Creative Commons Attribution 3.0 licence. Any further distribution of this work must maintain attribution to the author(s) and the title of the work, journal citation and DOI.
Published under licence by IOP Publishing Ltd

generalized statistical thermodynamic properties of the ideal monatomic gas in one, two and three dimensions. Although this statistical model is always considered in every modern textbook on statistical mechanics, the theory of ideal quantum gases is far from being a closed field for research [8,9]. Moreover, the proposed equations can be directly used to generalize other statistical models as well as to find new partition functions that could be useful in the treatment of some relevant thermodynamic applications.

2. Partition function

In a thermodynamic system at equilibrium, the canonical partition function *Z(T)*, which is given in terms of thermodynamic state variables, plays an important role in statistical physics due that all the thermodynamic functions associated to a statistical system, such as entropy, internal energy, free energy, heat capacity and others are obtained in terms of *Z(T)* and its derivatives [10]. Consequently, the *Z(T)* along with the particular properties of each system leads to the so-called equation of state that completely characterizes the system under study. The partition function for a classical discrete system is given by the equation

$$Z(T) = \sum_i e^{-\frac{E_i}{kT}} \tag{1}$$

where E_i is the energy of the system in a state i and k is the Boltzman constant. Thus defined, the partition function has a probabilistic sense because the factor $e^{-\frac{E_i}{kT}}$ is interpreted as the probability that the system is in a state i with energy E_i, i.e.

$$P_i = C\, e^{-\frac{E_i}{kT}}, \tag{2}$$

where C is a constant of proportionality. In the case we are studying, the partition function *Z(T)* plays the role of the normalization constant for the probability distribution in question Eq. (2). Indeed, if we apply the fact that the sum of all probabilities is the unity,

$$\sum_i P_i = 1, \tag{3}$$

then

$$\sum_i P_i = \sum_i C e^{-\frac{E_i}{kT}} = C \sum_i e^{-\frac{E_i}{kT}} = 1 \quad \Rightarrow \quad C = \frac{1}{\sum_i e^{-\frac{E_i}{kT}}} = \frac{1}{Z(T)} \tag{4}$$

such that the normalized probability distribution is

$$P_i = \frac{e^{-\frac{E_i}{kT}}}{\sum_i e^{-\frac{E_i}{kT}}} = \frac{1}{Z(T)}\, e^{-\frac{E_i}{kT}} \tag{5}$$

With this probability distribution, it is possible to calculate average values of various amounts of interest, such of the ones that we are going to show next and which will be useful for the purpose of this paper.

2.1. Calculations of average values for the energy.

As mentioned earlier, starting from the partition function and its derivatives, it is possible to deduce the thermodynamic functions of statistical mechanics. Indeed, based on the probability distribution of Eq. (5), the average value of energy $\langle E\rangle = \bar{E} = U$ is given by

$$U = \sum_i P_i E_i = \sum_i \frac{E_i e^{-\frac{E_i}{kT}}}{Z(T)} = \frac{1}{Z}\sum_i kT^2 \frac{\partial e^{-\frac{E_i}{kT}}}{\partial T} = \frac{kT^2}{Z(T)}\frac{\partial Z(T)}{\partial T} = kT^2 \frac{\partial \ln Z(T)}{\partial T}. \tag{6}$$

Similarly to the above equation, $\langle E^2\rangle = \overline{E^2}$, which is different of $\langle E\rangle^2 = (\bar{E})^2 = U^2$, becomes as

$$\langle E^2 \rangle = \sum_i P_i E_i^2 = \frac{1}{Z} \sum_i E_i^2 \, e^{-\frac{E_i}{kT}} . \tag{7}$$

However, by noticing that

$$\frac{\partial^2 Z(T)}{\partial T^2} = \left(\frac{1}{kT^2}\right)^2 \sum_i E_i^2 \, e^{-\frac{E_i}{kT}} - \frac{2}{kT^3} E_i \, e^{-\frac{E_i}{kT}} , \tag{8}$$

it can be rewritten as

$$\frac{\partial^2 Z(T)}{\partial T^2} = \left(\frac{1}{kT^2}\right)^2 Z(T) \left(\frac{\sum_i E_i^2 e^{-\frac{E_i}{kT}}}{Z(T)}\right) - \frac{2 Z(T)}{kT^3} \left(\frac{E_i e^{-\frac{E_i}{kT}}}{Z(T)}\right), \tag{9}$$

the expectation value of the square energy given in Eq.(7) will be

$$\langle E^2 \rangle = (kT^2)^2 \left(\frac{\frac{\partial^2 Z(T)}{\partial T^2}}{Z(T)}\right) + 2kT \langle E \rangle. \tag{10}$$

Also,

$$\frac{\frac{\partial^2 Z}{\partial T^2}}{Z} = \frac{1}{Z} \frac{\partial}{\partial T} \left(\frac{Z}{kT^2} kT^2 \left(\frac{\frac{\partial Z}{\partial T}}{Z}\right)\right) = \frac{1}{Z} \frac{\partial}{\partial T} \left(\frac{Z}{kT^2} \langle E \rangle\right), \tag{11}$$

leads to

$$(kT^2)^2 \left(\frac{\frac{\partial^2 Z(T)}{\partial T^2}}{Z(T)}\right) = kT^2 \frac{\partial \langle E \rangle}{\partial T} + \langle E \rangle^2 - 2kT \langle E \rangle. \tag{12}$$

Finally, we want to point out that the combination of Eqs. (10) and (12) allow us the calculation of fluctuations or dispersion of the energy $\overline{(\Delta E)^2}$, i.e.

$$\overline{(\Delta E)^2} = \overline{(E - \bar{E})^2} = \overline{E^2 - 2E \, \bar{E} + \bar{E}^2} = \langle E^2 \rangle - \langle E \rangle^2 \tag{13}$$

from where

$$\overline{(\Delta E)^2} = kT^2 \frac{\partial \langle E \rangle}{\partial T} = kT^2 \frac{\partial U}{\partial T}. \tag{14}$$

2.2. The canonical partition function and the Riccati equation.
As can be seen, Eq.(12) can be written as

$$\frac{1}{kT^2} \frac{\partial U}{\partial T} - \frac{2}{kT^3} U + \left(\frac{U}{kT^2}\right)^2 = \frac{Z(T)''}{Z(T)} , \quad with \quad Z(T)'' = \frac{\partial^2 Z(T)}{\partial T^2} \tag{15}$$

which means

$$\frac{\partial}{\partial T}\left(\frac{U}{kT^2}\right) + \left(\frac{U}{kT^2}\right)^2 = \frac{Z''}{Z} . \tag{16}$$

At this point, it is interesting to notice that the definition of a new thermodynamic function

$$R(T) = \frac{U(T)}{kT^2} \tag{17}$$

indicates that Eq. (16) becomes a Riccati equation

$$R'(T) + R^2(T) = \frac{Z''(T)}{Z(T)} \tag{18}$$

from where it is possible to generalize different statistical thermodynamic functions as will see next.

3. On the generalization of statistical thermodynamic functions

Let us consider the Riccati relationship

$$y'(x) + y^2(x) = f(x), \tag{19}$$

if a particular solution $y_p(x)$ is known, then the general solution is given by [11]

$$y_g(x) = y_p(x) + \frac{b}{e^{2\int y_p dx}\left(1+\mathrm{b}\int\frac{dx}{e^{2\int y_p dx}}\right)} = y_p(x) + \frac{d}{dx}\ln\left(1+\mathrm{b}\int\frac{dx}{e^{2\int y_p dx}}\right). \tag{20}$$

where the index g indicates general or generalized.

So, regarding with Eq.(18), by assuming $y_p(x) = R(T)$ this means that

$$R_g(T) = R(T) + \frac{b}{e^{2\int R(T)dT}\left(1+\mathrm{b}\int\frac{dT}{e^{2\int R(T)dT}}\right)}. \tag{21}$$

Moreover, due that according to Eq.(6) $U = kT^2\frac{\partial \ln Z(T)}{\partial T}$, or the equivalent $R(T) = \frac{Z'(T)}{Z(T)}$, there exist also $Z_g(T)$ such that

$$Z_g(T) = Z(T)\left(1+\mathrm{b}\int\frac{dT}{Z^2(T)}\right). \tag{22}$$

In fact Eq.(21) can be written as

$$R_g(T) = \frac{Z'(T)}{Z(T)} + \frac{b}{e^{2\int R(T)dT}\left(1+\mathrm{b}\int\frac{dT}{e^{2\int R(T)dT}}\right)} = \frac{d}{dT}\ln Z(T) + \frac{d}{dT}\ln\left(1+\mathrm{b}\int\frac{dT}{Z^2(T)}\right)$$

$$= \frac{d}{dT}\ln\left(Z(T)\left(1+\mathrm{b}\int\frac{dT}{Z^2(T)}\right)\right) = \frac{d}{dT}\ln(Z_g(T)), \tag{23}$$

that is to say

$$R_g(T) = \frac{Z'_g(T)}{Z_g(T)}, \tag{24}$$

leading to the generalization of Eq.(18)

$$R'_g(T) + R_g^2(T) = \frac{Z''_g(T)}{Z_g(T)}. \tag{25}$$

Consequently, this means that all the statistical thermodynamic functions associated with $R_g(T)$ can be straightforward generalized such as happen from Eq.(17)

$$U_g(T) = kT^2 R_g(T) = kT^2\frac{d}{dT}\ln Z_g(T) \tag{26}$$

that corresponds to the generalization of the internal energy of the system. In a similar way, other important statistical thermodynamic properties, related with the partition function, the internal energy and their derivatives, can be generalized. For example, the entropy $S(T)$, the Helmholtz free energy $F(T)$ and the heat capacity $C(T)$ are related with $Z(T)$ and $Z_g(T)$ as follow

$$S(T) = -\frac{d}{dT}Z(T) \dashrightarrow S_g(T) = k\ln Z_g(T) + \frac{U_g(T)}{T}, \tag{27}$$

$$F(T) = -kT\ln Z(T) \dashrightarrow F_g(T) = -kT\ln Z_g(T) \tag{28}$$

and

$$C(T) = \frac{d}{dT}U(T) = T\frac{d}{dT}S(T) = -T\frac{d^2}{dT^2}F(T) \dashrightarrow C_g(T) = \frac{d}{dT}U_g(T). \tag{29}$$

In short, by using Eq.(22) the above generalized statistical thermodynamic functions are written as

$$Z_g(T) = Z(T) + bZ(T)\int\frac{dT}{Z^2(T)}, \tag{30}$$

$$U_g(T) = U(T) + kT^2\frac{d}{dT}\ln\left(1 + \mathrm{b}\int\frac{dT}{Z^2(T)}\right), \tag{31}$$

$$S_g(T) = S(T) + \frac{d}{dT}\left(kT\ \ln\left(1 + \mathrm{b}\int\frac{dT}{Z^2(T)}\right)\right), \tag{32}$$

$$F_g(T) = F(T) - kT\ln\left(1 + \mathrm{b}\int\frac{dT}{Z^2(T)}\right) \tag{33}$$

and

$$C_g(T) = C(T) + T\frac{d^2}{dT^2}\left(kT\ \ln\left(1 + \mathrm{b}\int\frac{dT}{Z^2(T)}\right)\right). \tag{34}$$

As can be appreciated, the generalized statistical thermodynamic functions contain, algebraically, the corresponding standard functions plus an additional term which depends on temperature.

4. Applications

In this section, we will apply the above results to the system of an ideal gas with N particles confined in a container. In this case, the internal energy is given by

$$U(T) = \frac{DN}{2}kT \tag{35}$$

where D indicates the degrees of freedom of particles such that $D = 1, 2, 3$ corresponds to 1, 2 and 3 dimensions respectively. Thus, using Eqs.(6) and (17), the standard partition function becomes

$$Z(T) = e^{\int R(T)dT} = (AT)^{\frac{DN}{2}} \tag{36}$$

where A is a proper constant that leaves $Z(T)$ dimensionless. Accordingly to Eq.(22), the generalized partition function is given by

$$Z_g(T) = Z(T) + \frac{(Z(T))^{\frac{2}{DN}-1}}{DN-1}. \tag{37}$$

In consequence, the generalized statistical thermodynamic functions for the model of ideal gas in D dimensions are

$$Z_g(T) = (AT)^{\frac{DN}{2}} + \frac{(AT)^{1-\frac{DN}{2}}}{DN-1} \tag{38}$$

$$U_g(T) = \frac{DN}{2}kT - \frac{(DN-1)kT}{(DN-1)(AT)^{DN-1}+1} \tag{39}$$

$$S_g(T) = \frac{DN}{2}k + k\ln[(AT)^{\frac{DN}{2}}] + k\ln[1 + \frac{(AT)^{-(DN-1)}}{(DN-1)}] - \frac{(DN-1)k}{(DN-1)(AT)^{DN-1}+1} \tag{40}$$

$$F_g(T) = -kT\ln[(AT)^{\frac{DN}{2}}] - kT\ln[1 + \frac{(AT)^{-(DN-1)}}{(DN-1)}] \tag{41}$$

and

$$C_g(T) = \frac{DN}{2}k + k\frac{(DN-1)^2(DN-2)(AT)^{DN-1}-(DN-1)}{\left((DN-1)(AT)^{DN-1}+1\right)^2}. \tag{42}$$

From these results, it becomes evident that all the above generalized statistical thermodynamic functions tend asymptotically to the corresponding standard ones for a temperature beyond of which the generalized partition function reach its minimum. Namely, from Eq. (38) the minimum value of $Z_g(T)$ is obtained by solving the equation $Z'_g(T) = 0$ such that

$$Z_{gMin} = Z_g(T_{MinZg}) = \left(\frac{DN-2}{DN(DN-1)}\right)^{\frac{DN}{2(DN-1)}} + \frac{\left(\frac{DN-2}{DN(DN-1)}\right)^{\frac{2-DN}{2(DN-1)}}}{DN-1}, \tag{43}$$

where T_{MinZg} is the temperature at which $Zg(T)$ reach its minimum given by

$$T_{MinZg} = \frac{1}{A}\left(\frac{DN-2}{DN(DN-1)}\right)^{\frac{1}{3N-1}}. \tag{44}$$

Likewise, due that $U_g = kT^2 Z'_g/Z_g$ such that $U_g(T_{RootUg}) = 0$, it leads to $T_{RootUg} = T_{MinZg}$ where T_{RootUg} is the temperature at which U_g has a root, i.e. where U_g is null. Also, in the interval $(0,\ T_{RootUg})$ $U_g < 0$, reaching in this range a minimum value

$$U_{gMin} = \frac{k}{A}\left(\sqrt{\frac{DN(DN-2)+((DN-1)^2+1)^2}{(DN(DN-1))^2}} - \frac{(DN-1)^2+1}{DN(DN-1)}\right)^{\frac{1}{DN-1}}\left(\frac{DN}{2} - \frac{DN(DN-1)}{\sqrt{DN(DN-2)+((DN-1)^2+1)^2}-(DN-1)(DN-2)}\right) \tag{45}$$

at T_{MinUg} which is the temperature where $Ug(T)$ is minimal. That is $U'_g(T_{MinUg}) = 0$ leading to

$$T_{MinUg} = A^{-1}\left(\sqrt{\frac{DN(DN-2)+((DN-1)^2+1)^2}{(DN(DN-1))^2}} - \frac{(DN-1)^2+1}{DN(DN-1)}\right)^{\frac{1}{DN-1}}. \tag{46}$$

Besides, due that according to Eq. (29) $C_g(T) = U'_g(T)$, then $C_g(T_{MinUg}) = U'_g(T_{MinUg}) = 0$ this implies that $T_{MinUg} = T_{RootCg}$ which is the temperature at where C_g is null.

At this point, it is worth mentioning that by combining Eqs. (26) and (29) one has

$$C_g(T) = \frac{d}{dT}U_g(T) = \frac{d}{dT}\left(kT^2\frac{d}{dT}\ln Z_g(T)\right), \tag{47}$$

that permits to select the interval of validity of our equations. Finally, the heat capacity has a maximal value given by

$$C_{gMax} = C_g(T_{MaxCg}) = \frac{DN}{2}k + \left(\frac{DN}{2} - 1\right)^2 k \tag{48}$$

where T_{MaxCg} can be obtained from Eq. (42) after solving $C'_g(T) = 0$ as

$$T_{MaxCg} = A^{-1} \left(\frac{DN}{(DN-1)(DN-2)}\right)^{\frac{1}{DN-1}}. \quad (49)$$

In short, we want to notice that a similar study can be done directly from the proposed generalized partition or internal energy functions to other statistical thermodynamic properties different to those considered in this work as well as to other model systems [12].

5. Concluding remarks

Starting from the basic definitions of statistical mechanics, in this paper we propose a non-linear differential equation of Riccati-type involving the statistical thermodynamic functions of internal energy and partition function. To achieve this, it is used an *R(T)* function linking both the internal energy as the partition function through the logarithmic derivative *R(T)* = *Z* '/ *Z*. This same function is taken as a particular solution in order to find a general solution of the Riccati-type equation. So, this general solution leads to the generalization of the partition function $Z_g(T)$ and the internal energy $U_g(T)$. From there, the thermodynamic properties associated with these functions are directly generalized; specifically we consider the generalization of entropy, Helmholtz free energy and heat capacity in the case of the ideal monatomic gas in D-dimensions. According to our results, the statistical thermodynamic properties derived from the standard partition functions by means of ordinary statistical mechanics are incomplete, therefore should be corrected with the generalized equations that contain an extra term which is dominant at very low temperature. Besides, our proposal is general and can be directly applied to other thermodynamic models.

Acknowledgment

This work was partially supported by the projects UAM-A-CBI-2232004 and 009. We are grateful with SNI-Conacyt-México for the stipend received.

References

[1] K. Wang 1963 Statistical Mechanics, John Wiley & Sons, Inc. New York.
[2] F. W. Sears and G. L. Salinger 1980 Thermodynamics, Kinetic Theory and Statistical Thermodynamics, Addison Wesley, Massachusetts.
[3] J. J. Peña, M. A. Romero-Romo, J. Morales and J. L. López-Bonilla, *Int. Journ. Quantum Chem.*, **105** (2005) 731.
[4] F. Cooper, A. Khare and U. Sukhatme, *Phys. Rep.* **251** (1995) 297.
[5] J. J. Peña, G Ovando, D Morales-Guzmán and J Morales, *Int. Journ. Quantum Chem.*, **85** (2001) 244.
[6] J. Morales, J. J. Peña and A. Rubio-Ponce, *Theor. Chem. Accounts*, **110** (2003) 403.
[7] J. Morales and J. J. Peña, *Phys. Scr.* **74** (2006) 71.
[8] H. J. Schmidt.and J. Schanck, *Am. J. Phys.*, **70** (2002) 53.
[9] M. C. Gupta 2003 Statistical Thermodynamics (Second Edition), New Age International (P) Limited Publishers, New Delhi.
[10] R. E. Sonntag and G. J. Van Wylen 1971 Introduction to Thermodynamics:Classsical and Statistical, John Wiley & Sons, Inc. New York.
[11] H. T. Davis1962 Introduction to Nonlinear Differential and Integral Equations, Dover Publishers, New York.
[12] F. Reif 1965 Fundamentals of Statistical and Thermal Physics, Mc. Graw Hill International Book Co.Auckland.

Dynamics of a director reorientation and optical response of polymer films filled with a liquid crystal under strong electric field

S V Pasechnik, D V Shmeleva and A P Chopik

Moscow Technological University, Moscow 119454, Russia.

A A Vakulenko and A V Zakharov

Saint Petersburg Institute for Machine Sciences,

The Russian Academy of Sciences, Saint Petersburg 199178, Russia

E-mail: s-p-a-s-m@mail.ru

Abstract. The dynamics of the director reorientation in nematic liquid crystals (NLC) confined by cylindrical cavities of porous polymeric films under strong electric field E with has been investigated theoretically. The main attention was paid to the specific mode of field application characterized by abrupt changes of the applied voltage's polarity. In experiments with porous films filled with a liquid crystal 5CB such mode resulted in appearance of strong peak –like decreasing of an optical transparence of the films. Two mechanisms of such unusual response based on assumption of electrically induced motion of ions and overall motion of a liquid were considered and applied to explain experimental results.

1. Introduction

At present liquid crystals (LC), which play a key role in display industry are considered as very promising materials for non display applications, like E-paper, sensors and photonics [1, 2]. In the latter case there is a number of new requirements different from those applied for liquid crystal displays (LCD). In particular, decreasing of operating times for LC devices becomes of a great importance. Different ways of a solution of this problem, like usage of ferroelectric liquid crystals [2] were proposed. While FLC show very short operating times, their application in photonics devices meets the additional problems, like possibility of irreversible changes of a layer structure of FLC imposed by mechanical deformations.

The alternative way to decrease operation times is to use composite liquid crystal materials of different types [3, 4] where liquid crystals are confined by cavities of micro or nano sizes [5]. Among such materials, porous films filled with liquid crystals are of a special interest. These materials are characterized by well oriented pores of submicron or nanometer diameters. It provides an electric control of light propagation with operating times in microsecond range even in the case of nematic liquid crystals. For example, aluminum and silicon oxide porous films filled with LC were proposed as proper materials for fiber optics applications [6, 7]. Computer modeling of the electrically induced static changes of the initial orientational structure and optical properties was performed on the base of

Content from this work may be used under the terms of the Creative Commons Attribution 3.0 licence. Any further distribution of this work must maintain attribution to the author(s) and the title of the work, journal citation and DOI.
Published under licence by IOP Publishing Ltd

a minimization of the free energy [8] taking into account the dielectric contribution, which is proportional to the squared value of electric field intensity and does not depend on the field polarity. It was found, that application of strong enough electric field to the initially deformed escaped radial (ER) structure can induce a surface anchoring breaking which results in transition to the homogeneous orientation of LC along a pore axis and essential changes of an effective refractive indexes of LC.

Recently [9] we performed electro-optical investigations of polymeric (PET) porous films, filled with a standard nematic (5CB) and found a strong peak-like optical response arising after a very fast (about 10 ns) change of the electric field polarity. In this paper we consider electro-kinetic phenomena including both the electrically induced ion's motion and the overall motion of an anisotropic liquid as a base of mechanisms, describing peculiarities of the above mentioned optical response.

2. Experimental results

In our experiments we used PET films of thickness h = 23μm and different porosity P with cylindrical and randomly allocated pores (diameters d = 0.2, 0.3, 0.4, 0.5 and 0.6μm). Prepared samples were vacuumed and filled with well studied nematic 5CB characterized by a high positive value of a dielectric permittivity anisotropy (approximately $\Delta\varepsilon$ = 12 at room temperature). Samples were thoroughly wiped to remove parasite surface layers of LC and were placed between two glass substrates with ITO electrodes. The distance between the LC cell and the photodetector (22cm) was fixed in all experiments. Measurements were fulfilled at a room temperature via standard electro-optical setup with different position of input polarization.

It was established that the optical response of the samples on the applied voltage drastically depended on the frequency range and the application schema. In the case of video-pulses with changeable polarity at low frequencies ($f<f_c$) the optical response arises as an impulse of light intensity which decays with the time. In case of radio-pulse of high enough frequency ($f>f_c$) electric field induces an overall change in the light intensity, which is typical for an electro-optical response of LC layer in essentially shorter than the electric pulse duration.

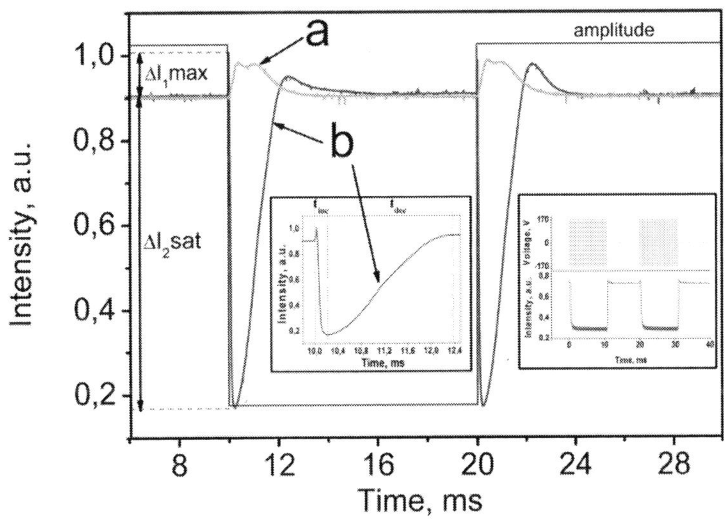

Figure 1. Electro-optical response of the sample with a pore diameter 0.3 μm (δ=0, without analyzer) on the applied square wave signals (f =50 Hz < f_c) for two values of voltage U: (a) U=80V<U^*, (b) U=170V>U^*. Left inset shows the (b) response in different time scale and right inset shows typical response on application of high frequency radio pulse (f=5 kHz > f_c, U=170V).

The description of the dynamic electro-optical response, made above, is based on a concept of change in the effective refractive index under the electric field, and does not include a mechanism of its influence on orientational structure of LC inside pores. In particular, the backward relaxation of Ω(r) to the initial state (Ω(r)= π/2) with the times essentially smaller than the pulse duration, cannot be explained by a traditional dielectric mechanism, describing reorientation of LC director inside pores.

This mechanism is based on an existence of dielectric torque produced by an electric field, with the value proportional to the square of the electric field strength. So, it is not changed at the change of polarity, except for the very fast (in order of nanoseconds) time intervals of the polarity inversion. The typical times of orientational relaxation of LC are essentially longer, and orientational structure has not to be changed during the change of polarity.

The well-known backflow effect, arising on a short time with fast switching on/off the electric field, could cause short peaks in optical response close to those shown above. Nevertheless in our case the polarity inversion time is much shorter (nanoseconds) than that, needed to form back flows.

In this paper we considered two mechanisms of such unusual optical response. The first one is based on a change of LC orientational structure caused by the integral shear flow (electro-osmotic flow). The second mechanism is connected with electrically induced motion of ions inside pores, filled with LC, which causes to the polarization of the system and adds a contribution to the total electric field.

3. Electro-osmotic flow

Electro-osmotic flows can arise due to electro-kinetic phenomena, observed in capillaries of different sizes and filled with polar isotropic liquids, such as a water and water solutions [10]. It is well known that the physical background of the electro-kinetic phenomena is connected with the absorption of polar molecules by the inner surface of the capillary, which leads to polarization of the surface and arising of the electrical double layer with thickness l_0 (of order some nanometers for typical water solutions). It produces non-homogeneous radial distribution of ions inside a capillary in the near surface diffuse layer with thickness l_D>l0. In turn, the application of DC (or low frequency) axial electric field E_z to the liquid induces a motion of ion's system which also results in arising of overall steady (or oscillating) electro-osmotic flow through a capillary. The flow direction depends on the field direction, so the flow oscillates with the same frequency as the AC electric field frequency. It is worth noting that the profile of a steady flow appears for a relatively short time τ_v, which can be determined as $\tau_v \approx \rho R^2/\eta$ [11], where ρ and η – are density and viscosity of a liquid. Estimations made for the water inside capillary with radius R=1μm gives $\tau_v \approx 10^{-6}$s. Increase of the viscosity to the typical for LC values $10^{-2}...10^{-1}$Pa.s and decrease of R should lead to the corresponding decrease of time τ_v, which is essentially smaller than the typical values of a period of E_z oscillations in the experiments mentioned above. In these experiments we used the samples of porous films with removed (after filling) parasitic layers of LC from the outer surfaces. It has to form the two meniscuses, which confine LC inside a cylindrical pore inside the pore (figure 2).

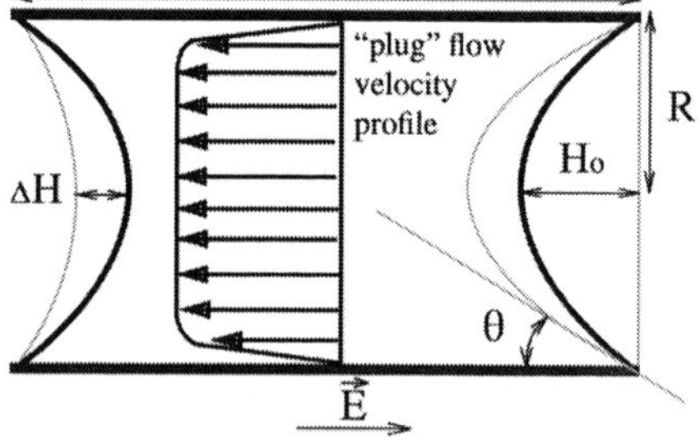

Figure 2. Schematic sectional view of a pore with two meniscuses on each side.

In this case electro-osmotic flow should lead to the deformation of meniscuses and arising of quasi elastic restore force F^{σ}, proportional to the surface tension coefficient σ, which can induce slowing of the electro-osmotic flow and it can even stop when the maximum value of the pressure gradient P_z^{σ}

exceeds the value of the electro-osmotic pressure gradient P_Z^E. Expression for P_Z^E can be obtained from the equation for the volumetric flow velocity Q induced by the pressure gradient P_Z and the electric field E_z [14]. In particular, for the "plug" flow (kR>>1) simplified expressions for Q can be written as:

$$Q = \frac{P_z R^2 S}{8\eta} - \left[\frac{\varepsilon_0 \varepsilon \zeta}{\eta}\right] S E_z \tag{1}$$

where $S=\pi R^2$ – cross section area of the pore, $k = l_D^{-1}$, ε – dielectric permittivity of a liquid, ζ – so-called zeta potential.

Assuming Q=0 (closed channel) one can easily get expression for the electrically induced pressure gradient:

$$P_z^E = 8\left[\frac{(\varepsilon_0 \varepsilon \zeta\ )}{R^2}\right] E_z \tag{2}$$

The estimated value of ζ = -12 mV for PET-5CB interface at the "plug" flow approximation was obtained from the direct measurements of volumetric velocities of the LC flow through porous films [12].

Maximum value (170V) of the voltage, applied to samples in the electro-optical experiments, mentioned above, corresponds to the value $E_z \approx 7$V/μm. For linear dependence P_z^E on E_z, predicted by eq. (2) it results in the maximal values of the electrically induced pressure gradient P_z^E and pressure difference $\Delta P^E_m = P_z^E h$ equal, correspondently to $2.6 \cdot 10^9$Pa/m and 6.10^4Pa for R = 150nm. The latter value can be compared with the pressure difference in capillary ΔP^σ arising due to deformation of two meniscuses induced by the electro-osmotic flow by using well-known expression for a capillary pressure P^σ:

$$P^\sigma = \frac{2\sigma cos\theta}{R} \tag{3}$$

where θ – the contact angle which depends on the wetting conditions. For the simplicity, we will consider the case of meniscuses with the initially flat surface (θ = π / 2 weak wetting), deforming to the ideal wetting state (θ = 0). In accordance with (3) it leads to the next expression for the maximum value of the capillary pressure difference ΔP^σ_m, acting in the direction, opposite to those for ΔP^E, $\Delta P^\sigma_m = 4\sigma / R$.

Assuming the value of $\sigma = 3{,}3\ .10^{-2}$Pa.m for 5CB [14] one can get $\Delta P^\sigma_m = 8.8.10^5$Pa, which is 15 times higher than previously estimated value of ΔP^E_m. So, the presented above estimations confirm the possibility of the significant slowing of an electro-osmotic flow slowing and stopping due to deformation of the meniscuses.

The equation, which describes the dynamics of such process, can be obtained by analysis of the forces, acting on the sample of a liquid confined inside a pore. For this aim, it is convenient to express the contact angle θ in terms of height H of the spherical segment corresponding to each meniscus $cos\theta = 2(H/R)[1+(H/R)^2]^{-1}$.

At the initial state (E_z=0, H = H_0) capillary forces are compensated, so there is no pressure gradient or flow. Under the influence of the electrically induced flow, both menisci are deformed, which causes the change of H from H_0 to H(t). The expression describing a connection between ΔH (t) = H(t) – H_0 and ΔV (t) = V(t)-V_0 (change of the spherical segment volume) in the case of weak deformations (ΔH<<R) can be written as:

$$\Delta V(t) = \left(\frac{\pi R^2}{2}\right)\left[1+\left(\frac{H_0}{R}\right)^2\right]\Delta H(t) \tag{4}$$

Thus volumetric velocity of the flow Q(t) is expressed as:

$$Q(t) = \frac{d\left[\Delta V(t)\right]}{dt} = \left(\frac{\pi R^2}{2}\right)\left[1+\left(\frac{H_0}{R}\right)^2\right]\frac{d\left[\Delta H(t)\right]}{dt} \quad (5)$$

The deformation of meniscuses results in increasing of the total area ΔS of the spherical segments' surface, defined for the case of small deformations (ΔH<<R) as $\Delta S = \Delta S_1 + \Delta S_2 = 2\pi(\Delta H)^2$ which, in turn, leads to the rise of the surface energy $\Delta W = \sigma\Delta S = 2\pi\sigma(\Delta H)^2$ and the appearance of the quasi elastic force F^σ = -(dW/d)(ΔH) = -4πσ(ΔH), which tries to return a system to the initial state. Expression for the viscous friction force Fv, which acts on the LC confined inside a pore, can be simplified by considering sliding of a quasi solid cylinder with the length L, separated from the inner surface of a pore by the layer of liquid of thickness, equal to the Debye length l_D with a linear velocity profile in the gap between the surfaces of the pore and the cylinder. Such approximation correlates with the "plug" type of the flow mentioned above for a liquid moving with the average velocity $v_0 \approx Q(t)/\pi R^2$ everywhere except for the thin diffuse layer, characterized by high viscosity losses due to large values of the velocity gradient $u = dv(r)/d$ which considered to be constant value, defined as $u = v_0/l_D$ in accordance with the simplified model.

Taking into account (5), one can get the next expression for the viscose friction force:

$$F^v = -\eta 2\pi R L u = -\eta\pi L\left(\frac{R}{l_D}\right)\left[1+\left(\frac{H_0}{R}\right)^2\right]\frac{d\left[\Delta H(t)\right]}{dt} \quad (6)$$

The expression for the electro-osmotic force F^E, which induces a motion of the cylinder, can be easily obtained from (2):

$$F^E = P_z^E \pi R^2 L = 8\pi(\varepsilon_0\varepsilon\zeta)E_z L \quad (7)$$

The cylinder motion, under the action of the forces described above, is described by the Newton's second law:

$$m\left(\frac{d^2x}{dt^2}\right) = F^E + F^\sigma + F^v \quad (8)$$

where x = ΔH and $m = \rho\pi R^2 L$ is the cylinder's mass. Equation (8) can be converted to canonical form, which describes vibrations of the pendulum in the presence of frictional forces:

$$\frac{d^2x}{dt^2} + 2\beta\left(\frac{dx}{dt}\right) + \omega_0^2 x = A^E(t) \quad (9)$$

where:

$$\beta = \left[\frac{\eta}{2\rho R l_D}\right]\left[1+\left(\frac{H_0}{R}\right)^2\right] \qquad \omega_0^2 = \frac{4\sigma}{\rho L R^2} \qquad A^E(t) = 8\left[\frac{\varepsilon_0\varepsilon\zeta}{\rho R^2}\right]E_z(t)$$

The solution of the equation (9) depends on the ratio:

$$\frac{\beta^2}{\omega_0^2} = \frac{\eta^2 L\left[1+(H_0/R)^2\right]}{16\,\sigma\rho l_D^{\,2}} \quad (10)$$

With the typical values for 5CB $\eta \approx 0.05$ Pa.s, $\sigma \approx 3{,}3\cdot 10^{-2}$ Pa.m, $\rho \approx 10^3$kg/m^{3}andconsidering $L \approx h$ = 23μm, $l_D \approx$ 15nm, $H_0 = 0$ it results in the next estimate $\beta^2/\omega_0^2 \approx 2.10^5 >>1$ which is in accordance with the motion of an over damped pendulum, with the characteristic time τ^σ written as:

$$\tau^\sigma = \frac{2\beta}{\omega_0^2} = \left(\frac{\eta L}{4\sigma}\right)\left(\frac{R}{l_D}\right)\left[1+\left(\frac{H_0}{R}\right)^2\right] \tag{11}$$

For typical values R=200nm, $\sigma=3.3\cdot 10^{-2}$ N/m, l_D=10nm and $\eta = 10^{-1}$Pa.s, the estimated value of $\tau^\sigma \approx$ 350μs which is essentially smaller than the typical values of t_{dec} shown in figure 3. Such deviation can be partly explained by the lag of a director relaxation, induced by the shear flow, relatively to the slowdown process by elastic energy.

Further analysis of obtained results requires additional consideration of the orientational relaxation processes under the action of the flow and the electric field. In turn, it depends on the type of the initial orientational structure of LC, realized at strong confinement inside pores. It NMR experiments [13] fulfilled with the similar system (polycarbonate porous membrane filled with LC), indicated on the possibility for LC to form two different types of an orientational structure – planar polar (PP) and escaped radial with point defects (ERPD) dependently on the radius R and anchoring strength W. The results of our dielectric investigations of PET-5CB system, which will be published elsewhere, correspond to PP structure, characterized by a director distribution in the plane of the film. For this case, one can expect electrically induced Freedericksz transition with the threshold intensity E_z^F and voltage U^F roughly defined for strong surface anchoring as $E_z^F = (\pi/d)[K_{33}/(\varepsilon_0\Delta\varepsilon)]^{1/2}$, $U^F = E_z^F h = (\pi h/d)[K_{33}/(\varepsilon_0\Delta\varepsilon)]^{1/2}$, where K_{33} – the Frank's module. The dependence of U^F on d obtained by using data for 5CB is presented in table 1.

Table 1. Calculated values of U^F for samples with different pore sizes.

d, μm	0.2	0.3	0.4	0.5	0.6
U^F,V	102	68	51	41	34

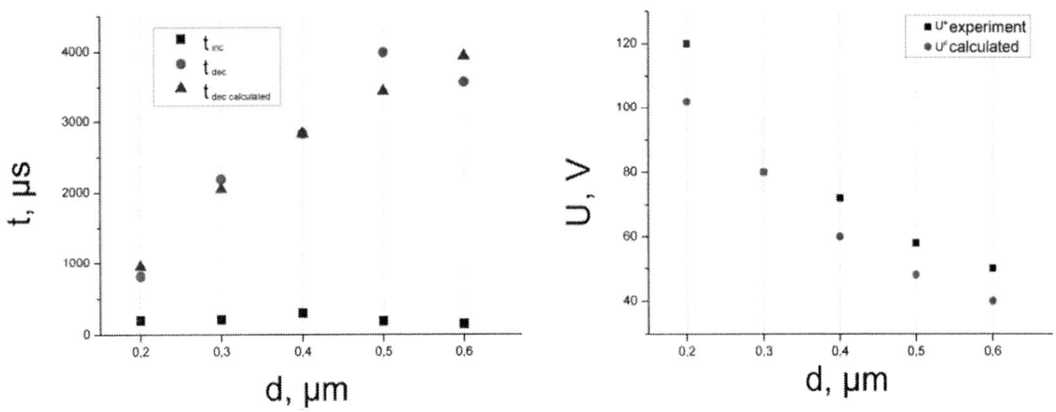

Figure 3. Characteristic times (t_{inc} and t_{dec}) of negative peak as functions of a pore diameter and corresponding dependence of threshold voltages U* and U^F.

The calculated values of U^F are slightly lower, than the experimentally determined threshold voltages, presented in figure 3, which correspond to the strong negative peak in the electro-optical response (figure 1). Moreover, the account of weak anchoring effects results in decreasing of U^F more pronounced for small values of *d* [14]. Nevertheless, the electro-osmotic flow can play the essential role in destabilization of the initial PP structure. Indeed, while the torques acting on a director from electric field and boundaries are equal to zero at $E_z<E_z^F$, the electro-osmotic flow produces the non zero torque, which tends to rotate a director towards the pore's axis. So, it can act as a trigger mechanism creating the escape of a director from the film plane with further enforcing due to arising of a dielectric torque. The initial PP structure can to be restored by the elastic torques on a stage of decay of the shear flow due to capillary forces. The relatively weak positive changes in the optical response, shown in figure 1 correspond to such scenario.

In more details, the influence of electric field on the orientational structure, induced by the electro-osmotic flow and the dielectric torque, can be obtained from the general equations of nematodynamics [14]. For the aim of simplicity, we will use the equation of a director motion for the escaped radial configuration, written in terms of angle Ψ (r,t) = $\pi/2 - \Omega$ (r,t), where Ω is the polar angle, which can be considered as a balance torque equation, written as:

$$M_K + M_d + M_{fl} = M_v \tag{12}$$

where

$$M_K = K \{[\frac{\partial^2\Psi}{\partial r^2} + \frac{1}{r}\frac{\partial\Psi}{\partial r} + \frac{1}{r^2} sin\Psi cos\Psi]\} \tag{13}$$

the Frank's type torque due to elastic curvature of a director field, where K – the effective Frank's constant,

$$M_E = \frac{[\varepsilon_0\varepsilon^{\perp}\varepsilon_{II}\Delta\varepsilon sin\Psi cos\Psi E_z^{\ 2}]}{(\varepsilon^{\perp} + \Delta\varepsilon cos^2\Psi)^2} \tag{14}$$

the dielectric torque, which is proportional to E_z^2, where $\varepsilon^{\perp}=\varepsilon$ (Ψ=0),$\varepsilon_{II} = \varepsilon$ (Ψ=0), $\Delta\varepsilon =\varepsilon$ ($\Psi=\pi/2$) -ε (Ψ=0) - the principal values and anisotropy of a dielectric permittivity,

$$M_{fl} = -\frac{1}{2}\frac{\partial v}{\partial r}(\gamma_2 cos2\Psi - \gamma_1) \tag{15}$$

the hydrodynamic torque, proportional to the velocity gradient, where $\gamma_1 = (\alpha_3 - \alpha_2) > 0$ – the rotational viscosity coefficient, $\gamma_2 = (\alpha_3+\alpha_2)<0$ – the reactive parameter arising due to connection between the director and the velocity field.

The approximate solution of equation (12) can be easily found in the case when the flow induced torque M_{fl} acting on a director is essentially larger, than the dielectric moment M_E, and elastic curvature moment M_K. It may occur at the intensive shear flow on the initial stage of a director relaxation (small values of the angle Ψ describing the deviation of a director field from the initial state Ψ (r,0) = 0). Indeed, in the latter case both M_E and M_K decrease to zero at Ψ (r,t) $\to 0$, whereas the moment M_{fl} reaches its maximal absolute value

$$M_{fl}^{max} = \frac{\alpha_2 \partial v}{\partial r} \tag{16}$$

This moment tends to induce an overall orientation of LC in the direction of a flow (along a capillary axis).

So for the mentioned above approximation the torques balance equation $M_{fl} = M_v$ includes only the flow induced torque, defined by eq. (15) and viscous torque:

$$M_v = \gamma_1 \frac{d\Psi}{dt} \tag{17}$$

Eq.(17) can be easily solved for a steady flow, when a velocity gradient $u = \partial v/\partial r$ does not depend on time. The corresponding analytic solution in this case is written as:

$$tg\Psi = \frac{tg\Psi_{fl}[1 - exp\frac{-t}{\tau_{fl}}]}{[1 + exp\frac{-t}{\tau_{fl}}]} \qquad (18)$$

where the flow alignment angle Ψ_{fl} and the relaxation time τ_{fl} are defined as:

$$tg\Psi_{fl} = \left(\frac{\alpha_2}{\alpha_3}\right)^{1/2} \qquad (19)$$

$$\tau_{fl} = \left[\frac{\alpha_3 - \alpha_2}{(\alpha_2\alpha_3)^{1/2}}\right]\left(\frac{-\partial v}{\partial r}\right)^{-1} \qquad (20)$$

By using the values of Leslie coefficients for 5 CB [14] $\alpha_2 = -0.0812$ Pa.s, $\alpha_3 = -0.0036$ Pa.s one can get $\tau_{fl} \approx 45\ u^{-1}$, where $u = (-\partial v/\partial r) > 0$

For the "plug" type flow, considered above, the maximal value of the velocity gradient u_{max} takes place in the boundary layer. In general, this value depends on the Debay length l_d [14]. For rough estimations we used the value $l_d \approx 15$ nm, which corresponds to variations of a parameter kR from 6.7 (for R = 100 nm) up to 20 (for R = 300 nm) . Calculations, made with the parameters mentioned above, show, that in the near surface layer ($0.95<kR<1$) the mean value of u_{max} varies from $2.0 .10^4\ s^{-1}$ to $1.75 .10^4\ s^{-1}$ at variation of R from 100 nm to 300 nm. So the corresponding value of τ_{fl} only slightly varies with a radius (from 220 μs to 250 μs). The obtained values of τ_{fl} are in a rough agreement with the results for t_{inc} shown in figure 3 and confirms that flow induced changes of orientation may play the main role in an electro-optical response arising on the first stage, which takes place after very fast (for time of about τ_v) initial increasing of the flow velocity rate up to the maximal value Q_{max}.

Nevertheless the estimates made above are valid only for very small values of Ψ at least for liquid crystals with large value of $\Delta\varepsilon$ (as 5CB). Dielectric moment M_d is proportional to E^2 and essentially exceeds the flow induced moment M_{fl} even at relatively small (some degrees) values of Ψ. So the alternative mechanism describing the intermediate and final stages of electro-optical response has to be proposed. It will be considered in the next paragraph.

4. Ion motion

To calculate the field $E_1(t)$ let's exchange the set of charged sectors of LC, formed on the surface of the chaotically allocated pores, by the homogeneous distribution of the surface charges with an average density ρ_e, localized on the meniscus's surfaces in pores.

In consideration of the lower and upper surfaces impact on the value $E_1(t)$, which takes place in case of the equal ion mobility, the expression can be written:

$$E_1(t) = \frac{\rho_e(t)}{\varepsilon_0} = \frac{q_1(t)n}{\varepsilon_0} \qquad (21)$$

where $q_1(t)$ – the surface charge value in a pore, equal for both positive and negative ions.

An upper surface charge is proportional to the number of ions M(t) reached that surface for the time *t* and can be written as:

$$q_1(t) = q_0^+ c\mu^+ \pi R^2 \int_0^t [E_0 - E_1(t')]dt' \tag{22}$$

Equations (21) and (22) lead to the next differential equation for $E_1(t)$:

$$\frac{dE_1(t)}{dt} + \alpha E_1(t) = \alpha E_0 \tag{23}$$

where

$$\alpha = \frac{\pi R^2 q_0^+ nc\mu^+}{\varepsilon_0} = \frac{Pq_0^+ c\mu^+}{\varepsilon_0} \tag{24}$$

The solution to the equation (23) corresponds to the simple law:

$$E_1(t) = E_0[1 - e^{-t/\tau_q}] \tag{25}$$

with the time of relaxation

$$\tau_q = \alpha^{-1} = \frac{\varepsilon_0}{Pq_0^+ c\mu^+} = \frac{\varepsilon_0}{P\sigma} \tag{26}$$

where σ - specific conductance of LC. Estimations made in accordance with (26) give us τ_q=4,4ms for typical values of the specific conductance of LC ($2.10^8 Om^1 m^{-1}$) and the film porosity (0.1), which corresponds to the typical values of experimentally measured decay times.

Obtained expressions enable us to explain qualitatively the dependence of the decay time for electro-optical response on the pore radius. It is known that the electric field intensity has to be increased with reducing the pore radius, to have a notable impact on the refractive index of LC. It corresponds to the low values of the shielding field E_1in comparison with the field E_0 and the small values of the decay times to the initial state, in the framework of the model mentioned above. At the same time, the rise of the pore sizes leads to the increasing of the intensity range where strong changes of orientation take place. It results in the rise of the ratio E_1/E_0, and increasing (accordingly to (26)) of a decay time, needed for relaxation of the system to the initial state.

The quantitative comparison of the theory with the experiment can be done by using the dependence of the threshold voltages $U^\bullet$ on a pore diameter, which results in an appearance of the pike-shaped response (figure 1).

The data from this graph can be used to calculate the field intensity $E_c = U^\bullet/h$, corresponding to the appearance (or disappearance) of a strong electro-optical response, to the initial intensity $E_0 = U_0/h$ which is constant for the data shown in figure 1. Taking into account the above-mentioned remarks one can get from equation (25) the next expressions:

$$E = E_0 - E_1(t) = E_0 e^{-t/\tau_q} \tag{27}$$

$$t_{dec}(d) = \tau_q \ln[\frac{U_0}{U^\bullet(d)}] \tag{28}$$

Due to the variety of the ion diffusion mechanisms, it is reasonable to consider the specific conductance of LC as a fitting parameter.

According to the data (figure 3) the dependence $U^\bullet(d)$ can be described by the law $U^\bullet(d) = A/d$

It results in the next expression for a decay time:

$$t_{dec}(d) = \tau_q \ln[\left(\frac{U_0}{A}\right)d] \tag{29}$$

The obtained dependence (29) is in an agreement with the results of experiment, as it shown in figure 3.

5. Conclusion

In this paper two physical mechanisms, which can be responsible for the specific electro-optical response of a liquid crystal, infiltrated into pores of a porous polymer film are considered. Both mechanisms are connected with electro-kinetic phenomena, namely motion of ions and overall electro-osmotic flow, arising when an axial electric field is applied to the liquid crystal confined by a pore's surface. The first mechanism describes a deformation of menisci, induced by electro-osmotic flow, which results in arising of capillary forces and slowing of the flow. Such mechanism can act as a trigger which stimulates orientational changes in the vicinity of the Freedericksz transition at the abrupt change of the field polarity. The second mechanism describes time variations of the effective value of the electric field's strength arising due to the periodical changes of direction of ion's motion at a periodical changes of the field polarity. This mechanism can be responsible for a breaking and restoring of surface anchoring leading to a sharp negative peak in the electro-optical response.

Acknowledgements

This work was partially supported by the Ministry of Education and Science of the Russian Federation [grant no. 3.1921.2014/K] and the RFBR [project no. 16-02-00041].

References

[1] Chigrinov V G 2004 *Liquid Crystal Photonics* (New York: Nova Science Publishers)
[2] Pasechnik S V, Chigrinov V G and Shmeliova D V 2009 *Liquid Crystals: Viscous and Elastic Properties in Theory and Applications* (New York: Wiley)
[3] Park M S, Lee Y and Kim J K 2005 Chem. Mater. **17** 3944
[4] Ito R, Kumagai T, Yoshida H, Takeya K, Ozaki M, Tonouchi M, and Nose T 2011 Mol. Cryst. Liq. Cryst. **543** 77
[5] Chychlowski M, Yaroshchuk O, Kravchuk R and Wolinski T *Liquid crystal alignment in cylindrical microcapillaries* Opt.–El. Rev. **20**(1) 0–0
[6] Bradac Z, Kralj S and Zumer S 1998 Molecular dynamics study of nematic structures confined to a cylindrical cavity. Phys. Rew. E **58**(6) 7447
[7] Marroum R.-M, Iannacchione G S, Finotello D and Lee M A 1995 *Numerical study of cylindrically confined nematic liquid crystals*. Phys. Rew. E. **51**(4) R2743
[8] Tkachenko A, Dyomin A A, Tkachenko G V, Abbate G, and Sukhoivanov I A 2008 J. Opt. A **10** 055301
[9] Chopik A P, Pasechnik S V, Semerenko D A, Shmeliova D V, Dubtsov A V, Srivastava A K, Chigrinov V G 2014 Electro-*optical effects in porous PET films filled with liquid crystal: new possibilities for fiber optics and THZ applications*. Optics Letters. **39** pp 1453–1456
[10] Masliyah J and, Bhattacharjee S 2006 *Electro-kinetic and Colloid Transport Phenomena* (New York: Wiley)
[11] Tichomolova K P and Kemp T J, *Electro-osmosis* 1993 (Ellis Horwood Limited)
[12] Pasechnik S V, Chopik A P, Shmeliova D V, Drovnikov E M, Semerenko D A, Dubtsov A V, Zhang W and Chigrinov V G 2015 *Electro-kinetic phenomena in porous PET films filled with liquid crystals* Liquid Crystals **44**(11) pp 1537-1542
[13] Allender D W, Crawford G P, Doane J W 1991 *Determination of the Liquid-Crystal Surface Elastic Constant* K_{24} Phys. Rev. Letters **67**(11) pp 1442-1445
[14] Stewart I W 2004 The Static and Dynamic Continuum Theory of Liquid Crystals (London: Tayor & Francis)

The relationship between trading volumes, number of transactions, and stock volatility in GARCH models

Tetsuya Takaishi[1] and Ting Ting Chen[2]
[1]Hiroshima University of Economics, Hiroshima 731-0192, JAPAN
[2]Faculty of Integrated Arts and Sciences, Hiroshima University, Higashi-Hiroshima 739-8521, Japan

E-mail: [1]tt-taka@hue.ac.jp

Abstract. We examine the relationship between trading volumes, number of transactions, and volatility using daily stock data of the Tokyo Stock Exchange. Following the mixture of distributions hypothesis, we use trading volumes and the number of transactions as proxy for the rate of information arrivals affecting stock volatility. The impact of trading volumes or number of transactions on volatility is measured using the generalized autoregressive conditional heteroscedasticity (GARCH) model. We find that the GARCH effects, that is, persistence of volatility, is not always removed by adding trading volumes or number of transactions, indicating that trading volumes and number of transactions do not adequately represent the rate of information arrivals.

1. Introduction

The empirical properties of asset returns have been intensively studied, and some universal properties are classified as "stylized facts" [1]. The notable stylized facts include (1) no significant autocorrelation in returns, (2) long autocorrelation in absolute returns, (3) fat-tailed return distributions, and (4) volatility clustering. The return dynamics explaining these stylized facts have been the subject of numerous studies. Assuming that the return dynamics can be described by a Gaussian random walk with time-varying volatility, one possible explanation is that $r_t = \sigma_t \epsilon_t$, where r_t is a return, σ_t^2 represents volatility, and ϵ_t is a random variable from $N(0,1)$ at time t. Several studies have verified this assumption [2]–[8] by examining whether r_t/σ_t is consistent with the random variable $\sim N(0,1)$.

Yet another unresolved issue relates to volatility dynamics. Under the mixture of distributions hypothesis (MDH) proposed by Clark [9], volatility dynamics is related to the rate of information arrivals to the market. Since the rate of information arrivals is latent and unobservable, Clark used trading volume as a proxy for the rate of information arrivals. Empirical evidence indicates the existence of a contemporaneous correlation between volatility and trading volume; see, for example, [10].

On the other hand, the dynamic behavior of volatility is well captured by the autoregressive conditional heteroscedasticity (ARCH) model [11] and its extension, the generalized ARCH (GARCH) model [12]. In particular, the GARCH model successfully captures the persistence of volatility variation, referred to as GARCH effects. In the GARCH model, the volatility process is described by a function of past volatilities and returns. The MDH also implies that the volatility process is described by a function of trading volume. Lamourex and Lastraps [13] inserted trading volume into the GARCH process by using individual stocks in the US market and found that the GARCH effects disappear, supporting the MDH. Some subsequent studies,

Content from this work may be used under the terms of the Creative Commons Attribution 3.0 licence. Any further distribution of this work must maintain attribution to the author(s) and the title of the work, journal citation and DOI.
Published under licence by IOP Publishing Ltd

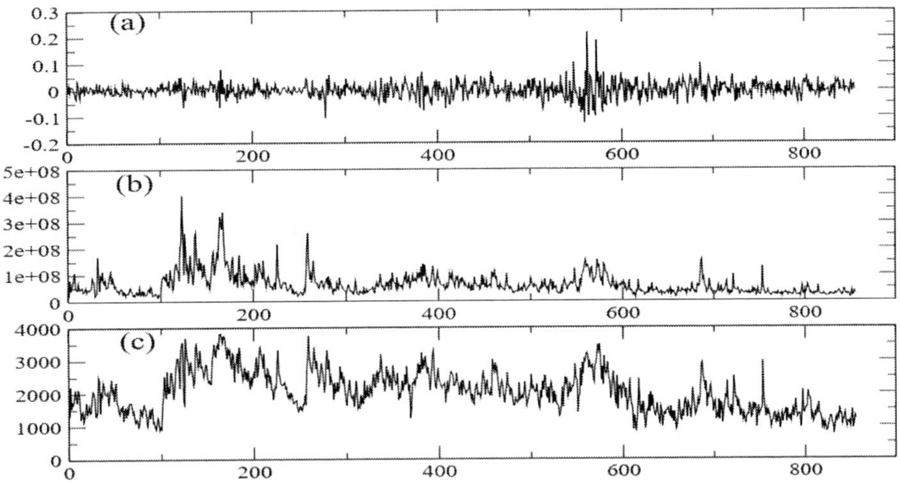

Figure 1. Time series of (a) stock returns, (b) trading volume, and (c) number of transactions for Nippon Steel Co.

for example, [14, 15, 16, 17], also support their finding, that is, that the inclusion of trading volume in the GARCH model reduces the GARCH effects. On the other hand, some other studies, such as [18, 19, 20, 21, 22], report that the inclusion of trading volume in the GARCH model does not completely remove the GARCH effects; thus, the MDH is not supported.

In order to elaborate the volatility dynamics, we examine the relationship between trading volume and stock volatility by using the daily stock data of the Tokyo Stock Exchange from June 3, 2006, to December 30, 2009. Specifically, by including trading volumes into the GARCH process, we can infer the GARCH parameters and examine whether the GARCH effects can be explained by trading volume. We also use the number of transactions as a proxy for the rate of information arrivals and examine its effect on GARCH volatility.

2. GARCH Test

We focus on the GARCH(1,1) model [12] described by $r_t = \sigma_t \epsilon_t$, and

$$\sigma_t^2 = \omega + \alpha r_{t-1}^2 + \beta \sigma_{t-1}^2, \tag{1}$$

where α, β and ω are the GARCH parameters to be determined. The magnitude of persistence of volatility, that is, the GARCH effects, is measured by $\alpha + \beta$, and for high persistence of volatility, we observe that $\alpha + \beta$ is close to 1. The effect of trading volume or the number of transactions is examined by adding a term to the GARCH process, as

$$\sigma_t^2 = \omega + \alpha r_{t-1}^2 + \beta \sigma_{t-1}^2 + \gamma N_t, \tag{2}$$

where N_t stands for either the trading volume or number of transactions at time t. We infer the GARCH parameters by the Bayesian inference conducted using the Markov Chain Monte Carlo (MCMC) method [23]–[28].

3. Empirical Study

Our analysis is based on four individual stock data, (1) Astellas Pharma Inc., (2) JFE Steel Co., (3) Nippon Steel Co., and (4) Seven & i Holdings Co., on the Tokyo Stock Exchange. The sample period of our data is from June 3, 2006, to December 30, 2009. The stock return is defined by the log-price difference: $r_t = 100 \times (\ln P_t - \ln P_{t-1})$, where P_t is the closing stock price at day t. Figure 1 shows the time series of (a) returns, (b) trading volume, and (c) number

Table 1. GARCH parameter results without trading volume and number of transactions. SD stands for standard deviation.

	α	β	ω	$\alpha+\beta$
Astellas Pharma Inc.	0.095	0.857	0.188	0.952
SD	0.018	0.026	0.066	
JFE Steel Co.	0.096	0.895	0.126	0.991
SD	0.019	0.020	0.065	
Nippon Steel Co.	0.159	0.825	0.202	0.984
SD	0.031	0.032	0.086	
Seven & i Holdings Co.	0.117	0.873	0.0761	0.990
SD	0.027	0.028	0.0324	

Table 2. GARCH parameter results with trading volume.

	α	β	ω	γ	$\alpha+\beta$
Astellas Pharma Inc.	0.232	0.251	0.043	1.97	0.483
SD	0.045	0.093	0.043	0.38	
JFE Steel Co.	0.108	0.851	0.046	0.373	0.959
SD	0.025	0.046	0.045	0.232	
Nippon Steel Co.	0.177	0.790	0.102	0.241	0.967
SD	0.037	0.043	0.081	0.131	
Seven & i Holdings Co.	0.166	0.786	0.0233	0.230	0.952
SD	0.045	0.064	0.0257	0.136	

of transactions for Nippon Steel Co. as a representative case. We find no strong correlation between the returns and trading volume or number of transactions. The correlation coefficient ρ between the returns and trading volume (number of transactions) is estimated to be 0.14 (0.02). On the other hand, we find a strong correlation between trading volume and number of transactions, $\rho \sim 0.84$.

For parameter estimations, we use the trading volume normalized by its average. Similarly, we use the number of transactions normalized by its average. We perform our GARCH parameter estimation in this study using the MCMC method based on the Bayesian inference. The MCMC method we use is the Metropolis–Hastings algorithm with a multivariate Student's t-proposal density, which has been shown to be particularly efficient for GARCH parameter estimations [23]-[28]. After the first 5000 Monte Carlo samples are discarded as "burn-in" or "thermalization" process, we collect 50000 samples for analysis.

Table 1 shows the GARCH parameter results. For all stocks, we find that $\alpha + \beta$ is close to 1, implying that a strong persistence of volatility, in other words, the GARCH effect, exists.

Table 2(3) shows the GARCH parameter results with trading volume (number of transactions). We find that even after including the trading volume or number of transactions in the GARCH process, the value of $\alpha + \beta$ does not change much, except for Astellas Pharma Inc. Moreover, we find that γ is always positive, indicating positive correlations between volatility and the trading volume or number of transactions.

4. Conclusions

We examined the relationship between stock volatility and trading volumes or number of transactions by using four individual stock data of the Tokyo Stock Exchange from June 3, 2006, to December 30, 2009. We find that including the trading volume or number of transactions in the GARCH process does not always reduce the value of $\alpha + \beta$, that is, the magnitude of the GARCH effects. Thus, the mixture of the distributions hypothesis using trading volumes or

Table 3. GARCH parameter results with number of transactions.

	α	β	ω	γ	$\alpha+\beta$
Astellas Pharma Inc.	0.194	0.588	0.052	0.827	0.782
SD	0.049	0.127	0.054	0.370	
JFE Steel Co.	0.100	0.872	0.050	0.233	0.972
SD	0.021	0.028	0.047	0.132	
Nippon Steel Co.	0.168	0.797	0.088	0.249	0.965
SD	0.034	0.041	0.079	0.150	
Seven & i Holdings Co.	0.132	0.846	0.0313	0.099	0.978
SD	0.033	0.037	0.0317	0.061	

number of transactions as a proxy for the rate of information arrivals is not completely verified. Since our findings are based on only four individual stock data, it might be interesting to further investigate the robustness of the volatility dynamics using other stock data.

Acknowledgment

Numerical calculations of this work were carried out at the Yukawa Institute Computer Facility and at the facilities of the Institute of Statistical Mathematics. This work was supported by JSPS KAKENHI Grant Number 25330047.

References

[1] Cont R 2001 *Quantitative Finance* **1** 223-236.
[2] Andersen T G, Bollerslev T, Diebold F X, and Labys P 2000 *Multinational Finance Journal* **4** 159-179.
[3] Andersen T G, Bollerslev T, Diebold F X, and Labys P 2000 Great Realizations *Risk* 105-108.
[4] Andersen T G, Bollerslev T, Diebold F X, and Labys P 2001 *Journal of the American Statistical Association* **96** 42-55.
[5] Andersen T G, Bollerslev T, Diebold F X, and Ebens H 2001 *Journal of Financial Economics* **61** 43-76.
[6] Andersen T G, Bollerslev T, Frederiksen F, and Nielsen M O 2010 *Journal of Applied Econometrics* **25** 233-261.
[7] Takaishi T, Chen T T, and Zheng Z 2012 *Progress of Theoretical Physics Supplements* **194** 43-54.
[8] Takaishi T 2012 *Procedia - Social and Behavioral Sciences* **65** 968-973.
[9] Clark P K 1973 *Econometrica* **41** 135-155.
[10] Karpoff J M 1987 *Journal of Financial and Quantitative Analysis* **22** 109-126.
[11] Engle R F 1982 *Econometrica* **50** 987-1007.
[12] Bollerslev T 1986 *Journal of Econometrics* **31** 307-327.
[13] Lamoureux C G and Lastrapes W D 1990 *Journal of Finance* **45** 221-229.
[14] Sharma J L, Mougoue M, and Kamath R 1996 *Applied Financial Economics* **6** 337-342.
[15] Brailsford T 1996 *Accounting and Finance* **35** 89-111.
[16] Miyakoshi T 2002 *Japan and the World Economy* **14** 215-231.
[17] Carroll R and Kearney C 2012 *Applied Financial Economics* **22** 1993-2008.
[18] Najand M and Yung K 1991 *Journal of Futures Markets* **11** 613-621.
[19] Bessembinder H and Seguin P J 1992 *Journal of Finance* **47** 2015-2034.
[20] Floros C and Vougas D V 2007 *International Research Journal of Finance and Economics* **12** 98-115.
[21] Ureche-Rangau L and de Rorthays Q 2009 *Journal of Applied Statistics* **36** 779-799.
[22] Bose S and Rahman H 2015 *Applied Economics* **47** 1899-1908.
[23] Takaishi T 2009 *Lecture Notes of the Institute for Computer Sciences, Social Informatics and Telecommunications Engineering. Complex Sciences* Vol. **5** 1424.
[24] Takaishi T 2009 *New Advances in Intelligent Decision Technologies, Studies in Computational Intelligence* Vol. **199** 635.
[25] Takaishi T 2009 *Proceedings of 8th IEEE/ACIS International Conference on Computer and Information Science* 525-529, doi:10.1109/ICIS.2009.173.
[26] Takaishi T 2010 *Journal of Physics: Conference Series* **221** 012011.
[27] Takaishi T and Chen T T 2012 *International Proceedings of Economics Development and Research* **29** 303-307.
[28] Takaishi T 2013 *Procedia Computer Science* **22** 1056-1064.

Dynamics of the director reorientation in confined nematic liquid crystals imposed by a strong orthogonal electric field

S V Pasechnik[1], **A A Vakulenko**[2], **and A V Zakharov**[2]

[1] Moscow Technological University, Vernadsky pr. 78, Moscow 119454, Russia
[2] Saint Petersburg Institute for Machine Sciences, The Russian Academy of Sciences, Saint Petersburg 199178, Russia

E-mail: alexandre.zakharov@yahoo.ca

Abstract. The dynamics of the periodic distortions in confined nematic liquid crystals (LCs) has been investigated theoretically basing on the hydrodynamic theory including the director motion with appropriate boundary and initial conditions. Analysis of the numerical results for the turn-on process provides an evidence for the appearance of the spatially periodic patterns in confined LC film, only in response to the suddenly applied strong electric field. It has been shown that there is a threshold value of the amplitude of the thermal fluctuations of the director over the LC sample which provides the nonuniform rotation mode rather than the uniform one, whereas the lower values of the amplitude dominate the uniform mode. During the turn-off process the reorientation of the director to the direction preferred by the surfaces is characterized by the complex destruction of the initially periodic structure to a monodomain state.

1. Introduction

In this paper we will particularly refer to the electrically driven director reorientation in a thin nematic liquid crystal (LC) film confined between two transparent electrodes and subjected to a strong orthogonal electric field. When the applied field **E** is much larger than a critical field, then the LC system is suddenly placed far from the equilibrium. It responds by creating a distortion which maximizes the rate at which the LC lowers its total free energy. In this case, the final form of deformation depends on viscous, elastic and electric torques, as well as the boundary and initial conditions. In this work we are focusing on such geometry where the electric field is orthogonal (or approximately orthogonal) to the horizontal boundaries. In this configuration the state of the system immediately becomes unstable after applying the orthogonal electric field. When the misalignment of the director with respect to the direction preferred by the surfaces is due to the thermal fluctuations with small amplitudes, the reorientation following the sudden application of a sufficiently strong and orthogonal electric field manifests itself by the growing of one particular Fourier mode. After switching- off the field the long-range elastic interactions ensure that the molecules reorient themselves in the direction preferred by surfaces.

In this work, switching will be driven by a series of a voltage pulses: V>0, for $0 < t < t_0$; V=0, for $t_0 \leq t \leq t_1$; and V<0, for $t_1 < t < t_2$. So, the aim of this paper is to shows how in the

Content from this work may be used under the terms of the Creative Commons Attribution 3.0 licence. Any further distribution of this work must maintain attribution to the author(s) and the title of the work, journal citation and DOI.
Published under licence by IOP Publishing Ltd

LC system of this geometry switching can be driven by the electric field and to demonstrate the kinetic pathway along which the switching proceeds.

2. Formulation of the relevant equations for dynamical reorientation of the director field

We consider a nematic system such as cyanobiphenyls which is delimited by two horizontal and two lateral surfaces at mutual distances 2d and 2L on a scale in the order of tens micrometers. According to this geometry, the LC system may be seen as two-dimensional, since the director is maintained within the xz-plane (or in the yz-plane) defined by the electric field and the unit vector $\hat{\mathbf{i}}$ directed parallel to the horizontal surfaces, $\hat{\mathbf{k}}$ is a unit normal vector which coincides with the direction of the electric field $\mathbf{E}$, and $\hat{\mathbf{j}} = \hat{\mathbf{k}} \times \hat{\mathbf{i}}$. We can suppose that the components of the director, $\hat{\mathbf{n}} = n_x\hat{\mathbf{i}} + n_z\hat{\mathbf{k}} = \cos\theta(\mathrm{x}, \mathrm{z}, t)\hat{\mathbf{i}} + \sin\theta(\mathrm{x}, \mathrm{z}, t)\hat{\mathbf{k}}$ depend only on x- and z-components and on time t. Here θ denotes the angle between the director and the unit vector $\hat{\mathbf{i}}$. In the case of the quasi-two-dimensional LC system the dimensionless torque balance equation describing the reorientation of $\hat{\mathbf{n}}$ to its equilibrium orientation $\hat{\mathbf{n}}_{\mathrm{eq}}$ can be written as (for details, see the Refs.[1, 2, 3])

$$\theta_{,\tau} = \delta_1 \left[\Delta_1(\theta)\theta_{,xx} + \Delta_2(\theta)\theta_{,zz} + \Delta_3(\theta)\left(-2\theta_{,xz} + \theta_{,x}^2 - \theta_{,z}^2\right) + \Delta_4(\theta)\theta_{,x}\theta_{,z}\right] + \frac{1}{2}\overline{E}^2(z)\sin 2(\alpha - \theta) + \frac{1}{2}(\psi_{,xx} + \psi_{,zz}) - \psi_{,z}\theta_{,x} + \psi_{,x}\theta_{,z} + \gamma\left[\sin 2\theta\psi_{,xz} + \frac{1}{2}(\psi_{,xx} - \psi_{,zz})\right], \quad (1)$$

where $\Delta_1(\theta) = \sin^2\theta + K_{31}\cos^2\theta$, $\Delta_2(\theta) = \cos^2\theta + K_{31}\sin^2\theta$, $\Delta_3(\theta) = \frac{1-K_{31}}{2}\sin 2\theta$, and $\Delta_4(\theta) = (K_{31} - 1)\cos 2\theta$ are four functions of the angle θ, whereas $\delta_1 = \frac{4K_1}{\epsilon_0\epsilon_a V^2}$, $\gamma = \frac{\gamma_2}{\gamma_1}$, and $K_{31} = \frac{K_3}{K_1}$ are three parameters of the system. Here γ_1 and γ_2 are two rotational viscosity coefficients (RVCs), and K_1 and K_3 are the splay and bend elastic constants of the nematic phase, d is the film thickness, ϵ_0 is the absolute dielectric permittivity of free space, and ϵ_a is the dielectric anisotropy of the nematic LC phase. The dimensionless Navier-Stokes equation for the velocity field $\mathbf{v} = u(x, z, \tau)\hat{\mathbf{i}} + w(x, z, \tau)\hat{\mathbf{k}}$ takes the form (for details, see the Refs.[1, 2, 3])

$$\delta_2\left[(\Delta\psi)_{,\tau} + \psi_{,z}(\Delta\psi)_{,x} - \psi_{,x}(\Delta\psi)_{,z}\right] = \hat{\mathcal{L}}\psi + \mathcal{F}, \quad (2)$$

where the dimensionless components of the velocity field vector $\mathbf{v} = u\hat{\mathbf{i}} + w\hat{\mathbf{k}}$ are expressed by the dimensionless stream function $\psi(\mathrm{x}, \mathrm{z}, \tau)$ as $u = \frac{\partial\psi}{\partial z}$ and $w = -\frac{\partial\psi}{\partial x}$, respectively, $\Delta\psi = \psi_{,xx} + \psi_{,zz}$, both the operator $\hat{\mathcal{L}}\psi$ and the function $\mathcal{F}$ are listed in the Ref.[4]), whereas $\overline{\mathrm{x}} = \mathrm{x}/\mathrm{d}$ and $\overline{\mathrm{z}} = \mathrm{z}/\mathrm{d}$ are the dimensionless space variables, d is the film thickness, $\tau = \frac{\epsilon_0\epsilon_a}{\gamma_1}\left(\frac{V}{2d}\right)^2 t$ is the dimensionless time, $\delta_2 = \frac{\rho\epsilon_0\epsilon_a V^2}{4\gamma_1^2}$ is an additional parameter of the system, whereas the electric field $\mathbf{E} = E_x\hat{\mathbf{i}} + E_z\hat{\mathbf{k}} = E(\mathrm{z})\cos\alpha\hat{\mathbf{i}} + E(\mathrm{z})\sin\alpha\hat{\mathbf{k}}$ makes the angle α with the horizontal surfaces, and its values are varied in the vicinity of $\frac{\pi}{2}$. Notice that the overbars in the space variables x and z have been (and will be) eliminated in the last as well as in the following equations.

The application of the voltage across the nematic film results in a variation of $E(z)$ through the film which is obtained from [1, 2, 3]

$$\frac{\partial}{\partial \mathrm{z}}\left[\left(\frac{\epsilon_\perp}{\epsilon_a} + \sin^2\theta\right)\overline{E}(z)\sin\alpha\right] = 0,$$
$$1 = \int_{-1}^{1}\overline{E}(z)dz, \quad (3)$$

where $\overline{E}(z) = \frac{2dE(z)}{V}$, and V is the voltage applied across the cell, ϵ_0 is the absolute dielectric permittivity of free space, and ϵ_a is the dielectric anisotropy of the nematic LC phase. Consider now the nematic film between two electrodes when the director is weakly anchored at the horizontal surfaces and the anchoring energy takes the form $W^{\rm an} = \frac{1}{2}A\sin^2(\theta_s - \theta_0)$, where θ_s and θ_0 are the angles corresponding to the director orientation on the solid surface, $\hat{\mathbf{n}}_s$, and easy axis, $\hat{\mathbf{n}}^0$, respectively. In order to elucidate the role of the thermal fluctuations in maintaining of the spatially periodic patterns in the LC sample under the influence of the strong orthogonal electric field, we have performed a numerical study of Eqs.(1)-(3) with the mixed boundary condition for the angle θ, which reads in the dimensionless form as (hereafter referred to as case A)

$$\theta_{,\rm z}(-10 < {\rm x} < 10, {\rm z} = \pm 1) = \pm\delta_3\theta(-10 < {\rm x} < 10, {\rm z} = \pm 1),$$
$$\theta({\rm x} = \pm 10, -1 < {\rm z} < 1) = 0, \quad (4)$$

where $\delta_3 = \frac{Ad}{K_1}$, and with the strong anchoring condition for the angle θ, which reads in the dimensionless form as (hereafter referred to as case B)

$$\theta\,(-10 < {\rm x} < 10, {\rm z} = \pm 1) = 0, \theta\,({\rm x} = \pm 10, -1 < {\rm z} < 1) = 0. \quad (5)$$

Moreover, we will assume the no-slip boundary conditions for the nematogenic molecules on these boundary surfaces, i.e.,

$$(\psi_{,\rm z})_{-10<{\rm x}<10,{\rm z}=\pm 1} = (\psi_{,\rm z})_{{\rm x}\pm 10,-1<{\rm z}<1} = 0,$$
$$(\psi_{,\rm x})_{-10<{\rm x}<10,{\rm z}=\pm 1} = (\psi_{,\rm x})_{{\rm x}\pm 10,-1<{\rm z}<1} = 0. \quad (6)$$

In order to observe the formation of the spatially periodic patterns developing spontaneously from homogeneous state and excited by the strong orthogonal electric field, we consider the initial condition in the form

$$\theta({\rm x}, {\rm z}, 0) = \theta_0\cos(q_{\rm x}{\rm x})\cos(q_{\rm z}{\rm z}), \quad (7)$$

which defines the thermal fluctuations of the director over the LC sample with the amplitude θ_0 and wavelengths $q_{\rm x}$ and $q_{\rm z}$ of an individual Fourier component of the modulation. In the case A, the wavelengths $q_{\rm x}$ and $q_{\rm z}$ of an individual Fourier components are described as $q_{\rm x} = \frac{\pi}{20}(2k+1)$, where ${\rm k} = 0, 1, 2, ...$ and ${\rm q_z} = \mp(\delta_3 \cot {\rm q_z})_{\rm z=\pm 1}$, whereas in the case B, as $q_\alpha = \frac{\pi}{2}(2k+1)$, where $\alpha = {\rm x}, {\rm z}$ and ${\rm k} = 0, 1, 2, ...$ In turn, the initial velocity field is $\mathbf{v}\,(x, z, 0) = 0$, so the initial condition for the dimensionless stream function has the form

$$\psi\,(x, z, 0) = 0. \quad (8)$$

In the following the reorientation of the director in the nematic film, under the influence of the strong electric field directed perpendicular to the horizontal bounding surfaces, will be obtained by solving the nonlinear differential equations (1) and (3), with appropriate bounding conditions for the angle θ (Eq.(4) (case A) or (Eq.(5) (case B)), and for the dimensionless stream function ψ (Eq.(6), together with the initial conditions both for the angle θ (Eq.(7)) and ψ (Eq.(8)), respectively. The optimal dimensionless wavelengths $q_{\rm x}$ and $q_{\rm z}$ are determined by the minimum condition of the total energy ${\rm W} = {\rm W_{elast}} + {\rm W_{el}}$ (for details, see the Refs.[1, 2, 3]). Thus we have obtained a closed system of nonlinear differential equations (1)-(3) with appropriate boundary (4)-(6) and initial conditions (7) and (8) which can provides the nonuniform rotation mode rather than the uniform one.

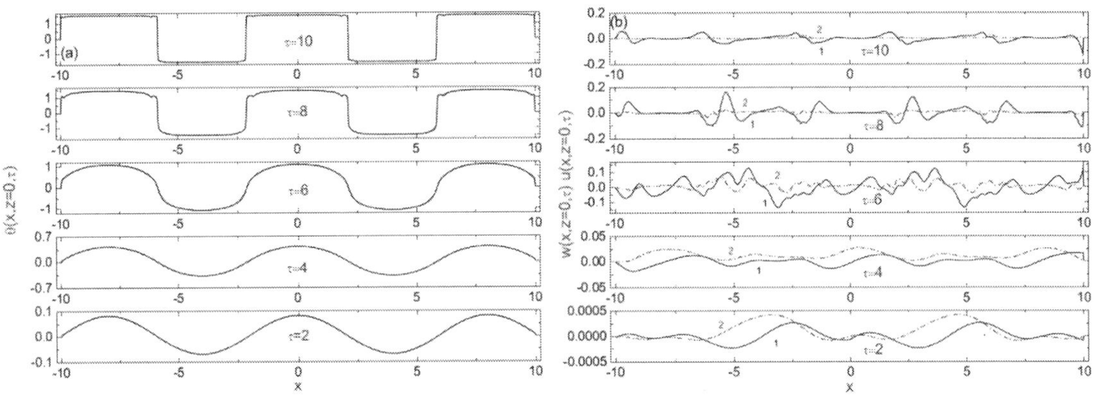

Figure 1. The evolution of the angle $\theta\,(\mathrm{x}, \mathrm{z}=0, \tau)$ (a) and the velocity field $\mathbf{v} = u\,(\mathrm{x}, \mathrm{z}=0, \tau)\,\hat{\mathbf{i}} + w\,(\mathrm{x}, \mathrm{z}=0, \tau)\,\hat{\mathbf{k}}$ (b) during the turn-on process ($\mathrm{E} > 0$) along the length of the dimensionless LC film ($-10 \leq \mathrm{x} \leq 10$), for a number of dimensionless times τ =2, 4, 6, 8, and 10, respectively. In part (b) the solid (curves 1) and the dash dotted (curves 2) lines denote the numerical results for the vertical $w\,(\mathrm{x}, \mathrm{z}=0, \tau)$ and horizontal $u\,(\mathrm{x}, \mathrm{z}=0, \tau)$ components of the velocity field $\mathbf{v} = u\hat{\mathbf{i}} + w\hat{\mathbf{k}}$, respectively.

2.1. Turn–on process in the positive sense (case A)

When a strong electric field $\mathbf{E}$ is applied in the positive sense at the angle α close to the right angle to the unit vector $\hat{\mathbf{i}}$, the director moves from being parallel to the direction preferred by the surfaces to being parallel to the electric field (the turn-on process), because dielectric anisotropy is positive for all cyanobiphenyls. For the case of 4-cyano-4′-pentylbiphenyl (5CB), at the temperature $300K$, corresponding to nematic phase and the density 10^3 kg/m^3, the set of δ-parameters, which is involved in Eqs.(1)-(4), takes the values [2, 3] $\delta_1 = 8.6 \times 10^{-6}$, $\delta_2 = 0.19$, and $\delta_3 = 19.5$. These nonlinear differential equations has been solved by the numerical relaxation method [4]. The relaxation criterion $\epsilon = |\left(\theta_{(\mathrm{n}+1)}(\tau) - \theta_{(\mathrm{n})}(\tau)\right)/\theta_{(\mathrm{n})}(\tau)|$ for calculating procedure was chosen equal to be 5×10^{-4}, and the numerical procedure was then carried out until a prescribed accuracy was achieved. The evolution both of the angle $\theta\,(\mathrm{x}, \mathrm{z}=0, \tau)$ (Fig.1(a)) and the velocity components $w(x, z=0, \tau)$ and $u(x, z=0, \tau)$ (Fig.1(b)) of the vector $\mathbf{v} = u\hat{\mathbf{i}}+w\hat{\mathbf{k}}$ along the x-axis ($-10 \leq x \leq 10$), for a number of times τ =2 (~12 ms), 4 (~24 ms), 6 (~36 ms), 8 (~48 ms), and 10 (~60 ms), are shown in Fig.1(a) and (b), respectively. The values of the dimensionless time $\left(\tau = \frac{\epsilon_0\epsilon_a}{\gamma_1}\left(\frac{V}{2d}\right)^2 t\right)$ is accounted for after switching- on the electric field. The time propagation of the angle $\theta\,(\mathrm{x}, \mathrm{z}=0, \tau)$ along the x-axis ($-10 \leq x \leq 10$) with the value of the amplitude θ_0 which is equal to 0.01 ($\sim 1.1°$) is characterized by the well developed periodic structure with the lattice points at $\mathrm{x} = \pm2.175$ and ±5.83 (case A). According to our calculations, the evolution process of the dimensionless velocities $w(x, z=0, \tau)$ (Fig.1(b), solid curves) and $u(x, z=0, \tau)$ (Fig.1(b), dash dotted curves) is characterized by the increase of these components upon increasing of τ up to 8 (~48 ms), before getting to the distributions of $w(x, z=0, \tau=8)$ and $u(x, z=0, \tau=8)$ along the x-axis ($-10 \leq x \leq 10$). That distribution of w is characterized by the maximum $w(x, z=0, \tau=8) \sim 0.15$ ($\sim 5\ mm/s$) in the vicinity of the lattice points at $\mathrm{x} = \pm2.175$ and ±5.83. In that case the distribution of the velocity field $\mathbf{v}$ in the LC sample under the influence of the strong electric field is characterized by maintaining the well developed vortices, in the framework of the cells delimited by the scaled distances $-10 < x < -5.83$, $-5.83 < x < -2.175$, $-2.175 < x < 2.175$, $2.175 < x < 5.83$,

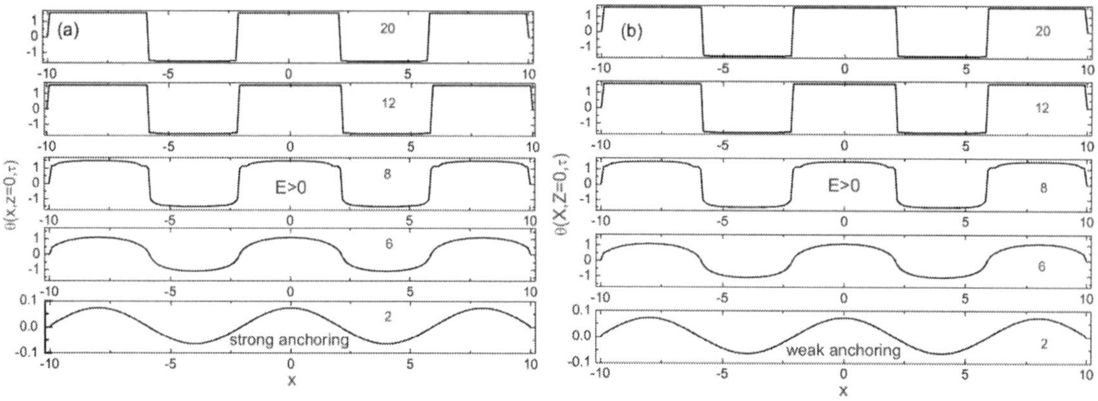

Figure 2. The evolution of the angle $\theta\,(\mathrm{x}, \mathrm{z}=0, \tau)$ during the turn-on process ($\mathrm{E}>0$ and $\alpha = 1.57$ ($\sim$ 89.96°)) along the length of the dimensionless LC film ($-10 \leq \mathrm{x} \leq 10$), for a number of dimensionless times τ =2 (∼12 ms), 6 (∼36 ms), 8 (∼48 ms), 12 (∼72 ms), and 20 (∼0.12 s), respectively. Part (a) shows the case of strong anchoring (case B), while (b) shows the case of weak anchoring (case A). In all these cases the amplitude of the thermal fluctuation θ_0 is equal to 0.01 ($\sim$ 1.1°).

and $5.83 < x < 10$, as shown in Fig.1(a). For both cases A and B, and for the values of $\theta_0 = 0.01$ ($\sim$ 1.1° and $\alpha = 1.57$), the time propagation of the angle $\theta\,(\mathrm{x}, \mathrm{z}=0, \tau)$ profile along the x-axis ($-10 \leq x \leq 10$) is characterized by well-developed periodic structure with the lattice points at $\mathrm{x} = \pm 2.175$ and ± 5.83 (case A) (Fig.2(a)) and $\mathrm{x} = \pm 2.19$, and ± 5.80 (case B) (Fig.2(b)), respectively. Physically, this means that the character of the molecules anchoring to the bounding surfaces does not influence on the final maintaining of the spatially periodic patterns (see Fig.2). At the value of the angle $\alpha = 1.57$ the optimal dimensionless wavelengths q_x and q_z provide the minimal values of the total energy [2, 3] $\mathrm{W} = \mathrm{W}_{\mathrm{elast}} + \mathrm{W}_{\mathrm{e}}$. In our case, the equilibrium $\theta_{\mathrm{eq}}^{\mathrm{ON}}(\mathrm{x}, \mathrm{z})$ value of the angle $\theta(\mathrm{x}, \mathrm{z}, \tau = 12)$ corresponding to the turn-on process is achieved after the dimensionless time term τ =12. It is also shown that only for the values of $\mathrm{q_x}$=0.785, $\mathrm{q_z}$=64.336, $\alpha = 1.57$, and θ_0 =0.01 ($\sim$ 1.1°) and higher the periodic structure may appear spontaneously from homogeneous nematic phase under the above- mentioned conditions. These values of $\mathrm{q_x}$ and $\mathrm{q_z}$ provide the minimal values of the total energy W.

2.2. *Turn-off process (case B)*

After the electric field is removed, the director relaxes back to the direction preferred by the surfaces (the turn-off process). Now the reorientation of the director in the nematic film under the influence of the long-range elastic interactions can be obtained by solving the nonlinear differential equations (1) and (3), for the case of $\mathrm{E} = 0$, and both with the mixed Eq.(4) (case A) or strong Eq.(5) (case B) boundary conditions for the angle θ. The initial condition is taken in the form $\theta(\mathrm{x}, \mathrm{z}, 0) = \theta_{\mathrm{eq}}^{\mathrm{ON}}(\mathrm{x}, \mathrm{z})$, where $\theta_{\mathrm{eq}}^{\mathrm{ON}}$ is defined as an equilibrium distribution of the director over the LC film, obtained during the turn-on process and at the value of the angle α being equal to 1.57. Figure 3 shows the evolution of the angle $\theta\,(\mathrm{x}, \mathrm{z}=0, \tau)$ during the turn-off process along the length of the dimensionless LC film ($-10 \leq \mathrm{x} \leq 10$), for two cases, A (Fig.3(a)) and B (Fig.3(b)), respectively, and for a number of times τ =2 (22) (∼12 ms), 6 (26) (∼36 ms), 8 (28) (∼48 ms), 12 (32) (∼72 ms) and 20 (40) (∼0.12 s). Here in our notation the first value denotes the dimensionless ($\tau = \frac{\epsilon_0 \epsilon_a}{\gamma_1}\left(\frac{V}{2d}\right)^2 t$) time after switching- off the electric field, whereas the second value is the total time after starting the process. It is shown that after

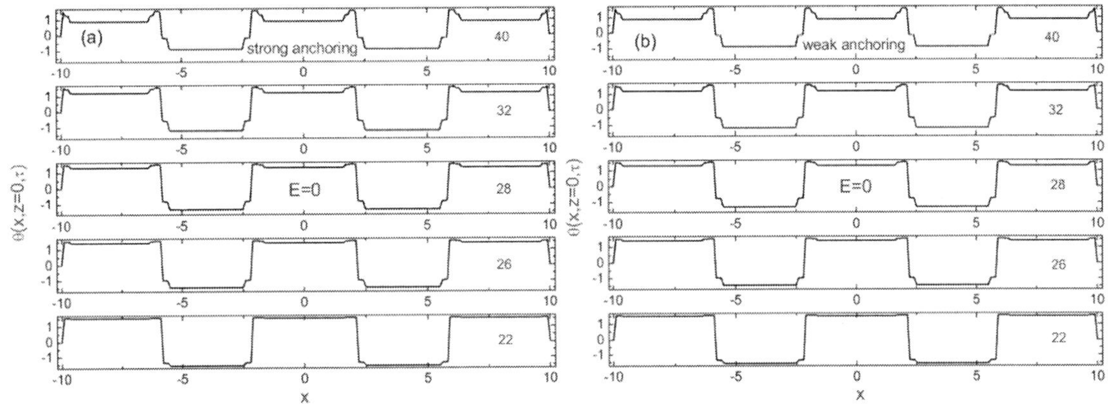

Figure 3. The evolution of the angle $\theta\,(\mathrm{x}, \mathrm{z}=0, \tau)$ during the turn-off process ($\mathrm{E}=0$) along the length of the dimensionless LC film ($-10 \le \mathrm{x} \le 10$), and for a number of dimensionless times τ =2 (22), 6 (26), 8 (28), 12 (32) and 20 (40), respectively. Part (a) shows the case of strong anchoring (case B), while (b) shows the case of weak anchoring (case A), respectively.

time τ ~18 (~0.108 s), when the electric field is removed, the director slowly relaxes back to the direction preferred by the surfaces and that process is characterized by the complex destruction of the initially periodic structure (see Fig.3), especially in the vicinity of the lattice points. Our calculations show that the director slowly relaxes back to the direction preferred by the surfaces and the angle $\theta \to 0$ only after time $\tau_R \sim 400$ (2.4 s). Physically, this means that the role of both the elastic and viscous forces becomes comparable.

2.3. Turn-on process in the negative sense (case C)

When the strong electric field $\mathbf{E}$ is applied again but in the negative sense at the angle $\alpha \sim -\frac{\pi}{2}$, the director moves from being parallel to the $\hat{\mathbf{n}}^{\mathrm{OFF}}$ to being parallel to the electric field. Here $\hat{\mathbf{n}}^{\mathrm{OFF}}$ is the final orientation of the director after the time term τ_{OFF}, when the electric field was removed. Now the reorientation of the director in the nematic film under the influence of the external forces can be obtained by solving the nonlinear differential equations (1) and (3), with the appropriate boundary Eqs.(4)-(5) and initial conditions. In that case the initial condition is taken in the form $\theta(\mathrm{x}, \mathrm{z}, 0) = \theta^{\mathrm{OFF}}(\mathrm{x}, \mathrm{z})$, where $\theta^{\mathrm{OFF}}(\mathrm{x}, \mathrm{z})$ is defined as the final distribution of the director over the LC film, obtained during the turn-off process when the electric field was removed. Two scenarios of the evolution of the director distribution across the LC film, under the influence of the strong electric field directed in the negative sense at the angle α =-1.57 to the horizontal bounding surfaces are shown in Fig.4. That evolution is shown for time sequences τ=0 (248), 2 (250), 4 (252), 6 (254) and 8 (256) (see Fig.4), respectively. Here the first value denotes the dimensionless time after switching-on the process in the negative sense, whereas the second value is the total time after starting the process. In that case, the electric field is switching- on again after the dimensionless time terms τ =248 (~1.488 s). So, in that case the electric field was removed during τ_{OFF} =228 ($20 \le \tau \le 248$) dimensionless units. The main result of these calculations is that the final maintaining of the spatially periodic patterns, at α =-1.57, is possible only when the electric field was removed during 228 dimensionless time units or $20 \le \tau_{\mathrm{OFF}} \le 248$ ($0.12\ s \le \mathrm{t}_{\mathrm{OFF}} \le 1.488\ s$) (see Fig.4). This result confirm our previous suggestion that there is a thereshold value of the amplitude θ_0 which provides the nonuniform rotation mode rather than the uniform one, whereas the lower value of θ_0 dominates the uniform mode [4]. Notice that the final maintaining of the spatially periodic (Fig.4) structure is achieved

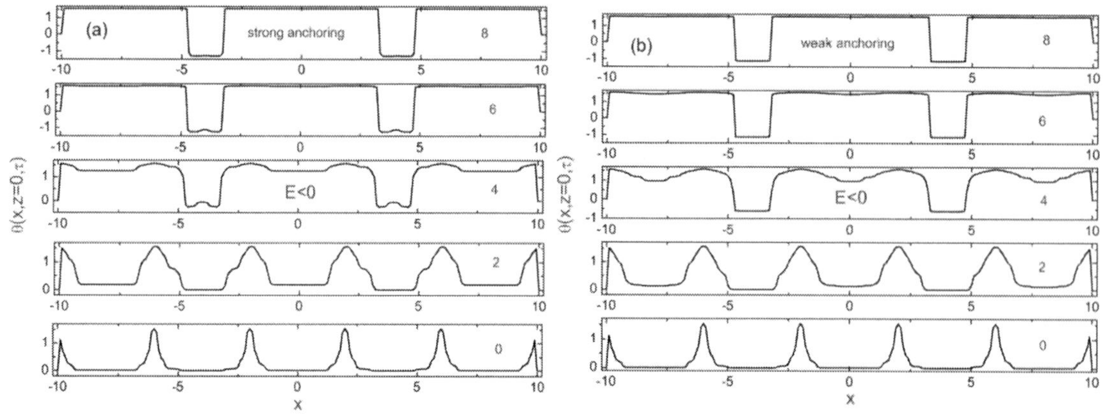

Figure 4. Formation of the nonuniform periodic evolution of the angle $\theta\,(\mathrm{x},\mathrm{z}=0,\tau)$ along the length of the dimensionless LC film $(-10 \leq x \leq 10)$, for the turn-on process in the negative sense $(E < 0)$, and for a number of dimensionless times τ=0 (248), 2 (250), 4 (252), 6 (254), and 8 (256). Part (a) shows the case of strong anchoring (case B), while (b) shows the case of weak anchoring (case A), respectively. Here $\alpha = -1.57$ ($\sim -89.96°$), and the value of the amplitude θ_0 is equal to 0.01 ($\sim 1.1°$).

after the same dimensionless time 8 (~48 ms).

3. Conclusion

In summary, we have numerically investigated the peculiarities in the director relaxation during both the turn-on and turn-off aligning processes in the nematic phase. Analysis of the numerical results for the turn-on process provides the evidence for the appearance of the spatially periodic patterns in confined 5CB LC film only in response to the suddenly applied strong electric field directed orthogonal (or approximately orthogonal) to the horizontal bounding surfaces. During the turn-off process (τ_{OFF}), when the electric field is removed, the director relaxes back to the direction preferred by the surfaces and that process is characterized by the complex destruction of the initially periodic structure to a monodomain state.

Acknowledgments

We acknowledge financial support of the Russian Funds for Fundamental Research (Grant No. 16-02-00041-a) and of the Ministry of Education and Science of the Russian Federation (Grant No. 3.1921.2014/K).

References

[1] Vakulenko A A and Zakharov A V 2013 *Phys. Rev.* E **88** 022505
[2] Zakharov A V and Vakulenko A A 2013 *J. Chem. Phys.* **139** 244904
[3] Zakharov A V and Vakulenko A A 2015 *Journal of Non-Newtonian Fluid Mechanics* **217** 23
[4] Berezin I S and Zhidkov N P 1965 *Computing Methods* (Oxford: Clarendon)

Vector potential quantization and the photon wave-particle representation

C Meis[1*] and P R Dahoo[2]

[1] CEA Saclay. National Institute for Nuclear Science and Technology, Université Paris Saclay – 91191 Gif-sur-Yvette, France.
[2] LATMOS /IPSL, UVSQ Université Paris-Saclay, UPMC Univ. Paris 06, CNRS, F-78280, Guyancourt, France.

E-mail: constantin.meis@cea.fr

Abstract. The quantization procedure of the vector potential is enhanced at a single photon state revealing the possibility for a simultaneous representation of the wave-particle nature of the photon. Its relationship to the quantum vacuum results naturally. A vector potential amplitude operator is defined showing the parallelism with the Hamiltonian of a massless particle. It is further shown that the quantized vector potential satisfies both the wave propagation equation and a linear time-dependent Schrödinger-like equation.

Introduction

We analyse first the fundamental link between the electromagnetic wave theory and quantum electrodynamics (QED) [1-3]. In the classical description issued from Maxwell's equations the energy density of an electromagnetic wave with electric and magnetic fields $\vec{E}(\vec{r},t)$ and $\vec{B}(\vec{r},t)$ respectively is

$$W_C(\vec{r},t)=\frac{1}{2}\left(\varepsilon_0\left|\vec{E}(\vec{r},t)\right|^2+\frac{1}{\mu_0}\left|\vec{B}(\vec{r},t)\right|^2\right) \qquad (1)$$

where ε_0 and μ_0 are the electric permittivity and magnetic permeability of the vacuum.
In the case of a monochromatic plane wave the electric and magnetic fields are proportional to the vector potential amplitude $A_0(\omega)$ and the energy density writes

$$W_C(\vec{r},t)=4\varepsilon_0\omega^2\left|A_0(\omega)\right|^2\sin^2(\vec{k}\cdot\vec{r}-\omega t) \qquad (2)$$

whose mean value over a period, that is over a wavelength, becomes time and space independent

$$W_C=2\varepsilon_0\omega^2\left|A_0(\omega)\right|^2 \qquad (3)$$

In the quantum theory the energy density for N photons with angular frequency ω in a volume V is

$$W_Q=\frac{N\hbar\omega}{V} \qquad (4)$$

where $\hbar=h/2\pi$ is Planck's reduced constant.

Content from this work may be used under the terms of the Creative Commons Attribution 3.0 licence. Any further distribution of this work must maintain attribution to the author(s) and the title of the work, journal citation and DOI.
Published under licence by IOP Publishing Ltd

In order to link the classical and quantum description it is generally imposed for $N = 1$, thus for a single photon state, the relations (3) and (4) to be equal. In this way, the vector potential amplitude is

$$|A_0(\omega)| = \sqrt{\frac{\hbar}{2\varepsilon_0 \omega V}} \tag{5}$$

It is worth noting that as a result of this procedure an external arbitrary parameter V has been introduced in the last equation which is supposed to express naturally the photon vector potential amplitude, an intrinsic physical property. Nevertheless, this equation is used to define the fundamental link relations between the classical and quantum theory of light through the definition of the vector potential amplitude operators for a photon

$$A_{k\lambda} = \sqrt{\frac{\hbar}{2\varepsilon_0 \omega_k V}} a_{k\lambda} \qquad A^*_{k\lambda} = \sqrt{\frac{\hbar}{2\varepsilon_0 \omega_k V}} a^+_{k\lambda} \tag{6}$$

where $a_{k\lambda}$ and $a^+_{k\lambda}$ are the annihilation and creation operators respectively for a k-mode and λ-polarization photon with angular frequency ω_k .

Vector potential in QED

It is useful to examine now the way the last relations are used in QED calculations [1-4]. The vector potential operator writes generally as a superposition of the vector potentials of all the k-modes and λ–polarization photons with polarization unit vector $\hat{\varepsilon}_\lambda$

$$\vec{A}(\vec{r},t) = \sum_{k,\lambda} \sqrt{\frac{\hbar}{2\varepsilon_0 \omega_k V}} \left[a_{k\lambda} \hat{\varepsilon}_\lambda e^{i(\vec{k}\cdot\vec{r} - \omega_k t + \theta)} + a^+_{k\lambda} \hat{\varepsilon}^*_\lambda e^{-i(\vec{k}\cdot\vec{r} - \omega_k t + \theta)} \right] \tag{7}$$

The discrete summation over the modes k is generally replaced by a continuous one over the angular frequencies ω following the transformation issued from the density of states theory

$$\sum_{k,\lambda} \rightarrow \frac{V}{2\pi^2 c^3} \sum_\lambda \int \omega^2 d\omega \tag{8}$$

where c is the velocity of light in vacuum and λ takes two values corresponding to the Left and Right hand circular polarizations. Consequently, for all the calculations involving the square of the amplitude of the vector potential this mathematical operation eliminates the volume parameter V. Notice also that the relation (8) has been obtained under the condition that all the photons wavelengths are much smaller than the characteristic dimensions of the volume V.

Single photon state vector potential and the quantum vacuum

The methodology presented above gives quite physical results when considering a system of photons within a volume whose dimensions are much bigger compared to their wavelengths. However, the experimental evidence [1-4] has shown that a single photon is an indivisible entity with definite energy and momentum. Despite of this, the relation (6) gives no precise information on its vector potential amplitude.

Now, the energy density of the electromagnetic waves in the classical description as well as that of photons in QED depends quartically on the angular frequency ω^4 [1,3]. Consequently, the relation (3) entails automatically that the vector potential amplitude is proportional to ω. Indeed, the unit analysis of the general solution of Maxwell's equations for the vector potential shows that it is inversely proportional to time, thus proportional to an angular frequency. Consequently, for a *k*-mode photon the vector potential amplitude α_0 can be written as [4,5]

$$\alpha_{0_k} = \xi \omega_k \tag{9}$$

where ξ is a constant.

Thus, the fundamental physical quantities characterizing both the wave and particle nature of a single photon: energy and momentum (particle), vector potential amplitude and wave vector (wave), are all related to the angular frequency as follows [4,5]

$$\frac{E_k}{\hbar} = \frac{|\vec{p}_k|}{\hbar / c} = \frac{\alpha_{0_k}}{\xi} = |\vec{k}| c = \omega_k \tag{10}$$

In the plane wave representation the vector potential for a *k*-mode and λ–polarization photon can be expressed over a period and repeated successively along the propagation axis as

$$\vec{\alpha}_{k,\lambda}(\vec{r},t) = \omega_k \left(\xi \hat{\varepsilon}_\lambda e^{i(\vec{k}\cdot\vec{r} - \omega_k t + \theta)} + cc \right) = \omega_k \vec{\Xi}_{k\lambda}(\omega_k, \vec{r}, t) \tag{11}$$

with *cc* the complex conjugate and has to satisfy the wave propagation equation

$$\vec{\nabla}^2 \vec{\alpha}_{k,\lambda}(\vec{r},t) - \frac{1}{c^2}\frac{\partial^2}{\partial t^2}\vec{\alpha}_{\kappa,\lambda}(\vec{r},t) = 0 \tag{12}$$

leading to

$$\left[\alpha_0^2 + \xi^2 c^2 \vec{\nabla}^2 \right] \vec{\alpha}_{k,\lambda}(\vec{r},t) = 0 \tag{13}$$

The last expression entails that the vector potential amplitude of the photon can be expressed as an operator

$$\tilde{\alpha}_0 = -i\xi c \vec{\nabla} \tag{14}$$

which is quite symmetrical to the relativistic Hamiltonian operator for a massless particle

$$\tilde{H} = -i\hbar c \vec{\nabla} \tag{15}$$

Applying the vector potential amplitude operator (14) upon the vector potential expression (11) we get finally

$$i\xi \frac{\partial}{\partial t}\vec{\alpha}_{k,\lambda}(\vec{r},t) = \tilde{\alpha}_0 \vec{\alpha}_{k,\lambda}(\vec{r},t) \tag{16}$$

This is an equation for the vector potential with quantization constant ξ analogue to Schrödinger's equation for the energy with quantization constant h. Hence, we get the coupled equation for a photon in a non-local representation with $\vec{\alpha}_{k,\lambda}(\vec{r},t)$ as a wave function that may be suitably normalized [3]

$$i\begin{pmatrix}\xi \\ \hbar\end{pmatrix}\frac{\partial}{\partial t}\vec{\alpha}_{k;\lambda}(\vec{r},t)=\begin{pmatrix}\widetilde{\alpha}_0 \\ \widetilde{H}\end{pmatrix}\vec{\alpha}_{k,\lambda}(\vec{r},t) \tag{17}$$

In a first approximation the value of the constant ξ has been evaluated to be [6,7]

$$|\xi| \propto \left|\frac{1}{(2\pi)^{3/2}}\sqrt{\frac{\hbar}{8\alpha_{FS}\varepsilon_0 c^3}}\right| = \left|\frac{\hbar}{4\pi e c}\right| = 1.747\ 10^{-25}\ V\ m^{-1}s^2 \tag{18}$$

where α_{FS} =1/137 is the Fine Structure constant and e is the electron charge.

Now, at very low frequencies the wavelength tends to infinity and the vector potential to zero but the function $\vec{\Xi}_{k\lambda}(\omega_k,\vec{r},t)$ composing the vector potential does not vanish and tends to a unique expression for all modes

$$\vec{\Xi}^{(0)} = \xi\hat{\varepsilon}_\lambda e^{i\theta} + cc \tag{19}$$

Consequently, in absence of photons $\vec{\Xi}^{(0)}$ is a real field with amplitude ξ having electric units and permeating all space. Thus, it can be characterized as a component of the quantum vacuum. This means that the electromagnetic waves, that is photons, are oscillations of the vacuum field $\vec{\Xi}^{(0)}$.

Conclusion and discussion

We have seen here that the quantization of the vector potential amplitude enhanced at a single photon state (9) complements the fundamental relations of the photon (10) and leads to the coupled equation (17) for which the vector potential (11) with the quantized amplitude $\xi\omega$ behaves as a real wave function. The quantization constant ξ of the photon vector potential amplitude has electric essence and derives from the quantum vacuum. Consequently, the vacuum is not a sea of photons, which leads to the well-known QED singularity of infinite vacuum energy [8] (so called *quantum vacuum catastrophe*), but the electromagnetic waves (photons) are waves of the quantum vacuum sea which is composed of a real potential field $\vec{\Xi}^{(0)}$. The relation (17) indicates that a vibration of the vacuum field at an angular frequency ω gives rise to a photon with vector potential amplitude $\xi\omega$ and energy $\hbar\omega$.

References

[1] Garrison J C and Chiao R, *Quantum Optics,* Oxford University Press (2008).
[2] Milonni P W, *The quantum vacuum,* Academic Press Inc. (1994).
[3] Grynberg G, Aspect A and Fabre C, *Quantum Optics,* Cambridge University Press (2010).
[4] Meis C, *Light and Vacuum*, World Scientific (2015).
[5] Meis C, *Found. Phys.*,**27**, (1997) 865.
[6] Meis C, *Phys. Essays*, **12**, 1 (1999).
[7] Meis C, *Phys. Res. Int.* Vol.2014, ID 187432 (2014).
[8] Weinberg S, *Rev. Mod. Phys.* **61**, (1989) 1.

Computer simulation of the mechanical properties of metamaterials

R A Gerasimov[1], V A Eremeyev[1], T O Petrova[1], V I Egorov[2], O G Maksimova[2] and A V Maksimov[2]

[1] Southern Federal University, 105/42 Bolshaya Sadovaya Str., Rostov-on-Don, 344006, Russian Federation

[2] Cherepovets State University, 5 Lunacharskii Av., Cherepovets, 162600, Russian Federation

E-mail: roman-gerasimoff@yandex.ru

Abstract. For a hybrid discrete-continual model describing a system which consists of a substrate and polymer coating, we provide computer simulation of its mechanical properties for various levels of deformations. For the substrate, we apply the elastic model with the Hooke law while for the polymeric coating, we use a discrete model. Here we use the Stockmayer potential which is a Lennard-Jones potential with additional term which describes the dipole interactions between neighbour segments of polymer chains, that is Keesom energy. Using Monte-Carlo method with Metropolis algorithm for a given temperature the equilibrium state is determined. We obtain dependencies of the energy, force, bending moment and Young's modulus for various levels of deformations and for different values of temperature. We show that for the increase of the deformations level the influence of surface coating on the considered material parameters is less pronounced. We provide comparison of obtained results with experimental data on deformations of crystalline polymers (gutta-percha, etc.)

1. Introduction

Nowadays the interest grows to design and control of surface properties of materials. In particular, surface properties related with wetting and dewetting are important in order to obtain super-hydrophobic or super-hydrophilic properties [1-6]. Among physical and chemical methods of surface enhancements, it is worse to mention the plasma treatment, electroplating, lithography, etching, polymers grafting [7-9]. Especially, the methods of grafting of polymeric chains are attractive. These methods give the possibility to change the effective surface properties of a substrate such as adhesion, wetting, tribological properties, etc, without changing of the bulk properties of substrate. In other words, the substrate surface may inherit the properties of attached polymer [1, 7]. Wide choice of existing polymers possesses the design of the surface properties and manufacturing of new hybrid materials. In addition, these methods may produce rough and porous surface to obtain self-cleaning coatings such as lotus leaves. In this case the wetting of the surface is determined by both chemical properties of polymer and microstructure of the coating [1].

Content from this work may be used under the terms of the Creative Commons Attribution 3.0 licence. Any further distribution of this work must maintain attribution to the author(s) and the title of the work, journal citation and DOI.
Published under licence by IOP Publishing Ltd

Particular attention is recently paid to polymeric brushes as a way of producing of smart surfaces [10]. A polymer brush is a monolayer of polymer chains connected with an impermeable surface (substrate) by terminal ends (figure 1).
As a substrate the aluminium, its alloys and oxides are widely used [1]. Modifications of the aluminium based substrates are performed in order to decrease contact friction, adhesion, and to improve corrosion resistance. Enhancements of textile fibres also may lead to new materials. Formed on a substrate surface nano-rough polymer film results in hydro- and oleophobic properties of fibres that may be used for manufacturing of protective and bactericidal clothes [11]. The aim of this paper is to simulation of such kind surface metamaterials.

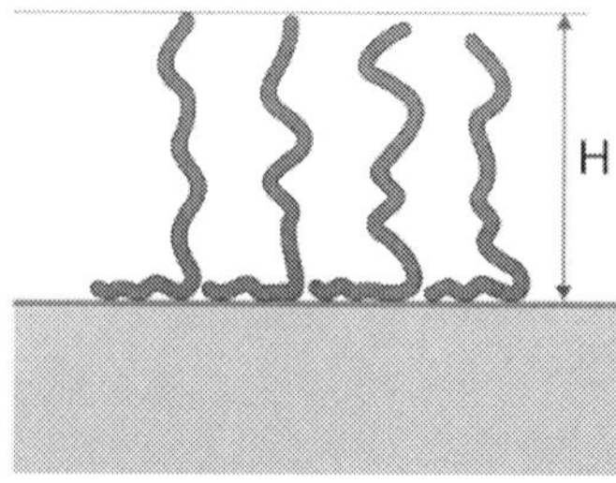

Figure 1. Plane polymer brush with chain height H.

2. Model

In figure 2 the hybrid discrete-continual system is shown. The system consists of plane rotators of length l attached to a substrate which is considered within the continuum model. The presented system is subjected by external tensile force F applied to both substrate and coating.

Within the framework of multiparticle models of polymers [12-17] $N_1N_2N_3$ rigid kinematical units, so-called rotators, of length l forms quasi-lattice $\vec{n} = (n_1, n_2, n_3)$, where indices n_1, n_2, n_3 take the following values: $n_1 = 1, \ldots, N_1$; $n_2 = 1, \ldots, N_2$; $n_3 = 1, \ldots, N_3$.

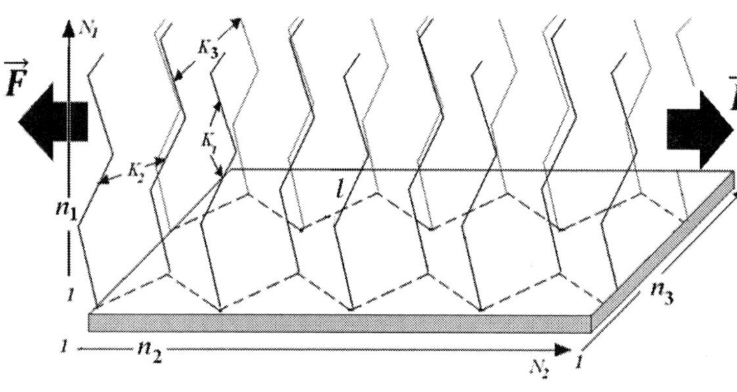

(a)

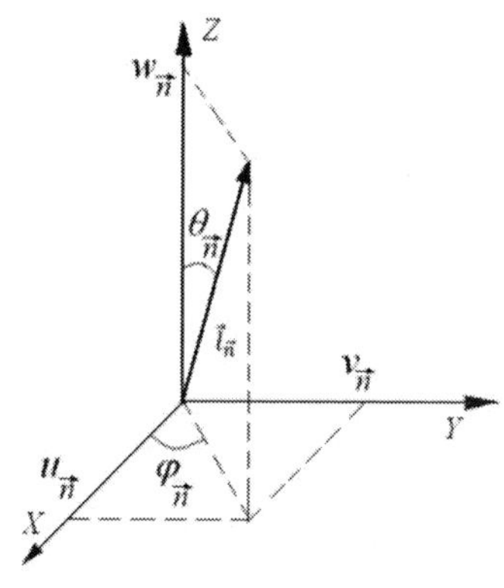

(b)

Figure 2. (*a*) Discrete-continuum model: model of 3D ordered N_2N_3 chains consisting of N_1 rigid kinematic elements, so-called rotators of length l connected with the substrate (here N_2 and N_3 are the numbers of chains along two orthogonal directions of cross-section of the system). The system is stretched by applied force $\vec{F}$.

(*b*) Vector $\vec{l}_{\vec{n}}$ of orientation of a chain segment located in the node $\vec{n}$ with Cartesian (x, y, z) and spherical (φ, θ) coordinates.

We assume that the energy of orientational interactions $H^{(i)}_{\vec{n},\vec{m}}$ between segments located in nodes $\vec{n}$ and $\vec{m}$ of quasi-lattice depends only on the angle of their relative orientation (Keesom's energy). For the potential of dipole type $H^{(i)}_{\vec{n},\vec{m}} = -K_i \cos\Phi_{\vec{n},\vec{m}}$ the energy is proportional to cosine of spatial angle $\Phi_{\vec{n},\vec{m}}$ between their axis as in Gotlib-Maksimov models for polymer systems [12]

$$H^{(i)}_{\vec{n},\vec{m}} = -K_i \frac{\left(\vec{l}_{\vec{n}}, \vec{l}_{\vec{m}}\right)}{l^2} = -K_i \frac{\left(u_{\vec{n}}u_{\vec{m}} + v_{\vec{n}}v_{\vec{m}} + w_{\vec{n}}w_{\vec{m}}\right)}{l^2}, \qquad (1)$$

where $u_{\vec{n}}$, $v_{\vec{n}}$ and $w_{\vec{n}}$ are projections on the coordinate axis of vector $\vec{l}_{\vec{n}}$ oriented along the kinematic unit located in the node $\vec{n}$ ($i = 1, 2, 3$).

Energetic-type parameter K_1 along "longitudinal" curvilinear direction n_1 of quasi-lattice describes the longitudinal orientational interactions and determines mean cosine of the angle between neighbour kinematic units [12-14]. Energetic-type parameters K_2 and K_3 characterize orientational interactions of neighbour kinematic units in transversal interactions n_2 and n_3 of quasi-lattice [13, 14]. Characteristic values for rigid as well as for flexible polymeric chains have been estimated in ref. [18]. During the deformations, parameter K_2 was calculated taking into account the dependence of chains interactions energy on relative displacements of chain segments [18]:

$$K_2 = K_3 \left(\frac{r}{r_0}\right)^{-3}. \qquad (2)$$

In equation (2) the parameter r_0 is the mean distance between neighbour segments of polymer in direction of force action that is along n_2 assuming equilibrium of configuration, $r = r_0 + \Delta r$, where Δr characterizes the change of segments' position during deformations. Energetic-type parameter K_3 keeps constant value during deformations.

Potential energy of the whole system taking into account elastic deformations of the substrate $H_{subs.}$ and interactions between nearest segments is described by the Stockmayer potential that is Lennard-Jones potential plus energy of dipole interactions is given by

$$H = \sum_{i=1}^{3} \sum_{\vec{n},\vec{m}} H^{(i)}_{\vec{n},\vec{m}} + 4\varepsilon\left[\left(\frac{\sigma}{r}\right)^{12} - \left(\frac{\sigma}{r}\right)^{6}\right] + H_{subs.}. \qquad (3)$$

In equation (3) ε is the depth of the Lennard-Jones potential pit, r is the distance between nearest segments of the polymer chains, σ is the distance where the energy of interactions is zero.

3. Simulation algoritm

Within the framework of Monte-Carlo method we generate random process consisting of sequence of the system configurations. Using ensembles with the large enough number of configurations, one can calculate mean values of almost all equilibrium physical quantities.

Input parameters for the simulation of the system shown in figure 2 are chosen the following parameters: dimensions N_1, N_2 and N_3 of quasi-lattice (number of segments in the directions n_1, n_2 and n_3), the interaction parameters K_1, K_2 and K_3, and the number of Monte-Carlo steps. For calculations, all parameters having dimension of energy were normalized with respect to the parameter $K_1 \sim 10^{-20}$ J [18], while all quantities determining the system dimensions were normalized using mean interatomic distance in the substrate $a \sim 10^{-10}$ m. The normalized temperature is defined as $T^* = k_B T/K_1 = 0.1$, where k_B is the Boltzmann constant, that corresponds to the value $T \sim 10^2$ K.

For the simulation, we use the following method:

- The initial configuration of the system is determined by the following parameters: position of segments attached to the substrate are given as well as at the system boundary in the direction n_2 are prescribed, at the upper boundary that is in the direction n_1 the segments are free. In the direction n_3 we assumed the periodic boundary conditions. So, in the n_3-direction the system is infinite. For a more rapid achievement of equilibrium state of the system at low temperatures, we assume parallel distributions of the chain segments, while for high temperatures the distribution is chaotic.

- The cycles of the Monte-Carlo steps with the Metropolis algorithm are given. We assume the quasistatic loading that is the system tends to equilibrium state at each step. Therefore, the calculated displacement of each segment under action of the external force F is independent on its place in the quasi-lattice.
- Taking into account random re-orientations of segments (rotators) at each step of deformation, we determine the equilibrium state and calculate the energy of polymer part of the system for given temperature. For the simplicity, the energy of elastic strains of the substrate is assumed to be $H_{subs.} = k(\Delta x)^2/2$, where k is the effective stiffness parameter.
- For small deformations, the applied force F is proportional to the displacement of chain segments from their initial positions, in other words $F \sim \Delta x$. The work of external loads during deformations transforms to the energy change of the system: $A = -\Delta H$, where

$$A = -\frac{F\Delta x}{2} = -\frac{E}{2}\cdot\left(\frac{\Delta x}{x_0}\right)^2 . \qquad (4)$$

In equation (4) E is Young modulus, x_0 is the initial length of the system along the action of force F. As a result, using (4) it is possible to calculate F and Young's modulus E of the system in dependence on the deformation Δx.

4. Results and discussion

The calculated dependencies of force and Young's modulus on stretching of the system are presented in figures 3 and 4. It is clear that with the increase of deformations the surface coating gradually ceases to have an effect on the investigated parameters. With the decrease of the value of chain interactions K_3 along the direction, that is not subjected to extension the influence of the coating disappears more quickly, see figures 3 and 4.

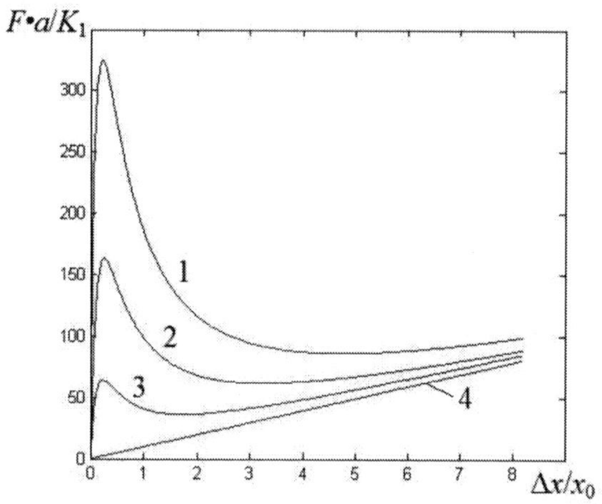

Figure 3. Applied normalized force $F{\cdot}a/K_1$ vs. relative strain $\Delta x/x_0$ for given values of parameters $K_1 = 1$ and: $K_3 = 0.1$ (curve *1*), 0.05 (curve *2*), 0.01 (curve *3*); force $F{\cdot}a/K_1$ vs. strain $\Delta x/x_0$ with the Hooke law (curve *4*).

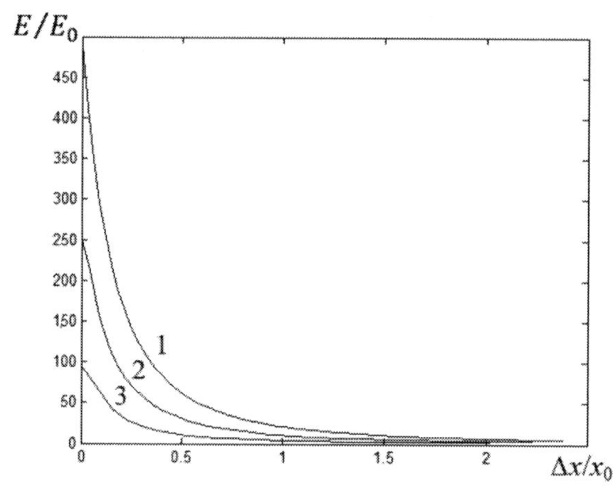

Figure 4. Normalized Young's modulus E/E_0 (E_0 is Young's modulus of the substrate) vs. relative strains $\Delta x/x_0$ for given values of the parameters $K_1 = 1$ and: $K_3 = 0.1$ (curve *1*), 0.05 (curve *2*), 0.01 (curve *3*).

Areas after the peaks in figure 3 can be explained by the break of bonds between adjacent polymer chains of the coating and increasing role during deformation of the substrate system. Similar curves "strain-stress" can be observed during the stretching of thin films of crystalline polymers such as gutta-percha and caoutchouc [19], where the behaviour is also related with the structural changes from ordered structure to amorphous one. The following coincidence of curves with lines is related with a dominant role of the substrate that is with Hooke's law and negligible influence of the coating. Effective Young's modulus of such systems decreases and tends to Young's modulus of the substrate (figure 4).

Changing the length of attached to substrate chains and their density it is possible to analyse the dependencies of elastic parameters on size of the system in one direction or on the number of the attached polymer chains per unit area of the surface. In the forthcoming works, we intend to analyse the system not only for tension and compression but also for the bending, that may be important for flexible substrates.

Acknowledgements

The work is performed within the framework of the project "Methods of microstructural nonlinear analysis, wave dynamics and mechanics of composites for research and design of modern metamaterials and elements of structures made on its base" (grant № 15-19-10008 of by the Russian Science Foundation).

References

[1] Boinovich L B, Emelyanenko A M 2008 *Russian Chemical Reviews* **77(7)** 583
[2] Vinogradova O I, Belyaev A V 2011 *J. Phys.: Condens. Matter* **23** 184104
[3] Zheng L, Wu X, Lou Z et al. 2004 *Chin. Sci. Bull* **49** 1779
[4] Bico J, Thiele U, Quere D 2002 *Colloid Surf. A: Physicochem. Eng. Asp.* **206(1-3)** 41
[5] Callies M, Quere D 2005 *Soft Matter* **1(1)**, 55
[6] Bhushan B, Jung Y 2011 *Progress in Materials Science* **56** 1
[7] Shirtcliffe N, McHale G, Atherton Sh et al. 2010 *Advances in Colloid and Interface Science* **161(1–2)** 124
[8] Zhang X, Shi F, Niu J et al. 2008 *J. Mater. Chem.* **18** 621
[9] Lisichkin G V, Fadeev A Yu 2003 *Chemistry of the grafted surface compounds* (Moscow: FIZMATLIT)
[10] Rigoberto C A et al. 2004 *Polymer Brushes: Synthesis, Characterization, Applications* (Weinheim: WILEY-VCH Verlag GmbH & Co. KGaA)
[11] Ageev A A, Aksenova I V, Volkov V A, Eleev A F 2012 *Fluorine notes* **4(83)**
[12] Gotlib Yu Ya, Maximov A V 1992 *Polymer Science* A **34(10)** 902
[13] Maksimov A V, Pavlov G M 2007 *Polymer Science* A **49(7)** 828
[14] Maksimov A V, Gerasimov R A 2009 *Physics of the Solid State* **51(7)** 1365
[15] Gerasimov R A, Maksimov A V, Petrova T O and Maksimova O G 2012 *Physics of the Solid State* **54(5)** 1002
[16] Petrova T O, Maksimova O G, Gerasimov R A and Maksimov A V 2012 *Physics of the Solid State* **54(5)** 937
[17] Maksimov A V, Gerasimov R A, Maksimova O G 2012 *Ferroelectrics* **432(1)** 32
[18] Maksimova O G, Maksimov A V 2003 *Polymer Science* A **45(9)** 859
[19] Askadskii A A 1973 *Deformation of polymers* (Moscow: Khimiya)

Journal of Physics: Conference Series **738** (2016) 012101 doi:10.1088/1742-6596/738/1/012101

Computer simulation of the process of absorption of electromagnetic field by liquid crystal systems

T O Petrova[1], R A Gerasimov[1], O G Maksimova[2] and A V Maksimov[2]

[1]Institute of Mathematics, Mechanics and Computer Sciences, Southern Federal University, Bolshaya Sadovaya 105/42, 344006, Rostov-on-Don, Russian Federation
[2]Institute of Informational Technologies, Cherepovets State University, Prospect Lunacharskii 5, 162600, Cherepovets, Russian Federation

E-mail: to_87@bk.ru

Abstract. The behavior of LC systems in alternating external fields is investigated; hysteresis curves are calculated. The dependencies of the area of the hysteresis curves on the frequency of the external field, at various values of temperature, amplitude of external field and of intermolecular interactions constants are discussed. The position of maximum and the peak height depending on the temperature and the of interactions constants are investigated.

1. Inrtoduction

Liquid crystal (LC) materials are of high potential for use in modern chemical and medical industry, instrument-making and machine-building. Therefore, currently they synthesize new low molecular LC and intensively research their properties by various theoretical and experimental methods. Besides, polymer systems are synthesized, which successfully combine unique properties of both low-molecular liquid crystals, and high molecular compounds with their ability to form films, fibers and coatings. Such polymers include a wide range of homologues of poly-n-acrylates, poly-n-acryl methacrylates, simple and complex poly-n-acryl-vinyl esters, poly-n-alkyl-styrenes, etc.[1-5].

The main features of physical characteristics of low molecular and polymer LC are associated primarily with geometrical anisotropy of structural units (molecules, fragments of polymeric chains, etc.) of these systems, as well as with the values of intra- and intermolecular orientational interactions. Therefore taking into account similar molecular structure of such systems, the processes occurring in them can be investigated via similar models and methods.

Under the influence of electric fields physico-chemical properties of LC systems change [6], such as color, transparency and the ability to rotate the plane of polarization of the transmitted light. This, in turn, determines the wide range of applications of LC materials in optoelectronics and optical information processing systems, laser physics, medicine, thermography, etc. [7, 8].

The aim of this research is the theoretical study of the absorption of external alternating electric field by ordered low molecular weight and polymer LC systems within a unified model via the Monte Carlo method. For this, we calculated the dependencies of the areas of hysteresis curves on the frequency of the external field at different values of temperature and constants of the model.

Content from this work may be used under the terms of the Creative Commons Attribution 3.0 licence. Any further distribution of this work must maintain attribution to the author(s) and the title of the work, journal citation and DOI.
Published under licence by IOP Publishing Ltd

2. Model

To describe the LC the lattice model is used, a single one for low molecular weight (Figure 1a) and polymer (Figure 1b) systems. Each lattice node $\vec{n}=(n_1,n_2,n_3)$ has anisotropic elongated particle - rotator having rotational (orientational) degrees of freedom. Orientation of the rotator in a spherical coordinate system is defined by the angles θ_{n_1,n_2,n_3} and ϕ_{n_1,n_2,n_3}. In this model, it is assumed that the potential of orientational interactions $U_{\vec{n},\vec{m}}$ between rotators located in the lattice nodes $\vec{n}$ and $\vec{m}$ depends only on their mutual orientation, i.e. the spatial angle $\Phi_{\vec{n},\vec{m}}$ between their axes.

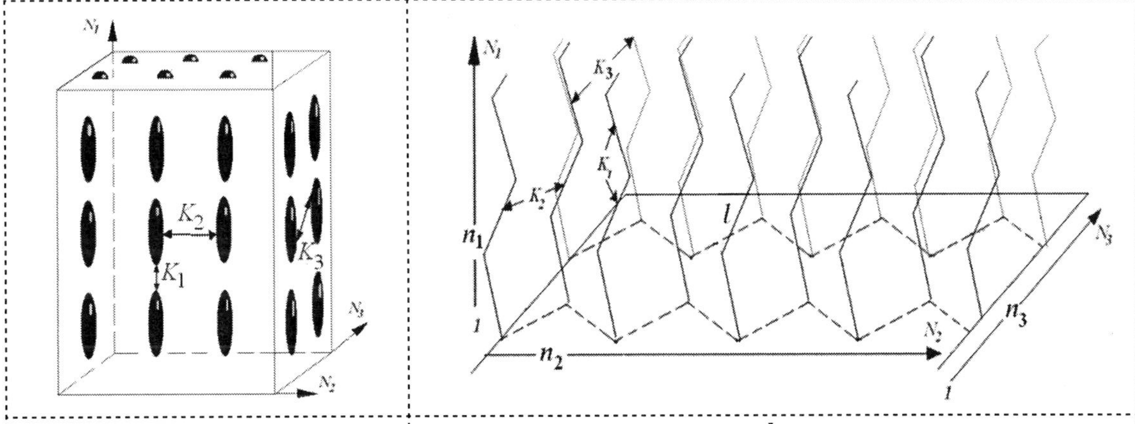

a b

Figure 1. Spatial models of the low molecular weight (a) and multichain polymer (b) LC systems with orientational interactions.

Quadrupole interaction energy of two nonpolar particles not possessing the electric dipole moment decreases with the increase of distance R proportionality $1/R^5$, while the energy of interaction of dipole moments, set due to their mutual polarizability, changes depending on the distance in proportion $1/R^8$. Therefore, the quadrupole interaction is dominant. To describe the orientational interactions between non-polar molecules capable of forming typical liquid crystal structures they usually apply quadrupole potential of interaction [9].

Therefore, in this study it is assumed, that potential energy of interaction of rotators with external electric field has the form:

$$U=-\sum_{n_1,n_2,n_3} K_1\cos^2(\Phi_{n_1,n_2,n_3,n_1+1,n_2,n_3})-\sum_{n_1,n_2,n_3} K_2\cos^2(\Phi_{n_1,n_2,n_3,n_1,n_2+1,n_3})- \\ -\sum_{i,j,k} K_2\cos^2(\Phi_{n_1,n_2,n_3,n_1,n_2,n_3+1})-fqE\sum_{n_1,n_2,n_3}\cos^2(\varphi_{n_1,n_2,n_3}) \tag{1}$$

where $\Phi_{n_1,n_2,n_3,n_1',n_2',n_3'}$ is the angle between the rotators located in the nodes n_1, n_2, n_3 and n'_1, n'_2, n'_3 of the lattice, q is the module of the quadrupole moment of the rotator, E is intensity of the external electric field, φ_{n_1,n_2,n_3} is the angle between the direction of orientation of the rotator and the vector of field intensity, f is coefficient of proportionality.

Energy constants characterize different types of interactions, depending on the considered chemical compound. For low molecular weight systems the constants K_i can be estimated by the multipole expansion of the interaction energy of two identical quadrupoles. For polymers energy constant K_1 along the longitudinal curved direction n_1 of the lattice describes intrachain orientational interactions and is associated with the chain bending ridity, which determines the elastic properties of the indidual chain. Its value can be estimated from the relation $K_1 = A \cdot k_B T / 4$, where A is Kuhn segment length (in numbers of rotators), T is temperature. The constants K_2 and K_3 characterize the orientational interactions between adjacent rotators (links) in different chains, in "transverse" directions of the lattice.

As a long-range quadrupole order parameter the Tsvetkov's factor is used:

$$S = \frac{3\langle \cos^2 \theta \rangle}{2} - \frac{1}{2} . \tag{2}$$

where θ is the angle of orientation of the rotator relative to the preferred direction.

The absorption of the energy of electromagnetic field in the liquid crystal system occurs due to periodic change of the molecular orientational order. Molecular mechanisms of formation of quadrupole order in the nematic phase in electric fields are associated with the turn of the rod-shaped molecules around the long and short axes [1]. The forces of intermolecular interactions are responsible for this type of long-range orientational order prevent such turns.

3. Results and discussion

In this work we calculated the dependencies of the order parameter (the Tsvetkov's factor S) in LC systems on the value of external harmonic electric field $E = E_0 \cdot \sin \omega t$ at different values of internal and external parameters. When simulating by the Monte Carlo method, the interaction potential was selected in the form (1) and dynamic interpretation of this method is given to averaging in the Markov process, i.e., the number k of successive configurations of the system is associated with the definite timescale, which allows to explore various dynamic characteristics of the system. Therefore, the frequency was selected inversely proportional to the number of Monte Carlo steps N_{MK}. We analyzed the behavior of LC systems in external alternating electric fields. As the standard we adopted the frequency value corresponding to the number of Monte Carlo steps $N_{MK} = 10^3$ when simulating 8,000 of rotators under periodic boundary conditions.

The calculated hysteresis curves for LC systems are asymmetric (Figure 2-3), their form is determined by the temperature, the value of the interaction constants, amplitude and frequency of the external field. The order parameter varies near its equilibrium value, which increases with the rise of amplitude of the external field.

The main characteristic of the hysteresis curve is its area, which is proportional to the energy expended on the periodic change of orientation order in the system. It is seen that with the increase of the amplitude of the external field E_0 at low and high temperatures the area of the hysteresis curve increases. It was found that the form of the hysteresis curve and its area is dependent on interaction constants of the model and temperature. Figure 4 shows the dependencies of the area of hysteresis curves (Ω) on the logarithm of the reduced frequency of the external field ($fqE/3K_1$) at various temperatures.

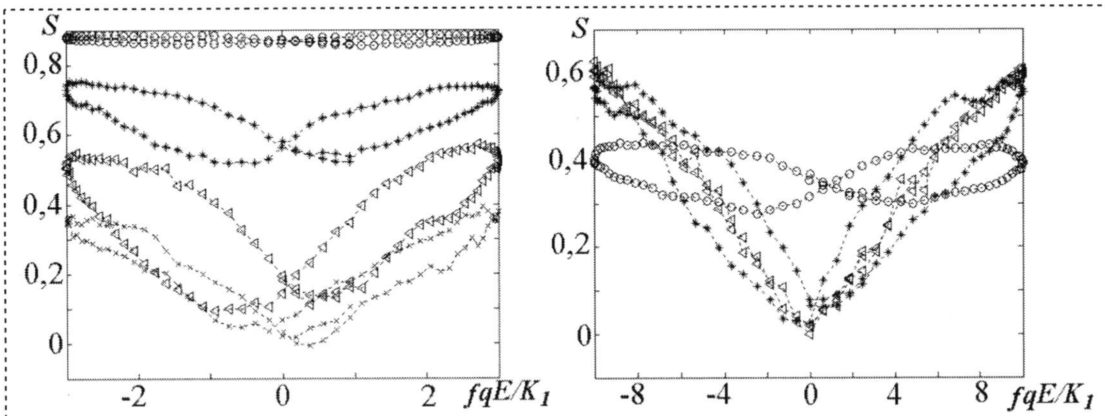

a *b*

Figure 2. Calculated hysteresis curves for LC systems vs. the reduced temperature $g = {k_B T}/{3K_1}$ and the frequency of the external field: ***a*** - $\omega = \omega_r$, $g = 0.5$(o); 1(*); 1,5(Δ); 2(x); ***b*** - $g = 0.5$, $\omega = 0.1\omega_r$(Δ); ω_r(*); $10\omega_r$(o).

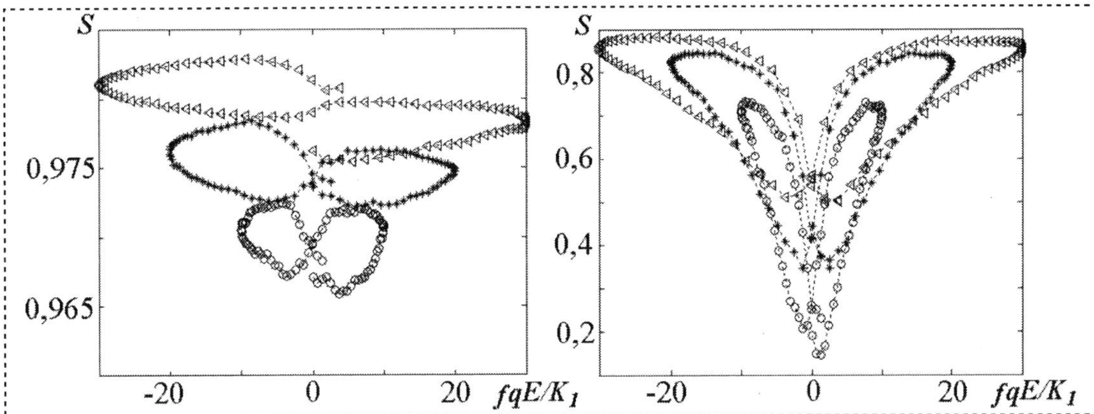

a *b*

Figure 3. Calculated hysteresis curves for LC systems vs. the reduced temperature $g = {k_B T}/{3K_1}$ and the amplitude of the external field E_0: ***a*** - $g = 0.2$, ${fqE_0}/{K_1} = 10$(o); 20(*); 30(Δ); ***b*** - $g = 2$, ${fqE_0}/{K_1} = 10$(o); 20(*); 30(Δ).

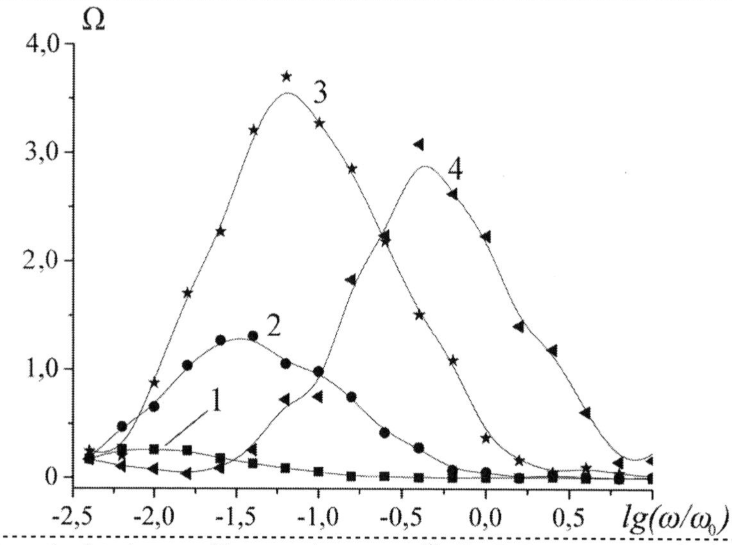

Figure 4. The calculated area of hysteresis curves Ω for LC systems vs. the logarithm of the reduced frequency ω/ω_0 of the external field at different values of the reduced temperature $g = k_B T / 3K_1$ = 0.2 (1); 0.8 (2); 1.5 (3); 9 (4).

4. Conclusion

At a certain frequency of the external field the phenomenon of resonance occurs and the maximum of hysteresis curve area is observed. The peak position and its value depend on the temperature and interaction constants and is defined by the relation $g = \frac{3k_B T}{K_1 + K_2 + K_3}$. With the increase of the value g the peak of the area of hysteresis curve is shifted towards higher frequencies. Moreover, with the increase of the reduced temperature the maximum value of the function $\Omega\left\{\lg\left(\omega/\omega_0\right)\right\}$ first increases, then decreases.

Acknowledgments

The reported study was funded by Russian Science Foundation; according to the research project No.15-19-10008.

References

[1] Formirovsky K A, Bragina N A, Mironov A F, Anan'Eva G A, Bykova V V, Usol'Tseva N V 2012 *Mendeleev Communications.* **22 (5)** 278

[2] Usol'Tseva N , Bykova V, Kudrik E et al. 2001 *Mol. Cryst. Lyq. Cryst.* **367** 509

[3] Shibaev V P 2014 *Polymer Science. Series A.* **56 (6)** P. 727-762

[4] Smirnova N N, Samosudova Y S, Markin A V, Shibaev V P, Boiko N I 2015 *Thermochimica Acta.* 2015 **614** 226

[5] Bobrovsky A, Boiko N, Shibaev V 2015 *Polymer.* **56** 263

[6] de Gennes P G 1974 *Clarendon. Oxford*

[7] Kreuzer F H et al. 1996 *Springer.* 218

[8] Witkiewicz Z, Oszczudlowski J, Repelewicz M 2005 *J.of Chromatography.* **1062 (2)** 155

[9] Birstein T M, Mercurieva A A, Prjamitsyn V A 1996 *Macromol.Theory Simul.* **5 (3)** 215

Mathematical modeling of a non-Newtonian fluid flow in the main fracture inside permeable porous media

A M Ilyasov, G T Bulgakova[1]

Ufa State Aviation Technical University, Ufa, Russian Federation

[1]E-mail: math@mail.rb.ru

Abstract. This paper describes a mathematical model of the main fracture isolation in porous media by water-based mature gels. While modeling injection, water infiltration from the gel pack through fracture walls is taking into account, due to which the polymer concentration changes and the residual water resistance factor changes as a consequence. The salutation predicts velocity and pressure fields of the non-Newtonian incompressible fluid filtration for conditions of a non-deformable formation as well as a gel front trajectory in the fracture.
The mathematical model of agent injection into the main fracture is based on the fundamental laws of continuum mechanics conservation describing the flow of non-Newtonian and Newtonian fluids separated by an interface plane in a flat channel with permeable walls. The mathematical model is based on a one-dimensional isothermal approximation, with dynamic parameters pressure and velocity, averaged over the fracture section.

Keywords: mathematical modeling, non-Newtonian fluid, the main fracture, porous media, the SIMPLE algorithm

1. Introduction

In current situation, one of the major oilfield problems is the water inflow restriction into the wells draining naturally fractured reservoirs and the isolation of a high-conductivity single fracture connecting production and injection wells or aquifer. The instant water-cut increase in the produced fluid is caused by formation and injected water breakthrough via high-permeability layers and fractures. Under such conditions, the objective to shut-off water inflow to production wells becomes extremely important [1-9]. The problem is particularly acute in the case of complex fractured oil reservoirs. For isolation of fractures acting as the main channels of water flow and breakthrough to production wells, a crosslinked (mature) gel is applied. Numerous experimental studies are carried out to investigate the mechanism of gel propagation through the fracture and determine the rheological and filtration properties of crosslinked polymer compositions.

The studies [3-6] shows that in the process of gel propagation through the fracture the gel dehydration, i.e., water removal, takes place. Water, separated from gel, infiltrates through the fracture walls into a porous matrix. Under the severe flow exposure conditions, gel concentration as well as gel stability increases. The gels, used for fracture water-shutoff, are crosslinked aqueous polymer solutions of low concentrations. According to rheological studies [7], a crosslinked gel is a non-Newtonian fluid which apparent viscosity decreases with a shear rate increase. The rheological characteristics of its composition are close to the pseudoplastic fluid model. Study [8] experimentally proves that the gel injection into a fracture forms two flows - gel flow and formation fluid flow-separated by a moving interface. The gel spreads in the fracture as a piston and the gravity force does not affect the shape of its front. Placement of immature and mature gels and their ability to block

Content from this work may be used under the terms of the Creative Commons Attribution 3.0 licence. Any further distribution of this work must maintain attribution to the author(s) and the title of the work, journal citation and DOI.
Published under licence by IOP Publishing Ltd

fractures during subsequent waterfloods were investigated in paper [9]. The immature gel and fully formed (mature) polymer gel show different behavior during placement in a fractured system, and the gels deposit differently in the fracture volume. Injection of different maturity gels at into a fracture may therefore influence the ability of gel treatment to block fractures, and hence its performance during conformance-control operations. At the same time, the problem of numerical modeling of non-Newtonian agent injection into the main fracture is very important for predicting the distribution of final injected volumes and creating stable gel barriers. This paper describes a mathematical model of the main fracture water-shutoff in the porous reservoir by the crosslinked gel. The mathematical model is based on the fundamental laws of continuum mechanics conservation describing the flow of non-Newtonian and Newtonian fluids separated by an interface in a flat channel with permeable walls. This study was targeted to optimize the design of gel placement in the fracture and to determine subsequently optimal technological parameters of the process and the size of a gel screens ensuring their stability under the excessive exposure to flow from Water-Shut-Off operations.

2 Mathematical model

Let us consider a rectangular zone in the oil reservoir of permeability k and porosity m, where a vertical main fracture of length L, width w_f and height h_f, connecting the injection and production wells, is symmetrically parallel to the reservoir boundaries. The relation between the width, height and length of the fracture can be described as $w_f \ll h_f \ll L$. The coordinate system is selected in such way that its origin coincides with the outer radius of the injection well and axis x goes along the fracture. In the main fracture, initially filled with formation water, gel injection forms two flow regions separated by moving interface $x_f(t)$. In the first region $0 \le x \le x_f(t)$, gel density is ρ_g, gel effective viscosity is μ_g. In the second region $x_f(t) \le x \le L$, formation water of constant density ρ_w and constant viscosity μ_w is displaced by gel in a piston-like manner. In the first region water infiltration from gel into the reservoir takes place, and in the second region - formation water losses or inflow, depending on the sign of pressure gradient in the direction perpendicular to the fracture walls.

The mathematical model of gel injection into a vertical fracture with permeable walls was built on the following assumptions: no mixing occurs on the formation water and gel interface; water stream lines from fracture to reservoir are straight lines in the reservoir region under review; as per rheology, the crosslinked gel is a non-Newtonian pseudoplastic liquid. The fluid flow inside the reservoir is isothermal and follows the Darcy's law; fluids in the fracture and reservoir are incompressible; and the reservoir rock matrix is rigid. The mathematical model is calculated under the hydraulic approximation with hydrodynamic parameters - pressure and velocity - averaged over the fracture cross section. It is assumed, that the main fracture is flushed out by previous flows, so its boundaries are affected only by fluid pressure and friction (the fracture walls are free of geomechanical stress).

2.1. Mass conservation equation.

The numerical modeling of gel injection into the main fracture with permeable walls requires to write down the basic laws of continuum mechanics conservation (CMC) in the hydraulic approximation describing a viscous incompressible fluid flow in the vertical fracture with permeable walls for the averaged fields:

$$\langle v\rangle = \frac{1}{S}\int_S v\,ds, \quad \langle v^2\rangle = \frac{1}{S}\int_S v^2\,ds = (1+\beta)\langle v\rangle^2, \quad \langle p\rangle = \frac{1}{S}\int_S p\,ds, \qquad (1)$$

where β is the Coriolis correction [10].

Let us review a specific fluid volume V in the fracture, which is bounded by a piecewise-smooth surface $\Sigma = S_0 + S_1 + \Sigma_l + \Sigma_r + \Sigma_{hd} + \Sigma_{hup}$, where: S_0, S_1 are inflow and outflow boundaries of the specific volume, respectively; Σ_l, Σ_r are left and right (flow-wise) porous boundaries of the specific volume; $\Sigma_{hd}, \Sigma_{hup}$ are lower and upper (reservoir base and top) impermeable boundaries of the specific volume. All boundaries in the reviewed volume are assumed as flat.

To derive the integral mass conservation law for the selected specific volume V of a rectangular parallelepiped shape we applies the known continuum mechanics theorem about the time-derivative of the tensor quantities integral over a mobile volume [11]:

$$\frac{d}{dt}\int_{V(t)}\rho dV = \int_{V(t)}\frac{\partial\rho}{\partial t}dV + \int_{S_1}\rho v_n d\Sigma + \int_{S_0}\rho v_n ds = J_m = -\int_{\Sigma_l}\rho u_l ds - \int_{\Sigma_r}\rho u_r ds, \tag{2}$$

where ρ is fluid density; v_n is the fluid particle velocity component, normal to the cross section; J_m is fluid mass flow into the reservoir through fracture sidewalls; u_r, u_l are the rate components of the outflow (inflow) from (into) the reservoir, normal to the left and right fracture walls. Equation (2) assumes flow absence at the reservoir top and base. For the one-dimensional flow of gel in the region $0 \le x \le x_f(t)$ and the symmetric infiltration $u_l = u_r = u_g$ equation (2) can be written as

$$\frac{\partial\rho_g}{\partial t} + \frac{\partial(\rho_g v)}{\partial x} = -\frac{2\rho_g u_g}{w_f}. \tag{3}$$

For water infiltration from gel, the rate is determined from the experimental data [3], approximation of which resulted in the power law of infiltration rate vs. time with a negative exponent. The infiltration rate in the International System of Units is defined as:

$$u_g = At^{-\alpha}, \quad A = 1{,}764\cdot 10^{-7}\ m\cdot c^{\alpha-1}, \quad \alpha = 0{,}55. \tag{4}$$

The next assumption is that the crosslinked gel is a binary water solution of polymer. Then, by definition, the "reduced" density of polymer ρ_p and water ρ_w in aqueous polymer solution is:

$$\rho_p = \frac{M_p(t)}{V(t)}, \qquad \rho_w = \frac{M_w(t)}{V(t)}, \qquad \rho_g = \rho_p + \rho_w, \tag{5}$$

where $M_p(t)$ and $M_w(t)$ are polymer and water masses in the gel material volume, respectively, and ρ_g is gel density. The experimental data [3] show that the crosslinked gel being injected behaves like a "sponge", from which into the reservoir only water infiltrates while polymer retains its mass in the specific volume of crosslinked gel: $M_p(t)$ = const.

Using the mass concentrations for the binary mixture components:

$$c_p = \frac{M_p}{M(t)} = \frac{\rho_p V(t)}{\rho V(t)} = \frac{\rho_p}{\rho_g}, \quad c_w = \frac{M_w}{M(t)} = \frac{\rho_w V(t)}{\rho V(t)} = \frac{\rho_w}{\rho_g}, \quad c_p + c_w = 1. \tag{6}$$

The reduced water and gel density can be described as:

$$\rho_w = (1-c)\rho_g(c), \quad \rho_p = c\rho_g(c), \quad c = c_p. \tag{7}$$

Using the reduced water and gel densities (7), the equation of continuity (3) for gel only in the flow region $0 \le x \le x_f(t)$ is reduced to an equivalent system for a binary mixture:

$$\frac{\partial \rho_g(c)}{\partial t} + \frac{\partial \rho_g(c) v}{\partial x} = -\frac{2(1-c)\rho_g(c) u_g}{w_f}, \tag{8}$$

$$\frac{\partial c\rho_g(c)}{\partial t} + \frac{\partial c\rho_g(c) v}{\partial x} = 0, \tag{9}$$

where equation (8) is the continuity equation for a binary mixture (gel), and equation (9) is the continuity equation for the polymer component of gel.

The application of continuity equation (2) to the specific volume, occupied by formation water in the flow region $x_f(t) \le x \le L$, leads to the equation:

$$\frac{\partial v}{\partial x} = -\frac{2u_w}{w_f}, \tag{10}$$

where water incompressibility in the fracture is $\rho_w =$ const, u_w is the rate of formation water infiltration through the flat channel walls.

The formation water infiltration rate through the fractures face is determined from the Darcy's law:

$$u_w(x,t) = \frac{k}{\mu_w L_k}(p(x,t) - p_k), \tag{11}$$

where k is reservoir permeability; μ_w is formation water viscosity; p_k is constant pressure at the external reservoir boundary at distance L_k from the fracture walls.

2.2. Equation of a fluid momentum conservation in a flat channel

Given that the momentum change in the specific liquid volume equal to the sum of all external surface forces and mass forces [10], we have

$$\int_{V(t)} \frac{\partial(\rho\vec{v})}{\partial t} dV + (1+\beta)\left[\int_{S_0} \rho\vec{v}(\vec{v}\cdot\vec{n})d\Sigma + \int_{S_1} \rho\vec{v}(\vec{v}\cdot\vec{n})d\Sigma\right] = \int_{S_0} \vec{p}_n d\Sigma + \int_{S_1} \vec{p}_n d\Sigma + \int_{\Sigma_l} \vec{p}_n d\Sigma + \int_{\Sigma_r} \vec{p}_n d\Sigma, \tag{12}$$

where $\vec{p}_n -$ tension vector on the surface Σ at the point with normal $\vec{n}$; $\vec{g}$ – gravitational acceleration.

The momentum equation (12) for the unsteady gel one-dimensional flow in a flat channel $0 \le x \le x_f(t)$ with permeable walls is reduced to a form:

$$\frac{\partial(\rho_g v)}{\partial t} + (1+\beta_g)\frac{\partial(\rho_g v^2)}{\partial x} = -\frac{\partial p}{\partial x} - \frac{(1-m)C_g \rho_g v^2}{w_f}. \tag{13}$$

Similarly, given the formation water incompressibility in the formation water flow region $x_f(t) \le x \le L$, the equation of motion is:

$$\frac{\partial v}{\partial t}+(1+\beta_w)\frac{\partial v^2}{\partial x}=-\frac{1}{\rho_w}\frac{\partial p}{\partial x}-\frac{(1-m)C_w v^2}{w_f}, \tag{14}$$

In equations (13) and (14), β_g, β_w are the Coriolis corrections for gel and water, C_g, C_w are hydraulic friction coefficients of gel and water, respectively. The viscous stress on the walls of the flat channel is determined by hydraulic friction coefficients.

2.3. Key model equations

The experimental data show [3] that if the initial polymer concentrations are $c = 0.001 \div 0.005$ then resulting from water infiltration from crosslinked polymer the gel concentration in aqueous solution can increase by an order of magnitude and achieve $c = 0.01 \div 0.05$. The aqueous polymer solution density varies in the range of $1-2\%$. If the initial solution concentration is $\rho_g = 1000$ kg/m^3, the solution density upon water infiltration changes by $\Delta\rho = 10-20$ kg/m^3. Therefore, it can be estimated that

$$\Delta\rho / \rho_g << 1, \tag{15}$$

allowing to apply an incompressible fluid approximation for describing gel injection. Thus, according to equation (15) for the crosslinked gel flow region $0 \le x \le x_f(t)$ in the fracture, equations (8), (9), (13) are reduced to the following form:

$$\frac{\partial v}{\partial x}=-\frac{2(1-c)At^{-\alpha}}{w_f}, \tag{16}$$

$$\frac{\partial c}{\partial t}+\frac{\partial (cv)}{\partial x}=0, \tag{17}$$

$$\frac{\partial v}{\partial t}+(1+\beta_g)\frac{\partial v^2}{\partial x}=-\frac{1}{\rho_g}\frac{\partial p}{\partial x}-\frac{(1-m)C_g v^2}{w_f} \tag{18}$$

For the formation water flow region $x_f \le x \le L$ in the fracture according to equation (15), equations (10) and (14) are reduced to the following form:

$$\frac{\partial v}{\partial x}=-\frac{2k(p(x,t)-p_k)}{w_f \mu_w L}. \tag{19}$$

$$\frac{\partial v}{\partial t}+(1+\beta_w)\frac{\partial v^2}{\partial x}=-\frac{1}{\rho_w}\frac{\partial p}{\partial x}-\frac{(1-m)C_w v^2}{w_f}. \tag{20}$$

When the crosslinked gel injecting into the fracture the regions, occupied by gel and formation water, change over time, which can be fully described by the law of gel - formation water interface movement. Therefore, the differential equation for the interface paths is defined to close the model using the following initial assumption at fracture inlet (walls of the injection well):

$$\frac{dx_f(t)}{dt}=v(x_f(t)). \tag{21}$$

2.4. Closing relations, initial and boundary conditions

The experimental studies on the crosslinked gel injection into the fracture [3-8] show that under the applicable gel injection rates the gel flow in fractures is generally characterized by the laminar flow regime.

To solve the gel flow equation (18), the hydraulic friction coefficient was calculated using equation [12]:

$$C_g(\mathrm{Re}) = \frac{16}{\mathrm{Re}_g} 2^{n-3} \left(\frac{3n+1}{n} \right)^n, \qquad \mathrm{Re}_g = \rho D_H^n v^{2-n} / K \tag{22}$$

where K is a consistency index for power fluid, $n \in (0,1)$ is the non-Newtonian degree of pseudoplastic fluid.

Equation (20) is closed by defining the friction coefficient C_w for the laminar flow regime $(R_e \le 2000)$ [13]:

$$C_w = \frac{24}{\mathrm{Re}_w}, \quad \mathrm{Re}_w = \frac{\rho_w v D_H}{\mu_w}, \quad D_H = \frac{4 w_f h_f}{2(w_f + h_f)} \approx 2 w_f \tag{23}$$

The approximate equality in equation (23) follows from the assumption that $w_f << h_f$.

In our study we assume that the production well is shut down and therefore the liquid in the main fracture initially is at rest, and the fracture pressure is equal to pore pressure. The initial conditions for the equations (16) - (21) are:

$$v(x,0) = 0, \qquad p(x,0) = p_k, \qquad c(x,0) = 0, \qquad x_f(0) = 0. \tag{24}$$

Gel of initial polymer concentration c_0 is injected into the fracture at a constant rate Q. The boundary conditions at the main fracture entry are:

$$v(0,t) = \frac{Q}{w_f h_f}, \qquad c(0,t) = c_0, \quad p(L,t) = p_k, \qquad \frac{\partial c(L,t)}{\partial x} = 0. \tag{25}$$

3. Results of numerical calculations

For the numerical solution of equations (16) - (25) by the finite volume approach the SIMPLE algorithm was updated [14] to be applied to the problem of non-stationary Newtonian fluid displacement by non-Newtonian power fluid in a flat channel with permeable walls. The numerical calculations used the following Coriolis parameters $\beta_g = \beta_w = 0$. The results of calculations for the reservoir of permeability $k = 32.7$ mD and porosity $m = 0.2$ are shown on Figures 1-4.

For the numerical calculations the accepted length of main fracture is $L = 300$ m, and its height is $h_f = 27$ m. The fracture is symmetrical with respect to the boundaries of the considered rectangular reservoir region and the distance from the fracture walls to the reservoir boundaries is $L_k = 50$ m. Reservoir pressures at the boundaries of the reviewed region, left and right of the fracture, are constant and equal $p_k = 7$ MPa. Formation water density is $\rho_w = 1010$ kg/m^3, its viscosity is taken as $\mu_w =$ 0.001 Pa·s. Gel consistency index is $K = 0.178$, non-Newtonian exponent is $n = 0.72$. The average fracture width is $w_f = 0.005$ m. The gel injection rate is $Q = 88.6$ m^3/day. The initial mass concentration of crosslinked polymer is $c_0 = 0.003$.

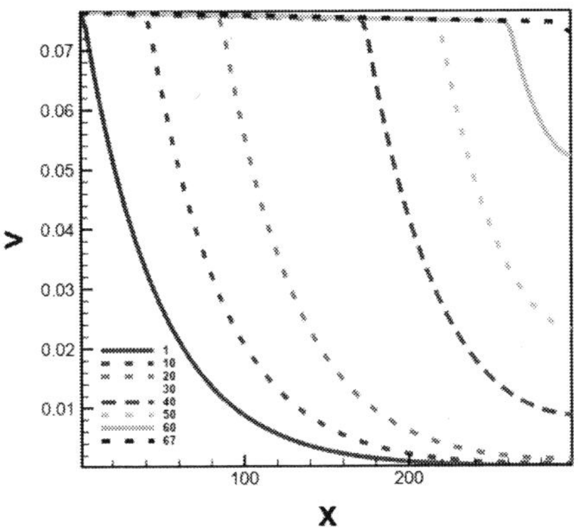

Fig. 1. Velocity (m/s) vs. fracture length (m) curves at time intervals $t = 10$ min.

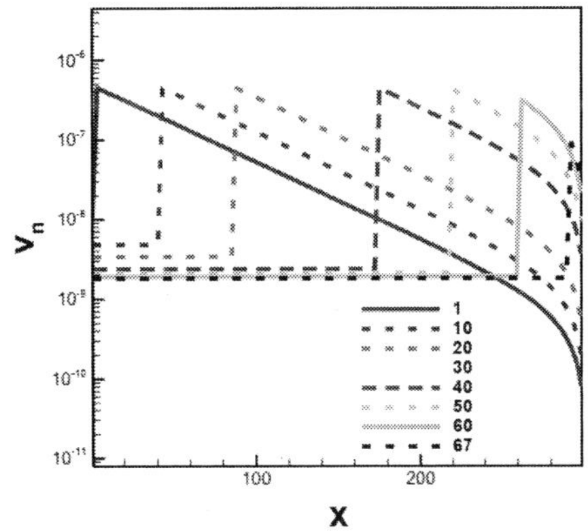

Fig. 2. Water infiltration velocity (m/s) vs. fracture length (m) at time intervals $t = 10$ min.

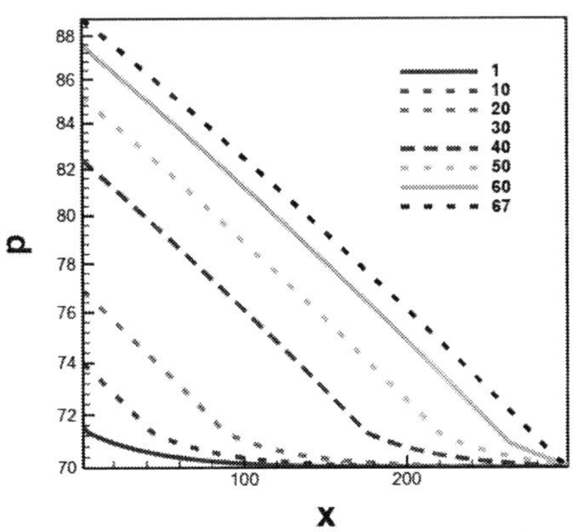

Fig. 3. Pressure (atm) vs. fracture length (m) at time intervals $t = 10$ min.

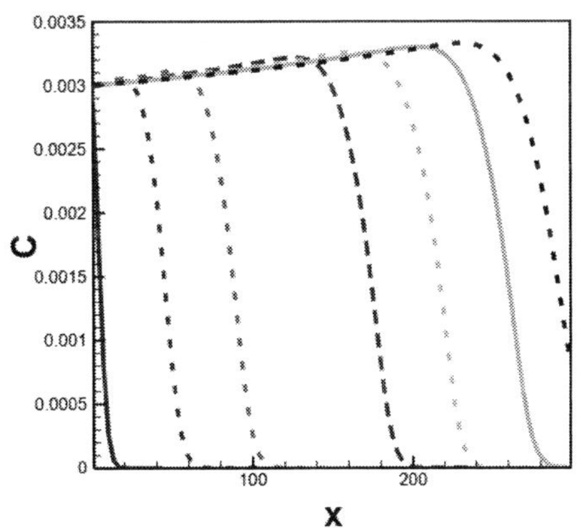

Fig. 4. Polymer concentration vs. fracture length (m) at time intervals $t = 10$ min.

Figure 1 shows the flow velocity behavior along the fracture length. In the gel flow region the rate is almost constant, but in the formation water region the rate drops due to its infiltration into the formation. The sharp bend of the velocity profile corresponds to the gel - formation water contact in the fracture. Figure 2 shows the rates of infiltration into formation both for water from gel and for formation water. Here, the shift of velocity gaps is also consistent with movement of the gel-water interface. It can be seen from the Figure 2 that in the process of fluid flow though the fracture the fluid loss velocity decreases. Figure 3 shows the profile of flow velocity vs. fracture length. At the first time points of the flow the pressure profile has a curved shape, but then, as gel fills the fracture, pressure profile is straightened and, when gel reaches the production well, it becomes almost linear. Figure 4 shows the crosslinked polymer concentration variation in gels. This calculation gave about 10% difference in initial and final concentrations of the crosslinked polymer. Even so, changes in crosslinked polymer concentrations depends on many factors including the injection rate, initial polymer concentration in the aqueous solution, crosslinker concentration, and rock permeability. Affecting by all these factors, the change in crosslinked polymer concentration with the flow time in the fracture can be significant, and in this case we must take into account the effect of gel rheological properties on the on-going crosslinked polymer concentration. The numerical simulation for the above-described mathematical model allows providing optimal technological parameters for successful gel placement in the main fractures.

4. Conclusion

We presented a mathematical model of pseudoplastic gel injection into the main fracture for the purpose of its isolation. The mathematical model can be easily generalized for considering the formation fluid and rock matrix compressibility, as well as the dependence of effective gel viscosity on the ongoing crosslinked polymer concentration and destruction of gel. The gel placement in the fracture is an important problem, based on it solutions the stability of gel barriers in the fracture can be solved if the limiting shear gradients are known for the applied gels.

Acknowledgments

This study was supported by the Ministry of Education and Science of the Russian Federation in 2016 as the base part of the state orders for higher educational institutions and the Russian Foundation for Basic Research (Project 14-01-97012 r_povolzh'e_a).

References

[1] Al-Assi A A, Willhite G P, Green D W and McCool C S 2009 *SPE J.* **14(3)** 450-61
[2] Hild G P, Wackowski R K 1999 *SPE Reservoir Evaluation & Engineering* **2(2)** 196-205
[3] Seright R S 2001 *SPE Production & Facilities* **16(4)** 225-31
[4] Seright R S 2003 *SPE Production & Facilities* **18 (1)** 65–72
[5] Seright R S 2003 Washout of Cr (III)-Acetate-HPAM gels from fractures *Presented at the International Symposium on Oilfield Chemistry* (Houston, USA, 5–7 February 2003) DOI: http://dx.doi.org/10.2118/ 80200-MS
[6] Seright R S, Lane R H, Sydansk R D 2003 *SPE Production & Facilities* **18(3)** 158–169
[7] Wang Y, Seright R S 2006 Correlating gel rheology with behavior during extrusion through fractures *Presented at the International Symposium on Improved Oil Recovery* (Tulsa, Oklahoma, USA, 22–26 April 2006) DOI: http://dx.doi.org/10.2118/ 99462-MS
[8] Hao Zhang, Baojun Bai 2011 *SPE J.* **16(2)** 388-400
[9] Bratteka B, Pedersen S G, Nistov H T et al 2015 *SPE Production & Operations* **30(2)** 99-109
[10] Charny I A 1975 *Un-steady flow of real liquid in pipes* (Moscow: Nedra) (in Russian)
[11] Serrin J 1959 *Mathematical principles of classical fluid mechanics* (Berlin: Springer- Verlag)
[12] Wilkinson W L 1960 *Non-Newtonian Fluids. Fluid Mechanics, Mixing and Heat Transfer* (London: Pergamon Press)
[13] Idelchik I E 2008 *Handbook of Hydraulic Resistance 4th Edition Revised and Augmented* (Redding, CT: Begell House)
[14] Patankar S V 1980 *Numerical Heat Transfer and Fluid Flow* (New York: McGraw-Hill Book Co.)

Cosmic rays intensity and atmosphere humidity at near earth surface

V V Oskomov, A N Sedov , N O Saduyev , O A Kalikulov, A Zh Naurzbayeva, N Sh Alimgazinova, I E Kenzhina

Institute of Experimental and Theoretical Physics of Al-Farabi Kazakh National University, 71 al-Farabi Ave., Almaty, Kazakhstan

E-mail: kenzhina_inesh@mail.ru

Abstract. Experimental studies of estimation the mutual influence of humidity and flux of cosmic rays in first approximation were carried out. Normalized cross-correlation function of time series of neutron monitors count rate and level of relative atmosphere humidity near cosmic rays registration point is studied. Corrected and uncorrected on pressure minute and hour data of 6NM64 neutron monitor count rate were used for the study. Neutron monitor is located in Al-Farabi Kazakh National University, at an altitude of 850 m above sea level. Also, data from NM64 neutron monitor of Tien Shan mountain research station of Institute of Ionosphere, located at an altitude of 3340 m above sea level were used. Uncorrected on pressure cosmic rays intensity better reflects the changes in relative atmosphere humidity. Average and sometimes strong relationship is often observed by time changes of atmosphere humidity near the point of cosmic rays detection and their intensity: the value of normalized cross-correlation function of respective signals, even in case of their long duration and a large number of data (eg, for minute changes at intervals of up to several months) covers 0.5 - 0.75 range, sometimes falling to ~ 0.4.

1. Introduction

Determination of atmosphere humidity at near Earth's surface is one of the most important tasks of hydro-meteorological and glaciological observations carrying out in agriculture, hydropower and water supply. Existing contact measurement methods are inefficient, time-consuming, and their accuracy is relatively low [1].

It is necessary to estimate the degree of mutual influence of humidity and flux of cosmic rays for experimental studies. In the first approximation it can be made by examining the normalized cross-correlation function of time series of neutron monitor count rate and the level of relative atmosphere humidity near the registration point of CR (cosmic rays) [2-3].

II. Method and calculations

Corrected and uncorrected on pressure minute and hour data of 6NM64 neutron monitor count rate were used for the study. Neutron monitor is located at al-Farabi Kazakh National University, at an altitude of 850 m above sea level. Also, data from NM64 neutron monitor of Tien Shan mountain research station of Institute of Ionosphere, located at an altitude of 3340 m above sea level were used. As well, data on changes in relative humidity were used, obtained from meteorological Tien Shan mountain station near Almaty, located at 43° 15'N, 76° 57'E. and at a height of 848 m above sea level [5].

Content from this work may be used under the terms of the Creative Commons Attribution 3.0 licence. Any further distribution of this work must maintain attribution to the author(s) and the title of the work, journal citation and DOI.
Published under licence by IOP Publishing Ltd

The majority of neutrons (over 90%) are produced in nuclear reactions such as "evaporative stars." In these reactions, the high-energy cosmic-ray hadrons collide with atmosphere nuclei, creating new unstable nuclei that emit secondary particles. The task is to calculate the average number of neutrons at different heights in the air to estimate their interaction with nuclei of material (elastic and inelastic scattering, absorption) and also to calculate the number of neutrons crossing the horizontal plane at different heights, that is neutron fluxes [6].

The flow of primary particles N depends on the height exponentially:

$$N = N_0 \cdot \exp(-\frac{x}{L}) \tag{1}$$

Here, x – thickness of layer (g/cm^2), passing by cosmic radiation is measured from top of atmosphere;

L – thickness of material layer, weakening the flow of e (L ≅ 140 g/cm^2);

N_0 - nominal flow of particles at the border of atmosphere.

The average number of neutrons, "evaporated" by unstable nucleus is denoted as N_k (for air k=1). The distribution of evaporation neutrons at energies is well described by expression:

$$\frac{\Delta N_k(E)}{\Delta E} = \frac{E}{\tau_k^2} \cdot \exp(-\frac{E}{\tau_k}) \tag{2}$$

Where τ_k is the average temperature (energy per nucleon) of unstable nuclei in the k substance.

These neutrons have energies of a few MeV order. The angular distribution of born neutrons is isotropic. The energy spectrum of neutrons is described satisfactorily by "evaporation" model. The generation of low-energy neutrons can be described by the following expression:

$$Q(x, E) = \int_{(E^*)}^{(E)} N_{ucn}^k (E^*) \frac{dn_{ucn}^k(E^*, E)}{dE} \cdot \mu_n^k \cdot \Phi_n(x, E^*) dE^* \tag{3}$$

Hadron component of cosmic rays at sea level consists mainly of high-energy neutrons, and its distribution can be written as:

$$\Phi_n(x, E^*) = \varphi(E^*) \cdot \exp(-\mu_1 \cdot x) \tag{4}$$

where:

$\varphi(E^*)$ - energy spectrum of neutrons at sea level at high energies (E > 10-100 MeV)

μ_1=I40 g/cm^2 – path of high energy neutron absorption in atmosphere.

"Evaporative" neutrons with energies of 1 MeV due to elastic and inelastic scattering on atoms nuclei are slowed down to thermal energies. Thermal neutrons are eliminated from the flow due to radiation absorption forming gamma rays, proton or other particles.

For the calculation of neutrons passing through the substance are given the interaction cross-section σ_{jm}^{tot}, σ_{jm}^{el}, σ_{jm}^{a} (full, elastic and absorption, respectively) with the atoms of the j element in the m neutron energy gradation (the average energy of gradation is E_m).

The calculations take into account the interaction of neutrons with the nuclei of atmosphere atoms.

III. Results and discussions

It was found that the uncorrected intensity of CR on pressure better reflects the changes in relative humidity of atmosphere: the value of normalized cross correlation function between the humidity and uncorrected counting rate of NM (neutron monitor) for all examined periods of time is considerably higher than in the case of corrected values of count rate. This is understandable when you consider that the air pressure depends on level of atmosphere humidity.

Average and sometimes strong relationship is often observed by time changes of atmosphere humidity near the point of cosmic rays detection and their intensity: the value of normalized cross-correlation function of respective signals, even in case of their long duration and a large number of data (eg, for minute changes at intervals of up to several months) covers 0.5 - 0.75 range, sometimes falling to ~ 0.4. The phase shift between these signals is typically about 10 hours.

Figures 1 - 3 shows correlation functions of time series of NM count rate and relative humidity for a period of time, confirming the above findings.

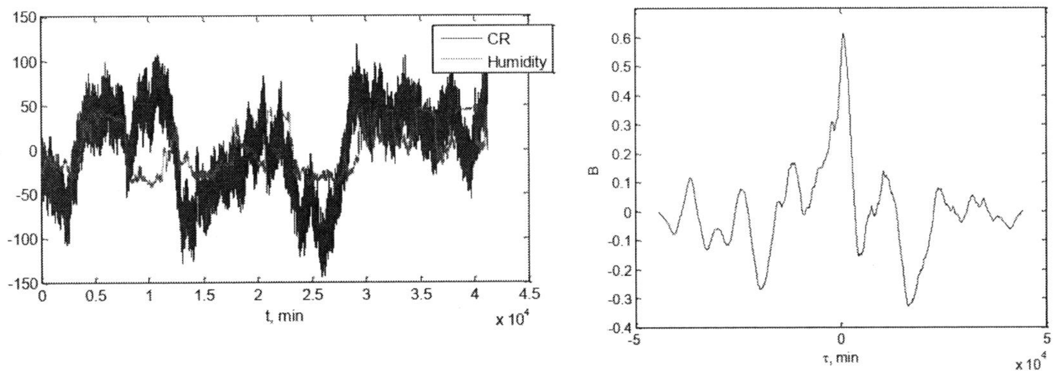

Figure 1. Time series of deviations from the average minute values of NM count rate and relative humidity at Tien Shan mountain station and normalized cross correlation function of the series for January 2014

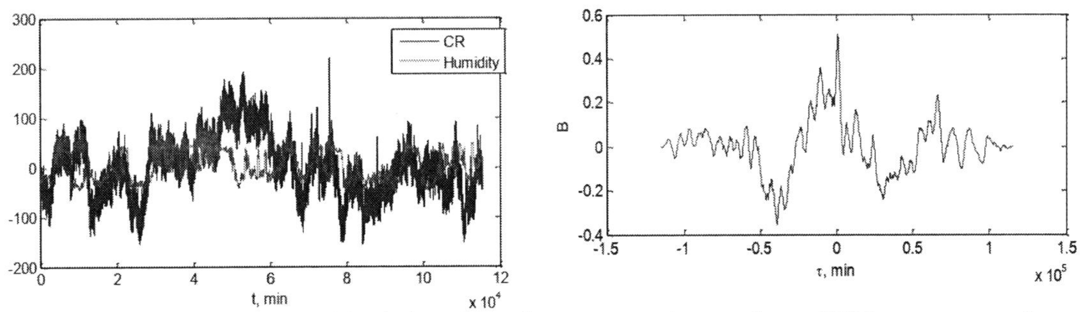

Figure 2. Time series of deviations from the average minute values of NM count rate and relative humidity at Tien Shan mountain station and normalized cross correlation function of the series for March 2014

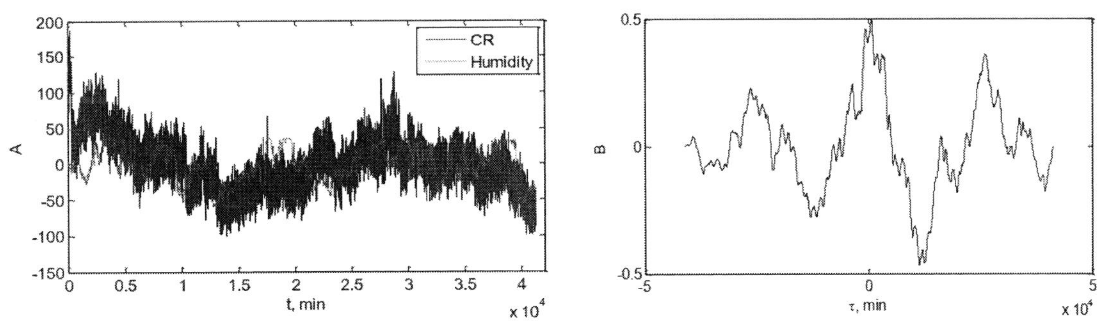

Figure 3. Time series of deviations from the average minute values of NM count rate and relative humidity at Tien Shan mountain station and normalized cross correlation function of the series for September 2014

III. Conclusion

Thus, the correlation between the intensity of cosmic rays and atmosphere humidity near the Earth was investigated.

It is planned to conduct further numerical study of mutual influence of humidity and the intensity of CR using the techniques such as wavelet analysis, analysis of phase synchronization, and others.

Acknowledgments

The study was carried out under the support of grant #3842/GF4 and #3838/GF4 of Ministry of Education and Science of Kazakhstan.

References

[1] Chubenko A P , Shepetov A L, Babaev M K, et al. 2005 Bulletin of the Russian Academy of Sciences: Physics **69** 3 433-436

[2] Dalkarov O D, Zhukov V V , Nam R A, et al 2013 Book Series: Journal of Physics Conference Series 409 012127

[3] Alexandrov K V, Ambrosio M, Ammosov V V et al. (The INCA Collaboration) Measurements of the neutron yield from a lead absorber for pion and proton projectiles Proc. 26th ICRC, Salt Lake City, 1999, v. 3, p. 195 – 198.

[4] Alexandrov K V, Ambrosio M, Ammosov VV et al. (The INCA Collaboration) 1999 Proc. 26th ICRC **3**, 195 – 198.

[5] Hatton C J, Carmichael H 1964 Canadian Journal of Physics **42** 2443.

[6] Le Couteur K J 1952 Proc. Phys. Soc **65(a)** 718-720.

Numerical optimization of a RIXS spectrometer using ray-tracing simulations

K Lieutenant[1], T Hofmann[1], C Zendler[3,1], C Schulz[1], E F Aziz[1,2], K Habicht[1]

[1] Helmholtz-Zentrum Berlin, Hahn-Meitner-Platz 1, 14109 Berlin, Germany

[2] Department of Physics, Freie Universität Berlin, Arnimallee 14, 14195 Berlin, Germany

[3] European Spallation Source ERIC, Tunavägen 24, 223 63 Lund, Sweden

E-mail: klaus.lieutenant@helmholtz-berlin.de

Abstract. At Helmholtz-Zentrum Berlin (HZB) the end-station PEAXIS (Photo Electron Analysis and X-ray resonant Inelastic Spectroscopy) combining Angle-dependent X-ray Photoelectron Spectroscopy (AdXPS) and Resonant Inelastic X-ray Scattering (RIXS) is currently built. The latter method uses a spherical variable line space (VLS) grating to focus the beam onto the detector. Working in first-order diffraction allows resolving photon energy by transferring the energy-dependent signal to a position-dependent focal spot on the detector. Focusing requires a precise combination of various parameters of the VLS grating and the geometry of the RIXS spectrometer. The VLS grating was optimized by calculating the geometry parameters for different photon energies, simulating the instrument and evaluating the pattern on the detector. As figure of merit we chose the intensity times the square of the resolving power averaged over the photon energies.

1. Introduction

Resonant Inelastic X-ray Scattering (RIXS) is a rapidly evolving technique to study inelastic processes using X-ray synchrotron beams. Due to the wide range of applications [1], spectrometers of this type have recently been built [2] or are under construction at most synchrotrons. At the BESSY II synchrotron, operated by the Helmholtz-Zentrum Berlin (HZB), the PEAXIS end-station offering RIXS measurements [3] is currently under construction. It combines Angle-dependent X-ray Photo-electron Spectroscopy (AdXPS) and RIXS measurements, both for solid and liquid samples, in a energy range of 200 – 1200 eV. The set-up allows measurements in a wide range of momentum transfer by varying the angle of detection between 30° and 150° relative to the incoming X-ray beam.

To perform the AdXPS measurements, a commercial photoelectron analyzer is part of the instrument. The RIXS spectrometer is designed by HZB and has to be optimized thoroughly. The basic idea is to use a grating with a high line density (> 1000 lines/mm) in first-order diffraction. For grazing incidence (87° < α < 90°, cf. figure 1), soft X-rays are reflected on the surface of the grating and scattered to a direction that depends on the photon wavelength. This transfers the energy dependent intensity distribution to a position dependent distribution on the detector surface and thus allows precise determination of the photon energy. To achieve a high energy resolution, good optical

Content from this work may be used under the terms of the Creative Commons Attribution 3.0 licence. Any further distribution of this work must maintain attribution to the author(s) and the title of the work, journal citation and DOI.
Published under licence by IOP Publishing Ltd

focusing from the sample to the detector surface is crucial. This is realized by state-of-the-art photon optics, a spherical variable line space (VLS) grating. Since its line density varies slightly along the beam, aberration effects can be significantly reduced. This optical system is described in the literature [4 - 6]; the analytical formulae are used here to find the best instrument parameters.

Most applications need a high energy resolution, but are also demanding in terms of intensity, because the processes studied by RIXS measurements have a low cross-section. Increasing the sample – detector distance improves the resolution, but reduces the intensity at the detector. So a good compromise for the instrument length as for other parameters has to be found. We decided to optimize the instrument for the highest value of intensity times the square of the resolving power.

As perfect imaging can only be realized for a single photon energy and aberrations increase with energy difference, two gratings are necessary to cover the large photon energy range. Based on existing knowledge, e.g. [2], instruments of different lengths and line densities were simulated using the program RAY [7] and performance compared. In this way preliminary parameter sets for the two gratings were found and published [3]. However, to find out if the set-ups found were really the best choice, the simulation of the RIXS instrument and its data evaluation was combined with numerical optimization using a swarm algorithm. This algorithm was transferred from the simulation package VITESS 3 [8] and adapted to the requirements for the optimization of a RIXS instrument using RAY.

2. Imaging using a VLS grating

The line density distribution of a variable lines space (VLS) grating can be described by a Taylor series

$$a(w) = a_0 + a_1 w + a_2 w^2 + a_3 w^3 + \dots \tag{1}$$

Imaging using a spherical concave VLS grating of radius R works perfectly only for a single wave-length λ_0 corresponding to a photon energy E_0. (The index 0 denotes parameters for this wavelength.) It can be described using the light path function F. The focal point is then given by the Fermat condition of vanishing derivatives $\partial F/\partial w=0$ and $\partial F/\partial l=0$ [4] (*w*, *l*: directions on the grating surface, *l* along the lines, *w* perpendicular to the lines). This leads to a set of equations relating the grating parameters R, a_0, a_1, a_2, etc. with the instrument parameters r_{10}, r_{20}, α_0, β_0 and λ_0. These are [5,6]:

$$a_0 = \frac{1}{k\lambda_0}\left[\sin\alpha_0 - \sin\beta_0\right] \quad \Leftrightarrow \sin\beta_0 = \sin\alpha_0 - k\lambda_0 a_0 \tag{2}$$

$$a_1 = \frac{1}{k\lambda_0}\left[\frac{\cos^2\alpha_0}{r_{10}} + \frac{\cos^2\beta_0}{r_{20}} - \frac{\cos\alpha_0 + \cos\beta_0}{R}\right] \tag{3}$$

$$a_2 = -\frac{3}{k\lambda_0}\frac{1}{2}\left[\frac{\sin\alpha_0}{r_{10}}\left(\frac{\cos^2\alpha_0}{r_{10}} - \frac{\cos\alpha_0}{R}\right) - \frac{\sin\beta_0}{r_{20}}\left(\frac{\cos^2\beta_0}{r_{20}} - \frac{\cos\beta_0}{R}\right)\right] \tag{4}$$

$$a_3 = \frac{4}{k\lambda_0}\frac{1}{8}\left[\begin{aligned}&\frac{4\sin^2\alpha_0}{r_{10}^2}\left(\frac{\cos^2\alpha_0}{r_{10}} - \frac{\cos\alpha_0}{R}\right) - \frac{1}{r_{10}}\left(\frac{\cos^2\alpha_0}{r_{10}} - \frac{\cos\alpha_0}{R}\right)^2 + \frac{1}{R^2}\left(\frac{1}{r_{10}} - \frac{\cos\alpha_0}{R}\right)\\ &+\frac{4\sin^2\beta_0}{r_{20}^2}\left(\frac{\cos^2\beta_0}{r_{20}} - \frac{\cos\beta_0}{R}\right) - \frac{1}{r_{20}}\left(\frac{\cos^2\beta_0}{r_{20}} - \frac{\cos\beta_0}{R}\right)^2 + \frac{1}{R^2}\left(\frac{1}{r_{20}} - \frac{\cos\beta_0}{R}\right)\end{aligned}\right] \tag{5}$$

Equation (2) is the so-called grating equation, equation (3) describes the focal condition and equations (4) and (5) minimize aberrations. An additional equation defines focusing along the detector surface, which is inclined by an angle γ relative to the beam direction ([9], see figure 1).

$$\tan\gamma_0 = -r_{20}\frac{\partial\beta}{\partial r_2} = \frac{\cos\beta_0}{2\sin\beta_0 - r_{20}(\tan\beta_0 / R + a_1 / a_0)} \tag{6}$$

Additionally, the total length L is the sum of r_1 and r_2. There is one more condition needed to solve the system of equations. This is chosen to have the same contributions from sample size and detector cell size to the line width broadening:

$$r_{10} = \frac{S_1\cos\alpha_0}{S_1\cos\alpha_0 + S_2\cos\beta_0\sin\gamma_0}L_0 \qquad r_{20} = L_0 - r_{10} \tag{7a,b}$$

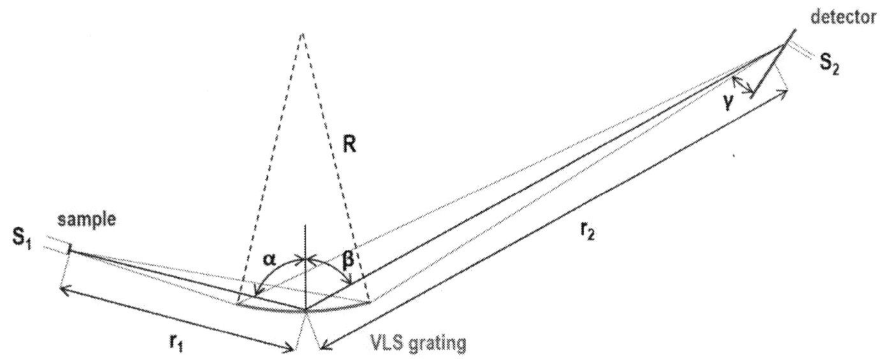

Figure 1. Sketch of the RIXS spectrometer layout showing the relevant instrument parameters

The first-order diffraction is used, i.e. k=1. The illuminated height S_1 of the sample and the (effective) detector cell size S_2 are specific for a certain beamline and instrument (cf. figure 1) and are given as input parameters. The total instrument length L_0, the standard photon energy E_0 and the standard values for incidence angle α_0, detector inclination γ_0 and the line density a_0 at the centre of the grating are now chosen. Equations (2) to (7) allow calculating the instrument parameters β_0, r_{10}, r_{20}, and R. Then the grating parameters a_1 to a_3 can be calculated using equations (3) to (5).

The system is now completely described for the standard photon energy E_0. However to achieve best imaging for photon energies $E \neq E_0$, the parameters α, β, γ, r_1 and r_2 all have to be changed. The equations (2) - (4) and (6) have to be applied again, but now for the parameter set α, β, …r_2. (Equation (5) cannot be fulfilled anymore.) This time the parameters R, a_1, a_2, a_3 are fixed and used to calculate the angles α, β, and γ and distances r_1 and r_2. However, with 4 equations for 5 parameters, there is one free parameter to choose. We choose α – an alternative is γ [10] – equation (2) delivers β then. Knowing α and β allows calculating r_1 and r_2 by combining equations (3) and (4) [3]:

$$r_1 = 1/x_+ \text{ and } r_2 = 1/y_+ \text{ if } x_+>0 \qquad r_1 = 1/x_- \text{ and } r_2 = 1/y_- \text{ if } x_->0 \tag{8}$$

using

$$x_{+,-} = \frac{a_1hcR + E(\cos\alpha + \cos\beta) - ERy_{+,-}\cos^2\beta}{ER\cos^2\alpha} \tag{9a}$$

$$y_{+,-} = \frac{2ER(E + a_1hcR/\cos\beta)\sin\alpha + (E^2R\cos\alpha\sin(\alpha-\beta) \pm \sqrt{W}/\cos\beta)/\cos^2\beta}{2E^2R^2(\sin\alpha\cos\beta - \cos^2\alpha\tan\beta)} \tag{9b}$$

with

$$W = E^2R^2\cos^2\alpha\cos^2\beta(8a_0hcR(B\sin\beta - a_2ER\cos^2\beta\sin\alpha/(3a_0)) + E^2\sin^2(\alpha+\beta))$$
$$B = a_2ER\cos^2\alpha/(3a_0) + a_1(a_1hcR + E(\cos\alpha + \cos\beta))\sin\alpha/(2a_0)$$

γ can then be calculated using equation (6).

3. Simulation and evaluation of a RIXS spectrum

3.1. Simulation of a RIXS spectrum

The spectrum is simulated using the RAY software package [7] in its new variant Ray-UI (version 2.7) containing a graphical user interface. Typically 1.5 Mio X-rays are generated in an area of 20 x 5 μm^2, which is the foreseen spot size of the beam at the sample position. They are directed towards the grating which is usually completely illuminated. Only 3 discrete energies E_m-½ΔE, E_m and E_m+½ΔE are generated which produce 3 lines at the detector that allow calculating the resolving power (see chapter 3.2). ΔE was usually 1 eV.

The grating component of RAY considers practically all physical effects that occur when the X-ray beam hits the grating: the reflectivity of the coating, the position-dependent line density, the profile of the lines and deviations from the ideal shape of the grating (the so-called slope error). Here we assume a so-called blaze grating, whose surface is inclined to have specular reflection into the direction of the first-order diffraction (for the standard energy E_0.) This type of grating has a better performance than others because it delivers higher intensity. However, up to now it can only be produced with a maximal line density of 2400 lines/mm. Rays that are not lost at the grating are propagated to the detector. The data is then stored as counts as a function of the position on the detector I(z).

3.2. Evaluation of the simulated RIXS spectra

The separation Δh_m of the 3 lines is calculated from the centers of mass of the outer peaks: $\Delta h_m=|h_3-h_1|$. The lines have a Gaussian shape. So the FWHM width δh_m can be determined from first and second moment of the intensity distribution as the average over the 3 lines. (The same formula is also used in cases where there is a significant deviation from a Gaussian shape.) The line broadening (compared to the delta function generated in the source) seen in the simulated spectrum arises from aberration effects, the size of the source and the slope error of the grating. The simulation does not include a broadening due to the effective cell size of the detector. So this contribution has to be added giving the total resolving power

$$R_{tot,m} = \frac{\Delta h_m \cdot E / \Delta E}{\sqrt{\delta h_m^2 + \kappa^2 h_{cell}^2 \sin^2 \gamma}} \qquad (10)$$

(κ is the ratio of effective detector cell size and physical cell size h_{cell}. A value of κ=1.5 is assumed.)

The relative intensity I_{rel} is defined as the ratio of the number of photons N_{det} counted in the detector to the number of photons emitted by the sample. In a simulation, photons are sent out only into a limited solid angle Ω_g so that I_{rel} can be calculated from the number N_{src} of photons sent as

$$I_{rel} = \frac{\Omega_g}{4\pi} \frac{N_{det}}{N_{src}} \qquad (11)$$

For small inclination angles γ of the detector, a part of the active area is shadowed by the detector frame, which reduces the detector count rate N_{det}. This effect is not included in the simulation and has to be considered in the data evaluation tool. Now the figure of merit FoM_m for one energy E_m can be calculated and averages over the simulated energies determined, either weighted with energy (13b) or not (13a).

$$FoM_m = I_{rel} R_{tot,m}^2 \qquad (12)$$

$$FoM_{av} = \frac{1}{M} \sum_{m'=1}^{M} FoM_{m'} \qquad FoM_{wt} = \sum_{m'=1}^{M} E_{m'} FoM_{m'} \Bigg/ \sum_{m'=1}^{M} E_{m'} \qquad (13a,b)$$

4. Numerical optimization of the RIXS spectrometer

4.1. Determination of grating and instrument parameters for a simulation

First, several parameter sets (L_0, E_0, α_0, γ_0, a_0) were simulated and evaluated. It was confirmed that low values of the detector inclination γ give best resolving power - cf. equation (10) - and thus highest figures of merit. On the other hand, shielding by the frame of the detector does not allow using angles below 17°, and the technical realization also prevents angles above 43.6°. Another limitation concerns the distance r_1: the technical realization requires that it is in the range 650 to 1250 mm.

So the angle of incidence α cannot be chosen arbitrarily. Instead, α is first set to α_0 and the system of equations solved - then it is increased until $\gamma(\alpha)$ equals γ_0. Now it is checked if the corresponding value of r_1 is within the given limits. If it is, this parameter set is used for the simulation; if it is not, α is varied until r_1 and γ are within the limits (see figure 2). This procedure is carried out for each parameter set (L_0, E_0, α_0, γ_0, a_0) and each photon energy E_m. In some cases, there are no values for r_1 and γ within the limits fulfilling the equations. In this case, the figure of merit is set to zero.

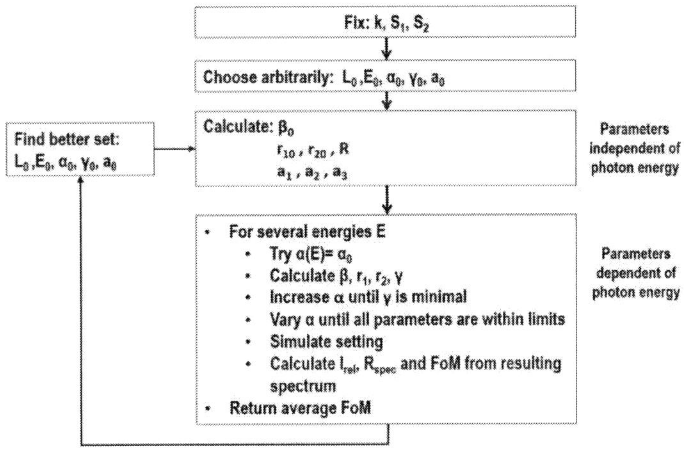

Figure 2. Flow chart of the procedure to find improved settings for the RIXS spectrometer by simulation and data evaluation

4.2. Combination of numerical optimization and simulations

In the new variant Ray-UI of the RAY simulation package [7], the complete instrument is described by an XML file. A simulation can be started by giving this file as a parameter to an executable. This allows running a new simulation by simply substituting the parameter values in the XML file and re-starting the simulation.

To find the best configuration, the particle swarm optimization (PSO) [11] is used: a number of individuals (bees) explores the parameter space, but they are influenced in their search by the results of other individuals. The implementation of this interaction follows the *time-varying acceleration coefficients* (TVAC) method described in [12]. The number of particles (bees) is set to 15 and the maximal number of steps to 80; after 20 steps without improvement the optimization is stopped. Other parameters determine the interaction of the bees. All these optimization parameters are defined in a parameter file.

Wide ranges were chosen for the allowed values in order to find the true optimum: L_0: 2000 – 6000 mm, E_0: 250 – 550 eV (low energy grating) and 500 – 1100 eV (high energy grating), α_0: 82.0° – 89.6°, γ_0: 20.0° – 43.6°, a_0: 600 – 3600 mm^{-1}. The optimization program defines all parameter sets to be tested. The whole simulation and data evaluation software has to deliver values of the figure of merit for each of the parameter sets. The optimization algorithm interprets this result and calculates new possible parameter sets to be tested. This is continued until the maximum number of steps or the maximum number of steps without improvement is reached.

5. Results

After strong variation during the first 20 to 30 steps, there is clear convergence of the parameters leading to an early completion of the optimization in some cases. Table 1 summarizes different parameter sets that fulfill the equations (2) to (7) for the standard photon energy E_0 and equations (2) to (4), (6) and (7b) for energies $E \neq E_0$ and yield grating distances r_1 and detector inclinations γ within the technical limits for the whole energy range needed. The upper part shows four settings of the low energy set-up (200 to 600 eV) and the lower part four settings of the high energy set-up (400 to 1200 eV). The first setting of each set-up (Lo-C and Hi-C) shows the best parameter set found without optimization. The other settings are the results of different runs of numerical optimization. In each case, 3 photon energies were simulated and evaluated, 223, 400 and 540 eV for the low energy set-up and 487, 800 and 1116 eV for the high energy set-up. The figure of merit was calculated for each energy according to equations (10) to (13).

The first numerical optimizations (Lo-OCV and Hi-OCV) used the same weight for all 3 energies (equation (13a)). This leads to low standard energies E_0 favoring the low energy end of the given energy range. So the total figure of merit was increased by achieving high performance for low energies while losing performance for high energies. As the aim is to have ideally the same performance over the full range, the following optimization runs were performed with a stronger weighting for the high energies (equation (13b). The resulting E_0 values are now much closer to the center of the range and the figures of merit, calculated according to equation (13b), are the highest of all settings simulated.

Table 1. Standard values of the instrument parameters as found by trial and error (Lo-C and Hi-C) and by different runs of numerical optimization, both for the low- and the high-energy set-up of the RIXS spectrometer. FoM-3 is the weighted average over the 3 energies used in the optimization, FoM-8 is the average over 8 energies. Fixed parameters are in italic.

Run	weight	R	FoM-8	FoM-3	L_0 [mm]	E_0 [eV]	α_0 [deg]	γ_0 [deg]	a_0 [mm^{-1}]	R [mm]
Lo-C	-	-	67.2	73.9	*4084*	*400.0*	*87.200*	*20.000*	*2400*	27157
Lo-OCV	const.	var	67.2	68.8	3985	325.8	87.253	20.000	2365	25422
Lo-OEV	E-dep.	var	68.8	77.3	4205	401.6	87.241	20.000	2747	27227
Lo-OEF	E-dep.	fix	69.2	76.1	*4040*	386.5	87.285	20.000	*2400*	27157
Hi-C	-	-	29.2	31.0	*4110*	*900.0*	*88.000*	*20.000*	*2400*	41513
Hi-OCV	const.	var	33.1	29.9	4174	571.4	87.813	20.000	2551	35278
Hi-OEV	E-dep.	var	35.2	36.3	4133	723.9	87.694	20.162	3357	33317
Hi-OEF	E-dep.	fix	34.3	34.3	*4149*	691.0	88.133	20.000	*2400*	41513

However, while these seem to be the ideal parameters, they could not be used for a practical realization: The first problem is that the maximal line density available for a blaze grating is 2400 lines/mm and the result of the optimization yielded higher values, especially for the high energy set-up. The second problem was that the substrates of the gratings were already ordered with the radii obtained for the best settings found without optimization, i.e. Lo-C and Hi-C.

Therefore, new optimization runs were started, keeping the line density a_0 fixed to 2400 lines/mm and the radius to 27157 and 41513 mm resp. Keeping the radius fixed practically fixes L_0 as well. So the last optimization had only 3 free parameters: E_0, α_0 and γ_0. However, the performance of the optimized setting was about as good as for the 5 parameter optimization.

The relative intensity, the resolving power and the figure of merit as a function of energy are shown in figure 3 in comparison to the settings found without numerical optimization. The comparison clearly shows a gain in performance, in particular a higher resolving power for the low energy end of each energy range.

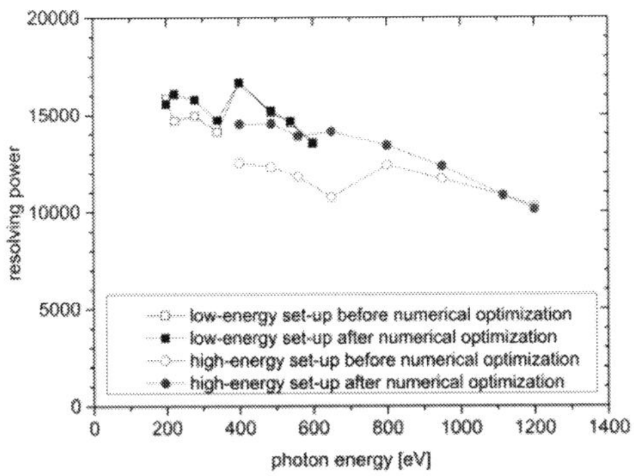

(a)

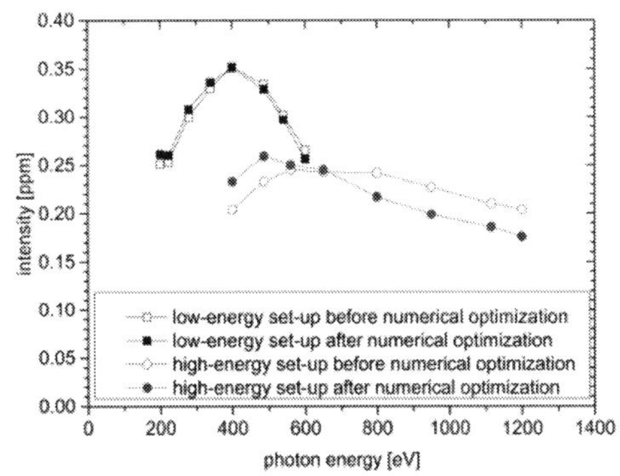

(b)

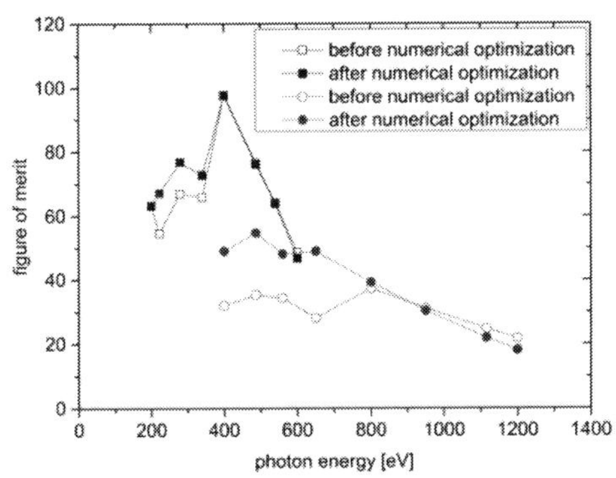

(c)

Figure 3. Comparison of the instrument performance before and after numerical optimization for both, low-energy (black squares) and high energy set-up (red circles): resolving power R (a), detector count rate I_{rel} (b) and figure of merit $FoM=I_{rel}\,R^2$ (c)

6. Discussion

From several simulations of the RIXS spectrometer [3] a range of the parameter space of good performance was known, but it was not clear if it is the global optimum. So initial values were chosen in a wide range for each parameter to explore the whole parameter space. The result was that all parameter settings of good performance and especially all optimization results were in the known region. However, the performance could be improved significantly, especially for the high energy set-up (see table 1).

The combination of ray-tracing simulations and numerical optimization had been introduced [13] and successfully applied [14] for neutron scattering instruments before. These results show that this combination also works for X-ray instruments. The swarm algorithm is well suited for this kind of optimization, especially if it is not clear where the global optimum is. In this case the swarm algorithm is superior to other algorithms tested in this context [15]; the only drawback is the high number of simulations needed.

The restriction to 3 energies during optimization seems justified: the figures of merit using the average of 3 energies are not very different from that for 8 energies (cf. table 1). It also appeared that using 15 bees was sufficient. On the other hand, a higher number of rays would have been useful, because the figure of merit for one parameter set varied from simulation to simulation. Unfortunately, the number of rays was limited in this version of RAY-UI and the simulation is rather slow. When these weaknesses are overcome in upcoming versions, RAY-UI will be a well suited program to combine with optimization, because the description of the instrument in an XML file makes this combination very easy.

Interestingly, about the same performance could be achieved with a reduced number of free parameters (cf. optimizations –OEV and –OEF in table 1). For the low energy grating the fixed parameters are quite close to the results of the other optimization; so it is not that surprising that the same figure of merit could be reached. But for the high energy grating, radius and line density were significantly different, while nearly the same figure of merit could be reached (see table 1). This means that there is not a single optimum giving a performance better than all others. Instead there seem to be many optima with comparable figures of merit.

It is important to mention that the limit for the total instrument length L_0 was never reached in any optimization. Of course a short instrument spoils the resolution, because the broadening caused by most parameters (like source size) decreases with increasing length making a long instrument a high resolution instrument. However the contribution of the slope error is independent of length, which limits the gain in resolving power with increasing length. On the other hand, the count rate is more and more reduced with instrument length. So an intermediate length yields the highest figure of merit. For the slope error of 0.1 arcsec, which could be achieved by our supplier [16], the ideal length L_0 is about 4 m, giving a maximal instrument length of about 4.7 m, which was (fortunately) in the range that can be realized at the foreseen beamline position.

7. References

[1] Simon M and Schmitt T 2014 *J. Electron Spectrosc Relat. Phenom.* **188** 1

[2] Ghiringhelli G et al. 2006 *Rev. Sci. Instrum.* **77** 113108

[3] Lieutenant K, Hofmann T, Schulz C, Yablonskikh M, Habicht K, Aziz E F 2016 (accepted for publication in *J. Electron Spectrosc. Relat. Phenom.*, DOI: 10.1016/j.elspec.2015.08.009)

[4] Namioka T 1959 *J. Opt. Soc. Am.* **49** (1959) 446

[5] Harada T and Kita T 1980 *Applied Optics* **19** 3987

[6] Howells M R 2009 Gratings and Monochromators *X-Ray Data Booklet* (*LBNL/PUB-490 Rev.3*), ed Thompson A C (Berkeley, California: LBNL) chapter 4.3 pp. 4-16–4-26 (http://xdb.lbl.gov/xdb-new.pdf)

[7] Schäfers F 2008 The BESSY raytrace program RAY *Modern Developments in X-ray and neutron Optics* (*Springer Series in Optical Sciences* vol 137) eds Erko A, Idir M, Krist Th and Michette A G (Berlin: Springer-Verlag) chapter 2 pp 9–41

[8] Zendler C, Lieutenant K, Nekrassov D and Fromme M 2014 *J. Phys.: Conf. Ser.* **528** 012036

[9] Osborn K D and Callcott T A 1995 *Rev. Sci. Instrum.* **66** 3131

[10] Strocov V N et al. 2011 *J. Synchrotron Rad.* **18** 134

[11] Kennedy J and Eberhart R 1995 *Proc. IEEE Int. Conf. Neural Networks* **1-6** 1942

[12] Ratnaweera A, Halgamuge S and Watson H C 2004 *IEEE Transactions on Evol. Comp.* **8** 240

[13] Lieutenant K and Zsigmond G 2004 *Physica B* **350** *Suppl. 1* E687

[14] Klenø K H, Lieutenant K, Andersen K and Lefmann K 2012 *Nucl. Instrum. Methods Phys. Res. A* **696** 75

[15] Lieutenant K 2005 *J. Phys.: Cond. Matter* **17** S167

[16] www.zeiss.com

Coherence of Physics and Chemistry Curricula in Terms of the Electron Concept

Ivanova Elena

The School of Natural Sciences, Far Eastern Federal University, 8, Sukhanova St. Vladivostok, 690950 Russian Fed.

E-mail: lena---iv@mail.ru

Abstract. One of the major contradictions in subject teaching is the contradiction between the unity of the world and the discrete separated generalized content of natural sciences that study natural phenomena. These are physics, chemistry, biology and more. One can eliminate the conflict if opens the content's interdisciplinary links set by the events that are studied by different disciplines. The corresponding contexts of the phenomenon content arise depending on the discipline, and they are not enough coordinated. Obviously, we need a mechanism that allows establishing interdisciplinary links in the content quickly and without losing the logic of the material and assess their coherence in academic disciplines. This article uses a quantitative method of coherence assessment elaborated by T.N. Gnitetskaya. The definition of the concept of the semantic state introduced by the authors is given in this article. The method is applied to coherence assessment of physics and chemistry textbooks. The coherence of two pairs of chemistry and physics textbooks by different authors in different combinations was calculated. The most cohered pairs of textbooks (chemistry-physics) were identified. One can recommend using the pair of textbooks for eighth grade that we offered that favors the development of holistic understandings of the world around us.

1. Introduction

Studying natural sciences such as high school courses of physics, chemistry and biology should facilitate forming an integral perception of the scientific picture of the world; understanding of connectedness and correlation of natural sciences. That's why it's significant to have a mechanism which allows establishing interdisciplinary links in different subjects.

Nowadays there are a lot of textbooks in which different authors describe a course of physics. Surely, these textbooks vary both in order of chapters and materials included. One could say the same about chemistry textbooks. Obviously, it raises the questions: how fully do contents of physics and chemistry textbooks correspond with each other and do the connectedness of contents of physics and chemistry textbooks by different authors differ?

The answers may be found if we calculate a degree of coherence for pairs of chemistry and physics textbooks in different combinations using the quantitative method of interdisciplinary links graph model worked out by T.N. Gnitetskaya. [6,7].

2. Coherence of high school courses of physics and chemistry on the basis of interdisciplinary links graph model.

Content from this work may be used under the terms of the Creative Commons Attribution 3.0 licence. Any further distribution of this work must maintain attribution to the author(s) and the title of the work, journal citation and DOI.
Published under licence by IOP Publishing Ltd

For analysis we select two physics courses for 7-9th grades: (Physics-1) by A.V. Peryshkin [2], (Physics-2) by S.V. Gromov, N.A. Rodina [3]; and two chemistry courses for 8-9th grades: (Chemistry-1) by O.S. Gabriyelyan [4], (Chemistry-2) by R.G. Ivanova [5], described in the appropriate textbooks. The chosen textbooks are on the federal list of textbooks recommended for use in the process of learning at the educational institutions by the Ministry of Education and Science of Russian Federation. At the first stage of analysis we build four spaces of interdisciplinary content: Physics-1 with Chemistry-1; Physics-2 with Chemistry-1; Chemistry-1 with Physics-1; Physics-2 with Chemistry-2.

At the second stage of analysis we calculate quantitative characteristics of interdisciplinary links on the basis of the method suggested in [8,9]. The values of interdisciplinary links length and strength obtained for these chemistry and physics courses build the hierarchical sequence. The hierarchy of physics course's concepts which are used in chemistry course includes, first of all, the fundamental concepts. It's meaningful that the list of fundamental concepts for physics is similar both for Chemisty-1 and Chemistry-2. Calculation of quantitative characteristics of the interdisciplinary links graph model (relative maximal length L and relative connection strength C) [6,7] enables to establish the hierarchical sequence of physical concepts where an interdisciplinary fundamental core of physics and chemistry composes physical concepts which relative maximal length *L* is equal L = 1 and relative connection strength *C* lies in the interval of $1{,}00 \leq C \leq 0{,}73$.

At the third stage of analysis we calculate a degree of coherence between the chosen chemistry and physics courses [6,7]. The coherence of two courses is regarded [1] as the sum of all interdisciplinary links graphs' strengths for a set of knowledge elements. This article deals with the coherence of physics and chemistry for the group of concepts which is regarded as the calculation of sum of all graphs' strengths of direct interdisciplinary links for the set of physics knowledge elements used at the chemistry course or the set of chemistry knowledge elements used at the physics course with the reverse direction of interdisciplinary links.

Table 1 Forest of orgraphs of interdisciplinary links for physics and chemistry established using the electron concept in chemistry course for 8-9th grades by O.S. Gabriyelyan

Structure of chemistry course / Elements of physics course	Atoms of chemical substances	Compounds of chemical substances	Chemical reaction rates. Chemical equilibrium.	Solubility. Solutions. Ion-exchange reactions and redox reaction	Introduction. Overview of chemical substances	Metals	Nonmetals of subgroups VI -VII A	Nonmetals of subgroups IV-V A	Organic substances	Relative max. length *L*	Connection relative strength *C*
Group of concepts	8th grade					9th grade					
	1	2	3	4	5	6	7	8	9		
Electron	◆	◆		◆		◆	◆	◆	◆	1,00	0,77

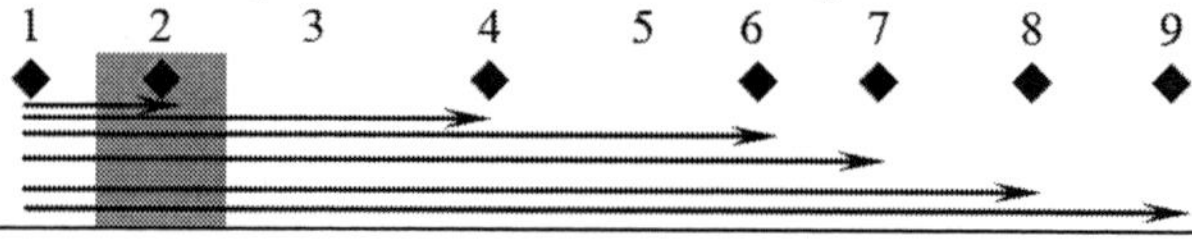

For instance, with the physical concept of electron from physics to structural elements of chemistry course for 8-9th grades by O.S. Gabriyelyan eight links are built between the following structural elements of chemistry: 1) Atoms of chemical substances; 2) Compounds of chemical substances; 3) Solubility. Solutions. Ion-exchange reactions and redox reaction; 4) Introduction. Overview of chemical substances; 5) Metals; 6) Nonmetals of subgroups VI -VII A; 7) Nonmetals of subgroups IV-V A; 8) Organic substances. They compose a forest containing eight orgraphs.

The described analysis of interdisciplinary content of high school chemistry and physics courses shows the most significant, relevant to both subjects' concepts, such as: an atom, a molecule, an electron, a substance and others.

It may be useful to take the obtained results into consideration while choosing chemistry and physics textbooks. Thus, for those who learn physics with A.V. Peryshkin' s textbook, chemistry textbook by O.S. Gabriyelyan may be recommended. When learning physics with the textbook by S.V. Gromov, N.A. Rodina, chemistry textbook by R.G. Ivanova may be useful for grades studying the general course of chemistry.

3. Semantic states of a physical concept.

One of the major goals of secondary school is to develop a holistic unity of the World around us. This goal may be achieved during the process of natural disciplines' studying on the interdisciplinary basis. The content of the concept expands and its meaning is transformed in the process of studying the courses. However, the volume of information in the concept increases while transferring from the previous meaning to the following one. Let us introduce and formulate the term "semantic state" of a term, which is identical to the "meaning" of a term.

Semantic state of a concept is the state of its content, which serves to form one of the concept's meanings completed at the given stage of studying, and possessing the required informational features. Then the process of concept formation can be viewed as a process of transfer from one semantic state to the other [8].

Two paragraphs of the 8th grade chemistry course contain eight semantic states of an electron (a particle of atom which transfers electricity, free electron, the lowest negative charge equal to "-1", a structural unit of atom, a part of the atom planetary model, an element of electronic shell, a structural unit of atom with defined energy, an electron of external atomic level). Moreover, the study period of these meanings of electrons in chemistry does not coincide with the time of studying this concept in physics. Electrons are studied 13 weeks later in the selected course of physics. Three paragraphs are devoted to forming of the concept "electron"; these paragraphs describe three semantic states of the term (a negatively charged particle, an atomic structural unit, free electron).

At first sight, it is possible to find corresponding semantic states of the term "electron" in chemistry to all three semantic states of this term in physics. However, their semantic contents differ.

- a negatively charged particle, a structural unit of atom, a free electron.
- an atom particle that transfers electricity, a free electron; the lowest negative charge equal to "- 1", a structural unit of atom; a part of atom planetary model; an element of electronic shell, a structural unit of atom with defined energy, an electron of external atomic level.

For example, the first semantic state of the concept "electron" as a negatively charged particle in physics has a similar definition in chemistry as well. However, the course of physics says that an electron has the lowest negative charge and provides its quantitative value and measurement units; the course of chemistry does not take that into consideration and describes an electron as having a negative charge equal to "-1". It is surely possible to explain this discrepancy by an attempt to ensure a more convenient description of the topics such as "Degree of oxidation", "Oxidation-reduction reactions" and others. Nevertheless, the ambiguity of concept definitions used in academic courses and lack of connections between these two definitions result in confusion thus complicating the process of studying these fundamental concepts.

This dramatic difference is caused by the fact that in chemistry the concept "electron" as an atomic structural unit is formed based on the experiments by Thomson and Rutherford who have their own arsenal of concepts, for example α- и β-particles, γ rays, radioactivity (introduced to explain the experiment of Becquerel). However, neither these experiments nor the concepts related to them are used in the chemistry course for 8-9th grades again, but they overload the content of the paragraph and there are doubts about the practicability of structuring the course in this manner.

Besides, the graph of structure of the concept "electron" from the chemistry course contains three additional semantic states of the concept "electron" (an atom particle, which transfers electricity, a free electron, the lowest negative charge equal to (-1)) that are obviously unnecessary as the physical structure shows (contains only one additional semantic state of the concept "electron" as a negatively charged particle).

In order to reduce the volume of information, a detailed description of the Rutherford's experiment could be removed from the chemistry course while forming the fourth semantic state of the concept "electron" as a structural atom unit. Some of the remaining semantic states in the course of chemistry and physics for 8th grade are not used again, for example the concept of electron as a part of the atom planetary model.

4. Summary

Therefore, the problem of breaking in knowledge which is one of the main reasons of complexity to percept physics and chemistry may be solved with involving preliminary coherence assessment of disciplines' courses on the basis of the quantitative method of interdisciplinary links graph model. Good coherence of physics and chemistry courses assure developing an integral perception of natural phenomena for pupils and in its turn facilitates more effective learning of natural sciences and the educational process in general. Structuring content of physics and chemistry courses by defining semantic states of fundamental concepts, for example, an electron, helps to build integral and obvious inter-subject links between the chosen connected courses.

References

[1] Gnitetskaya T.N.2006. Information models of intra- and interdisciplinary connections as a basis of training physics technology. Dr. dissert., Moscow, – 2006. - 321 p.

[2] Peryshkin A.V. Physics. 7-9th grades: For Secondary-Education Institutions. – 7th reprint edition. - Moscow: "Drofa", 2011.

[3] Gabriyelyan O.S. Chemistry. 8-9th grades: Textbook for Secondary-Education Institutions. – 18th reprint edition. - Moscow: 2011.

[4] Gromov S.V., Rodina N.A. Physics. 7-9 grades: Textbook for Secondary-Education Institutions. – 3rd reprint edition. – Moscow: "Prosveshcheniye", 2003.

[5] Ivanova R.G. Chemistry. 8-9th grades: Textbook for Secondary-Education Institutions. – 6th reprint edition. – Moscow: "Prosveshcheniye", 2003.–287 p.: illust.

[6] Gnitetskaya T. Graph Model of Interdisciplinary Links as a Tool to Optimize the Content of Physics and Mathematics Courses Procedia - Social and Behavioral Sciences, Volume 89, 2013, Pages 139-143.

[7] Gnitetskaya T. Modeling of Interdisciplinary Connections in Science Courses //Journal of Physics: Conference Series. Volume 490, Issue 1, 2014. **DOI:** 10.1088/1742-6596/490/1/012068

[8] Gnitetskaya Tatyana, N. , Ivanova Elena, B. _ Semantic status of chemical concepts (Conference Paper) Advanced Materials Research Volume 550-553, 2012, Pages 3429-3432/ **DOI:** 10.4028/www.scientific.net/AMR.550-553.3429.

Time Fractional Diffusion Equations and Analytical Solvable Models

Evangelos Bakalis[1,2] and Francesco Zerbetto[1,2]

[1]Dipartimento di Chimica "G. Ciamician", Università di Bologna, V. F. Selmi 2, 40126, Bologna, Italy

[2]ebakalis@gmail.com, francesco.zerbetto@unibo.it

Abstract. The anomalous diffusion of a particle that moves in complex environments is analytically studied by means of the time fractional diffusion equation. The influence on the dynamics of a random moving particle caused by a uniform external field is taken into account. We extract analytical solutions in terms either of the Mittag-Leffler functions or of the M-Wright function for the probability distribution, for the velocity autocorrelation function as well as for the mean and the mean square displacement. Discussion of the applicability of the model to real systems is made in order to provide new insight of the medium from the analysis of the motion of a particle embedded in it.

1. Introduction

Diffusion is a non-equilibrium phenomenon. The Einstein-Smoluchowski (ES) equation, $<\vec{r}^2(t)>= 2dDt$, describes a diffusive process in homogeneous and isotropic environments, which is called normal diffusion. D is the diffusion coefficient in units (cm^2/s) and d is the dimensionality of the space where the motion evolves. It says that the variance of the displacement of a diffusing particle grows linearly in time. The above description is made in the frame of statistical physics where the probability density function (pdf) and its moments are used to describe the dynamics. For normal diffusion, the pdf is Gaussian, its moments are finite and only the first two of them are necessary to capture the entire dynamics [1]. Experimental evidence, however, in a broad range of phenomena that take place in diverse fields, that is, from physics to biology, and even to social and to economical sciences, indicate deviations from normal diffusion.

Complexity either spatial [2] and/or temporal [3] leads to divergent probability distributions, resulted of different time scales. Heterogeneities, entrapment, crowding, confinements, to name a few phenomena, give rise to individual sequences, which potentially evolve at similar time scales competing each other. In other words, divergence of microscopic time scales may be connected to the anomalous behavior in longer time scales, which means that the system gets some memory over time. The overall outcome reflects on the dynamics that differ radically from normal diffusion. This kind of motion is called anomalous diffusion, and the variance of the displacement, $W(t) =<\vec{r}^2(t)> - <\vec{r}(t)>^2$, grows as a power law of time

$$W(t) = \frac{2dK_a t^a}{\Gamma(1+a)} \tag{1}$$

Content from this work may be used under the terms of the Creative Commons Attribution 3.0 licence. Any further distribution of this work must maintain attribution to the author(s) and the title of the work, journal citation and DOI.
Published under licence by IOP Publishing Ltd

K_a is a generalized diffusion coefficient in units of (cm^2/s^a), $\Gamma()$ is the gamma function and a is the exponent that classifies the motion. For $a = 1$ the motion is truly Brownian and the ES equation is recovered, for $0 < a < 1$ the motion is sub-diffusive, for $1 < a < 2$ it is super-diffusive, and for $a = 2$ it becomes ballistic.

Deviation from normal diffusion mirrored on the pdf, which cannot be described by the classical Gaussian. Finding of exact propagators that incorporate all the interactions with the environment is feasible only under certain circumstances [4-5]. Toy models based on random walks and associated to specific stochastic scenarios have been introduced and describe anomalous diffusion [6-7]. Fractional Brownian motion (FBM) [8-9], continuous time random walk (CTRW) [10-11], Levy flights (LFs), and Levy walks (LWs) [12-13] are the most known. Each of them, and according to stochastic scenario that obeys, may describe diffusion process within complex systems if and only the stochastic mechanism associated to it is the only operative cause, which gives rise to anomalous behavior. Detailed studies showed that more than one stochastic mechanism may govern the motion of a diffusing particle and potentially lead to multifractal behavior. Subordination [14], mixed noises [15-17], confinement either result of an external field or result of environmental constrains [18], are, just to name a few, cases where more than one stochastic mechanisms are present. The method of the generalized moments [19], which has successfully been applied in diverse fields [20-23], is a good diagnostic tool towards distinguishing multifractality [24-25]. One-sided Levy stable distributions, $l_a(x)$, may describe pdf's of multifractal processes [25], and its characteristic function in Laplace space, $L[l_a(x)](s) = \int_0^\infty e^{-sx} l_a(x)dx = e^{-s^a}$, $0 < a < 1$, $s > 0$, is the well-known Kohlraush-Williams-Watts function [26] or stretched exponential. These functions, $l_a(x)$, find applications in many fields and especially in fractional kinetics [27-28]. It has been pointed out that the function $l_a(x)$ is the solution of the fractional Fokker-Planck equation in the force free case [29]. Fractional calculus can describe spatial and temporal complexity; see [30], and references therein, for an introduction to modeling complex phenomena by a fractional point of view. The application of fractional calculus in diffusion is straightforward since deals with fractional derivatives and integrals, which by definition incorporate memory terms that are responsible for anomalous diffusion [13, 29, 31-34].

In the present study we consider the time fractional diffusion equation with an additive term (reaction term) [35-36], which stands for the influence of a uniform external field. Notice that the same methodology may be followed for generalization of non-fractional diffusion model [37]. In each case, we obtain the pdf, the normalized mean and mean square displacement as well as the normalized velocity autocorrelation function. We provide criteria of estimating the generalized diffusion coefficient and the exponent that classifies the motion. The paper is organized as follows; in Section 2 we give the general mathematical framework, in Section 3 we discuss the influence of a uniform external field on a particle's random motion within complex environments, and we conclude in section 4.

2. The Mathematical Framework

The time fractional reaction diffusion equation of order a, $a \in (0,1)$, in the Riemann-Liouville sense, see also Appendix A for definitions, reads

$$\frac{\partial^a P}{\partial t^a} = LP + N(x,t) \tag{2}$$

$P \equiv P(x,x_0;t)$ is the pdf to find the particle at x at time t given that it was at x_0 at t_0. L is the operator that acts on the pdf, and has the form $L = K_a \frac{\partial^2}{\partial x^2} - \frac{K_a}{k_B T} F(x) \frac{\partial}{\partial x}$, where $F(x)$ stands for the force that a potential exerts on the moving particle, $k_B T$ is the Boltzmann's constant times the temperature of the bath. $N(x,t)$ is the reactive term and models modification on the overall motion of the particle that caused either by an external stimulus or by a conformational change of the particle itself (reaction). Eq. (2) describes the dynamics of a particle within the domain, $-b_1 < x < b_2$ (boundaries), and under the initial condition $P(x,x_0;0) = f(x)$. We apply natural boundary conditions, which mean $b_{1,2} \to \infty$ and $\lim_{x\to\pm\infty} P(x,x_0;t) = 0$. By setting $F(x) = 0$ the solution of eq. (2) in integral form reads

$$P(x,x_0;t) = f(x) + \frac{K_a}{\Gamma(a)}\int_0^t (t-\tau)^{a-1}\frac{\partial^2 P(x,x_0;\tau)}{\partial x^2}d\tau + \frac{1}{\Gamma(a)}\int_0^t (t-\tau)^{a-1}N(x,t-\tau)d\tau \tag{3}$$

By applying consecutive the integral transforms, $L\{f(t)\} = \int_0^\infty e^{-st} f(t)dt$, for the Laplace, and $F\{f(x)\} = \int_0^\infty e^{-ikx} f(x)dx$, for the Fourier transform, eq. (3) takes the form

$$\bar{\hat{P}}(k,s) = \frac{s^{a-1}\hat{f}(k)}{(s^a + k^2 K_a)} + \frac{\bar{\hat{N}}(k,s)}{(s^a + k^2 K_a)} \tag{4}$$

where, the bar/hat stands for the Laplace and the Fourier transform respectively. Eq. (4) is the characteristic function and contains all the information about the dynamics, which the particle followed, by means of the analytical form of all the moments of the pdf in Laplace space. More specifically, the zeroth moment is the normalization factor of the pdf, the first moment describes the un-normalized mean displacement and provides information about directionality of the motion, that is, if it is zero then all directions are equi-probable. The second moment gives the mean square displacement of the particle and classifies the motion either as normal or as sub/super diffusive. Notice that if the zero moment is different than one, then proper normalization is needed. More information about the pdf may be extracted, for example, the skewness and the kurtosis can also be obtained, which means that the third and the fourth moment ought to be obtained.

By considering that $f(x) = \delta(x - x_0)$, we write the moments up to the second order by using eq. (4)

$$m_0(s) = \int \bar{P}(x,s)dx = \left.\bar{\hat{P}}(k,s)\right|_{k=0} \tag{5}$$

$$m_1(s) = <(x-x_0)>(s) = i\left.\frac{d\bar{\hat{P}}(k,s)}{dk}\right|_{k=0} - x_0 m_0(s) \tag{6}$$

and

$$m_2(s) = <(x-x_0)^2>(s) = i^2\left.\frac{d^2\bar{\hat{P}}(k,s)}{dk^2}\right|_{k=0} - 2x_0 i\left.\frac{d\bar{\hat{P}}(k,s)}{dk}\right|_{k=0} + x_0^2 m_0(s) \tag{7}$$

Furthermore, assuming that the second dissipation fluctuation theorem (DFT) holds true, $W(t) = \frac{k_B T}{M}\int_o^t dt_1 \int_0^t C_v(t_1,t_2)dt_2$, then an analytical derivation of the normalized velocity autocorrelation function in Laplace space is feasible through the relation [38]

$$C_v(s) = \frac{M}{2k_B T}s^2 W(s) \tag{8}$$

where M is the mass of the particle, see also [39] for scaling Green-Kubo relation applicable to aging systems. By assuming that $N(x,t) = 0$ and by using eqs (4-8) we take; a) $m_0(s) = 1/s$, which means that the pdf is normalized since $L^{-1}(s^{-1}) = 1$, b) $m_1(s) = 0$ and underlines that there is no preferred direction of motion, c) $m_2(s) = 2K_a s^{-1-a}$, and d) $C_v(s) = \dfrac{MK_a}{k_B T} s^{1-a}$. Inverting in real space we take $<(x-x_0)^2>(t) = 2K_a \dfrac{t^a}{\Gamma(1+a)}$ for the mean square displacement, and $C_v(t) = \dfrac{MK_a}{k_B T}\dfrac{t^{-2+a}}{\Gamma(-1+a)}$ for the velocity autocorrelation function. Since, $0<a<1$ the particle performs sub-diffusive motion with anti-correlated steps as the velocity autocorrelation function shows because it goes to zero from negative as t^{-2+a}. The pdf of the diffusing particle is obtained by the inversion of the Fourier transform in eq. (4), which reads $\bar{P}(x,s) = \dfrac{\exp\left(-\sqrt{\frac{s^a}{K_a}}\left|x-x_0\right|\right)}{2K_a^{1/2} s^{1-a/2}}$ and its inversion in real time domain is feasible giving $P(x,x_0;t) = \dfrac{1}{\sqrt{4K_a t^a}} M_{a/2}\left(\dfrac{\left|x-x_0\right|}{\sqrt{K_a t^a}}\right)$, where $M_{a/2}(\ )$ is the M-Wright function [40]. It should be noticed that the above pdf has also been expressed in closed form either in terms of a Fox function [41] or in terms of a one sided Levy stable distribution [29]. We prefer the use of the M-Wright function since it is the generalization of the Gaussian density function [42].

3. Influence of a uniform external field.

Considering $N(x,t) = uP(x,x_0;t)$ and $f(x) = \delta(x-x_0)$, eq.(2) reads

$$\frac{\partial^a P(x,x_0;t)}{\partial t^a} = K_a \frac{d^2 P(x,x_0;t)}{dx^2} - uP(x,x_0;t) \tag{9}$$

where, from dimensional analysis u stands for a generalized frequency term in units (s^{-a}) which alters uniformly the probability distribution, increase/decrease so it can take either negative or positive values and for zero value the original time fractional diffusion equation is recovered. The corresponding to eq. (9) characteristic function reads

$$\bar{\hat{P}}(k,s) = s^{a-1}\frac{\exp(-ikx_0)}{(s^a + u + K_a k^2)} \tag{10}$$

The inverse Fourier transform of eq. (10) delivers

$$\bar{P}(x,s) = \frac{\exp\left(-\sqrt{\frac{s^a+u}{K_a}}\left|x-x_0\right|\right)}{2K_a^{1/2} s^{1-a}\sqrt{s^a+u}} \tag{11}$$

By using eqs (5-7) we write

$$m_0(s) = \frac{s^{a-1}}{s^a + u} \tag{12}$$

$$<m_2>(s) = \frac{2K_a s^{a-1}}{(s^a+u)^2} \tag{13}$$

and $<(x-x_0)>(s) = 0$. The latter is expected since the uniform way that the external stimulus acts on the complex environment cannot produce any preferred direction of motion. Finally, by using eq. (8) we write for the velocity autocorrelation function

$$C_v(s) = \frac{MK_a}{k_B T}\frac{s^{1+a}}{(s^a+u)^2} \tag{14}$$

For u=0 the well-known results discussed in the previous section are recovered. For u different than zero a more attractive physical picture is revealed. For the inversion of eqs (12), (13) and (14) we make use of the relations (A.4) and (A.5) [42], see Appendix A, and we write

$$m_0(t) = E_a(-ut^a) \tag{15}$$

$$< m_2 > (t) = 2K_a t^a E^2_{a,a+1}(-ut^a) \tag{16}$$

and

$$C_v(t) = \frac{MK_a}{k_B T} t^{a-2} E^2_{a,a-1}(-ut^a) \tag{17}$$

where $E_a(x)$, $E_{a,b}(x)$ and $E^k_{a,b}(x)$ are the one, two and three parameter Mittag-Leffler functions, see Appendix A for the definitions. By dividing eq.(16) with eq.(15) we take, $< (x - x_0)^2 >_{norm} (t) = 2K_a t^a E^2_{a,a+1}(-ut^a) / E_a(-ut^a)$, which describes the normalized mean square displacement. The asymptotic behavior of this quantity can be found either by using eq. (A.7) of Appendix A and retaining up to the first two terms of the expansion or by expanding eqs (12) and (13) for $s^a \rightarrow 0$ and inverting in real time domain. Both ended up with exactly the same result, which reads

$$\lim_{t\to\infty} < (x - x_0)^2 >_{norm} (t) \approx \frac{2K_a}{u} - \frac{4K_a}{u^2}\frac{\Gamma(1-a)}{\Gamma(1-2a)} t^{-a} \tag{18}$$

By following the same procedure the normalized velocity autocorrelation function is defined as, $C_v^{norm}(t) = \frac{MK_a}{k_B T} t^{a-2} E^2_{a,a-1}(-ut^a) / E_a(-ut^a)$ and in the long time limit its behavior reads

$$\lim_{t\to\infty} C_v^{norm}(t) \approx \frac{1}{u}\frac{\Gamma(1-a)}{\Gamma(-1-a)} t^{-a} \left\{1 + \left(2\frac{\Gamma(-1-a)}{\Gamma(-1-2a)} - \frac{\Gamma(1-a)}{\Gamma(1-2a)}\right)\frac{t^{-a}}{u}\right\} \tag{19}$$

The leading term of eq. (18) describes the full confinement of the particle since the mean square displacement reaches a plateau. This behavior is already known [18]. Eq. (18) may be used to estimate the diffusion coefficient of a particle that moves within a complex environment. Before the plateau, eq. (18) changes according to a power law. Therefore, these two terms of eq.(18) describe two different regimes. On the one hand, the value of the plateau corresponds to the ratio K_a / u, generalized diffusion coefficient per generalized frequency imposed by the external field. On the other hand, the second regime can give the exponent that classifies the motion and its slope correlates the exponent with the parameter u. By exploiting both regimes the values of a, K_a, and u can be explicitly given. It should be noticed, however, that confinement could potentially be the result of many factors that affect the diffusing particle. For example, reflected boundaries can lead to confinement [5], which are not taken into account in the current analysis. Additionally, and in order to ensure that the confinement on the motion is result of the uniform external field, the normalized autocorrelation velocity function should satisfy eq. (19).

4. Conclusions

In conclusion, we study the erratic motion of a particle within a complex environment under the influence of a uniform external field. We provide criteria under which properties of the system could be extracted, specifically, the generalized diffusion coefficient as well as the exponent that classifies the motion and the frequency term installed by the field. The mean square displacement in the long time limit indicates the existence of two regimes; the first one goes as a power law before reaches a plateau, where the plateau is the second one. This condition is not enough to ensure the kind of the motion; therefore, experimental evidence should also fulfil the normalized velocity autocorrelation function, eq.(19). The results presented herein may be helpfully in trying to determine complex systems from experimental results.

Appendix A

The one parameter Mittag-Leffler (M-L) function is defined as [44-45]

$$E_a(x) = \sum_{n=0}^{\infty} \frac{x^n}{\Gamma(an+1)} \tag{A.1}$$

where $x \in \mathbb{C}$, $\mathbb{R}(a) > 0$. The two parameter M-L function has been introduced and studied quite later [46-47].

$$E_{a,\beta}(x) = \sum_{n=0}^{\infty} \frac{x^n}{\Gamma(an+\beta)} \tag{A.2}$$

where $x, \beta \in \mathbb{C}$, $\mathbb{R}(a) > 0$. At the early seventies Prabhakar introduced the three parameter M-L function [48]

$$E_{a,\beta}^{k}(x) = \sum_{n=0}^{\infty} \frac{(k)_n}{\Gamma(an+\beta)} \frac{x^n}{n!} \tag{A.3}$$

$x, \beta, k \in \mathbb{C}$, $\mathbb{R}(a) > 0$ and $(k)_n = \Gamma(k+n)/\Gamma(k)$ is the Pochhammer symbol. From equations (A.1) to (A.3) the following identity, $E_{a,1}^{1}(x) = E_{a,1}(x) = E_a(x)$, is easily established. The Laplace transform of the two-parameter M-L function [43]

$$L\{t^{\beta-1}E_{a,\beta}(\pm bt^a)\} = \int_0^{\infty} e^{-st}t^{\beta-1}E_{a,\beta}(\pm bt^a)dt = \frac{s^{a-\beta}}{s^a \mp b} \tag{A.4}$$

where $\mathbb{R}(s) > |b|^{1/a}$. The Laplace transform of the three-parameter M-L function [43,48]

$$L\{t^{\beta-1}E_{a,\beta}^{k}(\pm bt^a)\}(s) = \frac{s^{ak-b}}{(s^a \mp b)^k} \tag{A.5}$$

where $\mathbb{R}(s) > |b|^{1/a}$. The asymptotic expansion for the three-paramater M-L function satisfies the following relations [49]

$$E_{a,b}^{k}(x) = \sum_{n=0}^{\infty} \frac{\Gamma(k+n)}{\Gamma(k)\Gamma(an+b)} \frac{x^n}{n!}, \quad |x| < 1 \tag{A.6}$$

and

$$E_{a,b}^{k}(x) = \frac{(-x)^{-k}}{\Gamma(k)} \sum_{n=0}^{\infty} \frac{\Gamma(k+n)}{\Gamma(b-a(k+n))} \frac{(-x)^n}{n!}, \quad |x| > 1 \tag{A.7}$$

References

[1] Feller W 1967 An Introduction to Probability Theory and its Applications, 3nd Ed (John Wiley & Sons.
[2] Newman M E J 2010 Networks, An Introduction, Oxford University Press, Oxford, New York.
[3] Turalska M, Lukovic M and West B J 2009 *Phys. Rev. E*, **80**, 021110.
[4] Kosmas M and Bakalis E 2006 *Phys. Lett. A* **358**, 354.
[5] Bakalis E 2012 *Physica A* **391**, 3093.
[6] Metzler R and Klafter J 2000 *Phys. Rep.* **339**, 1.
[7] Sokolov I M 2012 *Soft Matter* **8**, 9043.
[8] Kolmogorov A N 1940 *Dokl. Acad. Sci. USSR* **26**, 115.

[9] Mandelbort B B and van Ness J W 1968 *SIAM Rev.* **10**, 422.
[10] Montroll E W and Weiss G H 1965 *J. Math.Phys.* **6**, 167.
[11] Havlin S and Ben-Avraham D 1987 *Adv. Phys.* **36**, 695.
[12] Shlesinger M F, Zaslavsky G M and Klafter J 1993 *Nature* **363**, 31.
[13] Eliazar I I and Shlesinger M F 2013 *Phys. Rep.* **527**, 101.
[14] Tabei S M A et. al. 2013 *PNAS* **110**, 4911.
[15] Jeon J-H, Barkai E and Metzler R 2013 *J. Chem. Phys.* **139**, 121916.
[16] Sandev T and Tomovski Z 2014 *Phys. Lett. A*, 2014.
[17] Bakalis E and Zerbetto F 2016 "Erratic motion under composite noises" to be submitted.
[18] Burov S, Jeon J-H, Metzler R and Barkai E 2011 *Phys.Chem.Chem.Phys.* **13**, 1800.
[19] Andersen K H, Castiglione P, Mazzino A and Vulpiani A 2000 *Eur. Phys. J. B.* **18**, 447.
[20] Gal N and Weihs D 2010 *Phys. Rev. E.* **81**, 020903(R).
[21] Seuront L and Stanley H E 2014 *PNAS* **111**, 2206.
[22] Bakalis E, Höefinger S, Venturini A and Zerbetto F 2015 *J. Chem. Phys.* **142**, 215102.
[23] Sändig N, Bakalis E and Zerbetto F 2015 *Chem.Phys.Lett.* **633**, 163-168.
[24] Bacry E, Delour J and Muzy J F 2001 *Phys. Rev. E* **64**, 026103.
[25] Muzy J F and Bacry E 2002 *Phys. Rev. E* **66**, 1-16.
[26] Andersen R S, Husain S A and Loy R J 2004 *ANZIAM J.*, **45**, C800.
[27] Sokolov I M, Klafter J and Blumen A 2002 *Phys. Today*, No 11, **55**, 48.
[28] Chechkin A V et. al. 2008 *Phys. Rev. E*, **78**, 021111.
[29] Barkai E, Metzler R and Klafter J 2000 *Phys.Rev.E* **61**, 132.
[30] West B J 2014 *Rev. Mod. Phys.* **86**, 1169.
[31] Lutz E 2001 *Phys. Rev. E* **64**, 051106.
[32] Stanislavsky A A 2003 *Phys. Rev. E*, **67**, 021111.
[33] Barkai E and Silbey R J 2000 *J. Phys. Chem. B* **104**, 3866-3874.
[34] Seki K, Wojcik M and Tachiya M 2003 *J. Chem. Phys.* **119**, 2165.
[35] Henry B I and Wearne S L 2000 *Physica A*, **276**, 448.
[36] Henry B I, Langlands T A M and Wearne S L 2006 *Phys. Rev. E*, **74**, 031116.
[37] Bakalis E, Vlahos C and Kosmas M 2006 *Physica A* **360**, 1-16.
[38] Kneller G R 2011 *J. Chem. Phys.* **134**, 224106.
[39] Dechant A, Lutz E, Kessler D A and Barkai E 2014 *Phys. Rev. X* **41**, 011022.
[40] Mainardi F et.al. 2010 *Inter. J. Diff. Equat.*, 104505/doi: 10.1155/2010/104505.
[41] Schneider W R and Wyss W 1989 *J.Math.Phys.* **30**,134.
[42] Pagnini G 2013 *Fract. Calc. Appl. Anal.* **16**, 436-453/ doi: 10.2478/s135540-013-0027-6.
[43] Podlubny I 1999 Fractional Differential Equations, Acad. Press, San Diego.
[44] Mittag-Leffler C 1903 *C. R. Acad. Sci. Paris*, **137**, 554.
[45] Winman A 1905 *Acta Math.* **29**, 191.
[46] Agarwal R P 1953 *C. R. Acad. Sci. Paris* **236**, 2031.
[47] Humbert P 1953 *C. R. Acad. Sci. Paris* **236**, 1467.
[48] Prabhakar T R 1971 *Yokohama Math. J.* **19**, 7.
[49] Saxena R K, Mathai A M and Haubold H J 2004 *Astrophys. Space. Sci.* **209**, 299.

Journal of Physics: Conference Series **738** (2016) 012107 doi:10.1088/1742-6596/738/1/012107

A model to describe potential effects of chemotherapy on critical radiobiological treatments

D. Rodríguez-Pérez, M.M. Desco and J.C. Antoranz

Dep. Física Matemática y de Fluidos, Universidad Nacional de Educación a Distancia. c/ Senda del Rey, 9, 28040 - Madrid, Spain

E-mail: {daniel,mmdesco,jcantoranz}@dfmf.uned.es

Abstract. Although chemo- and radiotherapy can annihilate tumors on their own, they are also used in coadjuvancy: improving local effects of radiotherapy using chemotherapy as a radiosensitizer. The effects of radiotherapy are well described by current radiobiological models. The goal of this work is to describe a discrete radiotherapy model, that has been previously used describe high radiation dose response as well as unusual radio-responses of some types of tumors (e.g. prostate cancer), to obtain a model of chemo+radiotherapy that can describe how the outcome of their combination is a more efficient removal of the tumor. Our hypothesis is that, although both treatments haven different mechanisms, both affect similar key points of cell metabolism and regulation, that lead to cellular death. Hence, we will consider a discrete model where chemotherapy may affect a fraction of the same targets destroyed by radiotherapy. Although radiotherapy reaches all cells equally, chemotherapy diffuses through a tumor attaining lower concentration in its center and higher in its surface. With our simulations we study the enhanced effect of combined therapy treatment and how it depends on the tissue critical parameters (the parameters of the non-extensive radiobiological model), the number of "targets" aimed at by chemotherapy, and the concentration and diffusion rate of the drug inside the tumor. The results show that an equivalent chemo-radio-dose can be computed that allows the prediction of the lower radiation dose that causes the same effect than a radio-only treatment.

1. Introduction

Chemo- and radiotherapy are the two most often used treatments to deal with cancer tumors. Although these two treatments can remove tumors on their own, they are also used in coadjuvancy: chemotherapy, administered globally, may improve the effects of local radiotherapy; this synergistic effect is actually exploited in the clinic, where patients who received combined therapy had a significant increase in survival [1]. This radiotherapy enhancement, known as radiosensitisation by chemotherapy, has been investigated *in vitro* using mammalian normal and cancer cells [2]. These studies show that this interaction is more effective in cell lines which are relatively sensitive to chemotherapy.

The effect of radiotherapy is well modeled by linear-quadratic (LQ) models currently in use that predict cell survival fractions for low and fractionated doses [3]. However, for higher doses, such as those used in hypofractionated therapy, better predictions are obtained using a model derived from non-extensive Tsallis entropy [4]. In a previous paper [5] the former model was discretized and new predictions emerged, describing unusual behaviors of some types of tumors (e.g. prostate cancer) under radiotherapy, not well predicted by LQ model without introducing further hypotheses.

Content from this work may be used under the terms of the Creative Commons Attribution 3.0 licence. Any further distribution of this work must maintain attribution to the author(s) and the title of the work, journal citation and DOI.
Published under licence by IOP Publishing Ltd

Experimental results describe radiosensitization numerically as a change in LQ parameters [2]. Two different behaviors can be observed depending on the type of cells and the chemopharm used in the culture: one in which chemotherapy largely enhances the effects of radiotherapy and one in which its effect is less significant. And, in the first case, there are cases in which the enhancement does depend on the chemopharm dosage and cases in which it does not. We hypothesize that all these behaviors can be reproduced within the framework of our former radiotherapy discrete model based on random damage of units in the cells that increase their risk of failure, that is, of random death. This could provide a way to generalize to chemo+radiotherapy the non-extensive formulae that described fractioned radiotherapy based on the Tsallis entropy model [6, 7].

Hence, our hypothesis is that, although both chemo- and radiotherapy have different mechanisms, both affect similar key points of cell metabolism and regulation, that may lead equally to cellular death. The goal of this paper is to model radiosensitization effects using the radiotherapy model in [5].

2. Methodology

We will consider a discrete model that starts with a population of N cells, each with identical T_0 units. Upon being hit by a damage event, each cell has a death risk inversely proportional to the number of remaining intact units T so that

$$\frac{\delta N}{N} = \gamma \frac{\delta T}{T} \tag{1}$$

being $\delta T = 1$, the target that will be efaced after the event. In [5] this model was introduced considering damage by radiation, so that taking the continuum limit the survival fraction F_r after a dose D (proportional to the average number of damage events per cell) could be obtained as [4]

$$F_r = \left(1 - \frac{D}{D_0}\right)^{\gamma} \tag{2}$$

As the number of damaged units, $T_0 - T$, is proportional to the dose given D, thus the critical dose D_0 is proportional to the number of units T_0 in the intact cell.

We model chemotherapy following the same rules, although considering the possibility that only a number T_q of the initial T_0 units can be affected by a given chemopharm, and that the rate at which units are removed may depend on the local concentration $C(r)$ of the compound, which may change across the tumor due to diffusion or angiogenesis [8]. We consider a radial exponential distribution of the form

$$C(r) = \frac{1}{K} \exp(-r \log K) \tag{3}$$

where K stands for the ratio of the concentrations on the boundary (where we assume $C(R) = 1$) and at the center of the tumor ($C(0) = 1/K$).

The simulation of chemo-sensitized radiotherapy will be carried out following these steps:

- chemotherapy sensitization: chemotherapy is simulated for the given γ and T_0 of the tumor cells and the $C(r)$ chemopharm distribution within the tumor, defined by K. The simulation is run starting from the intact tumor until the number of cells contained in it reaches a fraction F_q characteristic of the intensity of chemotherapy treatment.
- radiotherapy cycles: radiotherapy is simulated on the cells remaining after chemotherapy. The simulation is run continuously and the number of living cells is recorded after given dosages. In order to have a dimensionless dosage unit, it is taken as that number of events required to halve the number of cells in the intact tumor (i.e. prior to chemotherapy).

Hence, we will study the consequences that chemotherapy model parameters T_q, K and F_q may have on the radiotherapy dose-survival functions of a given tumor (described by γ and T_0).

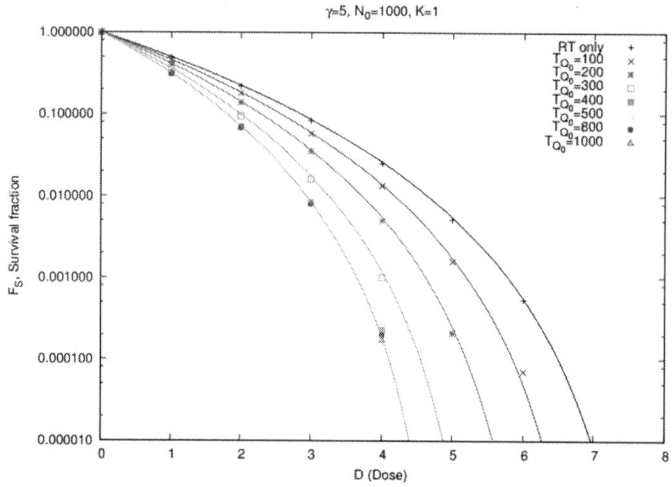

Figure 1. Simulated radiotherapy survival fractions after radiosensitization with a chemopharm, for homogenous tumours ($\gamma = 5$, $T_0 = 1000$, $K = 1$) with different fractions of chemosensitive units: $T_q/T_0 = 0.1, 0.2, 0.3, 0.4, 0.5, 0.8, 1$. Survival fractions of radio-only therapy are also shown.

3. Results

Let us first examine the effect of T_q, the number of units chemotherapy can harm in the cell, on radiotherapy dose-survival curves. Figure 1 shows the effects on a tumor totally permeable to the chemopharm ($K = 1$) under a chemotherapy reduction effects of $F_q = 0.10$. It can be seen that larger reductions (small F_q) do depend on T_q, whereas smaller reductions become independent of T_q.

The effect of the chemopharm distribution along the tumor is shown in Figure 2 for different values of $K = 1, 10, 100$: the larger the tumor permeability to chemopharm, the more acute the effect of radiotherapy.

Finally, the intensity of chemotherapy, measured from its reduction effect F_q, also increases the effectiveness of radiotherapy, according to Figure 3. As in the former cases, isoeffect doses can be obtained as shifts in the critical dose, that can be interpreted as an equivalent chemo-radio-dose administered through chemotherapy.

4. Discussion

A consequence of describing the two therapies through a unified chemo-radiotherapy model is that two apparently unrelated mechanisms can be described in the common language of therapy session combinations [6, 7], helping to interpret (and predict) the radiosensitizing effects of some chemopharms. This is in agreement with the fitted curves in Figures 4a,b : these curves are just the corresponding radiotherapy-only curves transformed to a rescaled dose. The experimental situation would be similar to that represented in Figures 3 for increasing concentrations of the chemopharm, causing a decrease of $F_q = 0.10, 0.25, 0.50$ of the initial tumor volume (after shrinkage). In terms of the model, a fraction of units equal to

$$\frac{\Delta T}{T_0} = 1 - F_q^{1/\gamma} \tag{4}$$

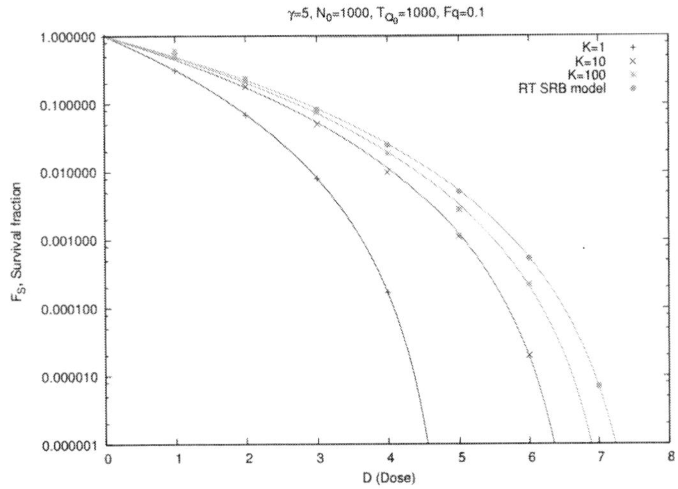

Figure 2. Simulated radiotherapy survival fractions after radiosensitization for tumours ($\gamma = 5$, $T_q/T_0 = 1$, $F_q = 1$) with varying chemopharm perfusion: $K = 1$, 10, 1000. Survival fractions of radio-only therapy are also shown.

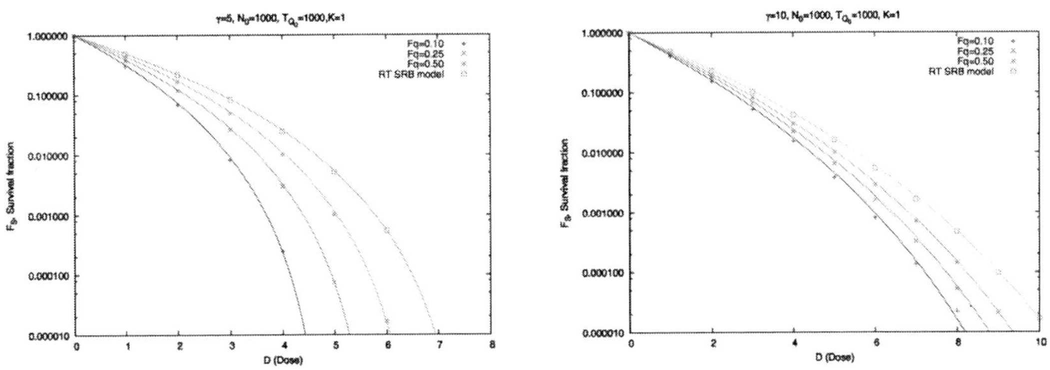

Figure 3. Simulated radiotherapy survival fractions after radiosensitization of tumors with different initial radiosensitivities ($\gamma = 5$ and 10, $T_q/T_0 = 1$, $K = 1$) and different chemotherapy intensities: $F_q = 0.1$, 0.25, 0.5. Survival fractions of radio-only therapy are also shown.

has been damaged by chemotherapy, thus the radiotherapy survival curves will behave as

$$F_s = \left(1 - \frac{D}{D_0 - D_q}\right)^{\gamma} \tag{5}$$

where

$$D_q = D_0 \frac{\Delta T}{T_0} \tag{6}$$

which is in good agreement with the simulations and provides for the data in [2] values of $D_q/D_0 = 0.17$ (Figure 4a, where the dose of the radiotherapy-only curve has been rescaled by 0.83 to adjust the cisplatin-radiosensitized curve) and $D_q/D_0 = 0.31$ (Figure 4b, where dose has

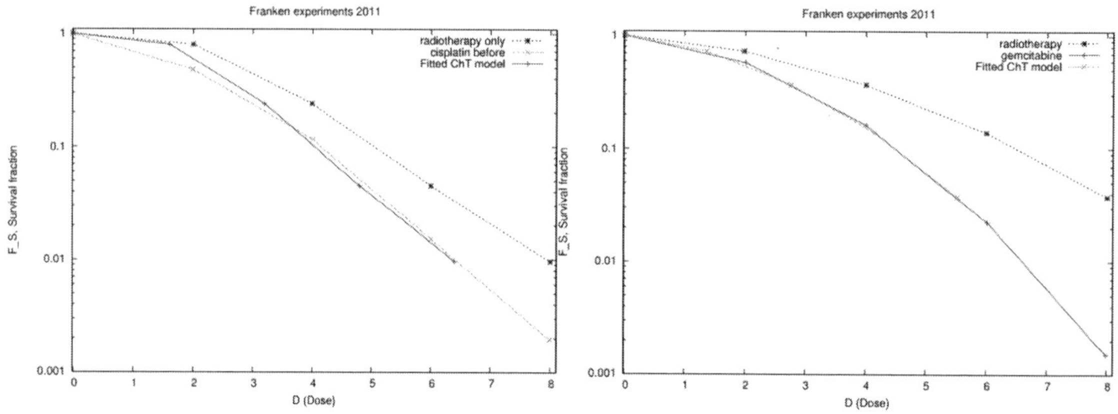

Figure 4. Experimental data of radiotherapy survival fractions extracted from [2]: (left) after treatment with cisplatin $1\,\mu$M during one hour, (right) emcitabine 100 nM during 24 h. Also shown are the radiotherapy only survival fraction and the dose-rescaled version of this curve that best matches the radiosensitized survival data (in accordance with, Eq. (5)).

been rescaled by 0.69 to match the gemcitabine-radiosensitized curve). That is, according to our model, those would be the chemopharm equivalent radio-doses of the cisplatin $1\,\mu$M acting during one hour, and gemcitabine 100 nM during 24 h.

Under these assumptions (homogeneous distribution of the chemopharm across the tumor), the chemopharm equivalent radio-dose D_q will be given only by its effect on the tumor size. E.g. given a reduction fraction F_q of the tumoral cells attained at a unit concentration ($C = 1$). However, according to our hypotheses this behavior can be limited by two other factors: the number of chemosensitive units T_q and the chemopharm distribution. The former must only be taken into account for very intensive chemotherapy. Figure 2 shows that above a given number of chemosensitive units (around 36% of the total in that particular case) chemotherapy effect is no longer dependent of T_q: there are enough radio-sensitive units to be damaged during the chemotherapy treatment, thus radiotherapy starts from a tumor that has a number of intact units given only by F_q. For tumors with less radio-sensitive units, radiosensitization will be imposed by that number T_q, for a given F_q.

On the other hand, when tomor perfusion is not homogeneous, an outer shell will be more affected by the chemopharm than the inner core, about K times more affected. That means that for a low dose of chemopharm about K times more units will be damaged in the outer cells than in the inner ones. For intense chemotherapy, the outer shell can be totally removed, while the inner core will have about $1/K$ times less unit damages than would a similar homogenous tumor. According to our model, the radiosensitized radiotherapy curves would behave as

$$F_s \simeq \left(1 - \frac{D}{D_0 - D_q/K}\right)^{\gamma} \tag{7}$$

Although this expression qualitatively explains why the curves having larger K in figure 2 get closer to the radiotherapy-only curve (the rightmost one), the expression is not straightforwardly applicable: the effective values of D_q/K that fit the simulation data are much larger than the ones expected from F_q alone; that is, radiosensitization is much larger than expected from chemopharm perfusion in the tumor. This can be explained by the rough approximation made in (7) and the assumption that all the outer shell is removed by chemotherapy, which is not true for less aggressive chemotherapies.

5. Conclusions

The simple discrete model described above seamlesly combines chemotherapy and radiotherapy in one single model, so that a equivalent chemo-radio-dose can be defined. This allows a simple derivation of the chemo+radiotherapy survival fraction and predicts the influence of other factors on radiosensitization, such as chemosensitivity or chemopharm perfusion (that is, its radial concentration profile) inside a tumor.

Acknowledgements

Authors acknowledge the financial support from the Spanish Ministerio de Economía y Competitividad under project FIS2012-37408.

6. References

[1] Herskovic A, Martz K, Al-Sarraf M *et al.* 1992 *Combined Chemotherapy and Radiotherapy compared with Radiotherapy Alone in Patients with Cancer of the Esophagus.* N. Engl. J. Med. **326** 1593-1598

[2] Franken N A P, Hovingh S, Rodermond H, Stalpers L, Barendsen G W and Crezee J 2011 *Radiosensitization with Chemotherapeutic Agents and Hyperthermia: Effects on Linear-quadratic Parameters of Radiation Cell Survival Curves.* J. Cancer Sci. Ther. **S5** 002

[3] Douglas B G and Fowler J F 1976 *The effect of multiple small doses of X rays on skin reactions in the mouse and a basic interpretation.* Radiat. Res. **66** 401

[4] Sotolongo-Grau O, Rodriguez-Perez D, Antoranz J C and Sotolongo-Costa O 2010 *Tissue radiation response with maximum Tsallis entropy.* Phys. Rev. Lett. **105** 158105

[5] Rodriguez-Perez D, Sotolongo-Grau O, Desco M M and Antoranz J C 2015 *Stochastic effects in a discrete RT model with critical behaviour.* J. Physics: Conference Series **633** 012089

[6] Sotolongo-Grau O, Rodríguez-Pérez D, Sotolongo-Costa O and Antoranz J C 2013 *Tsallis entropy approach to radiotherapy treatments.* Physica A **392** 2007

[7] Rodriguez-Perez D, Sotolongo-Grau O, Sotolongo-Costa O and Antoranz J C 2014 *Radiotherapy treatments by using Tsallis entropy statistical approach.* J. Physics: Conference Series **490** 012132

[8] Wu J, Long Q, Xu S and Padroni A R 2009. *Study of tumor blood perfusion and its variation due to vascular normalization by anti-angiogenic therapy based on 3D angiogenic microvasculature.* J. Biomechanics **42** 712

The generalized susceptibility of dislocation segment in nondissipative crystal

I L Bataronov and V V Dezhin
Voronezh State Technical University, Moskovsky prospect 14, Voronezh 394026, Russian Federation

E-mail: viktor.dezhin@mail.ru

Abstract. Oscillations of a dislocation segment dependence on the external forces are considered. Expressions are obtained for matrix elements which are expressed by means of the generalized susceptibilities matrix of dislocation oscillators. Elimination of non-physical features in the range of large wave numbers was made within the Peierls dislocation model. An expression was obtained describing the low-frequency asymptotic behavior of the diagonal and non-diagonal elements of the generalized susceptibilities matrix of dislocation oscillators.

1. Introduction

The important role of dislocations in the formation of mechanical, thermodynamic, electrical and other properties of crystals is a well known fact nowadays. Therefore, the vibrations of a crystal having dislocations are of considerable practical interest. For a crystals with a simple lattice the problem of vibrations of crystal dislocation and the phonon scattering by dislocations was solved in [1-3] using self-consistent dynamic theory of dislocations. In [4-7] generalized susceptibility endless dislocation for different types of crystals was found, which determines the response of dislocations to an external force. In this paper we describe the obtained the generalized susceptibility inverse matrix and its analysis.

2. Calculation of the generalized susceptibility dislocation oscillators

Neglecting the elastic fields of defects that perpetuate this dislocation, the equation of dislocation motion can be obtained by putting the respective boundary conditions on the displacement of the points in the dislocation line $\xi(l,t)$ in the pinning points, where l is the coordinate along the dislocation line, t is time. In the case of a dislocation pinning under consideration we require that $\xi(l,t)=0$ for $l\in(-\infty,-L/2]\cup[L/2,\infty)$, where L is the length of a dislocation segment. Under these conditions, the equation oscillations dislocation [5] after the conversion which eliminates the divergence, will take the form

$$\int_{-L/2}^{L/2} dz' G_0(z-z',\omega)\xi(z',\omega)=f(z,\omega). \qquad (1)$$

Here z is axis extending along of the dislocation line, ω is frequency. Function $\xi(z,\omega)$ and external disturbance $f(z,\omega)$ will be sought in the form of trigonometric series satisfying the boundary conditions at the points of attachment $z=\pm L/2$:

Content from this work may be used under the terms of the Creative Commons Attribution 3.0 licence. Any further distribution of this work must maintain attribution to the author(s) and the title of the work, journal citation and DOI.
Published under licence by IOP Publishing Ltd

$$\xi(z,\omega)=\sum_{n=1}^{\infty}Q_n(\omega)\varphi_n(z),\qquad f(z,\omega)=\sum_{n=1}^{\infty}F_n(\omega)\varphi_n(z),\tag{2}$$

where is indicated $\varphi_n(z)=\begin{cases}\cos(\pi nz/L), & n=1,3,5,\ldots\\ \sin(\pi nz/L), & n=2,4,6,\ldots\end{cases}$. Substituting expressions (2) to the equation (1), we obtain, after some transformations, the matrix equation

$$\sum_n B_{mn}(\omega)Q_n(\omega)=F_m(\omega),\tag{3}$$

where the matrix elements B_{mn} are defined as:

$$B_{mn}(\omega)=\frac{2}{L}\int_{-L/2}^{L/2}\mathrm{d}z\int_{-L/2}^{L/2}\mathrm{d}z'G_0(z-z',\omega)\varphi_n(z')\varphi_m(z).\tag{4}$$

According to the linear response theory, we write down

$$Q_n(\omega)=\sum_m \alpha_{nm}(\omega)F_m(\omega),\tag{5}$$

where $\hat{\alpha}$ is the generalized susceptibilities matrix of dislocation oscillators. Comparing equations (3) and (5), we obtain $\hat{\alpha}(\omega)=\hat{B}^{-1}(\omega)$. Integral (4) calculation result in

$$B_{mn}=(-1)^{[m/2]+[n/2]}\frac{\pi mn}{2L^2}\mu\left\{3b_s^2L_{mn}^{t1}+2\gamma(b_e^2-2b_s^2)L_{mn}^{l1}-b^2q_t^2L_{mn}^{t0}-\gamma b_e^2q_l^2L_{mn}^{l0}-\frac{4b_s^2-b_e^2}{q_t^2}L_{mn}^{(2)}\right\}.\tag{6}$$

Here μ is the shear modulus of the crystal, $\boldsymbol{b}$ is Burgers vector of the dislocation, b_e and b_s are edge and screw components of the Burgers vector, $q_l=\omega L/c_l$, $q_t=\omega L/c_t$, c_t and c_l are the transverse and longitudinal sound velocities, $\gamma=c_t^2/c_l^2$. For $m\neq n$ case we have

$$L_{mn}^{\alpha k}=\frac{1}{2}\frac{\mathrm{i}}{q_n^2-q_m^2}\Big\{q_n^{2k-1}\left[\mathrm{Ein}(-\mathrm{i}(q_\alpha-q_n))-\mathrm{Ein}(-\mathrm{i}(q_\alpha+q_n))\right]$$
$$-q_m^{2k-1}\left[\mathrm{Ein}(-\mathrm{i}(q_\alpha-q_m))-\mathrm{Ein}(-\mathrm{i}(q_\alpha+q_m))\right]\Big\}-(-1)^m\exp(\mathrm{i}q_\alpha)\delta_{k,2},$$

$L_{mn}^{(2)}=L_{mn}^{t2}-L_{mn}^{l2}+\mathrm{i}(q_t-q_l)$, where $q_m=\pi m$, $q_n=\pi n$, $\mathrm{Ein}(z)=\int_0^z\frac{1-\exp(-t)}{t}\mathrm{d}t$ is the integer part of the integral exponential function [8], δ_{ik} is Kronecker's symbol. For $m=n$ case we have

$$L_{nn}^{\alpha k}=\frac{q_n^{2k-2}}{4}\left\{2\mathrm{C}+2\ln(k_{\mathrm{D}}L)+2\mathrm{i}q_\alpha\frac{1-(-1)^n\exp(\mathrm{i}q_\alpha)}{q_n^2-q_\alpha^2}-\left[\mathrm{Ein}(-\mathrm{i}(q_\alpha+q_n))+\mathrm{Ein}(-\mathrm{i}(q_\alpha-q_n))\right]\right\}$$
$$+\frac{1-2k}{4}\mathrm{i}q_n^{2k-3}\left[\mathrm{Ein}(-\mathrm{i}(q_\alpha+q_n))-\mathrm{Ein}(-\mathrm{i}(q_\alpha-q_n))\right],$$

where k_{D} is the Debye wave number, $\mathrm{C}\approx 0{,}577$ is Euler's constant,

$$L_{nn}^{(2)}=L_{nn}^{t2}-L_{nn}^{l2}+\mathrm{i}(q_t-q_l)-(-1)^n\left[\exp(\mathrm{i}q_t)-\exp(\mathrm{i}q_l)\right].$$

3. Matrix elements in the Peierls model

Expressions (6) obtained in section 2 are characterized by the fact that they lead to nonphysical features in the linear dislocation model in the range of wave numbers in approximation to k_{D}. These features can be eliminated by means of calculations within the frame of Peierls dislocation model which can taken into account with the respective analytic continuation of generalized susceptibility expressions of the infinite dislocation beyond k_{D}. In the process of calculation of the matrix elements for dislocation oscillators it leads to the necessity to single out and specify the Δ amendments. For non-diagonal elements ($m \neq n$) we have

$$\Delta_{mn} = \mathrm{i} \int_{k_D L}^{\infty} \mathrm{d}(q_l + \mathrm{i}x)[1-(-1)^m \exp(\mathrm{i}(q_l + \mathrm{i}x))] \frac{\gamma[2(q_l + \mathrm{i}x)^2(2b_s^2 - b_e^2) + b_e^2 q_l^2] - (4b_s^2 - b_e^2)(q_l + \mathrm{i}x)^4 / q_l^2}{[(q_l + \mathrm{i}x)^2 - q_m^2][(q_l + \mathrm{i}x)^2 - q_n^2]}$$

$$+\mathrm{i} \int_{k_D L}^{\infty} \mathrm{d}(q_t + \mathrm{i}x)[1-(-1)^m \exp(\mathrm{i}(q_t + \mathrm{i}x))] \frac{b^2 q_t^2 - 3b_s^2(q_t + \mathrm{i}x)^2 + (4b_s^2 - b_e^2)(q_t + \mathrm{i}x)^4 / q_t^2}{[(q_t + \mathrm{i}x)^2 - q_m^2][(q_t + \mathrm{i}x)^2 - q_n^2]}. \quad (7)$$

Integrating the (7) and taking into account the main summands, we obtain

$$\Delta_{mn} \approx \frac{1}{2} \frac{1}{q_n^2 - q_m^2} \left[-\left((3-4\gamma)b_s^2 + 2\gamma b_e^2\right)\left(q_n \mathrm{arctg} \frac{2q_n}{k_D L} - q_m \mathrm{arctg} \frac{2q_m}{k_D L} \right) \right.$$

$$\left. + (b^2 + \gamma^2 b_e^2) q_t^2 \left(\frac{1}{q_n} \mathrm{arctg} \frac{2q_n}{k_D L} - \frac{1}{q_m} \mathrm{arctg} \frac{2q_m}{k_D L} \right) \right].$$

Thus, the expression for the elements of inverse matrix generalized susceptibility in case of $m \neq n$ will be shown here:

$$(\alpha^{-1})_{mn} = (-1)^{[m/2]+[n/2]} \frac{\pi mn}{4L} \mu \Big\{ 3b_s^2 L_{mn}^{t1} + 2\gamma(b_e^2 - 2b_s^2) L_{mn}^{l1} - b^2 q_t^2 L_{mn}^{t0} - \gamma b_e^2 q_l^2 L_{mn}^{l0}$$

$$- (4b_s^2 - b_e^2) L_{mn}^{(2)} / q_t^2 - \Delta_{mn} \Big\}. \quad (8)$$

Let us now consider the diagonal elements of the generalized susceptibility inverse matrix ($m = n$). In this case, the integrals should be calculated:

$$\Delta_{nn} = \mathrm{i} \int_{k_D L}^{\infty} \mathrm{d}(q_l + \mathrm{i}x)[1-(-1)^m \exp(\mathrm{i}(q_l + \mathrm{i}x))] \frac{\gamma[2(q_l + \mathrm{i}x)^2(2b_s^2 - b_e^2) + b_e^2 q_l^2] - (4b_s^2 - b_e^2)(q_l + \mathrm{i}x)^4 / q_t^2}{[(q_l + \mathrm{i}x)^2 - q_n^2]^2}$$

$$+\mathrm{i} \int_{k_D L}^{\infty} \mathrm{d}(q_t + \mathrm{i}x)[1-(-1)^m \exp(\mathrm{i}(q_t + \mathrm{i}x))] \frac{b^2 q_t^2 - 3b_s^2(q_t + \mathrm{i}x)^2 + (4b_s^2 - b_e^2)(q_t + \mathrm{i}x)^4 / q_t^2}{[(q_t + \mathrm{i}x)^2 - q_n^2]^2}. \quad (9)$$

Integrating (9) and taking into account the main components, we obtain

$$\Delta_{nn} \approx -\frac{1}{2} \left[\left((3-4\gamma)b_s^2 + 2\gamma b_e^2\right)\left(\frac{1}{2q_n} \mathrm{arctg} \frac{2q_n}{k_D L} + \frac{k_D L}{(k_D L)^2 + q_n^2} \right) \right.$$

$$\left. + (b^2 + \gamma^2 b_e^2) \frac{q_t^2}{q_n^2} \left(\frac{1}{2q_n} \mathrm{arctg} \frac{2q_n}{k_D L} - \frac{k_D L}{(k_D L)^2 + q_n^2} \right) \right].$$

Thus, the expression for the elements of the generalized susceptibility inverse matrix in case of $m = n$ takes the form shown here:

$$(\alpha^{-1})_{nn} = \frac{\pi n^2}{4L} \mu \left\{ 3b_s^2 L_{nn}^{t1} + 2\gamma(b_e^2 - 2b_s^2) L_{nn}^{l1} - b^2 q_t^2 L_{nn}^{t0} - \gamma b_e^2 q_l^2 L_{nn}^{l0} - \frac{4b_s^2 - b_e^2}{q_t^2} L_{nn}^{(2)} - \Delta_{nn} \right\}$$

$$+ \frac{\mu}{16\pi L} \Big\{ q_n^2 \left[(5-4\gamma) b_s^2 + (1-\gamma) b_e^2 \right] - q_t^2 \left[b_s^2 + (1-\gamma^2) b_e^2 / 2 \right] \Big\}. \quad (10)$$

The requirement to use the second summand in (10) was established in [9].

4. Matrix elements in the low-frequency limit

Is of practical interest to consider the low-frequency asymptotic formulas (8) and (10). For this purpose, we transform the expressions $L_{mn}^{\alpha k}$ and expand them to powers of q_α. Then, for the non-diagonal elements ($m \neq n$) we have:

$$L_{mn}^{\alpha 1} \approx \left(\frac{q_m \mathrm{Si}(q_m) - q_n \mathrm{Si}(q_n)}{q_n^2 - q_m^2} + \frac{(-1)^n q_\alpha^4}{4 q_n^2 q_m^2} \right) + \mathrm{i} \frac{\left(1 - (-1)^n\right) q_\alpha^3}{3 q_n^2 q_m^2},$$

$$q_t^2 L_{mn}^{\alpha 0} \approx \frac{q_t^2}{q_n^2 - q_m^2} \left(\frac{\mathrm{Si}(q_m)}{q_m} - \frac{\mathrm{Si}(q_n)}{q_n} \right) + \mathrm{i} \frac{1 - (-1)^n}{q_n^2 q_m^2} q_t^2 q_\alpha,$$

$$\frac{L_{mn}^{(2)}}{q_t^2} \approx \frac{(-1)^n (1 - \gamma^3)}{6 q_n^2 q_m^2} q_t^4 + \mathrm{i} \frac{(1 - (-1)^n)(1 - \gamma^{5/2})}{5 q_n^2 q_m^2} q_t^3.$$

Here in the real parts, summands $\sim q_\alpha^4$ can be neglected, and all imaginary parts have the same order of $\sim q_\alpha^3$. Therefore we may write down the expression (8) explicitly in the low-frequency limit:

$$B_{mn} = (\alpha^{-1})_{mn} = (-1)^{[m/2]+[n/2]} \frac{\mu}{4\pi L} \left(C_{mn} - M_{mn} q_t^2 + \mathrm{i} \Gamma_{mn} q_t^3 \right), \tag{11}$$

where we have introduced the reciprocal rigidity matrix C_{mn}, mass coefficient M_{mn} and damping coefficients Γ_{mn} of the dislocation oscillators, respectively:

$$C_{mn} = \left[(3 - 4\gamma) b_s^2 + 2\gamma b_e^2 \right] \frac{q_m q_n}{q_n^2 - q_m^2} \left[q_m \mathrm{Si}(q_m) - q_n \mathrm{Si}(q_n) + \frac{1}{2} \left(q_n \operatorname{arctg} \frac{2 q_n}{k_\mathrm{D} L} - q_m \operatorname{arctg} \frac{2 q_m}{k_\mathrm{D} L} \right) \right],$$

$$M_{mn} = (b^2 + \gamma^2 b_e^2) \frac{q_n q_m}{q_n^2 - q_m^2} \left[\frac{\mathrm{Si}(q_m)}{q_m} - \frac{\mathrm{Si}(q_n)}{q_n} + \frac{1}{2} \left(\frac{1}{q_n} \operatorname{arctg} \frac{2 q_n}{k_\mathrm{D} L} - \frac{1}{q_m} \operatorname{arctg} \frac{2 q_m}{k_\mathrm{D} L} \right) \right],$$

$$\Gamma_{mn} = -\frac{4}{5} \frac{1 - (-1)^n}{q_m q_n} b^2 \left(1 + \frac{2}{3} \gamma^{5/2} \right).$$

Now we shall define the low-frequency asymptotic behavior of diagonal elements ($m = n$). To do this, we use the expansions of functions with q_α powers. With these expansions we obtain:

$$q_\alpha^2 L_{nn}^{\alpha 0} = \frac{1}{4} \left\{ \left[2 \ln(k_\mathrm{D} L \mathrm{e}^\mathrm{C}) - 2 \mathrm{Cin}(q_n) + 2 \frac{\mathrm{Si}(q_n)}{q_n} \right] \frac{q_\alpha^2}{q_n^2} + 4 \mathrm{i} \frac{1 - (-1)^n}{q_n^4} q_\alpha^3 \right\},$$

$$L_{nn}^{\alpha 1} = \frac{1}{4} \left\{ 2 \ln(k_\mathrm{D} L \mathrm{e}^\mathrm{C}) - 2 \mathrm{Cin}(q_n) - 2 \frac{\mathrm{Si}(q_n)}{q_n} + \frac{q_\alpha^2}{q_n^2} + \frac{4}{3} \mathrm{i} \frac{1 - (-1)^n}{q_n^4} q_\alpha^3 \right\},$$

$$\frac{L_{nn}^{(2)}}{q_t^2} = \frac{1 - \gamma}{4} + \frac{1 - \gamma^2}{8} \frac{q_t^2}{q_n^2} + \mathrm{i} \frac{1 - (-1)^n}{5 q_n^4} (1 - \gamma^{5/2}) q_t^3.$$

Then the low-frequency asymptotic behavior of elements B_{nn} is expressed by formula (11) for the non-diagonal elements $B_{nn} = (\alpha^{-1})_{nn} = \frac{\mu}{4\pi L} \left(C_{nn} - M_{nn} q_t^2 + \mathrm{i} \Gamma_{nn} q_t^3 \right)$, where we have now

$$C_{nn} = \frac{1}{2} \left(2\gamma b_e^2 + (3 - 4\gamma) b_s^2 \right) \left(\mathrm{C} + \ln(k_\mathrm{D} L) - \mathrm{Cin}(q_n) - \frac{\mathrm{Si}(q_n)}{q_n} + \frac{1}{2 q_n} \operatorname{arctg} \frac{2 q_n}{k_\mathrm{D} L} + \frac{k_\mathrm{D} L}{(k_\mathrm{D} L)^2 + q_n^2} \right) q_n^2,$$

$$M_{nn} = \frac{1}{2} (b^2 + \gamma^2 b_e^2) \left(\mathrm{C} + \ln(k_\mathrm{D} L) - \mathrm{Cin}(q_n) + \frac{\mathrm{Si}(q_n)}{q_n} + \frac{1}{2 q_n} \operatorname{arctg} \frac{2 q_n}{k_\mathrm{D} L} - \frac{k_\mathrm{D} L}{(k_\mathrm{D} L)^2 + q_n^2} \right),$$

$$\Gamma_{nn} = -\frac{4}{5} \frac{1 - (-1)^n}{q_n^2} b^2 \left(1 + \frac{2}{3} \gamma^{5/2} \right).$$

It is obvious from the recorded expressions that C_{nn} and M_{nn} unlike C_{mn} and M_{mn} contain summands $\ln(k_{\mathrm{D}} L e^{\mathrm{C}})$, so the diagonal matrix elements exceed significantly the non-diagonal ones in their absolute value.

5. Conclusion

The results obtained in this paper can be used to find the oscillation eigenfrequencies of dislocation segment and their damping coefficients, for the calculation of dynamic characteristics of a dislocation segment and the dislocation internal friction.

References

[1] Ninomiya T and Ishioka Sh 1967 *J. Phys. Soc. Japan* **23** 361
[2] Ninomiya T 1968 *J. Phys. Soc. Japan* **25** 830
[3] Ninomiya T 1970 *Fundamental Aspects of Dislocations Theory*, ed J A Simmons, R de Wit and R Bullough (Nat. Bur. Stand. (U.S.) Spec. Publ. 317) p 315
[4] Dezhin V V, Nechaev V N and Roshchupkin A M 1989 *Ferroelectrics Letters* **10** 155
[5] Bataronov I L, Dezhin V V and Roshchupkin A M 1993 *Bulletin of the Russian Academy of Sciences - Physics*, **57** 1947
[6] Roshchupkin A M, Bataronov I L and Dezhin V V 1995 *Bulletin of the Russian Academy of Sciences - Physics*, **59** 1648
[7] Bataronov I L, Dezhin V V and Roshchupkin A M 1995 *Bulletin of the Russian Academy of Sciences - Physics*, **59** 1690
[8] *Handbook of Mathematical Functions with Formulas, Graphs and Mathematical Tables*, ed Abramowitz M and Stegun I A 1964 (Nat. Bur. Stand. Applied Mathematics Series 55)
[9] Bataronov I L, Nadeina T A, Dezhin V V and Roshchupkin A M 2002 *Interaction of Defects and Inelastic Phenomena in Solids: Proc. 10th Int. Conf.* (Tula: Tula State University) p 179

The dynamics of an edge dislocation in a ferromagnetic crystals

V V Dezhin and V N Nechaev
Voronezh State Technical University, Moskovsky prospect 14, Voronezh 394026, Russian Federation

E-mail: viktor.dezhin@mail.ru

Abstract. The system of equations describing the bending vibrations of the dislocation in the ferromagnetic crystal is written. Elastic and magnetostrictive properties of the ferromagnetic crystals are considered isotropic. The linearization of the resulting system produced a relatively small contribution to the magnetization from the influence of dislocation. In the linear approximation of the dislocation displacement system of equation describing vibrations of a ferromagnetic crystal with an edge dislocation is obtained. The equation of motion of an edge dislocation in a ferromagnetic crystal is found.

1. Introduction

In papers [1] and [2] distribution of electric and magnetic fields near the fixed dislocation has been investigated. Braking straight dislocation, moving with constant velocity, considered in paper [3]. However, without considering the bending vibrations of dislocation is impossible to give a correct description of the behavior of crystals in external fields. For crystals with a simple lattice the problem of crystal vibrations with a dislocation and of phonon scattering by dislocations is solved by application of self-consistent dynamic theory of dislocations [4]. In papers [5] and [6] a generalized susceptibility of dislocations is calculated in the vicinity of a structural phase transition point and in ferroelectrics. Effective mass and effective rigidity of the dislocation with respect to bending among them discussed in these papers. Of interest to consider the dynamics of dislocations in a ferromagnetic crystal study of the effect of the magnetic subsystem on the dynamic characteristics of dislocation – the effective mass and effective rigidity. In the present paper we give an accurate solution of the problem in the framework of linear theory of elasticity and Landau theory of phase transitions.

2. Set of Equations Describing Crystal Vibrations with a Dislocation

The set of equations includes, generalized for the ferromagnetic case, the dynamic Beltrami equation [7] (derived similar to papers [5] and [6]):

$$\frac{1}{c_t^2}\frac{\partial^2\sigma_{ij}}{\partial t^2}-\frac{1}{2(1+\nu)}\frac{1}{c_t^2}\delta_{ij}\frac{\partial^2\sigma_{ll}}{\partial t^2}-\nabla^2\sigma_{ij}-\frac{1}{1+\nu}\left(\frac{\partial^2\sigma_{ll}}{\partial x_i\partial x_j}-\delta_{ij}\nabla^2\sigma_{ll}\right)+\rho a_2\frac{\partial^2}{\partial t^2}\left(M_iM_j\right)$$

$$-\frac{1}{2}\rho\left(a_1+a_2\right)\delta_{ij}\frac{\partial^2}{\partial t^2}\left(M^2\right)+\mu\left(a_1+a_2\right)\delta_{ij}\nabla^2\left(M^2\right)-\mu\left(a_1+a_2\right)\frac{\partial^2\left(M^2\right)}{\partial x_i\partial x_j}+\mu a_2\frac{\partial^2\left(M_jM_l\right)}{\partial x_i\partial x_l}$$

Content from this work may be used under the terms of the Creative Commons Attribution 3.0 licence. Any further distribution of this work must maintain attribution to the author(s) and the title of the work, journal citation and DOI.
Published under licence by IOP Publishing Ltd

$$+\mu a_2 \frac{\partial^2 (M_i M_l)}{\partial x_j \partial x_l} - \mu a_2 \nabla^2 (M_i M_j) - \mu a_2 \delta_{ij} \frac{\partial^2 (M_l M_k)}{\partial x_l \partial x_k} = \rho \frac{\partial}{\partial t}(j_{ij} + j_{ji} - \delta_{ij} j_{ll}) - 2\mu \eta_{ij}, \quad (1)$$

the Landau-Lifshits equation, describing magnetization vector dynamics:

$$\frac{\partial \boldsymbol{M}}{\partial t} = \gamma (\boldsymbol{M} \times \boldsymbol{H}^e) \quad (2)$$

and magnetostatic equations:

$$\nabla \cdot (\boldsymbol{H} + 4\pi \boldsymbol{M}) = 0, \qquad \nabla \times \boldsymbol{H} = 0. \quad (3)$$

Here $c_t = \sqrt{\mu/\rho}$ is the sound velocity of transverse waves, μ is shear modulus, ρ is density of the crystal, σ_{ij} is stress tensor, t is time, ν is Poisson's ratio, x is coordinate, δ_{ij} is Kronecker delta,

$$a_{iklm} = a_1 \delta_{ik} \delta_{lm} + \frac{1}{2} a_2 (\delta_{il} \delta_{km} + \delta_{im} \delta_{kl}) \quad (4)$$

is magnetostriction coefficient tensor, $\boldsymbol{M}$ is magnetization vector, j_{ij} is the dislocation current density tensor which acquires the following form for a single dislocation [8]:

$$j_{ij} = -n_i b_k V \delta(\xi), \quad (5)$$

$\boldsymbol{n}$ is the unit vector of the normal to the dislocation slip plane, $\boldsymbol{b}$ is the Burgers vector of the dislocation, V is the velocity vector projection of the dislocation line at a given point to the direction of the normal to it in the slip plane, $\delta(\xi)$ is the two-dimensional delta function, ξ is the two-dimensional radius-vector counted from the dislocation axis within the plane normal to vector $\boldsymbol{\tau}$ at a given point to the dislocation line, $\boldsymbol{\tau}$ is the unit vector tangent to the dislocation line,

$$\eta_{ij} = e_{jkn} \frac{\partial}{\partial x_k} \left(\tau_i b_n - \frac{1}{2} \boldsymbol{\tau} \cdot \boldsymbol{b} \delta_{in} \right) \delta(\xi) \quad (6)$$

is the deformation incompatibility tensor [9] due to the presence of dislocations within the crystal, e_{jkn} is Levi-Civita symbol, $\boldsymbol{H}$ is the magnetic field vector,

$$H_i^e = \alpha \nabla^2 M_i + H_i + a_1 \sigma_{ll} M_i + a_2 \sigma_{ik} M_k - \beta M_{\perp i}, \quad (7)$$

$\boldsymbol{H}^e$ is the effective magnetic field applied to magnetization vector $\boldsymbol{M}$, α is exchange constant, β is the magnetic anisotropy constant,

$$\boldsymbol{M}_{\perp} = \boldsymbol{M} - \boldsymbol{l} \cdot (\boldsymbol{M} \cdot \boldsymbol{l}) \quad (8)$$

is the component of vector $\boldsymbol{M}$ normal to the axis of easy magnetization, whose position is given by unit vector $\boldsymbol{l}$, γ is gyromagnetic ratio. Elastic and magnetostrictive properties of the crystal, for simplicity are considered isotropic. Equations (1)–(8) together with equation of dislocation motion

$$f_i = e_{ikl} \tau_k \sigma_{lm} b_m = 0,$$

which represents the zeroing condition of Peach-Koehler force at any instant of time at any dislocation point, completely describe the dynamics of the ferromagnetic crystal within a dislocation.

Let us represent the magnetization vector as $\boldsymbol{M} = \boldsymbol{M}_0 + \boldsymbol{m}$, where $\boldsymbol{M}_0$ is the vector of spontaneous magnetization, $\boldsymbol{m}$ is a small contribution due to inhomogeneous deformations generated by a dislocation within the crystal. Now, we linearize the set of Equations (1)–(3) in respect to $\boldsymbol{m}$, remembering, that in this approximation $\boldsymbol{M}_0 \cdot \boldsymbol{m} = 0$ by virtue of conservation of the modulus of magnetization vector $\boldsymbol{M}$. We find:

$$\frac{1}{c_t^2} \frac{\partial^2 \sigma_{ij}}{\partial t^2} - \frac{1}{2(1+\nu)} \frac{1}{c_t^2} \delta_{ij} \frac{\partial^2 \sigma_{ll}}{\partial t^2} - \nabla^2 \sigma_{ij} - \frac{1}{1+\nu} \left(\frac{\partial^2 \sigma_{ll}}{\partial x_i \partial x_j} - \delta_{ij} \nabla^2 \sigma_{ll} \right)$$

$$+ \mu a_2 \left[\frac{1}{c_t^2} \frac{\partial^2}{\partial t^2} (M_{0i} m_j + M_{0j} m_i) - \nabla^2 (M_{0i} m_j - M_{0j} m_i) + M_{0i} \frac{\partial}{\partial x_j} \nabla \cdot \boldsymbol{m} + M_{0j} \frac{\partial}{\partial x_i} \nabla \cdot \boldsymbol{m} \right.$$

$$+\boldsymbol{M}_0\cdot\nabla\left(\frac{\partial m_i}{\partial x_j}+\frac{\partial m_j}{\partial x_i}-2\delta_{ij}\nabla\cdot\boldsymbol{m}\right)\Bigg]=\rho\frac{\partial}{\partial t}\left(j_{ij}+j_{ji}-\delta_{ij}j_{ll}\right)-2\mu\eta_{ij}, \quad (9)$$

$$\frac{\partial \boldsymbol{m}}{\partial t}=\gamma\left(\boldsymbol{M}_0\times\boldsymbol{H}^e(\boldsymbol{m})\right),\qquad \nabla^2\psi-4\pi\nabla\cdot\boldsymbol{m}=0,$$

where ψ is a scalar magnetic field potential ($\boldsymbol{H}=-\nabla\psi$).

Then from Equation (9), in a common way we derive the equation for the quantity $f_\perp=n_l\sigma_{lm}b_m$, which is a projection of Peach-Koehler force onto the slip plane, and the value of the hydrostatic pressure $p=-\sigma_{ll}/3$

$$\frac{1}{c_t^2}\frac{\partial^2 f_\perp}{\partial t^2}-\nabla^2 f_\perp+\frac{3}{1+\nu}n_i b_j\frac{\partial^2 p}{\partial x_i\partial x_j}+\mu a_2\left[\frac{1}{c_t^2}\frac{\partial^2}{\partial t^2}\left((\boldsymbol{M}_0\cdot\boldsymbol{n})(\boldsymbol{m}\cdot\boldsymbol{b})+(\boldsymbol{M}_0\cdot\boldsymbol{b})(\boldsymbol{m}\cdot\boldsymbol{n})\right)\right.$$

$$-\nabla^2\left((\boldsymbol{M}_0\cdot\boldsymbol{n})(\boldsymbol{m}\cdot\boldsymbol{b})+(\boldsymbol{M}_0\cdot\boldsymbol{b})(\boldsymbol{m}\cdot\boldsymbol{n})\right)-\left((\boldsymbol{M}_0\cdot\boldsymbol{n})(\boldsymbol{b}\cdot\nabla(\nabla\cdot\boldsymbol{m}))+(\boldsymbol{M}_0\cdot\boldsymbol{b})(\boldsymbol{n}\cdot\nabla(\nabla\cdot\boldsymbol{m}))\right)$$

$$\left.+\boldsymbol{M}_0\cdot\nabla\left(\boldsymbol{b}\cdot\nabla(\boldsymbol{m}\cdot\boldsymbol{n})+\boldsymbol{n}\cdot\nabla(\boldsymbol{m}\cdot\boldsymbol{b})\right)\right]=\rho b^2\frac{\partial}{\partial t}\left(V\delta(\xi)\right)+2\mu n_i\eta_{ij}b_j, \quad (10)$$

$$\frac{1}{c_l^2}\frac{\partial^2 p}{\partial t^2}-\nabla^2 p-\frac{2}{3}\mu a_2\frac{1+\nu}{1-\nu}\boldsymbol{M}_0\cdot\nabla(\nabla\cdot\boldsymbol{m})=-\frac{2}{3}\mu\frac{1+\nu}{1-\nu}\eta_{ll}. \quad (11)$$

3. The bending vibrations equation of edge dislocation

Expressions obtained in section 2 are characterized by the fact that they lead to nonphysical features in We consider small bending vibrations of edge dislocation lying along the axis Oz of the vector tangent to the dislocation line $\tau_0=(0,\,0,\,-1)$, the Burgers vector $\boldsymbol{b}=(b,\,0,\,0)$ and the vector normal to the slip plane of the dislocation $\boldsymbol{n}=(0,\,1,\,0)$. In the linear approximation of the dislocation displacement $u=u(z,\,t)$

$$\tau=(-\partial u/\partial z,\,0,\,-1),\qquad \delta(\xi)=\delta(x)\delta(y)-\delta'(x)\delta(y)u$$

and system of Equations (10), (11) are reduced to the form:

$$\frac{1}{c_t^2}\frac{\partial^2 f_\perp}{\partial t^2}-\nabla^2 f_\perp+\frac{3b}{1+\nu}\frac{\partial^2 p}{\partial x\partial y}+\mu a_2 b\left[\frac{1}{c_t^2}\left(M_{0x}\frac{\partial^2 m_y}{\partial t^2}+M_{0y}\frac{\partial^2 m_x}{\partial t^2}\right)-M_{0x}\left(\frac{\partial^2 m_y}{\partial y^2}+\frac{\partial^2 m_y}{\partial z^2}+\frac{\partial^2 m_z}{\partial y\partial z}\right)\right.$$

$$\left.-M_{0y}\left(2\frac{\partial^2 m_x}{\partial x^2}+\frac{\partial^2 m_x}{\partial z^2}+\frac{\partial^2 m_z}{\partial x\partial z}\right)+M_{0z}\left(\frac{\partial^2 m_y}{\partial x\partial z}+\frac{\partial^2 m_x}{\partial y\partial z}\right)\right]=-\mu b^2\left(\frac{1}{c_t^2}\frac{\partial^2 u}{\partial t^2}-\frac{\partial^2 u}{\partial z^2}\right)\delta(x)\delta(y), \quad (12)$$

$$\frac{1}{c_l^2}\frac{\partial^2 p}{\partial t^2}-\nabla^2 p-\frac{2}{3}\mu a_2\frac{1+\nu}{1-\nu}\boldsymbol{M}_0\cdot\nabla(\nabla\cdot\boldsymbol{m})=-\frac{2}{3}\mu b\frac{1+\nu}{1-\nu}\left[\delta(x)-u\delta'(x)\right]\delta'(y). \quad (13)$$

Here taken into account Equations (5) and (6). If the vector of spontaneous magnetization is parallel to the dislocation line: $\boldsymbol{M}_0=(0,\,0,\,M_{0z})$, then the system of equations (1), (2) is essentially simplified:

$$\frac{1}{c_t^2}\frac{\partial^2 f_\perp}{\partial t^2}-\nabla^2 f_\perp+\frac{3b}{1+\nu}\frac{\partial^2 p}{\partial x\partial y}+\mu a_2 bM_{0z}\left(\frac{\partial^2 m_y}{\partial x\partial z}+\frac{\partial^2 m_x}{\partial y\partial z}\right)=-\mu b^2\left(\frac{1}{c_t^2}\frac{\partial^2 u}{\partial t^2}-\frac{\partial^2 u}{\partial z^2}\right)\delta(x)\delta(y),$$

$$\frac{1}{c_l^2}\frac{\partial^2 p}{\partial t^2}-\nabla^2 p-\frac{2}{3}\mu a_2\frac{1+\nu}{1-\nu}M_{0z}\left(\frac{\partial^2 m_x}{\partial x\partial z}+\frac{\partial^2 m_y}{\partial y\partial z}\right)=-\frac{2}{3}\mu b\frac{1+\nu}{1-\nu}\left[\delta(x)-u\delta'(x)\right]\delta'(y).$$

Fourier transforms of the coordinates and the time of Equations (12) and (13) are

$$\left(q^2-\frac{\omega^2}{c_t^2}\right)\widetilde{f}_\perp-\frac{3b}{1+\nu}q_xq_y\widetilde{p}+\mu a_2 b\left[\widetilde{m}_x\left(2M_{0y}q_x^2+M_{0y}q_z^2-M_{0z}q_yq_z-M_{0y}\frac{\omega^2}{c_t^2}\right)\right.$$
$$\left.+\widetilde{m}_y\left(2M_{0x}q_y^2+M_{0x}q_z^2-M_{0z}q_xq_z-M_{0x}\frac{\omega^2}{c_t^2}\right)+\widetilde{m}_z\left(M_{0z}q_yq_z+M_{0y}q_xq_z\right)\right]=-\mu b^2\left(q_z^2-\frac{\omega^2}{c_t^2}\right)\widetilde{u}, \quad (14)$$

$$\left(q^2-\frac{\omega^2}{c_t^2}\right)\widetilde{p}+\frac{2}{3}\frac{1+\nu}{1-\nu}\mu a_2\left[\widetilde{m}_x\left(M_{0x}q_x^2+M_{0y}q_xq_y+M_{0z}q_xq_z\right)+\widetilde{m}_y\left(M_{0x}q_xq_y+M_{0y}q_y^2+M_{0z}q_yq_z\right)\right.$$
$$\left.+\widetilde{m}_z\left(M_{0x}q_xq_z+M_{0y}q_yq_z+M_{0z}q_z^2\right)\right]=-\frac{2}{3}\frac{1+\nu}{1-\nu}\mu b\left(q_xq_y\widetilde{u}+4i\pi^2q_y\delta(q_z)\delta(\omega)\right), \quad (15)$$

where $\widetilde{f}_\perp=\widetilde{f}_\perp(\boldsymbol{q},\omega)$ is the Fourier transform of the projection Peach-Koehler force onto the slip plane, $\boldsymbol{q}$ is wave vector, ω is oscillation frequency, $\widetilde{p}=\widetilde{p}(\boldsymbol{q},\omega)$ is the Fourier transform of the hydrostatic pressure, $\widetilde{\boldsymbol{m}}=\widetilde{\boldsymbol{m}}(\boldsymbol{q},\omega)$ is the Fourier transform of $\boldsymbol{m}$, $\widetilde{u}=\widetilde{u}(q_z,\omega)$ is the Fourier transform of the dislocation displacement. Solving the set of Equations (14), (15) the equation of bending vibrations of an edge dislocation in a ferromagnetic crystal we find:

$$\widetilde{f}_\perp=-\frac{2}{1-\nu}\mu bq_xq_y\frac{a_2(\boldsymbol{q}\cdot\widetilde{\boldsymbol{m}})(\boldsymbol{q}\cdot\boldsymbol{M}_0)+b\left(q_xq_y\widetilde{u}+4i\pi^2q_y\delta(q_z)\delta(\omega)\right)}{\left(q^2-\omega^2/c_t^2\right)^2}-\mu b\frac{a_2\boldsymbol{Q}\cdot\widetilde{\boldsymbol{m}}+b\left(q_z^2-\omega^2/c_t^2\right)\widetilde{u}}{q^2-\omega^2/c_t^2},$$

where

$$\boldsymbol{Q}=\left\{M_{0y}\left(2q_x^2+q_z^2-\omega^2/c_t^2\right)-M_{0z}q_yq_z,\ M_{0x}\left(2q_y^2+q_z^2-\omega^2/c_t^2\right)-M_{0z}q_xq_z,\ M_{0x}q_yq_z+M_{0y}q_xq_z\right\}.$$

If the vector of spontaneous magnetization is parallel to the dislocation line: $\boldsymbol{M}_0=(0,0,M_{0z})$, then the equation of bending vibrations an edge dislocation in a ferromagnetic crystal is simplified and takes the form

$$\widetilde{f}_\perp=-\frac{2}{1-\nu}\mu bq_xq_y\frac{a_2(\boldsymbol{q}\cdot\widetilde{\boldsymbol{m}})q_zM_{0z}+b\left(q_xq_y\widetilde{u}+4i\pi^2q_y\delta(q_z)\delta(\omega)\right)}{\left(q^2-\omega^2/c_t^2\right)^2}$$
$$+\mu b\frac{a_2q_zM_{0z}\left(q_y\widetilde{m}_x+q_x\widetilde{m}_y\right)-b\left(q_z^2-\omega^2/c_t^2\right)\widetilde{u}}{q^2-\omega^2/c_t^2}.$$

Further investigation of the dynamic properties of dislocations in ferromagnetics is similar to papers [5], [6]. To avoid awkward computations, associated, in particular, with increased complexity of the form of the response function to the external force of the spin subsystem and also due to weakness of magnetoelastic coupling, we shall give a qualitative description of spin effects. Taking into account elastic field-to-magnetic subsystem interaction leads to an increase in effective mass and rigidity of the dislocation, and also in bending wave velocity along its axis by a factor $1+\varepsilon$, where $\varepsilon\sim\mu a_2^2M_0^2m$. Using typical values for the evaluation, we find $\varepsilon\sim10^{-3}$. Nearly this order of magnitude is attributed to the relative contribution of spin wave attenuation to the total dislocation vibration damping. These effects may be considerably enhanced in the vicinity of magnetic transitions of the orientational type, due to an anomalous increase of magnetoelastic coupling.

4. Conclusion

The set of equations derived may be used for the evaluation of magnetic fields and magnetization distribution generated by vibrating dislocations. The results of the present study may also be used for the investigation of attenuation and scattering of ultrasonic and electromagnetic waves, light scattering, thermal conductivity and heat capacity of ferromagnetic crystals with dislocations. The most significant effects, associated with dislocations, will, evidently, be observed close to the point of magnetic phase transition, where the system response to any external influence, corresponding to the order parameter, is anomalously large.

References

[1] Kosevich A M and Feldman E P 1967 *Fizika Tverdogo Tela* **9** 3415
[2] Gann V V and Feldman E P 1976 *Fizika Nizkih Temperatur* **2** [9] 30
[3] Barjakhtar V G and Druinskij E I 1977 *Zhurnal Experimentalnoy i Teoreticheckoy Fiziki* **72** 218
[4] Ninomiya T 1970 *Fundamental Aspects of Dislocations Theory*, ed J A Simmons, R de Wit and R Bullough (Nat. Bur. Stand. (U.S.) Spec. Publ. **317**) p 315
[5] Dezhin V V, Nechaev V N and Roshchupkin A M 1990 *Fizika Tverdogo Tela* **32** 810
[6] Dezhin V V, Nechaev V N and Roshchupkin A M 1989 *Ferroelectrics Letters* **10** 155
[7] Bross H 1964 *Physica Status Solidi* **5** 329
[8] Landau L D and Lifshitz E M 1986 *Theory of Elasticity* (*Course of Theoretical Physics* vol 7) (Oxford: Butterworth-Heinemann)
[9] de Wit R 1970 *Fundamental Aspects of Dislocation Theory*, ed J A Simmons, R de Wit and R Bullough (Nat. Bur. Stand. (U.S.) Spec. Publ. **317**) p 651

Size effects on the open probability of two-state ion channel system in cell membranes using microcanonical formalism based on gamma function

Rıza Erdem[1], Ekrem Aydıner[2]
[1]Department of Physics, Akdeniz University, 07058 Antalya, Turkey
[2]Department of Physics, Istanbul University, 34134 Istanbul, Turkey

E-mail: rerdem@akdeniz.edu.tr

E-mail: ekrem.aydiner@istanbul.edu.tr

Abstract. Ion channel systems are a class of proteins that reside in the membranes of all biological cells and forms conduction pores that regulate the transport of ions into and out of cells. They can be investigated theoretically in the microcanonical formalism since the number of accessible states can be easily evaluated by using the Stirling approximation to deal with factorials. In this work, we have used gamma function ($\Gamma(n)$) to solve the two-state or open-close channel model without any approximation. New values are calculated for the open probability (p_0) and the relative error between our numerical results and the approximate one using Stirling formula is presented. This error $(p_0^{app} - p_0)/p_0$ is significant for small channel systems.

1. Introduction

The interest in ion channel systems of all biological cells has increased greatly. They are a class of proteins that reside in the membranes of the cells and forms conduction pores that regulate the transport of ions into and out of cells [1]. One of their important functions is the conduction of action potentials in neurons. The equilibrium value function or the probability of finding a channel in the open state at equilibrium is crucial in the channel kinetics for fitting the experimental data. The probability is generally described by the Boltzmann statistics [2] and shows a direct proportion with the membrane potential. Boltzmann curves are sigmoidal-shaped curves on linear plot of open probability against potential. They are proved to be adequate for describing the steady-state data from many experiments [3]. So far, a more detailed theoretical explanation of the curves based on the free energy minimization, generalized statistical mechanics, and Monte-Carlo simulation technique has already been published by us [4, 5, 6]. In the previous work [4], we have defined the entropy in the microcanonical ensemble in terms of open and closed channel populations and used the Stirling approximation to deal with the factorials. Then we have minimized the free energy to derive the voltage-dependent open state probability of a single channel.

In the present work, we propose a new methodology to define the equilibrium value function for fitting the experimental data found during the measurements on small number of specific channel clusters in excitable membranes [7]. The methodology is again based on the

Content from this work may be used under the terms of the Creative Commons Attribution 3.0 licence. Any further distribution of this work must maintain attribution to the author(s) and the title of the work, journal citation and DOI.
Published under licence by IOP Publishing Ltd

microcanonical formulation in statistical physics approach but it simply uses the definition of gamma function instead of Stirling approximation.

2. The Model and Method

We consider a two-state (or open-close) channel system in which each channel is opened by the movement of a single gating particle which carries a charge. At any moment, the particle is in one of two positions, 1 and 2, and these are associated with closed and open states, respectively. The positions 1 and 2 correspond to two wells (with energy levels ε_1 and ε_2) in energy profile, and there is a single energy barrier between them. A simple expression for the internal energy of such a system in the presence of a membrane potential (V) is given by

$$E = n\left(1 - p_0\right)\varepsilon_1 + np_0\varepsilon_2 + nqe_0\left(1 - p_0\right)V \tag{1}$$

where n is total number of channels, e_0 elementary electronic charge and q is the number of charges on gating particle. In Eq. (1), p_0 and 1-p_0 are the open and closing probabilities of single channel, respectively. The number of accessible states of the system is defined by

$$\Omega = \frac{n!}{\left[n\left(1 - p_0\right)\right]!\left(np_0\right)!} . \tag{2}$$

Based on the Stirling approximation ($\ln n! \cong n\ln n - n$), we obtain the familiar Boltzmann entropy ($S = k\ln\Omega$) as

$$S = -nk\left[\left(1 - p_0\right)\ln\left(1 - p_0\right) + p_0\ln p_0\right] \tag{3}$$

where k is the Boltzmann constant. In the following, we call Eq. (3) as the approximated entropy. The approximated Helmholtz free energy is given by the thermodynamic relation $F = E - TS$, where T is the temperature. Using the minimization condition ($\partial F/\partial p_0 = 0$) one obtains the channels open probability for the static properties as follows:

$$p_0^{app} = \left\{1 + \exp\left[-\frac{qe_0}{kT}\left(V - V_0\right)\right]\right\}^{-1} \tag{4}$$

where the superscript *app* denotes approximation and V_0 [equal to $-\left(\varepsilon_1 - \varepsilon_2\right)/qe_0$] is the voltage at which half of the channels are open. Eq. (4) gives a sigmoidal shaped curves which start from zero and increase gradually to one as the membrane potential rises. In the literature, most of the data available from experiments done at different temperatures and in different axons were fitted using Eq. (4). From the experimental recordings made at several bath temperatures, Boltzmann parameters (such as q and V_0) can easily be determined [2].

On the other hand, the use of Stirling approximation is generally justified in the thermodynamic limit ($n \to \infty$). But, it generates significant errors with respect to the solution using gamma function for small systems [8]. From the definitions of gamma function (GF) ($\Gamma\left(n + 1\right) = n!$), we can write a value for the number of accessible states as

$$\Omega_{GF} = \frac{\Gamma\left(n + 1\right)}{\Gamma\left[n\left(1 - p_0\right) + 1\right]\Gamma\left(np_0 + 1\right)} . \tag{5}$$

Then, entropy ($S_{GF} = k\ln\Omega_{GF}$) becomes

$$S_{GF} = k\left\{\ln\Gamma\left(n + 1\right) - \ln\Gamma\left[n\left(1 - p_0\right) + 1\right] - \ln\Gamma\left(np_0 + 1\right)\right\} . \tag{6}$$

The free energy is now calculated by the formula

$$F_{GF} = E - TS_{GF} . \tag{7}$$

Using the condition $\partial F/\partial p_0 = 0$ we can finally obtain the following steady-state equation for the channel opening:

$$\psi^0 (np_0 + 1) = \psi^0 (n - np_0 + 1) + \frac{qe_0}{kT} (V - V_0) \tag{8}$$

where $\psi^n (x) = \frac{d^{n+1}}{dx^{n+1}} \ln \Gamma (x)$ with $\psi^0 (x) = \frac{d}{dx} \ln \Gamma (x)$ as in Refs. [8]. It is now possible to establish a relationship between the value of open probability (p_0), and the number of channels in the membrane. From the solutions of Eq. (8) we have plotted p_0 vs. V for several number of channels in Fig. 1. Also, we have calculated the relative error $(p_0^{app} - p_0)/p_0$ as functions of the membrane voltage and channel number in Figs. 2 and 3, respectively. We will briefly discuss the figures in the next section.

3. Results and Discussion

In our calculations, we have considered the voltage-gated potassium channels from xenopus oocytes [9]. Using several values of n, the value of probability of channel opening as function of the applied voltage is shown in Fig. 1 for the experiment illustrated in Ref. [9] with the parameter values $V_0 = -42$ mV, $kT/qe_0 = 4.8$. From the figure one can see that equilibrium value function

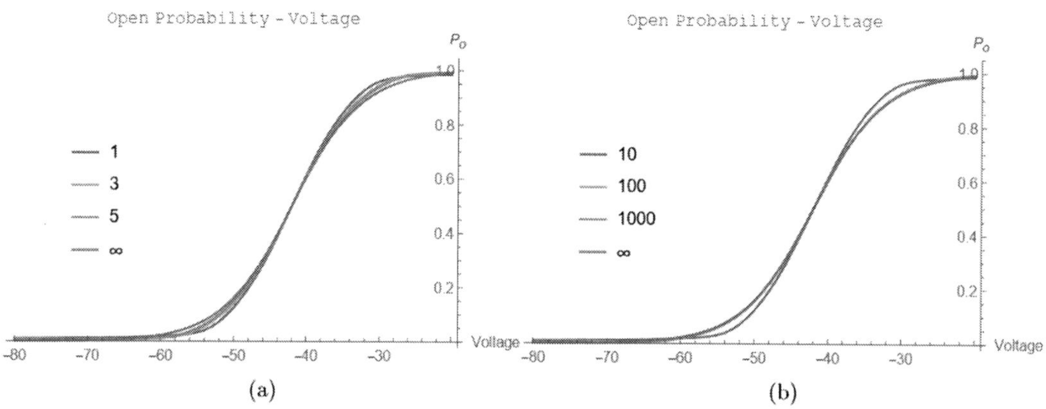

Figure 1: (a) Open probability vs. membrane voltage for the sizes $n = 1$, $n = 3$, $n = 5$ and $n = \infty$. (b) Same as Fig. 1 (a) but for $n = 10$, $n = 10^2$, $n = 10^3$.

gives a sigmoidal-shaped curve on linear plot of p_0 against V, i.e. it rises to 1 from zero as the value of V grows, as in the previous works [3, 4, 5, 6]. This figure also shows the effect of number of channels on p_0. An increase in the number of channels located on the membrane alters the steepness of the $p_0 - V$ relation at midpoint potential (V_0) being smaller with higher values of n and the solution of Eq. (8) reduces to one using Stirling approximation or Eq. (4) when $n \to \infty$ as expected. This corresponds to a sharp increase of the probability for opening in an ensemble of many potassium channels (or high channel densities) observed at potentials around $V \approx -60$ mV. In contrast, at low channel densities the chances for finding channels in open state shift to more negative potentials below $V = -60$ mV. The relative error $(p_0^{app} - p_0)/p_0$ is depicted in Fig. 2 as a function of V for four different sizes ($n = 5$, $n = 10$, $n = 10^2$, $n = 10^3$). For membranes with very small number of channels $n < 100$, the difference between the approximated and new open probability values is relevant in the voltages below $V = -40$ mV. For larger membranes with $n > 10^3$ channels the error is lower but it is still relevant for very low voltages. In Fig. 3, the relative error is also shown in terms of system size (n) for different values of voltages ($V = -50, -55, -60, -70, -75, -85, -90$ mV). It becomes clear from the figure that for sizes

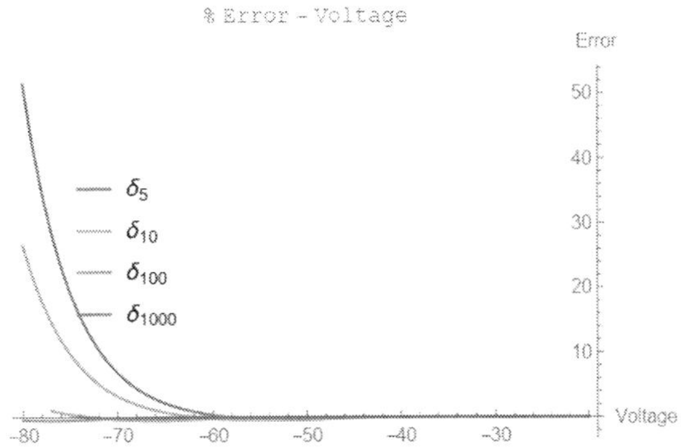

Figure 2: Open probability relative error of the two-state ion channel system in terms of the membrane potential. The diffferent curves are for four sizes.

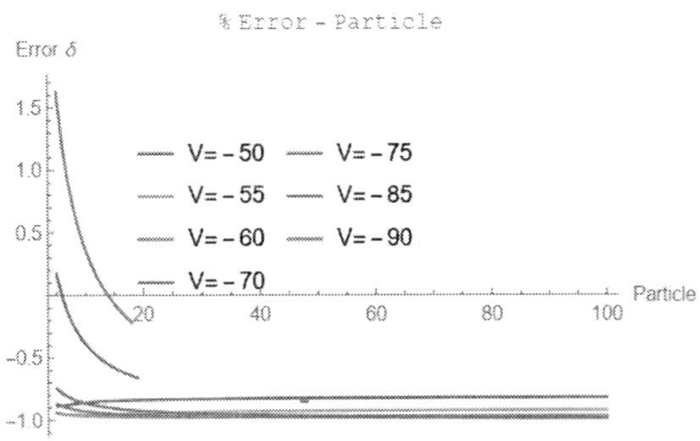

Figure 3: Open probability relative error of the two-state ion channel system in terms of the system size. The diffferent curves are for seven membrane voltages.

smaller than $n = 10^2$ the difference between the approximated and new open probabilities can be significant.

4. Conclusion

In this work, a two state (open-close) ion channel model of biological membranes has been solved. The same system has already been considered in microcanonical ensemble using Stirling approximation in many references. We have presented an analysis of the model in the microcanonical formalism without any approximation but using gamma function and its derivatives. From this analysis we conclude that we observe an excellent matching of our calculations and the one using Stirling approximation in the limit $n \to \infty$. Above result is also consistent with the experimentally observed data at high channel densities. According to the voltage-clamp data [9], the mechanism of membrane potential stabilization is determined at high channel densities by the activation-deactivation rates while it is only affected by the inactivation process at low densities with small number of channel populations. It other words, the characteristics of voltage fluctuations for small number of channels reflect the competition between voltage-dependent rates of transition and the voltage-independent one [10]. This leads to a noticeable difference between two approaches at lower potential regime for smaller sizes ($n < 100$). From the conceptual viewpoint, this is known as a lower bound to the Stirling approximation just as in Ref. [8]. Our results also develop a useful approach in the steady-state studies of channels that are not voltage-gated.

Acknowledgments
Authors thank to Dr. Ferhat Nutku for his kind assistance with this work. This work was supported by Scientific Research Projects Coordination Unit of Istanbul University. Project numbers: BEK-2016-20733.

References
[1] Hille B. (1992). *Ionic Channels of Excitable Membranes* Sinauer Associates, Sunderland, MA.
[2] Correa A. M., Bezanilla F., Latorre R., (1999). Biophys. J. 61, 1332.
[3] Willms A. R., Baro D. J., Harris-Warrick R. M., Guckenheimer J., (1999). J. Comp. Neurosci. 6, 145.
[4] Erdem R., Ekiz C., (2005) Physica A 349, 283.
[5] Erdem R., (2007) Physica A 373, 417.
[6] Erdem R, Aydiner E., (2009) Phys. Rev. E 79, 031919.
[7] Shuai J. W., Jung P., (2006) Chaos 16, 026104.
[8] Bertoldi D. S., Bringa E. M., Mirande E. N., (2011) Eur. J. Phys. 32, 1485.
[9] Marom S., Salmon H., Lyakhov V., Braun E., (1996) J. Membrane Biol. 154, 267-274.
[10] Salman H., Braun E., (1997) Phys. Rev. E 56, 852-864.

The distribution of the electric field around the vibrating edge dislocation in the ferroelectric crystal

V V Dezhin and V N Nechaev
Voronezh State Technical University, Moskovsky prospect 14, Voronezh 394026, Russian Federation

E-mail: viktor.dezhin@mail.ru

Abstract. The amplitude of the electric fields around the vibrating edge dislocation on the distance from the dislocation line in long-wave and short-wave cases is found. The laws of decrease modulus of the amplitude of the electric fields at the distance from the dislocation line are determined.

1. Introduction

The problem of the distribution of electric fields around the dislocations in the static case was solved in paper [1]. In papers [2-7] were further investigated electroelastic fields created by dislocations. In [8] the general expressions for the electroelastic fields of moving dislocations and disclinations in piezoelectric crystals are obtained. In this paper, based on the results of [9], [10] are expressions for the distribution of the electric field around oscillating edge dislocation in a ferroelectric is found. The case of small oscillations of the dislocation is considered. It was assumed that the active ferroelectric axis coincides with the coordinate axis Oz: $\boldsymbol{P}=(0,0,P)$ along the same axis located equilibrium position of the edge dislocation line: $\boldsymbol{\tau}_0=(0,0,-1)$, Burgers vector is directed along the axis Ox: $\boldsymbol{b}=(b,0,0)$. In this case, the electric fields associated with the dislocation are of a purely dynamic nature.

2. The distribution of electric fields in the long-wave case

Earlier [10], under the assumption that the dislocation makes bending vibrations according to the $u=u_0\exp(\mathrm{i}kz-\mathrm{i}\Omega t)$, for the components of the electric field intensity authors obtained expression

$$E_z(\boldsymbol{r},t)=4\mu bP_0u_0\chi_0k^2\left(\frac{1+\nu}{1-\nu}g_1\frac{\partial^2 J_l}{\partial x\partial y}+g_2\frac{\partial^2 J_t}{\partial x\partial y}\right)\exp(\mathrm{i}kz-\mathrm{i}\Omega t), \quad (1)$$

where

$$J_{l,t}=\frac{\eta}{\omega_0^2}K_0\left(\frac{r\omega_0}{\sqrt{\eta}}\right)-\frac{\pi}{2}\frac{1}{dk^2/\omega_0^2+\Omega^2/c_{l,t}^2}Y_0\left(r\sqrt{\Omega^2/c_{l,t}^2-k^2}\right)$$
$$-\frac{1}{dk^2/\omega_0^2+\Omega^2/c_{l,t}^2}K_0\left(rk\sqrt{1+d/\omega_0^2}\right). \quad (2)$$

After calculating derivatives [11], the expression for the electric field intensity components (1) takes the form:

Content from this work may be used under the terms of the Creative Commons Attribution 3.0 licence. Any further distribution of this work must maintain attribution to the author(s) and the title of the work, journal citation and DOI.
Published under licence by IOP Publishing Ltd

$$E_z(r,t) = 4\mu b P_0 u_0 \chi_0 k^2 \left(\frac{1+\nu}{1-\nu} g_1 M_l + g_2 M_t \right) \exp(ikz - i\Omega t), \tag{3}$$

where

$$M_{l,t} = \frac{\sqrt{\eta}}{\omega_0} \frac{xy}{r^2} \left[\frac{2}{r} K_1\left(\frac{r\omega_0}{\sqrt{\eta}}\right) - \frac{\omega_0}{\sqrt{\eta}} K_0\left(\frac{r\omega_0}{\sqrt{\eta}}\right) \right] - \frac{\pi}{2} \frac{\sqrt{\Omega^2/c_{l,t}^2 - k^2}}{dk^2/\omega_0^2 + \Omega^2/c_{l,t}^2} \frac{xy}{r^2} \left[\frac{2}{r} Y_1\left(r\sqrt{\Omega^2/c_{l,t}^2 - k^2} \right) \right.$$
$$\left. - \sqrt{\Omega^2/c_{l,t}^2 - k^2}\, Y_0\left(r\sqrt{\Omega^2/c_{l,t}^2 - k^2} \right) \right] - \frac{k\sqrt{1+d/\omega_0^2}}{dk^2/\omega_0^2 + \Omega^2/c_{l,t}^2} \frac{xy}{r^2}$$
$$\times \left[\frac{2}{r} K_1\left(rk\sqrt{1+d/\omega_0^2} \right) - k\sqrt{1+d/\omega_0^2}\, K_0\left(rk\sqrt{1+d/\omega_0^2} \right) \right]. \tag{4}$$

Here $Y_0(w)$, $Y_1(w)$ are Neumann functions of zero and first order, $K_0(w)$, $K_1(w)$ are MacDonald functions of zero and first order, $r = \sqrt{x^2 + y^2}$, c_t и c_l are the transverse and longitudinal sound velocities, μ is shear modulus of the crystal, ν is Poisson's ratio, g_1 and g_2 are striction coefficients, P_0 is polarization equilibrium value in a homogeneous crystal, $\omega_0^2 = \sqrt{2\alpha/m}$, $\eta = \delta/m$, $d = 4\pi/m$, h и m are damping coefficient and mass coefficient for polarization vibrations, respectively, δ is correlation constant, α и β are coefficients in Landau expansion for free energy, $\chi_0 = 1/(m\omega_0^2)$. In the formulas (1)-(4) satisfaction of the condition $k^2 < \Omega^2/c_{l,t}^2$ was assumed.

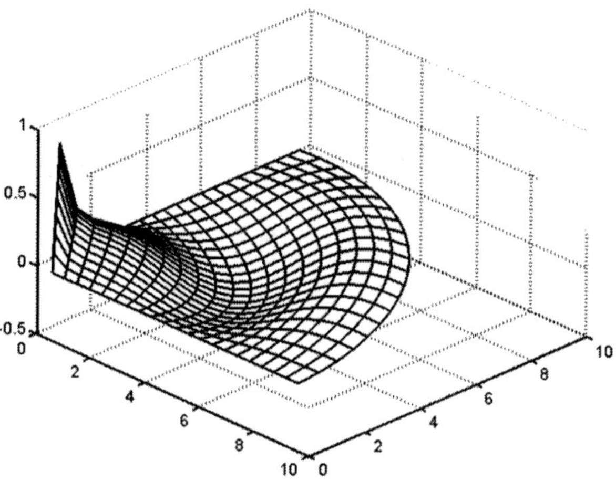

Figure 1. Dependence of the electric field intensity component of the distance to the dislocation line and the polar angle in the long-wave case.

Figure 1 in cylindrical coordinates shows the dependence of the amplitude of the electric field intensity component $E_z(r,t)$ of the distance from the dislocation line and the polar angle, based on the formulas (3) and (4) in normalized units in the long-wave case $k^2 < \Omega^2/c_{l,t}^2$. From formulas (3)-(4) it follows that for small values of the distance to the dislocation line $r < \left(\Omega^2/c_{l,t}^2 - k^2\right)^{-1/2}$ electric field amplitude decreases proportionally r^{-2}, and for large values $r > \left(\Omega^2/c_{l,t}^2 - k^2\right)^{-1/2}$ the modulus of the electric field amplitude decreases proportionally $r^{-1/2}$ (figure 2).

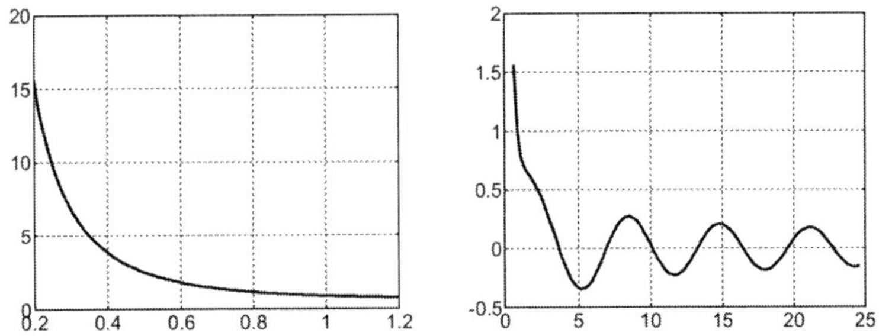

a) *b*)

Figure 2. Dependence of the electric field intensity amplitude of the distance from the dislocation line at small (a) and large (b) of values r in the long-wave case.

3. The distribution of electric fields in the short-wave case

In section 2 has been considered a long-wave case $k^2 < \Omega^2/c_{l,t}^2$. If $k^2 > \Omega^2/c_{l,t}^2$, in the expression (2) for $J_{l,t}$ the replacement $Y_0\left(r\sqrt{\Omega^2/c_{l,t}^2 - k^2}\right) \to -\frac{2}{\pi}K_0\left(r\sqrt{k^2 - \Omega^2/c_{l,t}^2}\right)$ needs to be done. Then for the components of the electric field we obtain

$$E_z(\boldsymbol{r},t) = 4\mu b P_0 u_0 \chi_0 k^2 \left(\frac{1+\nu}{1-\nu} g_1 T_l + g_2 T_t\right)\exp(\mathrm{i}kz - \mathrm{i}\Omega t)\,, \tag{5}$$

where

$$T_{l,t} = \frac{\sqrt{\eta}}{\omega_0}\frac{xy}{r^2}\left[\frac{2}{r}K_1\left(\frac{r\omega_0}{\sqrt{\eta}}\right) - \frac{\omega_0}{\sqrt{\eta}}K_0\left(\frac{r\omega_0}{\sqrt{\eta}}\right)\right] + \frac{\sqrt{k^2 - \Omega^2/c_{l,t}^2}}{dk^2/\omega_0^2 + \Omega^2/c_{l,t}^2}\frac{xy}{r^2}\left[\frac{2}{r}K_1\left(r\sqrt{k^2 - \Omega^2/c_{l,t}^2}\right)\right.$$

$$\left. - \sqrt{k^2 - \Omega^2/c_{l,t}^2}\,Y_0\left(r\sqrt{k^2 - \Omega^2/c_{l,t}^2}\right)\right] - \frac{k\sqrt{1 + d/\omega_0^2}}{dk^2/\omega_0^2 + \Omega^2/c_{l,t}^2}\frac{xy}{r^2}$$

$$\times\left[\frac{2}{r}K_1\left(rk\sqrt{1 + d/\omega_0^2}\right) - k\sqrt{1 + d/\omega_0^2}\,K_0\left(rk\sqrt{1 + d/\omega_0^2}\right)\right]. \tag{6}$$

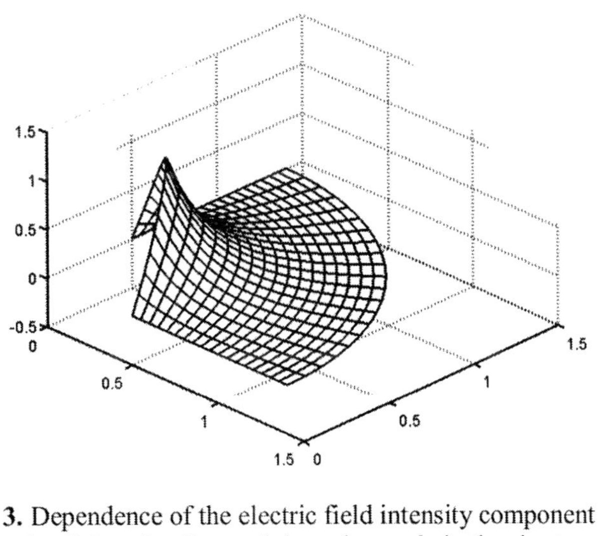

Figure 3. Dependence of the electric field intensity component of the distance to the dislocation line and the polar angle in the short-wave case.

Figure 3 in cylindrical coordinates shows the dependence of the amplitude of the electric field intensity component $E_z(\mathbf{r},t)$ of the distance from the dislocation line and the polar angle, based on the formulas (5) and (6) in normalized units in the long-wave case. From formulas (5), (6) it follows that for small values of the distance to the dislocation line $r < \left(k^2 - \Omega^2/c_{l,t}^2\right)^{-1/2}$ electric field amplitude decreases proportionally r^{-2}, and for large values $r > \left(k^2 - \Omega^2/c_{l,t}^2\right)^{-1/2}$ the modulus of the amplitude of the electric field decreases proportionally $r^{-1/2}\mathrm{e}^{-r}$ (figure 4).

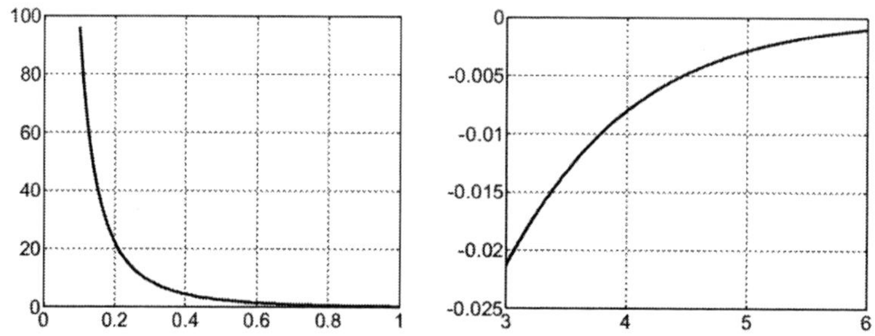

a) *b)*

Figure 4. Dependence of the electric field intensity amplitude of the distance from the dislocation line at small (a) and large (b) of values r in the short-wave case.

4. Conclusion

In this paper a quality form of the dependences of the electric field intensity amplitude component of the angle and the distance to the dislocation line is obtained. For the exact form the specific values of the constants of the crystal must be used. The results can be used to study the attenuation and scattering of ultrasound and electromagnetic waves in a ferroelectric with dislocations.

References

[1] Kosevich A M, Pastur L A and Feldman E P 1967 *Kristallografiya* **12** 916
[2] Minagawa S 1984 *Phys. Stat. Sol. (b)* **124** 565
[3] Smirnova I S 1984 *Phys. Stat. Sol. (b)* **126** 177
[4] Omelchenko S A, Bulanyi M F and Khmelenko O V 2003 *Fizika Tverdogo Tela* **45** 1608
[5] Tyalin Y I, Tyalina V A 2009 *Vestnik Tamb. Univ. (Estestv. i tehn. Nauki)* **14** 1105
[6] Tyalin Y I, Tyalina V A 2010 *Vestnik Tamb. Univ. (Estestv. i tehn. Nauki)* **15** 1086
[7] Tyalin Y I, Tyalina V A 2012 *Vestnik Tamb. Univ. (Estestv. i tehn. Nauki)* **17** 1410
[8] Hannanov Sh H 1999 *Fizika Tverdogo Tela* **41** 1210
[9] Dezhin V V, Nechaev V N and Roshchupkin A M 1989 *Ferroelectrics Letters* **10** 155
[10] Dezhin V V, Nechaev V N and Roshchupkin A M 1990 *Fizika Tverdogo Tela* **32** 1148
[11] *Handbook of Mathematical Functions with Formulas, Graphs and Mathematical Tables*, ed Abramowitz M and Stegun I A 1964 (Nat. Bur. Stand. Applied Mathematics Series 55)

Numerical simulation of steady and unsteady flow for generalized Newtonian fluids

Radka Keslerová, David Trdlička, Hynek Řezníček

Czech Technical University in Prague, Faculty of Mechanical Engineering, Dep. of Technical Mathematics, Czech Republic

E-mail: Radka.Keslerova@fs.cvut.cz, David.Trdlicka@fs.cvut.cz, Hynek.Reznicek@fs.cvut.cz

Abstract. This work presents the numerical solution of laminar incompressible viscous flow in a three dimensional branching channel with circle cross section for generalized Newtonian fluids. The governing system of equations is based on the system of balance laws for mass and momentum. Numerical solution is based on cetral finite volume method using explicit Runge-Kutta time integration. In the case of unsteady computation artificial compressibility method is considered.

1. Introduction

Branching of pipes occurs in many technical or biological applications. In [14] the effects of viscoelasticity on the pitchfork bifurcation using a numerical finite volume method was investigated. Results from both the upper-convected Maxwell and Oldroyd-B models show that the instability occurs at lower Reynolds numbers for viscoelastic fluids in comparison to the Newtonian base case. In [13] computational fluid dynamics simulations of steady viscoelastic flows through a planar two dimensional T-junction is considered and the influence of constitutive model and fluid elasticity upon the main recirculating flow characteristics formed at the junction and the shear stress fields is studied. In biomedical applications, it is the complex branching system of blood vessels in human body. The blood can be characterized by shear-thinning viscoelastic property and the blood flow can be described by generalized Oldroyd-B model. In [1] the new model to describe the rheological characteristics of blood (namely shear-thinning and deformation dependent viscoelasticity) in both steady and unsteady flows was developed. In [3] a comparative numerical study of non-Newtonian fluid models capturing shear-thinning and viscoelastic effects of blood flow in idealized and realistic stenosed vessels was presented. In [16] a novel, time-accurate approach for solving the unsteady incompressible NavierStokes equations on non-staggered grids is proposed. The approach modifies the standard, dual-time stepping artificial-compressibility iteration scheme by incorporating ideas from pressure-based, fractional-step formulations.

In prewious work we studied the numerical simulation of generalized Newtonian and Oldroyd-B fluids flow in 2D branching channel, [8], [9]. In this article the problem of the unsteady numerical simulation for the generalized Newtonian fluids flow is presented. The modelled domain is the three dimensional branching channel with T-junction.

Content from this work may be used under the terms of the Creative Commons Attribution 3.0 licence. Any further distribution of this work must maintain attribution to the author(s) and the title of the work, journal citation and DOI.
Published under licence by IOP Publishing Ltd

2. Mathematical model

The governing system of equations is the system of generalized Navier-Stokes equations, see [2], [12]. This system consists of the continuity equation and the momentum equation

$$\text{div } \boldsymbol{u} = 0 \tag{1}$$

$$\rho \frac{\partial \boldsymbol{u}}{\partial t} + \rho(\boldsymbol{u}.\nabla)\boldsymbol{u} = -\nabla P + \text{div } \mathbf{T}, \tag{2}$$

where P is the pressure, ρ is the constant density, $\boldsymbol{u}$ is the velocity vector. The symbol $\mathbf{T}$ represents the stress tensor.

The stress tensor $\mathbf{T}$ is defined by corresponding rheological model, Newtonian, see e.g. [1], [3], [4], [5], [8]

$$\mathbf{T} = 2\mu_s \mathbf{D} \tag{3}$$

where μ_s is the dynamic fluid viscosity and tensor $\mathbf{D}$ is the symmetric part of the velocity gradient, $\mathbf{D} = \frac{1}{2}(\nabla \boldsymbol{u} + \nabla \boldsymbol{u}^T)$.

For the numerical modelling of the generalized Newtonian fluids flow it is necessary to generalize the mathematical model. The constant viscosity coefficient μ_s is replaced by a shear rate dependent viscosity function $\mu(\dot{\gamma})$ where shear rate $\dot{\gamma}$ is defined by $\dot{\gamma} = 2\sqrt{\frac{1}{2}\text{tr } \mathbf{D}^2}$. This function can be written in the following general form (for more details see [15], [17])

$$\mu(\dot{\gamma}) = \mu_\infty + \frac{\mu_0 - \mu_\infty}{(1 + (\lambda\dot{\gamma})^b)^a}, \quad \dot{\gamma} = 2\sqrt{\frac{1}{2}\text{tr } \mathbf{D}^2}, \tag{4}$$

with special parameters $\mu_0 = 1.6 \cdot 10^{-1} Pa.s, \mu_\infty = 3.6 \cdot 10^{-3} Pa.s, a = 1.23, b = 0.64, \lambda = 8.2s$. For Newtonian flow, the viscosity μ_s is kept constant and equal to μ_∞.

3. Numerical solution

The mathematical models described above are solved numericaly the artificial compressibility approach combined with the finite-volume discretization. The artificial compressibility method [6], [8], [12] is used to obtain equation for pressure. It means that the continuity equation is completed by a pressure time derivative term $\frac{\partial p}{\beta^2 \partial t}$, where β is positive parameter, making the inviscid part of the system of equations hyperbolic. The parameter β for steady case is chosen equal to the maximum inlet velocity. This value ensures good convergence to steady state but is not large enough to make the transient solution accurate in time. Therefore it is suitable for steady flows only. In the case of unsteady simulation this parameter is 100x higher. The discretization is done by a cell-centered finite-volume method with hexahedral finite volumes. The system including the modified continuity equation and the momentum equations can be written

$$\tilde{R}_\beta W_t + F_x^c + G_y^c + H_z^c = F_x^v + G_y^v + H_z^v, \quad \tilde{R}_\beta = \text{diag}(\frac{1}{\beta^2}, 1, \cdots, 1), \tag{5}$$

where W is vector of unknowns, $W = (p, u, v, w)$, by superscripts c and v the inviscid and the viscous fluxes are denoted. Eq. (5) is discretized in space by the finite volume method and the arising system of ODEs is integrated in time by the explicit multistage Runge–Kutta scheme ([8], [11]).

The flow is modelled in a bounded computational domain where a boundary is divided into three mutually disjoint parts: a solid wall, an outlet and an inlet. At the inlet Dirichlet boundary condition for velocity vector is used and for the pressure homogeneous Neumann boundary condition is used. At the outlet parts the pressure value is prescribed and for the velocity vector homogeneous Neumann boundary condition is used. The no-slip boundary condition for the velocity vector is used on the wall. For the pressure homogeneous Neumann boundary condition is considered.

3.1. Unsteady computation

For numerical solution of unsteady flows the artificial compressibility method is applied. In this case the artificial compressibility parameter β is set to be 100x higher than for steady case. The question arises, how big the parameter β should be in numerical calculations. Numerical calculations show, that $\beta = 6\text{m s}^{-1}$ is the proper choice. The artificial compressibility approach used for unsteady incompressible flows is modifying the system of equations by adding an unsteady term to the continuity equation in the same way as for steady case, see e.g. [7].

The unsteady boundary conditions are defined as follows. In the inlet, in the solid wall and in one of the outlet part the steady boundary conditions are prescribed. In the second outlet part (the branch) new boundary condition is defined. The pressure value is prescribed by the function

$$p_{branch} = \frac{1}{4}\left(1 + \frac{1}{2}\sin(\omega t)\right), \tag{6}$$

where ω is the angular velocity defined as $\omega = 2\pi f$, where f is a frequency. In this work these values of the frequency are tested: $f = 2, 5, 10, 20, 50, 100$ Hz.

4. Numerical Results

This section deals with the comparison of the numerical results of Newtonian and generalized Newtonian fluids flow. Numerical tests are performed in an idealized branched channel with the circle cross-section. Fig. 1 (left) shows the shape of the tested domain. The computational domain is discretized using a structured, wall fitted mesh with hexahedral cells. The domain is divided to 19 blocks with 125 000 cells.

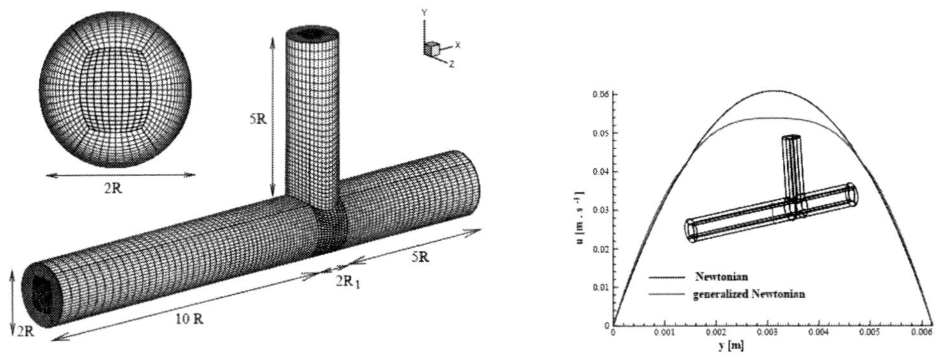

(a) structure of the domain (b) axial velocity profile

Figure 1. Structure of the tested domain (left) and axial velocity profile of tested fluids (right).

All numerical results presented in this section were computed using in-house software. The computational code was verified for the steady flow of an incompressible fluid in a straight tube by prescribing a constant velocity profile at the inlet. Fully developed velocity profiles for all tested fluids were used as the initial velocity profile for following computations in the branching channel with T-junction. At the outlet the constant pressure values are prescribed (0.0005 Pa (main channel) and 0.0005 Pa (branch)).

The computations were performed with the following model parameters: $R = 0.0031$ m, $R_1 = 0.0025$ m, $\mu_s = 0.0036$ Pa s, $U_0 = 0.0615$ m s^{-1}, $\rho = 1050$ kg m^{-3}. In Fig. 1 the axial velocity profile for tested types of fluids close to the branching is shown. The lines for Newtonian fluids are similar to the parabolic line, as was assumed. From this velocity profile is clear that the shear thinning fluids attain lower maximum velocity in the central part of the channel (close

Journal of Physics: Conference Series **738** (2016) 012112 doi:10.1088/1742-6596/738/1/012112

to the axis of symmetry) which is compensated by the increase of local velocity in the boundary layer close to the wall.

In Figs. 2 the velocity isolines and the cuts through the main channel and the small branch are shown.

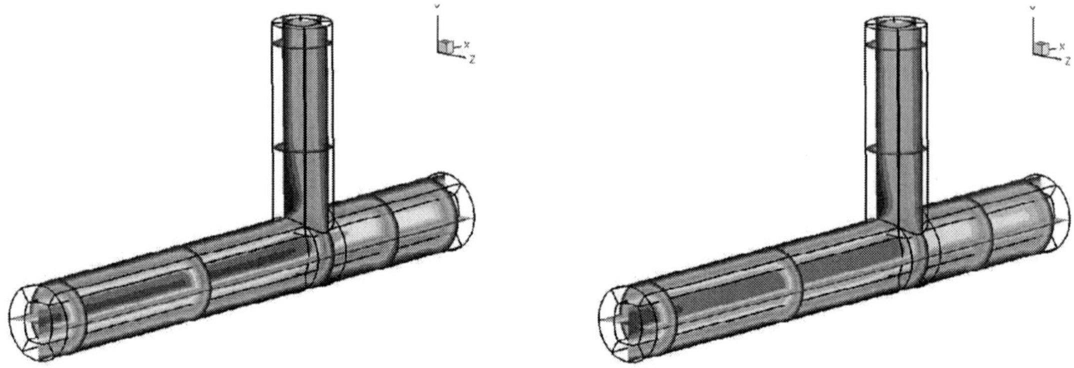

(a) Newtonian (b) generalized Newtonian

Figure 2. Velocity isolines of steady flows for generalized Newtonian fluids.

The axial velocity isolines in the center-plane area for all tested fluids are shown in the Figs. 3. It can be observed from these that the size of separation region for generalized Newtonian fluids is smaller than for Newtonian fluids, see in detail Fig. 4.

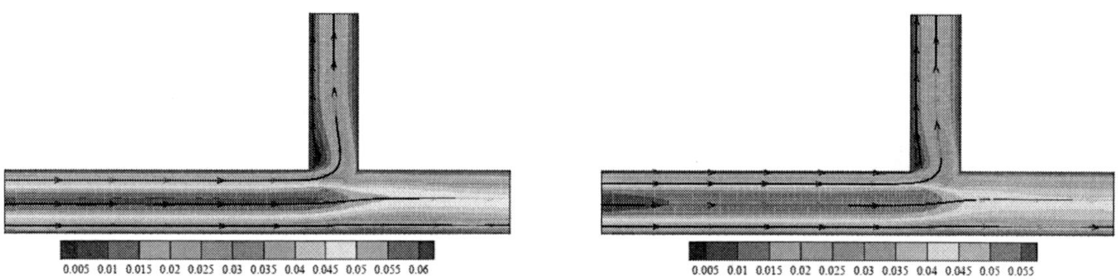

(a) Newtonian (b) generalized Newtonian

Figure 3. Axial velocity isolines in the center-plane area for generalized Newtonian fluids.

These steady numerical results are used as initial condition for unsteady numerical computation. The artificial compressibility method is used for the unsteady numerical simulation of generalized Newtonian fluids flow in the branching channel with T-junction with circle cross section. In this case the unsteady boundary condition are considered. It means that the pressure value in the branch outlet part is changing with given pressure function 6. Six values of frequency are tested, $f = 2, 5, 10, 20, 50, 100$ Hz. Parameter β is set to be 100x higher than for steady case, $\beta = 6.0$ m s^{-1}.

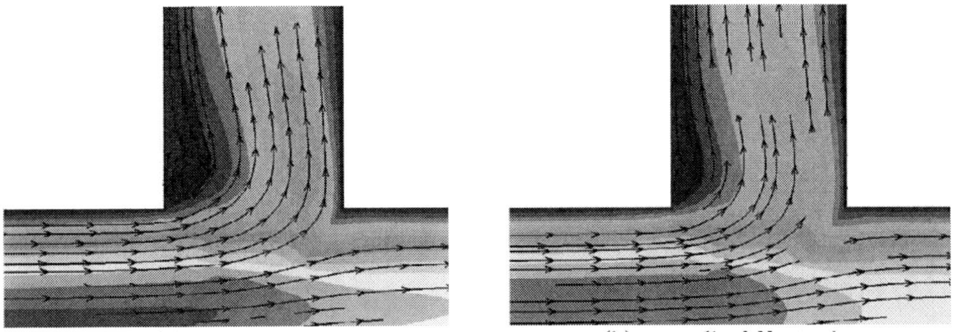

(a) Newtonian (b) generalized Newtonian

Figure 4. Axial velocity isolines in the center-plane area in the separation region.

In Fig. 5 and 6 graphs of velocity as the function of time for lower and higher values of the frequency are shown.

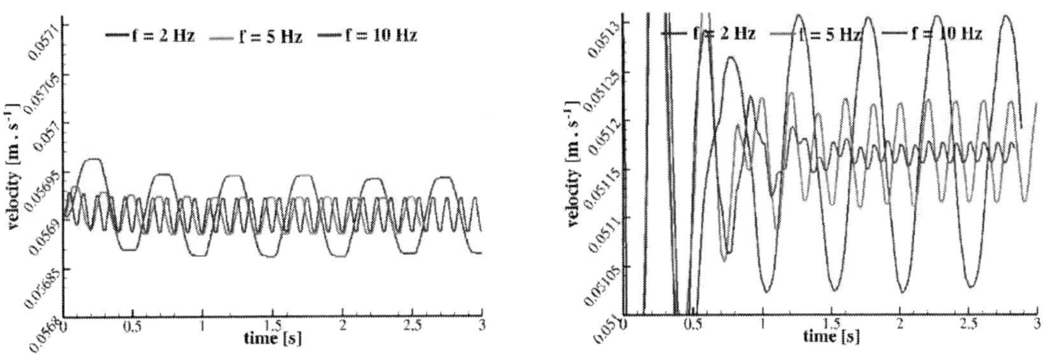

(a) Newtonian (b) generalized Newtonian

Figure 5. The graphs of the velocity as the function of time for lower values of frequency.

5. Conclusion

Classical Newtonian, as well as its generalized (shear-thinning) modification have been considered to model flow in the branching channel with T-junction, to investigate shear-thinning effects in steady flow simulations. From the presented velocity profile is clear that the shear thinning fluids (generalized Newtonian fluids) attain lower maximum velocity in the central part of the channel (close to the axis of symmetry) which is compensated by the increase of local velocity in the boundary layer close to the wall.

The numerical method used to solve the governing equations seems to be sufficiently robust and efficient for the appropriate resolution of the given class of problems.

One approach was considered for numerical solution of unsteady governing equations - an artificial compressibility approach. The artificial compressibility parameter β set to be $6ms^{-1}$. Several values for frequency were tested. Pressure value in the branch outlet was prescribed by periodic function. The numerical results given by graphs of the velocity as the function of the time were presented.

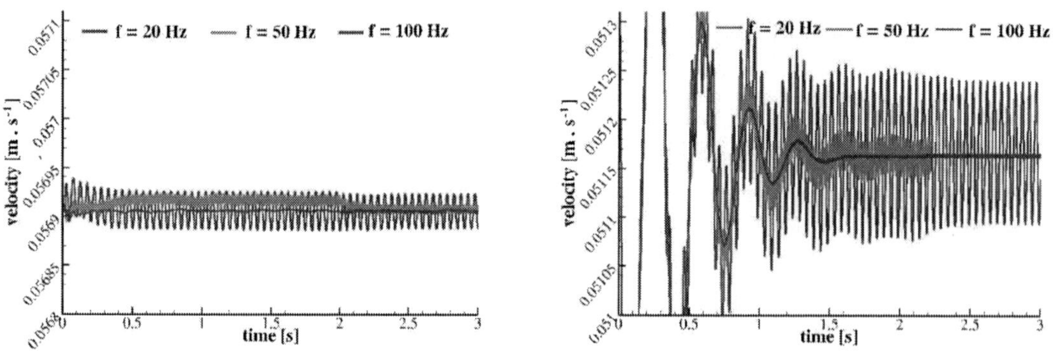

(a) Newtonian (b) generalized Newtonian

Figure 6. The graphs of the velocity as the function of time for higher values of frequency.

For the future work the extending this unsteady simulation for generalized Oldroyd-B fluids flow is considered. The new unsteady method, dual-time stepping method, will be used.

Acknowledgments

This work was supported by the grant SGS16/206/OHK2/3T/12.

References

[1] Anand M, Kwack J and Masud A 2013 *International Journal of Engineering Science* **72** 78–88
[2] Beneš L, Louda P, Kozel K, Keslerová R and Štigler J 2013 *Applied Mathematics and Computation* **219** 7225–7235
[3] Bodnar T, Sequeira A and Prosi M 2010 *Applied Mathematics and Computation* **217** 5055–5067
[4] Bodnar T and Sequeira A 2010 *Advances in Mathematical Fluid Mechanics* 83–104
[5] Bodnar T, Sequeira A and Pirkl L 2009 *In Numerical Analysis and Applied Mathematics* **2** 645–648
[6] Chorin AJ 1967 *Journal of Computational Physics* **135** 118–125
[7] Gaitonde AL 1998 *Intern. Journal for Numerical Methods in Engineering* **41** 1153–1166
[8] Keslerová R and Kozel K 2012 *Proc. 16th Sem. on Programs and Algorithms of Num. Math.* (Inst. of Math. Ac. of Sc. of CR) 100
[9] Keslerová R and Kozel K 2011 *Proc. 6th Intern. Symposium on Finite Volumes for Complex Applications* (Berlin: Springer) 589
[10] Leuprecht A and Perktold K 2001 *Computer Methods in Biomechanics and Biomechanical Engineering* **4** 149–163
[11] LeVeque R 2004 *Finite-Volume Methods for Hyperbolic Problems* (Cambridge University Press)
[12] Louda P, Kozel K, Příhoda J and Sváček P 2013 *International Journal of Heat and Fluid Flow* **43** 268–276
[13] Matos HM and Oliveira PJ 2014 *Journal of Non-Newtonian Fluid Mechanics* **213** 15–26
[14] Poole RJ, Haward SJ and Alves MA 2014 *Procedia Engineering* **79** 28–34
[15] Rabby MG, Razzak A and Molla MM 2013 *Procedia Engineering* **56** 225–231
[16] Tang HS and Sotiropoulos F 2007 *Computers and Fluids* **36** 974–986
[17] Vimmr J, Jonášová A and Bublík O 2010 *Mathematics and Computers in Simulation* **80** 1324–1336

Journal of Physics: Conference Series **738** (2016) 012113 doi:10.1088/1742-6596/738/1/012113

Use of ELVIS II platform for random process modelling and analysis of its probability density function

Yu.S. Maslennikova, I.S. Nugmanov

Kazan Federal University, Kremlevskaya str.18, Kazan 420018, Russia

E-mail: JSMaslennikova@kpfu.ru, nugm13@yandex.ru

Abstract. The problem of probability density function estimation for a random process is one of the most common in practice. There are several methods to solve this problem. Presented laboratory work uses methods of the mathematical statistics to detect patterns in the realization of random process. On the basis of ergodic theory, we construct algorithm for estimating univariate probability density distribution function for a random process. Correlational analysis of realizations is applied to estimate the necessary size of the sample and the time of observation. Hypothesis testing for two probability distributions (normal and Cauchy) is used on the experimental data, using $\chi 2$ criterion. To facilitate understanding and clarity of the problem solved, we use ELVIS II platform and LabVIEW software package that allows us to make the necessary calculations, display results of the experiment and, most importantly, to control the experiment. At the same time students are introduced to a LabVIEW software package and its capabilities.

1. Introduction

Probability density function (PDF) estimation is one of the major concern in areas such as signal detection and recognition, machine learning, neural networks, digital signal processing and computer vision. On the one hand, it offers a flexible way to investigate the properties of a given data set and provides a solid basis for efficient data mining tools. On the other hand, it is crucial in unsupervised learning tasks and Bayesian inference and classification [1].

Probability density estimation for random variables and stochastic processes is one of the key sections in any course of statistics and statistical physics, in particular. Recently, many mathematical tools provide special libraries for the empirical evaluation of probability density function in different ways. But often, the use of ready-made functions does not allow students to fully understand the essence of the matter. The most effective method of development of educational material for this section is to perform laboratory practicum, when students independently perform the measurement of stochastic processes dynamics and then estimate various statistical parameters of these processes.

The National Instruments Educational Laboratory Virtual Instrumentation Suite (NI ELVIS) is an educational design and prototyping platform based on NI LabVIEW. The NI ELVIS teaching platform takes students from discovering engineering theory to practical hands-on experience with industry relevant technology and can be used in laboratory practicum of statistical radiophysics [2].

In this paper, we propose an example of a laboratory practicum for a probability density function estimation using the ELVIS II platform capabilities platform and LabVIEW software package. This laboratory practicum is successfully used during statistical radiophysics course at the Kazan Federal University for many years.

Content from this work may be used under the terms of the Creative Commons Attribution 3.0 licence. Any further distribution of this work must maintain attribution to the author(s) and the title of the work, journal citation and DOI.
Published under licence by IOP Publishing Ltd

2. PDF function estimation for ergodic random processes

Below we describe theoretical backgrounds for the theory of ergodic stochastic processes and methods of probability density function estimations.

According to the ergodic theorem, mean and correlation function of the stationary process $\xi(t)$ can be calculated by averaging over time the values of a single realization of $v(t)$ for a continuous random process or a single realization $v(n)$ for a random sequence $\xi(n)$.

A sufficient condition for the applicability of the ergodic theorem is the lack of correlation between the two values a stationary random process (in broad sense) divided by an infinite length of time, i.e., $\lim_{T\to\infty}\frac{1}{T}R(\tau)=0$, where $R(\tau)$ is autocorrelation function of the process $\xi(t)$. Another sufficient condition for the applicability of the ergodic theorem is the Slutsky equation [3]:

$$\lim_{T\to\infty}\frac{1}{T}\int_0^T R(\tau)d\tau=0, \tag{1}$$

where T is the time of observation for the random process.

For ergodic process averaging value is equal to

$$\langle\xi(n)\rangle=\lim_{N\to\infty}\frac{1}{N}\sum_{n=0}^{N-1}v(\mathrm{n}), \tag{2}$$

where N is the number of elements in the observed series. Practically, number of the elements N in the observed sequence is limited.

Proposed method is based on the measurements of the time of the realization $v(t)$ for an ergodic stochastic process $\xi(t)$ between two fixed levels v_i and $v_i+\Delta v$ over a sufficiently large time interval of observation T (figure 1b).

Let us estimate the probability density function of noise with sufficiently large T. First of all, we write a new random process $\eta(t)$ (figure 1), depending on $\xi(t)$, with values:

$$y(\mathrm{t})=\begin{cases}1, \text{ if } v_i<\xi(\mathrm{t})\le v_i+\Delta v,\\ 2, \text{ if } v_i>\xi(\mathrm{t}) \ or\ \xi(\mathrm{t})>v_i+\Delta v.\end{cases} \tag{3}$$

Mathematical expectation of the process $\eta(t)$ in an arbitrary moment in time:

$$M[\eta(t)]=P(v_i<\xi(t)\le v_i+\Delta v). \tag{4}$$

For sufficiently small Δv, ignoring the change in the probability density function $w(v)$ in the interval $(v_i, v_i+\Delta v)$, we have $M[\eta(t)]\approx w(v)\Delta v$. The process $\eta(t)$ is also ergodic.

Mean value for the process $\eta(t)$:

$$\langle\eta(\mathrm{t})\rangle=\lim_{T\to\infty}\frac{1}{T}\int_0^T y(\mathrm{t})\,\mathrm{dt}=\lim_{T\to\infty}\frac{1}{T}\sum_{j=1}^{n}\tau_{ij}, \tag{5}$$

where $y(t)$ - realization of the random process, $\eta(t)$ - pulses with unit amplitude and a random duration of τ_{ij}, τ_{ij} – time spent in the interval $(v_i, v_i+\Delta v)$ for process $\xi(t)$, n – number of intervals of random length, when $v_i<\xi(t)\le v_i+\Delta v$ for an observation time T.

Values of τ_{ij} and n are random and depend on the one-dimensional density $w(v)$. Then suppose that the measurement time T so great that we can write:

$$\langle\eta(\mathrm{t})\rangle\approx\frac{1}{T}\sum_{j=1}^{n}\tau_{ij}. \tag{6}$$

Using the ergodic property of the process $\eta(t)$:

$$\langle \eta(\mathrm{t}) \rangle \approx E[\eta(\mathrm{t})] \Rightarrow w(v_i)\Delta v \approx \frac{1}{T}\sum_{j=1}^{n}\tau_{ij} \,. \tag{7}$$

Time of observation T, total duration of the pulses of unit amplitude $\sum_{j=1}^{n}\tau_{ij}$ voltage $(v_i, v_i+\Delta v)$ can registered using different devices. Thus, it is possible to experimentally estimate one-dimensional probability density function of the noise belonging to the interval $(v_i,\ v_i + \Delta v)$ and replace the exact density distribution $w(v_i)$ with its estimation $\hat{w}(v_i)$ in equation (7). Each level of quantization v_i corresponds estimation $\hat{w}(v_i)$, calculated according to the equation:

$$\hat{w}(v_i) \approx \frac{1}{\Delta v T}\sum_{j=1}^{n}\tau_{ij} \,. \tag{8}$$

The experimental estimations $\hat{w}(v_i)$ can be used to construct a histogram of the probability density function (figure 1a).

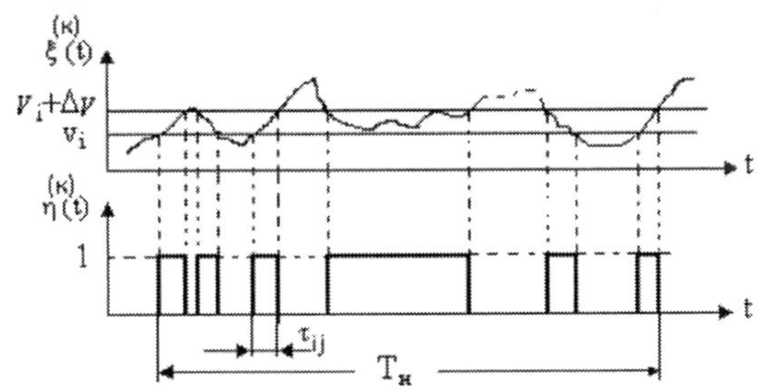

Figure 1a. Original stochastic process $\xi(t)$ and binary random process $\eta(t)$.

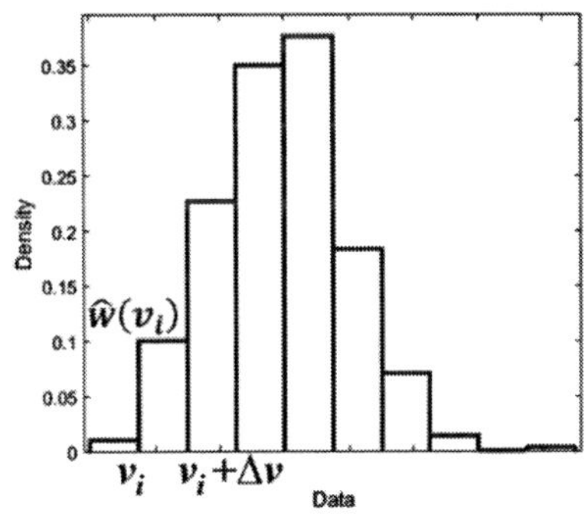

Figure 1b. Empirical estimation of a probability density function

For experimental evaluation of the density function $w(v_i)$ we must justify the choice of the observation time T (or the number of counts N during the observation T), the number of quantization levels of the process $\xi(t)$ and the size of the quantization interval Δv.

Statistical hypothesis testing of distribution law requires independent sampling in time. If the observations are dependent, then one-dimensional empirical distribution function $\hat{w}(v_i)$ will be distorted.

It is impossible to specify the exact value Δt for the test process used in this laboratory practicum. We need to select the time Δt from the condition of uncorrelated values of noise, separated by an interval of time Δt, though it is impossible for certain to state that these two values will be independent. However, the theory has proven that if the distribution of random variables is normal, then uncorrelated random variables are independent.

3. Random process modelling and PDF function estimation using ELVIS II platform and LabVIEW software package

In this laboratory practicum use a personal computer, NI ELVIS II laboratory platform and a test breadboard. For numerical analysis of the probability density function we use virtual apparatus, created with a universal programming system LabVIEW. The main view of this program is shown in the figure 2.

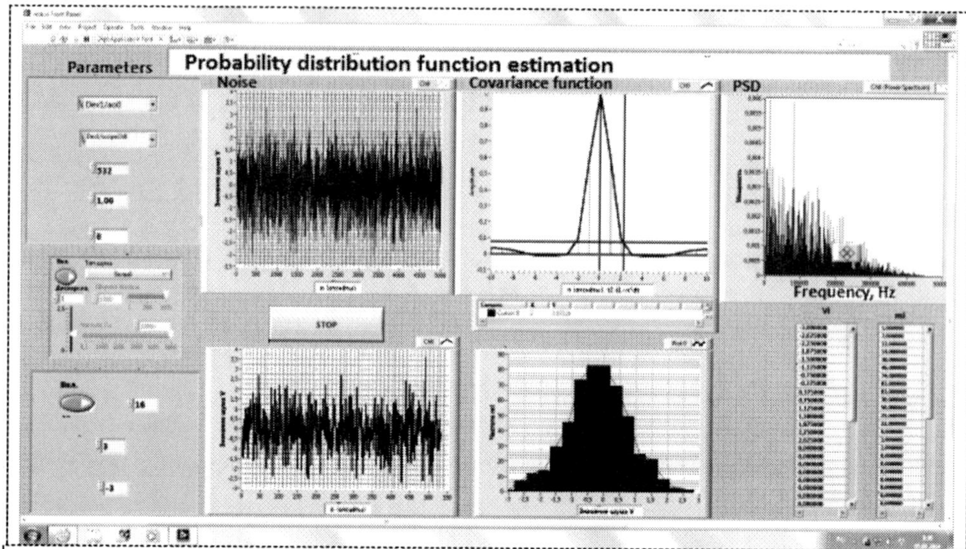

Figure 2. The main view of laboratory practicum for the PDF function estimation (in LabVIEW software package).

We use the noise generation in the LabVIEW environment and we output the noise from digital to analog devices (pin A0 0 connector) to the first channel oscilloscope CH 0 (Channel 0 SCOPE). The virtual apparatus has the following tasks:

- Generate a random process ("Noise" on the figure 2);
- Show the covariance function of the generated data ("Covariance function" on the figure 2);
- Show the power spectral density ("PSD" on the figure 2);
- Visualizes a data sample.
- Display a histogram of the distribution density for the noise values.

At the beginning of this work we need to calculate the observation time. Let's describe basics of this procedure.

If the generated set of noisy data consists of N samples from the general population with the distribution $w(v)$, then the difference between the probability density $\hat{p}_i$ for v_i interval and theoretical probability p_i should be smaller than ε and would be determined only by random errors. The probability of such deviation is written as:

$$P\left(\left|\hat{p}_i - p_i\right| > \varepsilon\right) = \alpha, \tag{9}$$

where ε is infinitesimal value.

According to the law of large numbers [2] the distribution of random variable $\psi = \hat{p}_i - p_i$ when $N \to \infty$ corresponds to the normal distribution law with a mean value $E[\Psi] = 0$ and variance $D[\Psi] = p_i (1 - p_i)/N$.

The probability of the case, that a random variable $\psi = \hat{p}_i - p_i$ is in range of $(-\varepsilon, \varepsilon)$, is equal to

$$P\left(-\varepsilon < \hat{p}_i - p_i < \varepsilon\right) = \frac{2}{\sqrt{2\pi}} \int_0^{2\pi} \exp\left(-\frac{x^2}{2}\right) dx = 1 - \alpha, \tag{10}$$

where $z_\alpha = \varepsilon \sqrt{\dfrac{N}{p_i(1-p_i)}}$.

Probability integral (10) is tabulated [4]. With known confidence level *(1-α)* and using tabulated values of the normal distribution or by numerical integration, we can find the value of z_α. Using z_α, we get $N = \frac{z_\alpha^2 p_i(1-p_i)}{\varepsilon^2} \le \frac{z_\alpha^2}{\varepsilon^2}$.

The value of ε is usually set based on practical needs.

After estimation of the number observation points *N*, we generate the random set of the white noise data. While generating we need to control the covariance function of the noise and its probability distribution function. It is needed to be sure that our measurements are not correlated therefore independent. After that, we can see the histogram of probability distribution function. If everything is good, we export generated values for the father analysis.

4. Testing hypothesis of distribution law

Experimentally found PDF function must be identified with one of the distribution law. We can make a hypothesis H_0 of the type of distribution law and the proposed distribution of $w(v)$ can be either true and false. We introduce a metrics *D* of the difference between the empirical distribution function and the hypothetical distribution.

Because of the empirical distribution function $w(v)$ corresponds to random variables then the metrics *D* is a random variable too. In this laboratory work we chosen a measure of deviation of the hypothetical distribution from the empirical as:

$$D = \chi^2 = \sum_{i=1}^{h} \frac{(m_i - Np_i)^2}{Np_i}, \tag{11}$$

where *h* is the number of quantization intervals for a noise, p_i - the hypothetical probability of the interval v_i, m_i – empirical mean value in the interval v_i.

From the literature [5] it is known that the measure of the metrics $D = \chi^2$ has χ^2 distribution with $h - 1$ degrees of freedom, where h is the number of quantization intervals for a noise. If we have number *L* of parameters of distribution function $w(v|H_0)$, then the number of degrees of freedom of χ^2 distribution is reduced by *L*. Therefore, χ^2 distribution degrees of freedom will be equal to $k = h - L - 1$.

If the confidence level *q* is specified, the boundary of the critical region χ^2_q is determined from the equation $P(\chi^2 > \chi^2_q|H_0) = q$, where H_0 is hypothesis that the sample belongs to the expected distribution law $w(v)$. The function $P(\chi^2 > \chi^2_q|H_0)$ is tabulated. [5]

Practice has shown that for the χ^2 test it is enough for the number of degrees of freedom to be in the range of (11–20). If during processing of the experiment results it occurs that $m_i < 10$, you need to combine intervals and add up m_i in corresponding united intervals. The resulting probability is equal to the sum of probabilities of united intervals. The number of degrees of freedom for χ^2 distribution is reduced by the number of intervals *j* merged and will be equal to $h = h - L - j - 1$.

Table of χ^2 distribution [1] makes it possible to determine the critical divergence (threshold) χ^2_q with the chosen permissible error probability *q* with accepted hypothesis H_0 about the law of distribution.

The procedure for the hypothesis testing is then reduced to the calculation of χ^2 and checking the following inequalities: if $\chi^2 > \chi^2_q$, H_0 hypothesis is rejected, if $\chi^2 \le \chi^2_q$, H_0 hypothesis is not rejected.

In the laboratory practicum we propose to check hypotheses about the normal distribution law and the Cauchy distribution. During hypothesis testing, students show empirical PDF function and compare it with hypothetical distributions on the graphics. After that they make a conclusion about hypothesis testing.

5. Conclusion

In this paper we proposed an example of laboratory practicum for the probability density function estimation and a random processes analysis. On the basis of ergodic theory, we realized the algorithm for estimating of probability distribution function for a random process. Use of ELVIS II platform and LabVIEW software package provides wide possibilities for students. They make necessary calculations, display results of the experiment and, most importantly, to control the experiment. At the same time students are introduced to a LabVIEW software package and its capabilities. Proposed laboratory practicum has been successfully used for many years at Kazan Federal University for statistical radiopysics studies.

References

[1] Parzen E 1962 On estimation of a probability density function and mode *Ann. Math. Stat.* **33** 1065–1076
[2] http://www.ni.com/ni-elvis/
[3] Nicholson W 2005 Microeconomic Theory (10th ed.) *Mason, Ohio: Thomson Higher Education*
[4] Moll Victor Hugo 2014 Special Integrals of Gradshteyn and Ryzhik: the Proofs *Series: Monographs and Research Notes in Mathematics I(1 ed.)* **Vol I**
[5] Box, Hunter and Hunter 1978 Statistics for experimenters *Wiley*

A computing model of ion energy distributions for dual frequencies capacitive couple discharges

Zhuwen Zhou[1,2,3,*], Bo Kong[1,3], Deliang Chen[1,2], Yuee Luo[1,3], and Yuansheng Wang[1,2]
[1]Physics and Electronic Science College, Guizhou Education University, Guiyang 550018, China
[2]Key Laboratory of Photoelectron Materials Design and Simulation in Guizhou Province, Guiyang 550018, China
[3]Scientific Research Innovation Team in Plasma and Functional Thin Film Materials in Guizhou Province, Guiyang 550018, China

E-mail: zzwwdxy@gznc.edu.cn

Abstract. We have developed an analytical dual-frequency model, which results in accurate dual-peaks of the ion energy distributions (IEDs) for the given control parameters. The model can analytically calculate and predict the IEDs dual peaks and energy width with necessary plasma parameters. This model can also predict the IEDs from dual frequencies' drives at any voltage, the associated results are compared with particle-in-cell (PIC) simulations. The IED at a surface is an important parameter for processing in dual radio frequencies driven capacitive couple discharges. The computing model is developed for the IED in a low pressure discharge based on a linear transfer function that relates the time-varying sheath voltage to the time-varying ion energy response at the surface. This model is in good agreement with PIC simulations over a wide range of dual frequencies driven capacitive couple discharge excitations.

1. Introduction

Capacitive radio frequency discharges are widely used for thin film deposition processing of silicon and surface modification in the microelectronics industry. The ion energy distribution (IED) at the substrate surface is an important parameter in these processes. Since the IED is essentially determined by the sheath response, it is necessary to establish the relationship between the control parameters and the sheath response. It is valuable to develop an analytical model and particle in cell (PIC) simulations to better understand the physics of the discharge and more accurately predict the IED. Multiple frequency drives are often used to control the width and the average of the IED. The physics of IEDs in low pressure single frequency capacitive discharges is fairly well understood.[1–3] In some approaches,[4–7] simulations regarding dual frequency ion energy distribution study sheath parameters and conclude that it is possible to control the IED by varying the voltage of the low frequency drive. In another approach,[8] an effective frequency concept has been introduced for dual

* To whom any correspondence should be addressed.

Content from this work may be used under the terms of the Creative Commons Attribution 3.0 licence. Any further distribution of this work must maintain attribution to the author(s) and the title of the work, journal citation and DOI.
Published under licence by IOP Publishing Ltd

frequency capacitive discharges to determine the width and average of the IED. The effective frequency takes the place of both frequencies and is dependent on the frequencies and voltages of the dual frequency drive. The IED can then be determined as for a single frequency discharge. Dual frequency discharges have drawn much attention due to their independent control of the ion fluxes and ion bombarding energies at the substrates. Semianalytical [9,10] models were developed to understand the mechanics in multi-frequency capacitive discharges to some extent. For both single- and dual-frequencies, research has been performed by experiments [11] and PIC simulations.[5] Analytical collisional sheath models with charge exchange collisions were developed.[12,13] A global model for capacitive discharges in the collisionless sheath model was proposed to understand bimodal for IED of single frequency capacitive discharge.[14] However, understanding the mechanics of multiple frequencies driven discharges is less than adequate. We have developed an analytical multi-frequency model, which results in accurate multi-peaks IEDs for the given control parameters. Incorporating Benoit-Cattin's model [2] our model can analytically predict the IEDs from multi-frequency drives at any voltage with necessary plasma parameters. PIC simulations are used to verify this model.

2. Theory analysis

Benoit-Cattin and Bernard2 analytically calculated the IEDs and DEi for single frequency drive, in the high frequency regime for a collisionless rf sheath, they assumed (i) a constant sheath width, (ii) a uniform sheath electric field, (iii) a sinusoidal sheath voltage, (iv) zero initial ion velocity at the plasma sheath boundary. For the high frequency regime, the ions take many rf cycles to cross the sheath and can no longer respond to the instantaneous sheath voltage. Instead, the ions respond only to an average sheath voltage and the phase of the cycle in which they enter the sheath becomes unimportant, resulting in a narrower IED. In this high-frequency regime, ΔE_i was calculated analytically for a collisionless sheath by Benoit- Cattin et al and found to be directly proportional to $\tilde{V}_s$、$\bar{V}_s$, and inverse proportional to ω_{hf}. Thus, as ω_{hf} increases, the IED width shrinks and the two peaks of the IED approach each other until, at some point, they can no longer be resolved (see figure 1). Therefore the expressions for the ion energy distributions function (IEDF- f (E)), and ΔE_{hf} at high- frequency are:

$$f(E) = \frac{2n_t}{\omega_{hf}\Delta E_{hf}}[1 - \frac{4}{(\Delta E_{hf})^2}(E - e\bar{V}_s)^2]^{-1/2} \quad (1)$$

$$\Delta E_{hf} = \frac{2e\tilde{V}_s}{\omega_{hf}\bar{s}_{hf}}(\frac{2e\bar{V}_s}{M})^{1/2} \quad (2)$$

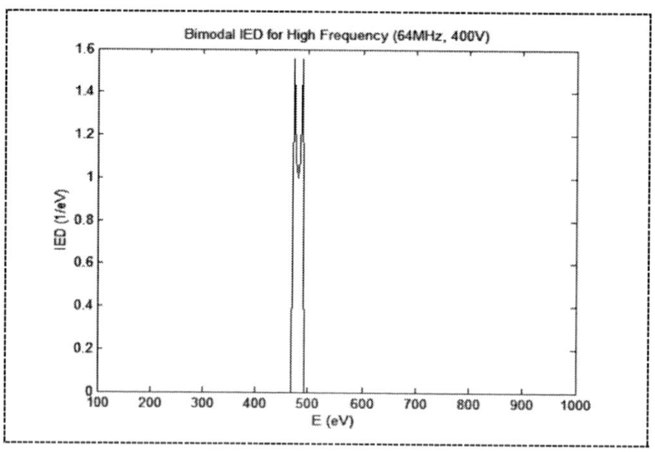

Figure 1. Bimodal IED high frequency (64MHz, 400V).

For the low-frequency regime, the ions cross the sheath in a small fraction of an rf cycle and respond to the instantaneous sheath voltage. Thus, their final energies depend strongly on the phase of the rf cycle in which they enter the sheath. As a result, the IED is broad and bimodal,
and the IED width ΔE_i approaches the maximum sheath drop. The two peaks in the distribution correspond to the minimum and maximum sheath drops (see figure 2). The resulting expressions for f (E) and the ΔE_{lf} at low frequency are

$$f(E) = \frac{2n_t}{\omega_{lf}\Delta E_{lf}}[1 - \frac{4}{(\Delta E_{lf})^2}(E - e\overline{V}_s)^2]^{-1/2} \tag{3}$$

$$\Delta E_{lf} = \frac{2e\widetilde{V}_s}{\omega_{lf}\overline{s}_{lf}}(\frac{2e\overline{V}_s}{M})^{1/2} \tag{4}$$

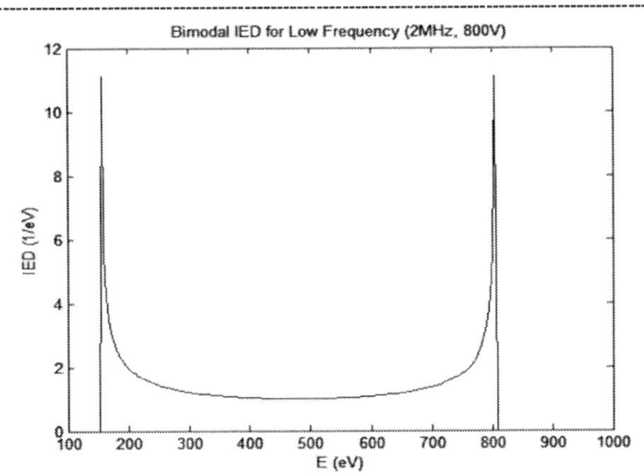

Figure 2. Bimodal IED low frequency (2MHz, 800V).

For the dual-frequency(both high and low frequency) regime, according to Benoit-cattin et al assumed a sinusoidal sheath voltage $V_s(t) = \overline{V}_s + \widetilde{V}_s \sin \omega t$, the sheath voltage functions at dual frequency (2MHz800V, 64MHz400V) showed in Figure 3, dual frequency (3MHz800V, 64MHz400V) showed in figure 4, these figures showed that low frequency voltage of the sheath was modulated by high frequency voltages, so the IED of low frequency sheath voltage was respectively modulated by of high frequency sheath voltage.

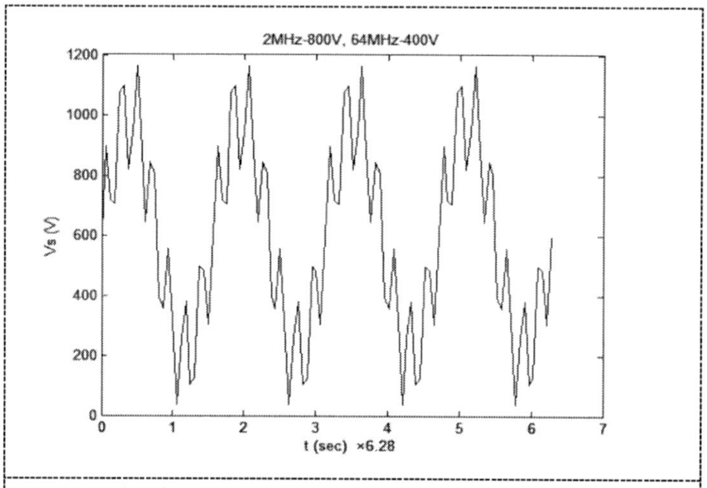

Figure 3. the sheath voltage functions at dual frequency (2MHz800V, 64MHz400V), low frequency (2MHz) voltage of the sheath was modulated by high frequency (64MHz) voltages.

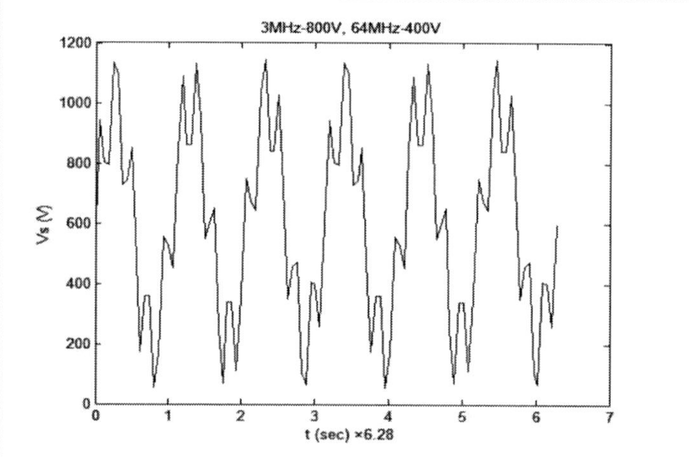

Figure 4. the sheath voltage functions at dual frequency (3MHz800V, 64MHz400V), low frequency (3MHz) voltage of the sheath was modulated by high frequency (64MHz) voltages.

From Eq. (4) we can see, for low frequency regime, the IED width ΔE_i is the max when sheath voltage approaches maximum of peak value, $\Delta E_i = \Delta E_{\max}$, and the bimodal of the IED max width $\Delta E_{\max}$ is in accord with upper and lower limit of IED. More peaks are distributed between the upper and lower limit due to high frequency IED width is small (see Eq. (2)), and its bimodal is repeatedly distributed on the low frequency most width bimodal. We obtain the expressions for the IED, $f(E)$ and the energy width ΔE_{lf}, for dual-frequency regime are:

$$f(E) = \frac{2n_t}{\omega_{lf}\Delta E_{lf}}[1-(\frac{2}{\Delta E_{lf}}(E-e\overline{V}_s))^2]^{-1/2} + \frac{2n_t}{\omega_{hf}\Delta E_{hf}}[1-\frac{4}{(\Delta E_{hf})^2}\sum_n(E-(e\overline{V}_s \pm n\Delta E_{hf}))^2]^{-1/2}$$

$$\Delta E_{lf} = \frac{2e\,\widetilde{V}_s}{\omega_{lf}\bar{s}_{lf}}(\frac{2e\overline{V}_s}{M})^{1/2},\quad \Delta E_{hf} = \frac{2e\,\widetilde{V}_s}{\omega_{hf}\bar{s}_{hf}}(\frac{2e\overline{V}_s}{M})^{1/2}. \tag{5}$$

Where ω_{lf} is low radian frequency of dual-frequency, $\omega_{hf}(>\omega_{lf})$ is higher radian frequency than ω_{lf}. ΔE_{lf} is the most width ion energy distribution relevant to low frequency, the width centre is at $e\overline{V}_s$ with upper limit $e\overline{V}_s + \Delta E_{lf}/2$ and lower limit $e\overline{V}_s - \Delta E_{lf}/2$. ΔE_{hf} is small width ion energy distribution relevant to high frequency, it distribute between upper limit and lower limit of IED. n represents integer ratio of $n = \Delta E_{lf}/\Delta E_{hf}$. n_t is the number of ions entering the sheath per unit time. $\bar{s}$ is the average sheath width, and $\overline{V}_s$ is the average sheath potential. We obtain the mean sheath width $\bar{s}$ in terms of the mean sheath voltage $\overline{V}_s$ by using the collisionless Child–Langmuir law

$$\bar{s} = \frac{2}{3}(\frac{2e}{M})^{1/4}(\frac{\varepsilon_0}{\overline{J}_i})^{1/2}\overline{V}_s^{3/4} \tag{6}$$

The ion current density in the sheath is given by

$$\overline{J}_i = en_s u_B \approx 0.61 n_0 u_B \tag{7}$$

Where n_0 is the bulk plasma density, n_s is the ion density at the pre–sheath boundary and $u_B = (kT_e/M)^{1/2}$ is the Bohm velocity. This implies [1]

$$\bar{s} \approx \frac{2}{3}(\frac{2e}{kT_e})^{1/4}(\frac{\varepsilon_0}{0.61 n_0 e})^{1/2}\overline{V}_s^{3/4} \tag{8}$$

As the voltage drop across the bulk plasma is relatively small at low pressures, we let, $\widetilde{V}_{rf}$ is the fundamental rf voltage amplitude applied to the electrodes. The time-averaged sheath voltage $\overline{V}_s$, which is also the ion kinetic energy per ion hitting the electrode, ε_i is given by,[3]

$$\varepsilon_i = \overline{V}_s = \widetilde{V}_s/2 \approx 0.8\widetilde{V}_{rf}/2 \tag{9}$$

We illustrated the multi-peaks of IEDs for multi-frequency using the computational analytical model from Eq. (5), (6), (7), (8), (9), for dual frequency excitation with ($\omega_{lf} = 2\pi \times 2$ MHz, $\widetilde{V}_{rf} = 800$ V, $\omega_{hf} = 2\pi \times 64$ MHz, $\widetilde{V}_{rf} = 400$ V, $\bar{s} = 0.0113$ m) in figure 5, and with ($\omega_{lf} = 2\pi \times 3$ MHz, $\widetilde{V}_{rf} = 800$ V, $\omega_{hf} = 2\pi \times 64$ MHz, $\widetilde{V}_{rf} = 400$ V, $\bar{s} = 0.0113$ m) in figure 6. In order to the model is more truthful, parameter 0.6 instead of 0.8 in Eq. (9). We also compared the model with semi-analytical [10], the IEDs peaks for the model are in accord with semi-analytical.

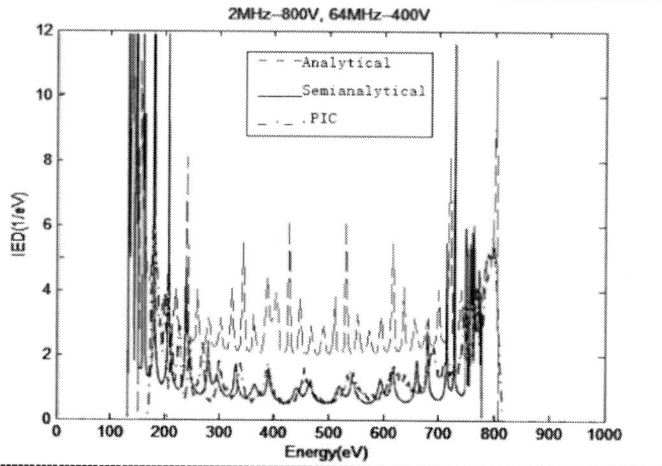

Figure 5. (Color online) Analytical, semianalytical, and PIC IEDs for dual frequency 2MHz - 800V, 64MHz - 400V.

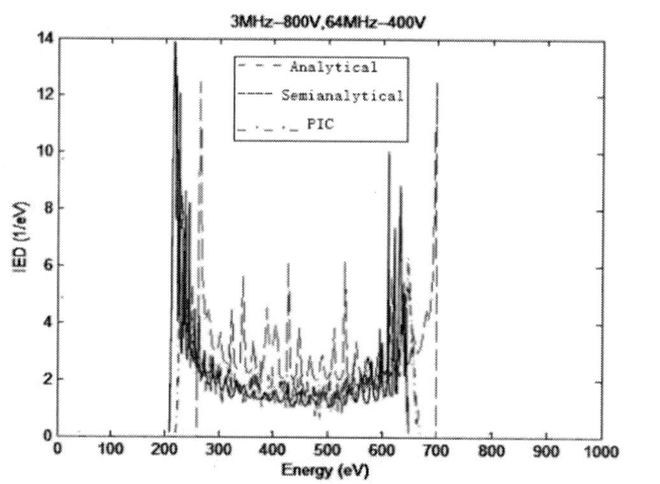

Figure 6. (Color online) Analytical, semianalytical, and PIC IEDs for dual frequency 3MHz - 800V, 64MHz - 400V.

3. PIC Simulations

The simulations are performed using a one-dimensional particle-in-cell code XPDP1. [16] Starting from the initial conditions, particle and field values are sequentially advanced in time. The particle equations of motion are advanced one time step, using fields interpolated from the discrete grid to the continuous particle locations. Next, particle boundary conditions such as absorption are applied. The Monte Carlo collision scheme is applied for electron-neutral collisions. [17] Ion-neutral and neutral-neutral collisions are not considered here. Source terms, charge density ρ , and current density J , for the field equations are accumulated from the continuous particle locations to the discrete mesh locations. The fields are then advanced one time step, and the time step loop starts over again. The cell size Δx , and time step Δt , must be determined from the Debye length λ_{D}, and the plasma frequency ω_p respectively, for accuracy and stability: $\Delta x < \lambda_D = \sqrt{\varepsilon_0 T_e / en}$ and

$(\Delta t)^{-1} >> \omega_p = \sqrt{e^2 n / \varepsilon_0 m}$. Here, ε_0 is the permittivity of free space, e is the elementary charge, and m is the mass of the lightest species (electrons). A violation of these conditions will result in inaccurate and possibly unstable solutions. After the simulation becomes steady state, the IED is collected over many cycles of the lowest frequency and normalized to unity. The PIC/MCC IED results are then compared with the computational analytical model. In the simulation, the discharge gap is L =0.03 m, consistent with typical processing applications. The gap width is divided into 500 cells, max $\Delta x / \lambda_D \approx 0.8$, and the simulation time step is $\Delta t = 7.63 \times 10^{-12}$, max $\Delta t \omega_p \approx 0.08$, Argon gas is used with p =30 mTorr, $np2c$=1E8, initial density of particles $n_0 = 1 \times 10^{15}\,\mathrm{m}^{-3}$. These parameters chosen are in order to resolve the Debye length and fulfill simulation stability criteria. The analytical result is shown as the dash line, semi-analytical result as the solid line and the PIC result is shown as the dash-dot line in figures 5, 6. As can be seen, the model predicts the 2 MHz/64 MHz (in figure 5) and the 3 MHz/64 MHz (in figure 6) dual frequency case reasonably well. We have similarly calculated the IEDs for other dual and triple frequency cases and find good agreement. The number of peaks and their energies are reasonably predicted.

4. Conclusions

We have developed the computational analytical model for multi-frequency that result in accurate multi-peaks of the ion energy distributions (IEDs) given the control parameters, which can analytically predict the IEDs multiple peaks and energy width with some plasma parameters provided. This model predicts the IEDs from dual frequencies drives at any voltage, the results are compared with particle-in-cell (PIC) simulations. The IED at a surface is an important parameter for processing in dual radio frequencies driven capacitive discharge. The analytical model is developed for the IED in a low pressure discharge based on a linear transfer function that relates the time-varying sheath voltage to the time-varying ion energy response at the surface. This model is in good agreement with PIC simulations over a wide range of single, dual, and even triple frequency driven capacitive discharge excitations. The present model neglects collisional effects in the sheath, which is the subject of future work.

5. Acknowledgments

The authors gratefully acknowledge the use of PTSG Software from the University of California, Berkeley. This project was supported by the Natural Science Research Project of the Key Laboratory of Guizhou Province Education Bureau Grant No. KY[2014]217, of the Scientific research innovation team in plasma and functional thin film materials of Guizhou Province Education Bureau Grant No.QJTD[2014]38, and of the Key Support Disciplines of Theoretical Physics of Guizhou Province Education Bureau Grant No. ZDXK[2015]38.

6. References

[1] Kawamura E, Vahedi V, Lieberman M A and Birdsall C K, 1999 *Plasma Sources Sci.Technol.* **8**, R45
[2] Benoit-Cattin P and Bernard L –C 1968 *J. Appl. Phys.* **39** 5723
[3] Lieberman M A and Lichtenberg A J 2005 *Principles of Plasma Discharges and Materials Processing*, 2nd ed. Wiley, New York, Chap. 11
[4] Georgieva V, Bogaerts A and Gijbels R 2004 *Phys. Rev. E* **69** 026406
[5] Lee J K, Manuilenko O V, Babaeva N Yu, Kim H C and Shon J W 2005 *Plasma Sources Sci. Technol.* **14** 89
[6] Guan Z Q, Dai Z L and Wang Y N 2005 *Phys. Plasmas* **12** 123502
[7] Wang L H, Dai Z L and Wang Y N 2006 *Chin. Phys. Lett.* **23** 668
[8] Kim H C and Lee J K 2005 *J. Vac. Sci. Technol.* A **23** 651
[9] Olevanov M, Proshina O, Rakhimova T and Voloshin D 2008 *Phys. Rev. E* **78** 026404
[10] Wu A C F, Lieberman M A and Verboncoeur J P 2007 *J. Appl. Phys.* **101** 056105

[11] Rakhimova T V, Braginsky O V, Ivanov V V, Kovalev A S, Lopaev D V, Mankelevich Y A, Olevanov M A, Proshina O V, Rakhimov A T, Vasilieva A N and Voloshin D G 2007 *IEEE Trans. Plasma Sci.* **35** 1229
[12] Lee J K, Manuilenko O V, Babaeva N Y, Kim H C and Shon J W 2005 *Plasma Sources Sci. Technol.* **14** 89
[13] Chen W C, Zhu X M, Zhang S and Pu Y K 2009 *Appl. Phys. Lett.* **94** 211503
[14] Israel D and Riemann K -U 2006 *J. Appl. Phys.* **99** 093303
[15] Wang Y, Lieberman M A, Wu A C F and Verboncoeur J P 2011 *J. Appl. Phys.* **110** 033307
[16] Verboncoeur J P, Alves M V, Vahedi V and Birdsall C K 1993 *J. Comput. Phys.* **104** 321
[17] Vahedi V and Surendra M 1995 *Comput. Phys. Commun.* **87** 179

Nonlinear Dynamic Analysis of Functionally Graded Timoshenko Beam fixed to a Rotating Hub

B Panigrahi[1], **G Pohit**[2,3]
[1,2]Department of Mechanical engineering, Jadavpur University, Kolkata-32, India
[2]E-mail: gpohit@gmail.com

Abstract. The present work accounts centrifugal stiffening effect on the nonlinear vibration response of an FGM Timoshenko beam. Analysis is carried out for a cantilever beam fixed with a rotating hub. Material is assumed to have a gradation relation along the depth of the beam. Centrifugal force and axial displacement raised due to the rotating hub is incorporated in the strain energy equations. Subsequent to this, an iterative technique is employed to obtain amplitude dependent vibration response of a rotating Timoshenko beam while material follows a gradation relation along the beam depth. Main objective of the work is to obtain the effects of rotational speeds, hub radius, and different gradation relations on the linear as well as nonlinear frequencies and mode shapes.

1. Introduction

Functionally graded materials (FGMs) have gained tremendous popularity in recent years. Use of these materials becomes more essential especially in situations where unevenly distributed thermal, chemical or mechanical loads are present. Such materials can be used in rotating machinery components for better performance. In order to avoid undesirable resonance phenomena and determination of suitable vibration control strategies, it becomes much more important to observe the vibration characteristics of such system. Numbers of investigations on vibration response of rotating beams have been proposed in recent times. Putter and Manor [1] and Kaya [2] found out natural frequencies of a rotating Timoshenko beam using differential quadrature method. Benerjee [3] obtained the effect of centrifugal stiffening on the dynamic response by deriving the dynamic stiffness matrix. Pohit et al. [4] used Perturbation technique to obtain free vibration response of a rotating beam with a nonlinear elastomeric constraint. Panigrahi and Pohit [5] proposed an iterative technique to study nonlinear free vibration of cracked FGM beams.

It is evident from the review that most of the literatures on the vibration of rotating beams have focused on linear vibration problem of isotropic beams. Literatures explaining the nonlinear effects and functionally graded material properties are rare. The present work accounts centrifugal stiffening effect on the nonlinear vibration response of a Timoshenko beam made of FGM. Main objective of the work is to obtain the effects of rotational speeds, hub radius and different gradation relations on the linear as well as nonlinear frequencies and mode shapes.

2. Solution methodology

Fig. 1 shows a cantilever beam of length L and thickness h, fixed end of which is attached to a hub of radius R and rotating with an angular speed of $\overline{\Omega}$ rad/sec. Midplane of the beam can be represented by z=0. Variation of the properties are assumed to vary exponentially in thickness direction as given in eq. 1, where E_I and ρ_I are effective young's modulus and mass density of beam at top surface

[3] Corresponding author

Content from this work may be used under the terms of the Creative Commons Attribution 3.0 licence. Any further distribution of this work must maintain attribution to the author(s) and the title of the work, journal citation and DOI.
Published under licence by IOP Publishing Ltd

respectively and k is the material index (E_2/E_1). Poisson's ratio v is taken to be constant throughout the analysis.

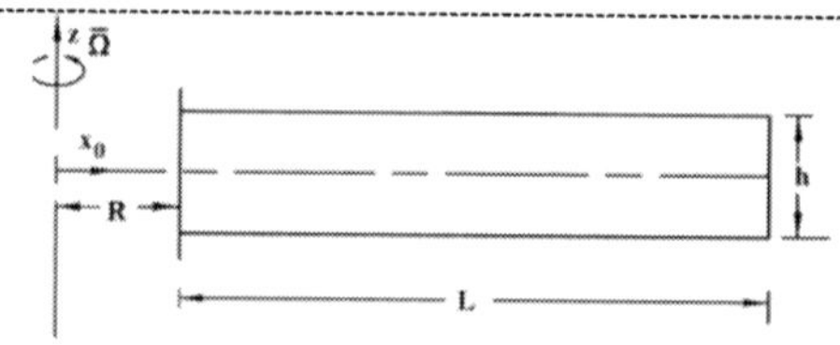

Figure 1. Cantilever beam fixed with rotating hub.

$$E(z) = E_1 \sqrt{k} e^{\left(\frac{z}{h}\ln(k)\right)}, \qquad \rho(z) = \rho_1 \sqrt{k} e^{\left(\frac{z}{h}\ln(k)\right)} \tag{1}$$

In present work neutral plane is considered for the analysis purpose, in order to avoid bending-extension coupling [5]. Position of the neutral plane from the mid plane of the beam is C distance apart. Value of C can be easily obtained [5].

2.1 *Timoshenko beam theory and energy equations*

In fig. 1, z is the coordinate in the thickness direction, x_0=R+x is the axial coordinate, where x is the span wise distance of the point from the hub edge. Displacement fields in space and time coordinates of Timoshenko beam can be given as follows.

$$U(x,z,t) = U(x,t) + z\{\Psi(x,t)\}, \quad W(x,z,t) = W(x,t) \tag{2}$$

In the eq. 2, U(x, t), W(x, t) and Ψ(x, t) represent longitudinal stretching, transverse and rotational displacement of neutral plane, respectively. Von Kerman type nonlinearity is incorporated in Strain-displacement relation. Motion is assumed to be harmonic motion of frequency θ rad/sec. Following the energy method as expressed in [5] and incorporating the effect of centrifugal force and axial displacement raised due to rotating condition as potential of load (PL), Maximum form of Kinetic energy (KE_{max}), Potential energy (PE_{max}) and potential of load (PL_{max}) can be derived as

$$KE_{\max} = \frac{\theta^2}{2}\int_0^L \left(M_1 U^2 + M_3 \Psi^2 + M_1 W^2\right) dx, \; PE_{\max} = \frac{1}{2}\int_0^L \left(\begin{array}{l} K_1\left(\frac{\partial U}{\partial x}\right)^2 + K_3\left(\frac{\partial \Psi}{\partial x}\right)^2 + K_4\left(\frac{\partial W}{\partial x} + \Psi\right)^2 \\ + K_1\left(\frac{\partial U}{\partial x}\right)\left(\frac{\partial W}{\partial x}\right)^2 + \frac{K_1}{4}\left(\frac{\partial W}{\partial x}\right)^4 \end{array} \right) dx$$

$$PL_{\max} = \frac{1}{2}\int_0^L M_1 \bar{\Omega}^2 L^2 \left\{\left(\frac{R}{L}\right) - \left(\frac{Rx}{L^2}\right) + \frac{1}{2} - \left(\frac{x^2}{2L^2}\right)\right\}\left(\frac{\partial W}{\partial x}\right)^2 dx \tag{3}$$

Here K_1, K_2, $K_3$and K_4 are stiffness parameter and M_1, M_2 and M_3 are inertial parameter as given by Panigrahi and pohit [5].Dimensionless parameters are taken as mentioned in eq. 4

$$\xi = \frac{x}{L}, u = \frac{U}{h}, w = \frac{W}{h}, \psi = \Psi, \alpha = \frac{L}{h} \; \delta = \frac{R}{L} \; r^2 = \frac{I}{AL^2} = \frac{h^2}{12L^2}, \quad \{m_3, k_3, k_4\} = \left\{\frac{M_3}{M_1 h^2}, \frac{K_3}{K_1 h^2}, \frac{K_4}{K_1}\right\}$$

$$\omega^2 = \frac{12 M_1 \theta^2 L^4}{K_1 h^2}, \Omega^2 = \frac{12 M_1 \bar{\Omega}^2 L^4}{K_1 h^2} \tag{4}$$

Using eq. 3 and eq. 4 and multiplying each expression of eq. 3 by $12\alpha^2 / K_1 h$ dimensionless form of energy expression can be derived separately for linear and nonlinear potential energy as

$$KE^*_{\max} = \frac{\omega^2}{2}\left[\int_0^1 \left(u_1^2 + m_3\psi_1^2 + w_1^2\right) d\xi\right]$$

$$PE^*_{linear} = \frac{12\alpha^2}{2}\int_0^1 \left\{\left(\frac{\partial u}{\partial \xi}\right)^2 + k_3\left(\frac{\partial \psi}{\partial \xi}\right)^2 + k_4\left(\frac{\partial w}{\partial \xi}\right)^2 + k_4\alpha^2\psi^2 + k_4\alpha\frac{\partial w}{\partial \xi}\psi\right\} d\xi \quad (5)$$

$$PE^*_{nl} = \frac{12\alpha^2}{2}\int_0^1 \left\{\frac{1}{\alpha}\left(\frac{\partial u}{\partial \xi}\right)\left(\frac{\partial w}{\partial \xi}\right)^2 + \frac{1}{4\alpha^2}\left(\frac{\partial w}{\partial \xi}\right)^4\right\} d\xi \qquad PL^*_{\max} = \frac{\Omega^2}{2}\int_0^1 \left(\delta + \frac{1}{2} - \delta\xi - \xi^2\right)\left(\frac{\partial w}{\partial \xi}\right)^2 d\xi$$

Energy functional can be obtained as - $\Pi = PE^*_{linear} + PE^*_{nonlinear} + PL^*_{\max} - KE^*_{\max}$ (6)

Displacements (admissible functions) are chosen as $u = \sum_{j=1}^{j=N} A_j \xi^j, w = \sum_{j=1}^{j=N} B_j \xi^j, u = \sum_{j=1}^{j=N} C_j \xi^j$. (7)

Minimizing the energy functional by taking derivative with unknown coefficients, sets of nonlinear governing equations can be derived in the form of $[K_{linear}]\{UC\} + [K_{nonlinear}]\{UC\} = \omega^2 [M]\{UC\}$ (8)
Here UC is unknown coefficients. Elements of nonlinear stiffness matrix contain terms of unknown coefficients as well. In order to solve nonlinear problem, an iterative technique is employed with the following steps: First all the nonlinear terms are neglected and a linear set of unknown coefficients are obtained. Linear set of unknown coefficients are then scaled up in such a manner that the maximum nonlinear displacement occurs at the free end and is equal to the assumed amplitude of vibration (w_{max}). Using the scaled unknown vectors nonlinear set of vectors are found out. Nonlinear set of unknown coefficient vectors are again scaled up. Repeating the above process until relative error between two consecutive values of the non-linear frequency is less than the tolerance.

3. Results and discussions

A matlab code is developed for analysis purpose. Number of Polynomial terms is taken as eight.

3.1 *Comparison study*

Results obtained from the present analysis are compared with the results available in the literatures. For comparison purpose geometric and material properties are taken as δ=0, $E_1/\kappa G_1$ =3.059, r=1/30, where κ is shear correction factor. Otherwise for all the results r is taken as 0.1 and other properties remain same until and unless specified. Table 1 show that the results from present analysis matched fairly well with that of Benerjee [3].

Table 1. Comparison study.

Ω	0	1	2	3	4
Dimensionless First linear frequency (present)	3.4798	3.6451	4.0995	4.7562	5.5383
Benerjee[3]	3.4798	3.6445	4.0971	4.7516	5.5314

3.2 *Effect of material gradient on linear frequencies*

Fig. 2 shows the variation of the linear frequency with respect to the variation in the material gradient index (k). Ordinate is taken as the ratio of linear natural frequency of FGM material to the same of isotropic material and abscissa contains the variation of k. Results for first three modes of vibration are shown.

3.3 *Effect of material gradient, rotational speed and hub radius on non-linear frequencies*

Fig. 3 represents the backbone curve for 1st mode, which shows the effect of k (1, 0.2 and 2) and rotational speeds. Nonlinear frequency ratios are comparatively higher for k=0.2, whereas a very small difference in nonlinear frequency ratio is observed for k=1 and k=2. Rotational speed causes the centrifugal stiffening effect. Stiffening effect makes the beam stiffer and linear as well as nonlinear frequency increases. Therefore Backbone curves become straighter as the rotational speed increases. Table 2 shows the variation in the nonlinear frequency ratios with the hub radius δ for Ω=4. With increase in hub radius nonlinear frequency ratios are found to be reduced.

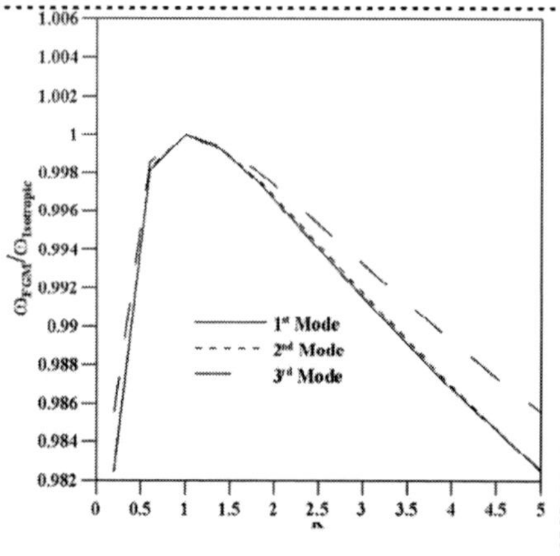

Figure 2. Variation of linear frequency with material index k.

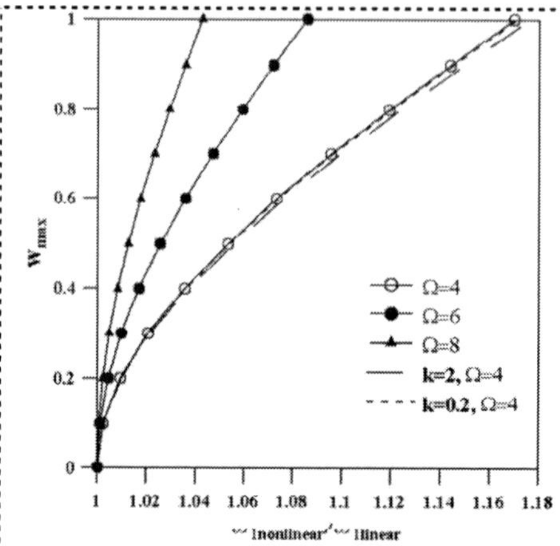

Figure 3. Variation in backbone curves for first mode with material index k and rotational speeds Ω.

Table 2. Effect of hub radius.

Hub radius	δ=0	δ=0.1	δ =0.2	δ =0.3	δ =0.4
$\omega_{1nl}/\ \omega_{11}$	1.072	1.015	1.008	1.005	1.002
$\omega_{2nl}/\ \omega_{21}$	1.099	1.057	1.043	1.036	1.032

4. Conclusion

An iterative technique is employed in order to capture the nonlinear dynamic response of a rotating FGM beam. Methodology is computationally efficient and accurate. Effects of centrifugal stiffening arise due to rotation of hub is incorporated and effects are captured effectively. Effects of different gradation relation on linear as well as nonlinear vibration response are captured successfully.

References

[1] Putter S and Manor H 1978 Natural frequencies of radial rotating beams *J. Sound Vib.* **56** 175–185.

[2] Kaya M O 2006 Free Vibration analysis of a rotating Timoshenko beam by differential transform method *Aircr. Eng. Aerosp. Tec* **78(3)** 194-203.

[3] Benerjee J R 2001 Dynamic stiffness formulation and free vibration analysis of centrifugally stiffened Timoshenko beams *J. Sound Vib.* **247(1)** 97–115.

[4] Pohit G, Mallik A K and Venkatesan C 1999 Free out-of-plane vibration of a rotating beam with non-linear elastomeric constraints *J. Sound Vib.* **220(1)** 1–25.

[5] Panigrahi B and Pohit G 2015 Nonlinear modelling and dynamic analysis of cracked Timoshenko functionally graded beams based on neutral surface approach *P I Mech Eng C-J Mec.*, online DOI: 10.1177/0954406215576560.

Adhesive friction based on finite element study and n-point asperity model

Prasanta Sahoo and Ajay K Waghmare

Department of Mechanical Engineering, Jadavpur University, Kolkata, India

Email: psjume@gmail.com, psahoo@mech.jdvu.ac.in

Abstract. The present work considers analysis of adhesive friction of rough surfaces using n-point asperity concept for statistical definition of surface roughness features, and accurate finite element analysis of elastic-plastic deformation of single asperity contact. Well defined adhesion index and plasticity index are used to study the prospective contact situations arising out of variation in material properties and surface roughness features. From the present results it is possible to locate the combinations of adhesion index and plasticity index that may yield very low coefficient of friction. Thus suitable choice of surface and material parameters for the contact of two rough surfaces can be made in order to minimize friction typically at low load and micro scale roughness situations.

1. Introduction

Friction is defined as resistance to movement of one body over the other. For smooth surfaces carrying nano to micro scale size of roughness features, adhesion at contact spots is the main cause of friction. Bowden and Tabor [1] in their classical theory, described the adhesive friction as a tangential force required to shear off adhesive bonds formed at the tip of contacting spots due to local plastic deformation. Tabor [2] emphasized that three basic elements are involved in dry friction of solids viz. the true area of contact, the nature and strength of the interfacial bonds formed at the region of contact, and the way in which the material around the contact regions is sheared during sliding. Chang et al. [3] developed a static friction model incorporating these basic elements. The model relates the start of slip to material properties and treats the first yielding of a single material point as a criterion for sliding inception. But, Kogut and Etsion [4] proved that such assumption underestimates the permissible tangential force as the first failed point remains surrounded by considerable volume of elastic material which can support additional tangential load. Roy Chowdhury and Ghosh [5] (henceforth called RG model) used slip and yield as two separate limiting criteria to get total tangential force. In their analysis, RG assumed purely elastic and plastic deformations of contact spots with normally applied load but they ignored the intermediate elastic-plastic deformation. Chang, Etsion, and Bogy [6] (hence forth called as CEB approach) used conservation of tip volume concept to account for behaviour of material under intermediate elastic-plastic deformation. But it is observed that the CEB analytical model suffers from a discontinuity in the contact load as well as in the first derivatives of contact load and contact area at elastic to elastic-plastic transition point. Literature shows some work [7-9] intended to remove the discrepancies found in CEB's model. But these works were either mere mathematical exercises without much relevance to physical contact phenomenon [7-8] or they lacked the accuracy [9]. Kogut and Etsion [10] (hence forth called as KE approach), based on their accurate FEA (Finite

Content from this work may be used under the terms of the Creative Commons Attribution 3.0 licence. Any further distribution of this work must maintain attribution to the author(s) and the title of the work, journal citation and DOI.
Published under licence by IOP Publishing Ltd

Element Analysis) results of single contact spot, developed analytical expressions for transition zone behaviour of load and contact area.

Surface roughness (asperity) plays significant role in the study of adhesion and friction. In conventional models the asperities are defined as peaks having some regular form at the tip. Also all such asperities on the surface are assumed to have identical tip radius with only variation in peak heights. Such definition of surface roughness features is far away from reality. The n-point asperity concept introduced by Hariri, Zu, and Ben Mrad [11] assumes that at particular level of separation between an interfering plane and a rough surface, there exist asperities of different sizes (i.e. different radii of curvatures and different heights). Also each asperity doesn't exist as a separate entity throughout the progression of contact but the earlier asperity gets merged into a new asperity with contact progression. Thus scope exists to study adhesive frictional contact by extending the RG model to cover intermediate transition zones of deformation and using n-point asperity model. To the best of authors knowledge no such study is found in the existing literature. In earlier two studies [12, 13], the present authors studied the frictional contact in n-point asperity frame work. The first study [12] considers adhesive friction without any consideration for transition type deformation; while the second study [13] though uses KE considerations for transition zone, it is for non-adhesive contact situation.

2. Modeling adhesive friction in n-point asperity frame work

Contact between two rough surfaces can be analysed as a contact between an equivalent rough surface and a rigid smooth plane [3-6, 11-14]. In present study, the equivalent rough surface is assumed to be carrying equivalent height ordinates, and existence of n-point asperities is defined statistically by using such height ordinates [11]. In modeling adhesive friction, first formulations are obtained for normally applied load and tangentially applied load (friction load) on a single n-point asperity and then by incorporating this into statistical multi asperity contact model, the total effect is quantified for the whole surface. As mentioned before, in present study the RG model of adhesive friction is extended to cover the transition zones of deformation also by incorporating CEB and KE expressions.

KE model, based upon the critical value of interference between flat plane and an asperity, $\delta^*_{(y)_n}$ (which marks the yield inception), demarcates the whole range of deformation into four distinct zones. Total non dimensional applied load for all asperities having specific 'n' value will be

$$P^*_n = P^*_{(e)_n} + P^*_{(ep-I)_n} + P^*_{(ep-II)_n} + P^*_{(p)_n} \tag{1}$$

Where the non-dimensional applied load values in elastic, elastic-plastic stage-I, elastic-plastic stage-II, and plastic zones of deformation are calculated as:

$$P^*_{(e)_n} = M_{(e)_n} E\left[\Delta P^*_{(e)_n}\right] \quad for \quad 0 < \delta^*_n < \delta^*_{(y)_n} \tag{2}$$

$$P^*_{(ep-I)_n} = M_{(ep-I)_n} E\left[\Delta P^*_{(ep-I)_n}\right] \quad for \quad \delta^*_{(y)_n} \le \delta^*_n \le 6\delta^*_{(y)_n} \tag{3}$$

$$P^*_{(ep-II)_n} = M_{(ep-II)_n} E\left[\Delta P^*_{(ep-II)_n}\right] \quad for \, 6\delta^*_{(y)_n} < \delta^*_n \le 110\delta^*_{(y)_n} \tag{4}$$

$$P^*_{(p)_n} = M_{(p)_n} E\left[\Delta P^*_{(p)_n}\right] \quad for \quad \delta^*_n > 110\delta^*_{(y)_n} \tag{5}$$

Here parameter $'n'$ indicates the number of height ordinates of which an asperity is comprised; presence of $'n'$ at subscript or sub-subscript position in a parameter indicates that the value of the parameter is for an asperity which is comprised of specific $'n'$ number of ordinates. M values are the number of n-point asperities lying in respective zones of deformation, $E[.]$gives expected values, and Δ terms are the values of single n-point asperity. δ^*_n is the non dimensional interference of an n-point asperity while $\delta^*_{(y)_n}$ marks non dimensional critical interference corresponding to yield inception. For brevity, Δ term for only elastic zone of deformation is provided here as below.

$$\Delta P^*_{(e)_n} = \frac{(n-1)}{(-\ln\rho)}\left\{\frac{0.67\delta_n^{*3/2}}{\psi\, c_n^{1/2}} - \frac{1.77(n-1)^{1/2}(-\ln\rho)^{1/2}\phi^{1/2}\delta_n^{*3/4}}{\psi\, c_n^{3/4}}\right\} \quad for \;\; 0 \le \delta^*_n \le \delta^*_{(y)_n} \qquad (6)$$

Where, c_n is the curvature coefficient of an n-point asperity; $\phi = (\gamma\beta^*)/(E\sigma^2)$ is the adhesion index [14]; and $\psi = (H\beta^*)/(E\sigma)$ is the plasticity index [11]. The terms σ, γ, E and H used in these indices are standard deviation of height ordinates (root mean square roughness), work of adhesion, composite elastic modulus of equivalent rough surface and hardness value of soft surface respectively.

Equation (1) as given above when solved for all values of $'n'$, viz. $n = 3 \; to \; \infty$ (minimum 3 height ordinates are required to define an n-point asperity), will give total applied load on the rough surface.

In RG model, asperities deforming in purely elastic zone are considered to contribute to friction force. But in present study while using KE to demarcate different zones of deformation, contribution to friction force (T^*) is obtained from the asperities deforming in first two zones (viz. purely elastic and elastic-plastic stage-I) by applying RG's slip-yield theory. The method adopted to obtain slip and yield parts of tangential resistance is given in author's earlier works [12].

Finally the same contact situation is formulated with CEB's considerations of transition zone deformation using n-point asperity surface roughness features.

3. Results and discussion

In order to study adhesive friction with two different approaches viz., CEB and KE, the corresponding formulations modeled for total applied load and total frictional force are solved by using MATLAB code. The applied load and tangential load formulations consists of different surface roughness and material properties related parameters like number of height ordinates (n), correlation coefficient (ρ), adhesion index (ϕ) and plasticity index (ψ).The different conditions of contact with respect to variation in material properties and surface roughness features is obtained by varying the indices, ϕ and ψ while variation in normally applied and frictional load is obtained by varying the mean surface separation $(h^* = h/\sigma)$ between contacting surfaces in the range of 0 to. $+3\sigma$. For standard normal distribution of height ordinates, this range of mean separation covers 99.7% asperities on the surface under consideration. Physical significance of the different parameters and the related data used in present study is the same as discussed in author's earlier study [12]. In present study three different values of adhesion index (viz., $\phi = 0.1, 0.5 \; and \; 0.9$) and three different values of plasticity index (viz., $\psi = 2.0, 0.9 \; and \; 0.3$) are used. Care is taken in selection of these indices for study so that we get significantly different zones with respect to adhesion effect and plasticity effect. Increase in adhesion index value from 0.1 to 0.9 represents increased adhesion effect while decrease in plasticity index from 2.0 to 0.3 represents increased plasticity effect. Physically, increase in adhesion effect is characteristic of smoother surfaces and/or surfaces possessing higher surface energy, while increase in plasticity effect is the characteristic of rough and/or soft surfaces. Figure 1 shows non-dimensional friction force against non-dimensional applied load. The general trend of behaviour observed is that the friction force is almost proportional to applied load. For the same applied load surfaces with higher ϕ value carry higher friction force. In other words, the smoother surfaces with higher tendency for adhesion yield higher resistance to tangential movement. Also it is observed that plots with combination of higher ψ and lower ϕ show small non linearity. Physically it means surfaces which have higher tendency to deform elastically and are more rough show non-linear behaviour of friction force with normally applied load. Figure 2 shows coefficient of friction versus non-dimensional applied load plots in high adhesion effect $(\phi = 0.9)$ zone. From the plots it is observed that plots with $\psi = 0.3$ and 0.9 have almost zero slope while that with $\psi = 2.0$ have some slope. Physically it means surfaces undergoing predominantly plastic type of deformation and having moderate to higher adhesion have constant coefficient of friction. Also it is observed that at a particular value of applied load, for smaller value of ψ , coefficient of friction values are high. Physically it means higher the tendency to deform plastically higher will be coefficient of friction. It is to be noted here that the rough surfaces have higher tendency to undergo plastic type of deformation. Comparison of results from the KE and CEB approaches show that there is qualitative agreement between the results but from quantitative agreement point of view, CEB approach formulation overestimates the results in predominantly elasto-plastic and plastic zones of deformations. From the present results it is possible to locate the combinations of adhesion index and plasticity index that may yield very low coefficient of friction. Thus suitable choice of surface and material parameters for the contact of two rough surfaces can be made in order to minimize friction typically at low load and micro scale roughness situations.

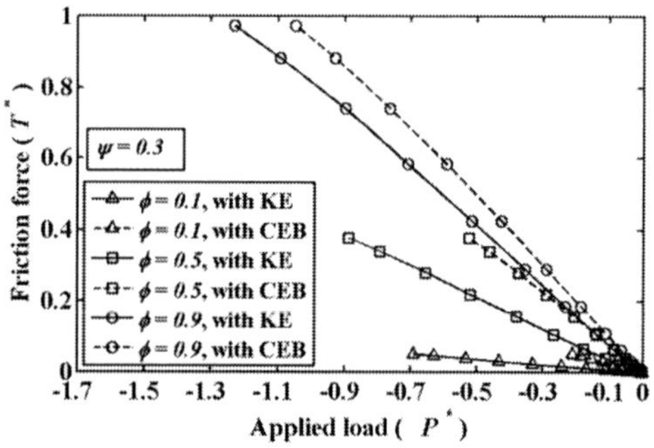

Figure 1 Typical variation of friction force with applied load

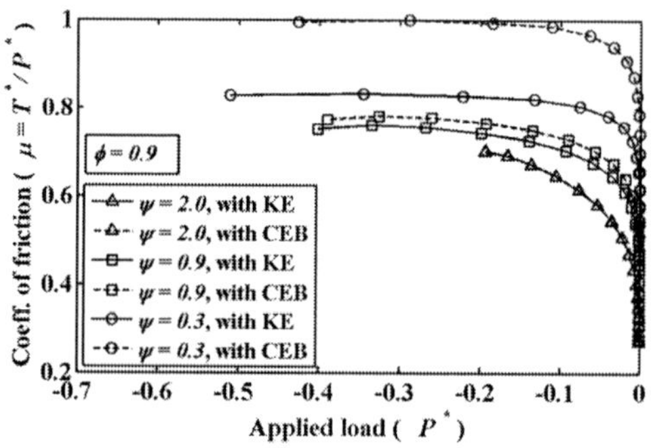

Figure 2 Typical variation of coefficient of friction with applied load

4. Conclusion

Use of n-point asperities to define surface roughness features represent more realistic picture than any other conventional method. Also incorporation of KE approach formulation for intermediate transition state of deformation gives complete solution. Qualitative trend of the results found is same as that observed in literature with conventional asperities. It is possible to locate the combinations of adhesion index and plasticity index that may yield very low coefficient of friction. Thus suitable choice of surface and material parameters for the contact of two rough surfaces can be made in order to minimize friction typically at low load and micro scale roughness situations.

References

[1] Bowden F P and Tabor D 1964 The friction and lubrication of solids Vol. **2** OUP
[2] Tabor D 1981 *Journal of Lubrication Technology* **103(2)** 169-179
[3] Chang W R, Etsion I and Bogy D B 1988 *Journal of Tribology* **110** 50-56
[4] Kogut L and Etsion I 2003 *Tribological Transactions* **46(3)** 383–390
[5] Roy Chowdhury S K and Ghosh P 1994 *Wear* **174** 9-19
[6] Chang W R, Etsion I and Bogy D B 1987 *Journal of Tribology* **109** 257-263
[7] Evseev D G, Medvedev B M and Grigoriyan G G 1991 *Wear* **150** 79–88
[8] Zhao Y, Maietta D M and Chang L 2000 *Journal of Tribology* **122(1)** 86-93
[9] Kucharski S, Klimczak T, Polijaniuk A and Kaczmarck J 1994 *Wear* **177** 1-13
[10] Kogut L and Etsion I 2002 *Journal of Applied Mechanics* **69(5)** 657-662
[11] Hariri A, Zu J W and Ben Mrad R 2006 *ASME Journal of Tribology* **128** 876-885
[12] Waghmare A K and Sahoo P 2015 *Engineering Science and Technology, an International Journal* **18** 463- 474
[13] Waghmare A K and Sahoo P 2016 *Journal of Engineering Tribology* (Online published, *doi:10.1177/1350650116632018*)
[14] Sahoo P, Mitra A and Saha K 2009 *Journal of Physics D Applied Physics* **42(6)** 1–13.

Journal of Physics: Conference Series **738** (2016) 012117 doi:10.1088/1742-6596/738/1/012117

Some Preliminary Results on an SEIARD Epidemic Model with Vaccination and Antiviral Treatment Controls and Dead-Infective Culling Action

M. De la Sen[1], R. Nistal, S. Alonso-Quesada, A. J. Garrido
Institute of Research and Development of Processes,
University of the Basque Country,
Bilbao, Spain

E-mail: manuel.delasen@ehu.es

Abstract. This paper studies the non-negativity and stability properties of the solutions of a newly proposed SEIADR model which incorporates asymptomatic and dead-infective subpopulations to those defining the standard SEIR model and, in parallel, it incorporates feedback vaccination and antiviral treatment controls.

1. Introduction

Relevant attention is being paid in the last two decades to the study of mathematical epidemic models, which are modelled by integro-differential equations and/ or difference equations. Those models describe the evolution of the various subpopulations considered as the disease under study progresses. Typically, the models have three essential subpopulations (namely, susceptible, infected and recovered by immunity) whose dynamics are mutually coupled. There are different degrees of complexity available in the statement of the models. The simpler models basically describe "susceptible" (S) and "infected" (I) subpopulations and are referred to as SI- models. A second degree of complexity adds a third one said to be the "recovered by immunity" subpopulation and those models are said to be SIR-models. A further complexity degree splits the infected into two subpopulations (or compartments), namely, the so-called "infected" or "exposed" (E) (those infective having the disease but do not present yet external symptoms) and the "infectious" or "infective" (those infective having external symptoms). The generic acronym used for this last category of models is SEIR, being referred to as SEIR epidemic models. A general description of epidemic models and some mathematical analysis on them is given in some classical books. See, for instance, [1-2]. More sophisticated models have been described and analyzed in the literature. See, for instance, [4-8] and references therein. On the other hand, it turns out, due to medical experience, that there are individuals who are infected but do not have significant external symptoms, the so-called the "asymptomatic" (A) subpopulation. This occurs even in the common known influenza disease. If such an asymptomatic subpopulation is considered in the model then it turns out that the exposed have potential distinct transitions to the infective and to the asymptomatic so that a part of the exposed become asymptomatic after some time while others become infective. On the other hand, it is well-known that in the Ebola disease case, the abandoned

---

[1] To whom any correspondence should be addressed. This research is supported by the Spanish Government through Grants DPI2012-30651 and DPI2015-64766-R

Content from this work may be used under the terms of the Creative Commons Attribution 3.0 licence. Any further distribution of this work must maintain attribution to the author(s) and the title of the work, journal citation and DOI.
Published under licence by IOP Publishing Ltd

lying dead corpses are infective [3, 4] what causes serious sanitary problems in third world tropical countries with low or scarce sanitary technical means when Ebola disease spreads thoroughly specially when it is transmitted from scarcely populated rural areas to high populated urban ones. In particular, it has been pointed out in some background literature that the simultaneous presence of asymptomatic infective population with dead-infective corpses is compatible in the Ebola disease propagation. The dead lying corpses can be considered in the model as a new infective subpopulation "D".

2. The model and its controlling actions

The proposed SEIADR model is the following:

$$\dot{S}(t)=b_1-(b_2+\beta I(t)+\beta_A A(t)+\beta_D D(t))S(t)+\eta R(t)-V(t)$$
$$\dot{E}(t)=-(b_2+\gamma)E(t)+(\beta I(t)+\beta_A A(t)+\beta_D D(t))S(t)$$
$$\dot{I}(t)=-(b_2+\alpha+\tau_0+\xi(t))I(t)+\gamma p E(t)$$
$$\dot{A}(t)=-(b_2+\tau_0)A(t)+\gamma(1-p)E(t)$$
$$\dot{D}(t)=-\mu D(t)+b_2(I(t)+A(t))+\alpha I(t)-\rho_D(t)D(t)\Sigma_{t_i\in ImpD}\delta(t-t_i)$$
$$\dot{R}(t)=-(b_2+\eta)R(t)+\tau_0(I(t)+A(t))+\xi(t)I(t)+V(t)$$

; $\forall t\in \boldsymbol{R}_{0+}$ with initial conditions satisfying $min\,(S(0),E(0),I(0),A(0),D(0),R(0))\geq 0$, where:
$\delta(t)$ is the Dirac distribution, $ImpD=\{t\in \boldsymbol{R}_{0+} : D(t)\neq D(t^-)\}=\cup_{t\in \boldsymbol{R}_{0+}} ImpD(t)$, with $\boldsymbol{R}_{0+}=\boldsymbol{R}_+\cup\{0\}=\{t\in \boldsymbol{R}: t\geq 0\}$, is the total set of impulsive ("culling") time instants where actions or removal of infective corpses are performed, (the notation for $f(t^+)$ is simplified to $f(t)$), and

$ImpD(t^-)=\{\sigma\in ImpD : \sigma<t\}$,

$ImpD(t)=\{\sigma\in ImpD : \sigma\leq t\}=ImpD(t^-)$ if $t\notin ImpD$, and

$ImpD(t)=\{\sigma\in ImpD : \sigma\leq t\}=ImpD(t^-)\cup\{t\}$ if $t\in ImpD$

and the (nonnegative) parameters and controls are the following:

b_1 is the recruitment rate,

b_2 is the natural average death rate,

β, β_A, β_D are the various disease transmission coefficients to the susceptible from the respective infective, asymptomatic and infective corpses subpopulations,

η is a parameter such that $1/\eta$ is the average duration of the immunity period reflecting a transition from the recovered to the susceptible,

γ is the transition rate from the exposed to all (i.e. both symptomatic and asymptomatic) infectious,

α is the average extra mortality associated with the symptomatic infectious subpopulation,

τ_0 is the natural immune response rate for the whole infectious subpopulation (i.e. $A+I$), respectively,

p is the fraction of the exposed, which become standard infectious,

$1-p$ is the fraction of the exposed which become asymptomatic infectious,

$1/\mu$ is the average period of infectiousness after death,

$V(t)$ and $\xi(t)$ are, respectively, the vaccination and antiviral treatment controls and $\rho_D(t_i)D(t_i)$ is the impulsive action of removal of corpses (or "culling") for all $t_i\in ImpD$ with $\rho_D(t)\in[0,1]$, $\rho_D(t)\in(0,1]$ if $t\in ImpD$, being the fraction of corpses removal. The controls can be of different types including constant and feedback actions.

The proposed time-varying feedback vaccination and antiviral treatment laws, eventually including feedback impulses on the susceptible and symptomatic infective subpopulations, are the following:

$V(t)=V_0+K_V(t)S(t)+\rho_V(t)S(t)\sum_{t_i\in ImpS}\delta(t-t_i)$

$\xi(t)=\xi_0+K_\xi(t)I(t)+\rho_\xi(t)I(t)\Sigma_{t_i\in Imp\,I}\,\delta(t-t_i)$

$;\forall t\in \boldsymbol{R}_{0+}$, where $V_0\geq 0,\xi_0\geq 0$, $K_V, K_\xi\in PC^0(\boldsymbol{R}_{0+},[0,1])$ and $ImpS$ and $Imp\,I$ are the sets of impulsive time instants for the vaccination action on the susceptible and antiviral action on the symptomatic infective, $\rho_V(t)\in[0,1]$ and $\rho_\xi(t)\in[0,1]$; $\forall t\in \boldsymbol{R}_{0+}$ give the fractions of vaccinated susceptible or antiviral infective treated subpopulations.

3. Some model properties

The proposed SEIADR model has the following properties which have been rigorously proved but the proofs are omitted by space reasons:

Property 1 (existence and uniqueness of the solution). The solutions of the SEIADR model exist and are uniquely defined for any given initial conditions and any given vaccination and antiviral controls and can be expressed with explicit formulas.

Property 2 (positivity of the solution). Assume that $V_0\in[0,b_1]$ and $\rho_D,\rho_V:\boldsymbol{R}_{0+}\to[0,1]$. Then, the SEIADR model is "positive" in the sense that its trajectory solution is nonnegative for all time under non-negative initial conditions:

$$[min\,(S(0),E(0),I(0),A(0),R(0),D(0))\geq 0]\Rightarrow[min\,(S(t),E(t),I(t),A(t),R(t),D(t))\geq 0;\ \forall t\in \boldsymbol{R}_{0+}]\,.$$

This property allows a potential validity of the model for its application to real epidemic disease spreads on populations in both presence and absence of eventually mixed vaccination-antiviral-corpses culling controls. The proof is organized by using contradiction arguments to some subpopulation value being the first one to reach the zero value at some nonzero time instant. It follows due to the chained coupling dynamics in-between sub-populations that this supposed zero value should have been reached at a former time instant.

Properties 3 (uniform boundedness of the solution):

(i) $\limsup_{t\to\infty} I(t)\leq b_1/\alpha$ and $\sup_{t\in \boldsymbol{R}_{0+}} I(t)<+\infty$,

(ii) $N(t)<+\infty$; $\forall t\in \boldsymbol{R}_{0+}$,

(iii) $max\left(\sup_{t\in \boldsymbol{R}_{0+}} S(t),\ \sup_{t\in \boldsymbol{R}_{0+}} E(t)\ \sup_{t\in \boldsymbol{R}_{0+}} I(t),\ \sup_{t\in \boldsymbol{R}_{0+}} A(t),\ \sup_{t\in \boldsymbol{R}_{0+}} R(t),\ \sup_{t\in \boldsymbol{R}_{0+}} D(t)\right)<+\infty$.

where $N(t)$ is the total population obtained by summing up the values of all the subpopulations at each time instant. The proof is made in two parts: a) It is proved that the whole subpopulation is bounded from its differential time equation $\dot{N}(t)=-b_2N(t)+b_1-\alpha I(t)$; $\forall t\in \boldsymbol{R}_{0+}$, b) This above property together with the non-negativity of all the subpopulations for all time leads to the boundedness of all of those subpopulations for all time.

Properties 4 (existence and stability of the unique disease-free equilibrium periodic solution or equilibrium point): Assume that

1) $K_V^*<min\,(\mu,b_2+min\,(\gamma,\tau_0,\eta))$,

2) the initial conditions of (1)-(6) to be non-negative, $V_0 \in [0, b_1]$, $0 < T_{min} \leq T_i = t_{i+1} - t_i \leq T_{max} < +\infty$; $\forall t_i \in ImpS \cup ImpI \cup Imp\, D$,

3) $\rho_D(t_i)(\in [0,1]) \to \rho_D^*$ as $t_i(\in ImpD) \to \infty$, $\rho_V(t_i)(\in [0,1]) \to \rho_V^*$ as $t_i(\in ImpS) \to \infty$, $\rho_\xi(t_i)(\in [0,1]) \to \rho_\xi^*$ as $(t_{i+1} - t_i) \to T_V^*$ $t_i(\in Imp\, I) \to \infty$ and $K_V(t) \to K_V^*$ as $t \to \infty$.

Then, the following properties hold:

(i) There exists a unique disease- free equilibrium periodic solution if $\rho_V^* \in [0,1)$ in the set $ImpS$ of impulsive vaccination of the susceptible subpopulation, which is independent on the antiviral control, and which is defined by:

$$x_{df}^*(T_V^* + \sigma) := \left(S_{df}^*(T_V^* + \sigma), E_{df}^*, I_{df}^*, A_{df}^*, D_{df}^*, R_{df}^*(T_V^* + \sigma)\right)^T = \left(S_{df}^*(T_V^* + \sigma), 0,0,0,0, R_{df}^*(T_V^* + \sigma)\right)^T$$

; $\forall \sigma \in [0, T_V^*]$, where

$$S(T_V^* + \sigma) = \lim_{t_i = nT_V^*(\in ImpS) \to \infty} S(t_i + \sigma) = \frac{b_2(b_1 - V_0) + \eta b_1}{b_2(b_2 + \eta + K_V^*)} + o\left(T_V^{*-1}\right) + o(\rho_V^*); \ \forall \sigma \in [0, T_V^*)$$

$$R(T_V^* + \sigma) = \lim_{t_i = nT_V^*(\in ImpS) \to \infty} R(t_i + \sigma) = \frac{b_2 V_0 + K_V^* b_1}{b_2(b_2 + \eta + K_V^*)} + o\left(T_V^{*-1}\right) + o(\rho_V^*); \ \forall \sigma \in [0, T_V^*)$$

$$S(T_V^*) = (1 - \rho_V^*) S(T_V^{*-}) \quad , \quad R(T_V^*) = (1 + \rho_V^*) R(T_V^{*-})$$

; $\forall t_i \in ImpS$ with an associated limit total population $N_{df}^* = S_{df}^*(\sigma) + R_{df}^*(\sigma) = \frac{b_1}{b_2} = \lim_{t \to \infty} N(t)$;

$\forall \sigma \in [0, T_V^*]$ and controls $\xi_{df}^* = \xi_0$, and

$$\lim_{n \to \infty} V(nT_V^* + \sigma) = V_{df}^*(T_V^* + \sigma) = V_0 + K_V^* S_{df}^*(T_V^* + \sigma) + \rho_V^* S_{df}^*(T_V^{*-}) \delta(\sigma - T_V^*)$$

; $\forall \sigma \in [0, T_V^*]$ where $\delta(.)$ is the Dirac distribution. If $\rho_V^* = 0$ or the impulsive vaccination ends in finite time then the periodic oscillation becomes a disease- free equilibrium point.

(ii) Assume that the extended disease transmission coefficient $\bar{\beta} = max(\beta, \beta_A, \beta_D)$ is small enough to satisfy $\bar{\beta} < \frac{1}{S_{df}^*} min\left(\mu, b_2 + min\left(K_V^*, \gamma, \tau_0, \eta\right)\right)$. Then, the linearized trajectory solution about the disease- free equilibrium point is locally asymptotically stable.

If $\bar{\beta} > \frac{1}{S_{df}^*} min\left(\mu, b_2 + min\left(K_V^*, \gamma, \tau_0, \eta\right)\right)$ then the linearized trajectory solution about the disease- free equilibrium point is locally unstable.

The main message of Properties 4 is the following: "If the extended disease transmission rate $\overline{\beta} = max(\beta, \beta_A, \beta_D)$ (typically $\overline{\beta} = \beta$ in practice) is small enough then the disease- free equilibrium point is globally asymptotically stable. As a result, the whole nonlinear model about such an equilibrium is also locally asymptotically stable". Note that under impulsive controls there is a jump from the left to the right in the values of the equilibrium periodic regime if the control impulses do not end after a finite time interval.

Property 5 (reproduction-like number): It is concluded from Property 4 (ii) that we can define $R_{df} = \frac{\overline{\beta} S_{df}^*}{min\left(\mu, b_2 + min\left(K_V^*, \gamma, \tau_0, \eta\right)\right)}$ as a disease- free reproduction-like number. If $R_{df} < 1$, or equivalently, if the infective disease transmission coefficient is small enough such that $\overline{\beta} < \frac{1}{S_{df}^*} min\left(\mu, b_2 + min\left(K_V^*, \gamma, \tau_0, \eta\right)\right)$, then the disease does not spread for small deviations from the disease-free equilibrium. If $R_{df} > 1$, equivalently, if $\overline{\beta} > \frac{1}{S_{df}^*} min\left(\mu, b_2 + min\left(K_V^*, \gamma, \tau_0, \eta\right)\right)$ then the equilibrium oscillation is locally asymptotically unstable.

Properties 6 (existence and uniqueness of the endemic equilibrium oscillation/point): (i) If the given assumptions for Properties 4 hold, then there is no endemic equilibrium point or limit periodic oscillation:

$$x_{end}^* := \left(S_{end}^*, E_{end}^*, I_{end}^*, A_{end}^*, D_{end}^*, R_{end}^*\right)^T$$

or equilibrium periodic solution , where $E_{end}^* \neq 0$, provided that $\underline{\beta} = min(\beta, \beta_A, \beta_D) < \beta_\delta$ for some existing some $\beta_\delta \in \boldsymbol{R}_+$. As a result, if $\overline{\beta} < min\left(\frac{1}{S_{df}^*} min\left(\mu, b_2 + min\left(K_V^*, \gamma, \tau_0, \eta\right)\right), \beta_m\right)$ for some $\beta_m (\leq \beta_\delta) \in \boldsymbol{R}_+$ then there only exists the disease-free equilibrium periodic solution or point in the particular of asymptotic removal of vaccination and antiviral impulsive controls, which is globally asymptotically stable.

(ii) Assume that $K_V(t) = K_\xi(t) = \rho_V(t) = \rho_\xi(t) = 0$; $\forall t \in \boldsymbol{R}_{0+}$. Then, the endemic equilibrium point, if it exists, is unique.

The endemic equilibrium periodic solution or endemic equilibrium point I in the case of asymptotic impulsive - free case) is characterized bi its exposed population to be nonzero, i.e. $E_{end}^* \neq 0$, what causes the asymptomatic and symptomatic infective endemic populations to be nonzero. Note that the endemic equilibrium periodic solutions or points make the disease to be permanent in the sense that the infective population cannot be asymptotically removed contrarily to the disease- free equilibrium points where the infection is asymptotically removed. This result establishes is that for small enough coefficient transmission rates only the disease- free equilibrium is relevant since the endemic one does not exist.

4. Example

It is now presented a set of numerical simulation work imnplemented in Matlab. The parameters of the

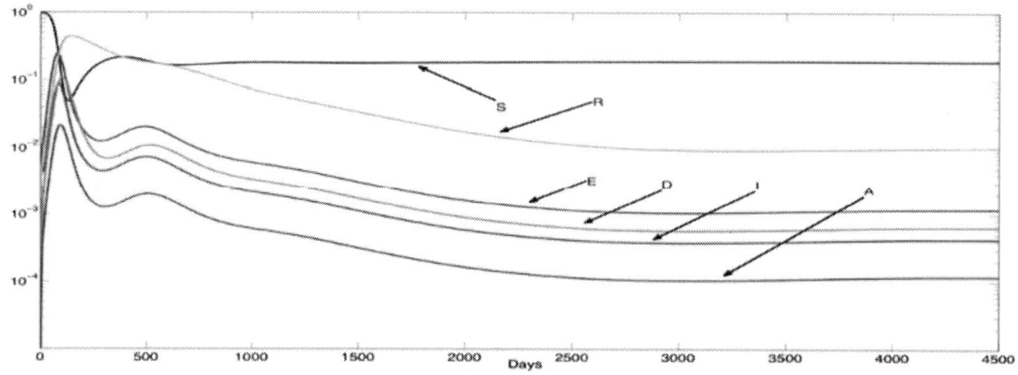

Figure 1. Endemic equilibrium reached for a dynamic without culling

model are obtained from real data from a study of Ebola disease [9]:
The recruitment rate and the natural average death rate are $b_1 = b_2 = 1/(70x365)$ days^{-1} so that the total population is normalized to unity while the disease transmission coefficients are $\beta = 0.16, \beta_A = 0.05, \beta_D = 0.5$ respectively. The average duration of the immunity period reflecting a transition from the recovered to the susceptible is determined by $1/\eta$=1000 days, γ=1/15.8 days^{-1}, α =1/13.3 days^{-1}, τ_0=1/12 days^{-1}, p=0.9, the average duration of infection $1/\mu$=20 days and $V(t)$ and $\xi(t)$ are, respectively, constant values V=0.2b_1 and ξ=0.02.

First, a study is concerned with initial conditions $S(0)= S_{DFE}$, $R(0)=R_{DFE}$, $A(0)=E(0)=D(0)=0$ and $I(0)$=0.01 without culling, or impulsive removal of the infectious diseases at regular times (fig.1). In Figure 1, it can be seen where the dynamic of the subpopulations is seen to evolve into a endemic equilibrium point, and the disease remains endemic. In Fig.2 the culling of a 95% of the infectious

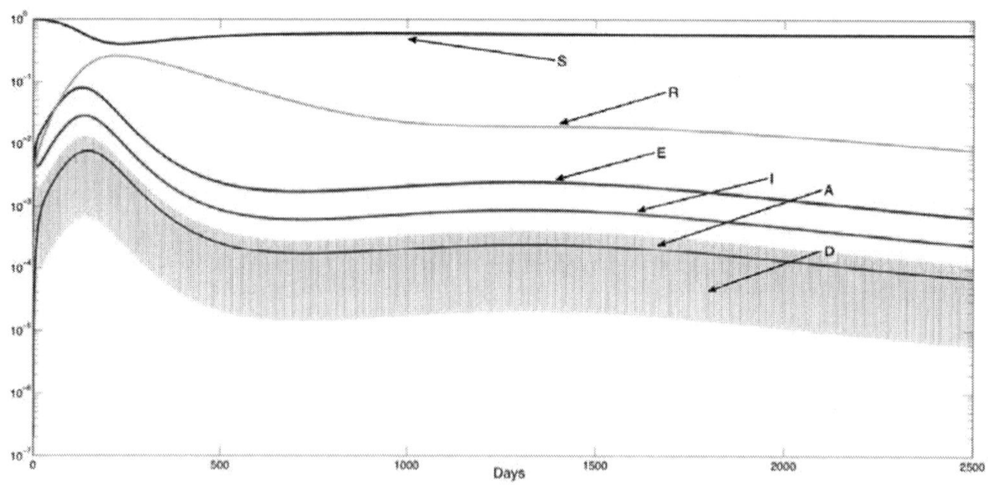

Figure 2. A dynamic with regular culling of 95% of D(t) subpopulation each 7 days

bodies performed regularly each week. In Figure 2 it can be seen how the values related to the sick subpopulations decrease exponentially, as the susceptible and the recovered subpopulation reach the disease-free equilibrium.

5. Concluding remarks

In this paper, the properties of a new proposed epidemic model which is referred to as an SEIADR epidemic model have been described. This model incorporates the asymptomatic infective and dead-infective populations and three class of optional controls (vaccination on the susceptible, antiviral treatment on the symptomatic infective- both of them can contain optional constant and feedback regular and impulsive actions- plus a culling action, which is impulsive by nature, on the lying effective corpses. The positivity and stability properties of the solution as well as the existence of the unique disease-free equilibrium point and the existence, non-existence and uniqueness (if it exists) of the endemic equilibrium point are investigated. The proposed model could be useful to deal with the propagation of the Ebola disease where lying un-recovered dead corpses are known to be also highly infective in urban areas with high- population density.

References

[1] Mollison D 1995 *Epidemic Models: Their Structure and Relation to Data, Publications of the Newton Institute, Cambridge University Press,* 1995, Denis Mollison Editor (transferred to digital printing 2003).

[2] Keeling MJ and P. Rohani P 2008 *Modeling Infectious Diseases in Humans and Animals*, Princeton University Press, Princeton and Oxford

[3] Bellan SE, Pulliam JRC, Dushoff J. and Meyers LA 2014, *The Lancet* **384** 9953 pp. 1499-1500

[4] Santermans E, Robesyn E, Ganiani T, Sudre B, Faes, C, Quinten, C, Van Bortel, W , Haber, T, Kovac F, Van Reeth F, Testa M, Hens N and Plachouras, D 2016 *Plos One* **11**(1), doi:101371/journal.pone.0147172

[5] Song XY, Jiang J and Wei HM 2009 *Applied Mathematics and Computation* **214**, pp. 381-390

[6] De la Sen M, Agarwal RP, Ibeas A and Alonso-Quesada S 2011 *Advances in Difference Equations*, **2011**, Article ID 748608, Vol. 2011, doi:10.1155/2011/748608

[7] De la Sen M, Agarwal RP, Ibeas A and Alonso-Quesada S 2010 *Advances in Difference Equations* **2010**, Article ID 281612, doi:10.1155/20101/281612

[8] De la Sen M and Alonso-Quesada S 2011 *Applied Mathematics and Computation* **218**, pp. 3888-3904

[9] Rivers CM, Lofgren ET, Marathe M, Eubank S, Lewis BL *PLOS Currents Outbreaks* **2014** Oct 16 . Edition 1. doi: 10.1371/currents.outbreaks.fd38dd85078565450b0be3fcd78f5ccf.

Rise-Time Distortion of Signal without Carrying Signal

N S Bukhman
Samara state university of architecture and civil engineering ,
Russian Federation, 443001, Samara, 194 Molodogvardeiskaya str.

E-mail: nik3141rambler@rambler.ru

Abstract. The article deals with one-dimensional problem of rise-time distortion signal without carrying signal, that appears in the starting point intermittently, that is signal distortion at front edge or one of its derivative. The authors show that front edge of signal isn't distorted in case of propagation in unrestricted (including absorbing) area (amplitude of starting signal step or of one of its derivatives doesn't change) and move with the accuracy of vacuum light speed. The paper proves that it is the time interval shortage that causes signal loss with the route extension, but not the reduction of its starting amplitude, during which front edge of signal retains its starting value. The research presents new values for this time interval.

1. Introduction

The matter of rise-time distortion of the signal as it propagates in a material medium has got a long history, but it is still researched today (see [1-7]). The papers quoted above deal with signal with "carrier frequency"; this paper deals with the rising edge of video signal, i.e. signal "without carrier frequency".

Let's consider the signal $E(t)$ as it propagates in a medium with complex dielectric constant $\varepsilon(\omega)$ and refractive index $n(\omega)=\sqrt{\varepsilon(\omega)}$. It's evident that $E(\omega)=\int_{-\infty}^{+\infty} E(t)\exp(-i\omega t)dt$, $E(t)=(2\pi)^{-1}\int_{-\infty}^{+\infty} E(\omega)\exp(i\omega t)d\omega$, where $E(\omega)$ is the frequency spectrum of the signal in the starting point ($z=0$). As the signal propagates, with increase of path length z, both its spectrum $E(\omega,z)$ and its time dependence $E(t,z)$ change.

After the signal goes a path the length of which is z in a homogeneous medium with wave number $k(\omega)=(\omega/c)n(\omega)$, we have the frequency spectrum of the signal: $E(\omega,z)=F(\omega,z)E(\omega)$ where $F(\omega,z)=\exp(-i(\omega/c)z)F_{def}(\omega,z)$ where $\exp(-i(\omega/c)z)$ is the frequency characteristic of the ideal delay line (for the time of "light delay"; further, to reduce writing, we usually imply this delay, but do not explicitly write it), and $F_{def}(\omega,z)=\exp\left(-i(\omega/c)(n(\omega)-1)\right)$ is the additional (with respect to the latter) part of frequency characteristic of the above-mentioned filter, that describes the deformation (distortion) of the signal as it propagates in the medium.

To be specific, let's consider the case of collisional plasma; as it turns out, basically the obtained results are of much more general character. For the plasma, we have the dielectric constant [8.9]

Content from this work may be used under the terms of the Creative Commons Attribution 3.0 licence. Any further distribution of this work must maintain attribution to the author(s) and the title of the work, journal citation and DOI.
Published under licence by IOP Publishing Ltd

$\varepsilon(\omega) = 1 - \omega_p^2/(\omega^2 + \nu^2) + i\,\omega_p^2(\nu/\omega)/(\omega^2 + \nu^2)$, where $\omega_p = (4\pi ne^2/m)^{1/2}$ - is the plasma frequency, ν - is the effective collision frequency.

The first three expansion terms $F_{def}(\omega, z)$ in series in powers of $1/\omega$ are $F_{def}(\omega, z) = 1 + i(k_p z/2)(\omega_p/\omega) - ((\nu/\omega_p)(k_p z/2) + (k_p z)^2/8)(\omega_p/\omega)^2$, where $k_p \equiv (\omega_p/c)$.

To understand further analysis it is necessary to recall that in the theory of complex variable functions there is a theorem on the uniqueness of analytical function (see [10]) which states that two analytic functions coinciding on any finite section of real axis, coincide on all real axis. In particular, the analytical function which is identically zero on some section of the real axis is identically zero on all real axis. In our case, this means that the signal $E(t)$, which is the analytical function of real variable t in point $z = 0$ and is not identically zero, can not become 0 on any finite section of real axis t - in particular, it can not satisfy the condition $E(t) = 0$ if $t \le t_0$. This means that such signal existed, exists and will always exist (if $-\infty < t < +\infty$), and basically it can not be used to transmit information.

Indeed, in any point in space z at any given time t the part of the signal received at all previous times is already available for analysis. By this part of the signal, in accordance with the above-mentioned theorem on the uniqueness of analytical function, its time dependence at any past or future point of time can be basically reconstructed. Therefore, no signal used to transmit information can be the analytical function on all real axis; it can coincide with the analytical function only if $t > t_0$, where t_0 - is the time of the signal's emergence. The point t_0 here is the point of breach of analyticity, which usually manifests itself as a breach of the time dependence of the signal and (or) its derivatives. The main purpose of this paper is to research, how exactly this breach of the time dependence of the signal in the starting point and its near neighborhood look after the signal goes a path the length of which is z in a substance.

2. Results

Let's suppose the signal $E(t)$ with sharp rising edge propagates in the medium, it emerges in the starting point $z = 0$ at the point of time t_0 ($E(t) = 0$ if $t < t_0$, $E(t) \ne 0$ if $t > t_0$). Then for the asymptotics (if $\omega \to \infty$) of the spectrum of such signal in the starting point $z = 0$, with third order accuracy in powers of $1/\omega$, we have (see Appendix) $E(\omega) = \exp(-i\omega t_0)(E(t_0)(i\omega)^{-1} + E'(t_0)(i\omega)^{-2} + E''(t_0)(i\omega)^{-3})$, where $E(t_0), E'(t_0), E''(t_0)$ - is the value of the signal and its first derivatives on the edge, fully describing (See Appendix) the relevant terms of its spectrum's high-frequency asymptotics. The result for the signal's spectrum in point z (with third order accuracy in powers $1/\omega$) is: $E(\omega, z) = \exp(-i\omega t_0)(E(t_0, z)(i\omega)^{-1} + E'(t_0, z)(i\omega)^{-2} + E''(t_0, z)(i\omega)^{-3})$, where $E(t_0, z) = E(t_0)$, $E'(t_0, z) = E'(t_0) - (1/2)E(t_0)(k_p z)\omega_p$,

$E''(t_0, z) = E''(t_0) - (1/2)E'(t_0)(k_p z)\omega_p + E(t_0)\omega_p^2((1/2)(\nu/\omega_p)(k_p z) + (1/8)(k_p z)^2)$ - are the values of the amplitude of the signal and its two first derivatives on the edge (if $t = t_0$) in point z. Let us discuss the characteristics of the changes of the rising edge of the signal, that emerges as a jump in point $z = 0$ at the point at time $t = t_0$ ($E(t) = 0$ if $t < t_0$, $E(t_0) \ne 0$), as it propagates in the medium, that is let us make some obvious conclusions from the above formulae.

1. In the case of signal amplitude jumping from 0 to final value $E(t_0) \ne 0$ at the point of time t_0 in starting point $z = 0$, its amplitude also jumps from 0 to the same final value $E(t_0) \ne 0$ in any other

point in space z at the point of time $t_0 + z/c$ corresponding to the light delay time of the signal's edge. In other words, the amplitude of the initial jump of the signal propagating in an arbitrary medium (including an absorbing or an amplifying one) does not change, and the delay time of its rising edge in any medium is exactly equal to the light delay time. We would like to emphasize that the delay time of the initial jump (that is, the rising edge of the signal) can not be less or more than "the light delay", that is, no "early" or "late" appearance of the rising jump is possible.

The fact that this conclusion does not refer to any particular medium (for example, collisional plasma), but it refers to any arbitrary material medium, is connected with the aspect that in the most general case (see [8]), if $\omega \to \infty$, the dielectric constant of an arbitrary medium with an accuracy up to terms of order ω_p^2/ω^2 ($\omega_p = \left(4\pi n e^2/m\right)^{1/2}$ where n - is the total content of free and bound electrons in the medium) is $\varepsilon(\omega) = 1 - \omega_p^2/\omega^2$ and, if $\omega \to \infty$, it tends to 1.

Of course, in any dispersive medium signal deformation occurs - it is distorted (usually stretched) and its amplitude decays; in an absorbing medium, in addition to this, the signal is attenuated (i.e., its energy is lost), but all this is not related to the amplitude of the signal's rising edge – the latter never changes. As an example, the figure below shows the numerical results for the time dependence of square wave signal (a) after it goes path $z = 10/k_p$ in collisionless plasma $\nu = 0$ (b) and in collisional plasma $\nu = \omega_p$ (c). "Light delay" is not shown. Line (d) shows the "linear" approximation, used to assess the time interval during which the initial amplitude of the signal remains the same.

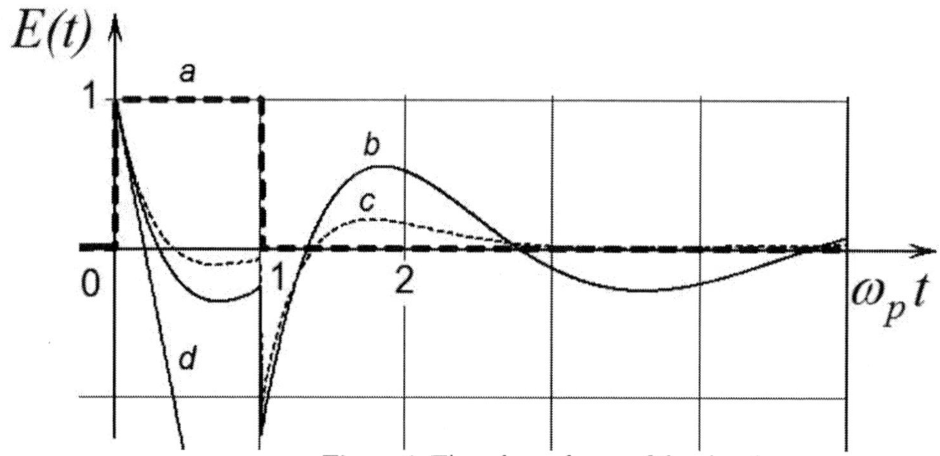

Figure 1. Time dependence of the signal

2. To estimate the duration of the time interval during which the signal which emerges as a jump maintains the amplitude close to that of the initial jump in any arbitrary medium, for the energy of the signal's rising pulse we have the following approximate estimates: $\tau_{fr} \approx 2/(\omega_p k_p z)$, $W_{fr} \approx E^2(t_0)\tau_{fr}/3 = 2E^2(t_0)/(3\omega_p k_p z)$. You can get evaluation formula $z_{def} = (2mc\varepsilon_0)/ne^2\tau_{pulse}$ for the characteristic weak distortion distance of the square-wave pulse.

3. Similarly let us consider the signals, on the rising edge of which there's a breach of time dependence not of the signal as such, but only of its derivatives of order k and higher: $E(t) = 0$ if $t < t_0$, $E(t) \neq 0$ if $t > t_0$, $E''(t_0) = 0, \ldots, E^{(k-1)}(t_0) = 0$, $E^{(k)}(t_0) \neq 0$. It is easy to show that in this case, as the signal propagates in the medium, it is the value of the jump of the derivative $E^{(k)}(t_0)$ that does not change; as a result in *the arbitrary medium* the amplitude of the rising pulse of the signal

slowly decays in accordance with the law $E_{fr} \sim E^{(k)}(t_0)/\left(\omega_p^k (k_p z)^k\right) \sim 1/\left(k_p z\right)^k$, and the duration is still determined by the law $\tau_{fr} \sim 1/\left(\omega_p k_p z\right)$.

Appendix

Let us research the Fourier spectrum $E(\omega)$ that exists for a limited period of time (within the range $[t_1, t_2]$) of smooth signal $E(t)$ with sharp rising edge t_1 and falling edge t_2 ($E(t) = 0$ if $t < t_1$ and $t > t_2$, $E(t) \neq 0$ if $t_1 < t < t_2$). Obviously $E(\omega) = \int_{t_1}^{t_2} E(t)\exp(-i\omega t)dt$, $E(t) = (2\pi)^{-1} \int_{-\infty}^{+\infty} E(\omega)\exp(i\omega t)d\omega$. After n-fold integration by parts ($n \to \infty$), the first of these formulae can be written as an asymptotic formula $E(\omega) = -\sum_{k=0}^{\infty} \left(\exp(-i\omega t)(i\omega)^{-k-1} E^{(k)}(t)\right)\Big|_{t_1}^{t_2}$. The latter formula demonstrates a direct link between the breaches of the signal on the rising and falling edges and the asymptotic behaviour (if $\omega \to \infty$) of its spectrum. Of course, the noted connection between the asymptotics of the signal spectrum, if $\omega \to \infty$, and the character of its breaches can be used "reverse" too - the presence of relevant terms in the asymptotics of the signal's spectrum indicates that the signal itself (or its derivatives of the corresponding order) are experiencing breaches of the corresponding rank and amplitude at the relevant points of time. This fact allows, while researching the rising edge of the signal, limiting yourself to the minimum amount of information about the properties of the medium, i.e. the first few expansion terms of its refractive index in series in powers of $1/\omega$.

References

[1] Sommerfeld A 1914 *Ann. Phys.* **44** 177
[2] Brillouin L 1914 *Ann. Phys.* **44** 203
[3] Maske B and Segard B 2012 *Phys. Rev.* A **86** 013837
[4] Maske B and Segard B 2013 *Phys. Rev.* A **87** 043830
[5] Akulshin A M, McLean R J 2010 *Journal of Optics* **12** 104001
[6] Malykin G B, Romanets E A 2012 *Optics and Spectroscopy* **112** 920
[7] Bukhman N 2007 *J. Commun. Technol. Electron.* **52** 555
[8] Landau L D and Lifshits E M 1960 *Electrodynamics of Continuous Media* (Oxford: Pergamon Press).
[9] Vinogradova M B, Rudenko O V and Sukhorukov A P 1979 *The Wave Theory* (Moscow: Nauka)
[10] Smirnov V I *The higher mathematics course* 1974 vol 3 part 2 (Moscow: Nauka) p 72

Modeling the dynamics of the chassis of construction machines

R L Sakhapov[1,2,3], R V Nikolaeva[1], M H Gatiyatullin[1] and M M Makhmutov[1]

[1]Kazan State University of Architecture and Engineering, 420043, Kazan, Zelenaya str., 1, Russia

[3]E-mail: rustem@sakhapov.ru

Abstract. The article presents the results of a study of the transfer functions of a construction machine as a complex dynamic system. Authors constructed a dynamic model of a construction machine. The paper formulates and solves a system of nonlinear differential equations of motion of the chassis system of a construction machine on the basis of the d'Alembert-Lagrange equation. The numerical values of the transfer function coefficients for the construction machines were determined from the experimentally obtained curves of acceleration, processed by area method. Authors determined the experimental curves of the transition process of chassis system of a construction machine. The results of the study show that the difference of source curves ordinates and calculation of transients is less than 4% on average, which indicates a fairly accurate description of the process. The resulting expressions of transfer system functions of the chassis with sufficient precision can be used for practical purposes in the design and development of new construction machines.
Keywords: mathematical modeling of the dynamic model, chassis system, motor construction machine, acceleration curves, Laplace transform.

1. Introduction

A construction machine is one of the most massive earth-moving machinery, employed in road construction. Modern market conditions require high-performance, high-speed and powerful heavy transport machine. Constant increase of available power leads to an increase in dynamic loads acting on both the human operator and the machine.

2. Construction of dynamic models

For the investigation of dynamics of the construction machine chassis we constructed its dynamic model according to its kinematic scheme [2].

On the basis of the d'Alembert-Lagrange system of equations we made non-linear differential equations of motion of the construction machines chassis system [1]:

1) a bent shaft of the engine and a power shaft of a reducer with the general moment of inertia of J_1;

2) output shaft of a distributing reducer and a gear box with J_2 moment of inertia;

3) shaft of running wheels with the given J_3 construction machine moment of inertia.

On the basis of it the system of the differential equations of running system has the following appearance:

[2] To whom any correspondence should be addressed.

Content from this work may be used under the terms of the Creative Commons Attribution 3.0 licence. Any further distribution of this work must maintain attribution to the author(s) and the title of the work, journal citation and DOI.
Published under licence by IOP Publishing Ltd

$$\left.\begin{aligned}&J_1\dot{\omega}_1+M_{c\theta 1}+M_{T1}+\frac{M_{1.2}}{i_{1.2}\eta_{1.2}}+\frac{M_{1.5}}{i_{1.5}\eta_{1.5}}=M_1(\omega_1);\\&J_2\dot{\omega}_2+M_{c\theta 2}+M_{T2}+\frac{M_{2.3}}{i_{2.3}\eta_{2.3}}=M_{1.2}(\omega_1\omega_2);\\&J_3\dot{\omega}_3+M_{c\theta 3}+M_{T3}+F_x(t)=M_{2.3}(\omega_2\omega_3);\end{aligned}\right\}\tag{1}$$

where J_1, J_2, J_3 – respectively the constant given moments of inertia on cranked to an engine shaft with a power shaft of a reducer, clutch with an output in bulk of a gear box, on a shaft of running wheels of the construction machine; ω_1, ω_2, ω_3 – angular speeds of rotation of the corresponding shaft (rad / s); $M_{c\theta 1}$, $M_{c\theta 2}$, $M_{c\theta 3}$ – air drag torques on the corresponding shaft, H·m.

$$M_{c\theta 1}=B_1\omega_1^2;\quad M_{c\theta 2}=B_2\omega_2^2;\quad M_{c\theta 3}=B_3\omega_3^2;$$

B_1, B_2, B_3 – the coefficients considering air and other resistance, N·m·s^2; MT_1, MT_2, MT_3 – the moments of powers of constant friction (N·m) on the corresponding shaft of reduction; $M_1(\omega_1)$ – the partial or external high-speed characteristic of the engine on a torque on cranked to a shaft which for this case is accepted [2]:

$$M_1(\omega_1)=A-C\cdot\omega_1,$$

$M_{1.2}(\omega_1,\omega_2)$, $M_{2.3}(\omega_2,\omega_3)$ – the moments arising in hydro mechanical and friction gears which are accepted in a look [4]:

$$M_{1.2}(\omega_1,\omega_2)=A_1-C_1\cdot(\omega_1/\omega_2)\cdot i_{1.2};\ M_{2.3}(\omega_2,\omega_3)=A_2-C_2\cdot(\omega_2/\omega_3)\cdot i_{2.3};$$

A, A_1, A_2, C, C_1, C_2 – constant coefficients; $i_{1.2}$, $i_{2.3}$, $i_{1.5}$, $\eta_{1.2}$, $\eta_{2.3}$, $\eta_{1.5}$ – reduction ratios and efficiency of drives; $F_x(t)$ – external loading (H·m), time-dependent.

We accept the moment $M_{1.5}$ arising in a drive gear of a distributing reducer to constants.

Let's lead the system of the differential nonlinear equations (1) relatively ω_1, ω_2, ω_3 to a linear look. For this purpose we decompose the components of the equations (1) in ranks and we are limited only to linear elements of increments $\Delta\omega_1$, $\Delta\omega_2$, $\Delta\omega_3$ of rather established values $\tilde{\omega}_1$, $\tilde{\omega}_2$, $\tilde{\omega}_3$ [5]:

$$\Delta M_{c\theta 1}=\left(\frac{\partial M_{c\theta 1}}{\partial\omega_1}\right)\Delta\omega_1=\Pi_1\Delta\omega_1;\tag{2}$$

$$\Delta M_{c\theta 2}=\left(\frac{\partial M_{c\theta 2}}{\partial\omega_2}\right)\Delta\omega_2=\Pi_2\Delta\omega_2;\tag{3}$$

$$\Delta M_{c\theta 3}=\left(\frac{\partial M_{c\theta 3}}{\partial\omega_3}\right)\Delta\omega_3=\Pi_3\Delta\omega_3;\tag{4}$$

$$\Delta M_1=-C\Delta\omega_1;\tag{5}$$

$$\Delta M_{1.2}=\frac{C_1 i_{1.2}}{\omega_2}\left[-\Delta\omega_1+\frac{\omega_1}{\omega_2}\Delta\omega_2\right];\tag{6}$$

$$\Delta M_{2.3}=\frac{C_2 i_{2.3}}{\omega_3}\left[-\Delta\omega_2+\frac{\omega_2}{\omega_3}\Delta\omega_3\right],\tag{7}$$

Where $\Delta M_{c\theta 1}$, $\Delta M_{c\theta 2}$, $\Delta M_{c\theta 3}$, ΔM_1, $\Delta M_{1.2}$, $\Delta M_{2.3}$ – increments of the moments; $\Pi_1=2B_1\tilde{\omega}_1$; $\Pi_2=2B_2\tilde{\omega}_2$; $\Pi_3=2B_3\tilde{\omega}_3$.

Having substituted expressions (2) – (7) in (1), we will receive the following system of the linear equations of the movement of a running gear of the construction machine:

$$\left.\begin{aligned} &J_1\Delta\dot{\omega}_1 + \left[II_1 - \frac{C_1}{\eta_{1.2}\tilde{\omega}_2} + C\right]\Delta\omega_1 + \frac{C_1\tilde{\omega}_1}{\eta_{1.2}\tilde{\omega}_2^2}\Delta\omega_2 = 0;\\ &J_2\Delta\dot{\omega}_2 + \left[II_2 - \frac{C_2}{\eta_{2.3}\tilde{\omega}_3} - \frac{C_1\tilde{\omega}_1 i_{1.2}}{\tilde{\omega}_2^2}\right]\Delta\omega_2 + \frac{C_2\tilde{\omega}_2^2}{\eta_{2.3}\tilde{\omega}_3^2}\Delta\omega_3 + \frac{C_{1.2}i_{1.2}}{\tilde{\omega}_2}\Delta\omega_1 = 0;\\ &J_3\Delta\dot{\omega}_3 + \left[II_3 - \frac{C_2\tilde{\omega}_2 i_{2.3}}{\tilde{\omega}_3^2}\right]\Delta\omega_3 + \frac{C_2 i_{2.3}}{\tilde{\omega}_3}\Delta\omega_2 = -\Delta F_x(t). \end{aligned}\right\} \tag{8}$$

We will enter designations:

$$G_{11} = II_1 - \frac{C_1}{\eta_{1.2}\tilde{\omega}_2} + C;\ G_{12} = \frac{C_1\tilde{\omega}_1}{\eta_{1.2}\tilde{\omega}_2^2};\ G_{21} = \frac{C_{1.2}i_{1.2}}{\tilde{\omega}_2};\ G_{22} = II_2 - \frac{C_2}{\eta_{2.3}\tilde{\omega}_3} - \frac{C_1\tilde{\omega}_1 i_{1.2}}{\tilde{\omega}_2^2};$$

$$G_{23} = \frac{C_2\tilde{\omega}_2^2}{\eta_{2.3}\tilde{\omega}_3^2};\ G_{32} = \frac{C_2 i_{2.3}}{\tilde{\omega}_3};\ G_{33} = II_3 - \frac{C_2\tilde{\omega}_2 i_{2.3}}{\tilde{\omega}_3^2}.$$

With the accounting of designations the system of the equations (8) looks like this:

$$\left.\begin{aligned} &J_1\Delta\dot{\omega}_1 + G_{11}\Delta\omega_1 + G_{12}\Delta\omega_2 = 0;\\ &J_2\Delta\dot{\omega}_2 + G_{22}\Delta\omega_2 + G_{23}\Delta\omega_3 + G_{21}\Delta\omega_1 = 0;\\ &J_3\Delta\dot{\omega}_3 + G_{33}\Delta\omega_3 + G_{32}\Delta\omega_2 = -\Delta F_X(t). \end{aligned}\right\} \tag{9}$$

Having $\Delta\omega_2$ expressed from the third equation of system, having substituted $\Delta\omega_2$in the second equation of system (9) we will receive:

$$-\frac{J_2}{G_{32}}\left[\Delta\dot{F}_x(t) + J_3\Delta\ddot{\omega}_3 + G_{33}\Delta\dot{\omega}_3\right] - \frac{G_{22}}{G_{32}}\left[\Delta F_x(t) + J_3\Delta\dot{\omega}_3 + G_{33}\Delta\omega_3\right] + G_{23}\Delta\omega_3 + G_{21}\Delta\omega_1 = 0, \tag{10}$$

From the received ratio we will express $\Delta\omega_1$and also we will find a derivative on time from $\Delta\omega_1$:

$$\begin{aligned} \frac{d\Delta\omega_1}{dt} &= \frac{J_2}{G_{21}G_{3.2}}\left[\frac{d^2\Delta F_x(t)}{dt^2} + J_3\frac{d^3\Delta\omega_3}{dt^3} + G_{33}\frac{d^2\Delta\omega_3}{dt^2}\right] +\\ &+ \frac{G_{22}}{G_{21}G_{32}}\left[\frac{d\Delta F_x(t)}{dt} + J_3\frac{d^2\Delta\omega_3}{dt^2} + G_{33}\frac{d\Delta\omega_3}{dt}\right] - \frac{G_{23}}{G_{21}}\frac{d\Delta\omega_3}{dt}. \end{aligned} \tag{11}$$

Having substituted expressions (10)-(11) in the first equation of system (9) we will receive the differential equation:

$$\begin{aligned} &\frac{J_1J_2}{G_{21}G_{3.2}}\left[\frac{d^2\Delta F_x(t)}{dt^2} + J_3\frac{d^3\Delta\omega_3}{dt^3} + G_{33}\frac{d^2\Delta\omega_3}{dt^2}\right] + \frac{G_{22}J_1}{G_{21}G_{32}}\left[\frac{d\Delta F_x(t)}{dt} + J_3\frac{d^2\Delta\omega_3}{dt^2} + G_{33}\frac{d\Delta\omega_3}{dt}\right] -\\ &-\frac{G_{23}J_1}{G_{21}}\frac{d\Delta\omega_3}{dt} + \frac{G_{11}J_2}{G_{21}G_{3.2}}\left[\frac{d\Delta F_x(t)}{dt} + J_3\frac{d^2\Delta\omega_3}{dt^2} + G_{33}\frac{d\Delta\omega_3}{dt}\right] + \frac{G_{11}G_{22}}{G_{21}G_{32}}\left[\Delta F_x(t) + J_3\frac{d\Delta\omega_3}{dt} + G_{33}\Delta\omega_3\right] -\\ &-\frac{G_{11}G_{23}}{G_{21}}\Delta\omega_3 - \frac{G_{12}}{G_{32}}\Delta F_X(t) - \frac{G_{12}J_3}{G_{32}}\frac{d\Delta\omega_3}{dt} - \frac{G_{12}G_{33}}{G_{32}}\Delta\omega_3 = 0. \end{aligned} \tag{12}$$

After transformations, we will receive the differential equation of a look:

$$a_3 \frac{d^3\Delta\omega_3}{dt^3} + a_2 \frac{d^2\Delta\omega_3}{dt^2} + a_1 \frac{d\Delta\omega_3}{dt} + a_0\Delta\omega_3 = -b_2 \frac{d^2\Delta F_x(t)}{dt^2} - b_1 \frac{d\Delta F_x(t)}{dt} - b_0\Delta F_x(t), \quad (13)$$

where a_3; a_2; a_1 a_0; b_2; b_1; b_0 – movement equation coefficients.

For the purpose of convenience we enter the following dimensionless sizes:

$$X = \frac{\Delta\omega_3}{\widetilde{\omega}_3};\ t_1 = \widetilde{\omega}_3 t;\ Y = \frac{\Delta F_x(t)}{\widetilde{F}_x};\ \widetilde{a}_3 = \frac{\varepsilon_3\widetilde{\omega}_3^4}{\widetilde{F}_x};\ \widetilde{a}_2 = \frac{\varepsilon_2\widetilde{\omega}_3^3}{\widetilde{F}_x};\ \widetilde{a}_1 = \frac{\varepsilon_1\widetilde{\omega}_3^2}{\widetilde{F}_x};\ \widetilde{a}_0 = \frac{\varepsilon_0\widetilde{\omega}_3}{\widetilde{F}_x};$$

$$\widetilde{b}_2 = b_2\widetilde{\omega}_3^2;\ \widetilde{b}_1 = b_1\widetilde{\omega}_3;\ \widetilde{b}_0 = b_0,$$

where $\widetilde{F}_x$ и $\widetilde{\omega}_3$ –chosen in advance values of entrance sizes.

We will write down the equation (13) in an operator form, having replaced each member of the equation with the image corresponding to it according to Laplace, under zero entry conditions:

$$[a_3P^3 + a_2P^2 + a_1P + a_0]\cdot Y = -[b_2P^2 + b_1P + b_0]\cdot X, \quad (14)$$

where X и Y – images according to Laplace entrance α (rad.) and output V (m/s) sizes; $P=d/dt$ – operator of differentiation.

According to (10) transfer function of running system of the construction machine $W_n(P)$ has an appearance:

$$W_n(P) = -\frac{b_2P^2 + b_1P + b_0}{a_3P^3 + a_2P^2 + a_1P + a_0}, \quad (15)$$

where $a_1, a_2, a_3, a_0, b_2, b_1, b_0$ – the equation coefficients determined by formulas: $b_2=\widetilde{b}_2/\widetilde{\varepsilon}_0$; $b_1=\widetilde{b}_1/\widetilde{\varepsilon}_0$; $b_0=\widetilde{b}_0/\widetilde{\varepsilon}_0$; $a_3=\widetilde{a}_3/\widetilde{a}_0$; $a_2=\widetilde{a}_2/\widetilde{a}_0$; $a_1=\widetilde{a}_1/\widetilde{a}_0$; $a_0=\widetilde{a}_0/\widetilde{a}_0=1$.

Numerical values of coefficients of transfer function for the construction machine (11) were determined by experimentally received curves of dispersal [4].

When determining curves of dispersal indignation was carried out by sharp change of entrance size – an angle of rotation of the lever steering productivity of the fuel pump of hydromechanical transmission (GMT) on the established corner $\Delta\alpha$.

Speed of progress got out in the range of the working speeds of 1,2 - 1,8 m/s [3]. Rotary speed measurement ω_3 was performed the tacho generator and an induction marker [6,9], and the entrance size α - the electrometric sensor of the PLP-21 brand.

Experimental curves of transition process of running system of the construction machine are given in figure 1.

With a practical accuracy it is possible to accept that transfer function of simpler look:

$$W_1(P) = \frac{1}{a_3P^3 + a_2P^2 + a_1P + a_0}, \quad (16)$$

where$a_3=F_3=0,197$c^3;$a_2=F_2=1,209$c^2;$a_1=F_1=1,87$c; $a_0=1$ – for the construction machine.

Then transfer function of running system of the construction machine taking into account time of delay will assume an air:

$$W_1(P) = \frac{K_1e^{-0,1P}}{0{,}197\,P^3 + 1{,}209\,P^2 + 1{,}87\,P + 1}, \quad (17)$$

where K = 3,13 – coefficient of strengthening of running system; τ = 0,2 c – delay time.

Journal of Physics: Conference Series **738** (2016) 012119 doi:10.1088/1742-6596/738/1/012119

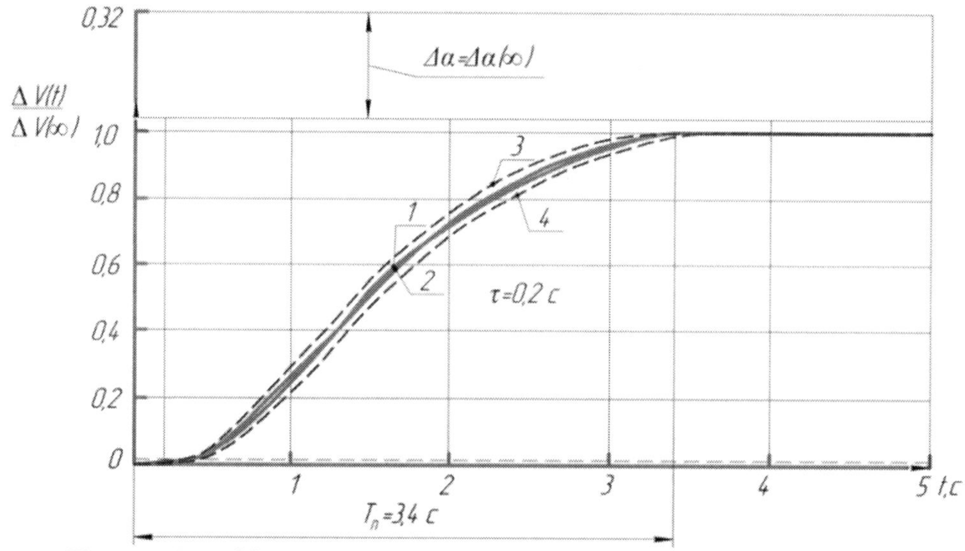

Figure 1. Transition processes of running system of the construction machine: 1 – the experimental; 2 – the settlement; 3, 4 – borders of confidential intervals atconfidential probability 0,95

Next we define an error of approximation of transfer functions. For this purpose we will find analytical expressions of transition processes $\Delta V(t)$construction machine [8].

We will write down the image of transition process according to Laplace:

$$L[\Delta V_п(t)]=L[\alpha(t)]W_1(P). \tag{18}$$

At single indignation $\alpha\ (t)=1;$ we have $L[\alpha\ (t)]=1/P$ and

$$L[\Delta V(t)]=\frac{K_1 e^{-\tau_1 P}}{(a_3P^3+a_2P^2+a_1P+a_0)P}, \tag{19}$$

where $L[\Delta V_п(t)]$ и $L[\alpha(t)]$ – images according to Laplace of a deviation of speed of progress of the construction machine and the revolting influence.

The roots of denominators of transfer functions of running system (19) found an iterative method [10] the following: $P_1=-4,141;\ P_{2,3}=-0,998\pm0,479i.$

We will write down the original of transition process in a general view:

$$\Delta V(t)=\frac{R(0)}{D(0)}+\frac{R(P_1)}{P_1D^I(P_1)}\text{K}\cdot e^{-P(t-\tau)}+2A_k e^{(t-\tau_1)}\cos\left[\beta_k(t-\tau_1)+\varphi_k\right], \tag{20}$$

where R и D – numerator and denominator of transfer functions; P_1– valid root of a denominator; $A_к$, $\varphi_к$ – amplitude and phase of fluctuations which to be determined by formulas:

$$A_k=\sqrt{\delta^2+\sigma^2};\ \varphi_k=\text{arctg}\sigma/\delta,$$

where α_k and β_k – respectively valid and imaginary parts of complex roots of a denominator; δ and σ – respectively valid and imaginary speak rapidly expressions $R(P_k)/D^I(P_k)$in a case when R_k is a complex root. After substitution of roots in (20), we will receive expressions of transition processes for the construction machine:

$$\Delta V_n(t)=\{1-0,121e^{-4,141(t-0,1)}-3,011e^{-0,998(t-0,1)}\cdot\cos[0,479(t-0,1)-1,275]\}/\Delta V(\infty), \tag{21}$$

where $\Delta V_n(t)$ – deviations of speed of the movement of the construction machine from the established value.

We will check correctness of calculations by comparison of the transition processes calculated according to (21) and received experimentally.

3. Conclusion

1. Analyzing curves in figure 1 it is possible to say that the toe-out of ordinates of initial curves of 1 and settlement curve 2 transition processes on average doesn't exceed 4% that points to rather exact description of process.

2. Probe of dynamics of chassis of the construction machine showed that its mathematical model is described by the differential equation of the third order. The received expressions of transfer functions of running system are confirmed with pilot studies and with a sufficient accuracy can be used in the form of (21) for practical calculations.

References

[1] Ayzerman M A 1958 *Lectures on the theory of automatic regulation*. Moscow publishing house of physical and mathematical literature p 520

[2] Alekseeva T V, Artemyev K A and Bromberg A A 1972 *Cars for earthwork.* chapter 3, reslave and additional Moscow Mechanical engineering. pp 490-504

[3] Brehman A I, Sakhapov R L and Absalamova C G 2014 Innovation model of integration science and business in road-building industry. *Naukaitechnica v doroshno-stroitelnoyotrasli* №4 pp 75-78

[4] Kryzhanovsky O M 1950 About an iterative method of determination of approximate values of roots of the equations. *Avtomatika i telemechanika* № 5 pp 390-404

[5] Makhmutov M M 2008 Analytical model of coefficient of by volume crumpling of soil. *Vestnik Kazgau*, № 4 (10) pp 298-302

[6] Sakhapov RL and Makhmutov MM 2014 The influence of the studied factors on the performance of road cutters. *Izvestia KGASU*, № 4(30) Kazan pp 497-502

[7] Simou M P 1957 Determination of coefficients of transfer functions of the linearized links and systems of regulation. *Avtomatika i telemechanika* № 6 100

[8] Slavkin V I 1992 On the determination of transfer and transition function chassis self-propelled potato harvester KCK-4-I. *In collection of scientific works MYISP chapter 1, vol.* 18, *Moscow MYISP* 6 pp 66-72

[9] Sakhapov R L and Absalyamova S G 2015 The use of telecommunication technologies in education network. *Proceedings of 2015 12th International Conference on Remote Engineering and Virtual Instrumentation, Bangkok, REV* pp 14-17

Free vibration of laminated composite stiffened hyperbolic paraboloid shell panel with cutout

Sarmila Sahoo

Department of Civil Engineering, Heritage Institute of Technology, Kolkata, India

Email: sarmila.sahoo@gmail.com, sarmila.sahoo@heritageit.edu

Abstract. Composite shell structures are extensively used in aerospace, civil, marine and other engineering applications. In practical civil engineering applications, the necessity of covering large column free open areas is often an issue and hyperbolic paraboloid shells are used as roofing units. Quite often, to save weight and also to provide a facility for inspection, cutouts are provided in shell panels. The paper considers free vibration characteristics of stiffened composite hyperbolic paraboloid shell panel with cutout in terms of natural frequency and mode shapes. A finite element code is developed for the purpose by combining an eight noded curved shell element with a three noded curved beam element. The size of the cutouts and their positions with respect to the shell centre are varied for different edge conditions to arrive at a set of inferences of practical engineering significances.

1. Introduction

Composite shell structures are extensively used in aerospace, civil, marine and other engineering applications. Civil engineering applications often require covering large column free open areas and in such cases, thin shells instead of flat plates are used as in airports, parking lots, hangers, and the like. Automobile industries and medical plants applications require entry of north light through the roofing units as an additional requirement. At times, to save weight and also to provide a facility for inspection, cutout is provided in shell panels. In practice the margin of the cutouts are stiffened to take account of stress concentration effects. In civil engineering construction, hyperbolic paraboloid shells are aesthetically appealing although they offer less stiffness than other doubly curved shells. Moreover, civil engineers use laminated composites to fabricate these shell forms due to high specific stiffness and strength properties. Realizing the importance of laminated composite doubly curved shells in the industry, several aspects of shell behaviour such as bending, buckling, vibration, impact etc. are being reported by different researchers. The present investigation is however, restricted only to the free vibration behaviour of composite stiffened hyperbolic paraboloid shell panels with cutout.

Free vibration study of doubly curved shells was done by Qatu [1], Liew and Lim [2], Chakravorty et al [3], Shin [4], Tan [5], and Kant et al [6]. Qatu et al. [7] reviewed the work done on the vibration aspects of composite shells during 2000 - 2009 and observed that most of the researchers dealt with closed cylindrical shells with little attention to other shell geometries. Hyperbolic paraboloid shells with cutout (stiffened along the margin) are far from complete in the existing literature. Accordingly, the present endeavor focuses on the free vibration behavior of composite hyperbolic parabolic shell with cutout (stiffened along the margin) with concentric and eccentric cutouts and considers the shells to have various boundary conditions.

Content from this work may be used under the terms of the Creative Commons Attribution 3.0 licence. Any further distribution of this work must maintain attribution to the author(s) and the title of the work, journal citation and DOI.
Published under licence by IOP Publishing Ltd

2. Mathematical Formulation

A laminated composite hyperbolic paraboloid shell of uniform thickness h (Fig.1) and radius of curvature R_{xx} and R_{yy} is considered. Keeping the total thickness the same, the thickness may consist of any number of thin lamina each of which may be arbitrarily oriented at an angle θ with reference to the X-axis of the co-ordinate system. An eight-noded curved quadratic isoparametric finite element is used. The five degrees of freedom taken into consideration at each node include two in-plane and one transverse displacement and two rotations about the X and Y axes. The detailed finite element formulation for doubly curved shells with cutout stiffened along the margin of the cutout is reported elsewhere [8]. The code developed can take the position and size of cutout as input.

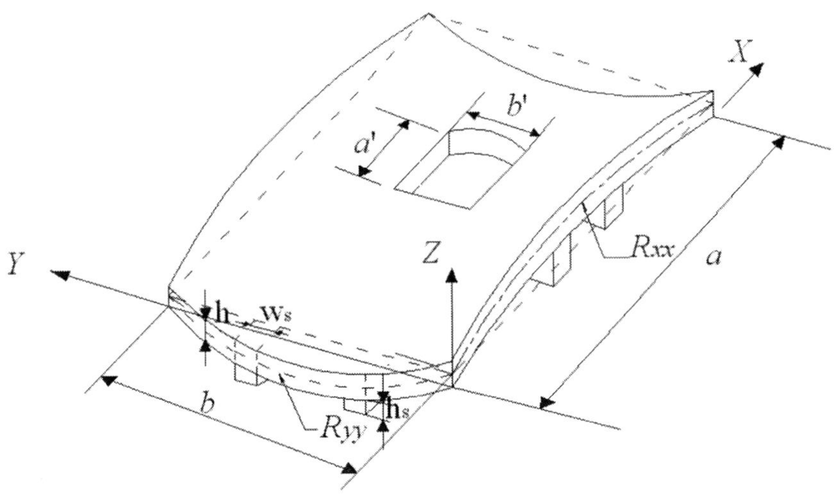

Fig.1 Hyperbolic paraboloid shell with a concentric cutout stiffened along the margins

3. Validation study

The present finite element approach is capable of modeling free vibration problem of composite stiffened hyperbolic paraboloid shells with cutout which is provided by solution of benchmark problems. The validation of stiffener formulation is reported elsewhere [8]. Free vibration of corner point supported, simply supported and clamped spherical shells of $(0/90)_4$ lamination with cutouts is also considered. The fundamental frequencies of spherical shell with cutout obtained by the present method agree well with those reported by Chakravorty et al. [3] as evident from Table 1, establishing the correctness of the cutout formulation in doubly curved shells.

Table 1: Non-dimensional fundamental frequencies ($\bar{\omega}$) for laminated composite spherical shell with cutout

a'/a	CS		SS		CL	
	Chakravorty et al.[3]	Present model	Chakravorty et al.[3]	Present model	Chakravorty et al.[3]	Present model
0.0	34.948	34.601	47.109	47.100	118.197	117.621
0.1	35.175	35.926	47.524	47.114	104.274	104.251
0.2	36.528	36.758	48.823	48.801	98.299	97.488
0.3	37.659	37.206	50.925	50.920	113.766	113.226
0.4	39.114	39.412	53.789	53.788	110.601	110.094

a/b=1, a/h=100, a'/b'=1, h/R_{xx}= h/R_{yy}=1/300, CS=Corner point supported, SS=Simply supported, CL=Clamped, non-dimensional fundamental frequency, $\bar{\omega} = \omega a^2 \left(\rho / E_{11} h^2\right)^{1/2}$

4. Results and Discussion

In order to study the effect of cutout size and position on the free vibration response additional problems for hyperbolic paraboloid shell panels with four layered symmetric cross ply and angle ply (0/90/90/0 and +45/-45/-45/+45) lamination and different boundary conditions have been solved. But for the sake of brevity, the results of only cross ply shells are presented here. The positions of the cutouts are varied along both of the plan directions of the shell for different practical boundary conditions to study the effect of eccentricity of cutout on the fundamental frequency.

Table 2 shows the results of non-dimensional frequency ($\bar{\omega}$) of 0/90/90/0 stiffened hyperbolic paraboloid shells with cutout. The shells considered are of square plan form ($a=b$) and the cutouts are also taken to be square in plan ($a'/b'=1$). The cutout sizes (i.e. a'/a) are varied from 0 to 0.4 and boundary conditions are varied along the four edges. Cutouts are concentric on shell surface. The boundary conditions are designated by describing the support clamped or simply supported as C or S taken in an anticlockwise order from the edge x=0. This means a shell with CSCS boundary is clamped along x=0, simply supported along y=0 and clamped along x=a and simply supported along y=b. The stiffeners are placed along the cutout periphery and extended up to the edge of the shell. The material and geometric properties of shells and cutouts are mentioned along with the Tables. From the tables it is seen that when a cutout is introduced to a stiffened shell the fundamental frequency increases in all the cases. Typical mode shapes corresponding to CSSC boundary condition for different sizes of the cutouts are shown in Fig. 2. It is seen that with increase in cutout size, mode shapes remain almost similar.

Table 2 Fundamental frequencies ($\bar{\omega}$) for 0/90/90/0 shell with central square cutout

Boundary conditions	Cutout size (a'/a)				
	0	0.1	0.2	0.3	0.4
CCCC	104.21	117.21	139.79	155.00	157.43
CSCC	79.32	94.17	111.86	129.18	131.77
CCSC	96.15	106.07	123.62	142.36	143.51
CCCS	78.99	92.31	111.13	128.37	131.51
CSSC	69.38	80.04	94.69	109.11	111.21
CCSS	69.32	79.82	94.65	109.07	111.19
CSCS	87.92	86.46	107.33	126.02	128.52
SCSC	100.44	104.16	122.09	139.55	140.44
CSSS	63.30	73.69	90.73	102.68	104.34
SSSC	67.72	77.92	93.63	103.87	104.39
SSCS	63.33	73.78	90.73	102.68	104.34
SSSS	64.63	71.16	91.13	96.79	97.42
Point supported	25.89	31.63	38.35	46.61	53.07

$a/h=100$, $h/R_{xx}=1/300$, $R_{xx}/R_{yy}=-1.5$; $E_{11}/E_{22}=25$, $G_{23}=0.2E_{22}$, $G_{13}=G_{12}=0.5E_{22}$, $\nu_{12}=\nu_{21}=0.25$.

To study the effect of cutout positions on fundamental frequencies, results are obtained for different locations of a cutout with a$'$/a =0.2. Each of the non-dimensional coordinates of the cutout centre ($\bar{x}=x/a$, $\bar{y}=y/b$) is varied from 0.2 to 0.8 along both the plan directions so that the distance of a cutout margin from the boundary is not less than one tenth of the plan dimension. The ratio of the fundamental frequency of a shell with an eccentric puncture to that of a shell with concentric puncture expressed in percentage is denoted by r (Table 3). The mode shapes for different eccentric positions of the cut-out are shown in Fig. 3 for CSSC shells only.

Table 3 Values of 'r' for 0/90/90/0 hyperbolic paraboloid shells

Edge conditions	$\bar{y}$	$\bar{x}$						
		0.2	0.3	0.4	0.5	0.6	0.7	0.8
CCCC	0.2	89.09	90.82	94.20	97.90	94.20	90.81	89.08
	0.3	87.70	89.45	93.11	97.48	93.11	89.45	87.70
	0.4	89.58	91.24	94.77	98.64	94.77	91.24	89.58
	0.5	91.95	93.63	96.96	100.00	96.96	93.63	91.95
	0.6	89.58	91.24	94.77	98.64	94.77	91.24	89.58
	0.7	87.70	89.45	93.11	97.48	93.11	89.45	87.70
	0.8	88.85	90.56	93.88	97.90	94.11	90.61	88.91
CSCC	0.2	95.76	98.63	101.84	103.62	101.83	98.63	95.75
	0.3	97.34	102.57	107.80	110.44	107.81	102.58	97.35
	0.4	96.74	101.33	108.80	118.09	108.96	101.36	96.65
	0.5	92.79	94.34	97.49	100.00	97.66	94.41	92.45
	0.6	87.32	87.05	88.41	98.82	88.60	87.12	86.84
	0.7	84.18	83.46	84.49	85.77	84.65	83.53	83.73
	0.8	83.37	82.72	83.75	84.96	83.90	82.81	83.00
CCSC	0.2	80.87	83.53	89.02	99.01	103.19	96.08	91.14
	0.3	80.35	83.00	88.42	98.24	102.22	95.45	90.92
	0.4	80.64	83.24	88.71	98.88	103.07	95.51	90.31
	0.5	81.17	83.80	89.42	100.00	104.29	95.94	90.09
	0.6	80.64	83.24	88.71	98.88	103.07	95.51	90.31
	0.7	80.35	83.00	88.42	98.24	102.22	95.45	90.92
	0.8	80.70	83.34	88.73	98.94	103.01	95.83	90.97
CCCS	0.2	83.55	83.26	84.19	85.14	84.19	83.26	83.55
	0.3	84.26	83.96	84.89	85.90	84.88	83.96	84.26
	0.4	87.41	87.57	88.80	89.91	88.80	87.57	87.41
	0.5	93.06	94.92	97.89	100.00	97.89	94.92	93.06
	0.6	97.27	101.96	109.54	123.05	109.35	101.96	97.27
	0.7	97.97	103.23	108.52	111.16	108.51	103.23	97.97
	0.8	95.72	99.06	102.48	104.26	102.47	99.07	95.91
CSSC	0.2	71.91	79.82	90.94	105.34	103.51	91.42	81.58
	0.3	76.06	84.77	97.05	111.76	109.58	97.04	86.44
	0.4	76.78	84.91	96.55	112.82	113.81	99.77	88.26
	0.5	71.96	77.67	86.27	100.00	104.79	92.31	82.45
	0.6	65.73	70.08	77.28	89.88	94.36	83.59	75.28
	0.7	61.75	65.95	73.15	85.93	89.06	78.68	70.97
	0.8	60.26	64.82	72.38	85.51	87.40	76.91	69.26
CCSS	0.2	60.29	64.88	72.46	85.55	87.13	76.82	69.24
	0.3	61.74	65.98	73.17	85.96	88.70	78.55	70.94
	0.4	65.73	70.11	77.31	89.89	93.92	83.45	75.26
	0.5	71.97	77.71	86.30	100.00	104.38	92.20	82.44
	0.6	76.81	84.94	96.58	112.86	113.77	99.75	88.26
	0.7	76.10	84.81	97.09	111.81	109.62	97.07	86.47
	0.8	71.48	79.44	90.68	105.35	103.53	91.32	81.41
CSCS	0.2	82.94	82.40	82.54	82.67	82.53	82.40	82.94

	0.3	84.82	84.47	85.00	85.48	85.00	84.47	84.82
	0.4	88.75	89.08	90.31	91.26	90.31	89.08	88.75
	0.5	93.70	96.63	98.42	100.00	98.42	96.63	93.70
	0.6	88.75	89.08	90.31	91.26	90.31	89.08	88.75
	0.7	84.83	84.47	85.00	85.48	85.00	84.47	84.82
	0.8	82.93	82.38	82.51	82.66	82.52	82.39	82.93
SCSC	0.2	79.29	83.00	89.32	99.59	89.32	83.00	79.29
	0.3	79.31	82.86	88.93	99.17	88.93	82.86	79.31
	0.4	78.97	82.72	89.07	99.16	89.07	82.72	78.97
	0.5	78.85	82.88	89.58	100.00	89.58	82.88	78.85
	0.6	78.97	82.72	89.07	99.87	89.07	82.72	78.97
	0.7	79.31	82.86	88.93	98.75	88.93	82.86	79.31
	0.8	79.12	82.81	89.03	99.17	89.24	82.84	79.20
CSSS	0.2	51.33	57.82	67.63	82.96	84.15	72.06	62.83
	0.3	55.42	61.59	70.76	85.37	88.03	76.33	67.22
	0.4	61.04	67.79	76.99	91.12	95.07	83.04	73.27
	0.5	65.13	74.29	87.10	100.00	104.47	89.74	78.06
	0.6	61.04	67.79	76.99	91.12	95.06	83.04	73.27
	0.7	55.43	61.59	70.76	85.37	88.03	76.33	67.22
	0.8	51.27	57.80	67.60	82.94	84.13	72.03	62.79
SSSC	0.2	69.80	79.08	91.01	102.97	91.01	79.08	69.80
	0.3	73.61	83.66	96.66	108.19	96.66	83.66	73.61
	0.4	73.41	83.03	95.76	110.20	95.75	83.03	73.40
	0.5	67.68	75.36	85.52	100.00	85.52	75.36	67.68
	0.6	61.54	67.96	76.68	90.04	76.79	67.96	61.54
	0.7	58.03	64.14	72.68	86.39	72.68	64.14	58.03
	0.8	56.92	63.18	71.96	85.77	72.00	63.20	56.92
SSCS	0.2	62.83	72.05	84.15	82.96	67.62	57.82	51.33
	0.3	67.22	76.33	88.03	85.37	70.75	61.59	55.42
	0.4	73.27	83.04	95.06	91.12	76.99	67.79	61.04
	0.5	78.04	89.74	104.47	100.00	87.10	74.29	65.13
	0.6	73.27	83.04	95.06	91.12	76.99	67.79	61.04
	0.7	67.22	76.33	88.03	85.37	70.75	61.59	55.42
	0.8	62.78	72.03	84.13	82.94	67.60	57.80	51.28
SSSS	0.2	46.61	55.16	66.25	81.99	66.24	55.16	46.61
	0.3	50.60	58.74	69.22	84.36	69.22	58.74	50.60
	0.4	56.16	64.73	75.23	89.84	75.23	64.73	56.16
	0.5	60.47	71.24	85.11	100.00	85.11	71.24	60.47
	0.6	56.16	64.72	75.22	89.84	75.22	64.72	56.16
	0.7	50.60	58.74	69.22	84.36	69.22	58.74	50.59
	0.8	46.56	55.14	66.21	81.97	66.21	55.14	46.57
CS	0.2	91.37	102.29	109.41	112.10	109.41	102.29	91.37
	0.3	93.48	100.29	104.82	106.65	104.82	100.29	93.48
	0.4	95.96	98.46	100.57	101.67	100.55	98.46	95.96
	0.5	97.29	97.68	98.93	100.00	98.93	97.68	97.29
	0.6	95.93	98.46	100.55	101.67	100.55	98.46	95.93
	0.7	93.46	100.29	104.82	106.65	104.82	100.29	93.46
	0.8	91.21	102.01	109.05	112.02	109.20	102.11	91.29

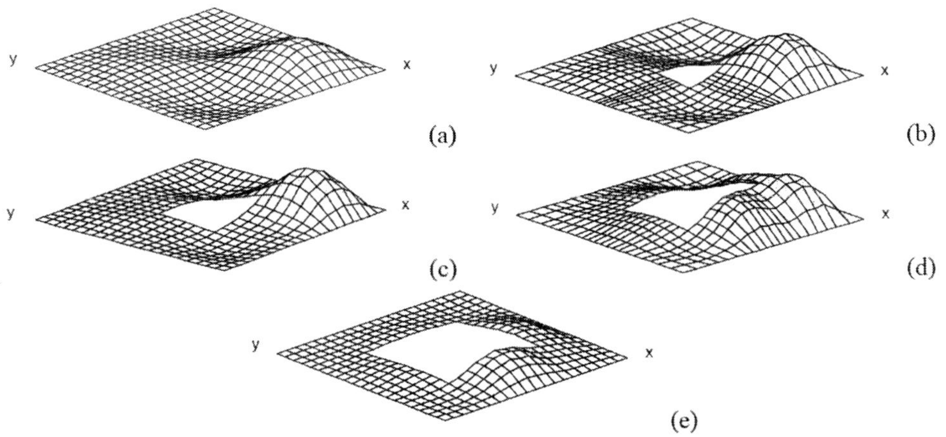

Fig. 2 Typical mode shapes of 0/90/90/0 shell for different sizes of central square cut-out and CSSC boundary (a) a'/a=0, (b) a'/a= 0.1, (c) a'/a=0.2, (d) a'/a=0.3, (e) a'/a=0.4

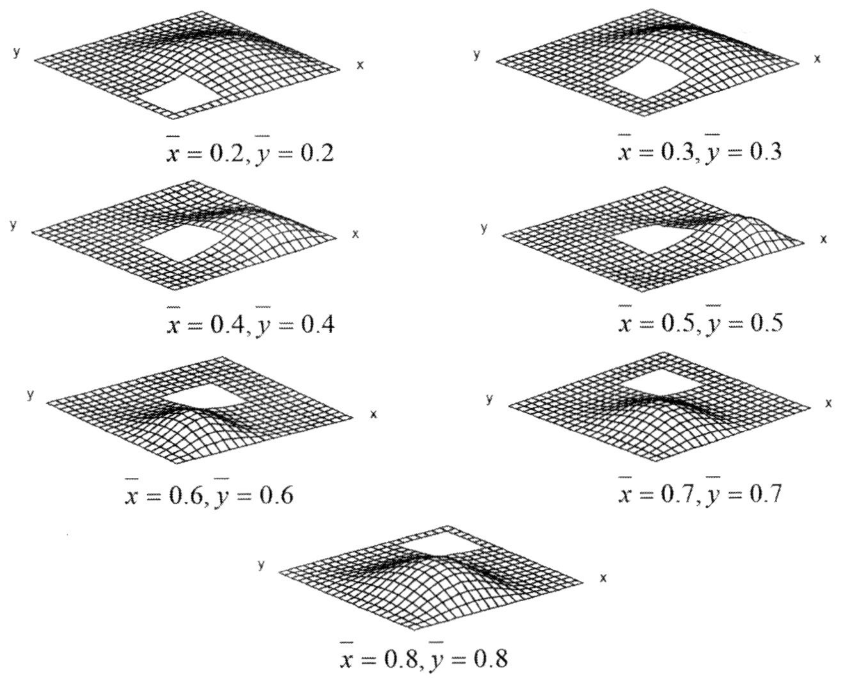

Fig. 3 Typical mode shapes of 0/90/90/0 shell for different position of the central square cutout and CSSC boundary condition

Table 4 provides the maximum values of r together with the position of the cutout. The table also shows the rectangular zones, within which r is always greater than or equal to 95. It is to be noted that at some other points r values may have similar values, but only the zone rectangular in plan has been identified. These tables indicate the maximum eccentricity of a cutout which can be permitted if the fundamental frequency of a concentrically punctured shell is not to reduce a drastic amount. So such tables will help practicing engineers.

Table 4 Maximum values of r with coordinates of cutout centres and zones for r≥95 for 0/90/90/0 shells

Boundary Condition	Maximum values of r	Co-ordinate of cutout centre	Area in which the value r≥95
CCCC	100.00	(0.5,0.5)	$\bar{x}$=0.5, 0.2 ≤ $\bar{y}$ ≤0.8
CSCC	118.09	(0.5,0.4)	0.2≤ $\bar{x}$ ≤0.8, 0.2≤ $\bar{y}$ ≤0.4
CCSC	104.29	(0.6,0.5)	0.5≤ $\bar{x}$ ≤0.7, 0.2≤ $\bar{y}$ ≤0.8
CCCS	123.05	(0.5,0.6)	0.3≤ $\bar{x}$ ≤0.7, 0.5 ≤ $\bar{y}$ ≤0.8
CSSC	113.81	(0.6,0.4)	0.5≤ $\bar{x}$ ≤0.6, 0.2 ≤ $\bar{y}$ ≤0.5
CCSS	113.77	(0.6,0.6)	0.5≤ $\bar{x}$ ≤0.6, 0.5 ≤ $\bar{y}$ ≤0.8
CSCS	100.00	(0.5,0.5)	0.3≤ $\bar{x}$ ≤0.7, $\bar{y}$ =0.5
SCSC	100.00	(0.5,0.5)	$\bar{x}$ =0.5, 0.2≤ $\bar{y}$ ≤0.8
CSSS	104.47	(0.6,0.5)	$\bar{x}$ =0.6, 0.4 ≤ $\bar{y}$ ≤0.6
SSSC	110.20	(0.5,0.4)	$\bar{x}$ =0.5, 0.2 ≤ $\bar{y}$ ≤0.5
SSCS	104.47	(0.4,0.5)	$\bar{x}$ =0.4, 0.4 ≤ $\bar{y}$ ≤0.6
SSSS	100.00	(0.5,0.5)	$\bar{x}$ =0.5, $\bar{y}$ =0.5
CS	112.10	(0.5,0.2)	0.3≤ $\bar{x}$ ≤0.7, 0.2 ≤ $\bar{y}$ ≤0.8

a/h=100, h/R_{xx}=1/300, R_{xx}/R_{yy}=-1.5; E_{11}/E_{22} = 25, G_{23} = 0.2E_{22}, G_{13} = G_{12} = 0.5E_{22}, ν_{12} = ν_{21} =0.25.

5. Conclusion

The present study shows that the finite element code used here is suitable for analyzing free vibration problems of stiffened hyperbolic paraboloid shell roofs with cutouts as this approach produce results in close agreement with those of the benchmark problems. Concentric cutouts may be provided safely on stiffened hyperbolic paraboloid shell surfaces for functional requirements upto d'/a=0.4. The information regarding behavior of stiffened hyperbolic paraboloid shell with eccentric cutouts for wide range of eccentricity may be used as design aids by structural engineers.

References

[1] Qatu M S 1991 *Composite Structures* **17** 227-255.
[2] Liew K M and Lim C W 1995 *Journal of Engineering Mechanics* **121** 1277-1283.
[3] Chakravorty D, Sinha P K and Bandyopadhyay J N 1998 *Journal of Engineering Mechanics* **124** 1-8.
[4] Shin D K 1997 *Computers and Structures* **62** 35-49.
[5] Tan D Y 1998 *Journal of Sound and Vibration* **213** 15-33.
[6] Kant T, Kumar S and Singh U P 1994 *Computers and Structures* **50** 135-146.
[7] Qatu M S, Sullivan R W and Wang W 2010 *Composite Structures* **93** 14-31.
[8] Sahoo S 2013 *Journal of Engineering* ID: **230120** 1-18.

Evaluation of Waveform Structure Features on Time Domain Target Recognition under Cross Polarization

M A Selver[1], M Seçmen[2], E Y Zoral[1]

[1]Department of Electrical and Electronics Engineering, Dokuz Eylul University, Kaynaklar Campus, 35160, Buca, Izmir, TURKEY

[2]Department of Electrical and Electronics Engineering, Yaşar University, Selçuk Campus, 35320, Bornova, İzmir, TURKEY

Email alper.selver@deu.edu.tr, mustafa.secmen@yasar.edu.tr, yesim.zoral@deu.edu.tr

Abstract. Classification of aircraft targets from scattered electromagnetic waves is a challenging application, which suffers from aspect angle dependency. In order to eliminate the adverse effects of aspect angle, various strategies were developed including the techniques that rely on extraction of several features and design of suitable classification systems to process them. Recently, a hierarchical method, which uses features that take advantage of waveform structure of the scattered signals, is introduced and shown to have effective results. However, this approach has been applied to the special cases that consider only a single planar component of electric field that cause no-cross polarization at the observation point. In this study, two small scale aircraft models, Boeing-747 and DC-10, are selected as the targets and various polarizations are used to analyse the cross-polarization effects on system performance of the aforementioned method. The results reveal the advantages and the shortcomings of using waveform structures in time-domain target identification.

1. Introduction

The classification of aircraft targets from scattered electromagnetic signals is a challenging problem which is highly altered by aspect dependency [1]. Thus, the scattered signal from a target is highly dependent on the polarization of the incident and reflected fields. An effective target classification strategy should eliminate the adverse effects of the aspect angle while elevating the signal differences due to different target geometries and material type [2]. Recently, waveform structure of time domain scattered signals, which include several hills and valleys, are used for triangular re-modeling of the data. Then, the geometrical information of each triangle is fed to a classifier, where edge lengths and slopes of the triangles as well as the time-interval between triangle peaks constitute a feature set. The effects of noise at low SNR levels are eliminated using a coarse-to-fine approximation strategy that relies on radial basis function networks [3]. Opposed to well-known wavelet multi-resolution analysis using the discrete wavelet transform [4], proposed approach is shown to perform more precise reconstruction of the scattered signal and enable better extraction of the structural features, which eventually result with higher classification rates.

Although the approach described above has been applied to several different target geometries and material types, the experiment conditions are set to some special cases, where only a single planar

Content from this work may be used under the terms of the Creative Commons Attribution 3.0 licence. Any further distribution of this work must maintain attribution to the author(s) and the title of the work, journal citation and DOI.
Published under licence by IOP Publishing Ltd

component of electric field is received without cross polarization at the observation point. In this incremental study, the cross-polarization effects are analyzed for the method mentioned above.

The target recognition methods based on polarization diversity generally requires the data of the polarization matrix, which is a 2×2 complex scattering operator that characterizes a target's scattering properties. This matrix is obtained from the scattered signals received on both horizontal and vertical polarizations, where Horizontal-Horizontal (HH) and Vertical-Vertical (VV) are both co-polar components and Horizontal-Vertical (HV) and Vertical-Horizontal (VH) are both cross-polar components. There have been many attempts to use the polarization matrix of the target's scattered signal and other polarization features for target recognition [5]-[7]; however, there has not been the desired success in applying these methods for practical target recognition [1]. Besides these optical region methods, there exist other approaches in literature that are based on neural networks (NN), support vector machines [8], [9] or probabilistic hidden-Markov models and the Bayesian algorithm [10], [11].

Based on the discussions above, in this study, a target identification system for cross-polarization applications is developed. The system initially approximates scattered signals in order to enable robust determination of the peaks and the lobes of the waveform. Then, a triangularization process is carried out to model the scattered signal. Finally, the synthesized model is used to extract several features prior to the target identification. The rest of the paper is organized as follows: Section 2 covers the properties of the targets, the scattered signal data set, methodology (i.e. signal approximation, feature extraction and classification stages) of the developed system. Section 3 presents the application results and draws the conclusions.

2. Scattered Signal Data Set and Methodology

In this study, two small-scale aircraft targets, Boeing 747 and DC-10, are used to test the proposed approximation, feature extraction and classification algorithms. Albeit the number of targets is only two, the polarization invariance property of the suggested target recognition technique can be analyzed since the scattered signal database includes measurements at various polarizations. The aircraft targets to be used in the classifier design are again small-scale models of the actual Boeing 747 and DC-10 aircraft targets with the scale factor of 3/500.The monostatic frequency response measurements of these aircraft targets were done at the ElectroScience Laboratory of the Ohio State University for different polarizations and aspect angles [12]. The list of measurement polarizations and aspect angles for the targets is given Table 1.

Table 1. The list of available scattered data for the model targets Boeing 747 and DC-10 measured at various polarizations and aspect angles (HH: horizontal horizontal, VV: vertical-vertical, X: cross polarization).

Polarization	HH		VV		X	
Aircraft Target	Boeing 747	DC-10	Boeing 747	DC-10	Boeing 747	DC-10
Aspect Angle (degrees)	0, 30, 45, 60, 90, 120, 150, 180	0, 30, 45, 60, 70, 90, 120, 150, 180	0, 30, 45, 60, 90, 120, 150, 180	0, 30, 45, 70, 90, 120,150	30, 60, 90, 120, 135, 150	0, 30, 45, 60, 70, 90, 150, 180

The operating frequency band of the measurements is 1-12 GHz with 50 MHz frequency resolution (221 points) except the Boeing 747's HH polarization measurement data. The frequency band of measurements for these exceptional cases is (1.5-12 GHz) with the same frequency resolution. The scattered time signals of both aircraft targets for VV polarization, 90 degree case is given in Figure 1.

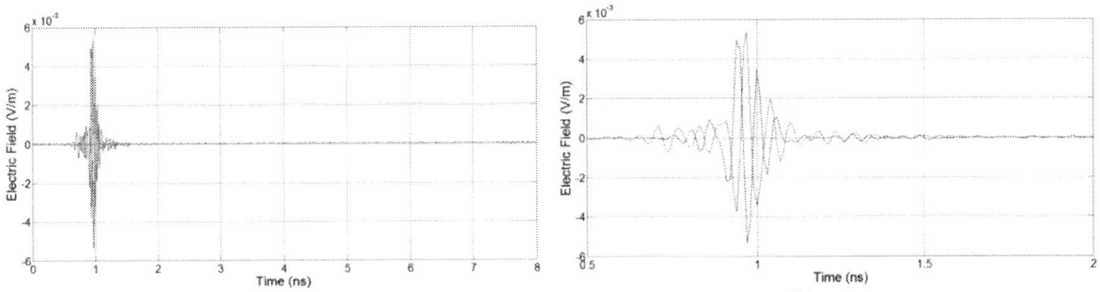

(a) (b)

Figure 1. The scattered time signals of the small-scale aircraft targets for VV polarization/90 degree case and the frequency band of (1-12 GHz) (blue solid Boeing 747, red dashed DC-10). (a) Full time span, (b) the time interval that contains most of the fluctuations.

The differences between the scattered signals from HH and X polarizations are given for DC10 in Figures 2 and 3, respectively.

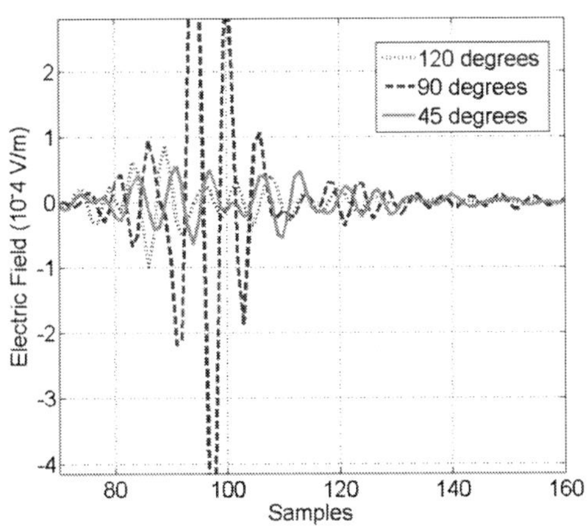

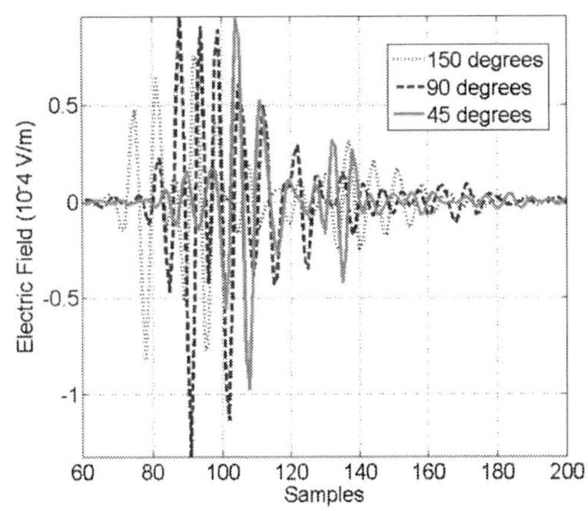

Figure 2. The scattered time domain signals for DC-10 at HH polarization and for aspect angles of 120° (blue), 90° (red) and 45° (green)

Figure 3. The scattered time domain signals for DC-10 at V polarization and for aspect angles of 120° (blue), 90° (red) and 45° (green)

Typical scattered signals in Figures 1, 2 and 3 show that these signals consist of fluctuating waveforms and they can be divided into sub-waves. Moreover, these sub-waves can be modeled by a triangularization process, which is based on determination of their amplitudes and durations. In previous studies, it is shown that the differences of these fluctuations can be used to eliminate the aspect angle dependency [3]. In order to do that, structural features, which characterize the waveform of each sub-wave, should be extracted. Specifically, these structural properties correspond to the 'peaks' (both negative and positive), hills and valleys of the scattered signals. Since each peak is assumed to have a lobe around it, these lobes should be found prior to the extraction of the features.

Here, it is important to point that a method utilized to find the peaks and the lobes associated with them should be robust to noise or other high frequency distortions (such as radar glint) such that the recognized peaks represent the main lobes of the signal waveform. Thus, a pre-processing step is needed to remove unwanted high-frequency components while preserving characteristic information of the signal waveform. Unfortunately, the use of smoothing filters is not possible, since they suppress the high frequency components on the entire data. This would also eliminate the highly variable frequency content of the fluctuating waveform and therefore, local analyses are needed. In this respect, the adaptive approach of MSA can be particularly effective. Hierarchical radial basis function (HRBF) networks offer a coarse-to-fine approach that particularly suits this need [3].

HRBF performs a coarse-to-fine multi-scale analysis by fitting low frequency components first, and then allocates more Gaussian units, where the data contain higher frequencies. The process continues until a pre-defined degree of accuracy is achieved. Since this approach relies on local operations and guarantees a uniform residual error in almost real time, it is used prior to peak detection and feature extraction. Figure 4 illustrates a scattered signal and its approximation created by HRBF, which allow much more robust determination of peaks and their associated lobes.

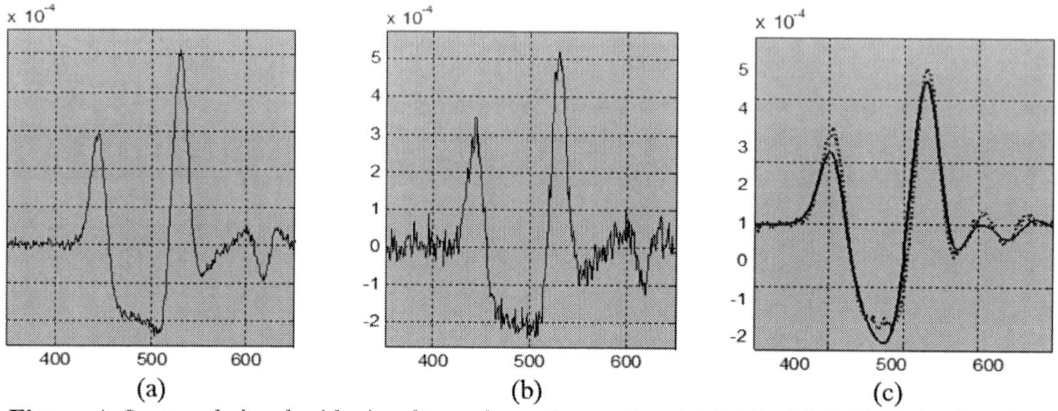

Figure 4. Scattered signal with signal to noise ratio equal to (a) 5 dB, (b) 0 dB. It is clear that peak detection gets harder as SNR decreases. (c) HRBF approximated scattered signal.

Instead of estimating the almost arbitrary shape of a lobe, it can be approximated by a simpler geometry while preserving important characteristic information. For this purpose, the coordinates of each lobe is determined such that it can be modeled by a triangle. The corners of the triangles are determined using the following three steps:

1) The maximum or minimum point found by using a peak detection algorithm constitutes the first corner of the triangle (See p_1 and p_2 at Figure 5).
2) The starting point of the lobe found by following the change in the sign of the slope or sign of the data value (i.e. whichever the first) at the left side of the peak constitute the second corner (See $[x_{11}, y_{11}]$ and $[x_{21}, y_{21}]$ at Figure 5)
3) The end point of the lobe found by following the change in the sign of the slope or sign of the data value (i.e. whichever the first) at the right side of the peak constitute the third corner (See $[x_{12}, y_{12}]$ and $[x_{22}, y_{22}]$ at Figure 5)

Next, the following five features are extracted from the triangular model:

1) The distance between the start and the end positions of the lobe constitute the first feature (See the arrows between $\{[x_{11}, y_{11}], [x_{12}, y_{12}]\}$ and $\{[x_{21}, y_{21}], [x_{22}, y_{22}]\}$ at Figure 5)
2) The distance between the start and the peak positions of the lobe constitute the second feature (i.e. $\{[x_{11}, y_{11}], p_1\}$ and $\{[x_{21}, y_{21}], p_2\}$ at Figure 5)
3) The distance between the end and the peak positions of the lobe constitute the third feature (i.e. $\{[x_{12}, y_{12}], p_1\}$ and $\{[x_{22}, y_{22}], p_2\}$ at Figure 5)
4) The slope between the start and the peak positions of the lobe constitute the fourth feature (See α_1 and α_2 at Figure 5)
5) The slope between the end and the peak positions of the lobe constitute the fifth feature (See β_1 and β_2 at Figure 5)

Finally, inter-peak relations are represented with an additional feature set, which consists of the distances between the peaks. It is clear that the size of this additional feature set depends on the number of peaks found. Moreover, the total number features also depend on the number of peaks. In this study, the number of peaks is limited, that is only the first fifteen peaks in terms of peak magnitude are considered such that the number of features for each signal become equal.

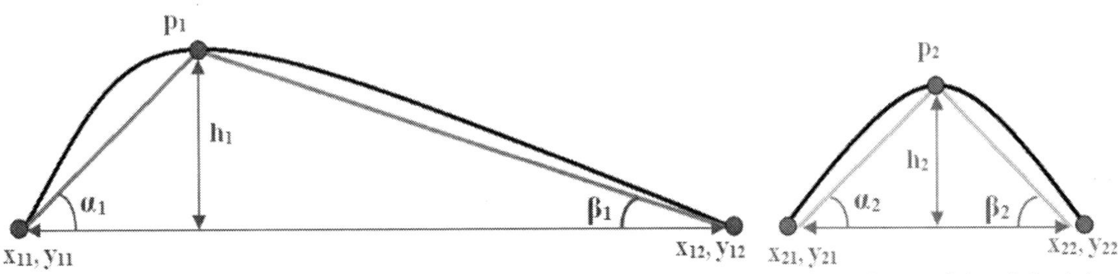

Figure 5. Two lobes determined around the two peaks (p_1 and p_2), the triangular models of the lobes and the features determined for each triangular model.

3. Application & Results

Two small scale aircraft models, Boeing 747 and DC-10, are used to test the method developed in this study. Feature vectors extracted for each target for the aspect angles given in Table 1, which generates a data set of 17 samples for HH polarization, 15 samples for VV polarization and 14 samples for X polarization. Since this is a small data set, cross-validation techniques, particularly K-fold analysis, which partitions the data set into K parts, are used [13]. K-fold performs K experiments, in which K-1 folds are used for training and the remaining fold is used for testing. The advantage of K-fold cross validation is that it prevents over-fitting by systematically using all the examples in the data set for both training and testing. The experiments are repeated for different K values including 2, 3 and 4. The best performance is obtained for K=3 and presented in Table 2.

For classification, a Multi-Layer Perceptron (MLP) network is chosen due to its universal approximation property [14]. MLP is trained using back-propagation with adaptive learning rate, which is initialized to 0.01, structured to have one hidden layer with 8 neurons with tangent sigmoid activation functions and an output layer with 2 neurons with linear activation functions. The network goal is chosen to be 0.001 and the maximum number of iterations is determined as 10000 epochs.

The results are shown in Table 2. Correct classification percentage (accuracy), selectivity and specificity metrics are used to measure the system performance. The results show that, although the developed method is successful for HH and VV polarizations, the correct classification percentage using the approximated data significantly decreases for cross polarization (i.e. X). It is observed that the classification performance for X polarization can be increased if more peaks and lobes are used for feature extraction since the scattered signal characteristics are represented in a better way. However, due to the small number of samples in the data set, the classification via supervised learning would not converge for such an extended feature vector.

Table 2. Target recognition results at different polarizations

	%	Accuracy	Selectivity	Sensitivity
HH	**Boeing 747**	94.25	90.82	92.13
	DC-10	91.10	87.96	89.94
VV	**Boeing 747**	93.55	91.24	92.82
	DC-10	90.42	88.35	89.77
X	**Boeing 747**	86.25	84.82	85.45
	DC-10	87.64	82.96	87.86

References

[1] Secmen M and Turhan-Sayan G 2009 Radar target classification method with reduced aspect dependency and improved noise performance using MUSIC algorithm *IET Radar Sonar and Navigation* **3** pp 583-595

[2] Turhan-Sayan G 2005 Real time electromagnetic target classification using a novel feature extraction technique with PCA-based fusion *IEEE Transactions on Antennas and Propagation* **53** 2 pp 766-776

[3] Selver M A, Zoral E Y and Dorak S 2013 Aspect independent spherical target recognition using structural features of scattered signals *Conf. on the Computation of Electromagnetic Fields* (*Budapest, Hungary 30 June – 4 July 2013*)

[4] Mallat S G 1999 *A Wavelet Tour of Signal Processing* San Diego CA Academic Press 2nd ed

[5] Li H J, Lane R Y 1995 Utilization of multiple polarization data for aerospace target identification *IEEE Trans. Antennas and Propagation* **43** 12 pp 1436-1440

[6] Garren D A, Odom A C, Osborn M K, Goldstein J S, Pillai S U, Guerci J R, 2002 Full-polarization matched-illumination for target detection and identification *IEEE Trans. Aerospace Elect. Syst.* **38** 3 pp. 824-837

[7] Dhai D H, Wu X F, Li Y Z, Wang X S, Xiao S P, 2005 Polarization target detection algorithm based on instantaneous Stokes sub-vector *IEEE International Symposium on Microwave, Antenna, Propagation and EMC Technologies for Wireless Communication* **1** pp 310-313

[8] Robinson M, Azimi-Sadjadi M R, Salazar J, 2005 Multi-aspect target discrimination using hidden Markov models and neural networks *IEEE Transactions on Neural Networks* **16** 2 pp 447-459

[9] Jouny I, 2005 On SVM for classification of real and synthetic radar signatures *IEEE Antennas and Propagation Society International Symposium* **1** pp 2-5

[10] Gader P D, Mystkowski M, Zhao Y, 2001 Landmine detection with ground penetrating radar using hidden Markov models *IEEE Transactions on Geoscience and Remote Sensing* **39** 6 pp 1231-1243

[11] Mei W, Shan G L, Li X R, An efficient Bayesian algorithm for joint tracking and classification *International Conference on Signal Processing (ICSP)* **3** pp 2090-2098

[12] Walton E K and Young J D 1994 The Ohio State University compact radar cross-section measurement range *IEEE Transactions on Antennas and Propagation*, **32** 11 pp 1218-1223

[13] Raudis S J and Jain A K 1991 Small sample size effects in statistical pattern recognition *IEEE Transactions on Pattern Analysis and Machine Intelligence* **13** 3 pp 252-264

[14] Duda R and Hart P 1973 *Pattern Classification and Scene Analysis* 1st ed

User Aligned Histogram Stacks for Visualization of Abdominal Organs via MRI

M Özdemir[1], O Akay[2], C Güzeliş[3], O Dicle[4], M A Selver[2]

[1]Institute of Natural and Applied Sciences, Dokuz Eylul University, Kaynaklar Campus, 35160, Buca, Izmir, TURKEY

[1]Department of Electrical and Electronics Engineering, Dokuz Eylul University, Kaynaklar Campus, 35160, Buca, Izmir, TURKEY

[2]Department of Electrical and Electronics Engineering, Yaşar University, Selçuk Campus, 35320, Bornova, İzmir, TURKEY

[1]Department of Radiology, Faculty of Medicine, Dokuz Eylul University, Balçova, Izmir, TURKEY

Email merveozdemir89@gmail.com, olcay.akay@deu.edu.tr, oguz.dicle@deu.edu.tr, cuneyt.guzelis@yasar.edu.tr , alper.selver@deu.edu.tr,

Abstract. Multi-dimensional transfer functions (MDTF) are occasionally designed as two-step approaches. At the first step, the constructed domain is modelled coarsely using global volume statistics and an initial transfer function (TF) is designed. Then, a finer classification is performed using local information to refine the TF design. In this study, both a new TF domain and a novel two-step MDTF strategy are proposed for visualization of abdominal organs. The proposed domain is generated by aligning the histograms of the slices, which are reconstructed based on user aligned majority axis/regions through an interactive Multi-Planar Reconstruction graphical user interface. It is shown that these user aligned histogram stacks (UAHS) exploit more a priori information by providing tissue specific inter-slice spatial domain knowledge. For initial TF design, UAHS are approximated using a multi-scale hierarchical Gaussian mixture model, which is designed to work in quasi real time. Then, a finer classification step is carried out for refinement of the initial result. Applications to several MRI data sets acquired with various sequences demonstrate improved visualization of abdomen.

1. Introduction

Visualization of abdominal organs (kidneys, liver, and spleen) plays critical role in several clinical procedures including but not limited to surgery planning, treatment follow-ups, and education. Although there are many segmentation techniques developed for abdomen, they mostly rely on explicit classification of the voxels belonging to the organ of interest ignoring the rest of the data. However, in most of the cases, what is interesting from the clinical point of view is not only an organ or tissue itself, but also its relationships with the neighboring organs and/or vessel systems. Thus, even if the segmentation is perfectly realized, it is not possible to render it together with the adjacent tissues/organs and related vascular trees unless another segmentation process is performed for each object of interest [1].

Content from this work may be used under the terms of the Creative Commons Attribution 3.0 licence. Any further distribution of this work must maintain attribution to the author(s) and the title of the work, journal citation and DOI.
Published under licence by IOP Publishing Ltd

This limitation of segmentation motivates the development of complicated TF specification techniques which allow enhancement of potentially important features during the rendering. In other words, TF design provides suppression of the redundant data, while facilitating intervention of physician for further modifications over the complete volume. Moreover, TF design is the only interactive step of the direct volume rendering pipeline and intuitive design user interfaces allow production of informative renderings [2].

TFs map data attributes (i.e. domain) to visual properties (i.e. range) such as color and opacity. In medical applications, domain should utilize enough features for discrimination of different tissues/organs. At the same time, complexity of user interaction and input dimensionality need to be considered carefully in order to preserve intuitiveness and better computational performance. In other words, tissues are usually indistinctive in low dimensional domains and the interactions become counterintuitive or infeasible in high dimensional ones. This trade-off between preserving intuitive user interaction and possessing a discriminative domain spurs ongoing research efforts in TF design.

There are three main approaches of TF design: Manual [3], image-centric [9-12], and data-centric. Data-centric methods have been gaining more importance since they increase the classification capabilities of TFs. In this manner, iso-value determination [4], edge detection concepts [5], topology analysis [6], stochastic properties [7], texture characteristics and 3-D filter responses [13] are some of the proposed effective techniques. Various graphical user interfaces (GUIs) [15] and interaction mechanisms [16-17] have also been developed using multi-dimensional transfer functions (MDTF) [18-19].

For medical visualization, recent studies show that incorporating medical domain knowledge through semantic expressions, anatomical/physiological information, and/or modality characteristics allow obtaining accurate renderings with a TF specification [20].

The method proposed in this study shares the same strategy (i.e. coarse classification plus refinement) with the above discussed MDTF techniques. Compared to those, our method performs further steps on TF specification in the following aspects: First of all, a new and very intuitive TF domain is constructed by histogram stacks, which are calculated from reconstructed images (i.e. resampled volume data) based on user aligned majority axis through an Multi-Planar Reconstruction (MPR) user interface (Section 3). Secondly, a multi-scale hierarchical Gaussian Mixture Model (GMM) strategy, which is able to locate and roughly classify tissue of interest automatically and in (quasi) real time, is developed (Section 4). Thirdly, refinement of classification results based on spatial and morphological relations of the tissues of interest are performed. Finally, the developed approach is tested using various clinical abdominal MRI data sets with different sequence protocols (Section 5).

2. Data Sets

The analyses in this study are applied to several MRI series, which are obtained by the Radiology Department of Dokuz Eylül University. The data sets are acquired using a 1,5T Philips MRI modality that produces 12 bit DICOM images with a resolution of 256 x 256. The data sets were retrospectively collected from the Picture Archiving and Communication System of the same department. Three different MRI sequences (i.e. T1-DUAL (in-phase), THRIVE, T1-WATS) are used to test the proposed method. Each of these sequences is routinely being used to scan the same part of the body, but they are obtained from different combinations of radiofrequency pulses and gradients [23]. The details of the characteristics of these data sets and the utilized sequences are given below and their sample images are illustrated in Figure 1.

T1-DUAL (in phase) is a fat suppression sequence which uses the time differences in the z-axis recoveries of fat and water protons. The signal is acquired twice: first, when water and fat protons are in phase; and second, when they are out of phase (while excited protons are returning to their first position). By choosing the suitable time of echo, fat suppression can be accomplished by subtracting corresponding frequencies of fat and water signals. Thus, this sequence is very useful to understand the fat content. Being T1-weighted, T1-DUAL is very effective to identify blood and tissues that are rich in protein and helps in determining the lubrication levels. In this study, 14 data sets, which have

ISDs that vary between 5.5 mm and 9 mm (average 7.84 mm), x/y spacing is between 1.44 - 1.89 mm (average 1.61 mm) and the number of slices between 26 and 50 (average 36), are used. THRIVE is a fast imaging sequence that is obtained by scanning the liver consecutively in a short time with thin slices to obtain the response of a known lesion to a contrast agent. THRIVE provides observation of the signal increase originating from the contrast agent instead of image quality. In total, 7 THRIVE data sets with 2.5 mm of ISD are used. Due to this small ISD, these data sets have slices between 80 and 110 (average 93). Although their ISDs are constant, x-y spacing changes between 1.56 and 1.97 mm (average 1.67 mm). In this study, 5 THRIVE data sets are used. Finally, WATS (WATer Selective) is a high resolution T1-weighted fat suppression sequence. 5 WATS data sets used in this study have x-y spacing between 5.5 - 9 mm (average of 7.9 mm). Their x-y spacing changes between 1.36 - 1.67 mm (average 1.43 mm) and the number of slices vary between 26 and 40 (average 37).

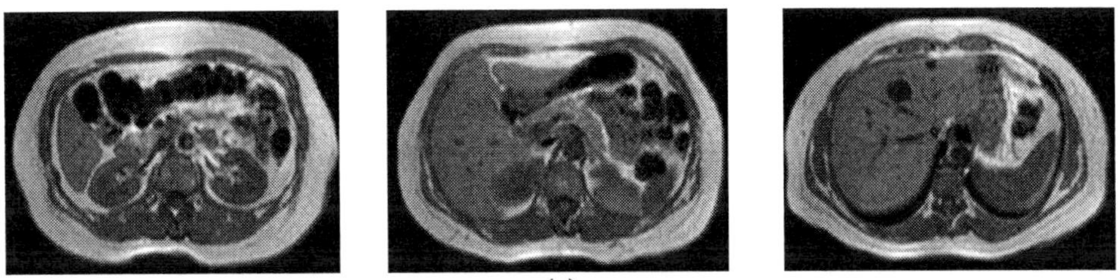

(a)

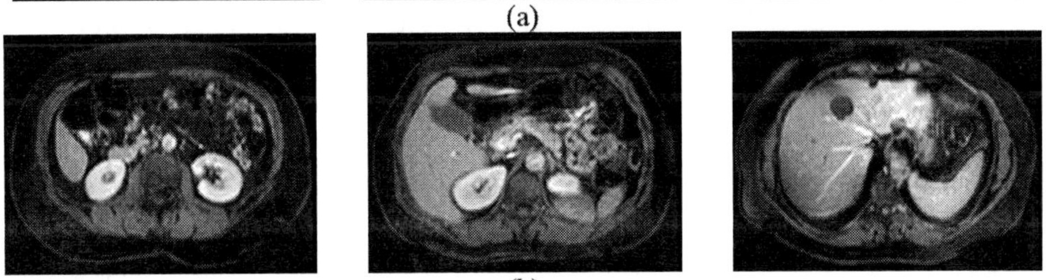

(b)

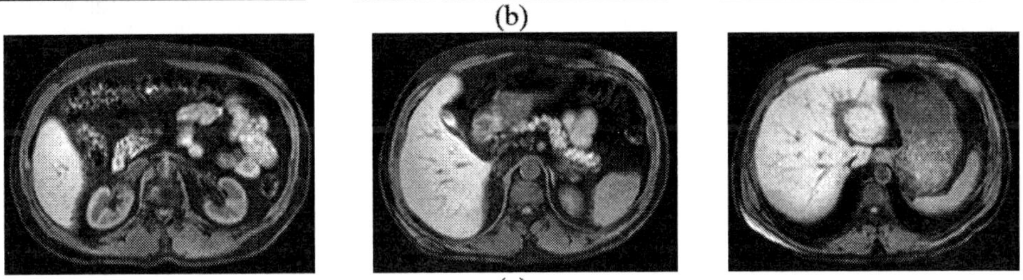

(c)

Figure 1. Sample abdominal images from sequences: a) T1-DUAL, b) THRIVE, c) WATS.

3. User Aligned Volume Histogram Stacks

Construction of a User Aligned Histogram Stack (UAHS) is a simple procedure for a physician who uses MPR interfaces and tools during his/her daily workflow. By using the developed MPR interface, the physician determines a majority axis or region for the organ of interest. The majority axis can be a spline or a (piecewise) line and majority region can be an ellipse or a rectangle. The selection of the region or axis should be consistent with the shape and size of the organ of interest. As an example, the geometry of the kidney is elliptical in axial slices. Therefore, applying an elliptical region would be more effective than the others (Fig 2.a-c). As a second example, selecting a rectangular region to reconstruct the liver is not sensible, because its complex size would require a very large area. Thus, a spline based majority axis is better suited to represent the liver (red line in Fig 2.h).

After this selection is done on a single MPR plane (axial, sagittal, or coronal), one of the remaining MPR planes is used to determine the slices, where the organ of interest is visible. This is a very simple interaction that only needs the user to draw a line over the organ of interest (see Fig 2.b for the kidney and Fig 2.a or 2.b for the liver).

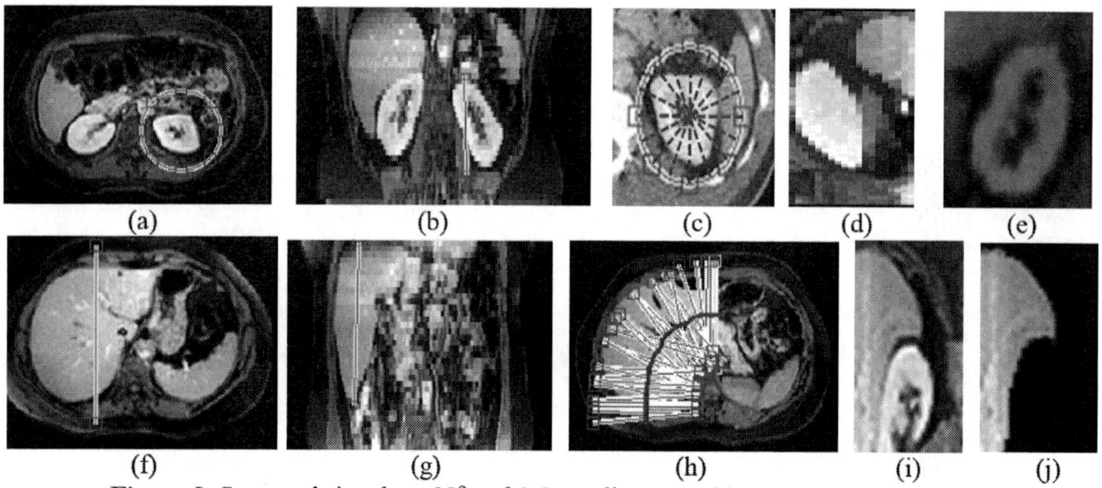

Figure 2. Scattered signals at 90° and 1.8 cm distance with 10, 0, and -10 dB SNR.

These two interactions are the only ones required for the construction of UAHS. Next, the developed software generates new reconstructed images using these user inserted planes and by dividing the selected areas into new sampling planes (See Fig 2.d and 2.e for the kidney and Fig 2.i and 2.j for the liver). By aligning the histograms of each reconstructed image (Fig. 3), UAHS, which provides a new data representing the object of interest in a more focused way, is obtained (Fig. 4).

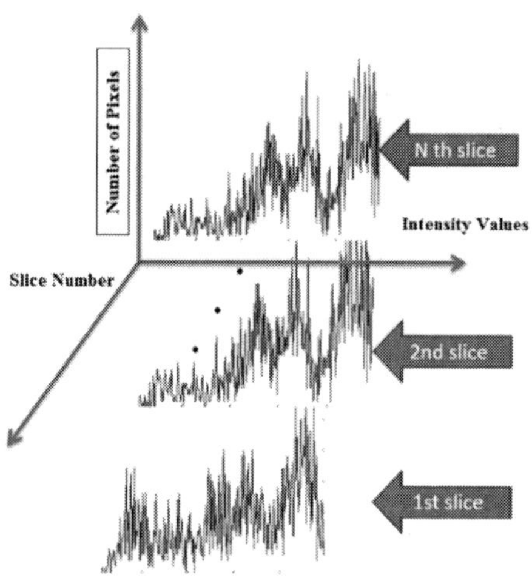

Figure 3. Aligning the histograms of the reconstructed image slices.

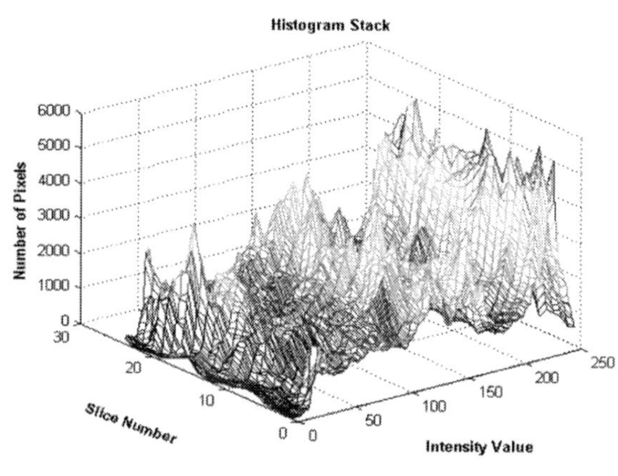

Figure 4. An example of UAHS.

4. Approximation of UAHS via Gaussian Functions

The gray values for organs of interest have varying intensity ranges and characteristics even for the same organ in different MRI acquisitions. However, by evaluating UAHS with respect to different MR sequences, some common characteristic properties are observed in UAHS. One of them is that UAHS becomes narrower at the beginning or end slices and widens in the middle slices. It should be noted that the use of DICOM image series, which supply large numbers of gray levels, is necessary to obtain a compact desired histogram. The other important point is that the desired histogram should be distinguishable from the unwanted components/tissues. This distinction in UAHS generates two valleys in both sides of the UAHS. Figure 5 illustrates three examples of UAHS data, which are obtained using the developed GUI and for liver visualization, along with their desired counterparts.

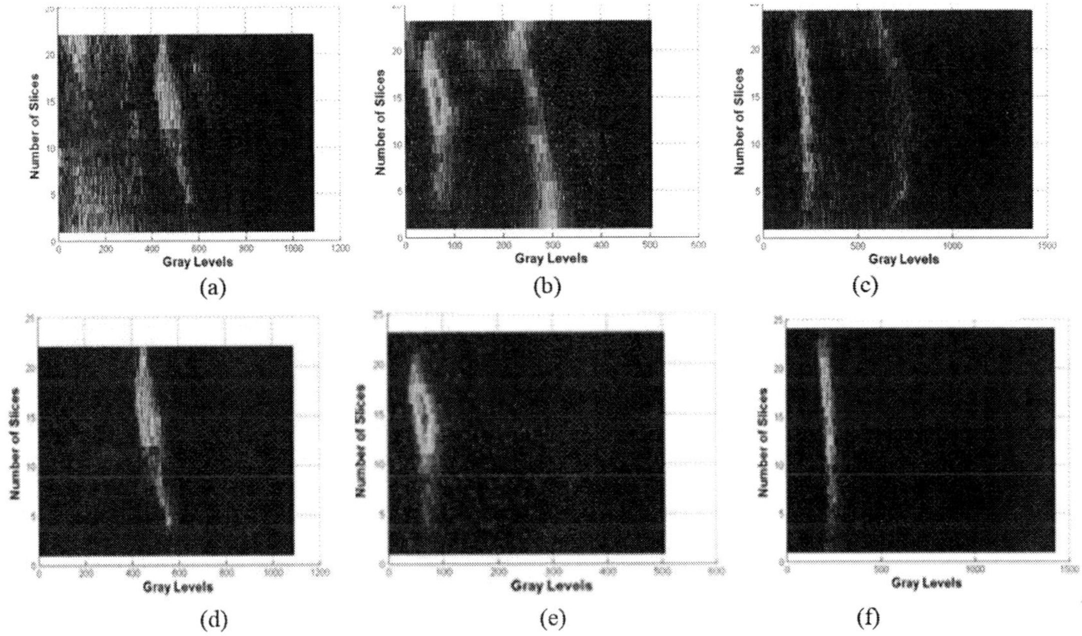

Figure 5. (a)-(c) Top view of resultant UAHS of the GUI constructed for liver visualization from different MRI sequences, (d)-(f) Top view of their respective desired counterparts.

The UAHS should be modeled in order to determine the valleys in a robust manner. Thus, the selection of the lobes from the UAHS requires an approximation to obtain the desired lobes. For modeling a UAHS, some important issues must be taken into account. One of them is the non-smooth structure of UAHS because of the intensity fluctuations of MRI. The second problem is the variability in the gray level range of the same organ among slices. The third problem is the slice thickness of the MRI data that causes a limited sampling in z-direction and results in a high inter-slice distance.

The shape of the UAHS is like a bell curve with a certain radius and intensity. In the center of the UAHS, there exist a lot of pixels that create high amplitude bells, each of which is called a "lobe". Gaussian function is a good approximation for fitting the lobes of UAHS. Besides, defining the UAHS in terms of Gaussian functions provides an intuitive initial TF design employing the parameters of Gaussian function. Various possible Gaussian basis functions can be created as a filter bank employing multiple orientation and scale values.

In this study, by considering the above mentioned criteria, a hierarchical and multi-scale modeling is proposed. In an iterative manner, the proposed model tries to find the best Gaussian basis to fit the corresponding lobe of the UAHS. The developed algorithm provides a procedure for capturing all the suppressed lobes of importance in a successive manner by associating the lobes with a suitable number of Gaussian basis functions.

Using different scales provides the user with the convenience of expressing different lobes in the right way. However, the characteristics of UAHS demonstrate that lobes can have curvy structures. The best approximation for these curvy parts is to integrate the rotation parameter to the algorithm as an extra degree of freedom. For a fixed scale parameter, the best rotation parameter is determined through the calculation of correlation between the Gaussian basis functions and UAHS. Correlation can be calculated in an easier way in the frequency domain by multiplying the Fourier transform of the UAHS by the conjugate of the Fourier transform of the Gaussian basis function.

A rotated and axis scaled Gaussian basis function can be expressed with respect to given a, θ, and ε parameters as follows;

$$g_{a,\theta}(x',y') = e^{-1/2(\varepsilon^{-1}(a^2(x'\cos\theta - y'\sin\theta)^2) + a^2(x'\sin\theta - y'\cos\theta)^2)} \quad (1)$$

where, variable a is the axis scaling parameter of the transformation. The rotation variable θ controls the orientation and has a critical importance on finding the appropriate basis function that fits the corresponding lobe of the UAHS. Figure 6 shows the top and side views of the Gaussian basis functions which have the same a and ε parameters, but different θ parameter values.

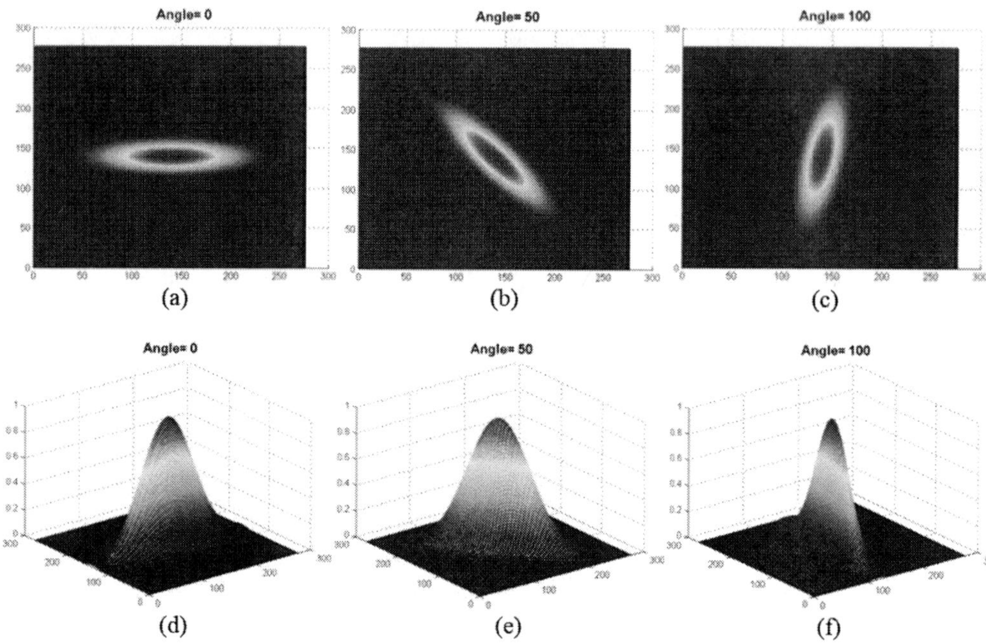

Figure 6 2-D Gaussian functions with respect to different angles (a)-(c) Top view (d)-(f) Side view.

Let us denote our final approximated UAHS as $UAHS(x,y)$ which is formed by combining all approximating Gaussian basis functions in an additive manner. However, before combining the suitable Gaussian basis functions, the best rotation angle among all possible choices of angles must be determined in an iterative manner. Here, each iteration corresponds to a different scale. Iterations start from the largest scale value and continue until the smallest scale value in a hierarchical manner. In our simulations, we used 8 different scale values. Thus, denoting the number of iterations by N, we have $N = 8$. For each scale, θ is varied from 0 up to a maximum of 170 degrees with an increment of 10 degrees. Thus, $\theta = \{0,10,\ldots,170\}$ defines the vector of rotations. Denoting the length of the vector, θ, by M, we have M=18. $Corr\{.\}$ is defined as the 2-D correlation operation between Gaussian basis function, $g_{a_i,\theta_j}(x',y')$, and the UAHS which is represented by *UAHS(x,y)*. For a fixed scale, the correlation is calculated for different rotations as:

$$\mathrm{Corr}_j = Corr\left\{g_{a_i,\theta_j}(x',y'), UAHS(x,y)\right\} \quad (2)$$

In Equation 3, we represent the best rotation angle as $\hat{\theta}$ which corresponds to the maximum correlation between UAHS and Gaussian basis functions.

$$\hat{\theta}=\arg\max_{\theta_j}\left(Corr\left\{g_{a_i,\theta_j}(x',y'),UAHS(x,y)\right\}\right)_{j=1}^{M} \tag{3}$$

We can express our approximated $UAHS(x,y)$ as a weighted summation of the rotated and axis scaled Gaussian basis functions, $g_{a_i,\theta_j}(x',y')$.

$$UAHS(x,y)=\sum_{i=1}^{N}\sum_{j=1}^{M}w_j g_{a_i,\theta_j}(x',y') \quad (4$$

where $w_j = \begin{cases} 1\ ; & \text{max correlation for scale } a_i \text{ among all possible } \theta_j \text{ angles} \\ 0\ ; & \text{for the rest of the } \theta_j \text{ angles} \end{cases}$

After approximating the UAHS, a method should be used to threshold the approximated UAHS by determining the two valleys so that upper and lower thresholds can be assigned correctly. Our main goal is to determine the two threshold values which correspond to the valleys on the left and right sides of the corresponding lobe of the organ of interest. As stated before, the main idea behind the Gaussian modeling is to find the basis functions that best fit the desired lobes and then to determine a threshold range to express the region of interest in the UAHS. Thus, after the Gaussian fitting step, as seen in Figure 7.a, the gray levels which belong to the organ of interest are grouped together. The center of the group is indicated by the mean of the gray levels, which are obtained via points through which the spline passes. The points, at which the tail of each Gaussian becomes zero, indicate the threshold range. As seen in Figure 7.b, a varying threshold range can be obtained using the Gaussian bases which are found by the multi-scale hierarchical model.

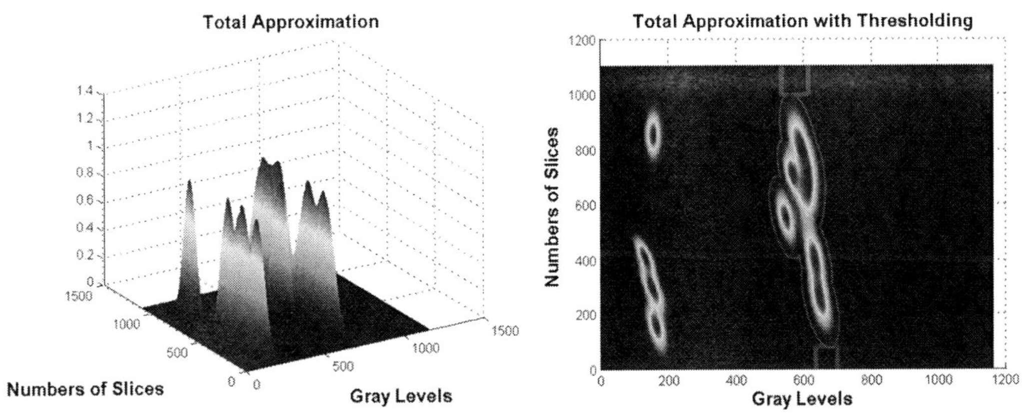

(a) (b)

Figure 7. (a) The resultant approximating Gaussian functions (b) Top view of the resultant approximation of Gaussian functions with a varying threshold range represented by red lines.

5. Application and Results

The developed method is evaluated using well established segmentation metrics including the area error rate (AER), which is defined as the area difference between the region segmented by the algorithm and the region segmented manually [21]. Volumetric overlap error (VOE) is similar to AER criterion where, 0 indicates perfect segmentation with no limitation for any upper bound [22]. Besides the difference in volume values, conformational characteristics of the obtained segmentation results are also important. For this purpose, Symmetric Surface Distance (SSD) metrics are also used for performance evaluations [22]. SSD allows the comparison of segmented 3-D organ with the reference 3-D organ by using surface voxels of a 3-D object. The surface voxels are defined according to at least one voxel that does not belong to the object in 18 possible neighboring. For each surface voxel in the reference organ, the distance to the surface voxel in the segmented organ is calculated according to the

Euclidean Distance Metric via different formulations that produce Average SSD (ASSD), Maximum SSD (MSSD), and RMS SSD (RSSD).

Some visual examples of the results are given in Figure 8 and calculated metrics for all the data sets are given in Table 1. Red pixels show True Positive pixels after the thresholding process. This means that the red pixels are classified as the organ of interest and these pixels are actual organ pixels. Therefore, it can be stated that by using Gaussian approximations and the thresholding process, the segmentation of the organ is accomplished substantially. Green pixels of the images represent the organ pixels which are not found by our algorithm. Finally, blue pixels are the pixels which belong to the unwanted pixels/neighboring tissues but are found as belonging to the organ of interest by our algorithm. Observing Figure 8, it can be stated that using the coarse approximation at the first and last slices is not enough for segmentation. In contrast to the first and last slices, in the middle of the image series our algorithm performs better.

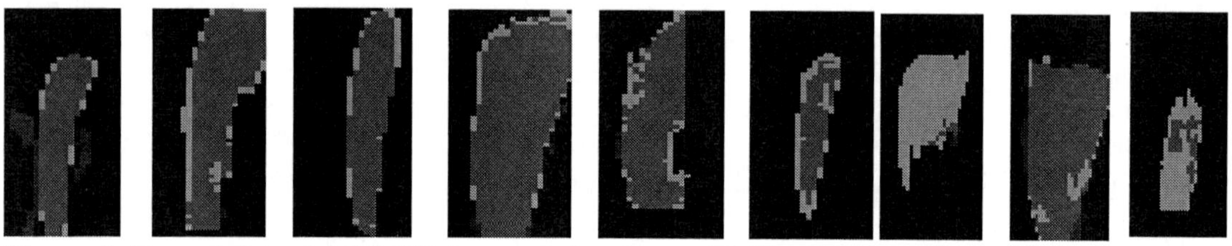

Figure 8 (From left to right) MPR images belonging to T1 WATS with slice numbers 3-11-21. T1-DUAL with slice numbers 12-18-24 and THRIVE with slice numbers 2-10-22.

The results show that the proposed approach can provide acceptable initial classification results by approximating the lobes of UAHS using proposed GMM strategy (i.e. see the first three rows of Table 1). On the other hand, these results must be enhanced for obtaining clinically usable visualizations. In this study, the finer classification steps include binary morphological operations and region growing using initial coarse classification results. The last three rows of Table 1 show that post-processing operations can significantly improve the results, since the initial model obtained by GMM is satisfactory. Future work could include using statistical shape models which would be initialized by the proposed GMM strategy.

Table 1. Comparison of the average metric results of all sequences before and after region growing.

	Dataset	*ASSD*	*RSSD*	*MSSD*	*AHO*	*VHO*
Without Post-Processing	**T1 DUAL**	3.69	6.72	34.30	53.89	44.69
	T1 WATS	6.95	11.37	46.47	54.04	52.46
	THRIVE	6.13	10.22	42.60	42.82	44.06
With Post-Processing (Region Growing)	**T1 DUAL**	3.88	6.84	35.21	59.93	42.03
	T1 WATS	1.73	2.61	18.75	33.61	28.70
	THRIVE	2.97	3.97	17.69	29.62	28.51

Acknowledgment
This study is supported by The Scientific and Technological Research Council of Turkey (TÜBİTAK) under grant number EEEAG 112E032.

References

[1] Fischer, F., Selver, M.A., Hillen, W., Güzeliş, C, Integrating Segmentation Methods from Different Tools into a Visualization Program Using an Object Based Plug-In Interface, IEEE Transactions on Information Technology in Biomedicine, Vol. 14(4), 923-934, 2010.

[2] R. A. Drebin, L. Carpenter, and P. Hanrahan. Volume rendering, In Proceedings of ACM Siggraph 1988, pages 65–74, 1988.
[3] H. Pfister, B. Lorensen, C. Bajaj, G. Kindlmann, W. Schroeder, and R. Machiraju, The Transfer Function Bake-Off, In Proceedings of IEEE Visualization 2000, 523-526.
[4] F. Shiaofen, B. Tom, and T. Mihran, Image-Based Transfer Function Design for Data Exploration in Volume Visualization, In Proceedings of IEEE Visualization 1998, 319-326.
[5] A. Konig, and E. Gröller, Mastering Transfer Function Specification by Using VolumePro Technology, In Proceedings of Computer Graphics'01, vol. 17, 279-286, 2001.
[6] J. Marks, B. Andalman, P.A. Beardsley, and H. Pfister, Design Galleries: A General Approach to Setting Parameters for Computer Graphics and Animation. SIGGRAPH, (1997) 389-400.
[7] F. Shiaofen, B. Tom, and T. Mihran, Image-Based Transfer Function Design for Data Exploration in Volume Visualization, In Proceedings of IEEE Visualization 1998, 319-326.
[8] Liang Zhou, Hansen, C., Transfer function design based on user selected samples for intuitive multivariate volume exploration, Visualization Symposium IEEE Pacific, pp. 73 – 80, 2013.
[9] C. L. Bajaj, V. Pascucci, D. R. Schikore, The Contour Spectrum, IEEE Visualization, 1997, 167-173.
[10] G. Kindlmann, and J. W. Durkin, Semi-Automatic Generation of Transfer Functions for Direct Volume Rendering, In Proceedings of IEEE Visualization 1998, 79-86.
[11] I. Fujishiro, T. Azuma, and Y. Takeshima, Automating Transfer Function Design for Comprehensible Volume Rendering Based on 3D Field Topology Analysis, In Proc. of IEEE Visualization'98, 467-470.
[12] T. He, L. Hong, A. Kaufman and H. Pfister, Generation of Transfer Functions with Stochastic Search Techniques, In Proceedings of IEEE Visualization 1996, 227-234.
[13] Jesus J. Caban, Penny Rheingans, "Texture-based Transfer Functions for Direct Volume Rendering," IEEE Transactions on Visualization and Computer Graphics, vol. 14(6), 1364-1371, Nov./Dec. 2008.
[14] Y. Sato, C.F. Westin, A. Bhalerao, S. Nakajima, N. Shiraga, S. Tamura, and R. Kikinis, "Tissue Classification Based on 3D Local Intensity Structures for Volume Rendering", IEEE Trans. On Visualization and Computer Graphics, Vol.6, No.2, pp. 160-180, 2000.
[15] C. Rezk Salama, M. Keller, P. Kohlmann, High-Level User Interfaces for Transfer Function Design with Semantics, IEEE Transactions on Visualization and Computer Graphics, Vol. 12(5), 1021 – 1028, 2006.
[16] J. Kniss, G. Kindlmann, C. Hansen, Interactive Volume Rendering Using Multi-Dimensional Transfer Functions and Direct Manipulation Widget, Visualization 2001, 255-262.
[17] Eric B. Lum, James Shearer, Kwan-Liu Ma: Interactive multi-scale exploration for volume classification, The Visual Computer, vol. 22(9-11): 622-630 (2006)
[18] J. Kniss, G. Kindlmann, and C. Hansen, Multi-Dimensional Transfer Functions for Interactive Volume Rendering, IEEE Transactions on Visualization and Computer Graphics, vol. 8(3), 270-285, (2002).
[19] Fan-Yin Tzeng, Eric B. Lum, and Kwan-Liu Ma: An Intelligent System Approach to Higher-Dimensional Classification of Volume Data. IEEE Trans. On Vis. and Computer Graphics 11(3): 273-284 (2005).
[20] P. Rautek, S. Bruckner, and M. E. Groller, Semantic Layers for Illustrative Volume Rendering, IEEE Transactions on Visualization and Computer Graphics, Volume 13, Issue 6, Page(s):1336 – 1343, 2007.
[21] Seo, K. S., Ludeman, L. C., Park, S. J., & Park, J. A. (2004) Efficient liver segmentation based on the spine. Lecture Notes in Computer Science, 3261, 400-409.
[22] Ginneken, B. V., Heiman, T., & Styner, M. (2007). 3-D segmentation in the clinic: A grand challenge. MICCAI Workshop Proceedings, 7-15.
[23] C. Westbrook, Handbook of MRI technique. (3rd edt.). Oxford, United Kingdom: Wiley-Blackwell, 2008

Fuzzy Logic Module of Convolutional Neural Network for Handwritten Digits Recognition

E.A. Popko and I.A. Weinstein

NANOTECH Centre, Ural Federal University, Mira Street, 19, Yekaterinburg, Russia

e.a.popko@urfu.ru

Abstract. Optical character recognition is one of the important issues in the field of pattern recognition. This paper presents a method for recognizing handwritten digits based on the modeling of convolutional neural network. The integrated fuzzy logic module based on a structural approach was developed. Used system architecture adjusted the output of the neural network to improve quality of symbol identification. It was shown that proposed algorithm was flexible and high recognition rate of 99.23% was achieved.

1. Introduction

Fuzzy logic is actively used in many technological problems related to the pattern recognition [1]. In particular, great attention is paid to accurate identification of handwritten characters in classification and analysis of documents, automatic serial number recording, and vehicle identification[1, 2]. One of the promising trends is the employment of artificial convolutional neural networks to imitate biological nervous systems with learning ability.

Convolutional neural networks are the extension of a multi-layer perceptron, optimized for two-dimensional pattern recognition (a character identification task), face detection, and so on[3]. The existence of neural layers of different types and functionality is a distinguishing attribute of CNNs. CNNs combine three architectural ideas: local receptive fields to extract elementary features from images; shared weights to extract the same set of elementary features from the entire input image and to lower computational costs; local averaging and sub-sampling to reduce the resolution of feature maps. The goal of this work is to recognize handwritten digits and improve the classification results by way of a newly devised version of convolutional neural network through integrating a logical correction module into their structure.

2. Model

Traditional character recognition methods can be divided into three main types - template, structural and feature ones [4, 5], each of which has its own advantages and restrictions (Table 1). At present, it can be boldly argued that the most rapidly developing algorithms are those of classification, including the strategy of learning by examples. Convolutional neural networks also possess the aforementioned peculiarity. As distinct from a classical multilayer perceptron, CNNs provide the invariance of image analysis to shift, rotation, scale, and distortions.

Content from this work may be used under the terms of the Creative Commons Attribution 3.0 licence. Any further distribution of this work must maintain attribution to the author(s) and the title of the work, journal citation and DOI.
Published under licence by IOP Publishing Ltd

Table 1. The main classes of the recognition methods

Methods	Advantages	Disadvantages
Template	High speed, ability to work under existence of character defects, provides noise immunity	Weak versatility, fails to identify the font, which differs from the input
Structural	It doesn't matter what size and font are	Complexity of defect-character recognition, works relatively slow
Feature	Good generalizing ability, resistance to changes in the shape of characters	Instability to various defects, loss of information part

In this work, as a prototype, we have used the network architecture described in [6]. The main difference of the implementation is to use six feature maps in a second layer, instead of five. The general functional scheme of CNN is shown in Figure 1. The input layer receives images of recognizable characters. This is followed by a layer of six convolutional feature maps of size 13 × 13. Each map element is connected to the receptive field of 5 × 5 on an input image. Further, there is an additional convolutional layer consisting of 50 maps of size 5 × 5. The next layer is fully-connected and contains 100 neurons. The fifth layer contains neurons corresponding to the digits from 0 to 9.

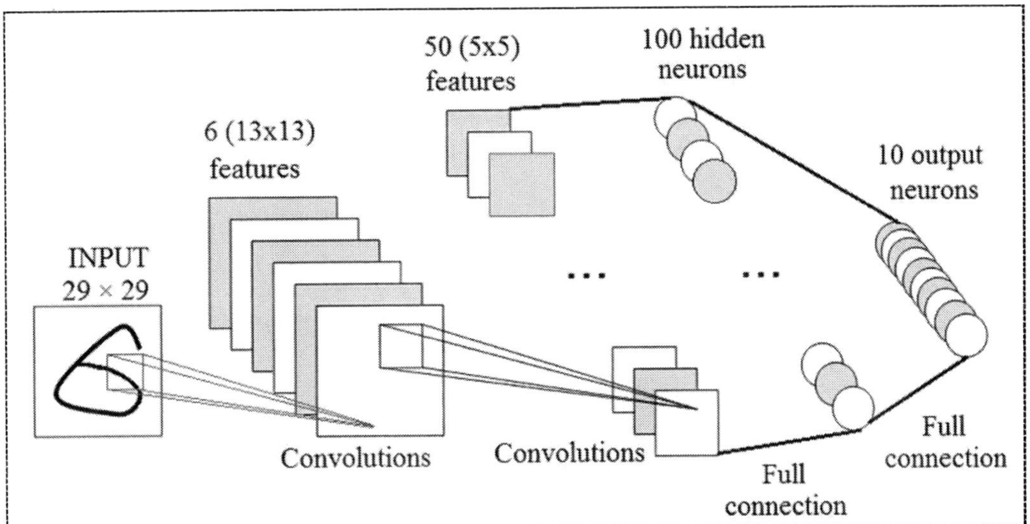

Figure 1. The neural network architecture being used

The connection weights of the convolutional neurons located in one feature map are shared, and the position of the local feature becomes less important. Such a way allows the shift invariance to be achieved. The output neural values Y_n^l of the n-th feature map of the convolutional layer l can be calculated as follows [3]:

$$Y_n^l(x,y) = f\left(\sum_{m \in M_n^l} \sum_{i=0}^{K^l-1} \sum_{j=0}^{K^l-1} \omega_{mn}^l(i,j) \cdot Y_m^{l-1}(x+i, y+i) + b_n^l \right), \tag{1}$$

where x, y are coordinates of a neuron inside the feature map; M_n^l is the set of feature maps of the previous layer l–1 that are associated with the n-th feature map of the layer l; ω_{mn}^l is the matrix of synaptic coefficients (the convolution kernel); K^l is the size of the receptive field of the neurons of the l-th layer; b_n^l being the bias term for the n-th feature map of the l-th layer.

As an activation function $f(z)$ of the neural network, the scaled hyperbolic tangent is used here:

$$f(z) = A \cdot \tanh(Sz), \qquad (2)$$

where z is an input value corresponding to the neuron activation value; A and S are parameters of the function. The work [3] proposes to employ the values of A = 1.7159 and S = 2/3.

This paper is based on the standard back-propagation algorithm to tune the neural network. As data for training and testing, we have utilized a set of images of the MNIST handwritten digit database [7].

3. Algorithm

As shown by preliminary tests, the major difficulty of CNNs is to identify gaps in specific areas of characters and individual structural elements, for the classification. The network errors mentioned above appear due to the omission of these features (Figure 2). For example, in the case when the character "2" was incorrectly identified as "7", the neural network was not able to determine clearly whether the bottom element as a line corresponds to two or seven (see the sample 321 in Figure 2).

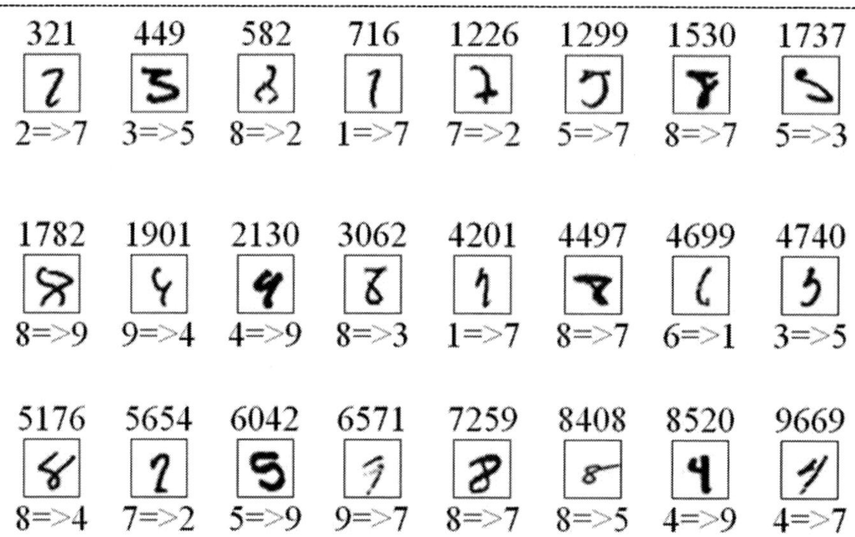

Figure 2. Examples of images of characters recognized incorrectly. The top of handwritten digit boxes indicate a serial number in the MNIST test database; the bottom shows the "desired value" => "the recognition result".

We have introduced a fuzzy logic module into the system. Its main intention is to ascertain the existence of certain structural elements of recognizable characters and to yield an appropriate correction penalty for the neural network outputs. The logical module includes a rule database for producing correction values. Each fuzzy rule can be represented as a vector model: P = <*Name*, *Alg*, *Rule*, *Outs*, *Val*>, where *Name* is the name of the rule; *Alg* is the algorithm for ascertaining the existence of the element; *Rule* is the definition of fuzzy rule; *Outs* is the set of the neural network outputs being corrected; *Val* is the maximum correction value.

For example, for more reliable identification of the characters "2" and "7" , the rule is formulated as follows: P_1 = <"correction for the character 7", "algorithm that determines the correspondence of the lower element (the line) to the digit "7" or "2", "if (element is wide) then apply penalty", {7}, -0.5>. In this case, the penalty -0.5 is assigned to the network output neuron, which is responsible for the digit "7" provided that the bottom of the symbol has a horizontal underline of sufficient size, corresponding to the digit "2". The corresponding membership function μ is shown in figure 3.

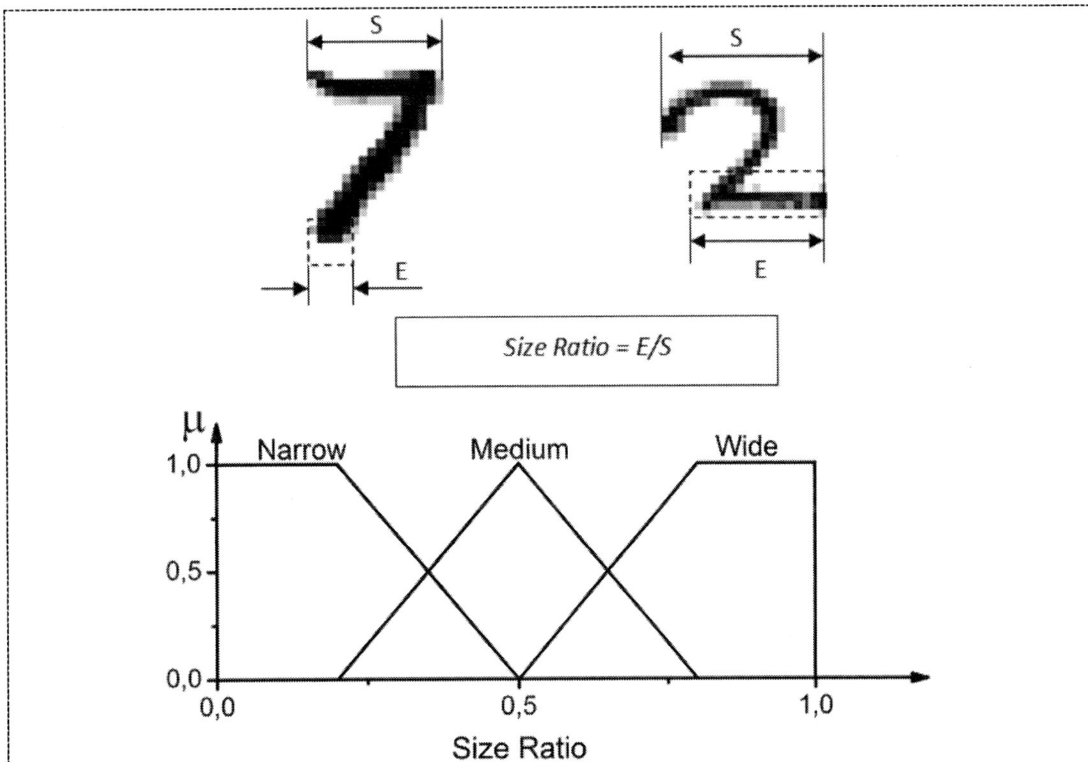

Figure 3. Example of fuzzy membership function for the characters "7" and "2"

4. Results

For testing, we have devised a software module in C++ with an open source cross-platform computer vision library (OpenCV) to implement the recognition method proposed [8]. Figure 4 displays the overall scheme of verifying the system's performance capability.

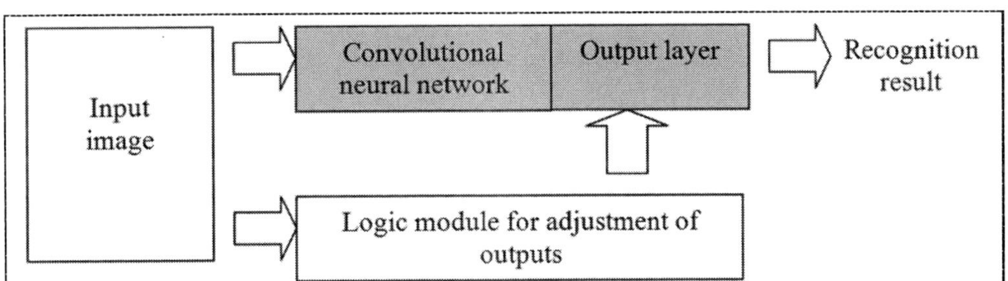

Figure 4. The system testing scheme

The verification of the system was first performed on a training set of 60000 characters, and then run on a test one of 10000 examples of the MNIST base. The training of the network was held for 50 epochs, the total tuning time amounted to 8 hours. The program was governed using both the logic correction module and without it. The findings of both program implementations are reported in Table 2.

Table 2. Recognition results

Implementation	Recognition accuracy for the training set	Recognition accuracy for the test set
Original CNN	97.35%	99.17%
Using the correction module	99.10%	99.23%

During the testing, it was found that the system equipped with the correction module allows the recognition quality to be enhanced. Over against the original CNN, the system succeeded in recognizing correctly digits from the test set with the numbers 321, 1299, 1530, 5654, 7259, 9669 (see. Figure 2). The achievable accuracy of the developed hybrid system is 99.23%, which is quite a good result [9, 10, 11]. Given the confident recognition, it can be claimed that one of the advantages of the proposed implementation is quite simple architecture of CNN as compared to analogs [9, 10]. The tuning of the configuration and internal network parameters for each particular case is a separate optimization problem. The traditional approach, as a rule, involves a series of preliminary experiments and the choice of structure. One of the most promising approaches uses evolutionary algorithms [12, 13, 14, 15].

5. Conclusion

Thus, in this study we have developed a modification of convolutional neural network using a fuzzy logic correction module. In this case, the embedded unit requires no implementation of a full recognition procedure, but merely modifies the network outputs by computer vision methods. It is shown that this simplified system architecture allows improving the quality of digit recognition and increases the accuracy to 99.23%.

The application of the convolution neural networks is quite a powerful tool for handwriting recognition owing to the possibility of their integrating into hybrid computer vision systems. Further investigations may be aimed at creating a learning technique of CNNs, advancing their optimal structure and fine-tuning of appropriate weights.

Acknowledgments
This work was supported by contract № 02.A03.21.0006, Act 211 Government of the Russian Federation.

References

[1] Ross T J 2010 *Fuzzy logic with engineering applications* (West Sussex, UK: John Wiley & Sons)
[2] Latoria A, Chauhan A S, Saxena A S 2012 *International J. of Computers&Technology* **3** 458
[3] LeCun Y 1999 *Proc. of IEEE* vol 86 2278
[4] Davies E R 2012 *Computer and Machine Vision: Theory, Algorithms, Practicalities* (Oxford, UK: Elsevier)
[5] Bishop C M 2006 *Pattern Recognition and Machine Learning* (New York: Springer)
[6] Simard P Y, Steinkraus D, Platt J C 2003 *Proc. 7th Int. Conf. on Document Analysis and Recognition* p 958
[7] The MNIST Database. Available from: http://yann.lecun.com/exdb/mnist/

[8] Popko E A, Weinstein I A 2016 *Submitted certificate on official registration of the computer program* (Moscow, Russia)
[9] Yang Z, Moczulski M, Denil M, et. al. 2015 *Proc. of the IEEE Int. Conf. on Computer Vision* p1476
[10] Lee H, Grosse R, Ranganath R, Ng A Y 2009 *ICML Proc. of the 26th Annual Int. Conf. on Machine Learning* p 609
[11] Thom M, Palm G 2013 *J. of Machine Learning Research* **14** 1091
[12] Oulette R, Browne M, Hirasawa K 2004 *Proc. of the 2004 IEEE Congress on Evolutionary Computation* **1** p 516
[13] Stanley K O, Miikkulainen R 2002 *Evolutionary Computation* **10** 99
[14] Weinstein I A, Popko E A 2008 *Radiat. Meas.* **43** 218
[15] Weinstein I A, Popko E A 2006 *Tech. Phys. Letters* **32** 534

Journal of Physics: Conference Series **738** (2016) 012124 doi:10.1088/1742-6596/738/1/012124

Liquid filtration properties in gravel foundation of railroad tracks

A Strelkov, S Teplykh, N Bukhman

Samara State University of Architecture and Civil Engineering, 194 Molodogvardeyskaya St, Samara, 443001, Russia

E-mail: kafvv@mail.ru

Abstract. Railway bed gravel foundation has a constant permanent impact on urban ecology and ground surface. It is only natural that larger objects, such as railway stations, make broader impact. Surface run-off waters polluted by harmful substances existing in railroad track body (ballast section) flow along railroad tracks and within macadam, go down into subterranean ground flow and then enter neighbouring rivers and water basins. This paper presents analytic calculations and characteristics of surface run-off liquid filtration which flows through gravel multiple layers (railroad track ballast section). The authors analyse liquids with various density and viscosity flowing in multi-layer porous medium. The paper also describes liquid stationary and non-stationary weepage into gravel foundation of railroad tracks.

1. Introduction

The Primary equations for liquid filtration in porous medium are considered [1, 2] to be the equation of continuity $\partial(\rho m)/\partial t + div(\rho\vec{u}) = 0$ and Darcy's law $-grad(p) + \rho\vec{g} - \mu\vec{u}/k = 0$ in which ρ is liquid density; m is medium porosity (that is relative proportion of open cells in a certain material); $\vec{u}$ is liquid filtration speed ($\vec{u} = m\vec{v}$ here $\vec{v}$ is liquid motion average speed in porous medium); t is time; p is liquid pressure; $\vec{g}$ is free-fall acceleration; μ is absolute viscosity of liquid; k is porous medium penetration (or penetration coefficient).

The optimal solution for this set of equations is $p = 0$, $\vec{u} = C(\vec{g}/g)$, where $C = k\rho g/\mu$ is liquid filtration coefficient in a certain medium. It is clear that this solution shows liquid free-flow in porous medium under the action of gravity. This solution makes clear the two physical meanings of the filtration coefficient. Firstly, C is speed of this liquid free-flow gravitational filtration in a certain porous medium. Secondly, C is the liquid maximum volume in a definite unit of time (discharge) on a definite area unit which is capable of free-flow filtration in this medium. Porosity and water filtration coefficient of different media typical value (measured by the researchers) are given in Table 1.

Table 1. Porosity and filtration coefficient of different media typical value

Material	m, %	C, m/s	Material	m, %	C, m/s
Broken granit 40x70 mm	46.0	0.01	River sand 1 mm	15.0	0.51×10^{-3}
Broken granit 20x40 mm	45.2	0.004	Fluvial soil	60.0	0.40×10^{-5}
Broken granit 5x20 mm	44.8	0.18×10^{-2}	Sand clay	75.0	0.22×10^{-5}
Quartz sand 2-3 mm	30.0	0.10×10^{-2}	Clay	50.0	0.23×10^{-7}

Content from this work may be used under the terms of the Creative Commons Attribution 3.0 licence. Any further distribution of this work must maintain attribution to the author(s) and the title of the work, journal citation and DOI.
Published under licence by IOP Publishing Ltd

2. Atmospheric precipitation filtration

Let us assume that this area precipitation rate is q_{20} [5]. It is evident that liquid accumulation (puddles) on porous medium surface can be non-existent only under the condition $q_{20} < C$. Let us compare q_{20} = 70 l/(s·ha) = $0.7 \cdot 10^{-5}$ m/s (typical for Samara region) with the data presented above in Table 1. We realize that no matter what real precipitation rate is there is no liquid accumulation on either sand or gravel surface. Puddling is possible only if the surface is fluvial soil, sand clay or clay. On the other hand, railroad track ballast section is multi-layer porous medium in itself. Its upper layer is gravel, the next under-layer is sand, then there is geofabric which is deposited above main soils [1] (usually these soils are fluvial soil, sand clay or just clay). It leads us to the assumption that rain water flows immediately through ballast section gravel and sand layers and comes down to the main soils.

The most likely scenario then can unfold in either if the following ways:

- if the main soils are water-permeable (sandy) rain water drains down to the nearest waterproof layer;
- if the main soils are not water-permeable enough (either clay or fluvial soil or sand clay) rain water starts accumulating on the border between sub-ballast and the main soils. This underground liquid accumulations slowly leak water down into the soils.

If we put water-permeable geofabric on the border between sub-ballast and the main soils it will change nothing. But if we put waterproof geofabric on the border between sub-ballast and the main soils it would prevent water from leaking further down. Thus, we make a kind of water-bearing layer. It makes liquid flow horizontally up to the geofabric edge. Then this liquid flows out of ballast section in lateral direction generating road-side puddles which later sink into soils.

3. Quasistationary anthropogenic (man-made) water discharge (leakages) filtration

Let us consider a situation when there is some stationary liquid weepage (that is usually very polluted water) in a certain area. It punctually drenches ballast section surface, its volume flow rate being Q ($m^3/(m^2 s)$=m/s). In this case there is a filtration channel developing gradually under the leakage point, its radius being $R = (Q/\pi C)^{1/2}$ The filtration channel radius depends on porous medium character. The channel in gravel is narrower than it is in sand, and the channel in sand is narrower than it is in soils: $R_{gr} < R_{sand} < R_{soils}$.

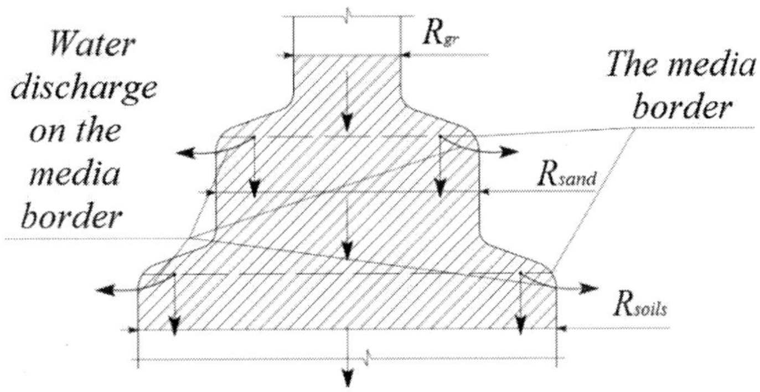

Fig. 1. Liquid filtration channel widening and water discharge on the media border

For this reason a filtration channel widens both on the border of gravel and sand and on the border of sub-ballast and soils (as seen on Fig. 1). When some part of the filtration channel goes beyond railway embankment it leads to liquid discharge in lateral direction (either on the border of gravel and sand or on the border of sand and soils).

Let us consider R_{max} to be a filtration channel the maximum permissible radius (that is a distance between the center of one 11.7 m railroad track, the railroad here being double track to the edge of the railway embankment, R_{max}=11.7/4 = 2.925 m [4, 6, 8]). In this case the maximum permissible liquid leakage up to the point of its discharge in the lateral direction is $Q_{max} = \pi C R_{max}^2$.

The researchers calculated the maximum permissible liquid leakage basing upon the filtration coefficient typical values. While calculating we assume the main soils to be sand clay, sand foundation to be of river sand, its upper layer to be broken granite of 40×70 mm. At $Q < Q_{soils}$ =0.059 l/s water gradually flows through gravel, sand and soils within the railroad embankment. At 0.059 l/s= $Q_{soils} < Q < Q_{sand}$ =14.0 l/s water flows through gravel, sand and soils within the railroad embankment, a lateral discharge of water originating on the border of sub-ballast and gravel (some water flowing farther into the soils outside the railroad embankment). At 14.0 l/s = $Q_{sand} < Q < Q_{gr}$ =269.0 l/s water flows through gravel within the railroad embankment. There is a lateral discharge of water originating on the border of gravel and sand. Some water goes on flowing farther through the sand. There is the second lateral discharge of water originating on the border of sand and soils. At $Q > Q_{gr}$ =269.0 l/s some water flows down the gravel and some – through the gravel within the railroad embankment. There is a lateral discharge of water originating on the border of gravel and sand. Some water goes on flowing farther through the sand. There is the second lateral discharge of water originating on the border of sand and soils. The process is schematically shown on Figure 2.

4. Petroleum products filtration

Now we consider petroleum products filtration (e.g. crude oil) when split on railroad tracks [7, 10]. All equations presented in the previous part of our research work here as well. There are, though, some quantitative changes, which are due to greater viscosity of petroleum products.

In fact, water viscosity (at 20° C) is approximately 10^{-3} Pa*s, its density being 10^3 kg/m^3. Crude oil viscosity is (100...200)*10^{-3} and more Pa*s, its density being (0.8 ...1.0)*10^3 kg/m^3. There is very littler difference in there density while their viscosity is 100 ... 200 times as different. One should also take into account that oil viscosity dramatically rises at low temperatures (that is in spring, autumn and winter) and that some petroleum products viscosity (e.g. industrial fuel oil) is even greater than that of crude oil.

Penetration coefficient of the media does not depend on the properties of liquid [1, 2]. That is why penetration coefficients co-relation for one and the same liquid in different media does not depend on the properties of liquid either. It means that the geometric formula for filtration channel widening (given above) are the same both for water and petroleum products. It is important, though, that time required for petroleum products filtration is μ_{oil} / μ_{water} approximately 100 ... 200 ... times longer. In other words, petroleum products filtration goes on the same way as water filtration but it takes 100 ... 200 times longer. As a result, critical values for leakage leading to a lateral discharge of petroleum products from ballast section are 100 ... 200 times less than those values for water discharge. In case of a stationary leakage, a lateral discharge of petroleum products originates on the border of sub-ballast and soils when $Q > Q_{soils}$ =(0.059 l/s)/(100...200)=(0.6...0,3) ml/s. In this case the sub-ballast ends up to be heavily polluted everywhere along its ballast section. When $Q > Q_{sand}$ =(14.0 l/s)/(100...200)=(140...70) ml/s there is another lateral discharge of petroleum products on the border of gravel and sand. In this case it is not only the sub-ballast which ends up to be heavily polluted everywhere along its ballast section but also the gravel layer. It is quite clear that in this case ballast section mechanical properties are seriously affected and it becomes dangerous to go on using this railway tracks.

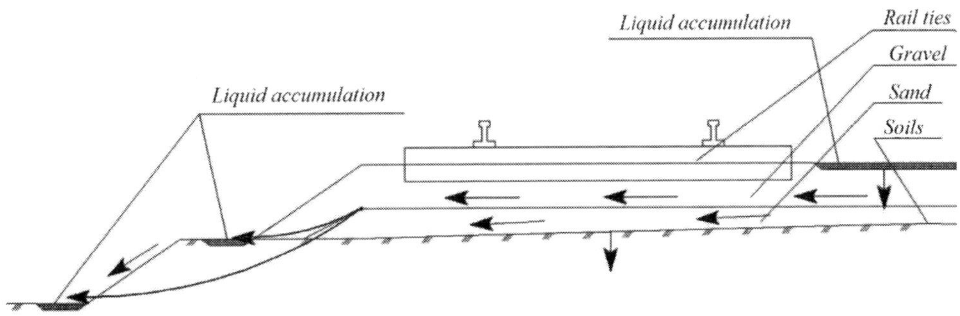

Fig. 2. Fluid weepage, outlet and discharge on different environment border

5. Conclusions

The research proves that any liquid which lands on railroad tracks goes on flowing down the railway embankment and then turns in the lateral direction. It happens either on the border of ballast section and sub-ballast or on the border of sub-ballast and soils.

Thus, we can conclude that if ballast section pollutants (discharged in the lateral direction) are properly collected and then utilized or disposed, environmental damage arising of railroads use can be reduced to a minimum. It is also very important that in this case ballast section reliable performance grows upward as well.

References

[1] Leontjev N V 2009 *Basics of filtration theory: Textbook* (Moscow: MSU)
[2] Muscat M 1937 *The Flow of Homogeneous Fluids in Porous Media* (McGrow-Hill, New York)
[3] Teplykh S Y and Strelkov A K 2015 Characteristics of railroad natural-technogenic complexes *Procedia Engineering* **111** 742-747
[4] Strelkov A K, Stepanov S V, Teplykh S Yu and Sargsyan A M 2016 Monitoring pollution level in railroad right-of-way *Procedia Environmental Sciences* **32** 147-154
[5] Balzannikov M I, Rodionov M V and Seliverstov M A 2011 Improving environmental security of underground hydraulic structures *Vestnik SGASU, Gradostroitelstvo i arhitektura* [Vestnik of SSUACE, Town Planning and Architecture] **1** 100-105
[6] Strelkov A K, Teplykh S Yu, Gorshkaljev P A and Sargsyan A M 2015 Water basins pollution by surface run-off *Water supply and sanitary technique* **12** 10-13
[7] Gorshkaljev P A and Teplykh S Yu 2011 *Railroad tracks run-off. Complex system of defining qualitative and quantitative parameters* LAP LAMBERT Academic Publishing GmbH & Co.
[8] Teplykh S Yu and Sargsyan A M 2012 Surface run-off from railway tracks influence on water facilities *Tracks and trackside* **15** 27-29
[9] Strelkov A K, Teplykh S Yu and Gorshkaljev P A 2014 Economic activity influence on surface run-off quality *Water supply and sanitary technique* **8** 21-26
[10] Strelkov A K, Teplykh S Yu, Gorshkaljev P A and Sargsyan A M 2015 Collecting and purifying surface run-off from railroads: contemporary state of the problem . *Science Review* **4** 123-129

Design of multiplier-less minimum-phase filters based on sharpening compensated comb filters

Gordana Jovanovic Dolecek

Department of Electronics, Institute INAOE, Puebla, Mexico

gordana@ieee.org

Abstract. Minimum-phase (MP) filters have all zeros inside and/or unit circle. As a consequence, the group delay of an MP system is always less than that of non-minimum phase systems, having the equal magnitude responses. Minimum-phase (MP) filters find applications where it is necessary to have a low group delay, like in communications, speech processing, and predictive coding, among others. This paper presents a novel simple method for the direct design of low-pass minimum-phase (MP) filters. Method is based on design of two compensated combs, using a multiplier-less minimum-phase compensator, and sharpening technique. The first comb defines the stop band and pass band of the MP filter, while the second comb decreases side lobes of the first comb, thus increasing attenuation of the resulting MP filter. Knowing that all zeros of comb filter are on the unit circle, the compensated comb is also a MP filter. Similarly, under the special condition, the sharpening of multiplier-less compensated comb may also result in a MP multiplier-less filter. The benefit of the proposed method is illustrated in the provided design examples.

1. Introduction

Minimum-phase (MP) filters find applications where it is necessary to have a low group delay, like in communications, speech processing, and predictive coding, among others. The zeros of minimum-phase filter are either on the unit circle, or inside the unit circle. Different methods are proposed for MP filter design. The methods can be divided into three principal groups:

- Methods based on the linear-phase proptotype filter [1-5],
- Methods based on complex cepstrum [6-8], and
- Direct methods [9-10].

We present here a novel direct method for MP filters design based on the compensated comb filters and sharpening technique. The rest of the paper is organized in the following way. Next section introduces brief review of comb filters, minimum-phase comb compensator, and sharpening technique. Section 3 elaborates the proposed method, and is illustrated with examples. Finally, Section 4 provides conclusion.

2. Brief reviews of comb, compensation filter and sharpening

In this section we present a brief review of comb filters, simple minimum-phase comb compensator recently proposed in literature [11], and sharpening technique [12].

2.1. Comb filters

Content from this work may be used under the terms of the Creative Commons Attribution 3.0 licence. Any further distribution of this work must maintain attribution to the author(s) and the title of the work, journal citation and DOI.
Published under licence by IOP Publishing Ltd

Comb filters are simple filters with all coefficients equal to unity [13]. The transfer function is given as:

$$H(z)=\left[\frac{1}{M}\frac{1-z^{-M}}{1-z^{-1}}\right]^K=\left[\frac{1}{M}\sum_{i=0}^{M-1}z^{-i}\right]^K, \tag{1}$$

where M is the length of the filter, and K is the number of the cascaded filters.

Comb filter has all zeros on the unit circle, and thus it is a minimum-phase filter. Its zeros are defined in the frequencies $2\pi i/M$, $i=1,\ldots,M$.

The magnitude characteristic is given as:

$$\left|H(e^{j\omega})\right|=\left|\frac{1}{M}\frac{\sin(\omega M/2)}{\sin(\omega/2)}\right|^K. \tag{2}$$

Comb filters are usually used as decimation filters [13]. However, here we will use comb filters as a low pass filters. The comb magnitude characteristic exhibits a pass band droop, and a low stop band attenuation. The goal here is to increase comb attenuations, and decrease the pass band droop. Increasing the comb parameter K both, the pass band droop and the attenuation increase. However, cascading comb filter with given parameter M and comb filter with a different parameter M_1 may result in an increasing of the resulting attenuation, without affecting significantly comb pass band. The value of comb parameter M_1 is chosen in such a way that the first zero of this comb filter be in the frequency band of the first side lobe of the comb with the parameter M:

$$\frac{2\pi}{M_1}\approx\frac{3\pi}{M}. \tag{3}$$

From here:

$$M_1=\text{nint}\left(\frac{2}{3}M\right), \tag{4}$$

where nint(*) is a closest integer to (*).

This idea is illustrated in the following example.

Example 1

We consider two values of M and in both cases is chosen $K_1=K_2=1$. Considering $M=9$, from (4) it follows $M_1=18/3=6$. Figure 1(a) shows overall magnitude responses of both combs, and their cascade. However, taking $M=7$, from (4) the value of M_1 is closest integer to $14/3=4.6667$, which is equal to 5. Similarly, Figure 1(b) shows the corresponding magnitude responses for this case. Note that in both cases, the first zero of the comb with the parameter M_1 is in the middle of the first lobe of the comb with the parameter M.

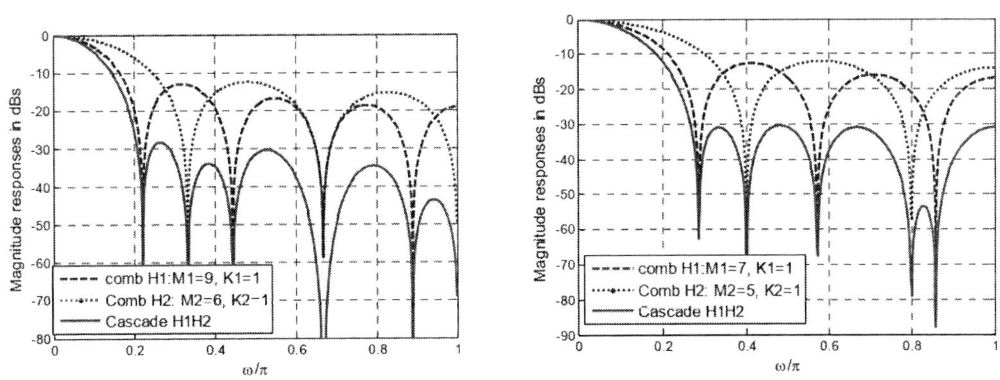

(a) Overall magnitude responses: $M_1=9$, $M_2=6$. (b) Overall magnitude responses: $M_1=7$, $M_2=5$.

Figure 1. Illustration of Example 1.

2.2. Minimum phase comb compensation filter

Here we adopt the simplest minimum phase comb compensator from [11] with the following transfer function:

$$G(z^M) = 1 + 2^{d-2} - 2^{d-2} z^{-M}, \tag{5}$$

where d is integer which depends on the comb parameter K as shown in Table 1 [11].

Table 1. The values of parameter d.

K	d
2,3	0
4,5	1
6	1 or 2

Example 2

Consider comb parameters: M=5 and K=2. From Table 1 it follows the parameter d=0. Figure 2(a) shows the zeros of compensated comb, thus confirming that the filter is a minimum-phase filter. Similarly, Figure 2(b) shows the magnitude responses of comb and compensated comb. Additionally, the pass band zoom is shown to demonstrate the decreasing of the comb pass band droop in the compensated comb.

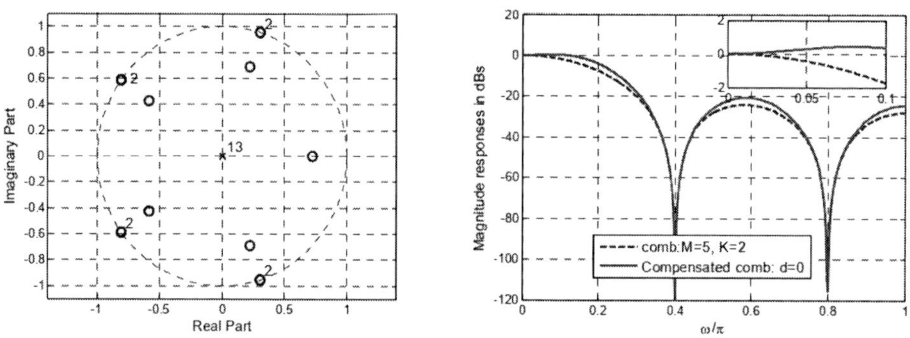

(a) Zeros of compensated comb. (b) Magnitude responses.

Figure 2. Illustration of comb compensation.

2.3. Sharpening

The sharpening technique is introduced in [12] for simultaneous improvements of both the pass-band and stop-band of a linear-phase FIR filters. In [9] is shown how this technique can also be used for minimum-phase filters. The sharpening technique uses a polynomial relationship of the form $H_{sh} = f(H)$ between the amplitudes of the sharpened and the prototype filters, H_{sh} and H, respectively. We will use here the simple sharpening polynomial: $Hsh=2H-H^2$.

3. Description of method

The comb parameter M is defined by the stop band edge frequency ω_s of the designed filter:

$$M = \mathrm{nint}(2/\omega_s), \tag{6}$$

where nint(*) is nearest integer of (*).

For a given M the pass band edge frequency ω_p is:

$$\omega_p = \pi / 2M. \tag{7}$$

The prototype filter has the following transfer function:

$$H_p(z) = H_1(z) G_1(z^{M_1}) H(z) G(z^M), \tag{8}$$

where $H_1(z)$ and $H(z)$ are combs with parameters M_1, $K_1$and M, K, respectively and $G_1(z^{M1})$ and $G(z^M)$ are the corresponding comb compensators given in (5).

Taking the simplest sharpening polynomial $2H\text{-}H^2$, the proposed MP sharpened filter is:

$$F(z) = Sh\{H_p(z)\} = H_p(z)[2 - H_p(z)] = H_p(z)F_1(z), \tag{9}$$

where Sh{*} means sharpening of {*}, and $H_p(z)$ is given in (8).

The filter (9) is a minimum-phase filter, because the filters $H_p(z)$ and $F_1(z)$ are minimum-phase filters. (In [9] is shown that the filter $F_1(z)$ is also a minimum-phase filter if the value of 2 is added to the coefficient with index 1 of the filter $- H_p(z)$). The attenuation and absolute value of pass band deviation depend on the values K_1 and K_1.

Table 2 shows the tentative values of minimum attenuation A and maximum absolute value of passband deviation δ in terms of the values K and K_2, obtained by MATLAB simulation.

Table 2. The values of A and δ in terms of the values K and K_1.

K	K_1	A [dB]	δ [dB]
2	2	45.5	0.1
2	3	55	0.03
3	2	62	0.01
2	4	68	0.2
3	3	75	0.04
4	2	75	0.2
3	4	85	0.05
4	3	87	0.01
4	4	100	0.4

Example 3

We design a minimum-phase filter with ω_s=0.3π, and minimum stop band attenuation of 60 dB and the absolute value of maximum pass band deviation of 0.01 dB. From (6) we get M=7. From (4) we get M_1=5. From Table 1 we chose K=3, and K_1=2. The magnitude response of the designed filter (9) and the pass band zoom are shown in Fig.3.

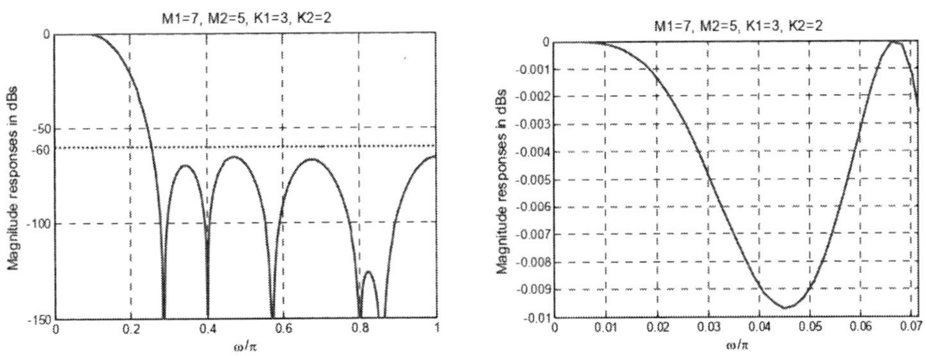

(a) Overall magnitude response. (b) Pass band zoom.

Figure 3. Magnitude response of the minimum-phase filter from Example 3.

Figure 4(a) shows pole-zero plot of the designed filter $F(z)$, while Figure 4 (b) shows pole-zero plot of the filter $F_1(z)$. Figure 4 thus confirms that the designed filter is a minimum-phase filter, i.e. has all its zeros inside the unit circle.

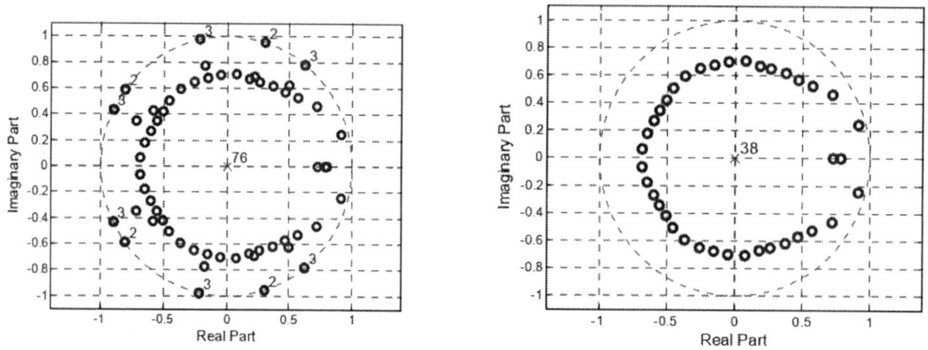

(a) Zeros of filter $F(z)$. (b) Zeros filter $F_1(z)$.

Figure 4. Minimum-phase property of filters from Example 3.

Example 4

In this example we design a MP filter with ω_s=0.23π, while the values of A and δ are the same as in Example 3 (A=60 dB and δ=0.01dB). From (6) and (4) we get M=9 and M_1=6. From Table 2 we have: K=3 and K_1=2. The corresponding magnitude response and pass band zoom are shown in Figure 5.

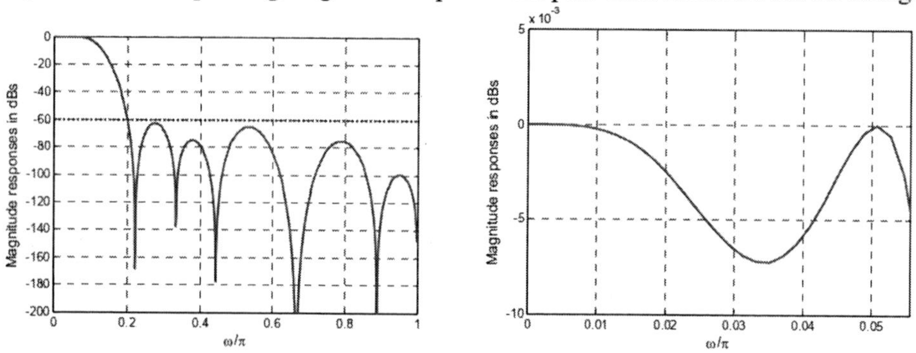

(a) Overall magnitude response. (b) Passband zoom.

Figure 5. Magnitude response of the filter in Example 4.

The pole zero plot of the designed MP filter is shown in Fig.6a. Similarly Figure 6b shows zeros of the filter $F_1(z)$ to demonstrate that the designed filter has all zeros inside the unit circle.

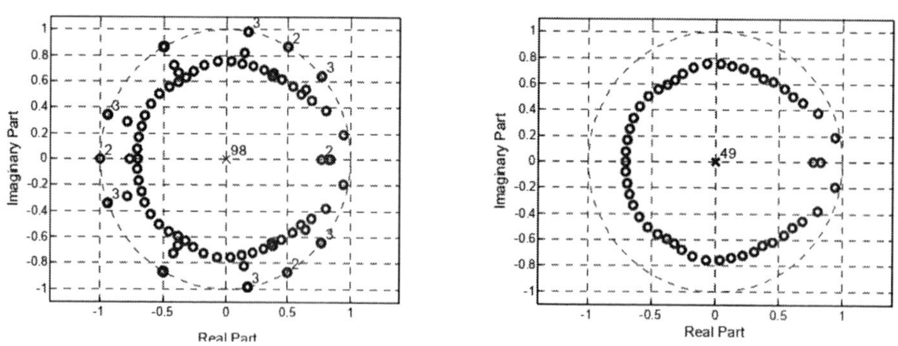

(a) Zeros of designed MP filter. (b) Zeros of the corresponding filter $F_1(z)$.

Figure 6. Pole-zero plots in Example 4.

4. Conclusion

This paper presents a novel method for minimum-phase filters design based on two compensated combs and sharpening technique. The designed parameters are: stop band edge frequency, minimum attenuation in the stop band and absolute value of the maximum pass band deviation. The pass band edge frequency is defined by the stop band edge frequency.

Acknowledgement
This work is supported by CONACYT Mexico under grant No.179587.

References

[1] Kale I, Cauin G.D and Morling R C S 1995 *Electronic Letters* Minimum-phase filter design from linear-phase start point via balanced model truncation **31** 1728-1729.
[2] Stathaki T and Fotinopoulos I 2001 *IEEE Trans. on Circuits and Systems-I: Analog and Digital Signal Processin* Equiripple minimum-phase FIR filter design from linear phase systems using root moments **6** 580-587.
[3] Huang Xiang-dong J Wang Zhao-hua, 2006 *Journal of Tianjin University* A novel technique to design filter with discontinuous frequency character **39** 614-620.
[4] G Jovanovic Dolecek & J Diaz Carmona 1997 *Electronic Letters* Lowpass minimum phase filter design using IFIR filters,", **33** 1933-1935.
[5] G Jovanovic Dolecek & V Dolecek 2001 *Electronic Letters* Method for narrowband minimum phase filter design **37** 324-325.
[6] S C Pei, & H S Lin 2006 *IEEE Transactions on Circuits and Systems-II: Express Briefs* Minimum-phase FIR filter design using real cepstrum **53** 1113-1117.
[7] H H Dam, S Nordebon & L Svensson 2003 *IEEE Transactions on Signal Processing* Design of minimum-phase digital filters as the sum of two allpass functions using the cepstrum technique **51** 726-731.
[8] T Nagarajan, K Prasad & H A Murthy 2003 *Electronic Letters* Minimum phase signal derived from root cepstrum **39** 941-942.
[9] G Jovanovic Dolecek & A Fernandez-Vazquez 2013 *Journal of Physics: Conference Serie* Sharpening minimum phase filters **410** 1-4.
[10] G Jovanovic Dolecek & A Fernández-Vazquez 2015 *Computacion y Sistemas* Sharpening minimum-phase interpolated finite impulse response filters **19** 805-815.
[11] D E Troncoso Romero, G Molina Salgado & G Jovanovic Dolecek 2015 *Electronics letters* Simple two-adders CIC compensator **51** 993-994.
[12] J F Kaiser & R W Hamming 1977 *IEEE Transactions on Acoustics, Speech, and Signal Processing,* Sharpening the response of a symmetric non-recursive filter **ASSP-25** 415-422.
[13] G Jovanovic Dolecek 2002 *Multirate Signal Processing: Design and Application* (Hershey: IGP).

Phase Diagram of the Antiferromagnetic Blume-Capel Model on Triangular Lattice

Sojeong Park, Wooseop Kwak
Department of Physics, Chosun University, Gwangju, 501-759, South Korea
E-mail: wkwak@chosun.ac.kr

Abstract. We perform Monte-Carlo simulations of the anti-ferromagnetic (AF) spin-1 Blume-Capel (BC) model and the AF Ising model on triangular lattice. We estimate the exact critical magnetic fields for both models at zero temperature using the Wang-Landau sampling method. We also show the phase diagrams and the critical lines for the models using the joint density functions. We find that the shapes of critical lines for the models are identical, but the phase transitions across the critical lines are different.

1. Introduction

Monte Carlo simulations currently play a major role in statistical physics to study phase transitions and critical properties. They are well-known in the case of anti-ferromagnetic (AF) Ising on square lattice. However, the studies of frustrated systems such as triangular lattice [1, 2, 3, 4, 5, 6, 7, 8, 9, 10, 11, 12, 13, 14, 15, 16, 17, 18] is also highlighted by the difficult tasks to understand the various phenomena. It has been reported that AF Ising models have three types of conjectured patterns of critical lines [19]. The Blume-Capel (BC) model [20, 21, 22, 23, 24, 25] has been studied to understand systems such as metamagents, ternary alloys, and multicompent fluids. This spin-1 Ising model shows first-order transitions and second-order transition. In this paper, we study the critical lines and properties of the AF BC model as well as those of the AF Ising model on triangular lattice using the Wang-Landau sampling method [26, 27, 28, 29] and microcanonical analysis [1].

2. Model and Simulation

The Hamiltonian of the AF Ising model defined on $L \times L$ square lattice in two dimensions is given by

$$\mathcal{H} = -JE - hM, \tag{1}$$

where $E = \sum_{\langle i,j \rangle} \sigma_i \sigma_j$, $M = \sum_i \sigma_i$, $\sigma_i = \pm 1$, J is a negative coupling constant, and h is an external magnetic field. The Hamiltonian of the AF BC model is as follows:

$$\mathcal{H} = -JE - hM + DM^2, \tag{2}$$

where D is the crystal field (also called the single-spin anisotropy parameter or the spin impurity chemical potential), $M^2 = \sum_i {\sigma_i}^2$, and $\sigma_i = -1, 0,$ or 1. Here, a negative coupling constant ($J < 0$) defines the AF BC model. The physical origin of the crystal field D arises from the non-central potentials for the metal atoms coordinated with various ligands in the crystal.

Content from this work may be used under the terms of the Creative Commons Attribution 3.0 licence. Any further distribution of this work must maintain attribution to the author(s) and the title of the work, journal citation and DOI.
Published under licence by IOP Publishing Ltd

Journal of Physics: Conference Series **738** (2016) 012126 doi:10.1088/1742-6596/738/1/012126

Using the WL algorithm, the $3d$ random walks are performed in the joint space [30, 31] $A = (E, M)$ for AF Ising model and $A = (E, M, M^2)$ for AF BC model by randomly changing the states of spins A, where the order parameter M is $\sum_i S_i$ and the square of order parameter M^2 is $\sum_i S_i^2$, but the state A associated with each spin configuration is only accepted with a probability proportional to the reciprocal of the joint density of states $g(A)$ [30, 31, 32, 33].

Therefore, the transition probability from state A to A' is

$$p(A \to A') = min\left(\frac{g(A)}{g(A')}, 1\right), \tag{3}$$

which indicates that if $g(A') \leq g(A)$, a state with spin configuration A' is always accepted, and that if $g(A') > g(A)$, it is accepted with the probability $g(A)/g(A')$.

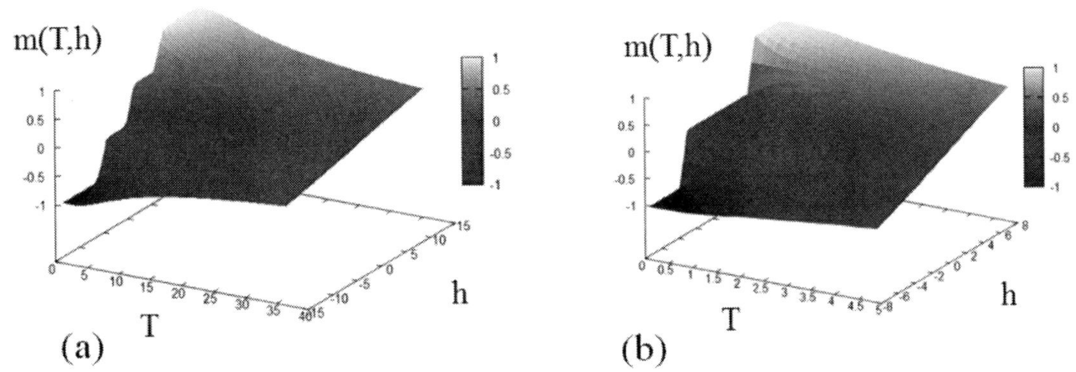

Figure 1. Magnetization m of (a) AF Ising model with $L = 12$ and (b) AF BC model with $L = 6$ as a function of energy, temperature, and h field on triangular lattice.

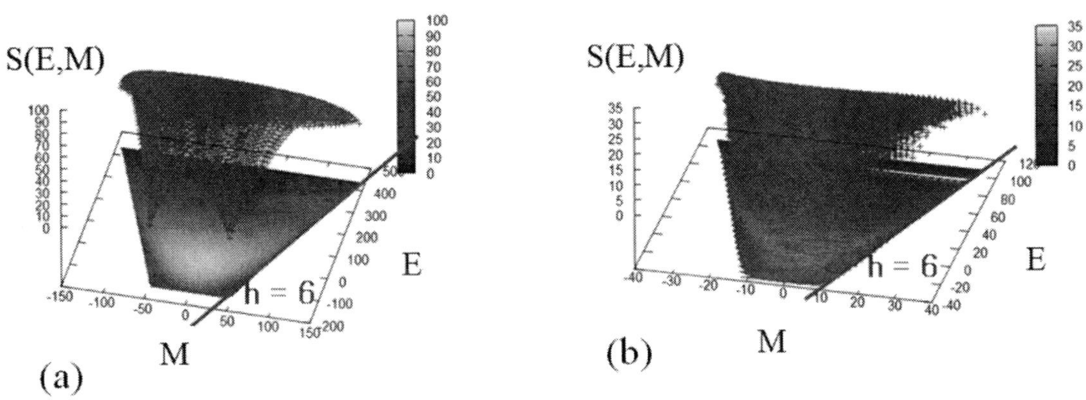

Figure 2. Microcanonical entropy of (a) AF Ising model with $L = 12$ and (b) AF BC model with $L = 6$ as a funcion of energy and magnetization on triangular lattice.

3. Results and Discussion

The canonical distributions of magnetization m for AF Ising model with linear dimension $L = 12$ and AF BC model with $L = 6$ on on triangular lattice are shown in figure 1. The line of m

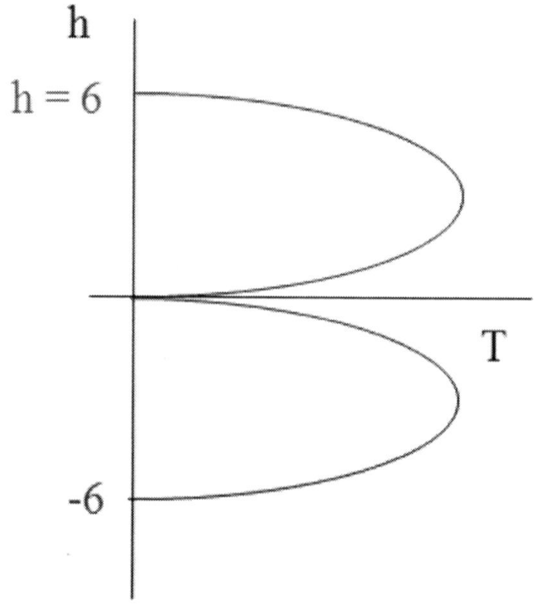

Figure 3. Schematic plot of three types of critical lines of AF Ising model with $L = 12$ and AF BC model with $L = 6$ on triangular lattice.

as a function of h in the vicinity of $T \sim 0$ shows discontinuity for AF BC model, but it does not show for AF Ising. Figure 2 shows microcanonical entropy $S(E, M)$ of (a) AF Ising model with $L = 12$ and (b) AF BC model with $L = 6$ as a funcion of energy and magnetization on triangular lattice. In Figure 2, the contour plot of ME diagram has four vertices. The top-left vertice corresponds to the state of all spins up. The locations of top-right vertices are as follows: $E = 432$ and $M = 144$ for AF Ising model, and $E = 102$ and $M = 35$ for AF BC model. In the contour plot of ME diagram in Figure 2, bottom line corresponds to the frustrated AF ground state without h. The locations of bottom-right vertices are $E = -144$ and $M = 48$ for AF Ising model, and $E = -36$ and $M = 12$ for AF BC model.

We can observe that at zero temperature the term with the minimum E_t is dominant, where $E_t = E - hM$ is the total energy. The total energy E_t can be interpreted as the intersection on the E axis of the linear line $E = hM + E_t$ in the ME diagram. Note that the slope of the line connecting the top right vertex and the bottom right vertex is six which is the critical magnetic field h_c, where $h_c = 6$ for AF Ising and AF BC models on triangular lattice.

Figure 3 shows the estimated critical lines of the AF Ising and AF BC models on triangular lattice. Although the patterns of critical lines for both models are the same, the transitions in the vicinity of $T \sim 0$ are different from each other. As shown in Figure 3, the critical line of AF BC model at $T \sim 0$ and $h = 6$ indicates the first order phase transition, whereas that of AF Ising at $T \sim 0$ and $h = 6$ indicates the second order phase transition.

4. Conclusions

In this paper, we study the AF BC and AF Ising model on triangular lattices using Wang-Landau sampling method and micro-canonical anaysis. We find critical lines of both models and find the critical magnetic field $h_c = 6$ at zero temperature. We also find that the critical line of AF BC model at $T \sim 0$ indicates the first order phase transition and that of AF Ising indicates the second order phase transition. Thus, our future intention is to extend the present investigations to the critical properties and phenomena of AF BC model on triangular lattice with non-zero D and non-zero h field.

Acknowledgment
This research was supported by Basic Science Research Program through the National Research Foundation of Korea (NRF) funded by the Ministry of Education Science and Technology (NRF-2013R1A1A2065043).

[1] C.-O. Hwang, S.-Y. Kim, D. Kang, J. M. Kim 2007 J.Stat.Mech. L05001
[2] C.-O. Hwang, S.-Y. Kim, D. Kang, J. M. Kim 2008 J. Korean Phys. Soc. **52** S203
[3] X. Yao 2010 Solid State Commun. **150** 160
[4] J.D. Noh, H. Rieger, M. Enderle, K. Knorr 2002 Phys. Rev. E **66** 026111
[5] Henk W. J. Blote, M. Peter Nightingale 1993 Phys. Rev. B **47** 15046
[6] C.-O. Hwang, S.-Y. Kim 2010 Physica A **389** 5650
[7] M. Zukovic, A. Bobak 2013 Phys. Rev. E **87** 032121
[8] O. Nagai, S. Miyashita, T. Horiguchi 1993 Phys. Rev. B **47** 202
[9] W. Li, S.-S. Gong, Y. Zhao, S.-J. Ran, S. Gao, G. Su 2010 Phys. Rev. B **82** 134434
[10] Y. L. Loh, D. X. Yao, E. W. Carlson 2008 Phys. Rev. B **77** 134402
[11] J.D. Noh, D. Kim 1992 Int. J. Mol. Sci. **6** 2913
[12] S.-H. Tsai, F. Wang, D.P Landau 2006 Braz. J. Phys. **36** 635
[13] M.A. Novotny, D.P. Landau 1981 Phys. Rev. B **24** 1468
[14] J.B. Santos-Filho, J.A. Plascak, D.P. Landau 2010 Physica A **389** 2934
[15] X. Qian, M. Wegewijs, Henk W. J. Blote 2004 Phys. Rev. E **69** 036127
[16] F.W.S. Lima, J.A. Plascak 2013 Eur. Phys. J. B **86** 1
[17] U. Yu 2015 Phys. Rev. E **91** 062121
[18] S. Hu, S.-H. Tsai, D.P. Landau 2014 Phys. Rev. E **89** 032118
[19] L.-H. Gwa 1990 Phys. Rev. B **41** 7315
[20] M. Blume 1966 Phys. Rev. **141** 517
[21] H.W. Capel 1966 Physica (Amsterdam) **32** 966
[22] M. Blume, V.J. Emery, R.B. Griffiths 1971 Phys. Rev. A **4** 1071
[23] A.K. Jain, D.P. Landau 1980 Phys. Rev. B **22** 22.
[24] C.-J. Liu, H.-B. Sch'uttler 2002 Phys. Rev. E. **65** 056103
[25] W. Kwak, J. Jeong, J. Lee, D.H. Kim 2015 Phys. Rev. E. **92** 022134
[26] F. Wang, D. P. Landau 2011 Phys. Rev. Lett. **86** 2050
[27] D. P. Landau, S.-H. Tsai, M. Exler 2004 Am. J. Phys. **72** 1294
[28] B. J. Schulz, K. Binder, M. Müller, D. P. Landau 2003 Phys. Rev. E **67** 067102
[29] C. Zhou, T. C. Schulthess, S. Torbrügge, D. P. Landau 2006 Phys. Rev. Lett. **96** 120201
[30] D.P. Landau, F. Wang 2002 Computer Physics. Commun. **147** 674
[31] S.-H. Tsai, F. Wang, D.P. Landau 2007 Phys. Rev. E. **75** 061108
[32] S. Ryu, W. Kwak 2013 J. Korean Phys. Soc. **62** 559
[33] S. Ryu, W. Kwak 2013 J. Korean Phys. Soc. **62** 861

Study on the C-terminal beta-hairpin of protein G in AB heteropolymer model

Seung-Yeon Kim

School of Liberal Arts and Sciences, Korea National University of Transportation, Chungju 27469, South Korea

E-mail: sykimm@ut.ac.kr

Abstract. The off-lattice AB heteropolymer model, consisting of the hydrophobic (A) and hydrophilic (B) polymers, is one of popular protein models. Its energy function includes the bending energy and the van der Waals interaction energy. The properties and the energy landscape of the C-terminal beta-hairpin of protein G are studied in the off-lattice AB heteropolymer model with conformational space annealing, a powerful global optimization method.

1. Introduction

Proteins [1] are fundamental components of all living organisms. There are about one hundred thousand kinds of proteins in human body, controlling human's biological activities. Information on the three-dimensional tertiary native structure of a protein is vital in unveiling its biological function and role. Understanding the folding of a protein into its three-dimensional tertiary native structure only with its one-dimensional amino-acid sequence information (the so-called primary structure) is one of the most important problems in modern science. The failure of a proper protein folding results in the malfunction of a biological system, leading to fatal diseases. Proteins are linear polymers, built of twenty different amino acids (or residues), with a defined residue sequence. An amino acid consists of the intrinsic side-chain and the common backbone. Because different amino acids have a common backbone structure, their side-chains determine their nature - hydrophilic or hydrophobic. Hydrophobic amino acids of an unfolded protein in water drive it to fold into its three-dimensional tertiary native structure. Ten residues, glycine (G, Gly), alanine (A, Ala), valine (V, Val), leucine (L, Leu), isoleucine (I, Ile), cysteine (C, Cys), methionine (M, Met), phenylalanine (F, Phe), tyrosine (Y, Tyr), and tryptophan (W, Trp), are usually classified as hydrophobic amino acids. The other ten residues, proline (P, Pro), serine (S, Ser), threonine (T, Thr), asparagine (N, Asn), glutamine (Q, Gln), aspartic acid (D, Asp), glutamic acid (E, Glu), lysine (K, Lys), arginine (R, Arg), and histidine (H, His), belong to hydrophilic amino acids, where aspartic acid and glutamic acids are negatively charged but lysine and arginine are positively charged.

Simplified model proteins have been popularly used to understand the physical principles of protein folding because studying the protein-folding problem is too difficult. The HP model [2, 3] is the most simplified model for protein folding, consists of two kinds of amino acids, the hydrophobic (A) and hydrophilic (B) polymers, and is configured as self-avoiding walks on lattices. A more generalized off-lattice AB heteropolymer model [4, 5, 6, 7, 8, 9] is based on

Content from this work may be used under the terms of the Creative Commons Attribution 3.0 licence. Any further distribution of this work must maintain attribution to the author(s) and the title of the work, journal citation and DOI.
Published under licence by IOP Publishing Ltd

a realistic potential energy including the bending energy and the van der Waals interaction energy. In this work, we investigate the properties and the energy landscape of the C-terminal beta-hairpin of protein G in the off-lattice AB heteropolymer model using conformational space annealing [10]. Protein G is an immunoglobulin binding protein (PDB ID: 2GB1). Its C-terminal β-hairpin fragment [11, 12, 13] with the amino-acid sequence, Gly-Glu-Trp-Thr-Tyr-Asp-Asp-Ala-Thr-Lys-Thr-Phe-Thr-Val-Thr-Glu, has been frequently studied to understand protein folding. This sequence corresponds to ABABABBABBBABABB in the off-lattice AB heteropolymer model.

2. Off-lattice AB heteropolymer model and global optimization

The energy function of the off-lattice AB heteropolymer model [4, 5, 6, 7, 8, 9], consisting of two energy terms (the bending energy and the van der Waals interaction energy), is defined by

$$E = \frac{1}{4}\sum_{i=2}^{N-1}(1+\cos\theta_i) + \sum_{i=1}^{N-2}\sum_{j=i+2}^{N}\left(\frac{4}{r_{ij}^{12}} - \frac{V_{ij}}{r_{ij}^{6}}\right). \tag{1}$$

Here, θ_i is the bond angle defined by three residues $i-1$, i, and $i+1$, and r_{ij} is the distance between residues i and j. In addition, the van der Waals constant V_{ij} is 4, 2, and -2 for, respectively, AA, BB, and AB pairs, thus, yielding a strong attraction between AAs, a weak attraction for BBs, and a weak repulsion between ABs. All bond lengths are fixed to the unity, and all physical quantities are unitless.

Global optimization is important in various fields of science and engineering. Global optimization problems are hard to solve because many of them belong to the NP-complete class. The protein-folding problem is one of the well-known classic examples. A powerful global optimization method called conformational space annealing (CSA) was proposed [10], and applied to the protein-folding problem [14, 15, 16, 17]. The CSA method unifies the essential ingredients of the three global optimization methods, simulated annealing, genetic algorithm, and basin-hopping algorithm. Here, we describe the concept of CSA briefly. The details of CSA are well explained in Ref. [10]. First, as in basin-hopping algorithm, we consider only the phase space of local minima, that is, all conformations are energy-minimized by a local minimizer. Secondly, as in genetic algorithm, we consider many conformations (called a bank in CSA) collectively, and we perturb a subset of bank conformations (seeds) using the information in other bank conformations. This procedure is similar to mating in genetic algorithm. Finally, as in simulated annealing, we introduce an annealing parameter D_{cut} (a cutoff distance in the phase space of local minima), which plays the role of temperature in simulated annealing. The diversity of sampling is directly controlled in CSA by introducing a distance measure between two conformations and comparing it with D_{cut}, whereas there is no such systematic control in simulated annealing. The value of D_{cut} is slowly reduced just as in simulated annealing, hence the name conformational space annealing.

3. Beta-hairpin in AB heteropolymer model

We study the C-terminal β-hairpin fragment of protein G in the off-lattice AB heteropolymer model using CSA, and obtain 300 distinct low-energy conformations between energies -9.20604 and -8.30429. The lowest-energy conformation with $E_1 = -9.20604$ can be called the (putative) global-minimum energy conformation. Figure 1 shows the stereographic view of the global-minimum energy conformation whose all hydrophobic residues constitute a well-formed hydrophobic core.

To investigate the energy landscape of the β-hairpin fragment, we measure the root-mean-square deviation (RMSD) of all low-energy conformations from the global-minimum energy conformation. Figure 2(a) depicts the distribution of the RMSD as a function of energy, showing

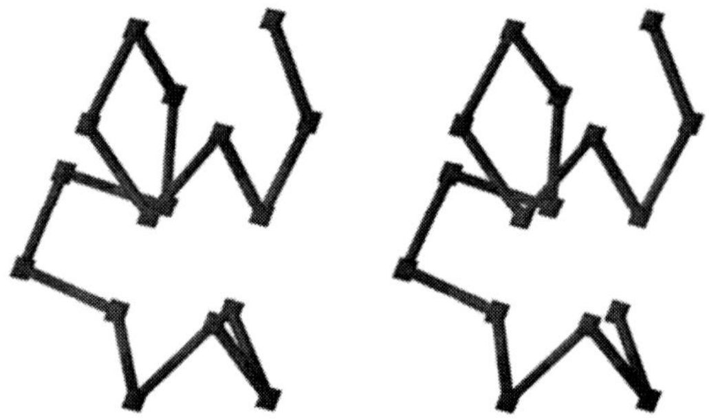

Figure 1. Stereographic view for the global-minimum energy conformation ($E_1 = -9.20604$) whose radius of gyration is 1.3428. The red and blue spheres indicate hydrophobic (A) and hydrophilic (B) polymers, respectively.

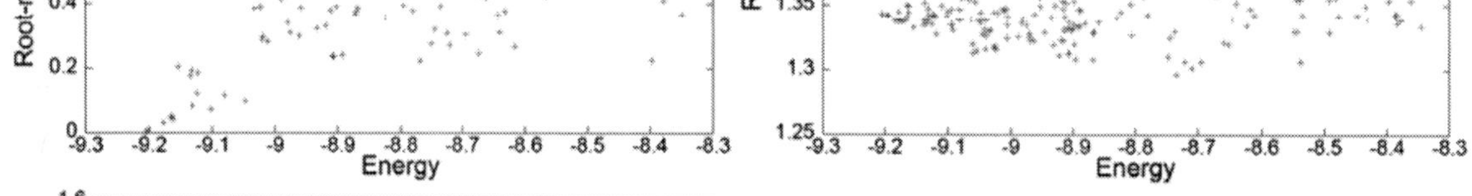

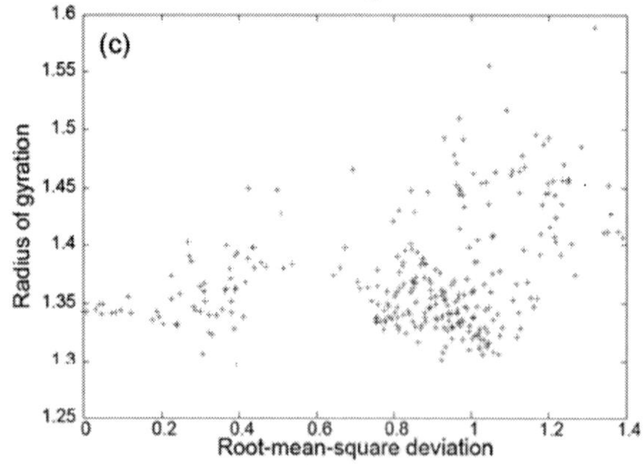

Figure 2. Distributions of (a) the root-mean-square deviation (RMSD) and (b) the radius of gyration as a function of energy for all 300 distinct low-energy conformations, estimated from the global-minimum energy conformation. The distribution of the radius of gyration versus RMSD is also shown in (c).

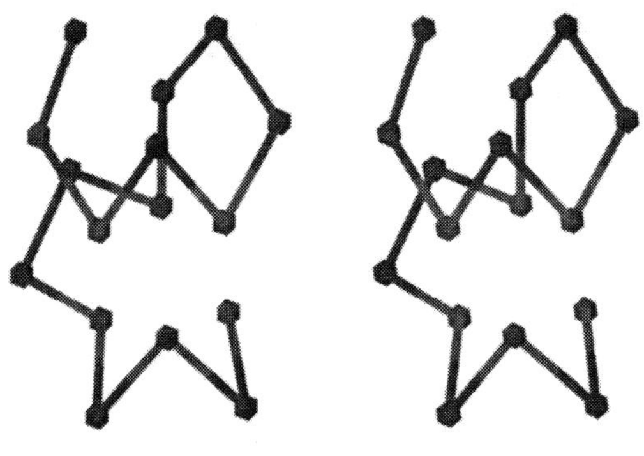

Figure 3. Stereographic view for the lowest-energy conformation ($E_3 = -9.19626$) of the second group whose RMSD is 0.9638 and radius of gyration is 1.3416.

that these low-energy conformations are classified as two groups: the first group with RMSD $\leq$ 0.5358 and the second group with RMSD $\geq$ 0.6410. Only 64 low-energy conformations belong to the first group. On the other hand, most of the low-energy conformations (amounting to 236 distinct conformations) belong to the second group whose minimum energy is $E_3 = -9.19626$ (the third lowest energy) with a tiny energy difference $\Delta E = E_3 - E_1 = 0.00978$, implying a rugged energy landscape.

The radius of gyration (R_g) for all 300 low-energy conformations are also estimated. Figure 2(b) shows the distribution of the radius of gyration as a function of energy, and figure 2(c) shows the distribution of the radius of gyration versus RMSD. The first group and the second group can not be distinguished in figure 2(b), but they are again distinguished in figure 2(c). As shown in figure 2(b), for $E \leq -8.6$, the low-energy conformations are relatively compact since $R_g \leq 1.4086$. However, for $E > -8.6$, the low-energy conformations show a wide range of R_g values ($1.3064 \leq R_g \leq 1.5554$).

Figure 3 shows the stereographic view of the lowest-energy conformation of the second group whose radius of gyration is 1.3416 and RMSD is 0.9638 from the global-minimum energy conformation. Also, it shows a well-formed hydrophobic core, similar to the global-minimum energy conformation.

Acknowledgments
This was supported by Korea National University of Transportation in 2016.

References

[1] Creighton W E 1993 *Proteins: Structures and Molecular Properties*, 2nd edition (New York: W H Freeman and Company)
[2] Lau K F and Dill K A 1989 *Macromolecules* **22** 3986
[3] Lee J H, Kim S-Y and Lee J 2015 *AIP Advances* **5** 127211
[4] Stillinger F H, Head-Gordon T and Hirshfeld C L 1993 *Phys. Rev. E* **48** 1469
[5] Kim S-Y, Lee S B and Lee J 2005 *Phys. Rev. E* **72** 011916
[6] Kim J, Straub J E and Keyes T 2006 *Phys. Rev. Lett.* **97** 050601
[7] Lee J W, Joo K H, Kim S-Y and Lee J 2008 *J. Comput. Chem.* **29** 2479
[8] Li B, Li Y and Gong L 2014 *Engineering Applications of Artificial Intelligence* **27** 70
[9] Kim S-Y 2016 *Chaos, Solitons and Fractals* in press (http://dx.doi.org/10.1016/j.chaos.2016.04.010)
[10] Lee J, Scheraga H A and Rackovsky S 1997 *J. Comput. Chem.* **18** 1222
[11] Munoz V, Thompson P A, Hofrichter J and Eaton W A 1997 *Nature* **390** 196
[12] Lee I-H, Kim S-Y and Lee J 2005 *Chem. Phys. Lett.* **412** 307
[13] Lee J 2013 *Phys. Rev. Lett.* **110** 248101
[14] Lee J, Kim S-Y, Joo K H, Kim I and Lee J 2004 *Proteins* **56** 704
[15] Kim S-Y, Lee W and Lee J 2006 *J Chem Phys* **125** 194908
[16] Kim S-Y 2010 *J Chem Phys* **133** 135102
[17] Joo K H, Joung I, Lee J H, Lee J W, Lee W, Brooks B, Lee S J and Lee J 2015 *Proteins* **83** 2251

Dynamic mesh for TCAD modeling with ECORCE

A Michez[1], J Boch[1], A Touboul[1], F Saigné[1]
[1]Université de Montpellier, IES – UMR UM/CNRS 5214, 860 Rue Saint Priest, 34095 Montpellier Cedex 5, France

E-mail: alain.michez@umontpellier.fr

Abstract. Mesh generation for TCAD modeling is challenging. Because densities of carriers can change by several orders of magnitude in thin areas, a significant change of the solution can be observed for two very similar meshes. The mesh must be defined at best to minimize this change. To address this issue, a criterion based on polynomial interpolation on adjacent nodes is proposed that adjusts accurately the mesh to the gradients of Degrees of Freedom. Furthermore, a dynamic mesh that follows changes of DF in DC and transient mode is a powerful tool for TCAD users. But, in transient modeling, adding nodes to a mesh induces oscillations in the solution that appears as spikes at the current collected at the contacts. This paper proposes two schemes that solve this problem. Examples show that using these techniques, the dynamic mesh generator of the TCAD tool ECORCE handle semiconductors devices in DC and transient mode.

1. Introduction

TCAD (Technology Computer Aided Design) modeling is widely used to design semiconductor devices and simulate their behaviour. The use of TCAD software is uneasy and most often at least one engineer of a team is dedicated to this work. The mesh design is the main problem since, to compute a true solution, the mesh must accurately follow the gradients of the 3 Degrees of Freedom (DF): potential and densities for electrons and holes. Indeed, carrier densities in semiconductors can change by several orders of magnitude inside depletion regions. These regions can be very thin depending on the doping level and need a highly refined mesh. Conversely, areas where doping are constant represent the main surface of the component and will only need a coarse mesh. Furthermore external events like varying bias or ionizing radiations induce fast and substantial changes in the spatial distribution of carriers.

However, for now, meshes in TCAD software are manually defined and can not be modified during modeling. Then the designer of a mesh must take into account all distributions of DF that will occur during modeling to create the fixed mesh. For this he adds nodes that will be unnecessary for most steps of the simulation and that will increase CPU time. Anyway, the task is so complex that a human operator can at best get an approximate result.

Tools that automatically create and modify a mesh according to gradients of DF during modeling are already used in others fields of physics like fluid and solid mechanics [1][2]. These dynamic mesh generators insure users to get the best balance between precision of solution and CPU time.

But, as explained previously, a dynamic mesh generator for TCAD modeling will face harsh numerical constraints. To correctly define gradients of DF that change by several orders of magnitude in thin areas, we need a reliable criterion that predicts when adding or when removing a node. Adding

Content from this work may be used under the terms of the Creative Commons Attribution 3.0 licence. Any further distribution of this work must maintain attribution to the author(s) and the title of the work, journal citation and DOI.
Published under licence by IOP Publishing Ltd

too late or removing too early a node triggers oscillations in the solution that result in numerical noise at the electrode currents. Furthermore DF of added nodes must be carefully calculated especially for transient modeling where time derivative of DF depends on previous steps.

We showed in a previous paper that ECORCE [3] is the first TCAD software that provides a dynamic mesh for transient analysis [4]. In this paper we describe the criterion used by ECORCE to drive the refinement and coarsening of the mesh and the calculation strategy for added nodes.

2. Overview of ECORCE

ECORCE is a TCAD tool that provides a Graphical User Interface (GUI) and a solver, including a dynamic mesh generator, integrated in one application. The whole software is developed in C++. The GUI is based on Qt [5] and offers a friendly interface for all steps of modeling: designing of model, execution of calculation, display and analyse of results. The main part of ECORCE is distributed under the GNU General Public License (GPL) [3].

ECORCE uses a well-known drift-diffusion model [4,6] coupled with heat equation. It generates structured meshes build with rectangular elements and groups of three triangular elements for connecting rectangular elements of different sizes as shown in Figure 1. Our meshing algorithm prevents the creation of obtuse triangles that are known to generate inaccurate results and numerical instabilities and ensures that all angles will be less than or equal to 90°.

To ease changes of the mesh, data are structured using 3 objects: .node, edge and face. Each object can access adjacent objects via a list of pointers as shown in Figure 2.

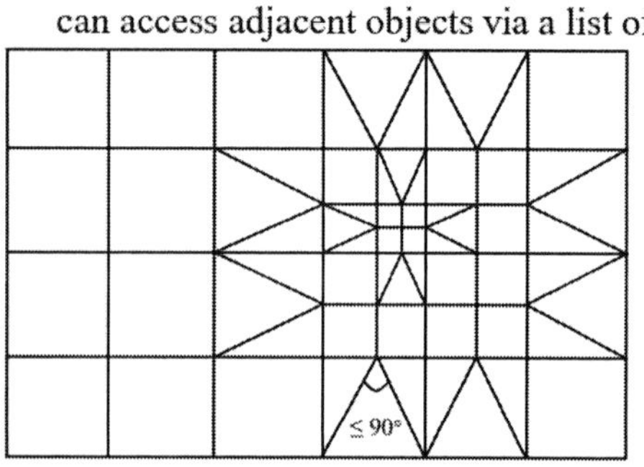

Figure 1. Example of mesh generated by ECORCE

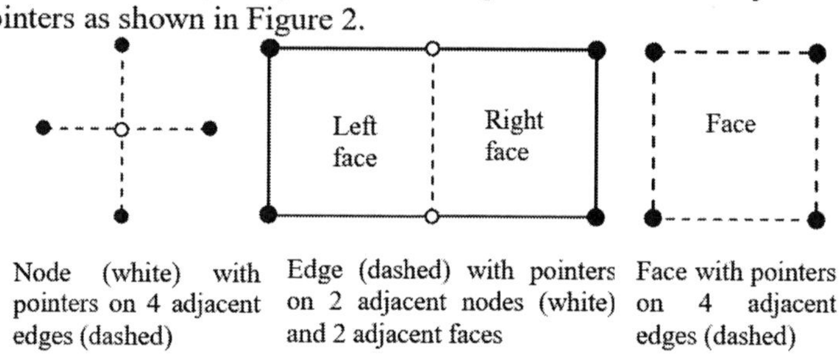

Figure 2. Connections between objects of the mesh

3. Criterion to refine and coarse the mesh

DF are assumed to vary linearly within an element when discretizing differential equations. This approximation is correct as long as the gradient is small enough between two consecutive elements.

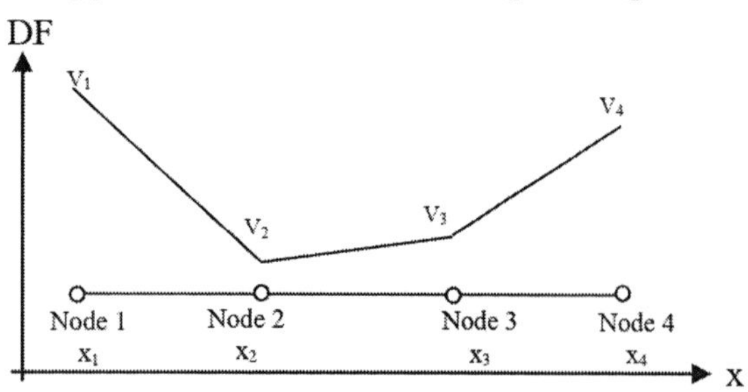

Figure 3. Example of adjacent edges and values of DF associated to each node.

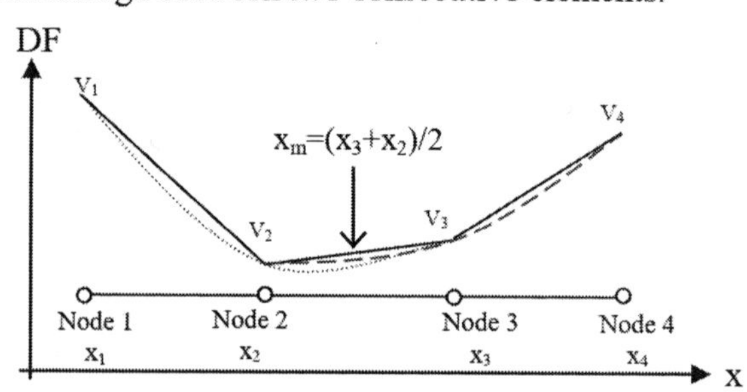

Figure 4. Quadratic interpolation on nodes 1, 2 3 (red dotted line) and on node 2, 3, 4 (green dashed line). Differences between linear and quadratic interpolations are computer at point $x_m=(x_3+x_2)/2$ (arrow).

Then, to correct the mesh, we need an estimate of the error induced by the linear interpolation on one edge. This estimation can be done by calculating the value in the middle of the edge using a quadratic interpolation and comparing with the same value calculated with a linear interpolation. For this we need to retrieve adjacent edges (Figure 3). Thank to our structured data, this can be easily done including two dimension modeling.

Interpolating value in the middle of the edge 32 assuming a linear interpolation between nodes 2 and 3 gives:

$$V_{32} = \frac{V_3 + V_2}{2} \tag{1}$$

Interpolating the same value assuming a quadratic interpolation between nodes 1, 2, 3 (Figure 4: red dotted line) and between nodes 2, 3, 4 (Figure 4: green dashed line) gives:

$$V_{321} = L_1(x_m)\, V_1 + L_2(x_m)\, V_2 + L_3(x_m)\, V_3 \qquad \text{with}\, L_i(x) = \prod_{j=1, j\neq i}^{3} \frac{x - x_i}{x_i - x_j} \tag{2}$$

$$V_{432} = N_2(x_m)\, V_2 + N_3(x_m)\, V_3 + N_4(x_m)\, V_4 \quad \text{with}\, N_i(x) = \prod_{j=2, j\neq i}^{4} \frac{x - x_i}{x_i - x_j} \tag{3}$$

$L_1(x)$, $L_2(x)$, $L_3(x)$ are the Lagrange polynomials of order 2 on nodes 1, 2, 3 and $N_1(x)$, $N_2(x)$, $N_3(x)$ are the Lagrange polynomials of order 2 on nodes 2, 3, 4.

Then the relative error of one DF on the edge 32 can be expressed by the differences:

$$E_{321}^{DF} = 2\frac{|V_{321} - V_{32}|}{|V_{321} + V_{32}|} \qquad \text{and} \qquad E_{432}^{DF} = 2\frac{|V_{432} - V_{32}|}{|V_{432} + V_{32}|} \tag{4}$$

and we keep the higher of these 2 errors:

$$E_{32}^{DF} = \max(E_{321}^{DF}, E_{432}^{DF}) \tag{5}$$

For the DF whose value changes slightly over the device, like potential (psi) and temperature (T), this formula can be used as is. But for the DF that can change by several orders of magnitude like electron and hole densities, the relative error is computed using the decimal logarithm of the DF. Then the formula is weighted to reduce the refinement generated by low values that have a negligible impact on the global solution. The weight function (hereinafter called W) is a spline function as shown on Figure 5.

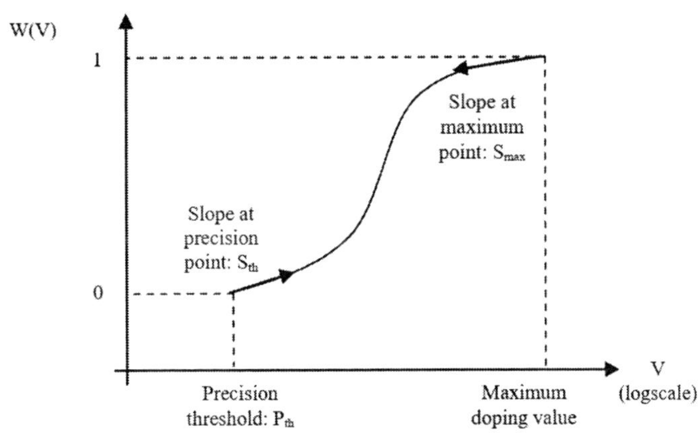

Figure 5. Weight function to reduce refinement for low values of electron and hole densities.

Three parameters of the weight function can be adjusted: the precision threshold (P_{th}) for n or p densities, the slope at the precision threshold point (S_{th}) and the slope at the maximum doping value (S_{max}). These values have been adjusted to optimize the refinement ratio between high carrier densities (at maximum doping value) and low carrier densities (which can be considered negligible):

$P_{th}=10^8 cm^{-3}$, $S_{th}=0$ and $S_{max}=0.5$

With the weight function, the relative error formulas for n and p densities are:

$$E_{32}^{n} = \max(E_{321}^{n}, E_{432}^{n})W(n) \qquad E_{32}^{p} = \max(E_{321}^{p}, E_{432}^{p})W(p) \tag{6}$$

and we keep the higher errors on all DF:

$$E_{32} = \max(E_{32}^{psi}, E_{32}^{T}, E_{32}^{n}, E_{32}^{p}) \tag{7}$$

Thus, the dynamic mesh generation in ECORCE is driven by four parameters:

- the minimum size of edges ($Size_{min}$)
- the maximum size of edges ($Size_{max}$)
- the relative precision for edge splitting (P_{refine})
- the relative precision for edge merging ($P_{release}$)

With these four parameters we define the conditions to refine one edge or merge two edges. An edge is refined if:

$$(Length > Size_{max})\, OR \left((E_{edge} > P_{refine})\, AND (Length > 2\, Size_{min})\right) \tag{8}$$

where Length and E_{edge} define the edge length and the higher relative error for this edge (on all DF).

Two edges (numbered 1 and 2) are merged if:

$$(E_{edge1} < P_{refine} * P_{release})\, AND (E_{edge2} < P_{refine} * P_{release})\, AND\ (Length1 + Lenght2 < Size_{max}) \tag{9}$$

where Length1, Length2, E_{edge1} and E_{edge2} define edge 1 length; edge 2 length, the higher relative error of edge 1 and the higher relative error of edge2.

Consistent values for the relative precisions are 10% to 0.1% for P_{refine} and 10% to 0.1% for $P_{release}$. Low values of P_{refine} result in coarse meshes and low CPU time. High values of P_{refine} give finer meshes but higher CPU time.

Too high values of $P_{release}$ (more than 10%) can lead to oscillations between mesh refinement and release, increasing the CPU time for the mesh calculation. Too low values of $P_{release}$ limit the removal of nodes, increasing the CPU time.

4. Calculation strategy for DF of new nodes

Adding one node, even in a device for which the mesh is properly designed, can bring a significant change on the entire solution. As an example consider the PN junction presented Figure 6 and the electron and hole densities in the junction for a -1V reverse bias presented Figure 7.

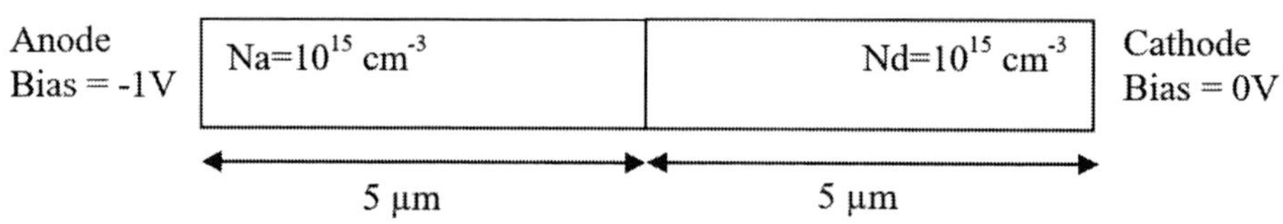

Figure 6. Example of a PN junction

For this modeling coarsening of the mesh is disabled and others parameters are: $Size_{min}=10^{-3}\mu m$, $P_{refine}=10^{-2}$ and $Size_{max}=10\mu m$, resulting in a 95 nodes mesh. Then performing the same modeling, adding only one node at the position $x=4.96\mu m$ (as shown in Figure 7), we get the change for electron and hole densities displayed on Figure 8. Since the node have been added inside the space charge region we can expect to see a change only in this region, namely for low electron and hole densities.

But we see on Figure 8 that this change is also significant for high electron and hole densities. This is highlighted on Figure 9 which presents the total electrical charge density and the difference of electrical charge density between the two modeling. The node is added at the position 4.96μm and the electric charge shows a significant change up to 2μm from that point. Thus adding one node in a device affects nodes over a wide area around the insertion point. This is not a problem for DC modeling where solution can be recomputed after each change of the mesh. But for transient modeling, derivative as a function of time of electron and hole densities are calculated from previous time steps. As shown on Figure 10 n steps are needed to calculate time derivatives at time t_0. n depends on the order of the interpolation polynomial [7]. Then DF of the n steps for each node of the affected area (Figure 9) become unsettled. If the solution is not corrected, the next steps of the modeling may be flawed.

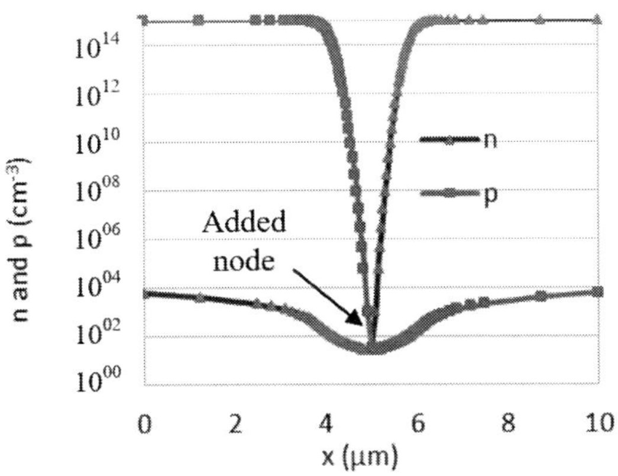

Figure 7. Electron and hole densities in the PN junction

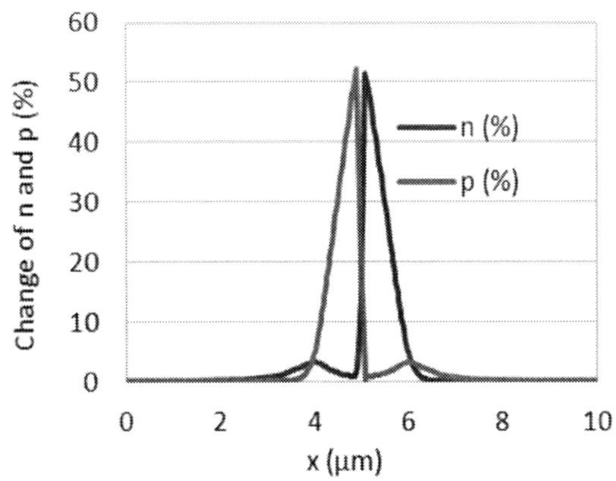

Figure 8. Change of electron and hole densities after adding one node at the position x=4.96μm

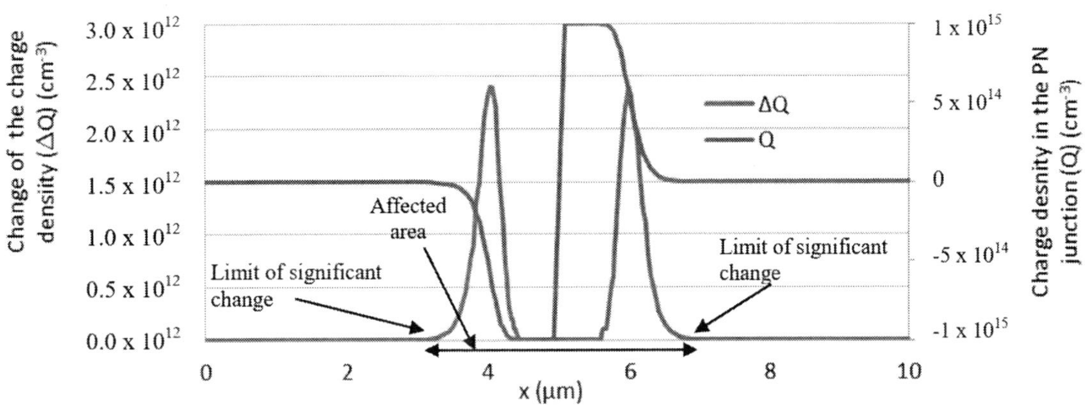

Figure 9. Charge density in the PN junction and change of this density when adding one node at position x=4.96μm

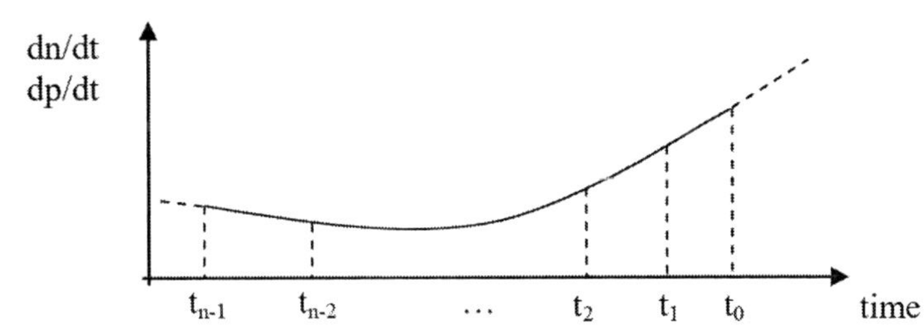

Figure 10. Time steps used to compute n and p densities derivative as a function of time.

4.1. First scheme

As a first method, after a mesh change in transient mode, DF of new nodes for steps 1 to n-1 are linearly interpolated using the DF of the split edge. Then, the complete solution for step 0 is recalculated, namely all DF for all nodes are recomputed for this step. With this method, if the mesh is not fine enough, adding nodes on steps 1 to n-1 brings a too high change in the solution in the affected area. Then the derivatives of electron and hole densities as a function of time show oscillations that appear as spikes in the current collected at the contact. Figure 11a displays the current crossing the PN junction for a sinusoidal bias applied on the Anode with a frequency of 20MHz, amplitude of 0.6V and a 3rd order polynomial interpolation for time derivative. At this frequency, the diode behaves roughly as a capacitor. Thus, the shape of the current is correct but spikes induced by added nodes are clear.

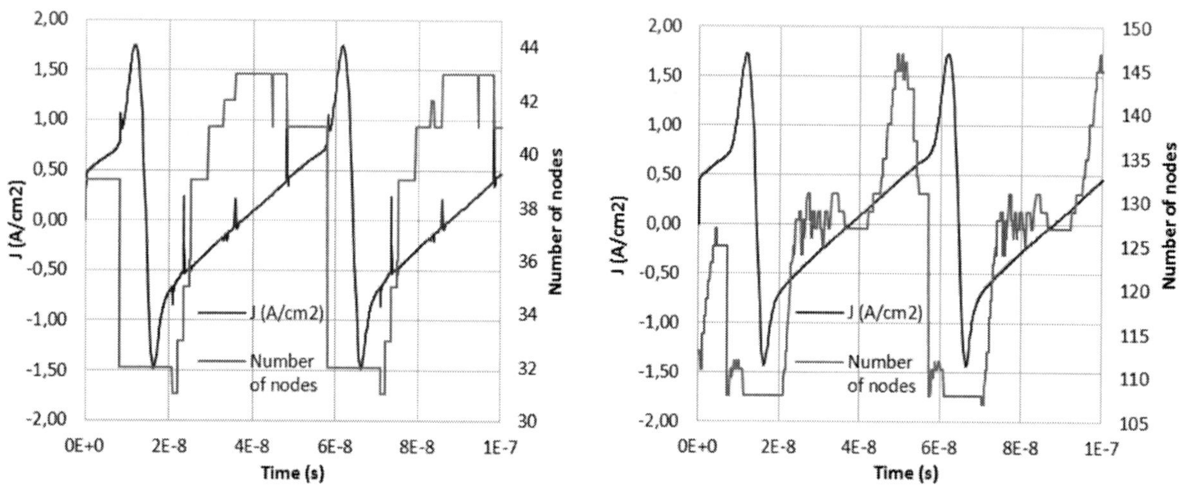

a) P_{refine}=0.2, $P_{release}$=0.2, $SIZE_{min}$=10^{-3}μm b) P_{refine}=0.005, $P_{release}$=0.05, $SIZE_{min}$=10^{-3}μm

Figure 11. Current density and number of nodes for a PN junction with a sinusoidal bias on the Anode, frequency: 20 MHz, amplitude: 0.6V, for the first scheme.

Applying tighter parameters to refine the mesh solves the problem through an increase of the node number and thus an increase of the CPU time. Spikes disappear on Figure 11b but we need 3 to 4 times more nodes than for results displayed Figure 11a.

4.2. Second scheme

Another way to solve this problem is to linearly interpolate DF of new nodes only for step n-1. Then the solution of step n-2 can be recalculated using step n-1 to compute the time derivative $\frac{dn}{dt}$ and $\frac{dp}{dt}$ by a first order interpolation. Step n-3 can then be recalculated using à 2nd order interpolation for time derivative using step n-1 and n-2. And so by degrees the full chain of steps needed to compute step 0 is recalculated.

With this scheme, for each mesh change, the DF for all nodes from step n-2 to 0 are recomputed, especially in the affected area. This method increases the CPU time, but much less nodes are needed to get an accurate solution. Figure 12a displays the current and number of node for this scheme for the same bias and mesh parameters than those of Figure 11a. On Figure 12a the spikes are hardly visible for the same number of nodes than Figure 11a. Furthermore by tightening slightly the mesh parameters (Figure 12b) the current shows no more defects for 2 to 3 times less nodes than for Figure 11b.

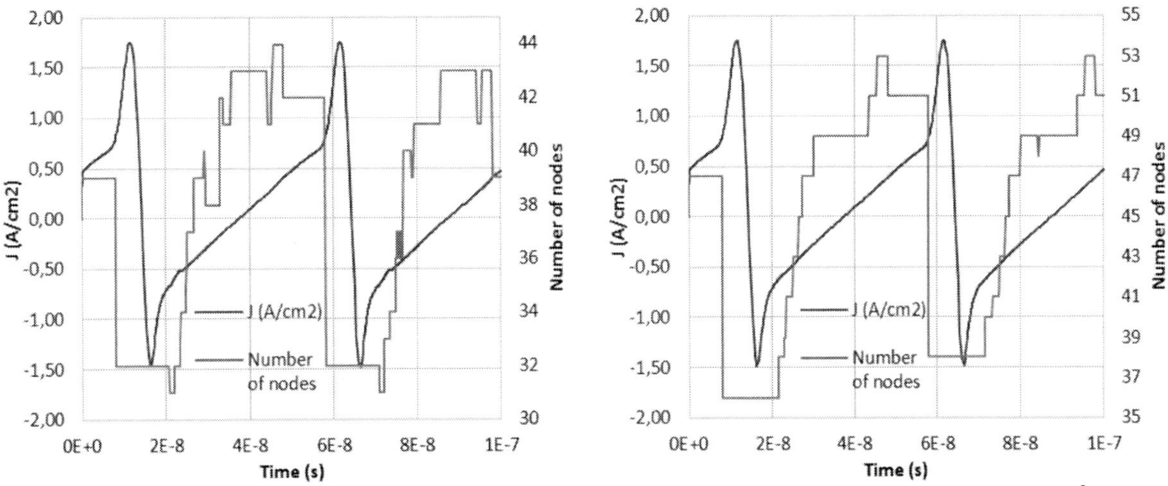

a) P_{refine}=0.2, $P_{release}$=0.2, $SIZE_{min}$=10^{-3}μm b) P_{refine}=0.1, $P_{release}$=0.1, $SIZE_{min}$=10^{-3}μm

Figure 12. Current density and number of nodes for a PN junction with a sinusoidal bias on the Anode, frequency: 20 MHz, amplitude: 0.6V, for the second scheme.

5. Conclusions

Meshing is a critical step of TCAD modeling. From the mesh quality depend the solution precision and the CPU time. But meshing in TCAD software is for now hand made and only one grid is used for all the steps of the simulation. Yet spatial distribution of DF change during modeling and the mesh should be adjusted to gradient of DF at each step. A dynamic mesh generator is a good solution to solve this problem. But driving the mesh generator needs a criterion that tells when to add nodes and when to remove nodes. Thank to the structured data of the mesh, ECORCE can easily retrieve adjacent nodes. Thus a criterion based on polynomial interpolation between 4 adjacent nodes gives a good idea of the DF gradients and allows a fine adjustment of the mesh.

Furthermore, use of dynamic mesh in transient mode with TCAD modeling is challenging. Because electron and hole densities can change by several orders of magnitude in depletion zone, adding only one node in such a place changes the solution in a large area around the insertion point. Then in transient modeling, changing the mesh induces oscillations in the solution that appears as spikes on the current collected at the contacts. Two schemes are proposed to solve this problem. One requires more nodes but less solving cycles, while the other one needs less nodes but more solving cycles.

Using the dynamic mesh generator, the criterion to adjust the mesh and the schemes to remove oscillations of the solution in transient modeling, ECORCE suppresses the hardest step of TCAD modeling.

References

[1] Hudson B 2007 Dynamic mesh refinement (Carnegie Mellon University)
[2] Lopez D, Bruno L , Dynamic Mesh Optimization for Free Surfaces in Fluid Simulation, presented at IMR - 21th international meshing roundtable - 2012, p. 1–5.
[3] *ECORCE Website* (http://www.ecorce.eu/)
[4] Michez A, Dhombres S and Boch J 2015 ECORCE: A TCAD Tool for Total Ionizing Dose and Single Event Effect Modeling *Nucl. Sci. IEEE Trans. On* **62** 1516–27
[5] *Qt Website* (http://www.qt.io/)
[6] Sze S M and Ng K K 2006 *Physics of semiconductor devices* (John Wiley & Sons)
[7] Michez A and Bordure G 1991 Transient simulation of silicon devices under high carrier injection. Comparison of various time stepping schemes. *COMPEL Int. J. Comput. Math. Electr.* **10** 231–40

A note on parallel efficiency of fire simulation on cluster

L Valasek[1], **J Glasa**
Slovak Academy of Sciences/Institute of Informatics, Dubravska cesta 9, 845 07 Bratislava, Slovakia

E-mail: lukas.valasek@savba.sk

Abstract Current HPC clusters are capable to reduce execution time of parallelized tasks significantly. The paper discusses the use of two selected strategies of cluster computational resources allocation and their impact on parallel efficiency of fire simulation. Simulation of a simple corridor fire scenario by Fire Dynamics Simulator parallelized by the MPI programming model is tested on the HPC cluster at the Institute of Informatics of Slovak Academy of Sciences in Bratislava (Slovakia). The tests confirm that parallelization has a great potential to reduce execution times achieving promising values of parallel efficiency of the simulation, however, the results also show that the use of increasing numbers of computational meshes resulting in increasing numbers of used computational cores does not necessarily decrease the execution time nor the parallel efficiency of simulation. The results obtained indicate that the simulation achieves different values of the execution time and the parallel efficiency in regard of the used strategy for cluster computational resources allocation.

1. Introduction

Current fire simulation systems are capable to utilize the knowledge of the CFD (Computational Fluid Dynamics) theory and model fires and their consequences involving many physical and chemical processes related to fire such as combustion, pyrolysis, heat transfer, thermal radiation, turbulence, fluid dynamics and so on. However, simulation of fires in large areas generally requires carrying out the calculation in parallel on HPC (high performance computing) systems.

The FDS (Fire Dynamics Simulator) system [1] is an advanced CFD-based fire simulator intended for fires in various environments. It was developed by NIST (National Institute of Standards and Technology, U. S. Department of Commerce) in cooperation with VTT (Technical Research Centre of Finland). FDS numerically solves a form of Navier-Stokes equations for low-speed fire induced flows with emphasis on transport of smoke and heat from fire. The system is capable to utilize various types of parallelization. To take advantage of computational resources available, FDS supports four programming models: the sequential model designed for sequential computers, the parallel MPI (Message Passing Interface) model designed for systems with distributed memory, the multithreading OpenMP (Open Multi-Processing) model designed for systems with shared memory and the hybrid MPI&OpenMP model designed for systems with distributed shared memory. In previous research we investigated applicability of FDS to simulate fires in various environments [2-4].

In this paper, we focus on the MPI model of parallelization of fire simulation on the HPC cluster at the Institute of Informatics of Slovak Academy of Sciences in Bratislava (Slovakia).

[1] To whom any correspondence should be addressed.

Content from this work may be used under the terms of the Creative Commons Attribution 3.0 licence. Any further distribution of this work must maintain attribution to the author(s) and the title of the work, journal citation and DOI.
Published under licence by IOP Publishing Ltd

There were 52 IBM dx360 M3 computational nodes available for testing purposes; each node consisted of 2 sockets with 6-core processors Intel E5645, 2.4GHz (12 cores per node). Within each node, 48 GB RAM was available (24 GB of RAM per processor). In the MPI model, the computational domain is divided into several computational meshes and the computation on each mesh is considered as a single MPI process assigned to a single computational core. The communication between MPI processes is provided by MPI. We use the 64-bit MPI version 6.3.2 of FDS for Linux and the open-source Open MPI version 1.10.0 of MPI. The script mpirun (a part of the Open MPI) allows determining exactly how the MPI processes representing a task will be mapped by individual computational nodes and then bound to individual sockets or cores. The aim of this paper is to compare the impact of 2 selected computational resources allocation strategies on parallel efficiency of a corridor fire simulation.

2. Simulation of a corridor fire

We consider a 10-second fire of a corridor with the dimensions of 7.2 m x 1.2 m x 2.7 m and an 100 kW rectangular fire source with the dimensions of 0.1 m x 1 m located about 1 mm above the floor at the 3.64 m distance from the left end of the corridor (figure 1, left). Its boundaries are 20 cm far from the corridor walls. The computational domain with the dimensions of the corridor is divided into 2916000 cube cells forming the computational mesh of the 2 cm resolution (360 cells in the x-direction, 60 cells in the y-direction and 135 in the z-direction). The corridor walls and the corridor ends are represented by INERT and OPEN boundary conditions on the domain borders. The fire source is represented by the corresponding VENT with the 1000 kW/m^2 HRRPUA (heat release rate per unit area) boundary condition. We assume the ambient temperature and initial temperature inside the corridor being equal to 20°C, the default dominant chemical reaction and the use of the LES (Large Eddy Simulation) method in regard of the corridor dimensions and mesh resolution. The chosen computational mesh fulfils the conditions required for usage of FFT for solving the pressure equation. The mesh sensitivity study realized (figure 1, middle) shows that the chosen 2 cm mesh corresponds to "fine mesh" in the sense of [5] already since the 2 s to the fire.

The simulation (denoted by 1M) was realized sequentially on a single computational core of the HPC cluster with the execution time of 17.45 hours. The simulated fire behaviour was as follows (figure 1). At first, smoke spread from the fire source upwards to the corridor ceiling. At the 2 s, the fire intensity reached the value of 100 kW and then fluctuated around the value (figure 1, right). At the 3 s to the fire, smoke hit the ceiling and began to spread under the ceiling in all directions from the hit point at the ceiling. At the 6th s, smoke reached both ends of the corridor.

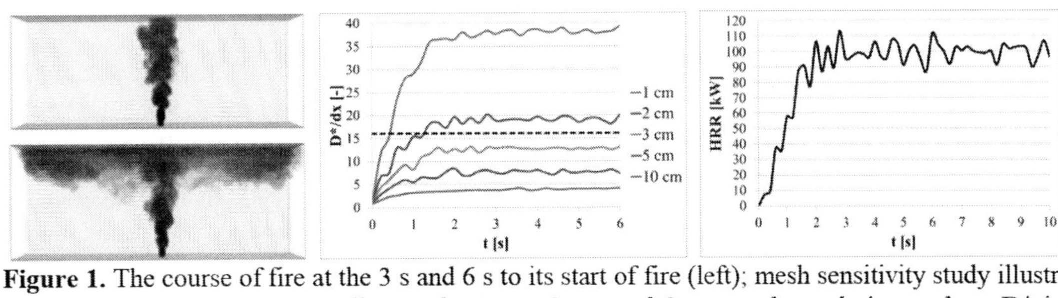

Figure 1. The course of fire at the 3 s and 6 s to its start of fire (left); mesh sensitivity study illustrated by the D*/dx curves corresponding to the 1 cm, 2 cm and 3 cm mesh resolutions, where D* is the characteristic fire diameter, dx is the nominal size of a mesh cell and t is time, and the horizontal dashed line represents the values corresponding to “fine mesh” in sense of [5] (middle); and the heat release rate (HRR) curve of fire (right)

3. Parallelization of the simulation and its realization

We consider a decomposition of the computational domain in x-direction onto m subdomains of the same size and 9 parallel simulations mM, where m = 2, 3, ..., 6, 8, 9, 10, 12 (figure 2). The decomposition for m = 7 and m = 11 does not fulfil the conditions required for solving a Poisson equation using FFT, therefore, we do not consider the simulations 7M and 11M. The tested simulations have the same 2 cm resolution with the total number of cells equal to 2916000 (2916000/m cells per individual mesh). Since the chosen mesh resolution is the same as in the sequential calculation and we simulate the same fire scenario, the mesh sensitivity study performed for the sequential calculation and analysis of the HRR curves of executed individual parallel simulations showed that we can suppose that also the computational meshes in parallel simulations correspond with "fine mesh" in the sense of the study [5].

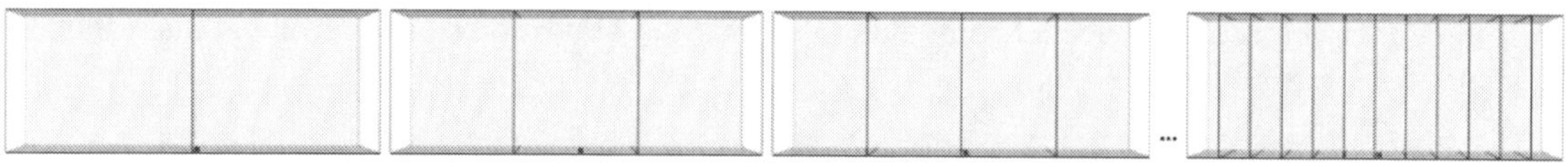

Figure 2. Decomposition of the computational domain for the chosen simulations

In the MPI model each simulation mM is represented by m MPI processes, each of them carried out on a single computational core. The mapping process of a given simulation on individual computational nodes of the HPC cluster and their subsequent binding to individual sockets or computational cores consists of three phases: mapping, ranking and binding [6]. In this paper, we use 2 allocation strategies: --map-by core, --bind-to core (denoted by CC) and --map-by socket, --bind-to socket (denoted by SS). The CC strategy binds the individual MPI processes to available individual computational cores in such way that it gradually binds individual MPI processes to corresponding individual computational cores of the socket 0 and then to corresponding individual cores of the socket 1. The SS strategy binds individual MPI processes alternately to the group of all cores of given socket the way that it binds the MPI process 0 and even MPI processes to the cores of the socket 0 and odd MPI processes to the cores of the socket 1. We reserved the whole computational node for calculation of each simulation in order to eliminate its interference with other tasks being resolved on the HPC cluster resulting in increase of the execution time of simulation.

Table 1. Comparison of considered parallel simulations

	2M	**3M**	**4M**	**5M**	**6M**	**7M**	**8M**	**9M**	**10M**	**11M**	**12M**
CC	12.35	8.12	7.72	6.03	6.39	-	5.02	4.11	4.32	-	3.74
SS	11.56	7.50	6.44	5.10	4.78	-	4.17	3.78	3.92	-	3.75
CC	70.65	71.63	56.51	57.88	45.51	-	43.45	47.17	40.39	-	38.88
SS	75.48	77.56	67.74	68.43	60.84	-	52.31	51.29	44.52	-	38.78
CC, SS	-4.83	-5.92	-11.23	-10.55	-15.33	-	-8.86	-4.12	-4.12	-	0.10

In Table 1, the values of execution times (in hours, rows 1 and 2) and parallel efficiency (in %, rows 3 and 4) of the considered simulations carried out with the CC and SS strategies, and their differences (in %, row 5) are shown (see figure 3). The parallel efficiency was calculated according to the number of computational meshes. The comparison of values of the parallel efficiency of simulations carried out with the CC and SS strategies shows that the tests do not confirm the original expectation that increasing number of computational meshes (or used computational cores) decreases the parallel efficiency of simulations (see the increase of the parallel efficiency of the couples of simulations 2M, 3M; 4M, 5M and 8M, 9M carried out with the CC strategy and the couples of simulations 2M, 3M and 4M, 5M carried out with the SS strategy).

Figure 3 also indicates that the values of parallel efficiency of the simulations realized by the CC strategy are smaller than the ones realized by the SS strategy (except the case of m = 12). Their difference varies from -15.33 to -4.12 (in the case of m = 12 it equals to 0.10). Thus, it can be concluded that the simulations carried out with the CC strategy reach substantially greater execution times and substantially smaller parallel efficiencies as compared with the ones realized by the SS strategy.

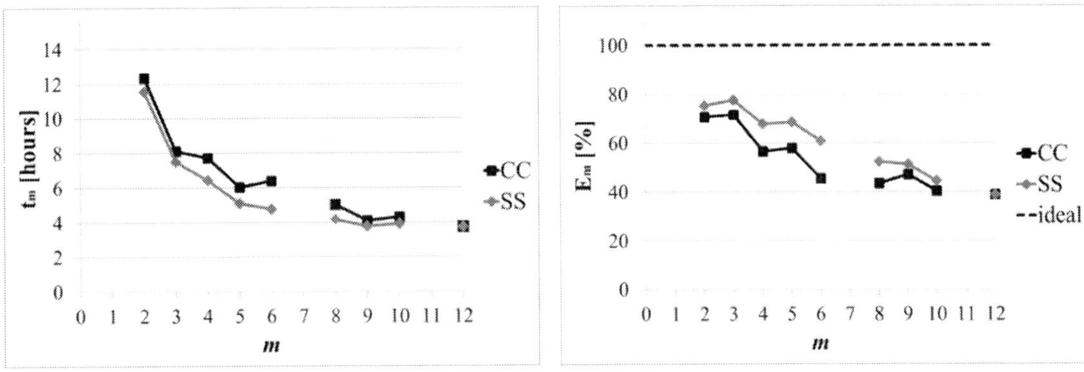

Figure 3. Execution times t_m (left) and parallel efficiency E_m of the simulations mM carried out with the CC and SS strategies, where m means the number of computational meshes. The dashed line represents values of ideal parallel efficiency of simulation.

4. Conclusion

In this paper, the effect of the 2 strategies for allocation of the MPI processes representing the given simulation carried out on computer cluster to individual cores or sockets on parallel efficiency of fire simulation was examined. The series of one sequential and 9 parallel simulations of simple corridor fire scenario was realized on the HPC cluster at the Institute of Informatics of Slovak Academy of Sciences in Bratislava using the parallel MPI programming model of FDS and the strategies --map-by core, --bind-to core and --map-by socket, --bind-to socket. The parallel efficiencies of parallel simulations were compared. The tests confirm a great potential of parallelization to reduce execution times and achieve promising values of parallel efficiency of simulation. The results indicate that the simulations carried out with the CC strategy reach markedly smaller values of parallel efficiency as compared with the simulations carried out with the SS strategy. Analysis of the simulation results also shows that parallelization of the simulation does not necessarily decrease the execution time nor the parallel efficiency of simulation with increasing number of computational meshes. The tests indicate that the problem of efficient parallel realization of fire simulation on computer clusters using FDS and the MPI programming model must be carefully considered.

Acknowledgement
This paper was supported by the Slovak Science Foundation VEGA (project No. 2/0184/14).

5. References

[1] McGrattan K, Hostikka S, McDermott R, Floyd J, Weinschenk C and Overholt K 2015 Fire Dynamics Simulator (Version 6): User's Guide, NIST Special Publication 1019-6, NIST, Washington

[2] Weisenpacher P, Glasa J, Halada L, Valasek L and Sipkova V 2014 Computing and Informatics 33 6 1237-1268

[3] Glasa J, Valasek L, Halada L and Weisenpacher P 2014 J. Phys.: Con. Ser. 490 012067 DOI: 10.1088/1742-6596/490/1/012067

[4] Valasek L 2015 J. Phys.: Con. Ser. 633 012083 DOI: 10.1088/1742-6596/633/1/012083

[5] Hill K, Dreisbach J, Joglar F, Najafi B, McGrattan K, Peacock R and Hamins A 2007 Veriffication and Validation of Selected Fire Models for Nuclear Power Plant Applications, NUREG 1824. United States Nuclear Regulatory Commission, Washington, DC.

[6] Documentation of OpenMPI, v1.10.0, Aug. 25, 2015. https://www.open-mpi.org/doc/v1.10/.

Design of order statistics filters using feedforward neural networks

Yu.S. Maslennikova, V.V. Bochkarev

Kazan Federal University, Kremlevskaya str.18, Kazan 420018, Russia

E-mail: JSMaslennikova@kpfu.ru, vbochkarev@mail.ru

Abstract. In recent years significant progress have been made in the development of nonlinear data processing techniques. Such techniques are widely used in digital data filtering and image enhancement. Many of the most effective nonlinear filters based on order statistics. The widely used median filter is the best known order statistic filter. Generalized form of these filters could be presented based on Lloyd's statistics. Filters based on order statistics have excellent robustness properties in the presence of impulsive noise. In this paper, we present special approach for synthesis of order statistics filters using artificial neural networks. Optimal Lloyd's statistics are used for selecting of initial weights for the neural network. Adaptive properties of neural networks provide opportunities to optimize order statistics filters for data with asymmetric distribution function. Different examples demonstrate the properties and performance of presented approach.

1. Introduction

In recent times, many linear filtering techniques were widely used in various fields of science and technology. At the same time, the use of linear filtering does not provide a satisfactory solution for many practical applications. It is known, for example, that the problem of optimal filtering can be solved in the class of linear filters when the additive noise is statistically independent and has a distribution function that is close to the normal distribution. In practice, the additive noise can be dependent on desired signal or have a distribution function the is different from the normal. In these cases, the optimal solution can be found in the class of non-linear filters, for example, using artificial neural networks [1].

Linear order statistics (or linear order estimation of some variable μ) is a combination $\hat{\mu}=\sum_{i=1}^{n}\alpha_i x^{(i)}$, where $x^{(i)}$ - i-th term in an ordered (ranked) series. Particularly, the arithmetic average can be obtained as an example of the linear order statistic by setting coefficients $\alpha_i = 1/n$ [2]. Some cases of order statistics play a major role in the various statistical studies, such as extrema, sampling range, sample median and other.

The problem of optimal weighting of observations (or finding the optimal linear statistics) was set in the middle of the last century, and soon was successfully solved by E.H. Lloyd, who found a common solution (on the assumption that the distribution function is known) for the evaluation of the two unknown parameters of the distribution function: mean and variance [3]. However, in practice, such linear estimates are insufficient. Therefore, in recent years nonlinear order statistics filters have been developed, especially for digital image processing. Authors of the paper [4, 5] present filters that is based on nonlinear order statistics using perceptron.

Content from this work may be used under the terms of the Creative Commons Attribution 3.0 licence. Any further distribution of this work must maintain attribution to the author(s) and the title of the work, journal citation and DOI.
Published under licence by IOP Publishing Ltd

In the present work non-linear analogues of order statistics based on feedforward neural networks is proposed. The main idea of the proposed technique is based on Lloyd's statistics, that are used for initialization of the neural network weights. Nonlinear activation function of neurons improves the quality of filtering results in cases with distribution different from the normal.

2. Lloyd's statistics

For Lloyd's statistics estimation it is assumed that the form of density function for observed values is known, but its parameters (like mean value μ and variance σ) can change. The density distribution function can be determined as following:

$$w(x) = \frac{1}{\sigma} f(\frac{x-\mu}{\sigma}), \tag{1}$$

If we know the density function (1), we can estimate mathematical expectation and variance of order statistics $\mu = 0, \sigma = 1$:

$$\xi_i = Mx^{(i)}, \tag{2}$$

$$B_{i,j} = M(x^{(i)} - Mx^{(i)})(x^{(j)} - Mx^{(j)}), \tag{3}$$

Theoretical equations for ξ_i and $B_{i,j}$ values are rather complicated, we do not present them here. For unknown random parameters μ and σ mathematical expectation of order statistics equals to [2]:

$$Mx^{(i)} = \mu + \sigma\xi_i, \tag{4}$$

where x is a vector of observed order statistics.

For estimation of the unknown parameters it is needed to select values that minimize standard deviation between conditional mathematical expectations (4) and current set of order statistics. Let write simpler form of these parameters for father calculations:

$$A = \begin{bmatrix} 1 & \xi_1 \\ 1 & \xi_2 \\ \vdots & \vdots \\ 1 & \xi_n \end{bmatrix} \text{ and } \Theta = \begin{pmatrix} \mu \\ \sigma \end{pmatrix}.$$

The expression for the least-squares criterion can be written taking into account the correlation of order statistics as: $\Phi = (x - A\Theta)^H B^{-1} (x - A\Theta)$.

Varying the parameter Θ and equating the result to zero, we found the unknown parameters in the following form $\hat{\Theta} = (A^H B^{-1} A)^{-1} A^H B^{-1} x$.

Matrix (5) of found coefficients gives us the required coefficients of linear order statistics [3].

$$(A^H B^{-1} A)^{-1} A^H B^{-1} = \begin{bmatrix} \alpha_1 & \alpha_2 & \cdots & \alpha_n \\ \beta_1 & \beta_2 & \cdots & \beta_n \end{bmatrix}, \tag{5}$$

Mathematical expectation and variance can be found as the following expressions:

$$\hat{\mu} = \sum_{i=1}^{n} \alpha_i x^{(i)}, \tag{6}$$

$$\hat{\sigma} = \sum_{i=1}^{n} \beta_i x^{(i)}, \tag{7}$$

If we do not know the analytic form of distribution, but have a sufficiently large set of empirical data, we can calculate the average value and covariance of order statistics, and then use equations (6) and (7) for finding of approximate values for coefficients α и β.

If $x^{(i)}$ is the set of ranged values, there are some obvious requirements for ordered series which are used to estimate the mean and variance, such as $\mu = f_\mu\left(x^{(1)}, x^{(2)}, \ldots x^{(n)}\right)$.

First of all, if we add some constant C to all values, estimations of the mean value should change on the same value. That's why the first condition on the function f:

$$f_\mu\left(x^{(1)} + C, x^{(2)} + C, \ldots x^{(n)} + C\right) = f_\mu\left(x^{(1)}, x^{(2)}, \ldots x^{(n)}\right) + C, \tag{8}$$

At the same time, for the function that provides estimation of the variance $\hat{\lambda} = f_\lambda\left(x^{(1)}, x^{(2)}, \ldots x^{(n)}\right)$, adding of additional constant C shouldn't change the estimation of the variance. That's why the second condition is:

$$f_\lambda\left(x^{(1)} + C, x^{(2)} + C, \ldots x^{(n)} + C\right) = f_\lambda\left(x^{(1)}, x^{(2)}, \ldots x^{(n)}\right), \tag{9}$$

Based on these conditions we can write following equations for coefficients of order statistics:

$$\sum_k \alpha_k = 1, \ \sum_k \beta_k = 0, \tag{10}$$

3. Nonlinear order statistics filter

Unlike the linear statistics, for neural network with nonlinear activation functions the conditions (8) can't be satisfied, because for some values of constant C activation functions will saturate and properties of estimations will change. Suppose that we know the probability distribution function of the estimated parameters μ и λ, or at least the range of their variations. Then we can create a representative set of training examples, and we need to achieve the results that the neural network gives a good performance for this prior distribution. In this case the exact conditions (8) are not guaranteed, but if the distribution of the estimated parameters corresponds to our expectations, then we will get satisfactory results (figure 1). Design of order statistics based on neural networks has obvious similarity to the estimates based on the criterion of maximum of a posteriori probability.

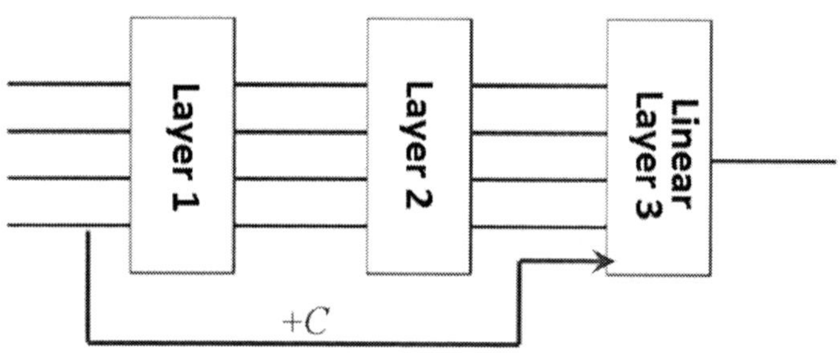

Figure 1. Neural network architecture for order statistics estimations.

It would be better to use information about Lloyd's order statistics for initial distribution of neurons weights [7]. Consider the simplest technique for solving this problem in the case when functions implemented by various network layers are evidently separated. The error that occurs due to the nonlinearity of neurons activation functions prior to the network training should be small. This fact imposes certain restrictions on the choice of activation functions for layers. For example, the hyperbolic tangent (the sigmoid) has a domain with a weak nonlinearity at the origin of coordinates.

While using neurons with sigmoidal activation functions, by properly scaling the training set we can make transformations of a signal by the network neurons close to linear.

Let us consider, as an example, the application of the proposed initialization algorithm to a three-layer feedforward neural network. The input layer of the network scales the data to make all values of the series to be located within the linear domain of the hyperbolic tangent. The intermediate layer performs a weak nonlinear transformation with the help of the weight matrix that contains coefficients of the Lloyd's statistics filter. The output layer performs the inverse data transform. By a proper choice of the α coefficient one can make the error stipulated by the nonlinearity of the activation function arbitrarily small. In this case the neural network with weights chosen in such a way will estimate the current value of order statistics in accordance with expressions (6, 7) [6].

$$A=\begin{bmatrix}0 & 0 & 0 & \dots & 1/\alpha\end{bmatrix}\begin{bmatrix}0 & 1 & 0 & \dots & 0\\ 0 & 0 & 1 & \dots & 0\\ \dots & \dots & \dots & \dots & \dots\\ 0 & 0 & 0 & \dots & 1\\ -a_1 & -a_2 & -a_3 & \dots & -a_p\end{bmatrix}\begin{bmatrix}\alpha & 0 & 0 & \dots & 0\\ 0 & \alpha & 0 & \dots & 0\\ 0 & 0 & \alpha & \dots & 0\\ 0 & 0 & 0 & \dots & 0\\ 0 & 0 & 0 & \dots & \alpha\end{bmatrix}$$

Note that the neural network output before training nearly coincide with the Lloyd's statistics. When training the network with an algorithm with a monotonic decrease of the error (e.g., gradient and quasi-Newton algorithms, conjugate gradients method, etc), the result obtained on any training step is the same or even better than that obtained with the help of the linear statistics. In this paper we use the Levenberg-Marquardt algorithm; it belongs to the class of quasi-Newton methods that guarantes a high convergence rate [1].

4. Results

Consider the synthesis of nonlinear order statistics on a model example with Generalized Extreme Value distribution (GEV) with the parameter k:

$$w_t(x)=\frac{1}{\sigma}\exp\left(-\left(1+k\frac{x-\mu}{\sigma}\right)^{-\frac{1}{k}}\right)\left(1+k\frac{x-\mu}{\sigma}\right)^{-1-\frac{1}{k}}, \tag{11}$$

This type of distribution is often used to simulate the smallest and largest value among a large set of independent and identically distributed random values representing measurements of some values or observations. The main feature of this distribution is that it is asymmetric (figure 2).

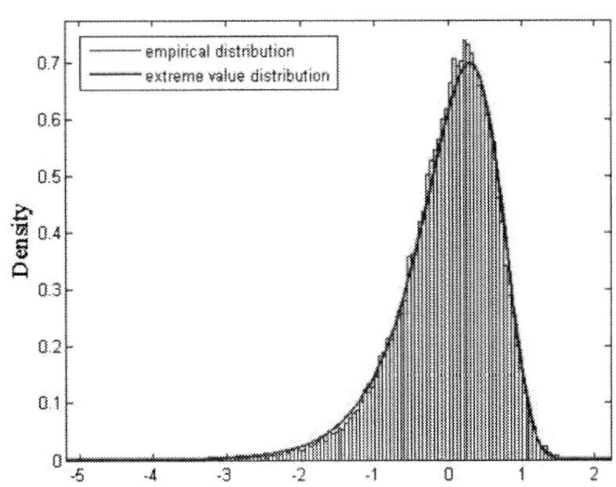

Figure 2. An example of GEV distributed data.

We modeled the set of N samples of the same type, consisting of n counts each. The purpose of this investigation is to compare the Lloyd's statistics coefficients with the neural networks statistics under the assumption that the distribution type is GEV.

The procedure involves the following stages:

1) With the help of a random number generator we create a matrix $N \times n$ of random numbers distributed according to the GEV distribution, then we add to the each line a constant C, distributed by the normal law with a mean of 0 and a variance σ_{shift};

2) Sorting the numbers in each lines. Thus, each line forms an ordered sample of n elements, and we have N number of such test samples.

3) Averaging the columns, we find estimates for the expectation of order statistics. Calculating the covariance between the columns, we find an estimation of the covariance matrix for order statistics;

4) According to the equations (4, 5) we calculate two sets of coefficients - for the expectation and variance. After that we can train the neural network. The example of such modelling is shown on the figure 3.

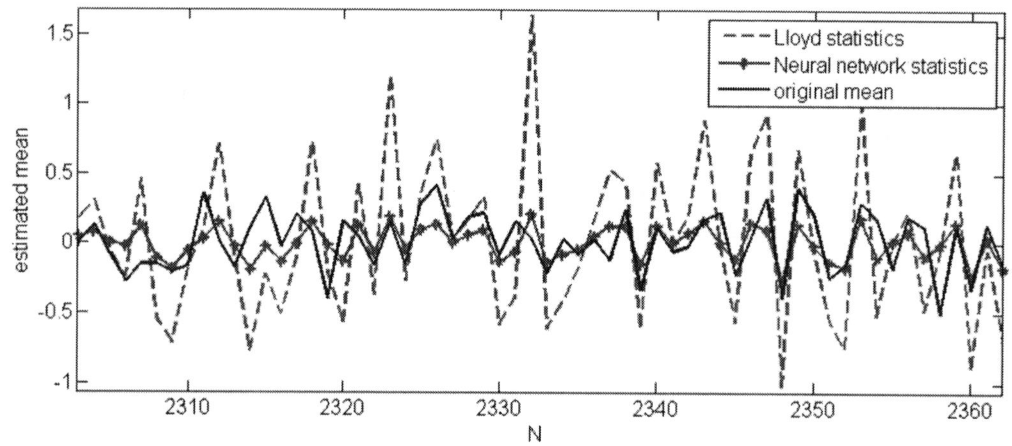

Figure 3. Comparison of expectation values based on Lloyd's and neural network statistics.

Table 1 shows the results of distribution parameters estimations for a sample of random variables with GEV-distribution. The table includes standard deviations for different statistics (Lloyd and neural network statistics) varying the distribution parameter k and the sample size n.

Table 1. Standard deviation of expected value estimations.

Method	n	Parameter k for the GEV-distribution					
		0	0.1	0.5	1	1.5	2
Lloyd's statistics	5	0.479	0.489	0.542	0.610	0.857	0.975
NN statistics	5	0.218	0.217	0.203	0.203	0.823	0.934
Lloyd's statistics	15	0.275	0.274	0.296	0.315	0.334	0.342
NN statistics	15	0.166	0.161	0.167	0.094	0.268	0.214

Figure 4a shows an example of a grayscale image. The same image with GEV-distributed noise is shown on the figure 4b. Figure 4c shows the results of the image filtering based on Lloyds statistics. And the figure 4d shows the results of the neural network statistics filter. It is clearly seen that nonlinear order statistics filter is better that linear order statistics filter.

Figure 4a. Original image.

Figure 4b. Image with a noise of GEV-distribution.

Figure 4c. The result of the Lloyd's statistics filter

Figure 4d. The result of the neural network statistics filter

However, when we work with real data, we cannot determine in advance the type of noise distribution in the processed signal. Therefore, the filtering of a set of real data will include the following stages: in the first stage the separation of noise from the original dataset with the median filtering or moving average, then an assessment of species distribution and its parameters. Then forming a test data with a given type of distribution, and the search for the coefficients of linear and neural network of order statistics, with the help of a network formed by filtering the original data. To assess the quality of filtering, you can repeat the procedure of filtering, starting with the first step.

References

[1] Haykin S 1999 Neural Networks: A Comprehensive Foundation, *Prentice Hall*
[2] David H A and Nagaraja H 2003 Order Statistics *Wiley Series in Probability and Statistics*
[3] Lloyd E H 1952 Least-Squares Estimation of Location and Scale Parameters Using Order Statistics *Biometrika* **39.1/2**
[4] Chang-Shing Lee, Yau-Hwang Kuo and Pao-Ta Yuhttp 1997 Weighted fuzzy mean filters for image processing *Fuzzy Sets and Systems*. **89** 157-180
[5] Yiming Pi and Shunji Huang 1995 Design of multistage weighted order statistic filters by a neural network *International Conference on Image Processing (ICIP'95)* **Vol. 1** 346-250.
[6] Maslennikova Yu S and Bochkarev V V 2011 Nonlinear order statistics using artificial neural network *MEPHI Neiroinformatics-2011 Book of papers* **2** 102-113 (In Russian).
[7] Maslennikova Yu S and Bochkarev V V 2010 Initialization of multilayer forecasting artifical neural networks *Uchenye Zapiski Kazanskogo Universiteta. Seriya Fiziko-Matematicheskie Nauki,* **152** 7-14 (In Russian). arXiv:1410.6413 (In English)

Approximation of an $M/M/s$ queue by the $M/M/\infty$ one using the operator mathod

A Bareche[1], **M Cherfaoui**[2] **and D Aïssani**[1]

[1] Research Unit LaMOS (Modeling and Optimization of Systems), University of Bejaia, 06000 Bejaia, Algeria
[2] Department of Mathematics, University of Biskra, 07000 Biskra, Algeria

E-mail: aicha_bareche@yahoo.fr

Abstract. In this paper, we provide an approximate analysis of an $M/M/s$ queue using the operator method (strong stability method). Indeed, we use this approach to study the stability of the $M/M/\infty$ system (ideal system), when it is subject to a small perturbation in its structure ($M/M/s$ is the resulting perturbed system). In other words, we are interested in the approximation of the characteristics of an $M/M/s$ system by those of an $M/M/\infty$ one. For this purpose, we first determine the approximation conditions of the characteristics of the perturbed system, and under these conditions we obtain the stability inequalities for the stationary distribution of the queue size. To evaluate the performance of the proposed method, we develop an algorithm which allows us to compute the various obtained theoretical results and which is executed on the considered systems in order to compare its output results with those of simulation.

1. Introduction

Multi-servers queues are analytically tractable queues of practical importance. However, the analytical results of such systems are only available in terms of Laplace transforms or/and generating functions which are often cumbersome and not useful in practice. To avoid and circumvent this problem, many authors use numerical and approximation methods to analyze such type of systems (see [5, 6, 7, 8]). On the other hand, many situations are modeled by a multi-server queue, but for which we seek to have the average number of waiting customers to tend to zero. This situation coincides with an infinite-servers queue that has the non waiting property.

It is why we focus, in this work, on the strong stability method (see [1, 11]) which allows us to make both qualitative and quantitative analysis helpful in understanding complicated models by more simpler ones for which an evaluation can be made. This method, also called "method of operators" can be used to investigate the ergodicity and stability of the stationary and non-stationary characteristics of the imbedded Markov chains (see [11]). In contrast to other methods, it supposes that the perturbations of the transition kernel are small with respect to some norms in the operators space. This stringent condition gives better stability estimates and enables us to find precise asymptotic expansions of the characteristics of the perturbed system.

The applicability of the strong stability method is well proved and documented in various fields and for different purposes. In particular, it has been applied to several proposals (see for example [2, 3, 4, 12]).

Content from this work may be used under the terms of the Creative Commons Attribution 3.0 licence. Any further distribution of this work must maintain attribution to the author(s) and the title of the work, journal citation and DOI.
Published under licence by IOP Publishing Ltd

This paper aims to study the strong stability of an $M/M/\infty$ system (ideal system) after a small perturbation of its structure (the $M/M/s$ is the resulting perturbed system). We first clarify the conditions for such an approximation, then we provide an upper bound for the norm of the difference between the two stationary distributions of the considered systems. Note that there is a first attempt to study this particular case of approximation of the $M/M/s$ queue by the $M/M/\infty$ one, using the theory of Markov chains and the special norm L_1 (see [9]). The authors of this latter work had produced two important results. The first one is summarized in the upper bound of the absolute difference (L_1 norm) between the stationary probabilities of the $M/M/s$ and $M/M/\infty$ systems, and in the second point the authors have proved that this difference tends to zero when the number of servers tends to infinity. Unlike [9], we use here the general weight norm ($\|.\|_v$) instead of the norm L_1 to approach the $M/M/s$ system by the $M/M/\infty$ one. Moreover, we provide the domain within which this approximation is valid.

2. Description of $M/M/s$ and $M/M/\infty$ models

Let us consider an $M/M/s$ system with s servers, where inter-arrival times are independently distributed with an exponential distribution $E_\lambda(t)$ and mean inter-arrival time $1/\lambda$, and service times are distributed with $E_\mu(t)$ (exponential with parameterμ).
Let X_k be the number of customers in the system just prior to the arrival of the kth customer. Therefore, $X = (X_k; k = 0, 1, \ldots)$ is an homogeneous Markov chain with a state space $\mathbf{N} = \{0, 1, 2, \ldots\}$ and with a transition operator $P = (P_{ij})_{i,j\geq 0}$ where:

$$P_{ij} = \begin{pmatrix} \frac{\mu}{\lambda+\mu} & \frac{\lambda}{\lambda+\mu} & 0 & 0 & \cdots & 0 & 0 & 0 & \cdots \\ \frac{\mu}{\lambda+\mu} & 0 & \frac{\lambda}{\lambda+\mu} & 0 & \cdots & 0 & 0 & 0 & \cdots \\ 0 & \frac{2\mu}{\lambda+2\mu} & 0 & \frac{\lambda}{\lambda+2\mu} & \ddots & 0 & 0 & 0 & \ddots \\ 0 & 0 & \frac{3\mu}{\lambda+3\mu} & 0 & \ddots & \ddots & 0 & 0 & \ddots \\ \vdots & \vdots & \ddots & \ddots & \ddots & \ddots & \ddots & \ddots & \ddots \\ 0 & 0 & 0 & \ddots & \frac{s\mu}{\lambda+s\mu} & 0 & \frac{\lambda}{\lambda+s\mu} & 0 & \ddots \\ 0 & 0 & 0 & 0 & \ddots & \frac{s\mu}{\lambda+s\mu} & 0 & \frac{\lambda}{\lambda+s\mu} & \ddots \\ \vdots & \vdots & \ddots & \ddots & \ddots & \ddots & \ddots & \ddots & \ddots \end{pmatrix} \quad (1)$$

Consider also an $M/M/\infty$ system, with the same distributions of arrivals and service times as in the previous system, and with no waiting room.
Let $\tilde{X}_k$ be the number of customers in the system just prior to the arrival of the kth customer. Therefore, $\tilde{X} = (\tilde{X}_k; k = 0, 1, \ldots)$ is an homogeneous Markov chain with a state space $\mathbf{N} = \{0, 1, 2, \ldots\}$ and with a transition operator $\tilde{P} = (\tilde{P}_{ij})_{i,j\geq 0}$ where:

$$\tilde{P}_{ij} = \begin{pmatrix} \frac{\mu}{\lambda+\mu} & \frac{\lambda}{\lambda+\mu} & 0 & 0 & \cdots & 0 & 0 & 0 & \cdots \\ \frac{\mu}{\lambda+\mu} & 0 & \frac{\lambda}{\lambda+\mu} & 0 & \cdots & 0 & 0 & 0 & \cdots \\ 0 & \frac{2\mu}{\lambda+2\mu} & 0 & \frac{\lambda}{\lambda+2\mu} & \ddots & 0 & 0 & 0 & \ddots \\ 0 & 0 & \frac{3\mu}{\lambda+3\mu} & 0 & \ddots & \ddots & 0 & 0 & \ddots \\ \vdots & \vdots & \ddots & \ddots & \ddots & \ddots & \ddots & \ddots & \ddots \\ 0 & 0 & 0 & \ddots & \frac{s\mu}{\lambda+s\mu} & 0 & \frac{\lambda}{\lambda+s\mu} & 0 & \ddots \\ 0 & 0 & 0 & 0 & \ddots & \frac{(s+1)\mu}{\lambda+(s+1)\mu} & 0 & \frac{\lambda}{\lambda+(s+1)\mu} & \ddots \\ \vdots & \vdots & \ddots & \ddots & \ddots & \ddots & \ddots & \ddots & \ddots \end{pmatrix} \quad (2)$$

Let π and $\tilde{\pi}$ be, respectively, the stationary distributions of the $M/M/s$ and $M/M/\infty$ systems.

3. The strong stability concept

3.1. Preliminary notations

Let $\mathcal{M} = \{\nu_j\}$ be the space of finite measures on $\mathbf{N}$, and let $\mathcal{N} = \{f(j)\}$ be the space of bounded measurable functions on $\mathbf{N}$. We associate with each transition kernel P the linear mapping:

$$(\mu P)_k = \sum_{j\geq 0} \mu_j P_{ik}, \tag{3}$$

$$(Pf)(k) = \sum_{i\geq 0} f(i) P_{ki}. \tag{4}$$

Introduce on $\mathcal{M}$ the v-norm of the form:

$$\|\nu\|_v = \sum_{j\geq 0} v(j)|\nu_j|, \tag{5}$$

where $v(k) = \beta^k$, for all $k \in \mathbf{N}$ and $\beta > 1$ is a real parameter. This norm induces in the space $\mathcal{N}$ the norm

$$\|f\|_v = \sup_{k\geq 0} \frac{|f(k)|}{v(k)}. \tag{6}$$

Moreover, for all $\nu \in \mathcal{M}$ and $f \in \mathcal{N}$, the symbols νf and $f \circ \nu$ denote respectively the summation and the kernel defined as below

$$\nu f = \sum_{k=0}^{+\infty} f(k)\nu_k, \tag{7}$$

$$(f \circ \mu)(k, j) = f(k)\,\mu_j, \text{ for all } (k, j) \in \mathbf{N} \times \mathbf{N}. \tag{8}$$

Let us consider $\mathcal{B}$, the space of linear operators, with the norm

$$\|Q\|_v = \sup_{k\geq 0} \frac{1}{v(k)} \sum_{j\geq 0} v(j) Q_{kj}. \tag{9}$$

3.2. Strong stability criteria

For more details on the strong stability method, see [1, 10, 11].

Let us give the definition of the strong stability (qualitative aspect) for an homogeneous Markov chain in the phase state $(\mathbf{N}, \mathcal{B}(\mathbf{N}))$ with respect to the v-norm. Here $\mathcal{B}(\mathbf{N})$ is the σ-algebra generated by the singletons $\{j\}$.

Definition 1 *(see [1, 11]) A Markov chain X with a transition kernel P and an invariant measure π is said to be strongly v-stable with respect to the norm $\|.\|_v$ if $\|P\|_v < \infty$ and each stochastic kernel Q on the space $(\mathbf{N}, \mathcal{B}(\mathbf{N}))$ in some neighborhood $\{Q : \|Q - P\| < \epsilon\}$ has a unique invariant measure $\mu = \mu(Q)$ and $\|\pi - \mu\|_v \to 0$ as $\|Q - P\|_v \to 0$*

4. Strong stability in the $M/M/\infty$ system

4.1. Strong stability conditions

Proofs of the results provided in this section, are based on theoretical results (theorems and corollaries) given in [1, 10].

Theorem 1 *Suppose that in the $M/M/\infty$ system, the condition $\lambda/\mu < 1$ is fulfilled. Then, there exists $\beta \in]1, \mu/\lambda[$ such that*

$$\rho = \left(\frac{1}{\beta}\left(\frac{\mu + \lambda\beta^2}{\lambda + \mu}\right)\right) < 1. \tag{10}$$

In addition, for all β such that $1 < \beta < \mu/\lambda$ the embedded Markov chain $\tilde{X}$ is $v-$strongly stable for the test function $v(k) = \beta^k$.

4.2. Strong stability estimates

Theorem 2 *Suppose that the conditions of the preceding theorem hold. Then, for all $1 < \beta < \mu/\lambda$, we have:*

$$\|P - \widetilde{P}\|_v = \frac{\alpha(1+\beta^2)}{\beta(s+\alpha)}, \tag{11}$$

and under the condition:

$$s \geq \alpha \left(\left(\frac{1+\beta^2}{\beta-1}\right)\left(\frac{1+\alpha}{1-\alpha\beta}\right)\left(1+e^{\alpha(\beta-1)}\right) - 1 \right), \tag{12}$$

we have

$$\|\pi - \tilde{\pi}\|_v \leq \frac{\alpha\,(1+\alpha)\,(1+\beta^2)\left(1+e^{\alpha(\beta-1)}\right)\left(e^{\alpha(\beta-1)}\right)}{(s+\alpha)\,(\beta-1)\,(1-\alpha\beta) - \alpha\,(1+\alpha)\,(1+\beta^2)\,(1+e^{\alpha(\beta-1)})} = B_\beta. \tag{13}$$

Where $\alpha = \lambda/\mu$.

5. Numerical application

To be able to put into practice the previous theoretical results concerning the approximation of the $M/M/s$ system by the $M/M/\infty$ one, we have developed the following algorithm:

5.1. Algorithm

Step 1. Introduce input parameters: inter-arrivals mean rate λ, number of servers s and service mean rate μ.

Step 2. Verify the existence of β_0
if $\mu > \lambda$ **then**
the system is stable and **goto** step 3
else *disp 'the system is not stable'* and **goto** step 7.

Step 3. Determine the constant $\beta_0 := \arg\max_{\beta} \rho < 1$.

Step 4. Determine the constant $\beta_{min} := \arg\min_{\beta}\{1 \leq \beta \leq \beta_0/$ the lowest value of β satisfying the condition **(12)**$\}$.

Step 5. Determine the constant $\beta_{max} := \arg\max_{\beta}\{1 \leq \beta \leq \beta_0/$ the highest value of β satisfying the condition **(12)**$\}$.

Step 6. Determine the constant β_{opt}:= the value of β minimizing the bound **(13)** where $\beta_{min} \leq \beta \leq \beta_{max}$.

Step 7. end.

For numerical applications, we consider the two following situations:

5.2. Situation 1: Variation of the number of servers s of the $M/M/s$ system

We fix the service rate $\mu = 10$ and some values for λ (for example $\lambda = 1$ (ie. the load $\alpha = \lambda/\mu = 0.1$) and $\lambda = 3$ (ie. $\alpha = 0.3$)), and for each fixed value of α we vary the values of the number of servers s. Then, for each value of s we compute the value of β minimizing the bound (13). For this value of β noted $\beta_{optimal}$, we compute the minimal bound $B_{\beta_{optimal}}$ (algorithmic error). The obtained results are listed in table 1 (case $\alpha = 0.1$) and table 2 (case $\alpha = 0.3$). The variations of the bound $B_{\beta_{Optimal}}$ in function of s for the case $\alpha = 0.1$ (respectively the case $\alpha = 0.3$) are represented in figure 1 (respectively figure 3). Figure 2 (respectively figure 4) compares between the algorithmic and simulation errors in function of s for the case $\alpha = 0.1$ (respectively the case $\alpha = 0.3$).

5.3. Situation 2: Variation of the load $\alpha = \lambda/\mu$ of the $M/M/s$ system

We fix the service rate $\mu = 20$, the number of servers $s = 20$, and we vary λ from 1 to 10 with step 0.5. Then, for each fixed value of $\alpha = \lambda/\mu$ we compute the value of β minimizing the bound (13). For this value of β noted $\beta_{optimal}$, we compute the minimal bound $B_{\beta_{optimal}}$ (algorithmic error). The obtained results are listed in table 3. The variations of the bound $B_{\beta_{Optimal}}$ in function of α are represented in figure 5. Figure 6 compares between the algorithmic and simulation errors in function of α.

Table 1. Values of $\beta_{optimal}$ and $B_{\beta_{optimal}}$ (case $\alpha = 0.1$).

s	$\beta_{optimal}$	Error ($B_{\beta_{optimal}}$)
1	×	×
2	1.9890	2.4503
3	1.9620	0.9668
4	1.9490	0.6020
5	1.9420	0.4370
6	1.9360	0.3430
7	1.9330	0.2823
8	1.9300	0.2398
9	1.9280	0.2084
10	1.9260	0.1843
11	1.9250	0.1652
12	1.9240	0.1497
13	1.9230	0.1369
14	1.9220	0.1260
15	1.9210	0.1168
16	1.9210	0.1088
17	1.9200	0.1019
18	1.9200	0.0958
19	1.9190	0.0903
20	1.9190	0.0855

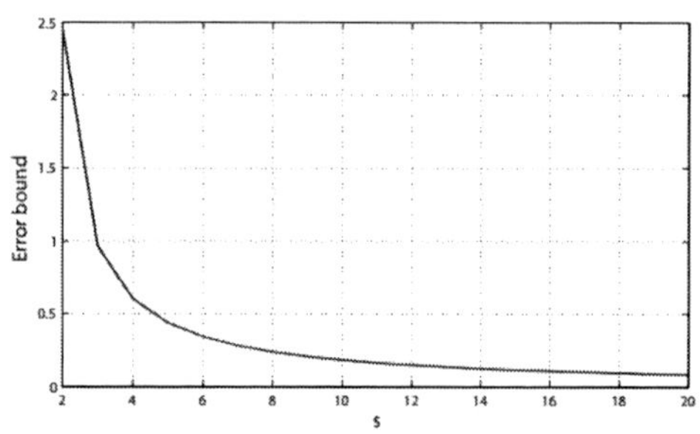

Figure 1. Variations of the bound $B_{\beta_{Optimal}}$ in function of s (case $\alpha = 0.1$).

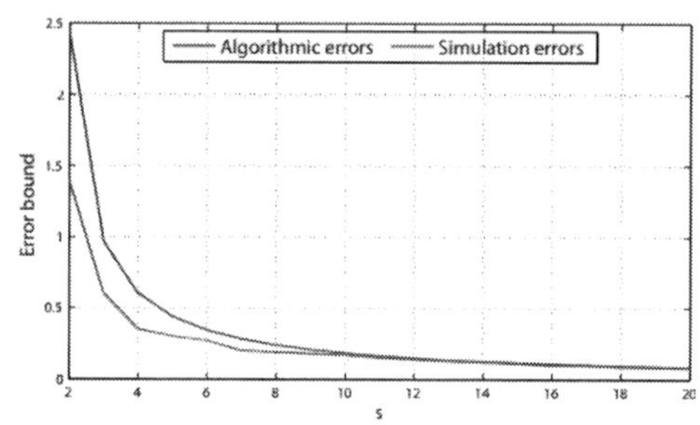

Figure 2. Comparison of algorithmic and simulation errors in function of s (case $\alpha = 0.1$).

Table 2. Values of $\beta_{optimal}$ and $B_{\beta_{optimal}}$ (case $\alpha = 0.3$).

s	$\beta_{optimal}$	Error ($B_{\beta_{optimal}}$)
≤ 9	×	×
10	1.6040	22.3974
11	1.5960	7.6876
12	1.5880	4.6363
13	1.5830	3.3175
14	1.5780	2.5822
15	1.5730	2.1133
16	1.5700	1.7883
17	1.5660	1.5498
18	1.5630	1.3674
19	1.5610	1.2233
20	1.5590	1.1067

Remark 1 *The symbol (×) in the the different tables indicates that the system is unstable for the fixed parameters and for the test function* $v(k) = \beta^k$.

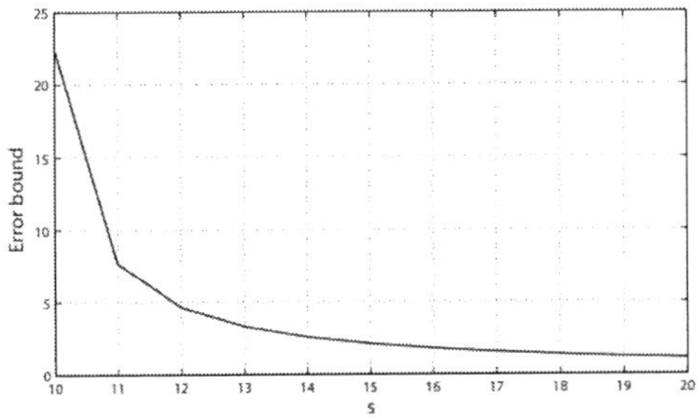

Figure 3. Variations of the bound $B_{\beta_{Optimal}}$ in function of s (case $\alpha = 0.3$).

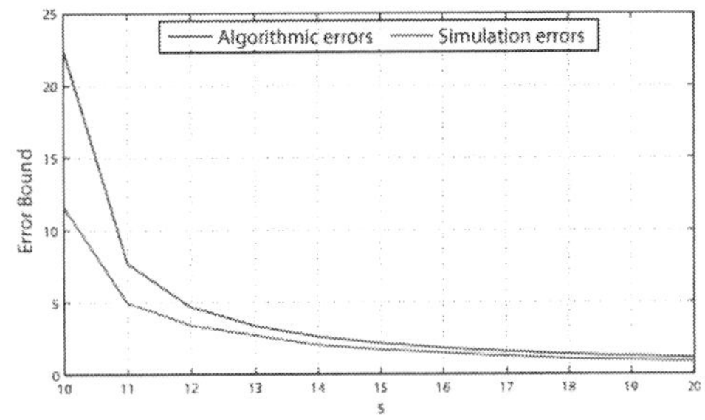

Figure 4. Comparison of algorithmic and simulation errors in function of s (case $\alpha = 0.3$).

Table 3. Values of $\beta_{optimal}$ and $B_{\beta_{optimal}}$ (case $\mu = 20$ and $s = 20$).

α	$\beta_{optimal}$	Error ($B_{\beta_{optimal}}$)
0.0500	2.1050	0.0322
0.0750	2.0020	0.0556
0.1000	1.9190	0.0855
0.1250	1.8490	0.1234
0.1500	1.7900	0.1718
0.1750	1.7380	0.2340
0.2000	1.6940	0.3152
0.2250	1.6540	0.4234
0.2500	1.6190	0.5716
0.2750	1.5870	0.7837
0.3000	1.5590	1.1067
0.3250	1.5330	1.6488
0.3500	1.5100	2.7282
0.3750	1.4890	5.8599
0.4000	1.4710	107.3026
$\alpha \geq 0.4250$	$\times$	$\times$

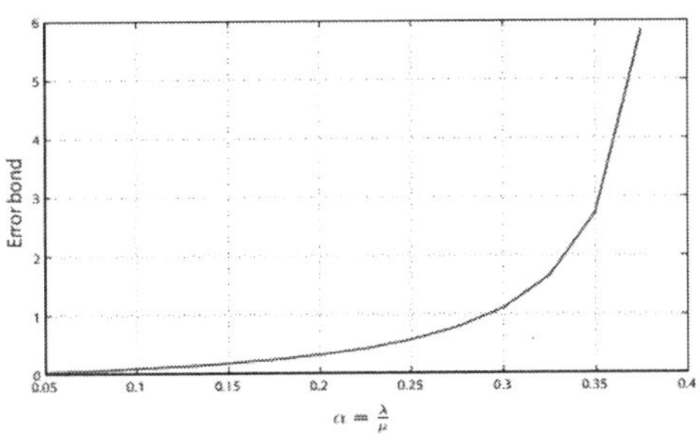

Figure 5. Variations of the bound $B_{\beta_{Optimal}}$ in function of α.

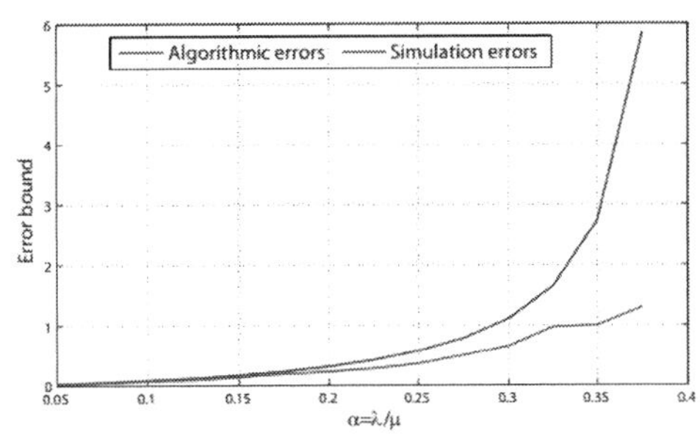

Figure 6. Comparison of algorithmic and simulation errors in function of α.

5.4. *Discussion of results*

- The error (B_β) decreases exponentially with the increase of the number of servers s (see figures 1 and 3). Moreover, this error converges to zero when s becomes large. This constatation confirms the tendency of the $M/M/s$ system to behave as an $M/M/\infty$ one, when s is rather large. Note also that the system becomes stable starting from the value $s = 2$ for the case $\alpha = 0.1$ (see table 1), whereas it becomes stable starting from the value $s = 10$ for the case $\alpha = 0.3$ (see table 2). This highlights the role of the load α of the system in its stability.

- The conditions and the stability bound of the system depend exponentially on the parameters of the system. Indeed, the error (B_β) increases exponentially with the increase of the load of the system α (see figure 5), where it passes from the stability state to the instability state starting from the value $\alpha = 0.4250$ (see table 3). This means that the stability of the system is strictly related not only to the number of its servers but also to its load.
- According to figures 2, 4 and 6, we notice that the simulation results are always lower than the algorithmic ones. This confirms that the bound (B_β) is an upper bound for the deviation $\|\pi - \tilde{\pi}\|_v$.

6. Conclusion

In this work, we applied for the first time the strong stability method on the $M/M/\infty$ queue which is subject to a small perturbation of its structure. The application of the method in question, allows us to determine the stability conditions of the $M/M/\infty$ system and to obtain stability bounds for the stationary characteristics of the $M/M/s$ system by those of the $M/M/\infty$ one. To validate and to illustrate the manner in which the theoretical results can be exploited in practice, we have presented numerical examples based on simulation studies. We may extend this analysis to the approximation case of the $GI/M/s$ system by the $GI/M/\infty$ one.

References

[1] Aïssani D and Kartashov N V 1983 Ergodicity and stability of Markov chains with respect to operator topology in the space of transition kernels *Compte Rendu Academy of Sciences U.S.S.R* A **11** 3–5.
[2] Bareche A and Aïssani D 2008 Kernel density in the study of the strong stability of the $M/M/1$ queueing system *Operations Research Letters* **36** 535–538.
[3] Benouaret Z and Aïssani D 2010 Strong stability in a two-dimensional classical risk model with independent claims *Scandinavian Actuarial Journal* **2** 83–92.
[4] Bouallouche-Medjkoune L and Aïssani D 2006 Measurement and performance of the strong stability method *Theor. Probab. Math. Statist.* **72** 1–9.
[5] Ferreira M A M 2013 *Infinite servers queue systems computational simulation* Proceedings: 12th Conference on Applied Mathematics APLIMAT 2013, At Bratislava, vol. 1, 1–16.
[6] Gross D, Shortie J F, Thompson J M and Harris C M 2013 *Bounds and Approximations. Fundamentals of Queueing Theory* John Wiley and Sons, Fourth edition.
[7] Guillemin F M, Mazumdar R R and Simonian A D 1996 On heavy traffic approximations for transient characteristics of $M/M/\infty$ queues *ournal of Applied Probability* **33(2)** 490–506.
[8] Halfin S and Whitt W 1981 Heavy traffic limits for queues with many exponenetial servers *Operation Research* **29(3)** 567–588.
[9] Helary J.M and Pedrono R 1983 *Recherche opérationnelle, Travaux Dirigés* Hermann.
[10] Kartashov N V 1986 Strong stability of Markov chains *Journal of Soviet Mathematics* **34** 1493–1498.
[11] Kartashov N V 1996 *Strong Stable Markov Chains* TbiMC Scientific Publishers, Utrecht, Kiev.
[12] Rabta B and Aïssani D 2005 Strong stability in an (R, s, S) inventory model *Int. J. Prod. Econ.* **4** 159171.

Integrable hierarchies of Heisenberg ferromagnet equation

G. Nugmanova, A. Azimkhanova

L.N. Gumilyov Eurasian National University, Astana, Kazakhstan

E-mail: nugmanovagn@gmail.com

Abstract.
In this paper we consider the coupled Kadomtsev-Petviashvili system. From compatibility conditions we obtain the form of matrix operators. After using a gauge transformation, obtained a new type of Lax representation for the hierarchy of Heisenberg ferromagnet equation, which is equivalent to the gauge coupled Kadomtsev-Petviashvili system.

1. Introduction

Korteweg-de Vries equation (KdV) is one of the most important equation in the theory of integrable systems , which have multi-soliton solution, an infinite number of conservation laws and many scientific applications [1].

In this paper we consider the expansion of the KdV equation - coupled Kadomtsev-Petviashvili equation (KP) with three potentials:

$$q_t = \frac{1}{4}(q_{xxx} - 6qq_x + 3\int q_{yy}dx + 6(pr)_x), \tag{1}$$

$$p_t = \frac{1}{2}(-p_{xxx} - 6qp_x + 3qp_x + 3p\int q_y dx - 3p_{xy}), \tag{2}$$

$$r_t = \frac{1}{2}(-r_{xxx} + 3qr_x - 3r\int q_y dx + 3r_{xy}), \tag{3}$$

where indexes represent partial derivatives. This system is also integrable and can assume multi-soliton solutions [2].

By using certain transformations [3], the system (1)-(3) can be reduced to a coupled system of the KdV equation [4], and the standard KP equation [5].

The main goal of this paper is to construct the equivalent spin system to the coupled KP system through gauge transformation.

2. Searching unknown elements of matrix operators for the Lax representation

In [6] presented decomposed form of this system by the first two terms into (1+1) dimensional hierarchy of Ablowitz-Kaup-Newell-Sigur:

$$u_y = -u_{xx} + 2u^2 v, \tag{4}$$

Content from this work may be used under the terms of the Creative Commons Attribution 3.0 licence. Any further distribution of this work must maintain attribution to the author(s) and the title of the work, journal citation and DOI.
Published under licence by IOP Publishing Ltd

$$v_y = v_{xx} - 2uv^2, \tag{5}$$

$$u_t = u_{xxx} - 6uvu_x, \tag{6}$$

$$v_t = v_{xxx} - 6uvv_x \tag{7}$$

and offers the following assumption: if (u, v) is the solution of system (4)-(7), then the functions (p, q, r), defined as $q = 4uv, p = -2u^2, r = -2v^2$ is a solution of (1)-(3) [7].

The Lax representation of the system (4)-(7) is defined as

$$\Psi_x = U\Psi, \tag{8}$$

$$\Psi_y = V\Psi, \tag{9}$$

$$\Psi_t = W\Psi, \tag{10}$$

where U, V, W - matrix operators. Let U looks like:

$$U = -\frac{\lambda}{2}\sigma_3 + Q, \tag{11}$$

where $\sigma_3 = \begin{pmatrix} 1 & 0 \\ 0 & -1 \end{pmatrix}$ and $Q = \begin{pmatrix} 0 & u \\ v & 0 \end{pmatrix}$. Matrix operator V given in the form

$$V = \lambda^2 V_2 + \lambda V_1 + V_0, \tag{12}$$

where $V_n = \begin{pmatrix} k_n & m_n \\ l_n & -k_n \end{pmatrix}$, $n = 0, 1, 2$. And operator W given by:

$$W = \lambda^3 W_3 + \lambda^2 W_2 + \lambda W_1 + W_0, \tag{13}$$

where $W_m = \begin{pmatrix} a_m & b_m \\ c_m & -a_m \end{pmatrix}$, $m = 0, 1, 2, 3$. Let's find the unknown elements of given matrices V and W. For this purpose, we will use the compatibility conditions of the system (4)-(7) and after this we obtain that

$$U_y - V_x + [U, V] = 0, \tag{14}$$

$$U_t - W_x + [U, W] = 0, \tag{15}$$

$$V_t - W_y + [V, W] = 0. \tag{16}$$

From (14) after substitution matrix (11) and (12) we have

$$Q_y - (\lambda^2 V_2 + \lambda V_1 + V_0)_x - \frac{1}{2}[\sigma_3, V] + [Q, V] = 0, \tag{17}$$

from which we obtain next system of equations in powers of λ:

$$\lambda^3 \;:\; -\frac{1}{2}[\sigma_3, V] = 0, \tag{18}$$

$$\lambda^2 \;:\; -V_{2x} - \frac{1}{2}[\sigma_3, V_1] + [Q, V_2] = 0, \tag{19}$$

$$\lambda^1 \;:\; -V_{1x} - \frac{1}{2}[\sigma_3, V_0] + [Q, V_1] = 0, \tag{20}$$

$$\lambda^0 \;:\; Q_y - V_{0x} + [Q, V_0] = 0. \tag{21}$$

Consistently solving the equations (18) - (21) we determine the unknown elements and form of the matrix operator V as following

$$V = \begin{pmatrix} -\frac{\lambda^2}{2} + uv & \lambda u - u_x \\ \lambda v + v_x & \frac{\lambda^2}{2} - uv \end{pmatrix}. \tag{22}$$

From (15) in terms of matrices U, V, W we have

$$Q_t - (\lambda^3 W_3 + \lambda^2 W_2 + \lambda W_1 + W_0)_x - \frac{\lambda}{2}[\sigma_3, W] + [Q, W] = 0. \tag{23}$$

Expanding in powers of λ, we obtain the following equation:

$$\lambda^4 \ : \ -\frac{1}{2}[\sigma_3, W_3] = 0, \tag{24}$$

$$\lambda^3 \ : \ -W_{3x} - \frac{1}{2}[\sigma_3, W_2] + [Q, W_3] = 0, \tag{25}$$

$$\lambda^2 \ : \ -W_{2x} - \frac{1}{2}[\sigma_3, W_1] + [Q, W_2] = 0, \tag{26}$$

$$\lambda^1 \ : \ -W_{1x} - \frac{1}{2}[\sigma_3, W_0] + [Q, W_1] = 0, \tag{27}$$

$$\lambda^0 \ : \ Q_t - W_{0x} + [Q, W_0] = 0. \tag{28}$$

From the system of equations (24)-(28) we obtain the unknown elements and form of the matrix W as

$$W = \begin{pmatrix} -\frac{\lambda^3}{2} + uv\lambda - vu_x + uv_x & u\lambda^2 - u_x\lambda - 2u^2v + u_{xx} \\ v\lambda^2 + v_x\lambda - 2uv^2 + v_{xx} & \frac{\lambda^3}{2} - uv\lambda + vu_x - uv_x \end{pmatrix}. \tag{29}$$

Thus, we found all unknown elements of the matrix operators V, W.

3. The new Lax representation form and coupled KP equivalent spin system

Now we are ready to obtain the Lax representation of a desired spin system by using a gauge transformation in form

$$\Phi = g^{-1}\Psi. \tag{30}$$

By definition, gauge equivalence is $g = \Psi|_{\lambda=0}$ and spin matrix is $S = g^{-1}\sigma_3 g$.

After using a gauge transformation, we get a new Lax representation for the hierarchy of Heisenberg ferromagnet equation, which is the gauge equivalence of the coupled KP system (1)-(3).

New Lax represenatation looks like:

$$\Phi_x = -\frac{\lambda}{2}S\Phi, \tag{31}$$

$$\Phi_y = -\frac{\lambda^2}{2}S\Phi + \frac{\lambda}{2}SS_x\Phi, \tag{32}$$

$$\Phi_t = -\frac{\lambda^3}{2}S\Phi + \frac{\lambda^2}{2}SS_x\Phi - \frac{3\lambda}{8}tr(S_x^2)S\Phi - \frac{\lambda}{2}S_{xx}\Phi. \tag{33}$$

Using the conditions of compatibility of the Lax representation we obtain a new spin system. From first condition $\Phi_{xy} = \Phi_{yx}$ after expanding by powers of λ we get the following equations:

$$S_x + \frac{1}{2}[SS_x, S] = 0, \tag{34}$$

$$S_y + \frac{1}{2}[S, S_{xx}] = 0. \tag{35}$$

From condition $\Phi_{xt} - \Phi_{tx} = 0$, expanding by powers of λ we have:

$$S_x + \frac{1}{2}[SS_x, S] = 0, \tag{36}$$

$$S_t - S_{xxx} - \frac{3}{4}tr(S_x^2)S_x - \frac{3}{4}[tr(S_x^2)]_x S = 0. \tag{37}$$

From the third conditions $\Phi_{yt} - \Phi_{ty} = 0$, expanding by powers of λ leads to the equations:

$$S_y + \frac{1}{2}[S, S_{xx}] = 0, \tag{38}$$

$$S_t + (SS_x)_y - \frac{3}{4}tr(S_x^2)S_x - \frac{1}{2}[S_{xx}, SS_x] = 0, \tag{39}$$

$$(SS_x)_t + S_{xxy} + \frac{3}{4}[tr(S_x^2)]_y S + \frac{3}{4}tr(S_x^2)S_y = 0. \tag{40}$$

Let us transform last two equations. After we consider (40)

$$S_t + (SS_x)_y - \frac{3}{4}tr(S_x^2)S_x - \frac{1}{2}[S_{xx}, SS_x] = 0. \tag{41}$$

In order to facilitate computations we divide this equation into the following two parts:

$$1) S_t - \frac{3}{4}tr(S_x^2)S_x = A, \tag{42}$$

$$2) (SS_x)_y - \frac{1}{2}[S_{xx}, SS_x] = B. \tag{43}$$

We consider the first term $(SS_x)_y$. Transform it considering the (35) and its derivative with respect to x ($S_{xy} = S_{yx}$):

$$(SS_x)_y = S_y S_x + SS_{xy} = \frac{1}{2}[S_{xx}, SS_x] - \frac{1}{2}S_{xxx} + \frac{1}{2}SS_{xxx}S. \tag{44}$$

Substituting this in (43) we find that

$$-\frac{1}{2}S_{xxx} + \frac{1}{2}SS_{xxx}S = B. \tag{45}$$

The second term $SS_{xxx}S$ we will express through the third derivative of the expression $S \cdot S = I$, multiplying it to S:

$$SS_{xxx}S = [S_{xx}S, S_x] - S_{xxx} - 2(S_x^2)_x S. \tag{46}$$

Substituting (46) into (45), we obtain

$$\frac{1}{2}[S_{xx}S, S_x] - S_{xxx} - (S_x^2)_x S = B. \tag{47}$$

Now we transform the first term of (47) $\frac{1}{2}[S_{xx}S, S_x]$:

$$\frac{1}{2}[S_{xx}S, S_x] = \frac{1}{2}(S_{xx}SS_x - S_x S_{xx}S) =$$

$$= \frac{1}{2}[(4uvg^{-1}\sigma_3 g + 2g^{-1}\begin{pmatrix} 0 & u_x \\ -v_x & 0 \end{pmatrix} g)g^{-1}\sigma_3 g \cdot 2g^{-1}\begin{pmatrix} 0 & u \\ -v & 0 \end{pmatrix} g -$$
$$-2g^{-1}\begin{pmatrix} 0 & u \\ -v & 0 \end{pmatrix} g \cdot (4uvg^{-1}\sigma_3 g + 2g^{-1}\begin{pmatrix} 0 & u_x \\ -v_x & 0 \end{pmatrix} g)g^{-1}\sigma_3 g]. \quad (48)$$

As a result, we find that

$$\frac{1}{2}[S_{xx}S, S_x] = -\frac{1}{4}[tr(S_x^2)]_x S. \quad (49)$$

Substituting this expression in (47)

$$-S_{xxx} - \frac{1}{4}[tr(S_x^2)]_x S - (S_x^2)_x S = B. \quad (50)$$

Considering the third term as the $(S_x^2)_x S = -(4uv)_x S = \frac{1}{2}[tr(S_x^2)]_x S$, we get in result B:

$$-S_{xxx} - \frac{1}{4}[tr(S_x^2)]_x S - \frac{1}{2}[tr(S_x^2)]_x S = -S_{xxx} - \frac{3}{4}[tr(S_x^2)]_x S = B. \quad (51)$$

Add the A and B, according to (41) we obtain

$$S_t - S_{xxx} - \frac{3}{4}tr(S_x^2)S_x - \frac{3}{4}[tr(S_x^2)]_x S = 0, \quad (52)$$

which corresponds to the equation (37).

Now let's consider the equation (40)

$$(SS_x)_t + S_{xxy} + \frac{3}{4}[tr(S_x^2)]_y S + \frac{3}{4}tr(S_x^2)S_y = 0. \quad (53)$$

We will transform each term. Let's start with the first, taking into account the expressions for S_x and its derivative with respect to t

$$(SS_x)_t = \frac{1}{2}S_t S_x + \frac{1}{2}SS_{xt} - \frac{1}{2}S_{xt}S - \frac{1}{2}S_x S_t, \quad (54)$$

$$\frac{1}{2}S_t S_x = \frac{1}{2}S_{xxx}S_x + \frac{3}{8}[tr(S_x^2)]_x SS_x + \frac{3}{8}tr(S_x^2)S_x^2,$$
$$\frac{1}{2}S_x S_t = \frac{1}{2}S_x S_{xxx} + \frac{3}{8}[tr(S_x^2)]_x S_x S + \frac{3}{8}tr(S_x^2)S_x^2. \quad (55)$$

Due to the validity of the expression $S_{xt} = S_{tx}$:

$$\frac{1}{2}SS_{xt} = \frac{1}{2}SS_{xxxx} + \frac{3}{4}[tr(S_x^2)]_x SS_x + \frac{3}{8}[tr(S_x^2)]_{xx} + \frac{3}{8}tr(S_x^2)SS_{xx}, \quad (56)$$

$$\frac{1}{2}S_{xt}S = \frac{1}{2}S_{xxxx}S + \frac{3}{4}[tr(S_x^2)]_x S_x S + \frac{3}{8}[tr(S_x^2)]_{xx} + \frac{3}{8}tr(S_x^2)S_{xx}S. \quad (57)$$

Combining the above, we find that the first term $(SS_x)_t$

$$(SS_x)_t = \frac{1}{2}[S_{xxx}, S_x] + \frac{1}{2}[S_x, S_{xxx}]\frac{9}{4}[tr(S_x^2)]_x SS_x + \frac{3}{8}tr(S_x^2)[S, S_{xxx}]. \quad (58)$$

Let's find the second $S_{xxy} = S_{yxx}$:

$$S_{yxx} = \frac{1}{2}[S_{xxxx}, S] + [S_{xxx}, S_x]. \tag{59}$$

By substituting the first two terms in the original equation (53), we obtain

$$\frac{3}{2}[S_{xxx}, S_x] + \frac{9}{4}[tr(S_x^2)]_x SS_x + \frac{3}{4}[tr(S_x^2)]_y S = 0. \tag{60}$$

We extend further transformation until we get in result the equation (37), or 0.

After the transformation of the first term, we have

$$\frac{3}{2}[S_{xxx}, S_x] = -\frac{9}{4}[tr(S_x^2)]_x SS_x + 6g^{-1}\begin{pmatrix} -vu_{xx} + uv_{xx} & 0 \\ 0 & -v_{xx}u + vu_{xx} \end{pmatrix} g. \tag{61}$$

Substituting the latest into (60) and obtain

$$\frac{3}{4}[tr(S_x^2)]_y S + 6g^{-1}\begin{pmatrix} -vu_{xx} + uv_{xx} & 0 \\ 0 & -v_{xx}u + vu_{xx} \end{pmatrix} g. \tag{62}$$

Transforming the first term, we find that

$$\frac{3}{4}[tr(S_x^2)]_y S = \frac{3}{2}(S_x S_x)_y S, \tag{63}$$

$$(S_x S_x)_y = S_{xy} S_x + S_x S_{xy} = -4g^{-1}\begin{pmatrix} -vu_{xx} + uv_{xx} & 0 \\ 0 & -v_{xx}u + vu_{xx} \end{pmatrix} g, \tag{64}$$

substituting the last expression into (63), and finally we find the following:

$$\frac{3}{4}[tr(S_x^2)]_y S = -\frac{3}{2}4g^{-1}\begin{pmatrix} -vu_{xx} + uv_{xx} & 0 \\ 0 & -v_{xx}u + vu_{xx} \end{pmatrix} g = -6g^{-1}\begin{pmatrix} -vu_{xx} + uv_{xx} & 0 \\ 0 & -v_{xx}u + vu_{xx} \end{pmatrix} g.$$

And finally substituting the last expression into (62), we obtain

$$-6g^{-1}\begin{pmatrix} -vu_{xx} + uv_{xx} & 0 \\ 0 & -v_{xx}u + vu_{xx} \end{pmatrix} g + 6g^{-1}\begin{pmatrix} -vu_{xx} + uv_{xx} & 0 \\ 0 & -v_{xx}u + vu_{xx} \end{pmatrix} g = 0. \tag{65}$$

Thus, in this paper was obtained integrable hierarchy of Heisenberg ferromagnet equation

$$S_t - S_{xxx} - \frac{3}{4}tr(S_x^2)S_x - \frac{3}{4}[tr(S_x^2)]_x S = 0, \tag{66}$$

which is the equivalent to a coupled KP system (1)-(3). Much more detailed study of the spin system (66) is the subject of our further research.

Acknowledgments

We express gratitude to Professor R. Myrzakulov for useful discussions and advices. The work is performed under the financial support of the grant of MES RK GPh4-2016.

4. References

[1] Wazwaz A 2011 *Pramana - journal of physics* **77** 233
[2] Isojima S, Willox R and Satsuma J 2002 *Journal of Physics A: Mathematical and General* **35** 6893
[3] Hirota R and Ohta Y 1991 *Journal of the Physical Society of Japan* **60** 798
[4] David D, Levi D and Winternitz P 1987 *Studies in Applied Mathematics* **76** 133
[5] Kakei S 2000 *Physics Letters A* **264** 449
[6] Geng X G 2003 *Journal of Physics A: Mathematical and General* **36** 2289
[7] Qi F-H, Tiana Bo, Liua W-J, Guoa R, Xua T and Zhanga H-Q 2011 *Verlag der Zeitschrift fur Naturforschung* Tubingen

A space-time geodesic approach for phase fitted variational integrators

Odysseas Kosmas and D.S. Vlachos

Department of Informatics & Telecommunications, University of Peloponnese, 22100, Greece

E-mail: odykosm@uop.gr, dvlachos@uop.gr

Abstract. The use of space-time geodesic approach of classical mechanics is investigated, in order to derive time adaptive high order phase fitted variational integrators. The proposed technique is employed for systems of which the Lagrangian is of separable form. To this end, first the unfolding of the standard Euler-Lagrange system to its space-time manifold is presented and then it is rewritten as a geodesic problem with zero potential energy. Preliminary simulation results (without optimizing the choice of step sizing) show that one can use the space-time geodesic formulation to generate an adaptive scheme that still preserves some underlying geometric structure.

1. Introduction

In recent years, in order to reduce computational cost, time adaptivity is an important ingredient. Therefore adaptive time integration schemes for ordinary differential equations (ODEs) have been well established. Although they perform extremely well for many applications, for problems involving the integration of Hamiltonian systems, there are good reasons for using symplectic integrators [1, 2].

On that end, studies have been done in order to derive and analyze symplectic integrators with variable time steps, but the early results were not promising [3, 4, 5]. Among these, there have been two types of time variation steps. In the first set of studies, the time step varies explicitly as the time is running resulting to problems. In the second set of studies, the time step was chosen using the dynamical variables of the system q, p, i.e. $\Delta t = \Delta(q, p)$. Using this choice of Δt, the equations are no longer in canonical Hamiltonian form (but rather Hamiltonian in non-canonical variables) and quite unreliable results are obtained. The results in both cases lead to the conclusion that if one needs an adaptive time step integrator is forced to use a high order non-symplectic scheme [6, 7].

In this paper, the Galerkin type, high order symplectic integrators of [8, 9, 10, 11, 12, 13, 14] are extended through the use of adaptive time stepping. To this end, in addition to the spacetime view point of [1, 2], the geodesic viewpoint [15, 16] is regarded in order to understand and overcome the problems appeared in the application of symplectic integrators with variable time steps. One of our purposes is to derive a method of optimal time-step adaptation scheme.

2. Geodesic approach

In order to construct time adaptive integrators, the Lagrangian function $L(x, \dot{x}) = \frac{1}{2}\dot{x}^2 - V(x)$ ($x \in \mathbb{R}$) must be considered. For the latter Lagrangian, the corresponding Euler-Lagrange

Content from this work may be used under the terms of the Creative Commons Attribution 3.0 licence. Any further distribution of this work must maintain attribution to the author(s) and the title of the work, journal citation and DOI.
Published under licence by IOP Publishing Ltd

equation is

$$\ddot{x} = -\frac{\partial V}{\partial x}. \tag{1}$$

Choosing initial conditions $x_0 = x(0)$ and $\dot{x}_0 = \dot{x}(0)$, $x(t)$ can be considered as a solution of (1) for some time interval $t \in [0, T]$.

We they consider the Lagrangian $\tilde{L} = \frac{1}{2}\acute{x}^2 + \frac{1}{2V}\acute{t}^2$, where the primes in $\acute{x}$ denote differentiation with respect to some parameter λ, see [15]. The corresponding Euler-Lagrange equations and initial conditions are

$$x'' = -\frac{1}{2V^2}\frac{\partial V}{\partial x}t'^2, \quad x_0 = x(0), \quad x_0' = \dot{x}_0 t_0' \tag{2a}$$

$$t'' = \frac{1}{V}\frac{\partial V}{\partial x}t'x', \quad t_0 = 0, \qquad t'(0) = \alpha V(x_0). \tag{2b}$$

Even though $\tilde{L}$ depends upon V and couples the space and time variables in a non trivial manner, the embedded evolution equations for x only depend on $\partial V/\partial x$. Of course one could add any constant to V without changing the x-dynamics [16].

It can be proved, see e.g. [15, 16], that if $\tilde{x}(\lambda), t(\lambda)$ solve (2b) because, some time interval $\lambda \in [0, \tilde{T}]$, then $\tilde{x}(\lambda) = x(2t/\sqrt{\alpha})$ for as long as both sides are defined i.e. the solutions for x and $\tilde{x}$ differ only by a constant that rescales the time. The discrete version of the above statement also holds true [16].

We now consider the following two Lagrangians [15, 16]

$$L_1 = \sqrt{\acute{x}^2 + f(x)\acute{t}^2}, \qquad L_2 = \frac{1}{2}\left(\acute{x}^2 + f(x)\acute{t}^2\right). \tag{3}$$

The action corresponding to L_1 is invariant under arbitrary reparametrization of λ, whereas the L_2 action is only affine reparametrization invariant, therefore, the Euler-Lagrange equations corresponding to L_2 are affine time reparametrization invariant [16]. The Euler-Lagrange equations corresponding to L_1 are

$$\frac{d}{d\lambda}\left(\frac{\acute{x}}{\sqrt{\acute{x}^2 + f(x)\acute{t}^2}}\right) = \frac{\acute{t}^2}{2\sqrt{\acute{x}^2 + f(x)\acute{t}^2}}\frac{\partial f}{\partial x} \tag{4a}$$

$$\frac{d}{d\lambda}\left(\frac{f(x)\acute{x}}{\sqrt{\acute{x}^2 + f(x)\acute{t}^2}}\right) = 0. \tag{4b}$$

Equations (4) are reparametrization invariant, see [16]. Notice also that the equations in (4) look like two evolution equations which should, in general, provide us not only the shape of the curve but also the parametrization of the curve.

3. Review of phase fitted variational integrators

As it is known, for the derivation of high order variational integrators, we need to apply discrete variational calculus [8, 9, 10]. As usually, a discrete Lagrangian, is a map $L_d : Q \times Q \to \mathbb{R}$ which may be considered as an approximation of a continuous action obtained through the Lagrangian $L : TQ \to \mathbb{R}$, i.e.

$$L_d(q_k, q_{k+1}, h_k) \approx \int_{t_k}^{t_{k+1}} L(q, \dot{q})dt. \tag{5}$$

The action sum $S_d : Q^{N+1} \to \mathbb{R}, N \in \mathbb{N}$ corresponding to the Lagrangian L_d is defined as

$$S_d(\gamma_d) = \sum_{k=0}^{N-1} h_k L_d(q_k, q_{k+1}, h_k), \tag{6}$$

with $\gamma_d = (q_0, \ldots, q_N)$ denoting the discrete trajectory. For any covector $\alpha \in T^*_{(x_1,x_2)}(Q \times Q)$ we have the decomposition $\alpha = \alpha_1 + \alpha_2$, where $\alpha_i \in T^*_{x_i}(Q \times Q)$. Thus, $dL_d(q_0, q_1) = D_1 L_d(q_0, q_1) + D_2 L_d(q_0, q_1)$, where the notation $D_i L_d$ indicates the slot derivative with respect to the i-argument of L_d. According to the discrete variational principle, as usually the solutions of the discrete system are determined from L_d by extremizing the action sum for given fixed points q_0 and q_N. Extremizing S_d over all the intermediate points of γ_d, the system of difference equations

$$h_{k-1} D_2 L_d(q_{k-1}, q_k, h_{k-1}) + h_k D_1 L_d(q_k, q_{k+1}, h_k) = 0, \qquad k = 1, \ldots, N-1, \tag{7}$$

are obtained which are commonly called the discrete Euler-Lagrange equations.

To derive high order methods, we approximate the action integral along the curve segment between q_k and q_{k+1} using a discrete Lagrangian that depends only on the end points. This way we obtain expressions for the configurations q_k^j and velocities $\dot{q}_k^j$, $j = 0, \ldots, S-1$, $S \in \mathbb{N}$, at time $t_k^j \in [t_k, t_{k+1}]$. Then, by expressing t_k^j as $t_k^j = t_k + C_k^j h_k$ for $C_k^j \in [0,1]$ such that $C_k^0 = 0$, $C_k^{S-1} = 1$ we write

$$q_k^j = g_1(t_k^j) q_k + g_2(t_k^j) q_{k+1}, \qquad \dot{q}_k^j = \dot{g}_1(t_k^j) q_k + \dot{g}_2(t_k^j) q_{k+1}, \tag{8}$$

where $h \in \mathbb{R}$ is the time step. We next choose the functions

$$g_1(t_k^j) = \sin\left(u - \frac{t_k^j - t_k}{h_k} u\right) (\sin u)^{-1}, \quad g_2(t_k^j) = \sin\left(\frac{t_k^j - t_k}{h_k} u\right) (\sin u)^{-1}, \tag{9}$$

to represent the oscillatory behavior of the solution [11, 12, 13, 14, 17]. For the sake of continuity the conditions $g_1(t_{k+1}) = g_2(t_k) = 0$ and $g_1(t_k) = g_2(t_{k+1}) = 1$ are required.

For any different choice of interpolation assumed we define the discrete Lagrangian by the weighted sum (see [11])

$$L_d(q_k, q_{k+1}, h_k) = \sum_{j=0}^{S-1} h_k w^j L(q(t_k^j), \dot{q}(t_k^j)), \tag{10}$$

where, as can be easily proved, $\sum_{j=0}^{S-1} w^j (C_k^j)^m = \frac{1}{m+1}$, with $m = 0, 1, \ldots, S-1$ and $k = 0, 1, \ldots, N-1$ [11, 12].

Applying the above interpolation technique with the trigonometric expressions of (9) and following the phase lag analysis of [11, 12], the parameter u must be chosen as $u = \omega h$. For problems including a definite frequency ω (such as the harmonic oscillator) the parameter u can be easily computed. For the solution of orbital problems of the general N-body problem, where no unique frequency is given, a new parameter u must be defined by estimating the frequency of the motion for any moving point mass [12, 13, 14].

4. Time adaptive phase fitted variational integrators

Using (10), for the length action given by L_1 in (3), the corresponding discrete Lagrangian reads

$$L_{1d}(q_k, q_{k+1}, h_k) = \sum_{j=0}^{S-1} h_k w^j \sqrt{\left(\acute{x}_k^j\right)^2 + f\left(x_k^j\right)\left(\acute{t}_k^j\right)^2}, \tag{11}$$

where x_k^j are defined using (8) and $\acute{x}_k^j$, $\acute{t}_k^j$ using the expression

$$\acute{q}_k^j = \frac{\partial q_k^j}{\partial \lambda} = \frac{\partial t}{\partial \lambda}\left(\dot{g}_1(t_k^j)q_k + \dot{g}_2(t_k^j)q_{k+1}\right) = \dot{g}_1(t_k^j)q_k + \dot{g}_2(t_k^j)q_{k+1}. \tag{12}$$

For the latter Lagrangian, the discrete Euler-Lagrange equations (7) give the discrete analogues of (4a) as

$$\sum_{j=0}^{S-1} w^j \frac{h_k}{2d_{k,k-1}}\left[2\dot{g}_2(t_k^j)\left(\dot{g}_1(t_k^j)x_{k-1} + \dot{g}_2(t_k^j)x_k\right) + \frac{\partial}{\partial x_k} f\left(g_1(t_k^j)x_{k-1} + g_2(t_k^j)x_k\right)\left(\dot{g}_1(t_k^j)t_{k-1} + \dot{g}_2(t_k^j)t_k\right)^2\right]$$
$$+\sum_{j=0}^{S-1} w^j \frac{h_{k+1}}{2d_{k+1,k}}\left[2\dot{g}_1(t_k^j)\left(\dot{g}_1(t_k^j)x_k + \dot{g}_2(t_k^j)x_{k+1}\right) + \frac{\partial}{\partial x_k} f\left(g_1(t_k^j)x_k + g_2(t_k^j)x_{k+1}\right)\left(\dot{g}_1(t_k^j)t_k + \dot{g}_2(t_k^j)t_{k+1}\right)^2\right] \tag{13}$$

and of (4b) as

$$\sum_{j=0}^{S-1} w^j \frac{h_k}{d_{k,k-1}}\left[\dot{g}_2(t_k^j)f\left(g_1(t_k^j)x_{k-1} + g_2(t_k^j)x_k\right)\left(\dot{g}_1(t_k^j)t_{k-1} + \dot{g}_2(t_k^j)t_k\right)^2\right]$$
$$+\sum_{j=0}^{S-1} w^j \frac{h_{k+1}}{d_{k+1,k}}\left[\dot{g}_1(t_k^j)f\left(g_1(t_k^j)x_k + g_2(t_k^j)x_{k+1}\right)\left(\dot{g}_1(t_k^j)t_k + \dot{g}_2(t_k^j)t_{k+1}\right)^2\right]. \tag{14}$$

In these equations $d_{k+1,k}$ is given by

$$d_{k+1,k} = \sqrt{\left[\dot{g}_1(t_k^j)x_k + \dot{g}_2(t_k^j)x_{k+1}\right]^2 + f\left(g_1(t_k^j)x_{k-1} + g_2(t_k^j)x_k\right)\left[\dot{g}_1(t_k^j)t_{k-1} + \dot{g}_2(t_k^j)t_k\right]^2}. \tag{15}$$

and similarly for $d_{k,k-1}$.

In accordance with the continuous case, the equations (13) and (14) are not independent. To solve the system above, we can choose arbitrary step sizes in either t or x direction and solve for the x or t, respectively.

Once we have solved the discrete Euler-Lagrange equations (13) and (14), we get a sequence of points $(x_0, t_0), \ldots, (x_N, t_N)$, where $t_0, \ldots, t_N$ does not necessarily present the physical time. Using this sequence of points, for the discrete Hamiltonian

$$H_d(x_0, x_1, h_0) = -h_0 D_3 L_d(x_0, x_1, h_0) - L_d(q_0, q_1, h_0) \tag{16}$$

and recalling that the energy expressed by the Hamiltonian is conjugate variable of the physical time, i.e.

$$H_d(x_0, x_1, h_0) = H_d(x_1, x_2, h_1), \tag{17}$$

we reconstruct the physical time.

References

[1] J.E. Marsden, G. Patrick, and S. Shkoller, Multisymplectic geometry, variational integrators, and nonlinear pdes, Comm. Math. Phys. 199 (1998) 351.
[2] C. Kane, J.E. Marsden, and M. Ortiz, Symplectic energy-momentum preserving variational integrators, J. Math. Phys. 40, (1999) 3353.

[3] R.D. Skeel, Variable step size destabilizes the Störmer/leapfrog/Verlet method, BIT Numerical Mathematics 33 (1993) 172.
[4] M.P. Calvo, and J.M. Sanz-Serna, The development of variable-step symplectic integrators, with application to the two-body problem, SIAM J. Sci. Comput. 14 (1993) 936.
[5] J.P. Wright, Numerical instability due to varying time steps in explicit wave propagation and mechanics calculations, Journal of Computational Physics 140 (1998) 421.
[6] E. Hairer, Variable time step integration with symplectic methods. Applied Numerical Mathematics 25 (1997) 219.
[7] S. Reich, Backward error analysis for numerical integrators, SIAM Journal on Numerical Analysis 36 (1999) 1549.
[8] J.M. Wendlandt, J.E. Marsden, Mechanical integrators derived from a discrete variational principle, Physica D 106 (1997) 223.
[9] C. Kane, J. Marsden, M. Ortiz, Symplectic-energy-momentum preserving variational integrators, Journal of Mathematical Physics 40 (1999) 3353.
[10] J.E. Marsden, M. West, Discrete mechanics and variational integrators, Acta Numerica 10 (2001) 357.
[11] O.T. Kosmas, D.S. Vlachos, Phase-fitted discrete Lagrangian integrators, Computer Physics Comm. 181 (2010) 562.
[12] O.T. Kosmas, S. Leyendecker, Phase lag analysis of variational integrators using interpolation techniques, PAMM Proc. Appl. Math. Mech. 12 (2012) 677.
[13] O.T. Kosmas, Charged particle in an electromagnetic field using variational integrators, ICNAAM Numerical Analysis and Applied Mathematics 1389 (2011) 1927.
[14] O.T. Kosmas, S. Leyendecker, Analysis of higher order phase fitted variational integrators, Advances in Computational Mathematics 42 (2016) 605.
[15] L.P. Eisenhart, Dynamical trajectories and geodesics, Annals of Mathematics 30 (1929) 591.
[16] S. Nair, Time adaptive variational integrators: A space-time geodesic approach, Physica D: Nonlinear Phenomena 241 (2012) 315.
[17] O.T. Kosmas, D.S. Vlachos, Local path fitting: a new approach to variational integrators, Journal of Computational and Applied Mathematics 236 (2012) 2632.

Finite difference computation of Casimir forces

Fabrizio Pinto

Laboratory for Quantum Vacuum Applications, Department of Physics, Faculty of Science, Jazan University, P.O. Box 114, Gizan 45142, Kingdom of Saudi Arabia.

E-mail: `fpinto@jazanu.edu.sa`

Abstract. In this Invited paper, we begin by a historical introduction to provide a motivation for the classical problems of interatomic force computation and associated challenges. This analysis will lead us from early theoretical and experimental accomplishments to the integration of these fascinating interactions into the operation of realistic, next-generation micro- and nanodevices both for the advanced metrology of fundamental physical processes and in breakthrough industrial applications. Among several powerful strategies enabling vastly enhanced performance and entirely novel technological capabilities, we shall specifically consider Casimir force time-modulation and the adoption of non-trivial geometries. As to the former, the ability to alter the magnitude and sign of the Casimir force will be recognized as a crucial principle to implement thermodynamical nano-engines. As to the latter, we shall first briefly review various reported computational approaches. We shall then discuss the game-changing discovery, in the last decade, that standard methods of numerical classical electromagnetism can be retooled to formulate the problem of Casimir force computation in arbitrary geometries. This remarkable development will be practically illustrated by showing that such an apparently elementary method as standard finite-differencing can be successfully employed to numerically recover results known from the Lifshitz theory of dispersion forces in the case of interacting parallel-plane slabs. Other geometries will be also be explored and consideration given to the potential of non-standard finite-difference methods. Finally, we shall introduce problems at the computational frontier, such as those including membranes deformed by Casimir forces and the effects of anisotropic materials. Conclusions will highlight the dramatic transition from the enduring perception of this field as an exotic application of quantum electrodynamics to the recent demonstration of a human climbing vertically on smooth glass.

1. Introduction

At the dawn of modern understanding of the existence of interatomic forces, made possible by the application of non-relativistic quantum mechanics [1], two issues became immediately central: the role played by the optical properties of interacting particles, which led to the use of the term *dispersion* by Eisenschitz and London [2] to characterize such forces, and the question of dispersion force *additivity*, quickly identified by London [3] as critical to the development of a theory of cohesion. Starting during World War II, it was suggested by Overbeek [4] and Verwey [5] that an additional element must be crucial to a satisfactory explanation of existing experimental data on the stability of lyophobic colloids: electrodynamical retardation [6].

This profound intuition was the inspiration for the work by Casimir and Polder towards a theory of interatomic forces fully consistent with quantum electrodynamics (QED) [7]. In a development fraught with immense consequences in the history of physics, Casimir continued his work in search for a deeper meaning for the result he had found with Polder and, after

Content from this work may be used under the terms of the Creative Commons Attribution 3.0 licence. Any further distribution of this work must maintain attribution to the author(s) and the title of the work, journal citation and DOI.
Published under licence by IOP Publishing Ltd

an illuminating conversation with Niels Bohr [8], he reformulated the problem in terms of the zero-pont energy of the electromagnetic field in the quantum vacuum [9]. An application of this approach to the problem of the force between two neutral, perfectly conducting parallel planes separated by an empty gap led Casimir, in 1948, to the famous equation for what is now referred to as the Casimir effect [10].

These momentous results successfully addressed the issue of retardation but the problem of the interaction between macroscopic bodies at very close range could only still be considered by assuming additivity and proceeding by integrations in the continuum approximation [6] as done for slabs and spherical particles by de Boer [11] and Hamaker [12], respectively. As pointed out by Milonni [13] with the benefit of hindsight, since the Clausius-Mossotti relation provides a link between atomic optical properties and the dielectric functions of the bulk material, a straightforward application of additivity to the retarded Casimir-Polder result could have shown that the Casimir effect between perfect conductors should be, in some sense, expected.

However, a full generalization of the equation for the Casimir force to the simplest realistic case of two semi-infinite dielectric slabs including retardation had to wait until 1955. This remarkable achievement was contributed by Lifshitz, who employed a classical stochastic approach [14], based on solving the Maxwell equations including a noise term introduced by Rytov [15] to describe the fluctuating field within the dielectric media [16]. The final expression, reached through a very significant mathematical effort, has been confirmed by several other approaches [13] and it stands, still today, as the standard solution to the problem. Importantly, as also noticed by Lifshitz, use of the theory requires knowledge of the complex dielectric function of the bulk interacting materials over the entire frequency range [17].

Direct experimental verification of the force between two plane slabs separated by a gap is exceedingly difficult due to the challenge to accurately ensure parallelism [18] and it has been successfully reported very rarely [19]. Much more typical since the very first attempts in this field is the use of a sphere-plane system, which entirely circumvents such difficulties [13, 20, 21, 22, 23, 24, 25, 26, 27, 28]. The connection between the Lifshitz expression and experiments involving sphere-plane or sphere-sphere setups is made by the proximity force approximation [20, 29] (PFA), although challenges remain [30].

A development fueling further great interest in Casimir forces is the growing recognition that applications of such interactions are likely to lead to novel micro- and nano-technology solutions impossible or impractical to achieve by any other means [31, 32, 33, 34, 35, 36]. One of the major challenges in executing this transfer of Casimir force-enabled technologies into marketable solutions is the fact that industrial devices will not at all conform to the strict idealizations of the Lifshitz theory, including its basic plane-plane geometry. Hence the problem of the computation of dispersion forces in realistic geometries has become timely for reasons beyond even the needs of fundamental physics and not dissimilar from those motivating the development of computer-aided design (CAD) tools [37].

In the rest of this paper, rather than to discuss extreme mathematical detail, we present the key issues with a few useful references in order to provide a fish-eye view of the computational program underway at our Laboratory including fundamental motivations, status and goals in several subfields. This material is complemented by further details and references in recent mini-reviews at Refs. [38, 39, 40].

2. Dispersion force computation: Challenges and opportunities

Unlike such fundamental interactions as gravitation and electromagnetism, field equations in the form of the Einstein of Maxwell equations are not known for the general case of dispersion forces. In the limit of rarefied materials, in which the additivity approximation is held to apply, it is possible to formulate the problem in terms of central forces inversely proportional to a power $p \geq 3$ of the interatomic distance, s, as already done by Newton and rediscovered by de Boer and Hamaker [41]. However, the results are completely unreliable in the general case and indeed counterexamples exist of systems in which the net force of atoms attracting each other as pairs is repulsive [42]. Under these circumstances, the natural option is to seek numerical solutions but, for quite some time, the very problem to solve could not be formulated.

The situation changed drastically in 2007 with the introduction of a numerical imaginary frequency Green's function computation (IFGFC) method by researchers at MIT working with their collaborators [43]. Not only has the theoretical foundation provided a path to the numerical computation of dispersion forces in any geometry but it has also shown that the algorithm can be built based, with relatively little modification, on procedures largely known from classical electromagnetism. In the last few years, implementations and variations of this algorithm, initially pursued by the finite-difference frequency domain (FDFD) method, have therefore grown exponentially providing the first glimpse of interacting systems that could never been previously treated [26]. Hence this is an extremely exciting time in the field of the numerical calculation of Casimir forces as the opportunity exists to explore systems for which no previous intuition can anticipate a reliable solution.

Of course, as with any numerical algorithm, the issue of computational facilities has emerged forcefully somewhat ameliorated by the fact that the intrinsic parallelization of the problem lends itself to supercomputing implementations.

3. Motivations for a finite-difference (FD) computation approach

Our program in this area has moved in a different direction than that of the mainstream [37]. Instead of pursuing development of code of increased complexity running on highly competitive machines, we have pursued the study of implementations based on the simplest possible approach, finite differencing (FD), to be run on readily available computers and a widely used computer algebra system [44] or on Fortran 90/95. Our goals with this approach are multifaceted. First, understanding the behavior of Casimir force solutions acquired by the FD method is of intrinsic interest not necessarily motivated by the need for competitiveness. Second, any solution represents a useful tool for developers working on other, possibly more effective algorithms and in need to obtain a check with results from completely different systems about problems that have never been previously solved and for which analytical solutions are probably unavailable. Third, the typical simplicity of FD algorithms translates into a relatively small code development time investment in exchange for access to problems of great importance. Finally, the FD approach is readily accessible to undergraduate students [45] thus offering opportunities for research projects that might otherwise be unaccessible at much higher levels of sophistication.

3.1. Standard FD approach

The first step into the standard FD implementation of Casimir forces was taken choosing the one-dimensional (1D) case for diagnostics [37]. In this model, both the electric and magnetic fields are assumed to fluctuate along fixed axes with respect to the two interacting slabs. In addition to demonstrating convergence to the available analytical solution, an important step forward was taken by considering dielectric boundaries not coinciding with mesh points. A second step was the generalization of this method to the case of 1D multilayers [46]. In order to carry out initial meaningful parallelization experiments, we have employed a computer algebra system capable to detect and take advantage of multiple cores on a legacy Cray CX1-LC (up to 8 cores in the present implementation) with Red Hat Enterprise Linux.

3.2. Non-standard FD schemes

Another important contribution of our initial effort was the identification of a non-standard FD scheme that exactly reproduces the necessary discrete Green's functions (DGFs). The motivation to pursue such an approach, beyond its intrinsic interest, is that it might shed light on more efficient FD schemes in higher dimensions.

3.3. Pedagogy
There exists a very rich literature of applications of the FD scheme in electrostatics due to its immediacy for undergraduate students [45]. Our goal is to develop FD code of Casimir force computations as part of activities for students carrying out research projects both in Fortran 90/95 and in a computer algebra system. Initial experience has already shown that students exposed to this curriculum will be well prepared to make transitions to other languages such as C and other computational methods such as FE, FDFD, BEM, etc on any platform. This process also presents unique opportunities to illustrate the role of Green functions (GF) by employing their discrete counterparts (DGF) for easier visualization.

4. Exemplary outstanding problems
Here we survey some interesting problems to be considered as both diagnostics and research by our numerical effort in dispersion forces.

4.1. Multilayers
No experiment is ever actually carried out between the two homogeneous dielectric slabs of the archetypal Lifshitz formulation since layers naturally form on any surface and also because such layers will often be part of the design. For instance, many experiments involve the interaction of substrata coated in gold (Au) or with even more complex architectures [23]. In addition, there is great interest in the use of multilayers, such as photonic crystals, as a means to engineer the Casimir force by altering its dependence as a function of the gap width between the two slabs. Although analytical solutions for such parallel plane geometry exist [17], research on multilayers in more complex geometries is relatively young.

4.2. Edge effects
In analogy to the standard capacitor problem of electrostatics, a typical assumption of the Lifshitz theory is that the gap width be much smaller than the later size of the interacting slabs. Numerical solutions can aid in developing or testing analytical expressions for Casimir force fringe effects.

4.3. Static deflection
One of the earliest considerations of the effects of Casimir forces in micro-electromechanical systems (MEMS) was as a leading cause of stiction [32]. In a later development, an effort was made to analytically determine the static deflection of microbridges and conditions for pull-in of such structures to the underlying substratum [47]. All such calculations are necessarily approximations due to the fact that, rigorously speaking, the microbridge and the substratum are no longer parallel. Numerical computations are necessary to explore the reliability of all conclusions drawn on the basis of such analytical methods specially in the case of 2D structures for which much less information is available.

As already mentioned, dispersion forces are making a rapid transition from being avoided as the cause of a failure mode – stiction – to being exploited as an enabling design component. One concept along these lines has been that of an adaptive optics system deformed not by mechanical pressure or electrostatically as typically is the case but by Casimir forces [48].

4.4. Free and driven anharmonic Casimir vibrations
Although the problem of a 1D anharmonic Casimir oscillator has been analyzed and also explored experimentally in MEMS [49], no generalization of such results to 2D membranes has been developed. Such systems could also serve as parametric resonators and amplifiers if the magnitude of the Casimir force between the substratum and the oscillating membrane can be modulated in time [50].

4.5. Non-trivial topologies, anisotropic materials and Casimir torque
A very active area for recent generalizations of the Lifshitz theory has been that of anisotropic materials both in the slabs and in the gap itself because, under specific circumstances, not only dispersion forces but also dispersion torques may appear [51, 52, 53, 54, 55, 56, 57, 58, 59, 60] and play obviously important technological roles. The behavior of such systems in more complex geometries as well as with elastic membranes is largely unexplored.

4.6. Nanomechanical oscillators
One of the environments in which dispersion forces completely dominate system dynamics due to the extremely small interlayer gap is that of multiwalled telescoping nanotubes with cores in oscillation or rapid ejection [35, 38]. In such structures, numerical methods would be beneficial to assess use of the proximity force approximation in very small tubes as well as the effect of illumination on the optical properties of the interacting layers [61].

4.7. Nanomechanical engine cycles
No realistic nanotechnological application of dispersion forces can be implemented without a means to transfer energy to and from the system. Typically, this limitation is addressed by mechanically actuating [23, 49] one of the two interacting surfaces so as to drive the other one into vibration. The problem is even more critical in nanotube applications in which it may be nearly impossible to access the oscillating shuttle [38, 62]. One of the most realistic proposals to implement nanoengines [63] is to use the dependence of dispersion forces on illumination as verified, with some open issues, by Arnold, Hunklinger and Dransfeld [64] and confirmed several years later by means of atomic force microscope (AFM) methods [65]. Numerical models are needed to create far more realistic models of the engine cycles so as to account for mutual spectral perturbations of moving interacting boundaries expected on energy conservation considerations [36] also in non-trivial geometries.

4.8. Acoustic Casimir effect
Although this is not often appreciated, the Casimir effect takes place not only in quantized but also in classical fields. This can be appreciated from the fact that, for instance, the radiation pressure interpretation of the Casimir force, suggested by Casimir himself [10], reiterated by Debye [66], and fully developed by Milonni, Cook, and Goggin, does *not* require that the radiation be quantized (Ref. [13], p. 293 and Ref. [67], p. 210). The most dramatic demonstration of such a classical effect is probably the acoustic Casimir effect, in which broadband noise is introduced by means of compression drivers according to a spectrum specifiable by the experimenter [68, 69]. Unlike the electrodynamical Casimir effect, most experimentation in this area has explored the parallel-plane case since, due to the vastly smaller value of the speed of sound with respect to the speed of light, typical gap widths will be $s \sim 1$ cm. For this reason, both theoretical calculations and experimentation on non-parallel plates and completely different geometries are unavailable.

5. Target applications

Let us consider the possible applications of knowledge developed, for instance, by FD schemes to evaluate Casimir forces and torques based on the assigned geometry and physics of any system possibly also under dynamical conditions.

5.1. Design tools

Dispersion forces are expected to become a key enabling technology on the micro- and nano-scales and therefore a sophisticated design tool to document the effect of such interactions in realistic market product designs is expected to be highly desirable. A possible outcome will be the evolution of dispersion force computation algorithms into an indispensable module for multiphysics analysis of such novel technologies [70]. Due to the different nature of these interactions, connected to this computing capability is also the need for enhanced visualization solutions in 3D so as to aid the designer take advantage of dispersion forces to achieve a specific technological goal.

5.2. Experiment modeling

Next generation dispersion force experimentation is poised to measure extremely small forces to unprecedented accuracy [71]. Yet, dispersion forces are now known to strongly interact with other environmental variables, such as temperature [72, 73], in possibly complex ways so that a quantitative understanding of the complex role of dispersion forces in yielding the data acquired will contribute to realistic estimates of experimental errors, still a very controversial topic [74, 75].

5.3. NOEMS and nanostructures

Beyond fundamental physics experimentation, a host of dispersion force based applications are expected to enter the marketplace in the near future [39, 40]. Perhaps the most stunning is the class of products generally referred to as "gecko glue," which represents the most convincing demonstration that forces once characterized as weak and exotic [36] have proven themselves as potentially dominant on the macroscale. Fascinatingly, these materials have more recently evolved into novel human climbing [76] and space grappling [77] tools. Because of the drastic dependence of dispersion forces on topology, such products can only be optimized by means of accurate mathematical modeling. Such considerations extend to all nanomachines with parts actuated by dispersion force modulation, including, for instance, nanodevices for various types of medical purposes [78, 79].

5.4. Energy storage and conversion

The possibility to employ the van der Waals interaction, an important cause of cohesion and adhesion [80], as a 'reservoir' for energy storage was first proposed by Robert Forward in 1984 [31]. At the time, the idea, though fascinating, appeared somewhat unworkable because the surface energy density of two ordinary surfaces separated by realistic gap widths is many orders of magnitude below values of practical interest when compared, for instance, to traditional chemical batteries. However, further research into the concept has revealed that, for instance, arrays of vertical telescoping nanotubes, due to their much higher specific surface and much smaller gap width (in this case, the distance between the outer wall and the sliding inner tube) could store energy at densities even higher than those of chemical storage [35]. Far more accurate computation tools, however, could dramatically expedite the exploration of the extremely vast parameter space of various geometries and compounds and identify the best route to this new storage technology.

6. Conclusions

The computation of dispersion forces is an exponentially growing field of research with unique connections to such theories as quantum electrodynamics (QED) and acoustics, numerical techniques including, as we have seen, traditional FD methods, and extending horizontally to software design, applied physics, engineering, product analysis and business development. This field will contribute to the development of breakthrough industrial products and to next generation scientific experiments, and, as such, it offers remarkable opportunities for discovery both for professional researchers and for students just beginning to explore the practice of science.

6.1. Acknowledgments

Partial funding was made available by a Jazan University Deanship of Scientific Research grant. Nurdoğan Can (Jazan University) provided additional indispensable logistical resources. Elias C. Vagenas (Kuwait University) kindly hosted an invited talk on the subject of this paper at the 5th International Conference on Mathematical Modeling in Physical Sciences (IC-MSQUARE) held in Athens, Greece.

References

[1] Wang S C 1927 *Phys. Z.* **28** 663–666
[2] Eisenschitz, R V and London F 1930 *Z. Phys.* **60** 491–527
[3] London F 1937 *Trans. Faraday Soc.* **33** 8–26
[4] Verwey E J W and Overbeek J T G 1946 *Trans. Faraday Soc.* **42** B117–123
[5] Verwey E J W 1947 *J. Phys. Colloid Chem.* **51** 631–636
[6] Verwey E J W and Overbeek J T G 1948 *Theory of the Stability of Lyophobic Colloids* (New York: Elsevier Publishing Company, Inc.)
[7] Casimir H B G and Polder D 1948 *Phys. Rev.* **73** 360–372
[8] Milonni P W and Shih M L 1992 *Contemp. Phys.* **33** 313–322
[9] Casimir H B G 1949 *J. Chim. Phys.* **46** 407–410
[10] Casimir H B G 1948 *Proc. Kon. Ned. Akad. Wetenshap* **51** 793—-795
[11] de Boer J H 1936 *Trans. Faraday Soc.* **32** 10–37
[12] Hamaker H 1937 *Physica* **4** 1058–1072
[13] Milonni P W 1994 *The Quantum Vacuum* (San Diego: Academic Press)
[14] Rosa F S S, Dalvit D A R and Milonni P W 2010 *Phys. Rev. A* **81** 033812
[15] Rytov S M 1959 Theory of Electric Fluctuations and Thermal Radiation Tech. rep. Armed Services Technical Information Agency, Rep. AFCRC-TR-59-162 (Air Force. Cambridge Research Center, Bedford, Mass.) Arlington, VA
[16] Lifshitz E M 1956 *Sov. Phys. JETP* **2** 73–83
[17] Parsegian V A 2006 *Van der Waals Forces* (Cambridge: Cambridge University Press)
[18] Sparnaay M J 1958 *Physica* 751–764
[19] Bressi G, Carugno G, Onofrio R and Ruoso G 2002 *Phys. Rev. Lett.* **88** 041804
[20] Derjaguin B V 1934 *Koll. Ztschr.* **69** 155–164
[21] Derjaguin B V, Abrikosova I I and Lifshitz E M 1956 *Quart. Rev. (London)* **10** 295–329
[22] Derjaguin B V and Abrikosova I I 1957 *Sov. Phys. JETP* **3** 819–829
[23] Hunklinger S, Geisselman H and Arnold W 1972 *Rev. Sci. Instrum.* **43** 584–587
[24] Lamoreaux S K 1997 *Phys. Rev. Lett.* **78** 5–8
[25] Mohideen U and Roy A 1998 *Phys. Rev. Lett.* **81** 4549–4552
[26] Dalvit D A R, Milonni P W, Roberts D C and Rosa F S S (eds) 2011 *Casimir Physics. Springer Lecture Notes in Physics 834* (Heidelberg: Springer)
[27] Bordag M, Mohideen U and Mostepanenko V M 2001 *Phys. Rep.* **353** 1–205
[28] Bordag M, Klimchitskaya G L, Mohideen U and Mostepanenko V M 2009 *Advances in the Casimir Effect* (Oxford: Oxford University Press)
[29] Blocki J, Randrup J, Swiatecki W J and Tsang C F 1977 *Ann. Phys.* **105** 427–462
[30] Milton K A 2011 *Am. J. Phys.* **79** 697–711
[31] Forward R L 1984 *Phys. Rev. B* **30** 1700–1702
[32] Serry F M, Walliser D and Maclay G 1995 *J Microelectromech. Sys.* **4** 193–205
[33] Pinto F 2008 *Space Technology and Applications Int. Forum (STAIF-2008), AIP Conf. Proc. 969* ed El-Genk M S (Melville, New York: AIP) pp 959–968

[34] Pinto F 2008 *J. Phys. A* **41** 164033
[35] Pinto F 2013 *Nanotube Superfiber Materials* ed Schulz M, Shanov V N and Zhangzhang Y (New York: Elsevier) chap 27, pp 789–806
[36] Pinto F 2014 *Am. Scie.* **102** 280–289
[37] Pinto F 2009 *Phys. Rev. A* **80** 042113
[38] Pinto F 2012 *48th AIAA/ASME/SAE/ASEE Joint Propulsion Conference (JPC) & Exhibit* July (Atlanta: AIAA) pp 1–31 10.2514/6.2012-3713
[39] Pinto F 2016 *5th International Science Congress & Exhibition APMAS2015, Lykia, Oludeniz, April 16-19, 2015. Acta Phys. Pol. A* vol 129 pp 819–825
[40] Pinto F 2016 *3rd International Workshop on Metrology for Aerospace* (Florence, Italy: IEEE) Paper ID: 4292221
[41] Sparnaay M J 1989 *Physics in the Making* ed Sarlemijn A and Sparnaay M J (Amsterdam: Elsevier Science Publishers) chap 9A, pp 235–246
[42] Axilrod B M and Teller E 1943 *J. Chem. Phys.* **11** 299–300
[43] Rodriguez A W, Ibanescu M, Iannuzzi D, Joannopoulos J D and Johnson S G 2007 *Phys. Rev. A* **76** 032106
[44] Pinto F 2008 *International Mathematica User Conference 2008* (Champagne, IL, USA) URL `http://library.wolfram.com/infocenter/Conferences/7274/`
[45] Pinto F 2007 *Am. J. Phys.* **75** 513–519
[46] Pinto F 2009 *Quantum Field Theory under the Influence of External Conditions (QFEXT09)* ed Milton K A and Bordag M (New Jersey: Wolrd Scientific)
[47] Serry F M, Walliser D and Maclay G 1998 *J. Appl. Phys.* **84** 2501–2506
[48] Pinto F 2008 *Optomechatronic Technologies 2008* vol 7266 ed et al Otani Y (San Diego: SPIE) p 726616
[49] Chan H, Aksyuk V, Kleiman R, Bishop D and Capasso F 2001 *Phys. Rev. Lett.* **87** 21–24
[50] Pinto F 2016 *TAUP 2015: Proceedings of the XIV International Conference on Topics in Astroparticle and Underground Physics. J. of Phys. Conf. Ser. 718* ed Fornengo N, Regis M and Zechlin H S p 072004
[51] Parsegian V A and Weiss G H 1972 *J. Adhes.* **3** 259–267
[52] Sarlah a and Zumer S 2001 *Phys. Rev. E* **64** 051606
[53] Podgornik R, Hansen P L and Parsegian V A 2003 *J.Chem. Phys.* **119** 1070–1077
[54] Deng G, Liu Z Z and Luo J 2008 *Phys. Rev. A* **78** 062111
[55] Rosa F S S, Dalvit D A R and Milonni P W 2008 *Phys. Rev. A* **78** 032117
[56] Deng G, Liu Z Z and Luo J 2009 *Phys. Rev. A* **80** 062104
[57] Veble G and Podgornik R 2009 *Phys. Rev. B* **80** 075422
[58] Sun J, Hua X K and Gao L 2012 *Solid State Comm.* **152** 1666–1669
[59] Kornilovitch P E 2013 *J. Phys: Condens. Matter* **25** 035102
[60] Deng G, Tan B h, Pei L, Hu N and Zhu J r 2015 *Phys. Rev A* **91** 062126
[61] Pinto F 2013 *J. Appl. Phys.* **114** 024310
[62] Kis A and Zettl A 2008 *Phil. Trans. R. Soc. A* **366** 1591–1611
[63] Pinto F 1999 *Phys. Rev. B* **60** 14740–14755
[64] Arnold W, Hunklinger S and Dransfeld K 1980 *Phys. Rev. B* **21** 1713–1713
[65] Chen F, Klimchitskaya G L, Mostepanenko V M and Mohideen U 2007 *Opt. Express* **15** 4823–4829
[66] Chu B 1967 *Molecular Forces, Based on the Baker Lectures of Peter J. W. Debye* (New York: John Wiley & Sons)
[67] Lamoreaux S K 2004 *Rep. Prog. Phys.* **68** 201–236
[68] Larraza A, Holmes C D, Susbilla R T and Denardo B 1998 *Phys. Lett. A* **248** 151–155
[69] Larraza A, Holmes C D and Susbilla R T 1998 *J. Acoust. Soc. Am.* **103** 2267–2272
[70] Lethbridge P *The Industrial Physicist* 28–31
[71] Almasi A, Brax P, Iannuzzi D and Sedmik R I P 2015 *Phys. Rev. D* **91** 102002
[72] Sushkov A O, Kim W J, Dalvit D A R and Lamoreaux S K 2011 *Nature Phys.* **7** 230–233
[73] Lamoreaux S K 2012 *Annu. Rev. Nucl. Part. Sci.* **62** 37–56
[74] Lamoreaux S K 2011 *Casimir Physics. Springer Lecture Notes in Physics 834* (Heidelberg: Springer) chap 7, pp 219–248
[75] Klimchitskaya G L and Mostepanenko V M 2011 *8th Alexander Friedmann International Seminar - Int. J. Phys. Conf. Ser.* vol 3 (World Scientific) pp 541–554
[76] Hawkes E W, Eason E V, Christensen D L and Cutkosky M R 2015 *J. R. Soc. Interface* **12** 20140675
[77] Jiang H, *et al.* 2015 *IEEE Int. Conf. on Robotics and Automation* (Seattle, Washington: IEEE) pp 1–8
[78] Yun Y H, *et al.* 2009 *NanoScience in Biomedicine* ed Shi D (Berlin: Tsinghua University Press, Springer) pp 451–484
[79] Schulz M J 2012 *Science* **338** 893–894
[80] Krupp H 1967 *Advan. Colloid Interface Sci.* **1** 111–239